Water Systems Analysis, Design, and Planning

Water Systems Analysis, Design, and Planning

Urban Infrastructure

Mohammad Karamouz

CRC Press is an imprint of the
Taylor & Francis Group, an **informa** business

MATLAB® is a trademark of The MathWorks, Inc. and is used with permission. The MathWorks does not warrant the accuracy of the text or exercises in this book. This book's use or discussion of MATLAB® software or related products does not constitute endorsement or sponsorship by The MathWorks of a particular pedagogical approach or particular use of the MATLAB® software.

First edition published 2022
by CRC Press
6000 Broken Sound Parkway NW, Suite 300, Boca Raton, FL 33487-2742

and by CRC Press
2 Park Square, Milton Park, Abingdon, Oxon, OX14 4RN

© 2022 Taylor & Francis Group, LLC

CRC Press is an imprint of Taylor & Francis Group, LLC

Reasonable efforts have been made to publish reliable data and information, but the author and publisher cannot assume responsibility for the validity of all materials or the consequences of their use. The authors and publishers have attempted to trace the copyright holders of all material reproduced in this publication and apologize to copyright holders if permission to publish in this form has not been obtained. If any copyright material has not been acknowledged please write and let us know so we may rectify in any future reprint.

Except as permitted under U.S. Copyright Law, no part of this book may be reprinted, reproduced, transmitted, or utilized in any form by any electronic, mechanical, or other means, now known or hereafter invented, including photocopying, microfilming, and recording, or in any information storage or retrieval system, without written permission from the publishers.

For permission to photocopy or use material electronically from this work, access www.copyright.com or contact the Copyright Clearance Center, Inc. (CCC), 222 Rosewood Drive, Danvers, MA 01923, 978-750-8400. For works that are not available on CCC please contact mpkbookspermissions@tandf.co.uk.

Trademark notice: Product or corporate names may be trademarks or registered trademarks and are used only for identification and explanation without intent to infringe.

Library of Congress Cataloging-in-Publication Data
Names: Karamouz, Mohammad, author.
Title: Water systems analysis, design, and planning : urban infrastructure / Mohammad Karamouz.
Description: First edition. | Boca Raton : CRC Press, [2022] | Includes index.
Identifiers: LCCN 2021028450 (print) | LCCN 2021028451 (ebook) |
ISBN 9780367528454 (hardback) | ISBN 9781032149189 (paperback) |
ISBN 9781003241744 (ebook)
Subjects: LCSH: Waterworks.
Classification: LCC TD485 .K368 2022 (print) | LCC TD485 (ebook) |
DDC 628.109173/2—dc23
LC record available at https://lccn.loc.gov/2021028450
LC ebook record available at https://lccn.loc.gov/2021028451

ISBN: 978-0-367-52845-4 (hbk)
ISBN: 978-1-032-14918-9 (pbk)
ISBN: 978-1-003-24174-4 (ebk)

DOI: 10.1201/9781003241744

Typeset in Times
by codeMantra

To the great power of mind and those who have striven to utilize it for the betterment of our hydro-planet.

Mohammad Karamouz

Contents

Preface xxv
Acknowledgments xxxi
Author xxxiii

Chapter 1 Introduction 1

1.1 Introduction 1
1.2 Urban Water Cycle 2
1.2.1 Components 3
1.2.2 Interdependencies 4
1.2.3 Impact of Urbanization 4
1.3 Interaction of Climatic, Hydrologic, Cultural and Esthetic Aspects 5
1.3.1 Climatic Effects—Rainfall Type 5
1.3.2 Hydrologic Effects 5
1.3.3 Urban Heat Islands 5
1.3.4 Cultural and Esthetic Aspects 6
1.4 Urban Water Infrastructure Management 7
1.4.1 Life Cycle Assessment 8
1.4.2 Environmental, Economic, and Social Performances 9
1.4.3 Urban Landscape Architecture 10
1.5 Systems Approach 12
1.5.1 General Systems' Characteristics 13
1.5.2 System Properties 15
1.6 Hydrologic Variability 17
1.6.1 Hydrologic Variables and Parameters 18
1.7 Representations, Statistical, and Simulation Models 18
1.8 Extreme Values, Vulnerability, Risk, and Uncertainty 19
1.9 Tools and Techniques 20
1.9.1 Systems Modeling 22
1.9.2 Model Resolution 22
1.10 The Hierarchy of Water for Life and Total Systems Approach 23
1.10.1 The Biosphere 24
1.10.2 System-Based Thinking 24
1.10.3 Natural Systems 25
1.10.4 Human and Institutional Systems 25
1.10.5 Built Environment—Infrastructure 25
1.10.6 Disasters and Interdependencies 25
1.11 People's Perception—Public Awareness 26
1.11.1 Integrated Water Cycle Management 28
1.12 Economics of Water 28
1.13 Clean Water Act 29
1.13.1 The Basis of State Water Laws in the United States 30
1.14 Concluding Remarks and Book's Organization 31
Problems 31
References 33

Chapter 2 Urban Water Cycle and Interactions 35

2.1 Introduction 35
2.2 Urban Water Cycle 37
2.2.1 Components—Water Movements 37
2.2.2 Impact of Urbanization—Water Distribution, Waste Collection 39
2.3 Interactions on Urban Components 40
2.3.1 Climatic Effects—Different Climates 40
2.3.2 Hydrologic Effects 41
2.3.3 Qualitative Aspects 42
2.3.4 Greenhouse Effect 42
2.3.5 Urban Heat Islands—Mitigation 43
2.3.6 Cultural Aspects 43
2.4 Remotely Sensed and Satellite Data 44
2.5 Water Balance Elements 44
2.5.1 Precipitation 44
2.5.1.1 Measurement by Standard Gauges 44
2.5.1.2 Various Types of Rain Gauges 46
2.5.1.3 Measurement by Weather Radar 47
2.5.1.4 Measurement by Satellite 48
2.5.1.5 The PERSIANN System 48
2.5.1.6 Estimation of Missing Rainfall Data 49
2.5.1.7 Station Average Method 51
2.5.1.8 Snowmelt Estimation 51
2.5.2 Evaporation and Evapotranspiration 54
2.5.2.1 Evaporation Evaluation 55
2.5.2.2 Water Budget Method 55
2.5.2.3 Mass Transfer Method 55
2.5.2.4 Pan Evaporation 56
2.5.2.5 Measurement of Evapotranspiration 56
2.5.2.6 Thornthwaite Method 57
2.6 Interception Storage and Depression Storage 58
2.7 Infiltration 59
2.8 Palmer Drought Severity Index (PDSI) 60
2.8.1 Agricultural Drought Indicators 60
2.8.2 Potential Climatic Values 61
2.8.3 Coefficients of Water Balance Parameters 61
2.8.4 Precipitation for Climatically Appropriate for Existing Condition, $\hat{P}_i$ 61
2.8.5 Drought Severity Index 62
2.9 Groundwater 63
2.10 Reservoirs and Lakes 65
2.11 Water Balance 65
2.11.1 Thomas Model (abcd Model) 66
2.11.2 A Case Study: Water Balanced-Based Sustainability 68
2.12 Interactions between the Urban Water Cycle and Urban Infrastructure Components 72
2.12.1 Interactions with the Wastewater Treatment System 73
2.12.2 Interactions between Water and Wastewater Treatment Systems 73
2.12.3 Interactions between Water Supply and Wastewater Collection Systems 74

2.12.4 Interactions between Urban Drainage Systems and Wastewater Treatment Systems 74
2.12.5 Interactions between Urban Drainage Systems and Solid Waste Management 75
2.12.6 Interactions between Urban Water Infrastructure and Urban Transportation Infrastructures 76
2.13 Livable Cities of the Future 77
2.13.1 Daniel Loucks View Points 77
2.13.1.1 Urban Water in the Larger Water Nexus 77
2.13.1.2 Gray Infrastructure 78
2.13.1.3 Green Infrastructure 79
2.13.1.4 Challenges of Future 80
2.13.2 David Miller's View Points 80
2.13.2.1 Setting the Stage 80
2.13.2.2 Cities Around the World Lead the Way 81
2.13.2.3 A Closer Look at Toronto's Strategies for Sustainability ... 82
2.13.3 Craig S. Ivey's View Points 83
2.13.3.1 Facts and Figures 83
2.13.3.2 Challenges 84
2.13.3.3 Solutions 84
2.14 Concluding Remarks 86
Problems 86
References 89

Chapter 3 Urban Water Hydrology 93

3.1 Introduction 93
3.2 Urban Watersheds 93
3.3 Watershed Geomorphology 94
3.4 Land Use and Cover Impacts 94
3.4.1 Urban Areas 96
3.4.2 Wetland Areas 96
3.5 Rainfall–Runoff Analysis in Urban Areas 97
3.5.1 Drainage Area Characteristics 98
3.5.2 Rainfall Losses 99
3.6 Travel Time 99
3.6.1 Definitions of Time of Concentration 99
3.6.2 Classifying Time Parameters 100
3.6.3 Velocity Method 101
3.6.4 Sheet Flow Travel Time 102
3.6.5 Empirical Formulas 103
3.7 Excess Rainfall Calculation 103
3.7.1 Interception Storage Estimation 104
3.7.2 Estimation of Infiltration 105
3.7.3 Green–Ampt Model 106
3.7.3.1 Ponding Time 108
3.7.3.2 Horton Method 110
3.7.3.3 Simple Infiltration Models 111
3.8 Rainfall Measurement 112
3.8.1 Intensity–Duration–Frequency Curves: Advantages and Disadvantages 113

3.8.1.1 Selection of Rainfall Duration ... 113
3.9 Estimation of Runoff Volume ... 114
3.9.1 Rational Method ... 114
3.9.2 SCS Method ... 117
3.10 Unit Hydrographs ... 122
3.10.1 UH Development ... 122
3.10.2 SCS UH ... 122
3.10.3 Application of the UH Method ... 123
3.10.4 S-Hydrograph Method ... 124
3.11 IUH–Convolution Integral–Nash Model ... 127
3.11.1 Convolution Integral ... 127
3.12 Instantaneous Unit Hydrographs ... 131
3.12.1 Nash Model ... 133
3.12.2 Laplace Transformation Model ... 138
3.12.2.1 Basin as a Linear Reservoir ... 138
3.12.2.2 Basin as a Channel ... 141
3.13 Routing Methods ... 142
3.13.1 Hydrologic Methods of River Routing ... 142
3.13.1.1 Muskingum Method ... 142
3.13.1.2 Determination of Storage Constants ... 144
3.14 Revisiting Flood Records ... 144
3.14.1 Urban Effects on Peak Discharge ... 145
3.14.2 Flood Record Adjusting ... 145
3.15 Test of the Significance of the Urban Effect ... 146
3.15.1 Spearman Test ... 147
3.15.2 Spearman–Conley Test ... 149
3.16 Time Series Analysis ... 151
3.16.1 ARMA(p, q) Model Identification ... 152
3.16.1.1 Autocorrelation Function ... 152
3.16.1.2 Partial Autocorrelation Function (PACF) ... 153
3.16.2 Autoregressive (AR) Models ... 154
3.16.3 Moving Average Process ... 157
3.16.4 Autoregressive Moving Average Modeling ... 158
3.16.5 Akaike's Information Criterion (AIC) ... 159
3.16.6 ARIMA Models Considerations ... 160
3.17 Concluding Remarks ... 162
Problems ... 162
References ... 164

Chapter 4 Urban Water Hydraulics ... 167
4.1 Introduction ... 167
4.2 Channel Geomorphology ... 167
4.2.1 Length of a Channel ... 168
4.2.2 Slope of a Channel ... 170
4.2.3 Law of Stream Slopes ... 174
4.2.4 Channel Cross Section ... 174
4.2.5 Channel Roughness ... 174
4.2.6 Urban Morphology Challenges ... 177
4.3 Travel Time ... 178

4.4 Open-Channel Flow in Urban Watersheds 179
4.4.1 Open-Channel Flow 179
4.4.1.1 Open-Channel Flow Classification 181
4.4.1.2 Hydraulic Analysis of Open-Channel Flow 181
4.4.2 Overland Flow 183
4.4.2.1 Overland Flow on Impervious Surfaces 184
4.4.2.2 Overland Flow on Pervious Surfaces 187
4.4.3 Urban Channel Routing 189
4.4.3.1 Muskingum Method 189
4.5 Hydraulics of Water Distribution Systems 192
4.5.1 Energy Equation of Pipe Flow 192
4.5.2 Evaluation of Head Loss Due to Friction 193
4.5.2.1 Darcy–Weisbach Equation 193
4.5.2.2 Hazen–Williams Equation for the Friction Head Loss 196
4.5.2.3 Minor Head Loss 198
4.5.2.4 Pipes in Series 203
4.5.2.5 Pipes in Parallel 206
4.5.2.6 Pipe Networks 208
4.6 Concluding Remarks 211
Problems 212
References 216

Chapter 5 Urban Stormwater Drainage Systems 217

5.1 Introduction 217
5.2 Urban Planning and Stormwater Drainage 217
5.2.1 Land Use Planning 220
5.2.1.1 Dynamic Strategy Planning for Sustainable Urban Land Use Management 220
5.2.1.2 Identification of System's Components 220
5.2.1.3 Identification of the Dynamic Relationships among the Components 220
5.2.1.4 DSR Dynamic Strategy Planning Procedure 222
5.2.2 Best Management Practices 222
5.2.2.1 Sediment Basins 224
5.2.2.2 Retention Pond 224
5.2.2.3 Bioretention Swales 226
5.2.2.4 Bioretention Basins 227
5.2.2.5 Sand Filters 227
5.2.2.6 Swales and Buffer Strips 230
5.2.2.7 Constructed Wetlands 231
5.2.2.8 Extended Detention Basin (EDB) 233
5.2.2.9 Ponds and Lakes 236
5.2.2.10 Infiltration Systems 236
5.2.2.11 Grass Buffer 237
5.2.2.12 Aquifer Storage and Recovery 237
5.2.2.13 Porous Pavement 237
5.3 Drainage in Urban Watersheds 238
5.3.1 Overland Flow 239
5.3.2 Channel Flow 239

5.4 Components of Urban Stormwater Drainage System.......239
5.4.1 General Design Considerations.......239
5.4.2 Flow in Gutters.......241
5.4.2.1 Gutter Hydraulic Capacity.......242
5.4.3 Pavement Drainage Inlets.......242
5.4.3.1 Inlet Locations.......243
5.4.4 Surface Sewer Systems.......244
5.4.5 Drainage Channel Design.......246
5.4.5.1 Design of Unlined Channels.......247
5.4.5.2 Grass-Lined Channel Design.......247
5.5 Combined Sewer Overflow.......248
5.5.1 Reduce Combined Sewer Overflows with Green Infrastructure.......249
5.5.2 Reduce Combined Sewer Overflows with High-Level Storm Sewers Citywide.......249
5.6 Culverts.......249
5.6.1 Sizing of Culverts.......250
5.6.2 Protection Downstream of Culverts.......251
5.7 Design Flow of Surface Drainage Channels.......251
5.7.1 Probabilistic Description of Rainfall.......251
5.7.1.1 Return Period and Hydrological Risk.......251
5.7.1.2 Frequency Analysis.......253
5.7.2 Design Rainfall.......255
5.7.2.1 Selecting Design Rainfall and Runoff.......255
5.7.3 Design Return Period.......256
5.7.4 Design Storm Duration and Depth.......257
5.7.5 Spatial and Temporal Distribution of Design Rainfall.......257
5.8 Stormwater Storage Facilities.......257
5.8.1 Sizing of Storage Volumes.......259
5.9 Risk Issues in Urban Drainage.......260
5.9.1 Flooding of Urban Drainage Systems.......260
5.9.1.1 Case Study: Improvement of Urban Drainage System Performance under Climate Change Impact.......262
5.9.2 DO Depletion in Streams—Discharge of Combined Sewage Effects.......264
5.9.3 Discharge of Chemicals.......265
5.10 Urban Floods.......265
5.10.1 Urban Flood Control Principles.......266
5.11 Overland Flow Models.......267
5.11.1 StormNET: Stormwater and Wastewater Modeling.......267
5.11.2 GSSHA.......267
5.11.3 LISFLOOD-FP.......267
5.12 Stormwater Infrastructure of Selected Cities.......267
5.12.1 Philadelphia, USA.......267
5.12.1.1 Characteristics of the system.......267
5.12.1.2 Improvement and Future Plans.......269
5.12.1.3 Recommendations.......269
5.12.2 Los Angeles, California.......269
5.12.2.1 Characteristics of the System.......269
5.12.2.2 Improvement and Future Plans.......270
5.12.2.3 Recommendations.......271
5.12.3 Chongqing, China.......271

5.12.3.1 Characteristics of the System 271
5.12.3.2 Improvement and Future Plans 271
5.12.3.3 Recommendations 272
5.12.4 London, England 272
5.12.4.1 Characteristics of the System 272
5.12.4.2 Improvement and Future Plans 272
5.12.5 Amsterdam, Netherlands 273
5.12.5.1 Characteristics of the System 273
5.12.5.2 Improvement and Future Plans 273
5.12.5.3 Recommendations 274
5.12.6 Stockholm, Sweden 274
5.12.6.1 Characteristics of the System 274
5.12.6.2 Improvement and Future Plans 274
5.12.6.3 Recommendations 276
5.13 Concluding Remarks 276
Problems 276
References 279

Chapter 6 Urban Water Supply Infrastructures 283

6.1 Introduction 283
6.1.1 History of Water Supply Development 284
6.1.2 Water Availability 284
6.1.3 Water Development and Share of Water Users 286
6.1.4 Natural Resources for Water Supply 287
6.1.5 Supplementary Sources of Water 288
6.2 Water Supply Infrastructures 288
6.2.1 Reservoirs and Water Supply Storage Facilities 290
6.2.2 Water Storage 293
6.2.2.1 Types of Dams 293
6.2.3 Planning Issues 294
6.2.3.1 Cascade Reservoirs 295
6.2.4 Parallel Reservoir 295
6.2.5 Reservoir Operation 297
6.2.6 Flood Control 297
6.2.7 Creative Thinking Examples of Supply Expansion 298
6.2.7.1 Curing a Dam—Bookan Reservoir: Increasing the Operational Efficiency 298
6.2.7.2 Curing Lar Dam in Iran (Karamouz et al., 2003b) 300
6.2.8 Groundwater Storage 302
6.2.8.1 Well Hydraulics 303
6.2.8.2 Confined Flow 304
6.2.8.3 Unconfined Flow 306
6.2.9 Urban Storage Reservoirs and Tanks 307
6.2.10 Water Transfers and Conveyance Tunnels 308
6.3 Water Treatment Plants 309
6.3.1 Water Treatment Infrastructure 309
6.3.2 Unit Operations of Water Treatment 311
6.3.2.1 Coagulation/Flocculation 312
6.3.2.2 Sedimentation 312
6.3.2.3 Filtration 312

6.3.2.4 Disinfection ... 312
6.4 Water Distribution System ... 314
6.4.1 System's Components ... 314
6.4.2 Hydraulics of Water Distribution Systems ... 316
6.4.3 Wáter Supply System Challenges ... 317
6.5 Urban Water Demand Management ... 317
6.5.1 Basic Definitions of Water Use ... 317
6.5.2 Water Supply Quantity Standards in Urban Areas ... 318
6.5.3 Water Demand Forecasting ... 319
6.5.4 Water Quality Modeling in a Water Distribution Network ... 319
6.5.4.1 Water Quality Standards ... 319
6.5.4.2 Water Quality Model Development ... 319
6.5.4.3 Chlorine Decay ... 321
6.5.5 Water Demand and Price Elasticity ... 323
6.6 Hydraulic Simulation of Water Networks ... 324
6.6.1 EPANET ... 324
6.7 Assessing the Environmental Performance of Urban Water Infrastructure ... 325
6.8 Life Cycle Assessment ... 325
6.9 Sustainable Development of Urban Water Infrastructures ... 326
6.9.1 Selection of Technologies ... 327
6.9.1.1 Further Development of Large-Scale Centralized Systems ... 328
6.9.1.2 Separation for Recycling and Reuse ... 328
6.9.1.3 Natural Treatment Systems ... 330
6.9.1.4 Combining Treatment Systems ... 331
6.9.1.5 Changing Public Perspectives ... 331
6.10 Leakage Management ... 331
6.10.1 Acceptable Pressure Range ... 333
6.10.2 Economic Leakage Index ... 333
6.11 Nondestructive Testing (NDT) ... 335
6.12 Water Supply Infrastructure of Selected Cities ... 336
6.12.1 Case 1: Philadelphia, USA ... 336
6.12.2 Case 2: Los Angeles ... 339
6.12.3 Case 3: Copenhagen, Denmark ... 341
6.12.4 Case 4: Amsterdam, Netherland ... 342
6.12.5 Case 5: ACCRA, Ghana ... 344
6.12.6 Case 6: Stockholm, Sweden ... 345
6.13 Concluding Remarks ... 346
Problems ... 348
References ... 356

Chapter 7 Wastewater Infrastructure ... 359

7.1 Introduction ... 359
7.2 The Importance of Wastewater Systems ... 359
7.3 Wastewater Management ... 360
7.4 Wastewater Treatment ... 360
7.4.1 Primary Treatment ... 362
7.4.2 Secondary (Biological) Treatment ... 363
7.4.2.1 Biological Treatment Processes ... 364
7.4.2.2 Aerobic Treatment ... 366

7.4.2.3 Anaerobic Treatment 367
7.4.2.4 Activated Sludge 368
7.4.2.5 Suspended Growth 378
7.4.3 Advanced Treatment 378
7.4.4 Technologies for Developing Region 378
7.4.5 Wetlands as a Solution 380
7.5 Satellite Wastewater Management 380
7.5.1 Satellite Wastewater Treatment Systems 380
7.5.2 Interception Type 381
7.5.3 Extraction type 381
7.5.4 Upstream Type 382
7.5.5 Decentralized Systems 382
7.5.6 Infrastructure Requirements 382
7.6 Collection System Alternatives 382
7.6.1 Conventional Gravity Sewers 382
7.6.2 Septic Tank Effluent Gravity (STEG) 382
7.6.3 Septic Tank Effluent Pumps (STEP) 382
7.6.4 Pressure Sewers with Grinder Pumps 383
7.6.5 Vacuum Sewers 383
7.7 Wastewater Package Plants 383
7.8 Examples of Wastewater Treatment Development 384
7.8.1 Caribbean Wastewater Treatment 384
7.8.2 The Lodz Combined Sewerage System 386
7.8.2.1 Upgrading the Old Sewerage System 387
7.9 Case Studies 388
7.9.1 Case Study 1: Reliability Assessment of Wastewater Treatment Plants Under Coastal Flooding 388
7.9.2 Case Study 2: Uncertainty Based Budget Allocation of Wastewater Infrastructures' Flood Resiliency 392
7.9.3 Case Study 3: Margin of Safety-Based Flood Reliability Evaluation of Wastewater Treatment Plants 398
7.10 Wastewater Collection and Treatment of Selected Cities 401
7.10.1 Amsterdam, Netherlands 401
7.10.1.1 Characteristics of the System 401
7.10.1.2 Improvement and Future Plans 401
7.10.2 Stockholm, Sweden 401
7.10.2.1 Characteristics of the System 401
7.10.2.2 Improvement and Future Plans 402
7.10.3 Philadelphia, USA 403
7.10.3.1 Characteristics of the System 403
7.10.4 Zaragoza, Spain 405
7.10.4.1 Characteristics of the System 405
7.10.5 Paris, France 405
7.10.5.1 Characteristics of the System 405
7.10.6 Copenhagen, Denmark 409
7.10.6.1 Characteristics of the System 409
7.10.6.2 Infrastructures Sustainability 409
7.10.6.3 Improvement and Future Plans 409
7.11 Standards and Planning Considerations 410
7.11.1 Standards on Water and Wastewater Services 410
7.12 Concluding Remarks 413

Problems 413
References 415

Chapter 8 Urban Water Economics—Asset Management 419

8.1 Introduction 419
8.2 Urban Water Systems Economics—Basics 419
8.2.1 Economic Analysis of Multiple Alternatives 423
8.2.2 Economic Evaluation of Projects Using Benefit-Cost Ratio Method 425
8.2.3 Economic Models 426
8.2.4 Financial Statement 427
8.2.4.1 Balance Sheet 428
8.2.4.2 Financial Analysis 429
8.3 Asset Management 430
8.3.1 Attributes of Asset Management 430
8.3.2 Asset Management Drivers 431
8.3.4 The Objectives in Asset Management 432
8.3.3 Asset Management Steps 432
8.3.3.1 Status and Condition 432
8.3.3.2 Level of Service 432
8.3.3.3 Risk Management 433
8.3.3.4 Life Cycle Cost Analysis 435
8.3.3.5 Case Study 1: Reliability-Based Assessment of Life Cycle Cost of Urban Water Distribution Infrastructures 437
8.3.4 Sustainable Service Delivery 439
8.3.5 Select AM Tools and Practices for Municipalities 440
8.3.5.1 Asset Management Strategic 440
8.3.5.2 Condition Assessment 440
8.3.5.3 Defining Levels of Service (LoS) 440
8.3.5.4 Software Trends 441
8.3.5.5 Conclusion for Municipalities 441
8.4 Performance Measures 442
8.5 Developing Asset Management Plans for Water and Sewer Utilities 442
8.5.1 Water Infrastructure Asset Management 443
8.5.1.1 Stages in Water System's Asset Management 445
8.5.2 Asset Management for Water Supply Infrastructures (Dams and Reservoirs) 445
8.5.3 Infrastructure for the Water Distribution System 446
8.5.3.1 Water Treatment and Water Mains 448
8.5.3.2 Design and Construction of the Water Main System 448
8.5.4 Asset Management Programs for Stormwater and Wastewater Systems 451
8.5.4.1 Scoring Assets 452
8.5.4.2 Costs of Wastewater Infrastructures 452
8.5.4.3 Cost of Stormwater Infrastructures 453
8.5.4.4 Wastewater Program Funding 454
8.5.4.5 Case Study 2: Asset Management-Based Flood Resiliency of Water Infrastructures 455
8.5.4.6 Stormwater Program Funding 456
8.5.5 Tools for Inspecting Water and Wastewater Linear Assets 458

8.5.5.1 Underground Infrastructure ... 459
8.5.5.2 Check-Up Program for Small Systems (CUPSS) ... 459
8.5.5.3 Benefits of Using CUPSS ... 460
8.6 Financing Methods for Infrastructure Development ... 460
8.6.1 Tax-Funded System ... 460
8.6.2 Service Charge-Funded System ... 461
8.6.3 Exactions and Impact Fee-Funded Systems ... 461
8.6.4 Special Assessment Districts ... 461
8.7 Assessing the Environmental Performance of Urban Water Infrastructure ... 463
8.8 Critical Infrastructure Interdependencies ... 465
8.8.1 Restoration of Interdependent Assets ... 467
8.9 Concluding Remarks ... 467
Problems ... 468
References ... 471

Chapter 9 Urban Water Systems Analysis and Conflict Resolution ... 475

9.1 Introduction ... 475
9.1.1 System Representation and Domains ... 475
9.1.2 Water Systems Analysis ... 477
9.2 Data Preparation Techniques ... 477
9.2.1 Regionalizing Hydrologic Data ... 477
9.2.1.1 Theoretical Semivariogram Models ... 478
9.2.1.2 Kriging System ... 480
9.2.1.3 Fitting Variogram ... 482
9.2.1.4 Cross-Validation ... 483
9.2.2 Multicriteria Decision-Making ... 484
9.2.2.1 Deterministic MCDM ... 485
9.2.2.2 Probabilistic MCDM ... 485
9.2.3 Fuzzy Sets and Parameter Imprecision ... 486
9.2.4 Fuzzy Inference System ... 488
9.3 Simulation Techniques ... 489
9.3.1 Probabilistic Distribution of the System's Characteristics ... 489
9.3.2 Stochastic Processes ... 492
9.3.3 Artificial Neural Networks "Data-Driven Modeling" ... 493
9.3.3.1 The Multilayer Perceptron Network (Static Network) ... 496
9.3.3.2 Temporal Neural Networks ... 500
9.3.4 Monte Carlo Simulation ... 500
9.3.4.1 Sequential Gaussian Simulation ... 500
9.3.5 Mathematics of Growth ... 501
9.3.5.1 Exponential Growth ... 502
9.3.5.2 Logistic Growth ... 503
9.3.5.3 Limits to Growth ... 505
9.3.5.4 Environmental Limits ... 505
9.3.5.5 Social Limits to Growth ... 507
9.3.6 System Dynamics ... 508
9.3.6.1 Modeling Dynamics of a System ... 509
9.3.6.2 Time Paths of a Dynamic System ... 511
9.4 Optimization Techniques ... 514
9.4.1 Linear Method ... 515
9.4.1.1 Simplex Method ... 518

9.4.2 Nonlinear Methods ... 520
9.4.3 Dynamic Programming ... 521
9.4.3.1 Stochastic DP ... 524
9.4.3.2 Markov Chains ... 524
9.4.4 Evolutionary Algorithms ... 526
9.4.4.1 Genetic Algorithms ... 526
9.4.4.2 Simulation Annealing ... 532
9.4.4.3 Ant Colony ... 533
9.4.4.4 Tabu Search ... 534
9.4.5 Multiobjective Optimization ... 536
9.5 Conflict Resolution ... 539
9.5.1 Conflict Resolution Process ... 539
9.5.2 A System Approach to Conflict Resolution ... 540
9.5.3 Conflict Resolution Models ... 541
9.6 Game Theory and Agent Based Modelling ... 546
9.6.1 Application of Game Theory in Multi-Objective Water Management ... 546
9.6.1.1 Non-Cooperative Stability Definitions ... 549
9.6.2 Agent Based Modelling ... 554
9.6.2.1 Agent Based Modelling for Water Management ... 555
9.6.2.2 Agents and Their Characteristics ... 555
9.7 Case Study ... 557
9.7.1 Reliability Evaluation of Wastewater Treatment Plants Using MCDM Approach and Margin of Safety Method ... 557
9.7.1.1 Probabilistic Load and Resistance Reliability ... 559
9.7.1.2 Margin of Safety (MOS) Method ... 560
9.8 Concluding Remarks ... 560
Problems ... 561
References ... 566

Chapter 10 Risk and Reliability ... 571

10.1 Introduction ... 571
10.2 Design by Reliability ... 572
10.3 Probabilistic Treatment of Hydrologic Data ... 573
10.3.1 Discrete and Continuous Random Variables ... 574
10.3.2 Moments of Distribution ... 575
10.3.3 Flood Probability Analysis ... 578
10.4 Common Probabilistic Models ... 580
10.4.1 The Binomial Distribution ... 580
10.4.2 Normal Distribution ... 580
10.4.3 The Exponential Distribution ... 581
10.4.4 The Gamma Distribution ... 582
10.4.5 The Log Pearson Type 3 Distribution ... 582
10.5 Return Period or Recurrence Interval ... 582
10.6 Classical Risk Estimation ... 584
10.7 Reliability ... 585
10.7.1 Reliability Assessment ... 590
10.7.1.1 State Enumeration Method ... 590
10.7.1.2 Path Enumeration Method ... 592
10.7.2 Reliability Analysis—Load-Resistance Concept ... 595

10.7.3 Direct Integration Method....596
10.7.4 Margin of Safety....598
10.7.5 Factor of Safety....598
10.8 Water Supply Reliability Indicators and Metrics....599
10.8.1 Risk Analysis Methods and Tools....600
10.8.2 Event Tree of Risk Assessment....600
10.8.3 Environmental Risk Analysis....602
10.9 Vulnerability....603
10.9.1 Vulnerability Estimation....603
10.9.1.1 Vulnerability Assessment Tools....606
10.9.2 Risk Reduction through Reducing Vulnerability....607
10.10 Resiliency....608
10.11 Sustainability Index....609
10.11.1 Case Study 1: Uncertainty Analysis of the Water Supply and Demand Indicators....609
10.12 Uncertainty Analysis....611
10.12.1 Implications of Uncertainty....615
10.12.2 Uncertainty of Hydrological Forecasting....615
10.12.3 Measures of Uncertainty....616
10.12.3.1 Uncertain Soil Moisture (SM) Estimation....616
10.12.3.2 Flood Inundation Maps, Machine Learning (Kalman Filter), and SMAP Soil Moisture....617
10.12.3.3 Inundation Probability Map....617
10.12.3.4 Load and Resistance Concept and Probabilistic Multicriteria Decision-Making....617
10.13 Entropy Theory....618
10.14 Probability Theory—Bayes' Theorem....620
10.15 Concluding Remarks....622
Problems....622
Appendix....624
References....627

Chapter 11 Urban Water Disaster Management....629
11.1 Introduction....629
11.2 Sources and Kinds of Disasters....629
11.2.1 Drought....630
11.2.2 Floods....631
11.2.2.1 Principles of Urban Flood Control Management....631
11.2.3 Widespread Contamination....632
11.2.4 System Failure....632
11.2.5 Earthquakes....632
11.3 What Is UWDM?....633
11.3.1 Policy, Legal, and Institutional Framework....633
11.4 Societal Responsibilities....634
11.5 Planning Process for UWDM....634
11.5.1 Taking a Strategic Approach....634
11.5.2 Scope of the Strategy Decisions....634
11.5.3 UWDM as a Component of a Comprehensive DM....635
11.5.4 Planning Cycle....635
11.6 Water Disaster Management Strategies....636

11.6.1 Disaster Management—Governance Perspective 637
11.6.2 Initiation 638
11.6.2.1 Political and Governmental Commitment 638
11.6.2.2 Policy Implications for Disaster Preparedness 639
11.6.2.3 Public Participation 640
11.6.2.4 Lessons on Community Activities 640
11.6.3 Steps in Drought Disaster Management 641
11.6.4 Drought Management Case—Georgia, USA 641
11.6.5 Flood Management Case—Northern California, USA 642
11.6.5.1 Flood Characteristics 642
11.6.5.2 Response 642
11.7 Situation Analysis 642
11.7.1 Steps in the Development of Situation Analysis 643
11.7.1.1 Approach 643
11.7.1.2 Objectives 643
11.7.1.3 Data Collection 643
11.7.2 Urban Disasters Situation Analysis 644
11.8 Disaster Indices 644
11.8.1 Reliability 644
11.8.1.1 Reliability Indices 647
11.8.1.2 Mean Value First-Order Second Moment (MFOSM) Method 648
11.8.1.3 AFOSM Method 648
11.8.2 Time-to-Failure Analysis 649
11.8.2.1 Failure and Repair Characteristics 649
11.8.2.2 Availability and Unavailability 649
11.8.3 Resiliency 649
11.8.4 Vulnerability 650
11.8.5 Sustainability Index 651
11.8.6 Drought Early Warning Systems 652
11.9 Uncertainties in Urban Water Engineering 652
11.9.1 Implications and Analysis of Uncertainty 653
11.9.2 Measures of Uncertainty 653
11.9.3 Analysis of Uncertainties 653
11.10 Risk Analysis: Composite Hydrological and Hydraulic Risk 656
11.10.1 Risk Management and Vulnerability 656
11.10.2 Risk-Based Design of Water Resources Systems 659
11.10.3 Creating Incentives and Constituencies for Risk Reduction 659
11.11 System Preparedness 660
11.11.1 Evaluation of WDS Preparedness 661
11.11.2 Hybrid Drought Index 668
11.11.3 Disaster and Scale 673
11.11.4 Disaster and Uncertainty 673
11.11.5 Water Supply Reliability Indicators and Metrics 674
11.11.6 Issues of Concern for the Public 675
11.12 Water Resources Disaster 676
11.12.1 Prevention and Mitigation of Natural and Man-Induced Disasters 676
11.12.2 Disaster Management Phases 676
11.13 Other attributes of Disaster Management 677
11.13.1 Disaster and Technology 678
11.13.2 Disaster and Training 678

11.13.3 Institutional Roles in Disaster Management 679
11.14 A Pattern of Analyzing System's Preparedness 679
11.14.1 A Monitoring System for the Water Supply and Distribution Networks 680
11.14.2 Organization and Institutional Chart of Decision Makers in a Disaster Committee 680
11.15 Concluding Remarks 683
Problems 683
References 686

Chapter 12 Urban Hydrologic and Hydrodynamic Simulation 689

12.1 Introduction 689
12.2 Mathematical Simulation Techniques 689
12.2.1 Stochastic Simulation 690
12.2.2 Stochastic Processes 690
12.2.3 Markov Processes and Markov Chains 690
12.2.4 Monte Carlo Technique/Simulation 691
12.3 Artificial Neural Networks 694
12.3.1 Probabilistic Neural Network 694
12.3.2 Radial Basis Function 697
12.4 Overland Flow Simulation 699
12.4.1 IHACRES 700
12.4.2 Hydrologic Modeling System (HEC-HMS) 700
12.4.2.1 Rainfall–Runoff Simulation 701
12.4.2.2 Parameters Estimation 701
12.4.3 StormNET 701
12.4.4 HBV 702
12.4.5 Distributed Hydrological Models 702
12.4.5.1 Watershed Modeling System 702
12.4.5.2 GSSHA Model 704
12.4.5.3 LISFLOOD Model 708
12.5 Hydrodynamic (Offshore) Modeling 709
12.5.1 Physical Models 709
12.5.2 Numerical Modeling 710
12.5.2.1 Open-Source Models 711
12.6 Hydraulic-Driven Simulation Models 715
12.6.1 EPANET 715
12.6.2 QUALNET 717
12.6.3 Event-Driven Method 717
12.7 Case Studies 718
12.7.1 Case Study 1: DEM Error Realizations in Hydrologic Modeling 718
12.7.1.1 Methodology 718
12.7.1.2 Results 719
12.7.2 Case Study 2: Simulation of Ungagged Coastal Flooding—Nearshore and Inland BMPs 719
12.7.2.1 Area Characteristics 720
12.7.2.2 Methodology 720
12.7.2.3 Results 723
12.7.2.4 Concluding Remarks 723

12.7.3 Case Study 3: Infrastructure Flood Risk Management—MCDM-Based Selection of BMPs and Flood Damage Assessment ... 724
12.7.3.1 Methodology ... 725
12.7.3.2 Results ... 727
12.7.3.3 Case Study Concluding Remarks ... 729
12.8 Summary and Conclusion ... 729
Problems ... 730
Appendix ... 739
References ... 748

Chapter 13 Flood Resiliency of Cities ... 751

13.1 Introduction ... 751
13.2 Setting the Stage—Flood Types and Formations ... 752
13.2.1 Inland and Coastal Flooding ... 752
13.2.1.1 Inland Flooding ... 752
13.2.1.2 Coastal Flooding ... 753
13.3 Flood Analysis ... 762
13.3.1 Flood Time Series ... 762
13.3.1.1 Peaks Over Threshold Series ... 763
13.3.2 Partial Frequency Analysis ... 763
13.3.2.1 Stationary Analysis ... 763
13.3.2.2 Non-stationary Analysis ... 765
13.3.2.3 Ungagged Flood Data ... 766
13.3.3 Testing Outliers ... 769
13.4 Flood Recurrence Interval ... 771
13.5 Flood Routing ... 772
13.5.1 Storage-Based Routing ... 772
13.6 Urban Floods ... 773
13.6.1 Urban Flood Control Principles ... 773
13.7 Understanding Flood Hazards ... 774
13.7.1 Climate Change and Flooding ... 774
13.7.2 Sea Level Rise and Storm Surge ... 775
13.8 Evacuation Zones ... 775
13.9 Interdependencies Role on Water Infrastructure Performance ... 777
13.9.1 Resiliency of New York City's Wastewater System ... 778
13.10 Flood Damage ... 780
13.10.1 Stage-Damage Curve ... 780
13.10.2 Expected Damage ... 781
13.10.2.1 Case Study 1: Coastal Flood Damage Estimator: An Alternative to FEMA's HAZUS Platform ... 783
13.11 Flood Risk Management ... 787
13.11.1 Resiliency and Flood Risk Management ... 788
13.11.1.1 Wastewater Treatment Plants of New York City ... 790
13.11.1.2 Case Study 2: Prioritizing Investments in Improving Flood Resilience and Reliability of Wastewater Treatment Infrastructure ... 791
13.12 Floodplain Management ... 794
13.12.1 Structural and Nonstructural Measures ... 794
13.12.2 BMPs and Flood Control ... 795

13.12.2.1 Case Study 3: Integration of Inland and Coastal Storms for Flood Hazard Assessment Using a Distributed Hydrologic Model 795
13.12.2.2 Case Study 4: Nonstationary-Based Framework for Performance Enhancement of Coastal Flood Mitigation Strategies 798
13.12.2.3 Case Study 5: Conceptual Design Framework for Coastal Flood Best Management Practices 800
13.12.2.4 Case Study 6: Improvement of Urban Drainage System Performance 803
13.12.3 Watershed Flood Early Warning System 806
13.12.4 Flood Insurance 807
13.13 Livable Cities of the Future 809
13.13.1 Mayor's Office Point of View 809
13.13.2 Infrastructure Renewal: An Agency View Point 810
13.13.2.1 The Big Picture 810
13.13.2.2 The Critical Role of Transportation 810
13.13.2.3 Sustainable Urban Renewal 810
13.13.2.4 Energy as the Core of NYC 811
13.13.2.5 Using Water for Urban Renewal 811
13.13.3 Fighting Climate Change—A Former Mayor View Points 811
13.13.4 Urban Challenges: The Way Forward—An IT Expert View Points ... 812
13.13.4.1 The Challenges 812
13.13.4.2 The Path to Success 812
13.14 Concluding Remarks 813
Problems 813
Appendix 818
References 819

Chapter 14 Environmental Visualization 823
14.1 Introduction 823
14.1.1 Sensed Water Infrastructure 823
14.2 Environmental Sensing 824
14.2.1 Introduction 824
14.2.2 Ubiquitous Environmental Sensor Technologies 825
14.2.3 Remote Sensing and Earth Observation 826
14.2.4 Soil Moisture Active Passive (SMAP) 828
14.2.4.1 Step-by-Step Procedure to Download SMAP Data for a Specific Date at a Specific Location 829
14.3 Pattern Recognition 830
14.3.1 Introduction 830
14.3.2 Parameter Estimation 832
14.3.2.1 Moments Method 832
14.3.2.2 Maximum Likelihood (MLE) 833
14.3.2.3 Maximum Posteriori (MAP) 835
14.3.2.4 Nonparametric Density Estimation—Parzen Method 836
14.3.3 Feature Extraction and Selection 841
14.3.3.1 Backward Elimination 841
14.3.3.2 Forward Selection 842
14.3.4 Discrimination Analysis 842

14.3.4.1 Fisher Discriminant Analysis (FDA) 842
14.3.4.2 Linear Discriminant Analysis 843
14.3.4.3 Principal of Component Analysis (PCA) 844
14.3.5 Supervised Classification 851
14.3.5.1 Bayes Decision Theory 851
14.3.5.2 Density Function Estimation 852
14.3.5.3 Parametric Density Estimation 852
14.3.5.4 k-Nearest Neighbor Estimation (k-NN) 853
14.3.5.5 Min-Mean Distance Classification 856
14.3.5.6 Support Vector Machines (SVM) 857
14.3.6 Unsupervised Classification (Clustering) 857
14.3.6.1 Sequential Clustering 857
14.3.6.2 Optimization Based Clustering 858
14.3.7 Image Processing 859
14.3.7.1 Image RGB Analysis 860
14.4 Data Assimilation 862
14.4.1 Kalman Filter (KF) 863
14.4.2 VIC Model Application with Data Assimilation 863
14.5 Environmental Visualization Attributes 866
14.5.1 Intelligent Visualization and Image Analysis Systems 866
14.5.2 Water-Related Environmental Visualization 867
14.5.3 Visualization to Control and Reduce Flood Risk 870
14.5.4 Environmental Effects and Protection—Signatures and Symbols 872
14.5.5 SLP as a Mean for Wet Front/Storm Movement 874
14.6 Map Resolution 876
14.6.1 DEM Resolution 876
14.6.2 Digital Terrain Model 878
14.6.3 Quality and Accuracy of DEM/DTM 878
14.6.4 Digital Surface Model (DSM) 878
14.6.5 Common Uses of DEMs 878
14.6.6 Effect of Map Resolution on Modeling 879
14.6.7 Kriging Interpolation 879
14.6.8 Variogram Modeling 880
14.6.9 Resampling 883
14.6.10 DEM Error 884
14.7 Case Studies 885
14.7.1 Case Study 1: A Satellite/Citizen Science-Based Soil Moisture Estimator 885
14.7.1.1 Clustering for Soil Moisture Applications 886
14.7.1.2 Data Estimation Platform—Error Analysis 886
14.7.1.3 Study Area and Data Collection 886
14.7.1.4 Digital Image Processing 889
14.7.1.5 Cross-Validation of Citizen Science with Satellite Data 889
14.7.2 Case Study 2: Data Assimilation for Flood Assessment 890
14.7.2.1 Soil Moisture Estimation Results 892
14.7.2.2 RVIC Model Results 893
14.8 Concluding Remarks 893
Problems 895
References 896

Index 903

Preface

This book has been assembled with the intent of exploring various theoretical, practical, and real-world applications of systems analysis, design, and planning of urban water infrastructures. Review and inclusion of some of the recent research done in each area has also been covered. Fourteen chapters of this book are organized in an integrated fashion in order to be used as a whole package, while each chapter can be utilized independently. Four chapters have been dedicated to background information regarding water science and engineering focusing on urban challenges. Subsequently, another three chapters cover stormwater, water supply, and wastewater and related infrastructures, followed by four chapters discussing a wide range of novel topics ranging from water assets management, water economics, systems analysis techniques, risk, reliability, and disaster management. Finally, modeling, flood resiliency, and environmental visualization chapters are a compilation of tools and emerging techniques that elevate us to a higher plateau in water systems' assessment. Another key feature of this book is the taking into account of the critical emerging urban and coastal issues such as satellite applications, citizen science, and digital data model (DEM) advancements in water-related issues.

We should ask what is precisely unique about the contents of this book that could make it distinctive. To address that, it could plausibly be argued that there are very few textbooks available containing such different and comprehensive teaching materials on urban water infrastructure. Furthermore, this book's uniqueness has been assured through: (a) transparency, (b) technical soundness, (c) containment of exciting materials, (d) practicality, and (e) being forward-looking.

These characteristics are further discussed in Chapter 1. The topics cover the state-of-the-art emerging technologies available and the main challenges we are facing such as satellite and digital data evolution, growing challenges of risk and uncertainty, and disaster preparedness.

It is emphasized that water conservation, better systems' operation, higher end use standards, and water allocation efficiencies are still the main instruments to offset the growing demand. But they are perhaps not enough for many societies that are struggling to bring supply and demand as well as storm and wastewater management to a sustainable level. More vigilant approaches are needed combined with political will to identify champions for water management in every region that faces scarce water and financial resources. Following some of the more recent initiatives in this book for water reuse, asset management, and continuous planning and performance evaluation equipped with the latest monitoring and environmental visualization techniques could be instrumental for bringing true water sustainability and resiliency to communities and for the livable cities of the future.

The materials covered in different chapters are described in a systematic and integrated fashion that are useful for undergraduate and graduate students and practitioners as follows:

Chapter 1 provides an introduction to different aspects of water science, engineering, planning, and management. It discusses the concepts of integration, sustainability, and public participation with an emphasis on laws, regulations, and public participation, and a brief reasoning on the absence of water governance in this book. Three distinct pillars for analysis, design, and planning are presented in this chapter: urban water cycle and variability as the state of water being; landscape architecture as the medium for built by design; and total systems as the planning approach.

Chapter 2 presents an overview of water balance in the hydrologic cycle, and interactions of climatic, hydrologic, and urban components are discussed. Losses or abstractions in hydrology (evaporation, evapotranspiration, and infiltration) are covered in detail in this chapter including physics-based methods such as Palmer drought severity index (PDSI). Inclusion of PDSI is particularly interesting as it is a widely used index to determine the state of water availability that agencies are relying on for short- and long-term planning. A case study of water balance-based water supply and demand sustainability is also included. The need for a simultaneous change in the cognitive, normative, and regulative conditions of the urban water management regime for sustainability transition is emphasized. Some attributes of livable cities of the future with emphasis on water drivers are also presented.

Chapter 3 discusses runoff and water accumulation quantity analysis and related issues such as excess rainfall estimation, rainfall–runoff analysis, calculation of peak flow, and its occurrence time as well as hydrograph analysis. Overland flow and water conveyance in urban area are more complicated than in undeveloped area because of the complexity of the paths that rain and stormwater are taking. Interdependencies of water infrastructure with other infrastructures make the urban hydrology and extreme value analysis more complicated and critical. Design values are key to many urban development issues and are analyzed by frequency analysis and the use of probability distributions. Time series analysis and modeling are discussed within a limited scope applicable to urban area. This chapter's unique feature is that it provides all the background needed on science and applied techniques in hydrology in general and urban water in particular to handle the engineering and planning issues presented in the other chapters.

Chapter 4 presents the technical aspects of developing and solving governing urban water hydraulics principles. Measuring the effect of geomorphic changes (land use driven) on urban water is essential to mitigate the potential effects. The urban water management is more effective if risk assessments included geomorphological changes to underpin nature-based management approaches. The land use alterations are addressed. Changes in geomorphic process regimes can also be triggered by extreme events. Implementing geomorphological adaptation strategies will enable communities to develop more resilient, less vulnerable socioeconomic systems fit for an age of climate extremes. The outcome of such approach will be of interest to landscape architects/planners and regulators because of the complexities related to stormwater collection and flood management. In this chapter, these issues and some basic concepts of hydraulic design of urban drainage and water distribution systems are introduced.

Chapter 5 describes different aspects of the design and planning of urban drainage systems. This chapter is particularly useful for developing urban areas. Many street and highway drainage issues and requirements are discussed that transportation engineers and contractors can effectively utilize. A dedicated and perhaps unique feature of this chapter is the inclusion of detailed land use planning, which is the essence of Integrated Urban Water Management, based on DSR (Driving force, State, and Response) dynamic strategy planning procedure. The following issues are also addressed: special characteristics of the urban storm; the complexity of urban watersheds; imperviousness and the maze of water pathways/channels; local ordinances and risk-based design values; streets/highways drainage (street gutters, stormwater inlets, and storm sewers); control structures/best management practices (BMPs) (with emphasis on green solutions); and urban flood, combined sewer overflow (CSO), and interdependencies (water, energy, and transportation). Perhaps the key to many growing urban stormwater management challenges is the lack of appropriate land use planning and excessive human intervention that exuberate the extreme events to disasters. Sections 5.3–5.7 are particularly useful for transportation and highway engineers and contractors.

In *Chapter 6* different components of urban water supply infrastructures have been presented and their interactions among UWC have also been discussed. Basic concepts on water storage and supply facilities are considered. Reservoir operation and supply and demand management are discussed. Some basic planning issue of the urban water supply infrastructures has also been discussed in this chapter. SIM (Structural Integrity Monitoring) of water mains in order to reduce their repair, rehabilitation, and replacement costs/issues and increase the reliability of supplying the demands are discussed. This requires monitoring the use of various methods for detecting leaks and predicting the impacts of alternative urban water systems on the life cycle including operation, maintenance, and repair policies of these systems. The relationship of head/pressure, discharge, and leakage as well as a brief head-driven modeling are presented for water distribution systems. Economic and financial analysis plays an important role in infrastructure development. Some examples are given. Environmental performance is investigated in the context of sustainable development. The interactions of systems' components are not often well integrated in the design, construction, rehabilitation, and maintenance of these systems. Inclusion of many elements of water supply system in this chapter could help the potential reader to better comprehend both analysis and design as well as operational issues of this vital lifeline.

Chapter 7 presents different issues and challenges in urban wastewater management. Wastewater treatment, an essential factor in urban wastewater planning, is discussed and followed by a detailed description of each part of the treatment plant. Both quantitative and qualitative aspects of wastewater treatment are addressed in this chapter. Then the concepts of wastewater planning are introduced. Afterward, the case studies related to this chapter as well as some selected cities' wastewater infrastructures are described. Many other case studies in other chapters are also related to wastewater treatment plants' vulnerability to coastal flooding. A good number of wastewater treatment facilities are located near water bodies and are prone to system failure with considerable consequences for the communities in their sewershed and the environment as the CSO impacts are getting more severe. Some planning and standard issues related to water utilities in general and wastewater are covered in the later part of this chapter. The main challenge of wastewater systems around the globe is in design, construction, operation, maintenance, funding, and standards of services. As these facilities are aging in many cities, the issue of the needed funds for rehabilitating or replacing them is in forefront of challenges many municipalities are facing.

Chapter 8 demonstrates how asset management (AM) can facilitate infrastructure performance corresponding to service targets over time. It helps to make sure risks are adequately managed, and that the corresponding costs of the life cycle are minimized. Lack of sound economic, regulatory frameworks and enforcement setup, and poor asset management practices, particularly underpriced water services, are common problems throughout the developing regions. The urban systems' governing bodies need plans to prioritize limited resource allocation. Therefore, this allocation should be in line with the current structure or reformed infrastructure asset management. In AM, effective decision-making requires a comprehensive approach that ensures the desired performance at an acceptable risk level, considering the costs of building, operating, maintaining, and disposing of capital assets over their life cycles. Sustainable management of the system's resources should respond to the growing need for financial stability and sound cash flow and investment strategies in capacity expansion and resource generation.

Chapter 9 discusses the system representation and domains with the essence of water system analysis. Data preparation and processing techniques are described followed by multicriteria decision-making (MCDM). Then, data-driven neural networks and fuzzy inference are introduced. Furthermore, mathematics of growth as a basis for systems dynamics is presented. Conventional and evolutionary optimization techniques are also introduced. Finally, the conflict resolution in the context of Nash bargaining theory, game theory, and agent base modelling is described. For considering integration and sustainability in water resources management, it is necessary to think over the social, economic, and environmental impacts of decisions. This integration in planning and management, especially in urban areas, needs a systematic approach, considering all the interactions among the elements of the system and with the outside world. Simulation of urban water dynamics will give the collective impacts of all possible water-related urban processes on issues such as human health, environmental protection, quality of receiving waters, and urban water demand.

Chapter 10 displays the topics needed to cover the objective of placing design and analysis of infrastructure into a reliability framework. To do that, the concept of probability, basic statistics, common probabilistic models; extreme value (flood) and frequency analysis—design values; and basic concept of risk and uncertainty are discussed. Then reliability in the context of serial and parallel components as well as load and resistant concept are described. The performance indicator with emphasis on vulnerability and resiliency is presented including a case study. The basis of uncertainty analysis is covered with a summary of how error and uncertainties have been quantified in the case studies of this book. The entropy theory including transinformation (measure of redundancy in information) is also discussed in the context of water resources issues to measure the information content of random variables and models, evaluate information transfer between hydrological processes, evaluate data acquisition systems, and design monitoring networks. The materials covered in this chapter allow a realistic assessment of how to characterize and manage risk. The adaptation and mitigation issues are further discussed in other chapters' case studies.

Chapter 11 provides information related to the nature of water disasters and the factors contributing to the formation and extent of changes caused by a disaster. The notion of water hazard (including water scarcity) as a "load" and our ability to withstand it as a "resistance" is discussed in the context of reliability and risk-based design. The elements of uncertainty and how risk management can be coupled with disaster management are presented. This chapter is divided into 11 sections with the following focus areas: First, an introduction to UWDM is presented. Then the planning process for UWDM is presented, followed by situation analysis, disaster indices, and risk and uncertainty elements of disaster. Guidelines for UWDM and preparedness planning are also presented.

Chapter 12 displays data-driven mathematical models' applications with accentuation on new information-based and machine learning models. In the subsequent part, various kinds of physical-based hydrologic models of rainfall–runoff including lumped models, semidistributed models like IHACRES (acronym for Identification of unit Hydrographs and Component flows from Rainfall, Evaporation, and Streamflow data), StormNET, and HBV (Hydrologiska Byråns Vattenbalansavdelning) are briefly discussed. Distributed models with applications of HEC-HMS (Hydrologic Modeling System), Gridded Surface/Subsurface Hydrologic Analysis (GSSHA), and LISFLOOD are presented that are widely used in urban flooding applications. A hydrodynamic model of Delft3D with two modules of FLOW and WAVE and hydraulic-driven models such as EPANET and QAULNET are also presented. The three case studies presented show how these models can help the engineers and the decision makers to better prepare for water-related incidents such as inland and coastal floods. There are many other applications of these data-driven and system dynamics models throughout this book, especially in Chapters 5, 6, 9, and 13.

Chapter 13 initiates flood resilient city that includes building smart communities, tools, models, and data processing and information management; flood hazard characterization and warning apparatus; inner and other links and interdependency characterization; infrastructure risk; sound asset and financial management; and performance measures. A number of these issues are discussed in this chapter and elaborated in the case studies. At the end, a gap analysis section discusses the remaining challenges and looking-forward perspective for a livable city of the future. The historical floods and real-world flood problems of water infrastructures which apply to one of the most crucial systems, New York City's WWTPs, are discussed. Factors contributing to flood hazards, evacuation zones, resiliency and flood risk management are also presented. The application of resiliency concept and how it can be used as a metric for performance evaluation and resource allocation are explored through a number of case studies.

Chapter 14 presents the significance of visualization for information representation. Its functionality to serve as a storage mechanism, a processing and research instrument, and a communication tool is increasing with the past of advances in optic and imagining technologies. Visualization technology is capable of transferring information into a simple image or animation. Environmental visualization (EV) is one of the hottest areas that many imaging technologies have been realized and utilized with many more potential for ground-breaking new developments. The generated image is typically a compilation of hundreds of pages of information from large and cumbersome reports. As a result, visualization acts as a data management tool that collates, organizes, and displays large volumes of information. It has special application in disaster management and flood inundation that can be utilized by a variety of users with low to very high technical capabilities. The water movement and accumulation and water infrastructures need to be continuously monitored. With the materials presented in this chapter ranging from sensors, to pattern recognition tools and techniques, to satellite technology and data collections almost all in imaging forms, to development of citizen science applications, we have realized many immerging opportunities. The future of water recourses assessment, protection, water hazard prevention, and effective visual communication lies on the advancement of environmental visualization and how we are prepared to utilize it.

This book incorporates feedback from my students in water system analysis courses and from my collaborators of many research and real-world projects in the national and international arena in the past 30 and more years. It is my hope that this book can add significant value to the application of systems analysis and design techniques for water infrastructure planning around the world.

Mohammad Karamouz
Tehran, Iran—Great Neck, New York
December 2021

MATLAB® is a registered trademark of The MathWorks, Inc. For product information, please contact:

The MathWorks, Inc.
3 Apple Hill Drive
Natick, MA 01760-2098 USA
Tel: 508-647-7000
Fax: 508-647-7001
E-mail: info@mathworks.com
Web: www.mathworks.com

Acknowledgments

Many individuals have contributed to the preparation of this book. The initial impetus for this book was provided by Professors Mark H. Houck and Jacques W. Delleur during the author's career at Purdue University. The authors acknowledge the significant contribution of Elham Ebrahimi, a PhD candidate, and Mohammad Reza Zare and Mohammad Movahhed, MS students at the University of Tehran. Dr. Karamouz's former PhD graduates, Drs. M.A. Olyaie, M. Fereshtehpoor, and D. Mahmoodzadeh reviewed and commented on several chapters. Many former and current graduate students in this university who contributed to different chapters of this book and who attended the author's System2 and System analysis and Planning of water infrastructure courses at the University of Tehran and urban systems management course at Pratt Institute and Polytechnic Institute of NYU were driving forces for publishing this book through completion of the class notes that formed the basis for and the significant extension, revision, and improvements of systems analysis components of first author's system analysis and other text books. In particular, the assistance of R. S. Alipoor, S. Mahmoodi, A. Zoghi, and F. Fooladi Mahani in Chapters 2, 12, 13, and 14, respectively, is hereby acknowledged; H. Meydani, M. Roohinia, and E. Elyasi helped with verifying the example problems. The seclusion experienced as a result of the COVID-19 pandemic provided the opportunity to expedite the formation and completion of this book. The artwork was done by the graduate students listed above.

Author

Mohammad Karamouz is a professor at the University of Tehran. He is an internationally known water resources engineer and consultant. He is licensed as a PE in the state of New York since 1985. He is the former dean of engineering at Pratt Institute in Brooklyn, New York. He is also a fellow of the American Society of Civil Engineers (ASCE) and a diplomat of the American Academy of Water Resources Engineers. Dr. Karamouz received his BS in civil engineering from Shiraz University, his MS in water resources and environmental engineering from George Washington University, and his PhD in hydraulic and systems engineering from Purdue University. He served as a member of the task committee on urban water cycle in UNESCO-IHP VI and was a member of the planning committee for the development of a 5-year plan (2008–2013) for UNESCO's International Hydrology Program (IHP VII). Among many professional positions and achievements, he also serves on a number of task committees for the ASCE. In his academic career spanning 35 years, he has held positions as a tenured professor at Pratt Institute (Schools of Engineering and Architecture in Brooklyn) and at Polytechnic University (Tehran, Iran). He was a visiting professor in the Department of Hydrology and Water Resources at the University of Arizona, Tucson, 2000–2003, and a research professor and Director of Environmental Engineering Program at Polytechnic Institute of NYU, 2009–2014. He was also the founder and former president of Arch Construction and Consulting Co. Inc. in New York City. His teaching and research interests include integrated water resources planning and management, flood resilient cities, groundwater and surface water hydrology and pollution, systems analysis and design, urban environmental systems management, DEM and Satellite data error analysis and downscaling, environmental visualization including image processing, pattern recognition, and data assimilation. He has more than 350 research and scientific publications, books, and book chapters to his credit, including four text books: *Water Resources System Analysis* published by Lewis Publishers in 2003 *Urban Water Engineering and Management* (2010), and *Hydrology and Hydroclimatology* (2012), and *Groundwater Hydrology* (1st ed. 2011 and 2nd ed. 2020) published by CRC Press. He also coauthored a book entitled *Urban Water Cycle Processes and Interactions* published by Taylor & Francis Group in 2008, and was the lead editor of the book *Livable Cities of the Future* (2014) published in collaboration with the National Academy of Engineers (NAE) by the National Academies Press in Washington, D.C. Professor Karamouz serves internationally as a consultant to both private and governmental agencies, such as UNESCO and the World Bank. In 2017, he received the year distinguished researcher award of the University of Tehran. Dr. Karamouz is the recipient of the 2013 ASCE Service to the Profession and 2018 ASCE Arid Land Hydraulic Engineering Awards.

During my academic career as a professor and in conviction of my personal life, I have received help and encouragement from so many people that it is not possible to name them all. To all of you, I express my deepest thanks. There were a few including some colleagues and former students that have tried to stop or discourage me. I am grateful to them too because without them I could not be so determined to improvise. Water resources system analysis has been a part of a personal journey that began years ago when I was a young boy with a love for water. Books are companions along the journey of learning, and I hope that you will be able to use this book in your own exploration of the field of water resources. Have a wonderful journey.

1 Introduction

1.1 INTRODUCTION

This book has been assembled with the intent of exploring various theoretical, practical, and real-world applications of system analysis, design, and planning of urban water infrastructures. Review and inclusion of some of the recent research done in each area has also been covered. To that end, 14 chapters of this book are organized in an integrated fashion in order to be used as a whole package, while each chapter can be utilized independently. Four chapters have been dedicated to background information regarding water science and engineering focusing on urban challenges. Subsequently, another three chapters cover stormwater, water supply, wastewater, and related infrastructures, followed by four chapters discussing a wide range of novel topics ranging from water assets, water economics, systems analysis, risk, reliability, and disaster management. Finally, modeling, flood resiliency, and environmental visualization chapters are a compilation of tools and emerging techniques that elevate us to a higher plateau in water systems' assessment. Another key feature of this book is the taking into account of the critical emerging urban and coastal issues such as satellite applications, citizen science, and digital elevation model (DEM) advancements in water-related issues.

One might think what is precisely unique about the contents of this book that could make it distinctive. To address that, it could plausibly be argued that there are very few textbooks available containing such different and comprehensive teaching materials on urban water infrastructure. Furthermore, this book's uniqueness has been assured through: (a) transparency, (b) technical soundness, (c) containment of exciting materials, (d) practicality, and (e) being forward-looking. Each of the mentioned attributes is further elaborated as follows:

a. *Transparency*: Arguments made in this book are supported by latest journal articles' citations and then illustrated through example problems. They are easy to follow and flow in a logical order. Chapters have a clear organizational structure with an introduction and a concluding remarks sections that sum up the most significant points of the chapter.
b. *Technical soundness*: Much experience and feedback have been derived from the last four textbooks of the author and have been put to use in gatherings in this book. The scientific phenomena discussed in this book are professionally based and have been tested in the included case studies.
c. *Excitement*: Each chapter has a distinctive characteristic and attributes ranging from science to engineering to a planning voyage, urban lifeline characteristics, system thinking, and dynamics all the way to asset/value-driven goals, to resiliency, pattern recognition, and environmental visualization.
d. *Practicality*: Targeting some of the most pressing real-world challenges through a practical approach by means of case studies that include engineering judgments and practice attributes.
e. *Forward-looking*: Covering the state-of-the-art technologies available and the emerging challenges topics such as satellite and digital data evolution; growing reliance on risk and uncertainty based solutions; and disaster preparedness planning.

Water and science of water (hydrology), urban water movement and services, infrastructures, and institutional supports are our domain in this book. Spatial and temporal variability; and social,

DOI: 10.1201/9781003241744-1

environmental, and economic issues are the states. The driving forces are natural hazards; human and anthropogenic interventions; water lifeline services; and health, safety, and preparedness attributes for planning purposes.

A watershed is the best hydrological unit that can be used to carry out water studies and planning. In urban areas, the term sewershed is often used that has watershed characteristics with man-made drainage elements. The urban setting alters the natural movement of water. Drastic land use changes in urban areas are a subset of urban and industrial development affecting natural landscapes and the hydrological response of watersheds. Although anthropogenic factors concerning waterways, pipes, abstractions, and built environment affect the elements of the natural environment, the main characteristics of the hydrological cycle remain the same in urban areas but are significantly altered by urbanization impacts of the services to the urban population, such as water supply, drainage, and wastewater collection and management.

As a conceptual way of looking at water balances in urban areas, the context of the urban water cycle is a total systems approach of natural, human and institutional, and built environment systems (see Section 1.10 for more details). Water balances studies are generally conducted on a different time scale, depending on the type of applications in a planning horizon. Among the planning objectives, water has to be distributed to growing populations and communities should be prepare to cope with storms from extreme weather and climatic variability and potential climate change.

Wastewater collection and treatment with a biological operation unit (which is climate sensitive) plays an essential role in a city's daily operation with many external elements such as stormwater that could threaten its safe operation by causing sewer overflow. Along with water supply and distribution, these threefold water-related services constitute a valuable asset for a city. Assets that are often poorly managed with their state of operation and maintenance. They should be consciously evaluated and planned to face/reduce the risk of failures as the natural disasters are getting more frequent, the customer dependencies are higher, and there are growing interdependencies with other infrastructures. These systems can be simulated with different models such as data-driven models that are subject to input, parameters, and model structure uncertainties. The climate-induced hazards and the sheer size of water systems, which are subject to many interdependencies and uncertainties, have brought new paradigms to measure performance and a new metric for resource allocation. It is called resiliency. In a number of case studies, it is demonstrated how resiliency is being used as a new norm for performance evaluation. Finally, new imaging and digital/satellite data technologies combined with pattern recognition and its machine learning attributes have brought new opportunities for utilizing many environmental visualization techniques. These emerging issues and opportunities have been realized through applications such as water-related land use and landscape, water and soil interactions, and flood hazard mitigation in this book. In the remaining of this chapter, three distinct pillars for analysis, design, and planning are presented: urban water cycle and variability as the state of water being; landscape architecture as the medium for built by design; and total system as the planning approach.

Water governance is not discussed here as the focus of this book is more on tools and techniques and less on institutional and administrative aspects of water infrastructure management. There are so much variability on the past practices, politics, organizational structure, even cultural aspects of water governance that is too difficult and sometimes too impractical to find and prescribe a good governance model for a region. We should hope that through the exercise of sound analysis and planning of water systems described in this book, the regional evolution of water governance happens. See Chapter 2 of Karamouz et al. (2010) on the water governance.

1.2 URBAN WATER CYCLE

The study of any phenomenon, process, and/or structure, such as climatic systems in urban areas or the urban water cycle, requires an integrated approach. The selected framework of study dictates the way an environment is dismantled for analysis. It also determines how to integrate the environmental components so that the analysis results can be incorporated holistically. For providing such a framework, an understanding of system concepts is needed.

Furthermore, the interactions and the variability in physical, chemical, and biological processes in and among urban system components need to be addressed. This understanding and knowledge can also be used to develop prediction and early warning systems widely used concerning the behavior of different components of the hydrologic cycle in general and urban water cycle in particular.

1.2.1 Components

Changes in the material and energy fluxes and the amount of precipitation, evaporation, and infiltration in urban areas result in changes in water cycle characteristics. The impacts of large urban areas on local microclimate have long been recognized and occurred due to changes in the energy regime such as air circulation patterns caused by buildings, transformation of land surfaces and land use planning, water transfer, waste generation, and air quality variations. These changes, which are depicted in Figure 1.1, can be summarized as follows:

- *Land use*: The transformation of undeveloped land into urban land, including transportation corridors
- *Demand for water*: Increased demand because of increased concentration of people and industries in urban and nearby suburban areas
- *Increased entropy*: The redundant use of unsustainable forms of energy
- *Waste production*: Substantial waste and industrial hazardous wastes and decreasing quality of different resources such as air, water, and soil
- *Water and food transfer*: Moved from other places to urban areas

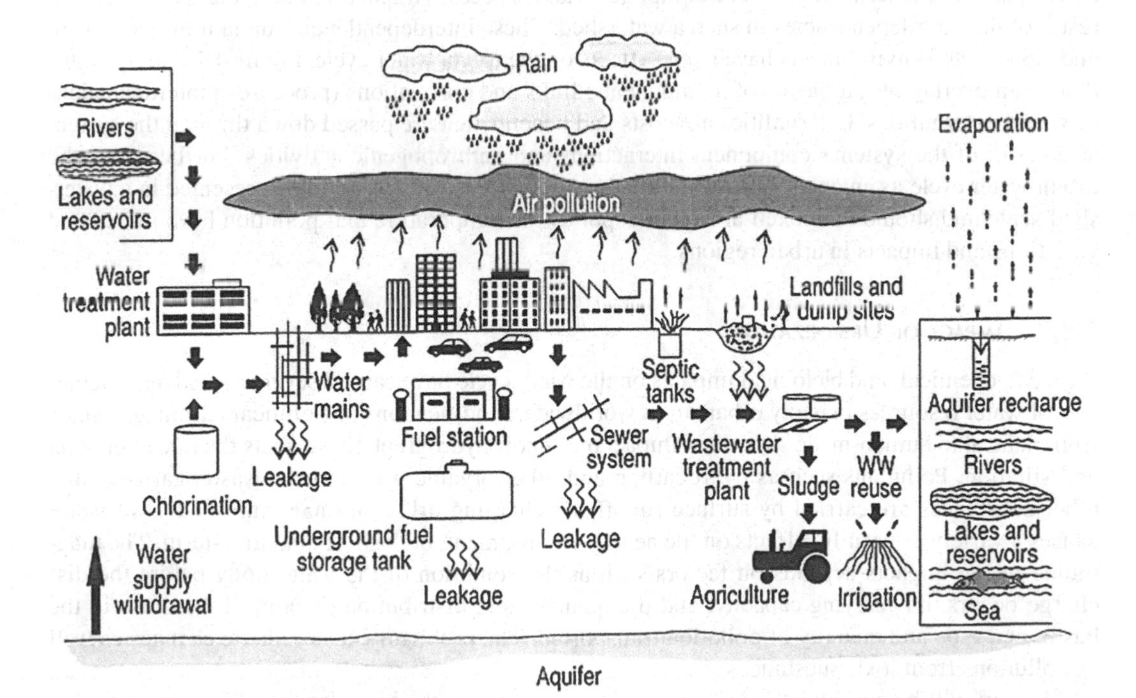

FIGURE 1.1 Changes in natural process of the hydrologic cycle due to urbanization. (From Marsalek, J., Cisneros, B.J., Karamouz, M., Malmquist, P.A., Goldenfum, J.A., and Chocat, B.. *Urban Water Cycle Processes and Interactions: Urban Water Series-UNESCO-IHP* (Vol. 2). CRC Press, Boca Raton, FL, (2008).)

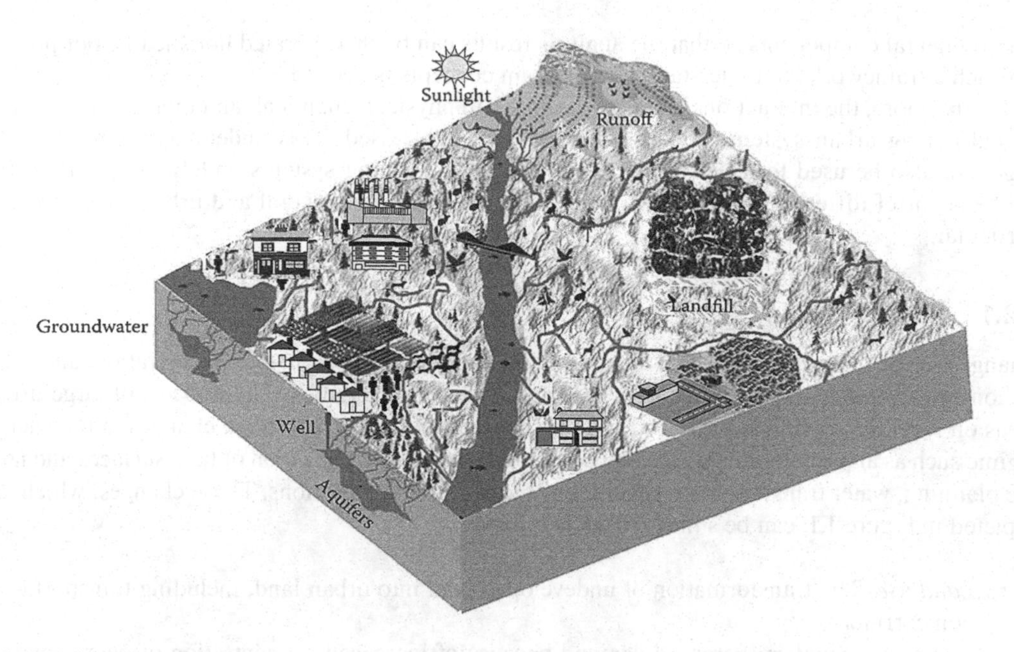

FIGURE 1.2 Interdependencies in natural, biological, and constructed environments in urban watershed scale.

1.2.2 Interdependencies

The urban water cycle may be depicted as a system on a watershed with varied land uses to show how it is impacted by external forces. The components and processes impacting this cycle are altered as a result of the interdependencies in such a watershed. These interdependencies in natural, biological, and constructed environments have ripple effects on the urban water cycle. Figure 1.2 demonstrates that as an overlay of four layers of: components; links and correlations (processes); interdependencies; and externalities. Externalities are costs and benefits that are passed down through the system as a result of the system's component interactions and anthropogenic activities. Furthermore, the urban water cycle as an open vapor/water/matter circulation system should be presented in a watershed scale and should be looked at as a hot spot as the temperature and pollution have intensified variations and impacts in urban regions.

1.2.3 Impact of Urbanization

Physical, chemical, and biological impacts on the water cycle have caused severe and adverse depletion of water resources in many urban areas worldwide. Modifications of significant drainage canals from natural to human-made structures impact the runoff hydrograph that affects the rate of erosion and siltation. Pollutants such as hydrocarbon and other organic wastes, food waste, garbage, and other substances are carried by surface runoff. Discharging urban drainage into bodies of water causes a variety of harmful effects on the nearby environment, both short- and long-term. The magnitude of the impacts depends on factors such as the condition of the water body before the discharge occurs, its carrying capacity, and the quantity and distribution of rainfall, land use in the basin, and type and quantity of pollution transported. The problems cause esthetic changes as well as pollutions from toxic substances.

The soil will be consolidated in metropolitan areas due to the high density of houses and other structures. As a result, soil porosity declines, reducing the quantity of water that can be held and released from urban aquifers. Another impact of urbanization on urban aquifers is the decreasing recharge to the aquifer because of decreasing infiltration. Certain urban areas with many absorption wells are subject to a high rate of wastewater infiltration into the aquifer. Groundwater resources have been polluted by urban activities in many cities of the world. For example, landfill leakages and leakage from wastewater sewers and septic tanks and gas stations are the most well-known point sources that cause groundwater pollution in urban areas.

The hydrological response of urban regions is determined by calculating surface runoff hydrographs, which are then routed via the drainage network's conduits and channels to create outflow hydrographs at the urban drainage outlets. In various areas of an urban region, the physical parameters of a catchment basin in terms of the rainfall–runoff process might change dramatically. The unit hydrograph is a traditional means of representing linear system response. However, it suffers from the limitation that the response function is lumped over the whole catchment and does not explicitly account for spatially distributed characteristics of the catchment's properties. Drought and flood severity and their impacts on urban areas are more significant. After urbanization, the peak of the unit hydrograph increases and occurs earlier, and the flooding condition is more severe. Also, because of high water demand, urban areas are more vulnerable during hydrological drought events. Due to the high rate of water use in urban areas, social, political, and economic issues related to water shortage in urban areas are intensified.

1.3 INTERACTION OF CLIMATIC, HYDROLOGIC, CULTURAL AND ESTHETIC ASPECTS

1.3.1 Climatic Effects—Rainfall Type

The interaction between large urban areas and local microclimate has long been recognized and occurred due to changes in the energy flux, air pollution, and air circulation patterns caused by buildings, land transformations, and the release of greenhouse gases. These factors contribute to changes in the radiation balance, precipitation, and evaporation, and consequently changes in the hydrologic cycle.

- Convective rainfall is more frequent, with high intensity and a short duration of time, covering small areas. This type of rainfall is more critical for an urban basin with a short time of concentration (high flow velocity due to gutter and pipe flows) and a small catchment area.
- Long periods of rainfall with high volumes of water result in water ponding in the streets. This situation is critical for detention systems. Since the wet periods are concentrated in only a few months (e.g., 500 mm in 15 days has a return period of about 15 years) and there is a storage system, its critical design condition is mainly based on rainfall volumes of a few days rather than on a short period of rainfall.

1.3.2 Hydrologic Effects

Urbanization increases surface runoff volumes and peak flows. Such excess rainfall may lead to flooding, sediment erosion and deposition, habitat washout (Borchardt and Statzner, 1990), geomorphologic changes (Schueler, 1992), and reduced recharge of groundwater aquifers. These effects may be divided into two categories: acute and cumulative. Flooding and stream channel incision go into the first category, whereas groundwater table lowering and morphological changes go into the second. There are more explanations about this heading in Chapter 2.

1.3.3 Urban Heat Islands

Urban heat islands (UHIs) have been forming over a while around the world. The UHI phenomenon occurs when the air in the urban city is 1°C–5°C (2°F–8°F) hotter than the surrounding rural area. Scientific data have shown that the maximum temperature of July during the last 30–80 years has been steadily increasing at a rate of 1.5°F every 10 years. Each city's UHI varies based on city layout, structure, infrastructure, and the range of temperature variations within the island. The urban area will have a higher temperature than the rural area due to the absorption and storage of solar energy by the urban environment and the heat released into the atmosphere from industrial and communal processes (Ytuarte, 2005). The UHI effect can adversely affect a city's public health, air quality, energy demand, and infrastructure costs (ICLEI, 2005). Attention should be paid to the following issues in UHI:

- *Poor air quality:* Hotter air in cities increases the frequency and severity of ground-level ozone (the primary component of smog) and can drive cities out of compliance. Smog is formed when air pollutants such as nitrogen oxides (NO_x) and volatile organic compounds (VOCs) are mixed with sunlight and heat. The rate of this chemical reaction increases when the temperature exceeds 5°C.
- *Risks to public health*: The UHI effect prolongs and intensifies heat waves in cities, making residents and workers uncomfortable and putting them at increased risk for heat exhaustion and heat stroke. In addition, high concentrations of ground-level ozone aggravate respiratory problems such as asthma, putting children and the elderly at particular risk.
- *High energy use:* Higher temperatures increase the demand for air conditioning, thus increasing energy use when demand is already high. This, in turn, results in power shortages and raises energy expenditures when energy costs are at their highest.
- *Global warming:* The combustion of fossil fuels to generate power for heating and cooling buildings contributes significantly to global warming. UHIs exacerbate global warming by increasing the demand for electricity to cool buildings. Depending on the fuel mix used in producing electricity in the region, each kWh of electricity consumed can produce up to 1.0 kg of carbon dioxide (CO_2), the main greenhouse gas contributing to global warming.

Mitigating UHI is a simple way of decreasing the risk to public health during heat waves while also reducing energy use, the emissions that contribute to global warming, and the conditions that cause smog.

Cities in cold climates may benefit from the wintertime warming effect of heat islands. Warmer temperatures can reduce heating energy needs and may help melt ice and snow on roads. In the summertime, however, the same city may experience the adverse effects of heat islands. Fortunately, communities can take a number of steps to lessen the impacts of heat islands. These "heat island reduction strategies" include the following:

- Reducing the high emission from transportation through traffic zoning and well-managed public transportation
- Installing ventilated roofs and utilizing passive sources of energy in buildings
- Planting trees and other vegetation

1.3.4 Cultural and Esthetic Aspects

Sustainable solutions to water-related problems must reflect the cultural (emotional, intellectual, and moral) dimensions of people's interactions with water. Culture is a powerful aspect of water resources management. Water is known to be a valuable blessing in most arid and semiarid countries and by most religions. Two cultural characteristics cause direct impacts on water resources management in urban areas: urban landscape architecture and people's lifestyle.

The climate characteristics of the area often reflect the practice of landscape architecture in urban areas. However, the traditional building-centered architecture in many large cities is going to be replaced by modern opens space-based architecture with water and plant adaptation features capable of mixing esthetic reflections with storm management practices. This may also cause many changes in urban settings of water infrastructures. The density of population and buildings, rainwater collection systems, the material used in construction, and wastewater collection systems are significant factors, among others, that alter the urban water cycle. The change in design paradigm has made significant changes in architecture and moves it towards ecological-based design.

Lifestyles in urban areas affect the hydrologic cycle through changes in domestic water demands. Water use per capita and water used in public centers such as parks and green areas are the main characteristics that define the lifestyle in large cities. Even though economic factors are important in determining these characteristics, the patterns of water use, tradition, and culture have more significant effects on the lifestyle in urban areas.

A turning point in exploring ideas and revisiting a combination of landscape architecture and stormwater management has been the occurrence of the notorious Superstorm Sandy in October of 2012. Sandy was a real wake-up call as far as the urgency for integration of critical infrastructures has been concerned but also for utilizing esthetically enhancing practices that could bring livelihood, mobility, and relief to the affected communities. There was an apparent lack of holistic/system-based thinking and an inadequate understanding of the region's vulnerability that had led to large-scale losses and casualties. Nationwide concerns led to the formation of a comprehensive effort through "Rebuild by Design." In December 2013, President Obama signed an executive order creating the Hurricane Sandy Rebuilding Task Force to ensure that the Federal government continues to provide appropriate resources to support affected state, local, and tribal communities to improve the region's resilience, health, and prosperity building for the future. The Task Force was commissioned to ensure cabinet-level, government-wide, and region-wide coordination to help communities make decisions about long-term rebuilding.

Following the presidential executive order, the US Department of Housing and Urban Development (HUD) initiated a competition with the collaboration of the Netherlands government called Rebuild by Design (RBD). It consisted of ten teams, including the finalists BIG U, New Meadowlands, Hudson River Project, and Hunts Point Lifelines among other groups made up of experts, landscape architects, and engineers to generate ideas and conceptual designs for flood risk management. The main objective was to find solutions suitable and adaptable for flood control infrastructure during both extreme events and normal weather conditions. The BIG U represented the notion of integrating a city park with floodwalls. The New Meadowlands team suggested an integrated linked system of embankments and wetlands to flood protection through the Meadowlands in New Jersey. The Hudson River Project proposed the green and gray infrastructure approach for reducing flood risk and achieving a more comprehensive flood management strategy such as landscape-based and engineered-based coastal defenses. Hunts Point Lifelines' proposal also included green and gray flood protection and measures to protect critical economic assets, including transportation in the region. See RBD (2014a–d) for more information. This competition provided a unique opportunity for landscape architects, planners, and engineers to explore many ideas generated for rethinking and rebuilding of flood resilient cities. Local cultural and esthetic reflections were core issues confronted by all the design groups involved. See Chapter 13 for more details.

1.4 URBAN WATER INFRASTRUCTURE MANAGEMENT

There are three main urban water infrastructures: water supply, sewerage, and stormwater drainage. Managing urban water infrastructures is a highly complex issue. While values and technology have changed, the nature of water infrastructure prevents the system from keeping pace with changes. Generally, in the last two decades, increased emphasis has been placed on the environmental outcomes of water infrastructure. Previously, social and economic outcomes dominated decisions on water infrastructure.

This change in focus has led to a perception that centralized, large-scale systems ought to be converted to alternative systems. This perception is not always correct, as the best results usually come from using a mix of systems, and this mix will change with location and time. Centralized, large-scale systems will still dominate water infrastructure in many regions for the next few decades, partly because they are there and because it is difficult to change them due to engineering, environmental, economic, and social reasons. For example, headwater dams and other facilities will be too difficult and costly to alter to any significant degree. The smaller alternative systems will have an increasingly important role, but their role will be limited by the source of supply in most cases. The more congested the site and the more the property interests involved, the harder it will be to replace centralized systems with other alternatives. In the following sections, narratives are presented for assessment, performance aspects, and more recent paradigm shifts to bring landscape architecture in the core water flow and conservation with many cultural, recreational, and esthetic exposures of urban life as an ultimate sustainability goal. Water Eco-Nexus System discussed in Section 1.4.3 could bring innovative and distributed treatment solutions to replace or enhance large-scale centralized systems for urban systems.

1.4.1 Life Cycle Assessment

The urban water system of a city is emphasized as a complex system defined by constant change and growth in an ecological way of thinking. Energy, natural resources, transportation, and solid waste can all be considered flows or processes in the urban water cycle, whether they are directly or indirectly involved. Building, maintaining, restoring, stimulating, and monitoring these lifeline flows and processes contribute to the sustainable development of an urban area. Urban water system performance measurement raises specific methodological problems, particularly concerns about physical and operational limits, time horizons, and uncertainties associated with functionality of unit operations. Conceptually, it is necessary to know the full environmental consequences of each decision or action made on water resource handling and usage to evaluate different performances or compare options. One of the primary methods for this is life cycle assessment (LCA). This serves as a foundation for calculating the medium- and long-term results of urban water resource planning and management, particularly those relating to the economic implications of these activities. Measures of the performance of urban water systems give decision makers a better foundation for evaluating their activities.

LCA can be defined as a systematic inventory and analysis of the environmental effect caused by a product or process starting from the delivery of raw materials, production, and use, up to waste treatment. For each of these steps, there will be an inventory of the use of materials and energy and emissions to the environment. With this inventory, an environmental profile will be set up, making it possible to identify weak points in the life cycle of the urban water system, including resources, water supply, treatment, distribution, wastewater collection and treatment, sludge disposal, and drainage systems. These critical stages of water movement and potential vulnerable states are the focal points for improving the performance of the urban water system from an environmental point of view and in the movement towards more sustainability.

LCA establishes the relationship between objectives and indicators, where one objective can relate to several indicators and one indicator can be used to assess the fulfillment of several objectives. Furthermore, LCA can enlarge traditional system limits in space, in time, and the number of concerned aspects. LCA can be directly and structurally related to life cycle cost as well as to other types of social and cultural impact assessment methods. LCA proceeds in four steps: (a) goal and scope definition, (b) inventory of extraction and emissions, (c) impact assessment, and (d) evaluation and interpretation.

The application of LCA to the urban water system is only relevant if situated within the larger conceptual framework of sustainable urban development. Furthermore, the traditional focus on environmental impacts has to be completed by considering aspects of the long-term water resource conservation of urban areas.

The application of LCA to water resources systems must respect and take into account cultural limitations to be effectively utilized. The sustainable development of accessible urban water resources is unquestionably at the start of a much longer process. Its goal cannot be achieved by using simple techniques to end the debate before a more in-depth discussion takes place. The physical and sociological elements of the urban water system are distinctive. It furthermore has a continuity that is both spatial and temporal. This continuity constitutes an essential value; it is a fundamental urban resource and must be protected. Protection is required to ensure sustainability through the continuity of development and the embedded social and physical values for urban residents (Hassler et al., 2004). See Chapter 8 for more explanation of LCA in the context of asset management and life cycle cost.

1.4.2 Environmental, Economic, and Social Performances

The environmental, economic, and social performances of urban water systems vary considerably in time and space, and their performances have been judged differently over time as community values change. Previously, water infrastructure development was seen to increase economic prosperity and social production of water resources, and environmental consequences were not seen as important. Such views have changed with the increase in environmental awareness. In addition, the necessity of greater efficiency in water usages due to social and environmental aspects has changed the perspective of what economic benefit of water infrastructure development delivers.

Given those values have changed since most water infrastructures were built, the performances of many aging water infrastructures are now viewed by many as unacceptable. While some problems such as the discharge of partly treated sewage or combined sewer overflow on the shoreline have been fixed, others such as the failure of some sewerage systems to meet their original technical specifications have not. Given that perspectives have changed over time, rather than lingering over the past, while we should preserve the desirable features of existing urban water systems, they should be transformed into more sound and functional systems. This includes improving and changing combined sewer system in many major cities such as New York City.

The cycle concept is a useful tool for stakeholder and community education and for consensus building around an action plan. The goal of restoring and maintaining the balance between the current demands of the community for water and the need to preserve the aquatic ecosystem for the benefit of future generations becomes understandable. Table 1.1 lists some of the enabling systems and practices that can be developed within an urban community.

The starting point for sound water stewardship is the understanding that urban water, from source to final disposition, flows through a series of four interrelated stages, listed in Table 1.1, in a continuous cycle. The cycle concept illustrates some important facts about urban water:

- Waste and contamination at any stage negatively affect the sustainability of the cycle as a whole and the health and safety of the community using that water.
- Urban planning, without considering the water cycle, results in water supply shortages, deteriorating aquifer water quality, groundwater infiltration into the distribution system, endemic health problems, and other symptoms of an unsustainable situation.
- Every citizen, institution, agency, and enterprise in the community has a contribution to make towards the goal of sustainability.

It should also be noted that all components of source control through management of the cycle at this level offers the opportunity to provide benefits for the consumer and the environment. The philosophy of source control is to minimize the cost of providing water and collection of stormwater and wastewater. Source control can be implemented through retention of roof rainwater (rainwater tanks), stormwater detention, on-site treatment of gray water (laundry, bathroom, and kitchen) and black water (toilet), use of water-efficient appliances and practices, and on-site infiltration.

TABLE 1.1
Examples of Community-Based Enabling Systems for a Sustainable Urban Water Cycle

1. Source
    - A long-term urban and watershed management master plan
    - A source water quality and quantity monitoring system
    - A geographic information and decision support system
    - An inspection and enforcement system to protect source water
    - A community education program
2. Use/reuse
    - A metering and billing system
    - An industrial discharge control program
    - Regulations and bylaws
    - An industrial incentive program
    - A community education program on water conservation
    - A network of supporting laboratories
    - A monitoring and control system
    - An emergency spill response system
3. Treatment/distribution
    - A potable water quantity and quality monitoring and control system
    - A utility operation and maintenance system, including training and accreditation of operators
    - A financial, administrative, and technical management structure
    - A flexible water treatment process
    - An operation, maintenance, leak detection, and repair system
    - Continuous pressurization
4. Treatment/disposition
    - An effluent quality monitoring and control system
    - A utility operation and maintenance system, including training and accreditation of operators
    - An environmentally sustainable biosolids management program
    - A financial, administrative, and technical management structure
    - A flexible treatment system
    - An end-user market

Source: United Nations University, International Network on Water, Environment and Health (UNU-INWEH). 2006. Four pillars. Available at http://www.inweh.unu.edu/inweh/.

1.4.3 Urban Landscape Architecture

To improve sustainable water management in urban water systems, enhancing water cycle with multiple blends of landscape architecture could expand water supply capacity, increase water use efficiency, and improve water environment quality. Moreover, the sustainable development goals by United Nations (UN SDG-6) further highlight the substantial increase of safe water reuse globally and implement integrated water resources management at all levels by 2030 (UN (United Nations), 2016). Consequently, a new Water Eco-Nexus Cycle System (WaterEcoNet) has been proposed by Chen et al. (2020). It aims to establish a safe and smart urban water system to enhance water quality and maximize water use efficiency while maintaining ecological, landscaping, and recreational attributes of urban architecture.

As can be seen from Figure 1.3, Water Eco-Nexus System generally follows a water cycle of "urban water use–wastewater drainage–water reclamation and reuse–urban water ecosystem supplement–urban water use." It can transform conventional urban water systems into innovative and distributed treatment which relies on distributed facilities rather than large-scale centralized systems for water ecosystems.

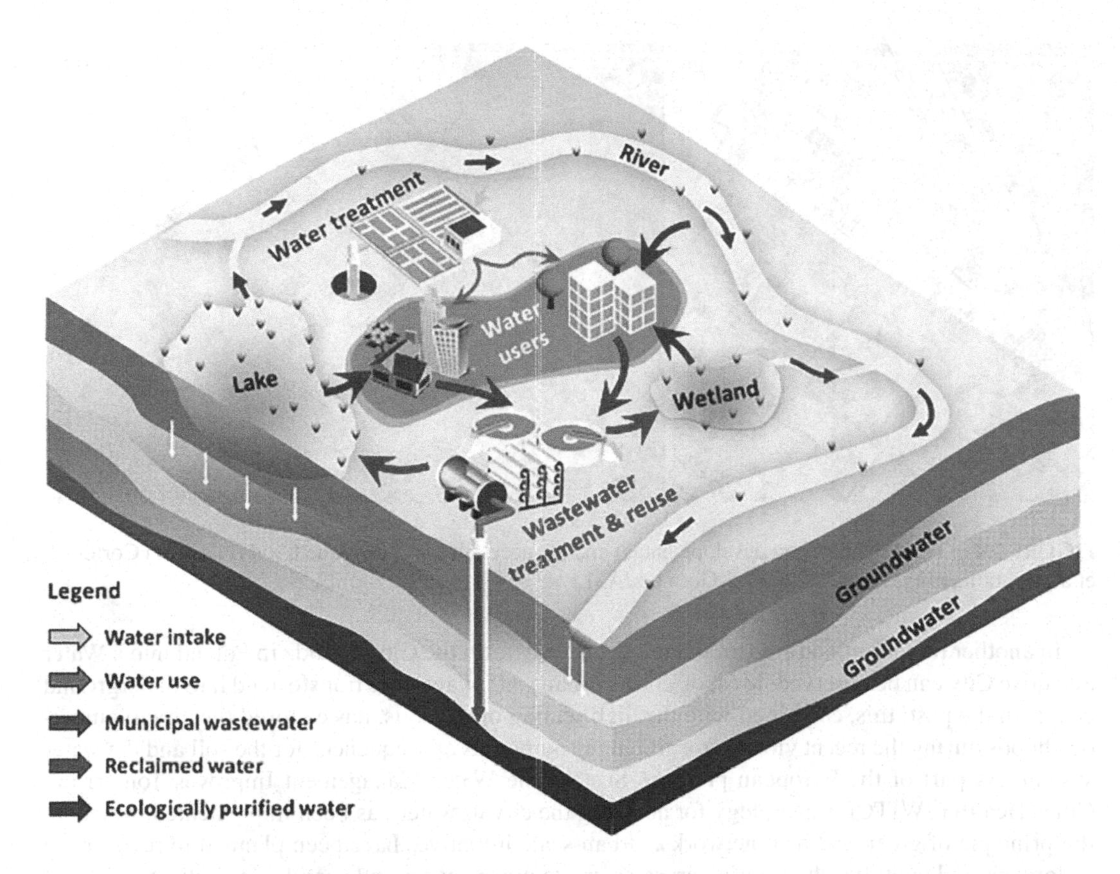

FIGURE 1.3 The concept of Water Eco-Nexus Cycle System (WaterEcoNet). (From Chen, Z. et al., Water Cycle, 2020. doi: 10.1016/j.watcyc.2020.05.004.)

It highlights the use of reclaimed water for water environment restoration and/or enlargement, such as river, lake, or wetland, while simultaneously storing and replenishing water. Afterward, cascading use can be implemented and the ecologically treated water will be applied for industrial, domestic, and agricultural uses to form an integral water ecological network in urban areas. Meanwhile, correlation and interaction among different water use applications are incorporated.

The landscape project as an integrated project has spread the seeds of a new approach to the consideration of the contemporary city in a water ecological manner.

In this global scenario, architecture and landscape projects are necessary means for the care of the physical world; while defending it, they understand and revise the causes of the transformation's phenomena. Architecture reaffirms a new ethical value and becomes responsible for a new relationship of stewardship between humans' actions and nature (Morgia, 2007). The architect is, according to Ian McHarg, an ecological planner since he/she leads the discipline "into a broad multidisciplinary tool of resource management and land use planning" (Zeunert, 2016).

Water and the related soil project are an essential component of context-based projects because they use existing resources that would otherwise be dispersed and, by introducing a time variable, allow the planning of an incremental evolution of places, this being the concept behind a new generation of landscape urbanism. In this part, we will therefore see how some interventions, namely the soil projects, manage to transform floods or overflow phenomena from risk factors to resources by creating new unstable urban landscapes.

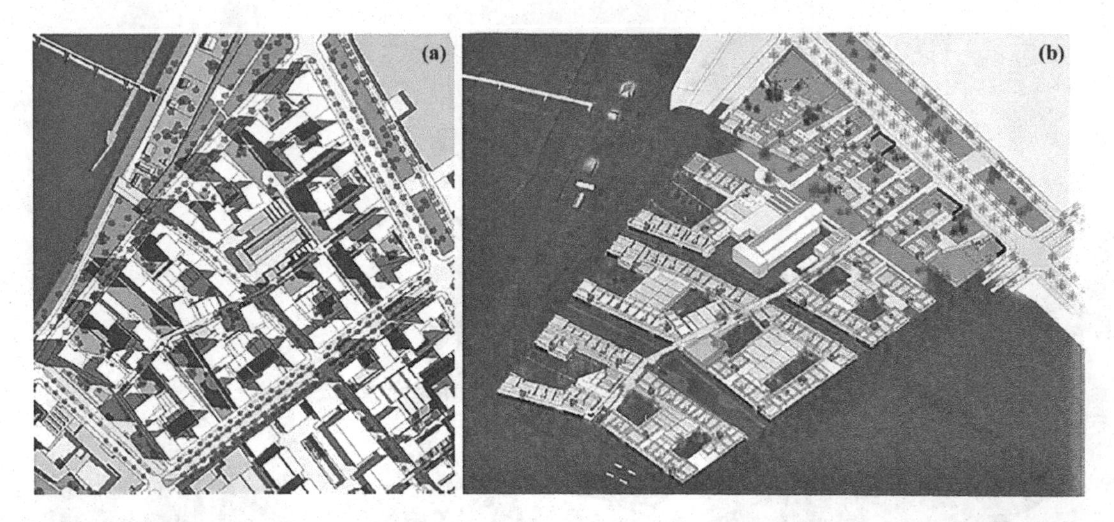

FIGURE 1.4 The neighborhood development (a) and the accessibility during the flood (b). (From Cortesi, I. et al., Sustainability, 12, 8840, 2020 © Germe&JAM.)

In another trend for flood control is a project to transform the City of Lodz in Poland into a Water Sensitive City can be observed. Most of Lodz's urban waterways were transformed into underground canals in the past; this, combined with the high sealing of the soils, has exposed the city to numerous floods during the recent violent storms having numerous consequences for the soil and the water system. As part of the European program Sustainable Water Management Improves Tomorrows Cities Health (SWITCH), a strategy for adapting the city to water has been implemented, based on the principle of green and blue networks. Urban-scale initiatives have been planned to recover the waterways and revitalize the riparian areas, to create green spaces and corridors to collect and manage rainwater, and to condition urban planning to the system of green and blue networks. Currently, in this context, there is a pilot project and an integrated approach to design led by a team, where all the actors involved in the management of water resources and ecosystem services are represented and have been carried out. The urban project for Vitry gare de Seine by Germe and Jam in Paris, France along the Seine River is more radical in building a water-resilient environment. It has as reference the flood of 100 years, and the issue of flooding risk is at the center of the design of the public space and the urban fabric (Figure 1.4). In Vitry, an artificial floor consisting of a network of elevated paths connected by walkways allows the connection of urban services and the entrances of residential buildings in the event of an extraordinary flood, while public spaces are designed to adjust to ordinary floods and violent rains.

The interesting element of this project is that by allowing the water to expand in the gardens and under the buildings, the designer managed to avoid the construction of blind walls at the urban routes level as happened in other projects, for instance, in HafenCity of Hamburg, Germany. The network of elevated paths can also be used by service vehicles in order to allow, in the event of a flood, an organized removal of the inhabitants.

1.5 SYSTEMS APPROACH

The study of any phenomenon, process, and/or structure such as climatic systems, urban water cycle, and heat islands requires an integrated approach. The selected framework of study dictates the way the structure of an environment is dismantled for analysis. Also, in order to incorporate the results of the analysis in a holistic fashion, the integration of environmental components should be done afterward. For providing such a framework, understanding of system concepts is needed.

Having a complete knowledge of the attributes of a system is often helpful to analyze the system properly. These attributes include goals or objectives, physical structure, operating rules of the system, system boundaries, environment or surroundings, condition, or performance of the system at any time. Furthermore, performance measures and the governing standards and criteria should be well defined. See Labi (2014) for more details.

In system analysis, the boundaries of a system need to be specified in order to determine which entities are internal to the system and which are external. For closed systems, the environment plays an intangible role in the analysis; but for open systems, the environment is often a critical factor in analyzing the system. In certain types of analysis, a system could be defined as being closed because the environment is considered an integral part of the system. The goals of a system are desired end states, and objectives are specific statements of goals. The objectives should be attainable and measurable. From the users' perspectives, system performance is often measured in terms of the direct benefits of the system in terms of delay, convenience, comfort, safety, or out-of-pocket fees, fares, or costs incurred when using the system. From the perspective of the system owner and sometimes a discerning general public, the system performance is viewed in terms of the system's physical condition. For an acceptable level of performance of a system, feedback and control are also essential.

1.5.1 General Systems' Characteristics

Although the word "system" is referred to in different fields, which, at first, may seem to be completely different, there are some common characteristics between different types of systems. The main similarities of systems based on White et al. (1996) are as follows:

1. All systems have a structured organization with distinct components. In other words, a system is a collection of things that are organized in a specific way and there are connections and links among them. For example, in a hydrologic system, precipitation produces the runoff that flows in rivers and rivers recharge to different water bodies. The evaporated water from these bodies forms clouds, and again they form precipitation.
2. The systems are generalizations, abstractions, or idealizations of what happens in the real world.
3. A function is defined for every system, and there are functional and structural relationships between the components of a system.
4. The flow and transfer of material through a system are defined by the function of the system. This material may be obvious, like the water in the hydrologic cycle, or it may be intangible, such as directives, decisions, ideas, and ideologies.
5. Certain driving forces or energy resources are needed for a system to function. In some systems, this motivation or stimuli is obvious, but in others, it is not clear what the impetus of the forces of supply and demand on people is. For example, in the water cycle, heat is needed for evaporation and the Earth's gravity is one of the driving forces acting during precipitation.
6. Each system has a level of integration.

However, the art and science of systems analysis have evolved through developments in separate disciplines of engineering, economics, and mathematics. Rapid developments have taken place, and the availability of high-speed computers has contributed to its widespread applications. Systems analysis has found extensive water resource planning applications as the science of systems analysis has progressed over the last several decades and as the scope of current water resource projects has risen. The origin of the activity may be said to be in the 1950s in the United States, and the pioneering work has been done by a group of engineers, economists, and political scientists at the Harvard University water program. Since then, the value of systems planning has been increasingly apparent, and progress has been achieved.

Different terms are used in relation to the system concepts. The main terms in describing systems are boundary, state, closed system, and open system. A system is separated from the surrounding environment by its boundary. Each system is identified with three kinds of property including elements, attributes, and relationships between elements and states or attributes in its boundary. For example, in a watershed system (as will be described in detail), the watershed boundary, the lower layer of the atmosphere, and the upper layer of the Earth can be considered as the system boundary. A watershed is composed of different elements such as precipitation, subbasins, drainage system, and groundwater where each of them has special attributes. These elements are related to each other in different types which will be described in future sections. The state of the system is defined when each of its properties (variables), that is, elements, attributes, and relationships, has a definite value. Several distinct types of systems can be distinguished according to the behavior of the system boundary:

1. *Isolated system*: In this kind of system, there is no interaction with the surroundings across the boundary, and they could only exist in the laboratory and be used for development of some concepts.
2. *Closed systems*: These systems are closed with respect to matter and with the exception of energy which can be transferred between the system and its surroundings. On the Earth, closed systems are rare, but it is often useful to treat complicated environmental systems as closed systems.
3. *Open systems*: Both matter and energy can be exchanged with the surroundings in the open system. All environmental systems are open systems and are characterized by the maintenance of structure in the face of continued throughputs of both matter and energy. For example, although there is a continuous throughput of water in a drainage system, it maintains the organization of the stream and river channel network and the contributing slopes.

The basis of any environmental system function is the laws of thermodynamics. So, any change in the system state is important. Although, in theory, the initial and final states of a system are considered, in practice, the pathway of system state is often of interest. A process is the way in which a change in system state happens. If a system returns to its initial state after a process, it is called a cycle. Regarding this definition, the hydrologic cycle is the way water vapor in the atmosphere moves through the Earth's system and returns to the atmosphere.

Environmental systems must preserve a stable state regarding their elements, attributes, and relationships through time. This characteristic state is called stationary or steady (Figure 1.5a). When a system is steady, at the macroscopic scale, the system state appears to be stable, but the steady state is completely dynamic, and only the average state is stable. For example, although there are considerable fluctuations in the rainfall pattern of a region even through several years and there are

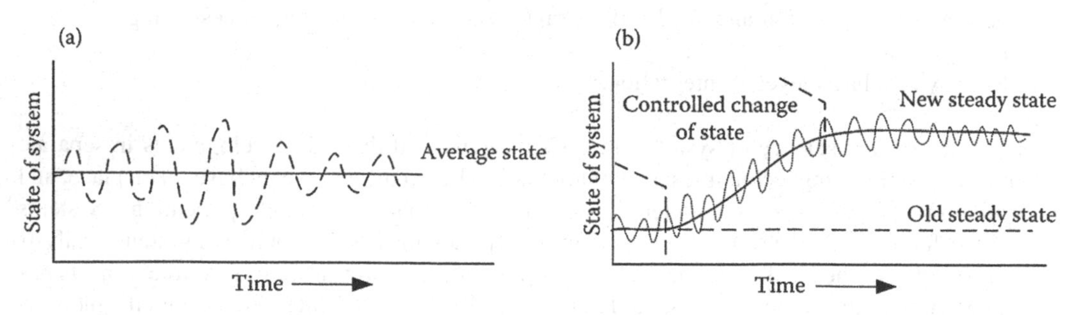

FIGURE 1.5 (a) System maintaining a steady (average) state through time. (b) Controlled change of state from an old to a new steady state effected by positive feedback. (From White, I.D. et al., *Environmental Systems*, Chapman and Hall, London, 1996. With permission.)

some very dry and very wet years, the mean rainfall of the region over the years is almost fixed in the long-term horizon.

When an open system has the capacity for self-regulation and it can be affected by some feedback from its surroundings, it can maintain its steady state. Negative-feedback mechanisms (homeostasis), which are control mechanisms, affect self-regulation in environmental systems. The result of the processes involved in negative feedback is returning the system to its initial state. For example, the detailed properties of a length of river channel may vary widely about the most effective state, but it will still transfer water and debris.

However, natural systems experience some changes, which are called positive feedback mechanisms (homeorhetic). The cumulative effects of these systems cause particular directions of change. Positive feedback has a deviation-amplification character that highly changes the initial state of the system; however, the mechanism of this change may be closely regulated (Figure 1.5b). An example of positive feedback is the climate change impact that has increased mean temperature over time. A composition of positive- and negative-feedback mechanisms is involved in the regulation of natural systems' states.

Some thresholds are related with positive-feedback processes due to their cumulative influence. Thresholds are state variables that, when given particular values, cause abrupt or drastic changes in state. For example, in a soil undergoing decalcification by leaching, when the calcium content reaches zero, the system will be transformed in state and the chemical characteristics of the soil will change rapidly. In other words, thresholds are extreme cases, and other certain state variables that play key roles in controlling the operation of the processes act as regulators.

1.5.2 System Properties

The main specifications of a system are as follows:

- Inputs
- Governing physical laws
- Initial and boundary conditions
- Outputs

Variables and parameters are used for definition of systems' inputs, outputs, and major characteristics. In system dynamics definitions, storage (stock), converters, and connectors are perceived as physical laws and conditions. Some system characteristics vary through time, these are called variables. Other system characteristics which normally do not change with time or can be considered fixed over a period of time are called parameters. Different values of variables can be viewed as possible states of nature or as alternative futures. Regarding the spatial scale of variables/parameters, they are classified as follows:

- Lumped, which are considered constant in space
- Distributed, which vary in one or more space dimensions

When the source of a contamination in a watershed scale is considered, point and nonpoint sources are examples of lumped and distributed variables. Memory, as an important characteristic of a system, shows the length of time in the past in which variations during that time could have an impact on the output. Different levels of memory are defined as follows:

- *Zero memory*: Only the input in the present time affects the system state and output.
- *Finite memory*: The history of the system for a specific time span (memory) can affect the state, output, and behavior of the system.
- *Infinite memory*: The state and output of the system are dependent on the entire system history.

Strictly, a random number and lag-one Markov process are examples of zero and finite/infinite memories, respectively. In another approach of system classification, systems are classified to be deterministic or stochastic. Because they contain one or more factors for which the connection between input and output is probabilistic rather than deterministic, deterministic systems always produce the same outcome for a given input, whereas stochastic systems may produce different outcomes for the same input.

The other detailed classifications of systems are as follows (Karamouz et al., 2003):

- *Continuous systems*: The system produces a continuous output.
- *Discrete systems*: The output changes after finite intervals of time.
- *Quantized systems*: The output values change only at certain discrete intervals of time and hold a constant value between these intervals.
- *Natural systems*: The system characteristics, including inputs, outputs, and other state variables, vary without any control on their behavior and only can be measured.
- *Devised systems*: The input of these systems may be both controllable and measurable.
- *Simple systems*: These systems do not include any feedback mechanism.
- *Complex systems*: There are complex loops and feedback in these systems' behavior.
- *Adaptive systems*: The performance of these systems is improved based on their history.
- *Causal systems*: An output cannot occur earlier than the corresponding input (cause and effect).
- *Simulation systems*: These systems are similar to causal systems providing a realization of environment.
- *Stable systems*: These systems have bounded output in case of bounded input and vice versa.
- *Damped systems*: The output of the system dies out without ever crossing the timescale.

Hydrologic systems are often stable and causal systems. Figure 1.6 shows examples of outputs of continuous, discrete, and quantized systems. The outputs for any system depend on both the nature of the components and the structure of the system according to which they are connected. However, in the system approach, concentration is on the system operations that depend on the physical laws and the nature of the system that is represented by the vertical components in Figure 1.7.

Considering the three basic components of a system (inputs, system operation, and outputs (Figure 1.7), systems analysts could face several different types of problems (Table 1.2):

- *Design problems*: In these problems, the system output must be quantified for specified inputs and system structure.
- *System identification problems*: The system should be specified for given inputs and outputs.
- *Detection problems*: System and outputs are known and inputs must be identified.
- *Synthetic problems (simulation)*: Inputs and outputs are known and the performance of models must be tested.

Hydrologists primarily deal with design and synthetic problems.

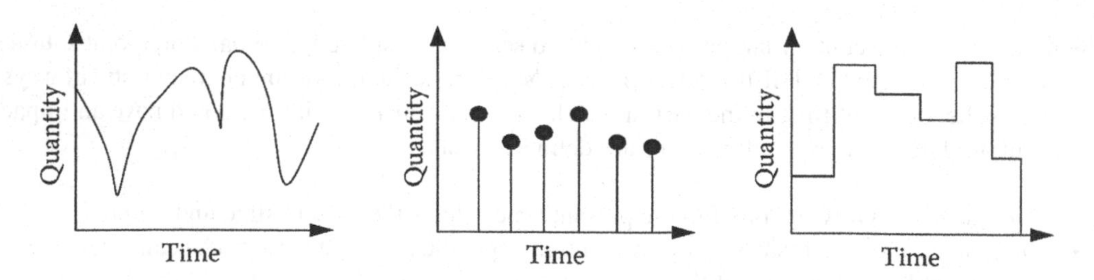

FIGURE 1.6 Outputs of continuous, discrete, and quantized systems. (From Karamouz, M. et al., *Water Resources Systems Analysis,* Lewis Publishers, New York, 2003. With permission.)

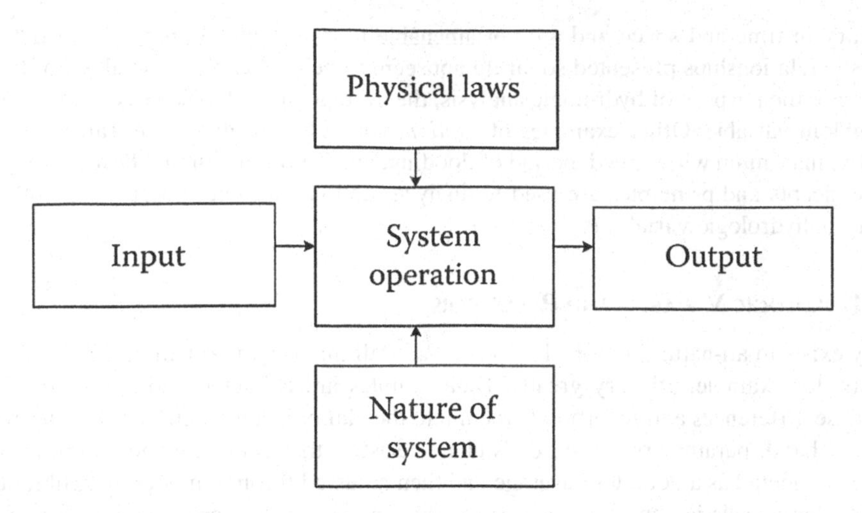

FIGURE 1.7 Specifications of a system. (From McCuen, R.H., *Hydrologic Analysis and Design* (2nd ed.), Prentice Hall, Upper Saddle River, NJ, 2005.)

TABLE 1.2
Classification for System Problems Based on What Is Given (G) and What Is Required (R)

Type of Problem		Input	System	Output
Analysis	Design	G	G	R
	Identification	G	R	G
	Detection	R	G	G
Synthesis	Simulation	G	G	R

Example 1.1

Consider that temperature prediction is needed for reservoir design. Describe the system concept of this problem in analysis and synthesis stages.

Solution:

In this simplified problem, evaporation is needed for reservoir design; thus, evaporation serves as the output variable. Temperature serves as the input. In the analysis phase, the problem is identification of the model for predicting evaporation; therefore, the model represents all of the hydrologic processes involved in converting temperature (e.g., heat energy) to evaporation. Obviously, temperature is only a single measurement of the processes that affect evaporation, but it is a surrogate variable for these processes in the conceptual representation of the system. The synthesis of the system output is used for the reservoir design problem. In summary, in the analysis phase, the temperature (input) and evaporation (output) are known, while in the synthesis phase, the temperature (input) and the model (transfer function) are known.

Note: The definition of system specifications is dependent on the purpose and applications.

1.6 HYDROLOGIC VARIABILITY

To investigate about hydrologic processes such as rainfall, snowfall, floods, and drought, usually their records of observations are analyzed in the first step. Many characteristics of these processes seem to vary in time and space and are not amenable to *deterministic analysis.* In other words, deterministic relationships presented so far do not seem to be applicable for analysis of these characteristics. For the purpose of hydrologic analysis, the average or peak discharge is then considered to be a random variable. Other examples of *random variables* are maximum rainfall, maximum temperature, maximum wind speed, period of flooding, and minimum annual flow. *Probability and statistics* concepts and principles are used to analyze random variables and to deal with risk and uncertainty in hydrologic variability.

1.6.1 Hydrologic Variables and Parameters

Variability exists in all-natural physical systems. Rainfall intensity, flood amplitude, and low flows in droughts, for example, all vary greatly. Data samples are collected and analyzed in order to research these differences and incorporate them into the planning and operation of water resources. On the other hand, parameters are system's characteristics that could change within a range, but their value is selected as a default or average and then adjusted through modeling calibration.

The hydrologic cycle is composed of various phenomena, such as precipitation, runoff, infiltration, evaporation, evapotranspiration, and abstraction. Different characteristic variables, which can simply be called hydrologic variables, have been defined to describe each of these phenomena. Depth or intensity of rainfall in different time steps of a rainstorm, monthly inflow discharge to a reservoir, and daily evaporation are some examples of hydrologic variables (Shahin, 2007). A dataset consists of a number of measurements of a phenomenon, and the quantities measured are variables.

When a variable is determined as continuous, it can have any value on its continuous domain; examples include volume of water flowing in a river or the amount of daily evaporation measured in a climatic station. A discrete variable represents an interval or the number of occurrences within each interval of time and space; the number of rainy days in a certain period of time (e.g., a year) is an example of a discrete hydrologic variable.

Hydrologic variables can also be classified as qualitative or quantitative. A qualitative variable can be expressed as a real number in a sensible and usable way; type of soil is an example of a qualitative hydrologic variable. The number of rainy days in a year and rainfall intensity in a day are examples of discrete and continuous quantitative hydrologic variables, respectively.

Hydrologic variables are often temporal and spatial variables. A time series is a collection of values arranged in chronological order. When analyzing the influence of numerous hydrologic factors on each other and modeling them throughout the decision-making process, conflicts might occur.

According to Nile (2018), the parameters used as input data for the hydraulic and hydrology models are as follows: subcatchment discharge (Q_c), discharge of pipe (Q_p), area of the subcatchment (A_c), cross-sectional area of pipe (A_p), slope of subcatchment (S_c), slope of pipe (S_p), velocity in subcatchment (V_c), velocity in pipe (V_p), time of concentration in subcatchment (T_{cc}), time of concentration in pipe (T_{cp}), length of subcatchment (L_c), and length of pipe (L_p). Other well-known parameters are Manning n, Chezy C, hydraulic conductivity K, and storage coefficient S_c.

In previous studies, the dynamics of hydrological model parameters are discussed including time-invariant parameters, "compensation" among parameters, the high dimensionality (too many, several parameters, and subperiods) of the parameters, and abrupt changes in the parameters. Compensation among parameters is for structural inadequacy and subperiods are identified based on different hydrological characteristics using a clustering technique. See Lan et al. (2019) for details.

1.7 REPRESENTATIONS, STATISTICAL, AND SIMULATION MODELS

For the analysis and planning of watersheds, simulations of hydrologic events are required. Hydrologic simulation models are utilized to give not just hydrologic predictions/projections but also a deeper knowledge of the processes that occur within the hydrologic cycle. A variety of modeling approaches are used for this purpose. These models are a simplified *conceptual representation* of a part/component of the water cycle. Different classifications are considered for hydrologic simulation. Two groups of *hydrologic models* are often defined:

1. *Mathematical models:* These models are usually black box models developed through tools such as regression, transfer functions, neural networks, and fuzzy inferences. Mathematical and statistical concepts are used in these models to develop a relationship between model input(s) such as rainfall and temperature and model output such as runoff.
2. *Physical (process)-based models:* These models attempt to simulate the physical processes that happen in the real world through the hydrological cycle based on identified physical and empirical relationships. In other words, in this model, different components of the hydrologic cycle are quantified using the relationships that interpret the relations between hydrologic cycle components. Typically, these models include representations of surface/subsurface runoff formation, evapotranspiration, infiltration, and channel flow/streamflow.

The main concepts that should be considered in models' development are data availability and quality. Sometimes, the type of the model utilized is determined by the limitations of the data at hand. Furthermore, the type and characteristics of the output data needed for analysis and decision making should also be considered in model selection and development. Data are represented in different forms: lumped and distributed. The application of distributed data enforces the application of models that are linked with a geographical information system. To evaluate a model performance and compare it with other models, some performance indices are defined. These indices should be selected regarding the model, data characteristics, and behavior. For example, in drought studies, indices are used to represent the severity of different types of droughts as explained in Chapter 12.

There are some powerful software in both kinds of simulation approaches that can be used for hydrologic studies and analyses. A general, powerful, user-friendly software that is commonly used in recent years for developing mathematical models is MATLAB® developed by MathWorks. The wide range of functions and toolboxes provided in this software can be used to develop different applications based on data-driven methods/models such as artificial neural network (ANN), fuzzy inference system (FIS), adaptive neuro fuzzy inference systems (ANFIS), and optimization techniques. There are a variety of physical models represented by different software, some of them explained in Chapters 9 and 12 for rainfall–runoff analysis in watershed scale. The performance of hydrologic models is highly dependent on data type and duration as well as on the watershed characteristics.

1.8 EXTREME VALUES, VULNERABILITY, RISK, AND UNCERTAINTY

Water can be utilized based on short-term or longer-term assessment of its supply, retention, and depletion and the nature of its use. Due to many social interdependencies and vulnerabilities associated with the need for water as urgent as water supply for domestic and industrial use, extreme low and high precipitations and extreme low and high temperatures are of major concern. Droughts and floods continuously devastate different regions in a variety of ways, targeting every essence of our ecosystem.

On the other hand, *floods* are still the most destructive natural hazard in terms of short-term damage and economic losses to a region. Flooding is caused primarily by hydrometeorological mechanisms, acting either as a single factor or as a combination of different factors. Measures taken to deal with the risks such as the repair of river embankments to make rivers capable of disposing rainfall of up to the conveyance capacity of the river and the improvement and expansion of reservoirs, diversion channels, and sewer systems to reduce/eliminate the danger of floods are available. Moreover, in order to promptly and correctly counter local changes in precipitation and the danger of high tides, an effective use of the information transmission system of a comprehensive flood prevention program is needed.

Therefore, vulnerability assessment and overlaying extreme hazards with vulnerability in risk management are in the forefront of today's hydrological and hydroclimatical challenges. Uncertainties range from

- Inherent natural uncertainties due to man's limited knowledge and complexity of comprehending principles and laws governing the formation of water and its migration through the Earth-atmosphere system.
- Parameters and model uncertainty due to assumptions, lack of enough reliable data, oversimplification of models, and/or error accumulation from too many parameters.
- Roughly 27% of the surface water runoff is floods, and this share of water is not considered a usable resource, even though it contributes to the recharge of mainly overexploited aquifers. However, floods are counted in the nation's TARWR (total actual renewable water resources) as part of the available annual water resource.
- Seasonal variations in precipitation, runoff, and recharge are not well reflected in annual figures. They are important for basin-scale decision-making and setting regional strategy.
- Many sizable countries have different climatic zones as well as scattered population and the TARWR does not reflect the ranges of these factors that can occur within nations.
- TARWR has no data to identify the volume of water that sustains ecosystems, that is, the volume that provides water for forests and directs rain-fed agriculture, grazing, and grass areas.

1.9 TOOLS AND TECHNIQUES

Water systems can be represented, in an integrated or component-by-component fashion, by different identification, simulation, and projection models of hydrology and hydroclimatology. The development and application of these models (the model-building process) consist of several interrelated stages. Hydrology is broad and complex, and due to the nature of its interactions in the Earth-atmosphere system with different elements including ecosystem, many tools are needed to analyze, assess, plan, and manage it. Figure 1.8 shows a framework for the combination of tools and methods needed for hydrology and hydroclimatic studies.

Guidelines are useful tools in the planning of the hydrologic cycle because they give certain guidelines for dealing with various components, reducing the need for trial-and-error and hastening the accomplishment of the optimal outcomes. *Algorithms* are commonly used for hydrologic cycle planning and management because they give a step-by-step method to achieving the intended aim in the planning of hydrologic components. Different modeling approaches through different software are highly developed in recent years for simulating hydrologic processes. *Heuristics* is a method of solving a problem for which no formula exists, based on informal methods or experience, and employing a form of trial-and-error iteration. They are often utilized for simulation purposes in hydrology when there is a high level of complexity, making it difficult to completely understand the nature and behavior of components and their interactions.

The process in which the model compares its responses or predictions with the historical data is called model calibration. Different error criteria as a function of the differences between predicted and observed variables can be used, such as mean square error (MSE), sum squared error (SSE), root mean square error (RMSE), and Nash–Sutcliffe, to judge the predictive capability of the model. See Chapter 11 for more details. If the predictions differ significantly from the historical conditions, then the model architecture, parameters, boundary, and initial conditions may be systematically varied to improve the model's performance. However, it is possible that although the model's performance may be acceptable, the parameter values may be physically unrealistic. In this event, the underlying assumptions of the model would have to be reexamined to determine their appropriateness in the context of the validation results. This is the feedback element of the model-building

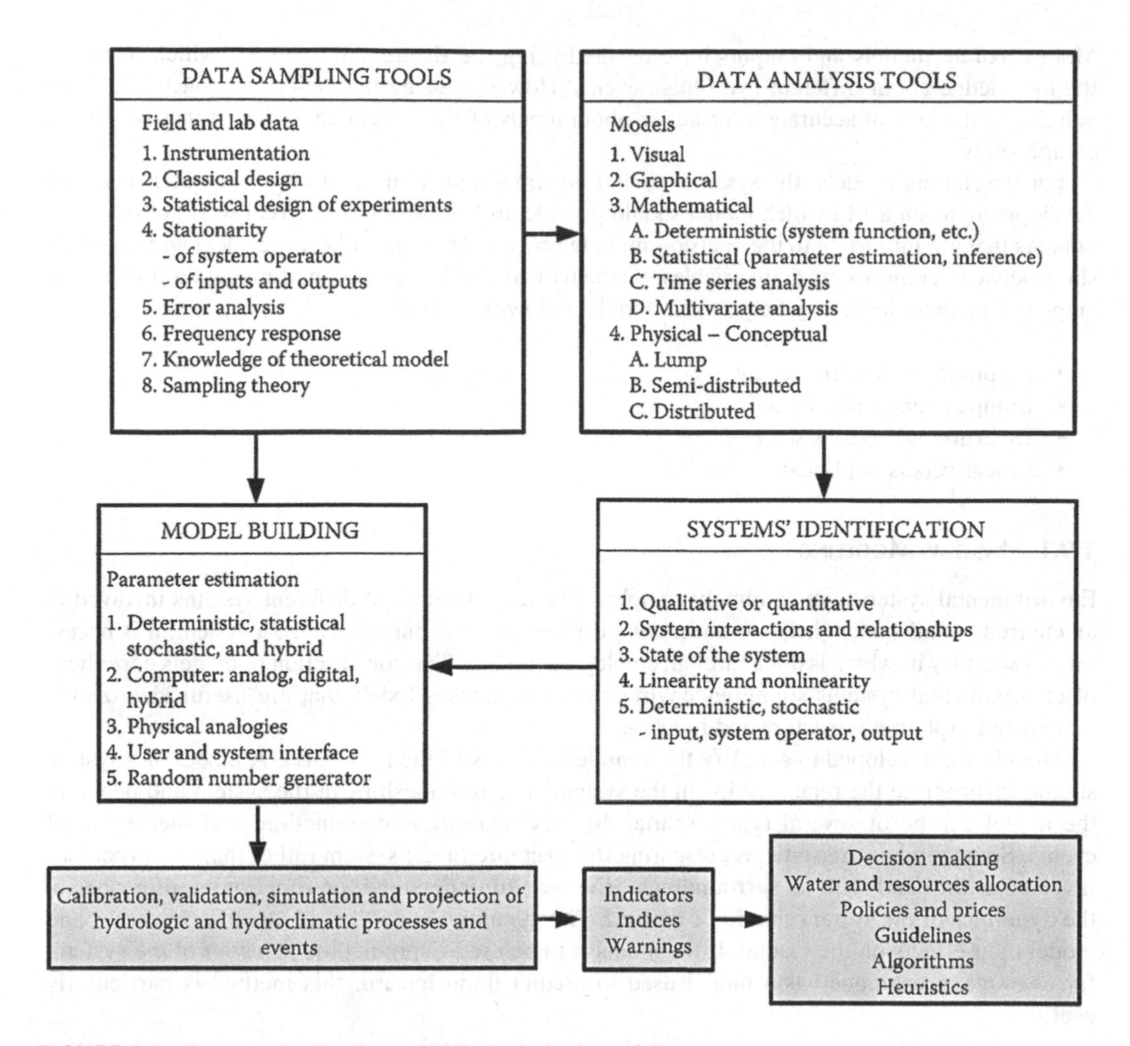

FIGURE 1.8 Framework of tools needed for hydrologic modeling.

process. In model validation, the model's response and performance are tested and verified based on the data that are not used to estimate the model parameters.

Following the calibration–validation process, the model can be used for the prediction (simulation) and/or projection of the hydrology and hydroclimatology systems. It is also worth remembering that simulation analysis provides only localized or limited information regarding the response properties of the system and the possible hydrologic and environmental trade-offs.

Complex systems are divided into subsystems, each having an input–output linkage as a component. The hydrologic systems are a subsystem of water resources that represents physical functioning of that system in a region. By analogy, a hydrologic system is defined as a structure or volume in space, surrounded by a boundary, that accepts water and other inputs, operates on them internally, and produces them as outputs (Chow et al., 2013).

Two approaches are considered for modeling hydrologic systems:

1. Theoretical approach, which models the physical components of a system
2. Empirical approach, which uses historical observations of different components of the hydrologic cycle to model its behavior

More accurate models and outputs have resulted using the theoretical approach which improves the knowledge about different hydrologic events. However, using theoretical approaches is difficult due to the lack of accurate information about inputs of the system and a wide range of natural complexities.

For developing models, the systems should be simplified. Loucks et al. (1981) refer to model development as an art in which model should provide an abstraction of the real world of the components that are important to the decision-making process. In terms of how they deal with relationships between components and variables, mathematical models, as a type of model built using the empirical methodology, may be grouped into the following categories:

- Empirical versus theoretical
- Lumped versus distributed
- Deterministic versus stochastic
- Linear versus nonlinear

1.9.1 Systems Modeling

Environmental systems are extremely complex. The identification of different systems involved in an environmental system helps to understand this complexity, but to analyze a system, it is necessary to simplify it, which is done through employing models. The construction of models or replicas of environmental systems simplifies environmental systems. Models that are useful idealize the system and explain its structure and functions.

Models are developed to simplify the complex systems of the real world. A model of a system should incorporate the relationships in the system. The relationships in the system and hence in the model can be of several types: spatial distance, causation, conjunction, and succession of events. Some models are static, representing the structure of the system rather than the processes involved within and with its surroundings. However, to understand the functioning of a system, the dynamics of the system should be defined. The dynamic models could identify processes and model their effects on the system. If the model is to be used to predict the behavior of the system, for example, a drainage basin model used to predict flood hazard, this method is particularly useful.

1.9.2 Model Resolution

The level of resolution of the models is dependent on the selected scale for study of the system. With fine resolution, reality is better resembled, which allows important small-scale processes to be simulated, and with coarse resolution, there is a gain in generality, but it increases the dependence on uncertain parameterizations. The summarized model of the system cannot be truly the same as reality because it lumps together elements and relationships and valuable information inside each part is ignored.

Egler (1964) called this the "meat grinder" approach and compartment models. Each part of these models is treated as a black box, which can be defined as any unit whose function may be evaluated without specifying the contents. Models with very low resolution are composed of a few number of relatively large black box compartments and the entire system can be treated as a black box. By increasing the resolution level of the model, the size of the black box compartments decreases and their number increases. Models with an intermediate level of resolution are called gray box models and provide a partial view of the system. When most of the elements, states, relationships, and processes of the system are identified and incorporated in the model, the gray box becomes a white box. It should be noted that because a model could not be a complete representation of reality, even in a white box model, there are some black box components (Figure 1.9). See Chapter 14 for more details.

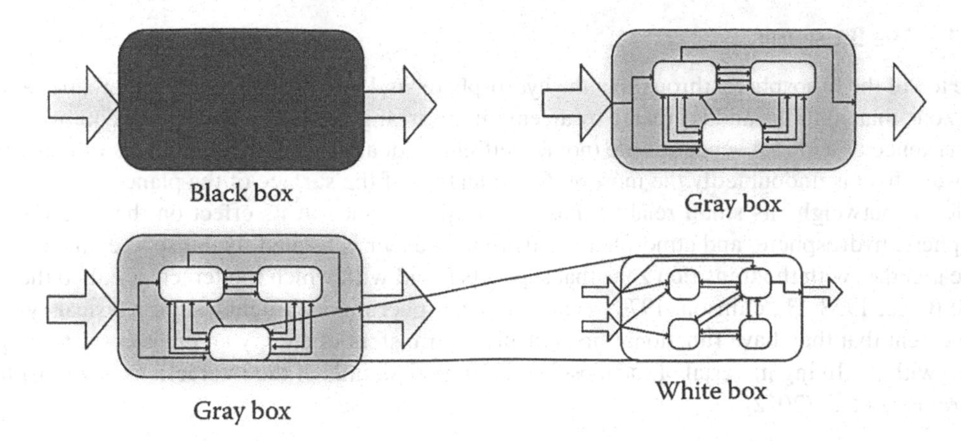

FIGURE 1.9 Levels of resolution in the modeling of systems. (From White, I.D. et al., *Environmental Systems,* Chapman and Hall, London, 1996. With permission.)

Example 1.2

Consider that simulation of produced runoff through a watershed is needed for a reservoir design. Discuss different modeling resolutions that could be applied for this purpose.

Solution:

Data availability and accessibility with different levels of resolution can be mentioned for this purpose. In cases with limited data and measurements of watershed properties, the black box models are preferred. For this purpose, different data-driven models such as artificial neural network, Fuzzy set theory, and statistical time series framework can be utilized. When some data on watershed characteristics and elements are available, some lumped rainfall–runoff models can be employed and the process of transformation of rainfall to runoff can be clarified. In this case, a gray box is developed. If there are enough detailed data about characteristics of watershed, complicated distributed models could be used to provide a level of white box models. However, it should be noted that it would not be a completely white box model. The details of the mentioned modeling approaches are given in the next sections.

Note: Data availability and accessibility as well as knowledge about the system function are the main factors in decision-making about the level of resolution of the system model.

With this general description of systems, we are trying to follow a system approach in the rest of this chapter (as well as in the rest of this book). This can promote system thinking in hydrologic analysis, design, prediction, and simulation.

1.10 THE HIERARCHY OF WATER FOR LIFE AND TOTAL SYSTEMS APPROACH

Different spheres that contain life and the flow of water and energy are characterized into a so-called "total systems" approach consisting of natural, human and institutional, and built environment (constructed systems). This is perhaps what we are trying to differentiate and elaborate with water as the focal point in this and other chapters. The fact that biosphere crosses through land (lithosphere), water (hydrosphere), and atmosphere gives livelihood and the notion of livable cities of the future a context that should be delivered in this book. So some discussion of biosphere could help to see the interaction between the triple systems.

1.10.1 The Biosphere

At the top of the lithosphere, throughout the hydrosphere, and into the lower atmosphere, lies a transition zone that contains and is created by an enigmatic arrangement of matter that we know as life. The existence of a global veneer of life (not forgetting the dead and decaying remains of that which once was alive) is undoubtedly the most profound feature of the surface of the planet.

Life far outweighs its small relative mass in the significance of its effect on the nature of the lithosphere, hydrosphere, and atmosphere. This living veneer is termed the biosphere, and the biosphere together with the transition zone that supports it and with which it interacts is called the ecosphere (Cole, 1958; Hutchinson, 1970). Hence, the ecosphere model includes the physical systems to the extent that they have functional links involving transfers of energy and matter (water in particular) with the living material of the biosphere. For more details of the interactions, see Chapter 2 of Karamouz et al. (2012).

1.10.2 System-Based Thinking

The success of total systems approach relies on how we position the three systems of natural, human, and constructed with respect to each other in water-related issues. Figure 1.10 shows this positioning. Natural system imposes load and sometimes hazards on the living environment. Even though, human and institutions provide resistance to the load imposed by the natural system, they are seldom exacerbating the effect of natural hazards due to many interventions including the constructed environment and the infrastructures that facilitates life and form the connectivity, mobility, and resources delivery, waste disposal, and other societal activities. System-based thinking for water provides realization of challenges, risks, linkages, setups, planning, strategies, tools, models, and gadgets to better manage and provide resistance to the loads inserted by the natural and the operational needs of supporting water infrastructures and their interdependencies with others. These elements are further discussed as they are related to each system with a focus on water related issues in the following sections.

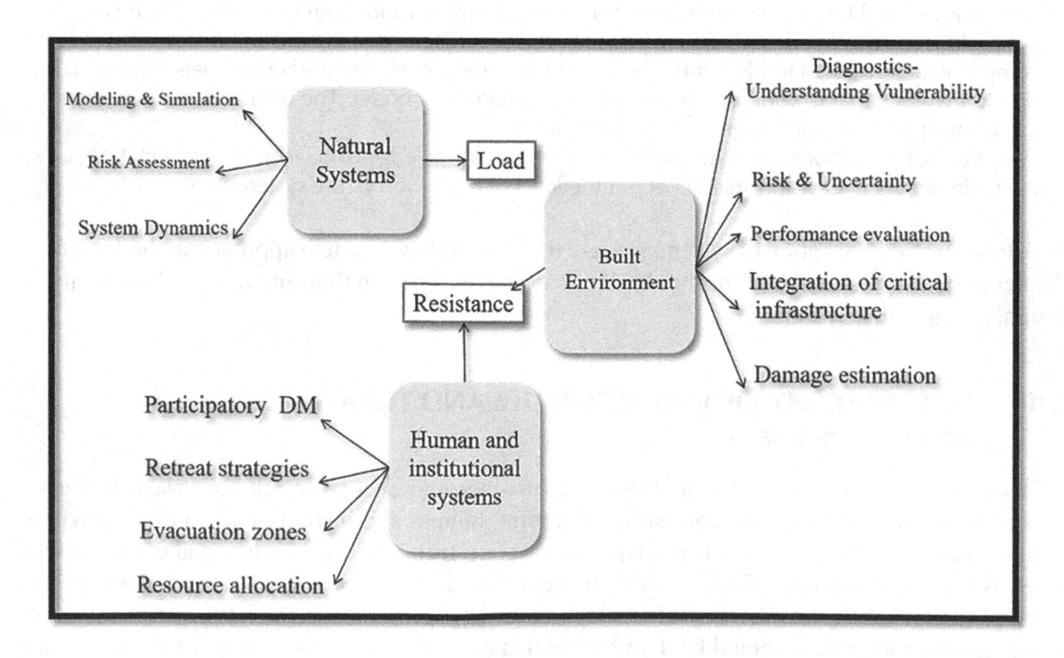

FIGURE 1.10 The human and institutional positioning, preparedness to manage built environment and face natural system load and water hazards.

1.10.3 Natural Systems

The focus is risk and preparation. Our concentration is on urban, coastal, and riparian systems that include weather events and climate change and variability, waterways and shoreline/coastal areas, environment effects, ecosystem dynamics, and land/bay interactions.

1.10.4 Human and Institutional Systems

The focus is response and strategy. Human shelters and utilities need to be built and protected. Adaptation to climate is main response of communities and societies. Therefore, preventive maintenance and state of preparation are the key strategies. Mitigation and recovery need to be planned. The institutional system is set up for investment and financing, insurance, law and societal expectations, government role and water governance for QA/QC, and setting standards and compliance. The policy criteria should have an overreaching characteristic for implementation and continuous revisitation. The overall objective often is economic welfare.

1.10.5 Built Environment—Infrastructure

The focus is resilience and reliability. There has been a change of paradigm in dealing with natural hazards such as flood from flood prevention and climate resiliency as standalone projects, focus on budget/costs, look at projects separately, focus on investment to interventions that can be seen as a function of other projects, focus on value (value creation and proposition), look at interrelationships, and focus on life cycle effects. Housing and commercial and industrial buildings are often targets of natural hazards that need to be protected; however, urban water systems have a dual function and vulnerabilities towards natural hazards such as flood. That is, their structural integrity and their state of operations should be protected. The thirst level of concern is the interdependencies to other infrastructures such as transportation and mobility, energy/utilities, and communications that could have a ripples effect on water infrastructure and essential services delivery.

1.10.6 Disasters and Interdependencies

The linkage and interaction of total systems components could be seen and examined in case of disasters such as floods. Figure 1.11 shows outer layer of interaction and shared activities between and among the three systems.

As can be seen in Figure 1.11, this structure is organized around three domains:

1. *Natural systems* affect, and are affected by, hydroclimatic risks such as floods and hurricanes. The main components of this domain are urban, coastal, and riparian systems, ocean and rivers, environment and climate change, ecosystem dynamics, and land/bay interactions.
2. *Constructed systems* (lifeline reliability and resiliency) provide the essential services in urban areas, as well as those used for flood management. These systems include housing, transportation, energy, communication, and urban water systems.
3. *Human and institutional systems* (governance, society and finance) are the basis for flood planning and management. Stakeholders in banking, insurance, and other industries, government agencies, law and societal expectation, policy criteria, economic welfare, standards, and public health all play a role in shaping these efforts.

Each of these domains has interactions and interdependencies with each other that can be measured and/or quantified to increase reliability and resiliency and decrease vulnerability of the coastal region. These interactions, in a finer and more complex structure, are presented in Figure 1.12 with the activities that will elucidate the inner and outer interactions and interdependencies within

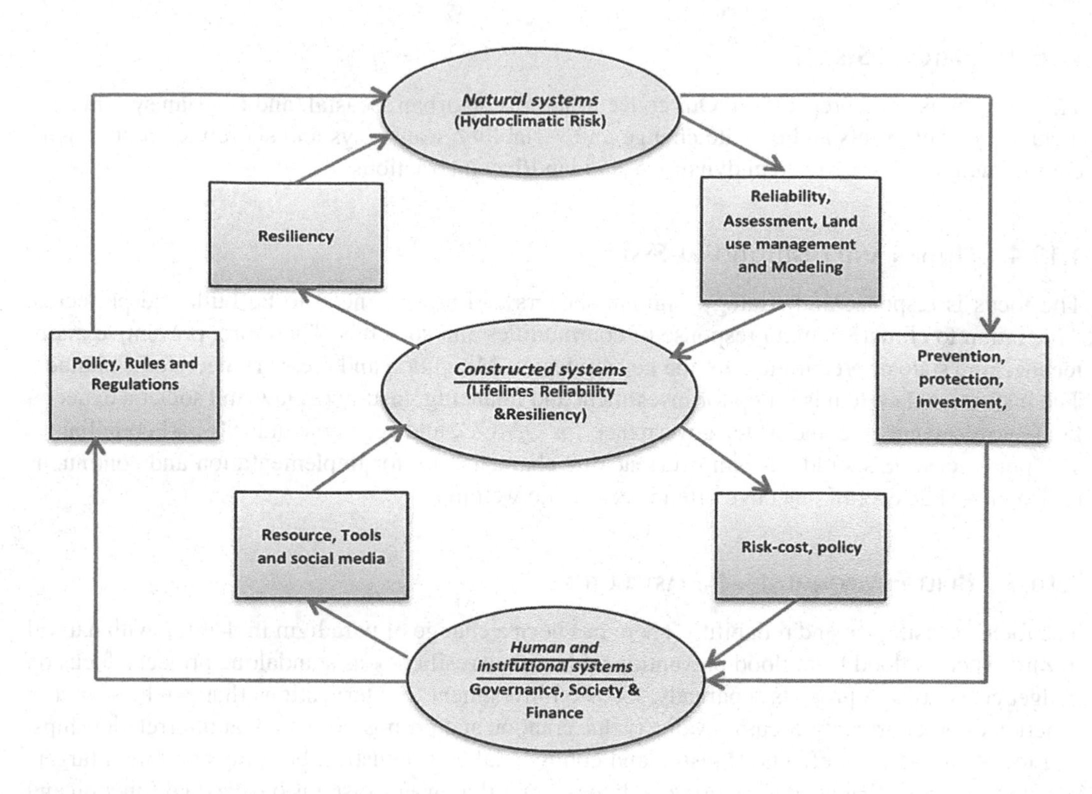

FIGURE 1.11 Disasters and interdependencies in the context of total systems approach.

and across the three domains. A better understanding of this feedback-based dynamic system can increase the adaptive capacity of the entire system and contribute to effective plans for the future.

1.11 PEOPLE'S PERCEPTION—PUBLIC AWARENESS

It may be claimed that having clean water was a luxury until recently, as delegations from impoverished nations frequently said in the United Nations. As they race towards economic growth, emerging nations should avoid making the same mistakes as industrialized nations and undermining environmental restrictions in their industrial and urban expansion.

In most of the developed countries, legislation for pollution control was introduced in the 1960s and 1970s. The United States Environmental Protection Agency (USEPA) was created in 1970 to administer environmental programs. It was an encouraging effort, but many details were left out at that time. The United Nations Conference on the Human Environment in 1972 in Stockholm focused on starting a global effort to bring the governments together to work on a shared vision about water and environment. Many more UN conferences followed, addressing topics such as population, food, women's rights, desertification, and human settlements, with a focus on environmental challenges.

Eckholm (1982) describes the enormous task faced by the developing countries, such as the provision of reasonable clean water and disposal facilities and the enforcement of sanitary standards. He estimated that, at that time, half of the people in the developing countries did not have reasonable access to safe water supplies and only one-fourth had access to adequate waste disposal facilities. Today, many people around the world still do not have access to clean water.

It will take extraordinary stewardship and wisdom at the national and international level to balance the economic development with the amount of sustainable water resources and the carrying capacity of rivers and aquifers to withstand quality degradation.

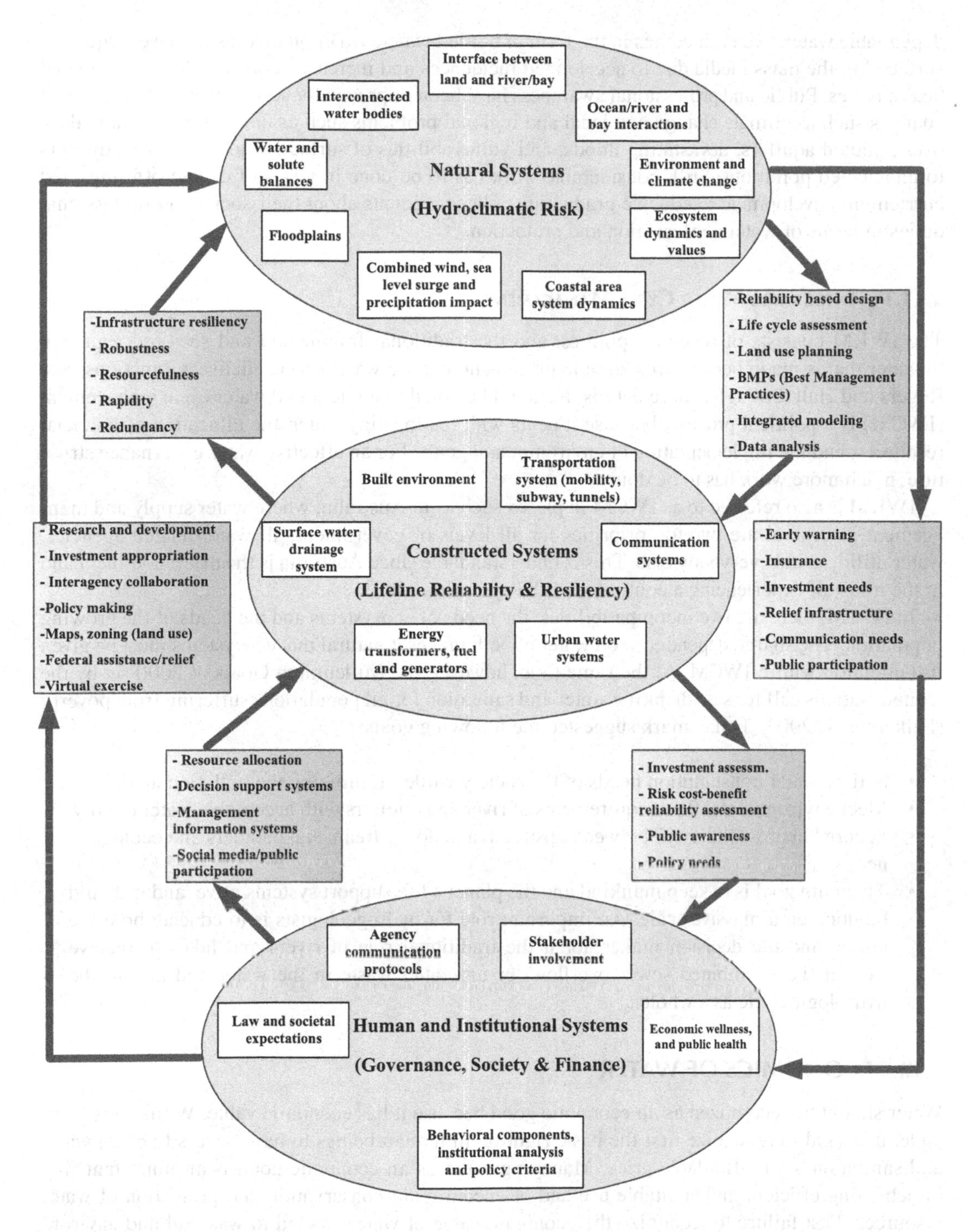

FIGURE 1.12 Interdependencies across and within domains.

When the social costs of welfare and unemployment have caused huge financial deficits for governments in the developed world and have brought many developing and underdeveloped countries to the brink of financial disaster, a high priority for environmental improvement will be difficult to impose.

Water scarcity, security, and regular droughts have modified people's perceptions of water as a God-given resource with no intrinsic monetary worth. People are now prepared to pay for safe, dependable water, even if it comes in the form of bottled water. Water quality issues have frequently surfaced in the mass media due to accelerated incidences and increased concerns for water-related health issues. Public and professional awareness have been improved for certain water-related global changes such as climate change and local and regional problems such as polluted rivers and lakes, over polluted aquifers, devastating floods, and vulnerabilities of surface waters and groundwaters to undetected pollutions. Still, considerable work has to be done in terms of course offerings and curriculum development to educate graduating college students about their social and professional duties in terms of water conservation and protection.

1.11.1 Integrated Water Cycle Management

The IWRM consists of schemes, politics, and the traditional fragmented and sectored approach to water that is made between resource management and the water service delivery functions. See Rogers and Hall (2003) for more details. It should be noted that integrated water cycle management (IWCM) is a political process, because it deals with reallocating water, the allocation of financial resources, and the implementation of environmental goals. For an effective water governance structure, much more work has to be done.

IWRM is also referred to as IWCM in places such as in Australia, where water supply and management in general are two top priorities for all levels of government, all water-related agencies, water utilities, and everybody else. This is understandable since Australia is the driest continent and in the midst of experiencing a continuous 1,000-year drought.

In IWCM, there are two incompatibilities: the needs of ecosystems and the needs of the growing population. The shared dependence on water of both makes it natural that ecosystems must be given full attention within IWCM. At the same time, however, the Millennium Goals of 2000 set by the United Nations call for safe drinking water and sanitation for all populations suffering from poverty (Falkenmark, 2003). Falkenmark suggested the following goals:

- Satisfy water consumption needs of the society while minimizing the pollution load.
- Meet environmental flow requirements of river and aquifers with acceptable water quality.
- Secure "hydro-solidarity" between upstream and downstream stakeholders and ecological needs.
- The main goal is to keep mankind and the planet's life support systems alive, and it should be done on a massive scale. An important role for hydrogeologists is to educate both the public and the decision maker about the traditional role of rivers and lakes to receive treated (i.e., combined sewer overflow) or untreated waste in the watershed and in the hydrologic cycle as a whole.

1.12 ECONOMICS OF WATER

Water should be recognized as an economic good because it has economic value. Within this principle, it is vital to recognize first the basic right of all human beings to have access to clean water and sanitation at an affordable price. Managing water as an economic good is an important way of achieving efficient and equitable use and of encouraging conservation and protection of water resources. Past failure to recognize the economic value of water has led to wasteful and environmentally damaging uses of the resource (Karamouz et al., 2011).

Van der Zaag and Savenije (2006) explained the overall concerns over the true meaning of "water as an economic good" based on two schools of thought. The first is the pure market-driven price. Its economic value would arise spontaneously from the actions of willing buyers and willing sellers. This would ensure that water is allocated to uses based on its highest value. The second school interprets the process of allocation of scarce resources in an integrated fashion, which may not involve financial transactions (McNeill, 1998). From this point of view, as water is a scarce resource, the allocation of the resource aims to achieve the balancing of demand and supply. Developing a water pricing framework based upon the cultural and social framework in a region, which could include subsidies for minimum consumption in low-income communities, is developed.

In a broader sense, a unique property of water is that it belongs to a system and it always affects many users. Any change in upstream water will affect the entire system downstream. Temporal and spatial variability of water resources is also constantly changing due to short- and long-term climate and land use changes. Water could have negative economic value in case of floods. All of these make it difficult to establish the value of external impacts on water use.

The price of water could be defined as the price water users are willing to pay per unit volume of water delivered per unit of time (e.g., cubic meters). In most cases, neither water users nor self-providers pay the full price (value) of water, which should be equal to the real value of water and should include the following:

- Capital cost of water withdrawal and distribution system
- Cost of operating and maintaining the system
- Investments for augmenting the existing and/or finding new sources of water and expanding the water transfer and distribution system
- Source protection cost reflecting its intrinsic value
- Environmental cost including the reduction of natural flows, water quality degradation, and loss of habitat
- Sustainability cost including the cost to bring the supply and demand to a level that could be managed over an extended period of time

Kresic (2008) described the full price and value of water. Among different costs, it is difficult to assess the environmental and sustainable costs. To consider the environmental costs, many intangible costs should be identified and quantified. This has some common attributes with sustainable cost when considering environmental sustainability. See Chapter 8 for more details.

1.13 CLEAN WATER ACT

Growing public awareness and concern for controlling surface water pollution resulted in the enactment of the Federal Water Pollution Control Act Amendments of 1972. It was amended in 1977, and this law became commonly known as the Clean Water Act (CWA). Major amendments were also enacted in the Water Quality Act (WQA) of 1987. The act established the basic structure for regulating discharges of pollutants into the surface waters of the United States. It gave the USEPA the authority to implement pollution control programs such as setting wastewater standards for industry. The CWA also continued requirements to set water quality standards (WQSs) for all contaminants in surface waters. The act made it unlawful for any person to discharge any pollutant from a point source into navigable waters, unless a permit was obtained under its provisions. It also funded the construction of sewage treatment plants under the construction grants program and recognized the need for planning to address the critical problems posed by nonpoint source pollution. In the WQA of 1987, the stormwater problem was considered. It made it a requirement for industrial stormwater dischargers and municipal dischargers to separate their storm sewer system. The permit exemption for agricultural discharges continued, but Congress created a nonpoint source pollution demonstration grant program at the USEPA to expand the research and development of nonpoint controls and management practices.

The key elements of the CWA are establishment of WQSs, their monitoring, and, if WQSs are not met, developing strategies for meeting them. To keep to the National Primary Drinking Water Regulations, which set enforceable maximum contaminant levels for particular contaminants, the policies and programs against degradation are employed, only if all WQSs are met, otherwise the water should be treated to reach the standards. Each standard includes requirements for water systems to test for contaminants in the water to make sure standards are achieved. Maximum contaminant levels are legally enforceable, which means that both the USEPA and individual states can take enforcement actions against water systems not meeting safety standards. For more details, refer to Copeland (2010). The USEPA and states may issue administrative orders, take legal actions, or fine water utilities. In contrast, the National Secondary Drinking Water Standards are not enforceable. The agency advises water utilities and professional bodies to implement compliance systems but does not require them. States of the union in the United States, on the other hand, may choose to embrace them as binding norms or to loosen them. The foundations of water law that regulate and are enforced in the states are presented in the following section.

1.13.1 The Basis of State Water Laws in the United States

Technological changes and population growth have stressed legal regimes for water in general and for managing surface waters in particular. Water law must become a tool for accomplishing water environmental sustainability rather than serving to perpetuate water pollution and water right issues (Dellapenna, 1999). Global climate disruption adds further complexities to the stresses facing water law regimes in the United States. Water will be at the center of adaptation to climate change because it is one of the most important resources for human existence and well-being. As water management systems struggle to adapt to global climate change, current water law frameworks are put under strain. Global climate change adds stress to existing water law regimes as water management systems struggle to adapt. Dellapenna (1999) stated that technological changes in particular have not eased the problems. It has put greater burden on rising water demand in the home and industrial sectors, while moderately reducing water demand in agricultural sectors through sophisticated irrigation techniques.

The basis of applicable laws in the United States goes back to the time of settlement of English colonies. They have been called the "riparian rights" following the Latin term *ripa* (the bank of a stream). They have limited the right to use water to those who owned or leased land having a water source. A riparian owner for the most part was free to make any use wanted so long as no one complained about direct interference with his or her water use. See Dellapenna (2009) for more details.

Riparian rights were in effect in 32 states as recently as 1950, and they continue to rule numerous governments today. Traditional riparian and traditional appropriative rights have been rendered obsolete by a combination of rising population, rising demand (including environmental demand), and climatic change. The challenge for both rights derives from the fact that water is utilized repeatedly by different users as it passes through the water cycle. According to Hardin (1998), lack of effective restraint on water use under riparian rights allows, and almost requires, users to exhaust the resource either by withdrawing and consuming the water or by polluting the water. This observation about water use still holds in many places in the United States and around the world.

As stated in detail by Dellapenna (2009), because of very different legal traditions, governments in the 32 riparian rights states and the 18 appropriative rights states have been faced with very different challenges. In riparian rights states, security of investment for water users as well as tightening protections for the public interest in the waters should be provided. In appropriative rights states, the needed flexibility into an increasingly sclerotic system of water rights is introduced while adequate protection to the public interest in waters should also be addressed. Devising proper responses in riparian rights states is easier than the other, because the water rights in riparian rights states are less firmly established than those in appropriative rights states.

The established systems of water rights have already been started to change by state governments. The 2009 list of current surface water law regime shows four classifications: 8 states such as Arizona, Colorado, Montana, and Utah, as most affected, follow appropriate rights, whereas 16 states such as Alabama, Indiana, Maine, Missouri, and West Virginia, as most affected, have riparian rights. The other 26 states have dual systems (ten states) or regulated riparian (16 states).

More than half of the states within the riparian tradition have abandoned traditional riparian rights in favor of publicly managed regulated riparianism. This transformation is likely to expand until traditional riparian rights in the 32 states that make up the riparian tradition have been completely altered by regulated riparianism. In the 18 appropriative rights states, many claims are being made that markets will solve water management problems. See Anderson and Snyder (1997) for more details. However, it has been demonstrated that true markets for water have only operated at small scales and within a short domain. Indeed, such changes as experienced in the western part of the United States have resulted from state intervention to reallocate water and not for market transitions. Concerns regarding ecosystem protection and property rights could also stymie the development of additional reservoirs or other infrastructure needed to deal with changing precipitation patterns and water storage requirements. Careful attention to the proper dimensions of legal rights might help alleviate these problems, but they are still more politically than legally motivated. See Dellapenna (2009) for more details.

1.14 CONCLUDING REMARKS AND BOOK'S ORGANIZATION

This chapter is the introduction to a *water system analysis, design, and planning* text book with concentration on urban infrastructure. The way components and effects of urbanization and the physical, chemical, and biological impacts have been *analyzed* is presented. Urbanization disturbs and changes different natural processes, so anything that is affected by urbanization and the complexity that offers should be *designed* differently. The effects of urban area developments in different aspects of climatic, hydrologic, cultural, and other special effects such as greenhouse and hot islands have been discussed in this chapter. These effects must be *planned* for future designs and decision-making about urban development, especially regarding the supply of water, storm management, and wastewater collection and treatment.

This chapter presents three distinct pillars for analysis, design, and planning: urban water cycle and variability as the state of water being; land scape architecture as the medium for built by design; and total systems as the planning approach. Even though these pillars are not exactly followed in the book chapters, many of their reflections could be seen in the way materials are presented and case studies are given. Each instructor of this book could design her or his way moving towards this emerging water analysis, design, and planning evolution.

The management of the urban water cycle and urban water infrastructures not only has been briefly mentioned in this chapter but also has been elaborated in many chapters of this book. Three chapters are dedicated to urban hydrology and hydraulics, three chapters are water infrastructures, four chapters to economics and asset management, systems analysis tools, risk and uncertainty, and disaster management. Final chapters are a journey through hydrologic and hydrodynamic simulation including offshore modeling, flood resiliency of cities, and environmental visualization, discussing a number of emerging issues with an eye on livable cities of the future.

PROBLEMS

1. Assess surface water and groundwater resources around the world and compare them with the US share of total surface water and groundwater resources. Hint: use UNESCO, World's water (www.worldwater.org), World Bank, and USGS sites, and publications.
2. Identify the federal and national agencies directly involved in the management of the hydroclimatic cycle, at least one in each part of the globe.
3. Name the main components of the urban water cycle and discuss their relationship.
4. Explain IWRM for the hydrologic cycle in the context of public awareness/participation and consideration of social impacts.
5. What is the estimated water cost for residential customers in the eastern (New York City) and western (Los Angeles) part of the United States? Compare it with the price of water

in five other major cities around the world. Is the price of groundwater in selected areas different from surface water resources?

6. To understand water conflict, briefly find and discuss a case of transboundary conflict over shared rivers, lakes, or aquifers.
7. Search and describe the current status of riparian rights and their transition to regulated water laws.
8. Identify and briefly explain factors affecting water supply schemes sustainability.
9. Identify three local (city or county) agencies in the United States or in other countries with the charter of protection and enforcement issues related to surface water resources.
10. Explain the different levels of modeling resolution that are commonly used in rainfall prediction.
11. Provide examples of different closed and open systems in the urban water cycle.
12. Explain the main difference between urban water cycle and hydrologic cycle in the context of systems approach.
13. Specify different components of groundwater system as black or white box. Explain how interactions with surface water can impact black and white box configuration.
14. Consider that river flow prediction is needed for a reservoir sizing. Describe the system concept of this problem at the design and synthesis stages.
15. Table 1.3 shows the measured rainfall and surface runoff from a 50 km^2 watershed. The given values are the averages over time intervals. Compute the accumulated storage (m^3) of the water within the watershed. Plot on the same coordination, the inflow, outflow, and cumulative storage as functions of time.
16. Identify the processes of the hydrologic cycle that affect flood runoff from a 1.5 km^2 forested lot. Discuss the relative importance of each of the processes. If the lot is cleared, what changes in the processes will occur? What will their importance be? If a single-family residence is constructed on the cleared lot, what processes will control flood runoff from the lot?
17. For a circular pipe flowing full, the head loss (h) can be computed using the Darcy–Weisbach equation:

$$h = f\left(\frac{L}{D}\right)\frac{V^2}{2g}$$

where f is the dimensionless friction factor that is a function of the Reynolds number and pipe roughness; L and D are the length and the diameter of the pipe, respectively; and V is the velocity of flow. A manufacturer intends to develop a pipe made of a new ceramic material for which f values are not available. Discuss the task in terms of analysis and synthesis.

TABLE 1.3
Measured Rainfall and Surface Runoff for Watershed in Problem 15

Time (minutes)	Rainfall (cm)	Runoff (m^3/s)
0	0	0
10	0.07	2.1
20	0.15	9.5
30	0.21	18.9
40	0.16	37.3
50	0.09	48.8
60	0.05	52.6

TABLE 1.4
Annual Precipitation and Evaporation of Catchment in Problem 21

Subarea	Area (m^2)	Annual Precipitation (mm)	Annual Evaporation (mm)
A	10.7	1,030	535
B	3.0	830	458
C	8.3	900	469
D	17.9	1,300	610

18. Estimate the constant rate of withdrawal from a 1,400 ha reservoir in a month of 30 days during which the reservoir level dropped by 0.7 m in spite of an average inflow into the reservoir of 0.5 MCM/day. During the month, the average seepage loss from the reservoir was 2.5 cm, the total precipitation on the reservoir was 18.5 cm, and the total evaporation was 9.5 cm.
19. A catchment has four subareas. The annual precipitation and evaporation from each of the subareas are given in Table 1.4. Assume that there is no change in groundwater storage on an annual basis and calculate for the whole catchment the values of annual average (a) precipitation and (b) evaporation.
20. A storm has an average rainfall depth of 9 cm over a 200 km^2 watershed. What size reservoir would be required to contain completely 20% of the rain?
21. A storm with a uniform depth of 3 cm falls on a 15 km^2 watershed. Determine the total volume of rainfall. If all of the water were collected in a storage basin having vertical walls and an area of 1,000 m^2, how deep a basin would be needed?

REFERENCES

Anderson, T.L. and Snyder, P. (1997). *Water Markets: Priming the Invisible Pump.* Cato Institute, Washington, DC.

Borchardt, D. and Statzner, B. (1990). Ecological impact of urban stormwater runoff studied in experimental flumes: Population loss by drift and availability of refugial space. *Aquatic Sciences*, 52(4), 299–314.

Chen, Z., Wu, G., Wu, Y., Wu, Q., Shi, Q., Ngo, H.H., Vargas Saucedo, O.A., and Hu, H.Y. (2020). Water Eco-Nexus Cycle System (WaterEcoNet) as a key solution for water shortage and water environment problems in urban areas. *Water Cycle*. doi: 10.1016/j.watcyc.2020.05.004.

Chow, V.T., Maidment, D.R., and Mays, L.W. (2013). *Applied Hydrology. Second Edition.* Tata McGraw-Hill Education, New York.

Cole, L.C. (1958). The ecosphere. *Scientific American*, 198, 83–92.

Copeland, C. (2010). Clean water act: A summary of the law. Congressional Research Service.

Cortesi, I., Ferretti, L.V. and Morgia, F. (2020). Soil and water as resources: How landscape architecture reclaims hydric contaminated soil for public uses in urban settlements. *Sustainability*, 12, 8840. doi: 10.3390/su12218840.

Dellapenna, J.W. (1999). Adapting the law of water management to global climate change and other hydropolitical stresses. *Journal of the American Water Resources Association*, 35, 1301–1326.

Dellapenna, J.W. (2009). In Beck, R.E., and Kelley, A.K. (eds.) *Waters and Water Rights* (3rd ed., Vol. 1). LexisNexis, Newark, NJ.

Eckholm, E. (1982). *Down to Earth: Environment and Human Needs.* Pluto Press, London.

Egler, F.E. (1964). Pesticides in our ecosystem. *American Scientist*, 52, 110–136.

Falkenmark, M. (2003). Water management and ecosystems: Living with change. Global Water Partnership Technical Committee, Global Water Partnership, Stockholm.

Hardin, G. (1998). Extensions of "the tragedy of the commons". *Science*, 280 (5364), 682–683.

Hassler, U., Algreen-Ussing, G., and Kohler, N. (2004). Urban life cycle analysis and the conservation of the urban fabric. SUIT Position Paper (6).

Hutchinson, G.E. (1970). The biosphere. *Scientific American*, 233 (3), 45–53.

International Council for Local Environmental Initiatives (ICLEI). (2005). Why should cities and counties care about urban heat Islands? Available at http://www.iclei.org/ (visited in December).

Karamouz, M., Szidarovszky, F., and Zahraie, B. (2003). *Water Resources Systems Analysis*. Lewis Publishers, Boca Raton, FL.

Karamouz, M., Moridi, A., and Nazif, S. (2010). *Urban Water Engineering and Management*. CRC Press, Boca Raton, FL, 595 p.

Karamouz, M., Ahmadi, A., and Akhabri, M. (2011). *Groundwater Hydrology: Engineering, Planning and Management*. CRC Press, Boca Raton, FL.

Karamouz, M., Nazif, S., and Falahi, M. (2012). *Hydrology and Hydroclimatology: Principles and Applications*. CRC Press, Boca Raton, FL.

Kresic, N. (2008). *Groundwater Resources: Sustainability, Management, and Restoration*. McGraw-Hill Professional, New York.

Labi, S. (2014). *Introduction to Civil Engineering Systems: A Systems Perspective to the Development of Civil Engineering Facilities*. Wiley. https://www.wiley.com/en-us p-9781118415306.

Lan, T., Lin, K., Xu, C., Tan, X., and Chen, X. (2019). Dynamics of hydrological model parameters: calibration and reliability. *Hydrology and Earth System Sciences*. doi: 10.5194/hess-2019-544.

Loucks, D.P., Stedinger, J.R., and Haith, D.A. (1981). *Water Resources Systems Planning and Analysis*. Prentice-Hall, Englewood Cliffs, NJ, 559 pp.

Marsalek, J., Cisneros, B.J., Karamouz, M., Malmquist, P.A., Goldenfum, J.A., and Chocat, B. (2008). *Urban Water Cycle Processes and Interactions: Urban Water Series-UNESCO-IHP* (Vol. 2). CRC Press, Boca Raton, FL.

McCuen, R.H. (2005). *Hydrologic Analysis and Design* (2nd ed.). Prentice Hall, Upper Saddle River, NJ.

McNeill, D. (1998). Water as an economic good. *Natural Resources Forum*, 22(4), 253–261.

Morgia, F. (2007). *Catastrofe istruzioni per l'uso* (1st ed.). Meltemi, Roma, Italy, pp. 96–98.

Nile, B.K. (2018). Effectiveness of hydraulic and hydrologic parameters in assessing storm system flooding. *Advances in Civil Engineering*, 2018, 17. doi: 10.1155/2018/4639172.

RBD (Rebuild by Design). (2014a). The big 'U' promoting resilience post Sandy through innovative planning, design, & programming. http://www.rebuildbydesign.org/data/files/675.pdf.

RBD (Rebuild by Design). (2014b). Hudson river project team's proposal. http://www.rebuildbydesign.org/data/files /672.pdf.

RBD (Rebuild by Design). (2014c). The hunts point lifelines team's proposal. http://www.rebuildbydesign.org/data/files/677.pdf.

RBD (Rebuild by Design). (2014d). The hunts point lifelines team's proposal. http://www.rebuildbydesign.org/data/files/677.pdf.

Rogers, P. and Hall, A.W. (2003). Effective water governance, TEC Background Papers No. 7, Global Water Partnership Technical Committee (TEC), Global Water Partnership, Stockholm.

Schueler, T.R. (1992). A Current Assessment of Urban Best Management Practices: *Techniques for Reducing Non-point Source Pollution in the Coastal Zone*. Metropolitan Washington Council of Governments, Washington, DC.

Shahin, M., (2007). Water resources and hydrometeorology of the Arab region (Vol. 59). Springer Science & Business Media.

UN (United Nations). (2016). The sustainable development agenda. Paris, France. Available at: https://www.un.org/sustainabledevelopment/development-agenda/.

Van der Zaag, P.V. and Savenije, H.H.G. (2006). Water as an economic good: The value of pricing and the failure of the markets. Value of Water Research Report Series.

White, I.D., Mottershead, D.N., and Harrison, S.J. (1996). *Environmental Systems*. Chapman and Hall, London.

Ytuarte, S.L. (2005). Urban hot island phenomenon of San Antonio Texas using time series of temperature from Modis images. *Geological Society of America Abstracts with Programs*, 37(3), 4–5.

Zeunert, J. (2016). *Landscape Architecture and Environmental Sustainability: Creating Positive Change through Design* (1st ed.). Bloomsbury, London, UK.

2 Urban Water Cycle and Interactions

2.1 INTRODUCTION

Water is the essence of life and is moving through Earth and its atmosphere in all three phases of gas/vapor, liquid, and solid. Evaporation of water from water bodies such as oceans and lakes, formation and movement of clouds, rain and snowfall, streamflow, and groundwater movement are some examples of the dynamic aspects of water. The various aspects of water related to the Earth can be explained in terms of a cycle known as the hydrologic cycle. The urban water cycle (UWC) is a subset of hydrologic cycle, and in order to better understand that, a brief description is presented in Figure 2.1. It includes the water interaction between the atmosphere and the Earth, namely, precipitation, infiltration, evaporation, and evapotranspiration, as well as water movement and storage on and under the surface of the Earth, namely, surface runoff, subsurface flow, groundwater flow, reservoirs, and lakes.

For water management in time and space, the hydrologic cycle is a model of holistic nature. There are different definitions of the hydrological cycle, but it is generally defined as a conceptual model describing the storage and circulation of water between the biosphere, atmosphere, lithosphere, and hydrosphere (Karamouz et al., 2012). Water can also be kept in nature in the atmosphere, ice and snowpacks, streams, rivers, lakes, groundwater aquifers, and oceans. Water cycle components are affected by processes such as temperature and pressure variation and condensation.

A watershed is the best hydrological unit that can be used to carry out water studies and planning in a systematic manner. The urban setting could alter the natural movement of water. Drastic land use changes in urban areas as a subset of urban and industrial development affect natural landscapes and the hydrological response of watersheds. Although anthropogenic factors concerning waterways, pipes, abstractions, and man-made infrastructures affect the elements of the natural environment, the main structure of the hydrological cycle remains the same in urban areas (McPherson and Schneider, 1974). But the characteristics of the hydrologic cycle are greatly altered by urbanization impacts of the services to the urban population, such as water supply, drainage, and wastewater collection and management.

As a conceptual way of looking at water balances in urban areas, the context of the UWC is the total system approach. To distribute water to growing populations and cope with extreme weather and climatic variations and potential climate change, Lawrence et al. (1999) emphasized the importance of integrated urban water management.

Liu and Jensen (2018) found out in the search for both immediate solutions and long-term transitions towards sustainability that green infrastructures (GIs) are increasingly linked to urban water management. In their study, the GI-based urban water management practices of five cities famous for their progressive approach to water management were investigated. Based on reviews of open-source city plans and strategies, supplemented with information obtained directly from city managers, the purpose was to share best practices for the transition to sustainable urban water management and to gain insight into the role, if any, of GI in urban water management. An analytical frame based on transition theory was adopted.

All five cities represented states of transition at the near end of a sustainable urban water management scale. Despite some overlap in challenges concerning water supply, environmental protection, and flood risk management, the development target of each city was unique, as were their solutions. GI has been applied as a way to reduce water footprints in Singapore and Berlin in Germany, to

DOI: 10.1201/9781003241744-2

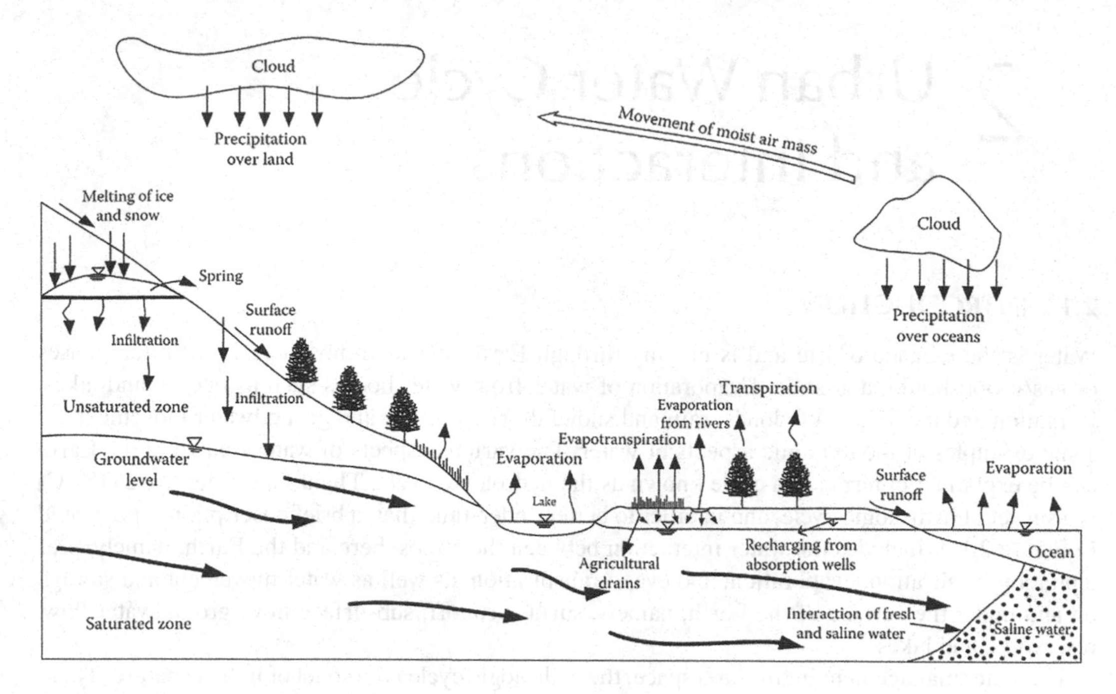

FIGURE 2.1 Hydrologic cycle and its components.

protect the environment in Philadelphia, USA, and to help save potable water for consumption in Melbourne in Australia and Sino-Singapore Tianjin Eco-city in China. Despite differences in scale, GI was, in all cases, applied as a supplement to the conventional water infrastructure. All five cities reveal a strong top-down approach towards sustainable urban water management and a strong mindset on GI's role for future development. However, all five cities point to similar challenges for GI implementation, including space and cost constrains as well as barriers for intersectorial and stakeholder collaboration, which limit the speed of city-wide upscaling of GI solutions and full realization of GI benefits.

This study was introduced here to indicate the need for a simultaneous change in the cognitive, normative, and regulative conditions of the urban water management regime for sustainability transition that is emphasized in this chapter.

Such a change requires a better balance between top-down and bottom-up planning to overcome barriers and foster innovation. The five cities jointly contribute to a noteworthy list of green solutions, city-wide strategies and guidelines, pilot project programs, regulations, and incentive programs, which may serve as inspiration for other cities' transition plans.

In this chapter, first, an estimation of precipitation, evaporation, evapotranspiration, and infiltration and excess rainfall calculations in the context of water balance are discussed. Analysis of surface runoff and subsurface flows is addressed in detail in Chapters 4 and 5, and details of groundwater flow are described in Karamouz et al. (2020). Furthermore, in this chapter, water balance in the hydrologic cycle and interaction of climatic, hydrologic, and urban components are discussed. Losses or abstractions in hydrology include evaporation, evapotranspiration, and infiltration are covered in detail in this chapter including physics-based methods such as the Penman evaporation equation and Green–Ampt infiltration. Snowmelt estimation, Palmer drought severity index (PDSI), and interactions between the UWC and urban infrastructure components are also presented. A case study of water balance-based water supply and demand sustainability is also included. Inclusion of PDSI is particularly interesting as it is perhaps the most widely used index to determine the state of water availability that agencies have to deliver to the customers. Finally, some attributes of livable cities of the future with emphasis on water elements are presented.

2.2 URBAN WATER CYCLE

A portrait of "A Man and His World" presents an ideal with all of the latest technological advancements and an abundance of facilities, utilities, and material. With this in mind, an urban population necessitates large amounts of resources and raw materials, as well as waste disposal. Environmental pollution is a byproduct of urban life. The main ingredients of urbanization and all key activities in modern cities are material and energy. The use of energy and material is evident in transportation, electric supply, water supply, waste management, heating, utilities, manufacturing, and so on. As a result, the concentration of people in urban areas changes material and energy fluxes significantly. The fluxes of water, sediment, chemicals, and microorganisms have been altered by drastic changes in landscape and land use, as well as the release of waste and heat. As a result of these changes, urban ecosystems, including urban waters, and natural resources and supporting facilities deteriorate. The number of these large cities continues to grow, especially in developing countries, exacerbating both human health and environmental issues on a regional scale.

2.2.1 Components—Water Movements

Changes in the material and energy fluxes in the urban water cycle are discussed in Chapter 1. Variations in the energy regime, such as air circulation patterns induced by buildings, transformation of land surfaces and land use planning, as well as water transfer, waste generation, and air quality variations, have long been recognized as having an effect on local microclimate. These changes are summarized in section 1.2.1 of Chapter 1.

Temperature, precipitation, evaporation, and infiltration in urban areas consequently result in changes in water cycle characteristics. Figure 2.2 shows the different components of the hydrologic cycle in urban areas. Each of these components is briefly explained in this section.

- *Temperature:* The temperature of the air over urban areas is normally 4°C–7°C higher than that of the surrounding areas. The higher evaporation rates (5%–20%) in urban areas are due to these thermal variations. Due to higher temperatures, convective storms can also be reported in urban areas.
- *Precipitation:* According to previous research, average annual precipitation in major industrialized cities is normally 5%–10% than in surrounding regions, and individual storms may have a % increase in precipitation. Because of much higher runoff coefficients, complexity in the conveyance system, congestion in transit corridors, and inadequate land use planning, urban areas are more vulnerable to storms.
- *Evaporation:* The mean temperature and evaporation are higher in cities due to the high rate of thermal and other sources of energy consumption. In rural areas, the presence of closed conduits and storage facilities can reduce evaporation rates as compared to open and surface flow and storage.
- *Transpiration:* Transpiration from trees and vegetation is normally reduced as a result of land use changes and the loss of open space and green areas.
- *Infiltration:* The following factors contribute to the decrease in infiltration rate:
  - Impervious areas (pavements, rooftops, parking lots, etc.)
  - Man-made drainage systems

Infiltration trenches and pervious pavements are examples of compensatory devices for urbanization effects that will improve infiltrated water flows in urbanized areas. The soil–water interaction is altered by pollutants from surface runoff, leaking underground tanks and absorption wells, as well as leaches from landfills. Therefore, the unsaturated zone in urban areas has different characteristics than other areas.

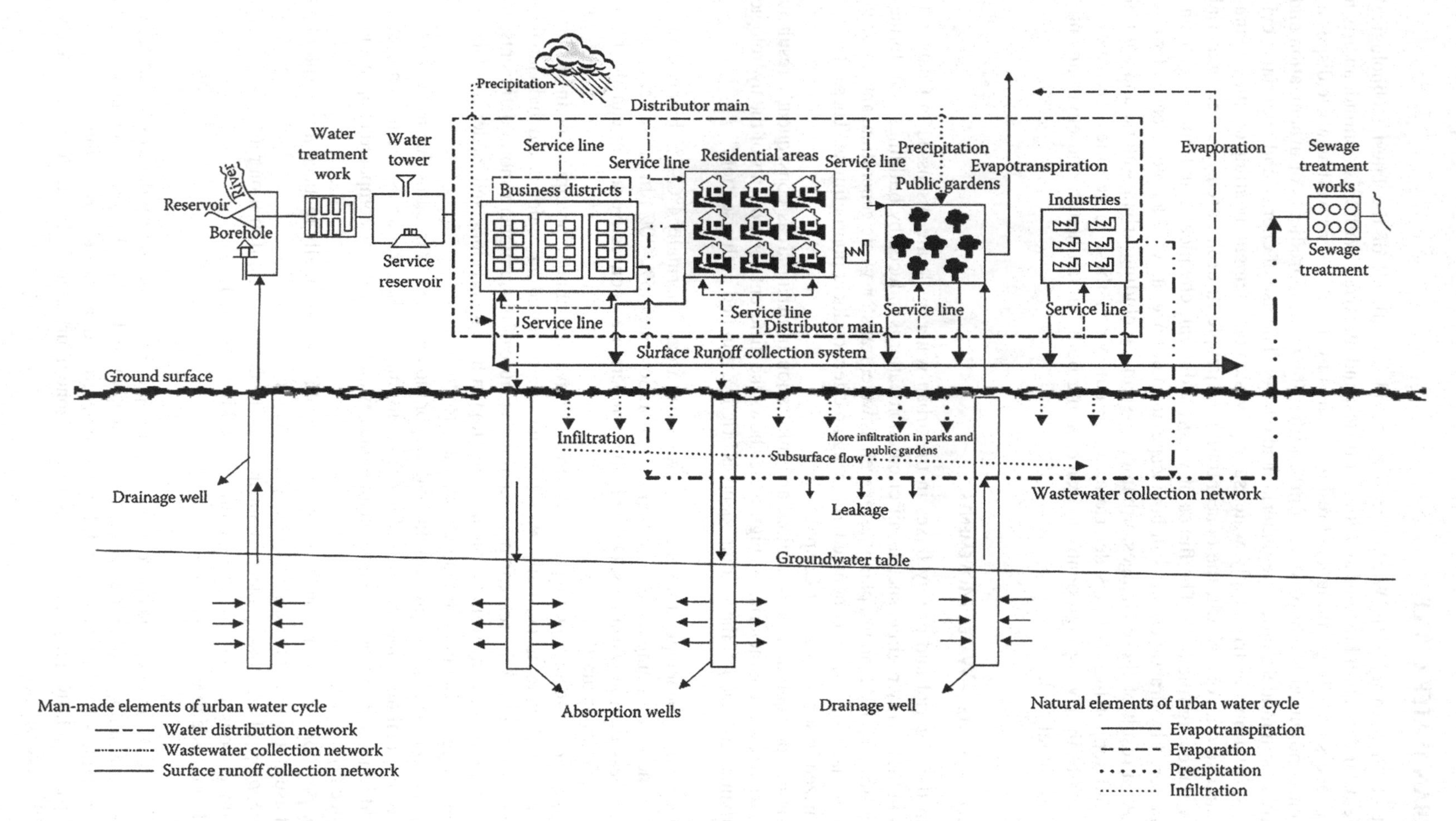

FIGURE 2.2 Different elements of the hydrologic cycle in urban areas. (From Karamouz, M., Moridi, A., and Nazif, S. *Urban Water Engineering, and Management*, CRC Press, Boca Raton, FL. (2010).)

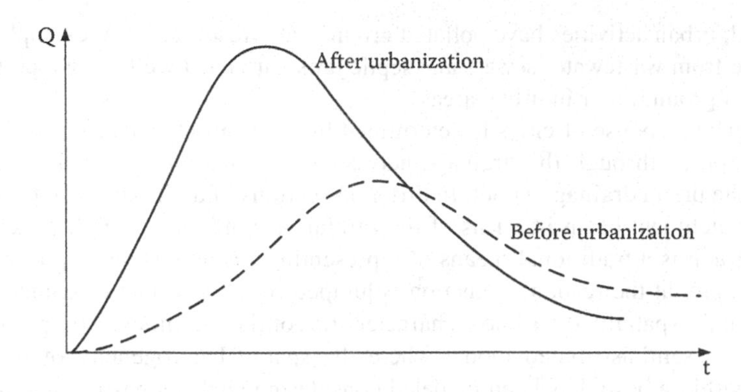

FIGURE 2.3 Unit hydrograph after and before urbanization.

- *Runoff*: In urban areas, the high proportion of impervious surfaces results in a higher runoff rate. Since man-made drainage systems make surface runoff collection and transfer easier, the peak of the unit hydrograph is higher and occurs earlier after urbanization. Figure 2.3 shows a comparison of an area's unit hydrograph before and after urbanization. After urbanization, the following change in order of magnitude could be expected for a drainage area of about 200 ha (1 ha = 10,000 m^2). The peak flow and runoff coefficient both increase by two to three times. The lag time between rainfall and runoff is reduced by three times, and the amount of runoff rises by three to four times.
- *Interflow*: Interflows in urban areas are often supplied by leakage from water distribution and wastewater collection networks, as shown in Figure 2.2, due to very low infiltration rates in urban areas. Increased interflows are also helped by higher infiltration rates in parks, green areas, and areas with absorption wells.
- *Water quality:* Urban activities have a huge impact on this. The efficiency of interflow water is also harmed by urbanization. As a result, the impact of this critical component of the hydrologic cycle has been dramatically changed.

2.2.2 Impact of Urbanization—Water Distribution, Waste Collection

The impacts on the water cycle have been briefly discussed in Chapter 1. The impacts can be categorized as physical, chemical, and biological. Physical effects include those that influence the amount of surface runoff and the UWC, such as those that affect soil consolidation. The above has an effect on aquifer storage in urban areas. In many major cities, air pollution from SO_2 and CO_2 is a common problem. When mixed with precipitation, it causes acid rains that destroy trees in urban areas and have adverse effects on surface and groundwater resources.

Many human infectious enteric diseases are transmitted to rivers and lakes through feces. In urban areas, these revelations happen faster. Oxygen plays a crucial role in most biological reactions. The dissolved oxygen in water decreases as the rate of biological reactions increases, affecting aquatic life. Disposal of industrial and residential wastewater into rivers and lakes degrades water quality and disrupts the hydrologic cycle in urban areas. For example, when water is polluted, the color darkens and it absorbs more energy from the sun. This increases the temperature of the water and changes the rate of evaporation.

In urban areas, because of the high density of buildings and other structures, soil will be consolidated and, therefore, the porosity of soil decreases and the amount of water that can be stored and discharged from urban aquifers decreases. Another effect of urbanization on urban aquifers is a reduction of aquifer recharge due to decreased infiltration. Certain urban areas with a large number of absorption wells experience high levels of wastewater leakage into the aquifer. In several cities around the world, urban activities have polluted groundwater resources. For example, landfill leakages and leakage from wastewater sewers and septic tanks are most well-known point sources that cause pollution of groundwater in urban areas.

The hydrological response of cities is determined by determining surface runoff hydrographs, which are then routed through the drainage network's conduits and channels to create outflow hydrographs at the urban drainage outlet. In different sections of an urban area, the physical characteristics of a catchment basin in terms of the rainfall–runoff process will differ significantly. The unit hydrograph is a traditional means of representing linear system response, but it suffers from the limitation that the response function is lumped over the whole catchment and does not explicitly account for spatially distributed characteristics of the catchment's properties. Schumann (1993) presented a semidistributed model where the spatial heterogeneity of the hydrological characteristics within a basin has been modeled considering each subbasin separately using geographical information system (GIS). Wang and Chen (1996) presented a linear spatially distributed model where the catchment is treated as a system that consists of a number of subcatchments, each assumed to be uniform in terms of rainfall excess and hydrological conditions. Using these approaches, different precipitation patterns and land covers and their effects on surface runoff can be modeled. These approaches are of significant value in urban areas because of the latter's complexity and land use.

- *Water distribution and wastewater collection systems:* A municipal water distribution system's goals are to provide clean, potable water for domestic use, appropriate amounts of water at sufficient pressure for fire safety, and industrial water for manufacturing. A typical network includes a source and a tank, as well as treatment, pumping, and distribution. Domestic or sanitary wastewater refers to liquid discharges from residences, business buildings, and institutions. Industrial wastewater is discharged from manufacturing, production, power, and refinery plants. Municipal wastewater is a general term applied to the liquid collected in sanitary sewers and treated in a municipal plant. Most communities collect storm runoff water in a separate storm sewer system with no established domestic or industrial connections and transport it to the nearest watercourse for discharge without treatment. Stormwater and wastewater collection systems are being mixed in many urban areas, resulting in low performance and a high rate of system failure.

2.3 INTERACTIONS ON URBAN COMPONENTS

2.3.1 Climatic Effects—Different Climates

Changes in energy flux, air pollution, and air circulation patterns, all of which are caused by buildings, land transformations, and the release of greenhouse gases, have long been recognized as a cause of interaction between large urban areas and local microclimate. These factors contribute to changes in the radiation balance, and the amounts of precipitation and evaporation, and consequently changes in the hydrologic cycle. The long-term behavior of weather over an area is referred to as climate. The most familiar features of a region's climate are probably average temperature and precipitation. Climate features also include windiness, humidity, cloud cover, atmospheric pressure, and fogginess. Latitude plays a huge factor in determining climate. Landscape can also help define regional climate. A region's elevation, proximity to the ocean or freshwater, and land-use patterns can all impact climate. Of course, no climate is uniform. Small variations, called microclimates, exist in every climate region. Microclimates are largely influenced by topographic features such as lakes, vegetation, and cities. Different parts of the world have different climates. Arid and semiarid, tropical, subtropical (continental), rain shadow, and cool coastal climates are the five types of climate in the world. Hydrologic processes in urban areas are influenced by hydrometeorological variables, which have varying ranges depending on the environment. Different climates are known under the following definitions:

- *Arid and semiarid areas* (e.g., arid areas of Asia and western USA): Seasonal temperatures in these areas range from very winters to very summers. Snow may fall, but its effectiveness may be limited. Summer rainfall is unreliable in this climate. Thompson (1975) lists high pressure, wind direction, topography, and precipitation as four main processes that explain aridity.
- *Tropical areas*: Climate interaction in tropical and subtropical areas can be classified by the amount of rain and rainy seasons, as well as regional water balance during the year. According to Tucci and Porto (2001), rainfall in tropical regions usually has either convective or long periods of rain that are discussed in section 1.3.1.

Climatic conditions in the humid tropics make the urban area more prone to frequent floods, with damage aggravated by socioeconomic factors. On the other hand, the larger mean volumes of precipitation and the greater number of rainy days lead to more complex management of urban drainage. Since larger volumes must be routed somewhere, the mean transported loads of solids are larger (because runoff continues for a longer time), there is less time for urban cleaning (more sediment and refuse remain on the streets) and for maintaining drainage structures, and there is more time to develop waterborne vectors or diseases (Silveira et al., 2001).

- *Subtropical areas* (e.g., Sahara, Arabia, Australia, and Kalahari): These areas are characterized by clear skies with high temperatures. The climate has hot summers and mild winters; hence, seasonal contrasts are evident with low winter temperatures due to freezing. Convective rainfall only develops when moist air invades the region (Marsalek et al., 2006).
- *Rain shadow areas* (e.g., mountain ranges such as the Sierra Nevada, the Great Dividing Range in Australia, and the Andes in South America): A rain shadow, or, more accurately, a precipitation shadow, is a dry region of land that is leeward of a mountain range or other geographic features, with respect to the prevailing wind direction. Mountains block the passage of rain-producing weather systems, casting a "shadow" of dryness behind them. The land gets little precipitation because all the moisture is lost on the mountains (Whiteman, 2000).
- *Cool coastal areas* (e.g., Namib and the Pacific coast of Mexico): These areas have reasonably constant conditions with a cool humid environment. When temperate inversions are weakened by moist air aloft, thunderstorms can develop. Cold ocean currents are onshore winds blowing across a cold ocean current close to the shore, which will be rapidly cooled in the lower layers (up to 500 m). Mist and fog may be resulted, as found along the coasts of Oman, Peru, and Namibia, but the warm air aloft creates an inversion preventing the ascent of air, and hence, there is little or no precipitation.

2.3.2 Hydrologic Effects

The urban surface absorbs heat in arid climates (as described in the previous section), which causes stormwater runoff to heat up and temperatures in receiving water bodies to increase by up to 10°C (Walker, 1987). In highly developed watersheds, these processes may lead to algal succession from cold-water species, impacts on invertebrates, and cold-water fishery succession by a warm-water fishery (Gallee et al., 1991).

Urban stormwater ponds, or small lakes, become chemically stratified during winter months (Marsalek et al., 2006), mostly by chlorides originating from road salting. High levels of dissolved solids and densimetric stratification, which prevent vertical mixing and transport of oxygenated water to the bottom layers and may also enhance the release of metal bounds in sediments, are among the environmental impacts.

In arid areas, the physical process of soil formation is active, resulting in heterogeneous soil types, with properties that do not differ greatly from the parent material and with soil profiles that retain their heterogeneous characteristics. Soils in arid lands may contain hardened or cemented horizons known as pans and are classified according to cementing agents such as iron and so on. The thickness and depth of formation of horizons influence infiltration and salinization to different degrees, as they behave as an obstacle to water and root penetration. Salt crusts can grow on the surface of salt media under certain conditions. The surface structure of salt-affected and sodic soils is very loose, making them vulnerable to wind and water erosion.

2.3.3 Qualitative Aspects

Urbanization is immediately associated with the pollution of water bodies due to untreated domestic sewage and industrial discharges. Recently, however, it was perceived that part of this pollution generated in urban areas also comes from surface runoff. Surface runoff carries pollutants such as organic matter, poisons, bacteria, and others. Thus, urban drainage discharge into water bodies introduces modifications that produce different negative impacts with short- and long-term consequences on the aquatic ecosystem.

Pollution caused by surface runoff of water in urban areas is referred to as diffuse pollution since it is dispersed through the river basin contribution region from pollution-depositing activities. The source of diffuse pollution is very diverse, and abrasion and wear of roads by vehicles, accumulated refuse on streets and sidewalks, organic wastes from birds and domestic animals, building activities, fuel, oil, and grease residues emitted by vehicles, air pollution, and so on contribute to it. The main pollutants carried in this manner are sediments, organic matter, bacteria, metals such as copper, zinc, magnesium, iron, and lead, hydrocarbons from petrol, and toxic substances such as pesticides and air pollutants that are deposited on surfaces. Storm events may raise the toxic metal concentrations in the receiving body to acute levels (Ellis, 1986).

When stormwater runoff washes off dispersed and diffused waste sources distributed through catchments, it becomes contaminated. In temperate climate zones, two main causes of stormwater runoff lead to soil erosion: raindrop impacts and shear stress action: (a) diffused sources originating primarily from atmospheric fallout and vehicle emission and (b) concentrated sources originating mostly from human activities, such as poor housekeeping practices, industrial wastes, and chemicals spread washoff by urban storm runoff.

Both processes result in the formation of soluble and suspended material. Depending on hydraulic conditions, settling and resuspension on the surface and in pipes, as well as biological and chemical reactions, occur during the transport process. These processes are often considered to be more intense in the initial phase of the storm (first flush effect); however, due to the temporal and spatial variability of rainfall and flowing water, first flush effects are more pronounced in pipes than on surfaces.

2.3.4 Greenhouse Effect

The sun's heat is partially absorbed by the earth, but a large portion of it is radiated out into space. The earth radiates the absorbed heat as infrared radiation. Greenhouse gases such as water vapor, carbon dioxide, methane, nitrous oxides, and others absorb this infrared radiation and in turn reradiate it in the form of heat. The amount of greenhouse gases in the atmosphere is increasing and has been scientifically established. Recent research has led us to believe that human activity is causing global climate change. Humans are directly contributing to global warming by emitting greenhouse gases. The most well-known effect of the greenhouse effect is an increase in average atmospheric heat of 2°C–5°C by 2050, which is anticipated. As the concentration of greenhouse gases increases, further climate change can be expected. In urban areas, some scientists suggest that the magnitude of the effects of changing climate on water supplies is more than in rural areas due to the added heat in flux in urban areas but may be much less important than changes in population, technologies, economics, or environmental regulations (Lins and Stakhiv, 1998).

2.3.5 Urban Heat Islands—Mitigation

Urban Heat Islands are described in Chapter 1. Heat islands form in urban and suburban areas because many common construction materials absorb and retain more of the sun's heat than natural materials in less-developed rural areas. There are two main reasons for this heating. First, most urban building materials are impermeable and watertight, so moisture is not readily available to dissipate the sun's heat. Second, dark materials in concert with canyonike configurations of buildings and pavement collect and trap more of the sun's energy. Temperatures of dark, dry surfaces in direct sun can reach up to 88°C (190°F) during the day, while vegetated surfaces with moist soil under the same conditions might reach only 18°C (70°F). Anthropogenic heat, or human produced heat, slower wind speeds and air pollution in urban areas also contribute to heat island formation. In colder cities at higher latitudes and elevations, the winter warming effects of the heat island are seen as beneficial. In some urban areas during the summer, shade around buildings can even create cooler areas for parts of the day. But in most cities throughout the world, the effects of the summer heat island are seen as a problem. Heat islands contribute to human discomfort, health problems, higher energy bills, and increased pollution. On top of the effects of global warming, heat islands are further reducing the habitability of urban and suburban areas.

Heat islands have air temperatures that are warmer than temperatures in surrounding rural areas. The difference between urban and rural air temperatures, also called the heat island strength or intensity, is often used to measure the heat island effect. This intensity varies throughout the day and night. In the morning, the urban–rural temperature difference is generally at its smallest. This difference grows throughout the day as urban surfaces heat up and subsequently warm the urban air.

Measuring a heat island's effects on regional climate is useful and interesting, but cannot tell us how effective mitigation measures would be at reducing a heat island's impacts. That is where modeling becomes necessary. Many different types of models are being used to predict how well mitigation measures can reduce urban temperatures, energy use and air pollution. Models have been developed to look at individual buildings, urban canyons and larger urban areas. Models of heat islands are used both to help understand how the heat island works and to estimate how effective it would be to apply different mitigation measures. There are five main types of models used to evaluate heat islands and their potential to be alleviated: building energy models, roof energy calculators, canyon and comfort models, ecosystem models, and regional models.

To evaluate the effects of heat island mitigation measures, the input for these models must be extensively manipulated. Temperature, energy use, emissions, and smog formation are very interdependent, so the two models must be run iteratively. In other words, modelling the effects of heat island mitigation is a huge undertaking. But as difficult as it is, researchers are making progress at seeing how heat island mitigation measures can reduce temperatures, smog formation and particulate pollution in different urban areas (Gartland, 2012).

2.3.6 Cultural Aspects

The cultural (emotional, intellectual, and moral) aspects of people's interactions with water must be reflected in sustainable solutions to water-related problems. Culture plays an important role in water resource management. Many arid and semiarid countries, as well as most religions, regard water as a precious blessing. There are two cultural characteristics that cause direct impacts on water resources management in urban areas: urban architecture and people's lifestyle.

The practice of architecture in urban areas is often reflected by the climate characteristics of the area. However, the traditional architecture in many large cities is going to be replaced by modern architecture because of population increase and globalization. This may also cause many changes in urban hydrology. The density of population and buildings, rainwater collection systems, material used in construction, and wastewater collection systems are major factors, among others, that alter the urban hydrologic cycle. The change in design paradigm has made major changes in architecture and moves it towards ecological-based design.

Changes in domestic water demands in urban areas have an effect on the hydrologic cycle. The key characteristics that define the lifestyle in large cities are water usage per capita and water used in public centers such as parks and green areas. Despite the importance of economic factors in deciding these characteristics, patterns of water use, tradition, and culture have a significant impact on urban lifestyles.

2.4 REMOTELY SENSED AND SATELLITE DATA

Big data in hydrology has become an ongoing challenge with the advancement of satellite and land surface technologies. Temperature, rainfall, evaporation, air pressure, wind speed, and streamflow data have been collected for many years and are available around the world. There are agencies with clear charter of collecting and disseminating this basic information. Different regions and localities have websites and public or restricted databases that are updated if not around the clock but at some intervals depending upon their needs and the cost associated with that. Urban-specific hydrologic data is also expanding by the use of more sensors and citizen science technologies. However, soil moisture data is scarce and snow pack data is not readily available globally. Some of the new advances in collecting data are discussed in the following sections. Here, two major efforts for snow/soil moisture information systems and soil climate analysis in the US are presented.

Snow Telemetry (SNOTEL) sensors provide the soil information, such as soil moisture and soil temperature, at five different depths (5, 10, 20, 50, and 100cm). The grid spacing of these sensors was determined so as to competently represent the US annual temperature and precipitation variance. The soil moisture and soil temperature for each station are obtained from three independent samples of soil located in a 5-m radius around the main instrument tower. This not only leads to an accurate measurement of such quantities but also yields a more realistic representation of soil properties in a specified location. The SNOTEL records the soil moisture profile at 5-minute intervals and transmits them as hourly data for satellite calibration/validation purposes (https://www.wcc.nrcs.usda.gov).

The Soil Climate Analysis Network (SCAN) sensors monitor soil moisture content at five different depths (5, 10, 20, 50, and 100cm approximately), air temperature, relative humidity, and some other climate variables mainly over the agricultural areas. The SCAN sensors measure the soil moisture values hourly to aid different applications such as drought assessment/monitoring and satellite soil moisture validation. Both SNOTEL and SCAN sensors report the soil moisture in the volumetric unit (m^3/m^3). Figure 2.4 demonstrates the SNOTEL and SCAN stations located at 158 sites across the Utah State. SNOTEL and SCAN stations are installed in specific areas to accommodate specific research needs. These two soil moisture networks have also been used for the validation of Soil Moisture Active Passive (SMAP) product that is describe later in this chapter and in Chapter 14 (https://www.wcc.nrcs.usda.gov). See also sections 2.5.1.2–2.5.1.5 for some details of radar and satellite measurement.

2.5 WATER BALANCE ELEMENTS

2.5.1 Precipitation

Precipitation is a major element of the hydrologic cycle. The measurement of precipitation is very important; there are essentially three different methods for precipitation measurements.

2.5.1.1 Measurement by Standard Gauges

This is the traditional and long-established method. Time series are developed using this method, and if the sampling location is moved, the time series should be adjusted. This method provides

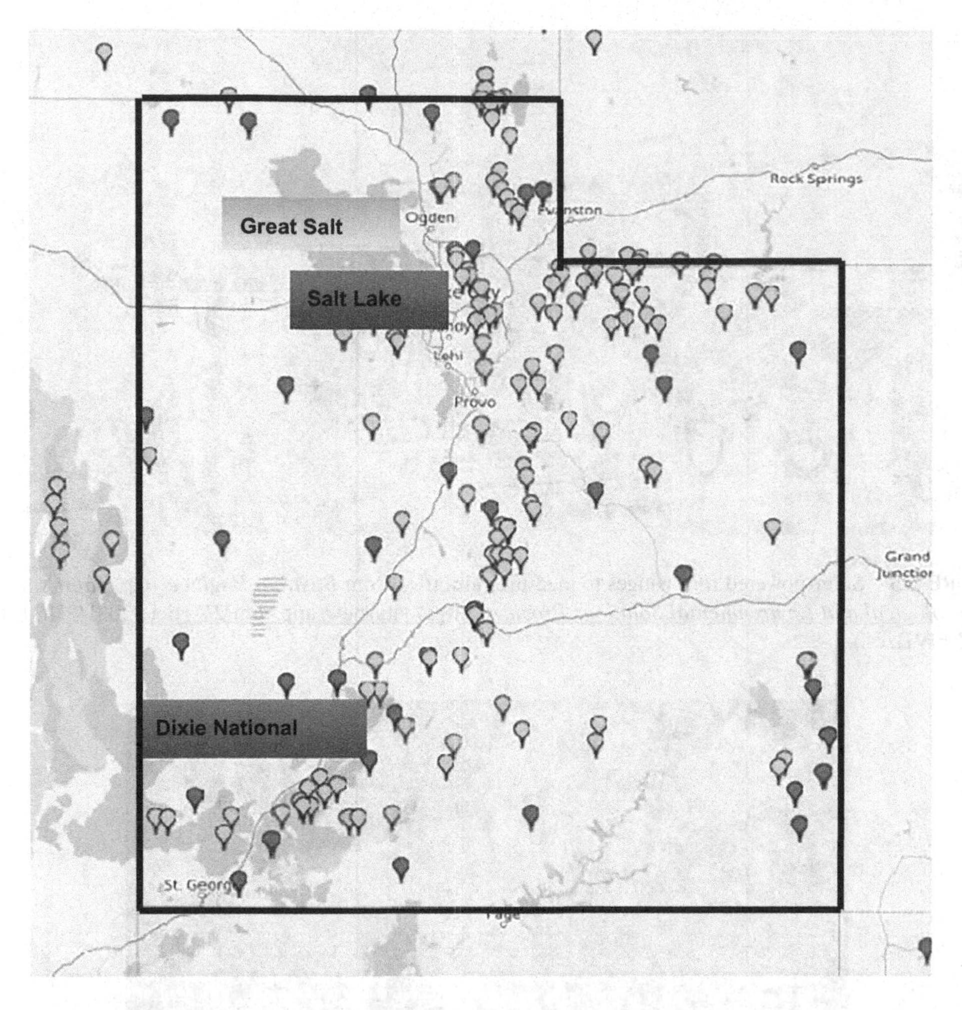

FIGURE 2.4 SNOTEL and SCAN stations located at 158 sites across the Utah State. (Modified from *Technical University of Wien*, https://www.geo.tuwien.ac.at/insitu/data_viewer/.)

point measurements; thus, it may not be representative even for a small area. There are sampling error and poor spatial averaging as well as recording and instrument errors.

Instruments for measuring precipitation include rain gauges and snow gauges. Various types are manufactured according to the purpose at hand. A solar-powered rain gauge to measure rainfall is illustrated in Figure 2.5. Rain gauges are classified into recording and nonrecording types. The latter include cylindrical and ordinary rain gauges, and measurement of precipitation with these types is performed manually by the observer. Some recording types, such as siphon rain gauges, have a built-in recorder, and the observer must physically visit the observation site to obtain data. Other types, such as tipping bucket rain gauges, have a recorder attached to them, and remote readings can be taken by setting a recorder at a site distant from the gauge itself to enable automatic observation.

As rain gauges measure the volume or weight of precipitation collected in a vessel with a fixed orifice diameter, the size of the orifice needs to be standardized. Commission for instruments and methods of observation (CIMO) provides that its area should be 200 cm^2 or more, and types with an orifice area of 200–500 cm^2 are widely used. (www.jma.go.jp).

A class of rain gauge in which the level of the collected rainwater can be measured by the position of a float resting on the water's surface is shown in Figure 2.6. This instrument is frequently

FIGURE 2.5 Solar-powered rain gauges to measure rainfall. (From *Bayanat Engineering: Qatar's leading Meteorological and Environmental Solutions Provider*, http://qatarnewsapp.com/2/Article/2031/211601848#.YWZ-hlVBzcs.)

FIGURE 2.6 The original floating rain gauge. (From *World's Coolest Rain Gauge* (2017). https://worldscoolestraingauge.com/.)

used as a recording rain gauge by connecting the float through a linkage to a pen that records on a clock-driven chart.

2.5.1.2 Various Types of Rain Gauges

a. **Cylindrical Rain Gauges and Ordinary Rain Gauges:** These instruments work according to a simple principle of measurement and also have a straightforward structure. They offer the advantage of having a low rate of problem occurrence.

b. Cylindrical Rain Gauges: As this type of rain gauge can also be used to measure snow, it is alternatively known as a cylindrical rain/snow gauge. It consists of a cylindrical vessel with a uniform diameter from top to bottom and an orifice at the top. It does not have a funnel. Rainwater enters through the orifice and accumulates in the cylindrical vessel, which is weighed at regular intervals with a precipitation scale. As the amount of precipitation is determined by subtracting the vessel weight from the total weight, the dry vessel is weighed before observation.

A rain-measuring glass may be used instead of a precipitation scale. To measure solid precipitation such as snow and hail with such a device, a known amount of warm water is added to melt the precipitation; the total amount is then measured with the measuring glass, and the amount of warm water added is subtracted from the total to obtain the precipitation amount. The precipitation scale is graduated in millimeters based on the size of the rain gauge orifice.

c. Ordinary Rain Gauges: Ordinary rain gauges are the type used at nonautomated observatories. With such devices, the observer takes measurements using a rain-measuring glass at regular intervals.

d. Siphon Rain Gauges: A siphon rain gauge enables automatic, continuous measurement and recording of precipitation.

e. Tipping Bucket Rain Gauges: This type of rain gauge generates an electric signal (i.e., a pulse) for each unit of precipitation collected and allows automatic or remote observation with a recorder or a counter. The only requirement for the instrument connected to the rain gauge is that it must be able to count pulses. Thus, a wide selection of configurations and applications is possible for this measuring system. Solid precipitation can also be measured if a heater is set at the receptacle.

f. Tipping Bucket Rain Gauge Recorder: This recorder counts and records pulses (signals) from a tipping bucket rain gauge, anemometer, etc. For each pulse counted, an electromagnet rotates a gear by one step, causing an eddy-type cam on the same shaft of the gear to drive a recording pen and mark a trace on the recording paper of a clock-driven drum.

2.5.1.3 *Measurement by Weather Radar*

This method can provide spatially distributed measurements in real time. A weather radar is a type of radar used to locate precipitation, calculate its motion, estimate its type (rain, snow, hail, etc.), and forecast its future position and intensity.

Heavy rainfall events often occur very locally, and with the use of rain gauges alone, there is a high probability that the rain gauges miss a rainfall event. To detect such events, continuous spatial monitoring such as rainfall radar monitoring is a better alternative. Radars measure circularly and via an antenna send out radio signals which may or may not hit raindrops falling from clouds. When the signal hits such raindrops it reflects and part of the reflected signal is then detected by the radar dish. The travel time of the radio signal and the diffusion of the signal are measures for distance and intensity of rainfall monitored by the radar.

The embedded software of the radar system converts the signal into a rainfall intensity on a certain place and converts it from dBZ (decibel reflectivity factor, Z) to mm of rainfall per minute. C-band radars which are most often applied for this purpose measure once every 5 minutes up to a distance of 100–200 km. Specific software such as SCOUT form hydro & meteo is used to combine the measurements of several radars and to transfer the radial measurements into a grid which can be used for water management analytics and operations.

The monitoring principle has a weakness in that radio signal is lost due to the collision with raindrops. This attenuates or lowers the intensity of the signal behind a rain front and consequently the radar underestimates rainfall intensity at such places.

The solution is to continually correct the radar signal using a correction factor field, which is determined by means of differences of the raw radar measurement with online point measurements of rain gauge. By doing so, the best of both worlds can be used: accuracy of local rainfall measurements with rain gauges and spatial measurement of the radar. Practice shows that quite accurate spatial rainfall monitoring can be made available by combining both products. An example is the corrected HydroNET rainfall product, available for the Netherlands and parts of Germany, South Africa, and Australia (www.hydrologic.com).

2.5.1.4 Measurement by Satellite

This is a speculative and unproven technology with the potential for global coverage. It has measurements for oceans and inaccessible land regions. A weather satellite is a type of satellite that is primarily used to monitor the weather and climate of the Earth and cloud systems. City lights, fires, effects of pollution, sand and dust storms, snow cover, ice mapping, boundaries of ocean currents, energy flows, etc., are other types of environmental information collected using weather satellites.

Many satellite services have two types of images: infrared and visible. Visible images are similar to those that a normal video camera (black and white) would see looking down at the Earth. The brightest clouds are usually the thicker ones low down in the atmosphere. Infrared images convert the temperature of the cloud, land, or sea (whichever the satellite can see at each point) to a shade of gray. The warmest points are shown as black at the ground surface and the coolest points are shown as white at a high level in the atmosphere. In between are shades of gray, which become brighter as clouds become colder (higher). Also, the differences in temperature of the ground can also be seen as different shades of gray. This type of information does not show actual rainfall; rather, it is inferred from the type of cloud cover.

Tropic Rainfall Measuring Mission (TRMM) was a research satellite in operation from 1997 to 2015, designed to improve our understanding of the distribution and variability of precipitation within the tropics as part of the water cycle in the current climate system. By covering the tropical and subtropical regions of the Earth, TRMM provided much needed information on rainfall and its associated heat release that helps to power the global atmospheric circulation that shapes both weather and climate. In coordination with other satellites in NASA's Earth Observing System, TRMM provided important precipitation information using several space-borne instruments to increase our understanding of the interactions between water vapor, clouds, and precipitation, that are central to regulating Earth's climate.

Since its launch in 1997, TRMM has provided critical precipitation measurements in the tropical and subtropical regions of our planet. The Precipitation Radar looked through the precipitation column and provided new insights into tropical storm structure and intensification. The TRMM Microwave Imager (TMI) measured microwave energy emitted by the Earth and its atmosphere to quantify the water vapor, the cloud water, and the rainfall intensity in the atmosphere. TRMM precipitation measurements have made critical inputs to tropical cyclone forecasting, numerical weather prediction, and precipitation climatologies, among many other topics, as well as a wide array of societal applications.

TRMM officially ended on April 15, 2015, after the spacecraft depleted its fuel reserves. TRMM was turned off and reentered Earth's atmosphere on June 15, 2015, over the South Indian Ocean. Originally designed for 3 years, TRMM continued to provide groundbreaking 3D images of rain and storms for 17 years. TRMM has helped spur additional precipitation measurement satellites that contain microwave radiometers such as the GPM Core Observatory.

2.5.1.5 The PERSIANN System

The current operational PERSIANN (Precipitation Estimation from Remotely Sensed Information Using Artificial Neural Networks) system uses neural network function classification/approximation procedures to compute an estimate of rainfall rate at each $0.25° \times 0.25°$ pixel of the infrared brightness temperature image provided by geostationary satellites. An adaptive training feature facilitates updating of the network parameters whenever independent estimates of rainfall are available. The PERSIANN system was based on geostationary infrared imagery and later extended to include the use of both infrared and daytime visible imagery. The PERSIANN algorithm used here is based on the geostationary long wave infrared imagery to generate global rainfall. Rainfall product covers 50°S to 50°N globally.

The system uses grid infrared (IR) images of global geosynchronous satellites (GOES-8, GOES-10, GMS-5, Metsat-6, and Metsat-7) provided by CPC, National Oceanic and Atmospheric Administration (NOAA) to generate 30-minute rain rates are aggregated to 6-hour accumulated rainfall. Model parameters are regularly updated using rainfall estimates from low-orbital satellites, including TRMM, NOAA-15, -16, -17, DMSP F13, F14, F15. This system was initially developed in the University of Arizona and it was moved to the University of California in Irvine.

2.5.1.5.1 Spectral Intervals and Applicable Satellites

- Long-wave infrared channel (10.2–11.2 μm) from GOES-8, GOES-10, GMS-5, Meteosat-6, and Meteosat-7.
- Instantaneous rainfall estimates from TRMM, NOAA, and DMSP satellites.
  - *Spatial scale*: 0.25° × 0.25° latitude/longitude scale
  - *Temporal scale*: 30 minutes accumulated to 6-hour accumulated rainfall

2.5.1.5.2 Operational Procedure

Since its early inception in 1997, PERSIANN has been a continuously evolving system. The current operational version of PERSIANN (link to appropriate decision support page) generates the above-described global precipitation maps product with 2 days delay time. The delay is due to international data access agreements, which allow us to access global IR composites 2 days after acquisition. The processing steps are as follows:

1. Global full-resolution IR composites are downloaded from the National Centers for Environmental Prediction (NCEP) server on daily basis, these are 15 minutes.
2. TRMM level 2A (TRMM-2A25) microwave imager data are downloaded twice a day also with 2 days delay.
3. The NOAA Satellite Active Archive system is polled every 30 minutes to acquire the most recent microwave-based precipitation estimate from six additional microwave instruments on board NOAA K, L, and M (15, 16, and 17, respectively) and DMSP 7, 8, and 9. Because these data are available in real-near time, they are archived for use whenever the corresponding global IR composites become available.
4. Each day, CHRS's Global IR data used to produce the intermediate 30-minute 4 km precipitation product with the neural network model is trained using all microwave-based precipitation estimates available for the given day.
5. The intermediate product is then aggregated to 0.25° six hourly precipitation maps 3. And the product is released as provisional PERSIANN precipitation estimates. At the same time, 24 hours, 3, 5, 7, 10, 15, and 30 days' total precipitation products are also created for distribution over the HyDIS-GWADI (Link) server.

2.5.1.6 Estimation of Missing Rainfall Data

Before using the rainfall records of a station, it is necessary to first check the data for continuity and consistency. The continuity of a record may be broken with missing data due to many reasons such as damage or fault in a rain gauge during a period. The missing data can be estimated by using the data of the neighboring stations. There are some methods for estimating missing rainfall data. In this calculation, the normal rainfall (the long-term mean annual rainfall) is used as a mean for comparison. The normal ratio method is the conceptually simplest one, which is based on the following equation:

$$\frac{P_x}{A_x} = \frac{1}{n}\left(\frac{P_1}{A_1} + \frac{P_2}{A_2} + \frac{P_3}{A_3} + \cdots + \frac{P_n}{A_n}\right). \quad (2.1)$$

where $P_1, P_2, \ldots, P_n$ are precipitations at neighboring stations 1, 2, …, n and $A_1, A_2, \ldots, A_n$ are areas at neighboring stations 1, 2, …, n. It should be noted that the missing annual rainfall, P_x, with area A_x at station x is not included in the above n stations.

A geostatical method called Kriging is now being used that can change the point data to regional data and therefore could be utilized to estimate rainfall at any ungagged location as well as estimating missing data at a gaged station. See Chapter 9 for details of Kriging methods.

Example 2.1

There are four rain gauges in a watershed named A, B, C, and D. A specific storm event for gauge A is missing. Data from other gauges for the storm event as well as the average annual rainfall of all gauges are given in Table 2.1. Estimate the missing data at gauge A.

Solution:

From Equation (2.1):

$$P_x = \frac{A_x}{n}\left(\frac{P_1}{A_1} + \frac{P_2}{A_2} + \cdots + \frac{P_n}{A_n}\right) = \frac{1{,}290}{3}\left\{\frac{123}{1{,}510} + \frac{148}{1{,}680} + \frac{119}{1{,}375}\right\}$$

$$= 110.01\,\text{mm} = 11.0\,\text{cm}.$$

For rainfall analyses in areas larger than a few square miles, it may be necessary to make estimates of average rainfall depths over subwatershed areas. There are different methods used to provide an estimation of areal rainfall. These methods are selected based on the available data and the characteristics of the study region. Three methods of extending point estimates to areal average are presented in this chapter. Each of the three methods provides a weighted average of measured catches. The station average method assumes equal weights. The weights for the Thiessen method are proportional to the size of the area geographically closest to each gauge. The isohyetal method assigns weights on the basis of storm morphology, the spatial distribution of the rain gauge, and orographic effect. As a general model, the average rainfall of a watershed is defined as follows:

$$\bar{P} = \sum_{i=1}^{n} w_i P_i, \quad (2.2)$$

TABLE 2.1
Data of Rainfall in Example 2.1

Gauge	Average Annual Rainfall (mm)	Total Storm Rainfall (cm)
A	1,290	Missing data
B	1,510	12.3
C	1,680	14.8
D	1,375	11.9

where $\overline{P}$ is the average rainfall, P_i is the rainfall measured in station i, n is the number of rain gauges, and w_i is the weight assigned to each station i. The mean $\overline{P}$ has the same units as P_i. Equation 2.2 is valid only when the following constraint holds:

$$\sum_{i=1}^{n} w_i = 1. \tag{2.3}$$

The general model of Equation 2.2 can be applied with all the methods discussed here. The difference in the application lies solely in the way that the weight is estimated.

2.5.1.7 Station Average Method

For this method, each gauge is given equal weight; thus, w_i equals $1/n$, and Equation (2.3) becomes

$$\overline{P} = 1/n \sum_{i=1}^{n} p_i. \tag{2.4}$$

The use of Equation (2.4) will provide a reasonably accurate estimate of $\overline{P}$ when there are no significant orographic effects, the gauges are uniformly spaced throughout the watershed, and the rainfall depth over the entire watershed is nearly constant. See Karamouz et al. (2012) for more details.

2.5.1.8 Snowmelt Estimation

Snowmelt and sequential streamflow form one of the most important phases of the hydrologic cycle. In hydrology, snowmelt is surface runoff produced from melting. In some cases, snowmelt is a high fraction of the annual runoff in a watershed. Because of the important role that the storage and melting of snow play in the hydrologic cycle of some areas, hydrologists must be able to reliably predict the contribution of snowmelt to overall runoff. Because snowmelt begins in the spring and the resulting runoff occurs before the peak water demand, control schemes such as building storage reservoirs should be a part of the planning and operation schemes for water supply. Under certain conditions, snowmelt can also contribute to flooding problems; snowmelt predictions are also valuable to power companies that generate hydroelectricity and to irrigation districts.

Snow mapping and monitoring are necessary for studies about snow melting. Furthermore, it is now widely recognized that satellite remote sensing has useful potential for snow mapping and monitoring. For repetitive ice and snow mapping over large areas, data from weather satellites, especially of the NOAA family, are more widely used. NOAA imagery is analyzed by the US National Weather Satellite Service to provide its customers with a range of medium- to small-scale ice and snow reports. Additionally, fully objective (automatic) techniques based on NOAA high-resolution picture transmissions (HRPTs) multispectral data are being developed for routine snow area and snow surface characteristics (surface temperature, melt accumulation statues, etc.), mapping, and monitoring. Evaluating snow depth is more difficult, as assessment from visible/infrared data is readily available only within a very few depth categories. Furthermore, cloud cover is still a nuisance, though much less so than in the case of the infrequently viewing Earth resources satellite systems. As with rainfall monitoring, new hope is emerging as passive microwave data will provide helpful supplementary information both spectrally and temporally (Figure 2.7).

Because of the range of uses for estimating the contribution of snowmelt to streamflow and the variation of condition applicable to each case, many methods have been developed for computing snowmelt as it affects streamflow. These methods have found that the timing and amount of runoff depend mainly on five factors: sensible heat condition from moist air to the snow surface, latent heat of condensation, solar radiation, heat transmitted by rainfall on the snowpack, and heat condition from underlying ground. Furthermore, it must be emphasized that snowmelt at a point involves a change of state requiring heat, whereas the production of snowmelt runoff from watershed is a

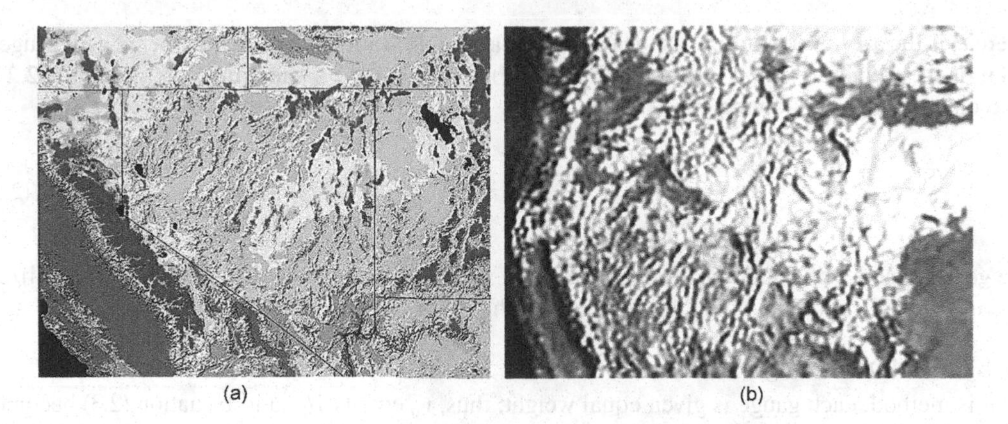

FIGURE 2.7 Digital snow maps for Great Basin in the United States on March 26, 2020: (a) NOAA-AVHRR satellite imagery. (From *NOAA's National Climatic Data Center (NCDC)*. (2020). Hursat AVHRR. HURSAT-Hurricane Satellite Data. https://www.ncdc.noaa.gov/hursat/index.php?name=hursat-avhrr.); (b) MODIS satellite imagery. (From Masuoka, E. *Modis: Snow and Sea Ice Global Mapping Project*. NASA, (2000). https://modis-snow-ice.gsfc.nasa.gov/?c=pap_intercomp.)

much more complicated process. The simple method of snowmelt runoff estimation is presented as follows:

$$M = 4.57 \times C(T - T_B), \tag{2.5}$$

where M is the snowmelt runoff (cm/day), C is snowmelt coefficient (between 0.015 and 0.2), T is daily mean temperature or maximum temperature (°C), and T_B is base temperature (°C). The base temperature is considered to be 0°C when mean daily temperature is used, and it would be 4.4°C when the maximum daily temperature is utilized. The suggested relations of snowmelt estimation by the US Army Corps of Engineers are given in Table 2.2.

In Table 2.2, M is snowmelt rate (cm/day); F is average basin forest-canopy cover shading of the area from solar radiation (decimal fraction); W is wind speed at 15.2 m above the snow surface temperature; A is average snow surface albedo, which is equal to $0.75/D^{0.2}$, where D is the number of days from the last snow; K is basin convection–condensation melt factor expressing average exposure to wind, which is equal to 1 for open site and 0.3 for heavily forested site; K' is basin shortwave radiation melt factor between 0.9 and 1.1; T' is difference between the air temperature measured at 3 m and the snow surface temperature (°C); T'_D is the difference between the dew point temperature measured at 3 m and the snow surface temperature (°C); R is rainfall (cm/day); and S is solar radiation on a horizontal surface (langleys/day).

TABLE 2.2
US Army Corps of Engineers' Suggested Snowmelt Equations

No.	Situation	Forest Cover Application	Equations
1	Rain in snow	Open site (0%–60% cover)	$M = 2.54\,[0.09 + (0.029 + 0.005KW + 0.0027R)\,(1.8T)]$
2		Heavily forested (60%–100% cover)	$M = 2.54\,[0.05 + (0.074 + 0.0027R)\,(1.8T)]$
3	Rain free	Open site (0%–60% cover)	$M = 2.54\big[K'(1-F)(0.004S)(1-A) + K(0.005W)$ $(0.396T' + 1.4047T'_D + 32) + F(0.052T' + 0.928)\big]$
4		Heavily forested (60%–100% cover)	$M = 2.54\big[0.074\ (0.953T' + 0.845T'_D + 32)\big]$

Source: US Army Corps of Engineers, *Snow Hydrology*, North Pacific Division, Portland, OR, 1965. With permission.

Example 2.2

The properties of a mountainous catchment are presented in Tables 2.3 and 2.4. Considering $K = 0.6$, $K' = 1.0$, and $F = 0.5$, estimate the snowmelt for each day. Assume the last snowfall occurred on January 31.

Solution:

a. *Area 1*:

From Table 2.4, the temperature for a height of 912 m is 4.4 and that for 1,216 m is 2.8 in area 1, then the mean temperature is (4.4+2.8)/2 = 3.6 and $K = 0.6$, $W = 16.9$, and $R = 0.127$. Equation 1 of Table 2.2 is used to estimate the snowmelt in this area because there is rain in snow and it is an open site.

$$M = 2.54\{0.09 + [0.029 + 0.005(0.6)(16.9) + 0.0027(0.127)](1.8 \times 3.6)\} = 1.54 \text{ cm/day}$$

Area 2:

From Table 2.4, the temperature for a height of 1,216 m is 2.8 and that for 1,520 m is 1.1 in area 2, then the mean temperature, T, is (2.8+1.1)/2 = 1.95, $K = 0.6$, $W = 16.9$, and $R = 0.127$. Equation 1 of Table 2.2 is used to estimate the snowmelt in this area because there is rain in snow and it is an open site.

$$M = 2.54\{0.09 + [0.029 + 0.005(0.6)(16.9) + 0.0027(0.127)](1.8 \times 1.95)\} = 0.94 \text{ cm/day}$$

b. Weighted average of snowmelt for February 3 is as follows:

$$M = \frac{1.54(815) + 0.94(1{,}890)}{815 + 1{,}890} = 1.120 \text{ cm/day.}$$

c. Snow storage at the end of February 3 is presented in Table 2.5.
d. There is no rainfall on February 4, and areas 1 and 2 are an open site. Equation 3 of Table 2.2 is then used to estimate the snowmelt in this area. In this equation, $F = 0.5$, $S = 576$, and $W = 177$.

TABLE 2.3
Catchment Properties for Example 2.2

Property	Area 1	Area 2
Height (m)	912–1,216	1,216–1,520
Area (km²)	815	1,890
Snow height (cm)	45	33

TABLE 2.4
Weather Information for Example 2.2

	Temperature (°C)			Dew Point (°C)					
Date	912 m Height	1,216 m Height	1,520 m Height	912 m Height	1,216 m Height	1,520 m Height	Rainfall (cm)	Wind Speed (km/h)	Solar Radiation (Langleys/day)
February 3	4.4	2.8	1.1	–0.5	–1.6	–3.3	0.127	16.9	400
February 4	8.9	7.2	5.5	–1.5	–3.2	0.1	0	17.7	567

TABLE 2.5
Estimation of Snow Storage at Day 1 of Example 2.2

	Area 1	Area 2
Snow storage at beginning of day (cm)	45	33
Snowmelt in first day (cm)	1.54	0.94
Snowfall during day (cm)	0	0
Snow storage at end of day (cm)	43.46	32.06

Area 1:

$$D = 4 \quad A = \frac{0.75}{4^{0.2}} = 0.57$$

$$T' = \frac{(8.9 + 7.2)}{2} = 8.05$$

$$T'_D = \frac{(-1.5 + (-3.2))}{2} = -2.35$$

$$M = 2.54\big[1(1-0.5)(0.004 \times 567)(1-0.57) + 0.6(0.005 \times 17.7)\big(0.396 \times 8.05 + 1.404 \times (-2.35 + 32)\big)$$
$$+ 0.5(0.052 \times 8.05 + 0.928)\big] = 7.24 \text{ cm/h}$$

Area 2:

$$T' = \frac{(7.2 + 5.5)}{2} = 6.35$$

$$T'_D = \frac{(-3.2 + 0.1)}{2} = -1.55$$

$$M = 2.54\big[1(1-0.5)(0.004 \times 567)(1-0.57)$$
$$+ 0.6(0.005 \times 17.7)\big(0.396 \times 6.35 + 1.404 \times (-1.55 + 32)\big)$$
$$+0.5(0.052 \times 6.35 + 0.928)\big] = 3.5 \text{ cm/h}$$

Weighted average of snowmelt for catchment in day 2 is as follows:

$$M = \frac{7.24(815) + 3.5(1{,}890)}{815 + 1890} = 4.62 \text{ cm/day.}$$

2.5.2 Evaporation and Evapotranspiration

Evaporation and evapotranspiration are important links in the hydrologic cycle in which water is transferred to the atmosphere as water vapor. Evaporation is the process in which a liquid changes to a gaseous state at the free surface, below the boiling point through the transfer of heat energy. While transportation takes place, plants lose moisture by the evaporation of water from soil and water bodies in the land area. In hydrology and irrigation practice, it is found that evaporation and transportation can be considered under one process—evapotranspiration.

For the hydrologist, the loss of water by evaporation must be considered from two main aspects. First, evaporation from an open water surface, E, is the direct transfer of water from lakes, rivers, and reservoirs to the atmosphere. This can be easily assessed if the water body has known capacity and does not leak by evaluating the changes in storage volume.

The second form of evaporation loss occurs in the form of the transpiration from vegetation, E_t. This is sometimes called evapotranspiration since loss by direct evaporation of intercepted precipitation and transpired water on plant surfaces is also included. Thus, E_t is usually thought of as the total loss by both transpiration and evaporation from a land surface and its vegetation. The value of E_t varies according to the type of vegetation, its ability to transpire, and the availability of water in the soil. It is much more difficult to quantify E_t than E since transpiration rates can vary considerably over an area and the source of water from the ground for plants requires careful definition.

Both forms of evaporation, E and E_t, are influenced by the general climatic conditions. Although the instrumental measurements of evaporation are not as simple as rainfall, it is a compensating factor that evaporation quantities are less variable from one season to another and, therefore, can be more easily predicted than rainfall amounts. Evaporation is one of the most consistent elements with unlimited supplies of water in the hydrologic cycle.

2.5.2.1 Evaporation Evaluation

There are different major approaches adopted in calculating evaporation from open water, E. Four primary methods are used to estimate evaporation from a water surface, that is, the water budget method, the mass transfer method, pan evaporation, and the Penman equation.

2.5.2.2 Water Budget Method

The water budget method for lake evaporation is based on the hydrologic continuity equation. Assuming that change in storage ΔS, surface inflow I, surface outflow O, subsurface seepage to groundwater flow GW, and precipitation P can be measured, evaporation E can be computed as

$$E = -\Delta S + I + P - O - GW. \tag{2.6}$$

The approach is simple in theory, but evaluating seepage terms can make the method quite difficult to implement. The obvious problems with the method result from errors in measuring precipitation, inflow, outflow, change in storage, and subsurface seepage.

2.5.2.3 Mass Transfer Method

In the mass transfer method of computing evaporation, the mass transfer coefficients are determined using energy-budget evaporation as an independent measure of evaporation.

The simple equation used for measuring evaporation by mass transfer method is as follows:

$$E = f(u)(e_s - e_d), \tag{2.7}$$

where E is the evaporation rate, $f(u)$ is a function of wind speed u, e_s is saturated vapor pressure of air at water surface (mb), and e_d is the saturated vapor pressure of the air at T_d, the dew point (mb). From a study of numerous reservoirs of different sizes up to 12,000 ha, an additional factor, surface area, can be incorporated into the equation to determine evaporation loss from a reservoir; thus,

$$E = 0.291A^{-0.05}u_2(e_s - e_d), \tag{2.8}$$

where E is evaporation (mm/day), A is area (m^2), u_2 is speed at 2 m (m/s), e_s is saturated vapor pressure of air at water surface (mb), and e_d is the saturated vapor pressure of the air at the dew point, T_d (mb).

Example 2.3

Calculate the annual water loss from a 5 km² reservoir, when u_2 is 10.3 km/h, e_s is 14.2, and e_d is 11.0 (mm Hg).

Solution:

$$A = 5\,\text{km}^2 = 5\times 1{,}000^2\ \text{m}^2$$

$$u_2 = 10.3\ \text{km/h} = \frac{10.3\times 1{,}000}{60\times 60} = 2.86\,\text{m/s}$$

$$e_s = 14.2\,\text{mm Hg} = 14.2\times 1.33 = 18.9\,\text{mb}$$

$$e_d = 11.0\,\text{mm Hg} = 11.0\times 1.33 = 14.6\,\text{mb}$$

$$E = 0.291\left(5\times 1{,}000^2\right) - 0.05\times 2.86(18.9-14.6) = \frac{0.291\times 2.86\times 4.3}{2.16} = 1.66\,\text{mm/day}$$

$$= 1.66\times 365 = 0.605.9 \cong 606\,\text{mm/year}$$

Total annual water loss from reservoir = $0.606\times 5\times 1{,}000^2 = 3.03\times 10^6\,\text{m}^3$.

2.5.2.4 Pan Evaporation

An evaporation pan is used to hold water during observations for the determination of the quantity of evaporation at a given location. Such pans are of varying sizes and shapes, the most commonly used being circular or square. The best known of the pans is the "Class A" evaporation pan. In Class A pan, an open galvanized iron tank 4 ft in diameter and 10 inches depth is used mounted 12 inches above the ground. To estimate evaporation, the pan is filled to a depth of 8 inches and must be refilled when the depth has fallen to 7 inches. The water surface level is measured daily, and evaporation is computed as the difference between observed levels, adjusted for any precipitation measured in a standard rain gauge. Alternatively, water is added each day to bring the level up to a fixed point. Pan evaporation rates are higher than actual lake evaporation and must be adjusted to account for radiation and heat exchange effects. The adjustment factor is called the pan coefficient, which ranges from 0.64 to 0.81 with an average of 0.70. However, the pan coefficient varies with exposure and climate conditions and should be used only for rough estimates of lake evaporation.

2.5.2.5 Measurement of Evapotranspiration

Evapotranspiration, E_t, can be measured by (a) tanks and lysimeters, (b) field plots, and (c) studies of groundwater fluctuations. A tank is a watertight container that is set into the ground with its rim nearly flush with the ground surface. The size should be sufficient to simulate natural growing conditions for the type of plants being studied. The tank is mounted on a scale to assist in necessary moisture measurements. Each plant requires certain moisture conditions for optimum growth, and these conditions are maintained during consumptive use measurements. E_t is determined by measuring the quantity of water necessary to maintain constant, optimum moisture conditions in the tank.

A lysimeter is essentially a tank with a pervious bottom. The bottom arrangement is such that excess soil moisture will drain through the soil, which can be collected and measured.

Specially designed field plots are used to determine E_t under field conditions. These plots are designed so that surface runoff water from the plot can be collected and measured. Deep percolation is captured by underground drain tiles. To determine E_t, water input in the form of precipitation or irrigation is measured.

Evapotranspiration equations have been developed to predict consumptive use for different conditions and different crops. These are empirical relations that have been useful to replace the difficult and expensive measurements. Only Thornthwaite method is explained here. For other methods refer to Karamouz et al. (2012).

2.5.2.6 Thornthwaite Method

Thornthwaite (1948) derived an equation to be used for limited water conditions. This equation produces monthly estimates of E_t using assumptions similar to those of the Blaney–Criddle method and is written as follows:

$$E_t = 1.62\left(\frac{10T}{I}\right)^a, \tag{2.9}$$

where E_t is evapotranspiration (cm), T is the mean monthly temperature (°C), and a can be estimated as follows:

$$a = \left(6.75\times10^{-8}\right)I^3 - \left(77.1\times10^{-6}\right)I^2 + 0.0179I + 0.492, \tag{2.10}$$

where I is estimated as follows:

$$I = \sum_{j=1}^{12}\left(\frac{T_j}{5}\right)^{1.514}, \tag{2.11}$$

where T_j is the mean temperature of the jth month.

Example 2.4

Estimate the potential evapotranspiration for the data in Table 2.6.

Solution:

From Equation (2.11):

$$I = \sum_{j=1}^{12}\left(\frac{T_j}{5}\right)^{1.514},$$

TABLE 2.6
Data of Example 2.4

Month	Precipitation (mm)	T (°C)
January	112.8	3.3
February	94	5.6
March	118.9	10
April	86.6	15
May	116.9	20
June	101.9	23.9
July	100.3	26.1
August	111.3	25
September	110.7	21.1
October	94.2	15
November	85.9	8.9
December	87.1	5

$$I=\sum_{j=1}^{12}\left(\frac{T_j}{5}\right)^{1.51}=\left(\frac{3.3}{5}\right)^{1.51}+\left(\frac{5.6}{5}\right)^{1.51}+\left(\frac{10}{5}\right)^{1.51}+\left(\frac{15}{5}\right)^{1.51}+\left(\frac{20}{5}\right)^{1.51}+\left(\frac{23.9}{5}\right)^{1.51}+\left(\frac{26.1}{5}\right)^{1.51}$$

$$+\left(\frac{25}{5}\right)^{1.51}+\left(\frac{21.1}{5}\right)^{1.51}+\left(\frac{15}{5}\right)^{1.51}+\left(\frac{8.9}{5}\right)^{1.51}+\left(\frac{5}{5}\right)^{1.51}=69.84.$$

From Equation 2.10:

$$a=\left(6.75\times10^{-8}\right)I^{\,3}-\left(77.1\times10^{-6}\right)I^{\,2}+0.0179I+0.492=1.598.$$

From Equation 2.8 and T_{mean} = 14.91:

$$E_t=1.62\left(\frac{10\times14.91}{69.84}\right)^{1.598}=5.44\,\text{cm}.$$

Example 2.5

If the monthly mean temperature is 30°C, estimate the potential evapotranspiration using the Thornthwaite method.

Solution:

From Equations 2.9–2.11:

$$I=\sum_{j=1}^{12}\left(\frac{T_j}{5}\right)^{1.514}=\sum_{j=1}^{12}\left(\frac{30}{5}\right)^{1.51}=12\times6^{1.51}=179.55$$

$$a=6.75\times10^{-8}(179.55)^3-\left(77.1\times10^{-6}\right)(179.55)^2+0.0179\times179.55+0.492=1$$

$$E_t=1.62\left(10\times\frac{30}{179.55}\right)^{10.1}=289.2\text{ cm}.$$

2.6 INTERCEPTION STORAGE AND DEPRESSION STORAGE

The process of interrupting the movement of water in the chain of transportation events leads to stream interception. By vegetation or cover depression storage in puddles and formations of land such as rills and furrows, the interception can take place. It is difficult to separate these sources of rainfall abstraction and they are usually dealt with using an overall figure.

When rain first begins, the water striking leaves and other organic materials spread over the surfaces in a thin layer or it is collected at points or edges. When the maximum surface storage capability on the surface of the material is exceeded, the material stores additional water in growing drops along its edges. Eventually, the weight of the drops exceeds the surface tension and water falls to the ground. Wind and the impact of raindrops can also release the water from the organic material. The water layer on organic surfaces and the drops of water along the edges are also freely exposed to evaporation. Additionally, interception of water on the ground surface during freezing and subfreezing conditions can be substantial. The interception of falling snow and ice on vegetation also occurs. The highest level of interception occurs when it snows on forests and hardwood forests that have not yet lost their leaves.

Estimating the losses due to interception is done by empirical methods. Horton (1919) suggested the following expression:

$$L_i=a+bP_T^n, \tag{2.12}$$

TABLE 2.7
Parameter Values for Interception Equations

	Equation 2.12			Equation 2.13	
Tree Type	***a***	***b***	***n***	***c***	***m***
Apple	0.04	0.18	1.0	0.25	0.73
Ash	0.015	0.23	1.0	0.26	0.88
Beech	0.02	0.23	1.0	0.21	0.65
Chestnut	0.04	0.20	1.0	0.30	0.77
Elm	0.0	0.23	0.5	0.15	0.48
Hemlock	0.0	0.20	0.5	0.32	0.74
Maple	0.03	0.23	1.0	0.29	0.77
Oak	0.03	0.22	1.0	0.24	0.66
Pine	0.0	0.20	0.5	0.30	0.70
Willow	0.02	0.40	1.0	0.43	0.85

Source: Kibler, D.F., *Urban Stormwater Hydrology*, America Geophysical Union, Washington, DC, 1982. With permission.

where L_i is interception (in), P_T is total rainfall depth (in.), and *a, b,* and *n* are constants.

Kibler (1982) reanalyzed Horton's data (collected from summer storms) and developed the following expression for estimation of interception storage:

$$L_i = cP_T^m, \tag{2.13}$$

where *c* and *m* are parameters that are different for different types of crops. Values of *a, b, n, c,* and *m* for different types of crops are given in Table 2.7.

Viessman et al. (1989) suggested that while estimates of losses due to interception can be significant in annual or long-term models, accounting for interception could be negligible for heavy rainfalls during individual storm events.

2.7 INFILTRATION

The term *infiltration* refers to the process by which rainwater passes through the ground surface and fills the pores of the soil on both the surface and subsurface. This process is an important part of rainfall losses.

The infiltration capacity or infiltration rate is defined as the maximum rate at which water can infiltrate. The actual rate of infiltration will be equal to the rate of rainfall if the rainfall rate is less than the infiltration capacity. Otherwise, the actual rate of infiltration will be equivalent to the infiltration capacity, and the rainwater that does not infiltrate will flow over the ground surface after filling the surface depressions. In other words,

$$f = i \quad \text{if } f_p > i \tag{2.14}$$

and

$$f = f_p \quad \text{if } f_p < i, \tag{2.15}$$

where f_p is the infiltration capacity, f is actual rate of infiltration, and i is rate of rainfall.

If infiltration is the only (or the dominating) type of abstraction:

$$i_e = i - f, \tag{2.16}$$

where i_e is rate of excess rainfall. An accurate evaluation of the infiltration capacity is a difficult task. Here, some simple methods to calculate the infiltration capacity, which are extensively used in engineering practices, are covered. The infiltration capacity depends on the soil characteristics and the humidity of the soil. Therefore, a credible infiltration capacity model should take into account the initial moisture conditions of the soil and the amount of water that has already infiltrated into the soil after the start of rainfall.

For water management in time and space, the hydrologic cycle is a model of holistic nature. There are different definitions of the hydrological cycle, but it is generally defined as a conceptual model describing the storage and circulation of water between the biosphere, atmosphere, lithosphere, and hydrosphere (Karamouz et al., 2012). Water can also be captured in nature in the atmosphere, ice and snow packs, streams, rivers, lakes, groundwater aquifers, and oceans. Water cycle components are affected by processes such as temperature and pressure variation and condensation. The means of water movement is through precipitation, snowmelt, evapotranspiration, percolation, infiltration, and runoff.

A watershed is the best hydrological unit that can be used to carry out water studies and planning in a systematic manner. The urban setting could alter the natural movement of water. Drastic land use changes in urban areas as a subset of urban and industrial development affect natural landscapes and the hydrological response of watersheds. Although anthropogenic factors with respect to waterways, pipes, abstractions, and man-made infrastructures affect the elements of the natural environment, the main structure of the hydrological cycle remains the same in urban areas (McPherson and Schneider, 1974). But the characteristics of the hydrologic cycle are greatly altered by urbanization impacts of the services to the urban population, such as water supply, drainage, and wastewater collection and management.

As a conceptual way of looking at water balances in urban areas, the context of the UWC is the total system approach. Water balances and budget studies are generally conducted on a different time scale, depending on the type of applications we are looking at in a planning horizon. For distributing water to growing populations and for coping with extreme weather and climatic variations and potential climate change, Lawrence et al. (1999) emphasized the importance of integrated urban water managements. Pressures, temperatures, and water quality buildups in urban areas have a pronounced effect on the UWC, which should be attended to through best management practice (BMP) schemes. See Chapter 3 for more details. In the next section, a unique indicator of water scarcity called PDSI is presented that is perhaps the most widely used index to determine the state of water availability that agencies have to deliver to the customers and water users.

2.8 PALMER DROUGHT SEVERITY INDEX (PDSI)

2.8.1 Agricultural Drought Indicators

The following steps have been taken to calculate the moisture anomaly index Z, which is necessary for calculating the PDSI. A soil moisture deficit occurs when the precipitation is not enough to supply the potential evapotranspiration. The soil moisture deficit, D_i is calculated as follows:

$$D_i = \Phi - S_i. \tag{2.17}$$

where Φ is soil moisture capacity; and S_i soil moisture in month i.

2.8.2 Potential Climatic Values

Besides the potential evapotranspiration, other potential values are estimated using the following definitions: potential recharge (PR) is the moisture needed by the soil to reach its moisture capacity; potential moisture loss (PL) is the amount of moisture loss of the soil to supply the evapotranspiration demand; and potential runoff (PRO) is the maximum runoff that can occur. PRO is a function of the difference between the precipitation and the soil moisture gain. The nature of the PRO is much more complicated than that of other potential values. Different investigators have assumed values as high as three times the amount of precipitation and as ambiguous as the difference between the soil moisture capacity and potential recharge. With the absence of any prior investigation of the relation between PRO and other hydrologic parameters, the value of PRO for each month is assumed to be the highest runoff experienced during the historical record for that month.

2.8.3 Coefficients of Water Balance Parameters

The evapotranspiration coefficient α is used to estimate the expected climatically appropriate evapotranspiration for the existing condition (CAFEC) from its potential figure. This terminology has been used frequently in drought studies. It simply implies the expected normal condition. The recharge coefficient β is the proportion of the average recharge for month i to the average potential recharge. The runoff coefficient γ is the ratio of the average runoff for month i to the potential runoff. The moisture loss coefficient δ is the ratio of the average moisture losses in month i to the potential loss. The amount of soil moisture losses, which is calculated by this coefficient, could be subtracted from the sum of the other three factors to obtain the amount of expected precipitation of a region for the normal condition.

2.8.4 Precipitation for Climatically Appropriate for Existing Condition, $\hat{P}_i$

Using the foregoing coefficients, the amounts of evapotranspiration, $\hat{ET}_i$, recharge, $\hat{R}_i$ runoff, $\hat{RO}_i$, moisture loss, $\hat{L}_i$, and expected precipitation, $\hat{P}_i$, at each month for the normal condition can be calculated as follows:

Each component of CAFEC precipitation has an average equal to the average historical observations.

$$\hat{P}_i = \hat{ET}_i + \hat{R}_i + \hat{RO}_i - \hat{L}_i. \quad (2.18)$$

where

$$\hat{ET}_i = \alpha \cdot PE_i \quad (2.19a)$$

$$\hat{R}_i = \beta \cdot PR_i \quad (2.19b)$$

$$\hat{RO}_i = \gamma \cdot PRO_i \quad (2.19c)$$

$$\hat{L}_i = \delta \cdot PL_i \quad (2.19d)$$

PDSI is an index for evaluating the severity of a drought. After calculating CAFEC precipitation and comparing it with the observed precipitation, the difference, d, can be calculated as:

$$d = P - \hat{P} \quad (2.20)$$

Average moisture supply is not always dependent on the precipitation in a period. In some occasions when there is insufficient precipitation, storage of moisture in the previous period is used. Using the average of moisture supplied, a climatic character, k_i, for each month, i, is calculated.

$$k_i = 17.67 \hat{K}_i \Big/ \sum_{j=1}^{12} \bar{D}_i \times \hat{K}_j \quad i = 1,\ldots,12 \tag{2.21}$$

where

$$\hat{K}_j = 1.5 \log_{10}\left(\frac{T_j + 2.8}{\bar{D}_j}\right) + 0.5 \quad j = 1,\ldots,12 \tag{2.22}$$

and

$$T_j = \left(\overline{PE}_j + \bar{R}_j + \overline{RO}_j\right) / \left(\bar{P}_j + \bar{L}_j\right) \quad j = 1,\ldots,12 \tag{2.23}$$

T_j is a measure of the ratio of moisture demand to moisture supply for a month j. $\hat{K}_j$ is a regional climatic character. $\overline{PE}_j$, $\bar{R}_j$, $\overline{RO}_j$, $\bar{P}_j$, and $\bar{L}_j$ are average monthly historical potential evapotranspiration, recharge, runoff, precipitation, and soil moisture losses, respectively. To estimate the Z index for a specific region, the coefficients of the above equations must be calculated based on the climatic character of the region. The monthly constants (k) are used as weighting factors of monthly deviations during dry spells. Then, a moisture anomaly index, Z, for month i is defined as follows:

$$Z_i = k_i \times d_i \quad i = 1,2,\ldots,N \tag{2.24}$$

2.8.5 Drought Severity Index

To obtain a regional drought index, all subbasins of the region must be investigated. Z-index series are calculated for all subbasins. For all negative values in the Z-index series, cumulative Z_s are calculated. The periods representing the maximum of the cumulated negative values are selected and plotted. Then the scale on the Z-axis is adjusted to –1 for the lowest and –4 for the highest cumulative absolute value of Z, according to the drought classification of the PDSI. This new coordinate is called the drought severity index, PDSI. Then the best line is fitted to the lower envelope of the graph PDSI versus t duration and a relation between the drought severity index, and Z is obtained for each month i. The following relation was obtained by Palmer (1965):

$$\text{PDSI}_i = 0.897\, \text{PDSI}_{i-1} + (1/3) Z_i \quad i = 2,\ldots,N \tag{2.25}$$

$$\text{PDSI}_1 = (1/3) Z_1, \tag{2.26}$$

where N is the number of time periods.

The coefficients should be adjusted for different region. Karamouz et al. (2004) applied this procedure for Zayandeh Rud River basin in Esfahan, Iran, and adjusted the two coefficients for Equation 2.25 as 0.679 and 1/16, respectively. The coefficient in Equation 2.26 was adjusted as 1/16. Based on Palmer's definition, dry and wet spells are categorized by PDSI. Table 2.8 shows the drought classification based on the PDSI value.

In order to obtain drought and wet spells using PDSI, one of the three different indices (X_i^k) must be used:

$$\begin{aligned} X_i^k &= 0.897 \times X_{i-1}^k + (1/3) \times Z_i k = 1,2,3 \\ i &= 1,2,3,\ldots,n \end{aligned} \tag{2.27}$$

TABLE 2.8
Drought Categories Based on PDSI

PDSI	Drought Category
≤ –4	Most severe drought
–1 to 1	Severe drought
–1 to 1	Medium drought
–1 to 1	Nearly drought
–1 to 1	Normal
1 to 2	Nearly wet
2 to 3	Medium wet
3 to 4	Severe wet
≥4	Most severe wet

Source: From Palmer, W.C., Meteorological Drought (Vol. 30), US Department of Commerce, Weather Bureau, 1965.

X_i^1 = Severity index for a wet spell in month i
X_i^2 = Severity index for a dry spell in month i
X_i^3 = Severity index for a spell that cannot be classified as either dry or wet in month i

To determine the drought severity index, the following steps should be taken:

If X_i^2 is ≤–1, then month i is in a drought spell. If X_i^k is ≥1, then month i is in a wet spell. During a period of drought, X_i^3 is equal to X_i^2, and during a wet period, X_i^3 is equal to X_i^1. In this situation, when the indicators (X_i^1 or X_i^2) are between −0.5 and 0.5, then X_i^3 is equal to zero and it is the termination time of a dry or wet spell, respectively. Often, there is only one indicator that is not equal to zero, and this non-zero indicator will be the PDSI. The advantage of this method is that once a dry or wet period is observed by either X_i^2 or X_i^1, then the other resets to zero. Therefore, the value of PDSI does not grow without any limits, which could be the case if only Equation 2.27 is used (Alley, 1984).

As stated in the methodology of calculating the PDSI, the coefficients of the PDSI are dependent on the regional climate and may change with the different climatic characteristics of different regions. Hence, it is necessary to derive new coefficients for the PDSI equation in different study areas. See Karamouz et al. (2004, 2012) for more details.

2.9 GROUNDWATER

The water below the surface of the Earth primarily is groundwater, but it also includes soil water. Movement of water in the atmosphere and on the land surface is relatively easy to visualize, but the movement of groundwater is not.

Groundwater moves along flow paths of varying lengths from areas of recharge to areas of discharge. The generalized flow paths start at the water table, continue through the groundwater system, and terminate at the stream or at the pumped well. The source of water to the water table (groundwater recharge) is infiltration of precipitation through the unsaturated zone. In the uppermost, unconfined aquifer, flow paths near the stream can be tens to hundreds of meters in length and have corresponding travel times of days to a few years. The longest and deepest flow paths may be thousands of meters to tens of kilometers in length, and travel times may range from decades to millennia. In general, shallow groundwater is more susceptible to contamination from human sources and activities because of its close proximity to the land surface. Therefore, shallow, local patterns of groundwater flow near surface water are emphasized in this section.

The principle hydrologic properties of aquifers are permeability and specific yield. Permeability shows the material ability to transmit water while specific yield indicates the volume of water that the aquifer yields during drainage. The materials that form the crust of the Earth have a lot of voids called interstices. These voids hold the water that is found below the surface of the land and that is recoverable through springs and wells. The occurrence of water under the ground is related to the character, distribution, and structure of the voids in soil. If the interstices are connected, the water can move through the rocks by percolating from one interstice to another and therefore the groundwater flow forms.

The porosity is considered as the percentage of the total volume of the rock that is occupied by interstices. The porosity of soil depends on the shape and arrangement of its particles, the particle sorting, the degree of cementation and compacting, and the fracturing of the rock resulting in joints and other openings. It should be mentioned that there is not definitely a correlation between the porosity and permeability of a material; a rock may contain many large but disconnected interstices and thus have a high porosity yet a low permeability. Clay may have a very high porosity (higher than some gravel) but a very low permeability.

Just a part of stored water in the interstices is recovered through wells and the remainder will be retained by the rock formations. Specific yield corresponds to the part that will drain into wells, and the part that is retained by the rocks is called the specific retention. Both specific yield and specific retention are expressed as the percentages of the total volume of material and their summation is equal to the porosity.

The water table fluctuates in response to recharge and discharge from the aquifer. The shape and slope of the water table, which are determining factors in the movement of groundwater, are affected by many factors such as the differences in permeability of the aquifer materials, topography, and the configuration of the underlying layer, amount, and distribution of precipitation, as well as the method of discharge of the groundwater. However, as a general concept, the slope of the water table varies inversely with the permeability of the aquifer material. This is shown by Darcy's law governing the movement of groundwater:

$$Q = KiA, \tag{2.28}$$

where Q is the quantity of water moving through a given cross-sectional area, K is the permeability or hydraulic conductivity coefficient, i is the hydraulic gradient, and A is the total cross-sectional area through which the water is moving.

Streams, lakes, and reservoirs interact with groundwater in all types of landscapes. The interaction takes place in three basic ways: streams gain water from inflow of groundwater through the streambed (gaining stream), they lose water to groundwater by outflow through the streambed (losing stream), or they do both, gaining in some reaches and losing in other reaches. For groundwater to discharge into a stream channel, the altitude of the water table in the vicinity of the stream must be higher than the altitude of the stream water surface. Conversely, for surface water to seep to groundwater, the altitude of the water table in the vicinity of the stream must be lower than the altitude of the stream water surface. Contours of water table elevation indicate gaining streams by pointing in an upstream direction, and they indicate losing streams by pointing in a downstream direction in the immediate vicinity of the stream. These interactions play an important role in surface and groundwater resource volume changes over time that should be considered in water supply schemes. The concepts used in the study of groundwater resources and their interactions with the surface water resources are explained in detail in Chapter 7 of Karamouz et al. (2012).

2.10 RESERVOIRS AND LAKES

Lakes and reservoirs are sites in a basin where surface water storage needs to be modeled. Thus, variables defining the water volumes at those sites must be defined. Let S_t^S be the initial storage volume of a lake or reservoir at site s in period t. Omitting the site index s for the moment, the final storage volume in period t, S_{t+1} (which is the same as the initial storage in the following period $t+1$), will equal the initial volume, S_t, plus the net surface and groundwater inflows, Q_t, less the release or discharge, R_t, and evaporation and seepage losses, L_t. All models of lakes and reservoirs include this mass balance equation for each period t being modeled.

$$S_t + Q_t - R_t - L_t = S_{t+1}. \tag{2.29a}$$

The release from a natural lake is a function of its surrounding topography and its water surface elevation. It is determined by nature, and unless it is made into a reservoir, its discharge or release is not controlled or managed. The release from a reservoir is controllable and is usually a function of the reservoir storage volume and time of year. Reservoirs also have fixed storage capacities, K. In each period t, reservoir storage volumes, S_t, cannot exceed their storage capacities, K.

$$S_t \leq K \text{ for each period } t. \tag{2.29b}$$

Equations (2.29a) and (2.29b) are the two fundamental equations required when modeling water supply reservoirs. They apply for each period t.

The primary purpose of all reservoirs is to provide a means of regulating downstream surface water flows over time and space. Other purposes may include storage volume management for recreation and flood control and storage and release management for hydropower production. Reservoirs are built to alter the natural spatial and temporal distribution of the streamflows. The capacity of a reservoir together with its release (or operating) policy determines the extent to which surface water flows can be stored for later release.

The use of reservoirs for temporarily storing streamflows often results in a net loss of total streamflow due to increased evaporation and seepage. Reservoirs also bring with them changes in the ecology of a watershed and river system. They may also displace communities and human settlements. When considering new reservoirs, any benefits derived from regulation of water supplies, from floodwater storage, from hydroelectric power, and from any navigational and recreational activities should be compared to any ecological and social losses and costs. The benefits of reservoirs can be substantial but so may the costs. Such comparisons of benefits and costs are always challenging because of the difficulty of expressing all such benefits and costs in a common metric.

Reservoir storage capacity can be divided among three major uses: (a) the active storage used for downstream flow regulation and for water supply, recreational development, or hydropower production; (b) the dead storage required for sediment collection; and (c) the flood storage capacity reserved to reduce potential downstream flood damage during flood events. The distribution of active and flood control storage capacities may change over the year. For example, there is no need for flood control storage in seasons that are not going to experience floods. Often, these components of reservoir storage capacity can be modeled separately and then added together to determine total reservoir storage capacity.

2.11 WATER BALANCE

In the previous sections, the mathematical concepts representing hydrologic components including evapotranspiration, infiltration, runoff, and groundwater flow were presented. In water resources related fields of hydrologic studies, in addition to evaluation of hydrologic components, it is useful to evaluate the changes in water resources. For this purpose, the water balance concept is employed. The water balance can be used to show the flow of water in and out of a system. This system can be considered as a watershed, an aquifer, a lake, or a global hydrologic cycle, depending on the desired region of study and corresponding boundaries.

The general form of the hydrologic balance for a specific time period is represented as follows:

$$P - R - G - E - T = \Delta S, \tag{2.30}$$

where P is precipitation, R is surface runoff, G is groundwater flow, E is evaporation, T is transpiration, and ΔS is change in storage. Since the runoff is a portion of rainfall, it can be estimated using the runoff coefficient. The runoff coefficient is defined as the ratio R/P.

Due to difficulty in estimation of hydrologic components, some parametric water balance models are developed. In these models, coefficients are defined to investigate the relationship between different components. In the next section, a simple water balance model called the Thomas model is described.

Example 2.6

In a given year, a watershed with an area of 2,500 km² received 130 cm of precipitation. The average rate of flow measured in a river draining the watershed was 30 m³/s. Estimate the amount of water lost due to the combined effect of evaporation, transpiration, and infiltration to groundwater. How much runoff reached the river for the year (in cm)? What is the runoff coefficient?

Solution:

The water balance equation (Equation 2.24) can be arranged to produce

$$E + T + G = P - R - \Delta S.$$

Assuming that the water levels are the same for $t = 0$ and $t = 1$ year, then $\Delta S = 0$ and

$$E + T + G = 130 - \frac{30 \times 86400 \times 365 \times 100}{2{,}500 \times 1000 \times 1000} = 130 - 37.9 = 91.1.$$

The runoff coefficient is divided by precipitation.

$$R / P = 37.8 / 130 = 0.29.$$

2.11.1 Thomas Model (ABCD Model)

The Thomas model is a parametric water balance model that includes two main variables. The first variable is the available water, W_i, at the beginning of time interval i, which is equal to the summation of the total precipitation in period i, P_i, and the soil moisture at the end of time interval $i-1$, S_{i-1}. This variable is formulated as follows:

$$W_i = P_i + S_{i-1}. \tag{2.31}$$

The second variable is Y_i, which is calculated as follows:

$$Y_i = E_i + S_i, \tag{2.32}$$

where E_i is actual evapotranspiration.

Thomas developed a nonlinear equation for relating these two variables as follows:

$$Y_i = \frac{W_i + b}{2a} - \left[\left(\frac{W_i + b}{2a}\right)^2 - \frac{W_i b}{a}\right]^{0.5}, \tag{2.33}$$

where a and b are model parameters. The parameters show the produced runoff before soil saturation. In cases of $a < 1$, it will result in $W_i < b$. For determining the percentage of evapotranspiration and soil moisture at the end of interval i from Y_i, it is assumed that the moisture loss due to evapotranspiration is directly related to soil moisture content and potential evapotranspiration as follows:

$$S_i = Y_i e^{-ET_{pi}/b}, \tag{2.34}$$

The difference between W_i and Y_i is equal to the summation of the direct runoff, DR_i, and groundwater recharge, GR_i. A part of infiltrated water increases the soil moisture. The direct runoff and groundwater recharge are estimated as follows:

$$GR_i = c(W_i - Y_i) \tag{2.35}$$

$$DR_i = (1 - c)(W_i - Y_i), \tag{2.36}$$

where c is the mode parameter indicating the part of runoff that is supplied from groundwater resources. If G_i shows the groundwater storage at the end of time interval i, it can be calculated as follows:

$$G_i = \frac{GR_i + G_{i-1}}{d + 1}. \tag{2.37}$$

The provided runoff from groundwater resources is also estimated as follows:

$$(QG)_i = dG_i, \tag{2.38}$$

where d is the model parameter showing the part of groundwater that is transformed to the surface runoff. The surface runoff at the end of time interval i is equal to

$$DR_i + QG_i. \tag{2.39}$$

The model parameters a, b, c, and d are estimated during model calibration based on measured water balance components in a specified region and specified time period. The estimation of initial values of soil moisture content, S_0, and groundwater storage, G_0, is needed for model application. The potential evapotranspiration can be estimated using different available models such as the Thornthwaite model.

Example 2.7

The average monthly precipitation and potential evapotranspiration (ETP) are given in Table 2.9.

The monthly precipitation and potential evapotranspiration data of Example 2.10 are used.

Estimate the average monthly soil moisture content, S_i, groundwater recharge, GR_i, and groundwater storage, G_i. Assume an initial moisture content of 8 cm, and an initial groundwater storage of 2 cm. a, b, c, and d are given as 0.98, 25, 0.1, and 0.35, respectively.

TABLE 2.9
Precipitation and Potential Evapotranspiration (ETP) of Example 2.7

Month	1	2	3	4	5	6	7	8	9	10	11	12
Precipitation (cm)	1.0	1.8	0	0.6	0	0.1	8.78	9.8	4.8	0.74	0.4	6.88
ETP (cm)	1.75	1.78	2.1	2.2	3.0	3.1	3.5	3.6	2.7	1.9	1.75	1.75

Solution:

For the first month, the following is obtained:

$$W_1 = P_1 + S_0 = 1.0 + 8.0 = 9.0 \text{ cm}.$$

The value of Y for the first month is calculated as follows:

$$Y_1 = \frac{W_1 + b}{2a} - \left[\left(\frac{W_1 + b}{2a}\right)^2 - \frac{W_1 b}{a}\right]^{0.5}$$

$$= \frac{9+25}{2(0.98)} - \left[\left(\frac{9+25}{2(0.98)}\right)^2 - \frac{(9)(25)}{0.98}\right]^{0.5} = 8.9.$$

Soil moisture content is estimated as

$$S_1 = Y_1 e^{-ETP_1/b} = 8.9e^{-1.75/25} = 8.3.$$

Groundwater recharge is estimated as

$$GR_1 = c(W_1 - Y_1) = 0.1(9 - 8.9) = 0.01.$$

Direct runoff is equal to

$$DR_1 = (1-c)(W_1 - Y_1) = (1-0.1)(9-8.9) = 0.09.$$

Finally, groundwater storage is obtained as

$$G_1 = \frac{(GR)_1 + G_0}{d+1} = \frac{0.01+2}{1+0.35} = 1.49.$$

The runoff produced by groundwater will be

$$QG_1 = dG_1 = 0.35 \times 1.49 = 0.52.$$

The total runoff is estimated as

$$TR_1 = DR_1 + QG_1 = 0.09 + 0.52 = 0.61.$$

The calculations for the succeeding months are given in Table 2.10.

2.11.2 A Case Study: Water Balanced-Based Sustainability

Karamouz et al. (2017) proposed a methodology for quantifying water supply–demand sustainability using water balance as summarized in Figure 2.8. Based on the proposed approach, first, a simulation model for this area has been developed. Calculations of a method called input–output water

TABLE 2.10
Summary of Calculations of Example 2.7

Month No.	Precipitation (cm)	ETP (cm)	*W* (cm)	*Y* (cm)	*S* (cm)	*GR* (cm)	*DR* (cm)	*G* (cm)	*QG* (cm)	*TR* (cm)
1	1	1.75	9.00	8.90	8.30	0.01	0.09	1.49	0.52	0.61
2	1.8	1.78	10.10	9.97	9.28	0.01	0.12	1.11	0.39	0.51
3	0	2.1	9.28	9.18	8.44	0.01	0.10	0.83	0.29	0.39
4	0.6	2.2	9.04	8.94	8.18	0.01	0.09	0.62	0.22	0.31
5	0	3	8.18	8.11	7.19	0.01	0.07	0.47	0.16	0.23
6	0.1	3.1	7.29	7.23	6.39	0.01	0.05	0.35	0.12	0.18
7	8.78	3.5	15.17	14.74	12.82	0.04	0.38	0.29	0.10	0.48
8	9.8	3.6	22.62	20.65	17.88	0.20	1.77	0.36	0.13	1.89
9	4.8	2.7	22.68	20.69	18.58	0.20	1.79	0.41	0.15	1.94
10	0.74	1.9	19.32	18.31	16.97	0.10	0.90	0.38	0.13	1.04
11	0.4	1.75	17.37	16.70	15.57	0.07	0.60	0.33	0.12	0.72
12	6.88	1.75	22.45	20.55	19.16	0.19	1.71	0.39	0.14	1.84

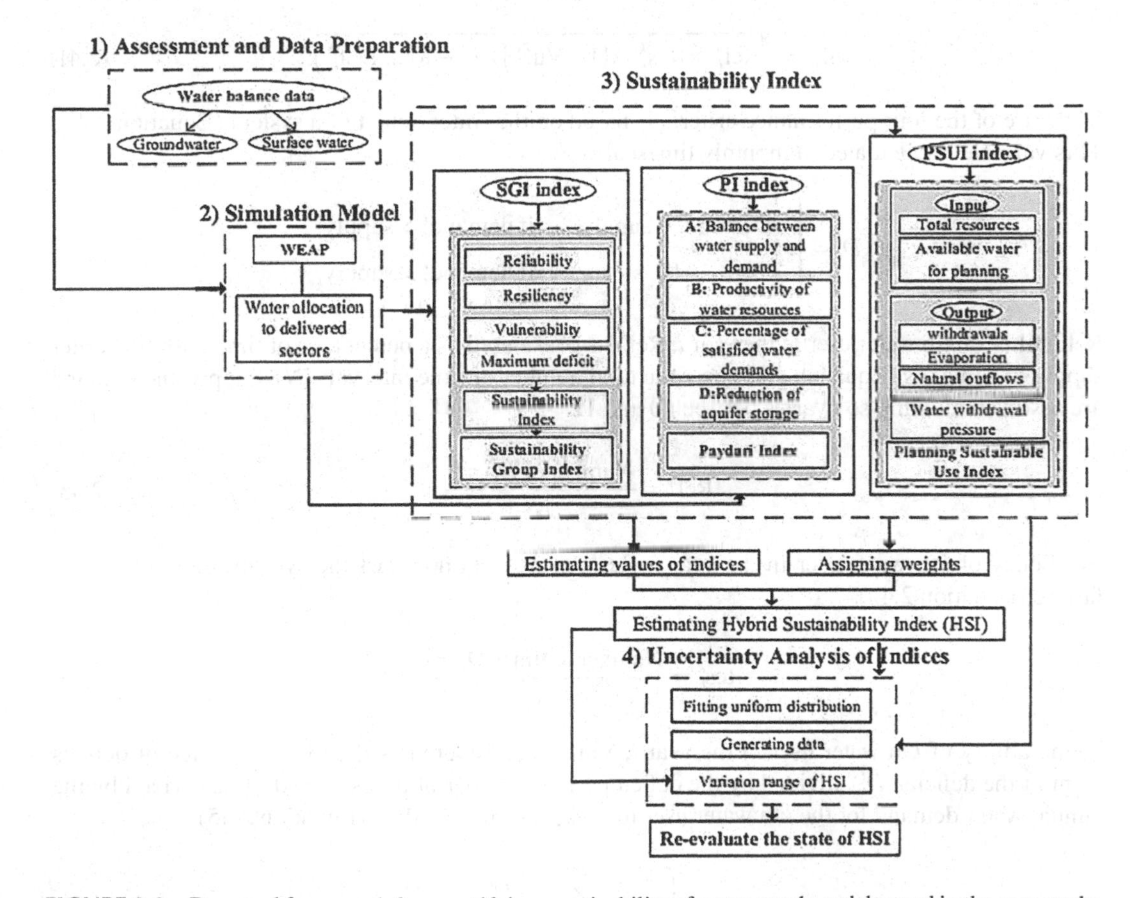

FIGURE 2.8 Proposed framework for quantifying sustainability of water supply and demand in the case study.

balance were done to estimate the Planning for Sustainable Use Index (PSUI) based on the water balance data released by the regional water board authority. Then, calculations of performance indicator (PI) (*P* is the indigenous term for sustainability in Farsi) and Sustainability Group Index (SGI) were obtained based on the outputs of the simulation model. These three indices are aggregated for the determination of a hybrid index called the Hybrid Sustainability Index (HSI). Indices' weights are estimated based on engineering judgment and expert's opinions. Finally, uncertainty analysis has been performed to determine the variation of each indices and HSI index based on uncertain variables.

The SGI developed by Sandoval-Solis et al. (2011) is one of the indicators used by Karamouz et al. (2017, 2021) to assess the water supply and demand sustainability. SGI can assess the performance of the system in satisfying demand by measuring four performance criteria of reliability, resiliency, vulnerability, and max deficit. Each of these performance criteria defines an important characteristic of a system as described in the following section. The structure of this indicator is as follows:

$$\mathrm{SGI}_i = \sum_{k=1}^{4} w_i^k + \mathrm{SI}_i^k \tag{2.40}$$

where w_i^k is the weight of the kth water user (domestic, agricultural, industrial, and environmental) in year i. SI_i^k is the sustainability index of kth water user in year i, and SGI_i is Sustainability Group Index in year i. SI is a geometric average of four criteria to quantify the system's performance in satisfying demands of k different water users. The values of these criteria are between 0 and 1 (Equation 2.41).

$$\mathrm{SI}_i^k = \sqrt[4]{\mathrm{Rel}_i^k \times \mathrm{Res}_i^k \times \left(1 - \mathrm{Vul}_i^k\right) \times \left(1 - \mathrm{Max\ Def}_i^k\right)} \tag{2.41}$$

Structure of the four performance criteria is based on the water deficit of a system (Equation 2.42). This variable is calculated in monthly timescale.

$$D_i^t = \begin{cases} \mathrm{Demand}_i^t - \mathrm{supply}_i^t & \text{if demand}_i^t > \text{supply}_i^t \\ 0 & \text{if demand}_i^t < \text{supply}_i^t \end{cases} \tag{2.42}$$

Reliability of kth water user in the year i, Rel_i^k, is considered as a percentage of time with full water supply to demands (Equation 2.43) over the total number of time intervals (N). Supply and demand are assessed monthly so t varies between 0 and 12.

$$\mathrm{Rel}_i^k = \frac{\text{Number of } D_i^t = 0}{N} \tag{2.43}$$

Resiliency of kth water user in the year i, Res_i^k, represents how fast the system recovers from a failure (Equation 2.44).

$$\mathrm{Res}_i^k = \frac{\text{Number of times } D_i^t = 0}{N} \tag{2.44}$$

Vulnerability of kth water user in the year i, Vul_i^k, is considered as the expected value of deficits (sum of the deficits, D_t^i), divided by the deficit period (number of times $D_i^t > 0$). It is divided by the annual water demand for the kth water user to make it dimensionless (Equation 2.45):

$$\mathrm{Vul}_i^k = \frac{\left(\dfrac{\sum_{D_i^t > 0} D_i^t}{\text{Number of time } D_i^t > 0}\right)}{\sum_{t=1}^{N} \mathrm{Demand}_i^t} \tag{2.45}$$

If the maximum deficit, Max Def_i^k, occurs, it is the worst-case annual deficit for the kth water user in the year i. Dividing the maximum annual deficit by the annual water demand represents the dimensionless maximum deficit (Equation 2.46):

$$\text{Max Def}_i^k = \frac{\text{Max}\left(D_{\text{Annual}}^k\right)}{\sum_{t=1}^{t=12} \text{Demand}_t^k} \tag{2.46}$$

where D_{Annual}^k is the annual water deficit, and time intervals are considered monthly.

PSUI is another indicator used in these studies. It is based on input–output water balance (Gascó et al., 2005)

$$\text{PSUI} = 1 - \frac{\text{PW}}{\text{GAA}} \tag{2.47}$$

where PW is primary water withdrawals, and GAA is gross annual available water.

The last indicator used for developing HSI is called PI that was introduced by Ahmadi (2011). The indicator PI-Plus as a sustainability index is expressed as follows. The P stands for Paydari which corresponds to the sustainability term in Farsi. This indicator is expressed as below

$$PI - \text{Plus}_i = w_A \times A_i + w_B \times B_i + w_C \times C_i + w_D \times D_i \tag{2.48}$$

where A_i indicates the balance between supply and demand in year i. B_i is defined as the portion of productivity of water resources in the year i over the maximum potential for productivity of water resources in the year i. The value of this parameter is always between zero and one. C_i represents the percentage of satisfied water demands for year i, and the value of this parameter is always between zero and one. D_i represents the compliment of the difference between actual and allowable withdrawals from groundwater divided by the aquifer capacity for year i. The ratio is considered as zero when it is negative. In this study, parameters of the PI index have been adjusted by considering total agricultural products in the parameter of B_i, and also the percentage of satisfied water demands in the agricultural sector is substituted by the percentage of satisfied water demands for all water demands in the parameter of C_i. Parameters of A_i, B_i, C_i, and D_i are defined as follows.

$$A_i = \frac{(\text{AGW} + I_i + S - S_{\min}) - (E_i + \text{WR}_i)}{S_{\max} + S_{\text{aquifer}}} \tag{2.49}$$

$$B_i = \frac{\dfrac{Y_i}{\text{AWA}_i}}{\dfrac{Y_m}{\text{CWR}}} \tag{2.50}$$

$$C_i = \frac{\text{SW}_i}{\text{WR}_i} \tag{2.51}$$

$$D_i = 1 - \frac{\text{GWW}_i - \text{AGW}}{S_{\text{aquifer}}} \tag{2.52}$$

where S is reservoir initial storage volume, $S_{\min}$ is inactive storage, $S_{\max}$ is storage capacity of the reservoir, AGW is allowable withdrawal from aquifer, I_i is inflows to the reservoir in year i, E_i is evaporation of reservoir in year i, WR_i is total water requirement in year i, S_{aquifer} is storage capacity of aquifer, Y_i is actual crop yield in year i, AWA_i is allocated water to agriculture in year i, Y_m

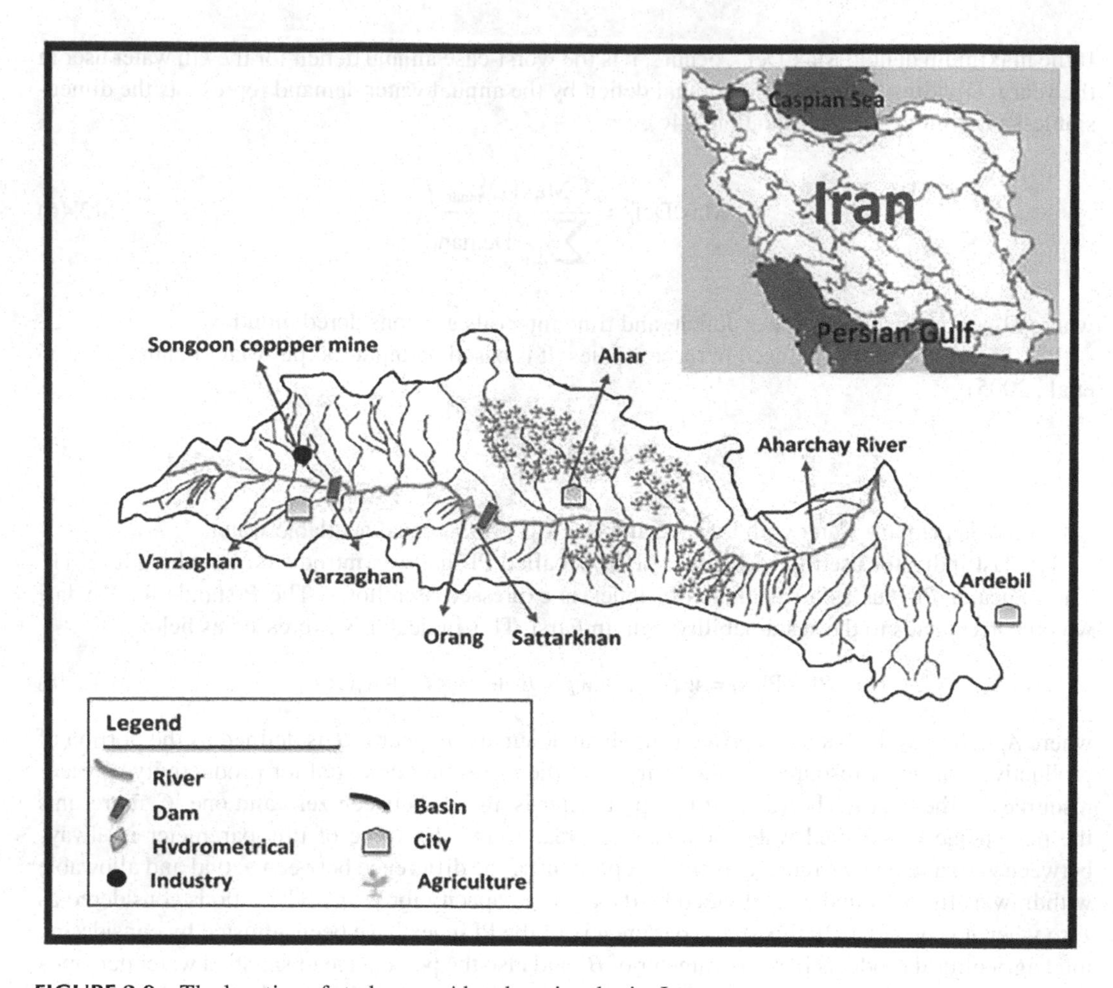

FIGURE 2.9 The location of study area, Aharchay river basin, Iran.

is maximum crop yield, CWR is agricultural water requirement, SW_i is supply delivered to all demands in year i, and GWW_i is withdrawals from groundwater in year i.

The Aharchay river basin in the northwestern part of Iran has been selected as the case study with the area of 2,232 km. This is a major subbasin of Aras river basin far northwest of Iran. The basin is located between 47° 20′ and 47° 30′ north longitude and 38° 20′ and 38°45′ east latitude. Figure 2.9 shows location of the Aharchay river basin. The main available surface water storage facility in the study area is Sattarkhan dam with 78.2 MCM effective capacity. Orang hydrometrical station is located just upstream of the dam. The average height of the basin is 1,430 m above sea level and mean basin slope is 22%. In this region, based on long-term data, the average rainfall is about 291.5 mm/year and the annual average minimum and maximum temperatures are 0.30°C and 35.8°C, respectively. Observed annual precipitation shows a decreasing trend over the last 20 years (1986–2007). It includes more than 3,000 ha of cultivated area with irrigation network with an efficiency of about 45%.

2.12 INTERACTIONS BETWEEN THE URBAN WATER CYCLE AND URBAN INFRASTRUCTURE COMPONENTS

There are many interactions between the UWC and urban infrastructure components. On one hand, wastewater treatment, being somewhere at the end of such a cycle, is regularly "suffering" from the "upstream" components, while on the other hand, it has itself a prominent impact on several "downstream" components. This section provides an overview of the interactions between wastewater treatment using water treatment, water transport and distribution, drainage, sewerage, solid waste, and transport. The overview of interactions concerning (1) drinking water quality and quantity; (2) disposal of exhausted adsorption materials from treatment processes; (3) discharge of filter backwash water; (4) disposal of sludge generated during water purification into drains and sewerage; (5) design, construction, and operation of septic tanks and possible groundwater pollution; (6) quality and quantity of treated effluents and their discharge into ground or surface waters; (7) unaccounted for water losses such as leakage, linked with infiltration and exfiltration in the sewage systems; (8) type of sewerage; (9) sewage characteristics; (10) sewerage as a bioreactor—transformation processes in the sewer system; (11) sewer design; (12) the role of combined sewer outflows (CSOs); (13) the effect of the "first flush"; (14) separation at source; (15) dry sanitation; (16) impact of sewage works on the hydraulics of sewage collectors; (17) solid waste deposition on streets, drains, and collectors; (18) leachate from dumps and landfills; (19) impacts of fecal sludge on sewerage and works; (20) codigestion of sludge with solid waste; (21) design and maintenance of roads; (22) the "road as a drain" approach; and so on. All of the abovementioned factors and issues urge for a truly integrated approach towards urban water systems management by which all major interactions will be taken properly into account. This should also include latest thinking on the development of cites that calls for an integrated approach that encompasses all aspects of urban planning.

2.12.1 Interactions with the Wastewater Treatment System

Wastewater treatment is an important component of both UWC and urban infrastructure (Figure 2.10). Looking at it from a rather conventional perspective, it can be said that it is located somewhere at the end of this cycle, although purely by the definition of "cycle," such a statement is incorrect.

Nevertheless, it is important to understand that wastewater treatment is influenced by both upstream and downstream components of the cycle and that it may have an impact on various components of urban infrastructure. In general, wastewater treatment is subject to interactions among water treatment, water distribution, drainage and sewerage, solid waste, and transport.

2.12.2 Interactions between Water and Wastewater Treatment Systems

Water treatment may affect wastewater treatment by several means, such as (a) drinking water quality and quantity, (b) disposal of exhausted adsorption materials from treatment processes, (c) discharge of filter backwash water, and (d) disposal of sludge generated during water purification into drains and sewerage. On the other hand, wastewater treatment practices can have a significant impact on water treatment via (a) the design, construction, and operation of septic tanks and possible groundwater pollution and (b) the quality and quantity of treated effluents and their discharge into ground or surface waters.

From the perspective of conservation of water resources, wastewater effluent reuse is increasingly considered as a feasible option, especially in cases where water is scarce and reuse option is one of the few possibilities left. There are numerous examples where the effluent of the sewage works makes up a large part of the rivers, creeks, or channels. At many locations worldwide, it is not uncommon to have the intake of raw water for drinking water treatment below the discharge point of the wastewater treatment plant (WWTP) effluent. These, among other factors, recommend that wastewater effluent be recognized as a resource and not waste. Because in the (urban) water cycle every drop of water becomes wastewater two or three times before it enters the global hydrological cycle as clean water, the effluent from WWTPs has a role in water supply from both the quality and the quantity perspective.

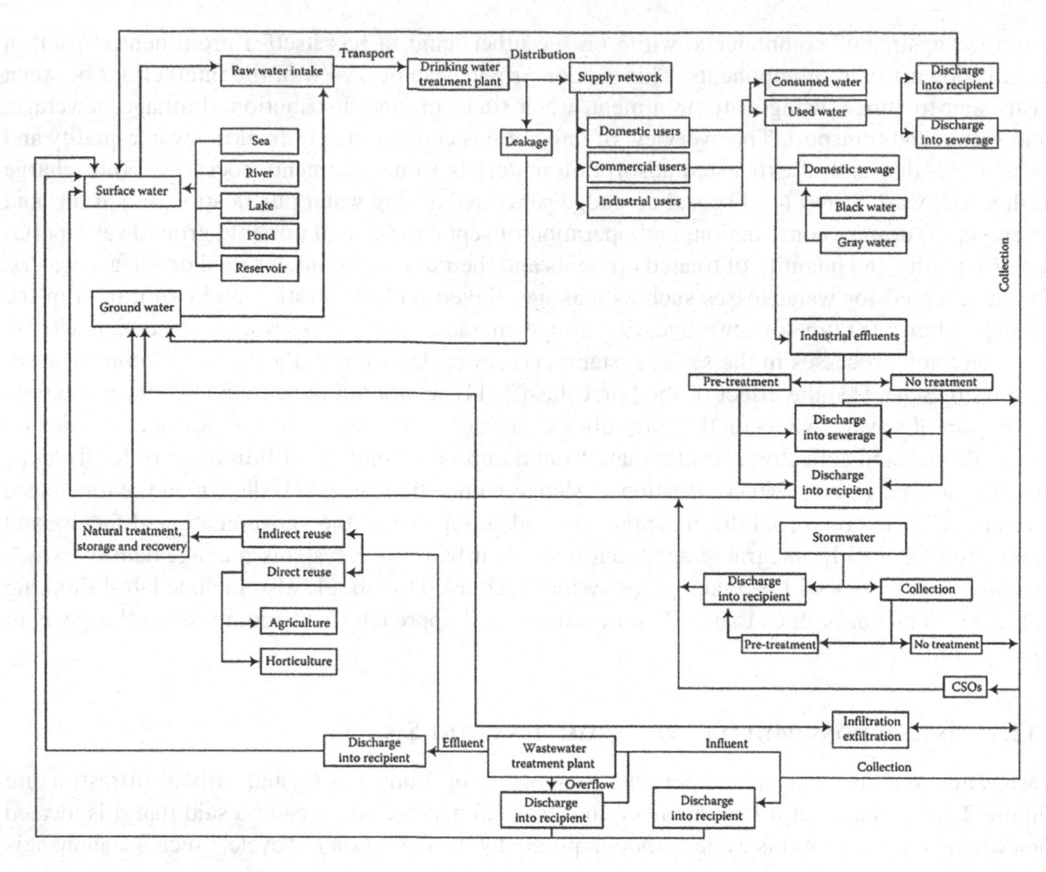

FIGURE 2.10 Schematic representation of UWC–infrastructure interactions.

2.12.3 Interactions between Water Supply and Wastewater Collection Systems

The most obvious interaction between water supply and wastewater treatment is resulting from the fact that, without exceptions, water supply networks are not fully efficient systems, where unaccounted for water losses such as leakage present the major reason for it. Even in new networks, physical water losses could exceed 10%, while in regular practice, especially in developing countries, it is between 30% and 50%, and sometimes even higher.

Leakage management has often an important impact on the groundwater level which may, depending on local hydrogeological conditions, increase infiltration into sewer collection systems. It is a generally known fact that infiltration is to be minimized in the sewers as it further dilutes the wastewater, increases hydraulic load to sewers and consequently sewage works, helps earlier activation of combined sewer overflows, and induces additional problems to plant operators.

Similar to water supply networks, sewerage systems are prone to exfiltration in cases where the groundwater table is below sewers, which is undoubtedly a problem by itself, but under certain situation can also cause contamination in the water supply network in case of loss of system pressure for whatever reason. The issue becomes more exaggerated in case of natural disasters (such as hurricanes or floods).

2.12.4 Interactions between Urban Drainage Systems and Wastewater Treatment Systems

Interaction among stormwater drainage, sewerage, and wastewater treatment is unavoidable as these components of the UWC should be generally considered as a single integrated system. Obviously, the primary role of sewerage is to transport pollution from one point to another and ultimately to the WWTP. Interactions between sewerage and sewage works depend very much on the type of the sewer system (separate or combined). In some countries, especially in newly developed and modern cities, one finds even three separate systems in place (sanitary sewers, industrial sewers, and stormwater drainage) (Saito et al., 2004).

The role of CSOs in stormwater and wastewater management is very important, because a decision on their position and number in the system, their design and moment of activation/spilling, and the pollution load allowed to be released into the environment through them has to be taken into account. Another stormwater-related issue that is closely related to CSOs and has direct consequences for plant operation is the effect of the "first flush." Furthermore, the impact of the first flush on a treatment plant depends very much on other components of urban infrastructure, such as transportation and solid waste management. For example, pollution accumulation and washoff collected by the first flush contain dry deposition of atmospheric pollutants on roofs; deposition of pollutants from traffic on streets, highways, and parking places; accumulation of dust, dirt, and larger residues on streets; solids' generation by roadway deterioration; highway surface contamination by vehicle and tire wear, fluid spills, and leaks; road surface contamination by salting and sanding in winter conditions; washoff by the energy of storm-generated runoff (washoff and erosion); washoff by dissolution and elutriation due to acidity of rainfall; and so on.

In general, small sewer systems are more sensitive to the first flush relative to large systems where dilution effect is more prominent. The first flush, usually considered to be 15–20 minutes' duration, removes the major part of pollution from the sources described above. It is also indicated that the first 25 mm of rainfall is responsible for 90% of such pollution. Furthermore, it is surprising that the effects of street cleaning using water are very low (contributes <10% to the reduction of pollution entering rainwater drainage system) (USEPA, 1993).

Based on the literature (USEPA, 1993), sewer networks are still the most efficient transport system, and despite the fact that separate sewers are 50% more expensive than combined sewers (data for the Netherlands), there is an increasing trend towards connection of separate sewers at the expense of combined sewers, at least in the developed parts of the world (Korving and Clemens, 2005). Associate costs per capita served are still low in comparison with individual wastewater treatment systems in remote locations where sewer connections are not feasible (Wilsenach, 2006). Regardless of alternatives currently available, it is expected that sewer networks will remain a sanitation backbone for urban areas, especially to those densely populated. However, contrary to the opinion of those who are constantly working on improving conventional wastewater systems, there is an emerging movement supporting the philosophy of alternative sanitation and urban water drainage, which promote pollution prevention rather than control, separation at source rather than end of the pipe treatment, and reuse of valuable resources rather than wasting by discharge into the environment. It is believed that source separation of rainwater and urine has the best prospect of improving urban water management (Wilsenach, 2006).

2.12.5 Interactions between Urban Drainage Systems and Solid Waste Management

Inappropriate solid waste management is a large and emerging problem especially in developing countries, sometimes being of more concern than wastewater-related problems. Interactions of solid waste with urban water infrastructure components are numerous and only the most prominent are briefly discussed below. Uncontrolled deposition of solid waste within urban areas, especially on streets, not only presents a major problem for public health but also has a direct impact on the functioning of the urban drainage system, as part of the waste can enter the sewerage system, be deposited/retained in the system, or eventually be discharged into the environment or reach the entrance to a sewage works.

As sewer networks and treatment plants are in general not designed to transport solid waste, such a practice has a negative impact on the efficiency, operation, and maintenance of urban wastewater infrastructure. There are numerous examples of dumping solid waste into open drains, which is accumulated at the banks of the channels, is deposited at the bottom, and is often accumulated at the entry of pumping stations, gates, weirs, and other infrastructure facilities within the system. Such a practice presents a huge obstacle to maintaining the sustainability of investments and is a continuous problem for system operators.

Another large problem caused by uncontrolled disposal of solid waste into rainwater drains is that with time these drains become loaded with waste, and their efficiency becomes drastically reduced. This is, in many cases, the major reason for the flooding during heavy rainfall and malfunctioning of the drainage system, which often has a consequence of flooding the urban areas.

In some cases of organized deposition of solid waste at designated locations, very often, and especially in developing countries, most of the sites are in fact nothing else than dumpsites rather than proper sanitary landfills. This consecutively means that most of the sites are characterized by uncontrolled leachate generation and its release into the environment. Such a practice has usually tremendous negative effects on the quality of receiving waters, especially groundwater. Taking into account that as a result of rapid population growth and often uncontrolled urbanization, urban and most frequently preurban agglomerations are getting more and more close to dumping sites, this issue becomes a very serious threat to public health affecting most of the world's poorest.

Fecal sludge is usually considered as a solid waste and its management can have a direct impact on water supply and sanitation infrastructure. Link with water supply is reflected by the risk of pollution of groundwater, which is often used as a source of on-site water supply where no piped water supply is available. In the case of leaking or overflowing septic tanks, there is a direct threat of bacteriological contamination of those humans who either get in contact with a septic tank content or drink contaminated water. Fecal sludge management practices have different impacts on sanitation infrastructure. In some cases, collected sludge is dumped to the nearest sewage manhole and transported with wastewater or is discharged at the entrance to the treatment plant—both practices are unwanted by treatment plant operators. Because fecal sludge represents a high load of organic matter and nutrients, such discharge practices often have a negative impact on sewage works (shock loads). In order to cope with such variations in load, some treatment plants have specially designed receiving tanks for fecal sludge, which is mixed with incoming wastewater and introduced to the treatment plant during periods of low load.

Another interaction within the urban context is joint treatment of organic fraction of solid waste and of the sludge generated from wastewater treatment at sewage works. This process is called codigestion and has distinctive advantages over separate handling of solid waste and sludge concerning both financial and environmental aspects.

In addition, due to its high organic content, sewage sludge is sometimes used as a supplement fuel in solid waste incineration plants to improve combustion process. It is important here to obtain sludge with as little as possible water content which is directly linked with the efficiency of its pretreatment (dewatering), usually taking place at sewage works.

2.12.6 Interactions between Urban Water Infrastructure and Urban Transportation Infrastructures

Transportation is a very important segment of infrastructure whose elements, such as roads, often host the other urban infrastructure systems (water supply pipes, sewerage pipes, urban drainage, telecommunication cabling, electrical cabling, etc.). There is a strong link between the design and maintenance of roads, solid waste management, and sewer systems (especially in the case of combined sewers). In addition, in very difficult cases where urban planning failed (often the situation in developing countries), due to unavailability of space, roads can be retrofitted to serve as a part of the urban drainage system (the "road as a drain" approach).

Another important factor is that road renovation/construction takes place more frequently in comparison with replacement of the old sewers beneath them. Because of the fact that old (brick) sewers were not designed to sustain traffic load of modern and heavy vehicles, there are numerous cases of sewer pipes collapsing under the impact of heavy traffic load. Such unwanted interactions have negative effects on both the traffic and the integrity and functioning of the sewer pipes and drains.

2.13 LIVABLE CITIES OF THE FUTURE

Smart Cities and Livable Cities of the Future are what city planners with engineers, social scientists, and IT experts strive to design. There are many initiatives and models. In late October 2012, a symposium to honor the legacy of George Bugliarello was hosted jointly by the Polytechnic Institute of New York University (NYU-Poly) and National Academy of Engineers (NAE). This event brought more than 200 policy makers, engineers, civic leaders, educators, and futurists to discuss how George Bugliarello's works and his technical point of view manifest themselves in innovating urban planning for livable cities of tomorrow.

The presentations and speeches of the symposium were recorded and transcribed into a proceeding of the symposium by Karamouz and Budinger (2014). Some designated parts of the proceedings were selected, titled as viewpoints, to be presented in this chapter and some in Chapter 13. They would paint a more precise picture of the challenges faced and the much needed attributes of the livable cities of the future.

2.13.1 Daniel Loucks View Points

The needs for clean water in all aspects of urban development and maintenance can be met through the integration of multiple decentralized schemes for capturing and collecting precipitation, wastewater sanitation management, and modernization of infrastructure maintenance technologies, as stated by the Professor at Cornell University. New York City has had success in using natural systems to provide clean drinking water and manage storm runoff. The city has saved billions of dollars through integration of diverse methods for controlling water quality, distribution, use, and reuse.

2.13.1.1 *Urban Water in the Larger Water Nexus*

Humans depend on water for life, which is obvious, but also for almost everything we see or make. Everything you see while reading a document required water to create, including the electricity and bulbs that provide the light you may be seeing it with. Humans are completely dependent not only on water but also on the fact that there are no substitutes.

Water is a critical input to all sectors of our economy and environment (Figure 2.11). All the components shown in this figure are impacted by each other, in part by how water is allocated

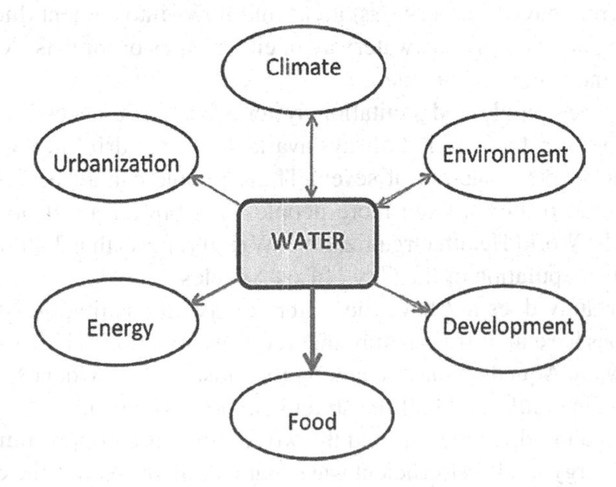

FIGURE 2.11 Water is a critical input to all sectors of our economy and to our built and natural environments.

to them and how climate, which seems to be changing, plays a role in determining the available supplies of water.

But what is amazing is that everyone living on this planet could not only survive but thrive on the small percentage of the total freshwater supply that is actually available to use globally. How much is that? If a half-liter (one pint) water bottle represents all the water on this planet, only a teaspoon of that water is available for human use. That is less than what you can put in the cap of that bottle. The trouble is that this water is not always where and when and of the quantity and quality needed. And then sometimes there is too much of it. Cities have to consider both the reliability and quality of their freshwater supplies as well as protection against too much of it in any given time period. Urban areas need stormwater management.

A region's demands for freshwater are a function of the need to provide clean drinking water and sanitation, to ensure public health in growing urban centers, to create electric and liquid fuel energy, to maintain a healthy environment and well-functioning ecosystems, to grow and process food, and to support the industries and economic development that provide jobs and welfare.

In developing regions meeting these needs is even more urgent, and often more difficult, especially in urban slum environments. But even in such cities, the options or opportunities can be improved given sufficient funding, effective governance, and appropriate technical expertise.

Let's focus on the water urbanization link. How do we provide the right amount of water at the right places, times, pressures, qualities, and costs? And how can water be used to enhance the urban environment and esthetics? After all, according to the *New York Times* (October 7, 2012), 80% of all Americans now live in urban areas. So this issue is important to most of us.

Urban renewal is a primary approach to building new infrastructure, attracting job-producing industry, stabilizing communities, and improving residents' quality of life. It almost always involves investments in infrastructure.

2.13.1.2 *Gray Infrastructure*

In the past, public works engineers had a major role, if not the only role, in the planning, design, development, installation, and operation of water supply and stormwater management infrastructures. These engineers are trained to use concrete and steel, so they do, and the concrete and steel infrastructure they build is called "hard" or "gray" (because of its color) construction.

Water supply systems typically pump natural water through pipes or canals to water treatment plants and then through storage and distribution systems to the tap. Wastewater systems typically pump wastewater through collection sewers to wastewater treatment plants, and the resulting effluent may be reused or released into natural water bodies. Many creeks, streams, and rivers now flow through cities underground in pipes and tunnels. Rainfall that used to soak into the ground now becomes runoff from paved (impervious) areas and flows into cement ditches and stormwater drains, again becoming an underground waterway in either pipes or tunnels. All of this water is out of sight, out of mind, and benefiting no one.

Reliable and safe water supply and sanitation systems are basic necessities of urban areas. In developing regions, however, they are not always available. Clean drinking water is still not available to about a billion people—one out of seven. These people cannot be fully productive members of their communities or cities. Even more people—2.4 billion, most of them in cities—lack adequate sanitation. The World Health Organization (WHO) reports that 3.4 million of these people die each year, about the population of the City of Los Angeles.

New York City certainly does not have the water supply and sanitation issues that developing regions have. But it does have an infrastructure that requires attention. The 15- to 35-million-gallon daily leak in the Delaware Aqueduct may be among the most visible evidence of this, except for the street where a water main breaks and half the street instantly disappears.

In New York City and in other cities around the world, there is an opportunity to manage stormwater runoff in more energy- and cost-efficient ways that will also enhance the environment of those who live and work in the city.

2.13.1.3 *Green Infrastructure*

The NYC Department of Environmental Protection (DEP) is widely recognized for successfully using natural systems to provide clean drinking water and manage stormwater. DEP estimates that such efforts have saved ratepayers billions of dollars—by eliminating the need for construction of hard infrastructure such as storm sewers and filtration plants—while preserving large tracts of natural areas. The department's *Green Infrastructure Plan* lays out how the city will improve the water quality in New York Harbor, for example, by capturing and retaining stormwater runoff before it enters the sewer system, and from there the harbor, through the use of StreetSide swales, tree pits, and blue and green rooftop detention techniques to absorb and retain stormwater (Figure 2.12). This hybrid approach will reduce combined sewer overflows by 12 billion gallons a year—over 2 billion gallons a year more than the current all-gray strategy—while saving New Yorkers $2.4 billion (NYCDEP, 2012).

Like other older urban centers, New York City is primarily serviced by a combined sewer system in which stormwater and wastewater are transported together through a single pipe. Treatment plants are designed to treat and disinfect twice the dry-weather flow, but the system can exceed its capacity during heavy storms. When this happens, a mix of stormwater and wastewater—called combined sewer overflow (CSO)—is discharged into New York Harbor. DEP has committed to reducing the annual volume of CSOs by more than 8 billion gallons over the next 20 years—10% of the runoff from the city's impervious surfaces.

Rather than build additional large storage tanks or tunnels to temporarily store stormwater at the end of the sewer system, DEP determined that it was more cost-effective to first construct source controls and "soft" infrastructure (e.g., bioswales, blue and green roofs, and subsurface detention systems) to control and reduce stormwater runoff from impervious spaces such as roofs, sidewalks, and parking lots. Together with conservation measures and operational improvements,

"And they said green roofs would never catch on."

FIGURE 2.12 Green roofs are an increasingly widespread and effective way to reduce stormwater runoff. (Courtesy of Grayman, W.M. et al., *Toward a Sustainable Water Future.* American Society of Civil Engineers, Reston, VA, 2012.)

the widespread adoption of such soft infrastructure can reduce CSOs at less cost than second-tier hard or gray infrastructure. Moreover, green infrastructure provides many quality-of-life benefits, by improving air quality, increasing shading, contributing to higher property values, and enhancing streetscapes.

The department is also implementing lots of other innovative measures such as giving people rain barrels, installing automated meter-reading devices, and developing an energy strategy that will (a) reduce the carbon footprint, including emissions of greenhouse gases; (b) reduce electricity demand, the cost of which is expected to almost double every 5 years in the absence of aggressive energy efficiency investments; and (c) explore clean energy options.

2.13.1.4 *Challenges of Future*

Urban populations are projected to rise, nearly doubling from the current 3.4–6.4 billion by 2050, with the number of people living in slums rising even faster, from 1.0 to 1.4 billion in just a decade. Already, half of the world's population lives in cities, and 80% of Americans do.

Providing adequate water supply and sanitation, particularly in urban areas, is a challenging task for governments throughout the world. Many, especially in Africa and Asia, have virtually no or only inadequate infrastructure and limited resources to address water and wastewater management in an efficient and sustainable way. Because of inadequate infrastructure, almost 85% of all wastewater is discharged to water bodies without treatment, resulting in one of the greatest health challenges, restricting development, and increasing poverty through costs to healthcare and lost labor productivity (UN, 2012).

This task is made even more difficult by predicted climate changes, which are associated with significant alterations in precipitation patterns, both spatially and temporally, affecting the availability and variability of water supplies.

In addition, technological and financial constraints are challenges in maintaining and upgrading infrastructure assets to deliver water to all sectors while maintaining the quality of water distributed to various users. Furthermore, population growth, urbanization, and industrial activities are leading to a dramatic increase in water use and wastewater discharge.

Cities are facing difficult strategic decisions. Do they continue business as usual, following a conventional technical, institutional, and economic approach to water and sanitation? Do they tinker, following the conventional approaches while trying to optimize and fine-tune them? Or do they look for a new paradigm that considers interventions over the entire UWC to provide security through diversification of water sources, reconsideration of the ways water is used (and reused), wastewater as a valuable resource, governance structures covering the entire UWC, and the resiliency of water and sanitation to global change pressures?

2.13.2 David Miller's View Points

Cities around the world are acting to both fight climate change and adapt to it as stated by the former Meyer of Toronto. They have strategic plans, like Toronto's Clean Air and Climate Change Action Plan, and in executing those plans are lowering carbon emissions and building new resilient infrastructures. The role of cities is critical, because as of 2008 most residents of the world lived in cities, which is where most jobs are and most carbon emissions occur. By addressing carbon emissions in three sectors—heating and cooling buildings, energy generation, and transportation—emissions can be dramatically reduced, the livability and resilience of cities improved, and new jobs created.

2.13.2.1 *Setting the Stage*

Cities around the world are facing economic challenges. One of the simplest ways to put people back to work is by rebuilding the infrastructure of cities—including infrastructure that helps mitigate and adapt to climate change. Large cities are seeing the impact of climate change through increased frequency and severity of storms, sea level rise, and other natural disasters (Figure 2.13), and they

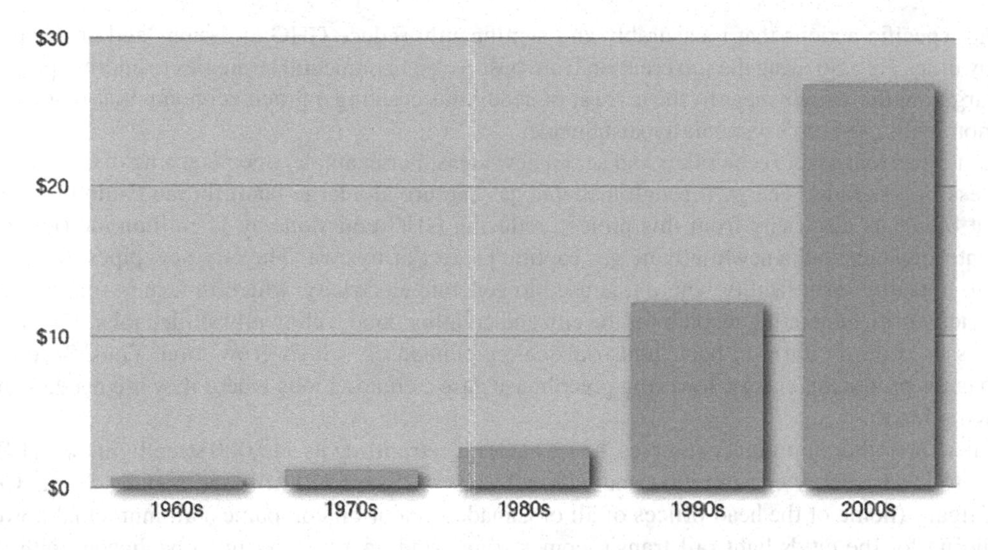

FIGURE 2.13 Rise in global disaster damage, based on annual insurance disaster claims, in billions of US dollars, adjusted for inflation, 1960s–2000s.

are responding to those challenges in ways that not only address environmental challenges but also create jobs.

2.13.2.2 *Cities Around the World Lead the Way*

The C40 group of cities (Figure 2.14) are leading the fight against climate change, partnered with the Clinton Climate Initiative.

About 80% of greenhouse gas (GHG) emissions come from activities that take place in or are needed to support cities. Most of these emissions are from three sources: generation of energy, use of energy to heat and cool buildings, and transportation. In each of those sectors, C40 cities are

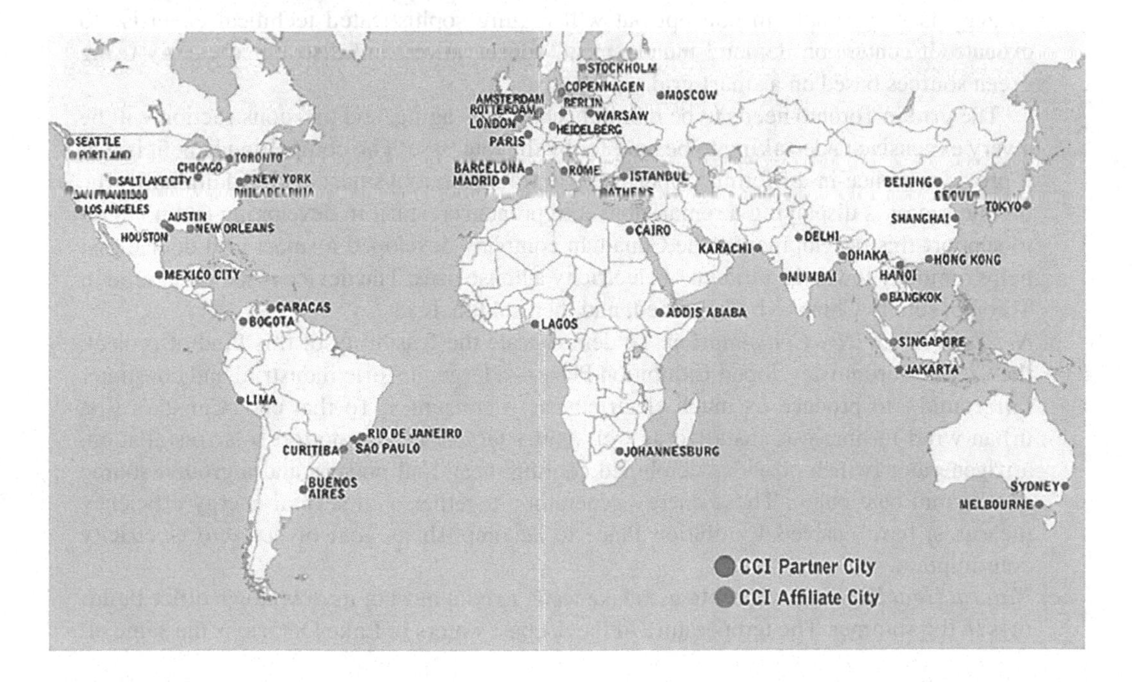

FIGURE 2.14 C40 partner cities and 18 affiliate cities. CCI = Clinton Climate Initiative.

taking specific actions that measurably and significantly reduce GHG emissions and create jobs. Many cities are also using the job creation from positive environmental strategies to address poverty by targeting the employment to those most in need, thus creating a green economy that is socially, economically, and environmentally sustainable.

Cities can learn from each other and act on new ideas. For example, upon learning of Sao Paulo's success in generating energy through methane gas capture at a large landfill (Sao Paulo generates about 8% of its electricity from this project, reducing GHG emissions by 11 million metric tons), Toronto implemented a new methane gas capture project of its own. The city now pipes the methane to a nearby farm facility where it is used to generate electricity, which in turn is sold into the electricity grid, generating income for the city and creating good technically skilled jobs. The waste heat is used on the farm to heat industrial-scale greenhouses, which grow food. Thus, Toronto's garbage is producing energy, lowering greenhouse gases, creating jobs where they are needed, and growing food!

There are other important examples. Los Angeles is retrofitting its 140,000 streetlights with LED bulbs to lower energy consumption and within 7 years will save $10 million annually. In the City of Calgary (home of the head offices of all of Canada's major oil companies) all municipal power, including for the city's light rail transit, comes from wind energy. And in Copenhagen, with the help of 80 turbines, 160 MW of electricity were generated from offshore wind, supplying energy to 150,000 homes.

2.13.2.3 *A Closer Look at Toronto's Strategies for Sustainability*

Toronto has adopted a variety of strategies to enhance energy generation, buildings, and transportation, dramatically lower greenhouse gases, and create an economically and socially sustainable community. The first step is a plan with specific targets and timetables. The city's 2007 Climate Change Action Plan, "Change Is in the Air" (updated in 2008 and 2009), requires it to meet the Kyoto targets for GHG emission reductions.

a. *A SMART Grid*: The city owns the local electricity utility (Toronto Hydro) and is working to develop a strategy to determine how to meet the city's GHG reduction targets. The strategy is very simple in concept but will require sophisticated technical expertise to execute. It centers on demand management, conservation, and distributed energy using green sources based on a smart grid.

   The grid in Toronto needs to be rebuilt because it is aging, and this construction will be a very expensive undertaking to be paid for by the rate base. The compelling benefit is that it provides a once-in-a-lifetime opportunity to transition to a smart grid. Additionally, the private sector is displaying a remarkable entrepreneurial spirit in developing technologies to support this development. One Canadian company developed a smart grid device that helps remotely lower the amount of electricity a house uses. The device is manufactured in Toronto, sent to China to be upgraded, and then sold in Texas.
b. *Net Zero Electricity Consumption*: To demonstrate the feasibility of this kind of project, the City of Toronto developed Exhibition Place—a large, historic industrial and consumer fairground—to produce as much electricity as it consumes. To that end, Canada's first urban wind turbine was installed as well as the largest urban rooftop solar installation, a trigenerator (which provides combined cooling, heat, and power), and a ground source geothermal heat pump. These energy generators, together with several energy efficiency measures, have enabled Exhibition Place to accomplish its goal of net zero electricity consumption.
c. *Natural Heat Exchange*: Toronto uses lake water to cool most of its downtown office buildings in the summer. The temperature of the deepest waters of Lake Ontario is the same all

year round, so a heat exchange system was installed. Energy demand peaks in the summer, as is the case in many cities, when on hot days everybody is using air-conditioning. Anything that can be done to lower the peak demand translates into large financial savings for the electricity system, as the cost of building new plants to accommodate the peak demands can be eliminated.

d. *Energy Retrofits*: Toronto has aggressively invested in energy retrofits for buildings. Its Better Buildings Partnership with the private sector and the (private) gas company has resulted in energy retrofits for millions of square feet of public, private, and institutional buildings. These activities are critical because the energy used to heat and cool buildings in Toronto creates 60% of the city's GHG emissions. In New York City, it is 80%, hence it is important for the commercial buildings to post their energy consumption, which helps to create awareness for energy efficiency initiatives.

   Toronto has the second highest concentration of concrete slab apartment buildings in North America after New York City. These buildings offer important benefits because they often were built with larger units. Furthermore, because they are concrete, their lifespan is least 100 years and proper maintenance can add to their durability. However, concrete has no insulating power, which translates into significant energy inefficiency. Engineers from the University of Toronto developed a strategy to reskin the buildings with aluminum and insulation, dramatically reducing energy consumption at a low cost (payback over just 7–10 years). Studies show that, if all of its buildings were insulated, Toronto would reduce its GHG emissions by 5%–6%—the first Kyoto target.

e. *Transportation*: In most cities, transportation sector emissions as a percentage of total emissions are a significant concern. Toronto has responded by developing, securing funding for, and starting construction of the Transit City Light Rail. The plan is effective because it creates a network of rapid transit to all neighborhoods of the city, meets transportation demands now and for the future, creates the opportunity for inner city development (rather than sprawl), and is cost-effective, allowing for rapid construction.

   Toronto is also adding capacity to its subway lines not by building new lines but by changing the cars, making it possible for them to run at shorter intervals because of technical advances in smart communication.

2.13.3 Craig S. Ivey's View Points

There are five critical components to the livable cities of the future: public safety, reliability, affordability, reduced environmental impact, and smarter and more secure facilities as stated by the former President of Con Edison. Cities must also prepare for both a growing demand for power and the effects of increasingly severe weather patterns that threaten the grid. This became dramatically clear with Superstorm Sandy, which hit the New York City region 3 days after this symposium, on October 29, 2012. Utilities, urban planners, climate experts, government leaders, and regulators must all collaborate to determine the best approaches to fortify the city's infrastructure and protect residents and businesses from future threats. New York City's largest energy provider is taking steps to address challenges and meet needs in order to ensure delivery of these critical components.

2.13.3.1 *Facts and Figures*

Con Edison (Con Ed) and its 14,000 employees support one of the most active and densely populated areas in the country. New York City and Westchester County are home to 9 million people, and more than 50 million visitors come to the city each year, based on 2011 data. The company serves 3.3 million electric customers and 1.1 million gas customers (Figure 2.15). The intense energy

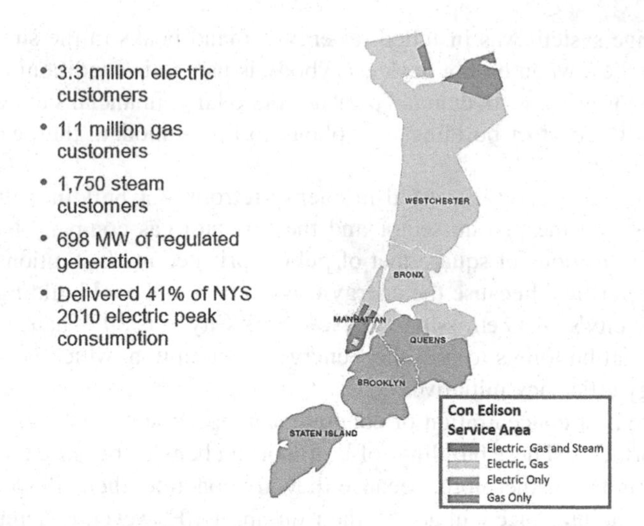

• 3.3 million electric customers
• 1.1 million gas customers
• 1,750 steam customers
• 698 MW of regulated generation
• Delivered 41% of NYS 2010 electric peak consumption
WESTCHESTER
BRONX
MANHATTAN
QUEENS
BROOKLYN
STATEN ISLAND
Con Edison Service Area
Electric, Gas and Steam
Electric, Gas
Electric Only
Gas Only

FIGURE 2.15 Energy provision in New York City and environs.

demand of the area requires a reliable energy infrastructure, and Con Ed substations are designed to safely meet the community's energy needs.

Sandy's relentless winds and unprecedented storm surge caused damage across the region unlike anything we've ever seen. Catastrophic flooding and corrosive salt water destroyed electrical equipment and downed trees ravaged our overhead system, making repairs difficult and time-consuming. Now and in the future, thoughtful, forward-thinking construction will help keep our systems reliable for the "new normal" that we must design and prepare for in the wake of Sandy.

Our investment in smart grid technologies and other innovations allow greater flexibility and reliability during extreme weather. For example, for the past 7 years, Con Ed has implemented a policy requiring any new business in a flood zone to either install submersible electrical equipment or locate its electrical equipment at higher elevations.

2.13.3.2 *Challenges*

Efforts to minimize construction in the streets are challenging and costly because of numerous underground structures that compete for limited space to accommodate growing demands not only for electricity but also for the distribution of communications, water, and natural gas (Figure 2.16).

In addition, demand is at peak—above 12,000 MW—for only 36 hours per year. Peak demand occurs when many users across an energy system simultaneously increase their energy use—for example, in the afternoon or evening of a day of extreme or record-breaking heat, when both homes and offices turn up the air-conditioning and households also turn on televisions, computers, washers and dryers, and other appliances. In the absence of storage mechanisms, energy must be produced when it is demanded, so the infrastructure as a whole must be ready to meet peak demand even if it is idle for the balance of the year.

2.13.3.3 *Solutions*

Smart grids enable two-way communication between our facilities and our customers' equipment (e.g., smart meters, distributed generators, plug-in vehicles), and switches enhance flexibility in the network, thus increasing reliability (Figure 2.17). Smart grid technology, which relies on underground auto-loop and wireless-controlled switches, reduces the likelihood and severity of service disruption caused by a network event.

During Sandy, Con Ed was able to use remote sensors on the distribution system and remotely operated switches to reduce the damage to the system and speed up repairs. New sensors allowed

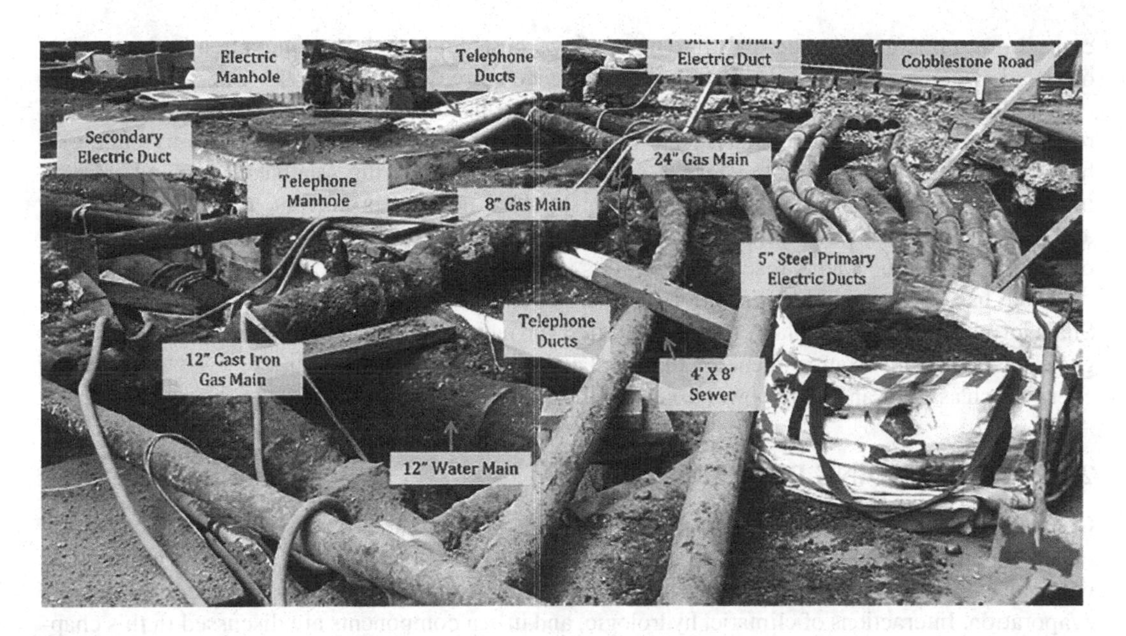

FIGURE 2.16 Limited underground space for pipes and conduits. (From Karamouz and Budinger, 2014.)

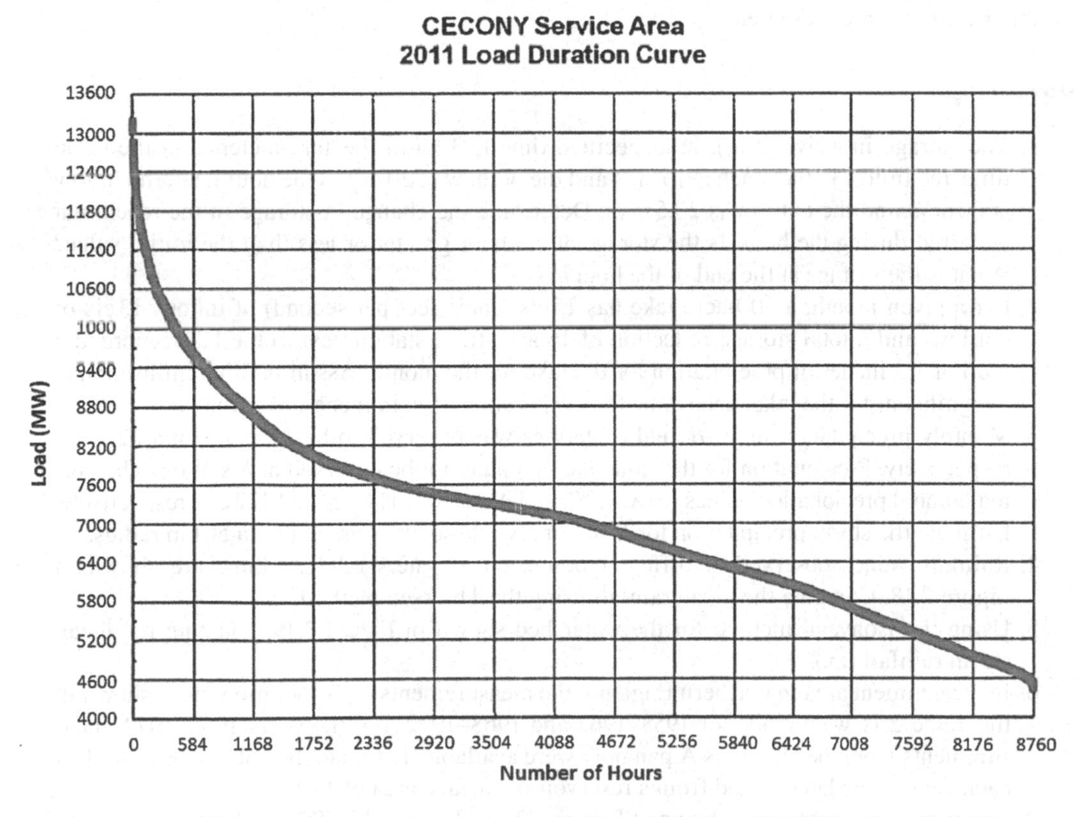

FIGURE 2.17 Utilities must invest to support a very short peak period. CECONY—Consolidated Edison Company of New York.

control room operators to see real-time power flows on feeders and, in conjunction with remotely operated switches, made it possible to reinstate service more quickly.

Con Ed was also able to sectionalize overhead lines ahead of the storm to improve both outage restoration times and public safety. New underground switches designed by company engineers have been installed, allowing greater flexibility and reliability during weather events and enhancing the ability both to monitor underground transformers, network protectors, and other equipment and to isolate problems. In addition, recently installed flood detectors in low-lying substations alert operators when flood waters reach critical levels.

There has also been a shift in sources of energy generation. With development of the Marcellus Shale formation, some coal plants have been retired and new gas plants have been established. Natural gas is a much cleaner source of energy than coal and oil. Furthermore, in 2010, renewable sources of energy such as solar and wind power exceeded the amount of oil-based energy in the region.

2.14 CONCLUDING REMARKS

Chapter 2 has covered the water cycle in urban areas on the basis of hydrology principles including precipitation, evaporation, evapotranspiration, and infiltration. Solar radiation and atmosphere water phase changes provide the main energy inputs and result in the generation of precipitation and evaporation. Interactions of climatic, hydrologic, and urban components are discussed in this chapter. Losses or abstractions in hydrology include evaporation, evapotranspiration, and filtration and are covered in detail in this chapter, including physics-based methods. Snowmelt estimation, PDSI, and water balance topics are also covered with a case study on the use of a water balance-based supply and demand sustainability index. Interactions between the UWC and urban infrastructure components are also presented. Finally, some attributes of livable cities of the future with emphasis on water elements are presented.

PROBLEMS

1. The storage in a river reach at a specified time is 3 ha-m (hectare-meters). At the same time, the inflow to the reach is 15 m³/s and the outflow is 20 m³/s. One hour later, the inflow is 20 m³/s and the outflow is 20.5 m³/s. Determine the change in storage in the reach that occurred during the hour. Is the storage at the hour greater or less than the initial value? What is the storage at the end of the hour?
2. For a given month, a 300-acre lake has 15 cfs (cubic feet per second) of inflow, 13 cfs of outflow, and a total storage reduction of 16 acre-ft. A station next to the lake recorded a total of 1.3 inches of precipitation for the lake for the month. Assuming that infiltration is insignificant for the lake, determine the evaporation loss, in inches, over the lake.
3. Monthly precipitation in *W*, *B*, and *A* stations was observed to be 11.5, 9.0, and 12.4 cm, respectively. Precipitation for the same month could not be observed at *X* station. The normal annual precipitation values for *X*, *W*, *B*, and *A* are 102, 114, 95, and 122 cm, respectively. Estimate the storm precipitation for *X* station. All these cities are within a 50 km radius.
4. Rainfall values observed at different points on a watershed (12.5 km²) are shown in Figure 2.18. Compute the mean rainfall using the Thiessen method.
5. Using the isohyetal method for the watershed shown in Figure 2.19, compute the mean storm rainfall (in.).
6. In a catchment area in northern England, the measurements of evaporation (mm) shown in the Table 2.11 were made for 1958–1962 and 1968–1972. For the years 1968–1972, measurements from the US Class A pan only were available. Estimate the volume of water lost each year in the later period from a reservoir of surface area of 1.4 km².
7. The rainfall hyetograph is shown in Figure 2.20 and is listed in Table 2.12. It is subject to a depression storage loss of 0.15 cm and Horton infiltration with parameters $f_0 = 0.45$ cm/h, $f_c = 0.05$ cm/h, and $k = 1$ h. Calculate the hyetograph of excess rainfall.

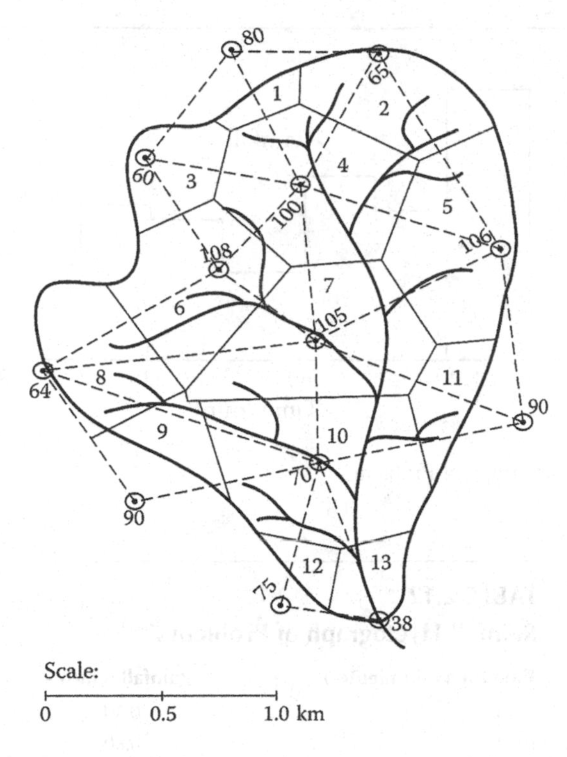

FIGURE 2.18 Thiessen polygon method for computing the mean areal rainfall for Problem 4.

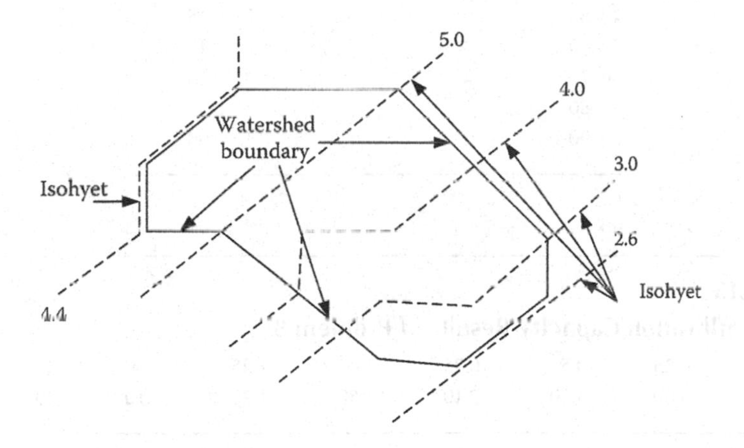

FIGURE 2.19 Watershed boundaries for Problem 5.

TABLE 2.11
Measurements of Evaporation for 1958–1962 and 1968–1972 for Problem 6

Year	U.K. Tank	U.S. Class A Pan	Year	U.S. Class A Pan
1958	351	491	1968	621
1959	536	713	1969	581
1960	502	653	1970	687
1961	437	586	1971	624
1962	486	612	1972	568

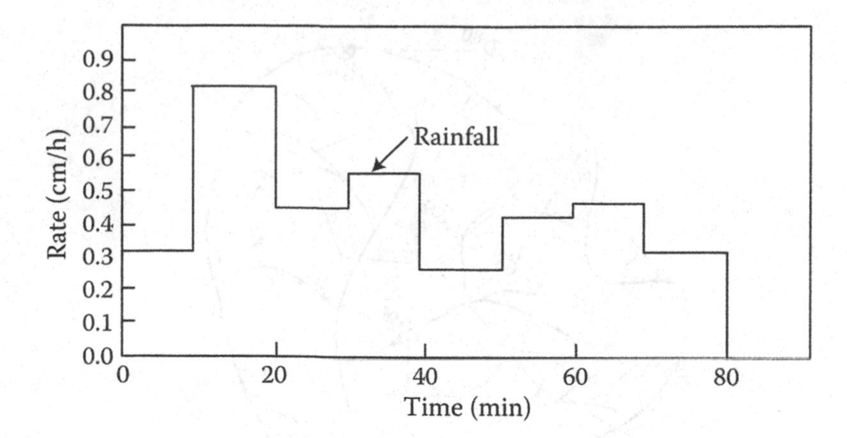

FIGURE 2.20 Rainfall hyetograph of Problem 7.

TABLE 2.12
Rainfall Hyetograph of Problem 7

Time Interval (minutes)	Rainfall (cm/h)
0–10	0.30
10–20	0.80
20–30	0.45
30–40	0.55
40–50	0.25
50–60	0.40
60–70	0.45
70–80	0.30
80–90	0.00

TABLE 2.13
Horton Infiltration Capacity Results of Problem 8

Time (h)	0.25	0.5	0.75	1.00	1.25	1.50	1.75	2.00
f_{ct} (cm/h)	5.60	3.20	2.10	1.50	1.20	1.10	1.00	1.00

8. Results to determine the Horton infiltration capacity in the exponential form are tabulated in Table 2.13. Determine the infiltration capacity exponential equation.
9. The data measured in 30 rain gauges and the corresponding coordinates are shown in Table 2.14.
   a. Draw the scattering of gauges.
   b. Estimate and draw the empirical variogram.
   c. Fit a theoretical variogram to an empirical variogram.
   d. Specify and draw the spatial distribution of data at the network with one dimension.
   e. Estimate rainfall at points with (3, 19), (5, 8), and (9, 9) coordinates using the developed Kriging model.

TABLE 2.14
Data Measured in 30 Rain Gauges for Problem 10

Rainfall (mm)	y-Coordinate	x-Coordinate	Rain Gauge
1	1	1	230
2	1	3	260
3	1	4	310
4	1	7	420
5	2	11	350
6	2	13	290
7	2	20	260
8	3	3	230
9	3	6	250
10	3	20	280
11	5	5	380
12	5	6	450
13	5	15	480
14	5	19	490
15	6	2	510
16	6	4	520
17	6	9	640
18	6	10	510
19	6	13	450
20	6	18	380
21	6	20	450
22	7	3	510
23	7	14	530
24	8	2	650
25	8	15	630
26	8	20	680
27	9	6	690
28	9	18	540
29	10	5	555
30	10	19	520

REFERENCES

Ahmadi, B. (2011). A climate driven model for increased in water productivity agricultural sector. Master thesis, University of Tehran.

Alley, W.M. (1984). The palmer drought severity index: Limitations and assumptions. *Journal of Applied Meteorology and Climatology*, 23(7), 1100–1109.

Ellis, J.B. (1986). Pollutional aspects of urban runoff. In H.C. Torno, J. Marsalek, and M. Desbordes (eds.) *Urban Runoff Pollution* (pp. 1–38). Springer, Berlin, Heidelberg.

Gallee, H., Van Ypersele, J.P., Fichefet, T., Tricot, C., and Berger, A. (1991). Simulation of the last glacial cycle by a coupled, sectorially averaged climate—ice sheet model: 1. The climate model. *Journal of Geophysical Research: Atmospheres*, 96(D7), 13139–13161.

Gartland, L.M. (2012). *Heat Islands: Understanding and Mitigating Heat in Urban Areas*. Routledge, England, UK.

Gascó, G., Hermosilla, D., Gascó, A., and Naredo, J.M. (2005). Application of a physical input–output table to evaluate the development and sustainability of continental water resources in Spain. *Environmental Management*, 36(1), 59–72. doi: 10.1007/s00267-004-0004-2.

Grayman, W.M., Loucks, D.P., and Saito, L. (eds.) (2012). *Toward a Sustainable Water Future*. American Society of Civil Engineers, Reston, VA.

Horton, R.E. (1919). An approach toward physical interpretation of infiltration capacity. *Soil Science Society Proceedings*, 5, 399–417.

International Council for Local Environment Initiatives (ICLEI). (2005). Case Studies (www3.iclei.org/iclei/casestud.htm), Case Reference & Cities Database (www.iclei.org).

Karamouz, M., Torabi, S., and Araghinejad, S. (2004). Analysis of hydrologic and agricultural droughts in central part of Iran. *Journal of Hydrologic Engineering*, 9(5), 402–414.

Karamouz, M., Hosseinpour, A., and Nazif, S. (2011). Improvement of urban drainage system performance under climate change impact: Case study. *Journal of Hydrologic Engineering*, 16(5), 395–412.

Karamouz, M., Nazif, S., and Falahi, M. (2012). *Hydrology and Hydroclimatology: Principles and Applications*. CRC Press, Boca Raton, FL.

Karamouz, M. and Budinger, T. F. Eds. (2014), *Livable Cities of the Future, National Academy of Engineering, Proceedings of a Symposium Honoring the Legacy of George Bugliarello*. National Academies Press, Washington, DC. https://doi.org/10.17226/18671.

Karamouz, M., Mohammadpour, P., and Mahmoodzadeh, D. (2017). Assessment of sustainability in water supply-demand considering uncertainties. *Water Resources Management*, 31(12), 3761–3778.

Karamouz, M., Ahmadi, A., and Akhbari, M. (2020). *Groundwater Hydrology: Engineering, Planning, and Management*. CRC Press, Boca Raton, FL, 750 p.

Karamouz, M., Rahimi, R., and Ebrahimi, E. (2021). Uncertain water balance-based sustainability index of supply and demand. *Journal of Water Resources Planning and Management*, 147(5), 04021015. doi: 10.1061/(asce)wr.1943-5452.0001351.

Kibler, D.F. (1982). *Urban Stormwater Hydrology*. America Geophysical Union, Washington, D.C.

Korving, H. and Clemens, F.H.L.R. (2005). Impact of dimension uncertainty and model calibration on sewer system assessment. *Water Science and Technology*, 52(5), 35–42.

Lawrence, A.I., Bills, J.B., Marsalek, J., Urbonas, B., and Phillips, B.C. (1999). Total urban water cycle based management. *In Proceedings of the 8th International Conference Urban Storm Drainage*, 30 August–3 September 1999, Sydney, Australia, pp. 1142–1149.

Lins, H.F. and Stakhiv, E.Z. (1998). Managing the nation's water in a changing climate 1. *JAWRA Journal of the American Water Resources Association*, 34(6), 1255–1264.

Liu, L. and Jensen, M.B. (2018). Green infrastructure for sustainable urban water management: Practices of five forerunner cities. *Cities*, 74, 126–133.

Marsalek, J., Watt, W.E., and Anderson, B.C. (2006). Trace metal levels in sediments deposited in urban stormwater management facilities. *Water Science and Technology*, 53(2), 175–183.

McPherson, M.B. and Schneider, W.J. (1974). Problems in modeling urban watersheds. *Water Resources Research*, 10(3), 434–440.

NYCDEP. (2012). New York City's Wastewater Treatment System, New York City Department of Environmental Protection, New York, NY

Palmer, W.C. (1965). *Meteorological Drought* (Vol. 30). US Department of Commerce, Weather Bureau, Washington, DC.

Saito, T., Brdjanovic, D., and Van Loosdrecht, M.C.M. (2004). Effect of nitrite on phosphate uptake by phosphate accumulating organisms. *Water Research*, 38(17), 3760–3768.

Sandoval-Solis, S., McKinney, D.C., and Loucks, D.P. (2011). Sustainability index for water resources planning and management. *Journal of Water Resources Planning and Management*, 137(5), 381–390.

Schumann, U. (1993). On the effect of emissions from aircraft engines on the state of the atmosphere.

Silveira, G.D., Borenstein, D., and Fogliatto, F. (2001). Mass customization: Literature review and research directions. *International Journal of Production Economics*, 72(1), 1–13.

Thompson, R.D. (1975). *The Climatology of the Arid World*. University of Reading, Reading.

Thornthwaite, C.W. (1948). An approach toward a rational classification of climate. *Geographical Review*, 38(1), 55–94.

Tucci, C.E. and Porto, R.L. (2001). Storm hydrology and urban drainage. Tucci, C. Humid Tropics Urban Drainage.

UN, (2012). Population connected to wastewater collecting system. http://unstats.un.org/unsd/environment/wastewater.htm

U.S. Army Corps of Engineers. (1965). *Snow Hydrology*. North Pacific Division, Portland, OR.

USEPA (1993), Combined Sewer Overflow Control (Manual), Environmental Protection Agency, Washington, DC

Viessman, W., Lewis, G.L., and Knapp, J.W. (1989). *Introduction to Hydrology* (3rd ed.). Harper and Row, New York.

Walker Jr, W.W. (1987). Phosphorus removal by urban runoff detention basins. *Lake and Reservoir Management*, 3(1), 314–326.

Wang, G.T. and Chen, S. (1996). A linear spatially distributed model for a surface rainfall-runoff system. *Journal of Hydrology*, 185(1–4), 183–198.

Whiteman, C.D. (2000). *Mountain Meteorology: Fundamentals and Applications*. Oxford University Press, Oxford.

Wilsenach, J. (2006). Source separation of urine: Sanitation. *Water Sanitation Africa*, 1(3), 22–25.

Ytuarte, S. (2005). Time series MODIS/aqua image based urban hot Island changes of San Antonio Texas using ArcMap 9.0 and ArcGIS 9.0.

3 Urban Water Hydrology

3.1 INTRODUCTION

Urbanization has adverse effects on the characteristics of local hydrology and stormwater quality. These effects have been realized in past two decades and different approaches have been utilized to address and mitigate them. UNESCO devoted the sixth phase (2002–2007) of International Hydrology Program (IHP 6) to the role of water in urban areas that was a turning point on the focus of hydrology communities on the urban water cycle. See Marsalek et al. (2008) for the details. In urban hydrology, considerable adjustment is needed to the principles that govern the water runoff initiation, ponding, drainage, and flood analysis. Increasing imperviousness and complexity in flow path and interaction with sanitary sewer are the overarching issues. The effect of controlling flow is more visible as the urban utilities are trying to curb, divert, and store accumulated water and construct best management practices (BMPs) to remediate flooding effects and improve stormwater quality. Urban water quality concepts rely on empirical techniques and limited available data. The emphasis of this chapter is on runoff and water accumulation quantity analysis and related issues such as excess rainfall estimation, rainfall–runoff analysis, calculation of peak flow, and its occurrence time as well as hydrograph analysis. Overland flow and water conveyance in urban area is more complicated than undeveloped area because of the complexity of the paths that rainwater and stormwater are taking. Interdependencies of water infrastructure with other infrastructures make the urban hydrology and extreme value analysis more complicated and critical. Design values are key to many urban development issues and are analyzed by frequency analysis and the use of probability distributions. Time series analysis and modeling are discussed within a limited scope applicable to urban area. The urban water hydraulics principles are discussed in Chapter 4 and the characteristics of storm water, water supply, and wastewater including hydraulic structures of different applications are discussed in Chapters 5–7.

3.2 URBAN WATERSHEDS

Determination of watershed characteristics is an important concept in all hydrologic designs. The watershed is defined in terms of a point named "outlet." The designs in the watershed are being made with regard to the outlet. The watershed can be well defined as the area that can be considered a bounded hydrologic system, within which all live components are linked to each other regarding their water supply resources. Each watershed includes many smaller subwatersheds that transfer water to the outlet during a rainstorm. Watersheds come in all shapes and sizes. They cross county, state, and national boundaries; for example, the continental United States includes 2,110 watersheds.

A watershed has three primary functions. First, it captures water from the atmosphere. Ideally, all moisture received from the atmosphere, whether in liquid or solid form, has the maximum opportunity to enter the ground where it falls. The water infiltrates the soil and percolates downward. Several factors affect the infiltration rate, including soil type, topography, climate, and vegetative cover. Percolation is also aided by the activity of burrowing animals, insects, and earthworms.

Second, a watershed stores rainwater once it filters through the soil. Once the watershed's soils are saturated, water will either percolate deeper or runoff the surface. This can result in freshwater aquifers and springs. The type and amount of vegetation, and the plant community structure, can greatly affect the storage capacity in any one watershed. The root mass associated with healthy vegetative cover keeps soil more permeable and allows the moisture to percolate deep into the soil for storage. Vegetation in the riparian zone affects both the quantity and quality of water moving through the soil. Finally, water moves through the soil to seeps and springs and is ultimately

DOI: 10.1201/9781003241744-3

released into streams, rivers, and the ocean. Slow release rates are preferable to rapid release rates, which result in short and severe peak instream flow. Storm events which generate large amounts of runoff can lead to flooding, soil erosion, and siltation of streams. Ultimately, the moisture will return to the atmosphere by way of evaporation. The hydrologic cycle (the capture, storage, release, and eventual evaporation of water) forms the basis of watershed function.

A watershed should be managed as a single unit. Each small piece of the landscape has an important role in the overall health of the watershed. Paying attention primarily to the riparian zone, an area critical to a watershed's release function, will not make up for lack of attention to the watershed's uplands. They play an equally important role in the watershed, the capture, and storage of moisture. It is seamless management of the entire watershed, and an understanding of the hydrologic process, that ensures watershed health (www.gdrc.org).

3.3 WATERSHED GEOMORPHOLOGY

Watershed geomorphology in urban area consists of land use characteristics and changes that could change from one urban to block to next. It also includes channel morphology where channels replace the streams and natural canal/rivers. Land use attributes are explained in the following section, and channel morphology is explained in Chapter 4. Basin area is briefly explained here.

Basin area (A) is hydrologically important because it directly affects the size of the storm hydrograph and the magnitude of mean and peak flows. The amount of sediment that eroded from the drainage basin is also related to the basin area. In fact, since almost every watershed characteristic is correlated with area, the area is the most important parameter in the description of form and processes of urban drainage basins.

The first step in determining the drainage area of a watershed is specifying the watershed boundary, which is determined based on a watershed topography map. A commonly used tool in delineating watershed boundaries is geographic information systems. There are also some software packages that can determine the watershed boundaries based on the introduced watershed outlet and given slope map. For manual computation of the watershed area, the planimeter can be used. The accuracy of the estimated value of the drainage area must be ensured because it is very important in hydrological design.

3.4 LAND USE AND COVER IMPACTS

When raining begins, people use trees as a temporary shield to protect themselves from the rain during the initial part of the storm. As the trees act like a barrier, it indicates that a forested watershed decreases the intensity of the flood runoff in comparison to a watershed without tree cover.

Another example of the impacts of land cover on runoff volume and rates is rooftop. You can see that the flow from the downspouts on houses starts very shortly and then increases gradually. This remarkable situation is enhanced due to impervious rooftops, planar surfaces, and steep slopes; therefore, there is delay in the flow. On the other hand, on the grassy hill that has the same size as the rooftop, after similar flow over a rooftop, flow down will begin. The sum of the water infiltrates into the topsoil and the grass, causing delay, which is hydraulically rougher than the shingles on the roof. These examples illustrate that flow from surfaces that are impervious would have smaller travel times and greater volumes than flow over surfaces that are pervious under conditions having similar shape, size, and slope.

Land cover could affect the runoff characteristics of a watershed. These two conceptual examples explain these effects even if all characteristics of watersheds stay the same except for land cover; the runoff characteristics such as the volume and timing of runoff and maximum flood flow rate would change significantly. Consequently, cover and use of land are among the primary problems in the hydrologic analysis and design phase.

Many descriptors of land cover/use are used in hydrologic design. A frequently used quantitative description is the index of runoff potential. As presented in Chapter 5, C is used in the rational method to reflect the runoff potential of a watershed. When a runoff coefficient C takes a high value, it reflects an increase in the runoff potential. For instance, the runoff potential values for commercial properties is $C = 0.75$; this value for residential areas is $C = 0.3$ and that for forested areas is $C = 0.15$. A transformed land cover/use index is used by the Soil Conservation Service (SCS), which is similar to that used in the rational method in their models. SCS also uses the runoff curve number *(CN)* in order to combine land use with the hydrologic effects of soil type and antecedent moisture.

In hydrology, how one considers urban land covers (impervious surfaces) is significant. Urban development can be considered as a source of many hydrologic design problems. The percentage of imperviousness is commonly used. Percentage of imperviousness ranges characteristic of "high-density residential areas" and "commercial and industrial areas" are from 40% to 70% and from 70% to 90%, respectively. Impervious covers in urban areas are not bounded to the watershed surface; for example, channels and pipes are used.

In a spatial analysis of the vulnerability and risk as shown in Figure 3.1, the areas exposed to the considered hazard are determined first represented by inundation map. Vulnerability map could be represented by the floodplain that is affected by human interventions reflected in land use and land cover that affects imperviousness. The spatial distribution of vulnerable socio-economic and environmental conditions are the dominating factors when combining the hazard and vulnerable maps to develop the risk map. The study region can be divided into some zones with different levels of vulnerability. Regarding the severity of vulnerability in different zones, the mitigation practices can be selected and prioritized for any specific application.

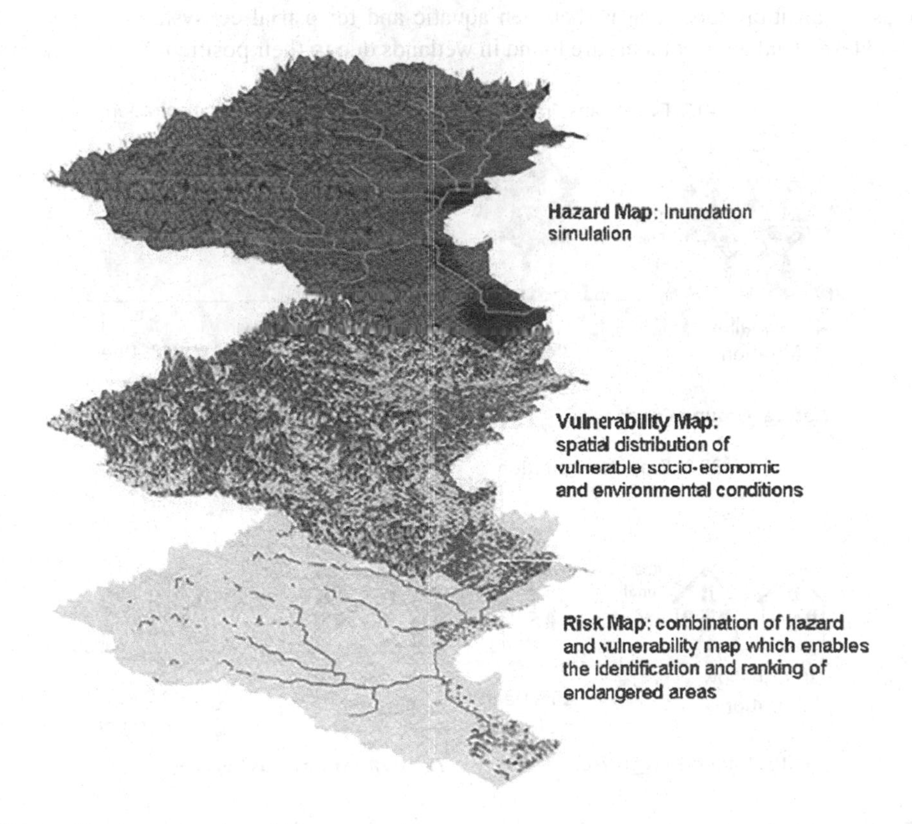
Hazard Map: Inundation simulation
Vulnerability Map: spatial distribution of vulnerable socio-economic and environmental conditions
Risk Map: combination of hazard and vulnerability map which enables the identification and ranking of endangered areas

FIGURE 3.1 The hazard, vulnerability, and risk maps.

3.4.1 Urban Areas

Land development influences how water naturally travels through the watershed. As mentioned above, about 50% of rainfall infiltrates into the ground with a natural ground cover, evaporation or plant transpiration make up 40% of rainfall (these together are called evapotranspiration), and only about 10% actually runs off the surface. Roads, houses, parking lots, sidewalks, and driveways are structures added onto the surface, as we develop the land, all of which are impervious surfaces. Water cannot pass through them as it can through soil; thus, instead of infiltrating, it is forced to either evaporate or run off. As shown in Figure 3.2, the rate of change in runoff is determined by the amount of impervious surface within a watershed. At 10%–20% impervious (i.e., medium-density residential areas), runoff is doubled, but the amount of water infiltrating is reduced. At 30%–50% impervious (similar to high-density residential developments), runoff is tripled, and at 75%–100% impervious (such as commercial areas), the majority of rainfall becomes runoff, and infiltration is less than one-third of what it was prior to development.

The results of increased runoff and reduced groundwater are twofold. First of all, much higher flows than naturally would occur in streams due to the large amount of excess runoff, besides the flow rate increases much more rapidly and drops off more rapidly after the storm. Second, as infiltration volumes are reduced, less water is available to be released slowly into the stream over time, which results in lower water levels between rainfall events. In summary, under natural conditions, a certain amount of water would infiltrate into the ground and would slowly make its way into nearby creeks; however, under urbanized conditions, that amount of water would run off and enter the stream all at once.

3.4.2 Wetland Areas

A distinctive hydrologic property of a landscape is wetland areas; this is mainly because of their position as a transition zone, that is, between aquatic and terrestrial ecosystems. Aspects of both aquatic and terrestrial environments are found in wetlands due to their position. Most of the available

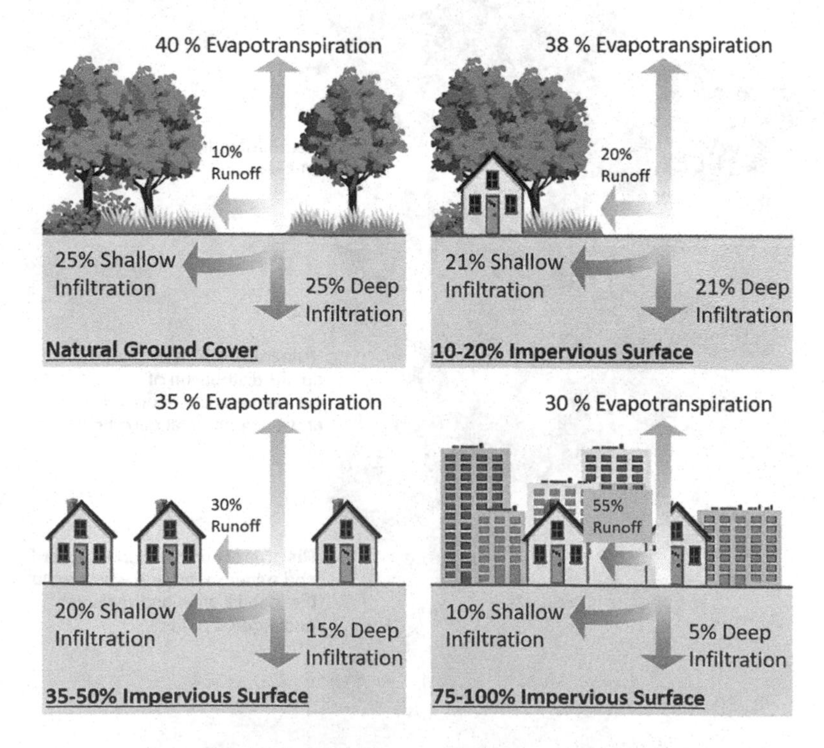

FIGURE 3.2 Effect of land use on runoff volume.

freshwater and marine aquatic environments, such as lakes, rivers, estuaries, and oceans, are characterized as having permanent water. However, terrestrial environments are generally characterized as having drier conditions, with an unsaturated (vadose) zone present for most of the annual cycle. As a result, the transition zone between predominantly wet and dry environments is occupied by wetlands.

One feature of wetlands that is very diagnostic is the proximity of the water surface (or water table below the surface) relative to the ground surface. In freshwater and marine aquatic habitats, the water surface lies well above the land surface; however, in terrestrial environments, it lies some distance below the root zone as a water table or zone of saturation. Unique biogeochemical conditions are created by the shallow hydrologic environment of wetlands that distinguish it from aquatic and terrestrial environments.

The water level, hydro pattern, and residence time are three hydrologic variables that are useful in characterizing wetland hydrologic behavior. A brief introduction of these concepts follows next. The general elevation of wetland water levels relative to the soil surface is considered as a hydrologic descriptor. Open water usually occurs in deeper areas where few, if any, emergent macrophytes exist. Any vegetation in these areas is not usually bonded to the wetland bottom, but vegetation may be floating on the surface of the water. However, it is possible for other wetlands to have large areas of exposed, saturated soil that is generally covered with macrophytic vegetation. Therefore, the water level can be used as an indicator of the vegetation types that probably occur in each of these zones. In wetland hydrology, the second descriptor is the temporal variability of water levels. Wetland hydro pattern incorporates the timing, duration, and distribution of water levels into the duration and frequency of water level perturbations. However, there are other more static wetland systems that may not display substantial short- or long-term variability. A function of the net difference between inflows and outflows from the atmosphere, groundwater, and surface water constitutes the wetland hydro pattern.

A third descriptor of wetland hydrology is the residence, or travel time, of water movement through the wetland. Water may be exchanged quickly in some wetland systems, with water remaining within the wetland for only a short duration of time, while water may travel very slowly through other wetland systems. The ratio of the volume of water within the wetland to the rate of flow through the wetland defines the residence time. When the flow through the wetland is large compared to its volume, the residence time is short, and when the flow is small compared to its volume, residence time will be long. The residence time is often related to the hydro pattern of that wetland; a wetland with large water level fluctuations may have shorter residence times, such as in tidal marshes. On the other hand, fluctuations of some wetlands might be rapid due to large changes in inflow, yet they have very long residence times due to slow loss rates.

3.5 RAINFALL–RUNOFF ANALYSIS IN URBAN AREAS

In order to estimate the effective rainfall (rainfall less losses), two main issues should be highlighted in the urban area, namely, change in infiltration and travel time. Change in evaporation is also important due to warmer temperature and hotter surfaces of streets and pavements. But the change in evaporation is often neglected due to short time of translation. Because of high imperviousness in urban areas, the mechanism of rainfall conversion to runoff is different from the other regions. For example, the accurate estimation of infiltrated rainfall into pervious surfaces in residential regions is very complicated and considerably affects the rainfall–runoff analysis. Travel time as often reflected in the time of concentration is greatly affected by the maze of pathways that rain water has to take to reach to a point of runoff accumulation. It is much quicker process in urban area that has to be realized in design of preventive measures specially for urban flooding. It is further discussed in Section 3.2.3.

The small variations in rainfall characteristics highly and quickly affect the response of urban areas, but these short-term variations do not affect the outflow of undeveloped regions. For precise estimation of a runoff hydrograph, the rainfall data should be available at time steps of 5 minutes or less. Also the peak discharge of urban areas is considerably higher because of faster movement of flow in these regions through channels and drainage systems. This means that for rainfall–runoff analysis in urban regions, one rain gauge is not sufficient. For explanation of the dynamics of moving storms, at least three rain gauges and a wind measurement are necessary (James and Scheckenberger, 1984; James et al., 2002).

3.5.1 Drainage Area Characteristics

An urban drainage area is characterized by the area, shape, slope, soil type, land use pattern, and percent of imperviousness, roughness, and different natural or man-made storage systems. The main factors in the estimation of runoff volume in a residential region are area and imperviousness. Note that, for accurate runoff estimation, the hydraulically connected impervious areas should be considered. The areas that are connected directly to a drainage system and drain into it are called hydraulically connected impervious areas. A street surface with curbs and gutters, which collects runoff from the surface and drains into a storm sewer, is an example of a hydraulically connected impervious area. The areas where the runoff drains to pervious areas and does not directly enter the storm drainage system are called non-hydraulically connected areas. Rooftops or driveways are examples of this kind of areas.

Often the imperviousness is estimated based on the land use. Typical values are available in texts for the estimation of imperviousness, such as Table 3.1. The population density can also be used for imperviousness estimation in large urban areas, such as Equation 3.1 developed by Stankowski (1974):

$$I = 9.6\text{PD}^{(0.479-0.017\ln \text{PD})}, \tag{3.1}$$

where I is the percent imperviousness and PD is the population density (persons/km^2).

The use of population density to estimate impervious cover is attractive since it provides a rapid technique for generating a quantitative estimation of present and projected land surface cover.

A network of channels and ducts that form a sewer system are the main part of an urban drainage system that normally flows with gravity in open channels. The geometry and hydraulic characteristics of the drainage system are described by a set of parameters. For the design of a

TABLE 3.1
Impervious Cover (%) for Various Land Uses

		Source				
Land Use	**Density (Dwelling Units/km^2)**	**Northern Virginia (NVPDC, 1980)**	**Olympia (COPWD, 1995)**	**Puget Sound (Aqua Terra Consultants, 1994)**	**NRCS (USDA, 1986)**	**Rouge River (Kluiteneberg, 1994)**
Low-density residential	<12.5	6	–	10	–	19
	12.5	–	–	10	12	
	25	12	–	10	20	
Medium- density residential	50	18	–	10	25	
	75	20	–	–	30	
	100	25	40	40	38	
High-density residential	125–175	35	40	40	–	38
Multifamily	Townhouse (>175)	35–50		60	65	–
Industrial	–	60–80	86	90	72	76
Commercial	–	90–95	86	90	85	56
Roadway	–	–	–	–	–	–

new drainage system, all of design parameters are determined considering the regional characteristics and the design needs. Evaluation of the life cycle and aging effect of system components and alterations made to an available drainage system is highly important for system development and improvement and needs a major site and future needs assessment.

3.5.2 Rainfall Losses

All of the possible losses during the rainfall-to-runoff conversion process are subtracted from the total rainfall to determine the excess rainfall. Rainfall loss includes depression storage (or "interception storage") on planted and other surfaces, infiltration, and evaporation. For a specific storm that happens in a short period, evaporation can be neglected, but in long-term analysis of the urban water budget, the evaporation amount should be considered.

Losses due to depression storage and infiltration cannot be separated over pervious areas. The amount of depression storage is highly variable in different areas; for example, it is about 16 mm for impervious surfaces and 63 mm for well-graded pervious surfaces (Tholin and Kiefer, 1960). In highly impervious areas, the depression storage is also related to the slope of the drainage area. Since it is very difficult to estimate depression storage, it is often determined during the rainfall–runoff model calibration process.

3.6 Travel Time

Most hydrologic designs involve some measure of flood discharge. The volume of flood runoff can be reflected by various watershed and hydrometeorological characteristics; for example, the product of the drainage area and the depth of rainfall give a volume of water that is potentially available for runoff, but volume alone is not adequate for many design problems. As the dimensions of discharge indicate (L^3/T), time is an important element in hydrologic design. However, a given volume of water may or may not present a flood hazard; in fact, the hazard will depend on the time distribution of the flood runoff. Flood damage occurs if a significant portion of the total volume passes a given location at about the same time.

Most hydrologic models require a watershed characteristic that reflects the timing of runoff due to the importance of the timing of runoff. A number of time parameters have been developed. As the runoff timing is a watershed characteristic, time parameters are formulated as a function of other watershed characteristics and, in some cases, rainfall intensity. Hydrologic and hydraulic models commonly use several time parameters such as the time of concentration, the time lag, and reach travel time.

The importance of time in hydrologic design is dependent on the type of hydrologic design problem. Unfortunately, most of the empirical formulas have been based on very limited data; therefore, lack of diversity in the data constrains their applicability. Considerable caution must be taken in the usage of an empirical estimation method either for watersheds having characteristics different from those of the watersheds used to calibrate the method or for watersheds in other geographic regions. Data extrapolation is represented in both uses. Therefore, a design engineer faced with a design problem requires an estimation of a time parameter and must choose from among many of the alternatives and often apply the time parameter with little or no knowledge of its accuracy.

3.6.1 Definitions of Time of Concentration

The time of concentration has two commonly accepted definitions. First, required time for a water particle to flow hydraulically from the most distant point in the watershed to the outlet or design point defines t_c; on this definition, watershed characteristics and sometimes a precipitation index such as the 2-year, 2-hour rainfall intensity are used in methods of estimation. Several empirical equations based on this definition will be discussed herein.

Some of the required terms have not been previously introduced; therefore, the second definition is introduced here only for completeness. In the second definition, t is based on a rainfall hyetograph (defined in Chapter 4) and the resulting runoff hydrograph (defined in Chapter 6). The rainfall excess and direct runoff are computed from the actual hyetograph and hydrograph. The time between the center of mass of rainfall excess and the inflection point on the recession of the direct runoff hydrograph defines the time of concentration. As an alternative, the time difference between the end of rainfall excess and the inflection point is sometimes accepted as t_c.

Both methods of estimating t_c do not provide either the true value or reproducible values of t_c. That is because the methods based on watershed characteristics may consider, for example, that Manning's equation is always valid and that the roughness coefficient applies throughout the flow regime. Even though one may ignore both the difficulties in selecting a single value of the roughness coefficient for a seemingly homogeneous flow regime and the assumption that the hydraulic radius remains constant, the design engineer must still contend with the problem of the input reproducibility; that is, different values of the input variables are selected by different users of a method even if on the same drainage area. In summary, in hydrologic design, an important input is the time of concentration; however, it is neither a highly accurate input nor highly reproducible.

There are some difficulties in estimation of t_c from rainfall and runoff data. There are no universally accepted methods of separating either base flow from direct runoff or losses from rainfall excess. Significant variation in estimated t values may be introduced by both of these separation requirements. In estimation of t_c using rainfall and runoff data, some factors must also be identified, such as antecedent soil moisture, intermittent rainfall patterns, nonlinearities in the convolution process, and variation in the recurrence intervals of the storm, and these require t_c to be adjusted. In summary, it is not possible to have a single correct method for estimating t_c, and therefore, the true value can never be determined.

3.6.2 Classifying Time Parameters

Sheet flow and concentrated flow methods are two classes for overland flow.

Input requirements and dominant flow regime can be used for classifying time parameters. The system could be further classified on the basis of the other time parameters such as time lag, time of concentration, and time to peak; however, in this chapter, the time of concentration is considered.

The runoff coefficient of the rational method, C, Manning's roughness coefficient, n, the percentage of imperviousness, I, the SCS runoff curve number, CN, and a qualitative descriptor of the land cover type are involved in measures of the overland flow resistance. The length and slope of overland flow can be represented by one of a number of size or slope parameters. A rainfall parameter is sometimes used as input to show the impact of the surface runoff availability on the time of travel.

Methods for estimating t_c, which include channel characteristics as input variables, should be used when channel flow makes an indicative contribution to the total travel time of runoff (see Table 3.2). The most widely used index of channel flow resistance is Manning's roughness coefficient, n. Espey and Winslow developed a coefficient that could also be an indicator of flow resistance to show the degree of channelization. Several slope and size parameters are used as variables. The precipitation intensity, the volume of surface runoff, and the hydraulic radius for bank full flow could be used as input factors for a channel.

This classification system can be used for separating time parameter prediction methods. A significant number of others will have to be identified as a "mixed" method when some prediction methods will fall into one of the four classes based on flow regime (i.e., one that includes variables reflecting different flow regimes). Although the term *overland flow* could be applied to a number of flow regimes, it is worthwhile to classify this into sheet flow and concentrated flow. Sheet flow occurs usually over very short flow paths in the upper reaches of a basin. The common method that is used to estimate travel times of concentrated flow is Manning's equation or the velocity.

TABLE 3.2
Criteria for Classifying Time Parameters and Variables Commonly Used

Flow Regime	Flow Resistance	Watershed Size	Slope	Inflow
Sheet flow	n	L	S	I
Concentrated flow	n, C, CN	L, A	S	I
Channel	n	L_{10-85}, L	S, S_{10-85}	R_h, i Q
Pipe	n	L	S	R_h

n = Manning's roughness coefficient; L = watershed length; S = average slope; i = rainfall intensity; C = runoff coefficient; CN = runoff curve number; A = drainage area; $L_{10-8}5$ = length of channel within 10% and 85% points; $S_{10-8}5$ = channel slope between 10% and 85% points; R_h = hydraulic radius; Q = channel discharge rate.

Manning's equation is used to estimate velocities of overland flow, both sheet flow and concentrated flow. It can be represented that the kinematic wave equation for estimating sheet flow travel time is based on Manning's equation. Manning's equation must have an assumption that the hydraulic radius equals the product of the travel time and the rainfall intensity. The velocity method for concentrated flow uses Manning's equation with an assumed Manning's n and depth. Frequently used curves of velocity versus slope are valid only as long as the assumed depth and n are accurate.

3.6.3 Velocity Method

The travel time is the basic concept of the velocity method. Travel time (T_t) for a particular flow path is a function of the length of flow (L) and the velocity (V):

$$T_t = \frac{L}{V}. \tag{3.2}$$

The travel time is computed for the principal flow path. Where the principal flow path consists of segments that have different slopes or land cover, the principal flow path should be divided into segments and Equation 3.3 should be used for each flow segment. The time of concentration is then the sum of the travel times:

$$t_c = \sum_{i=l}^{k} T_{t_i} \sum_{i=1}^{k} \left(\frac{L_i}{V_i} \right), \tag{3.3}$$

where k is the number of segments and the subscript i refers to the flow segment.

The velocity is a function of the type of flow (sheet, concentrated flow, gully flow, channel flow, pipe flow), the roughness of the flow path, and the slope of the flow path. Velocity can be estimated by various methods. Flow velocities in pipes and open channels can be computed using Manning's equation (Table 3.3):

$$V = \frac{1}{n} R_h^{2/3} \cdot S^{1/2}, \tag{3.4}$$

where V is the velocity (m/s), n is the roughness coefficient, R_h is the hydraulic radius (m), and S is the slope. Equation 3.4 can be simplified so that V is only a function of the slope by assuming values for n and R_h. This gives a relationship between the velocity and the average slope of the surface as follows:

$$V = k \cdot S^{0.5} \quad \text{and} \quad k = \frac{R_h^{2/3}}{n}, \tag{3.5}$$

TABLE 3.3
Parameters Used in Velocity–Slope Relationship

Land Use/Flow Regime	n	R_h	$K = \frac{R_h^{2/3}}{n}$
Forest			
Dense underbrush	0.8	0.83	1.11
Light underbrush	0.4	0.73	2.03
Heavy ground litter	0.20	0.67	3.82
Grass			
Bermuda grass	0.41	0.50	1.54
Dense	0.24	0.40	2.26
Short	0.15	0.33	3.20
Short grass pasture	0.025	0.13	10.44
Conventional Tillage			
With residue	0.19	0.20	1.80
No residue	0.09	0.17	3.37
Agricultural			
Cultivated straight row	0.04	0.40	13.57
Contour or strip cropped	0.05	0.20	6.84
Trash fallow	0.045	0.17	6.73
Rangeland	0.13	0.13	2.01
Alluvial fans	0.017	0.13	15.35
Grassed waterway	0.095	3.33	23.49
Small upland gullies	0.04	1.67	35.14
Paved area (sheet flow)	0.011	0.20	31.09
Paved gutter	0.11	0.07	14.95

where V is the velocity (m/s) and S is the slope. For Equation 3.4, the value of k is a function of the land cover with the effect measured by the value of n and R_h. After short distances, runoff tends to concentrate in rills and then gullies of increasing proportions. Manning's equation can be used where roughness coefficients exist for such flow.

3.6.4 Sheet Flow Travel Time

Runoff does not concentrate into well-defined flow paths such as gullies or swales at the upper reaches of a watershed. Instead, it flows over the surface at reasonably uniform, shallow depths. It is evident on long, sloping streets during rainstorms. For impervious surfaces, the sheet flow length can be several 100 ft for steep slopes. For shallow slopes or for pervious surfaces, concentrated flow will begin after relatively short sheet flow lengths.

In the upper reaches of a watershed, sheet flow runoff during the intense part of the storm will flow as a shallow layer with a reasonably constant depth. An equation, referred to as the kinematic wave equation for the equilibrium time, can be developed using Manning's equation with the assumption that the hydraulic radius equals the product of the rainfall intensity and the travel time, for example, $R_h = iT_t$. Using the velocity equation with the travel time equal to the time of concentration, Manning's equation becomes

$$V = \frac{L}{60T_t} = \frac{1}{n} R_h^{2/3} \cdot S^{1/2} = \frac{1}{n}\left(\frac{iT_i}{60(1{,}000)}\right)^{2/3} \cdot S^{1/2}, \tag{3.6}$$

where i is in millimeters per hour, T is in minutes, and L is in meters. Solving for the travel time yields

$$T_t = \frac{7.0}{i^{0.4}}\left(\frac{nL}{\sqrt{S}}\right)^{0.6}. \tag{3.7}$$

Since T_t is not initially known, it is necessary to assume a value of T_t to obtain i from a rainfall intensity–duration–frequency (IDF) curve and then compute T_t. A new estimate of i is obtained from the IDF curve using the computed value of T_t. If the initial assumption for T_t was incorrect, the iterative process should be repeated until the value of T_t does not change. For details refer to Karamouz et al. (2012).

3.6.5 Empirical Formulas

There are a variety of empirical methods for estimation of time of concentration; not all the methods were originally presented as equations for computing the time of concentration; therefore, it was necessary to adjust the empirical equations so that they would compute t in hours. For those methods designed to predict the lag time, the computed lag values were multiplied by a constant; the value of the constant depended on the definition of the lag. A value of 1.417 was used for the lag defined as the time difference between the centers of mass of rainfall excess and direct runoff, determined on the basis of the relationship between the time lag and the time of concentration for an SCS triangular hydrograph. This assumption is probably unimportant; that is, the results of comparisons are insensitive to this assumption. A conversion factor of 1.67 was used for methods in which the lag was defined as the time difference between the center of mass rainfall excess and the peak discharge; this constant was also based on analysis of a triangular hydrograph. Again, the use of this conversion factor would not be expected to affect the accuracy of the methods. The empirical formulas suitable for urban areas include Federal Aviation Agency (FAA) equation; Kerby-Hathaway formula; and Kirpich methods that are suitable for small watersheds. For details on these and others see Chapter 5 of Karamouz et al. (2012).

3.7 EXCESS RAINFALL CALCULATION

In this section, various methods used for the determination of the runoff quantity produced by a given storm event are investigated. Considerable parts of rainfall do not convert to runoff due to different lost sources. Plants intercept a part of rainfall. Some rainfall will remain in surface puddles. A part of rainfall infiltrates into the soil and a small amount of it evaporates before reaching the ground. The remaining rainfall after these losses will produce surface runoff.

The main sources of rainfall abstractions can be summarized as

- Interception
- Surface depression storage
- Evaporation
- Transpiration
- Infiltration

The evaporation and transpiration abstractions can be eliminated under design storm conditions, in an urban area. The relation between rainfall abstractions and overland flow (runoff) is presented in Figure 3.3. As mentioned before, the excess (effective) rainfall is equal to the total rainfall minus summation of all abstractions. The depth of excess rainfall produced per unit time is considered as the rate of excess (effective) rainfall. The excess rainfall rate is also equal to the rate of rainfall

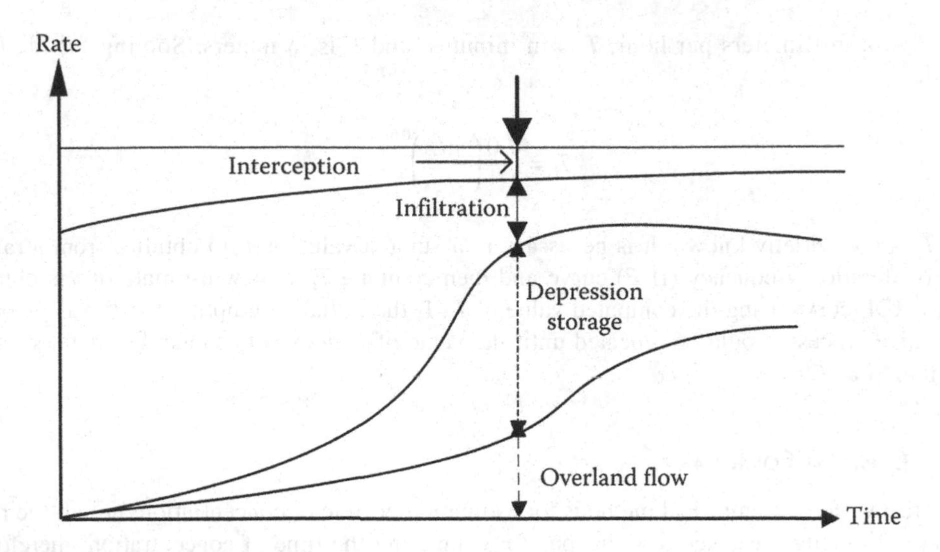

FIGURE 3.3 Relation between the interception, depression storage, infiltration, and overland flow.

minus the rate of loss in the situation where the excess rainfall is the only source of storm runoff in the considered urban watershed. Thus, the total volume of excess rainfall and the total volume of produced runoff are equal.

3.7.1 Interception Storage Estimation

The portion of the rainfall that is intercepted by flora before reaching the ground is named interception storage. The intercepted water is held by plant surfaces such as branches and leaves and, finally, evaporates without reaching the ground. Most of the interception occurs at the beginning of the storm, because the maximum water-holding capacity of leaves and branches is low and maximum interception is reached very soon.

The main factors in the determination of the amount of interception are the type and density of the plants as well as the amount of rainfall. Empirical methods are developed for the estimation of the losses due to interception. The following relationship is suggested by Horton (1919) for the estimation of interception storage:

$$\text{IS} = \alpha + \beta R_t^{\gamma}, \tag{3.8}$$

where IS is the interception storage depth (m), R_t is the total rainfall depth (m), and α, β, γ are the empirical constants.

Horton's equation is developed based on the data collected from summer storms, so Kibler et al. (1996) reanalyzed Horton's data and developed the following expression for the estimation of interception storage:

$$\text{IS} = kR_t^{n}, \tag{3.9}$$

where k and n are constant parameters that should be specified for each region. Values of α, β, γ, k, and n are presented in Table 3.4. It should be noted that Equations 3.8 and 3.9 are only applicable to the part of the urban area that is covered completely by plants.

Considering the interception loss during individual storm events is useless and estimation of losses due to interception is only significant in annual or long-term rainfall–runoff models (Viessman et al., 1989).

TABLE 3.4
Values of Constant Parameters in Interception Equations for Different Types of Trees

	Equation 3.7			Equation 3.8	
Tree type	α (×10⁻⁴)	β (×10⁻⁴)	γ	k	n
Apple	10.16	1.17	1.0	0.093	0.73
Ash	3.81	1.47	1.0	0.167	0.88
Beech	5.08	1.47	1.0	0.058	0.65
Chestnut	10.16	1.30	1.0	0.129	0.77
Elm	0.00	9.32	0.5	0.022	0.48
Hemlock	0.00	8.10	0.5	0.123	0.74
Maple	7.62	1.47	1.0	0.125	0.77
Oak	7.62	1.42	1.0	0.069	0.66
Pine	0.00	8.10	0.5	0.100	0.70
Willow	5.08	2.59	1.0	0.248	0.85

Source: Kibler, D.F. et al., *Urban Hydrology* (2nd ed.), ASCE Manual of Practice, No. 28, New York, 1996. With permission.

3.7.2 Estimation of Infiltration

The process of passing rainwater through the ground surface and filling the pores of the soil on both the surface (the availability of water for infiltration) and subsurface (the potential of the available water for infiltration) is called infiltration. In other words, infiltration is defined as the passage of water through the air–soil interface. Infiltration rates are affected by factors such as time since the rainfall event began, soil porosity and permeability, antecedent soil moisture conditions, and the presence of vegetation. Urbanization usually decreases infiltration with a resulting increase in runoff volume and discharge.

The maximum possible rate of water infiltration is considered as infiltration capacity or the potential infiltration rate, which shows the capability of available water to infiltrate. For rainfall rates less than the infiltration capacity, the real rate of infiltration is equal to the rate of rainfall. If the rate of rainfall exceeds the infiltration capacity, the actual rate of infiltration will be equivalent to the infiltration capacity. In this situation, the rainwater that does not infiltrate to the soil fills the surface depressions and, finally, flows over the ground surface.

$$f = i \quad \text{if} \quad f_f > i,$$

and

$$f = f_f \quad \text{if} \quad f_f < i,$$

where f_f is the infiltration capacity, f is the actual rate of infiltration, and i is the rainfall rate.

In the situation where infiltration is the only or the dominant source of rainfall abstraction, the rate of excess rainfall, i_e, is calculated as

$$i_e = i - f. \tag{3.10}$$

The infiltration capacity estimation is a complicated process, but some theoretical methods are developed for the determination of the infiltration capacity. The Richard equation is a sophisticated method for infiltration capacity estimation. This equation is nonlinear and its solving needs complex numerical algorithms. But in engineering practices, much simpler methods are used for infiltration capacity estimation. The application of the Richard equation in different regions reveals the dependency between the infiltration capacity and the soil characteristics, especially the available humidity in the soil before and during the infiltration. Therefore, for realistic modeling of the soil infiltration capacity variation during the infiltration process, the initial moisture condition of the soil and the amount of infiltrated water into the soil from the start of rainfall should be considered. Green–Ampt and Horton models are discussed here. For details of other models/methods to estimate infiltration refer to Karamouz et al. (2012).

3.7.3 Green–Ampt Model

The most physically based numerical infiltration model that is commonly used is the Green and Ampt (1974) model. All parameters incorporated into this model are determined from soil characteristics regarding their definition and physical basis. An illustration of a simplified model proposed by Green and Ampt is shown in Figure 3.4. It is assumed that the moisture of the soil underlying the pervious area is uniform and equals θ_i when rainfall starts.

But the basic assumption behind the Green and Ampt equation is that water infiltrates into (relatively) dry soil as a sharp wetting front. The soil is saturated on top of the wetting front, and the initial degree of saturation is maintained below it. In other words, the volumetric water contents remain constant above and below the wetting front as it advances. By this assumption, two distinctive zones are formed below the ground surface during the infiltration process. A saturated zone is developed below the surface a little after the start of the rain. As more water infiltrates into the soil, the depth of this zone increases. The other zone with the initial moisture content is below the wetting front, and it is assumed that this zone has unlimited depth (or, alternatively, the water table or the impervious bed rock is deep enough not to interfere with the infiltration process).

Darcy's law is applicable for the estimation of the infiltration capacity in the saturated zone adjacent to the soil surface:

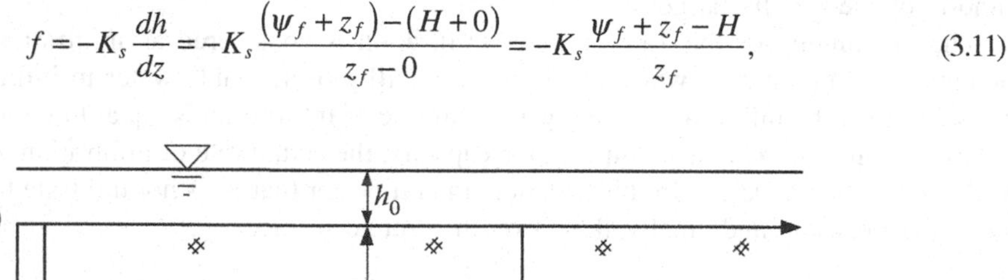

$$f = -K_s \frac{dh}{dz} = -K_s \frac{(\psi_f + z_f) - (H + 0)}{z_f - 0} = -K_s \frac{\psi_f + z_f - H}{z_f}, \tag{3.11}$$

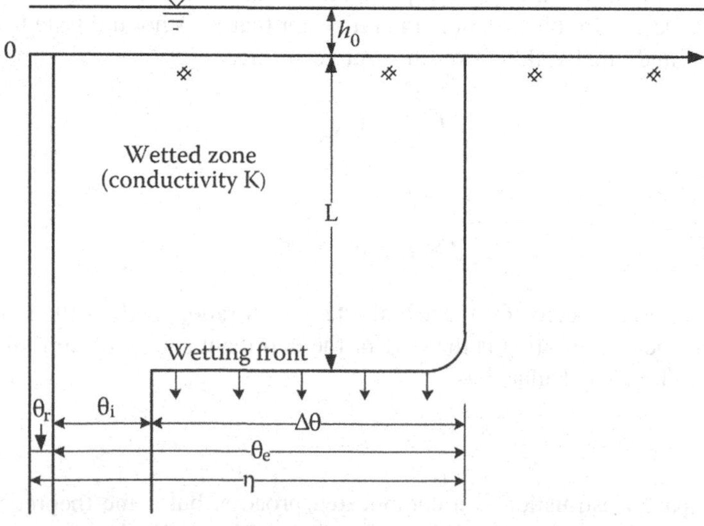

FIGURE 3.4 Typical profile during the infiltration process.

where H is the depth of ponding, K_s is the saturated hydraulic conductivity, f is the infiltration rate and is negative, ψ_f is the suction at the wetting front (negative pressure head), and z_f is the depth of the wetting front.

The Green–Ampt model is the most physical-based algebraic infiltration model available. The earliest equation was proposed by Green and Ampt (1974). All the parameters involved in this model have a physical basis, which are determined based on soil characteristics. The typical profile of infiltration considered in this method is shown in Figure 3.4. In this figure, the vertical axis is the distance from the soil surface; the horizontal axis is the moisture content of the soil. In this method, the wetting front is considered to be a straight line separating the lower unsaturated layer of soil with the moisture content of θ_i, from the upper saturated soil with moisture content of η. The wetting front has reached the depth, L, in time t since the start of filtration. Water is ponded to a small depth, h_0, on the soil surface.

To formulate the Green–Ampt equations, a cylindrical control volume with unit cross section and depth of L is considered. Hence, the increase in the water stored within the control volume as a result of infiltration is $L(\eta - \theta_i)$, which is equal to the cumulative depth of water infiltrated into the soil, F.

$$F(t) = L(\eta - \theta_i) = L\Delta\theta. \tag{3.12}$$

Darcy's law is applied to the saturated zone adjacent to the soil surface to determine the infiltration capacity as follows:

$$q = -K\frac{\delta h}{\delta z}. \tag{3.13}$$

In this case, the Darcy flux q is constant throughout the depth and is equal to $-f$, because q is positive upward while f is positive downward. If points 1 and 2 are located, respectively, at the ground surface and just on the dry side of the wetting front, the infiltration rate can be approximated by

$$f = K\left[\frac{h_1 - h_2}{z_1 - z_2}\right]. \tag{3.14}$$

The head h_1 at the surface is equal to the ponded depth h_0. The head h_2 in dry soil below the wetting front is equal to $-\psi - L$. Darcy's law for this system is written as follows:

$$f = K\left[\frac{h_0 - (-\psi - L)}{L}\right] \approx K\left[\frac{\psi + L}{L}\right]. \tag{3.15}$$

The second part is true when the ponded depth, h_0, is negligible compared to ψ and L. This assumption is usually appropriate for surface water hydrology problems because it is assumed that ponded water becomes surface runoff.

By the wetting front depth is $L = F/\Delta\theta$;

$$f = K\left(\psi\frac{\Delta\theta}{F} + 1\right). \tag{3.16}$$

In the case when the ponded depth, h_0, is not negligible, the value of $-\psi - h_0$ is substituted for ψ.

Substituting $f = \dfrac{dF}{dt}$ will result in

$$F(t) = Kt + \psi\Delta\theta \ln\left(1 + \frac{F(t)}{\psi\Delta\theta}\right). \tag{3.17}$$

TABLE 3.5
Soil Parameters for Green–Ampt Model

Soil Type	Porosity, η	Effective Porosity, θ_e	Wetting Front Soil Suction Head, ψ	Hydraulic Conductivity, K (cm/h)
Sand	0.437 (0.374–0.500)	0.417 (0.354–0.480)	4.95 (0.97–25.36)	11.78
Loamy sand	0.437 (0.363–0.506)	0.401 (0.329–0.437)	6.13 (1.35–27.94)	2.99
Sandy loam	0.453 (0.351–0.555)	0.412 (0.283–0.541)	11.01 (2.67–45.47)	1.09
Loam	0.463 (0.375–0.551)	0.434 (0.334–0.534)	8.89 (1.33–59.38)	0.34
Silt loam	0.501 (0.420–0.572)	0.486 (0.394–0.578)	16.68 (2.92–95.39)	0.65
Sandy clay loam	0.398 (0.332–0.464)	0.330 (0.235–0.425)	21.85 (4.42–108.0)	0.15
Clay loam	0.464 (0.409–0.519)	0.309 (0.279–0.501)	20.88 (4.79–91.10)	0.10
Silty clay loam	0.471 (0.418–0.524)	0.432 (0.347–0.517)	27.30 (5.67–131.50)	0.10
Sandy clay	0.430 (0.370–0.490)	0.321 (0.207–0.435)	23.90 (4.08–140.2)	0.06
Silty clay	0.479 (0.425–0.533)	0.423 (0.334–0.512)	29.22 (6.13–139.4)	0.05
Clay	0.475 (0.427–0.523)	0.385 (0.269–0.501)	31.63 (6.39–156.5)	0.03

Given K, t, ψ, and $\Delta\theta$, a trial value of F is substituted on the right-hand side (a good trial value is $F = Kt$), and a new value of F is calculated on the left-hand side, which is substituted as a trial value on the right-hand side, and so on, until the calculated values of F converge to a constant. The final value of cumulative infiltration F is substituted to determine the corresponding potential infiltration rate f.

An important issue in application of the Green–Ampt model is estimation of soil characteristics including the hydraulic conductivity K, the porosity η, and the wetting front soil suction head (ψ). Based on the laboratory tests of many soils, ψ can be expressed as a logarithmic function of an effective saturation, S_e. If the residual moisture content of soil after it has been thoroughly drained is denoted by θ_r, the effective saturation is the ratio of the available moisture $\theta - \theta_r$ to the maximum possible available moisture content $\mu - \theta_r$:

$$S_e = \frac{\theta - \theta_r}{\mu - \theta_r}, \tag{3.18}$$

where $\mu - \theta_r$ is called the effective porosity, θ_e. The effective saturation has the range $0 \le S_e \le 1.0$, provided $\theta_r \le \theta \le \mu$.

For the initial condition, when $\theta = \theta_i$, gives $\theta_i - \theta_r = S_e\,\theta_e$, and the change in the moisture content when the wetting front passes is $\Delta\theta = \mu - \theta_i = \mu - (S_e\,\theta_e + \theta_r)$; therefore,

$$\Delta\theta = (1 - S_e)\theta_e. \tag{3.19}$$

The value ranges and average values of the Green–Ampt parameters μ, θ_e, ψ, and K for different soil classes are given in Table 3.5.

3.7.3.1 Ponding Time

If rainfall intensity is constant and eventually exceeds infiltration rate, then at some moment the surface will become saturated and ponding starts when the infiltration rate is equal to the precipitation rate. The depth infiltrated at that moment, F_s is given by setting $f = i$, then:

$$F_s = \frac{[(\theta_s - \theta_i)\psi]}{[(i/K - 1)]}, \tag{3.20}$$

in which i must be greater than K. Time of ponding, t_p, is defined as follows:

$$t_p = \frac{F_s}{i}. \tag{3.21}$$

Example 3.1

For the following soil properties, determine the amount of infiltrated water when ponding occurs and the time to ponding. $K = 1.97$ cm/h, $\theta_i = 0.318$, $\theta_s = 0.518$, $i = 7.88$ cm/h, and $\psi_f = 9.37$ cm.

Solution:

The infiltrated water until the ponding time is obtained as

$$F_s = \psi_f \frac{\theta_s - \theta_i}{\frac{f}{K} - 1} = 9.37 \frac{0.518 - 0.318}{\frac{7.88}{1.97} - 1} = 0.625 \text{ cm}.$$

then

$$t_p = (0.625 / 7.88) = 0.079 \text{ hours} = 4.74 \text{ minutes} = 284 \text{ seconds}.$$

However, in the case of rainfall with variable intensity, determining the ponding time and infiltration depth is rather complicated. Here, a useful flow chart for estimating infiltration and ponding time by the Green–Ampt method proposed by Chow et al. (1988) is described.

Consider a time interval from t to $t + \Delta t$. The rainfall intensity during this interval is denoted i_t and is constant throughout the interval. The potential infiltration rate and cumulative infiltration at the beginning of the interval are f_t and F_t, respectively, and the corresponding values at the end of interval are $f_{t+\Delta t}$ and $F_{t+\Delta t}$. It is assumed that F_t is known from the given initial conditions or from previous computation.

There are two cases to be considered:

1. Ponding occurs throughout the interval.
2. There is no ponding throughout the interval.

The infiltration rate is always either decreasing or constant with time; thus, once ponding is established under a given rainfall intensity, it will continue. Hence, it is assumed that ponding cannot occur in the middle of an interval, but at the end of it, when the value of the rainfall intensity changes.

The first step is to calculate the current potential infiltration rate f_t from the known value of cumulative infiltration F_t. For the Green–Ampt method, one uses

$$f_t = K\left(\psi \frac{\Delta\theta}{F_t} + 1\right), \tag{3.22}$$

where f_t is infiltration rate, $\Delta\theta$ is difference in dry versus wet soil moisture, K is soil conductivity (cm/h), F_t is cumulative infiltration (cm), and ψ is suction head (cm). The Green–Ampt parameters can be estimated according to the soil texture and land use practices.

The resulting f_t is compared to the rainfall intensity i_t. If f_t is less than or equal to i_t, case (1) arises and there is ponding throughout the interval. In this case, the Green–Ampt equation, the cumulative infiltration at the end of the interval, $F_{t+\Delta t}$, is calculated from

$$F_{t+\Delta t} - F_t - \psi\Delta\theta \ \ln\left[\frac{F_{t+\Delta t} + \psi\Delta\theta}{F_t + \psi\Delta\theta}\right] = K\Delta t. \tag{3.23}$$

Both cases (2) and (3) have $f_t > i_t$ and no ponding at the beginning of the interval. Assume that this remains so throughout the interval; then, the infiltration rate is i_t and a tentative value for cumulative infiltration at the end of the time interval is

$$F'_{t+\Delta t} = F_t + i_t\Delta t. \tag{3.24}$$

Next, a corresponding infiltration rate $F'_{t+\Delta t}$ is calculated from $F'_{t+\Delta t}$. If $F'_{t+\Delta t}$ is greater than i_t, case (2) occurs and there is no ponding throughout the interval. Then $F_{t+\Delta t} = F'_{t+\Delta t}$ and the problem is solved for this interval.

If $F'_{t+\Delta t}$ is less than or equal to i_t, ponding occurs during the interval [case (3)]. The cumulative infiltration F_p at ponding time is found by setting $f_t = i_t$ and $F_t = F_p$ and solving for F_p to give, for the Green–Ampt equation,

$$F_p = \frac{K\psi\Delta\theta}{i_t - K}. \tag{3.25}$$

The ponding time is then $t + \Delta t$, where

$$\Delta t' = \frac{F_p - F_t}{i_t}, \tag{3.26}$$

and the cumulative infiltration $F_{t+\Delta t}$ is found by substituting $F_t = F_p$ and $\Delta t = \Delta t - \Delta t'$. The excess rainfall values are calculated by subtracting cumulative infiltration from cumulative rainfall and then taking successive differences of the resulting values. For details refer to Karamouz et al. (2012).

3.7.3.2 *Horton Method*

An exponential decay function for the evaluation of infiltration during a storm event is suggested by Horton (1940) based on experimental data:

$$f_p = f_f + \left(f_0 - f_f\right)e^{-mt}, \tag{3.27}$$

where f_p is the rate of infiltration into the soil at time t, f_f is the final infiltration capacity, f_0 is the initial infiltration capacity, m is the decay constant, and t is the time from the beginning of rainfall.

TABLE 3.6
Calculations of Example 3.2

Time (h)	Infiltration (cm/h)	Time (h)	Infiltration (cm/h)
0	2.900	4	1.283
0.1	2.834	5	1.092
0.2	2.769	6	0.947
0.3	2.707	7	0.838
0.4	2.646	8	0.756
0.5	2.586	9	0.693
0.6	2.529	10	0.646
0.7	2.473	15	0.536
0.8	2.418	20	0.509
0.9	2.365	25	0.502
1	2.314	30	0.501
2	1.871	35	0.500
3	1.536	40	0.500

Example 3.2

Given an initial infiltration capacity f_0 of 2.9 cm/h and a time constant m of 0.28 h^{-1} for a homogeneous soil, derive an infiltration capacity versus time curve if the ultimate infiltration capacity is 0.50 cm/h. For the first 8 hours, estimate the total volume of water infiltrated in cm over the watershed.

Solution:

Substituting the appropriate values into Equation 3.16 yields

$$f_p = 0.5 + (2.9 - 0.5)e^{-0.28t}.$$

For the times shown in Table 3.6, values of f are computed and entered into the table and, finally, the curve of Figure 3.5 is derived.

To find the volume of water infiltrated during the first 8 hours, Equation 3.16 can be integrated over the range of 0–8:

$$V = \int \left[0.5 + (2.9 - 0.5)e^{-0.28t}\right] dt,$$

$$= \int \left[0.5t + \left(2.4/(-0.28)e^{-0.28t}\right]_0^8 = 11.65 \text{ cm}.$$

Another method is Holton which is formed based on the concept that the infiltration capacity is related to the available storage for holding water in the surface layer of the soil. Because of the increase of water infiltration into soil, the infiltration capacity decreases because of less storage availability. The Holtan method is applicable for agricultural lands, calculating the rainfall losses in wooded parts of urban areas and the areas covered by grass and plants. For the details of this method that has limited applicability in the urban area refer to Karamouz et al. (2012).

3.7.3.3 *Simple Infiltration Models*

Different models for the calculation of infiltration rate and losses from rainfall are available in the literature. The simplest infiltration model is referred to as the Φ-index method. In this method, as shown in Figure 3.6, the infiltration capacity that is equal to the index Φ has been considered constant. Consequently, initial rates are underestimated and final rates are overstated if an entire

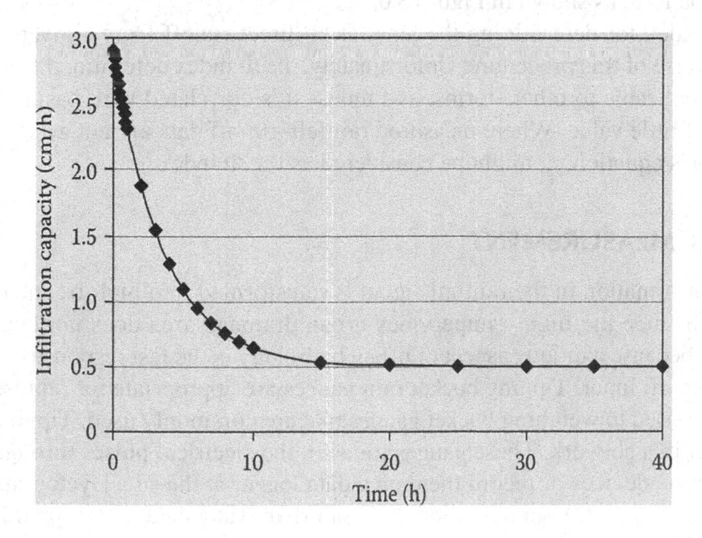

FIGURE 3.5 Infiltration curve for Example 3.2.

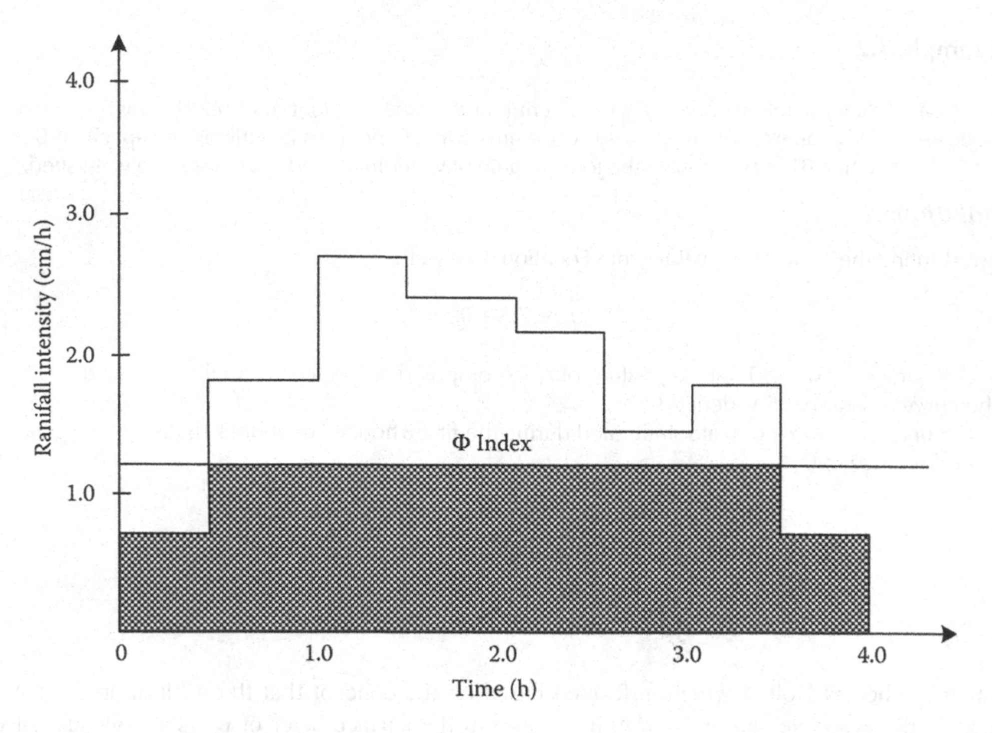

FIGURE 3.6 Representation of the Φ-index method.

storm sequence with little antecedent moisture is considered. The best application is to large storms on wet soils or storms where infiltration rates may be assumed to be relatively uniform.

In this method, the total volume of the storm period loss is estimated and distributed uniformly across the storm pattern. This index is estimated using measured rainfall–runoff data. The Φ-index model is a crude approximation to represent the losses due to infiltration. To determine the Φ index for a given storm, the amount of observed runoff is determined from the hydrograph, and the difference between this quantity and the total gauged precipitation is then calculated. The volume of loss (including the effects of interception, depression storage, and infiltration) is uniformly distributed across the storm pattern, as shown in Figure 3.6.

Use of the Φ index for determining the amount of direct runoff from a given storm pattern is essentially the reverse of this procedure. Unfortunately, the Φ index determined from a single storm is not generally applicable to other storms, and unless it is correlated with basin parameters other than runoff, it is of little value. Where measured rainfall–runoff data are not available, the ultimate capacity of Horton's equation, f_f, might be considered as the Φ index.

3.8 RAINFALL MEASUREMENT

High-frequency information in the rainfall signal is transformed into high-frequency pulses in the runoff hydrograph since the highly impervious urban drainage area does not considerably damp such fluctuations, because a unique aspect of urban hydrology is the fast response of an urban drainage area to the rainfall input. Tipping bucket rain gauges are appropriate for representation of high frequencies, as opposed to weighing bucket gauges that are commonly used. Tipping bucket gauges could easily be set in a network. These gauges transmit the electrical pulses through a telephone or other communication devices or record them on a data logger at the site. Hyetographs in 5 minutes or less time intervals are necessary for modeling an urban watershed, although this may be rather relaxed for larger basins.

3.8.1 Intensity–Duration–Frequency Curves: Advantages and Disadvantages

The total storm rainfall depth at a point, for a given rainfall duration and average return period, is a function of the local climate. Rainfall depths can be further processed and converted into rainfall intensities (intensity = depth/duration), which are then presented in IDF curves. Such curves are particularly useful in stormwater drainage design because many computational procedures require rainfall inputs in the form of average rainfall intensity. The three variables, frequency, intensity, and duration, are all related to each other. The data are normally presented as curves displaying two of the variables, such as intensity and duration, for a range of frequencies. These data are then used as the input in most stormwater design processes. IDF curves, which are the graphical symbols of the probability, summarize frequencies of rainfall depths or average intensities (Table 3.7).

An important point that should be considered in the application of IDF curves is that the intensities obtained from these curves are averaged over the specified duration and do not show the actual precedent of rainfall. The developed contours for different return periods are the smoothed lines fitted to the results of different storms so that the IDF curves are completely hypothetical. Furthermore, the obtained duration is not the actual length of a storm; rather, it is only a 30-minute period, say, within a longer storm of any duration, during which the average intensity was of the specified value. A detailed description of the method of developing IDF curves and the risks of inappropriate usages is given by McPherson (1978). An IDF curve for an urban area in logarithmic scale is presented in Figure 3.7.

3.8.1.1 *Selection of Rainfall Duration*

The main disadvantage of the application of IDF curves in practice is that they do not assign a unique return period to a storm with specific depth or average intensity or vice versa. Many combinations of average intensity and return period can result in a given depth, and as the duration decreases, the return period for a given depth increases. Since the durations on IDF curves are not the actual representative of the storm durations, the curves can only be used to make an approximation of, say, the 30-year rainfall event on the basis of total depth. Instead of implication of IDF curves, a frequency analysis of storm depths on the time series of independent storm events could be performed, which is more time-consuming.

The IDF curves are used in conjunction with the rational method (which is introduced in the next section) to calculate the runoff from a particular watershed. Although the IDF data are commonly used in this case, it should be reiterated that

1. Since in the development of IDF curves the intensities are averaged over the indicated duration, these curves do not represent actual time histories of real storms.
2. A single curve is developed by using the gathered data from several different storms.
3. The duration is not representative of the total duration of an actual storm and most likely refers to a short period of a long storm.
4. The IDF curves cannot be used to estimate a storm event volume because the duration should be assumed at first.

TABLE 3.7
Estimation of Constant Rate of Rainfall Loss Due to Hydrological Soil Group

Soil Group	Range of Loss Rate (cm/h)
A	0.75–1.13
B	0.38–0.75
C	0.125–0.375
D	0.0–0.125

Source: Adapted from Rawls, W.J., *Encyclopedia of Soil Science*, Marcel Decker, New York, 2006.

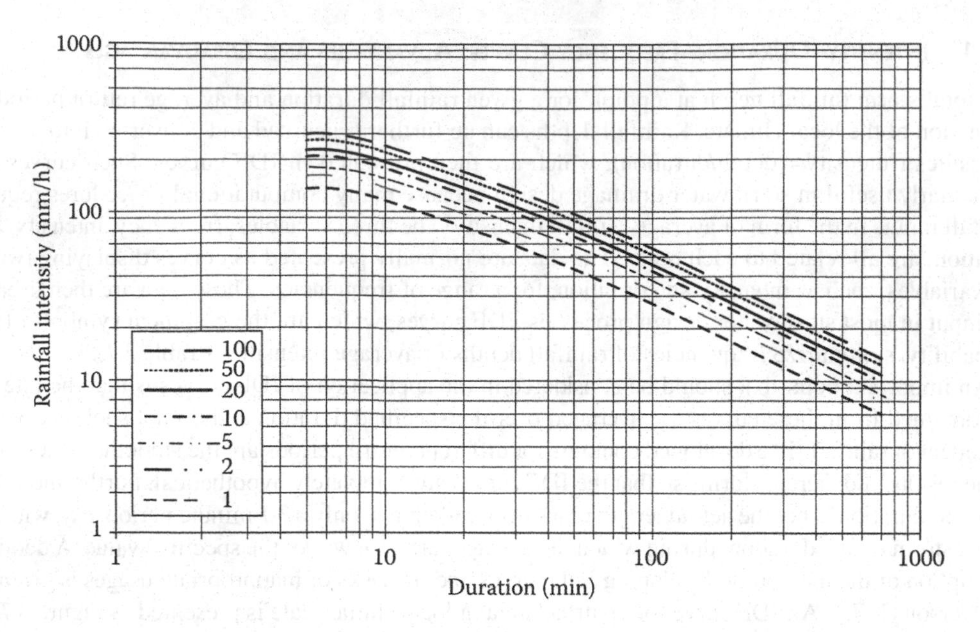

FIGURE 3.7 IDF curves for Kuala Lumpur. (From www2.water.gov.my/division/river/stormwater/Chapter_13.htm.)

3.9 ESTIMATION OF RUNOFF VOLUME

Total runoff volume and peak discharge are primary design variables in hydrology. Total runoff volume is used in the design of reservoirs and any other storage facilities. Peak discharge corresponds to the maximum volume flow rate passing a particular location during a storm event. Peak discharge is needed for the design of surface water drainage facilities such as pipe systems, storm inlets and culverts, and small open channels. Also, in some hydrologic planning, such as design and sizing of detention facilities in urban areas, peak discharge is used. It is an acceptable design variable for designs where the time variation of storage is not a primary factor in the runoff process. In this section, SCS and rational methods used for estimating runoff volume and/or peak discharge are discussed.

3.9.1 Rational Method

The rational method is an empirical relation between rainfall intensity and peak flow, which is widely accepted by hydraulic engineers. It is used to predict the peak runoff from a storm event. Despite being one of the oldest methods, it is still commonly used especially in the design of storm sewers, because of its simplicity and popularity, although it contains some limitations that are not often treated. The peak runoff is calculated using the following formula:

$$Q_p = k \cdot C \cdot i \cdot A, \tag{3.28}$$

where Q_p is the peak flow (m^3/s), C is the runoff coefficient, i is the rainfall intensity (mm/h), A is the drainage area (ha), and k is the conversion factor equal to 0.00278 for the conversion of ha mm/h to m^3/s.

The rational method makes the following assumptions:

- Precipitation is uniform over the entire basin.
- Precipitation does not vary with time or space.

- Storm duration is equal to the time of concentration.
- A design storm of a specified frequency produces a design flood of the same frequency.
- The basin area increases roughly in proportion to increases in length.
- The time of concentration is relatively short and independent of storm intensity.
- The runoff coefficient does not vary with storm intensity or antecedent soil moisture.
- Runoff is dominated by overland flow.
- Basin storage effects are negligible.

It is important to note that all of these criteria are seldom met under natural conditions. In particular, the assumption of a constant, uniform rainfall intensity is the least accurate. So, in watersheds with an area larger than about 2.5 km², the rational method should not be applied to the whole of the watershed. In these cases, the watershed is divided into some subdrainage areas considering the influence of routing through drainage channels. The rational method results become more conservative (i.e., the peak flows are overestimated) as the area becomes larger, because actual rainfall is not homogeneous in space and time.

The runoff coefficient *C*, which is usually given as a function of land use, considers all drainage area losses (Table 3.8).

TABLE 3.8
Runoff Coefficients for the Rational Formula versus Hydrologic Soil Group (A, B, C, and D) and Slope Range

	A			B			C			D		
Land Use	**0%–2%**	**2%–6%**	**6%+**	**0%–2%**	**2%–6%**	**6%+**	**0%–2%**	**2%–6%**	**6%+**	**0%–2%**	**2%–6%**	**6%+**
Cultivated	0.08[a]	0.13	0.16	0.11	0.15	0.21	0.14	0.19	0.26	0.18	0.23	0.31
land	0.14[b]	0.18	0.22	0.16	0.21	0.28	0.2	0.25	0.34	0.24	0.29	0.41
Pasture	0.12	0.2	0.3	0.18	0.28	0.37	0.24	0.34	0.44	0.3	0.4	0.5
	0.15	0.25	0.37	0.23	0.34	0.45	0.3	0.42	0.52	0.37	0.5	0.62
Meadow	0.1	0.16	0.25	0.14	0.22	0.3	0.2	0.28	0.36	0.24	0.3	0.4
	0.14	0.22	0.3	0.2	0.28	0.37	0.26	0.35	0.44	0.3	0.4	0.5
Forest	0.05	0.08	0.11	0.08	0.11	0.14	0.1	0.13	0.16	0.12	0.16	0.2
	0.08	0.11	0.14	0.1	0.14	0.18	0.12	0.16	0.2	0.15	0.2	0.25
Residential	0.25	0.28	0.31	0.27	0.3	0.35	0.3	0.33	0.38	0.33	0.36	0.42
lot size 500 m²	0.33	0.37	0.4	0.35	0.39	0.44	0.38	0.42	0.49	0.41	0.45	0.54
Residential	0.22	0.26	0.29	0.24	0.29	0.33	0.27	0.31	0.36	0.3	0.34	0.4
lot size 1,000 m²	0.3	0.34	0.37	0.33	0.37	0.42	0.36	0.4	0.47	0.38	0.42	0.52
Residential	0.19	0.23	0.26	0.22	0.26	0.3	0.25	0.29	0.34	0.28	0.32	0.39
lot size 1,300 m²	0.28	0.32	0.35	0.3	0.35	0.39	0.33	0.38	0.45	0.36	0.4	0.5
Residential	0.16	0.2	0.24	0.19	0.23	0.28	0.22	0.27	0.32	0.26	0.3	0.37
lot size 2,000 m²	0.25	0.29	0.32	0.28	0.32	0.36	0.31	0.35	0.42	0.34	0.38	0.48
Residential	0.14	0.19	0.22	0.17	0.21	0.26	0.2	0.25	0.31	0.24	0.29	0.35
lot size 4,000 m²	0.22	0.26	0.29	0.24	0.28	0.34	0.28	0.32	0.4	0.31	0.35	0.46
Industrial	0.67	0.68	0.68	0.68	0.68	0.69	0.68	0.69	0.69	0.69	0.69	0.7
	0.85	0.85	0.86	0.85	0.86	0.86	0.86	0.86	0.87	0.86	0.86	0.88
Commercial	0.71	0.71	0.72	0.71	0.72	0.72	0.72	0.72	0.72	0.72	0.72	0.72
	0.88	0.88	0.89	0.89	0.89	0.89	0.89	0.89	0.9	0.89	0.89	0.9

(Continued)

TABLE 3.8 *(Continued)*
Runoff Coefficients for the Rational Formula versus Hydrologic Soil Group (A, B, C, and D) and Slope Range

	A			B			C			D		
Land Use	**0%–2%**	**2%–6%**	**6%+**	**0%–2%**	**2%–6%**	**6%+**	**0%–2%**	**2%–6%**	**6%+**	**0%–2%**	**2%–6%**	**6%+**
Streets	0.7	0.71	0.72	0.71	0.72	0.74	0.72	0.73	0.76	0.73	0.75	0.78
	0.76	0.77	0.79	0.8	0.82	0.84	0.84	0.85	0.89	0.89	0.91	0.95
Open space	0.05	0.1	0.14	0.08	0.13	0.19	0.12	0.17	0.24	0.16	0.21	0.28
	0.11	0.16	0.2	0.14	0.19	0.26	0.18	0.23	0.32	0.22	0.27	0.39
Parking	0.85	0.86	0.87	0.85	0.86	0.87	0.85	0.86	0.87	0.85	0.86	0.87
	0.95	0.96	0.97	0.95	0.96	0.97	0.95	0.96	0.97	0.95	0.96	0.97

Source: McCuen, R., *Hydrologic Analysis and Design* (3rd ed.), Pearson Prentice Hall, Upper Saddle River, NJ, 2005. With permission.

[a] For storm return periods <25 years.

[b] For storm return periods more than 25 years.

A better estimate of rainfall coefficient is obtained from site measurements. There is often a considerable variation in C values in the analysis of actual storm data. But since a rational method gives a coarse estimate of rainfall losses, it could be useful. One of the major problems in the application of the rational method is considering a constant rate of loss during the rainfall, regardless of the total rainfall volume or initial conditions of the watershed. By an increase of imperviousness of the drainage area, this assumption becomes less important and the estimated runoff becomes closer to the observed value.

The rainfall intensity *(i)* is determined using an IDF curve attributed to the studied watershed for a specified return period and duration equal to the time of concentration of the watershed. The time of concentration is equal to the watershed equilibrium time, which means that in this time the whole drainage area contributes to flow at the watershed end point. Thus, for times less than the watershed time of concentration, the total drainage area, A, should not be used in the estimation of watershed outflow. But after reaching the watershed equilibrium, a higher rainfall intensity, that occurred earlier, should be used. The appropriate estimation of t_c is the main issue in the application of the rational method. The time of concentration has an inverse relation with the rainfall intensity, so that the kinematic wave equation could be used for its estimation. Because of the relation between t_c and the unknown intensity of the rainfall, an iterative solution, which combines Equation 3.28 with the IDF curves is utilized. This iterative procedure could be followed as described in Example 3.3.

The time of concentration is often approximated by constant overland flow inlet times, for more simplicity in the iterative process. These times vary between 5 and 30 minutes which are commonly used. For highly developed, impervious urban areas with closely spaced stormwater inlets, a value of 5 minutes is suitable for the time of concentration that increases up to 10 or 15 minutes for less developed urban areas of fairly lesser density. Values between 20 and 30 minutes are appropriate for residential regions with broadly spaced inlets. Longer values are appropriate for flatter slopes and larger areas.

The effect of channels in large drainage areas should also be considered in t_c estimation. For this purpose, the wave travel times in channels and conduits, t_r (based on the wave speed, using an estimate of the channel size, depth, and velocity), along the flow pathways, should be added to overland flow inlet times to yield an overall time of concentration. Since sometimes the objective is the design of drainage channels, an iterative procedure is used to find the corresponding values of t_c and t_r.

TABLE 3.9
Characteristics of the Watershed Considered in Example 3.3

Land Use	Area (ha)	Slope (%)
Open space	12	1.5
Streets and driveways	28	1
Residential lot size 1,000 m²	47	2.1
Pasture	13	3

Residential lot size 1,000 m²: 0.29.
Pasture: 0.28.

Example 3.3

Determine the 20-year peak flow at a stormwater inlet for a 100-ha urban watershed. Assume that the inlet time equals 30 minutes, the watershed soil belongs to group B, and the land use pattern and slope are as given in Table 3.9.

Solution:

At first the runoff coefficients of different land uses are determined using Table 3.9, as shown below:

Open space: 0.08
Streets and driveways: 0.71

The area-weighted runoff coefficient of the whole watershed is estimated as

$$\bar{C} = \frac{\sum A_i C_i}{\sum A_i} = \frac{12 \times 0.08 + 28 \times 0.71 + 47 \times 0.29 + 13 \times 0.28}{100} = 0.38.$$

For a 30-minute storm with a 20-year return period, the average intensity is estimated to be equal to 150 mm/h. The peak flow of this storm is calculated as

$$Q_p = 0.00278 \times 0.38 \times 150 \times 100 = 15.85 \text{ m}^3/\text{s}.$$

3.9.2 SCS Method

This empirical procedure, which is also called the runoff curve number (CN) method, is presented by the SCS (1986) to calculate excess rainfall (runoff). The rainfall excess in this method is quantified using a CN. This is a basin parameter that varies from 0 to 100, and its value is dependent on the hydrologic soil group, the soil cover type and condition, the percentage of impervious areas in the considered area, and the initial moisture condition of the soil. If the considered area is composed of several subareas with different CNs, a weighted average (based on area) or composite CN should be estimated for the entire area. The recommended CN values for various land use types are given in Table 3.10. CNs for four soil groups (A to D) are given in this table. These groups are defined by SCS (1986) according to their minimum infiltration rate. The soil of group A has the maximum infiltration capacity, whereas the soil of group D has the lowest infiltration capacity. The soil textures of these groups are as follows:

- Group A includes sand, loamy sand, and sandy loam.
- Group B includes silt loam and loam.
- Group C includes sandy clay loam
- Group D includes clay loam, silty clay loam, sandy clay, silty clay, and clay.

TABLE 3.10
Runoff CNs for Different Types of Land Uses

Land Use		CNS for Hydrologic Soil Group			
Land Cover Type and Condition	**Average Percent of Impervious Area**	**A**	**B**	**C**	**D**
	Rural Areas				
	Fallow				
Straight row with poor condition		77	86	91	94
	Row Crops				
Straight row with poor condition		72	81	88	91
Straight row with good condition		67	78	85	89
Contoured with poor condition		70	79	84	88
Contoured with good condition		65	75	82	86
Contoured and terraced with poor condition		66	74	80	82
Contoured and terraced with good condition		62	71	78	81
	Small Grain				
Straight row with poor condition		65	76	84	88
Straight row with good condition		63	75	83	87
Contoured with poor condition		63	74	82	85
Contoured with good condition		61	73	81	84
Contoured and terraced with poor condition		61	72	79	82
Contoured and terrace with good condition		59	70	78	81
	Closed-Seeded Legumes or Rotation Meadow				
Straight row with poor condition		66	77	85	89
Straight row with good condition		58	72	81	85
Contoured with poor condition		64	75	83	85
Contoured with good condition		55	69	78	83
Contoured and terraced with poor condition		63	73	80	83
Contoured and terraced with good condition		51	67	76	80
	Pasture or Range				
Poor condition		68	79	86	89
Fair condition		49	69	79	84
Good condition		39	61	74	80
Contoured with poor condition		47	67	81	88
Contoured with fair condition		25	59	75	83
Contoured with good condition		6	35	70	79
	Meadow				
Good condition		30	58	71	78
Woods					
Poor condition		45	66	77	83
Fair condition		36	60	73	79
Good condition		25	55	70	77
Farmsteads		59	74	82	86
	Fully Developed Urban Areas (Vegetation Established)				
	Open Space (Lawns, Parks, Golf Courses, Cemeteries, etc.)				
Poor condition (grass cover <50%)		68	79	86	89
Fair condition (grass cover 50%–75%)		49	69	79	84
Good condition (grass cover >75%)		39	61	74	80

(Continued)

TABLE 3.10 (*Continued*)
Runoff CNs for Different Types of Land Uses

Land Use		CNS for Hydrologic Soil Group			
Land Cover Type and Condition	**Average Percent of Impervious Area**	**A**	**B**	**C**	**D**
	Impervious Areas				
Paved parking lots, roofs, driveways, etc. (excluding right-of-way)		98	98	98	98
	Streets and Roads				
Paved; curbs and storm sewers (excluding right-of-way)		98	98	98	98
Paved; open ditches (including right-of-way)		83	89	92	93
Gravel (including right-of-way)		76	85	89	91
Dirt (including right-of-way)		72	82	87	89
	Urban Districts				
Commercial and business	85	89	92	94	95
Industrial	72	81	88	91	93
Residential areas by average lot size <500 m²	65	77	85	90	92
1,000 m²	38	61	75	83	87
2,800 m²	30	57	72	81	86
2,000 m²	25	54	70	80	85
4,000 m²	20	51	68	79	84
8,000 m²	12	45	65	77	82
	Developing Urban Areas				
Newly grades areas (pervious areas only, no vegetation)		77	86	91	94

Source: SCS, *Urban Hydrology for Small Watersheds* (2nd ed.), U.S. Department of Agriculture, Springfield, VA, 1986. With permission.

The given CNs in Table 3.10 are for average moisture conditions before a rainfall. The urban CNs for commercial, industrial, and residential districts are composite estimated based on the average percent imperviousness displayed in the second column. Antecedent soil moisture is a determining factor in the volume and rate of runoff. Regarding this important factor, SCS has considered three classes of antecedent soil moisture conditions labeled as AMC I, AMC II, and AMC III. The soil conditions of each class are defined as follows (Table 3.11):

AMC I: Dry soils, but before reaching the wilting point, satisfactory cultivation has taken place.
AMC II: Average conditions.
AMC III: The soil is saturated due to a heavy rainfall or light rainfall and low temperatures in the last few days.

TABLE 3.11
Total 5-Day Antecedent Rainfall (mm)

AMC	Dormant Season	Growing Season
I	<1.2	<3.6
II	1.2–2.75	3.6–5.25
III	More than 2.75	More than 5.25

TABLE 3.12
Modification of CN for Different Soil Moisture Conditions

AMC I (Dry)	AMC II (Normal)	AMC III (Wet)
100	100	100
87	95	98
78	90	96
70	85	94
63	80	91
57	75	88
51	70	85
45	65	82
40	60	76
35	55	74
31	50	70
26	45	65
22	40	60
18	35	55
15	30	50
12	25	43
9	20	37
6	15	30
4	10	22
2	5	13

In cases where the soil moisture is under condition AMC I or AMC III, the CN values are modified based on Table 3.12. The given CN values for AMC I and AMC III correspond to the envelope curves for real measurements of rainfall (P) and runoff (Q). This is because for a given watershed, similar storm events with almost equal values of P may result in different values of Q.

The CN values result from this variation in Q, and the variation was assumed to result from variations in antecedent soil moisture conditions. The given CN values for AMC I and AMC III show likely CN values, but what is needed for design purposes is the "most likely" value rather than an extreme value that is much less probable. The resulting uncertainty because of CN estimation should be incorporated in risk analysis.

Consider an urban area with a composition of commercial and pasture in good condition land uses. The CN values for commercial land use are based on an imperviousness of 85% and CN values of 89, 92, 94, and 95 are used for hydrologic soil groups A, B, C, and D, respectively. The CN values of pasture in good condition are also obtained from Table 3.10 and the weighted CN (CN_w) for this region is computed as follows:

$$CN_w = CN_p(1-f) + f\,CN_{im}, \tag{3.29}$$

where f is the proportion (not percentage) of impervious areas, CN_{im} is the CN for the impervious area, and CN_p is the CN for the pervious area (39, 61, 74, or 80). To show the use of Equation 3.29, the weighted CN value for this region considering soil group A is estimated as follows: $39(0.15) + (0.85)89 = 81.5$.

Many local drainage policies collect runoff from certain types of impervious land uses such as rooftops, driveways, and patios and direct it to pervious surfaces without any connection to storm drain systems. The decision makers who use these policies believe that disconnecting such impervious areas results in smaller and less costly drainage systems and, furthermore, increases groundwater recharge and improves water quality. However, before applying these policies, how much this disconnecting will affect the peak rates and volumes in the design of the drainage system should be evaluated.

Three variables are needed for estimating CNs for regions with some unconnected imperviousness including the CN of the pervious area, the percentage of the impervious area, and the percentage of the impervious area that is disconnected. The adjusted CN (CN_c) is estimated as follows:

$$CN_c = CN_p + f\left(CN_{im} - CN_p\right)\left(1 - 0.5r_{ut}\right), \tag{3.30}$$

where f is the fraction of impervious cover, CN_p is CN of the pervious area, CN_{im} is the CN for the impervious area, and r_{ut} is the ratio of the unconnected impervious area to the total impervious area.

Example 3.4

Consider the case of a drainage area having 25% imperviousness, a pervious area with a CN of 61%, and 50% of unconnected imperviousness. Compute the adjusted CN for this basin for the current soil condition.

Solution:

The CN of impervious area is determined as 98 for the impervious areas for all soil conditions. The adjusted CN (CN_c) is calculated as follows:

$$CN_c = 61 + 0.25(98 - 61)\left(1 - 0.5(0.5)\right) = 68.$$

If the unconnected impervious area were not important for this region, the composite CN would be 70. This means that in this region, by disconnecting 50% of the impervious cover, the CN value will be reduced by 2. This reduction can result in peak discharge and runoff volume reduction.

Note: The proportion of impervious areas over the basin is a determining factor of runoff volume and should be well managed to avoid flash floods over the urban areas.

After estimating the basin CN, the runoff is calculated as follows:

$$R = \frac{(P - I_a)^2}{(P - I_a) + S_r}, \tag{3.31}$$

where R is runoff (excess rainfall) in millimeters, P is rainfall in millimeters, I_a is initial abstraction in millimeters, and S_r is potential maximum soil moisture retention in millimeters at the time runoff (as opposed to rainfall) begins.

The initial abstraction is composed of different parts including the intercepted water by vegetation, the surface depressions, evaporation, and infiltration before runoff formation. It is often assumed that $I_a = 0.2S_r$. Substituting this into Equation 3.31 results in

$$R = \frac{(P - 0.2S_r)^2}{P + 0.8S_r}. \tag{3.32}$$

This equation is valid if $P > 0.2S_r$. In other cases, R is considered to be equal to $0.1\ I_a$. The empirical relationship is used for estimating S_r.

$$S_r = \frac{25{,}400 - 254\text{CN}}{\text{CN}}. \tag{3.33}$$

For given CN and P values, at first, S_r (mm) is calculated from Equation 3.32. Then R is determined. The family of curves represents a graphical solution to Equation 3.33. The CN method can be used to determine the total excess rainfall given the total rainfall. If the hyetograph of a rainfall is given, it can also be used to determine the rates of excess rainfall.

Example 3.5

A 400 km^2 watershed includes 350 km^2 of open area with 80% grass cover and 50 km^2 of an industrial district that is 72% impervious. The watershed is subjected to a 24-hour rainfall with a total depth of 6.8 cm. Determine the total amount of excess rainfall. Assume that the soil of the watershed belongs to the hydrologic soil group B.

Solution:

From Table 3.10, CN = 61 for the open area and CN = 88 for the industrial district are used. The area-weighted composite CN representing the whole watershed is estimated as follows:

$$\text{CN} = \frac{\left(350\ \text{km}^2\right)(61) + \left(50\ \text{km}^2\right)(88)}{400\ \text{km}^2} = 65.$$

Substituting $P = 6.8$ cm and CN = 65, $R = 9.3$ mm is obtained.

3.10 UNIT HYDROGRAPHS

A flood hydrograph for a basin can be simulated using a unit hydrograph (UH), defined as the direct runoff from a storm that produces one unit of rainfall excess. Using the UH method, the direct runoff hydrograph at the watershed outlet for given excess rainfall resulting from a particular storm event is calculated. As the spatial variations of physical characteristics of the watershed are not directly entered in the runoff calculations of UH, it is classified as a lumped method *T.*

The principal concept underlying the application of a UH is that each basin has one UH that does not change (in terms of its shape) unless the basin characteristics change. Because the physical characteristics (such as drainage area, slope) of a basin typically remain unchanged, changes in the UH usually reflect changes in land use patterns or urbanization. Given that the UH does not change in shape and represents stream flow response to 1 unit of runoff (excess rainfall) within a basin, flood hydrographs for actual storms can be simulated by multiplying the discharge ordinates from a UH by the excess rainfall computed from the observed rainfall record.

3.10.1 UH Development

A UH is a theoretical direct runoff hydrograph resulting from excess rainfall equal to unit depth with constant intensity for a specific watershed. A unit hydrograph is abbreviated as UH, and a subscript is used to indicate the duration of the excess rainfall. For a gauged watershed, the UH is developed by analyzing the simultaneous records of rainfall and runoff. In ungauged watersheds, the synthetic UH methods are used to develop a UH. In the development of a UH, different physical characteristics of the watershed are considered.

The derivation of a dimensionless hydrograph for all streams in the study area begins with the development of a station-average UH for each study site. The criteria for selecting a storm to develop a station-average UH for a site consisted of, to the extent possible, (a) concentrated storm rainfall that was fairly uniform throughout the basin during one period and (b) an observed hydrograph that had one peak. Storms in which the rainfall occurred over a long period punctuated by shorter periods of no rainfall resulted in complex, multipeak hydrographs that should generally be avoided.

3.10.2 SCS UH

The SCS curvilinear dimensionless UH procedure is one of the most well-known methods for deriving synthetic UHs in use today. The dimensionless UH used by the SCS is derived based on a large number of UHs from basins that varied in characteristics such as size and geographic location. The UHs are averaged and the final product is made dimensionless by considering the

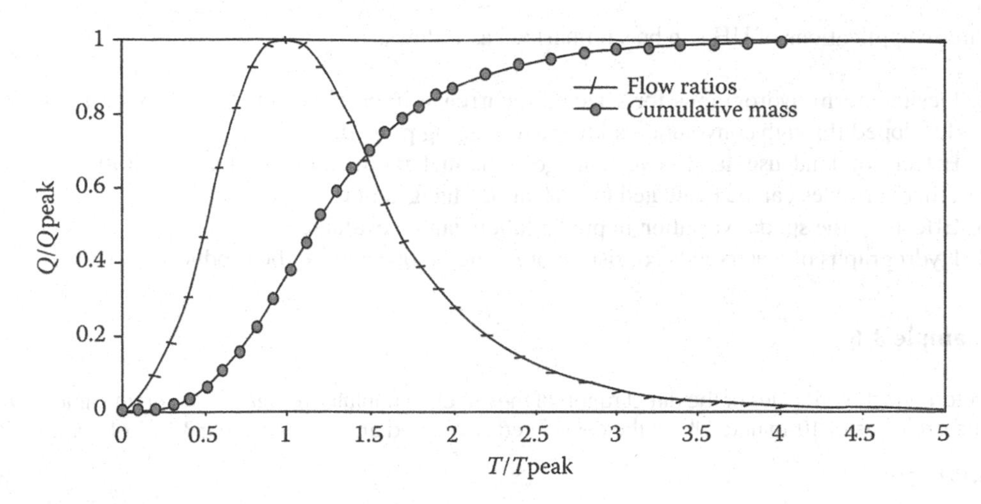

FIGURE 3.8 SCS dimensionless UH and mass curve (S-hydrograph).

ratios of q/q_p (flow/peak flow) on the ordinate axis and t/t_p (time/time to peak) on the abscissa, where the units of q and q_p are flow/cm of runoff/unit area. This final, dimensionless UH, which is the result of averaging a large number of individual dimensionless UHs, has a time-to-peak located at ~20% of its time base and an inflection point at 1.7 times the time to peak. The dimensionless UH is illustrated in Figure 3.8 which also illustrates the cumulative mass curve for the dimensionless UH.

3.10.3 Application of the UH Method

The UH method is formed based on the assumption of a linear relation between the excess rainfall and the direct runoff rates. The following are the fundamental assumptions of the UH theory.

1. The duration of direct runoff is always the same for uniform-intensity storms of the same duration, regardless of the intensity. This means that the time base of the hydrograph does not change and that the intensity affects only the discharge.
2. The direct runoff volumes produced by two different excess rainfall distributions are in the same proportion as the excess rainfall volume. This means that the ordinates of the UH are directly proportional to the storm intensity. If storm A produces a given hydrograph and storm B is equal to storm A multiplied by a factor, then the hydrograph produced by storm B will be equal to the hydrograph produced by storm A multiplied by the same factor.
3. The time distribution of the direct runoff is independent of the concurrent runoff from antecedent storm events. This implies that direct runoff responses can be superimposed. If storm C is the result of adding storms A and B, the hydrograph produced by storm C will be equal to the sum of the hydrographs produced by storms A and B.
4. Hydrological systems are usually nonlinear due to factors such as storm origin and patterns and stream channel hydraulic properties. In other words, if the peak flow produced by a storm of certain intensity is known, the peak corresponding to another storm (of the same duration) with twice the intensity is not necessarily equal to twice the original peak.
5. Despite this nonlinear behavior, the UH concept is commonly used because, although it assumes linearity, it is a convenient tool to calculate hydrographs and it gives results within acceptable levels of accuracy.
6. An alternative to the UH theory is the kinematic wave theory and distributed hydrological models.

The main applications of UH can be summarized as follows:

1. Design storm hydrographs for selected recurrence interval storms (e.g., 50 years) can be developed through convolution adding and lagging procedures.
2. Effects of land use–land cover changes, channel modifications, storage additions, and other variables can be evaluated to determine changes in the UH.
3. Effects of the spatial variation in precipitation can be evaluated.
4. Hydrographs of watersheds consisting of several subbasins can be produced.

Example 3.6

Determine the ordinates of the direct runoff of the effective rainfall hyetograph shown in Figure 3.9. The ordinates of 10-minute UH of the considered watershed are given in Table 3.13 (column 2).

Solution:

At first, the depth of the excess rainfall produced during each of the 10-minute periods is determined by multiplying the corresponding rainfall intensity by 1/6 hours. Considering the UH assumptions, the direct runoff hydrograph is calculated as

$$\text{DRH} = 0.17\text{UH} + 0.25\text{UH(lagged 10 minutes)} + 0.38\text{UH(lagged 20 minutes)}$$

$$+ 0.17\text{UH (lagged 30 minutes)} + 0.085\text{UH(lagged 40 minutes)}.$$

The calculations are summarized in Table 3.13.

3.10.4 S-Hydrograph Method

The characteristics of the obtained UH from the measured storm data depend on the effective duration of the rainfall excess. This is a result of coordination of rainfall excess duration with the watershed time of concentration. The S-hydrograph method is used to change the excess rainfall duration of the unit hydrograph to represent a different effective duration of the rainfall excess. The basic idea of the S-hydrograph method is that though an infinite number of effective durations are possible, as opposed to the UH, only one S-hydrograph exists for a watershed. The S-hydrograph is also referred to as the S-graph or summation hydrograph. For details refer to Karamouz et al. (2012).

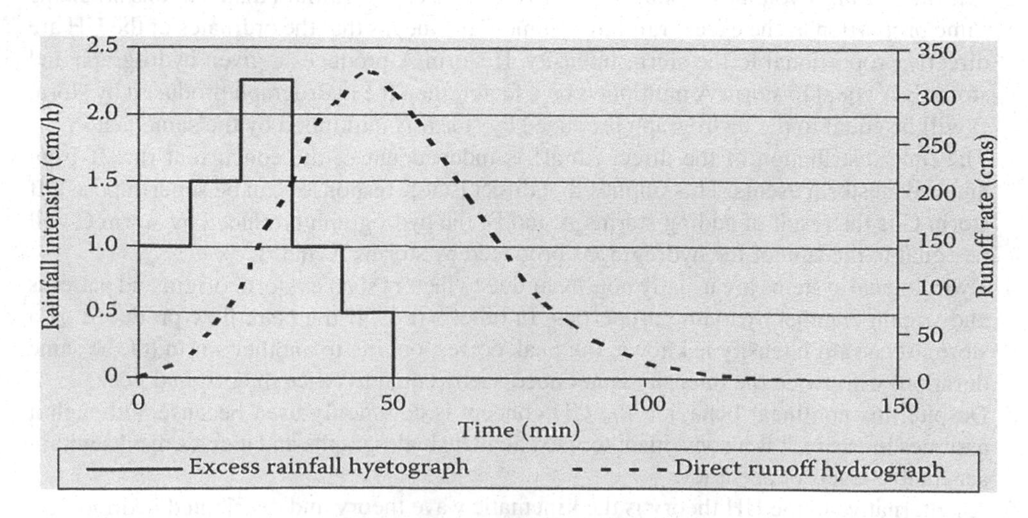

FIGURE 3.9 Characteristics of the rainfall event of Example 3.6.

TABLE 3.13
Application of the UN Method in Example 3.6

T (min)	UH (m³/s/cm)	0.17UH (m³/s)	0.25UH Lagged 10 minutes (m³/s)	0.38UH Lagged 20 minutes (m³/s)	0.17UH Lagged 30 minutes (m³/s)	0.085UH Lagged 40 minutes (m³/s)	Direct Runoff Hydrograph (m³/s)
0	0	0	0	0	0	0	0
10	50	8.5	0	0	0	0	8.5
20	110	18.7	12.5	0	0	0	31.2
30	170	28.9	27.5	19	0	0	75.4
40	240	40.8	42.5	41.8	8.5	0	133.6
50	150	25.5	60	64.6	18.7	4.25	173.05
60	70	11.9	37.5	91.2	28.9	9.35	178.85
70	30	5.1	17.5	57	40.8	14.45	134.85
80	15	2.55	7.5	26.6	25.5	20.4	82.55
90	5	0.85	3.75	11.4	11.9	12.75	40.65
100	0	0	1.25	5.7	5.1	5.95	18
110	0	0	0	1.9	2.55	2.55	7
120	0	0	0	0	0.85	1.275	2.125
130	0	0	0	0	0	0.425	0.425
140	0	0	0	0	0	0	0

For a given UH with effective duration of T hours, the S-curve is computed by adding an infinite series of T-hour unit hydrographs, each of them lagged by T hours, which is unique for each watershed. The shown S-graph is a result of the summation of the triangular unit hydrographs. It can be stated that the ith ordinate of the S-graph, S_i, is the sum of all UH ordinates, U_j, from the first ordinate to the ith ordinate as follows:

$$S_i = \sum_{j=1}^{i} U_j \tag{3.34}$$

$$S(t) = \int_0^t U(\tau)d\tau. \tag{3.35}$$

This computation applies when the time interval of the unit hydrograph is equal to the unit duration of the unit hydrograph (Figure 3.10).

When an S-curve is used to derive a UH with durations different from the duration of the UH used to develop the S-curve, the ordinates should be recorded on the time interval of the shorter of the two durations. For example, if the given S-curve corresponds to effective duration of 30- and 50-minute UH is needed, the ordinates must be recorded on a 30-minute interval, but when a 10-minute UH is needed, the ordinates must be recorded on a 10-minute interval. The intermediate points are derived by interpolation of the S-curve. After interpolation, it should be checked that the sum of the ordinates is equivalent to one area depth. This is true if linear interpolation between points on the S-graph is used. However, linear interpolation from the S-graph is not recommended because it always results in a peak discharge for the shorter-duration UH that is smaller than the peak of the longer-duration UH (See Example 3.7.).

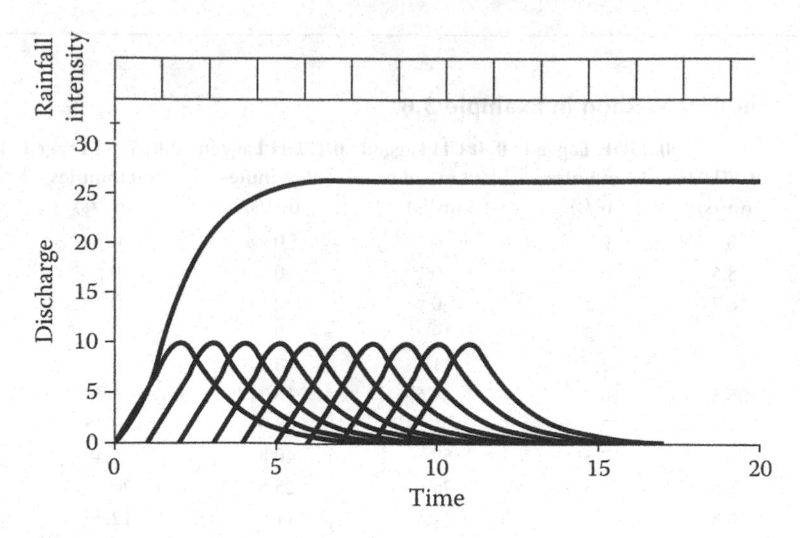

FIGURE 3.10 Derivation of an S-hydrograph.

Example 3.7

Given the 2-hour UH in Table 3.14, use the S-curve to develop the ordinates of a 3-hour and a 1-hour UHs.

Solution:

At first, the S-hydrograph is developed and then converted to a 3 and 1-hour UH. The results are given in Tables 3.15 and 3.16.

TABLE 3.14
Two-Hour Unit Hydrograph in Example 3.7

Time (h)	0	1	2	3	4	5	6
Q (m³/s)	0	200	500	400	200	100	0

TABLE 3.15
Results for Example 3.7

Time (h)	Q (m³/s)	S-Curve Additions	S-Curve	Legged S-Curve	Difference	Three-Hour UH × 2/3
0	0	–	0		0	0
1	200	–	200		200	133.3
2	500	–	500		500	333.3
3	400	200	600	0	600	400
4	200	500	700	200	500	333.3
5	100	400 + 200	700	500	200	133.3
6	0	200 + 500	700	600	100	66.7
7		100 + 400 + 200	700	700	0	0
				700		

TABLE 3.16
Results for Example 3.7 (1-hour UH)

Time (h)	Q (m³/s)	S-Curve Additions	S-Curve	Legged S-Curve	Difference	Two-Hour UH (6) × 2 (7)
0	0		0		0	0
1	200		200	0	200	400
2	500	0	500	200	300	600
3	400	200	600	500	100	200
4	200	500	700	600	100	200
5	100	400	500	700		
6	0	200	200	500		
7		100	100	200		
		0		100		

3.11 IUH–CONVOLUTION INTEGRAL–NASH MODEL

3.11.1 Convolution Integral

The same convolution process described earlier is also used for continuous processes. For a continuous process, all of the issues included in the convolution process including multiplication, translation, and addition are done using convolution integral:

$$Q(t)=\int_0^t x(\tau)U(t-\tau)d\tau, \tag{3.36}$$

where $U(t)$ is the continuous function of UH often called IUH and $Q(t)$ is the continuous function of direct runoff. $x(\tau)$ is the computed time distribution of rainfall excess, and τ is the time lag between the beginning times of rainfall excess and the IUH of direct runoff.

In practice, the convolution integral is treated in discrete form, which relates the time distributions of direct runoff $Q(t)$ with time distribution of rainfall excess $x(\tau)$ and the unit hydrograph $U(t-\tau)$:

$$Q(t)=\sum_{\tau=0}^{t} x(\tau)U(t-\tau). \tag{3.37}$$

This representation of the convolution process clearly indicates the multiplication, translation, and addition operations. The number of ordinates in the direct runoff distribution, the time base of the runoff, is computed (Figure 3.11).

Example 3.8

For the given rainfall excess hyetograph and 1-hour UH, compute the storm hydrograph for the corresponding watershed. Assume no losses, no infiltration, and no evaporation.

$$P_n=\{0.3,0.5,1.5,0.2,1\}\ \text{cm}$$

$$U_n=\{0,200,350,270,200,125,0\}\ \text{m}^3/\text{s}.$$

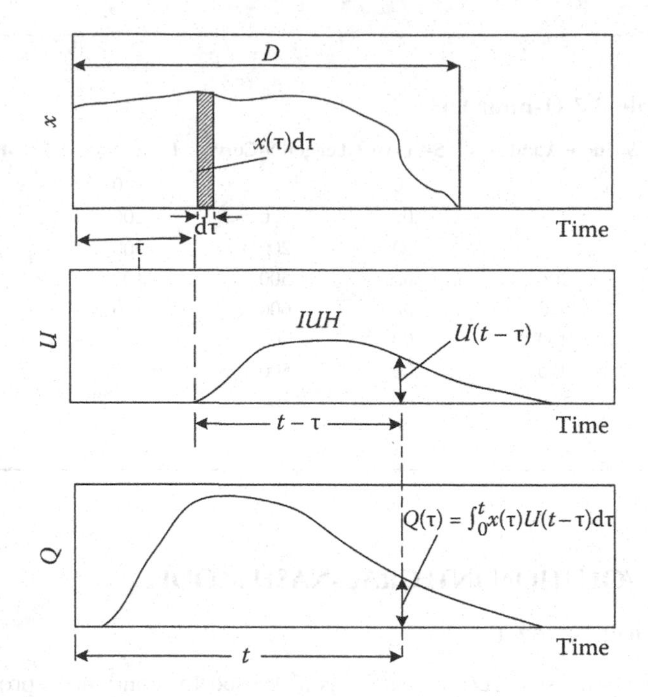

FIGURE 3.11 Schematic diagram of the convolution integral.

TABLE 3.17
Results for Example 3.8

Time (h)	P_1U_n	P_2U_n	P_3U_n	P_4U_n	P_5U_n	Q_n
0	0					0
1	60	0				60
2	105	100	0			205
3	81	175	300	0		556
4	60	135	525	40	0	760
5	37.5	100	405	70	200	812.5
6	0	62.5	300	54	350	766.5
7		0	187.5	40	270	497.5
8			0	25	200	225
9				0	125	125
10					0	0

Note: To develop the hydrograph of some successive rainfall events, at first, the corresponding hydrographs of individual rainfall events are produced by multiplying the unit hydrograph in depth of direct runoff. The developed hydrographs are summed with the appropriate lag times to generate the final rainfall hydrograph. The important point is the duration of individual rainfall events that should be considered in the selection of unit hydrograph as well as delaying hydrographs before their summation.

Solution:

The ordinates of direct hydrograph are computed as follows (Table 3.17):

$$Q_n = P_nU_1 + P_{n-1}U_2 + P_{n-2}U_3 + \cdots + P_1U_n.$$

Therefore,

$$Q_1 = (0.3)(0) = 0 \text{ m}^3/\text{s},$$

$$Q_2 = (0.5)(0)(0.3)(200) + 60 \text{ m}^3/\text{s},$$

$$Q_3 = (1.5)(0) + (0.5)(200) + (0.3)(350) = 205 \text{ m}^3/\text{s}, \ldots$$

Example 3.9

Construct a UH from the hydrograph depicted in Figure 3.12 and given in Table 3.18 for a 2-hour rainfall. The catchment area is 4.5 km². Use the constant value separation and construct a hydrograph representing the following complex rain pattern. The given values represent the excess rainfall depth (cm) in 2 hours.

$$P = \{1.1, 2.1, 1.8\}$$

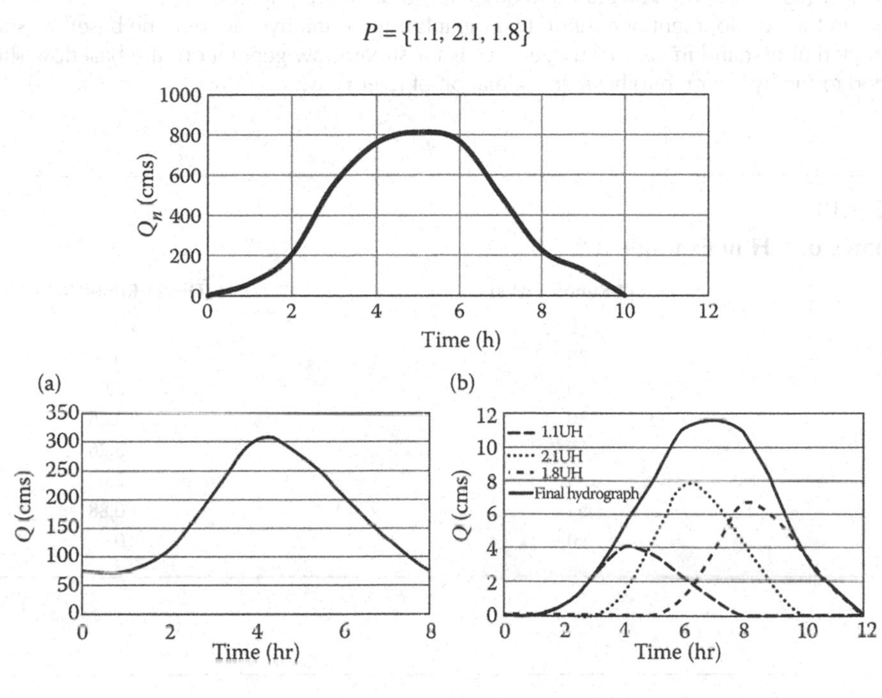

FIGURE 3.12 (a) Hydrograph in Example 3.9. (b) Development of hydrograph for a complex rainfall.

TABLE 3.18
Calculation of Direct Runoff for Example 3.9

t (h)	*Q* (m³/s)	Baseflow (m³/s)	Direct Runoff (m³/s)
1	75	75	0
2	110	75	35
3	205	75	130
4	305	75	230
5	280	75	205
6	205	75	130
7	130	75	55
8	75	75	0

Solution:

Volume of direct runoff is obtained as follows:

$$0.5\big[(0+35)+(35+130)+(130+230)+(230+205)$$
$$+(205+130)+(130+55)+(55+0)\big]\times(3{,}600\text{ s/h}) = 2.826\times10^6\text{ m}^3$$

$$\text{Effective runoff} = \left(2.826\times10^6\text{ m}^3\right)/\left(4.5\times10^6\text{ m}^2\right) = 0.628\text{ m.}$$

The ordinates of UH are calculated by dividing the ordinates of the hydrograph by 62.8 cm. The hydrograph of the complex rain is found by multiplying the ordinates of UH by the effective rain (rainfall minus the losses) and shifting the resulting hydrographs by 2 hours and adding ordinates (Tables 3.19 and 3.20).

Baseflow (75 m^3/s) should be added to ordinates to obtain full hydrograph.

Note: In the development of a runoff hydrograph from a unit hydrograph, the baseflow should be extracted at first and in case of usage of UHs for streamflow generation, the baseflow should be added to the hydrograph to have an estimation of river flow.

TABLE 3.19
Ordinates of UH in Example 3.9

t (h)	Direct Runoff (m^3/s)	UH (Direct Runoff/62.8) (m^3/s)
1	0	0
2	35	0.56
3	130	2.07
4	230	3.66
5	205	3.26
6	130	2.07
7	55	0.88
8	0	0

TABLE 3.20
Ordinates of Complex Rainfall in Example 3.9

t(h)	UH	UH × 1.1	UH × 2.1	UH × 1.8	Hydrograph Ordinates
1	0.00	0	0	0	0.61
2	0.56	0.61	0	0	0.61
3	2.07	2.28	0	0	2.28
4	3.66	4.03	1.17	0	5.20
5	3.26	3.59	4.35	0	7.94
6	2.07	2.28	7.69	1.00	10.97
7	0.88	0.96	6.86	3.73	11.54
8	0.00	0	4.35	6.59	10.94
9		0	1.84	5.88	7.71
10		0	0	3.73	3.73
11		0	0	1.58	1.58

3.12 INSTANTANEOUS UNIT HYDROGRAPHS

When the duration of rainfall excess approaches zero with a constant quantity (unit depth) of rainfall, a new form of UH is developed. The runoff produced by this instantaneous rainfall is called the IUH. In other words, IUH is a response function for a particular watershed to a unit impulse of rainfall excess. Each watershed has a unique function for the IUH, which is independent of time or antecedent conditions. The output function or total storm discharge, $y(t)$, is produced by the summation of the outputs due to all instantaneous input $x(t)$.

For a continuous process, all of the issues included in the convolution process including multiplication, translation, and addition are done using the convolution integral:

$$Q(t) = \int_0^t x(\tau)U(t-\tau)d\tau, \tag{3.38}$$

where $U(t)$ is the continuous function of UH and $Q(t)$ is the continuous function of direct runoff. $x(\upsilon)$ is the computed time distribution of rainfall excess, and υ is the time lag between the beginning times of rainfall excess and the unit hydrograph of direct runoff. A practical application of convolution integral is in the development of IUH. Rainfall contributes in the calculation of direct runoff as an input of the system, in instant time with a depth of $I(\tau)d\,\tau$. $U(t-\tau)$ is considered as a weighting function of rainfall intensities. Hence, multiplying $I(\tau)d\,\tau$ by $U(t-\tau)$, the runoff depth will be calculated as an output of the system. As a result, if the IUH of the watershed is determined, the direct runoff of the watershed would be evaluated.

The IUH can be derived from the S-curve hydrograph. It can be assumed that an S-curve is formed by a continuous rainfall. This rainfall is an infinite series of time units each separated by D. Given the summation of all unit hydrographs of duration D, the corresponding S-curve ordinates are derived as follows:

$$S(t) = \left[U(t) + U(t-D) + U(t-2D) + \cdots\right]. \tag{3.39}$$

The ordinate of a UH corresponding to a rainfall event with Δt duration, $U(t, \Delta t)$, is obtained as follows:

$$U(t,\Delta t) = \frac{D}{\Delta t}\left[S(t) - S(t-\Delta t)\right] = D = \frac{\Delta S(t)}{\Delta t}. \tag{3.40}$$

Regarding the IUH definition, when Δt approaches zero, $U(t)$ will be equal to IUH(t):

$$\text{IUH}(t) = \lim_{\Delta t \to 0} U(t,\Delta t) = D\frac{dS(t)}{dt}. \tag{3.41}$$

Therefore, the IUH in any time t can be calculated by multiplying the slope of the S-curve by D (the duration of the UH used for developing the S-curve).

In the same way, the S-curve can be derived from the IUH. The direct runoff depth is calculated from the convolution integral. If the input rainfall of system continues for the infinite period of time with the uniform intensity of ($1/D$), the resulting hydrograph would be the S-curve.

$$S(t) = \frac{1}{D}\int_0^t U(t-\tau)d\tau. \tag{3.42}$$

If $z = (t-\tau)$ is substituted and considering $d\tau = -dz$, it will result in

$$S(t) = -\frac{1}{D}\int_0^t U(z)dz. \tag{3.43}$$

If $U(t, D)$ is considered as the ordinate of a D-hour direct unit hydrograph (DUH) at time t, and $S(t)$ is the resulting S-curve from that DUH in time t, it will result in

$$u(t,\Delta t) = S(t) - S(t-\Delta t) = \frac{1}{D}\int_0^t U(z)\,dz - \frac{1}{D}\int_0^{t-D} U(z)\,dz$$

$$u(t,\Delta t) = \frac{1}{D}\int_{t-D}^{t} U(z)\,dz. \tag{3.44}$$

Suppose that IUH is linear between t and $(t - D)$, then

$$u(t,\Delta t) = U(t) = U\left(t - \frac{D}{2}\right). \tag{3.45}$$

Even though the IUH concept has some disadvantages and limitations especially when rainfall–runoff responses are nonlinear or are dependent on antecedent moisture conditions, it is widely used in practice.

Example 3.10

From the 1 cm/h S-curve in Table 3.21, determine the IUH and then use it to estimate a 2-hour UH.

Solution:

Here,

$$U(t) = D\frac{dS}{dt},$$

where $D = 2$ hours based on the data given in the example (1 cm/h S-curve is given) and dS/dt is approximated as $(S_{t+1} - S_{t-1})/\Delta t$, where $\Delta t = 1$ hour (Tables 3.22 and 3.23).

TABLE 3.21
S-Curve in Example 3.10

Time (h)	0	1	2	3	4	5	6	7	8	9
S-Curve (cm/s)	0	5	200	450	500	650	700	750	800	800

TABLE 3.22
IUH in Example 3.10

Time	S-Curve	IUH	
0	0	0	
1	50	50	
2	200	150	
3	450	250	This value is found as (450 – 200) × 1/1 = 250
4	500	50	
5	650	150	
6	700	50	
7	750	50	
8	800	50	This value is found as (800 – 750) × 1/1 = 50
9	800	0	

TABLE 3.23
Two-Hour UH in Example 3.10

Time	IUH	IUH_{t-1}	IUH_{t-2}	Two-Hour UH = 1/2[0.5IUH$_t$ + IUH$_{t-1}$ + 0.5IUH$_{t-2}$]	
0	0	0	0	0	
1	50	0	0	12.5	This value is found as (0.5 × 50 + 0 + 0)/2 = 12.5
2	150	50	0	62.5	
3	250	150	50	150	
4	50	250	150	175	
5	150	50	250	125	
6	50	150	50	100	This value is found as (50 × 0.5 + 150 + 50 × 0.5)/2 = 100
7	50	50	150	75	
8	50	50	50	50	
9	0	50	50	37.5	
10	0	0	50	12.5	

Since the UH duration is 2 hours and $\Delta t = 1$ hours, the 2-hour UH is obtained from the equation $Q = \frac{1}{2}[0.5\text{IUH}_t + \text{IUH}_{t-1} + 0.5\text{IUH}_{t-2}]$, and it is assumed that IUH is linear in each Δt.

Note: IUHs are a more general form of the UH that can be used to generate UHs of rainfalls with different duration.

3.12.1 Nash Model

The Nash model is a method for deriving the IUH. In the theoretical model of Nash (1959), a watershed is modeled as a cascade of n reservoirs in series. It has been assumed that the storage–discharge relation for each of the hypothetical reservoirs is linear, $S = KQ$, where K is the average delay time for each reservoir. Then, the continuity equation is combined with the storage relation as follows:

$$I - Q = \frac{dS}{dt} = K\frac{dQ}{dt}, \tag{3.46}$$

or

$$Q + K\frac{dQ}{dt} = I \tag{3.47}$$

Multiplying it by $e^{t/k}$, the above differential equation is solved as follows:

$$\frac{d}{dt}\left(Qe^{t/k}\right) = \frac{1}{K}e^{t/k}, \tag{3.48}$$

or

$$Qe^{t/k} = \frac{1}{K}\int \left(Ie^{t/k}\right)dt + C_1. \tag{3.49}$$

Considering an instantaneous inflow of unit volume to the reservoir, the reservoir discharge will be equal to the IUH dimension so

$$Q = e^{-t/k}\left[\frac{1}{K}\int e^{t/k}\delta(0)dt + C_1\right] = \frac{1}{K}\left(e^{-t/k}\right). \quad (3.50)$$

The above integral is a Laplace transform of the δ-function (Dirac delta function) that simply picks out the value of the function at $t = 0$:

$$L(\delta(0)) = \int \delta(0)e^{-pt}dt = e^{po} = 1. \quad (3.51)$$

These equations show that the response of a linear reservoir to the pulse input is a sudden jump at the moment of inflow followed by an exponential decline. Since it is assumed that there are n serial reservoirs in the basin, the outflow of the first reservoir flows into a second reservoir, where Q_1 is inflow and Q_2 is outflow, then

$$Q_1 - Q_2 = K\frac{dQ_2}{dt}, \quad (3.52)$$

and its solution is

$$Q_2 = \frac{1}{K}\left(\frac{1}{K}\right)e^{t/K}. \quad (3.53)$$

By continuing this process, the outflow of the nth reservoir (basin outflow) is estimated as follows:

$$Q_n = \frac{1}{K(n-1)!}\left(\frac{t}{K}\right)^{n-1} e^{-t/K}. \quad (3.54)$$

This equation is true for positive integer values of n, and for other real values of n, $(n - 1)!$ is substituted by the gamma function ($\Gamma(n)$) as follows:

$$Q_n = \frac{1}{K\Gamma(n)}\left(\frac{t}{K}\right)^{n-1} e^{-t/K}. \quad (3.55)$$

It is the probability density function of the gamma distribution. The general integral form of the gamma distribution is as follows:

$$\Gamma_n = \int_0^\infty e^{-x} \cdot x^{n-1}\,dx. \quad (3.56)$$

Although Nash (1959) suggests that the model represents the general equation form of the IUH, Gray (1962) has developed a UH model using the two-parameter gamma distribution. It can be used to derive the S-curve from the Nash model as follows:

$$S(t) = \int_0^t \mathrm{IUH}(t)\,dt = \frac{1}{K}\int_0^t \left(\frac{t}{K}\right)^{n-1} \cdot \frac{e^{t/k}}{\Gamma(n)}dt. \quad (3.57)$$

For a given instantaneous unit rainfall that spreads uniformly over the basin, the IUH can be interpreted as the frequency distribution of arrival times of water particles at the basin outlet. The time lag between centroids of the IUH and $x(\tau)$, the input rainfall hyetograph, is the expected value or average time. $E(t)$ also represents the first moment of the IUH about $t = 0$.

$$E(t) = \int_0^\infty u(t)t dt = nK \int_0^\infty \left(\frac{t}{K}\right)^n \frac{e^{-t/K}}{n!} d(t/K),$$

or

$$E(t) = nK. \tag{3.58}$$

The variance, Var(t), can also be derived as follows:

$$\text{Var}(t) = E\left(t^2\right) - [E(t)]^2 = \int_0^\infty u(t)t^2 dt - (nk)^2 = K^2 n(n+1) - n^2K^2 = K^2 n. \tag{3.59}$$

Example 3.11

For a catchment, the effective rainfall hyetograph of an isolated storm and the corresponding direct Runoff hydrograph (DRH) are given. Determine the coefficients n and k of Nash model IUH.

Solution:

As the first step, the M_{I1} and M_{I2} of rainfall hyetograph are estimated about the time origin as follows (Tables 3.24 and 3.25):

$$M_{I1} = \frac{4.3 \times 1 \times 0.5 + 3.2 \times 1 \times 1.5 + 2.4 \times 1.0 \times 2.5 + 1.8 \times 1.0 \times 3.5}{4.3 \times 1.0 + 3.2 \times 1.0 + 2.4 \times 1.0 + 1.8 \times 1.0} = 1.23 \text{ hours}$$

TABLE 3.24
Effective Rainfall Hyetograph

Time from Start of Storm (h)	Effective Rainfall Intensity (cm/s)
0–1.0	4.3
1.0–2.0	3.2
2.0–3.0	2.4
3.0–4.0	1.8

TABLE 3.25
Coordinates of DRH

Time from Start of Storm (h)	Direct Runoff (m³/s)	Time from Start of Storm (h)	Direct Runoff (m³/s)
0	0	9	32.7
1	6.5	10	23.8
2	15.4	11	16.4
3	43.1	12	9.6
4	58.1	13	6.8
5	68.2	14	3.2
6	63.1	15	1.5
7	52.7	16	0
8	41.9		

(For first-moment estimation, the numerator of the above relation, the area of each part of the hyetograph is multiplied by its center distance from the time origin.)

$$M_{12} = \frac{4.3\times1\times0.5^2 + 3.2\times1\times1.5^2 + 2.4\times1.0\times2.5^2 + 1.8\times1.0\times3.5^2}{4.3\times1.0 + 3.2\times1.0 + 2.4\times1.0 + 1.8\times1.0} = 2.89 \text{ hours}^2$$

(For second-moment estimation, the numerator of the above relation, the area of each part of the hyetograph is multiplied by the square of its center distance from the time origin.)

The same procedure is followed to estimate M_{Q1} and M_{Q2} based on the given runoff hydrograph. The calculations are summarized in Table 3.26.

$$M_{Q1} = \frac{\sum \Delta Q \cdot t}{\sum \Delta Q} = \frac{405.3}{68.2} = 5.94 \text{ hours.}$$

$$M_{Q2} = \frac{\sum \Delta Q \cdot t^2}{\sum \Delta Q} = \frac{3389.35}{68.2} = 49.69 \text{ hours}^2.$$

Nash IUH parameters are determined based on calculated values of M_{I1}, M_{I2}, M_{Q1}, and M_{Q2} as follows:

$$nK = M_{Q1} - M_{I1} = 5.94 - 1.23 = 4.71 \text{ hours.}$$

$$M_{Q2} - M_{I2} = n(n+1)K^2 + 2nKM_{I1} \Rightarrow 49.69 - 2.89 = n(n+1)K^2 + 2\times4.6\times1.23.$$

By solving the above equations, n is determined to be 2.82 and K is determined to be 1.67 hours.

TABLE 3.26
First and Second Moments of Outflow Hydrograph

Time from Start of Storm (h)	Direct Runoff (m³/s)	ΔQ	Distance of Time Interval Center from Time Origin, t	t^2	$\Delta Q \times t$	$\Delta Q \times t^2$
0	0	–	–	–	–	–
1	6.5	3.25	0.5	0.25	1.625	0.8125
2	15.4	4.45	1.5	2.25	6.675	10.0125
3	43.1	13.85	2.5	6.25	34.625	86.5625
4	58.1	7.5	3.5	12.25	26.25	91.875
5	68.2	5.05	4.5	20.25	22.725	102.2625
6	63.1	2.55	5.5	30.25	14.025	77.1375
7	52.7	5.2	6.5	42.25	33.8	219.7
8	41.9	5.4	7.5	56.25	40.5	303.75
9	32.7	4.6	8.5	72.25	39.1	332.35
10	23.8	4.45	9.5	90.25	42.275	401.6125
11	16.4	3.7	10.5	110.25	38.85	407.925
12	9.6	3.4	11.5	132.25	39.1	449.65
13	6.8	1.4	12.5	156.25	17.5	218.75
14	3.2	1.8	13.5	182.25	24.3	328.05
15	1.5	0.85	14.5	210.25	12.325	178.7125
16	0	0.75	15.5	240.25	11.625	180.1875
Sum	–	68.2	–	–	405.3	3389.35

Example 3.12

Develop an IUH and a 2-hour UH in the catchment with an area of 200 km² with Nash model parameters of $n = 3$ and $K = 5$ hours.

Solution:

$$Q_n = \frac{1}{K\Gamma(n)}\left(\frac{t}{K}\right)^{n-1} e^{-t/K},$$

$$K = 5 \text{ hours}, \quad n = 3, \quad \Gamma n = (n-1)!;$$

$$\Gamma_3 = 2! = 2, \quad K\Gamma n = 10, \qquad n-1 = 2.$$

The calculation of the IUH and 2-hour UH is shown in Table 3.27. For example, the calculation for time = 2 hours is as follows:

$$t = 2 \text{ hours} \rightarrow \frac{t}{K} = 0.4 \rightarrow e^{-t/k} = 0.67 \rightarrow \left(\frac{t}{K}\right)^{n-1} = 0.16$$

$$Q_n = \frac{1}{5\times 2}\left(\frac{t}{K}\right)^{n-1} e^{-t/K} = 0.01 \text{ cm/h}$$

$$\text{IUH}(2) = 2.778 \times A \times U(0,2) = 5.96 \text{ m}^3/\text{s}$$

$$\text{UH}(t) = \left(\text{IUH}(t) + \text{IUH}(t-1)/2 = 2.98 \text{ m}^3/\text{s.}\right)$$

TABLE 3.27
IUH and 2-Hour UH Calculation Using NASH Model

Time (h)	t/K	$e^{-t/K}$	$(t/K)^{n-1}$	$U(0, t)$	IUH	Two-Hour UH
0	0	1.00	0	0	0	0
2	0.4	0.670	0.16	0.010	5.96	2.98
4	0.8	0.449	0.64	0.028	16	10.98
6	1.2	0.301	1.44	0.043	24	20
8	1.6	0.201	2.56	0.051	28.4	26.2
10	2	0.135	4	0.054	30.1	29.3
12	2.4	0.090	5.76	0.052	29.1	29.6
14	2.8	0.060	7.84	0.047	26.7	27.9
16	3.2	0.040	10.24	0.041	23.4	25.1
18	3.6	0.027	12.96	0.035	19.4	21.4
20	4	0.018	16	0.029	16.3	17.8
22	4.4	0.012	19.36	0.023	13.1	14.7
24	4.8	0.008	23.04	0.018	10.5	11.8
26	5.2	0.005	27.04	0.014	8.3	9.4
28	5.6	0.003	31.36	0.011	6.45	7.38
30	6	0.002	36	0.008	5	5.73
32	6.4	0.001	40.96	0.006	3.83	4.42
34	6.8	0.002	46.24	0.005	2.82	3.32

3.12.2 Laplace Transformation Model

Laplace transformation can also be used for deriving the UH. Two approaches are considered in this method's application. In the first approach, the basin is simulated as a linear reservoir, and in the second, the basin is simulated as a channel. Each of these cases is described in the next sections.

3.12.2.1 *Basin as a Linear Reservoir*

The continuity equation for a linear reservoir is as follows:

$$I - Q = K\frac{dQ}{dt}, \tag{3.60}$$

where I is the basin input (rainfall), Q is the basin output (discharge), and K is the basin lag time. If the differential operator $D = d/dt$ is used,

$$Q = \frac{1}{1+KD} I(t). \tag{3.61}$$

This mathematically equals

$$Q = \frac{1}{k} e^{-t/k} Idt. \tag{3.62}$$

If the input is equal to 1, the resulting hydrograph would be an S-hydrograph as follows:

$$U(t) = 1 - e^{-t/k}. \tag{3.63}$$

As mentioned before, the slope of the S-curve is proportional to the ordinates of instantaneous hydrograph,

$$h(t) = \frac{e^{-t/k}}{K}. \tag{3.64}$$

If $1/K$ is replaced with a, then the Laplace transformation function is:

$$h(s) = a\left(\frac{1}{s+a}\right). \tag{3.65}$$

Using the above equation, the IUH of the basin in the form of the Laplace function and based on its lag time is obtained. The runoff–rainfall analysis is done using the following equation:

$$Q(s) = h(s)I(s), \tag{3.66}$$

where $I(s)$, $h(s)$, and $Q(s)$ are rainfall in the form of the Laplace function, UH, and runoff, respectively. The actual runoff values are determined by applying an inverse Laplace function on $Q(s)$.

For calculating $I(s)$, rainfall values should be defined in the form of a continuous function. Therefore, unit cascade function ($u_1(t)$) is used for definition of rainfall components.

$$I(t) = \sum_{j=0}^{m} w_j u_1(t - jD). \tag{3.67}$$

The unit cascade function is defined as follows:

$$u_1(t)=\begin{cases} 0, & t<0 \\ 1, & t\geq 0 \end{cases}. \tag{3.68}$$

D is rainfall measurement interval and w values are determined as follows:

$$w_0 = I_0, \quad w_j = I_j - I_{j-1} \quad j = 1,2,3,\ldots,m. \tag{3.69}$$

Example 3.13

Consider a basin as a linear reservoir with 6-hour lag time. Determine the resulting runoff from the rainfall given in Table 3.28.

Solution:

The weights and rainfall components are determined as follows:

$$w_j = I_j - I_{j-1}$$

$$w_0 = 5 - 0 = 5$$

$$w_1 = 2 - 5 = -3$$

$$w_2 = 1 - 2 = -1$$

$$w_3 = 2 - 1 = 1$$

$$w_4 = 0 - 2 = -2$$

and

$$I(t) = 5u_1(t) - 3u_1(t-0.5) - u_1(t-1) + u_1(t-1.5) - 2u_1(t-2).$$

Using the Laplace transformation function, the above equation is changed as follows:

$$I(s) = \frac{1}{s}\left[5 - 3\exp(-0.5s) - \exp(-s) + \exp(-1.5s) - 2\exp(-2s)\right],$$

The IUH Laplace function for this basin is

$$h(s) = \frac{1}{6}\left(\frac{1}{s+\frac{1}{6}}\right),$$

TABLE 3.28
Hyetograph in Example 3.13

Time (h)	0	0	0.5	1	1.5	2
***I* (cm/h)**	5	2	2	1	2	0

and runoff is calculated as follows:

$$Q(s)=\frac{1}{6}\left(\frac{1}{s+\frac{1}{6}}\right)\left[5-3\exp(-0.55s)-\exp(-s)+\exp(-1.5s)-2\exp(-2s)\right]\frac{1}{s}$$

$$=\frac{1}{6s\left(s+\frac{1}{6}\right)}\left[5-3\exp(-0.5s)-\exp(-s)+\exp(-1.5s)-2\exp(-2s)\right].$$

Using the equation of $\frac{1}{s(s+a)}=\frac{1}{sa}-\frac{1}{a(s+a)}$, it is simplified as follows:

$$Q(s)=5\left(\frac{1}{s}-\frac{1}{s+\frac{1}{6}}\right)-3\exp(-0.5s)\left(\frac{1}{s}-\frac{1}{s+\frac{1}{6}}\right)-\exp(-s)\left(\frac{1}{s}-\frac{1}{s+\frac{1}{6}}\right)$$

$$+\exp(-1.5s)\left(\frac{1}{s}-\frac{1}{s+\frac{1}{6}}\right)-2\exp(-2s)\left(\frac{1}{s}-\frac{1}{s+\frac{1}{6}}\right).$$

Using the inverse Laplace transformation, runoff is determined as follows:

$$Q(t)=5\times u_1(t)\left[1-\exp(-t/6)\right]-3u_1(t-0.5)\left[1-\exp(-(t-0.5/6))\right]$$

$$-u_1(t-1)\left[1-\exp(-(t-1)/6)\right]+u_1(t-1.5)\left[1-\exp(-(t-1.5)/6)\right]$$

$$-2u_1(t-2)\left[1-\exp(-(t-2)/6)\right].$$

The calculated runoff in different time steps is given in Table 3.29. For example, the calculations for the time = 0.5 and 1 hour are as follows:

$$Q(0.5)=5\times u_1(0.5)\left[1-\exp(-0.5/6)\right]=0.4$$

$$Q(1)=5\times u_1(1)\left[1-\exp(-1/6)\right]-2\times u_1(0.5)\left[1-\exp(-1/6)\right]=0.53.$$

TABLE 3.29
Runoff in Example 3.13

Time (h)	Q (cm/h)	Time (h)	Q (cm/h)
0	0	5.5	0.38
0.5	0.4	6	0.35
1	0.53	6.5	0.32
1.5	0.57	7	0.30
2.0	0.68	7.5	0.27
2.5	0.63	8	0.25
3	0.58	9	0.21
3.5	0.53	12	0.13
4.0	0.49	16	0.07
4.5	0.45	20	0.03
5.0	0.41		

3.12.2.2 *Basin as a Channel*

In this case, by assuming the basin as a linear channel, the UH is defined using the cascade function as follows:

$$h(t) = \frac{1}{T}\left[u_1(t) - u_1(t-T)\right], \tag{3.70}$$

where T is channel travel time. Considering the above equation and knowing the rainfall amount, the runoff is calculated. The Laplace transfers of commonly used functions are given in this chapter's Appendix.

Example 3.14

Consider a linear channel with 6-hour travel time. Using the effective rainfall hyetograph of Example 3.13, develop the DRH.

Solution:

The IUH of the given example is calculated, when $T = 6$ hours:

$$h(t) = \frac{1}{6}\left[u_1(t) - u_1(t-6)\right].$$

The Laplace form of this equation is

$$h(s) = \frac{1}{6s}\left[1 - \exp(-6s)\right].$$

Using the effective rainfall hydrograph given in the previous example, the direct runoff is estimated as follows:

$$\begin{aligned} Q(s) &= \frac{1}{6s}\left[1-\exp(-6s)\right]\left[5-3\exp(-0.5s)-\exp(-s)+\exp(-1.5s)-2\exp(-2s)\right]\frac{1}{s} \\ &= \frac{1}{6s^2}\left[5-3\exp(-0.5s)-\exp(-s)+\exp(-1.5s)-2\exp(-2s)-5\exp(-6s)+3\exp(-6.5s)\right. \\ &\quad \left. +\exp(-7s)-\exp(-7.5s)+2\exp(-8s)\right]. \end{aligned}$$

Applying the inverse Laplace results in Table 3.30

$$\begin{aligned} Q(t) = \frac{1}{6}&\left[5t - 3u_1(t-0.5) - u_1(t-1)(t-1) + u_1(t-1.5)(t-1.5) - 2u_1(t-2)(t-2)\right. \\ &- 5u_1(t-6)(t-6) + 3u_1(t-6.5)(t-6.5) + u_1(t-7)(t-7) \\ &\left. - u_1(t-7.5)(t-7.5) + 2u_1(t-8)(t-8)\right]. \end{aligned}$$

TABLE 3.30
DRH in Example 3.14

Time (h)	Q (cm/h)	Time (h)	Q (cm/h)
0	0	4.5	0.833
0.5	0.417	5	0.833
1	0.583	5.5	0.833
1.5	0.667	6	0.833
2	0.833	6.5	0.415
2.5	0.833	7	0.250
3	0.833	7.5	0.167
3.5	0.833	8	0.0
4	0.833		

The results are tabulated in Table 3.30. For example, the calculations for time = 0.5 and 1 hour are as follows:

$$Q(0.5)=\frac{1}{6}[5\times 0.5-3u_1(0)]=0.417$$

$$Q(1)=\frac{1}{6}[5-3u_1(0.5)(1-0.5)-u_1(0)(1-1)=0.583].$$

3.13 ROUTING METHODS

3.13.1 Hydrologic Methods of River Routing

The St. Venant equations are used to simulate the flow condition in unsteady and nonuniform rivers. Since the dynamic nature of the system is explained by these models, they are referred to as "dynamic models." However, there is no analytical solution for the St. Venant equations and complicated computer programs are needed to find the numerical solutions. Therefore, some methods with some simplifications are developed for river routing. The simplified methods are classified into two groups: hydraulic and hydrologic models. The St. Venant equations are used in hydraulic models; however, in developing hydrologic models, the momentum and continuity equations are replaced by some assumed relationships. These simpler models have some disadvantages, the most important of which is that some phenomena that affect the flow, such as downstream backwater, are not considered.

When a flood wave passes through a river reach, the peak of the outflow hydrograph is usually attenuated and delayed due to channel resistance and storage capacity. Regarding the quantity equation, the difference between input and output hydrographs is equal to the rate of storage change in the considered river reach. This is formulated as follows:

$$\frac{\Delta S}{\Delta t}=I-Q. \tag{3.71}$$

The sign of $\Delta S/\Delta t$ is positive when storage is increasing and negative when storage is decreasing.

$$\frac{S_2-S_1}{\Delta t}=\frac{I_1+I_2}{2}-\frac{Q_1+Q_2}{2}, \tag{3.72}$$

where Δt is the routing time period and subscripts 1 and 2 denote the beginning and end of the time period, respectively (Figure 3.13).

By plotting storage variations against outflow for a river reach, it will result in a loop-form curve. Therefore, the storage volumes during falling and rising stages are different and are bigger in the falling stage. The concept of prism and wedge storage can be used to evaluate water surface profile variations at various times during the passage of the flood wave from a river reach. During rising stages, before increasing the outflows, a large volume of wedge storage exists. During falling stages, the inflow decreases more rapidly than outflow, and the wedge storage becomes negative. Therefore, for hydrologic routing in rivers and channels, a storage relationship that allows for wedge storage is required. This is provided in the Muskingum method of flood routing by considering storage as a function of both inflow and outflow (McCarthy, 1938). The main disadvantage of this method is the use of the uniform flow rating curve instead of the loop curve.

3.13.1.1 *Muskingum Method*

In this method, the outflow hydrograph at the downstream of a river reach is calculated for a given inflow hydrograph at the upstream. The uniform short branches of the river are modeled as a single reach, but the long branches with various characteristics in length are often divided into several

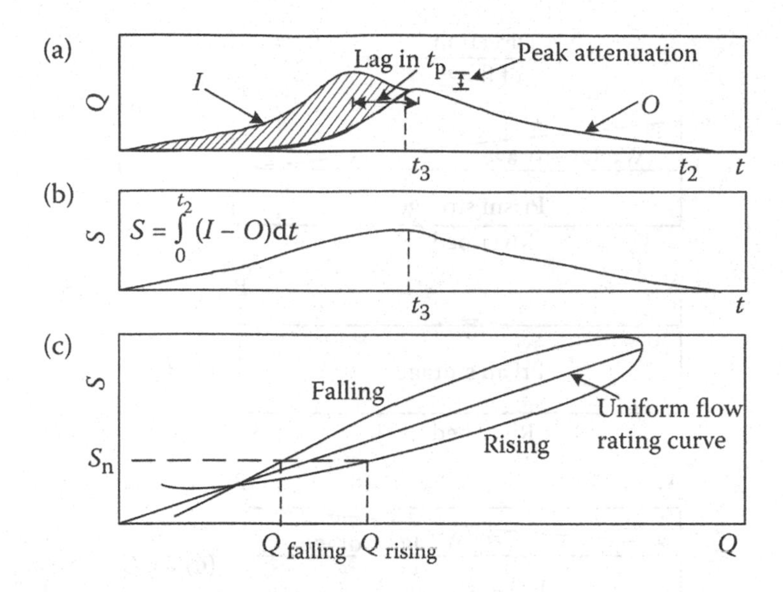

FIGURE 3.13 River routing. (a) Inflow and outflow hydrographs of a river reach. (b) Variation of storage in the river reach. (c) Variation of storage against outflow of the river reach.

reaches and each is modeled separately. When more than one reach is considered, the calculations start from the most upstream reach and then proceed in the direction of the downstream. The outflow from the upstream reach is used as the inflow for the next downstream reach.

As there are two unknown variables in the continuity equation including S and Q, a second relationship should be developed to solve the flood routing problem. In the Muskingum method, a linear relationship is assumed between S, I, and Q as follows:

$$S = K\left[XI + (1-X)Q\right], \tag{3.73}$$

where K is the travel time constant and X is the weighting factor, which varies between 0 and 1. The parameters K and X are constant for a river and do not have physical meaning. These parameters are determined during model calibration. For the case of linear reservoir routing where S depends only on outflow, $X = 0$. In smooth uniform channels, $X = 0.5$ yields equal weight to inflow and outflow, which theoretically results in pure translation of the wave. A typical value for most natural streams is $X = 0.2$ (Figure 3.14).

By substituting the obtained relation, the following equation is obtained:

$$Q_2 = C_0 I_2 + C_1 I_1 + C_2 Q_1, \tag{3.74}$$

where

$$C_0 = \frac{\left(\Delta t/K\right) - 2X}{2(1-X) + \left(\Delta t/K\right)} \tag{3.75}$$

$$C_1 = \frac{\left(\Delta t/K\right) - 2X}{2(1-X) + \left(\Delta t/K\right)} \tag{3.76}$$

$$C_2 = \frac{2(1-X) - (\Delta t/K)}{2(1-X) + (\Delta t/K)}. \tag{3.77}$$

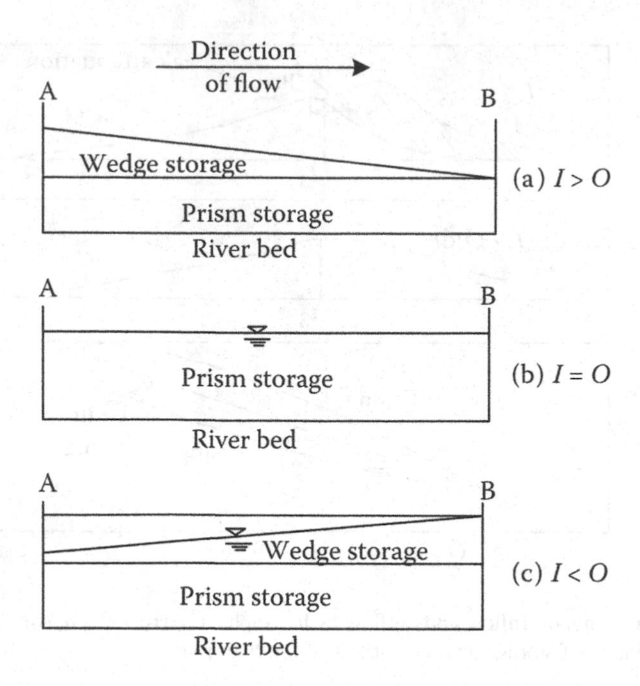

FIGURE 3.14 Prism and wedge storage concepts comparing inflow with outflow: (a) rising stage ($I > O$), (b) $I = O$, and (c) falling stage ($I < O$).

Note that $C_0 + C + C_2 = 1.0$ and K and Δt must have the same units.

Q_2 is the only unknown variable. The I_2 and I are obtained from the given inflow hydrograph and Q_1 is obtained from the initial condition of the river reach or from the previous time-step computation.

3.13.1.2 *Determination of Storage Constants*

For a natural stream, an average of 0.2 is considered for the Muskingum X and the travel time for a flood wave through the reach is used to estimate K. When both inflow and outflow hydrograph records are available, the graphical methods are preferred to better estimate K and X. In the graphical method considering different values of X, storage, S, is plotted against weighted discharge, $XI + (1 - X)O$. The best value for X corresponds to the plot that yields the most linear single-valued curve. In the Muskingum method, it is assumed that this curve is a straight line with reciprocal slope K. Thus, in the Muskingum method, storage is considered as a single-valued function of weighted inflow and outflow.

To achieve more accurate results from the Muskingum routing method, a river is divided into several reaches with slow flow changes over time. Thus, this method works quite well in ordinary streams with small slopes where the storage–discharge curve is approximately linear. In rivers with very steep or even mild slopes, different phenomena such as backwater effects, abrupt waves, or dynamic effects of flow affect the flow condition, and hydraulic routing methods are preferred. As another option in these cases, the Muskingum-Cunge method may be used. See Chapter 6 of Karamouz et al. (2012) for more details on this and on storage routing.

3.14 REVISITING FLOOD RECORDS

The annual flood records are not homogeneous because of the nonhomogeneity of the flood characteristics in an urban area due to land use variability. Thus, the assumption of homogeneity that is essential in a statistical flood frequency analysis is violated. Therefore, prior to making a frequency analysis on flood records, the results of inaccurate estimates of flood for any return period and the effect of nonhomogeneity in the analysis results should be investigated.

There are no specific well-known procedures suggested for adjustment of flood records in the literature. Often multiparameter watershed models are used for adjustment of flood records, but there is no single model or procedure widely accepted by the professional community and the results of different methods are not compared.

3.14.1 Urban Effects on Peak Discharge

As different methods are used for the estimation of peak discharge, the intensity of urban effects could be different. In the rational method, which is more popular in the estimation of urban peak discharge, urban development (increasing the impervious areas) affects the time of concentration and also the watershed runoff coefficient. Thus, the rate of increase in the peak discharge for a specific return period cannot be measured as imperviousness is different.

Some regression-based equations are suggested for the estimation of peak discharge using the percentage of imperviousness. Sarma et al. (1969) provided a general equation:

$$q_p = 629.34 A^{0.723} (1+U)^{1.516} P_E^{1.113} T_R^{-0.403}, \tag{3.78}$$

where A is the watershed area (km^2), U is the fraction of imperviousness, P_E is the depth of excess rainfall (mm), T_R is the duration of excess rainfall (hour), and q_p is the peak discharge (m^3/s).

The effect of urbanization on the peak discharge could be evaluated through the relative sensitivity (S_R) of Equation 3.79, which is calculated as

$$S_R = \frac{\partial q_p}{\partial U} \cdot \frac{U}{q_p}. \tag{3.79}$$

For instance, the peak discharge will be increased by 1.52% in response to 1% change in U. Since peak discharge is independent of both the value of U and the return period, this method estimates the average influence of urbanization on peak discharge.

Also, the following equation is suggested in the literature for evaluating the effect of urbanization on peak discharge:

$$f = 1 + 0.015U, \tag{3.80}$$

where f is the relative increase in peak discharge for percentage of imperviousness of U. See Karamouz et al. (2012) for more details.

3.14.2 Flood Record Adjusting

This method is developed based on the general trends of the data. The peak adjustment factor is determined using Figure 3.15 as a function of the exceedance probability for percentages of imperviousness <60%. Using this method, the greatest adjustment factors are obtained for more frequent events and the highest percentage of imperviousness.

The discharge must be adjusted from a partly urbanized watershed to a discharge for another urban condition. For this purpose, first the rural discharge is obtained by dividing the peak discharge in the existing condition to the peak adjustment factor. This way, the representative discharge of a nonurbanized condition is obtained. Then the resulting "rural" discharge is multiplied by the peak adjustment factor to obtain the peak discharge for the desired watershed condition. The exceedances probability is also necessary in using the proposed adjustment method of Figure 3.15. A plotting position formula could result in the best estimate of the probability for a flood record.

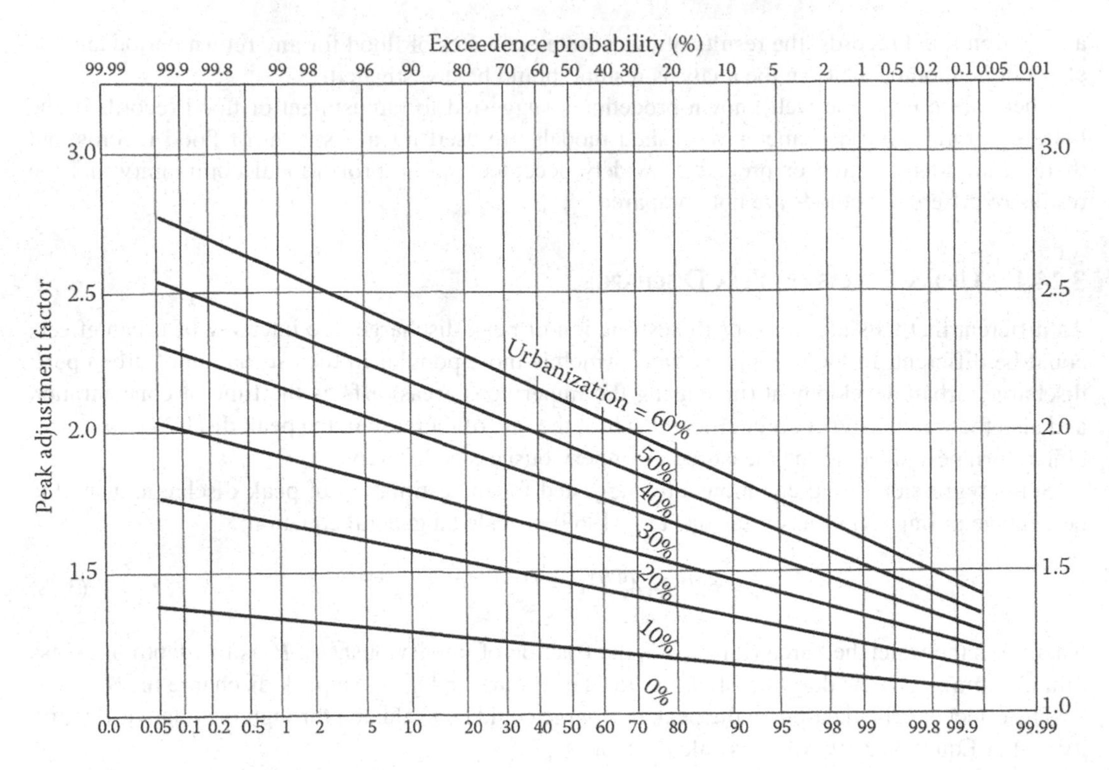

FIGURE 3.15 Peak adjustment factors for urbanizing watersheds. (From McCuen, R.H., *Hydrological Analysis and Design*, Prentice-Hall, Englewood Cliffs, NJ, 2005. With permission.)

In a watershed that is experiencing a continuous variation in the level of urbanization, a procedure could be utilized for adjusting a flood record (Karamouz et al., 2012).

3.15 TEST OF THE SIGNIFICANCE OF THE URBAN EFFECT

A series of data used for frequency analysis should be sampled from a unique population. This assumption in hydrological terms means that during data gathering the watershed should not significantly change. But urban watersheds are continuously changing due to developmental practices. The variation of some factors such as land cover and soil moisture can highly affect the hydrological processes in the watershed. The variations of these factors bring about the variation in the annual maximum flood series which are close to the natural variations of storms. For solving this problem, these factors are usually considered as random parameters and, in this way, the assumption of watershed homogeneity is not violated.

However, it should be noted that when major changes happen in land use, the watershed situation cannot be considered homogenous. The flood magnitudes or probabilities estimated from a flood frequency analysis are highly affected by these significant variations. Classical flood frequency analysis cannot be used for the determination of the accurate indicators of flooding for the changed watershed condition. For example, for a developing watershed, whose imperviousness increases from 5% to 45%, the duration and magnitude of the flood record are computed from a frequency analysis for 45% imperviousness. The results of this frequency analysis are not representative of this watershed, because many of the flood records that are used to derive the frequency curve correspond to a condition of much less urban development. Therefore, prior to making a flood frequency analysis, the homogeneity of data should be tested. The result of both hydrological and statistical assessments should suggest the necessity for adjustment. Among the available statistical methods for testing the nonhomogeneity of data, the most appropriate one should be selected due to watershed and data characteristics.

If the results of statistical analysis show the nonhomogeneity of data, then the adjustment of flood magnitudes is necessary before any frequency analysis. The results of the adjusted and unadjusted series could be significantly different. Two statistical tests that are commonly used to test the nonhomogeneity of flood records are introduced in the following sections. Both of these methods could be applied to records that have undergone a gradual change. Some other tests are also available for detecting a change in a flood record due to a rapid watershed change.

3.15.1 Spearman Test

If the annual measures of the gradual watershed change are available for each year of the flood record, the Spearman test could be applied. It is a bivariate test and needs both flood records and imperviousness data. For example, in a watershed with a 40-year flood record and the annual variations of percentage of impervious cover, the Spearman test could be used. The Spearman test includes the six general steps of a hypothesis test as follows (McCuen, 2005):

1. *State the hypotheses*: For testing the homogeneity of flood records in regard to urbanization, the null (H_0) and alternative (H_A) hypotheses are defined as follows:
   H_0: The urbanization does not affect the magnitude of the flood peaks.
   H_A: The magnitudes of the flood peaks are increased due to urbanization.
   The test is applied as a one-tailed test, as the statement of the alternative hypothesis indicates. The hypothesis statement also shows the expectation of a positive correlation.
2. *Determine the test statistic*: The Spearman test statistic is calculated using Equation 3.81. This statistic is named as the Spearman correlation coefficient (R_S) as shown below:

$$R_S = 1 - \frac{6\Sigma_{i=1}^{n} d_i^2}{n^3 - n}, \tag{3.81}$$

   where n is the sample size and d_i is the ith difference in the ranks of the two series, which in this case are the series of annual flood peaks and the series of the average annual percentage of imperviousness. After ranking the measurements of each series from largest to smallest, the difference in ranks d_i is computed.
3. Specify the desired level of significance. Often the value of 5% is considered, but in specific cases, values of 1% and 10% can also be used. The level of significance is usually determined based on expert judgment rather than through a rational analysis of the situation.
4. Estimate the test statistic R_S for the available sample through the following steps:
   a. Find the rank of each event in each series, while keeping the two series in chronological order.
   b. Calculate the difference in the ranks of the two series for each record i.
   c. Using Equation 3.63, compute the value of the Spearman test statistic, R_S.
5. Determine the critical value of the test statistic from Table 3.31. The values in the table are for the upper tail. As the distribution of R_S is symmetric, the critical values for the lower tail could be obtained by multiplying the value obtained from Table 3.31 by –1.
6. Make a decision. For an upper one-tailed test, if the computed value in step 4 is larger than the critical value obtained in step 5, the null hypothesis is rejected. For a lower one-tailed test, the null hypothesis is rejected if the computed value is less than the critical value.

For a two-tailed test, if the computed value lies in either tail, the null hypothesis is rejected, but the level of significance is twice that of the one-tailed test. If the null hypothesis is rejected, it could

TABLE 3.31
Critical Values for the Spearman Correlation Coefficient for the Null Hypothesis H_0: $|\rho_S|$ and Both the One-Tailed Alternative H_A: $|\rho_S| > 0$ and the Two-Tailed Alternative H_A: $\rho_S \neq 0$

	Level of Significance for a One-Tailed Test			
	0.050	**0.025**	**0.010**	**0.005**
	Level of Significance for a Two-Tailed Test			
Sample Size	**0.100**	**0.050**	**0.020**	**0.010**
5	0.900	1.000	1.000	1.000
6	0.829	0.886	0.943	1.000
7	0.714	0.786	0.893	0.929
8	0.643	0.738	0.833	0.881
9	0.600	0.700	0.783	0.833
10	0.564	0.648	0.745	0.794
11	0.536	0.618	0.709	0.818
12	0.497	0.591	0.703	0.780
13	0.475	0.566	0.673	0.745
14	0.457	0.545	0.646	0.716
15	0.441	0.525	0.623	0.689
16	0.425	0.507	0.601	0.666
17	0.412	0.490	0.582	0.645
18	0.399	0.476	0.564	0.625
19	0.388	0.462	0.549	0.608
20	0.377	0.450	0.534	0.591
21	0.368	0.438	0.521	0.576
22	0.359	0.428	0.508	0.562
23	0.351	0.418	0.496	0.549
24	0.343	0.409	0.485	0.537
25	0.336	0.400	0.475	0.526
26	0.329	0.392	0.465	0.515
27	0.323	0.385	0.456	0.505
28	0.317	0.377	0.448	0.496
29	0.311	0.370	0.440	0.487
30	0.305	0.364	0.432	0.478

be concluded that the peak flood records are not independent of the urbanization and the watershed change highly affects the hydrologic situation of the watershed as well as the annual maximum flood series.

Example 3.15

Eleven annual maximum discharges and the corresponding imperviousness for a watershed are given in Table 3.32. Test the homogeneity of these flood records.

Solution:

The homogeneity of data is tested using the Spearman test because of the gradual change in watershed imperviousness. Columns 4 and 5 of Table 3.32 present the ranks of the series, and the rank differences are calculated in column 6. Then the value of R_S is calculated as

$$R_S = 1 - \frac{6(288)}{11^3 - 11} = -0.310.$$

TABLE 3.32
Spearman Test for Example 3.15

Year	Annual Maximum Discharge Y_i (m³/s)	Average Imperviousness X_i (%)	Rank of $Y_i\ r_{yi}$	Rank of $X_i\ r_{xi}$	Difference $d_i = r_{yi} - r_{xi}$
1	123	40	11	5	6
2	453	35	5	10	−5
3	246	41	8	4	4
4	895	36	1	9	−8
5	487	37	4	8	−4
6	311	45	6	1	5
7	285	33	7	11	−4
8	206	39	9	6	3
9	803	42	2	3	−1
10	568	38	3	7	−4
11	189	44	10	2	8
					$\Sigma d_i^2 = 288$

Since the critical value of R_S is 0.536 for significance level of 0.05, from Table 3.31, which is more than the computed value of R_S, the null hypothesis is rejected. This means that urbanization has had a considerable effect on the annual maximum discharge series.

3.15.2 Spearman–Conley Test

If the annual data of watershed imperviousness change are not available, which is a common situation, the Spearman test cannot be used. In these cases, the Spearman–Conley test (Conley and McCuen, 1997; Eslamian et al., 2011) can be used to test for serial correlation. For applying the Spearman–Conley test, the following steps should be passed (McCuen, 2005):

1. *State the hypotheses*: For this test, the null and alternative hypotheses are defined as follows:
   H_0: There is no temporal relation between annual flood peaks.
   H_A: There is a significant correlation between consequence values of the annual flood series.
   The alternative hypothesis can be also expressed as a one-tailed test with an indication of positive correlation for a flood series supposed to be affected by watershed development. Since urban development increases the peak floods, a positive correlation coefficient is expected.
2. *Determine the test statistic*: The same statistic as the Spearman test (Equation 3.81) can be used for the Spearman–Conley test. However, the statistic is denoted as R_{sc}. In applying Equation 3.81, n indicates the number of pairs, which is 1 less than the number of annual maximum flood magnitude records. For the calculation of the R_{sc} value, a second series, X_t, is formed as $X_t = Y_{t-1}$. The value of R_{sc} is calculated with the differences between the ranks of the two series by using Equation 3.81.
3. *Specify the desired level of significance*: Again, this is usually determined by convention and a 5% value is commonly used.
4. Compute the test statistic R_{sc} for the available sample through the following steps:
   a. Create a second series of flood magnitudes (X_t) by offsetting the actual series (Y_{t-1}).
   b. Identify the rank of each event in each series, while keeping the two series in chronological order. The series are ranked in decreasing order so that a rank of 1 is used for the largest value in each series.
   c. Calculate the difference in ranks for each i.
   d. Compute the value of the Spearman–Conley test statistic R_{sc} using Equation 3.81.

TABLE 3.33
Critical Values for Univariate Analysis with the Spearman–Conley One-Tailed Serial Correlation Test

	Lower-Tail Level of Significance (%)					Upper-Tail Level of Significance (%)				
Sample Size	**10**	**5**	**1**	**0.5**	**0.1**	**10**	**5**	**1**	**0.5**	**0.1**
5	–0.8	–0.8	–1.0	–1.0	–1.0	0.4	0.6	1.0	1.0	1.0
6	–0.7	–0.8	–0.9	–1.0	–1.0	0.3	0.6	0.8	1.0	1.0
7	–0.657	–0.771	–0.886	–0.943	–0.943	0.371	0.486	0.714	0.771	0.943
8	–0.607	–0.714	–0.857	–0.893	–0.929	0.357	0.464	0.679	0.714	0.857
9	–0.572	–0.667	–0.810	–0.857	–0.905	0.333	0.452	0.643	0.691	0.809
10	–0.533	–0.633	–0.783	–0.817	–0.883	0.317	0.433	0.617	0.667	0.767
11	–0.499	–0.600	–0.746	–0.794	–0.867	0.297	0.406	0.588	0.648	0.745
12	–0.473	–0.564	–0.718	–0.764	–0.836	0.291	0.400	0.573	0.627	0.727
13	–0.448	–0.538	–0.692	–0.734	–0.815	0.287	0.385	0.552	0.601	0.706
14	–0.429	–0.516	–0.665	–0.714	–0.791	0.275	0.369	0.533	0.588	0.687
15	–0.410	–0.495	–0.644	–0.692	–0.771	0.270	0.358	0.521	0.574	0.671
16	–0.393	–0.479	–0.621	–0.671	–0.754	0.261	0.350	0.507	0.557	0.656
17	–0.379	–0.462	–0.603	–0.650	–0.735	0.256	0.341	0.491	0.544	0.641
18	–0.365	–0.446	–0.586	–0.633	–0.718	0.250	0.333	0.480	0.530	0.627
19	–0.354	–0.433	–0.569	–0.616	–0.703	0.245	0.325	0.470	0.518	0.613
20	–0.344	–0.421	–0.554	–0.600	–0.687	0.240	0.319	0.460	0.508	0.601
21	–0.334	–0.409	–0.542	–0.586	–0.672	0.235	0.313	0.451	0.499	0.590
22	–0.325	–0.399	–0.530	–0.573	–0.657	0.230	0.307	0.442	0.489	0.580
23	–0.316	–0.389	–0.518	–0.560	–0.643	0.226	0.301	0.434	0.479	0.570
24	–0.307	–0.379	–0.506	–0.549	–0.632	0.222	0.295	0.426	0.471	0.560
25	–0.301	–0.370	–0.496	–0.537	–0.621	0.218	0.290	0.419	0.463	0.550
26	–0.294	–0.362	–0.486	–0.526	–0.610	0.214	0.285	0.412	0.455	0.540
27	–0.288	–0.354	–0.476	–0.516	–0.600	0.211	0.280	0.405	0.448	0.531
28	–0.282	–0.347	–0.467	–0.507	–0.590	0.208	0.275	0.398	0.441	0.523
29	–0.276	–0.341	–0.458	–0.499	–0.580	0.205	0.271	0.392	0.434	0.515
30	–0.270	–0.335	–0.450	–0.491	–0.570	0.202	0.267	0.386	0.428	0.507

5. *Obtain the critical value of the test statistic*: It should be noted that the distribution of R_{sc} is not symmetric and is different from that of R_S. The critical values for the upper and lower tails of the Spearman–Conley test statistic R_{sc} are given in Table 3.33. In this table, the number of pairs of values is used to compute R_{sc}.
6. *Make a decision*: For a one-tailed test, if the computed R_{sc} is greater in magnitude than the value of Table 3.33, the null hypothesis is rejected. It could be concluded that the annual maximum floods are serially correlated if the null hypothesis is rejected. This correlation reflects the impact of urbanization on the peak flood discharges.

Example 3.16

Assuming the imperviousness data of Example 3.15 (column 3 of Table 3.32 is not available), test the homogeneity of the flood data.

Solution:

The flood series from the second year are tabulated consequently in column 1 of Table 3.34. The offset values are given in column 2. As the record should be offset, one value is lost, and thus, for

TABLE 3.34
Summary of the Spearman–Conley Test for Example 3.16

Annual Maximum Discharge t Y_t (m³/s)	X_t = Offset Y_t (m³/s)	Rank of Y_i r_{yi}	Rank of Xi r_{xi}	Difference $d_i = r_{yi} - r_{xi}$
453	123	5	10	−5
246	453	8	5	3
895	246	1	8	−7
487	895	4	1	3
311	487	6	4	2
285	311	7	6	1
206	285	9	7	2
803	206	2	9	−7
568	803	3	2	1
189	568	10	3	7
				$\Sigma d_i^2 = 200$

this test, $n = 10$. The ranks of the two series are given in columns 3 and 4, and the differences in ranks are listed in column 5. The sum of the squares of the d_i values is equal to 86. The test statistic value is computed as

$$R_{sc} = 1 - \frac{6(200)}{10^3 - 10} = -0.212.$$

The critical value is 0.433 from Table 3.33 for a 5% level of significance and a one-tailed upper test. As the calculated value of R_{sc} is less than the critical value, the null hypothesis is rejected and it can be concluded that the flood data are temporally correlated. This serial correlation could be the result of urbanization.

3.16 TIME SERIES ANALYSIS

The concept of random variables has been widely used for over a century in the field of hydrology and water resources analysis and modeling. Great improvements have been made in recent years in the following fields: understanding the stochastic nature of hydrologic variables such as streamflows and rainfall; modeling stochastic hydrological procedures; developing new statistical models; improving the parameter estimation techniques; proposing new model evaluation and fitness tests; and quantifying uncertainty and imprecision. Forecasting the future state of the resources is necessary for application of operating policies of water resources systems in real-time decision-making. For example, in a reservoir, which supplies water for different purposes, the amount of scheduled releases depends on the probable range of inflow to the reservoir.

Due to lack of enough knowledge about physical processes in the hydrologic cycle, the application of statistical models in forecasting and generating synthetic data is highly expanded. Furthermore, generation of synthetic data helps to incorporate the uncertainties and probable extreme events in hydrological analyses. The new advancements in this area have been mainly focused on coupling physical characteristics of the water resources systems and the effects of large-scale climate signals on water resources and early prediction of rainfall and streamflow. In this chapter, basic principles of hydrologic time series modeling and different types of statistical models as well as their application in water resources systems analysis are discussed.

A systematic approach to hydrologic time series modeling could consist of the following main steps (Salas et al., 1988):

- *Data preparation*: The first step is to remove trends, periodicity, and outlying observations and fit the data to a specific distribution by applying the proper transformations.

- *Identification of model composition*: The next step is to decide upon the use of a univariate or multivariate model, or a combination, with disaggregation models. This decision can be made based on the characteristics of the water resources system and existing information.
- *Identification of model type*: Various types of models, such as AR (autoregressive), ARMA (autoregressive moving average), and ARIMA (autoregressive integrated moving average), can be selected in this step; details of these models are presented in this chapter. Statistical characteristics of the time series and the modeler input and knowledge about the types of models are the key factors in identification of model type.
- *Identification of model form*: The form of the selected model should be defined based on the statistical characteristics of the time series such as seasonal and nonseasonal characteristics. The periodicity of the data and how it can be considered in the structure of the selected model is the main issue in this step.
- *Estimation of model parameters*: Several methods, such as method of moments and method of maximum likelihood, can be used for estimating the model parameters.
- *Testing the goodness of fit of the model*: In this step, assumptions such as test of independence and normality of residuals should be checked. The statistics used for verifying these assumptions are briefly explained in this chapter.
- *Evaluation of uncertainties*: When evaluating uncertainties, the model, parameter, and natural uncertainties in data should be analyzed separately. Model uncertainty may be evaluated by testing whether there are significant differences between the statistics of actual data and forecasted and generated data by alternative models. Parameter uncertainty may be determined from the distribution of estimated parameters.

The concepts and principles of development and application of regression-based models used for statistical simulation of hydrologic time series are introduced. The most general form of these models is called autoregressive integrated moving average modeling, ARIMA(p, d, q), where p is the autoregressive order, d is the order of differencing, and q is the moving average order. These models are, in theory, the most general class of models for forecasting a time series by assumption of data stationarity and normality. In this section, ARMA(p, q) family is described with AR(p) and MA(q) series of models, ARIMA(p, d, q) is discussed. It should be noted that trend removing and transformation should be performed before model development, if it is necessary.

3.16.1 ARMA(p, q) Model Identification

The first step in ARMA model development is to recognize the appropriate values of p and q. For this purpose, autocorrelation and partial autocorrelation functions (PACFs) are evaluated as follows:

3.16.1.1 Autocorrelation Function

The autocorrelations of a time series, $z_1, z_2, \ldots, z_k$, are defined as follows:

$$\rho_j = \frac{\text{cov}\left(z_t, z_{t-j}\right)}{\sigma^2(z_t)} = \frac{\gamma_j}{\gamma_o} \tag{3.82}$$

and

$$\gamma_j = \frac{1}{N}\sum_{t=j+1}^{N}(z_t - \bar{z})(z_{t-j} - \bar{z}), \tag{3.83}$$

where N is sample size and $\bar{z}$ is sample mean value. The plot of ρ_j against j is called the autocorrelation function (ACF).

Autocorrelation plots can be used for checking the randomness of a time series. For random series, the autocorrelations with different lags are near zero, and for nonrandom series, one or more of the autocorrelations will be significantly more than zero. Also, these plots can be used for determining the order of ARMA(p, g) models as discussed later.

There are some suggestions for generating confidence bands (determining the significance of autocorrelations). If an ARMA(p, g) model is assumed for the data, the following confidence bands are considered:

$$\left\{-u_{1-\alpha/2}\sqrt{\frac{\left(1+2\sum_{i=1}^{k} z_i^2\right)}{N}};+u_{1-\alpha/2}\sqrt{\frac{\left(1+2\sum_{i=1}^{k} z_i^2\right)}{N}}\right\}, \tag{3.84}$$

where $u_{1-\alpha/2}$ is the $1-\alpha/2$ quantile of the standard normal distribution, N is the sample size, α is the significance level, and k is the lag. In this case, the confidence bands increase as the lag increases.

3.16.1.2 Partial Autocorrelation Function (PACF)

PACF shows the time dependence structure of the series in a different way. The partial autocorrelation lag k can be considered as the correlation between z_t and z_{k+1}, adjusted for the intervening observations $z_2, \ldots, z_k$. The partial autocorrelation coefficient $\phi_p(p)$ in an ARMA(p, q) process of order p is a measure of the linear association between ρ_j and ρ_{j-p} for $j \le p$. $\phi_p(p)$ for $p = 1, 2, \ldots$ is the PACF. The pth autoregressive coefficient of the AR(p) model fitted to the autocorrelation coefficients is as follows:

$$\rho_j = \phi_1(p)\rho_{2-j} + \cdots + \phi_p(p)\rho_{p-j} \; j = 1, \ldots, p, \tag{3.85}$$

where $\phi_j(p)$ is the jth autoregressive coefficient of the ARIMA (p, q) model. The partial autocorrelation is given by the last coefficient $\phi_p(p)$, ($p = 1, 2, \ldots$).

$$\phi_1(p)\rho_0 + \phi_2(p)\rho_1 + \cdots + \phi_p(p)\rho_{p-1} = \rho_1, \tag{3.86}$$

$$\phi_1(p)\rho_1 + \phi_2(p)\rho_2 + \cdots + \phi_p(p)\rho_{p-2} = \rho_2, \tag{3.87}$$

$$\phi_1(p)\rho_2 + \phi_2(p)\rho_3 + \cdots + \phi_p(p)\rho_{p-3} = \rho_3, \tag{3.88}$$

$$\phi_1(p)\rho_{p-1} + \phi_2(p)\rho_{p-2} + \cdots + \phi_p(p)\rho_0 = \rho_p, \tag{3.89}$$

which can also be written as it is known as Yule Walker equations:

$$\begin{bmatrix} 1 & \rho_1 & \rho_2 & \cdots & \rho_{p-1} \\ \rho_1 & 1 & \rho_1 & \cdots & \rho_{p-2} \\ \rho_2 & \rho_1 & 1 & \cdots & \rho_{p-3} \\ \vdots & \vdots & \vdots & & \vdots \\ \rho_{p-1} & \rho_{p-2} & \rho_{p-3} & \cdots & 1 \end{bmatrix} \begin{bmatrix} \phi_1(p) \\ \phi_2(p) \\ \phi_3(p) \\ \vdots \\ \phi_p(p) \end{bmatrix} = \begin{bmatrix} \rho_1 \\ \rho_2 \\ \rho_3 \\ \vdots \\ \rho_p \end{bmatrix} \tag{3.90}$$

or

$$P_1\phi_p + \psi_p \rightarrow \phi_p = P_p^{-1}\psi_p,\ p = 1,2,\ldots \tag{3.91}$$

Thus, the PACF $\phi_p(p)$ is determined successively.

The PACF $\phi_p(p)$ may be also obtained recursively by means of Durbin's (1960) relations

$$\phi_1(1) = \rho_1,\ \ \phi_1(2) = \frac{\rho_1(1-\rho_2)}{(1-\rho_2^2)}\ \ \ \phi_2(2) = \frac{\rho_2 - \rho_1^2}{(1-\rho_1^2)}\ , \tag{3.92}$$

$$\phi_1(p) = \frac{\rho_p - \sum_{j=1}^{p-1}\phi_j(p-1)\rho_{p-j}}{1 - \sum_{j=1}^{p-1}\phi_j(p-1)\rho_j} \tag{3.93}$$

$$\phi_j(p) = \phi_{j-1}(p)\phi_{p-j}(p-1). \tag{3.94}$$

On the hypothesis that the process is AR(p), the estimated $\phi_p(p)$ for $k > p$ is asymptotically normal with a mean of zero and a variance $1/N$. Hence, the $1 - \alpha$ probability limits for zero partial autocorrelation may be determined by Box and Jenkins (1970):

$$\left\{-u_{1-\alpha/2}\big/\sqrt{N};\ \ +u_{1-\alpha/2}\big/\sqrt{N}\right\}, \tag{3.95}$$

where $u_{1-\alpha/2}$ is the $1 - \alpha/2$ quantile of the standard normal distribution, N is the sample size, and α is the significance level. The limits of Equation 3.95 may be used to give some guide as to whether theoretical partial autocorrelations are practically zero beyond a particular lag.

3.16.2 Autoregressive (AR) Models

Autoregressive (AR) models have been widely used in hydrologic time series modeling. They incorporate the correlation between time sequences of variables. These models are the simplest models and their development goes back to the application of pioneering Thomas Fiering and Markov lag one models. They can be classified into the following subsets:

- AR models with constant parameters, which are typically used for modeling of annual series
- AR models with timely variable parameters, which are typically used for modeling of seasonal (periodic) series

The basic form of the AR model of order p(AR(p)) with constant parameters is

$$z_t = \sum_{i=1}^{p}\phi_i(p)z_{t-i} + \varepsilon_t, \tag{3.96}$$

where z_t is time-dependent normal, and standardized series $N(0,\ 1)$, $\phi_i(p)$ are autoregressive coefficients, ε_t is the time-independent variable (white noise), and p is the order of autoregressive model. In order to identify whether an AR(p) is an appropriate model for a specific time series, it is necessary to estimate and investigate the behavior of the PACF of the series. It can be shown that the partial correlogram of an autoregressive process of order p has peaks at lags 1 through p and then cuts off. Hence, the PACF can be used to identify the p of an AR(p) model. The AR model with periodic parameters has the following form:

$$z_{v,\tau} = \sum_{i=1}^{P} \phi_{i,\tau}(p) z_{v,\tau-i} + \sigma_{\tau}(\varepsilon)\zeta_{v,\tau}, \tag{3.97}$$

where $z_{v\tau}$ is the normal and standardized value in year v and season τ, $\phi_{i,\tau}(p)$ are periodic autoregressive coefficients, and $\sigma_{\tau}(\varepsilon)$ is the periodic standard deviation of residuals. ζ_t is the standardized normal random variable. $z_{v,\tau}$ is estimated using seasonal mean and variances as follows:

$$z_{v,\tau} = \frac{x_{v,\tau} - \mu_{\tau}}{\sigma_{\tau}}, \tag{3.98}$$

where μ_{τ} and σ_{τ} are the mean and standard deviation of x in season τ. The parameter set of the model can be summarized as

$$\left\{\mu_{\tau}, \sigma_{\tau}, \phi_{1,\tau}(p), \ldots, \phi_{P,\tau}(p) = \sigma_{\tau}^{2}(\varepsilon), \tau = 1, \ldots, \eta\right\}, \tag{3.99}$$

where η is the total number of seasons.

$$\rho_k = \sum_{i=1}^{P} \hat{\phi}_{i,\tau}(p)\rho_{|k-i|,\tau-\min(k,i)}, \quad k \geq 0, \tag{3.100}$$

where ρ_k is the sample correlation coefficients of lag k. ^ stands for estimated value. The residual variance can be estimated using the following relation:

$$\hat{\sigma}_{\tau}^{2}(\varepsilon) = 1\sum_{i=1}^{P} \hat{\phi}_{j,\tau}(p)\hat{\rho}_{j,\tau}. \tag{3.101}$$

For AR(2) model,

$$\begin{cases} \rho_1 = \phi_1(2) + \phi_2(2)\rho_1 \\ \rho_2 = \phi_2(2) + \phi_2(2)\rho_1 \end{cases} \tag{3.102}$$

By solving the above equations,

$$\phi_1(2) = \frac{\rho_1(1-\rho_2)}{1-\rho_1^2} \tag{3.103}$$

and

$$\phi_2(2) = \frac{\rho(1-\rho_2)}{1-\rho_1^2} \tag{3.104}$$

which is the same as what was obtained. As can be seen, the above formulation is based on the standardized series, which can be obtained as follows:

$$z_t = \frac{x_t - \mu}{\sigma}, \tag{3.105}$$

where μ and σ are the mean and standard deviation of the series x_t. The parameter set of the model is

$$\{\mu, \sigma, \phi_1(p) = \phi_p(p)\sigma^2(\varepsilon)\}, \tag{3.106}$$

where $\sigma^2(\varepsilon)$ is the variance of the time-independent series. The model parameters can be estimated by solving the following linear equations, which are called Yule–Walker equations, simultaneously as was mentioned before:

$$\rho_i = \hat{\phi}_1(p)\rho_{i-1} = \hat{\phi}_2(p)\rho_{i-2} + \cdots + \hat{\phi}_p(p)\rho_{i-p}, \quad i \geq 1, \tag{3.107}$$

where ρ_i are the sample correlation coefficients of lag i. The parameter $\sigma^1(\varepsilon)$ can also be estimated using the following relation:

$$\hat{\sigma}^2(\varepsilon) = \left(1 - \sum_{i=1}^{P} \hat{\phi}_i(p)\rho_i\right), \tag{3.108}$$

where N is the number of data and $\hat{\sigma}^2$ is the sample variance.

The stationary condition must be met by the model parameters. For this purpose, the roots of the following equation should lie inside the unit circle (Yevjevich, 1972):

$$u^p - \hat{\phi}_1(p)u^{p-1} - \hat{\phi}_2(p)u^{p-1} - \cdots - \hat{\phi}_p(p) = 0. \tag{3.109}$$

In other words, we must have $|u_i|$. In order to forecast or generate annual AR models, the following relation can be used:

$$\hat{z}_t = \hat{\phi}_1(p)\hat{z}_{t-1} + \cdots + \hat{\phi}_P(p)\hat{z}_{t-P} + \hat{\sigma}(\varepsilon)\zeta_t. \tag{3.110}$$

Example 3.17

For an AR(2) model, the parameters have been estimated as $\phi_1(1) = 0.3$ and $\phi_2(2) = 00.4$. Check the parameters' stationary condition.

Solution:

The following expressions can be written:

$$u^2 - 0.3u - 0.4 = 0 \Rightarrow \begin{cases} u_1 = 0.8 \\ u_2 = -0.5 \end{cases}$$

The roots lie within the unit circle; therefore, the parameters pass the stationary condition.

Example 3.18

For a sample of 100-year normal and standardized annual inflows to a reservoir, an AR(2) model is selected. The first and second correlation coefficients are estimated as $\rho_1 = 0.65$ and $\rho_2 = 0.3$. Estimate the model parameters if the variance of normal and standardized inflow series is estimated as 1.8.

Solution:

It can be calculated that

$$\phi_1(2) = 0.79$$

and

$$\phi_2(2) = -0.21.$$

The variance of model residuals can also be estimated

$$\hat{\sigma}^2(\varepsilon) = \frac{100 \times 1.8}{(100-2)}(1.065 \times 0.79 + 0.3 \times 0.21) = 0.96$$

3.16.3 Moving Average Process

The autoregressive models can be used as an effective tool for modeling hydrologic time series such as streamflow in low-flow season, which is mainly supplied from groundwater and has low variations. However, previous studies have shown that the streamflows in high-flow season can be better formulated by adding a moving average component to the autoregressive component (Salas et al., 1988).

If the series, z_t, is dependent only on a finite number of previous values of a random variable, ε_t, then the process can be called a *moving average* (MA) process. The MA(q) process (the moving average model of order q) is formulated as follows:

$$z_t = \varepsilon_t - \theta_1\varepsilon_{t-1} - \theta_2\varepsilon_{t-2} - \theta_3\varepsilon_{t-3} - \cdots - \theta_q\varepsilon_{t-q}. \quad (3.111)$$

It can also be written as

$$z_t = -\sum_{j=0}^{q}\theta_j\varepsilon_{t-j}, \quad (\theta_0 = -1), \quad (3.112)$$

where $\theta_1,\ldots, \theta_q$ are q orders of MA(q) model parameters. The parameter set of the model can be summarized as

$$\{\mu, \theta_1, \ldots, \theta_q, \sigma^2(\varepsilon)\}. \quad (3.113)$$

The parameters of the model should satisfy the invertibility condition. For this purpose, the roots of the following polynomial should lie inside the unit circle:

$$u^q - \hat{\theta}_1 u^{q-2} - \hat{\theta}_2 u^{q-2} - \cdots - \hat{\theta}_q = 0. \quad (3.114)$$

Example 3.19

For an MA(2) model, the parameters have been estimated as $\theta_1 = 0.65$ and $\theta_2 = 0.3$. Check whether the parameters pass the invertibility condition.

Solution:

It can be written that

$$u^2 - 0.5u - 0.2 = 0. \Rightarrow \begin{cases} u_1 = 0.76 \\ u_2 = -0.26. \end{cases}$$

The roots lie within the unit circle; therefore, the parameters pass the invertibility condition.

3.16.4 Autoregressive Moving Average Modeling

A combination of an autoregressive model of order p and a moving average model of order q forms an ARMA model of order (p, q). The ARMA(p, q) model is formulated as follows:

$$z_t - \phi_1(p) z_{t-1} - \cdots - \phi_p(p) z_{t-p} = \theta_1 \varepsilon_{t-1} + \cdots + = \theta_q \varepsilon_{t-q}. \quad (3.115)$$

The ARMA(p, q) model can also be shown in the following compact form:

$$\phi(B) z_t = \theta(B) \varepsilon_t, \quad (3.116)$$

where $\phi(B)$ and θB are the pth and qth degree polynomials:

$$\phi(B) = 1 - \phi_1(p)B, - \cdots - \phi_p(p)\ B^p \quad (3.117)$$

$$\theta(B) = 1 + \theta_1 B + \cdots + \theta_q B^p. \quad (3.118)$$

The parameter set of ARMA(p, q) model can be summarized as

$$\{\mu, \theta_1, \ldots, \theta_q \phi_1(p), \ldots, \phi_1(p), \sigma^2(\varepsilon)\}. \quad (3.119)$$

The parameters of the ARMA(p, q) model should satisfy both the conditions of invertibility and stationarity. Some relationships between parameters and correlations ρ_k of ARMA processes of low order are given. After estimation of correlation coefficients with different lags, regarding the order of the selected ARIMA model, these equations can be used to estimate model parameters.

Example 3.20

For a time series with given autoregressive and partial autoregressive coefficients estimate the parameters of model ARMA(1, 1).

Solution:

Regarding Table 3.35, for model ARMA(1, 1), the following are obtained (Table 3.36):

$$\rho_1 = \frac{(1 - \theta_1 \phi_1(1))(\phi_1(1) - \theta_1)}{1 + \theta_1^2 - 2\phi_1(1)\theta_1}$$

$$\rho_k = \rho_1 (\phi_1(1))^{k-1}, \quad k \geq 2 \Rightarrow \rho_2 = \rho_1(1)$$

Using the above, the following is obtained:

$$\phi_1(1) = \frac{\rho_2}{\rho_1} = \frac{0.034}{0.157} = -0.22.$$

Replacing these values in the first equation, the following second-order equation based on θ_1 is obtained:

$$(\rho_1 - \phi_1(1))\theta_1^2 + (\phi_1(1)^2 - 2\rho_1\phi_1(1) + 1)\theta_1 + \rho_1 - \phi_1(1) = 0.$$

By solving this equation, two values (−0.3891 and −0.8282) are obtained, from which the value of $\theta_1 = -0.8282$ is finally selected. This value satisfies the invertibility condition (being between 1 and −1).

TABLE 3.35
Some Relationships between Parameters and Correlations ρ_k of ARMA Processes

ARMA (p, q)	Relationships
(1, 0)	$\rho_k = \left(\phi_1(1)\right)^k$
(2, 0)	$\rho_1 = \dfrac{\phi_1(2)}{1-\phi_2(2)}$
	$\rho_2 = \phi_2(2) + \dfrac{\left(\phi_1(2)\right)^2}{1-\phi_2(2)}$
	$\rho_k = \phi_2(2)\,\rho_{k-1} + \phi_1(2)\,\rho_{k-2}, \quad k \geq 2$
(1, 1)	$\rho_1 = \dfrac{\left(1-\theta_1\phi_1(1)\right)\left(\phi_1(1)1-\theta_1\right)}{1-\theta_1^2-2\phi_1(1)\theta_1}$
	$\rho_k = \rho_1\left(\phi_1(1)\right)^{k-1}, \quad k \geq 2$
(0, 2)	$\rho_1 = \dfrac{-\theta_1(1-\theta_2)}{1+\theta_1^2+\theta_2^2}$
	$\rho_2 = \dfrac{-\theta_2}{1+\theta_1^2+\theta_2^2}$
	$\rho_k = 0, \ k \geq 3$
(0, 1)	$\rho_1 = \dfrac{-\theta_1}{1+\theta_1^2}$
	$\rho_k = 0, \quad k \geq 2$

TABLE 3.36
Characteristic Behavior of AR, MA, and ARMA

Process	Autocorrelation	Partial Autocorrelation
AR (p)	Damped and infinite in extent exponentials and/or waves	Peaks at lags 1 through p and then cuts off
MA (q)	Peaks at lags 1 through q and then cuts off	Damped and infinite in extent exponentials and/or waves
ARMA (p, q)	Irregular in first q–p lags and then damped and infinite in extent exponentials and/or waves	Irregular in first p–q lags and then damped and infinite in extent exponentials and/or waves

3.16.5 Akaike's Information Criterion (AIC)

This criterion, which was first proposed by Akaike (1998), is usually used as the primary criterion for model selection. It considers the parsimony (using the least parameters in model development) in model building. The Akaike Information Criterion (AIC) among competing ARMA (p, q) models is defined as follows:

$$\text{AIC}(p,q) = N \cdot \ln\left(\hat{\sigma}^2(\varepsilon)\right) + 2 \tag{3.120}$$

where N is the sample size and $\hat{\sigma}^2(\varepsilon)$ is the maximum likelihood estimation of the residual variance. Based on this criterion, the model with minimum AIC is selected.

The goodness-of-fit tests should be applied to the model with minimum AIC to make sure the residuals are consistent with their expected behavior. If the model residuals do not pass the test, the models with higher values of AIC should be checked.

The concepts and principles of development and application of regression-based models used for statistical simulation of hydrologic time series are introduced. The most general form of these models is called autoregressive integrated moving average modeling, ARIMA(p, d, q). These models are, in theory, the most general class of models for forecasting a time series by assumption of data stationarity and normality. In this section, first, ARMA(p, q) family is described with AR(p) and MA(q) series of models, ARIMA(p, d, q) is discussed. It should be noted that trend removing and transformation should be performed before model development, if it is necessary. For details refer to Karamouz et al. (2012).

3.16.6 ARIMA Models Considerations

In ARIMA modeling, a model with one AR term, a first difference, and one MA term would have order (1, 1, 1). For the last model, ARIMA (1, 1, 1), a model with one AR term and one MA term is being applied to the variable $Z_t = X_t - X_{t-1}$. A first difference might be used to account for a linear trend in the data. The differencing order refers to successive first differences. For example, for a difference order = 2, the variable analyzed is $z_t = (x_t - x_{t-1}) - (x_{t-1} - x_{t-2})$, the first difference of first differences. This type of difference might account for a quadratic trend in the data.

Some well-known special cases arise naturally or are mathematically equivalent to other popular forecasting models. For example:

- An ARIMA(0, 1, 0) model (or $I(1)$ model) which is simply a *random walk.*
- An ARIMA(0, 1, 0) with a constant, which is a random walk with drift.
- An ARIMA(0, 0, 0) model is a *white noise* model.
- An ARIMA(0, 1, 2) model is a Damped Holt's model.
- An ARIMA(0, 1, 1) model without constant is a *basic exponential smoothing* model.
- An ARIMA(0, 2, 2) model which is equivalent to Holt's linear method with additive errors, or *double exponential smoothing.*

It is impossible to know the exact mathematical models of hydrologic time series, and only approximations are made for the inferred population. Since the exact model parameters in hydrology must be estimated from limited data, they will never be known precisely. Theoretically, time series modeling or stochastic modeling of hydrologic series refers to the estimations of models and their parameters from available data. Even though the improvement of time series modeling in hydrology has reached a level of complexity, unfortunately, simple methods are the basis of most time series modeling in practice. Here, there is an attempt to describe up-to-date advances in modeling in a systematic step-by-step approach, including various details and examples of modeling hydrologic time series. Generally, an important step towards modeling a hydrologic time series is the identification of the model composition, which could be a univariate or multivariate model, or a combination of a univariate and a disaggregation model, or a combination of a multivariate and a disaggregation model. After identifying the model composition, the type of the model(s) must be selected. Types of models could be AR, ARMA, ARIMA, or any other model that is available in stochastic hydrology. An important deciding factor in the selection of the type of model is the statistical characteristics of the samples of hydrologic series. For example, ARMA models are required rather than AR models for a series with low decaying correlograms (long memory). Lastly, other factors such as time limitation to solve a particular problem, funds for computer time, and availability of readymade programs of alternative models also contribute in the decision of selecting the type of model. Furthermore, the identification of the form of model must be done. The statistical characteristics of the historic time series are important for such model identification. When the model identification is done, the estimation of the parameters of the model must be made. An appropriate method of estimation, such as the method of moments and the (approximate) method of maximum likelihood, should be selected. The estimated parameters must be checked in order to comply with certain conditions of the model. An alternative form of the model must be selected if these conditions are not met. Refer to Salas et al. (1986) or Karamouz et al. (2012) for details of ARIMA model and periodic ARMA (seasonal) models.

Example 3.21

Find the appropriate ARIMA(p, d, q) model for discharge data of a reservoir (Table 3.37):

Solution:

The time series is not stationary and differencing is as such needed ($d > 0$). We start by finding out the order of differencing, d, using autocorrelation:

The time series is stationary at $d = 1$ where only the first lag is above the significance level. If your series is slightly underdifferenced, try adding an additional AR term and if it is slightly overdifferenced, maybe add an additional MA term. Knowing we should difference once, we go on to find out the order of AR, p. We get it by counting the number of lags above the level of significance in partial autocorrelation. The first lag is the only one vastly above the significance level and so $p = 1$. The autocorrelation function can tell the order of MA terms, q, needed to remove autocorrelation in the stationary series (Figure 3.16).

One lag can be found above the significance level and thus $q = 1$. Hence, the model is ARIMA (1, 1, 1).

TABLE 3.37
Data for Example 3.21

Date	Discharge (MCM)
2017.01.08	0.193
2017.01.15	0.174
2017.01.22	0.225
2017.01.29	0.141
2017.02.05	0.148
2017.02.12	0.126
2017.02.19	0.144
2017.02.26	0.141
2017.03.03	0.132
2017.03.10	0.102
2017.02.17	0.144
2017.03.24	0.138
2017.04.01	0.134
2017.04.08	0.166
2017.04.15	0.238
2017.04.22	0.157
2017.04.29	0.151
2017.05.06	0.233
2017.05.13	0.132
2017.05.20	0.172
2017.05.27	0.189

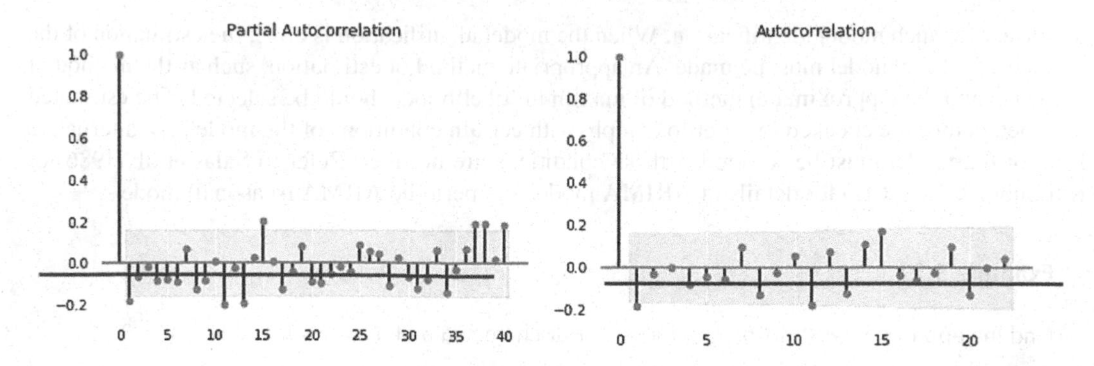

FIGURE 3.16 Autocorrelation and partial autocorrelation of Example 3.21.

3.17 CONCLUDING REMARKS

Specific characteristics of urban water hydrology are discussed in this chapter including the key factors of land use and land cover impacts, increased imperviousness that affects runoff coefficient, curve number, and even Manning coefficient; flow and drainage path complexities, travel time; interaction of stormwater with sanitary sewer; and the drawback of sewer overflow as a result of heavy rains. Losses or abstractions in hydrology include evaporation, evapotranspiration, and infiltration and are covered in Chapter 2 with more details in this chapter through the calculation of excess rainfall and rainfall runoff analysis. The classical issues and models of discrete and instantaneous unit hydrographs, total runoff hydrograph including base flow, S-curve, and the governing issues and methods of rainfall and runoff analysis are described. This includes the use of Nash model and Laplace transformation to estimate runoff hydrograph. The overland flow and routing methods that are combined in flooding and flood inundation analysis are also presented.

In this chapter, a practical overview of time series analysis and development of a well-known statistical ARMA models are discussed. Due to the stochastic nature of hydrologic events and variables, these models are successfully employed for simulation of these models' behaviors. However, in these models' applications, great attention should be given to the compatibility of the selected model and the data behavior. For this purpose, different tests are used to ensure these models' performance. The calibrated models can then be used for simulation of hydrologic events and generation of synthetic data. ARIMA models are also briefly discussed as a special class of ARMA models that can be used for forecasting behavior of hydrologic events.

This chapter's unique feature is that it provides all the background needed on science and applied techniques in hydrology in general and urban water in particular to handle the engineering and planning issues presented in the other chapters. The related water quality issues are discussed in Chapters 5–7.

PROBLEMS

1. Which one of the following is not a factor in controlling times of concentration of watershed runoff? (a) The drainage density, (b) the roughness of the flow surface, (c) the rainfall intensity, (d) the flow lengths, (e) all of the above are factors (Figure 3.17).
2. Using the least-squares UH, fit a gamma UH to the least squares UH. Area = 630 ha. Time increment = 30 minutes. Compare the least-squares and gamma synthetic UHs.
3. The characteristics of a given watershed are as follows:
   - Area: 800 km^2
   - Length of main channel: 35 km
   - Length of basin center to outlet point: 15 km
   - $C_t = 1.6$ and $C_p = 0.16$

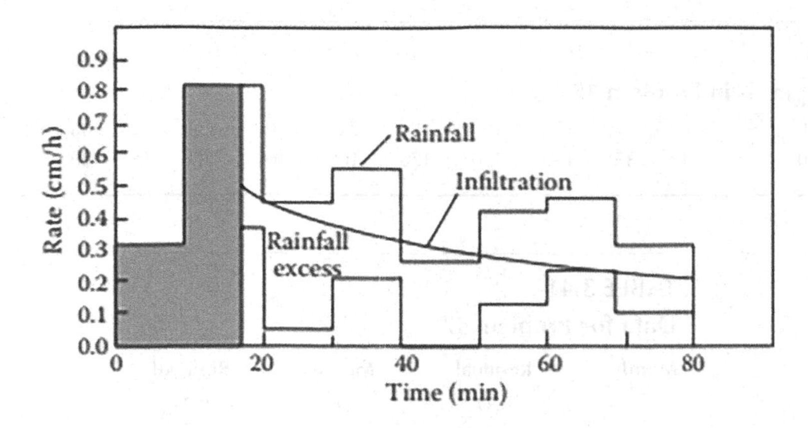

FIGURE 3.17 Rainfall and excess rainfall hyetographs of Problem 1.

4. Construct and draw an S-hydrograph using the 2-hour UH of the above exercise.
5. For an AR(2) model, the parameters have been estimated as $\phi_1(2) = 0.35$ and $\phi_2(2) = 0.5$. Check the parameters' stationary condition. If it is used for 100-year normal and standardized annual inflows to a reservoir and the first and second correlation coefficients are estimated as $\rho_1 = 0.6$ and $\rho_2 = 0.35$, estimate the model parameters if the variance of normal and standardized inflow series is estimated as 1.6.
6. For an MA(2) model, the parameters have been estimated as $\theta_1 = 0.8$ and $\theta_2 = 0.45$. Check whether the parameters pass the invertibility condition.
7. For a given month, a 300-acre lake has 15 cfs of inflow, 13 cfs of outflow, and a total storage reduction of 16 acre-ft. A station next to the lake recorded a total of 1.3 inches of precipitation for the lake for the month. Assuming that infiltration is insignificant for the lake, determine the evaporation loss, in inches, over the lake.
8. Compute the infiltration and cumulative infiltration after 1 hour of infiltration into a silt loam soil that initially had an effective saturation of 30% (*Note*: $\Delta\theta = (1 - S_e)\theta_e$).
9. Calculate the time of concentration of the watershed where the length of the main river is 3,000 m and the average slope is 0.02. If the curve number is estimated at 80, use the SCS equation for calculation.
10. A 6-hour UH of a catchment is triangular in shape with a base width of 53 hours and a peak ordinate of 25 m^3/s. Calculate the equilibrium discharge of an S-curve obtained by this 6-hour unit hydrograph.
11. A 2-hour UH is given by a rectangle whose base is 4 hours and height is 0.32 cm/h. Derive a 4-hour UH using the given 2-hour UH.
12. The ARMA(1, 1) model is fitted to the time series of streamflow at a station with parameters $\phi_1(1) = -0.6$ and $\theta_1 = -0.4$. Plot the autocorrelation functions of the time series from lags 1 to 5.
13. The peak of flood hydrograph due to a 3 h duration isolated storm in a catchment is 300 m3/s. The total depth of rainfall is 6.3 cm. Assuming an average infiltration loss of 0.3 cm/h and a constant baseflow of 18 m3/s, estimate the peak of the 3-hour unit hydrograph (UH) of this catchment. If the area of catchments is 420 km2, determine the base width of the 3-hour unit hydrograph by assuming it to be triangular.
14. Write the mathematical expression of order 1 and order 2 nonseasonal ARIMA. Why is the seasonal differencing done? Write the mathematical expression of the second-order seasonal differencing for monthly time series (assume 12 for the number of seasons).
15. The ordinates of the 6-hour unit hydrograph of a catchment are given in Table 3.40. Calculate the ordinate of the DRH due to a rainfall excess of 2.5 cm occurring in 6 hours.

TABLE 3.40
Unit Hydrograph in Problem 15

Time (h)	0	3	6	9	12	15	18	24	30	36	42	48	54	60	66
UH (m^3/s)	0	25	60	85	120	150	170	160	100	55	35	20	15	10	0

TABLE 3.41
Data for Problem 17

Month	Residual	Month	Residual
1	0.87	11	0.78
2	0.28	12	0.82
3	0.32	13	0.14
4	0.34	14	0.22
5	0.33	15	0.65
6	0.15	16	0.48
7	0.18	17	0.32
8	0.17	18	0.07
9	0.21	19	0.24
10	0.22	20	0.25

16. Develop ARIMA (1,1,1) (0,1,2)4. Consider that this model is used to predict the rainfall during a year, what would be the scale of the model outputs (daily, monthly, or seasonally)? At most, how many data are effectively used in order to predict the present rainfall values?
17. A series of residuals of a model fitted to precipitation data in a station in 20 time intervals is given in Table 3.41. Use the turning point test and comment on the randomness of the residuals.

REFERENCES

Akaike, H. (1998). Information theory and an extension of the maximum likelihood principle. In Akaike, H., Parzen, E., Tanabe, K., and Kitagawa, G. (eds.) *Selected Papers of Hirotugu Akaike* (Vol. I, pp. 199–213). Springer, New York.

Box, G.E.P. (1970). GM Jenkins' *Time Series Analysis.* Forecasting and Control'Holden-Day'San Francisco.

Chow, V.T., Maidment, D.R., & Mays, L.W. (1988). *Applied Hydrology.* McGraw-Hill Series in Water Resources and Environmental Engineering.

City of Olympia Public Works Department (COPWD). (1995). Impervious surface reduction study, Olympia, WA.

Conley, L.C. and McCuen, R.H. (1997). Modified critical values for Spearman's test of serial correlation. *Journal of Hydrological Engineering ASCE,* 2(3), 133–135.

Durbin, J. (1960). Estimation of parameters in time-series regression models. *Journal of the Royal Statistical Society: Series B (Methodological),* 22(1), 139–153.

Eslamian, S.S., Gilroy, K.L., and McCuen, R.H. (2011). Climate change detection and modeling in hydrology. In Blanco, J. and Kheradmand, H. (eds.) *Climate Change: Research and Technology for Adaptation and Mitigation* (pp. 87–100). IntechOpen, London, 504 pp.

Gray, D.M. (1962). Derivation of hydrographs for small watersheds from measurable physical characteristics. *Research Bulletin (Iowa Agriculture and Home Economics Experiment Station),* 506, 514–570.

Green, W.H. and Ampt, G.A. (1974). Studies on soil physics. 1: The flow of air and water through soils. *Journal of Agriculture Science,* 9, 1–24.

Horton, R.E. (1919). Rainfall interception. *Monthly Weather Review,* 47, 603–623.

Horton, R.E. (1940). An approach toward physical interpretation of infiltration capacity. *Soil Science Society Proceedings,* 5, 399–417.

James, W. and Scheckenberger, R. (1984). RAINPAK: A program package for analysis of storm dynamics in computing rainfall dynamics. *Proceedings of Stormwater and Water Quality Model Users Group Meeting*, Detroit, Michigan. EPA-600/9-85-003 (NTIS PB85-168003/AS), Environmental Protection Agency, Athens, GA, April.

James, W., Nimmrichter, P., James, R., and Scheckenberger, R.B. (2002). Robustness of the Rainpak algorithm for storm direction and speed. *Journal of Water Management Modeling*, 208–206. doi: 10.14 796/JWMM.R208-06.

Karamouz, M., Nazif, S., and Falahi, M. (2012). *Hydrology and Hydroclimatology: Principles and Applications*. CRC Press, Boca Raton, FL.

Kibler, D.F., Akan, A.O., Aron, G., Burke, C.B., Glidden, M.W., and McCuen, R.H. (1996). *Urban Hydrology* (2nd ed.). ASCE Manual of Practice, No. 28, New York.

Kluiteneberg, E. (1994). Determination of impervious area and directly connected impervious area. Memo for the Wayne County Rouge Program Office, Detroit, MI.

Marsalek, J., Cisneros, B. J., Karamouz, M., Malmquist, P. A., Goldenfum, J. A., & Chocat, B. (2008). *Urban Water Cycle Processes and Interactions: Urban Water Series-UNESCO-IHP* (Vol. 2). CRC Press, Boca Roten, FL.

McCarthy, G.T. (1938). The unit hydrograph and flood routing. *In Proceedings of Conference of North Atlantic Division*, US Army Corps of Engineers, Washington, DC, pp. 608–609.

McCuen, R.H. (2005). *Hydrologic Analysis and Design* (3rd ed.). Pearson Prentice Hall, Upper Saddle River, NJ.

McPherson, C.B. (1978). *Mainstream and Critical Position,* University of Toronto Press, Toronto.

Nash, J.E. (1959). Systematic determination of unit hydrograph parameters. *Journal of Geophysical Research*, 64(1), 111–115.

Northern Virginia Planning District Commission (NVPDC). (1980). *Guidebook for Screening Urban Nonpoint Pollution Management Strategies*. Northern Virginia Planning District Commission, Falls Church, VA.

Rawls, W.J. (2006). Infiltration properties. In Lal, R. (ed.) *Encyclopedia of Soil Science* (pp. 689–692). Marcel Decker, New York.

Salas, J.D., Delleur, J.W., Yevjevich, V., and Lane, W.L. (1988). *Applied Modeling of Hydrological Time Series*. Water Resources Publications, Littleton, CO, 484 pp.

Sarma, P.G.S., Delleur, J.W., and Rao, A.R. (1969). A program in urban hydrology. Part II. Technical Report No. 9, Waster Research Center, Purdue University.

Soil Conservation Service (SCS). (1986). *Urban Hydrology for Small Watersheds*, Technical Release 55 (2nd ed.). U.S. Department of Agriculture, Springfield, VA. (Microcomputer version 1.11. NTIS PB87-101598).

Stankowski, S.J. (1974). Magnitude and frequency of floods in New Jersey with effects of urbanization. Special Report 38, USGS, Water Resources Division, Trenton, NJ.

Tholin, A.L. and Keifer, C.J. (1960). Hydrology of urban runoff. *Transactions, ASCE*, 125, 1308–1379.

USDA. (1986). *Urban Hydrology for Small Watersheds*, Technical Release 55 (2nd ed.). Natural Resources Conservation Service (NRCS), United States Department of Agriculture, Washington, DC.

Viessman, W., Jr., Lewis, G.L., and Knapp, J.W. (1989). *Introduction to Hydrology*. Harper & Row, New York.

Yevjevich, V.M. (1972). *Structural analysis of hydrologic time series*. Doctoral dissertation, Colorado State University, Libraries.

4 Urban Water Hydraulics

4.1 INTRODUCTION

Water conveyance in urban area is more complicated than undeveloped area because of the complexity of the paths that rain and stormwater have taken and many interdependencies of water infrastructure with other infrastructures. The urban water hydraulics principles are discussed in this chapter, and the characteristics of hydraulic structures of different applications are discussed in Chapters 5–7. Measuring the effect of geomorphic changes (land use driven) on urban water is essential to mitigate the potential effects. The urban water management will be more effective if risk assessments included geomorphological changes to underpin nature-based management approaches and land use alterations are addressed. Changes in geomorphic process regimes can also be triggered by extreme events. Implementing geomorphological adaptation strategies will enable communities to develop more resilient, less vulnerable socioeconomic systems fit for an age of climate extremes. The outcome of such approach will be of interest to landscape architects/planners and regulators because of the complexities related to stormwater collection and flood management.

The terms river or channel morphology are used to describe the change in the shapes of river and channels and how their direction is changing. Analytical approach to river morphology are based on the physical principles for the hydraulics of flow and sediment transport processes. River morphology and changes have a long history in hydraulic engineering mostly due to movement of river channels affected by scouring and sedimentation mostly in flooding seasons in undeveloped watersheds. Land use change was not often considered as the governing factor. In urban area with overland flow moving to drainage channels in most areas, morphological changes are mainly due to change in land use. A continuous study of relationships between channels properties, their state of repair, and potential stormwater hazards should be undertaken in urban area.

There is a unique opportunity to assess the characteristics of each channel that can be greatly assisted with digital elevation models (DEMs) of high resolution such as LiDAR (Light Detection and Ranging). A terrain analysis can be done to find the relationships between channel form and flooding hazard. This will open the avenue for rapid risk assessment of ungauged areas. It gives the basis for a better understanding of the human role in changing urban water pathways. It allows developing a vulnerability assessment of hydrologic response through hydraulic components and structure to climate and human drivers on stormwater hazard and to identify increased flood frequency additional strain on the existing water body banks, infrastructure, and communities. DEM applications as well as DEM error analysis are discussed in Chapter 13 and in a number of case studies.

In this chapter, some basic concepts of hydraulic design of urban drainage and water distribution systems are introduced. This chapter is particularly useful for developing urban areas. The presented design procedures complement the hydraulic structure issues discussed in Chapters 5–7.

4.2 CHANNEL GEOMORPHOLOGY

In this section, the main attributes of channel morphology are explained with emphasis on the differences between their urban characteristics with emphasis on land use and landscape morphology. Two main questions to be answered are:

DOI: 10.1201/9781003241744-4

1. Which are the critical flow attributes that dominate the range and extent of geomorphic adjustments in urban settings?
2. What is the influence of these changes in channel morphology on flood risk assessment?

For more details of river and channel morphology refer to Chapter 5 of Karamouz et al. (2012).

4.2.1 Length of a Channel

Besides drainage area and watershed length that are discussed in Chapter 3, channel length is used frequently in hydrologic and hydraulics computations. The channel length is computed in the following two computational schemes:

1. The distance is measured from the watershed outlet to the end of the channel, along the main channel as indicated on a map, which is denoted as Le.
2. It is measured as the distance between two points located at the 10% and 85% of the distance along the main channel from the outlet, which is denoted as L10–85.

Figure 4.1 illustrates these definitions along with the watershed length. An extension of a line on the map from the end of the main channel to the divide, L, is required to obtain the length of the watershed, which requires some subjective assessment and is often a source of inaccuracy.

A measure of subjectivity is also involved in the definitions for channel length because the endpoint of the channel is dependent on the way that map was drawn; the location of the channel end depends on the level of flow at the time the map was compiled. The final design bears an unknown degree of inaccuracy due to these subjective assessments. Knowing exactly which definition was used in the development of a design aid is very important. For example, a bias may be introduced into the design if one of the definitions was used in developing the design method and the other is used in computing the channel length.

A stream channel may not be included in very small watersheds such as a section of overland flow or a section where the flow is in a swale or gully. As the size of the watershed increases, channel flow is in control and the watershed and channel lengths are essentially the same. Depending on the type of design problem, the appropriate measure of length can be chosen. Many urban channels are following street directions going through the complexity of urban landscape.

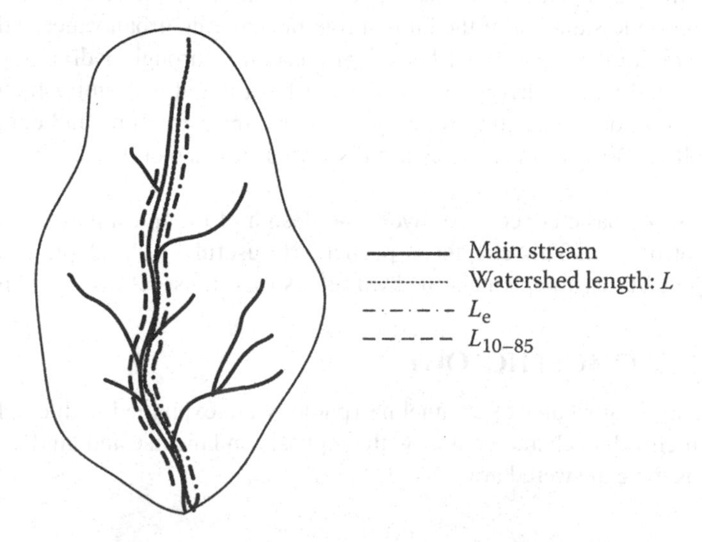

FIGURE 4.1 Delineation of watershed length and main stream length.

Example 4.1

The channel system shown in Figure 4.2 can be viewed as consisting of three reaches. Subareas 1, 2, and 3 have reaches of lengths 4,940, 2,440, and 3,670 m, respectively. As it is shown in Figure 4.2, the watershed lengths are defined as subwatershed lengths, and the length measurements for subareas 1 and 2 must be extended to the basin divide. For subwatersheds 1 and 2, the lengths are 6,810 and 4,875 m, respectively. The length of subwatershed 3 is defined as the flow path from the most distant point on the watershed divide to the outlet, and this point is the northernmost point, which is on the boundary separating subareas 2 and 3. This flow path is made up of 2,550 m of channel and 2,450 m of overland and gully flow for a total watershed length of 5,000 m. Determine the total basin length.

Solution:

Discussing time parameters reveals the importance of distinguishing between the channel length and the length of the subwatersheds. In the case of the demand for the subwatershed being indistinguishable from the need to estimate the total watershed length, the length would be the sum of the channel length for the reach through subarea 3 (e.g., 3,670 m) and the watershed length of subarea 1 (i.e., 6,810 m), or 10,480 m. This value provides the maximum length from the watershed boundary to its outlet.

Example 4.2

The total area of a watershed, divided into three subwatersheds as shown in Figure 4.3, is 26 km^2. The lengths of subbasins 1, 2, and 3 are 1,304, 342, and 252 m, respectively. If the lengths of the main rivers in these subbasins are 1,200, 311, and 220 m, respectively, calculate the elongation ratio for this basin.

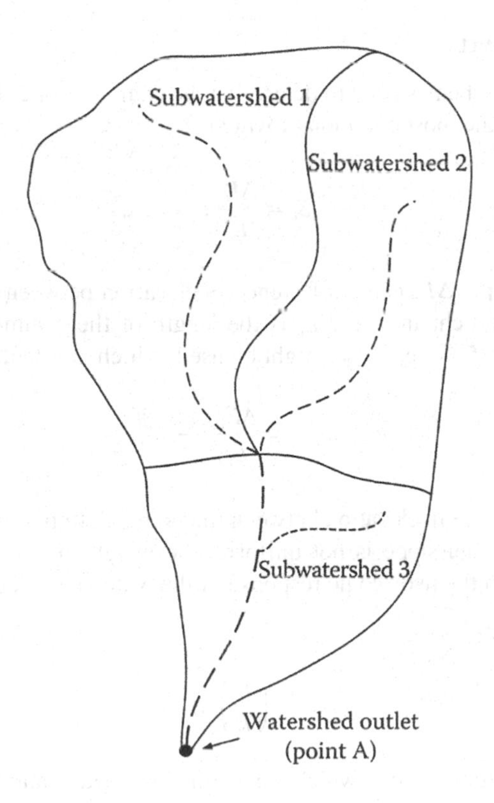

FIGURE 4.2 Watershed in Example 4.1.

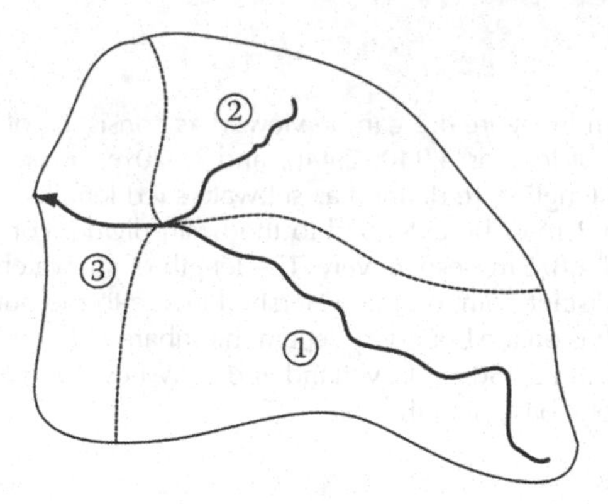

FIGURE 4.3 Watershed in Example 4.2.

Solution:

The length of subbasin 1 is 1,304 m, and it is used because the main river in subbasin 1 is less than the real length of the subbasin; moreover, the length of the main river in subbasin 3 along the length of the basin is 220 m. Hence, the sum of the length of subbasins 1 and 3 (1,524 m) is greater than the sum of the length of subbasins 2 and 3, which is 562 m. Consequently, R_e is calculated as follows:

$$R_e = \frac{2}{L_m} \times \left(\frac{A}{\Pi}\right)^{0.5} = \frac{2}{1{,}304 + 220} \times \left(\frac{26}{\Pi}\right)^{0.5} \rightarrow R_e = 3.78.$$

4.2.2 Slope of a Channel

Most of the computational schemes need to determine the channel slope. There are different definitions of the channel slope, the most common of which is

$$S_c = \frac{\Delta E_c}{L_c}, \tag{4.1}$$

where S_c is the channel slope, ΔE_c is the difference in elevation between the points that define the upper and lower ends of the channel, and L_c is the length of the channel between the same two points. In addition, the 10–85 slope, S_{10-85}, might be used, which is calculated as follows:

$$S_{10-85} = \frac{\Delta E_{10-85}}{L_{10-85}}, \tag{4.2}$$

where ΔE_{10-85} is the difference in elevation between the points defining the channel length L_{10-85}.

In cases where the channel slope is not uniform, the weighted slope index is used to better reflect the effect of slope on the hydrologic response of the watershed. A channel slope index, S_i, is defined as

$$S_i = \left(\frac{n}{k}\right)^2, \tag{4.3}$$

where n is the number of segments into which the channel is divided and k is given by

$$k=\sum_{i=1}^{n}\frac{1}{\left(\Delta e_i/l_i\right)^{0.5}}, \tag{4.4}$$

where Δe_i is the difference in elevation between the endpoints of channel segment i, and l_i is the length of segment i. The slope is considered relatively constant over the segmented channel.

Example 4.3

For the given watershed in Figure 4.2, calculate the channel and watershed slope based on the altitude data given in Table 4.1.

Solution:

For computing the channel slope of the watershed in Figure 4.2, the channel reach in each subwatershed is used. Channel and watershed slopes are given in Table 4.1 for both the subdivided watershed and the watershed as a whole. Generally, as a rule, the watershed slopes are greater than the channel slopes, because the side slopes of the watershed are almost always steeper than the channel.

Channel slope has a profound effect on the velocity of flow in a channel and, consequently, on the flow characteristics of runoff from a drainage basin. The importance of channel slope lies in two areas: (a) it helps in the determination of discharge and velocity using the Manning or Chezy equation and (b) it can be used as a variable in multivariate analysis to determine the amount of influence accounted for by the channel slope. Because the slope varies longitudinally, an average value of slope must be determined.

Example 4.4

Estimate the average slope of the watershed shown in Figure 4.4 regarding Table 4.2 if the watershed is divided into five intervals.

Solution:

$$K=\sum_{j=1}^{n}\frac{1}{\left(\frac{\Delta e_i}{L_i}\right)^{0.5}}=\left(\frac{1}{0.091^{0.5}}+\frac{1}{0.035^{0.5}}+\frac{1}{0.015^{0.5}}+\frac{1}{0.110^{0.5}}+\frac{1}{0.042^{0.5}}\right)=24.72$$

$$S_i=\left(\frac{n}{K}\right)^2=\left(\frac{5}{24.72}\right)^2=0.041.$$

Example 4.5

Estimate the average slope of Figure 4.4 without any interval.

Solution:

$$\Delta E=3{,}758-2{,}994=764$$

$$S_e=\frac{\Delta E}{L_e}=\frac{764}{3{,}200+1{,}700+2{,}000+2{,}300+3{,}000}=0.062.$$

TABLE 4.1
Length and Slope of the Watershed in Figure 4.2

	Length (m)		Channel Elevation (m)			Watershed Elevation (m)			Slope (%)	
Area	**Channel**	**Watershed**	**Upper**	**Lower**	**Difference**	**Upper**	**Lower**	**Difference**	**Channel**	**Watershed**
Sub 1	4,940	6,810	450	340	110	608	340	268	0.022	0.039
Sub 2	2,440	4,875	400	340	60	545	340	205	0.025	0.042
Sub 3	3,670	5,000	340	295	45	545	295	250	0.012	0.050
Total	8,610	10,480	450	295	155	608	295	313	0.018	0.030

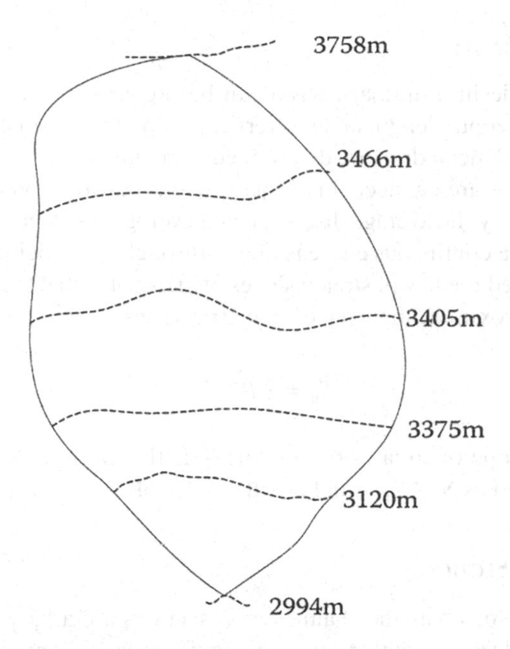

FIGURE 4.4 Watershed in Example 4.4.

TABLE 4.2
Characteristics of Watershed in Example 4.4

Interval Length (m)	Interval Number
3,200	1
1,700	2
2,000	3
2,300	4
3,000	5

TABLE 4.3
Calculated Slopes for Watershed in Example 4.4

Interval Number	Δe_i	Interval Slope
1	292	0.091
2	61	0.035
3	30	0.015
4	255	0.110
5	126	0.042

The method that is used in Example 4.4 is more accurate because it is divided into some intervals and so the slope of the basin for each interval is calculated independently. On the other hand, in Example 4.5, the difference in elevation in the entire basin is calculated. As it is shown in Table 4.3, the slope in some intervals is sharper than the others. Hence, the result of the first method is more reliable than the second one.

4.2.3 Law of Stream Slopes

A composite stream profile in a drainage basin can be prepared in the following way. For each first-order stream, the horizontal length and the vertical drop of the segment are determined. Then, mean horizontal length and mean drop are determined. Streams of all orders follow this procedure. The triangles for each order are connected in sequence to produce a composite profile. Since each segment slope is governed by the average discharge and average sediment load, a segmented profile looks logical even though a continuous curve is drawn through these points.

Horton (1945) introduced the law of stream slopes, which states that the average slope of streams of each order tends to approximate an inverse geometric series.

$$\bar{S}_w = \bar{S}_1 R_s^{W-w}, \tag{4.5}$$

where $\bar{S}_w$ is the average slope of streams of order w; $\bar{S}_1$ is the average slope of first-order streams; R_s is the slope ratio, defined as $\bar{S}_w / \bar{S}_{w-1}$; and w is the order of the basin. The value of R_s is ~0.55.

4.2.4 Channel Cross Section

There are a number of reasons why the channel cross sections are a very important part of hydrologic analysis and design. Design problems usually require cross-section information, including the cross-sectional area. Other important characteristics are the wetted perimeter, slope, roughness, and average velocity. Stream cross sections can take on a wide variety of shapes and sizes, which are very apparent if one walks along a river or stream, even for short distances.

V-shaped cross sections are often in upland areas for small streams. As the stream drains larger areas, past floods have sculptured out larger, often rectangular or trapezoidal sections. Relatively small V- or U-shaped channels are frequent in many rivers that carry the runoff from small storms and the runoff during periods between storm events. This small channel area is often part of a much larger flat area on one side or on both sides of the channel; this flat area is called the floodplain, which is the area that is covered with water during times of higher discharges.

The low-flow cross section is approximately trapezoidal; however, it represents a small portion of the total cross section that is used to pass large flood events. The floodplain could easily be represented as a rectangular section due to its low side slopes. Cross sections can change shape during flood events on account of channel instability and erosional processes, and where such changes take place, they should be accounted for in developing a discharge rating table. Development within the floodplain may also change the channel cross sections.

The width of the section was decreased due to the abutments on each side and the four twin piers that support the bridge; in the meantime, the low-flow section remains unchanged. From the continuity equation, we know that for a given discharge, either the average velocity will have to increase because of the reduced cross-sectional area created by the bridge or the flood surface elevation will increase.

In practice, it is required to calculate the flood surface elevations along the entire section of the channel affected by the structure in design problems. For example, at the bridge, the cross section changes and it affects the flood profile both above and below the bridge, which should be considered in flood management practices.

4.2.5 Channel Roughness

The characteristics of runoff are affected by the roughness of a surface, whether the water is on the surface of the watershed or in the channel. The surface roughness retards the flow with respect to the hydrologic cycle. In case of overland flow, increasing roughness delays the runoff and should increase the potential for infiltration. The erosion amount decreases due to reduced velocity caused by an increase in roughness. The general effects of roughness on a channel flow are similar to those for overland flow.

A number of hydraulic computations require Manning's roughness coefficient (n), and it is a necessary input in floodplain delineation and a number of runoff time estimation methods. Moreover, it is used in the design of stable channel systems. Hence, it is an important input.

There are a number of methods in estimating the roughness coefficient. First, a series of pictures of stream channels are provided in books with a recommended n value for each picture; therefore, by comparing the roughness characteristics of the channel with the pictures that appear most similar, a value of n can be obtained for any natural channel. Second, typical or average values of n for various channel conditions are available in tables similar to Table 4.4. However, the degree of homogeneity of channel conditions and the degree of specificity of the table determine the accuracy of n values from such tables. Due to their simplicity and not being highly inaccurate, the picture comparison and tabular look-up methods are frequently used.

TABLE 4.4
Recommended Design Values of Manning's Roughness Coefficient, *n*, Used for Channel Flow Analysis

	Manning *n* Range
a. Unlined Open Channels	
A. Earth. Uniform section	
1. Clean recently completed	0.016–0.018
2. After weathering	0.018–0.020
3. With short grass, few weeds	0.022–0.027
4. In graveled soil, uniform section, clean	0.022–0.025
B. Earth. Fairly Uniform Section	
1. No vegetation	0.022–0.025
2. Grass. Some weeds	0.025–0.030
3. Dense weeds or aquatic plants in deep channels	0.030–0.035
4. Sides clean gravel bottom	0.025 0.030
5. Sides clean cobble bottom	0.030–0.040
C. Dragline Excavated or Dredged	
1. No vegetation	0.028–0.033
2. Light brush on banks	0.035–0.050
D. Rock	
1. Based on design section	0.035
2. Based on actual mean section	
a. Smooth and uniform	0.035–0.040
b. Jagged and irregular	0.040–0.045
E. Channels Not Maintained, Weeds and Brush Uncut	
1. Dense weeds, high as flow depth	0.08–0.12
2. Clean bottom, brush on sides	0.05–0.08
3. Clean bottom brush on sides, highest stage of flow	0.07–0.11
4. Dense brush, high stage	0.10–014

(*Continued*)

TABLE 4.4 (*Continued*)
Recommended Design Values of Manning's Roughness Coefficient, *n*, Used for Channel Flow Analysis

	Manning *n* Range
b. Roadside Channels and Swales with Maintained Vegetation (Values Shown Are for Velocities of 0.6 and 1.8 m/s)	
A. Depth of Flow, up to 0.2 m	
1. Bermuda grass, Kentucky bluegrass, buffalo grass	
a. Mowed to 5 cm	0.07–0.045
b. Length 10–15 cm	0.09–0.05
2. Good stand, any grass	
a. Length about 30 cm	0.18–0.09
b. Length about 60 cm	0.30–0.15
3. Fair stand, any grass	
a. Length about 30 cm	0.14–0.08
b. Length about 30 cm	0.25–0.13
B. Depth of Flow, 0.2–0.45 m	
1. Bermuda grass, Kentucky bluegrass, buffalo grass	
a. Mowed to 5 cm	0.05–0.035
b. Length 10–15 cm	0.06–0.04
2. Good stand, any grass	
a. Length about 30 cm	0.12–0.07
b. Length about 60 cm	0.20–0.10
3. Fair stand, any grass	
a. Length about 30 cm	0.10–0.06
b. Length about 60 cm	0.17–0.09
c. Natural Stream Channels	
A. Minor Streams (Surface Width at Flood Stage <35 m)	
1. Fairly regular section	
a. Some grass and weeds, little or no brush	0.030–0.035
b. Dense growth of weeds, depth of flow materially greater than weed height	0.035–0.05
c. Some weeds. Light brush on bank	0.04–0.05
d. Some weeds. Heavy brush on bank	0.05–0.07
e. Some weeds. Dense willows on bank	0.06–0.08
f. For trees within channel, with branches submerged at high stage	0.01–0.10
2. Irregular sections, with pools, slight channel meander	0.01–0.02
3. Mountain streams, no vegetation in channel, bank usually steep, trees and brush along banks submerged at high stage	
a. Bottom of gravel, cobbles, and few boulders	0.04–0.05
b. Bottom of cobbles, with large boulders	0.05–0.07
B. Floodplains (Adjacent to Natural Streams)	
1. Pasture, no brush	
a. Short grass	0.030–0.035
b. High grass	0.035–0.05
2. Cultivated areas	
a. No crop	0.03–0.04

(*Continued*)

TABLE 4.4 (*Continued*)
Recommended Design Values of Manning's Roughness Coefficient, *n*, Used for Channel Flow Analysis

	Manning *n* Range
b. Mature row crops	0.032–0.045
c. Mature field crops	0.04–0.05
3. Heavy weeds, scattered brush	0.05–0.07
4. Light brush and trees	
a. Winter	0.05–0.06
b. Summer	0.06–0.08
5. Medium to dense brush	
a. Winter	0.07–0.11
b. Summer	0.10–0.16
6. Dense willows, summer, not bent over by current	0.15–0.20
7. Cleared land with tree stumps, 0.4–0.6 per km^2	
a. No sprouts	0.04–0.05
b. With heavy growth of sprouts	0.06–0.08
8. Heavy stand of timber, a few down trees, little undergrowth	
a. Flood depth below branches	0.10–0.12
b. Flood depth reaches branches	0.12–0.16
C. Major Stream (surface width at flood stage more than 35 m: Roughness coefficient is usually less than that for minor streams of similar description on account of less effective resistance offered by irregular banks or vegetation on banks; values of n may be somewhat reduced. The value of n for larger streams of most regular sections, with no boulders or brush, may be in the range shown.)	0.028–0.033

Source: Adapted from McCuen, R., *Hydrologic Analysis and Design* (3rd ed.), Prentice-Hall, London, UK, 2005.

4.2.6 Urban Morphology Challenges

In urban area with overland flow in most areas, morphological changes are mainly due to change in land use. There are more channels than rivers that are stabilized but still experience changes as the city grows. Therefore, a continuous study of relationships between channel properties and stormwater hazards should be undertaken in urban area.

This provides a unique opportunity to assess the physiographic characteristics of each channel that can be greatly assisted with DEMs of high resolution such as LiDAR. Hence, if the relationships between channel form and stormwater hazard can be directly derived from terrain analysis products, it will open the avenue for rapid risk assessment of ungauged areas. It also provides a framework to disentangle climate and human drivers on stormwater hazard. It gives the basis for a better understanding of the human role in changing urban water channels. It allows developing a vulnerability assessment of hydrologic response to climate and human impact and to identify floods whose increased frequency could put additional strain on the existing riverbanks, infrastructure, and human settlements. Many of these issues, challenges, and use of digital data breakthroughs are discussed in Chapters 12–14 with snapshots of case studies.

As for answering the questions raised earlier in this chapter, that is, the critical flow attributes that dominate the range and extent of geomorphic adjustments in urban settings; and the influence of these changes in channel morphology on flood risk assessment, refer to different topics related to environmental visualization and digital data in this book and specifically in Chapter 14. Also find related topics to risk and uncertainty in Chapter 10 and partial frequency analysis and understanding flood hazard in Chapter 13.

4.3 TRAVEL TIME

The volume of flood runoff can be reflected by various watershed and hydrometeorological characteristics; for example, the product of the drainage area and the depth of rainfall give a volume of water that is potentially available for runoff, but volume alone is not adequate for many design problems. As the dimensions of discharge indicate (L^3/T), time is an important element in hydrologic and hydraulic design. However, a given volume of water may or may not present a flood hazard; in fact, the hazard will depend on the time distribution of the flood runoff. Flood damage occurs if a significant portion of the total volume passes a given location at about the same time.

Conversely, flood damage will be minimal in case of distribution of the total volume over a relatively long period of time. Instantaneous inundation causes major damage such as when a car is submerged, but an agricultural crop field does not usually experience major damage in short-duration inundation. Therefore, duration of flooding is an important factor.

In most hydraulic design, watershed characteristics are required that reflect the timing of runoff and time to peak value. A number of time parameters have been developed. Hydrologic and hydraulic models commonly use several time parameters such as the time of concentration, the time lag, and reach travel time based on average velocity.

The importance of time in hydraulic design is dependent on the type of problem. On small watershed designs, such as for the design of either rooftop drains or street-drainage inlets, and cutters, the time parameter may be an indicator of three parameters: the intensity and volume of rainfall, average velocity, and the degree to which the rainfall will be attenuated. In small watersheds, since the storages of the hydrologic processes have minimal effects, a short time period would suggest little attenuation of rainfall intensity. On larger watershed designs, time parameters may be representative of watershed storage and the effect of storage on the time distribution of runoff. The shape and the time distribution of the runoff hydrograph are directly affected by the watershed storage. For designs where the time variation of flood runoff is routed through channel reaches, the effect of channel storage on the attenuation of the flood discharge is reflected by the reach travel time.

Errors in a time parameter will cause errors in designs. Errors in the estimated value of the time parameter are responsible for 75% of the total error in an estimate of the peak discharge. Numerous methods for estimating the various time parameters have developed due to the importance of time parameters in hydraulic design and evaluation. Unfortunately, most of the empirical formulas have been based on very limited data; therefore, lack of diversity in the data constrains their applicability. Considerable caution must be taken in the usage of an empirical estimation method (discussed in Chapter 3) either for watersheds having characteristics different from those of the watersheds used to calibrate the method or for watersheds in other geographic regions. Data extrapolation is represented in both uses. Therefore, a design engineer faced with a design problem requires an estimation of a time parameter and must choose from among many of the alternatives and often apply the time parameter with little or no knowledge of its accuracy.

The time of concentration, the lag time, the time to peak, the time to equilibrium, and the time area curve are the most frequently used time parameters. They are usually defined in terms of either the physical characteristics of a watershed or the distribution of rainfall excess and direct runoff. The most widely used time parameter in hydrology is the time of concentration and in hydraulics is the time estimated based on average velocity that are discussed in Chapter 3.

4.4 OPEN-CHANNEL FLOW IN URBAN WATERSHEDS

The detailed procedure of estimation of rainfall excess (or runoff) resulted from a rainfall event is discussed in Chapter 3. The processes for carrying the produced excess rainfall to the watershed outlet and/or desired point in the watershed, which often occurs in the form of open-channel flow, is discussed in this section.

The flow in most of the components of the urban drainage system is in the form of open-channel flow. The main characteristic of this type of flow is a free surface at atmospheric pressure. At first, the excess rainfall flows as "overland flow" with very small depth over the roofs, lawns, driveways, and street pavements. Then the rainfall excess is drained into a channel such as a street gutter, a ditch, a drainage channel, or a storm sewer, which are usually partly full. The excess rainfall finally reaches the watershed outlet in the form of a channel flow. The driving force in all types of open-channel flows is the gravitational force (Figure 4.5).

4.4.1 Open-Channel Flow

The open-channel flow is classified owing to the criterion considered in the evaluation of flow characteristics. If the flow characteristics at a specific point do not change with time, the flow is steady; otherwise, it is considered as an unsteady flow. When the considered criterion is space, if the flow characteristics do not change along the channel, the flow is considered uniform; otherwise it is called a nonuniform flow. It is also classified into gradually varied and rapidly varied flows depending on whether the variations along the channel are gradual or rapid (Figure 4.6). The hydraulic and geometric elements of the open-channel flow are depicted in the channel cross section of Figure 4.7.

The total energy head in each flow section is equal to the total energy per unit weight of the flowing water. The total energy head is measured vertically from a fixed horizontal datum (Figure 4.8). By assuming a hydrostatic distribution for the flow pressure, the total energy head of the flow is formulated as

FIGURE 4.5 Different open channel paths in urban areas. (From *Storm Water Management Model (SWMM), Helps predict runoff quantity and quality from drainage systems*, USEPA, Durham, NC. https://www.epa.gov/water-research/storm-water-management-model-swmm.)

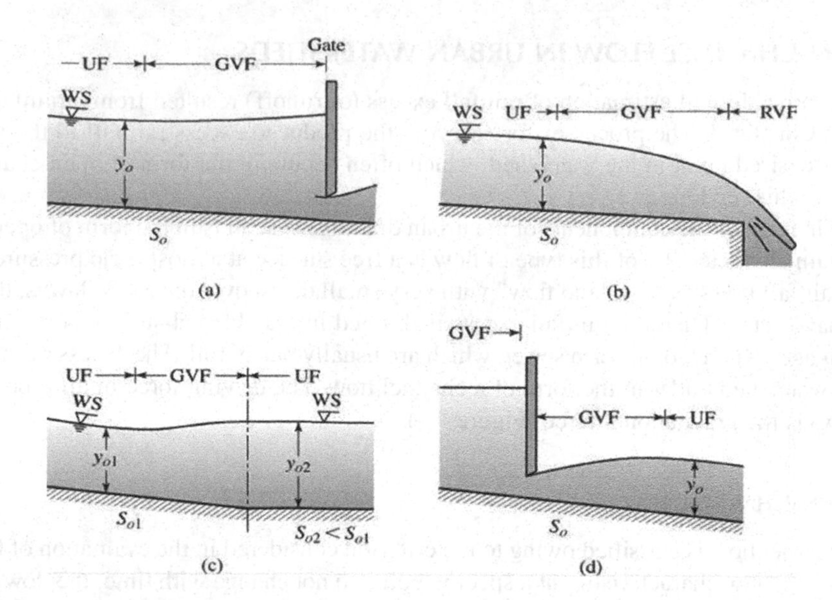

FIGURE 4.6 Examples of uniform (UF) and gradually varied flow (GVF) in all (a-d), and rapidly varied flow RVF) in (d) in open channels. (From "Class lectures Ch 14. CVE 341 – Water Resources," Presentation transcript, *Osborne Felix Fletcher*. https://slideplayer.com/slide/5946626/.)

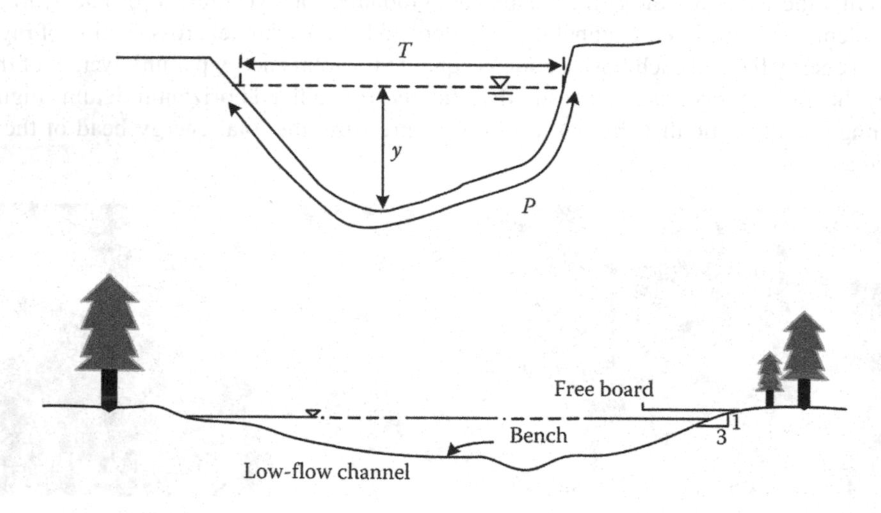

FIGURE 4.7 Elements of a channel section.

$$H = Z + y + \frac{V^2}{2_g}, \tag{4.6}$$

where H is the total energy head, Z is the elevation head, y is the flow depth, g is the gravitational acceleration, and $V^2/2_g$ is the velocity head.

The elevation head Z, which corresponds to the potential energy, is measured as the vertical distance between the channel bottom and the assumed datum. Flow depth y is called the pressure energy, and the kinetic energy of the flow is quantified by the velocity head equal to $V^2/2_g$. The energy grade line shows the variations of total energy head along the channel. The hydraulic head is equal to the summation of the elevation head and the pressure energy head, and in an open channel, the hydraulic head corresponds to the water surface.

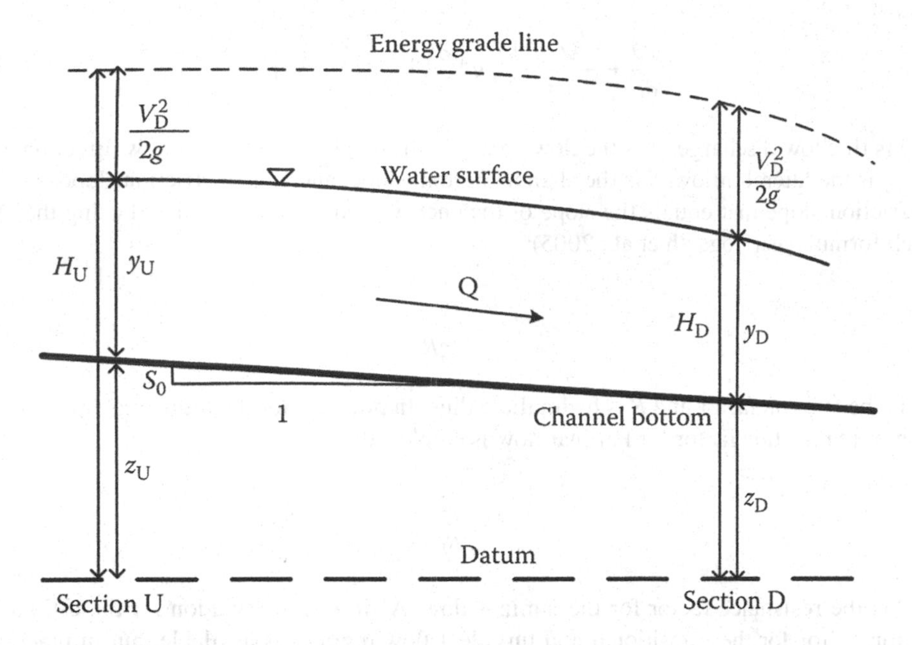

FIGURE 4.8 Definition of energy head and energy grade line.

4.4.1.1 Open-Channel Flow Classification

Various forces such as viscous, inertial, and gravitational forces affect the flow condition in open channels. The Reynolds number, *Re,* is defined as the ratio of inertial forces to viscous forces. The *Re* is calculated for open channels as

$$Re = \frac{VR}{\nu}, \tag{4.7}$$

where ν is the kinematic viscosity of water, V is the flow velocity, and R is the hydraulic radius.

Same as the pipes, for $Re < 500$, the flow is considered in the laminar state, and when $Re > 4{,}000$, the flow is turbulent. If $500 < Re < 4{,}000$, then a transitional flow occurs. Another dimensionless parameter that is used in determination of the flow state is the Froude number, *Fr,* which is equal to the ratio of the inertial to gravitational forces and is calculated as

$$Fr = \frac{V}{\sqrt{gy}}, \tag{4.8}$$

where y is the flow depth.

The flow is at the critical, subcritical, and supercritical states when Fr is equal to 1.0, less than 1.0, and more than 1.0, respectively. The flow in an open channel is significantly affected by the flow state as just described.

4.4.1.2 Hydraulic Analysis of Open-Channel Flow

The St. Venant continuity and momentum equations that are used for the description of unsteady, gradually varied flow in open channels are formulated as

$$\frac{\partial Q}{\partial x} + \frac{\partial A}{\partial t} = q_{LAT} \tag{4.9}$$

and

$$\frac{\partial Q}{\partial t}+\frac{\partial\left(Q^2/A\right)}{\partial x}+gA\frac{\partial y}{\partial x}=gA\left(S-S_f\right), \tag{4.10}$$

where Q is the flow discharge, A is the flow area, x is the displacement in the flow direction, t is the time, q_{LAT} is the lateral inflow, S is the channel bottom slope, and S_f is the friction slope.

The friction slope that equals the slope of the energy grade line is calculated using the Darcy–Weisbach formula as (Hogarth et al., 2005):

$$S_f=\frac{fV^2}{8gR}, \tag{4.11}$$

where f is the friction factor and R is hydraulic radius. In practice, for the following equation, determination of the friction factor for laminar flow is employed:

$$f=\frac{C}{Re}, \tag{4.12}$$

where C is the resistance factor for the laminar flow. A similar formulation for the estimation of the friction factor for the transitional and turbulent flow regimes is available, but in practice, the Manning formula is commonly used for evaluation of the friction slope in these situations as

$$S_f=\left(\frac{nQ}{A}\right)^2\frac{1}{R^{4/3}}, \tag{4.13}$$

where n is Manning's coefficient.

The values of Manning's coefficient are dependent on the channel material and also on flow depth as presented in Table 4.5. In turbulent flow, the dependency of Manning's coefficient on flow depth decreases and could be neglected. In this situation, the roughness factor is constant for a given channel material.

The St. Venant equations are valid for unsteady, gradually varied flow conditions. In steady flows, as the flow characteristics remain constant with time, all of the partial differential terms with respect to time t are zero. Also, by assuming that $q_{LAT}=0$, Equation 4.9 could be simplified to

$$\frac{\partial Q}{\partial x}=0, \tag{4.14}$$

TABLE 4.5
Manning's Coefficient, *n*, for Different Depths of Flow over Different Materials

	Depth Range		
Channel Material	**0–0.15 m**	**0.15–0.60 m**	**>0.6 m**
Concrete	0.015	0.013	0.013
Grouted riprap	0.010	0.030	0.028
Soil cement	0.025	0.022	0.0211
Asphalt	0.019	0.016	0.016
Rare soil	0.023	0.020	0.020
Rock cut	0.045	0.035	0.025

Equation 4.14 means that the discharge is constant under the steady-state flow conditions without lateral inflow. Similarly, Equation 4.10 is simplified to

$$\frac{d\left(Q^2/A\right)}{dx} + gA\frac{dy}{dx} = gA\left(S - S_f\right). \tag{4.15}$$

The term Q/A is substituted with V, and finite difference discretization is performed on Equation 4.15 to obtain Equation 4.16, which is commonly used in practice for the evaluation of steady-state flow conditions:

$$\left(y_1 + \frac{V_1^2}{2g}\right) - \left(y_2 + \frac{V_2^2}{2g}\right) = \Delta x\left(S - S_f\right), \tag{4.16}$$

where subscripts 1 and 2 correspond to the upstream and downstream sections, respectively, and A_x is the distance between the two considered sections. In Equation 4.16, S_f is the average friction slope between the upstream and downstream sections. Considering that the channel bottom slope, S, is equal to $(z_1 - z_2)/A_x$ and also head loss, h_f, is obtained by $(A_x)(S_f)$, Equation 4.16 is rearranged as

$$h_f = H_1 - H_2. \tag{4.17}$$

If the steady flow is also uniform (the flow characteristics are constant along the channel), Equation 4.15 can be simplified further by dropping all the differential terms with respect to x. What finally remains is $S_f = S$. This state of flow (steady, uniform) is called normal flow. In this condition, the energy grade line, the water surface, and the channel bottom are parallel to each other, and the Manning formula (Equation 4.18) can be used for the description of the water surface profile.

$$Q = \frac{1}{n}AR^{2/3}S^{1/2}. \tag{4.18}$$

However, it should be noted that ideally normal flow can occur only in long, prismatic channels without any flow control structures. But in practice, for the design of stormwater channels, the flow is considered normal with the constant design discharge equal to the peak discharge.

For the description of the relationship between the flow area, A, and the discharge, Q, at a given channel section, rating curves are developed. The field measured data should be used for the development of these curves, but in the cases where the field data are not available or accessible, Equation 4.18 can be used for the estimation of a rating curve. Considering that n and S are constant for a given channel, and R and A are related to each other, the channel rating curve can be formulated as

$$Q = eA^m, \tag{4.19}$$

where e and m are constants, which are determined analytically. It should be noted that e and m are dependent on the channel flow and their average values should be used for flow analysis.

4.4.2 Overland Flow

The overland flow occurs on impervious and pervious surfaces in urban watersheds such as roofs, driveways, and parking lots as well as lawns. The driving force in this kind of flow is gravitational, and flow over these surfaces occurs in the form of a sheet with very low depth.

Overland flow can be modeled as an open-channel flow with a very shallow depth so that the same equation as for the open-channel flow can be applied. The flow depth in overland flow is very small, and this results in a low *Re*; therefore, the overland flow can be classified as laminar when only *Re* is considered. But some other factors should be considered in the determination of the flow state. For instance, different surface coverage such as rocks, grass, and litter cause some disturbances in overland flow, which cannot be considered as laminar flow despite the low *Re* value (Akan, 2011; Tejada-Martínez et al., 2020). In this way, it is more logical to consider overland flow as turbulent flow. Different equations are employed for the description of the overland flow in laminar or turbulent flow conditions.

If the overland flow is turbulent, the Manning equation can be employed. Manning's coefficients used in overland flow analysis are different from those for channel flow because of the small depth and the turbulent nature of the flow. All the factors affecting the flow resistance in overland flow are aggregated into the Manning's coefficient that is referred to as the effective Manning's roughness factor. In Table 4.6, the effective n values for different types of land surfaces are given (Singh and Woolhiser, 2002; Engman, 1986; Sadeh et al., 2018).

4.4.2.1 Overland Flow on Impervious Surfaces

For the analysis of the overland flow on impervious surfaces, the basin soil is considered to be dry at the start of rainfall. At $t = 0$, excess rainfall and infiltration are produced. It is assumed that the rainfall would continue with a constant rate so that the discharge at the basin outlet would increase until basin equilibrium time (time of concentration) is reached, and then it would remain constant. The time of concentration for the overland flow is estimated as

$$t_c = \frac{L^{1/m}}{\left(\alpha i_0^{m-1}\right)^{1/m}}, \tag{4.20}$$

where t_c is the time of concentration, L is the length of the overland flow on the watershed, and $i_0 = i - f$ is the constant rate of rainfall excess. α and m are constants which are dependent on the flow condition that is either laminar or turbulent. For laminar flow, $m = 3.0$ and

TABLE 4.6
Effective Manning n Values for Overland Flow

Surface Type	n
Concrete	0.01–0.013
Asphalt	0.01–0.015
Bare sand	0.01–0.016
Graveled surface	0.012–0.3
Bare clay loam (eroded)	0.012–0.033
Fallow, no residue	0.008–0.012
Conventional tillage, no residue	0.06–0.12
Conventional tillage, with residue	0.16–0.22
Chisel plow, no residue	0.06–0.12
Chisel plow, with residue	0.10–0.16
Fall disking, with residue	0.30–0.50
No till, no residue	0.04–0.10
No till (20%–40% residue cover)	0.07–0.17
No till (60%–100% residue cover)	0.17–0.47
Sparse vegetation	0.053–0.13
Short grass prairie	0.10–0.20

$$\alpha = \frac{8gS}{Cv}, \tag{4.21}$$

where C is the laminar flow resistance factor, v is the kinematic viscosity of the water, and S is the basin slope.

For turbulent flow, $m = 5/3$ and

$$\alpha = \frac{S^{1/2}}{n}. \tag{4.22}$$

As the actual rainfall occurs in a finite duration, t_d, the shape and the peak discharge of the runoff hydrograph are dependent on the relation between t_d and t_c.

1. If $t_c < t_d$ (Figure 4.9, part a), the rising limb of the runoff hydrograph before reaching the equilibrium time is calculated as

$$q_L = \alpha(i_0 t)^m, \tag{4.23}$$

where q_L is the discharge produced per unit width at $x = L$, the basin outlet. The discharge will remain constant between the equilibrium time and the end of rainfall and is equal to

$$q_L = i_0 L, \tag{4.24}$$

By the end of rainfall at t_d, the discharge gradually reduces. The flow variations with time are calculated as

$$t = t_d + \frac{L/\left(\alpha y_L^{m-1}\right) - (y_L / i_0)}{m}, \tag{4.25}$$

where y_L is the flow depth at the downstream of the basin and the corresponding discharge is calculated as

$$y_L = \left(\frac{q_L}{\alpha}\right)^{1/m}. \tag{4.26}$$

2. When the time of concentration is greater than the rainfall duration (Figure 4.9, part b), the basin does not reach the equilibrium condition. The rising limb is calculated using Equation 4.23 before the rainfall ends. After that, the runoff hydrograph becomes flat with a constant discharge until t_p. The discharge is calculated as

$$q_L = \alpha(i_0 t_d)^m, \tag{4.27}$$

where

$$t_p = t_d - \left(\frac{t_d}{m}\right) + \frac{L}{m\alpha(i_0 t_d)^{m-1}}. \tag{4.28}$$

The runoff gradually decreases during the period when $t > t_p$, and Equations 4.25 and 4.28 can be applied again to compute the falling limb of the runoff hydrograph. It should be noted that in this case, the flat portion of the runoff hydrograph between t_d and t_p does not mean reaching the equilibrium condition. In this period of time, the depths and discharges

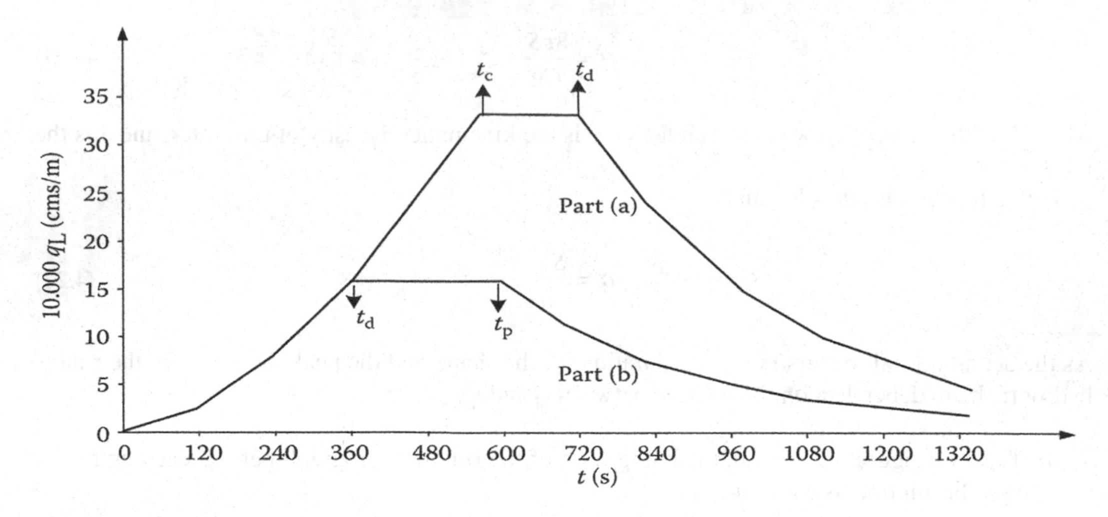

FIGURE 4.9 Overland flow hydrographs.

are decreasing at the upstream sections, while a constant flow rate is observed at the downstream of the watershed. A detailed description of this phenomenon can be provided using a characteristic method, which is beyond the scope of this text.

Example 4.6

Consider a parking lot with length 50 m and slope 0.005. The effective Manning's coefficient of the parking lot is 0.020. Determine the runoff hydrograph for a constant rainfall excess rate equal to 15 mm/h at the downstream end of the parking lot. Solve the example for two rainfall durations of 15 and 5 minutes.

Solution:

Considering that the flow is turbulent, α and t_e are calculated from Equations 4.20 and 4.22, respectively, using the excess rainfall of 15 mm/h = 4.167×10^{-6} m/s, $m = 5/3 = 1.667$, $1/m = 0.6$, and $(m - 1)/m = 0.4$,

$$\alpha = \frac{(0.005)^{0.5}}{0.020} = 3.54 \text{ m}^{1/3}/\text{s},$$

$$t_C = \frac{50^{0.6}}{(3.54)^{0.6}(4.167\times10^{-6})^{0.4}} = 695 \text{ seconds} = 11.59 \text{ minutes}.$$

1. For t_d = 15 minutes = 900 seconds, $t_c < t_d$. Thus, between time 0 and 695 seconds, Equation 4.23 is used to calculate the rising limb of the hydrograph. The resulting q_L values are given in column 2 of Table 4.7, part (a). The discharge at equilibrium time (695 seconds) is equal to 20.83×10^{-5} m³/s/m and remains constant until the rain ceases at $t = t_d = 900$ seconds. The q_L values in column 3 of Table 4.7 are arbitrarily chosen from the peak flow to zero to calculate the falling limb. Then, columns 4 and 5 are calculated using Equations 6.26 and 6.27.
2. For t_d = 5 minutes = 300 seconds, $t_c > t_d$. The rainfall ceases before reaching the basin equilibrium. Therefore, the peak discharge occurs at t = 300 seconds.

   Equation 4.23 is used for calculating the rising limb for $0 < t < 300$ seconds. As shown in column 2 of part (b) of Table 4.7, the maximum discharge is 5.14×10^{-5} m³/s/m. Then the t_p is estimated using Equation 4.28 as

TABLE 4.7
Kinematic Wave Model Example

	Rising Limb	Falling Limb		
t (s)	q_L (10^{-5} m³/s/m)	q_L (10^{-5} m³/s/m)	y_L (10^{-3}m)	t (s)
		(a)		
0	0.00	20.83	2.90	695
150	1.62	20.83	2.90	900
300	5.14	15	2.38	1,033
450	10.01	10	1.86	1,088
600	16.30	5	1.23	1,461
695	20.83	0.5	0.31	2,710
		(b)		
0	0.00	5.14	1.25	850
100	0.82	3	0.91	1,071
200	2.61	2	0.71	1,263
300	5.14	1	0.47	1,634
		0.5	0.31	2,106

$$t_P = 300 - \frac{300}{1.667} + \frac{50}{(1.667)(3.54)\left(4.167\times10^{-6}\right)^{0.667}(300)^{0.667}} = 850 \text{ seconds.}$$

The discharge will remain constant at 5.14×10^{-5} m³/s/m between 300 and 850 seconds. After 850 seconds, flow rates decrease. Equations 6.20 and 6.21 are used to calculate the falling limb, and the results are tabulated in part (b) of Table 4.7. The q_L values in column 3 are used to calculate the falling limb. Then columns 4 and 5 are calculated using Equations 4.26 and 4.25.

4.4.2.2 Overland Flow on Pervious Surfaces

The analytical solutions governed for impervious surfaces can also be used for pervious surfaces when there is a constant rate of rainfall excess. But for variable excess rainfall rates, an analytical solution cannot be governed. Akan and Houghtalen (2003) have developed a graphical solution for the determination of the peak overland flow discharge over pervious surfaces (Figure 4.10). Before using this chart, the following dimensionless parameters should be calculated:

$$q_p^* = \frac{q_P}{iL}, \tag{4.29}$$

$$\alpha^* = \frac{L^{0.6}}{i^{0.4} t_d \alpha^{0.6}}, \tag{4.30}$$

$$K^* = \frac{K}{i}, \tag{4.31}$$

$$P^* = \frac{P_{f\phi}(1 - S_i)}{i t_d}, \tag{4.32}$$

where q_p is the peak discharge per unit width at the basin outlet, i is the constant rate of rainfall, L is the length of the overland flow path, K is the hydraulic conductivity of soil, P_f is the soil suction

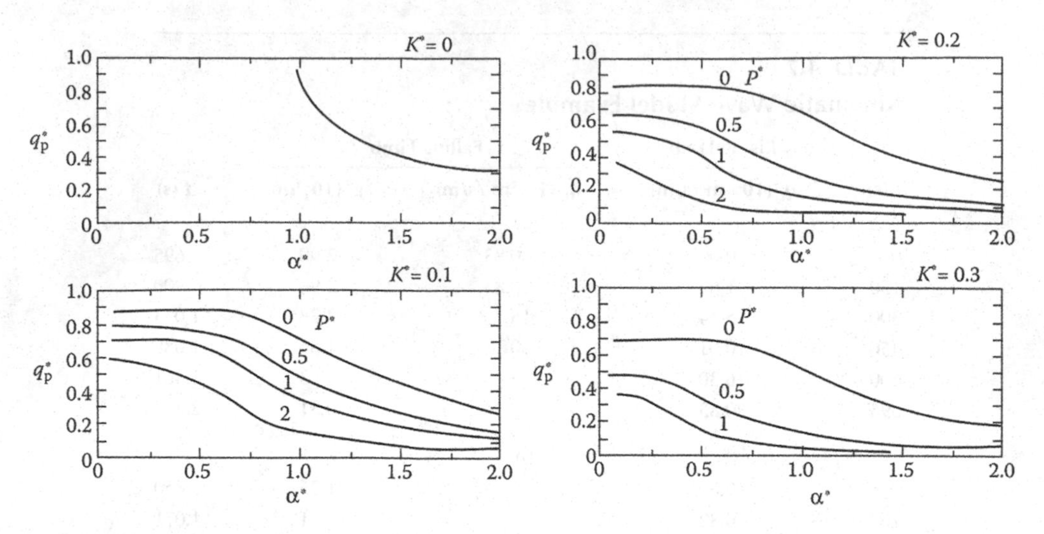

FIGURE 4.10 Overland flow from pervious surfaces. (From Akan, A.O. and Houghtalen, R.J., *Urban Hydrology, Hydraulics, and Stormwater Quality: Engineering Applications and Computer Modeling*, Wiley, Hoboken, NJ, 2003.)

head, ϕ is the effective soil porosity, S_i is the initial degree of saturation, t_d is the duration of rainfall, and

$$\alpha = \frac{1}{n} S^{0.5}. \tag{4.33}$$

It should be noted that Figure 4.10 is only applicable for basins with uniform characteristics and when the only source of rainfall loss is infiltration.

Example 4.7

Determine the peak discharge resulting from a 1-hour rainfall with a constant intensity equal to 1.5×10^{-5} m/s that happens on a pervious rectangular basin with the following characteristics: $L = 120$ m, $S = 0.002$, and $n = 0.02$. The infiltration parameters are $9 = 0.414$, $K = 1.5 \times 10^{6}$ m/s, and $P_f = 0.25$ m. At the beginning of the rainfall, the basin soil is 0.65% saturated.

Solution:

For utilizing Figure 4.10, variables α, α*, K*, and P* are calculated using Equations 4.33 and 4.30–4.32, respectively.

$$\alpha = \frac{(0.002)^{0.5}}{0.02} = 2.24 \text{ m}^{1/3}/\text{s}.$$

$$\alpha^* = \frac{(120)^{0.6}}{\left(1.5\times10^{-5}\right)^{0.4}(3,600)(2.23)^{0.6}} = 0.26,$$

$$K^* = \frac{1.5\times10^{-6}}{1.5\times10^{-5}} = 0.1,$$

$$P^* = \frac{(0.25)^{0.5}(0.414)(1.0-0.65)}{\left(1.5\times10^{-5}\right)(3,600)} = 0.67.$$

Then $q_p^* = 0.74$ using Figure 4.10. Finally, q_p is calculated as

$$q_p = (0.74)(1.5\times10^{-5})(120) = 1.33\times10^{-3}\ \text{m}^3/\text{s/m}.$$

4.4.3 Urban Channel Routing

Channel flow occurs in different elements of an urban watershed including gutters, ditches, and storm sewers. The flow in these structures is unsteady and nonuniform during and after a rainstorm. The St. Venant equations could be employed for simulation of this flow condition. The models that use these equations are referred to as "dynamic models," as they could explain the dynamic nature of the system. There is no analytical solution for St. Venant equation and complicated computer programs are developed for solving them. But as this procedure takes a lot of time and is costly, in practice less complex methods are employed.

The simplified methods are categorized into two groups including hydraulic and hydrological models. In hydraulic models, St. Venant equations are employed, but hydrological models use an assumed relationship instead of the momentum and continuity equations. The main disadvantage of simpler models is that some phenomena that affect the flow such as downstream backwater are not considered. Unsteady open-channel flow calculations are referred to as flood routing or channel routing. Muskingum flood routing described in this section is often used in conjunction with urban hydraulic structure despite of being classified as a hydrological routing.

4.4.3.1 Muskingum Method

In this method, the outflow hydrograph at the downstream of a channel reach is calculated for a given inflow hydrograph at the upstream. The uniform short channels are modeled as a single reach, but the long channels with various characteristics in length are often divided into several reaches and each is modeled separately. When more than one reach is considered, the calculations start from the most upstream reach and then proceed in the direction of the downstream. The outflow from the upstream reach is used as the inflow for the next downstream reach.

The continuity equation for a channel reach is written as

$$\frac{dS}{dt} = I - Q, \tag{4.34}$$

where S is the storage water volume in the channel reach, I is the upstream inflow rate, and Q is the downstream outflow rate.

As there are two unknown variables including S and Q in Equation 4.34, a second relationship should be developed to solve the flood routing problem. In the Muskingum method, a linear relationship is assumed between S, I, and Q as

$$S = K\left[XI + (I - X)Q\right], \tag{4.35}$$

where K is the travel time constant and X is the weighing factor that varies between 0 and 1.0.

Parameters K and X are constant for a channel and do not have physical meaning. These parameters are determined during the model calibration. Equation 4.34 can be rewritten in the finite difference form over a discrete time increment as

$$\frac{S_2 - S_1}{\Delta t} = \frac{I_1 + I_2}{2} - \frac{Q_1 + Q_2}{2}, \tag{4.36}$$

where $\Delta t = t_2 - t_1$, in which the variables with subscripts 1 and 2 are related to the beginning and the end of the time increment. Equation 4.35 can be written for times t_1 and t_2 to provide a relationship between S and Q in these times. By substituting the obtained relations into Equation 4.31, the following equation is obtained:

$$Q_2 = C_0 I_2 + C_2 Q_1, \tag{4.37}$$

where

$$C_0 = \frac{(\Delta t / K) - 2X}{2(1-X) + (\Delta t / K)}, \tag{4.38}$$

$$C_1 = \frac{(\Delta t / K) - 2X}{2(1-X) + (\Delta t / K)}. \tag{4.39}$$

$$C_2 = \frac{2(1-X) - (\Delta t / K)}{2(1-X) + (\Delta t / K)}, \tag{4.40}$$

Note that $C_0 + C_1 + C_2 = 1.0$.

Q_2 is the only unknown variable in routing Equation 4.37. I_2 and I_1 are obtained from the given inflow hydrograph, and Q_1 is obtained from the initial condition of the channel reach or from the previous time step computation.

Example 4.8

The inflow hydrograph for a channel reach with an initial steady flow equal to 45 m³/s is given in Figure 4.11. Calculate the outflow hydrograph in 1-hour time increments. The Muskingum parameters are determined as $K = 3$ hours and $X = 0.15$.

Solution:

Using Equations 4.38–4.40, the weighting factors are estimated as being $C_0 = 0.016$, $C_1 = 0.311$, and $C_2 = 0.673$. The routing calculations are given in Table 4.8. I_1 and I_2 values for each time are calculated using the inflow hydrograph presented in Figure 4.11. The beginning and end time of each 1-hour time increment are tabulated in columns 2 and 3 of Table 4.8. Because the initial

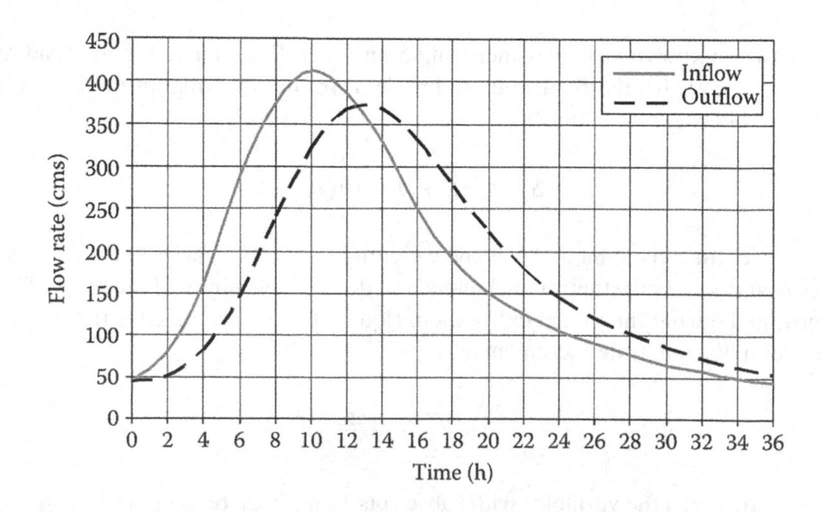

FIGURE 4.11 Inflow and outflow hydrographs of Example 4.8.

TABLE 4.8
Summary of Example 4.8 Calculations

Time Step	t_1 (h)	t_2 (h)	I_1 (cm)	I_2 (cm)	Q_1 (cm)	Q_2 (cm)
1	0	1	45	62	45	45
2	1	2	62	80	45	51
3	2	3	80	120	51	61
4	3	4	120	160	61	81
5	4	5	160	225	81	108
6	5	6	225	290	108	147
7	6	7	290	333	147	195
8	7	8	333	375	195	241
9	8	9	375	394	241	285
10	9	10	394	412	285	321
11	10	11	412	399	321	351
12	11	12	399	385	351	367
13	12	13	385	358	367	372
14	13	14	358	330	372	367
15	14	15	330	290	367	354
16	15	16	290	250	354	332
17	16	17	250	220	332	305
18	17	18	220	190	305	277
19	18	19	190	170	277	248
20	19	20	170	150	248	222
21	20	21	150	138	222	198
22	21	22	138	125	198	178
23	22	23	125	115	178	160
24	23	24	115	105	160	145
25	24	25	105	98	145	132
26	25	26	98	90	132	121
27	26	27	90	84	121	111
28	27	28	84	77	111	102
29	28	29	77	71	102	94
30	29	30	71	65	94	86
31	30	31	65	61	86	79
32	31	32	61	57	79	73
33	32	33	57	53	73	68
34	33	34	53	49	68	63
35	34	35	49	47	63	58
36	35	36	47	45	58	54

flow in the channel is equal to 45 m³/s, Q_1 of the first time period has been set as 45 m³/s. The routed outflow hydrograph is also presented in Figure 4.11. Three main points can be observed in Figure 4.11 by a comparison of the inflow and outflow hydrographs:

1. The peak outflow rate is less than the peak inflow rate.
2. The peak outflow occurs later than the peak inflow.
3. The volumes of inflow and outflow hydrographs are equal.

In any unsteady channel-flow situation, these observations are true. The only exceptions are channel flows with very high Froude numbers or channels with lateral flows.

4.5 HYDRAULICS OF WATER DISTRIBUTION SYSTEMS

In a water distribution system, almost all of the principal components work under pressure. When the pipes are full and the water in the system is moving under gravitation force, it is also considered as pressure flow. The sewage force mains that receive discharge from a pumping station also carry flows under pressure. There is a range of minimum to maximum pressure in which these components operate. Due to frictional resistance by the pipe walls and fittings, the pressure is reduced as water flows through the system. This is measured in terms of the head/energy loss. The principles of hydraulics that are used in water distribution modeling are described in the next section.

4.5.1 Energy Equation of Pipe Flow

A pipeline segment is illustrated in Figure 4.12. The total energy at any point consists of potential or elevation head, pressure head, and velocity head. The hydraulic grade line demonstrates the elevation of pressure head along the pipe (i.e., it is a line connecting the points to which the water will rise in piezometric tubes installed at different sections of a pipeline). This concept is similar to the water surface in open-channel flow. The total head at different points of a pipe section is represented by the energy grade line. In a uniform pipe, the velocity head is constant. Thus the energy grade line is parallel to the hydraulic grade line.

Applying the energy equation between points 1 and 2 yields

$$Z_1 + \frac{P_1}{\gamma} + \alpha\frac{V_1^2}{2g} = Z_2 + \frac{P_2}{\gamma} + \frac{P_2}{\gamma} + \alpha\frac{V_2^2}{2g} + h_f. \tag{4.41}$$

In Equation 4.41, h_f is the head loss along the pipeline due to friction. The energy gradient S_f is equal to h_f/L. Additional losses resulting from valves, fittings, bends, and so on are known as the minor losses, h_m, and have to be included when present. Then, in Equation 4.41, the term h_f will be replaced by the total head loss, h_{loss}, which is equal to the summation of h_f and h_m. Since the minor losses are localized, the energy grade line, represented by h_f/L, will have breaks wherever the minor losses occur. If a mechanical energy is added to the water by a pump or taken off by a turbine between the two points of interest, it should be added or subtracted from the left side of Equation 4.41.

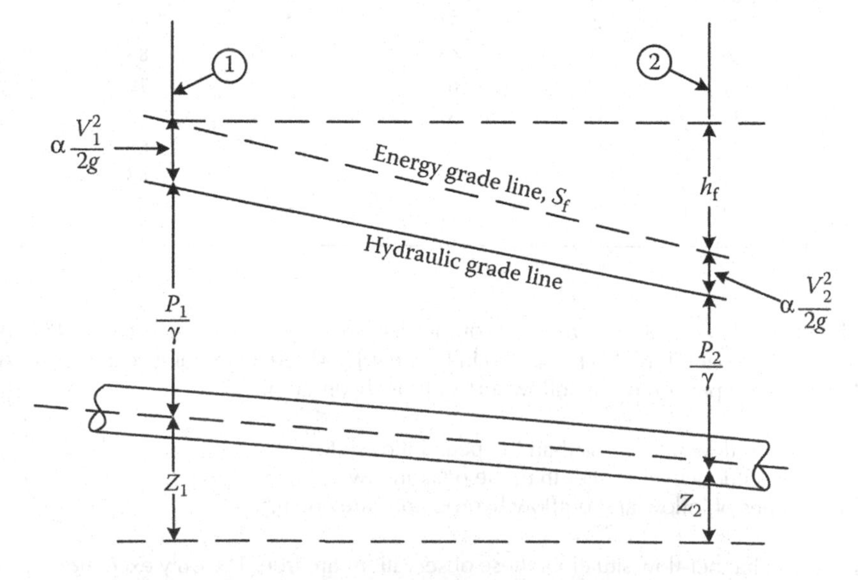

FIGURE 4.12 Hydraulic grade line and energy grade line in a pipe flow.

In a uniform pipe, $V_1 = V_2$, and elevations Z_1 and Z_2 are generally known. To ascertain the pressure reduction, it is necessary to evaluate the head loss (and minor losses if present).

Example 4.9

The diameter of a pipe varies between 200 and 100 mm from section A to section B, respectively. The pressure at section A is 8.15 m, and there is a negative pressure equal to 2.5 m at section B. The velocity in section A is 1.8 m/s. If section B is 7 m higher than section A, calculate (a) the flow rate, (b) the velocity at section B, (c) the flow direction, and (d) the head loss in the system.

Solution:

a. The pipe flow rate is calculated as

$$Q = A_A V_A = \frac{\pi}{4}(0.2)^2\, 1.8 = 0.057\ \text{m}^3/\text{s}.$$

b. Since the flow rate in the pipe is constant, the flow velocity at section B is

$$V_B = \frac{Q}{A_B} = \frac{0.057}{(\pi/4)(0.1)^2} = 7.26\ \text{m/s}.$$

c. For determining the flow direction, at first the total head at sections A and B should be calculated. For this purpose, it is assumed that the datum line corresponds with section A. Total head at section A:

$$H_A = Z_A + \frac{P_A}{\gamma} + \frac{V_A^2}{2g} = 0 + 8.15 + \frac{1.8^2}{2\times 9.81} = 8.32\ \text{m}.$$

Total head at section B:

$$H_B = Z_B + \frac{P_B}{\gamma} + \frac{V_B^2}{2g} = 7 + (-2.5) + \frac{7.26^2}{2\times 9.81} = 7.19\ \text{m}.$$

Since the total head in section A is more than that in section B, the flow direction is from section A to section B.

d. The system head loss is equal to the difference between the total head at sections A and B and is equal to

$$h_f = 8.32 - 7.19 = 1.13\ \text{m}.$$

4.5.2 Evaluation of Head Loss Due to Friction

4.5.2.1 Darcy–Weisbach Equation

This equation is the most general formula in the pipe flow application (see Karamouz et al. (2011) for more details). It was obtained experimentally as

$$h_f = \frac{fL}{d}\frac{V^2}{2g}, \tag{4.42}$$

where h_f is the head loss due to friction in the pipe (m), f is the friction factor, dimensionless, L is the length of the pipe (m), d is the internal diameter of the pipe (m), and V is the mean velocity of flow in the pipe (m/s).

The solution of Equation 4.42 requires the interim step of ascertaining an appropriate value of the friction factor, f, to be used in the equation.

4.5.2.1.1 Friction Factor

The friction factor relation depends on the Reynolds number which is the state of flow. The pipe diameter is a characteristic dimension and the Reynolds number is given by

$$Re = \frac{Vd}{\upsilon}, \tag{4.43}$$

where V is the average velocity of flow (m/s), d is the internal diameter of the pipe (m), and υ is the kinematic viscosity of the fluid (m^2/s).

The Reynolds number for laminar flow is given by the following expression as a function of the friction factor:

$$f = \frac{64}{Re}\left[\text{For laminar flow}\left(Re < 2{,}000\right)\right]. \tag{4.44}$$

Any friction factor relation cannot be applied in the critical region of Re between 2,000 and 4,000 the flow alternates between the laminar and turbulent flows ($Re > 4{,}000$). In turbulent flow, the friction factor is a function of the Reynolds number and the relative roughness of the pipe surface. The roughness is characterized by a e/d parameter, where e is the average height of roughness along the pipe. The turbulent flow is further categorized into three zones as follows:

1. Flow in a smooth pipe, where the relative roughness e/d is very small.
2. Flow in a fully rough pipe.
3. Flow in a partially rough pipe, where both the relative roughness and viscosity are significant.

From the obtainable implicit relations for calculation of the friction factor for the Darcy–Weisbach equation, Moody diagram has been prepared based on of the friction factor versus the Reynolds number and the relative roughness. The Moody diagram is illustrated in Figure 4.13. For the application of the Moody diagram, the velocity of flow and the diameter of the pipe should be known so that the Reynolds number can be determined. When the velocity or diameter is unknown, the procedure has been discussed in the next section.

Example 4.10

Determine the friction factor for water flowing at a rate of 0.028 cm in a cast iron pipe 50 mm in diameter at 80°F. The pipe roughness is $e = 2.4 \times 10^{-4}$ m.

Solution:

Since the area of the pipes cross section is $A = (\pi/4)5^2 \times 10^{-4} = 0.002\ m^2$, the velocity of flow is calculated as $V = Q/A = 0.028/0.002 = 14$ m/s.

At 80°F, the kinematic viscosity is $0.86 \times 10^{-6} m^2/s$ and the Reynolds number is estimated as

$$Re = \frac{Vd}{\upsilon} = \frac{14 \times 50.0 \times 10^{-3}}{0.86 \times 10^{-6}} = 8.1 \times 10^5.$$

Since $Re > 4{,}000$, the flow is turbulent. The equivalent roughness is $e = 2.4 \times 10^{-4}$ and the relative roughness is calculated as $e/d = 2.4 \times 10^{-4}/50 \times 10^{-3} = 0.005$. On the Moody diagram (Figure 4.13), the point of intersection of $Re = 8.1 \times 10^5$ and $e/d = 0.005$ is projected horizontally to the left to read $f = 0.03$.

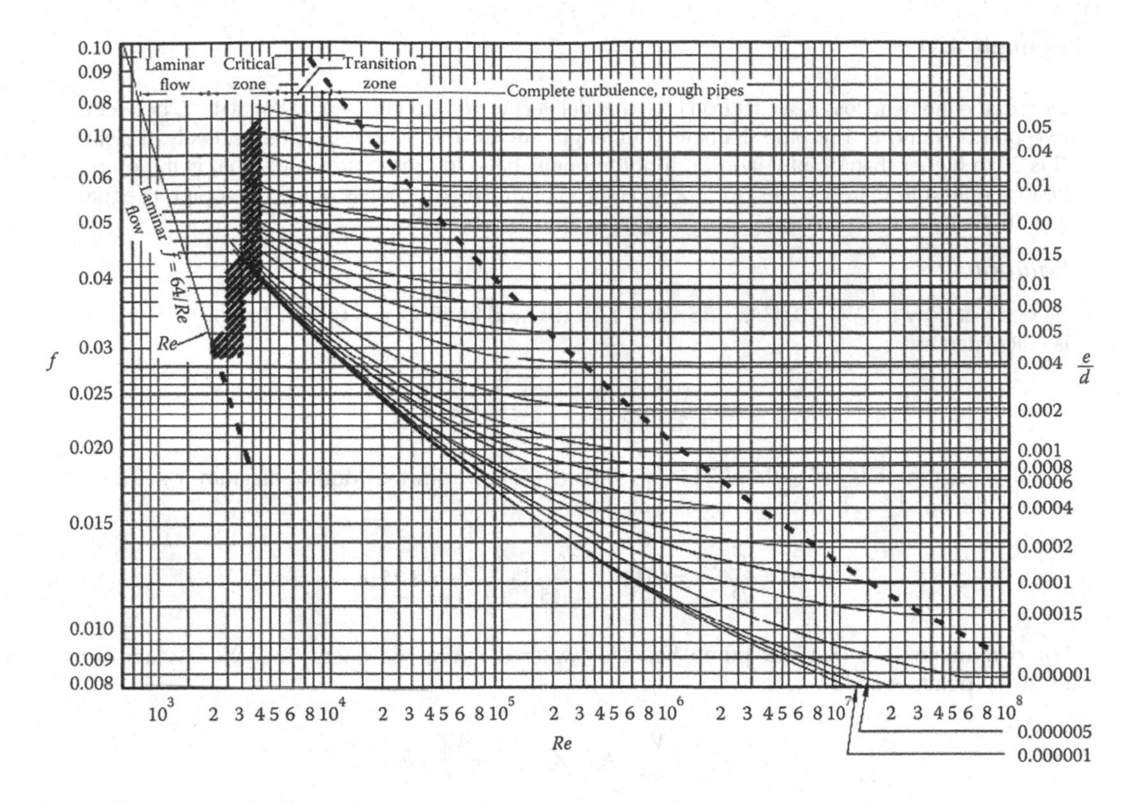

FIGURE 4.13 Moody diagram.

4.5.2.1.2 *Applications*

The Darcy–Weisbach Equation 4.42 has three applications (see Karamouz et al. (2011) for more details):

1. It is used to compute the head loss, h_f, in a pipe of given size, d, that carries a known flow, V or Q.
2. It is used to determine the flow, V (or Q), through a pipe of given size, d, in which the head loss, h_f, is known.
3. It is used to establish the pipe size, d, to pass a given rate of flow, Q, within a known limit of head loss, h_f.

In the first case, the application of Equation 4.42 is direct. From known V and d values, Re can be computed and then f can be determined by the Moody diagram. Thus, Equation 4.42 can be solved for h_f. In the second and third cases, Re, and f, cannot be determined. Since Re is unknown, many investigators have prepared particular diagrams between certain groups of variables in nondimensional form, instead of Re versus f, which enable direct determination of pipe size or flow. There is a trial-and-error procedure as follows:

1. A value of f is assumed near the rough turbulence if e/d is known.
2. Using the Darcy–Weisbach Equation 4.42, either V or d can be analyzed.
3. Re and revised f are determined.
4. Steps 2 and 3 are repeated until a common value of f is obtained.

Example 4.11

Determine the head loss in a cast iron pipe that delivers water at a rate of 0.03 cm at 45°C between two points *A* and *B*. The pipe diameter and length are 150 mm and 300 m, respectively. If point *B* is 25 m higher than point *A* and both of the points have the same pressure, what is the pump head required to deliver water from point *A* to point *B*? The roughness of cast iron pipe is equal to 0.000244 m.

Solution:

Velocity of flow is $Q/A = 0.03/(\pi(0.15)^2/4) = 1.70$ m/s, and at 45°C, υ is 0.86×10^{-6} m²/s. Thus, *Re* is calculated as

$$Re = \frac{Vd}{\upsilon} = \frac{1.70(0.150)}{0.86 \times 10^{-6}} = 3.0 \times 10^5.$$

Relative roughness is $e/d = 0.000244/0.15 = 0.0016$ and, from the Moody diagram, $f = 0.028$. Hence, the head loss in the pipe is calculated as

$$h_f = \frac{fL}{D}\frac{V^2}{2g} = (0.028)\left(\frac{300}{0.15}\right)\frac{1.7^2}{2(9.81)} = 8.25 \text{ m}.$$

For determining the required pump head for water delivery, the energy equation is applied between points *A* and *B*:

$$Z_A + \frac{P_A}{\gamma} + \frac{V_A^2}{2g} + h_p = Z_B + \frac{P_B}{\gamma} + \frac{V_B^2}{\gamma} + h_f.$$

Since the pressure and velocity of both points are the same, the equation is summarized as $Z_A + h_p = Z_B + h_f$, and then

$$0 + h_p = 25 + 8.25 \Rightarrow h_p = 33.25 \text{ m}.$$

4.5.2.2 Hazen–Williams Equation for the Friction Head Loss

The Hazen–Williams equation is another common formula for head loss in pipes. The Hazen–Williams formula is used in pipe designs, although it is accurate within a certain range of diameters and friction slopes. The Hazen–Williams equation in SI units is

$$V = 0.849CR^{0.63}S^{0.54}, \quad (4.45)$$

where V is the mean velocity of flow (m/s), C is the Hazen–Williams coefficient of roughness, R is the hydraulic radius (m), and S is the slope of the energy gradient and is equal to h_f/L.

Jain et al. (1978) and Valiantzas (2005) signified that an error of up to 39% can be involved in the evaluation of the velocity by the Hazen–Williams formula over a wide range of diameters and slopes. Two sources of error in the Hazen–Williams formula are the following:

1. The multiplying factor 0.849 should change for different R and S values for the same value of C.
2. The Hazen–Williams coefficient C is considered to be related to the pipe material only. Similar to the friction factor of Darcy–Weisbach, the Hazen–Williams coefficient also depends on pipe diameter, velocity, and viscosity.

A nomogram based on Equation 4.45 is given in Figure 4.14 by Gupta (2016) to facilitate the solution. Equation 4.6 and the nomogram provide a direct solution to all types of pipe problems:

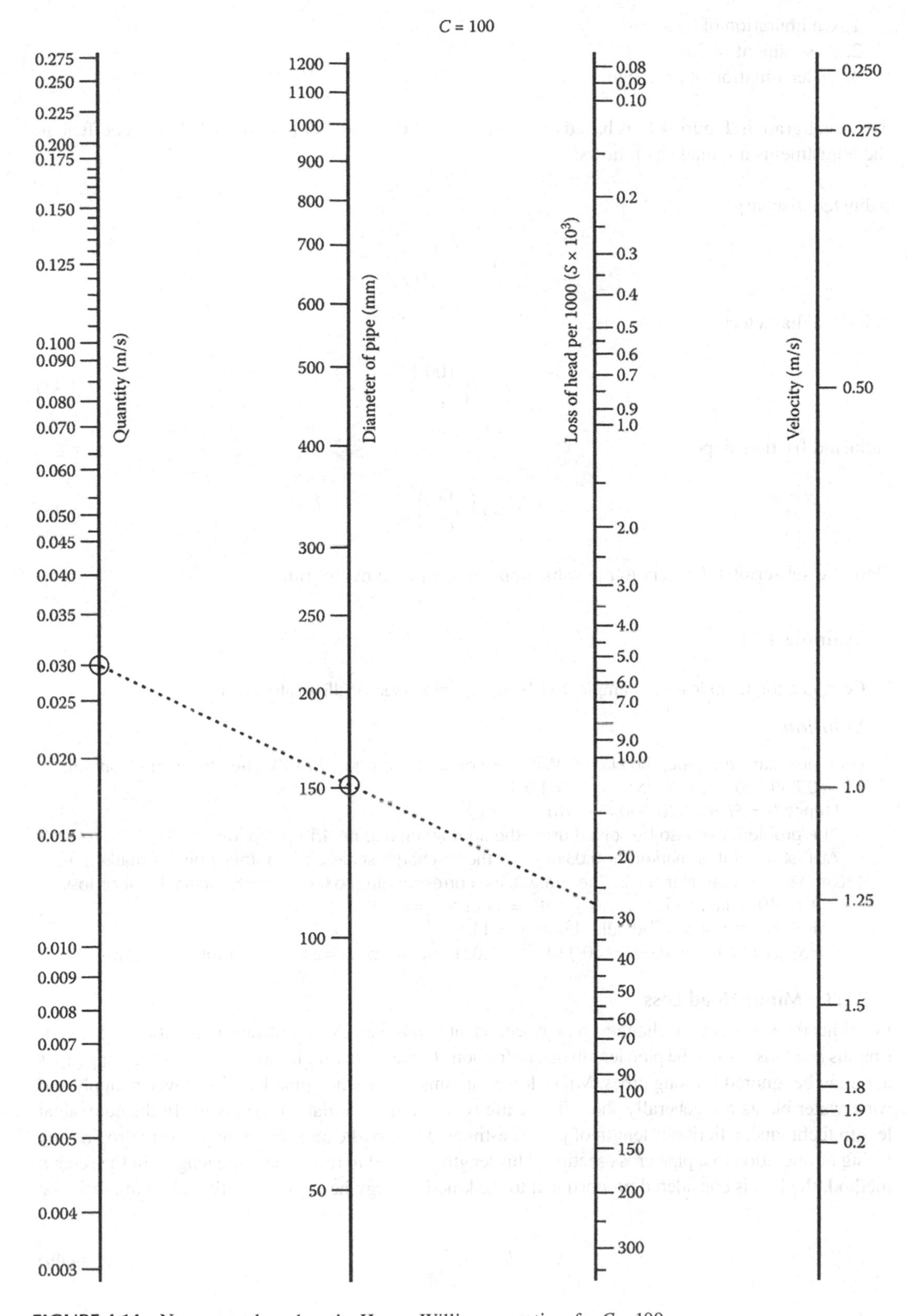

FIGURE 4.14 Nomogram based on the Hazen–Williams equation, for $C = 100$.

1. Computation of head loss
2. Assessment of flow
3. Determination of pipe size

The nomogram in Figure 4.14 is based on the coefficient $C = 100$. For pipes of a different coefficient, the adjustments are made as follows:

Adjusted discharge:

$$Q = Q_{100}\left(\frac{C}{100}\right). \tag{4.46}$$

Adjusted diameter:

$$d = d_{100}\left(\frac{100}{c}\right)^{0.38} \tag{4.47}$$

Adjusted friction slope:

$$S = S_{100}\left(\frac{100}{C}\right)^{1.85}. \tag{4.48}$$

Here the subscript 100 refers to the value obtained from the nomogram.

Example 4.12

Compute the head loss in Example 4.11 by using the Hazen–Williams formula.

Solution:

For a new cast iron pipe, the Hazen–Williams coefficient is equal to 130. Then from Equation 4.6, $1.7 = 0.849(130)(0.15/4)^{\circ.63}S^{\circ.54}$ or $S = 0.020$.

Hence $h_f = SL = 0.020(300) = 6.0\,\text{m}$.

The problem can also be solved using the nomogram (Figure 4.14) as follows:

At first a point is marked at $0.03\,\text{m}^3/\text{s}$ on the discharge scale and another point is marked at 150 mm on the diameter scale. The straight line corresponding to these points meets the head loss for the 1,000 scale at 35. Thus, $S_{100} \times 10^3 = 35$ or $S_{100} = 0.035$.

This value of S should be adjusted for $C = 130$.

$S = S_{100}(100/C)^{1.85} = (0.035)(100/130)^{1.85} = 0.021$ and hence $h_f = SL = 0.021(300.0) = 6.3\,\text{m}$.

4.5.2.3 Minor Head Loss

Local head losses occur at changes in pipe section, at bends, valves, and fittings in addition to the continuous head loss along the pipe length due to friction. These losses are important for <30-m-long pipes and may be ignored for long pipes. Minor losses are important when pipe lengths in water supply and wastewater plants are generally short. There are two ways to calculate these losses. In the equivalent length technique, a fictitious length of pipe is estimated that will cause the same pressure drop as any fitting or alteration in a pipe cross section. This length is added to the actual pipe length. In the second method, the loss is considered proportional to the kinetic energy head given by the following formula:

$$h_m = K\frac{V^2}{2g}, \tag{4.49}$$

where h_m is the minor head loss (m), K is the loss coefficient, and V is the mean velocity of flow (m/s).

TABLE 4.9
Minor Head Loss Coefficients

Item	Loss Coefficient (K)
Entrance Loss from Tank to Pipe	
Flush connection	0.5
Projecting connection	1.0
Exit loss from pipe to tank	1.0
Sudden Contraction	
$d_1/d_2 = 0.5$	0.37
$d_1/d_2 = 0.25$	0.45
$d_1/d_2 = 0.10$	0.48
Sudden Enlargement	
$d_1/d_2 = 2$	0.54
$d_1/d_2 = 4$	0.82
$d_1/d_2 = 10$	0.90
Fittings	
90° bend-screwed	0.5–0.9
90° bend-flanged	0.2–0.3
Tee	1.5–1.8
Gate valve (open)	0.19
Check valve (open)	3.00
Glove valve (open)	10.00
Butterfly valve (open)	0.30

Source: Gupta, R.S., *Hydrology and Hydraulic Systems* (4th ed.), Waveland Press, Long Grove, IL, 2016.

Some characteristic values of the minor head loss coefficient are given in Table 4.9.

The analysis consists of determination of the head loss or the rates of flow through a pipeline of a given size. Selection of a pipe size that will carry a design discharge between two points with specified reservoir elevations or a known pressure difference forms the design situation. The problems can be solved by the Darcy–Weisbach equation. If minor losses are neglected, the Hazen–Williams equation leads to the direct solution of both the analysis and design problems.

Example 4.13

Two reservoirs are connected by a 200-m-long cast iron pipeline, as shown in Figure 4.15. If the pipeline is proposed to convey a discharge of $0.2\ m^3/s$, what is the required size of the pipeline ($\upsilon = 1.3\ mm^2/s$)?

Solution:

The energy equation is applied between points 1 and 2:

$$Z_1 + \frac{P_1}{\gamma} + \frac{V_1^2}{2g} = Z_2 + \frac{P_2}{\gamma} + \frac{V_2^2}{2g} + h_{loss}$$

$$90 + 0 + 0 = 80 + 0 + 0 + h_{loss}$$

$$h_{loss} = 10\ \text{m}.$$

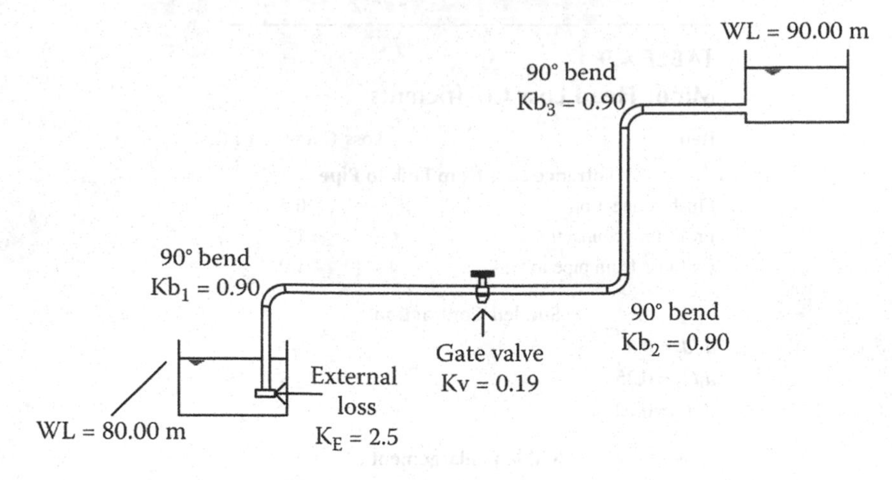

FIGURE 4.15 Pipe system connecting two reservoirs in Example 4.13.

Head loss is the summation of friction loss and minor losses ($h_{loss} = h_f + h_m$), which are calculated as follows:

Friction loss:

$$h_f = \frac{fL}{d}\frac{V^2}{2g} = \frac{fL}{d}\frac{Q^2}{\left[(\pi/4)d^2\right]\times 2g} = \frac{fLQ^2}{12.1d^5} = 0.66\frac{f}{d^5}.$$

Minor losses: The minor losses coefficients are listed in Table 4.10.

$$h_m = \sum\frac{KV^2}{2g} = \sum K\frac{Q^2}{\left[(\pi/4)d^2\right]^2\times 2g} = 0.003\frac{\sum K}{d^4} = \frac{0.016}{d^4},$$

$$h_{loss} = h_f + h_m \Rightarrow 0.066\frac{f}{d^5} + \frac{0.016}{d^4} = 10,$$

$$10d^5 - 0.016d - 0.66f = 0.$$

The above equation should be solved by trial and error. In the first trial, $f = 0.03$ and it has been calculated that $d = 0.30$ m. Thus

$$V = \frac{4Q}{\pi d^2} = \frac{4(0.2)}{\pi(0.30)^2} = 2.83 \text{ m/s},$$

TABLE 4.10
Minor Losses Coefficient for Example 4.13

Item	K
Three 90° bends	2.7
Gate valve	0.19
External loss	2.5
Total	5.39

$$Re = \frac{Vd}{\upsilon} = \frac{2.83(0.30)}{1.3\times10^{-6}} = 6.5\times10^{5},$$

$$\frac{e}{d} = \frac{2.44\times10^{-4}}{0.30} = 0.0008.$$

$f = 0.022$ (from the Moody diagram) is substituted in equation $10d^5 - 0.016\,d - 0.66\,f = 0$. Solved by trial and error: $d = 0.29$ m. Thus

$$V = \frac{4Q}{\pi d^2} = \frac{4(0.2)}{\pi(0.29)^2} = 3.03 \text{ m/s},$$

$$Re = \frac{Vd}{\upsilon} = \frac{3.03(0.29)}{1.3\times10^{-6}} = 6.8\times10^{5},$$

$$\frac{e}{d} = \frac{2.44\times10^{-4}}{0.29} = 0.0008.$$

$f = 0.022$ (from the Moody diagram). Since the amount of f has not changed, the required pipe diameter is 290 mm.

For lifting the water level and at intermediate points for boosting the pressure, the pumps are commonly used in waterworks and wastewater systems. A situation of water supplied from a lower reservoir to an upper-level reservoir is indicated in Figure 4.16.

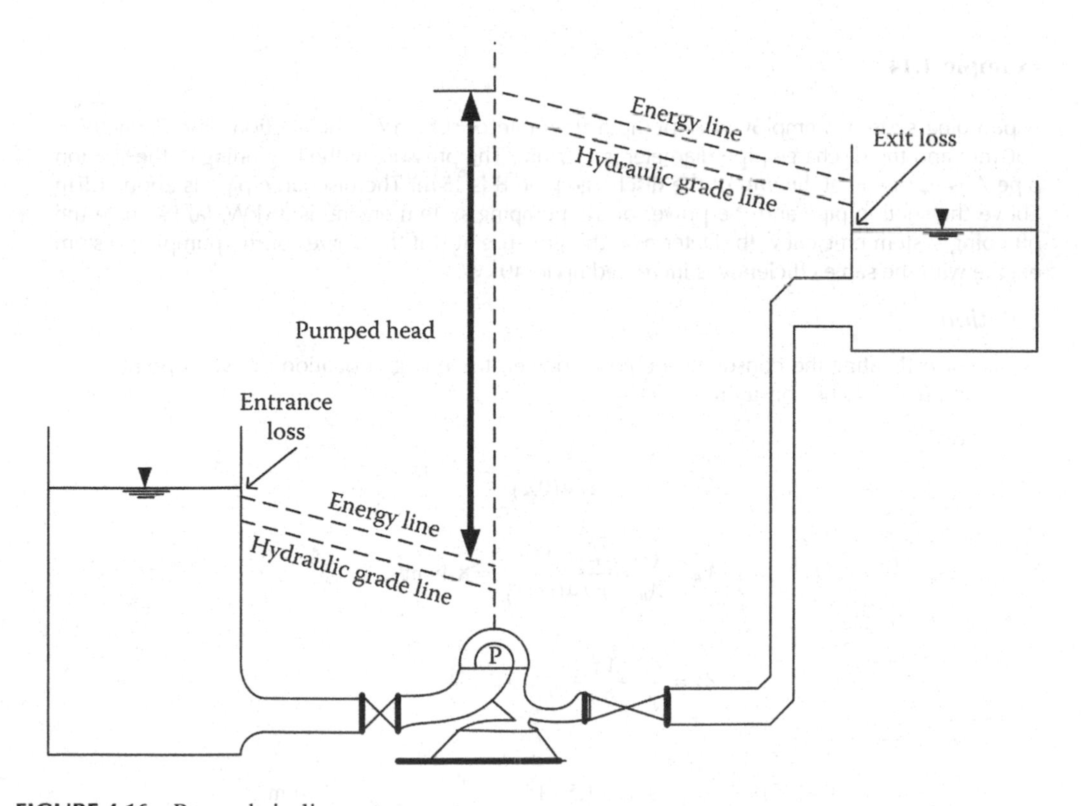

FIGURE 4.16 Pumped pipeline systems.

To analyze the system, the energy equation is applied between upstream and downstream ends of the pipe:

$$Z_1 + \frac{P_1}{\gamma} + \frac{V_1^2}{2g} + h_p = Z_2 + \frac{P_2}{\gamma} + \frac{V_2^2}{2g} + h_f + h_m. \quad (4.50)$$

Treating $V_1 = V_2$ yields

$$h_p = \left(Z_2 + \frac{P_2}{\gamma}\right) - \left(Z_1 + \frac{P_1}{\gamma}\right) + h_f + h_m, \quad (4.51)$$

$$h_p = \Delta Z + h_{\text{loss}}, \quad (4.52)$$

where h_p is the energy added by the pump (m), ΔZ is the difference in downstream and upstream piezometric heads or water levels, or total static head (m), h_f is the friction head loss and is equal to $(fL/D)(V^2/2g)$ (m), h_m is the minor head loss and $\sum KV^2/2g(\text{m})$, and h_{loss} is the total of friction and minor head losses.

The energy head, h_p, and the brake horsepower of the pump are related as

$$\text{BHP} = \gamma \frac{Qh_p}{\eta}, \quad (4.53)$$

where BHP is the brake horsepower (kW), Q is the discharge through the pipe (m^3/s), h_p is the pump head (m), and n is the overall pump efficiency.

Example 4.14

A pumping system is employed to pump a flow rate of $0.15\,m^3/s$. The suction pipe diameter is 200 mm and the discharge pipe diameter is 150 mm. The pressure at the beginning of the suction pipe A is –2 m and at the end of the discharge pipe B is 15 m. The discharge pipe is about 1.5 m above the suction pipe and the power of the pumping system engine is 35 kW. (a) Estimate the pumping system efficiency. (b) Determine the pressure at B if the power of the pumping system engine with the same efficiency is increased up to 40 kW.

Solution:

a. For estimating the consumed pumping power, the energy equation between points *A* and *B* should be considered.

$$V_A = \frac{Q}{A_A} = \frac{0.15}{\pi/4(0.2)^2} = 4.77 \text{ m/s},$$

$$V_B = \frac{Q}{A_B} = \frac{0.15}{\pi/4(0.15)^2} = 8.49 \text{ m/s},$$

$$Z_A = \frac{P_A}{\gamma} + \frac{V_A^2}{2g} + h_p = Z_B + \frac{P_B}{\gamma} + \frac{V_B^2}{2g},$$

$$0 + (-2) + \frac{(4.77)^2}{2\times 9.81} + h_p = 1.5 + 15 + \frac{(8.49)^2}{2\times 9.81} \Rightarrow h_p = 21.01 \text{ m}.$$

Pumping system power, $P = \gamma Q h_p = 9.81 \times 0.15 \times 21.01 = 30.91$ kW.
Pumping system efficiency = 30.91/35 = 0.88%.

b. At first, the effective pumping system power is calculated:

$$P = 0.88 \times 40 = 35.2 \text{ kW}.$$

Then the pumping head is calculated:

$$h_p = \frac{P}{\gamma Q} = \frac{35.2}{9.81 \times 0.15} = 23.92 \text{ m}.$$

Again the energy equation between points A and B is considered:

$$0 + (-2) + \frac{(4.77)^2}{2 \times 9.81} + 23.92 = 1.5 + P_B + \left(\frac{(8.49)^2}{2 \times 9.81} \Rightarrow P_B = 17.91 \text{ m.} \right)$$

4.5.2.4 Pipes in Series

Several pipes of different sizes could be connected together to form pipes in series or a compound pipeline as illustrated in Figure 4.17. According to the continuity and the energy equations, the following relations apply to the pipes in series:

$$Q = Q_1 = Q_2 = Q_3 = \ldots \tag{4.54}$$

$$h = h_1 = h_2 = h_3 + \ldots \tag{4.55}$$

For analysis purpose, different pipes are replaced by a pipe of uniform diameter of equivalent length that will pass a discharge, Q, with the total head loss, h_f. This is known as the equivalent pipe. The procedure is shown by an example.

Example 4.15

For a flow rate of 0.04 m³/s, determine the pressure and total heads at points A, B, C, and D for the series pipes shown in Figure 4.18 and Table 4.11. Assume a fully turbulent flow for all cases and the pressure head at point A to be 40 m. Determine the length of an equivalent pipe having a diameter of 1,000 mm and f equal to 0.025.

Solution:

The pressure and total heads are computed using the energy equation along the path beginning at point A. Given the pressure head and elevation, the total head at point A is

$$H_A = \frac{P_A}{\gamma} + Z_A = 40 + 20 = 60 \text{ m}.$$

Note that the velocity and the velocity head are

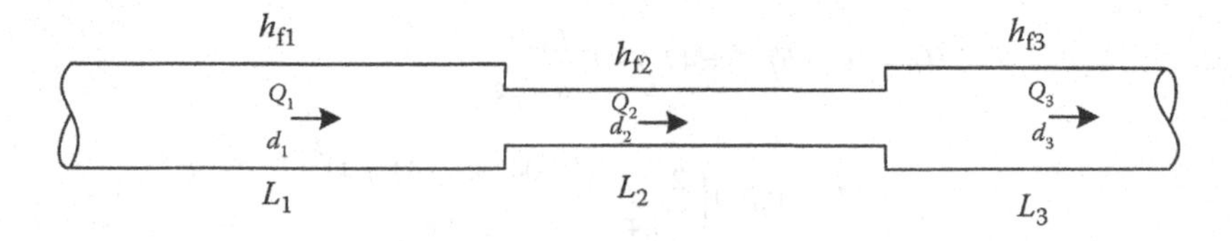

FIGURE 4.17 Compound pipeline.

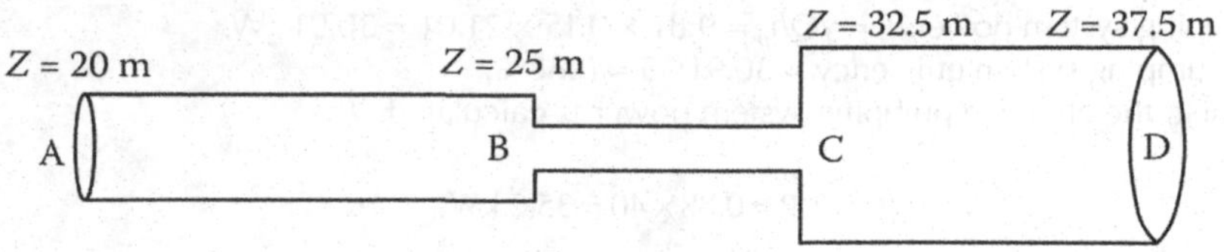

FIGURE 4.18 Compound pipe system of Example 4.15.

TABLE 4.11
Computation for Pipes in Series

Pipe	Pipe Diameter (mm)	Discharge (m^3/s)	Friction Slope	Pipe Length (m)	Head Loss (m)
1	300	0.04	0.022	2,000	2.4
2	200	0.04	0.025	1,000	10.3
3	400	0.04	0.021	2,000	0.5
	300 (Selected)	0.04	0.025 (Selected)	9,703 (Determined)	13.2

$$V_1 = \frac{Q}{A} + \frac{0.04}{\pi(0.3)^2/4} = 0.57 \text{ m/s}.$$

and

$$\frac{V_1^2}{2g} = \frac{0.57^2}{2(9.81)} = 0.017 \text{ m}.$$

The velocity head is four orders of magnitude less than the static head, so it can be neglected. Neglecting the velocity head is a common assumption in pipe network analysis.

All energy loss in the system is due to friction. So, following the path of flow, the total heads at A, B, C, and D are

$$H_B = H_A - h_f^{A-B} = 60 - f_1 \frac{L_1}{d_1}\frac{V_1^2}{2g} = 60 - 0.022\left(\frac{2000}{0.3}\right)\frac{0.57^2}{2(9.81)}$$

$$= 60 - 2.4 = H_B = 57.6 \text{ m}.$$

$$H_C = H_B - h_f^{B-C} = 57.6 - f_2 \frac{L_2}{d_2}\frac{V_2^2}{2g}$$

$$= 57.6 - 0.025\left(\frac{1,000}{0.2}\right)\frac{\left(0.04/\pi(0.2)^2/4\right)^2}{2(9.81)}$$

$$= 57.6 - 10.3 \Rightarrow H_C = 47.3 \text{ m}.$$

$$H_D = H_C - h_f^{C-D} = 47.3 - f_3 \frac{L_3}{d_3}\frac{V_2^2}{2g}$$

$$= 47.3 - 0.021\left(\frac{2,000}{0.4}\right)\frac{\left(0.04/\pi(0.4)^2/4\right)^2}{2(9.81)}$$

$$= 47.3 - 0.5 \Rightarrow H_D = 46.8 \text{ m}.$$

The pressure heads are

At point *B*:

$$H_B = \frac{P_B}{\gamma} + Z_B = 57.6 \; \frac{P_B}{\gamma} + 25 \Rightarrow \frac{P_B}{\gamma} = 32.6 \text{ m}$$

At point *C*:

$$H_C = \frac{P_C}{\gamma} + Z_C = 47.3 \; \frac{P_C}{\gamma} + 32.5 \Rightarrow \frac{P_C}{\gamma} = 14.8 \text{ m}$$

At point *D*:

$$H_D = \frac{P_D}{\gamma} + Z_D = 46.8 \; \frac{P_D}{\gamma} + 37.5 \Rightarrow \frac{P_D}{\gamma} = 9.3 \text{ m}$$

Total head loss of the equivalent pipe is $h_{\text{loss}} = H_A - H_D = 60 - 46.8 = 13.2$.

$$h_{\text{loss}} = f_{\text{eq}} \frac{L_{\text{eq}}}{d_{\text{eq}}^5} \frac{Q_{\text{eq}}^2}{12.1} \Rightarrow 13.2 = 0.025\left(\frac{L_{\text{eq}}}{0.3^5}\right)\frac{0.04^2}{12.1} \Rightarrow L_{\text{eq}} = 9{,}703 \text{ m}$$

The calculations are summarized in Table 4.11.

Example 4.16

For the series pipe system in Example 4.15, find the equivalent roughness coefficient and the total head at point *D* for a flow rate of 0.03 m³/s.

Solution:

The equivalent pipe loss coefficient is equal to the sum of the pipe coefficient or

$$K_{\text{eq}}^s = \sum_{l=1}^{3} k_l = \sum_{l=1}^{3} \frac{8 f_l L_l}{g\pi^2 d_l^5},$$

For this problem,

$$K_{\text{eq}}^s = K_1 = K_2 = K_3 = \frac{8f_1L_1}{g\pi^2 d_1^5} + \frac{8f_2L_2}{g\pi^2 d_2^5} + \frac{8f_3L_3}{g\pi^2 d_3^5}$$

$$= \frac{8(0.022)(2{,}000)}{9.81\pi^2(0.3)^5} + \frac{8(0.025)(1{,}000)}{9.81\pi^2(0.2)^5} + \frac{8(0.021)(2{,}000)}{9.81\pi^2(0.4)^5}$$

$$= 1{,}496 + 6{,}455 + 339 \Rightarrow K_{\text{eq}}^s = 8{,}290$$

Note that pipe 2 has the largest loss coefficient since it has the smallest diameter and highest flow velocity. As seen in the previous example, although it has the shortest length, most of the loss occurs in this section. The head loss between nodes A and D for $Q = 0.03\,\text{m}^3/\text{s}$ is then

$$h_f^{A-D} = K_{\text{eq}}^s Q^2 K_1 Q^2 + K_2 Q^2 + K_3 Q^2 = 8{,}290(0.03)^2 \Rightarrow h_f^{A-D} = 7.5 \text{ m}.$$

We can also confirm the result in the previous example by substituting $Q = 0.04\,\text{m}^3/\text{s}$ in which case

$$h_f^{A-D} = K_{eq}^s Q^2 = 8,290(0.04)^2 = 13.26 \text{ m}.$$

and

$$H_D = H_A - h_f^{A-D} = 60 - 13.26 \Rightarrow H_D = 46.74 \text{ m}.$$

That would be equivalent to the earlier result if the previous example was carried to two decimal places.

4.5.2.5 Pipes in Parallel

For the parallel or looping pipes of Figure 4.19, the continuity and energy equations provide the following relations:

$$Q = Q_1 + Q_2 + Q_3 + \ldots + Q_n. \quad (4.56)$$

$$h_f = h_{f1} = h_{f2} = h_{f3} = \ldots = h_{fn}. \quad (4.57)$$

A procedure similar to that used for pipes in series is also used in this case, as illustrated in the following example.

Example 4.17

Given the data for the three parallel pipes in Figure 4.20, compute (1) the equivalent parallel pipe coefficient, (2) the head loss between nodes A and B, (3) the flow rates in each pipe, (4) the total head at node B, and (5) the diameter of an equivalent pipe with length 100 m and $C_{hw} = 100$.

Solution:

(1) The equivalent parallel pipe coefficient allows us to determine the head loss that can then be used to disaggregate the flow between pipes. The loss coefficient for the Hazen–Williams equation for pipe 1 is

$$K_1 = \frac{10.66 L_1}{C_1^{1.85} d_1^{4.87}} = \frac{10.66(91.5)}{120^{1.85}(0.356)^{4.87}} = 21.253.$$

Similarly, K_2 and K_3 equal 6.366 and 9.864, respectively.

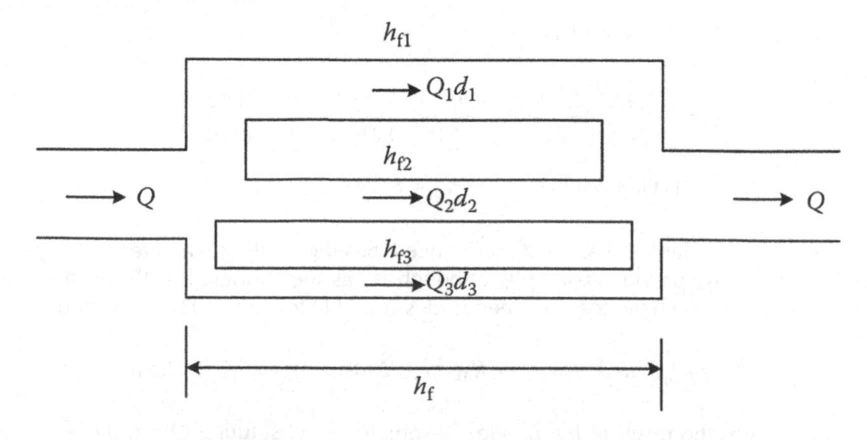

FIGURE 4.19 Parallel pipe system.

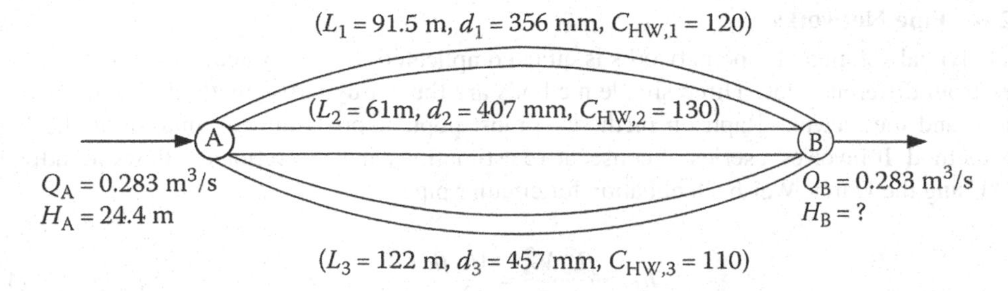

FIGURE 4.20 Pipes in parallel for Example 4.17.

The equivalent loss coefficient is

$$\left(\frac{1}{K_1}\right)^{1/n}+\left(\frac{1}{K_2}\right)^{1/n}+\left(\frac{1}{K_3}\right)^{1/n}=\left(\frac{1}{21.253}\right)^{0.54}+\left(\frac{1}{6.366}\right)^{0.54}+\left(\frac{1}{9.864}\right)^{0.54}$$

$$=0.192+0.368+0.291=\left(\frac{1}{K_{eq}^{p}}\right)^{0.54}\Rightarrow K_{eq}^{p}=1.348.$$

(2) and (4) The head loss between nodes A and B is then

$$h_L=K_{eq}^{p}\left(Q_{total}\right)^{n}=1.348(0.283)^{1.85}=0.131\text{ m}.$$

(4) So the head at node B, H_b, is

$$H_A=H_B-h_L=24.4-H_B=0.131\text{ m}\Rightarrow H_B=24.269\text{ m}.$$

(3) The flow in each pipe can be computed from the individual pipe head loss equations since the head loss is known for each pipe (h_L = 0.131 m),

$$Q_1=\left(\frac{1}{K_1}\right)^{1/n}h_L^{1/n}=\left(\frac{1}{21.253}\right)^{0.54}0.131^{0.54}=0.64\text{ m}^3/\text{s}.$$

The flows in pipes 2 and 3 can be calculated by the same equation and are 0.123 and 0.097 m³/s, respectively. The sum of the three pipe flows equals 0.284 m³/s, which is approximately same as the inflow to node A. The results are tabulated in Table 4.12.

(5) The diameter of the equivalent pipe is estimated as

$$K_{eq}=\frac{10.66L_{eq}}{C_{eq}^{1.85}d_{eq}^{4.87}}\Rightarrow 1.348=\frac{10.66(100)}{100^{1.85}\left(d_{eq}\right)^{4.87}}\Rightarrow d_{eq}=0685\text{ m}.$$

TABLE 4.12
Computation of Parallel Pipe System

Pipe	Pipe Diameter (cm)	Head Loss (m)	Pipe Length (m)	Chw	Discharge (m³/s)
1	356	0.131	91.5	120	0.064
2	407	0.131	61	130	0.123
3	457	0.131	122	110	0.097
	685	0.131	100	100	0.284
	(Computed)		(Assumed)	(Computed)	

4.5.2.6 Pipe Networks

The analytical solution of pipe networks is quite complicated. In this system, the flow to an outlet comes from different sides. Three simple methods are the Hardy Cross method, the linear theory method, and the Newton–Raphson method. A most popular procedure of analysis is the Hardy Cross method. It involves a series of consecutive estimations and corrections to flows in individual pipes. Using the Darcy–Weisbach equation for circular pipes,

$$h_f = \frac{fL^2}{d}\frac{V^2}{2g} = \frac{16}{\pi^2}\frac{fL}{d^5}\frac{Q^2}{2g}. \tag{4.58}$$

From the Hazen–Williams equation for circular pipes,

$$Q = 0.278\ Cd^{2.63} = \left(\frac{h_f}{L}\right)^{0.54}. \tag{4.59}$$

or

$$h_f = \frac{10.7L}{C^{1.85}d^{4.87}}Q^{1.85}. \tag{4.60}$$

Both h_f equations can be expressed in the general form

$$h_f = KQ^n, \tag{4.61}$$

where K is given in Table 4.13 and n is 2.0 for the Darcy–Weisbach equation and 1.85 for the Hazen–Williams equation.

The sum of head losses around any closed loop is zero (energy conservation), that is,

$$\sum h_f = 0.$$

Consider that Q_a is an assumed pipe discharge that varies from pipe to pipe of a loop to satisfy the continuity of flow. If S is the correction made in the flow of all pipes of a loop to satisfy the above equation, then by substituting $h_f = KQ_n$ in $\Sigma h_f = 0$,

$$\sum K(Q_a + \delta)^n = 0.$$

Expanding the above equation by the binomial theorem and retaining only the first two terms yields

TABLE 4.13
Equivalent Resistance, *K*, for the Pipe

Formula	Units of Measurement	*K*
Hazen–Williams	Q, m^3/s, L, m, d, m, h_f, m	$\frac{10.7L}{C^{1.85}d^{4.87}}$
Darcy–Weisbach	Q, m^3/s, L, m, d, m, h_f, m	$\frac{fL}{12.1\ d^5}$

TABLE 4.14
Iteration 1 of the Hardy Cross Procedure

Loop	Pipeline	K	Q_a (m³/s)	$h_f = KQ_a^{1.85}$	$1.852\lvert h_f / Q\rvert$	$Q_a + \delta$
1	1	5	1.8	14.850	15.279	+1.698
	2	8	−1.2	−11.213	17.305	−1.302
	3	8	−0.1	−0.112	2.083	−0.224
	Sum			3.525	34.667	–
2	3	8	0.1	0.112	2.083	−0.224
	4	6	1.1	7.158	12.052	+0.674
	5	5	−0.2	−0.254	2.350	−0.626
	Sum			7.016	16.485	–

$$\delta = -\frac{\sum KQ_a^n}{\sum KQ_a^{n-1}} \tag{4.62}$$

or

$$\delta = -\frac{h_f}{n\sum \lvert h_f / Q_a\rvert}. \tag{4.63}$$

Equations 4.60 and 4.62 are used in the Hardy Cross procedure. The values of n and K are obtained based on the Darcy–Weisbach or Hazen–Williams equations from Table 4.13. The procedure is summarized as follows:

1. Divide the network into a number of closed loops. The computations are made for one loop at a time.
2. Compute K for each pipe using the appropriate expression from Table 4.13 (column 3 of Table 4.14).
3. Assume a discharge, Q_a, and its direction in each pipe of the loop (column 4 of Table 4.14). At each joint, the total inflow should be equal to total outflow. Consider the clockwise flow to be positive and the counterclockwise flow to be negative.
4. Compute h_f in column 5 for each pipe by Equation 4.60, retaining the sign of column 4. The algebraic sum of column 5 is h_f.
5. Compute h_f/Q for each pipe without considering the sign. The sum of column 6 is $\sum \lvert h_f / Q\rvert$.
6. Determine the correction, δ, by Equation 4.62. Apply the correction algebraically to the discharge of each member of the loop.
7. For common members among two loops, both S corrections should be made, one for each loop.
8. For the adjusted Q, steps 4–7 are repeated until δ becomes very small for all loops.

Example 4.18

Find the discharge in each pipe of the pipe network shown in Figure 4.21. The head loss in each pipe is calculated by using $h_f = KQ1.852$, where the values of K for each pipe are given in Figure 4.21. The pressure head at point 1 is 100 m. Determine the pressure at different nodes.

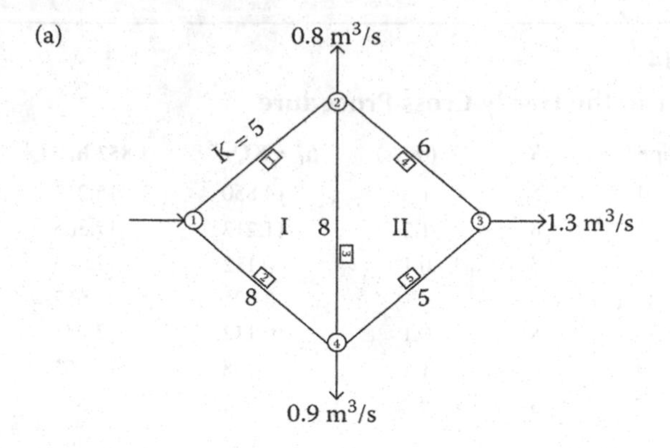

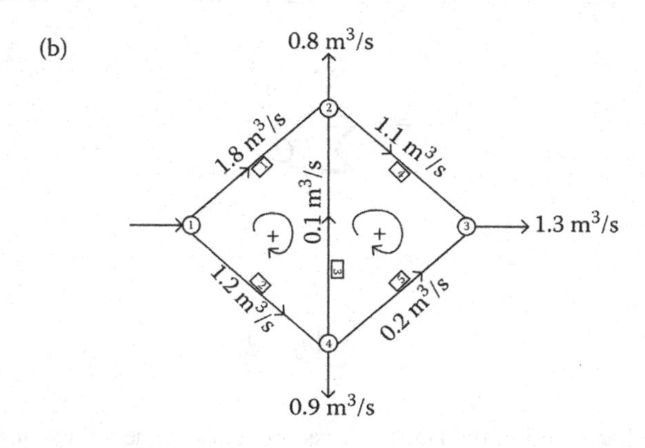

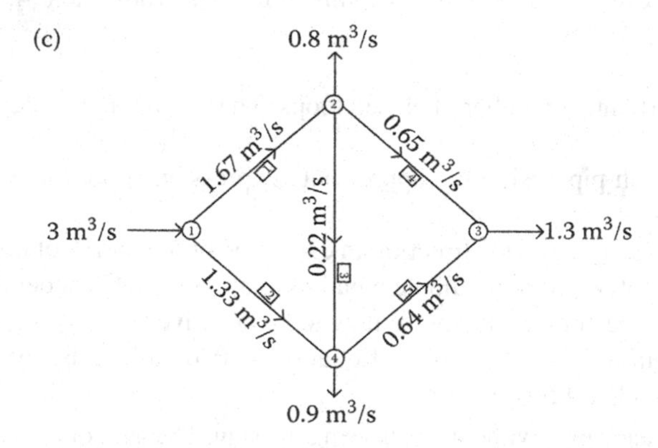

FIGURE 4.21 (a) Pipe network of Example 4.18; (b) assumed discharges; (c) discharge after the final iteration.

Solution:

Two first Hardy Cross iterations are summarized in Tables 4.14 and 4.15. In the first iteration, the flow in each pipe is assumed to be indicated in Figure 4.21.

Iteration 1:
For loop 1,

$$\delta = (-3.525 / 34.667) = -0.102, \quad \text{adjusted } Q_1 = +1.8 + (-0.102) = 1.698.$$

TABLE 4.15
Iteration 2 of the Hardy Cross Procedure

Loop	Pipeline	K	Q_a (m³/s)	$h_f = KQ_a^{1.85}$	$1.852\|h_f / Q\|$	Q Corrected = $Q_a + \delta$
1	1	5	1.698	13.330	14.538	1.677
	2	8	−1.302	−13.042	18.551	−1.323
	3	8	0.224	0.501	4.141	0.219
	Sum			0.789	37.230	–
2	3	8	−0.224	−0.501	4.141	−0.219
	4	6	0.674	2.890	7.940	0.658
	5	5	−0.626	−2.100	6.213	−0.642
	Sum			0.289	18.294	–

TABLE 4.16
Final Flows and Pressure Heads

Pipe	Flow (m³/s)	Head Loss (m)	Node	Pressure Head (m)
1	1.675	13.00	1	100 m (given)
2	1.326	13.49	2	$100 - h_1 = 87$
3	−0.222	0.49	3	$h_2 - h4 = 87 - 2.73 = 84.27$
4	0.653	2.73	4	$100 - h_2 = 86.51$
5	0.647	2.23	–	–

For loop 2,

$$\delta = (-7.016 / 16.485) = -0.426, \quad \text{adjusted } Q_3 = 0.1 - (-0.102) + (-0.426) = -0.224.$$

Iteration 2:
For loop 1,

$$\delta = (-0.789 / 37.230) = -0.102, \text{ adjusted } Q_1 = +1.698 + (-0.021) = 1.677.$$

For loop 2,

$$\delta = (-0.289 / 18.294) = -0.016, \text{ adjusted } Q_3 = -0.224 - (-0.021) + (-0.016) = -0.219.$$

The Hardy Cross process converges to the final results after five iterations. The final results are tabulated in Table 4.16.

4.6 CONCLUDING REMARKS

In this chapter, the hydraulic design of urban drainage and water distribution systems are introduced. This chapter is particularly useful for developing urban areas. The presented design procedures are based on fundamental hydrological and hydraulic design concepts. Water conveyance in urban area is more complicated than undeveloped area because of the complexity of the paths that rainwater and stormwater take and many interdependencies of water infrastructure with other infrastructures.

Measuring the effect of geomorphic changes (land use driven) on urban water is essential to mitigate the potential effects. The urban water management will be more effective if risk assessments include uncertainties associated with geomorphological changes to underpin nature-based management approaches and land use alterations. Changes in geomorphic process regimes can also be triggered by extreme events. Implementing geomorphological adaptation strategies will enable communities to develop more resilient, less vulnerable socioeconomic systems fit for an age of climate extremes. The outcome of such approach will be of interest to landscape architects/managers and regulators because of the criticalities related to stormwater collection and flood management.

This chapter can serve as an essential tool for anyone who aspires to become a civil or hydraulic engineer. Classic hydraulics that are presented in this chapter are the foundation upon which many other civil and environmental engineering and water resources planning and management assumptions are built. This chapter provides valuable insight into the governing equations and underlying concepts of channel geomorphology, open-channel flows and pipelines which then together can be used in carrying out flood plain, drainage systems and water distribution systems' design and analysis issues. The characteristics of hydraulic structures of different applications are discussed in Chapters 5–7.

PROBLEMS

1. Water having a temperature of 15°C is flowing through a 150 mm ductile iron main at a rate of 19 L/s. Is the flow laminar, turbulent, or transitional?
2. Does conservation of energy apply to the system represented in Figure 4.22. Based on the conservation of energy, the summation of head losses in a loop should be equal to zero. Data describing the physical characteristics of each pipe are presented in Table 4.16 and neglect the minor losses in this loop.
3. Manually find the discharge through each pipeline and the pressure at each junction node of the rural water system shown Figure 4.23. Physical data for this system are given in Table 4.17.
4. In the network depicted in Figure 4.24, determine the discharge in each pipe. Assume that $f = 0.015$. Physical data of the pipeline have been given in Table 4.17.
5. A cast iron pipe is employed to deliver a flow rate of 0.1 m^3/s between two points that are 800 m apart. Determine the pipe size if the allowable head loss is 10 m and the Hazen–Williams roughness coefficient is 130.
6. Using the Muskingum method, route the inflow hydrograph given in Table 4.18, assuming (a) $K = 4$ hours and $X = 0.12$ and (b) $K = 4$ hours and $X = 0.0$. Plot the inflow and outflow hydrographs for each case, assuming the initial outflow equals the inflow.

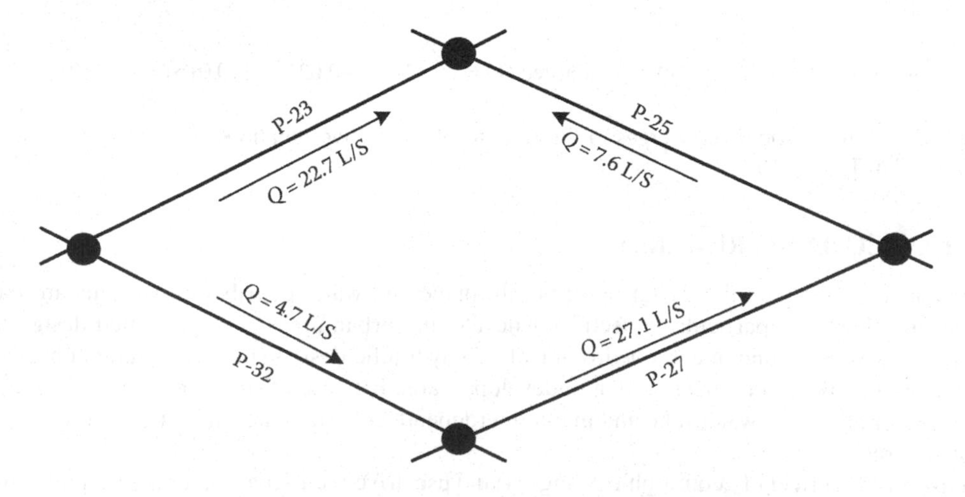

FIGURE 4.22 Pipeline of Problem 2.

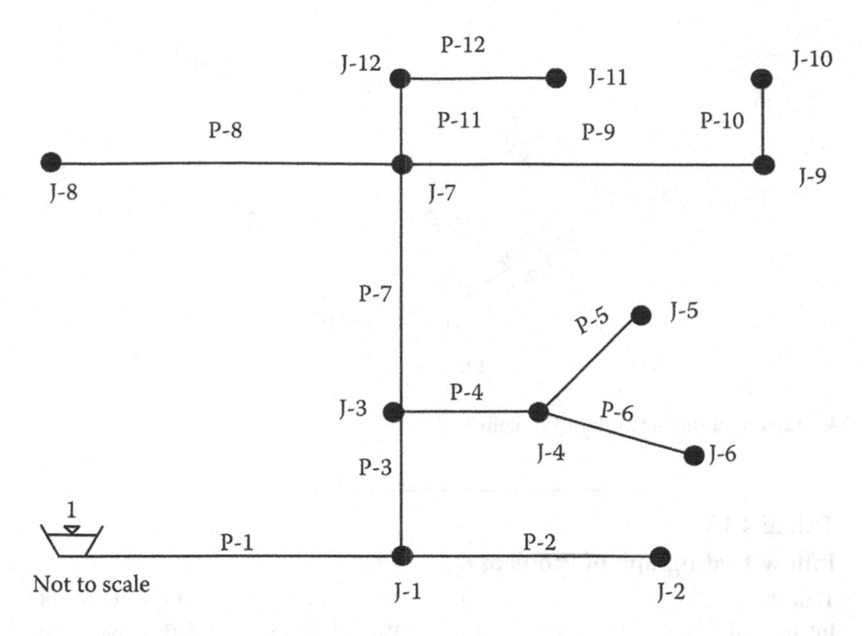

FIGURE 4.23 Pipeline of Problem 3.

TABLE 4.17
Physical Data of the Pipeline System of Problems 3 and 4

Pipe Label	Length (m)	Diameter (mm)	Hazen-Williams Roughness Coefficient (C)	Node Label	Elevation (m)	Demand (L/s)
P-1	152.4	254	120	R-1	320.0	N/A
P-2	365.8	152	120	J-1	262.1	2.5
P-3	1280.2	254	120	J-2	263.7	0.9
P-4	182.9	152	110	J-3	265.2	1.9
P-5	76.2	102	110	J-4	266.7	1.6
P-6	152.5	102	100	J-5	268.2	0.3
P-7	1585.0	203	120	J-6	269.7	0.8
P-8	1371.6	102	100	J-7	268.2	4.7
P-9	1676.4	76	90	J-8	259.1	1.6
P-10	914.4	152	75	J-9	262.1	0
P-11	173.7	152	120	J-10	262.1	1.1
P-12	167.6	102	80	J-11	259.1	0.9
				J-12	257.6	0.6

7. Determine the flow in a composite gutter with $W = 0.5\,\text{m}$, $S = 0.015$, $S_X = 0.03$, and $a = 0.08\,\text{m}$. Assume that T is equal to 2 m and the Manning coefficient is 0.01.
8. Consider an impervious area with the length of 40 m, slope of 0.03. The effective Manning coefficient is 0.03. Determine the runoff hydrograph for a constant rainfall excess rate of 20 mm/h with 10-minute duration at the exit point of the area.
9. Determine the peak discharge resulting from a 2-hour rainfall with a constant intensity equal to 2.5×10^{-5} mm/s that happens on an impervious rectangular basin with the following characteristics: $L = 200\,\text{m}$, $S = 0.005$, and $n = 0.04$. The infiltration parameters are $\Phi = 0.5$, $K = 3 \times 10^{-6}$ m/s, and $P_f = 0.25\,\text{m}$. At the beginning of the rainfall, the basin soil is 0.50% saturated.

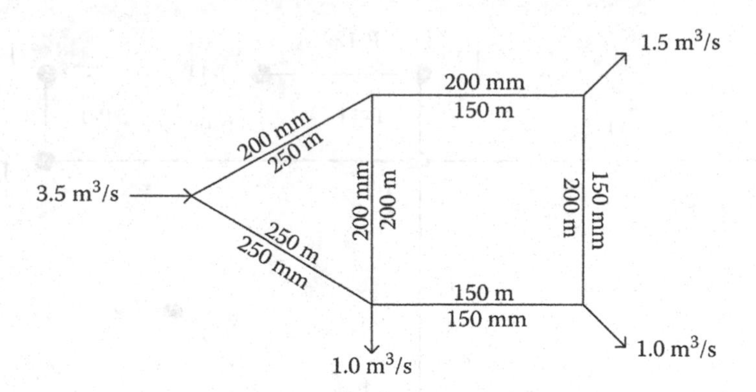

FIGURE 4.24 Layout of the network of Problem 4.

TABLE 4.18
Inflow Hydrograph of Problem 6

Time (h)	0	2	4	6	8	10	12	14	16	18
Inflow (L/s)	0	135	675	1350	945	567	351	203	68	0

10. Determine the flow depth and spread of a triangular gutter with $S = 0.02$, $S_X = 0.03$, and $n = 0.02$ where the peak flow is estimated as $0.1\,m^3/s$.
11. Two adjacent subwatersheds have characteristics as follows: Subarea 1: length = 2,500 m; elevation drop = 52 m. Subarea 2: length = 1,200 m; elevation drop = 65 m. Compute the slope of each subarea. Can you legitimately conclude that the average watershed slope for the entire area is the average of the two subarea slopes? Explain.
12. Derive expressions for computing the shape parameters (LC, L, L_i, R_e) of an elliptical watershed with lengths for the major and minor axes $2a$ and $2b$, respectively. Assume the watershed outlet is located at one end of the major axis.
13. Compute the channel slope between sections 1 and 6 and for each of the five reaches of the given data in Table 4.19. Compute the average of the computed slopes for the five reaches and compare it to the estimate for the entire reach. Discuss the results.
14. Compute the average channel slope using the station data from Problem 5.
15. Find the drainage density for the given watershed in Figure 4.25. The reach lengths, in miles, are shown in the figure (area = $80\,km^2$).
16. In a watershed, the number of streams of orders 1–6 is 42, 56, 19, 9, 3, and 1, respectively. Compute the bifurcation ratio.

TABLE 4.19
Characteristics of River Reach in Problem 13

Section	Elevation (m)	Distance from Outlet (m)
1	150	0
2	168	4,100
3	170	5,020
4	190	8,100
5	220	11,500
6	234	12,340

TABLE 4.20
Stream Data of Watershed in Problem 17

Stream Orders	Stream Number	Stream Length (km)	Stream Area (km²)
1	44	2.12	17.5
2	15	3.02	8.6
3	4	7.15	5
4	1	13.4	1.6

17. The number, length, and area of each stream order are shown in Table 4.20. Compute the bifurcation ratio and the relations for estimation of stream number, total length, and area of different orders in this watershed.
18. Calculate the time of concentration of the watershed where the length of the main river is 3,000 m and the average slope is 0.02. If the curve number is estimated at 80, use the Soil Conservation Service (SCS) equation for calculation.

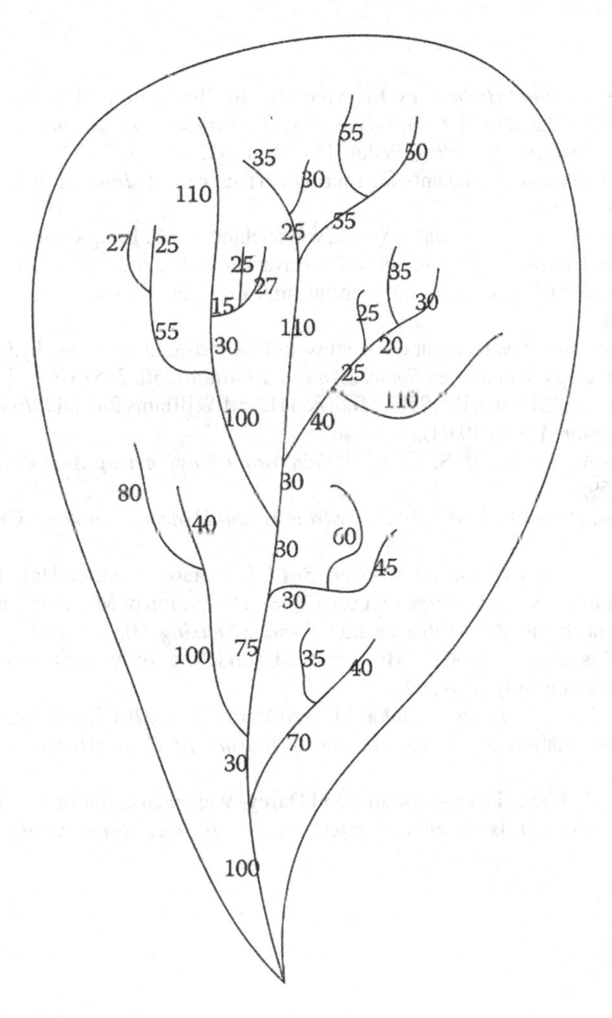

FIGURE 4.25 Watershed in Problem 15.

19. Using the velocity method, find the concentration time of a 1.6 ha drainage area which is primarily forested, with a natural channel having a good stand of high grass. The data of Table 3.3 can be used to estimate the time of concentration. The length and slope of the forested area are 100 km and 0.2% and the channel length is 20 km with a slope of 0.5%.
20. Consider a smooth concert surface of 30 m in length at a slope of 0.15%. If the rain intensity is 125 mm/h, calculate the estimated travel time.
21. Calculate the time of concentration of the dense grass watershed where the length of the main river is 2,000 m, the length of drainage is 1,200 m, and the average slope is 2%. If the curve number is estimated at 70, use the equations below for calculation.
    a. Kerby-Hathaway formula
    b. Kirpich's methods
    c. Van Sickle equation
22. Which one of the following characteristics is not a measure of basin shape? (a) The length to the center of area, (b) the elongation ratio, (c) the drainage density, and (d) the circularity ratio.
23. Which one of the following is not a factor in controlling times of concentration of watershed runoff? (a) The drainage density, (b) the roughness of the flow surface, (c) the rainfall intensity, (d) the flow lengths, and (e) all of the above are factors.

REFERENCES

Akan, A.O. (2011). *Open Channel Hydraulics.* Elsevier. Oxford, Burlington, MA, 384 pp.

Akan, A.O. and Houghtalen, R.J. (2003). *Urban Hydrology, Hydraulics, and Stormwater Quality: Engineering Applications and Computer Modeling.* Wiley, Hoboken, NJ, 392 pp.

Engman, E.T. (1986). Roughness coefficients for routing surface runoff. *Journal of Irrigation and Drainage Engineering,* 112(1), 39–53.

Gupta, R.S. (2016). *Hydrology and Hydraulic Systems.* Waveland Press, Long Grove, IL, 888 pp.

Hogarth, W.L., Parlange, J.Y., Rose, C.W., Fuentes, C., Haverkamp, R., and Walter, M.T. (2005). Interpolation between Darcy–Weisbach and Darcy for laminar and turbulent flows. *Advances in Water Resources,* 28(10), 1028–1031.

Horton, R.E. (1945). Erosional development of streams and their drainage basins, hydrophysical approach to quantitative morphology. *Geological Society America Bulletin,* 56, 275–370.

Jain, A.K., Mohan, D.M., and Khanna, P. (1978). Modified Hazed-Williams formula. *Journal of Environmental Engineering Diversion ASCE,* 104(1), 137–146.

Karamouz, M., Moridi, A., and Nazif, S. (2011). *Urban Water Engineering and Management.* CRC Press, Boca Raton, FL, 595.

Karamouz, M., Nazif, S., and Fallahi, M. (2012). *Hydrology and Hydroclimatology.* CRC Press, Boca Raton, FL, 716.

McCuen, R. (2005). *Hydrologic Analysis and Design, 3rd Ed.* Pearson, Prentice-Hall, London.UK

Sadeh, Y., Cohen, H., Maman, S., and Blumberg, D.G. (2018). Evaluation of Manning's n roughness coefficient in arid environments by using SAR backscatter. *Remote Sensing,* 10(10), 1505.

Singh, V.P. and Woolhiser, D.A. (2002). Mathematical modeling of watershed hydrology. *Journal of Hydrologic Engineering,* 7(4), 270–292.

Tejada-Martínez, A.E., Hafsi, A., Akan, C., Juha, M., and Veron, F. (2020). Large-eddy simulation of small-scale Langmuir circulation and scalar transport. *Journal of Fluid Mechanics,* 885, DOI: 10.1017/Jfm.2019.802

Valiantzas, J.D. (2005). Modified Hazen–Williams and Darcy–Weisbach equations for friction and local head losses along irrigation laterals. *Journal of Irrigation and Drainage Engineering,* 131(4), 342–350.

5 Urban Stormwater Drainage Systems

5.1 INTRODUCTION

As key infrastructures in urban areas, stormwater drainage systems play an essential role in communities' health and safety. The best practices of these systems are related to their design, maintenance, and different modes of operation. The urban is affected the most by problems related to inadequate drainage in cities of developing countries. Low-value land with steep-sided hillsides, prone to flooding, is often inhibited by poor communities that are unattractive for decision makers to improve. Although the consequences of flooding can be destructive, the disadvantages of the substandard drainage system are balanced by the benefits of living near sources of employment and urban services (Parkinson, 2002).

An accurate estimate of runoff and peak discharge is very important in urban drainage system planning and management. The estimation of the peak rate of runoff, runoff volume, and the temporal and spatial distribution of flow is the basis for the planning, designing, and construction of a drainage system. Urban drainage is usually designed based on the concept of draining water from urban surfaces (small watersheds) as quickly as possible through pipes and channels. However, the peak flow and the cost of the drainage system are increased by this approach (Tucci, 2007). There are also certain peak increases at minor drainage levels, which could highly impact the condition of the major drainage that should be controlled for the safety of highways, street intersections, and buildings with basements and underground parking. Any inaccurate analysis could result in the undersized or oversized design of the conveyance and control structures. In this chapter, different aspects of the design and planning of urban drainage systems are introduced. This chapter is particularly useful for developing urban areas. Many street and highway drainage issues and requirements are discussed that transportation engineers and contractors can effectively utilize. The design procedures presented here are based on the basic concepts explained in Chapters 3 and 4.

A dedicated and perhaps unique feature of this chapter is the inclusion of detailed land use planning, which is the essence of integrated urban water management, based on DSR (Driving force, State, and Response) dynamic strategy planning procedure. The following issues are addressed in the remaining sections: Special characteristics of the urban storm; the complexity of urban watersheds; imperviousness and the maze of water pathways/channels; local ordinances and risk-based design values; streets/highways drainage; control structures/best management practices (BMPs); urban flood, combined sewer overflow (CSO); and interdependencies (water, energy, and transportation). Perhaps the key to many growing urban stormwater management (SWM) challenges is the lack of appropriate land use planning and excessive human intervention that exuberates to disasters such as what was witnessed in some extreme urban flooding in New Orleans in 2005 and in New York City in October 2012. The issues related to storms in coastal cities are explained in Chapter 13.

5.2 URBAN PLANNING AND STORMWATER DRAINAGE

Urban stormwater drainage systems are one of the fundamental components of the urban infrastructure. Appropriate design of these systems minimizes flood damage and disruptions in urban areas during storm events and protects the environment of urban water resources. The primary objective of drainage systems is to collect excess stormwater from street gutters, convey the excess stormwater through storm sewers and along the street, and discharge it into a detention basin or the nearest receiving

DOI: 10.1201/9781003241744-5

water body (FHWA, 1996). Water quality BMP should also be exercised throughout the collection, treatment, and disposal processes. There are also some other objectives considered in the planning and management of drainage systems. These objectives could be listed as follows:

- Preserving the operation of the natural drainage system
- Providing safe passage of all means of transportation during storms
- Safeguarding the public and managing floods during storms
- Protecting the urban stream and local rivers environment
- Design and operation of these systems, at least capital and maintenance costs

Pollution and a wide range of problems connected to waterborne diseases in poorly drained areas are due to urban runoff that is mixed with sewage from overflowing sewers. Focally contaminated wet soils provide ideal circumstances for the spread of intestinal worm infections, and flooded septic tanks and leach pits provide breeding sites for mosquitoes. Infiltration of polluted water into low-pressure water distribution systems, contaminating drinking water supplies, could cause outbreaks of diarrhea and other gastrointestinal diseases (Parkinson, 2002).

Poor solid waste management causes many problems associated with the operation of stormwater drainage systems. Municipal agencies that are responsible for solid waste management usually do not have adequate resources and equipment for cleaning of drain inlets. There is poor communication and coordination among different urban authorities responsible for operating and maintaining the diverse components of the drainage network.

The urban drainage problems are amplified by residential housing and commercial developments that increase the impervious areas, which intensifies urban runoff. There is often little control over new developments that can restrict runoff drainage by downstream flow constrictions such as encroachment into the floodplain. Urban drainage engineers do not consider the existence of waterways and wetlands and are usually insensitive to the use of drainage through natural systems.

Hurricane Ida ripped across more than 1,500 miles (2,400 km) in the United States in late August 2021, setting numerous records and leaving a trail of destruction. Ida was most destructive while both entering and exiting the United States. In addition to strong winds, flooding was a huge concern for residents of Louisiana. Even before the storm hit, New Orleans had received over 65 inches of rain so far this year, making it the second wettest year on record to that date. The yearly average for the City of New Orleans is 62 inches. During moments of crisis, deferred and inadequate maintenance suddenly catches up with a city. These circumstances are cast in high relief when a single event—like Ida—reveals just how fragile a system really is. Ahead of Ida, which eventually triggered the first-ever flash-flood emergency in New York City, the city activated its Flash Flood Emergency Plan. (Chinchar and Gray, 2021).

A more modern, sustainable, and cost-effective approach that utilizes natural channels and flood basins for drainage and flood control could be developed based on the evaluation of conventional approaches to drainage system design and the advantages and disadvantages of traditional approaches. This approach requires improved catchment planning and greater coordination among relevant institutions responsible for drainage, irrigation, and land use in the urban and suburban areas.

Design, operation, and maintenance of urban drainage systems are major challenges for urban authorities and engineers. The effectiveness of SWM systems and the efficiency of urban management are directly linked. Many issues should be emphasized even though improvements in technology have provided effective tools for planning and for operationally sustainable and more cost-effective urban drainage systems. Parkinson (2002) summarizes these issues as follows:

1. Coordination among responsible urban authorities and agencies
2. Collaboration among private and public organizations as well as non-governmental organizations (NGOs) to promote effective partnership with civil society and the private sector
3. Capacity building for improved planning, design, and operation of urban drainage systems

Drainage systems, wastewater treatment plants (WWTPs), and their related facilities as water infrastructures are essential facilities for every community. These facilities need maintaining, repairing, or replacing due to their life cycle in order to act properly. Stormwater is one of the most important causes of water pollution in urban environments. Therefore, community leaders and decision makers need to make decisions that could lead to the best SWM in their communities.

Traditional "gray" stormwater includes gutters, drains, curbs, piping, and collection systems intended to move urban stormwater away from the constructed parts. Generally, traditional gray infrastructure collects and conveys stormwater from impervious surfaces, such as parking lots, roadways, and rooftops, into a series of piping that eventually releases untreated stormwater into a local water body, whereas "green" stormwater infrastructure is designed to emulate nature and capture rainwater where it falls. Green infrastructure decreases and treats stormwater at its source while also providing numerous advantages such as:

- Reducing localized flooding
- Improving community aesthetics
- Encouraging more neighborhood socialization
- Improving economic health by increasing property values and providing jobs and opportunities for small businesses
- Decreasing the economic and community impacts of flooding
- Delivering environmental, social, and financial benefits

As shown in Figure 5.1, green projects tend to be mostly self-maintaining and include low or zero energy usage. For permitting and selecting costs and materials, natural solutions can be far more affordable than conventional concrete-and-steel approaches that accomplish similar results.

Stormwater infrastructures in coastal areas have their own characteristics that are classified into three particular groups: (a) traditional coastal measures (gray infrastructure) have their own unique characteristics, including revetment, seawalls, levees, bulkhead, breakwaters, groins, and jetties

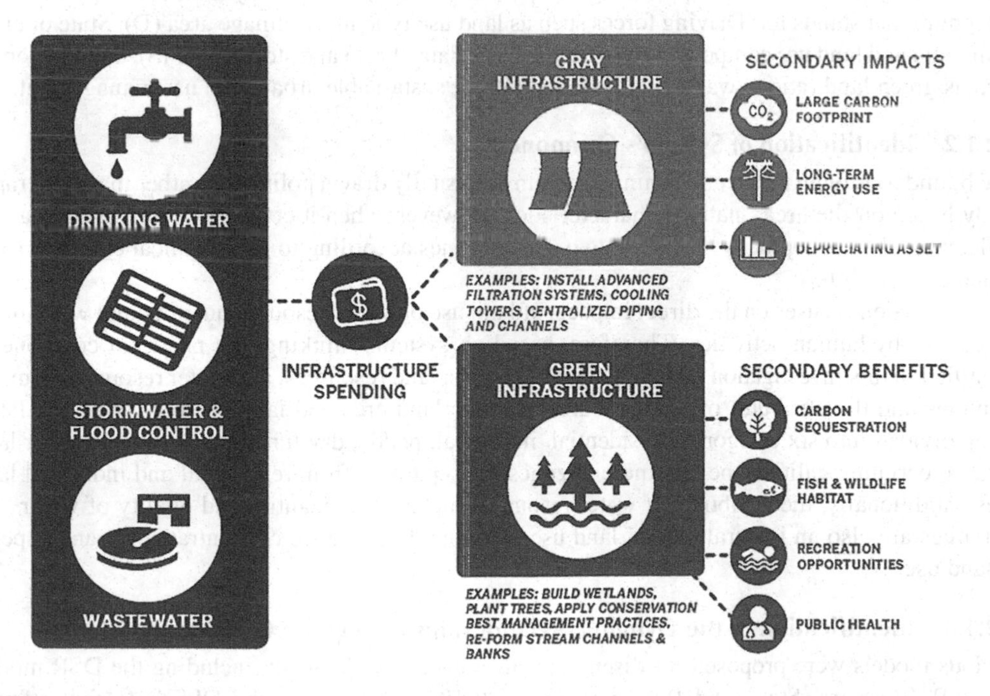

FIGURE 5.1 Green versus gray infrastructures' impacts and benefits. (*Source*: https://www.thefreshwatertrust.org/infographic-green-vs-gray-infrastructure/.)

(Marfai et al., 2015); (b) soft defense measures such as beach fills, sand bypassing, and sand dunes' stabilization (Masria et al., 2015); and (c) natural protection works (green infrastructure) such as constructed wetlands (CWs), submerged breakwaters, perched beaches, and artificial headlands (Tunji et al., 2012). These systems are utilized to significantly reduce the risk possibilities by integrating conventional flood protection mechanisms and flood mitigation, inhibition, and improvement tools (Sharif et al., 2010). Depending on the flood risk requirements related to the system support, the paired activities of individuals and organizations are combined to manage the flood risks. These activities include maintaining an interconnection between urbanization, flood protection, and financial support (Sayers et al., 2002). For instance, the US Department of Housing and Urban Development (HUD) held a national competition with a key objective of creating novel solutions to the flood protection management of New York City after Superstorm Sandy, called Rebuild by Design (RBD) (Rebuild by Design organization, 2014). RBD employed an innovative methodology to visualize the change of the watershed-scale affected area towards the improved resilient region. For further information, refer to Karamouz et al. (2018) and Šakić Trogrlić et al. (2018) and see Chapter 14 as well as the Department of Environmental Protection (DEP) website.

5.2.1 Land Use Planning

Many localities and states have stated that implementing sound land use planning is the essence of integrated water resources management. However, due to many financial, social, and political constraints, it is not fully exercised in many developing regions. To quantify, the attributes of driving forces in the context of "dynamic strategy planning" are examined here (Chen et al., 2005; Karamouz et al., 2010).

5.2.1.1 Dynamic Strategy Planning for Sustainable Urban Land Use Management

The dynamic strategy planning for sustainable urban watersheds' land use management includes the following: (a) identification of sustainable urban land use components and ordinances, (b) identification of the dynamic relationships among components, and (c) the DSR dynamic strategy planning procedure that stands for: Driving forces such as land use type in a drainage area (D); State of environmental and land use components such as the basic data of rain and stormwater (S); and Responses such as green land ratio to water accumulation (R) for sustainable urban land use management.

5.2.1.2 Identification of System's Components

The boundaries of an urban environment system are usually drawn politically rather than geographically based on the area's natural characteristics. However, when it comes to water, even at a city scale, it can be conceptually divided into drainage zones according to geographical characteristics (Chen et al., 2000).

This section focuses on the direct impacts of land use on water resource quantity and water quality caused by human activities. Therefore, based on systems thinking, the principal components identified in this investigation include human activities, land resources, and water resources. Human activities and the attributes of land resources include land area and land use type, with the latter being divided into six categories: residential, industrial, paddy, dry farming, forest, and other land uses. In certain localities, the commercial zones are separated from residential and industrial land uses. Additionally, the attributes of water resources include the quantity and quality of water. Air resources are also an integral part of land use planning, but here we concentrate on water aspects of land use.

5.2.1.3 Identification of the Dynamic Relationships among the Components

Various models were proposed for discussing sustainable development, including the DSR model, the PSR (Pressure, State, and Response) model (OECD, 1993), and the DPSIR (Driving force, Pressure, State, Impact, and Response) model (Denisov et al., 2000). Each model has its own characteristics and is suitable for different problem-solving situations. Since there are DSR dynamic interactions among human activities, land, water, and air resources for top land use management, the concept of the DSR framework and the theory of system dynamics are applied to address the cause-and-effect relationships. The dynamic relationships of the components in the sustainable urban land use management system are identified as shown in Figure 5.2. Among the components of the urban land use management system, human activities are the initial forces influencing the sustainability of land resources, which are identified as the driving forces.

Land use needs water resources and produces pollutants of different quality affecting the receiving water and air. Therefore, land use areas, land use types, water quantity, water quality, air quality, benefits, and populations are identified as states. Land use causes resource consumption and pollution production that influence the characteristics of an urban life system and benefits.

Decision makers may decide to respond based on the states' elements to modify driving forces affecting the pollution abatement, resource allocation, and water tariff as acceptable and equitable responses. In the responses, land resource allocation, water resource allocation, and water pollution abatement are implemented based on the available quantity of those resources in line with water and air quality standards. Water tariff is determined based on the land and water resource allocations which are affected by available resources to get more benefits through optimal use of them. Furthermore, land resource allocation, water resource allocation, and water pollution abatement influence one another. Land resource allocation also influences air pollution abatement, which could impact water quality as well.

Furthermore, there are dynamic interactions among the DSRs in an urban land use management system. Human activities use land, water, and air resources between driving forces and states and alter the states of land use type, a land use area, water quality, water quantity, air quality, and population density. If the states of land area and land use type are changed between states and responses, it will lead to a reallocation of land resources by the decision makers.

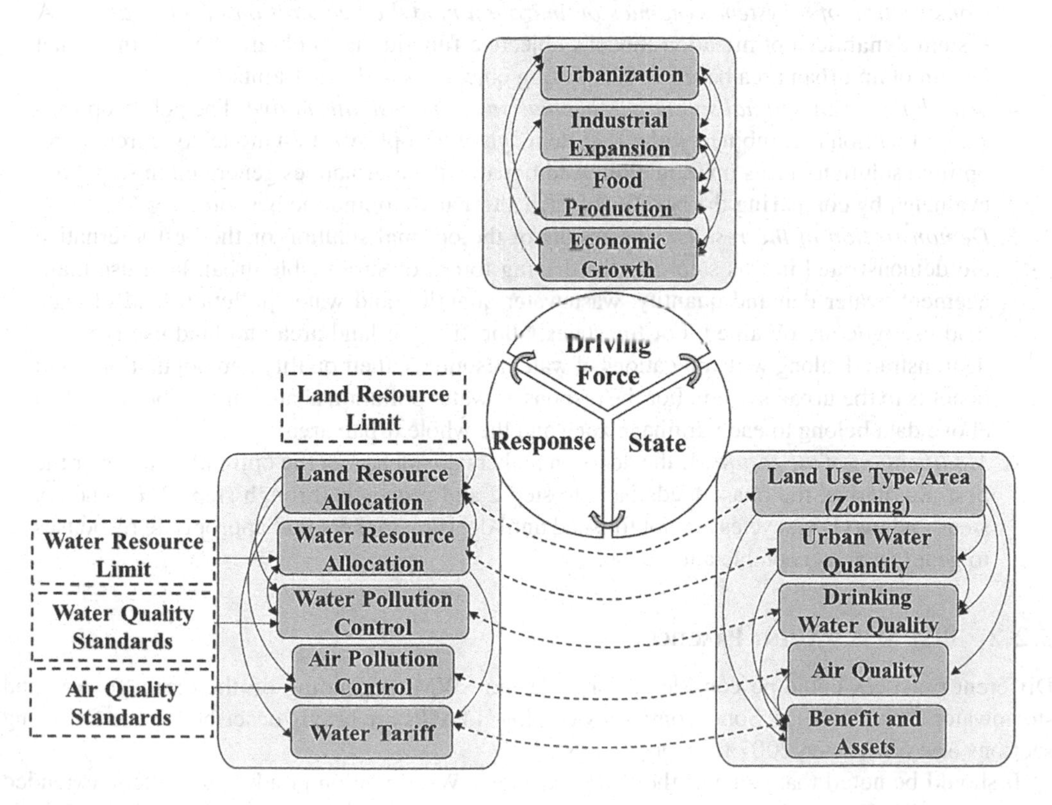

FIGURE 5.2 Dynamic relations among the sustainable urban land use management system components in the conceptual Driving force, State, and Response (DSR) framework (Adapted from Chen, C.H. et al., 2000. *Water Sci. Technol.*, 42, 389–396.)

5.2.1.4 DSR Dynamic Strategy Planning Procedure

As shown in Figure 5.3, the procedure includes six phases and two DSR cycles. Through continuous and iterative modification of the data of the two DSR cycles, an optimal solution or a satisfactory alternative for sustainable urban land use management can be obtained based on the DSR dynamic strategy planning procedure. See Chen (2005) for more details.

1. *Identification of an urban system's drainage zones*: The boundaries of the drainage environment system should be drawn geographically. The drainage system consists of mainstream/channels and branches.
2. *Identification of objectives and constraints*: The objective function of sustainable land use management for an urban setting is to maximize the net benefit of urban land use within its water resources' carrying capacity. The green cover ratio, forest land use ratio, the total available quantity of water resources, and mean available quantity of water resources are used as constraints for controlling land and water resource uses. The water and air quality standards are used as constraints to control pollution loads to meet water and air quality requirements. The constraints used for ensuring urban development can be kept within its natural resources, including water-carrying capacity.

   Because each human activity requires land resources and produces air and water pollution, each land use type can be assumed to be a related type of human activity. The loads, costs, and benefits of each land use type of each drainage zone are calculated by the loads, costs, and benefits per unit area of each land use type multiplying the land area of each land use type of each drainage zone. For the states, the area of each land use type of each drainage zone, available quantity of external water resources, and basic lithosphere, hydrosphere, and atmosphere data should be inputted here.
3. *Construction of a system dynamics optimization model or generation of alternatives*: A system dynamics optimization model's objective function is to obtain the maximum net benefit of an urban area based on identifying objectives and constraints.
4. *Searching for an optimal solution or evaluation of the best alternative*: The policy optimization function is combined with the system dynamics optimization model to search for an optimal solution in this investigation. Additionally, the alternatives generated in step 3 are evaluated by comparing the benefits of each alternative to find the best one.
5. *Demonstration of the results*: The results of the optimal solution or the best alternative are demonstrated in this step. For the driving forces of sustainable urban land use management, water demand quantity, wastewater quantity, and water pollution load of each land use type are obtained. For the states, allocations of land area and land use type are demonstrated, along with allocations of water resources, their quality and populations, and benefits to the urban system. For the responses, water pollution abatement is obtained. The above data belong to each drainage zone and the whole urban area.
6. *Modification of the results*: If the decision makers do not accept the optimal solution or the best alternative, the phase feeds back to step 2 and proceeds through steps 3–6. The six steps and two DSR cycles are not finished until decision makers find appropriate responses to generate an acceptable alternative.

5.2.2 Best Management Practices

Different practices could be considered for collected SWM, depending on the drainage area and stormwater characteristics. Some commonly employed BMPs are briefly described in the following sections (see McAlister, 2007 for more details).

It should be noted that some of the BMPs such as CWs, retention ponds, sand filters, extended detention basin (EDB), and grass buffers are discussed in more detail.

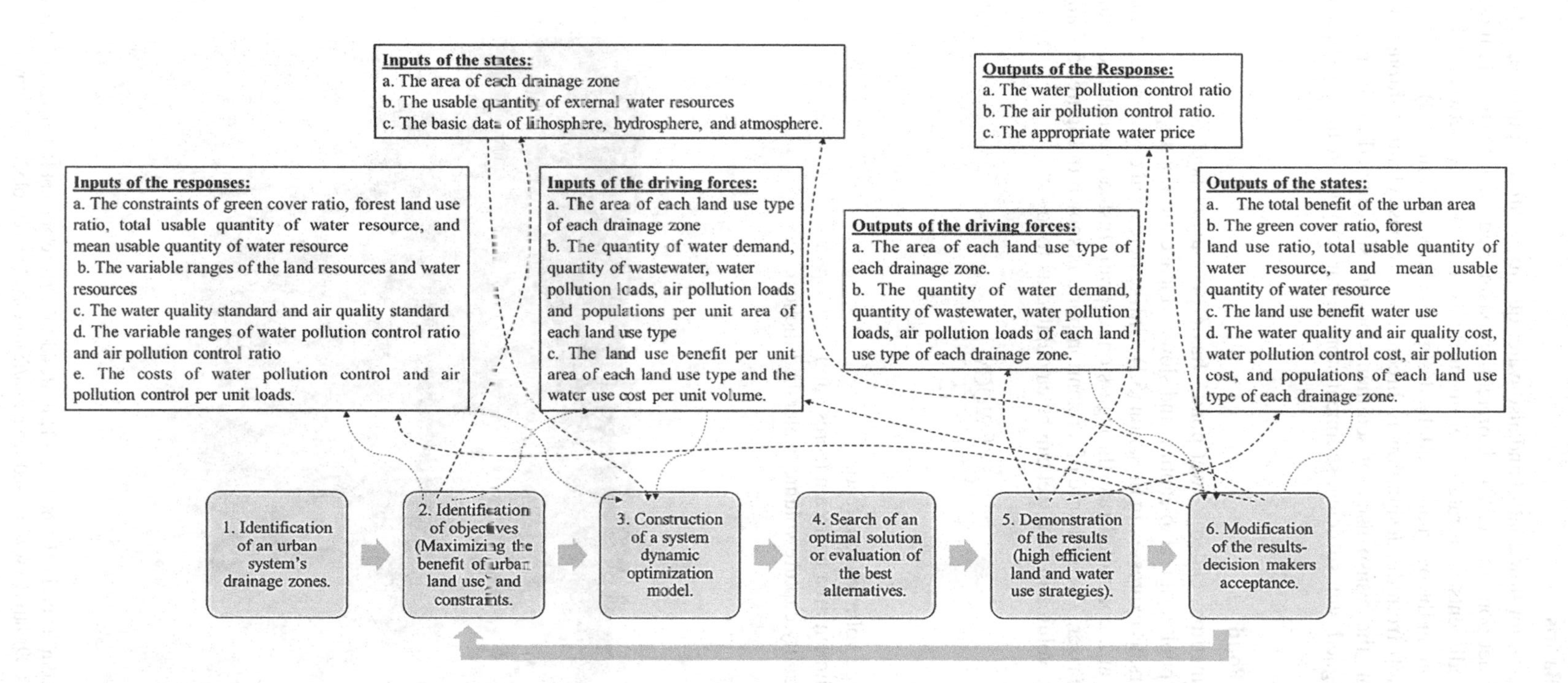

FIGURE 5.3 DSR dynamic strategy planning procedure for sustainable urban land use management. (Adapted from Chen, C.T. et al., 2005. *Sci. Tot. Environ.*, 346, 17–37. With permission.)

5.2.2.1 Sediment Basins

Sediment basins are used for flow control and also water quality treatment purposes. They are usually employed as an inlet pond to a CW or bioretention basin. Sediment basins are more effective in removing coarse sediments. The rate of sediment removal is usually between 70% and 90%; however, it is dependent on the basin area and the design discharge (Figure 5.4).

The exceeding runoff from the design flow is directed to a bypass channel through a secondary spillway. This prevents the resuspension of sediments previously trapped in the basin. Sediment basins should be designed with sufficient sediment storage capacity to ensure acceptable frequencies of desalting.

5.2.2.2 Retention Pond

A retention pond, sometimes called a "wet pond," has a permanent pool of water with capacity above the permanent pool designed to capture and slowly release the water quality capture volume (WQCV) over 12 hours, as expressed in Equation (5.1). Unlike detention dry ponds, retention ponds do not release water at a controlled rate through stone riprap. Instead, water accumulates till it reaches the top of the riser. The structure of a retention pond is shown in Figures 5.5 and 5.6. The required BMP storage volume in acre-feet can be calculated as follows:

$$V = \left[\frac{\text{WQCV}}{12}\right] A \tag{5.1}$$

where

V is required storage volume (acre-feet),
A is tributary catchment area upstream (acres), and
WQCV is water quality capture volume (watershed inches).

FIGURE 5.4 Typical sediment basin. (From "Loess Hills Residential Development Retains Rain," *Land and Water Magazine*. Vol. 52. (2). (2009). http://www.landandwater.com/features/vol52no2/vol52no2_1.html.)

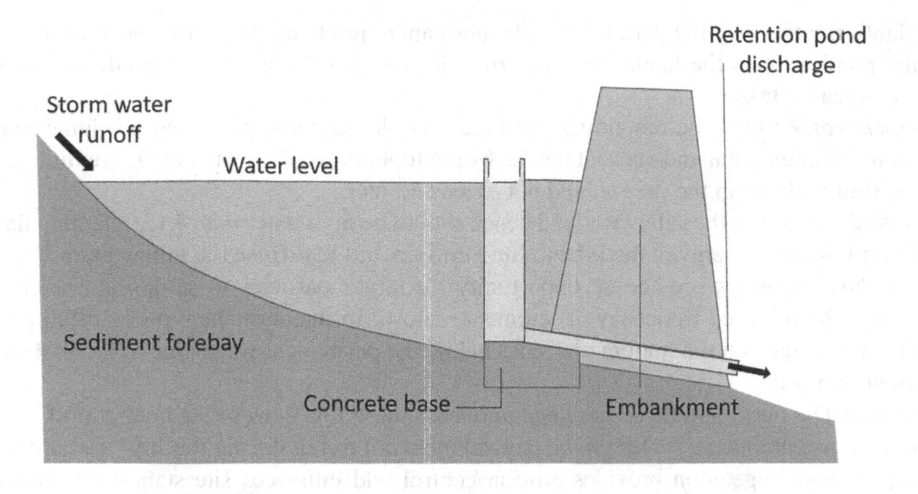

FIGURE 5.5 Typical retention pond.

FIGURE 5.6 Structure of a typical retention pond.

The pond characteristics are as follows:

a. *Site selection*: Retention ponds require groundwater or a dry-weather base flow if the permanent pool elevation is to be maintained year-round. High exfiltration rates can initially make it difficult to maintain a permanent pool in a new pond, but the bottom can eventually seal with fine sediment and become relatively impermeable over time. However, it is best to seal the bottom and the sides of a permanent pool if the pool is located on permeable soils and to leave the areas above the permanent.
b. *Basin shape*: Always maximize the distance between the inlet and the outlet. A basin length–to-width ratio between 2:1 and 3:1 is recommended to avoid short-circuiting.
c. *Permanent pool*: The permanent pool provides stormwater quality enhancement between storm runoff events through biochemical processes and continuing sedimentation. Aquatic

plant growth along the perimeter of the permanent pool can help strain surface flow into the pond, protect the banks by stabilizing the soil at the edge of the pond, and provide biological uptake.

d. *Open water zone*: The remaining pond area should be open, providing a volume to promote sedimentation and nutrient uptake by phytoplankton. To avoid anoxic conditions, the maximum depth in the pool should not exceed 12 feet.
e. Side slopes above the safety wetland bench should be no steeper than 4:1, preferably flatter.
f. *Inlet*: Dissipate energy at the inlet to limit erosion and to diffuse the inflow plume.
g. *Forebay*: Forebays provide an opportunity for larger particles to settle out, which will reduce the required frequency of sediment removal in the permanent pool. Install a solid driving surface on the bottom and sides below the permanent water line to facilitate sediment removal.
h. *Outlet*: The outlet should be designed to release the WQCV over a 12-hour period.
i. *Overflow embankment:* Design the embankment not to fail during the 100-year storm.
j. *Vegetation*: Vegetation provides erosion control and enhances site stability. Berms and side-sloping areas should be planted with native grasses or irrigated turf, depending on the local setting and proposed uses for the pond area.

Typical effectiveness to targeted pollutants is very good for sediment and solids but moderate for nutrients, total metals, and bacteria. It is a cost-effective BMP for larger tributary watersheds, creates wildlife and aquatic habitat, provides recreation, esthetics, and open space opportunities, and can increase adjacent property values. An open water environment could be a safety concern. Sediment, floating litter, and algae blooms can be difficult to remove or control. Table 5.1 lists the criteria for the Level 1 and 2 designs.

5.2.2.3 Bioretention Swales

Bioretention swales provide both flow conveyance and storage in the swale and water quality treatment through the bioretention area in the base of the swale (Figure 5.7). A high rate of water quality treatment can be provided through bioretention areas for small to modest flow rates. A limited flow detention capacity can be provided in cases where the cross section of the swale

TABLE 5.1
Levels 1 and 2 Wet Pond Design Guidance

Level 1 Design (RR:O[a]; TP:50[b]; TN:30[b])	Level 2 Design (RR:O[a]; TP:75[b]; TN:40[b])
Tv=[(1.0) (Rv) (A)/12)—volume reduced by upstream BMP	Tv=[1.5 (Rv) (A)/12]—volume reduced by upstream BMP
Single pond cell (with forebay)	Wet ED[c] (24 hours) and/or a multiple cell design[d]
Length of shortest flow path/overall length[e]=0.5 or more	Length of shortest flow path/overall length[e]=0.8 or more
Standard aquatic benches	Wetlands more than 10% of pond area
Turf in pond buffers	Pond landscaping to discourage geese
No internal pond mechanisms	Aeration (preferably bubblers that extend to or near the bottom or floating islands)

[a] Runoff volume reduction can be computed for wet ponds designed for water reuse and upland irrigation.

[b] Due to groundwater influence, slightly lower Total Phosphorous and Nitrogen (TP and TN) removal rates in coastal plain and CSN Technical Bulletin No. 2 (2009).

[c] Extended detention may be provided to meet a maximum of 50% of the treatment volume; refer to Design Specification 15 for Extended Detention (ED) design.

[d] At least three internal cells must be included, including the forebay.

[e] In the case of multiple inflows, the flow path is measured from the dominant inflows (that comprise 80% or more of the total pond inflow).

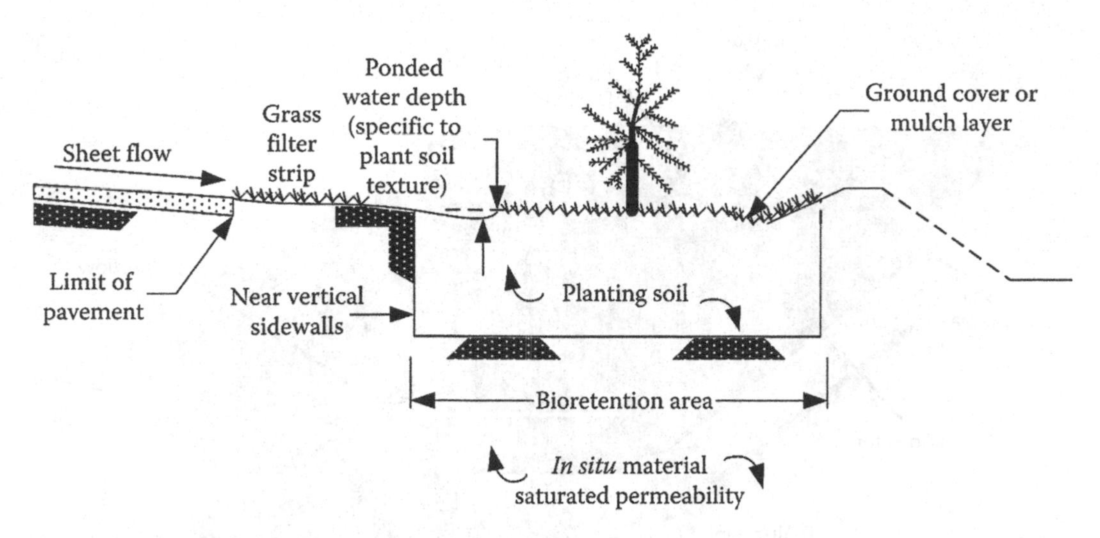

FIGURE 5.7 Structure of a typical bioretention swale.

is large relative to the flow rate. The recommended longitudinal slope for bioretention swales is between 1% and 4%. In these slopes, the flow capacity is maintained without creating high velocities, potential erosion of the bioretention or swale surface, and safety hazards. Pollutant removal is achieved through sedimentation and filtration through the filtration media and biological processes. The pollutant removal efficiency in a bioretention swale is dependent on the filter media, landscape planting species, and the hydraulic detention time of the system.

5.2.2.4 Bioretention Basins

The objectives of bioretention basins are flow control and water quality treatment. Bioretention basins can be used on lots where there are several buildings and the lot is under single ownership, and larger basins are frequently used as part of urban development plans. The pollutant removal efficiency is dependent on the employed filtration media and the relative magnitude of the extended detention component of the basin. The maximum annual pollutant removal efficiency simultaneously as flow control can be achieved if the extended detention component of the basin is used for small-to-medium runoff events. Pollutant removal is achieved through sedimentation and water filtration through the filter media and biological processes.

5.2.2.5 Sand Filters

A sand filter is filtering or infiltrating BMP that consists of a surcharge zone underlain by a sand bed with an underdrain system. During a storm, accumulated runoff collects in the surcharge zone and gradually infiltrates into the underlying sand bed, filling the void spaces of the sand. The absence of vegetation in a sand filter allows for active maintenance at the surface of the filter (i.e., raking for removing a layer of sediment). For this reason, sand filter criteria allow for a larger contributing area and greater depth of storage.

The operation of sand filters is similar to bioretention systems except that they do not usually support vegetation owing to the filtration media being free draining (Figure 5.8). Sand filters are usually employed in confined spaces without sustainable vegetation, such as below the ground. Sand filters typically include three separate chambers for sedimentation, sand filtration, and overflow. The medium-to-coarse sediments and pollutants are removed in the sedimentation chamber and the sand filter chamber then removes much of the medium-to-coarse sediment and some of the finer particulate and dissolved pollutants. To maintain removal efficiency, the sand filter should be regularly maintained to prevent crust formation.

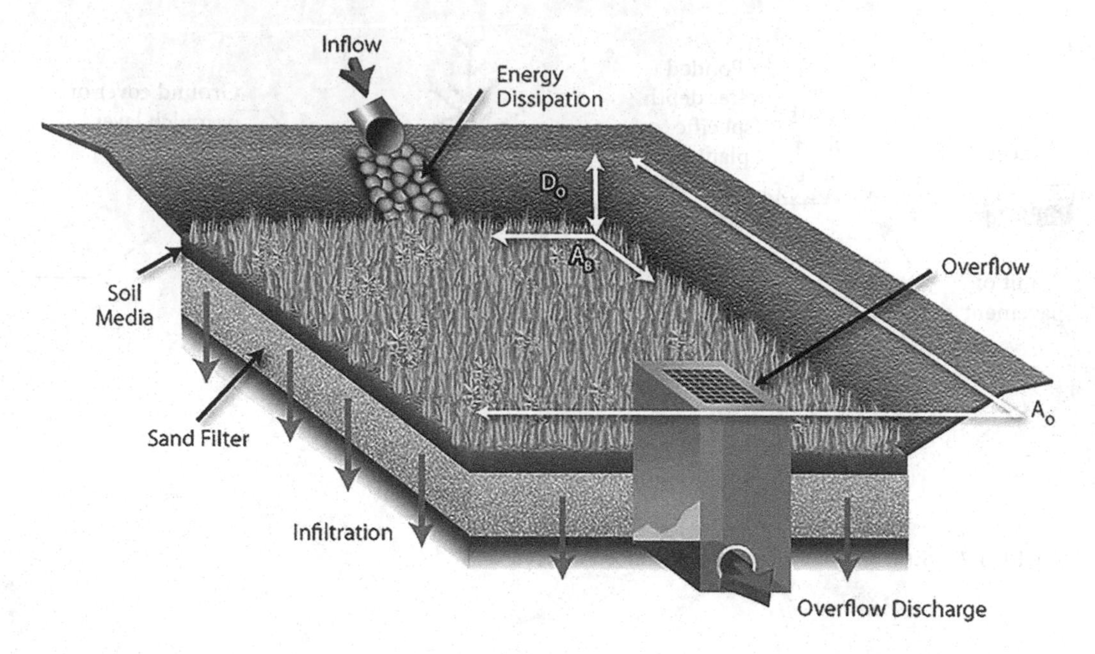

FIGURE 5.8 Structure of a typical sand filter.

The elements of a sand filter are as follows:

a. *Site selection*: Sand filters require a stable watershed. When the watershed includes phased construction, sparsely vegetated areas, or steep slopes in sandy soils, consider another BMP or provide pretreatment before runoff from these areas reaches the rain garden. When sand filters (and other BMPs used for infiltration) are located adjacent to buildings or pavement areas, protective measures should be implemented to avoid adverse impacts to these structures.
b. *Basin storage volume*: Provide a storage volume above the sand bed of the basin equal to the WQCV based on a 12-hour drain time.
c. *Basin geometry*: Use the below equation to calculate the minimum filter area, which is the flat surface of the sand filter.

$$A_F = 0.0125AI \tag{5.2}$$

A_F = minimum filter area (flat surface area) (ft^2)
A = area tributary to the sand filter (ft^2)
I = imperviousness of area tributary to the sand filter (percent expressed as a decimal)

d. *Inlet works*: Provide energy dissipation and a forebay at all locations where concentrated flows enter the basin.
e. *Outlet works*: Slope the underdrain into a larger outlet structure. Use an orifice plate to drain the WQCV over approximately 12 hours.

Construction considerations for a successful project include the following:

1. Protect area from excessive sediment loading during construction. The portion of the site draining to the sand filter must be stabilized before allowing flow into the sand filter. When using an impermeable liner, ensure enough slack in the liner to allow for backfill, compaction, and settling without tearing the liner.

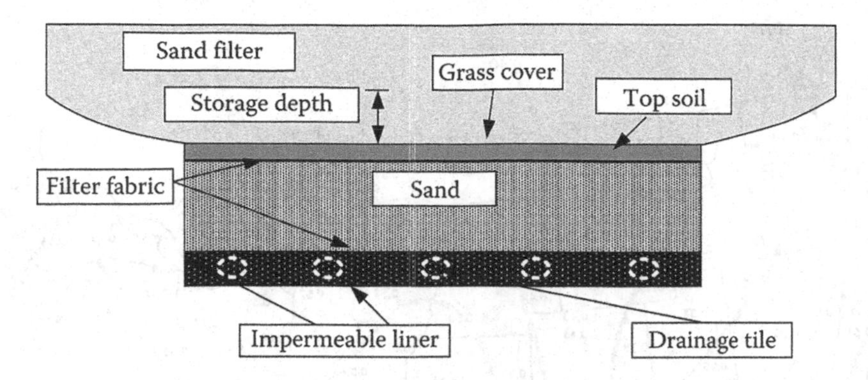

FIGURE 5.9 Schematic structure of a typical sand filter. (From "A holistic approach to stormwater management." *Stormwater Management* (2021) http://www.civil.ryerson.ca/Stormwater/menu_1/index.htm.)

2. Filtering BMPs provide effective water quality enhancement including phosphorus removal. This BMP may clog and require maintenance if moderate-to-high level of silts and clays are allowed to flow into the facility. This BMP should not be located within 10 feet of a building foundation without an impermeable membrane. See "Bioretention" section for additional information. The sand filter should not be put into operation while construction or major landscaping activities are taking place in the watershed. The schematic structure of a typical sand filter is shown in Figure 5.9.

5.2.2.5.1 Calculation of the WQCV

The first step in estimating the magnitude of runoff from a site is to estimate the site's total imperviousness. The WQCV is calculated as a function of imperviousness and BMP drain time

$$\text{WQCV} = \frac{a\left(0.91I^3 - 1.91I^2 + 0.78I\right)}{0.43} \times d_6 \tag{5.3}$$

where

WQCV = water quality capture volume (watershed inches),
a = coefficient corresponding to WQCV drain time,
I = imperviousness (%), and
d_6 = depth of average runoff producing storm (from Figure 5.10).

Values for drain time coefficient, a, for WQCV calculations are listed in Table 5.2.

Once the WQCV in watershed inches is found, the required BMP storage volume in acre-feet can be calculated as expressed in Equation (5.1).

Example 5.1

Calculate the WQCV for a 1.0-acre subwatershed in New York with a total area-weighted imperviousness of 50% that drains to a rain garden:

Solution:

As the first step, determine the appropriate drain time for the type of BMP. For a rain garden, the required drain time is 12 hours. The corresponding coefficient, a, from Table 5.2 is 0.8. Figure 5.10 can be also used to solve this problem.

$$\text{WQCV} = \frac{a\left(0.91I^3 - 1.91I^2 + 0.78I\right)}{0.43} \times d_6$$

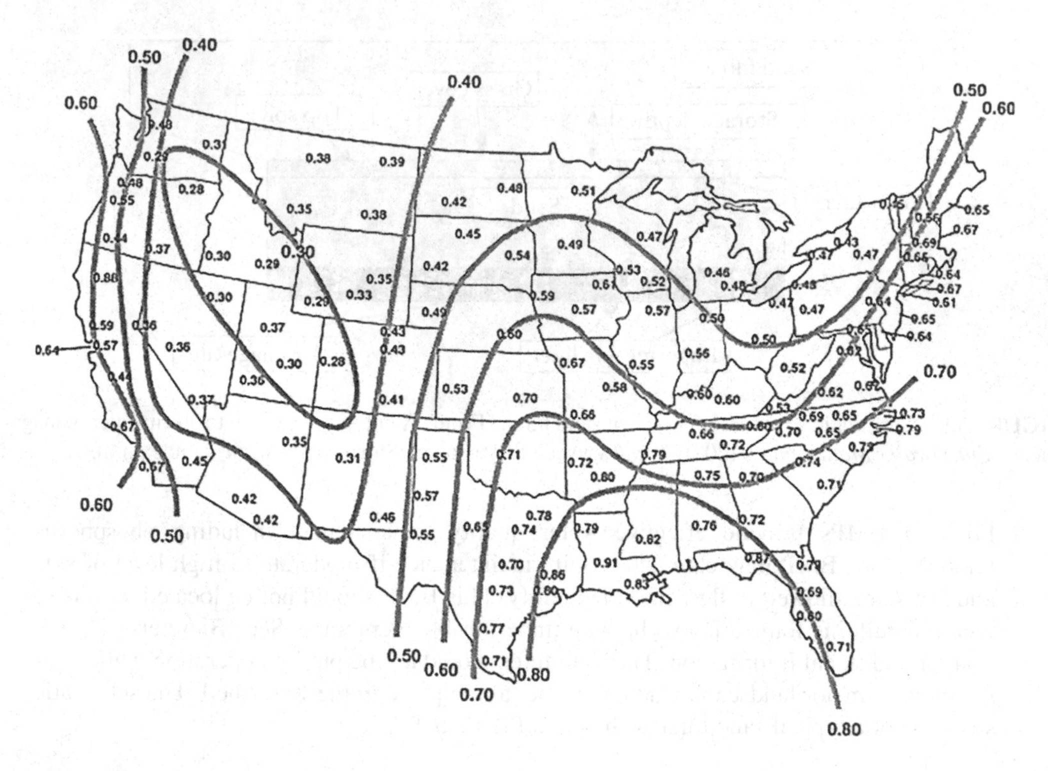

FIGURE 5.10 Map of the average runoff producing storm's precipitation depth in the United States in inches. (*Source*: Driscoll, E.D. et al., *Analysis of Storm Events Characteristics for Selected Rainfall Gauges throughout the United States*. Representative. U.S. Environmental Protection Agency (EPA), Washington, D.C., 1989.)

TABLE 5.2
Values for Drain Time Coefficient

Drain Time	Coefficient, *a*
12 hours (filtration BMPs and retention ponds)	0.8
24 hours (constructed wetland ponds)	0.9
40 hours (extended detention)	1.0

$$\text{WQCV} = \frac{1.0\left(0.91(0.50)^3 - 1.91(0.50)^2 + 0.78(0.50)\right)}{0.43} \times 0.65 = 0.40 \text{ (w.s.in)}$$

Calculate the WQCV in cubic feet using the total area of the subwatershed and appropriate unit conversions:

$$\text{WQCV} = 0.40 \text{ w.s.in} \times \frac{1 \text{ feet}}{12 \text{ inches}} \times \frac{43{,}560 \text{ ft}^2}{1 \text{ acre}} \approx 1{,}452 \text{ ft}^3$$

5.2.2.6 Swales and Buffer Strips

Flow conveyance function is done along the swale, and then water quality treatment is provided through sedimentation and contact of flowing water with swale vegetation. The most significant effect of water quality treatment functionality is on small-to-modest flow rates. A limited flow detention capacity can also be used when the cross section of the swale is large relative to the flow rate. The pollutant removal efficiency is dependent on the longitudinal slope, the vegetation height, and the area and length of the swale. The recommended longitudinal slope is the same as bioretention swales. Swales cannot effectively treat fine pollutants but can provide pretreatment for downstream measures.

5.2.2.7 Constructed Wetlands

CW systems are shallow retention pond designed, extensively vegetated water bodies (Figure 5.11), to permit the growth of wetland plants such as rushes, willows, and cattails. It slows down runoff and allows time for sedimentation, filtering, and biological uptake. They can also be used effectively in series with other flow/sediment-reducing BMPs that reduce the sediment load and equalize incoming flows to the CWs. Extended detention, fine filtration, and biological pollutant uptake processes are employed in these systems for pollutant removal. It can be designed as either an online or offline facility. Wetlands are composed of three major parts: an inlet zone (sedimentation basin), a macrophyte zone, and a high-flow bypass channel. The depth of the macrophyte zone varies between 0.25 and 0.5 m, and the theoretical detention time is between 48 and 72 hours. Wetlands store runoff during the rainfall and slowly release it after finishing the event and in this way provide a flow control function. When flows exceed the design flow of a wetland for protecting wetland vegetation and also avoiding resuspension of trapped pollutants, excess water is directed around the macrophyte zone via a bypass channel.

CWs are often organized into four groups:

a. Shallow Wetlands are large surface area CWs that primarily accomplish water quality improvement through the displacement of the permanent pool.
b. Extended Detention Shallow Wetlands are similar to Shallow Wetlands but use extended detention as another mechanism for water quality and peak rate control.

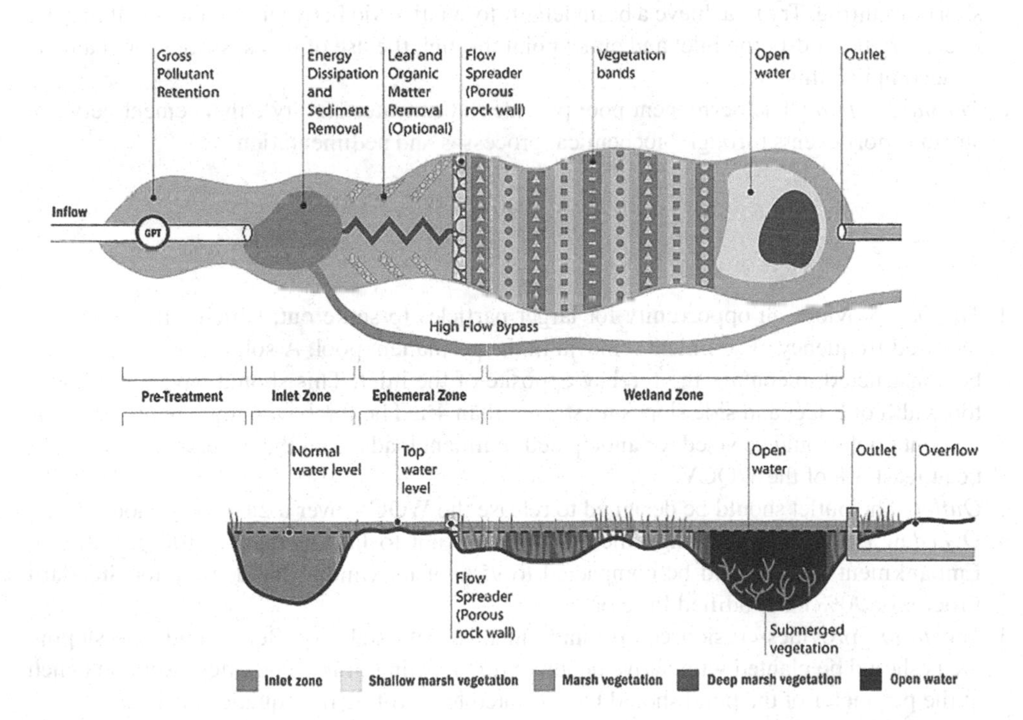

FIGURE 5.11 Schematic of a typical constructed wetland system. (From "Wetland principle © Melbourne Water. Urban wetlands: Urban green-blue grids." *Urban green-blue grids for sustainable and resilient cities* (2021). https://www.urbangreenbluegrids.com/measures/urban-wetlands/.)

c. Pocket Wetlands are smaller CWs that serve drainage areas between approximately 5 and 10 acres (1 acre ≈ 4,000 m^2) and are constructed near the water table.
d. Pond/Wetland systems are a combination of a wet pond and a CW.

The BMP characteristics are as follows:

a. *Site selection*: A CW pond requires a positive net influx of water to maintain vegetation and microorganisms. This can be supplied by groundwater or a perennial stream. An ephemeral stream will not provide adequate water to support this BMP. The following steps outline the design procedure for a CW pond.
b. *Baseflow*: Unless the permanent pool is established by groundwater, a perennial baseflow that exceeds losses must be physically and legally available. Net influx calculations should be conservative to account for significant annual variations in hydrologic conditions. Low inflow in relation to the pond volume can result in poor water quality.
c. *Surcharge volume*: Provide a surcharge storage volume based on a 24-hour drain time.
   - Determine the imperviousness of the watershed.
   - Find the required storage volume WQCV using Equation (5.1).
   - In order to maintain healthy wetland growth, the surcharge depth for WQCV above the permanent water surface should not exceed 2 feet.

The CW conjugations are as follows:

d. *Basin shape*: Always maximize the distance between the inlet and the outlet. Shape the pond with a gradual expansion from the inlet and a gradual contraction to the outlet to limit short-circuiting. Try to achieve a basin length-to-width ratio between 2:1 and 4:1. It may be necessary to modify the inlet and outlet point through the use of pipes, swales, or channels to accomplish this.
e. *Permanent pool*: The permanent pool provides stormwater quality enhancement between storm runoff events through biochemical processes and sedimentation

$$V_p \geq 0.75\left[\frac{\text{WQCV}}{12}\right]A \tag{5.4}$$

f. *Forebay* provides an opportunity for larger particles to settle out, which will reduce the required frequency of sediment removal in the permanent pool. A soil riprap berm should be constructed to contain the forebay opposite of the inlet. This should have a minimum top width of 8 feet and side slopes no steeper than 4:1. The forebay volume within the permanent pool should be sized for anticipated sediment loads from the watershed and should be at least 3% of the WQCV.
g. *Outlet*: The outlet should be designed to release the WQCV over a 24-hour period.
h. *Overflow embankment*: Design the embankment not to fail during the 100-year storm. Embankment soils should be compacted to 95% of maximum dry density for Standard Proctor or 90% for Modified Proctor.
i. *Vegetation* provides erosion control and enhances site stability. Berms and side sloping areas should be planted with native bunch or turf-forming grass. The safety wetland bench at the perimeter of the pond should be vegetated to facilitate the aquatic species.

Different components of CW are shown in Figure 5.12 with more emphasis on planting and vegetation.

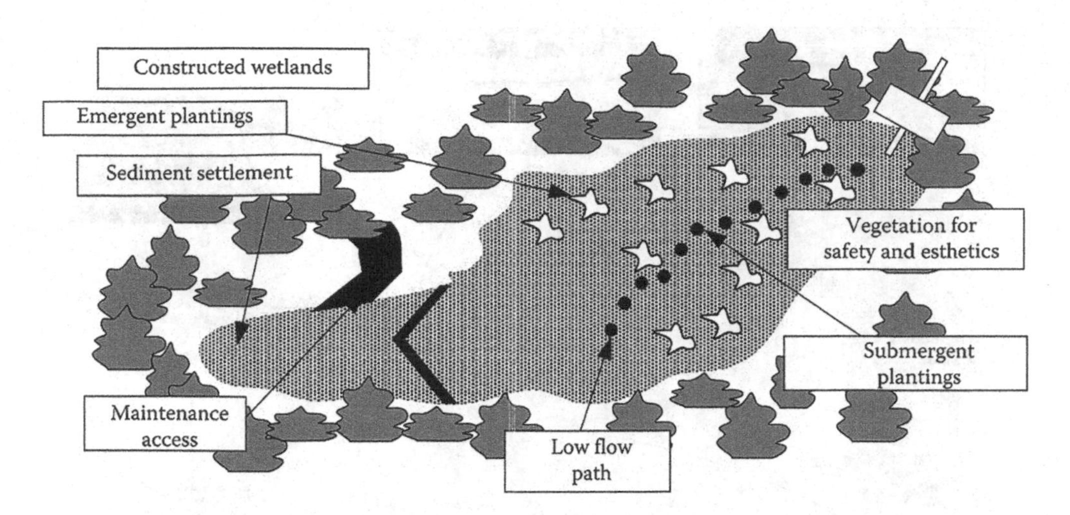

FIGURE 5.12 Different components of a typical constructed wetland. (From "A holistic approach to stormwater management." *Stormwater Management* (2021). http://www.civil.ryerson.ca/Stormwater/menu_1/index.htm.)

5.2.2.8 Extended Detention Basin (EDB)

An extended detention basin (EDB) is a sedimentation basin designed to detain stormwater for many hours after storm runoff ends. The EDB's 40-hour drain time for the WQCV is recommended to remove a significant portion of total suspended solids (TSS). Soluble pollutant removal is enhanced by providing a small wetland marsh or "micropool" at the outlet to promote biological uptake. A typical image of different components of the facility is shown in Figure 5.13.

Designers may choose to go with the baseline design (Level 1) or choose an enhanced design (Level 2) that maximizes nutrient removal and runoff reduction. To qualify for the higher nutrient reduction rates associated with the Level 2 design, EDB must be designed with a treatment volume equal to 1.25 (Rv) (A). Table 5.3 lists the criteria for the Level 1 and 2 designs.

The BMP characteristics of EDB are as follows (Table 5.4):

a. *Site selection*: EDBs are well suited for watersheds with at least five impervious acres up to approximately one square mile of the watershed. Smaller watersheds can result in an orifice size prone to clogging. Larger watersheds and watersheds with baseflows can complicate the design and reduce the level of treatment provided.
b. *Basin shape*: Always maximize the distance between the inlet and the outlet. A longer flow path from inlet to outlet will minimize short-circuiting and improve the reduction of TSS.
c. *Basin side slopes*: Basin side slopes should be stable and gentle to facilitate maintenance and access. Slopes that are 4:1 or flatter should be used.
d. *Inlet*: Dissipate flow energy at concentrated points of inflow. This will limit erosion and promote particle sedimentation.
e. *Forebay design*: The forebay provides an opportunity for larger particles to settle out in an area that can be easily maintained. The length of the flow path through the forebay should be maximized, and the forebay outlet should be sized to release 2% of the undetained peak 100-year discharge.
f. *Trickle channel*: Convey low flows from the forebay to the micropool with a trickle channel. The trickle channel should have a minimum flow capacity equal to the maximum release from the forebay outlet.
g. *Micropool and outlet structure*: Locate the outlet structure in the embankment of the EDB and provide a permanent micropool directly in front of the structure. Submerge the well screen to the bottom of the micropool.

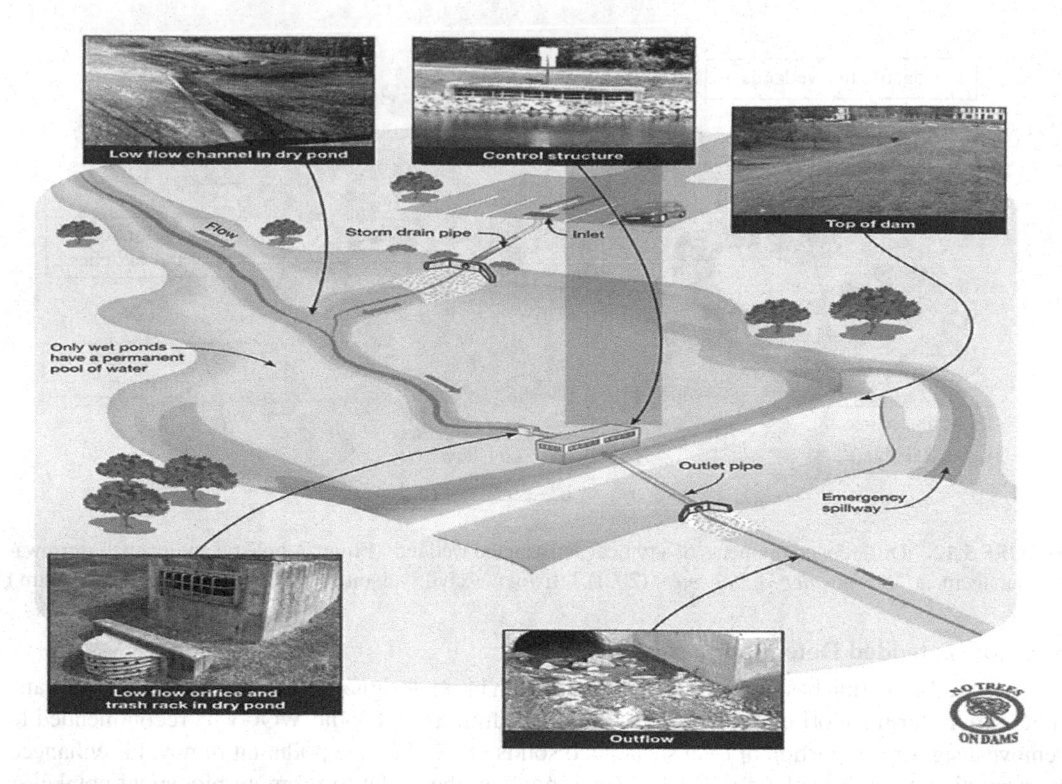

FIGURE 5.13 A typical image of different components of an extended detention basin. (From "Stormwater Ponds explained - retention or detention? - dry or wet?" *Platinum Ponds*. (2015). https://www.platinumlake-management.com/blog/stormwater-ponds-explained.)

TABLE 5.3
Extended Detention (ED) Pond Criteria

Level 1 Design (RR:O; TP:15; TN :10)	Level 2 Design (RR:15; TP:15; TN:10)
$Tv = [(1.0)(Rv)(A)]/12$—the volume reduced by an upstream BMP	$Tv = [(1.25)(Rv)(A)]/12$—the volume reduced by an upstream BMP
A minimum of 15% of the *Tv* in the permanent pool (forebay, micropool)	A. minimum of 40% of *Tv* in the permanent pool (forebay, micropool, or deep pool, or wetlands)
Length/width ratio or flow path = 2:1 or more	Length/width ratio or flow path = 3:1 or more
Length of the shortest flow path/overall length = 0.4 or more	Length of the shortest flow path/overall length = 0.7 or more
Average *Tv ED* time = 24 hours or less	Average *Tv ED* time = 36 hours
Vertical *Tv ED* fluctuation exceeds 4 feet	Maximum vertical *Tv ED* limit of 4 feet
Turf cover on floor	Trees and wetlands in the planting plan
Forebay and micropool	Includes additional cells or features (deep pools, wetlands, etc.). Refer to Section 5.5
Contributing drainage area (CDA) is less than 10 acres	CDA is greater than 10 acres

Typical effectiveness to targeted pollutants is good for sediment and solids but moderate for nutrients, total metals, and poor bacteria. It is best suited for tributary areas of 5 impervious acres or more, and it is not recommended for sites less than 2 acres of impervious area.

EDBs create wildlife and aquatic habitat; provide recreation, esthetics, and open space opportunities; and increase adjacent property values. Open water environment and depth could be a safety concern. Maintenance requirements are straightforward.

TABLE 5.4
BMP Characteristics of EDBs

	On-site EDBs for Watersheds up to 1 Impervious Acres	EDBs with Watersheds between 1 and 2 Impervious Acres	EDBs with Watersheds up to 5 Impervious Acres	EDBs with Watersheds over 5 Impervious Acres	EDBs with Watersheds over 20 Impervious Acres
Forebay release and configuration	EDBs should not be used for watersheds with less than 1 impervious acre	Release 2% of the undetained 100-year peak discharge by way of a wall/notch configuration	Release 2% of the undetained 100-year peak discharge by way of a wall/notch configuration	Release 2% of the undetained 100-year peak discharge by way of a wall/notch configuration	Release 2% of the undetained peak 100-year disc by way of a wall/no berm/notch or pipe configuration
Minimum forebay volume		1% of the WQCV	2% of the WQCV	3% of the WQCV	3% of the WQCV
Maximum forebay depth		12 inches	18 inches	18 inches	30 inches
Trickle channel capacity		The maximum possible forebay outlet capacity	The maximum possible forebay outlet capacity	The maximum possible forebay outlet capacity	The maximum possible forebay outlet capacity
Micropool		Area > 10 ft²	Area > 10 ft²	Area > 10 ft²	Area > 10 ft²
Initial surcharge volume		Depth > 4 inches	Depth > 4 inches	Depth > 4 in volume > 0.3% WQCV	Depth > 4 in volume > 0.3% WQCV

Example 5.2

The water quality volume (WQV) is determined to be 1.44 acres-feet for an area of 38 acres with 60% imperviousness.

Size the sediment forebay and the permanent pool volume for a wet pond and an extended detention wet pond.

Solution:

A sediment forebay has the same volume for wet ponds and extended detention wet ponds. It is calculated as 0.1 inches/impervious acre of drainage area. It can also be subtracted from the total WQV requirement.

For this example the following sediment forebay calculation could be calculated as:

$$\text{Sediment forebay volume} = 0.1 \text{ inches} \times \text{impervious acre age}$$

$$= \left(\frac{0.1}{12}\right)(38 \times 0.60) = 0.20 \text{ acre} - \text{feet}$$

You can also convert the volume to cubic feet by multiplying by 43,560, giving you 8,276 ft³.

The permanent pool volume for a wet pond is equal to the WQV. For an extended detention pond, the permanent pool volume is half of the WQV; however, the WQV should be increased by 15% for extended detention ponds. This is to count not only as a safety factor but also to account for the change in the curve used to generate the permanent pool requirements when extended detention is used.

Also, don't forget that the sediment forebay volume can be subtracted from the WQV when determining the permanent pool requirements.

The permanent pool volume equals the WQV with the sediment forebay volume subtracted for the wet pond.

$$\text{Permanent pool volume} = WQV - \text{Sediment forebay volume} = 1.44 - 0.20 = 1.24 \text{ acre-feet}$$

For the extended detention wet pond, the WQV must be increased by 15%. The permanent pool volume can then be calculated as half of the WQV with the sediment forebay volume subtracted.

$$\text{Permanent pool volume} = 0.5 \times \left[\left(WQV + (WQV \times 0.15)\right) - \text{Sediment forebay volume}\right]$$

$$= 0.5 \times \left[\left(1.44 + (1.44 \times 0.15)\right) - 0.20\right] = 0.728 \text{ acre} - \text{feet}$$

5.2.2.9 Ponds and Lakes

Ponds and lakes are usually formed by a simple dam wall with a weir outlet structure or are created by excavating below the natural surface level. Ponds and lakes collect water for stormwater reuse schemes and also act as part of a flood detention system. They also participate in pollutant removal through sedimentation, absorption of nutrients, and ultraviolet disinfection. Ponds cannot be considered as "stand-alone" stormwater treatment measures and require pretreatment via CWs or other measures. The historical runoff and/or predevelopment flows for a range of flood events are considered in the design of pond outlets. The minimum average turnover period between 20 and 30 days should be considered in designing a pond or lake to prevent water quality problems.

5.2.2.10 Infiltration Systems

Stormwater infiltration systems infiltrate stormwater into surrounding soils so that their performance efficiency is dependent on local soil characteristics. The best application of infiltration systems is in sandy loam soils with deep groundwater. Stormwater infiltration systems effectively decrease the volume and magnitude of peak discharges from impervious areas. A vital component in employing infiltration systems is pretreatment that avoids clogging, which deteriorates the infiltration effectiveness over time. Figure 5.14 shows the structure of a constructed infiltration bed, which could be used for stormwater drainage.

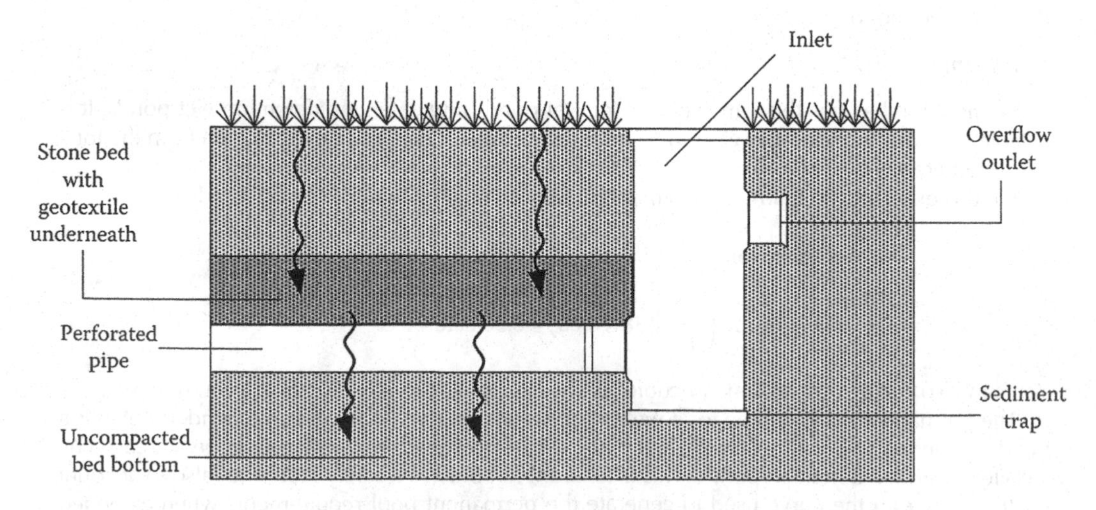

FIGURE 5.14 Structure of a constructed infiltration bed for stormwater drainage. (From "Pennsylvania Stormwater Best Management Practices Manual." Department of Environmental Protection Bureau of Watershed Management. Chapter 6. p.6 (2006).)

5.2.2.11 Grass Buffer

Grass buffers are densely vegetated strips of grass designed to accept sheet flow from upgradient development.

The BMP characteristics are as follows:

a. *Site selection*: Runoff can be directly accepted from a parking lot, roadway, or the roof of a structure, provided the flow is distributed uniformly over the width of the buffer. This can be achieved by using flush curbs, slotted curbs, or level spreaders where needed. Grass buffers are often used in conjunction with grass swales.
b. *Design discharge*: Determine the 2-year peak flow rate ($Q2$) of the area draining to the grass buffer.
c. *Minimum width*: The width (W), normal to the flow of the buffer, is typically the same as the contributing basin:
   W is the width of buffer (feet), and $Q2$ is 2-year peak runoff (cfs).
d. *Length*: The recommended length (L), the distance along the sheet flow direction, should be a minimum of 14 feet.
e. *Buffer slope*: The design slope of a grass buffer in flow direction should not exceed 10%. Generally, a minimum slope of 2% or more in turf is adequate to facilitate positive drainage. For slopes, less than 2%, consider including an underdrain system to mitigate nuisance drainage.
f. *Vegetation*: This is the most critical component for treatment within a grass buffer.

Durable, dense, and drought-tolerant grasses to vegetate the buffer should be selected. Also, consider the size of the watershed as larger watersheds will experience more frequent flows. The goal is to provide a dense mat of vegetative cover. Grass buffer performance falls off rapidly as the vegetation coverage declines below 80% (Barrett et al., 2004).

5.2.2.12 Aquifer Storage and Recovery

In an aquifer storage and recovery (ASR) system, water recharge to underground aquifers is enhanced through either pumping or gravity feed. Usually, treated stormwater or recycled water is used for this purpose to protect groundwater from deterioration of quality or aquifer properties. Therefore, ASR systems typically consolidate a CW, detention pond, dam, or tank, part or all of which act to remove pollutants and provide a temporary storage role. The stored water could provide a low cost compared to large surface water storages because it can be pumped during dry periods for subsequent reuse. The level of pretreatment of stormwater is dependent on the quality of the groundwater and its current use. The suitability of an ASR scheme is dependent on local hydrology, the underlying geology of an area, and the presence and nature of aquifers. If the salinity of an aquifer is greater than the injection water, then this may affect the viability of recovering water from the aquifer. A typical threshold ASR system using the collected rainfall from the rooftop is illustrated in Figure 5.15.

5.2.2.13 Porous Pavement

Porous paving surfaces allow stormwater to be filtered by a coarse subbase and may allow infiltration to the underlying soil (Figure 5.16). A variety of water management objectives could be promoted using porous paving. These objectives are mentioned as below:

- Reducing peak stormwater discharges from paved areas
- Increasing the groundwater recharge
- Improving the quality of stormwater
- Reducing the allocated area for SWM

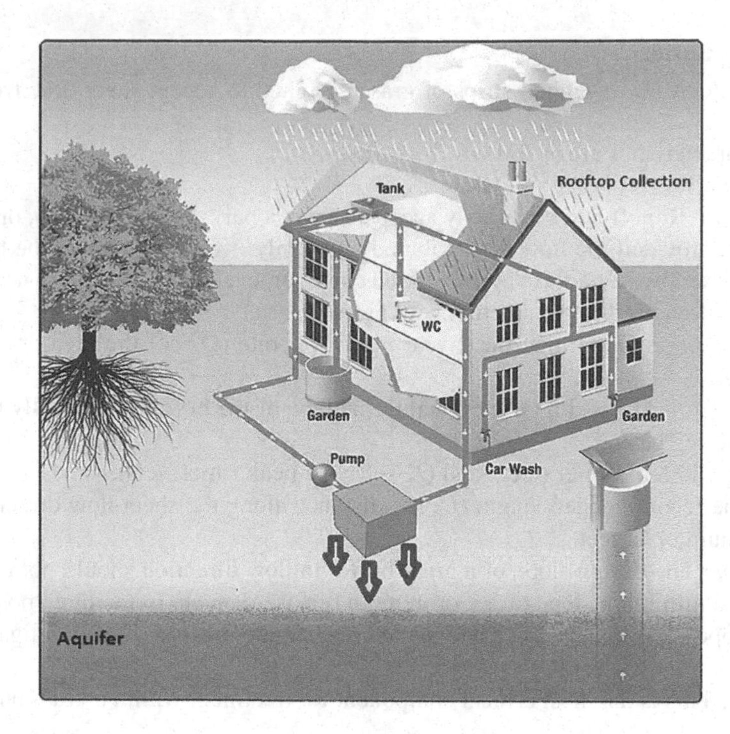

FIGURE 5.15 Schematic of aquifer storage and recovery system using collected rainfall from the roof (Adopted from https://www.civilclick.com/wp-content/uploads/2020/06/Rooftop-Rainwater-Harvesting.png?ezimgfmt=ng:webp/ngcb3).

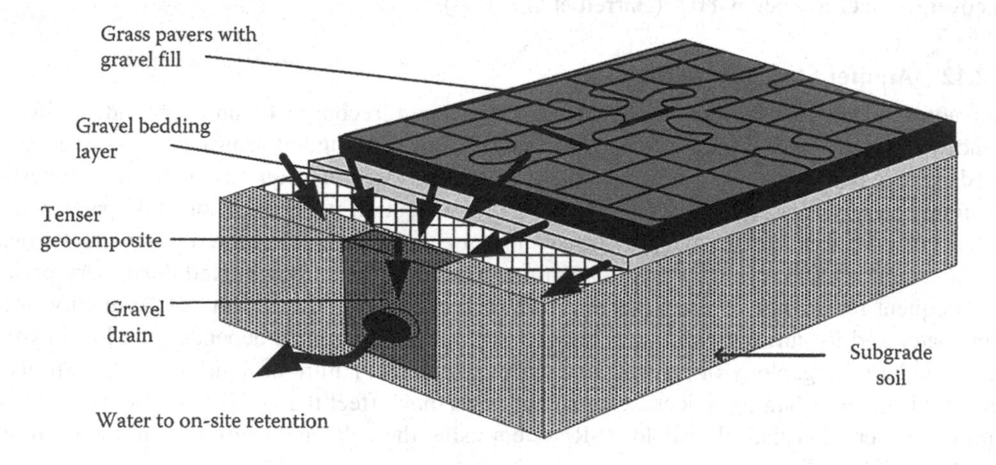

FIGURE 5.16 Structure of a typical porous pavement. (From http://www.tensar.co.uk/images/graphic4.gif.)

5.3 DRAINAGE IN URBAN WATERSHEDS

The flow in most of the components of the urban drainage system is in the form of surface and/or open-channel flow. The driving force is the gravitational force. At first, the excess rainfall flows as "overland flow" with very small depth over the roofs, lawns, driveways, and street pavements. Next, the excess rainfall is drained into a channel such as a street gutter, a ditch, a drainage channel, or a storm sewer under an impervious or porous sidewalk/pavement, which is usually partly full. The excess rainfall finally reaches the watershed outlet or passes through abstractions (i.e., highways) in the form of a channel flow discussed in Chapter 4 except the flow-through culverts discussed later in this chapter. The detailed procedure of estimation of excess rainfall and the procedure of carrying excess rainfall to the watershed outlet could be found in Karamouz et al. (2010).

5.3.1 Overland Flow

The overland flow occurs on impervious and pervious surfaces in urban watersheds such as roofs, driveways, and parking lots, as well as lawns. Overland flow can be modeled as an open-channel flow with a very shallow depth so that the same equation as for the open-channel flow can be applied. The low-flow depth in overland flow results in a low *Re*, so the overland flow can be classified as laminar when only *Re* is considered. However, some other factors, such as surface coverage, should be considered in determining the flow state. Therefore, when different surface coverage causes some disturbances in overland flow, it is more logical to consider overland flow as turbulent flow. Different equations are employed for the description of the overland flow in laminar or turbulent flow conditions.

If the overland flow is turbulent, the Manning equation can be employed. Manning's coefficients used in overland flow analysis are different from those for channel flow because of the small depth and the turbulent nature of the flow. All the factors affecting the flow resistance in overland flow are aggregated into Manning's coefficient that is referred to as the effective Manning's roughness factor. The details of overland flow are presented in Karamouz et al. (2010).

5.3.2 Channel Flow

Channel flow occurs in different elements of an urban watershed, including ditches, gutters, and storm sewers. During and after a rainstorm, the flow in these structures is unsteady and nonuniform. Dynamic models could explain the dynamic nature of the system and simulate this flow condition utilizing the Saint-Venant equations. As solving the Saint-Venant equations takes a lot of time and is costly, less complex methods are employed in practice.

The simplified methods are categorized into two groups, including hydraulic and hydrological models. In hydraulic models, the Saint-Venant equations are employed, but hydrological models use an assumed relationship instead of the momentum and continuity equations. The main disadvantage of simpler models is that some phenomena that affect the flow, such as backwater effects, are not considered. Unsteady open-channel flow calculations are referred to as flood routing or channel routing. The details of open channel and flood routing are described in Chapter 4.

5.4 COMPONENTS OF URBAN STORMWATER DRAINAGE SYSTEM

Urban stormwater drainage systems within a district are comprised of three primary components: (a) street gutters and roadside swales, (b) stormwater inlets, and (c) storm sewers (and appurtenances such as manholes and junctions).

Street gutters and roadside swales gather the runoff from the street (and adjacent areas) and convey the runoff to a stormwater inlet while maintaining the street's level of service. Inlets transfer the flow into storm sewers after collecting stormwater from streets and other land surfaces and often provide maintenance access to the storm sewer system. Storm sewers convey stormwater over a street's or a swale's capacity and discharge it into a SWM facility or a nearby receiving water body.

In this section, the basic and important concepts related to designing different curbs, gutters, and inlets of urban drainage systems' components are discussed.

5.4.1 General Design Considerations

Keeping the water depth and spread on the street below a permissible value for a design storm return period is the key issue in designing street drainage systems. Rainfall events vary greatly in magnitude and frequency of occurrence. As major storms rarely occur, stormwater collection

TABLE 5.5
Minimum Design Frequency and Spread for Different Types of Roads

Road Classification	Design Frequency	Design Spread
Arterial	10 years	Shoulder
Collector streets	50 years (sag vertical curve)	Shoulder + 90 cm
Residential street	10 years	$\frac{1}{2}$ Driving lane

Note: Check 100 years' storm for all road classifications. The water level should remain less than 45 cm.

and conveyance systems are not normally designed to pass the peak discharge during major storm events. Stormwater collection and conveyance systems are designed to pass the peak discharge of a storm event with minimal disruption to street traffic. As the traffic volume in the streets increases, storms with larger return periods should be considered. Typical design return periods and permissible spread for different kinds of roads are given in Table 5.5.

Inlets must be strategically placed to pick up the excess gutter or swale flow once the limiting spread of water is reached. The inlets direct the water into storm sewers, which are typically sized to pass the peak flow rate from the minor storm without any surcharge. Local ordinances or criteria establish the magnitude of the minor storm, and the 2-, 5-, or 10-year storms are most commonly used.

Sometimes, storms will occur that surpass the magnitude of the design storm event. When this happens, the spread of water on the street exceeds the allowable spread and the capacity of the storm sewers. Street flooding occurs, and traffic is disrupted. However, proper design requires that public safety be maintained and the flooding be managed to minimize flood damage. Thus, local ordinances also often establish the return period for the major storm event, generally the 100-year storm. For this event, the street becomes an open channel and must be analyzed to determine whether or not the flood consequences are acceptable with respect to flood damage and public safety.

Besides traffic flow, streets serve another important function in the urban stormwater collection and conveyance system. The street gutter or the adjacent swale collects excess stormwater from the street and adjacent areas and conveys it to a stormwater inlet. According to FHWA (1984), proper street drainage is essential to

- Maintain the street's level of service.
- Reduce skid potential.
- Maintain good visibility for drivers (by reducing splash and spray).
- Minimize inconvenience/danger to pedestrians during storm events.

Streets geometry is another important factor in the design of drainage systems. The longitudinal slope of curbed streets should not be <0.3% and should be >0.5%. An appropriate cross slope is necessary to provide efficient drainage without affecting driving comfort and safety. In most situations, a cross slope of 2% is preferred. Higher cross slopes could be used for multilane streets, but it should not be >4%.

An additional design component of importance in street drainage is the gutter (channel) shape. Most urban streets contain curb and gutter sections. Various types exist, which include spill shapes, catch shapes, curb heads, and roll gutters. The shape is chosen for functional, cost, or esthetic reasons and does not dramatically affect the hydraulic capacity. Swales are common along some urban and semiurban streets, and roadside ditches are common along rural streets. Their shapes are important in determining hydraulic capacity.

5.4.2 Flow in Gutters

Street slope can be divided into two components: longitudinal slope and cross slope. The longitudinal slope of the gutter essentially follows the street slope. The hydraulic capacity of a gutter increases as the longitudinal slope increases. The allowable flow capacity of the gutter on steep slopes is limited in providing for public safety. The cross (transverse) slope represents the slope from the street crown to the gutter section. A compromise is struck between large cross slopes that facilitate pavement drainage and small cross slopes for driver safety and comfort. Usually, a minimum cross slope of 1% for pavement drainage is recommended. For increasing the gutter capacity, composite sections are often preferred because the gutter cross slopes are steeper than street cross slopes.

For evaluating the drainage capacity of the street gutters, four major steps should be followed:

1. Calculating the theoretical gutter flow capacity to convey the minor storm based on the allowable spread.
2. Repeating step 1 based on the allowable depth.
3. Calculating the allowable street gutter flow capacity by multiplying the theoretical capacity (calculated in step 2) by a reduction factor used for safety considerations. The minimum of the capacities is calculated in step 1, and this step is the allowable street gutter capacity.
4. Calculating the theoretical major storm conveyance capacity based on the road inundation criteria. Reducing the major storm capacity by a reduction factor to determine the allowable storm conveyance capacity.

The flow should be routed through gutters to determine the flow depth and spread of water on the shoulder, parking lane, or pavement section under design flow conditions. The design discharge is calculated using the rational method. For more simplicity in calculations, the flow in a gutter is assumed to be steady and uniform at the peak design discharge. This approach usually results in more conservative results.

Triangular, V, and composite sections are three regular section types of gutters (Figure 5.17). Since the gutter flow is assumed to be uniform for design purposes, Manning's equation is utilized with a slight modification to account for the effects of a small hydraulic radius in designing gutters. Gutters with composite cross slopes are often used to increase the gutter capacity. In composite gutter sections, such as those illustrated on the right-hand side of Figure 5.17, the gutter flow is divided into two fractions: discharge in the depressed section and discharge in the nondepressed section to each other. For safety and maintenance reasons, the depths and side slopes of V-shaped sections should be as shallow as possible. Street-side sections collect the initial runoff and transport it to the nearest inlet or major drainage way. For effective performance, the velocity, depth, and cross-slope geometries of these sections should be limited. The usually considered limitations include a maximum 2-year flow velocity equal to 0.9 m/s, a maximum flow depth equal to 0.3 m, and a maximum side slope of each side equal to 5H:1V. In practice, flatter side slopes are usually used. The readers are suggested to refer to Karamouz et al. (2010) for more detailed information about the designing methodology of each section.

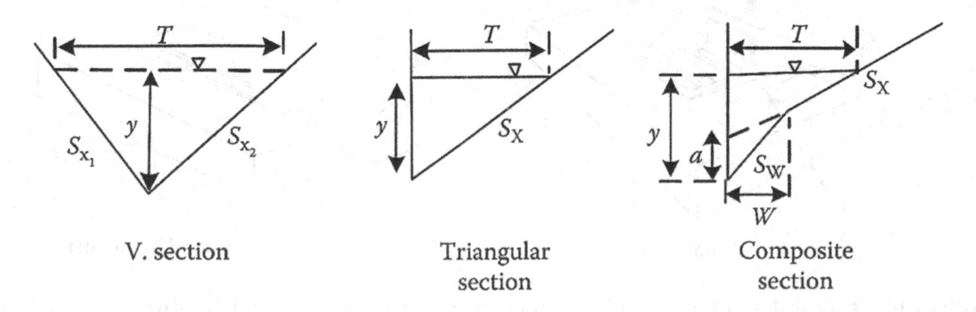

FIGURE 5.17 Various street gutter sections.

5.4.2.1 Gutter Hydraulic Capacity

Stormwater flowing along the streets exerts momentum forces on cars, pavements, and pedestrians. To limit the hazardous nature of heavy street flows, it is necessary to set limits on flow velocities and depths. As a result, the allowable sheet gutter hydraulic capacity Q_A is determined as the smaller value of Q_T or RQ_F.

Q_T is the street hydraulic capacity limited by the maximum water spread. Q_F is the gutter capacity when the flow depth equals allowable depth, and R is the reduction factor.

For example, Guo (2000) has defined two sets of reduction factors: minor storm events and the other for major events. The reduction factor is equal to one for a street slope less than 1.5% and then decreases in a nonlinear behavior as the street slope increases.

It is important for street drainage designs that the allowable street hydraulic capacity be used instead of the calculated gutter full capacity. Thus, wherever the accumulated stormwater flow rate on the street is close to the allowable capacity, a street inlet shall be installed.

5.4.3 Pavement Drainage Inlets

Stormwater inlets are a vital component of the urban drainage system. Inlets collect street surface runoff from the street, drain into an underground drainage system, and can provide maintenance access to the storm sewer system. They can be made of cast iron, steel, concrete, and/or precast concrete and are installed on the edge of the street adjacent to the street gutter or at the end of a swale.

Roadway geometrical features often dictate the location of pavement drainage inlets. In general, inlets are placed at all low points (sumps or sags) in the gutter grade, median breaks, intersections, and crosswalks. In other words, the drainage inlets are spaced so that the spread under the design (minor) storm conditions will not exceed the allowable flow spread (Akan and Houghtalen, 2003). Inlets used for the drainage of streets are divided into four major classes (Figure 5.18):

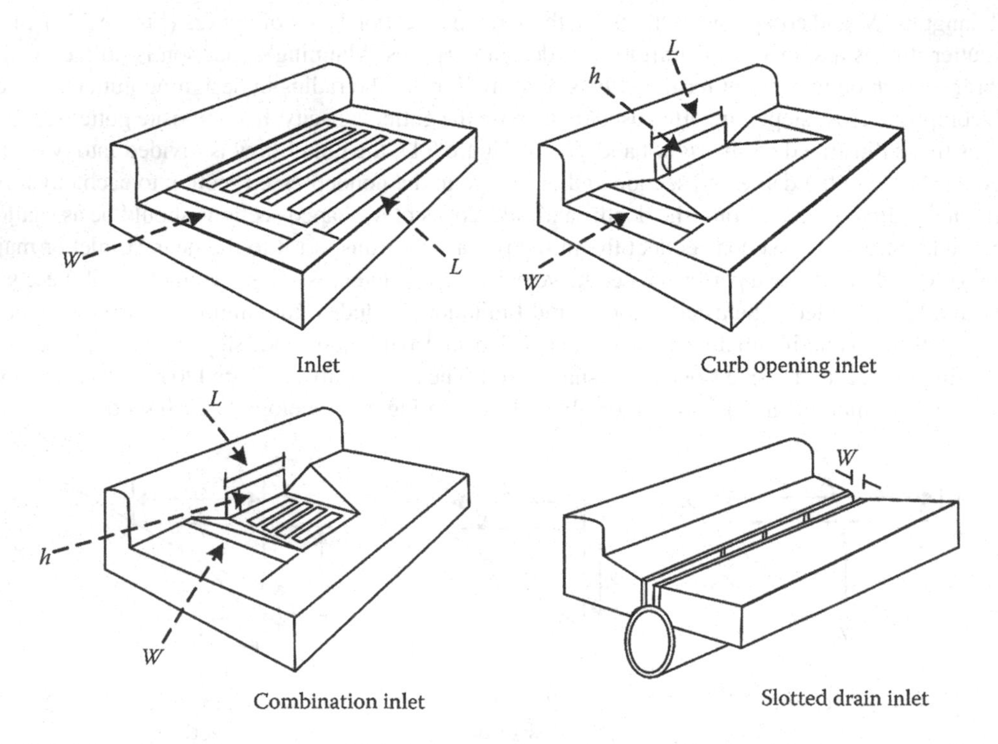

FIGURE 5.18 Four different types of inlets. (From Johnson, F.L., and Chang, F.M., *Drainage of Highway Pavements.* Hydraulic Engineering Circular, No. 12, Federal Highway Administration, McLean, VA, 1984.)

- Curb opening inlets.
- Grate inlets.
- Slotted drains, in which the performance is the same as curb opening inlets, can be considered as weirs lateral inflow.
- Combination inlets usually consist of some combination of three types of inlets. When curb is involved in a combination, its opening may extend upstream of the grate, and when the grate is involved, it is usually placed at the downstream end.

Each of these inlets can be used with or without a gutter depression. Inlets placed on continuous grades rarely intercept all of the gutter flow during the minor (design) storm. The efficiency of an inlet performance is calculated as

$$E = \frac{Q_i}{Q}, \tag{5.5}$$

where
E is the efficiency of inlet performance,
Q is the total gutter flow rate, and
Q_i is the intercepted flow rate.

The nonintercepted flow by an inlet, which is called bypass or carryover, is estimated as

$$Q_b = Q - Q_i, \tag{5.6}$$

where
Q_b is the carryover (bypass) flow rate.

The ability of an inlet to intercept flow (i.e., hydraulic capacity) on a continuous grade generally increases with the increasing gutter flow, but the capture efficiency decreases. In other words, even though more stormwater is captured, a smaller percentage of the gutter flow is captured. The inlet capacity depends on the inlet type and geometry, the flow rate (depth and spread of water), the cross (transverse) slope, and the longitudinal slope.

5.4.3.1 Inlet Locations

The inlet location may be dictated either based on physical demands, hydraulic requirements, or both. In all instances, the inlet location is coordinated with physical characteristics of the roadway geometry, utility conflicts, and feasibility of underground pipe layout.

Logical locations for inlets include sag configurations, near street intersections, at gore islands, crosswalks, entrance and exit ramp gores, and super elevation transitions. Inlets with locations not established by physical requirements should be located on the basis of hydraulic demand. In other words, the distance between drainage inlets should be determined in such a way that the spread does not exceed the allowable value under the design storm condition.

As mentioned before, the design discharge is calculated using the rational method. The main issue in using the rational method is an accurate estimation of the time of concentration. Because of the small drainage areas related to each inlet, the time of concentration is usually <5 minutes. But at most sites, rainfall data for durations <5 minutes are not recorded. Therefore, a time of concentration of 5 minutes is considered in the design of municipal drainage systems. This assumption leads to equal spacing for similar inlets. The carryover from the upstream inlet should be added to the stormwater runoff generated over the street section between the two adjacent inlets. The resulting runoff volume should be considered as the design discharge of the downstream inlet. The bypass flow from the upstream inlet is added to the peak flow rate obtained for the street section between two adjacent inlets for more simplicity. It should be noted that this approximated situation is completely different from the actual situation since the actual process is unsteady, and the peak street runoff and the bypass flow from upstream do not necessarily happen at the same time.

Example 5.3

Curb-opening inlets are placed along a 10-m-wide street. The length of the inlets is 0.8 m, and the inlets are placed on a continuous grade, and the runoff coefficient is 0.85. The inlets drain stormwater to a triangular gutter with S_X=0.025, S=0.015, and n=0.015. Determine the maximum allowable distance between inlets considering that the rainfall intensity is equal to 10 mm/h and the allowable spread is 1.8 m.

Solution:

First, the location of the most upstream inlet is specified. This inlet collects the runoff from a surface of $A_i = W_s L_d$, where W_s is the street width and L_d is the distance from the roadway crest to the inlet. As it is the first inlet, there is no carryover discharge from the upstream. The collected surface water at this inlet is calculated using the rational method as

$$Q = (0.85)(0.000003)(10)L_d = 0.00003L_d.$$

The adopted Manning formula for simple triangular gutter sections is

$$Q = \frac{0.56}{n} S_X^{5/3} S^{1/2} T^{8/3},$$

If S_X=0.025, S=0.015, n=0.015, and T=1.8, then

$$0.00003 \times L_d = \frac{(0.56)(1.8)^{8/3}(0.025)^{5/3}(0.015)^{1/2}}{(0.015)} = 0.047\text{m}^3/\text{s}.$$

Finally, L_d=0.047/(0.00003)=1,567 m, which is the distance of the initial inlet from the crest. This inlet drains a flow of Q=0.047m³/s, which corresponds to a spread of 1.8 m. Since the flow rate of next inlets is more than 0.047 m³/s, because of bypass flow from upstream inlets, their distance is less than 1,567 m. Therefore, the maximum distance between inlets is equal to 1,567 m.

5.4.4 Surface Sewer Systems

Buried pipes convey the collected stormwater through inlets to a point where stormwater is discharged to the receiving water body. In addition to pipes, different relevant structures such as inlets, manholes, junction chambers, transition structures, flow splitters, and siphons are included in a sewer system. An introduction to appurtenant structures is given here. For detailed descriptions of these components, see Brown et al. (1996).

For easy and convenient access to sewer systems for maintenance and repair, manholes (or access chambers) are used. Furthermore, manholes provide the appropriate air circulation in the sewer system. Manholes also serve as flow junctions and provide flow transitions for changes in pipe size, slope, and alignment. Often precast or cast-in-place concrete are used for manhole construction. A variety of configurations are used for manhole construction (Figure 5.19).

The diameter of the chamber is determined by the number and diameter of the sewer pipes coming into the manhole and the working space required. For deep manholes, the chamber should be large enough to provide benching or a landing adequate for two persons to stand upon. When a manhole is sited on a curve or additional pipes enter at the sides, a larger size may be required. However, the depth and diameter of usual manholes vary between 1.5 and 4.0 m and 1.2 and 1.5 m, respectively. Manholes are recommended:

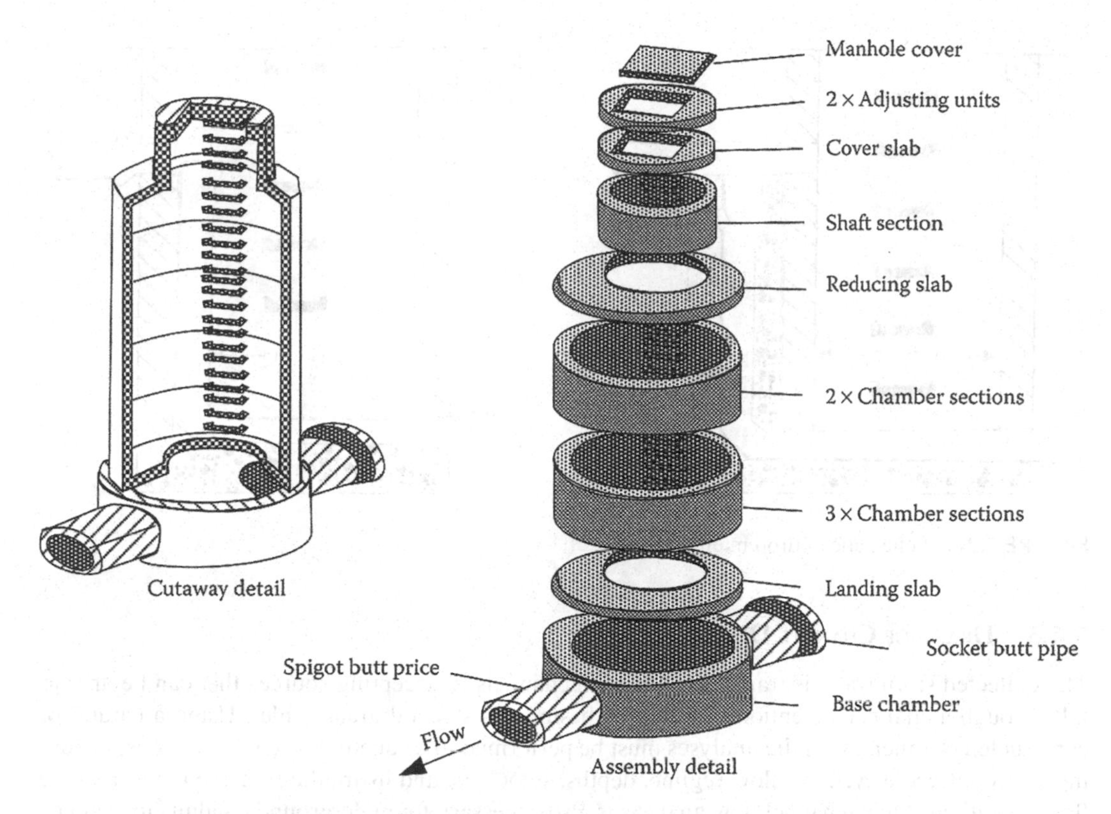

FIGURE 5.19 Typical manhole configuration.

- At intervals of up to 90 m, or 200 m, if man entry pipe runs are required
- Whenever there is a significant change of direction in a sewer
- Where another sewer is connected with the main pipe of a sewer
- Where there is a change of size or gradient of the pipelines

In the cases where the altitude of the incoming pipe is significantly higher than the outgoing pipe, drop manholes are utilized (Figure 5.20). For very large joining storm sewers that standard manholes cannot accommodate, a junction chamber is utilized, making the same material as manholes. To streamline the joining flows, channels are typically built into these chambers. A riser structure is placed in some junction chamber utilizations. The riser is extended to the street surface for more accessibility and interception of the surface flow.

Transition sections are used in the cases where the transition of pipe sizes is made without a manhole. As a significant amount of energy loss occurs where the pipe size suddenly changes, gradual expansion from one size to another is provided in these sections.

Flow splitters are appurtenant junction structures that divide and derive an incoming flow into two or more downstream storm sewers. Other appurtenant called flow deflectors are used for minimizing the energy losses in flow splitters. The deposition of suspended material in stormwater makes it difficult and costly to maintain the flow splitters. The flow is carried under an obstruction, like a stream, through inverted siphons (or depressed sewers). Commonly, at least two barrel siphons are constructed with each other. In systems with high potential for back-flood, flap gates are placed near storm sewer outlets because of high tides or high stages in the receiving system. These gates could be made of different materials that are available and economical on the construction site.

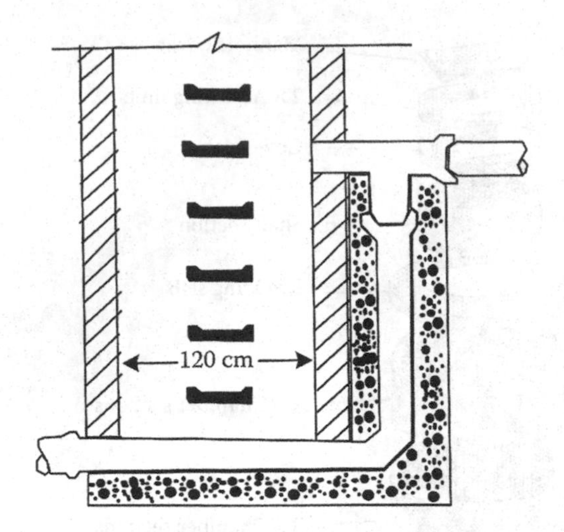

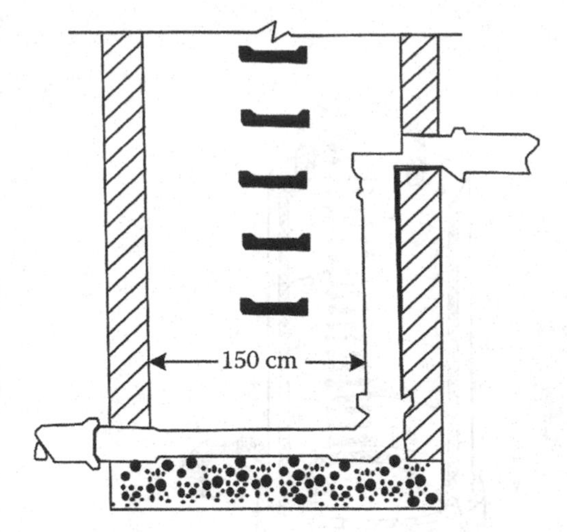

FIGURE 5.20 Schematic of drop manholes.

5.4.5 Drainage Channel Design

The collected stormwater is transmitted to open channels to accepting sources that can be an outfall through a channel, detention or retention basin, or a storm drainage inlet. Using a natural or constructed channel, hydraulic analyses must be performed to evaluate flow characteristics, including water surface elevations, flow regime, depths, velocities, and hydraulic transitions for multiple flow conditions. Open-channel flow analysis is also necessary for underground conduits to evaluate hydraulics for less-than-full conditions.

The maximum allowed velocity method or the tractive force method is commonly used in the design of channel liners. The primary assumption in using the maximum allowed velocity method is that until the average flow velocity of the channel is less than the maximum allowed velocity, erosion will not occur. In using the tractive force, the channel section is designed so that the resulting tractive force (shear force) that could move the particles on the channel bed does not exceed the values resisting the particle motion.

Trapezoid cross sections are commonly used in man-made open channels (Figure 5.21). The main geometrical parameters in designing open channels include the longitudinal bottom slope, the bottom width b, side slopes that should be <0.5, and the section depth y. The site topography usually governs the longitudinal bottom slope, and the characteristics of in situ soil dictate the section sides' slope. The bottom width and depth of the section are calculated after determining the longitudinal slope and the side slopes to convey the resulting runoff under the design situation without exceeding the allowed shear forces on the channel bed.

Similar to gutters, the open channels are designed for steady flow equal to the peak discharge, although in the actual situation, the surface drainage channel flow is not steady or uniform. A freeboard should be added to the obtained flow depth under design conditions to consider the deviations from the usual flow condition. The amount of appropriate freeboard is site-dependent, but a 15 cm freeboard is usually used to design roadside channels (Young and Stein, 1999).

On the other hand, the flow regime is determined by the balance of the effects of viscosity and gravity relative to the inertia of the flow. The Froude number (Fr) is considered in the design of channels for a definition of flow regime. In surface drainage channels, it is preferred that flow be subcritical so that the Froude number is <1.0 under design conditions. In Froude number close to one (critical flow), the flow condition may fluctuate between subcritical and supercritical states. In this situation, instabilities occur because of considerable changes in the discharge value.

5.4.5.1 Design of Unlined Channels

At first, the cross-sectional characteristics of a trapezoidal channel are formulated using the parameters illustrated in Figure 5.21 as follows:

$$A = (b + zy)y, \tag{5.7}$$

$$A = b + 2zy, \tag{5.8}$$

$$D = \frac{(b + zy)y}{b + 2zy}, \tag{5.9}$$

$$D = \frac{(b + zy)y}{b + 2y\sqrt{1 + z^2}}, \tag{5.10}$$

where

D is hydraulic depth and
R is hydraulic radius.

These expressions are substituted into the Manning formulation and rearranged to obtain

$$\frac{nQ}{S^{1/2}y^{8/3}} = \frac{(b/y + z)^{5/3}}{\left(b/y + 2\sqrt{1 + z^2}\right)^{2/3}}. \tag{5.11}$$

When applying Manning's equation, the choice of the roughness coefficient, n, is the most subjective parameter and is varying for different channel materials.

5.4.5.2 Grass-Lined Channel Design

The tractive force approach is usually used to design grass-lined channels, and the procedure is similar to unlined channels. The resulting shear stresses on the channel bed of the designed channel section should not exceed the maximum allowed value. The hydraulic characteristics of a grass lining depend on the type, maturity, and condition of the grass, and these factors should be incorporated into the design procedure.

The degree of retardance of grass lining and the hydraulic radius govern the Manning roughness coefficient of the channel according to the following equation (Chen and Cotton, 1988):

$$n = \frac{(3.28R)^{1/6}}{C_n + 19.97\log\left[(3.28R)^{1.4} S^{0.4}\right]}, \tag{5.12}$$

where

C_n is the dimensionless retardance factor.

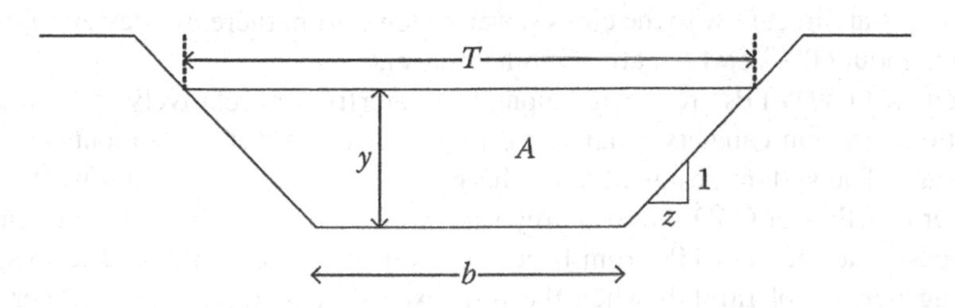

FIGURE 5.21 Schematic of a trapezoidal open channel section.

To determine this factor, first, the retardance class of the vegetal cover (A–E) is determined, and then the value of C_n and permissible shear stress are estimated.

The design procedure of grass-lined channels and that of unlined channels in cohesive soils are the same because of tractive forces on the channel bottom so that the maximum allowed flow depth (y_{allow}) can be calculated as

$$y_{allow} = \frac{C_s \tau}{\gamma S}, \tag{5.13}$$

C_s is the corrosion factor for the channel curvature, which is 1.0, 0.90, 0.75, and 0.6 for straight, slightly sinuous, sinuous, and sinuous channels, respectively. See Karamouz et al. (2010) for more details.

5.5 COMBINED SEWER OVERFLOW

When flooding, the combined volume of wastewater and flood may exceed the capacity of the combined sewer systems. Therefore, it may cause the plants not to function for a pretty long time fully. Flooding around the treatment plants may cause damage to the biological treatment unit, electronic, and pumping stations that support the plant's operation. In this situation, the plants bypass the wastewater with partial or no treatment into the close by waterbody. CSOs are a shock loading to the environment that could cause large amounts of pollutants entering water bodies, such as rivers, streams and harbors, and are a danger to human and aquatic life. A high tangible or intangible cost will be required to get rid of pollutions after sewer overflow. In New York City, with combined sewer, some CSO is happening after any heavy rain from many outfalls around the city.

Also, climate change impacts may affect the frequency of extreme rainfalls result in increasing the CSO's frequency of discharge. Schroeder et al. (2011) analyzed CSO datasets from four catchments of Berlin's in Germany combined sewer system. Comparison of rainfall characteristics (e.g., duration, maximum hourly intensity, total depth) with CSO volumes indicated that the total depth of rainfall was the best determinant of CSO occurrence. They proposed the definition of a critical rainfall depth for each catchment, above which CSO will occur.

New York's water and wastewater system is an engineering marvel of massive scale with a complex network of aqueducts and tunnels, some dating back more than 150 years. Water that enters the city's drains is conveyed through 7,500 miles of sewers and returned to New York City waterways. The DEP manages a complex system that begins with reservoirs located over 125 miles away from the city and ends at the city's 14 wastewater treatment plants with the release of treated effluent into New York Harbor. Although the system is integrated, it is best explained by separating it into two primary components: its water supply and distribution system and its collections and treatment system.

Every day, the city treats 1.3 billion gallons of wastewater and helps restore and maintain water quality in New York Harbor. The city is primarily served by a combined sewer system where stormwater and sanitary waste are carried through a single pipe. Sanitary waste enters the sewer system through direct connections from buildings. Stormwater enters the collections and treatment system from catch basins that direct flow to the city's sewer system. From there, wastewater flows by gravity through sewers, about 60% of which are combined sewers.

All of the city's 14 WWTPs are located along the waterfront at relatively low elevations. Under normal conditions, system capacity is adequate to perform complete treatment for the combined volume of sewage. The system designed to discharge a mix of stormwater and wastewater—called combined sewer overflow or CSO—into nearby waterways to quickly drain the city and prevent the biological processes at the WWTPs from becoming compromised could lead to extended service outages. During periods of rainfall when the flow exceeds two times dry weather capacity, the combined volume of sewage and stormwater quickly can exceed the capacity of the WWTPs.

In response to these CSO events, the city has invested billions of dollars. Recently innovative strategies have been implemented to absorb rain before it can enter sewers. In September 2010, Mayor Bloomberg launched the NYC Green Infrastructure Plan, a comprehensive 20-year effort to meet water quality standards. In March 2012, the plan was incorporated into a consent order with the state that will eliminate or defer $3.4 billion in traditional investments and result in approximately 1.5 billion gallons of CSO reductions annually by 2030.

Increased rainfall and heavy downpours may contribute to increases in street flooding, sewer backups, and CSOs. Improving the city's sewer systems will enhance the ability of the existing infrastructure to cope with environmental changes. To this end, NYC-DEP will continue to implement a number of its programs that are already underway. Two of these programs are explained briefly in the following section. Further details are given in chapter 13.

5.5.1 Reduce Combined Sewer Overflows with Green Infrastructure

As rainfall volume to the New York area increases due to climate change, shifts in the frequency and volume of CSOs will occur. The city will continue to implement its Green Infrastructure Plan and CSO Long-Term Control Plans (LTCPs) to reduce such CSOs. For this purpose, DEP, working with the Department of Parks and Recreation and New York City Department of Transportation (NYCDOT), will continue to pursue its plan to capture the first inch of runoff in 10% of impervious surfaces citywide in areas within the combined sewer system by 2030. At the same time, DEP will continue developing LTCPs to evaluate long-term solutions to reduce CSOs and improve water quality in New York City's waterways.

5.5.2 Reduce Combined Sewer Overflows with High-Level Storm Sewers Citywide

While the development of new, green infrastructure is an effective solution to manage rainfall and reduce CSOs in some locations, in other areas, it will be more cost-effective to enhance the city's existing sewer system. The city will enlarge existing combined sewers with high-level storm sewers in certain areas near the water's edge around the city. These high-level storm sewers sit on top of the combined sewer and accept stormwater from the street before diverting it to a nearby waterway, with the combined sewer below it sending wastewater and a reduced amount of stormwater to a treatment plant. Such high-level storm sewers are able to capture 50% of rainfall before it enters combined sewers. Mitigation of CSOs and the potential to reduce street flooding are the advantages of high-level storm sewers. To this end, DEP will continue to pursue approximately 15 high-level storm sewer projects that will be completed by 2023 and will continue to seek additional opportunities near the water's edge for additional cost-effective high-level storm projects.

5.6 CULVERTS

A culvert is a short, closed (covered) conduit that conveys stormwater runoff under an embankment, usually a roadway. The primary purpose of a culvert is to convey surface water, but it may also be used to restrict flow and reduce downstream peak flows if properly designed. In addition to the hydraulic function, a culvert must also support the embankment and/or roadway and protect traffic and adjacent property owners from flood hazards to the extent practicable. In detention reservoirs, culverts are also used as outlet conduits.

Different materials such as concrete, corrugated aluminum, corrugated steel, reinforced concrete pipes (RCPs) are used to construct culverts. RCPs are recommended for all flowing streams under roadways in which pipe slopes are <1%. RCP must be used for culverts designed for a 100-year storm if the culverts lie in public lands or easements. High-density polyethylene (HDPE) pipes may also be used where permitted by the Director. To reduce corrosion and flow resistance of the culverts, they are sometimes lined with another material, such as asphalt.

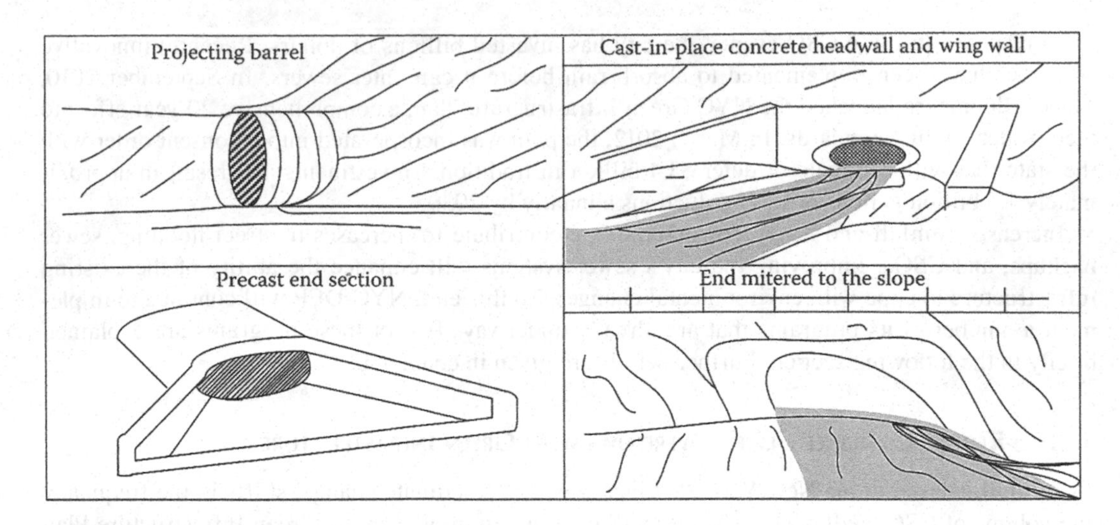

FIGURE 5.22 Standard inlet types. (From Normann, J.M. et al., *Hydraulic Design of Highway Culverts.* Hydraulic Design Series, No. 5, Federal Highway Administration, McLean, VA, 1985. With permission.)

An essential factor in the hydraulic performance of culverts is the inlet configuration. Different types of prefabricated and constructed-in-place inlet facilities are commonly used (see Figure 5.22). The culvert inlet comprises projecting culvert barrels, concrete headwalls, end sections, and culvert ends mitered to conform to the fill slope. Headwalls may be used for various reasons, including increasing the efficiency of the inlet, providing embankment stability and protection against erosion, providing protection from buoyancy, and shortening the length of the required structure. Headwalls are required for all metal culverts and where buoyancy protection is necessary. If high headwater depths are to be encountered, or the approach velocity in the channel will cause scour, a short channel apron shall be provided at the toe of the headwall. Wing walls must be used where the side slopes of the channel adjacent to the entrance are unstable or where the culvert is skewed to the normal channel flow. Where inlet conditions control the amount of flow that can pass through the culvert, improved inlets can significantly increase the hydraulic performance of the culvert.

The flow characteristics in a culvert depend on upstream and downstream conditions, inlet geometry, and barrel characteristics. It should also be considered that the flow condition for a specific culvert may vary over time. The inlet (upstream) or the outlet (downstream) condition controls the flow variations in a culvert. If the conveyance capacity of a culvert barrel is higher than the entrance flow of the inlet, the inlet controls the flow condition. Otherwise, the flow condition depends on the outlet situation. Different flow situations are discussed in Karamouz et al. (2010).

5.6.1 Sizing of Culverts

The determination of flow type in a culvert usually is not simple. So, designing (sizing) a culvert is a very complicated and difficult procedure.

In mild-slope culverts and steep culverts, the flow condition is often controlled by the outlet and inlet conditions. In these situations, only when $TW>D$, the full flow will occur. Various types of outlet control flow in the culvert are shown in figure 5.23.

Drainage culverts are often designed (sized) to avoid exceeding the headwater elevation from allowable tailwater elevation in the design discharge condition. The "minimum performance" approach is commonly used for sizing drainage system culverts in which the identification of the flow in the culvert is not needed and the culvert situation under both inlet and outlet control conditions is checked. As this method acts conservatively, it sometimes leads to a culvert size larger than what is needed.

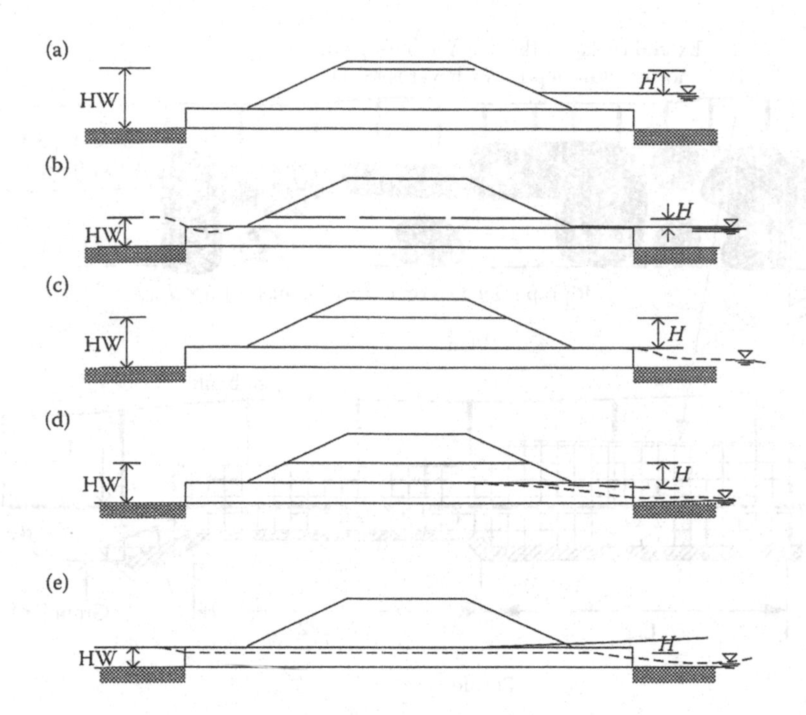

FIGURE 5.23 Types of outlet control flow in culverts. (From Normann, J.M. et al., *Hydraulic Design of Highway Culverts*. Hydraulic Design Series, No. 5, Federal Highway Administration, McLean, VA, 1985. With permission.)

5.6.2 Protection Downstream of Culverts

Inadequate protection at conduit and culvert outlets has long been a major problem. Scour resulting from the highly turbulent, rapidly decelerating flow is a common problem at conduit outlets. The riprap protection design is suggested for the conduit and culvert outlet where the channel and conduit slopes are parallel with the channel gradient. The conduit outlet invert is flushed with the riprap channel protection. Figure 5.24 shows typical riprap protection of culverts and significant drainage way conduit outlets. The rock size and length of riprap protection are determined based on the flow and culvert characteristics.

5.7 DESIGN FLOW OF SURFACE DRAINAGE CHANNELS

5.7.1 Probabilistic Description of Rainfall

Deterministic models do not result in accurate rainfall predictions because rainfall events are highly uncertain, and their characteristics vary in time and space. For incorporating these uncertainties in hydrological studies, rainfall events are treated as random processes, and probabilistic methods are employed for the determination of their occurrence and magnitudes.

5.7.1.1 Return Period and Hydrological Risk

The probability of equaling or exceeding a rainfall event with a specified duration and depth in any 1 year is called the exceedance probability, *p*. The return period is defined as the inverse of the return period T_r as

$$P = \frac{1}{T_r}. \tag{5.14}$$

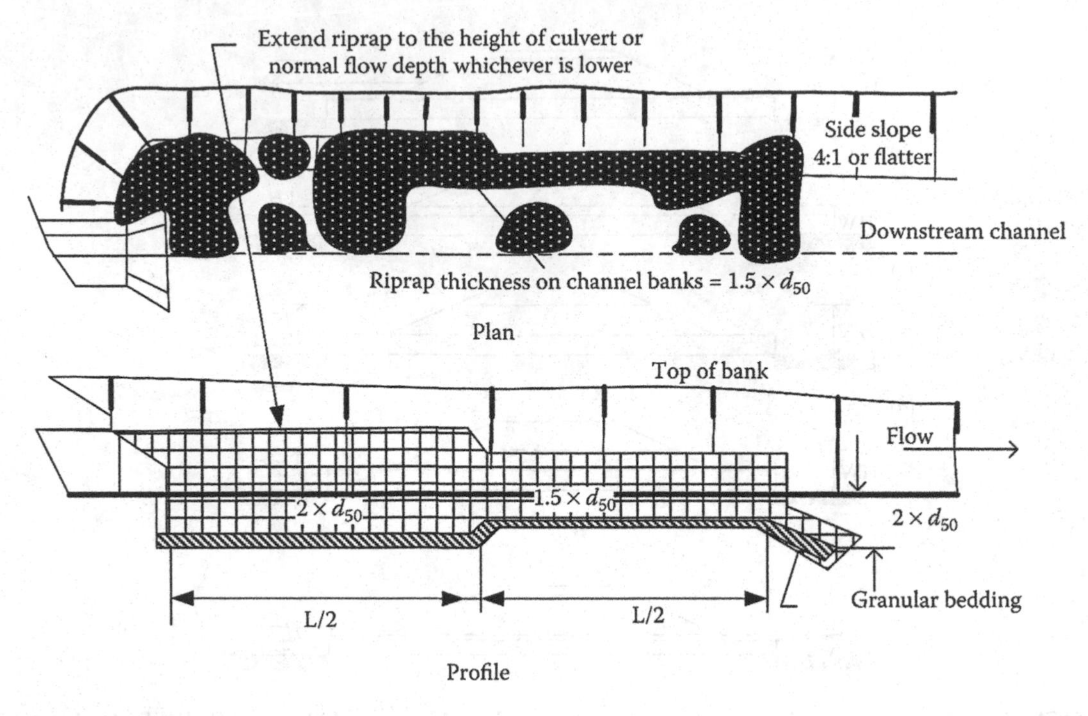

FIGURE 5.24 Culvert and pipe outlet erosion protection.

For example, if the 50-year, 18-hour rainfall is 10cm, then the probability that a depth of 10cm or higher will be produced over an 18-hour rainfall event in any given year is 2%. It should be noted that rainfall is considered a completely random process. This means that if a 50-year event is exceeded this year, the probability of its exceedance in the next year is still 2%.

Often, the design capacity of a stormwater structure is determined based on the rainfall events, and because of the random nature of the rainfall, the design capacity might be exceeded with a low probability at any given time. It can be concluded that there is always a hydrological risk (HR) the design should correspond to. The probability that, during the service life of the structure, the design event will be exceeded one time or more is considered an HR, which is calculated as

$$HR = 1 - \left(1 - \frac{1}{T_r}\right)^n, \tag{5.15}$$

where

T_r is the return period of the design event, and n is the life span of the considered structure.

Example 5.4

Determine the HR of a roadway culvert with 25 years' expected service life designed to carry a 50-year storm.

Solution:

Using Equation (5.15),

$$HR = 1 - \left(1 - \frac{1}{50}\right)^{25} = 0.40 = 40\%.$$

Example 5.5

Determine the design return period for the culvert of the previous example if the acceptable HR is 0.15 (or 15%).

Solution:

Equation 5.15 is employed again for the determination of the return period of the design rainfall event as

$$0.15 = 1 - \left(1 - \frac{1}{T_r}\right)^{25}.$$

T_r has been calculated as 154 years, whereas in practice, a return period of 150 years is considered to determine culvert size.

Fewer return periods may be considered because of economic reasons that lead to much higher HRs. In other words, flooding from a stormwater structure is not necessarily the result of poor structural design.

5.7.1.2 Frequency Analysis

Frequency analysis derives meaningful information from the available historical data. For instance, a frequency analysis should be carried out on the rainfall records at a specific rain gauge to determine the corresponding return period of rainfall with known depth and duration.

At first, for frequency analysis, the annual maximum series of rainfall depths for the specified duration is developed. It should be noted that "duration" is not necessarily equal to the entire duration of historic storms. For example, a historical storm may last for 50 minutes, but it can be used in the frequency analysis of 40-minute storms. If the rainfall in a 40-minute portion of this rainfall is more significant than all of the 40-minute rainfall events, the rainfall during this 40-minute portion is considered representative of the 40-minute rainfall year.

The best theoretical probability distribution that fits the rainfall annual maximum series is specified in the next step. The best-fitting probability distribution is determined based on the goodness of fit, which is quantified using statistical tests such as the chi-square test. However, experience shows that the "Gumbel distribution" (the extreme value type I distribution), most of the time, fits the maximum rainfall data well so that in practice, this distribution is often used for frequency analysis of rainfall data. The cumulative probability function of the Gumbel distribution is

$$p = 1 - e^{-e^{-b}}, \tag{5.16}$$

where

b is the distribution parameter.

The maximum rainfall depths at a specific return period, T, for a given rainfall duration, t_d, could be expressed as

$$R_T = \overline{R} + K\,S \tag{5.17}$$

where

R_T is the rainfall depth for a specified return period (T_r),
$\overline{R}$ is the mean annual maximum rainfall depths,
S is the standard deviation of annual maximum rainfall depths, and
K is the frequency factor.

TABLE 5.6
Frequency Factor *K* for the Gumbel Probability Distribution

	T_r (Years)				
N	5	10	25	50	1M
15	0.967	1.703	2.632	3.321	4.005
20	0.919	1.625	2.517	3.179	3.836
25	0.888	1.575	2.444	3.088	3.729
30	0.866	1.541	2.393	3.026	3.653
35	0.851	1.516	2.354	2.979	3.598
40	0.838	1.495	2.326	2.943	3.554
45	0.829	1.478	2.303	2.913	3.520
50	0.820	1.466	2.293	2.889	3.491
75	0.792	1.423	2.220	2.912	3.400
100	0.779	1.401	2.187	2.770	3.349
∞	0.719	1.305	2.044	2.592	3.137

Source: Haan, C.T., *Statistical Methods in Hydrology.* The lowa State University Press, Ames, IA, 1977.

The frequency factor, *K*, is determined based on the fitted probability distribution characteristics, the assigned design period, and the length of the employed annual maximum rainfall series. The *K* values of the Gumbel probability distribution for different combinations of return periods and the length of time series are shown in Table 5.6.

Example 5.6

Determine the 15-minute storm depths and the average intensities with return periods of 5, 10, 25, 50, and 100 years for a 25-year annual maximum series of 15-minute storm depths given in the first and third columns of Table 5.7. Suppose that the extreme value type I distribution (the Gumbel distribution) fits the annual maximum series.

Solution:

The mean and the standard deviation of the annual maximum depths are calculated as

$$\bar{P} = \frac{\sum P_i}{n} = \frac{40.752}{25} = 1.630 \text{ cm}$$

and

$$s = \sqrt{\frac{\sum \left(P_i - \bar{P}\right)^2}{n}} = 0.690 \text{ cm.}$$

For the return periods of T_r=5, 10, 25, 50, and 100 years, the frequency factors are determined as 0.888, 1.575, 2.444, 3.088, and 3.729, respectively, from Table 5.6. Then, Equation (5.16) is used to determine the 15-minute, 5-year depth as

$$P_T = 1.63 + (0.888)(0.690) = 2.243 \text{ cm.}$$

Likewise, for T_r=10 years,

$$P_T = 1.63 + (1.575)(0.690) = 2.717 \text{ cm.}$$

TABLE 5.7
Mean and Standard Deviation Example

P_j (cm)	$P_i - \bar{P}$	P_j (cm)	$P_i - \bar{P}$
2.742	1.112	1.548	−0.082
2.671	1.041	1.264	−0.366
2.657	1.027	1.201	−0.429
2.601	0.971	1.196	−0.434
2.526	0.896	1.105	−0.526
2.394	0.763	1.014	−0.616
2.156	0.526	1.009	−0.621
2.116	0.486	0.860	−0.770
1.990	0.360	0.849	−0.781
1.985	0.355	0.779	−0.851
1.636	0.006	0.708	−0.922
1.610	−0.020	0.585	−1.045
1.550	−0.081		

The rainfall depths corresponding to return periods of 25, 50, and 100 years are calculated in the same way and are equal to 3.32, 3.76, and 4.20 cm, respectively.

The rainfall depth is divided by the duration to find an average intensity. In this example, the duration is 15 minutes = 0.25 hours so that the calculated P_t is divided by 0.25 hours. The average intensities of 8.97, 10.87, 13.28, 15.04, and 16.80 cm/h are obtained for the return periods of 5, 10, 25, 50, and 100 years, respectively.

5.7.2 Design Rainfall

The design runoff event that a stormwater structure must carry out should be specified as the first step in sewer system development. It is assumed that the drainage system performs in its full capacity under design runoff so that the system may fail if the actual runoff is more than the designed value. Note that the system failure may correspond to physical damages to the system or a hydraulic failure in which the system does not perform as it is designed. An example of hydraulic failure is that water backs up and overtops the embankment if the storm runoff exceeds the design discharge of the associated culvert.

Often the runoff data are not available because stream gauges are too expensive and difficult to be used in urban areas. Therefore, instead of runoff data, rainfall data are used for the design of drainage systems. Thus, for estimating the design runoff, the rainfall data should be converted to runoff data through an appropriate rainfall–runoff model.

5.7.2.1 Selecting Design Rainfall and Runoff

The approaches used to select a design runoff event from historical rainfall data are divided into two groups: continuous simulation methods and single-event methods. These approaches have fundamental differences, but frequency analysis of hydrological data and application of a rainfall–runoff model is included in both of them. General procedures of these methods are shown in Figure 5.25. The top procedure in this figure is more acceptable than others. The application of rainfall–runoff models is discussed in Chapter 3, and in the preceding sections, a discussion on frequency analysis is given.

In applying the continuous simulation approach, the input of the rainfall–runoff model is a chronological series of rainfall, and the model output is a chronological series of runoff. Frequency analysis is performed on the resulting runoff series to determine the return period corresponding

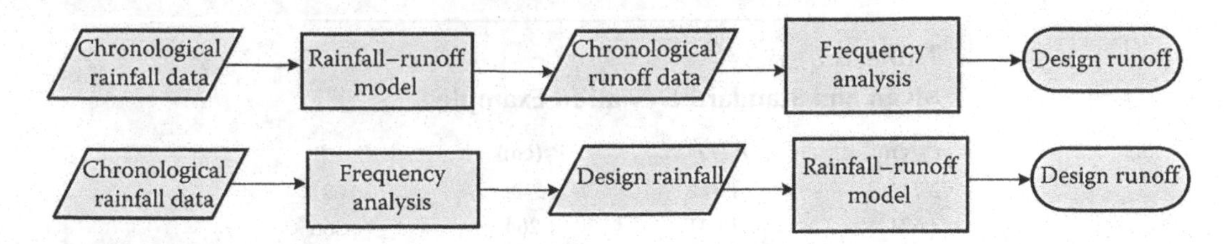

FIGURE 5.25 Design runoff formulation methods.

to different magnitudes of runoff events. The design runoff is selected based on the results of this frequency analysis. A subsurface flow factor should be incorporated into the rainfall–runoff model to consider the effects of water movement in the soil even during dry periods (zero rainfall). After the initial startup of the model, it would determine the initial basin soil moisture condition before each storm event. This issue plays an important role in determining design runoff by converting rainfall data because the resulting runoff from a rainfall event is significantly affected by the initial soil moisture condition.

The continuous simulation approach is preferred in detention basin projects because the sequence of rainfall events and the intervened time is important. For example, the filling capacity of a detention basin can be considered when the next storm occurs in this approach. But to benefit from the advantages of this approach, a sophisticated rainfall–runoff model should be employed. These kinds of rainfall–runoff models need big data, and their application is not easy. Furthermore, as there are large amounts of data, the simulation is often too time-consuming, especially when several decades, including hundreds of storms, are analyzed. The difficulty in applying this approach increases by decreasing the computational time increments to achieve more accurate results.

The Intensity–Duration–Frequency (IDF) curves or relationships are considered in the single-event approach to select a design rainfall. The selected design rainfall is entered into a rainfall–runoff model to estimate the design runoff. In this approach, only a single rainfall–runoff process for a single storm event is simulated; therefore, it is simpler than the continuous simulation approach. Also, a simple rainfall–runoff model can be employed, and it is not necessary to take into account the soil moisture condition. But the initial condition of the basin preceding the design rainfall occurrence should be specified.

Continuous rainfall records obtained synthetically using selected single rainfall events are also used in some recently developed hybrid methods. As the continuous simulation approach is too time-consuming and costly and needs extensive watershed data to yield accurate results, the single-event design storm method is more common in practice because of its simplicity and fewer data needed. A single-event design storm is identified by the corresponding return period, storm duration, depth (or average intensity), and its spatial and temporal distribution. The selection procedure of these characteristics is described in the following subsections.

5.7.3 Design Return Period

Structural or hydraulic failures of urban stormwater structures may cause serious difficulties for the public, flood damages of different magnitudes, and safety concerns. But as limited financial resources are available for sewer system construction, sonic risk of failure associated with the considered design return period is allowed in the design of stormwater collection infrastructures. The HR increases as design return periods decrease, but this results in larger sewer structures that are not economical.

There are a wide range of issues such as the importance of the structure, the cost, the level of protection provided by it, and the consequence of its failure that should be considered in selecting a design return period. The ideal design return period could be determined through a cost–benefit analysis. However, in practice, for establishing appropriate return periods (standards) for various systems, past experiences of failures and the resulting costs in the project location are considered. The design return periods of culverts are highly dependent on the traffic volume of the street, and usually, typical values of 5–10, 10–25, and 25–50 years are used for streets carrying low, intermediate, and high traffic volumes, respectively. By increasing the carrying capacity of the sewer structures, the design return period increases as design return periods of 2–5, 2–25, and 10–100 years are used for street gutters, storm sewers, and detention basins, respectively, and 50- to 100-year return periods are used for the design of major highway bridges. The design return periods of specific projects are determined based on the site characteristics and the local drainage regulations. In coastal regions, the joint occurrence of storm surges and rainfall is considered for a design period. See Chapter 13 for more details.

5.7.4 Design Storm Duration and Depth

The design storm duration is dependent on the project type. The design storm duration should result in the largest peak discharge for a specified return period because the storm sewers are designed for conveying the peak flow. Similarly, the duration that causes the largest detention volume is considered in the design of detention ponds. The design storm duration is specified by considering different values and evaluating their effects on the peak discharge and/or the detention volume.

IDF curves show the relationships between the average intensity (or depth), duration, and the return period as mentioned in Chapter 3. The average intensity is obtained from the local IDF curves after selecting the design return period and duration. Alternatively, the rainfall intensity for the area within the district can be approximated by the following equation:

$$I = \frac{28.5P_1}{(10+t_c)^{0.786}}, \tag{5.18}$$

where

I is the rainfall intensity (mm/h),
P_1 is the 1-hour point rainfall depth (mm), and
t_c is the time of concentration (minutes).

5.7.5 Spatial and Temporal Distribution of Design Rainfall

The spatial distribution of rainfall over a watershed is not uniform, which should be considered in rainfall–runoff analysis, especially in large watersheds. For incorporating the effect of watershed size on rainfall distribution, a reduction factor is applied to the design of rainfall. The rainfall intensity also varies during a single storm, and these variations affect the produced runoff rate. So, for a complete description of a design storm, the temporal variations of its intensity should also be considered. Different methods such as Soil Conservation Service (SCS) rainfall patterns are used to specification temporal variations in the design rainfall.

5.8 STORMWATER STORAGE FACILITIES

Detention and retention basins are used for stormwater runoff quantity control to mitigate the effects of urbanization on runoff flood peaks. As is the case with major drainage systems, the development of multipurpose, attractive detention facilities is recommended. These structures are safe, maintainable, and viewed as community assets rather than liabilities.

The storage facilities are divided into two groups. On-site runoff storage facilities are planned on an individual-site basis. Larger facilities that have been identified and sized as a part of some overall regional plan are categorized as "regional" facilities. In addition, the regional definition can also be applied to storage facilities that address moderately sized watersheds to encompass multiple land development projects.

On-site storage facilities are usually designed to control runoff from a specific land development site and are not located or designed with the idea of reducing downstream flood peaks along with a major drainage system. The total volume of runoff detained in the individual on-site facility is quite small, and the detention time is relatively short. Therefore, unless design (i.e., sizing and flow release) criteria and implementation are applied uniformly throughout the urbanizing or redeveloping watershed, their effectiveness diminishes rapidly along the downstream reaches of waterways. The application of consistent design and implementation criteria is of paramount importance if large numbers of on-site detention facilities effectively control peak flow rates along with major drainage systems (Glidden, 1981; Urbonas and Glidden, 1983).

The principal advantage of on-site facilities is that developers can be required to build them as a condition of site approval. Major disadvantages include the need for a larger total land area for multiple smaller on-site facilities than larger regional facilities serving the same tributary catchment area. If the individual on-site facilities are not properly maintained, they can become a nuisance to the community and a basis for many complaints to municipal officials. It is also difficult to ensure adequate maintenance and long-term performance at the design levels. Prommesberger (1984) inspected approximately 100 on-site facilities built or required by municipalities to be built, as a part of land developments over about a 10-year period. He concluded that a lack of adequate maintenance contributed to a loss of continued function of the facilities. He also concluded that a lack of local institutional structures contributed to the facilities no longer being in existence after their initial construction.

Some facilities are designed during the watershed planning process. These facilities, which are developed in a staged regional plan, are called regional facilities. These are often planned and located as part of the district's master planning process. They are typically much larger than on-site facilities. The main disadvantage of the regional facilities is the lack of an institutional structure to fund them. Another disadvantage of regional facilities is that they can leave substantial portions of the stream network susceptible to increased flood peaks, and plans must be developed to take this condition into account. In addition, to promote water quality benefits, some form of on-site SWM is necessary upstream of the regional facilities. Examples include minimized directly connected impervious areas (MDCIA) that promote flow across vegetated surfaces utilizing "slow-flow" grassed swales and grading lots to contain depressions.

More economical and hydrologically reliable results can be achieved through SWM planning for an entire watershed that incorporates the use of regional facilities. Regional facilities also potentially offer greater opportunities for achieving multiobjective goals such as recreation, wildlife habitat, enhanced property value, open space, and others.

There are several types of stormwater storage facilities, where they are classified as on-site or regional, namely

1. *Detention:* Detention facilities provide temporary storage of stormwater that is released through an outlet that controls flows to preset levels. Detention facilities typically flatten and spread the inflow hydrograph, lowering the peak to the desired flow rate. Often these facilities also incorporate features designed to meet water quality goals.
2. *Retention*: Retention facilities store stormwater runoff without a positive outlet or with an outlet that releases water at very slow rates over a prolonged period. These differ in nature and design from "retention ponds" that are used for water quality purposes.
3. *Conveyance (channel) storage*: Conveyance, or channel routing, is an often-neglected form of storage because it is dynamic and requires channel storage routing analysis. Slow-flow and shallow conveyance channels and broad floodplains can retard the buildup of flood peaks and alter the time response of the tributaries in a watershed.
4. *Infiltration facilities*: Infiltration facilities resemble retention facilities in most respects. They retain stormwater runoff for a prolonged period of time to encourage infiltration into

the groundwater. These facilities are difficult to design and implement because so many variables come into play.

5. *Other storage facilities*: Storage can occur at many locations in urban areas such as in random depressions and upstream of railroad and highway embankments. Special considerations typically apply to the use and reliance upon such conditions.

Detention and retention facilities can be further subdivided into

1. *In-line storage*: A facility located in line with the drainage channel/system and captures and routes the entire flood hydrograph. A major disadvantage with in-line storage is that it must be large enough to handle the total flood volume of the entire tributary catchment, including off-site runoff if any.
2. *Offline storage*: A facility located offline from the drainage way and depends on the diversion of some portion of flood flows out of the waterway into the storage facility. These facilities can be smaller and potentially store water less frequently than in-line facilities.

Irrespective of which type of storage facility is utilized, the designer is encouraged to create an attractive, multipurpose facility that is readily maintainable and safe for the public under different weather conditions (dry or wet). Designers are also encouraged to consult with other specialists such as urban planners, landscape architects, and biologists during planning and design.

5.8.1 Sizing of Storage Volumes

The reservoir routing procedure is usually employed for the sizing of detention storage volumes. This method is more complex and time-consuming than the use of empirical equations or the rational method. Its use requires the designer to develop an inflow hydrograph to the facility. This is an iterative procedure that is described as follows (Guo, 1999). It should be noted that some storage facilities/ BMPs sizing were discussed earlier in this chapter and what is presented here is general sizing and design consideration that complements the BMPs specific issues:

1. *Site selection*: The facility's location should be based on criteria developed for the specific project. Regional storage facilities are typically placed where they provide the greatest overall benefit. The location, geometry, and nature of these facilities are determined regarding multiuse objectives such as using the detention facility as a park or for open space, preserving or providing wetlands and/or wildlife habitat, or other uses and community needs.
2. *Hydrological determination*: The storage basin inflow hydrograph and the allowable peak discharge from the basin for the design storm events should be determined. The allowable peak discharge is limited by the local criteria or by the requirements spelled out in the available master plans.
3. *Initial sizing*: Based on the inflow hydrograph and basin characteristics, the initial size of the detention storage volume is estimated.
4. *Initial configuration of the facility*: The initial configuration of the facility should be based on site constraints and other goals for its use. Then a stage–storage–discharge relationship is developed for the facility.
5. *Outlets design*: The outlet's initial design entails balancing the facility's initial geometry against the allowable release rates and available volumes for each stage of the hydrological control. This step requires the sizing of outlet elements such as a perforated plate for controlling the releases, orifices, weirs, the outlet pipe, spillways, and so on.
6. *Preliminary design*: The results of steps 3–5 provide a preliminary design of the overall detention storage facility. In the preliminary design step, which is an iterative procedure,

the size and shape of the basin and the outlet works design are checked using a reservoir routing procedure and then modified based on the design goals. The modified design is again checked, and further modifications are made if needed. At the end of this step, the outlet works' storage volume and nature and sizes are finalized.

7. *Final design*: The final design phase of the storage facility is completed after the hydraulic design has been finalized. This phase includes structural design of the outlet structure, embankment design, site grading, a vegetation plan, accounting for public safety, spillway sizing, assessment of dam safety issues, and so on.

For reservoir routing procedure that is also used for storage sizing, see chapter 6 of Karamouz et al. (2012).

5.9 RISK ISSUES IN URBAN DRAINAGE

There are three typical urban drainage risk issues suggested by Hauger et al. (2005), including (a) flooding of urban drainage systems, (b) dissolved oxygen (DO) depletion in streams, and (c) discharge of chemicals to receiving waters. Table 5.8 shows these key issues.

5.9.1 Flooding of Urban Drainage Systems

The primary concern when sewers are flooded in urban areas is economic losses, including damage to buildings, damaged infrastructure, and delayed traffic. Injuries and infections are of secondary importance compared to the damage and other economic losses of floods in urban areas. The main source of risk is stormwater, which is the consequence of high-intensity precipitation. The total impact cannot be evaluated easily, because there are associated uncertainties. It is also difficult to get insurance information about the extent of the damage.

Traditional indicators of risk are directed towards the hazard. For many years, IDF curves were used as indicators because of the extended flexibility in utilizing them (Marsalek et al., 1993). These days, with computational power and sophisticated simulation models, more elaborate indicators have been introduced, such as design storm hyetographs and historical rainfall series (DWPC, 2005). The water level is another widely used indicator obtained from the simulation models. The advantage of water level is that it is highly correlated with damages. While water stays below the street level, the damage is limited, but the economic loss increases rapidly when the water level rises. The water level is also used as the accepted service level of the system, and a return period for exceeding a given water level is defined.

Water level and rain intensity are both indicators of the hazards. There are certain advantages concerning risk reduction when indicators of vulnerability are used. Design, building material, and travel access are indicators of vulnerability and are therefore more suitable for damage reduction. Detention basins and ponds are interventions that establish a barrier to keep a part of the storm or wastewater away from the receiving water.

In many European cities, sewer systems are built 50 years ago or more, and it is time to rehabilitate and upgrade them (Hauger et al., 2005). Climate change also adds to the need to upgrade the systems that have significant economic consequences. Evaluation of climate change effects on precipitation reveals different changes of behavior, which are different from region to region (see also Ashley et al., 2004; Grum et al., 2005).

Generally speaking, the pipe diameters have to be enlarged, and more storage is needed, which involves major capital investment. This follows what is called "hazard strategy." But the urban planners are more geared towards what is called "vulnerability strategy." It manages the increased volumes of stormwater by determining the areas with less flooding damages and then allowing them to be flooded in a controlled manner. This strategy is similar to a traditional strategy used in managing fluvial flooding of floodplains.

TABLE 5.8
Three Typical Risks in Urban Drainage with Respect to Indicators and Intervention Options

	Pluvial Flooding of Urban Drainage Systems	Sewage Discharge into a Stream (DO Depletion in a Stream)	Discharge of Chemicals to Receiving Water
Primary concern/ consequence	Economic loss (building interface structure, productive time)	Loss or degradation of habitat	Degradation of environmental quality Loss of recreational value
Risk source	Precipitation, high-intensity rain events	Organic matter in unwanted stormwater and wastewater	Chemicals that are dispersed after use
Indicator	IDF curves Return period for exceeding a specified water (service) level	Number of overflows Overflow volume Overflow duration	Inherent properties of chemicals Predicted environmental concentrations (PEC)
Risk object	Building, interface structure	Aquatic life in a stream, a stream as a recreational area	Receiving water, aquatic ecosystem
Indicator	Building design Building materials Location of the main interface structure routes	Minimum oxygen concentration Dissolved oxygen (DO) demand curves as a function of duration and return period	Predicted number of effect concentrations (PNEC) Fauna index
Analysis	Simulated water level for a rain event with a defined return period Model simulation with historical series	Long rain series used to calculate minimum oxygen concentration as a function of return period and duration	Calculate PEC and comparison with PNEC
Definition of acceptance	Return period for exceeding a given water level must be higher than a defined value	Demand for daily and 1-year minimum oxygen concentration interpolated to a curve showing oxygen concentration as a function of return period	PEC (multiplied with an appropriate safety factor) must be lower than PNEC Fauna index must live up to a target classification
Risk reduction options for hazards	Expected for surveillance and forecasting, improved early warning of hazards that cannot be reduced	Decentralized cleaning before discharge, keeping the first and most polluted part of water (first flush) away from streams	Reduce usage of chemicals, substitute hazardous chemicals with harmless ones
Vulnerability	Adjusted construction methods Use of robust materials Design (no cellar, no toilets in cellar) and space management point out the areas that are allowed to be flooded (urban flood plans)	Retrofit the stream to increase robustness Increase reaeration of the stream	Improve dilution rate contaminated sediment management (cleanup)
Barrier	Prevent sewer from flooding (increase capacity and storage; source control; keep stormwater out of the sewer system; infiltration reuse)	Keep wastewater and stormwater away from the stream (basins, infiltration, and increased capacity of the pipe system)	Treatment before discharge; source control; keeping water commissioning hazardous chemicals away from recipient

In recent years, climate change and its consequences have affected the total components of the water cycle as well as floods. These effects are intensified in urban areas because of the anthropogenic effects they have on the water cycle, such as reducing the infiltration capacity of basins, construction regardless of the channel's right of way, and disposal of sediment and solid wastes into channels that will decrease the channels' safe carrying capacity. In this way, incorporating climate change impact on urban water studies could help achieve more reliable results to be applied in real-time planning of urban areas by selecting BMPs. It is common to examine different BMPs in dealing with urban water drainage systems to find better alternatives for reducing probable urban flash flood damages. The BMPs may include structural or nonstructural solutions or a combination, with varying costs and benefits.

Karamouz et al. (2011) developed an algorithm to find effective BMPs for improving the performance of an urban drainage system in the time of flooding. In their algorithm, the effects of climate change and anthropogenic changes on the urban flood regime and characteristics are evaluated. Long-lead rainfall series under climate change effects are developed using down-scaling methods. A rainfall–runoff model is also developed to simulate the surface runoff and to provide more reliable results considering the uncertainties in the model parameters, the input data, and the modeling procedure. The BMPs are evaluated based on their effectiveness in reducing the flood volume and peak with the least cost. See more details about this study in the following case study.

5.9.1.1 Case Study: Improvement of Urban Drainage System Performance under Climate Change Impact

Stormwater collection and control in urban areas is an important and multifaceted job because of its effects on transportation and residents' life, especially in high-density urban areas. In flash-flood events when a high volume of water has to be collected in a very short period of time, this problem is intensified. An acceptable level of flood risk is defined to mitigate this problem, and the system is designed to control the corresponding flood. Different BMPs are employed to improve the system performance in flood transitions. The possible changes in rainfall and flood characteristics, such as climate change and anthropogenic effects, are incorporated into system performance evaluation. A part of the case study is explained in the case study 6 of chapter 13.

Karamouz et al. (2011) selected a case study in Tehran, the capital of Iran. This city has been rapidly developing without considering the diverse impacts on the environment, especially on the water cycle. This has resulted in a wide range of challenges and obstacles in the water and sanitation-related infrastructures in this area. The lack of a systematic approach to runoff management in Tehran has led to the frequent overflow of channels and some health hazard problems in rainy seasons. The north-eastern part of Tehran, between 51°22′ and 51°30′ E longitude, and 35°42′ and 35°53′ N latitude, has been considered as the main focus in this study as shown in Figure 5.26. It is a mountainous area with a high population density. Therefore, urban floods can cause considerable damage, and the evaluation and simulation of the flood effects in this area are very important. The drainage system of the study area is composed of natural and artificial channels. These channels carry large amounts of sediment, especially during floods, which drastically decrease their safe carrying capacity. These channels are used as combined surface and wastewater drainage facilities as well as snowmelt discharges.

- *Proposed methodology*: In this study, an algorithm is proposed to evaluate the urban drainage system performance in flood transmission using an optimal scheme of BMPs as follow:
    1. Data gathering, preparation, and analysis: In this step, the data needed for different study parts are collected and analyzed. These data are classified into four groups: hydrologic data, climate data, climate signals, and downscaled data.
    2. The joint probability of rainfall and sediment: The effect of sediment load should be considered in the development of flood hydrograph to provide a more realistic

FIGURE 5.26 Overall view of the study area. (From Karamouz, M., Hosseinpour, A., and Nazif, S.. "Improvement of urban drainage system performance under climate change impact: Case study." *Journal of Hydrologic Engineering*, 16(5), 395–412, (2011).)

configuration of the urban drainage system performance. For this purpose, the copula theory has been used to calculate the joint probability of rainfall and sediment.

3. Rainfall–runoff model development: StormNET model has been employed to simulate the flood and drainage system performance in the study area. To highlight the effect of sediment in flood characteristics, model performance results without considering sediment effect are compared with cases with sediment.
4. System performance evaluation: At this step, the system's performance is evaluated and compared with predefined criteria to decide on the application of the BMPs. If the system performance criteria are satisfied, the algorithm terminates. Otherwise, the algorithm continues by selecting a new group of BMPs.
5. BMP initiation: The application of BMPs in dealing with floods could help to solve soil erosion problems and improve the drainage system performance in transmitting the surface runoff. Such solutions result in capacity expansion and/or less runoff volume. Site geometry and structural characteristics of BMPs should be considered in selecting BMPs for urban area.
6. BMP selection: An optimization model is utilized to select the best composition of different BMPs to improve the system performance. The objective of this model is to minimize the system costs, considering BMPs' construction and flood damage costs.

- *Results and discussion*: Evaluation of climate change effects shows a significant increase in the number of rainfall events with high intensity, especially in dry seasons (7% increase in the average of maximum rainfalls of four seasons from 2007 to 2017 and 133% increase in maximum rainfall of summer from 2007 to 2017), which will result in more flash floods in the urban area and the need to revise drainage management strategies. The flood hydrograph characteristics are changed even more because of changes in the drainage system structure and urbanization. Some changes in the drainage system structure, such as blocking or changing the path of channels, have also intensified the flooding events in the study area.

The urban flood management practices considered in this case study are as follows:

1. Increase the green space;
2. Build diversion channels;
3. Double the closed channels;
4. Increase the open channel capacity;
5. Increase the detention pond capacity; and
6. Build a detention pond with a natural bed and stone wall.

As the case study has limited choices for the application of BMPs, the optimal composition of BMPs is determined by evaluating all possible situations regarding the objective function and constraints of the proposed optimization model. Using the optimal combination of BMPs has resulted in a significant decrease in flood volume and flooded areas, especially in the downstream subbasins. The results show that the damage resulting from extreme events will decrease in the future under climate change effects. This implies that adding some detention ponds and small channels downstream of the system could decrease the peak flood inundations in the present drainage system situation. Further investigations on identifying and mitigating problems and weaknesses of the drainage system are needed to ensure the success of flood management schemes in the study area.

5.9.2 DO Depletion in Streams—Discharge of Combined Sewage Effects

Often overflow structures are provided in combined sewer systems to minimize flooding inside the urban area. This solution causes other problems when a mixture of stormwater and wastewater is discharged directly without any treatment into receiving waters. The risk source is the organic matter and other compounds present in the storm or wastewater that cause water quality changes in the risk object (the ecosystem). Changes can be decreased in the complexity of the food web and substitution of sensitive species with more tolerant ones and a reduction of the recreational function of the area.

Hazard indicators such as the overflow frequency, overflow volume, and overflow duration are widely used (Marsalek et al., 1993). However, the concentration of easily degradable organic matter in overflowing water is rarely used. Classification of water quality using fauna indices or the DO concentration is used as indicators for vulnerability. DO concentration has been used to define the requirements on water quality in Denmark for many years (DWPC, 2005). This principle is similar to the Predicted No Effect Concentration (PNEC) that is widely used in chemical risk assessment. Concerning risk reduction, the situation is opposite to the previous example. The risk object is a natural system that shall remain as undisturbed as possible as per tradition. Unless the risk object is heavily modified by urban activity, the risk reduction potential in changing the physical conditions of the risk object is very limited. The risk reduction potential lies in reducing the risk source, and the possibilities are either avoiding that the combined sewage reaches the stream or reducing the pollution level in the combined sewage. Retrofitting a water body towards the original state results in much better conditions including increased possibilities for breeding in the stream and that the stream can receive a larger volume of stormwater considering the consequences related to water quality. Part of the retrofitting can be increased by natural reaeration, increasing the potential for the degradation of organic substances in the stream (Hauger et al., 2005).

BOD_5 test is a laboratory simulation of the microbial processes which measures the DO consumed in a sample diluted in a 300 mL bottle during a specified incubation period (usually 5 days at a temperature of 20°C in darkness). The DO is used by microorganisms as they break down organic material and certain inorganic compounds. Thus,

$$BOD_5 = (C_{DOI} - C_{DOF})/p \tag{5.19}$$

where

p is sample dilution (volume of sample/volume of bottle),
C_{DOI} is initial DO concentration (mg/L),
C_{DOF} is final DO concentration (mg/L).

BOD_5 is a very common parameter used in the control of treated wastewater and stormwater effluent quality through the setting and monitoring of discharge consent standards. Rivers are considered to be polluted if their BOD_5 exceeds 5 mg/L and DO concentration is <4 mg/L (5mg/L is widely accepted as a threshold).

Example 5.7

A laboratory test for BOD_5 is carried out by mixing a 10 mL sample with distilled water into a 300 mL bottle. Prior to the test, the DO concentration of the mixture was 7.45 mg/L, and after 5 days, it had reduced to 1.45 mg/L. What is the BOD_5 concentration of the sample?

Solution:

$$\text{Dilution, } p = 10/300 = 0.033$$

$$BOD_5 = (7.45 - 1.45)/0.033 = 181.82 \text{ mg/L}$$

5.9.3 Discharge of Chemicals

The discharge of chemicals into receiving waters is a major risk to human and aquatic ecosystems. They end up in the receiving waters via wastewater and surface runoff (Chocat and Desbordes, 2004). Indicators for the chemicals' hazard include toxicity, carcinogenicity, and bioaccumulation as Predicted Environmental Concentrations (PEC).

Vulnerability is used as PNEC. Also, the indicators PEC and PNEC define the acceptance level, which is a requirement that PEC must be lower than PNEC. A safety factor is used in estimating these factors. The major options include efficient treatment before discharge, reduction of usage of hazardous chemicals or less harmful substitutes, and controlling the fate of chemicals in receiving waters such as increasing the dilution rate or contaminated sediments the cleanup (Hauger et al., 2005).

5.10 URBAN FLOODS

In developing new urban areas, infiltration, reductions, and water retention characteristics of the natural land should be considered. The increasing quantity and the short duration of drainage produce high flood peaks that exceed the conveyed secure capacities of urban streams. The total economic and social damages in the urban areas tend to impact the national economies as flood frequencies increase significantly. Recent events in North America have been devastating in urban areas such as New Orleans.

With uncontrolled urbanization in the floodplains, a sequence of small flood events can be managed, but with higher flood levels, damage increases, and the municipalities' administrations have to invest in population relief. Structural solutions have higher costs but better chances of being implemented when damages are greater than their development or due to intangible social aspects. Nonstructural measures have lower costs but have less chances of being implemented because they are not politically attractive in developing regions (Tucci, 2007).

5.10.1 Urban Flood Control Principles

Certain principles of urban drainage and flood control are proposed by Tucci (2007) as follows:

- Promoting an urban drainage master plan.
- Involvement of the whole basin in evaluating flood control.
- Any city developments should be within urban drainage control plans.
- Priority should be given to source control and keeping the flood control measures away from downstream reaches.
- The impact caused by urban surface washoff related to urban drainage water quality should be reduced.
- Improvement of nonstructural measures for floodplain control such as real-time flood forecasting, flood zoning, and flood insurance.
- Public participation in urban drainage management.
- Consideration of full recovery investments.

A set of principles for urban drainage practices is not established in developing countries due to fast and unpredictable developments (Dunne, 1986). Furthermore, the local regulations are neglected in the urbanization of preurban areas, such as unregulated developments and invasion of public areas.

In many developing urban areas, the above principles are not incorporated because of the lack of the following:

- Sufficient funds.
- Appropriate garbage collection and disposal that decreases the water quality and the capacity of the urban drainage network due to filling.
- Preventive program for risk area occupation.
- Adequate knowledge on how to deal with floods.
- Institutional organization in urban drainage at a municipal level.

Urban floods have adverse impacts on the performance of urban infrastructures and the life of residents. The floods cause heavy damages and perturbation in the serviceability of urban infrastructures as well as transportation. Therefore, different factors affecting the urban water flood characteristics should be considered in urban development planning, especially in metropolitan areas.

Flooding also has significant impacts on urban infrastructures such as WWTPs. These impacts can mainly be seen by untreated or partially treated sewage flows that discharge into water bodies. These effects in regions with combined sewer systems are more severe, with a higher risk of occurrence. The world is experiencing extreme events (such as floods) much more than in the past, which calls for paying more attention to coastal risk management, which can also produce major social challenges for the coastal cities.

In addition, farmlands, water areas, and construction lands are the three types of land use that are most vulnerable to storm surge and inundation in coastal areas. The significant disasters caused by coastal floods in New York City reveal the importance of a better understanding of our vulnerabilities and the lack of integration in our vital infrastructure through flood risk management.

Flood risk management has many attributes with various interests for stakeholders (Kenyon, 2007). The first step to finding solutions for at-risk areas is to understand the hazard. For this purpose, observed data of the study area should be analyzed to find the frequency and intensity of the probable floods. For flood frequency analysis, a suitable probability distribution can be fitted to the observed data to estimate the occurrence frequency of extreme events (Khaliq et al., 2006; Machado et al., 2015). The next step is to build models and define the scenarios. New modeling techniques in integrating models based on geographic information systems (GIS) utilizing high-resolution digital elevation models (DEM) have provided ample opportunities for flood risk mapping. See Chapters 12 and 13 for case studies related to flood resiliency.

5.11 OVERLAND FLOW MODELS

To evaluate the impacts of the storms, different types of hydrodynamic models have been conducted, including the Gridded Surface-Subsurface Hydrological Analysis (GSSHA) (Karamouz et al., 2019) developed by the Engineer Research and Development Center of the United States Army Corps of Engineers and Lisflood developed by the floods group of the Natural Hazards Project of the Joint Research Centre (JRC) of the European Commission (Karamouz and Fereshtehpour, 2019) for flood resiliency and coastal flood inundation studies.

5.11.1 StormNET: Stormwater and Wastewater Modeling

StormNET is a powerful and comprehensive stormwater and wastewater modeling package available for analyzing and designing urban drainage systems, stormwater sewers, and sanitary sewers. StormNET is the only model that combines complex hydrology, hydraulics, and water quality in a completely graphical, easy-to-use interface. See Gong et al. (1996) and Chapter 12 for details of this software.

5.11.2 GSSHA

GSSHA is a two-dimensional, physically based watershed model. It simulates surface water and groundwater hydrology, erosion, and sediment transport. The GSSHA model is utilized for hydraulic engineering and research and is on the Federal Emergency Management Agency (FEMA) list of hydrologic models accepted for use in the national flood insurance program for flood hydrograph estimation. Input is best prepared by the watershed modeling system interface, which effectively links the model with GIS. GSSHA uses a square grid with constant size which represents characteristics and topography of watershed. Relevant model parameters are assigned to the model grids using index maps, which are often derived from soils, land use/land cover, or other physiographic maps. More details of these models are presented in Chapter 12.

5.11.3 LISFLOOD-FP

LISFLOOD-FP is a raster-based flood inundation model (Bates and De Roo, 2000; Horritt and Bates, 2001). The application of LISFLOOD-FP was to produce the simplest physical representation of dynamic flood spreading (Horritt and Bates, 2002). One advantage of a simple flood inundation scheme (such as that used in LISFLOOD-FP) over finite element models is computational efficiency, with approximately 40 times fewer floating-point operations per cell, per time step (Bates and De Roo, 2000). Despite modern high-speed computers, this becomes important when multiple simulations are conducted as part of sensitivity analysis or uncertainty assessment. Further, using a raster data structure makes the incorporation of multiple datasets in the model relatively straightforward. More details of these models are presented in Chapter 12.

5.12 STORMWATER INFRASTRUCTURE OF SELECTED CITIES

5.12.1 Philadelphia, USA

5.12.1.1 Characteristics of the system

Philadelphia is located in the Delaware River Watershed, which begins in New York State and extends 330 miles south to the mouth of the Delaware Bay. This city is renowned for implementing advanced SWM policies prior to most cities in the United States to reduce CSOs required by Federal

Clean Water Act (CWA). Thirty percent of the city's paved area is connected to traditional combined systems, compared to 70% connected to separate systems that discharge stormwater directly into the surface water. Stormwater runoff from the Philadelphia region, whether served by separate stormwater sewers or combined sewers, impairs the streams and rivers of the city. The combined sewer system covers almost two-thirds of the sewer service area in Philadelphia. There are 164 CSOs along the Delaware and Schuylkill rivers (Figure 5.27). The purpose of the Philadelphia Combined Sewer Overflow Public Notification System (CSO cast) is to alert the public of possible CSOs from combined sewer system outfalls.

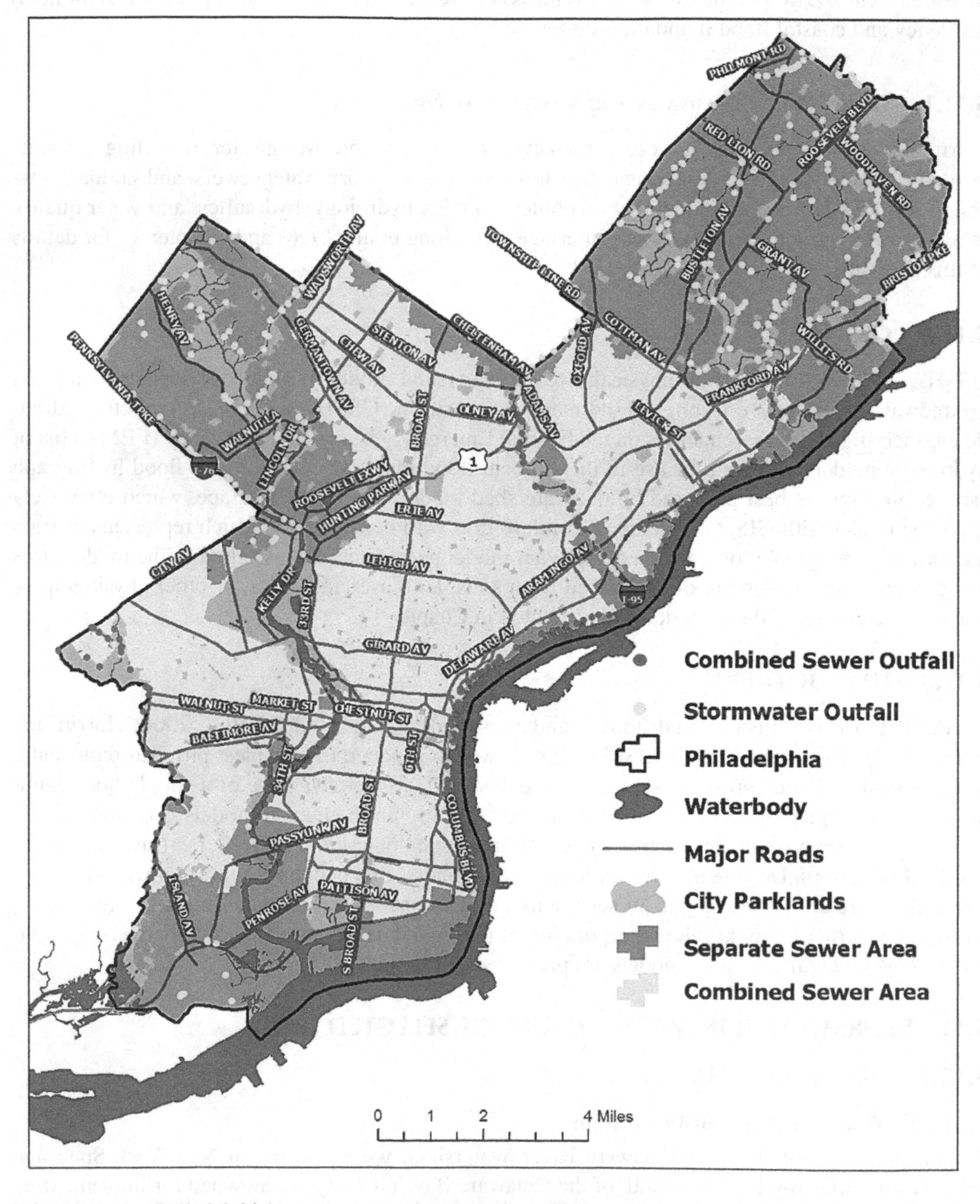

FIGURE 5.27 Philadelphia's sewer area and stormwater outfall.

5.12.1.2 Improvement and Future Plans

Philadelphia is the first city in the United States to undertake a major experiment in climate action planning—implementing a green SWM plan based almost exclusively on green infrastructure. The stormwater goals are to reconnect the natural links between land and water and "that green infrastructure—trees, vegetation, and soil—become the city's preferred stormwater management system."

The Office of Watersheds (OOW) was created within Philadelphia Water Department (PWD) to manage the city's CSO plan in 1999. Based on the success of this approach, in September 2009, OOW released Green City, Clean Waters, a 25-year, $2.48 billion updates to the original plan that calls for controlling runoff from 35% of the city's land and capturing 85% of existing CSOs by replacing a minimum of one-third of impervious surfaces with permeable pavement, expanding parks and green spaces, and various green stormwater infrastructure (GSI) techniques. The plan, officially adopted in June 2011, will reduce the city's CSO from 16 to 8 billion gallons/year. This approach is practical and less expensive than the cost of a gray "tunnel and tank" approach.

Green City, Clean Waters is a comprehensive but decentralized SWM strategy realized through eight programs: Green Schools; Green Streets; Green Parks; Green Public Facilities; Green Parking; Green Industry, Business, Commerce and Institutions; Green Alleys, Driveways, and Walkways; and Green Homes. The goal is to convert about 9,600 of the impervious surface into "greened acres," equivalent to an inch of managed stormwater per acre of impervious drainage area.

By implementing GSI projects, such as rain gardens and stormwater planters, the city can reduce water pollution impacts while improving our essential natural resources and making our neighborhoods more beautiful. To complement Philadelphia's Green City, Clean Waters plan, PWD is also working to restore the city's urban streams and keep them free from trash and debris.

5.12.1.3 Recommendations

Philadelphia has been surpassing its goals for reducing stormwater runoff and increasing GSI. However, to achieve its long-term goals for Green City, Clean Waters, it will need help from the community. Recommendations for Green City, Clean Waters going forward include:

- Dynamic public outreach;
- Upgrading methods of communication with the community;
- Community group networking;
- Connecting with peer organizations;
- Self-evaluation of progress;
- Large-scale habitat restoration;
- Building momentum for citywide sustainability.

5.12.2 Los Angeles, California

5.12.2.1 Characteristics of the System

In the Los Angeles Department of Water and Power (LADWP)'s recently completed Stormwater Capture Master Plan (SCMP), the areas tributary to the city were divided into 17 subwatersheds (15 of which contain city area). Figure 5.28 shows the stormwater system in the City of Los Angeles.

- There are approximately 2,500 miles of stormwater conveyance network identified in the City.
- For almost 80 years, the primary considerations for the design of stormwater projects in the city were flood control and water supply.
- Following the passage of the Porter-Cologne Act of 1969 and the CWA of 1972, stormwater began to be regulated from a water quality perspective.

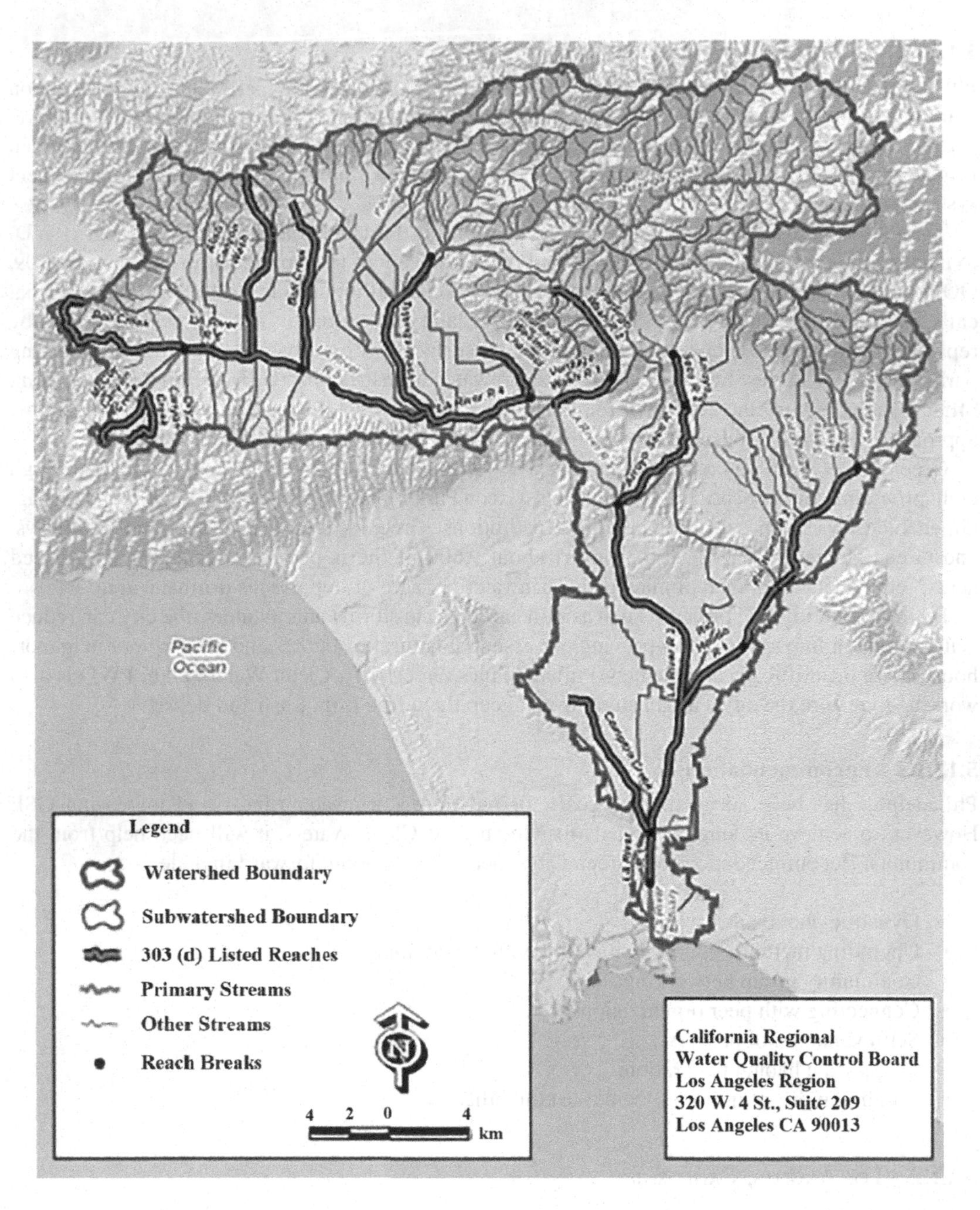

FIGURE 5.28 Stormwater system in the City of Los Angeles.

- Now, the city is leading the way as one of the most proactive cities in the nation regarding stormwater quality protection and enhancement.
- The stormwater infrastructure system within the city works collectively to provide multiple benefits to the public and includes both gray and green infrastructure.

5.12.2.2 Improvement and Future Plans

The gray infrastructure network includes some of the oldest stormwater assets in the city. The goal of the gray infrastructure network, which began to be installed in the city in the 1930s and 1940s, is to avoid flooding and route collected water away from urban areas and to the ocean as quickly as possible. On average, approximately 764 million gallons (MG) of total inflow to the city is estimated to occur per day. Approximately 92,000 acre-feet per year (AFY) of stormwater is captured for direct use, environmental and habitat supply, and groundwater recharge.

5.12.2.3 Recommendations

The following recommendations are for the city to pursue:

- Continuous cooperation with LA County, evaluating possible special taxes on parcels to, help pay for SWM.
- Continue to explore potential sources of funding and monitor legislative developments that may open new avenues for funding.
- Innovation in partnerships with other public agencies and the private sector that can help fund and implement SWM projects.
- Refined project cost estimates and value engineer individual projects as development proceeds.
- Developing budgets for SWM that are consistent with the strategy developed by the city. These budgets should match future revenues and consider potential costs from fines and sanctions on the city.
- An increased need for resources. As infrastructure grows, particularly green infrastructure, more funding is required.
- An increased demand for monitoring data. More and more projects are being constructed with a requirement for performance to be tracked via monitoring (e.g., water quality monitoring, flow monitoring).
- The need for an improved system to evaluate and assess project performance. With strict regulatory requirements in place, a deviation from performance for certain green infrastructure projects may imply immediate noncompliance.

5.12.3 Chongqing, China

5.12.3.1 Characteristics of the System

Chongqing has an annual rainfall of over 1,000mm, the bulk of which falls in summer and autumn.

The city is a pioneer in developing a solution for the optimization of sewer and stormwater networks. Chinese 'sponge cities' seek to turn water challenges into opportunities. During heavy rainfall, the water is soaked up by the porous bricks and flood-tolerant plants to prevent flooding, then used for irrigation and cleaning.

5.12.3.2 Improvement and Future Plans

Figure 5.29 shows the location of Chongqing city. The sponge city program was launched at the end of 2014, under the direct guidance and support of the Ministry of Housing and Rural-Urban Development (UHURD), Ministry of Finance (MOF), and Ministry of Water Resources (MWR).

These three ministries are responsible for reviewing, evaluating, and selecting candidate cities recommended by their respective provincial governments, based on a series of criteria concerning the rationality and feasibility of pilot goals, financing mechanisms, and the effectiveness of supporting measures from local governments.

The three ministries are also responsible for the assessment of pilot city performance.

In April 2015, the first group of 16 cities was selected as the pilot sponge cities; 1 year later in April 2016, the pilot program was expanded to another 14 cities.

The central government allocated to each pilot city between 400 and 600 million Chinese Yuan (CNY) each year for 3 consecutive years. Pilot cities are encouraged to raise matching funds through public–private partnership (PPP) and other financial ventures. The money will be used to implement innovative water and wastewater management measures that would transform these cities into sponge cities

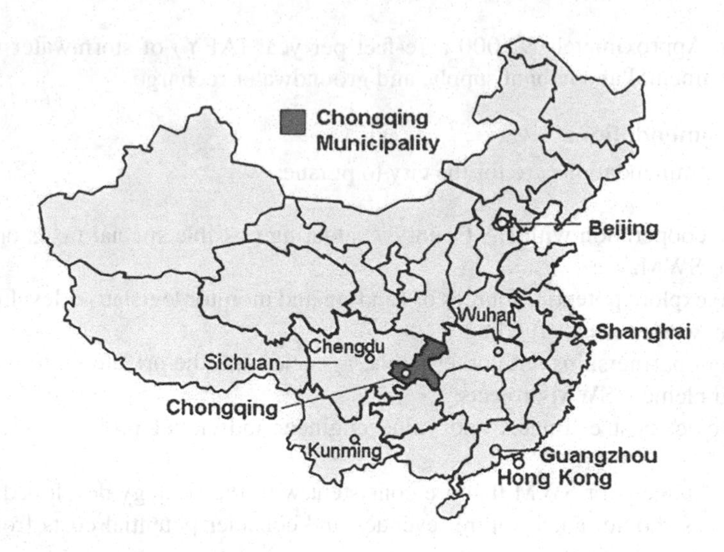

FIGURE 5.29 Location of Chongqing, China.

5.12.3.3 Recommendations

- Broad and diverse coalitions are necessary for discovering the benefits, exploring the possibilities, piloting the projects, and probing system-wide changes;
- Increased research efforts into the techniques, levels of performance, range of multiple benefits, life cycle analysis of costs, and other key areas of sponge city implementation are needed;
- Greater coordination is needed among agencies and communication among stakeholders, government officials, and staff.
- The development of green solutions.

5.12.4 London, England

5.12.4.1 Characteristics of the System

Three main flood risks are facing London: tidal surges, river water, and surface water. Figure 5.30 shows the critical drainage areas in London. Most parts of London are at risk of one or more of these flooding categories. The largest concentrations of risk are around rivers, especially River Thames.

5.12.4.2 Improvement and Future Plans

As of 2008, the City of London has approximately 100 SWM systems constructed within the city's boundaries. The Engineering and Environmental Services Department (EESD) anticipates that future development and capital work allocations will lead to constructing an additional 118 SWM facilities over the next 10–20 years. The total estimated cost for existing and proposed storm drainage and SWM infrastructures represents approximately $600 million. The design of SWM facilities is also based on the MOE's "Stormwater Management Planning and Design Manual." These documents are developed following Provincial and Federal Acts to ensure that each SWM facility will protect the public and property and meet the necessary environmental targets.

There are eight types of SWM systems constructed within the City of London boundaries:

- Wet ponds
- Dry ponds
- Energy dissipation systems

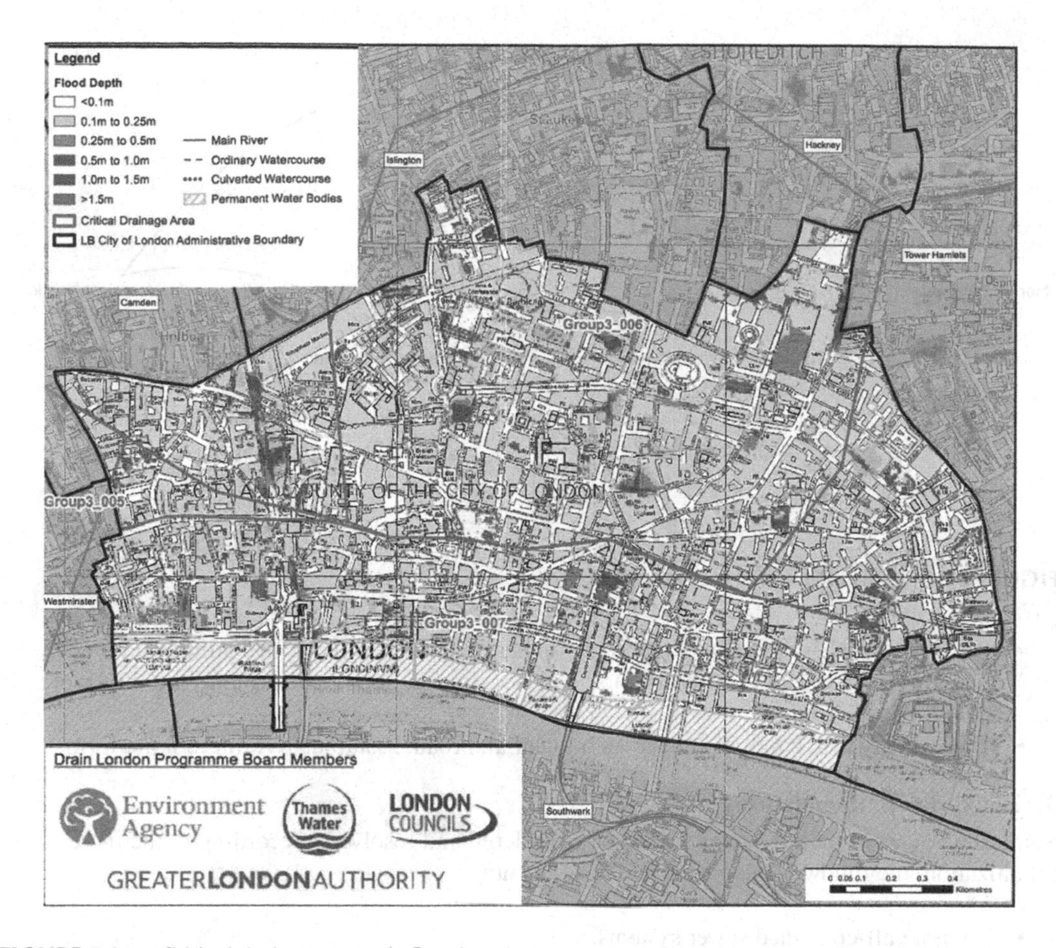

FIGURE 5.30 Critical drainage areas in London.

- Surface storage areas
- Wetlands
- Detention/retention channels
- Oversized pipes
    - Wet ponds represent the majority of new regional SWM systems in London. However, the city's SWM infrastructure constructed in the 1980s and the early 1990s includes some surface storage areas, dry ponds, channels, and large pipes.

5.12.5 Amsterdam, Netherlands

5.12.5.1 Characteristics of the System

Amsterdam has a 25%–40% impervious surface. More than 10% of the total area is covered by water, and it has a high groundwater level (ca. 60 cm—surface). Seventy-five percent of the drainage area has separated stormwater and sewerage systems. There are 1,670 km stormwater sewers and 524 km combined sewer for 800,000 inhabitants.

5.12.5.2 Improvement and Future Plans

- Amsterdam Rainproof is a platform that activates and stimulates different stakeholders to create a more resilient city for dealing with extreme rainfall. The goals of the Rainproof platform are mentioned in Figure 5.31.

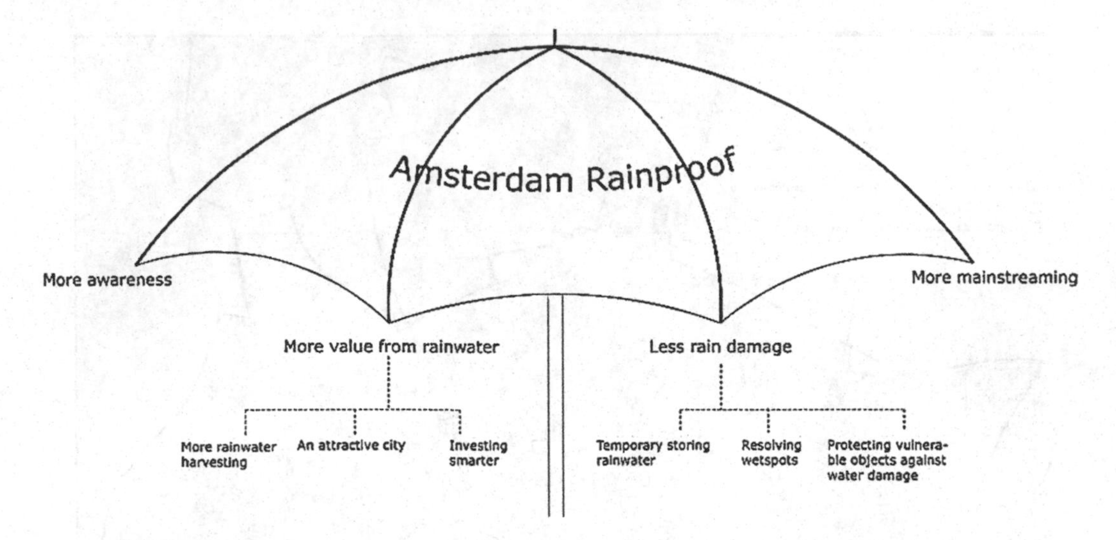

FIGURE 5.31 Amsterdam rainproof platform targets.

- The network of Amsterdam Rainproof wants to activate, connect, and stimulate citizens, city builders, officials, entrepreneurs, and housing corporations to make the city more rainproof.
- The strategy is to build and create an influential, broad, sustainable Rainproof platform.

5.12.5.3 Recommendations

Some challenges in these systems should be considered and resolved. According to the increase in urbanization, the following activity could be efficient:

- Separate all-combined sewer systems.
- Prepare a precise plan for dealing with climate change events.

5.12.6 Stockholm, Sweden

5.12.6.1 Characteristics of the System

In the past years, the City of Stockholm had to deal with surface water and with flooding events due to heavy rains. It had combined systems to mid-1900s leading away from both wastewater (sewage) and rainwater collected but has had a dual system from 1950 with separate pipes for stormwater and culvert creeks, and ditches are utilized. Figure 5.32 shows the blueprint of an urban area in Stockholm. The existing stormwater system in the city is designed for 2- to 10-year rainfall.

Today, the aim is to find more sustainable SWM that considers water quality, urban environment, and capacity. Figure 5.33 shows the sustainability factors implementable in Stockholm SWM.

5.12.6.2 Improvement and Future Plans

In order to maintain the city's functions, it is essential to take care of the stormwater in the most efficient way possible, concerning managing its quantity and quality. Rainwater contains pollutants that originate from the atmospheric emissions that occur in the region where the rain is formed. Intense rains causing the sewer network to be overloaded and causing basement floodings, surface flooding, and large overflows are the leading causes. Some problems with overflowing streams like in 2015, an 80-year rain event had caused an industrial area and surrounding houses to be flooded for 12 hours.

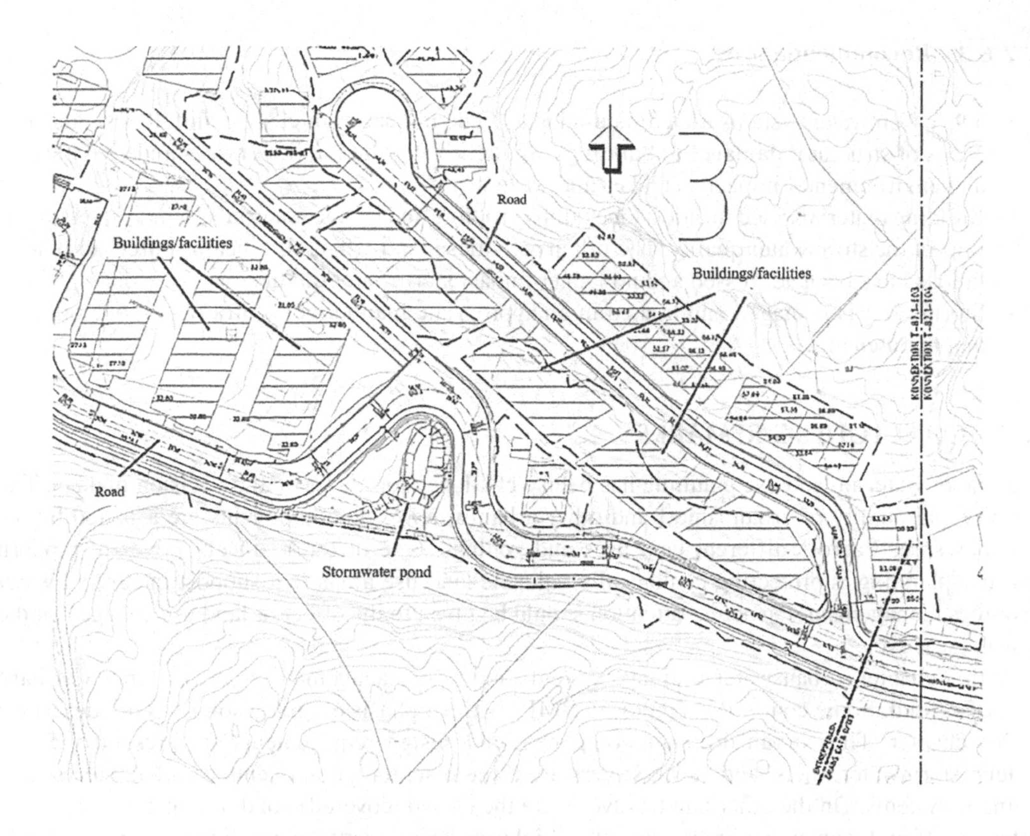

FIGURE 5.32 Blueprint of an urban area located in Nacka, Stockholm (Jansson, J., Nilsson, J., Modig, F., & Hed Vall, G.. Commitment to sustainability in small and mediumsized enterprises: The influence of strategic orientations and management values. *Business Strategy and the Environment*, 26(1), 69–83, (2017)).

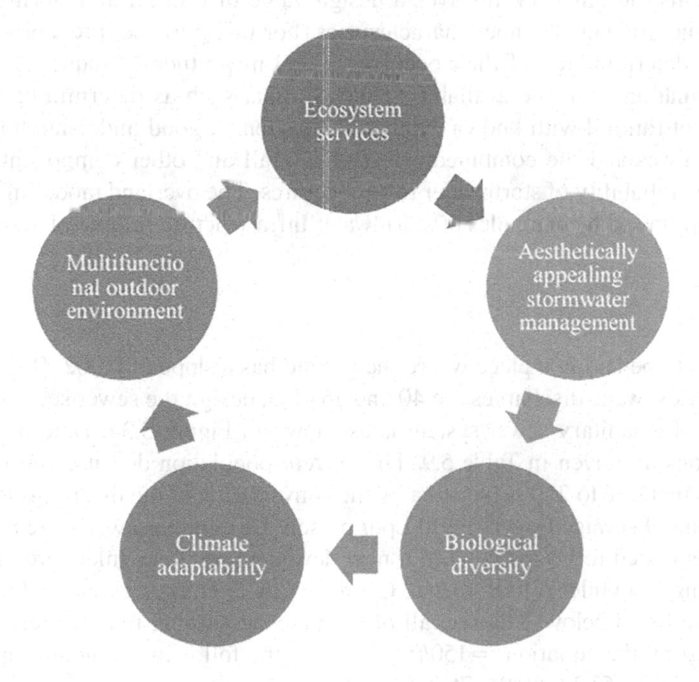

FIGURE 5.33 Sustainability factors implementable in stormwater management—City of Stockholm, Sweden. (From Jansson, J., et al., *Bus. Strategy Environ.*, 26(1), 69–83, 2017.)

5.12.6.3 Recommendations

- The urban water systems are climate-adapted. The impact of intensive rainfall is small in terms of structural damages to buildings, disruption of vital public services and activities, and environmental impacts on lakes and waterways.
- Building materials used in cities, for example, roofing and facade materials, do not adversely impact the stormwater quality. The urban activities identified as significant sources of pollutants have been addressed and no longer impact stormwater quality.
- Stormwater is managed and treated in an appropriate manner concentrating on operation and maintenance.

5.13 CONCLUDING REMARKS

The role of water in land use planning has been a challenge for planners and decision makers. The quantity and quality of urban runoff and other urban water cycle components are affected by the activities spread across different land uses and zonings. One of the main objectives of modern land use planning is protecting ecologically valuable land use areas that support integrated water resources management. Particular attention should be given to the effect of land use changes on the hydrological cycle.

In this chapter, urban water drainage systems and best management practices for SWM have been discussed. Some commonly employed BMPs with emphasis on green solutions are described in this chapter. Then urban drainage component and design requirements are presented. Street gutters, stormwater inlets, and storm sewers are three essential components of urban stormwater drainage systems. On the other hand, culverts are the closed (covered) conduit that conveys stormwater runoff under an embankment, usually a highway. These components should be appropriately designed to achieve the stormwater drainage system's objectives. Sections 5.3–5.7 are particularly useful for transportation and highway engineers and contractors.

Drainage systems' design work involves a design value of rainfall and storms for coastal cities which affect the drainage channel characteristics. For this purpose, probabilistic methods are employed for the determination of their occurrence and magnitudes. Frequency analysis derives meaningful information from the available historical data, such as determining the corresponding return period of rainfall with known depth and duration. A good understanding of nonstationary frequency analysis and the combined effect of rainfall and other components such as storm surge improves the reliability of stormwater infrastructures. The overland modeling models are also briefly presented followed by examples of stormwater infrastructure in several major cities.

PROBLEMS

1. A sewer has to be laid in a place where the ground has a slope of 0.002. If the present and ultimate peak sewage discharges are 40 and 165 L/s, design the sewer section.
2. The layout of a sanitary sewer system is as shown in Figure 5.34. Data on area, length, and elevations are given in Table 5.9. The current population density, 100 persons/ha, is expected to increase to 250 persons/ha by the conversation of the dwellings to apartments. The peak rate of sewage flow is 1,600 Lpd/ person. Design the sewer system.
3. Storm sewers need to be installed in a new development. Four inlets are proposed with pipes running from inlet A to B, then to C, and finally to D. Data associated with each pipe and inlet are listed below. The size all of the pipes, assuming that rainfall intensity, can be described by the equation $i=150/(t_d+20)$. Use the following standard pipe sizes: 38.1 (minimum), 45.72, 53.34, 60.96, 76.2, 91.44 cm, and so on.

   Inlet A: drainage area $=12{,}000\,m^2$, $t_0=10\,min$, $C=0.30$

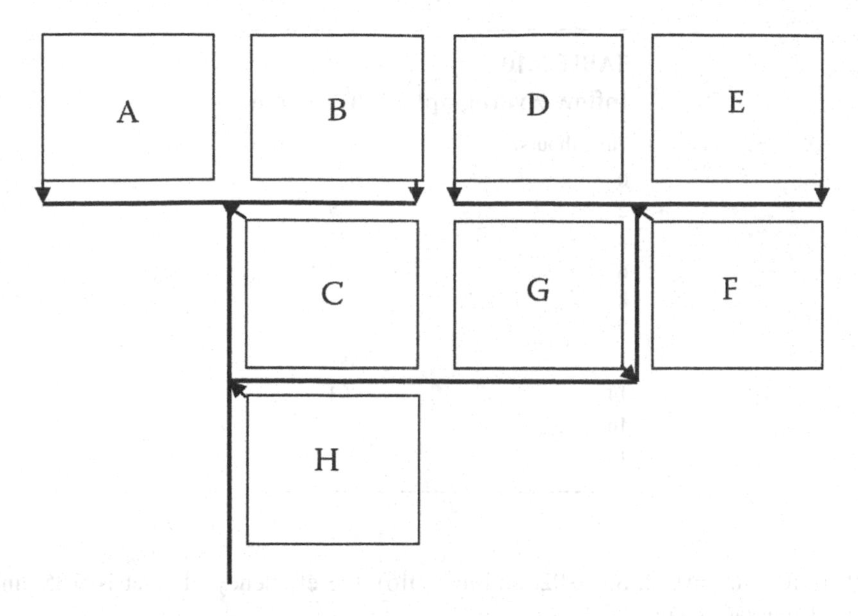

FIGURE 5.34 Layout of the sanitary sewer system of Problem 2.

TABLE 5.9
Characteristics of the Drainage Area of the Sanitary System of Problem 2

Block	Area (m²)	Length (m)	Elevation (m) Upstream	Downstream
A	8,000	118.87	30.94	29.62
B	10,000	106.68	30.68	29.62
C	6,000	100.58	29.62	28.43
D	5,200	70.10	30.08	29.73
E	4,800	89.92	30.63	29.73
F	8,400	91.44	29.73	28.74
G	22,800	198.12	28.74	28.43
H	14,000	167.64	28.43	26.34

Inlet B: drainage area = 16,000 m², t_0 = 12 min, C = 0.40
Inlet A: drainage area = 8000 m², t_0 = 10 min, C = 0.50
Inlet A: drainage area = 8000 m², t_0 = 10 min, C = 0.30
Pipe AB: length = 100 m, slope = 0.01, n = 0.013
Pipe BC: length = 135 m, slope = 0.02, n = 0.013
Pipe CD: length = 100 m, slope = 0.07, n = 0.013
Pipe DE: length = 165 m, slope = 0.01, n = 0.013

In your computations, take into consideration the various flow paths that lead to the determination of the time of the concentration. For instance, in determining the time of concentration for pipe BC, one flow path is from local inflow to inlet B and the other flow path is the local inflow time to inlet A plus the flow time through pipe AB. The entire drainage area only contributes to flow for the longest time of concentration.

4. Does a storm sewer pass its greatest flow rate when it is just flowing full? Explain.
5. Determine the spacing to the first and second inlets that will drain a section of highway pavement if the runoff coefficient is 0.9 and the design rainfall is 15 cm/h. The pavement

TABLE 5.10
Inflow Hydrograph of Problem 6

Time (hours)	Inflow (L/s)
0	0
2	135
4	675
6	1,350
8	945
10	567
12	351
14	203
16	68
18	0

width is 10 m (S_X=0.02, S_L=0.02, and n=0.016), the efficiency of inlet is 0.35, and the allowable spread is 2 m.

6. Using the Muskingum method, route the inflow hydrograph given in Table 5.10, assuming (a) K=4 hours and X=0.12 and (b) K=4 hours and X=0.0. Plot the inflow and outflow hydrographs for each case, assuming the initial outflow equals the inflow.
7. A storm event occurred in a watershed that produced a rainfall pattern of 5 cm/h for the first 10 minutes, 10 cm/h in the second 10 minutes, and 5 cm/h in the next 10 minutes. The watershed is divided into three subbasins (Figure 5.35) with the 10-minute unit hydrographs (UHs) as specified in Table 5.11. Subbasins A and B have a loss rate of 2.5 cm/h for the first 10 minutes and 1.0 cm/h afterward. Subbasin C has a loss rate of 1 cm/h for the first 10 minutes and 0.0 cm/h thereafter. Determine the storm hydrograph at point 2. Assume a lag time of 20 minutes.
8. Consider an impervious area with a length of 40 m, the slope of 0.03. The effective Manning coefficient is 0.03. Determine the runoff hydrograph for a constant rainfall excess rate of 20 mm/h with a 10-minute duration at the exit point of the area.
9. Determine the HR of a roadway culvert flooded with 50 years of expected service life designed to carry a 100-year storm.
10. Curb-opening inlets are placed along a 20 m wide street. The length of the inlets is 1 m that are placed on a continuous grade and the runoff coefficient is 0.9. The inlets drain the stormwater into a triangular gutter with S_X=0.03, S=0.02, and n=0.03. Determine the maximum allowable distance between inlets considering that the rainfall intensity is equal to 15 mm/h and the allowable spread is 3 m.

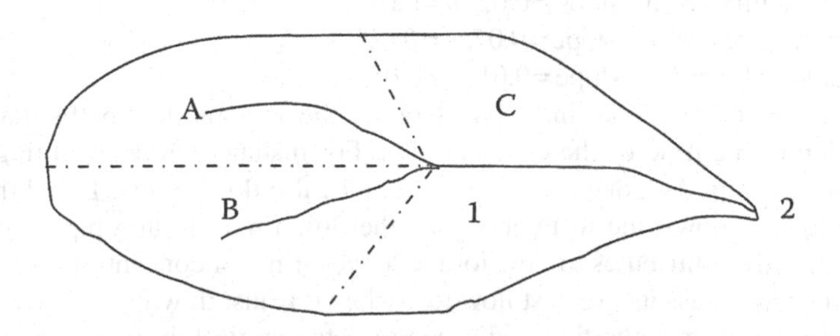

FIGURE 5.35 Schematic of watershed of Problem 7.

TABLE 5.11
Inflow Hydrographs of Problem 7

Time (min)		0	10	20	30	40	50	60	70	80	90	100
Inflow (L/s)	A	0	40	70	110	75	50	30	20	10	5	0
	B	0	30	65	100	75	50	35	20	7	3	0
	C	0	50	80	105	90	75	50	35	20	10	0

11. Determine the depth and width of a straight trapezoidal channel ($C_s = 1.0$) lined with weeping love grass which is expected to carry $Q = 0.75\,m^3/s$. The channel bottom slope is $S = 0.02$ and its side slopes are the same and equal to $z = 1.5$.
12. Determine the headwater depth for a culvert that conveys a flow of $0.5\,m^3/s$ under inlet control conditions. The culvert is circular and has a square edge inlet which is mitered with a headwall and a diameter of 2 m and a slope of 0.01.

REFERENCES

Akan, A.O. and Houghtalen, R.J. (2003). *Urban Hydrology, Hydraulics and Stormwater Quality: Engineering Applications and Computer Modeling*. Wiley, Hoboken, NJ. ISBN-10: 0-471-43158-3.

Ashley, R.M., Blamforth, D.J., Saul, A.J., and Blanksby, J.D. (2004). Flooding in the future—Predicting climate change, risks and responses in urban areas. *Proceeding of 6th ICUD Modeling*, Dresden, pp. 105–113.

Barrett, M., Lantin, A., and Austrheim-Smith, S. (2004). Storm water pollutant removal in roadside vegetated buffer strips. *Transportation Research Record*, 1890(1), 129–140.

Bates, P.D. and De Roo, A.P.J. (2000). A simple raster-based model for flood inundation simulation. *Journal of Hydrology*, 236(1–2), 54–77.

Brown, S.A., Stein, S.M., and Warner, J.C. (1996). *Urban Drainage Design Manual*. Hydraulic Engineering Circular 22. Federal Highway Administration, Washington, DC.

Chen, Y.H. and Cotton, B.A. (1988). *Design of Roadside Channels with Flexible Linings*. Hydraulic Engineering Circular No. 15 (HEC-15), FHWA, Publication No. FHWA-IP-87-7. USDOT/FHWA, McClean, VA.

Chen, C.H., Liu, W.L., Liaw, S.L., and Yu, C.H. (2005). Development of a dynamic strategy planning theory and system for sustainable river basin land use management. *Science of the Total Environment*, 346(1–3), 17–37.

Chen, C.H., Wu, R.S., Liaw, S.L., Suc, W.R., and Chiou, I.J. (2000). A study of water–land environment carrying capacity for a river basin. *Water Science and Technology*, 42, 389–396.

Chinchar, A. and Gray, J. (2021). *The Impacts of Hurricane Ida*. Royal Meteorological Society. https://www.rmets.org/metmatters/impacts-hurricane-ida

Chocat, B. and Desbordes, M. (2004). *Proceedings of Novatech 2004: Sustainable Techniques and Strategies in Urban Water Management*, Lyon, France.

Danish Water Pollution Committee (DWPC). (2005). Danish Water Pollution Committee Publication No. 27. Functional practice for sewer systems during rain (in Danish), Final draft, Dansk Ingeniør Forening.

Denisov, N., Grenasberg, M., Hislop, L., Schipper, E.L., and Sørensen, M. (2000). *Cities Environment Reports on the Internet: Understanding the CEROI Template*. UNEP/GRID-Arendal, Arendal.

Driscoll, E.D., Palhegyi, G.E., Strecker, E.W., and Shelley, P.E. (1989). *Analysis of Storm Events Characteristics for Selected Rainfall Gauges throughout the United States*. Representative. U.S. Environmental Protection Agency (EPA), Washington, D.C.

Dunne, T. (1986). Urban hydrology in the Tropics: Problems solutions, data collection and analysis. *Urban Climatology and its Application with Special Regards to Tropical Areas. Proceedings of the Mexico Technology Conference, November 1984 World Climate Programme*, WMO, Mexico.

Federal Highway Administration (FHWA). (1984). *Drainage of Highway Pavements*. Hydraulic Engineering Circular 12. Federal Highway Administration, McLean, VA.

Federal Highway Administration (FHWA). (1996). *Urban Drainage Design Manual*. Hydraulic Engineering Circular 22. Federal Highway Administration, Washington, DC.

Glidden, M.W. (1981). The effects of stormwater detention policies on peak flows in major drainage ways. Master of Science thesis, Department of Civil Engineering, University of Colorado.

Grum, M., Jørgensen, A.T., Johansen, R.M., and Linde, J.J. (2005). The effects of climate change in urban drainage: An evaluation based on regional climate model simulations. *10th ICUD*, Copenhagen, Denmark.

Gong, N., Ding, X., Denoeux, T., Bertrand-Krajewski, J.L., and Clément, M. (1996). StormNet: a connectionist model for dynamic management of wastewater treatment plants during storm events. *Water Science and Technology*, 33(1), 247–256. doi:10.2166/wst.1996.0024

Guo, J.C.Y. (1999). *Storm Water System Design*. University of Colorado, Denver, CO.

Guo, J.C.Y. (2000). Street storm water conveyance capacity. *Journal of Irrigation and Drainage Engineering*, 136(2), 119–124.

Haan, C.T. (1977). *Statistical Methods in Hydrology*. The Iowa State University Press, Ames, IA.

Hauger, M.B., Mouchel, J.M., and Mikkelsen, P.S. (2005). Indicators of hazard, vulnerability and risk in urban drainage. *Proceedings of 10th International Conference on Urban Drainage*, Copenhagen, Denmark, pp. 21–26.

Horritt, M.S., & Bates, P.D. (2001). Predicting floodplain inundation: Raster-based modelling versus the finite-element approach. *Hydrological Processes*, 15(5), 825–842.

Horritt, M.S., & Bates, P.D. (2002). Evaluation of 1D and 2D numerical models for predicting river flood inundation. *Journal of Hydrology*, 268(1–4), 87–99.

Johnson, F.L. and Chang, F.M. 1984. *Drainage of Highway Pavements*. Hydraulic Engineering Circular, No. 12. Federal Highway Administration, McLean, VA.

Jansson, J., Nilsson, J., Modig, F., & Hed Vall, G. (2017). Commitment to sustainability in small and medium-sized enterprises: The influence of strategic orientations and management values. *Business Strategy and the Environment*, 26(1), 69–83.

Karamouz, M., & Fereshtehpour, M. (2019). Modeling DEM errors in coastal flood inundation and damages: A spatial nonstationary approach. *Water Resources Research*, 55(8), 6606–6624.

Karamouz, M., Hossainpour, A. Nazif, S. (2011). Improvement of urban drainage system performance under climate change impact: Case study. *Journal of Hydrologic Engineering*, 16(5), 395–412.

Karamouz, M., Moridi, A., and Nazif, S. (2010). *Urban Water Engineering*. CRC Press, Boca Rotan, FL.

Karamouz, M., Nazif, S., and Falahi, M. (2012). *Hydrology and Hydroclimatology: Principles and Applications*. CRC Press, Boca Raton, FL.

Karamouz, M., Taheri, M., Khalili, P., and Chen, X. (2019). Building infrastructure resilience in coastal flood risk management. *Journal of Water Resources Planning and Management*, 145(4), 04019004.

Karamouz, M., Taheri, M., Mohammadi, K., Heydari, Z., and Farzaneh, H. (2018). A new perspective on BMPs' application for coastal flood preparedness. *World Environmental and Water Resources Congress 2018*, American Society of Civil Engineers, Reston, VA, pp. 171–180.

Kenyon, W. (2007). Evaluating flood risk management options in Scotland: A participant-led multi-criteria approach. *Ecological Economics*, 64(1), 70–81.

Khaliq, M.N., Ouarda, T.B.M.J., Ondo, J.C., Gachon, P., and Bobée, B. (2006). Frequency analysis of a sequence of dependent and/or non-stationary hydro-meteorological observations: A review. *Journal of Hydrology*, 329(3–4), 534–552.

Machado, M.J., Botero, B.A., López, J., Francés, F., Díez-Herrero, A., and Benito, G. (2015). Flood frequency analysis of historical flood data under stationary and non-stationary modelling. *Hydrology and Earth System Sciences*, 19(6), 2561–2576.

Marsalek, J., Barnwell, T.O., Geiger, W., Grottker, M., Huber, W.C., Saul, A.J., Schilling, W., and Torno, H.C. (1993). Urban drainage systems: Design and operation. *Water Science and Technology*, 27(12), 31–70.

Marfai, M.A., Sekaranom, A.B., and Ward, P. (2015). Community responses and adaptation strategies toward flood hazard in Jakarta, Indonesia. *Natural Hazards*, 75(2), 1127–1144.

Masria, A., Iskander, M., and Negm, A. (2015). Coastal protection measures, case study (Mediterranean zone, Egypt). *Journal of Coastal Conservation*, 19(3), 281–294.

McAlister, T. (2007). *National Guidelines for Evaluating Water Sensitive Urban Developments*. BMT WBM Pty Ltd, Queensland.

Normann, J.M., Houghtalen, R.J., and Johnston, W.J. (1985). *Hydraulic Design of Highway Culverts*. Hydraulic Design Series, No. 5. Federal Highway Administration, McLean, VA.

OECD. (1993). *Core Set of Indicators for Environmental Performance Reviews*. Environmental Monograph No. 83. OECD, Paris.

Parkinson, J. (2002). Stormwater management and urban drainage in developing countries. Available at http://www.sanicon.net/titles/topicintro.php3?topicId=5

Prommesberger, B. (1984). Implementation of stormwater detention policies in the Denver Metropolitan Area. *Flood Hazard News*, 14(1), 10–11.

Rebuild by Design Organization. (2014). http://www.rebuildbydesign.org/our%0A-work/sandyprojects (Accessed November 7, 2017).

Šakić Trogrlić, R., Rijke, J., Dolman, N., and Zevenbergen, C. (2018). Rebuild by design in Hoboken: A design competition as a means for achieving flood resilience of urban areas through the implementation of Green infrastructure. *Water*, 10(5), 553.

Sayers, P.B., Hall, J.W. and Meadowcroft, I.C. (2002, May). Towards risk-based flood hazard management in the UK. *Proceedings of the Institution of Civil Engineers-Civil Engineering*, 150(5), 36–42.

Sharif, H.O., Sparks, L., Hassan, A.A., Zeitler, J., and Xie, H. (2010). Application of a distributed hydrologic model to the November 17, 2004, flood of bull creek watershed, Austin, Texas. *Journal of Hydrologic Engineering*, 15(8), 651–657.

Schroeder K., Riechel, M., Matzinger, A., Rouault, P., Sonnenberg, H., Pawlowsky-Reusing E., Gnirss, R. (2011). Evaluation of effectiveness of combined sewer overflow control measures by operational data, combined sewer overflow control. *Water Science and Techology*, 63(2): 325–330.

Tucci, C.E.M. (2007). Urban flood management. World Meteorological Organization, Cap-Net International Network for Capacity Building in Integrated Water Resources Management. WMO/TD No. 1372.

Tunji, L.A.Q., Hashim, A.M., and Wan Yusof, K. (2012). Shoreline response to three submerged offshore breakwaters along Kerteh Bay coast of Terengganu. *Research Journal of Applied Sciences, Engineering and Technology, Maxwell Scientific Organization*, 4(16), 2604–2615.

Urbonas, B. and Glidden, M.W. (1983). Potential effectiveness of detention policies. *Flood Hazard News*, 13(1): 9–11.

Young, G.K. and Stein, S.M. (1999). Hydraulic design of drainage for highways. In L.W. Mays (ed.) *Hydraulic Design Handbook* (pp. 3.1–3.35). McGraw-Hill, New York.

6 Urban Water Supply Infrastructures

6.1 INTRODUCTION

A major part of the sustainable approach to the urban water infrastructure process is the provision of continuous service. Historically, water supply, stormwater collection and drainage, and sewage collection, treatment, and disposal are the main components of urban water systems that should be integrated into the provision of providing water services. Natural functions, in-stream uses, and withdrawals (e.g., for water supply) are the driving forces that control urban drainage, and wastewater effluents are the policies affecting the usage of receiving waters.

One of the major challenges is reaching consensus among various stakeholders on the environmental, social, and economic goals and values of urban water systems. More extensive community input and greater understanding of water management options will improve the sustainability of current systems and will make its development faster.

The water storm management is discussed in Chapter 5. Water supply is the theme of this chapter and wastewater is discussed in Chapter 7. But in all three chapters, there are many common grounds, and as such, there are issues and concerns related to all three topics in each chapter. Service-related issues are more common among the three as they are the most important and pressing issues. A shared vision for tripled services needs to be developed in the communities for utilizing the resources and exchanging information for the benefit of end users and the public.

There are some factors that cause the reduction of degraded urban water supply such as lower water consumption, conservation of natural drainage, water reuse and recycling, water contamination reduction, and preservation and/or enhancement of the receiving water ecosystem. This has led to the advocacy of the sustainable urban water system. Urban water systems should specifically meet the following basic objectives:

1. Supply of safe and good-tasting drinking water to the inhabitants on a continuous basis
2. Reclamation, reuse, and recycling of water and nutrients for use in farming, parks, or households in case of water scarcity
3. Water supply infrastructure, their state of operation, maintenance, and asset management
4. Collection and treatment of wastewater in order to protect the inhabitants against diseases as well as the environment from harmful impacts
5. Control, collection, transport, and quality enhancement of stormwater to protect the environment and urban areas from flooding and pollution

The first, second, and part of the third objectives are addressed in this chapter. The maintenance and asset management issues of objective 3 are discussed in Chapter 8. The fourth objective is presented in Chapters 7. The fifth objective is described in Chapter 5. Most of the sustainability issues have been addressed or are within reach in North America and Europe but are still far from being achieved in the developing parts of the world. The two specific goals related to water supply are (a) providing service to a high proportion of people who are incapable of accessing or affording safe drinking water and sanitation and (b) stopping the unsustainable exploitation of water resources by developing sound water management strategies at local, regional, and national levels, which promote both equitable access and adequate supplies.

DOI: 10.1201/9781003241744-6

The populations of the urban areas are expected to increase dramatically, especially in Africa, Asia, and Latin America. The population of urban areas in Africa is expected to rise more than two times over the next 25 years. Consequently, to meet the fast-growing necessities, the urban services will face great challenges over the coming decades. Water management with the provision of urban water supply services, including the basic requirements on urban water infrastructure, is the main focus of this chapter.

A balanced set of objectives of urban water demand management (UWDM) for the management and allocation of water resources are efficiency, equity, and sustainability. UWDM covers a wide range of technical, economic, capacity-building, and policy measures that are useful for municipal planners, water supply agencies, and consumers. Successful implementation of UWDM as a component of Integrated Water Resources Management (IWRM) plays a significant role in the reduction of poverty in societies through more efficient use of the available water resources and by facilitating water service through municipal water supply agencies in a region.

In the remaining part of this chapter, first an overview of water supply history and its occurrence variabilities in the recent years are given. Increasing the efficiency of the operation and maintenance of the water supply infrastructure in water districts where the service is provided is an important issue. Then different issues and challenges in the urban water supply and demand are presented. Water demand projection which is an important factor in the urban water supply planning is discussed followed by a detail description of hydraulic analysis of water distribution networks. Both quantity and quality aspects of water distribution networks are addressed in this chapter.

6.1.1 History of Water Supply Development

In formulating modern and unified drinking water systems, some valuable milestones are used. In about 3,000 BC, drinking water was distributed through lead and bronze pipes in Greece. In 800 BC, the Romans built aqueduct systems that provided water for drinking, street washing, public baths, and latrines. In 500 BC, the Persians had an integrated water and waste water system in Persepolis and built water supply/flood control reservoirs/dams as well as Qanat systems for groundwater supply management. At the beginning of the nineteenth century, the first public water supply systems in cities such as Philadelphia, USA, and Paisley, Scotland were constructed. In the middle of the nineteenth century, water quality became a problem, and hence, filter systems were introduced in some cities. The spread of cholera necessitated the use of disinfection, and the first chlorination plants were installed around 1900 in Belgium and in New Jersey, USA. During the twentieth century, advanced treatment of centrally supplied drinking water, including physical, biological, and chemical treatment, was presented by all large cities in North America and Europe. This was mainly done for surface waters; however, groundwater supplies in many countries have required and even now require minimal treatment or no treatment at all (e.g., in Slovenia and Denmark). An innovative treatment process called microfiltration of raw drinking water was introduced at the end of the twentieth century. Water treatment and water distribution were effectively governed by various laws and regulations, such as the US Safe Drinking Water Act (SDWA) in the United States (introduced in 1974, amended in 1986 and 1996) and the 1998 European Union Drinking Water Directive 98/83/EC. More extensive, globally developed guidelines on drinking water quality can be obtained from the World Health Organization (WHO, 2017).

6.1.2 Water Availability

A considerable uncertain environmental destruction appeared as a result of progress despite improvement in the quality of life. Since the beginning of nineteenth century, the population of the world has tripled, nonrenewable energy consumption has increased by a factor of 30, and the industrial production has multiplied by 50 times.

According to World Water Institute (2011), the world's urban population growth, which is a key issue in urban areas, is at four times the rate of the rural population. The urban population is projected to become more than 5 billion by the year 2025, and two-thirds of the world's population will be living in towns and cities. Furthermore, the rate of urban population growth is much higher in the developing countries. Developed regions include North America, Japan, Europe, Australia, and New Zealand. Developing regions include Africa, Asia (excluding Japan), South America and Central America, and Oceania (excluding Australia and New Zealand). The population projections in the future decades in the developed and developing countries show completely different behaviors. The results show about 2 billion population increase in the developing countries in the next 25 years. But the population remains constant in the developed countries in the same period. This indicates that the developed countries have reached their population growth limits. The rate of urbanization in different regions is also considerably different due to their rate of development. As an example, the average annual urban growth rate is about 4%/year in Asia, and Africa has the highest urban growth rate of about 5%/year (World Water Institute, 2011).

Example 6.1

Estimate the population of a town in the year 2010 based on the past population data given in Table 6.1.

Solution:

Since actual populations between the 10-year census intervals are not available, the data points by an assumed linear modification can be connected among each census. One can speculate on the reasons for some of the changes:

- The drop in population during the depression of the 1930s, most likely caused by people retreating significantly due to the drop in employment.
- The rapid augmentation in the war and postwar boom periods of the 1940s and 1950s.
- A slowdown in growth during the difficult time of the 1970s.

Are the 1980s being a replica of the 1930s? Does the drive to achieve more jobs through new industries succeed? Could the rate of growth in the coming years be as high as during the 1940 1960 rapid growth periods?

Answers to these questions will be available only with more information on the specific town and region. In fact, no one, even with broad information, can make more than an informed, educated guess. The best we can do are the following projections:

High projection: Assume that the current growth rate in 1950 will continue for 5 years, followed by an increasing rate of about four-fifths of the maximum previously experienced growth rate to the year 2000:

$$P_{(1990-2000)} = \frac{66{,}200 - 59{,}400}{10} = 680 \text{ persons/year},$$

$$P_{(1950-1960)} = \frac{45{,}050 - 32{,}410}{10} = 1{,}264 \text{ persons/year}.$$

TABLE 6.1
The Population Data of Example 6.1

Year	1900	1910	1920	1930	1940	1950	1960	1970	1980	1990	2000
Mid-Year Population	10,240	12,150	18,430	26,210	22,480	32,410	45,050	51,200	54,030	59,400	66,200

Therefore, the high population for 2010 is $66{,}200+5\times680+5\times4/5\times1{,}264=74{,}656=74{,}700$ persons.

Medium projection: The current growth rate in 1950 will continue for the next 17 years. Therefore, the medium projection for 2010 is $66{,}200+10\times680=73{,}000$ persons.

Low projection: Assume that the current growth rate in 1950 will continue for 5 years, followed by a decreasing rate of about four-fifths of the maximum previously experienced drop rate to the year 2010. The drop rate of 1930–1940 is

$$\frac{22{,}480-26{,}210}{10}=-373.$$

Therefore, the low population for 2010 is $66{,}200+5\times680-5-4/5-373=68{,}108=68{,}100$ in persons. Therefore, the estimated range of population growth is 4,900 or about 7% of the present population.

6.1.3 Water Development and Share of Water Users

Answers to the following questions could set the ground for water development framework: How development could be done in a way that is economically and ecologically sustainable? Are vision of the future and our planning schemes environmentally responsible and sensitive to the major elements of our physical environment, namely air, water, and soil? There is no single answer to these questions; however, each region could place these questions in forefront of their development plans.

Among these development attributes, water is of particular importance and should be considered for sustainable development (SD) in three distinct areas of water resources development, water conservation, and waste and leakage prevention as follows:

1. Improving the efficiency of water systems
2. Improving the quality of water
3. Downstream environmental flow consideration and water withdrawal and usage within the limits of the system
4. Water pollution reduction considering the carrying capacity of the streams
5. Water discharge from groundwater considering the safe yield of the system

The total world freshwater supply is estimated as 42,810 billion cubic meters (World Resources Institute, 2011). Only 8% of the total freshwater supply on earth has been used, with agriculture having the highest rate of 69% among the water uses. Eight percent of the total water consumption has been used in the domestic sector. Although the percentage of water consumption in the domestic sector is not high, the concentration of population in the urban areas has amplified the water shortages in this sector.

Water use in the world has increased by six folds between the years 1900 and 1995, which is more than twice the rate of the population growth. As mentioned earlier, only 8% of all the available freshwater, that is, about 220 L/d on average per capita, remains for all other domestic uses. In developing countries, especially in mega cities, drinking water demand growth is faster than urban population growth. Despite the fact that the urban population uses only small amounts of the available water for consumption, delivery of sufficient water volumes is becoming a difficult logistic and economical problem. In spite of efforts made during the past several decades, about 1.2 billion people in underdeveloped and developing countries do not have access to safe drinking water supplies. The population of water-short areas will be approximately 65% of the world population by the year 2050 (Milburn, 1996). More recent studies have shown that the pace of population growth is slowing down but the total population is increasing whereas the state of renewable water resources remains the same or even decreases due to quality issues.

Example 6.2

Consider an urban area with approximately 348,000 residents. Because of the water supply and distribution system, only 38% of the water supplied is used and the remaining is lost. To alleviate this problem, a rehabilitation and improvement program is considered.

(a) If the rehabilitation and improvement program reduced consumption losses by 1%, calculate the volume of water that would be saved in 1 year. The maximum water demand per capita is 350 m^3/year.
(b) Calculate the size of a city that could be served with the water saved, if the water demand is the same.

Solution:

(a) Current consumptive loss = 62% of the water used. Water saved by a 1% reduction of consumptive losses is

$$1\% \times 348{,}000(\text{person}) \times 350\left(\text{m}^3/\text{person} \cdot \text{year}\right) \times 0.62 = 755{,}160\left(\text{m}^3/\text{year}\right).$$

(b) Population that could be supplied by the reduction of loss is

$$\frac{755{,}160\left(\text{m}^3/\text{year}\right)}{350\left(\text{m}^3/\text{person} \cdot \text{year}\right)} = 2{,}158 \text{ person.}$$

6.1.4 Natural Resources for Water Supply

There is a constant expansion in the gap between society's needs and the natural capacity of water supply. The supply that bridges this gap is an important element of the total artificial water cycle in the urban areas. Groundwater or surface resources such as lakes, reservoirs, and rivers are the origins of natural water in urban areas. It is called untreated or raw water and is usually transferred to a water treatment plant. The water distribution network starts after the treatment facilities. The degree of treatment will be determined by the raw water quality and the intended use of water. In the past decades, different water quality standards for municipal purposes were developed and used.

Quantity, quality, time variation, and price are the four major characteristics of water supply. In almost all urban areas, the time variation of available water resources does not follow demand variations. However, when the quantity and time variation of water resources conform to the water use patterns in an urban area, storing or controlling water by man-made structures or tools is not needed. Certain facilities, therefore, should be implemented to store the excess water in the high-flow seasons for it to be consumed in the low-flow periods.

Planning for water development requires an assessment of the initial investment and operation costs and maintenance costs that should be used in the economic studies. In the case of treating water quality for different water uses, the costs should also be incorporated into the urban water resources development studies. Generally, the following categories represent water supply methods:

- Large-scale and conventional methods of surface and groundwater resources development
- Nonconventional methods

Some of the conventional methods of water supply are the construction of large-scale facilities such as dams, water transfer structures, and well fields. Dams are the most important elements of urban water resources systems in many large cities around the world.

6.1.5 Supplementary Sources of Water

Rainwater harvesting, bottled water, and wastewater reclamation and reuse are three important additional sources of water in urban areas. Rainwater harvesting, especially in places with relatively extreme rainfall and limited surface waters (e.g., small islands), is a supplementary or even primary water source at the household or small community level. In Water sensitive urban design (WSUD), the use of urban runoff for watering the open-space lawn and landscape or other uses in urban area parks is highly recommended. This method supports environmental sustainability by reducing water supply demands for those purposes and reducing urban runoff and its impacts. The most common collecting surfaces are roofs of buildings and natural and artificial ground collectors. Falkland (1991) suggested the following issues in the design of rainfall harvesting systems.

In the quantity issues, insufficient storage tank volumes or collector areas often affect the rainwater collection systems. Leaching from tanks due to poor design, selection of materials, construction, or a combination of these factors is a major problem of the rainwater collection systems. In the quality issues, the rainwater quality in many parts of the world is good, but water quality problems may arise within the collection systems or air quality. Physical, chemical, and biological pollution of rainwater collection systems occurs where inappropriate construction materials have been used or where maintenance of roofs and other catchment surfaces, gutters, pipes, and tanks is lacking.

In many parts of the world, rainwater tanks or other rainwater collection devices are used for centuries. Currently, this approach is used widely in Australia and India. Many farms in rural areas of Australia are not connected to water supply systems nor do they have adequate well water. These systems are encouraged by local water authorities in urban areas and in some cases by offering rebates to customers who use rainwater for subpotable uses. The most possible reuse of rainwater in urban areas is for gardening, which accounts for 35%–50% of domestic water use in many large cities of the world. A fairly simple system, with very low environmental risks, is used for reusing rainwater in the garden, and it is therefore encouraged by many water authorities (Karamouz et al., 2003a). An example of a rooftop rainwater harvesting system is presented in Figure 6.1. More details on application and design of rainwater harvesting systems are given in Chapter 5.

The collected rainwater can be used for toilet flushing (about 20% of domestic water use), the laundry, kitchen, bathroom, pools, and washing cars, so some potable water can be further saved. In some situations (e.g., in some rural areas), it may be possible to use rainwater for most domestic uses, without relying on the public water supply. In all these cases, strict regulations for reclaimed water quality must be followed and safety systems employed, particularly in connection with drinking water, which should be protected by the so-called "multiple barrier system." In these cases, multiple barriers are used to control the microbiological pathogens and contaminants that may enter the water supply system, thereby ensuring clean, safe, and reliable drinking water.

6.2 WATER SUPPLY INFRASTRUCTURES

These infrastructures for major cities include one or more reservoirs/dams, transfer tunnels, local aquifers, water treatment plants, and distribution network with all their appurtenances. A schematic of a water supply infrastructure system as a part of other water infrastructures is shown in Figure 6.2.

A real example of a water supply system could be seen for Tehran metropolitan area in Iran as shown in Figure 6.3 that includes three major dams, three transfer channels/tunnels, and five water treatment facilities. It is located on the southern slopes of the Alborz Mountains and includes the basins and dam reservoirs of Karaj, Lar, and Latian. Other than Lar basin dam, the remaining study area is located on the southern slopes of the Alborz Mountains. Several dams are constructed on the rivers in Tehran to control the surface water flow and for supplying a part of urban drinking water and agriculture requirements. Three most important dams are Karaj, Lar, and Latian dams. More recently, Taleghan and Mamlu dams are built. There are two other means of water supply that are added to this system including a transfer channel from Taleghan dam in the northwest and Mamloo

FIGURE 6.1 An example of a rainwater collection system.

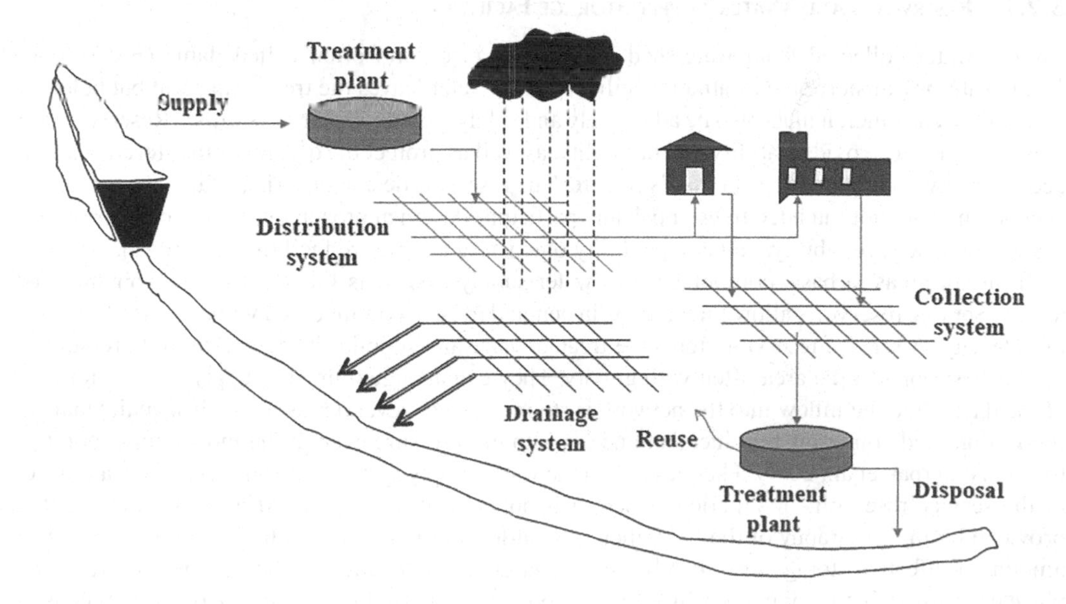

FIGURE 6.2 The water supply infrastructure system as a part of total water infrastructure.

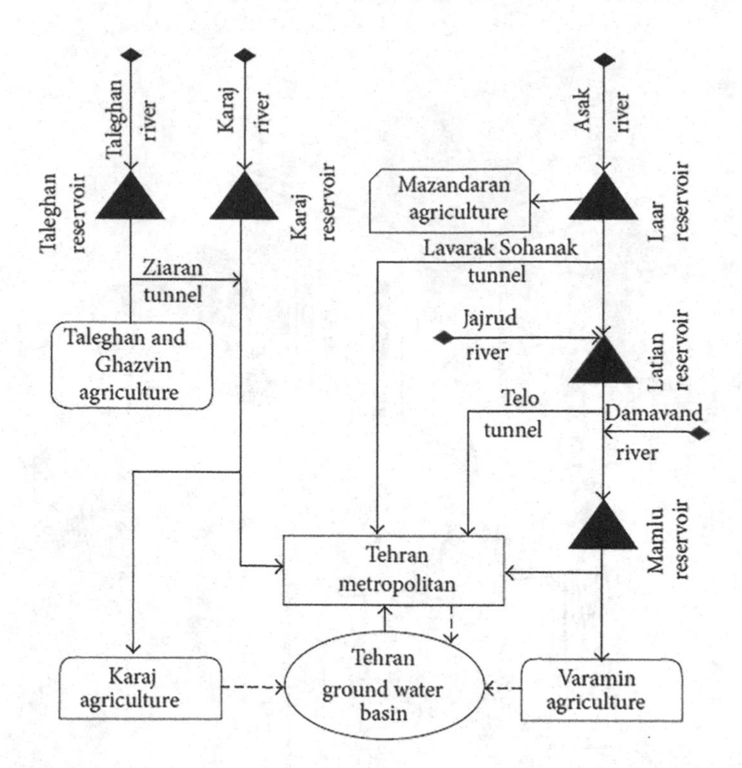

FIGURE 6.3 Schematic of water supply in Tehran Metropolitan Area (Karamouz, M., Goharian, E., and Nazif, S.. Development of a reliability based dynamic model of urban water supply system: A case study. *World Environmental and Water Resources Congress 2012: Crossing Boundaries*, 20–24 May 2102, Albuquerque, NM, pp. 2067–2078, (2012)).

dam in the east side of the city. Groundwater from Tehran aquifer supplies between 25% and 30% of water demand.

6.2.1 Reservoirs and Water Supply Storage Facilities

Natural water collected from watershed is stored in reservoirs often called dams (Figure 6.4). Then, water is transferred to treatment facilities. After water leaves the treatment plant but before it reaches the customer, it must also be adequately and safely stored in reservoirs/tanks. Reservoirs are to be designed to provide stability and durability, as well as protect the quality of the stored water in accordance with engineering standards of care. The reservoir design criteria as far as water supply is concerned are not intended to establish any particular design approach but rather to ensure water system adequacy, reliability, and compatibility of existing and future facilities (Bhardwaj, 2001).

In urban areas, a basic component of a water supply system is the service reservoir (tank or tower). Service reservoirs allow fluctuations in demand to be accommodated without loss of hydraulic integrity. A service reservoir stores the treated water and supplies it at the required pressure to the farthest point in the area often with gravity. They can also guarantee a supply, at least for part of the day, while the inflow into the network is stopped due to several reasons such as maintenance, renovation, and contamination incident and a minimum pressure even at the most remote point in the area (Værbak et al., 2019). The pressure in the water supply system depends upon the water level in the service reservoir. In situations where it is not critical to provide extra pressure above that provided by the geography of the land, then ground-level or underground tanks are sufficient. The amount of effective storage may also be dependent upon the location of the storage relative to the place of its use (whether or not it is in a different pressure zone and what distance the water needs to be conveyed) (Bhardwaj, 2001). Towers provide the extra benefit of increasing pressure head to the

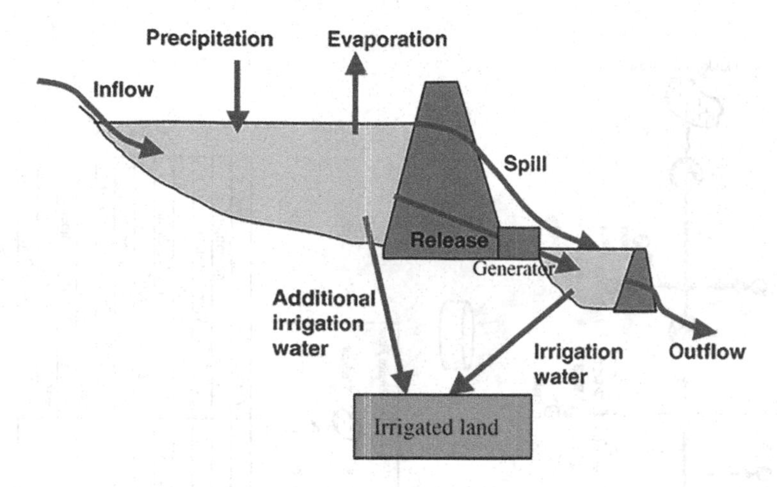

FIGURE 6.4 Schematic of reservoirs and water supply storage facilities.

downstream network, which is useful in flat regions. Total tank volume, as measured between the overflow and the tank outlet elevations, may not necessarily equal the effective volume available to the water system. Effective volume is equal to the total volume less any dead storage built into the reservoir. If a water system's source (well or booster pump) is not capable of delivering a design rate of flow above a certain water surface elevation, then that part of the volume of the tank is considered unavailable to the system and is not a part of the effective storage (Karamouz et al., 2003a).

Water supply and demand structures could be complicated and require a saved water accommodate structure. Figure 6.6 shows this structure for Zayandeh Rud River in central part of Iran. It is a unique setting of water planning to industrial, agricultural, and cosmetic sectors from the reservoir to the terminate part of Gavkhuni wetland (Figure 6.6).

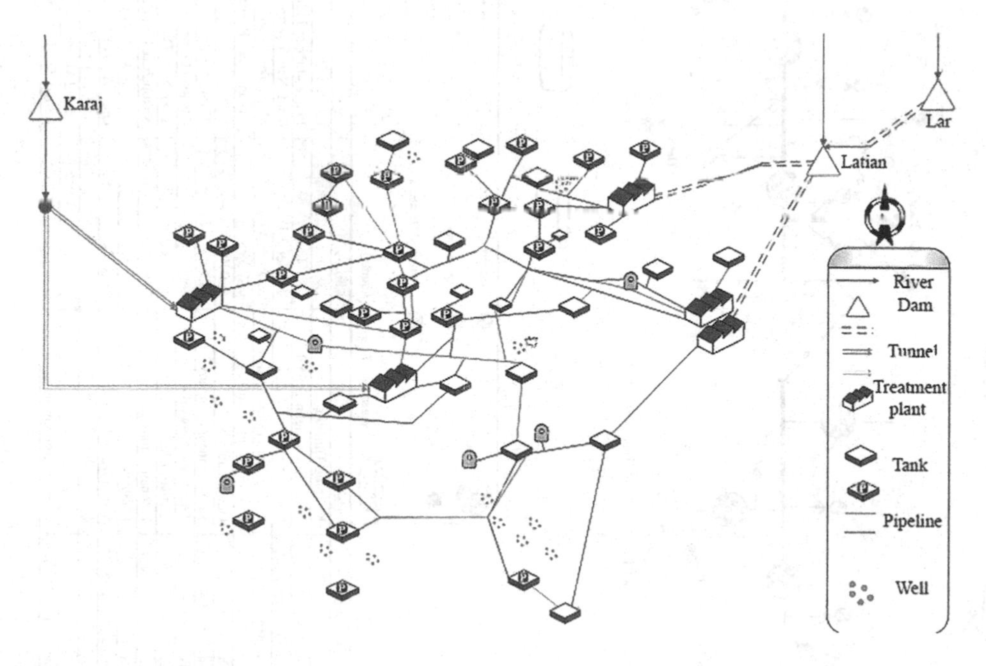

FIGURE 6.5 Surface water supply storage facilities for Tehran and related components. (From Karamouz, M., Zahraie, B., and Khodatalab, N. (2003b). Reservoir operation optimization: A nonstructural solution for control of seepage from Lar Reservoir in Iran. *IWRA – Water International*, 28(1), 19–26.)

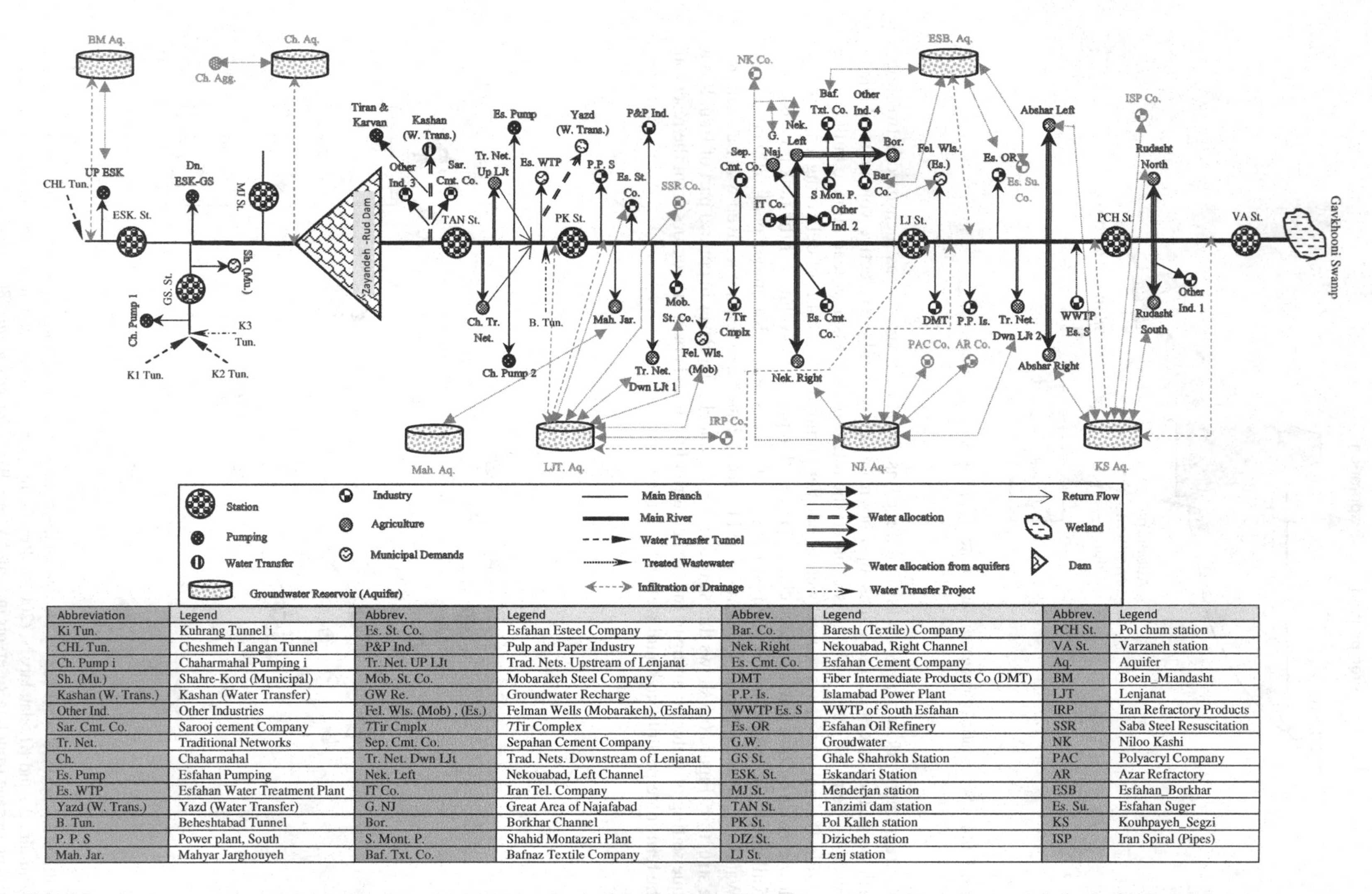

Abbreviation	Legend	Abbrev.	Legend	Abbrev.	Legend	Abbrev.	Legend
Ki Tun.	Kuhrang Tunnel i	Es. St. Co.	Esfahan Esteel Company	Bar. Co.	Baresh (Textile) Company	PCH St.	Pol chum station
CHL Tun.	Cheshmeh Langan Tunnel	P&P Ind.	Pulp and Paper Industry	Nek. Right	Nekouabad, Right Channel	VA St.	Varzaneh station
Ch. Pump i	Chaharmahal Pumping i	Tr. Net. UP LJt	Trad. Nets. Upstream of Lenjanat	Es. Cmt. Co.	Esfahan Cement Company	Aq.	Aquifer
Sh. (Mu.)	Shahre-Kord (Municipal)	Mob. St. Co.	Mobarakeh Steel Company	DMT	Fiber Intermediate Products Co (DMT)	BM	Boein_Miandasht
Kashan (W. Trans.)	Kashan (Water Transfer)	GW Re.	Groundwater Recharge	P.P. Is.	Islamabad Power Plant	LJT	Lenjanat
Other Ind.	Other Industries	Fel. Wls. (Mob) , (Es.)	Felman Wells (Mobarakeh), (Esfahan)	WWTP Es. S	WWTP of South Esfahan	IRP	Iran Refractory Products
Sar. Cmt. Co.	Sarooj cement Company	7Tir Cmplx	7Tir Complex	Es. OR	Esfahan Oil Refinery	SSR	Saba Steel Resuscitation
Tr. Net.	Traditional Networks	Sep. Cmt. Co.	Sepahan Cement Company	G.W.	Groudwater	NK	Niloo Kashi
Ch.	Chaharmahal	Tr. Net. Dwn LJt	Trad. Nets. Downstream of Lenjanat	GS St.	Ghale Shahrokh Station	PAC	Polyacryl Company
Es. Pump	Esfahan Pumping	Nek. Left	Nekouabad, Left Channel	ESK. St.	Eskandari Station	AR	Azar Refractory
Es. WTP	Esfahan Water Treatment Plant	IT Co.	Iran Tel. Company	MJ St.	Menderjan station	ESB	Esfahan_Borkhar
Yazd (W. Trans.)	Yazd (Water Transfer)	G. NJ	Great Area of Najafabad	TAN St.	Tanzimi dam station	Es. Su.	Esfahan Suger
B. Tun.	Beheshtabad Tunnel	Bor.	Borkhar Channel	PK St.	Pol Kalleh station	KS	Kouhpayeh_Segzi
P. P. S	Power plant, South	S. Mont. P.	Shahid Montazeri Plant	DIZ St.	Dizicheh station	ISP	Iran Spiral (Pipes)
Mah. Jar.	Mahyar Jarghouyeh	Baf. Txt. Co.	Bafnaz Textile Company	LJ St.	Lenj station		

FIGURE 6.6 Water supply and demand in Zayandeh Rud River Basins. (Safavi, H. R., Golmohammadi, M. H., & Sandoval-Solis, S. (2015). Expert knowledge based modeling for integrated water resources planning and management in the Zayandehrud River Basin. *Journal of hydrology*, 528, 773–789.)

6.2.2 Water Storage

Large dams are usually multipurpose structures. The purposes include providing water for domestic, agriculture, and industrial uses (the main objectives of reservoir planning and operation); hydropower electric production; flood control; and damage reduction A reservoir reduces the peak flow of a flood hydrograph to an amount lower than the river carrying capacity (Karamouz et al., 2003a). High efficiency, lower costs, and the specific capabilities of hydropower plants for controlling the frequency of power networks have made hydropower plants a necessary component of power systems.

6.2.2.1 Types of Dams

Dams are usually classified in terms of materials and forms. Topography and geomorphology are primary factors in weighting the comparative benefits of dam types. Common types are homogenous or zoned earthfills, rockfills with an earth core or concrete face, and concrete dams (Figure 6.7).

Embankment dams are constructed of earth and/or rock with a provision for controlling seepage. These dams have relatively poor resistance to overflow; therefore, the spillway capacity must be determined conservatively. A gravity dam is an essentially solid concrete structure that resists imposed forces principally by its own weight. Although they are usually straight in plan, these dams are sometimes curved or angled to accommodate site topography. Arch dams can carry large loads,

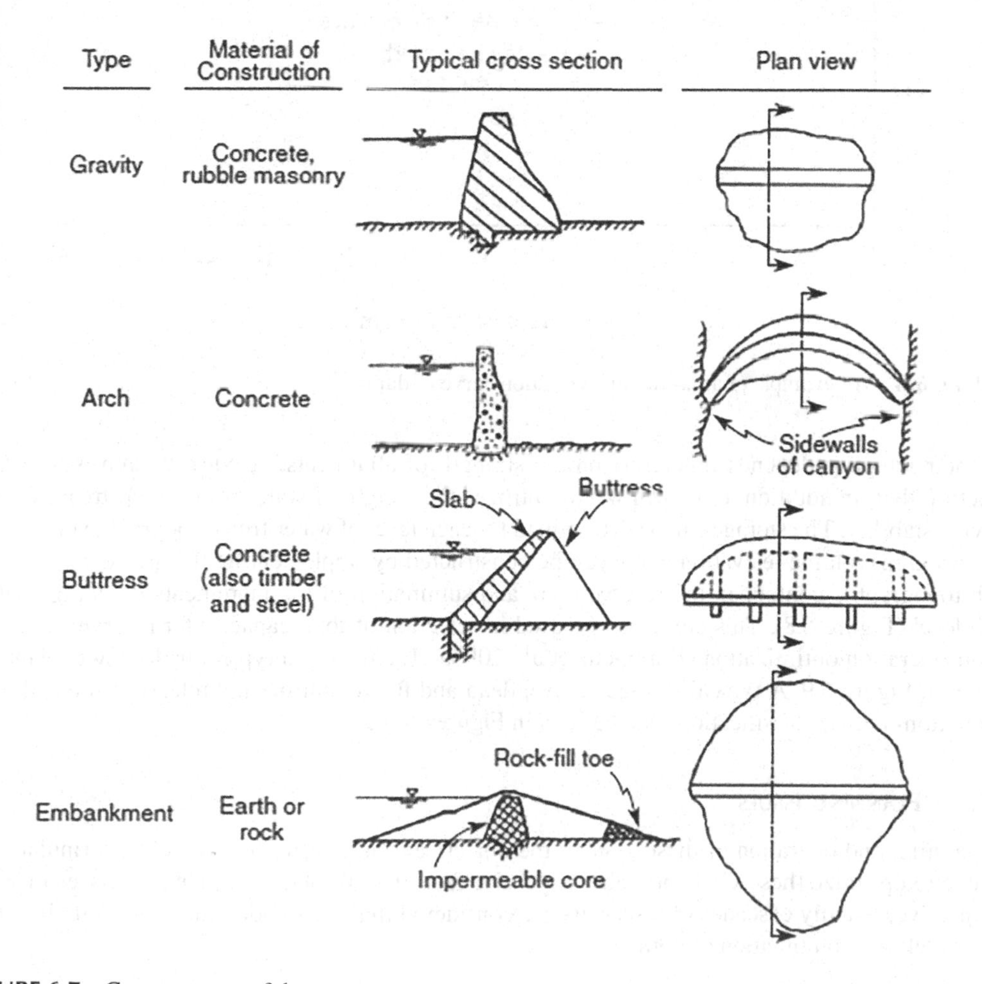

FIGURE 6.7 Common type of dams.

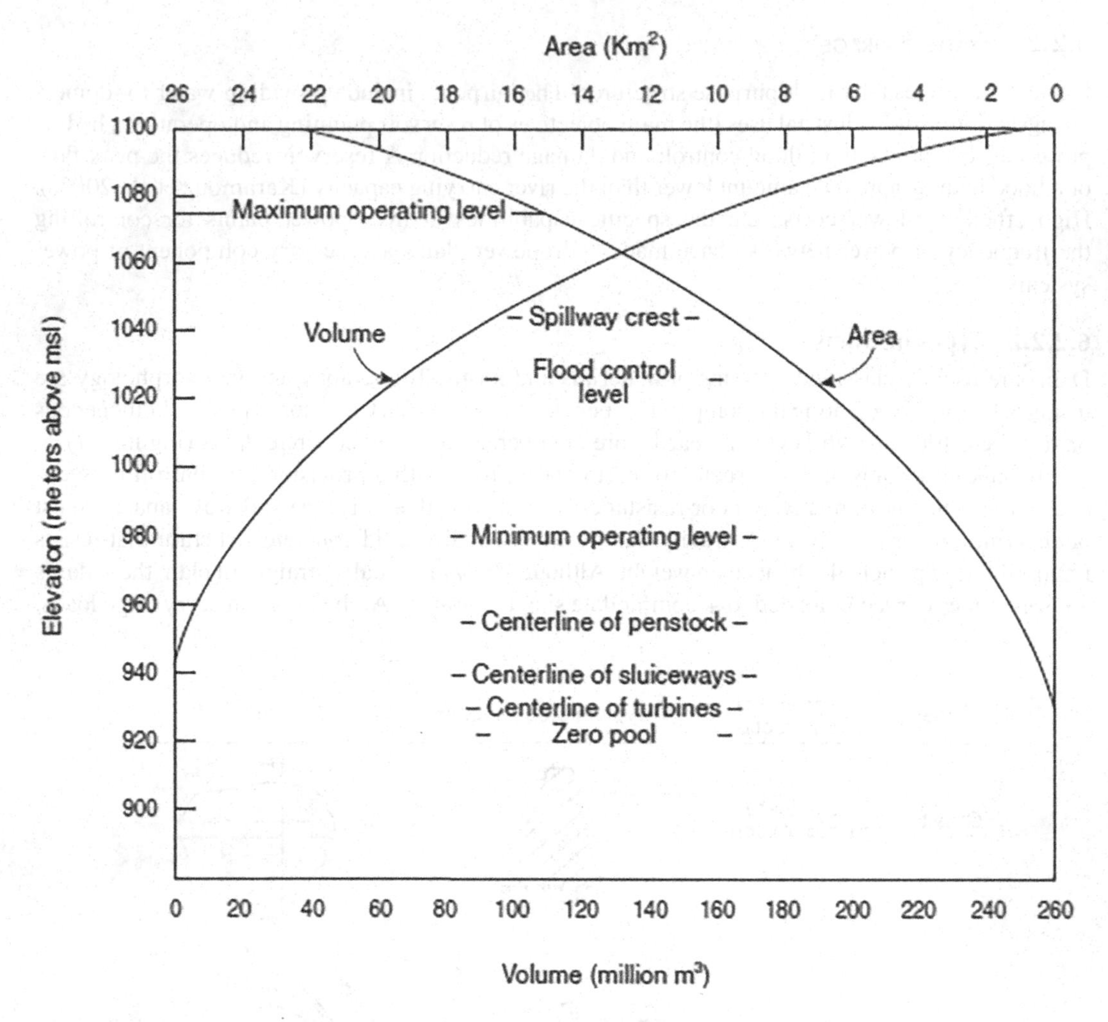

FIGURE 6.8 An example of area–volume–elevation curve of dams.

but their integrity depends inherently on the strength of abutments. A buttress dam is a gravity structure that, in addition to its own weight, utilizes the weight of water over the upstream face to provide stability. The storage can be determined for each level of water from topographic map of the site. An area–volume–elevation curve can be constructed by implementing the area enclosed with each topographic contour in the reservoir site and summation of the increments of storage below each level (Figure 6.8). This curve can be used in selection of total capacity for reservoir and reservoir operation optimization (Karamouz et al., 2003a). Element of a typical hydropower plant are shown in Figure 6.9. A typical storage (active, dead and flood control) and release (from spillway, and bottom outlet) classifications can be seen in Figures 6.10.

6.2.3 Planning Issues

In planning and operation of these system, the objectives and constraints should be formulated to be able to optimize these vital and rather expensive facilities. To place the planning issues in some perspectives usually cascades of reservoirs are considered that are divided into reservoirs in series or parallel or a combination of both.

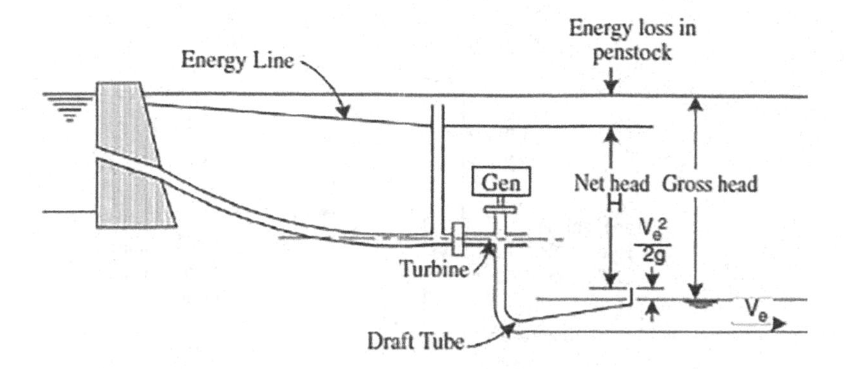

FIGURE 6.9 Element of a typical hydropower plant.

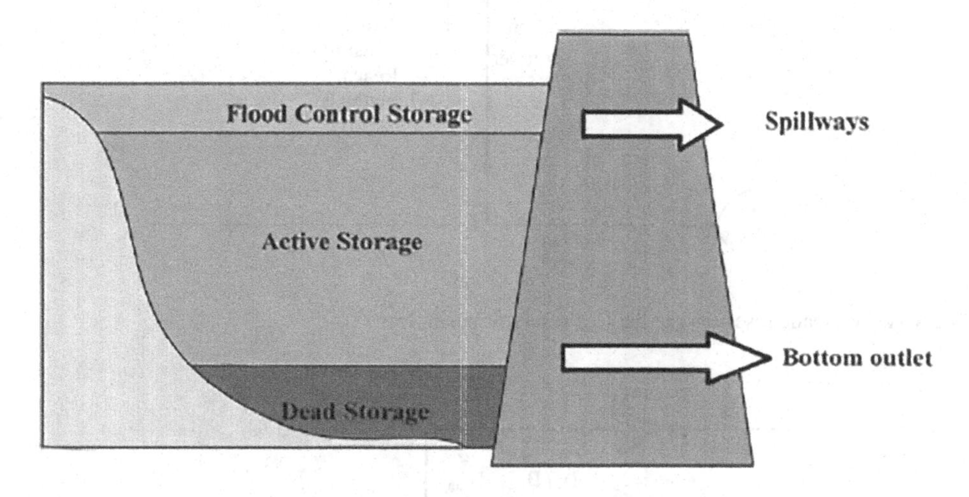

FIGURE 6.10 Reservoir storage—active, dead, and flood control storages.

6.2.3.1 Cascade Reservoirs

Cascade systems have more than one reservoir on the same river (Figure 6.11, Dez reservoir basin). Demand points are located along the river downstream of the reservoirs. Also, local rivers join the main river at various reaches.

6.2.4 Parallel Reservoir

Parallel reservoirs have a junction point with other rivers in the system. Demand points are located downstream of each reservoir and after the junction point. Reservoirs in parallel require special modeling when demand points downstream of the junction point should be supplied from upstream reservoirs (Karamouz et al., 2003a). Formulation of the optimization problem can be as follows (Figure 6.12):

$$\text{Minimize} \quad z = \sum_{i=1}^{n} \sum_{i=1}^{NR+1} \sum_{j=1}^{m_i} \text{loss}_{i,j,t}\left(RA_{i,j,t}, D_{i,j,t}, S_{i,t}\right) \tag{6.1}$$

$$\text{Subject to: } S_{i+1,t+1} = S_{i,t} + I_{i,t} - R_{i,t} - E_{i,t} - L_{i,t} \quad (\text{t} = 1,\ldots,n)\,(i = 1,\ldots,NR) \tag{6.2}$$

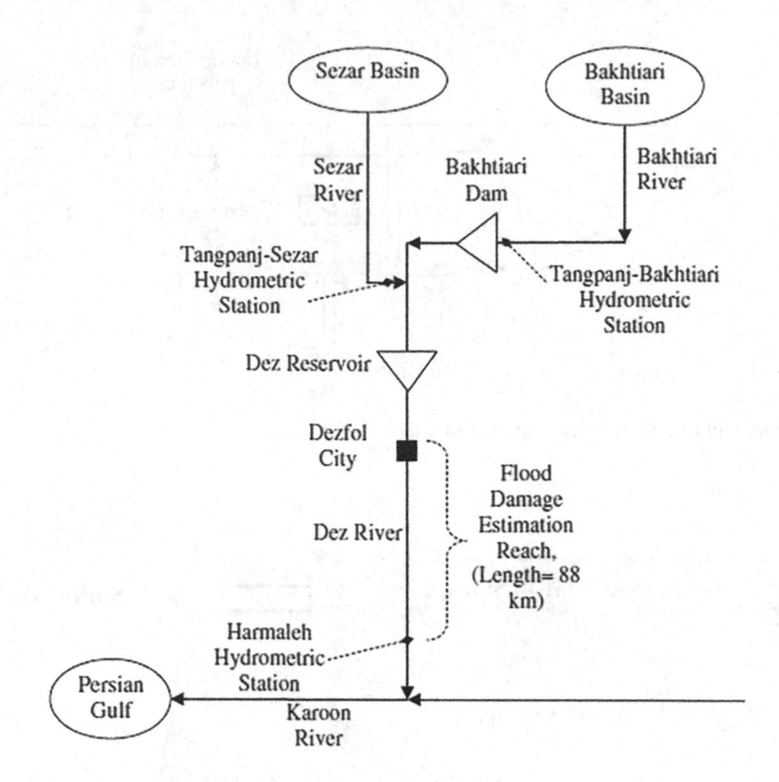

FIGURE 6.11 Cascade reservoirs in the Dez reservoir basin, Iran.

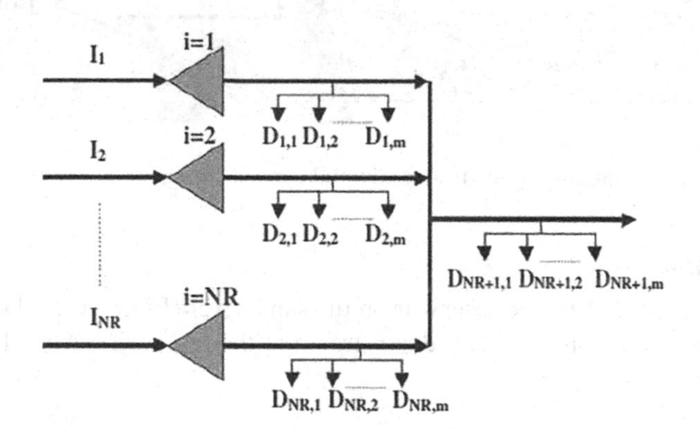

FIGURE 6.12 Parallel reservoir system (Karamouz, M., Szidarovszky, F, and Zahraie, B. *Water Resources Systems Analysis* (600 pp.). Lewis Publisher, Boca Raton, FL, (2003a)).

$$I_{NR+1,t} = \sum_{i=1}^{NR} R_{i,t} - \sum_{j=1}^{m_i} RA_{i,j,t} \quad (t = 1,\ldots,n) \tag{6.3}$$

$$S_{i,\min} \le S_{i,t} \le Cap_i \quad (t = 1,\ldots,n)\,(i = 1,\ldots,NR) \tag{6.4}$$

$$0 \le R_{i,t} \le R_{i,\max,t} \quad (t = 1,\ldots,n)\,(i = 1,\ldots,NR) \tag{6.5}$$

$$S_{i,j}, E_{i,t}, L_{i,t}, R_{i,t} \ge 0 \quad (t = 1,\ldots,n)\,(i = 1,\ldots,NR) \tag{6.6}$$

Thus, the contribution of each reservoir towards supplying the demand points along the $NR+1$ reach is determined.

Where, S, R, and I are storage, release from and I inflow to the reservoir (with Cap as total capacity), respectively. E and L are evaporation and other losses. The cost function is usually shown as a Loss function. RA is release allocated to demand points.

The objective function can be simplified for supplying water for an urban area from a multi-reservoir system:

$$\min Z = \sum_{t=1}^{T} \text{Loss}\left(\sum_{s=1}^{X} R_{s,t}\right), \tag{6.7}$$

where

T is the time horizon,

X is the total number of reservoirs in the system,

Loss is the cost of operation based on the ratio of supplied water ($R_{s,t}$) from reservoir s in month t to the monthly urban water demand.

6.2.5 Reservoir Operation

Reservoir operation optimization aims to determine release and transfer decisions that maximize water management objectives such as ensuring a reliable water supply, hydropower production, and mitigation of downstream floods. Reservoir operation consists of several control variables that define strategies for guiding a sequence of releases to meet downstream demands. In recent years, applying optimization techniques to reservoir operation by mathematical tools has become a major focus of water resource engineers. Optimization of reservoir operations is therefore more relevant than ever, both as a complement to the efficient design of new dams and for the revision of operations in existing ones. Here, we would define reservoir operation as the determination of how much water to abstract from sources (e.g. rivers), to transfer between reservoirs, and to release from reservoirs to points of consumption (e.g., for irrigation, domestic or industrial consumption) or use (e.g., hydropower production). Reservoir operation is a challenging decision-making problem because it requires finding a balance between decisions conflicting in time (e.g., whether to accept a cost in the short term in order to avoid a larger, but more uncertain, cost in the mid term) and across uses (e.g., between irrigation, hydropower, and municipal supply). Reservoir operation is discussed in detail in Karamouz et al. (2003a)

6.2.6 Flood Control

The flood storage capacity of a reservoir is one of the important components of many flood control systems. For this purpose, part of the active storage of a reservoir is kept empty to store potential floods and gradually release excess water at rates not exceeding the capacity of the downstream river. Methods of assessing the volume required for flood storage are based on minimizing expected damages. Flood routing procedures, along with knowledge of flood control operating policies and channel storage characteristics, can be used to predict the impact of flood peaks on downstream of the reservoir. The important step for determining flood control storage for different months is to define the hydrographs that represent the response of the watershed to floods with varying return periods. The next step would be to determine the ability of a reservoir to withstand floods of various magnitudes. Physical characteristics of the reservoir, discharge outlets, and the carrying capacity of the river downstream of the reservoir should be taken into account.

6.2.7 Creative Thinking Examples of Supply Expansion

6.2.7.1 Curing a Dam—Bookan Reservoir: Increasing the Operational Efficiency

Bookan dam was constructed on Zarrineh Rud, one of the important rivers of West Azerbaijan in Iran in the early 1970s. Since the dam construction, full operation of the dam has encountered some problems with sever noise and vibration whenever the gates (outlets) were opened more than 35%. This problem was resolved in April 1994. The successful experts' efforts in solving a 22-year problem, the scientific justification of the tunnel's hydraulic problem, the outlet valves of the dam, the proposed solution and its implementation have been described.

When opening the water release valves for agriculture exceeded 35%, vibrations and noise were created in the valves' gates and the intake tower which was located 200 m away from the valves. An international consultant engineer, who designed the dam, believed that the cause of the abovementioned problems was due to cavitation. In the presence of the main contractor of the dam and the contractor who constructed the hydromechanical equipment, experiments were carried out with changing the percentage of the valves' openings. To solve the problem, the valves were subjected to aeration tests and mounting of the speed bump at the branching point on the physical model. In a site visit by local experts headed by the author of this book, it became apparent that the cause of the sound and vibration is the short distance of the branching of the Howell Bunger gates and the end of the blocked tunnel and the occurrence of water hammer. Figure 6.13 shows the position of the tunnel and the water movement from the main tunnels of the Bookan dam. The tunnel closures in two of the four branches of tunnel were intended for supply of hydropower turbines that were not added and were sealed off. Two others were extended and used as outlets.

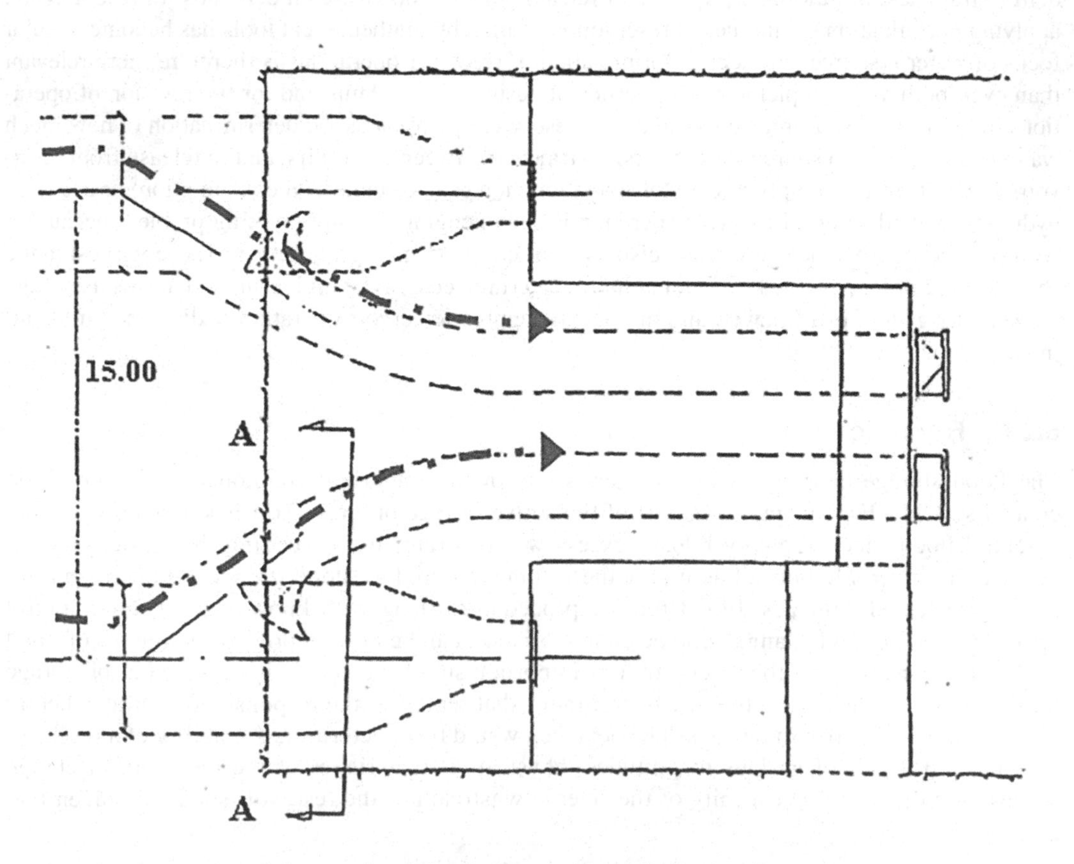

FIGURE 6.13 The plan of the tunnel and the discharge gates of the Bookan Dam.

The length of this section, branched off to the end of the tunnel (with 5 m diameter), has been about 6 m. Therefore, it was suggested to extend it the way the other two parts with gates that were originally constructed (with 2 m diameter). It was believed that this change would allow full operation of the discharge capacity of the valves. It is also proposed to install the internally made butterfly valves to the end of the tunnel so that the discharge capacity could reach to 220 m^3/s (from earlier about 35 m^3/s safe operation). Fortunately, by accepting this proposal, the project was executed without any further deliberations, spending less than half a million US dollars. This resulted in resolving the noise and vibration problems and doubling the outlet capacity. Figure 6.14 shows the final schematic for this modified plan.

In the branching place, corrosion in the body of the branching pipes was repaired a few years ago, and thus, no further corrosion was observed in that part. Thus, contrary to the initial assumption, cavitation was not the cause of the noise and vibration. Extending the two blocked branches of the main tunnel and reduce/remove the adverse effects of water hammer.

A summary of the operations that were planned and implemented within a month is as follows:

- Construction and installation of two conduits with a length of 10.25 m and an internal diameter of 2,000 mm
- Construction and installation of two reducing diameter pipes from 2,000 mm to 1.3 m (due to the lack of availability of internally made 2,000 mm butterfly valves)
- Construction and installation of two pipes with a length of 1 m and an internal diameter of 1,600 mm at the end of the butterfly valves
- Implementation of the project area, through the construction of a cofferdam and installation of a pump motor for discharging of water
- Electrical work between the valves and the command room

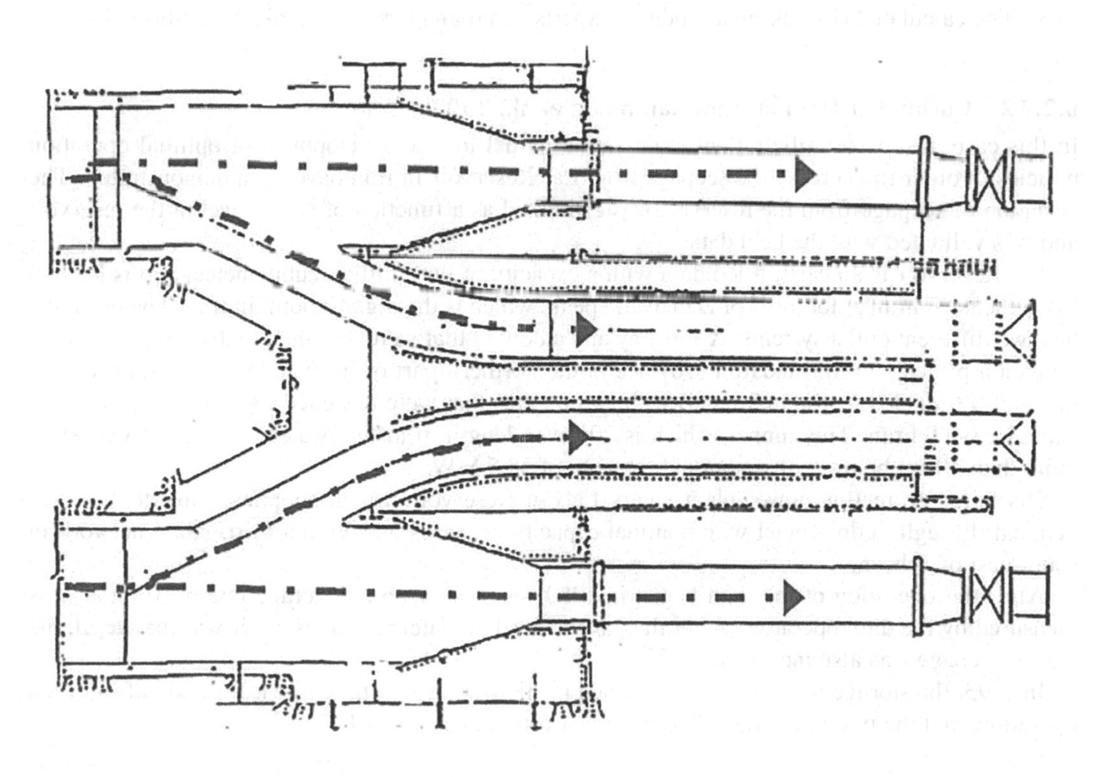

FIGURE 6.14 The schematic of modified plan.

It should be mentioned that these operations were at a cost of less than 500 million Rials (about $300k) this was a fraction of what an international consultant asked to run some tests. Its implementation results are summarized as follows:

- Eliminating the vibration of original two valves and allowing them have completed opening if (100%) needed
- Tripling of the release capacity in the dam by installing new valves and eliminating the problems in the existing valves (the safe discharge capacity of the valves was increased from 65 m^3/s to about 220 m^3/s)
- Eliminating the threats to the lands and installations of the nearby city, caused by the flooding in the Zarrineh Rud River as a result of inability to timely flood control included.

What happened at the Bookan dam in February and March of 1994 was an example of professional conduct with strict observance of the principles of science and engineering over a short period of time to solve a major problem. The following lessons can be learned from the Bookan dam experience:

- The noise and vibration in the tunnel and the outlet valves of the Bukan dam were not due to the cavitation phenomenon and there was no need for aeration to overcome this problem.
- The water hammer phenomenon can occur whenever a change in the amount and direction of flow occurs or if there is a short distance sudden blockage after a pipe convergence.
- Solving problems such as this requires self-confidence in expressing comments and professional opinion.
- The calculated risk-taking attitude by experts and managers could make breakthroughs.

6.2.7.2 Curing Lar Dam in Iran (Karamouz et al., 2003b)

In this case, results of utilizing an optimization model for the development of optimal operation policies in order to decrease the seepage from Lar Reservoir in Iran have been demonstrated. The variation of seepage from the reservoir was estimated as a function of water level in the reservoir and was validated with the field data.

Lar Reservoir is an earth-filled dam with a capacity of 960 million cubic meters and is located in northeast Tehran at foothills of Damavand peak, which is the highest point in Iran. The reservoir has two different outlet systems: A spillway and an outlet that were designed to discharge water for irrigation projects in Mazandaran Province in the northern part of Iran; and bottom outlet system includes a tunnel (Lar-Kalan Tunnel) with two intakes that were designed to transfer water to the capital city, Tehran. This tunnel, which is 20 km in length, transfers water to Kalan Power Plant more than 400 m below with a nominal capacity of 38.5 MW.

Discharge from this power plant enters Latyan Reservoir, which supplies water to Tehran's demand through Tello Tunnel with nominal capacity of 9 m^3/s and Varamin irrigation network in south-east of Tehran.

After the operation of the dam began in 1980, seepage, with an average rate of 3.3 m^3/s, was measured by the dam operators. After that, as the level of water in the reservoir was increased, the rate of seepage was also increased.

In 1993, the storage in the reservoir reached its highest level in the historical period of Lar dam operation, and the maximum rate of seepage was reported as about 12.1 m^3/s.

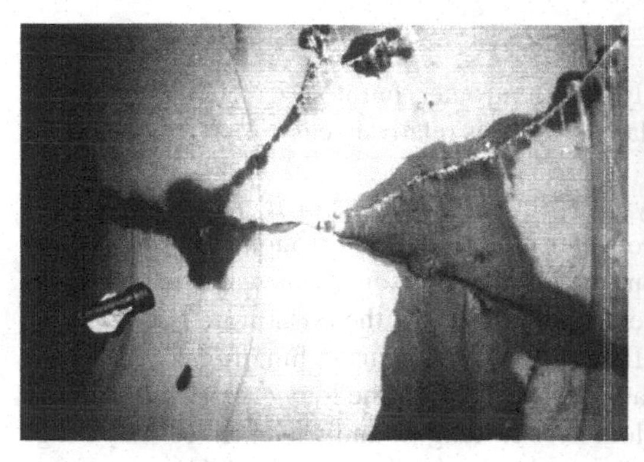
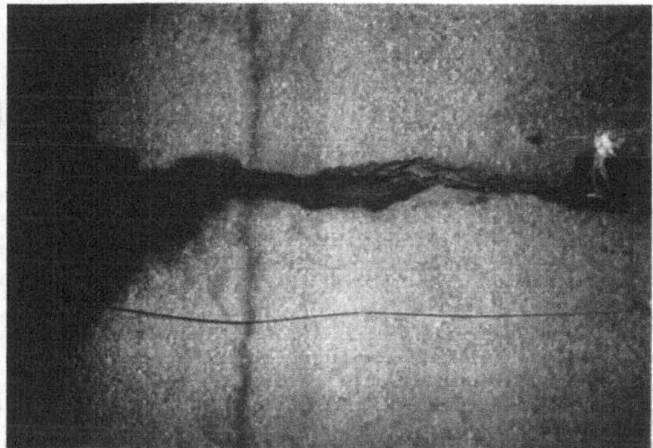

FIGURE 6.15 The state of repair of the water transfer tunnel - Lar Reservoir, Tehran, Iran.

Separate investigations were performed in order to find the sources and route of the seepage using isotopes and different tracers.

Results of these studies have shown that:

- Sources of seepage are the presence of different sinkholes at the bottom of the reservoir.
- A good portion of seepage returns to the river that supplies water to Mazandaran agricultural fields via Galoogah and Haraz Springs.
- There is a close correlation between rate of seepage and the water level in the reservoir.
- Because of large amount of seepage, the spillway and more than half of the total capacity of the dam have never been used.

However, even though the dam has not provided the amount of water that was expected from the project, it still plays an important role in supplying the consumption demand of more than 7 million people at that time in Tehran metropolitan area.

Different structural methods for controlling the seepage have been studied and implemented, such as the injection of a concrete mix in the soil layers under the foundation of the dam and the bottom of the lake (sink holes), which had no significant impact on the reduction of seepage.

Generally, results of the optimization model have shown that by reducing the water storage in the reservoir, the seepage could be significantly decreased. Results of various scenarios have shown that so much as 19% of average annual seepage can be reduced by better operating policies obtained from the model developed, compared with historical operation of the reservoir. Different scenarios also have been defined in order to estimate the rate of seepage and water supply for different demands with changing the location of Lar-Kalan Tunnel intake.

In a site visit in the winter of 1994, it became apparent that the relatively large dead storage of 100 MCM in this reservoir is the main storage volume that is subject to seepage; it was proposed to pump the water from the dead storage to the lower inlet at approximately 10 m distance. Above that, a rate of 6 m^3/s was selected as it is the minimum rate in which power can be generated at the power plant. In the visit's report, the state of tunnels repair was also stated (Figure. 6.15).

The results of having different inflow to the reservoir in a given year, have shown that 20%–40% of average annual seepage can be reduced. The elevation of the power plant is 431 m lower than the bottom tunnel intake in the Lar Reservoir, which is a constant head for power generation in that plant. Therefore, the cost of pumping station is justified by the added benefit of power generation by more than a 1:20 factor while utilizing more water for the consumption of Tehran, the capital of Iran (see Karamouz et al., 2003b for more details). Results of this study have shown that a crisis can be changed into an opportunity for utilizing resources with considerable economic gain.

6.2.8 Groundwater Storage

The water below the surface of the earth primarily is groundwater, but it also includes soil water. Movement of water in the atmosphere and on the land surface is relatively easy to visualize, but the movement of groundwater is not.

Groundwater moves along flow paths of varying lengths from areas of recharge to areas of discharge. The generalized flow paths start at the water table continue through the groundwater system and terminate at the stream or at the pumped well. The source of water to the water table (groundwater recharge) is infiltration of precipitation through the unsaturated zone. In the uppermost, unconfined aquifer, flow paths near the stream can be tens to hundreds of meters in length and have corresponding travel times of days to a few years. The longest and deepest flow paths may be thousands of meters to tens of kilometers in length, and travel times may range from decades to millennia. In general, shallow groundwater is more susceptible to contamination from human sources and activities because of its close proximity to the land surface. Therefore, shallow, local patterns of groundwater flow near surface water are emphasized in this section.

The principal hydrologic properties of aquifers are permeability and specific yield. Permeability shows the material ability to transmit water while specific yield indicates the volume of water that the aquifer yields during drainage. The materials that form the crust of the earth have a lot of voids called interstices. These voids hold the water that is found below the surface of the land and that is recoverable through springs and wells. The occurrence of water under the ground is related to the character, distribution, and structure of the voids in soil. If the interstices are connected, the water can move through the rocks by percolating from one interstice to another and therefore the groundwater flow forms.

The porosity is considered as the percentage of the total volume of the rock that is occupied by interstices. The porosity of soil depends on the shape and arrangement of its particles, the particles' sorting, the degree of cementation and compacting, and the fracturing of the rock resulting in joints and other openings. It should be mentioned that there is not definitely a correlation between the porosity and permeability of a material; a rock may contain many large but disconnected interstices and thus have a high porosity yet a low permeability. Clay may have a very high porosity (higher than some gravel) but a very low permeability.

Just a part of stored water in the interstices is recovered through wells and the remainder will be retained by the rock formations. Specific yield corresponds to the part that will drain into wells, and the part that is retained by the rocks is called the specific retention. Both specific yield and specific retention are expressed as the percentages of the total volume of material and their summation is equal to the porosity.

The water table fluctuates in response to recharge and discharge from the aquifer. The shape and slope of the water table, which are determining factors in the movement of groundwater, are affected by many factors such as the differences in permeability of the aquifer materials, topography, and the configuration of the underlying layer, amount, and distribution of precipitation, as well as the method of discharge of the groundwater. However, as a general concept, the slope of the water table varies inversely with the permeability of the aquifer material. This is shown by Darcy's law governing the movement of groundwater:

$$Q = KiA, \tag{6.8}$$

where

Q is the quantity of water moving through a given cross-sectional area,
K is the permeability or hydraulic conductivity coefficient,
i is the hydraulic gradient, and
A is the total cross-sectional area through which the water is moving.

Streams, lakes, and reservoirs interact with groundwater in all types of landscapes. The interaction takes place in three basic ways: streams gain water from inflow of groundwater through the streambed (gaining stream), they lose water to groundwater by outflow through the streambed (losing stream), or they do both, gaining in some reaches and losing in other reaches. For groundwater to discharge into a stream channel, the altitude of the water table in the vicinity of the stream must be higher than the altitude of the stream water surface. Conversely, for surface water to seep to groundwater, the altitude of the water table in the vicinity of the stream must be lower than the altitude of the stream water surface. Contours of water table elevation indicate gaining streams by pointing in an upstream direction, and they indicate losing streams by pointing in a downstream direction in the immediate vicinity of the stream. These interactions play an important role in surface and groundwater resource volume changes over time that should be considered in water supply schemes. The concepts used in the study of groundwater resources and their interactions with the surface water resources are explained in detail in Karamouz et al. (2020).

6.2.8.1 Well Hydraulics

Well drilling aids groundwater exploration with an objective to discover aquifers in different hydrogeological conditions and determination of hydraulic parameters. Groundwater exploitation from the well leads to decline in the water level that limits the yield of the basin. Water is removed from the aquifer surrounding the well during water pumping and the piezometric surface decreases. Hence, prediction of hydraulic-head drawdowns in aquifers under proposed pumping schemes is one of the goals of groundwater resource studies. The pumping well creates an artificial discharge area by drawing down (lowering) the water table around the well.

Figure 6.16 shows the creation stages of a cone of depression immediately surrounding the well when water is pumped out. At the initial stage in pumping an unconfined aquifer, water begins to flow towards the well screen. Most water follows a path with a high vertical component from the water table to the screen (Figure 6.16a). At the intermediate stage in pumping an unconfined aquifer, although dewatering of the aquifer materials near the well bore continues, the radial component of the flow becomes more pronounced (Figure 6.16b). At the approximate steady-state stage in pumping an unconfined aquifer, the profile of the cone of depression is established (Figure 6.16c).

If pumping continues, more water must be derived from the aquifer storage at greater distances from the bore of the well and the following are observed:

1. Expansion of the cone of depression,
2. Increasing the radius of influence of the well due to the expansion of the cone,
3. Increasing the drawdown at any point in the depth of the cone to provide the additional head required to move the water from a greater distance.
4. Expansion of the cone and deepening more slowly with time since an increasing volume of stored water is available with horizontal expansion of the cone.

The expansion of the cone of depression during equal intervals of time is much like that shown in Figure 6.17. Calculations of the volume of each cone would show that cone 2 has twice the volume as that of cone 1, and cone 3 has three times the volume as that of cone 1. It is because a constant pumping rate discharges the same volume of water from the well. As, if the aquifer is homogeneous and the well is being pumped at a constant rate, the increase in the volume of the cone of depression is constant over time, deepening or expansion of the cone during short intervals of pumping is barely visible and it becomes stabilized. The cone of depression continues to enlarge until the flow in the aquifer with the source of surface water, vertical recharge from precipitation, is intercepted to equal the pumping rate and the leakage that occurs through overlying or underlying formations. Equilibrium occurs when continued pumping results in no further drawdown and the cone stops expanding.

As steady-state groundwater problems are relatively simpler, steady-state radial flow into a well, under both confined and unconfined aquifer conditions, is presented.

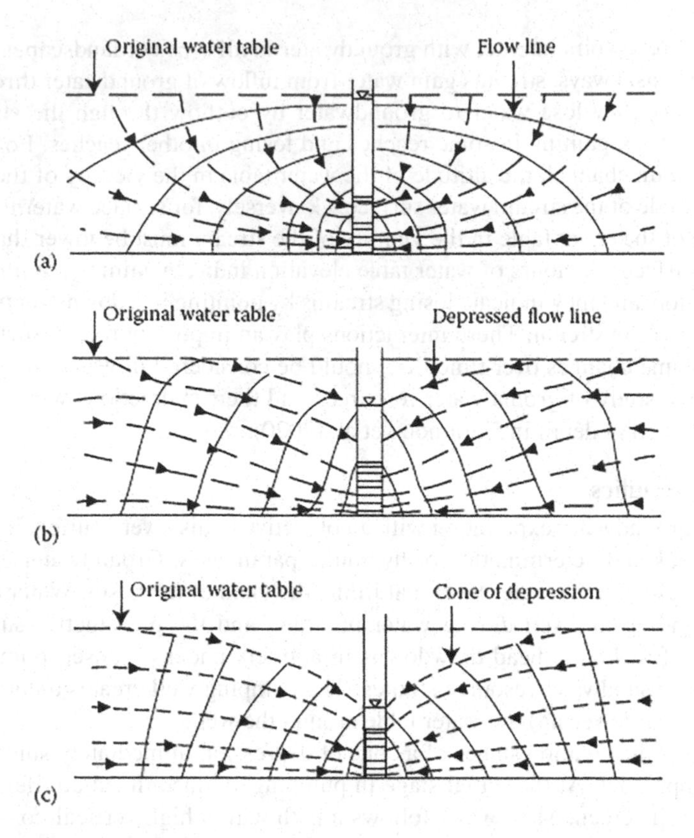

FIGURE 6.16 Development of flow distribution around a discharging well in an unconfined aquifer: (a) initial stage, (b) intermediate stage, and (c) steady-state stage. (From Karamouz, M. et al., *Groundwater Hydrology: Engineering, Planning, and Management*. CRC Press, Boca Raton, FL, 2020.)

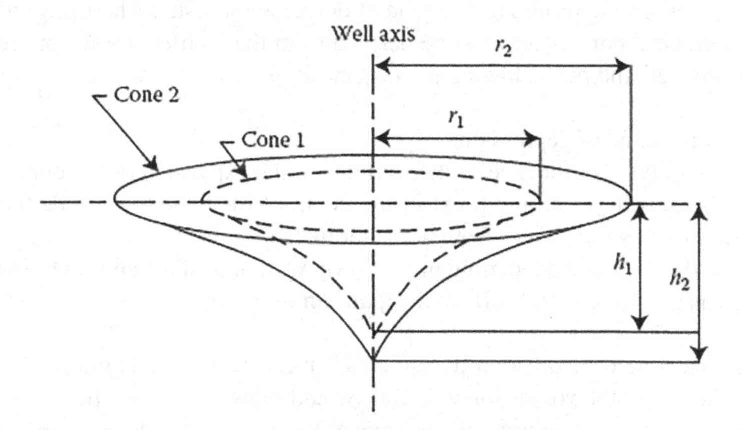

FIGURE 6.17 Changes in radius and depth of cone of depression (Karamouz, M., Ahmadi, A., and Akhbari, M.. *Groundwater Hydrology: Engineering, Planning, and Management* (p. 750) 2nd Ed., CRC Press, Boca Raton, FL, (2020)).

6.2.8.2 Confined Flow

A confined aquifer with steady-state radial flow to the fully penetrating well being pumped is shown in Figure 6.18. Thus, farther away from the well, the flow velocity decreases, the surface is flatter, and piezometric surface is almost static. For a homogeneous, isotropic aquifer shown in Figure 6.19, the well discharge at any radial distance r from the pumped well is

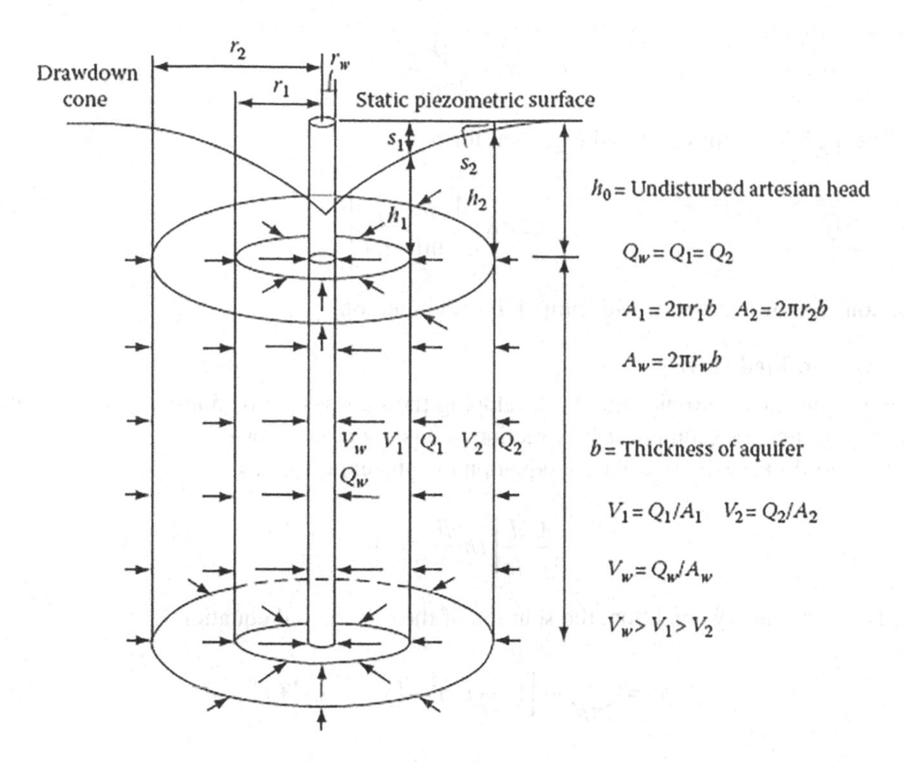

FIGURE 6.18 Flow distribution to a discharging well in a confined aquifer. (From U.S. Bureau of Reclamation, *Groundwater Manual*, U.S. Government Printing Office, Denver, CO, 1981.)

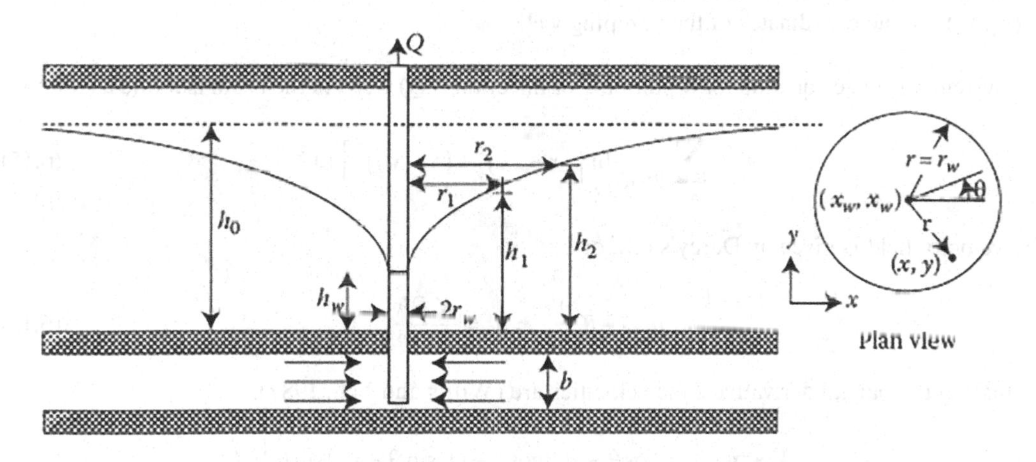

FIGURE 6.19 Well hydraulics for a confined aquifer (Karamouz, M., Ahmadi, A., and Akhbari, M.. *Groundwater Hydrology: Engineering, Planning, and Management* (p. 750) 2nd Ed., CRC Press, Boca Raton, FL, (2020)).

$$\sum_{CS} \rho V dA = Q = 2\pi K r b \frac{dh}{dr} \tag{6.9}$$

Equation (6.10) can be presented for boundary conditions of $h = h_1$, at $r = r_1$, and $h = h_2$, at $r = r_2$:

$$\int_{h_1}^{h_2} dh = \frac{Q}{2\pi K b} \int_{r_1}^{r_2} \frac{dr}{r} \tag{6.10}$$

$$h_2 - h_1 = \frac{Q}{2\pi Kb} \ln \frac{r_2}{r_1} \tag{6.11}$$

Solving Equation (6.12) for Q gives the general form of:

$$Q = 2\pi Kb \left[\frac{h - h_1}{\ln(r/r_1)} \right] \tag{6.12}$$

The equation is known as the equilibrium or Theim equation.

6.2.8.3 Unconfined Flow

In a homogeneous and isotropic aquifer, developing the steady-state response equation relating the velocities (q_r, q_θ) and the groundwater extraction rate (s), Q is as follows.

In radial coordinates, the Boussinesq equation may be expressed as

$$\frac{1}{r}\frac{d}{dr}\left[rh \frac{dh}{dr} \right] = 0 \tag{6.13}$$

With the Darcy boundary condition, the solution of the differential equation is:

$$h^2 = \frac{Q}{2\pi K} \ln \left[(x - x_w)^2 + (y - y_w)^2 \right] + C \tag{6.14}$$

where,

C is the constant of integration and
(x_w, y_w) are the coordinates of the pumping well.

The system response equation for w pumping or injection ($-Q$) wells in the basin is found as follows:

$$h^2 = \sum_w \frac{Q_i}{2\pi K} \ln \left[(x - x_w)^2 + (y - y_w)^2 \right] + C \tag{6.15}$$

The velocity field is given by Darcy's law, or

$$q_r = -K\frac{\partial h}{\partial r}, \quad q_\theta = \frac{K}{r}\frac{\partial h}{\partial \theta} \tag{6.16}$$

Evaluating the partial derivatives, the velocities are (Willis and Yeh, 1987):

$$q_r = -\frac{K}{\pi} \sum_w Q_i \left\{ \frac{(r\cos\theta - x_w)\cos\theta + (r\sin\theta - y_w)\sin\theta}{(r\cos\theta - x_w)^2 + (r\sin\theta - y_w)^2} \right\} \Big/ (2h) \tag{6.17}$$

$$q_\theta = \frac{K}{\pi} \sum_w Q_i \left\{ \frac{(r\cos\theta - x_w)(-r\sin\theta) + (r\sin\theta - y_w)r\cos\theta}{(r\cos\theta - x_w)^2 + (r\sin\theta - y_w)^2} \right\} \Big/ (2rh) \tag{6.18}$$

The velocities, in this case, are nonlinear functions of the pumping or injection rates.

By rearranging Equation (6.19),

$$hdh = \frac{Q}{2\pi K}\frac{r}{dr} \tag{6.19}$$

Integrating between lines r_1 and r_2 where the water table depths are h_1 and h_2, respectively, and rearranging,

$$Q = \frac{\pi K\left(h_2^2 - h_1^2\right)}{\ln\left(r_2/r_1\right)} \tag{6.20}$$

In addition, for a confined groundwater system can be derived as follows:

$$Q = \frac{\pi K\left(h_2^2 - h_1^2\right)}{\ln\left(r_2/r_1\right)} = \frac{\pi K\left(h_2 - h_1\right)\left(h_2 + h_1\right)}{\ln\left(r_2/r_1\right)} = \frac{2\pi Kb\left(h_2 - h_1\right)}{\ln\left(r_2/r_1\right)} \tag{6.21}$$

For further information and details, see Chapter 4 of Karamouz et al. (2020).

6.2.9 Urban Storage Reservoirs and Tanks

Storage tanks are the main components of urban water distribution networks and are necessary to supply variable water demand, to provide fire protection, and for emergency needs. Surface reservoirs, standpipes, and elevated tanks are three types of reservoirs used in the urban water distribution networks. Surface reservoirs are situated where they could provide adequate water pressure. They are usually covered to avoid contamination. Sufficient pressure will be provided by the natural elevation on a hill or through the use of pumps. Standpipes are basically tall cylindrical tanks. The upper portion of these tanks is used for storage to produce the necessary pressure head and the lower portion serves to support the structure. Standpipes over 15 m in height are uneconomical, and above this height, elevated storage tanks are the preferred choice.

High-lift pumps at the treatment facilities are not normally designed to meet variation of residential water demand; the usual practice is to pump water into the distribution system at a fixed rate for a given period and allow a reservoir to either supply extra water if the demand exceeds this rate or receive water if the demand is less than the pumping rate. In this way, reservoirs accomplish their regulating function by hydrostatic pressure alone. In large cities, reservoirs could be located in the middle of several distribution areas. How the location of a reservoir affects its capability to balance operating pressures throughout a distribution system is shown in Figure 6.16. Note how high water use and the accompanying friction losses increase the slope of the pressure profile so that water starts to flow from the reservoir to the surrounding area.

The slope of the hydraulic grade line from the pump to the tank decreases by decreasing the demand and allowing water to enter the tank and recharge the storage. Recently, elevated tanks have become less popular, partially due to their increased cost and the availability of relatively inexpensive variable speed pumps and controls, which make it possible to adjust pumping rates to varying demand.

The type and location of storage and storage size must be selected and determined based on the population (water demand) and the purpose of the storage (Figure 6.20). The sum of the three volumes for balancing, fire, and emergency is the storage capacity provided in a municipal water supply system. It will normally be about one day's average consumption. Volumes for the three purposes are calculated separately according to the required time period. Operating/equalizing storage is used to meet variable water demands while maintaining adequate pressure on the system. Where information on water demand is accessible, storage volume can be calculated or found graphically (e.g., from a mass diagram). When no information is available, operating storage is taken to be 15%–25% of maximum daily consumption.

Fire storage is calculated by taking the product of fire flow and fire duration. The values suggested by the National Fire Protection Association (NFPA) are given in Table 6.2. Fire flow capacity may be elevated or lowered depending on the reliability of the water supply source. For instance, a

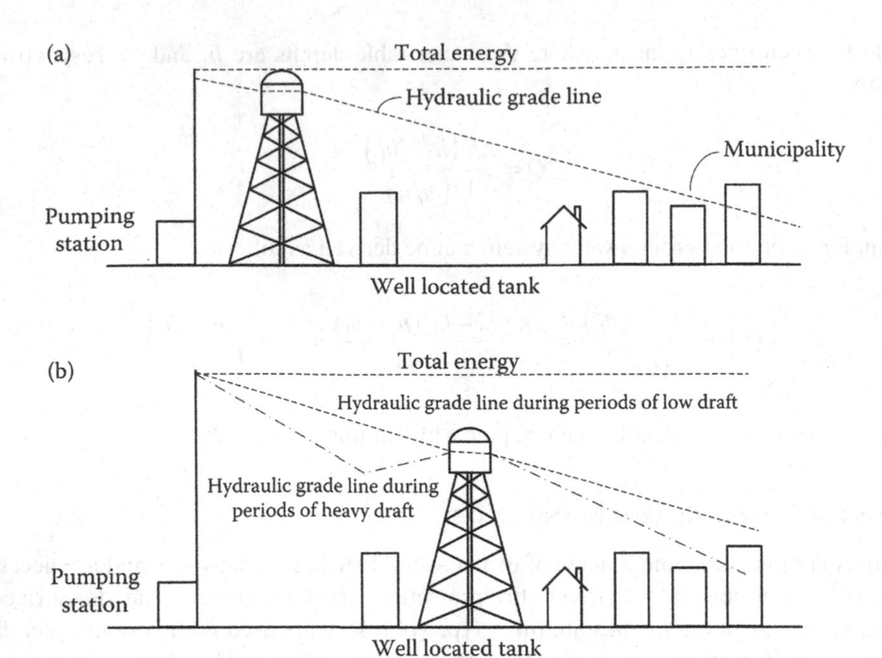

FIGURE 6.20 Effect of water storage reservoir location on pressure distribution.

TABLE 6.2
Duration of Required Fire Flow

Required Fire Flow (L/s)	Duration (hours)
160 or less	2
190	3
220	3
250	4
280	4
320	5
350	5
380	6
440	7
550	8
570	9
630 or more	10

municipality may increase its flow storage capacity if a water source such as a single well is used. Emergency storage of up to five times the maximum daily demand is suggested by the Insurance Advisory Organization to provide water during shutdowns for maintenance or repair to the system. This is rarely done in practice, and emergency storage is usually estimated to be one-quarter to one-third of the sum of the operating and fire capacity requirements (Gupta, 1989).

6.2.10 Water Transfers and Conveyance Tunnels

Compounded by issues associated with aging infrastructure and growing water needs, there is an increasing need for transfer tunnels requiring connections to new and existing water sources.

Water conveyance tunnels require special considerations regarding friction losses, drop shafts for vertical conveyance, air removal, control of infiltration and exfiltration, tunnel linings, and connections to existing/under operation installations/tunnels (Westfall, 1996). For potable water transmission, the engineering of water conveyance tunnels requires an appreciation of a range of construction and operational considerations, including:

- Subsurface tunneling conditions
- Groundwater hydrology
- In situ ground stresses as related to conveyance leakage
- Hydraulics of confined and unconfined flow
- Transient pressures
- Lining requirements for long-term serviceability and maintainability
- Head losses related to tunnel surface roughness

The most common finished interior surfaces of water conveyance tunnels may be categorized as follows: Case I. The tunnel is excavated by drill-and-blast methods and is left unlined (Manning coefficient: 0.038). Case II. The tunnel is excavated by the use of a full-face tunnel boring machine (TBM) and is left unlined (Manning coefficient: 0.018). Case III. The tunnel is lined with precast concrete segments (Manning coefficient: 0.016). Case IV. The tunnel is lined with cast-in-place concrete (Manning coefficient: 0.013). Designers of water conveyance tunnels should consider the economic value of presenting alternative tunnel liners and, hence, tunnel diameters in their designs (Westfall, 1996). Water conveyance tunnels are designed with a flow velocity of about 10 ft/s (3 m/s) and a maximum velocity not exceeding 20 ft/s (6 m/s).

6.3 WATER TREATMENT PLANTS

A water treatment system is operated to bring raw water up to drinking water quality. Producing safe water free of microbial pathogens is critical for any water source, but surface water has a much greater chance of microbial contamination. The particular type of treatment equipment required to meet these standards depends largely on the source of water. Most large cities rely more heavily on surface water, whereas most small towns or communities depend more on groundwater. Typically surface water treatment focuses on particle removal, and groundwater treatment focuses on removal of dissolved inorganic contaminants and reduction of water hardness (Masters and Ela, 2008).

6.3.1 Water Treatment Infrastructure

The raw water from surface and groundwater resources should be treated before they can be used in a municipal distribution. Many smaller communities are using groundwater only, but most of the larger cities are utilizing both. The treatment elements are called unit operators with three distinct functions of removing the floated, suspended and inorganic dissolved materials, disinfections, and sludge removal. A typical treatment of surface water includes the following as shown in Figure 6.21:

1. *Screening* and *grit removal* take out relatively large floating and suspended debris and the sand and grit that settle very rapidly which may damage equipment.
2. *Primary sedimentation* (also called *settling* or *clarification*) removes the particles that will settle out by gravity alone within a few hours.
3. *Rapid mixing* and *coagulation* use chemicals and agitation to encourage suspended particles to collide and adhere into larger particles.
4. *Flocculation*, which is the process of gently mixing the water, encourages the formation of large particles of floc that will more easily settle.
5. *Secondary settling* slows the flow enough so that gravity will cause the flocto settle.

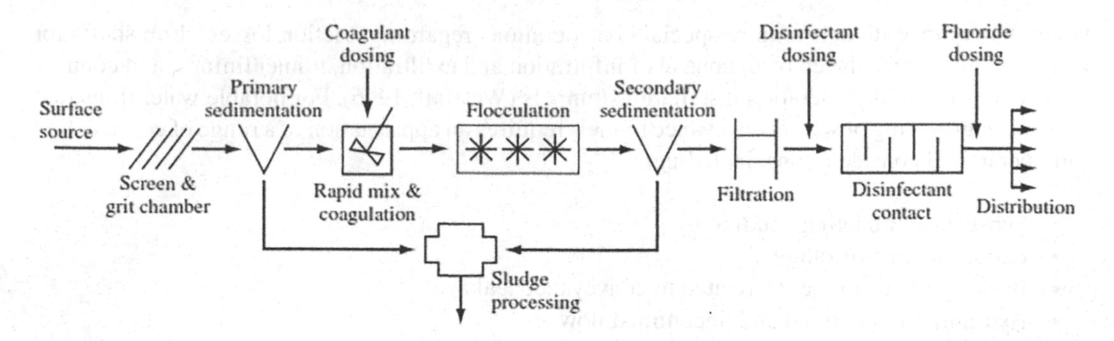

FIGURE 6.21 Schematic of a typical surface water treatment plant. (From Masters, G.M., and Ela, W.P., *Introduction to Environmental Engineering and Science* (No. 60457). Prentice Hall, Englewood Cliffs, NJ, 2008.)

6. *Filtration* removes particles and floc that are too small or light to settle by gravity.
7. *Sludge processing* refers to the dewatering and disposing of solids and liquids collected from the settling tanks.
8. *Disinfection contact* provides sufficient time for the added disinfectant to inactivate any remaining pathogens before the water is distributed.

Groundwater has less particles and pathogens than surface water, and in some instances, it is delivered after disinfection. However, because groundwater often moves through the soils mixed with minerals before withdrawal, it usually contains some levels of dissolved material and or volatile gases. The most common dissolved solid contaminants are calcium and magnesium, which are called hardness. They can be removed through some softening steps similar to the particle removal in surface water treatment. The following unit operations as shown in Figure 6.22 are typical groundwater treatment process:

1. *Aeration* removes excess and objectionable gases.
2. *Flocculation* (and *precipitation*) follows chemical addition, which forces the calcium and magnesium above their solubility limits.
3. *Sedimentation* removes the hardness particles that will now settle by gravity.
4. *Recarbonation* readjusts the water pH and alkalinity and may cause additional precipitation of hardness-causing ions.
5. *Filtration*, *disinfection*, and *solids processing* serve the same purposes as for surface water treatment.

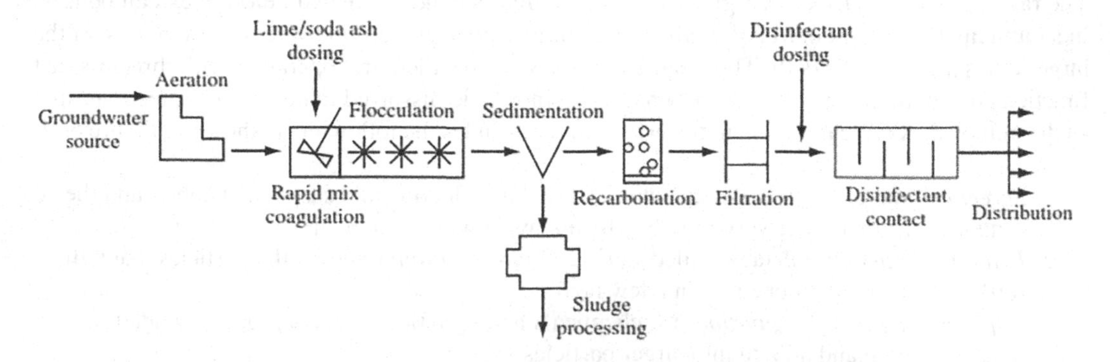

FIGURE 6.22 Schematic of a typical groundwater treatment plant. (From Masters, G.M., and Ela, W.P., *Introduction to Environmental Engineering and Science* (No. 60457). Prentice Hall, Englewood Cliffs, NJ, 2008.)

In particular, the *CT* (*CT* value is the product of the concentration of disinfectant agent in the water and the time of contact) values for 99% (2-log) and 99.99% (4-log) inactivation of bacteria and viruses for various disinfectant agents are reported. The concept of disinfectant concentration and contact time is important to the understanding of disinfection kinetics and the practical application of *CT* concept (which is defined as the product of the residual disinfectant concentration *C*, expressed in mg/L, and the contact time *T*, expressed in minutes, that residual disinfectant is in contact with water) is important. *CT* value represents an operative parameter, and it is an indicator of the effectiveness of the disinfection process.

Comparing the above unit operations, there is a major difference between surface and ground water treatment as illustrated in Figures 6.21 and 6.22. More details of different unit operations are provided in the next section.

6.3.2 Unit Operations of Water Treatment

An alternative system delivers raw source water, or just primarily treated water, directly to the consumer, who further treats this water in small, local treatment units near the point of consumption. The avoided deterioration of the water quality during transport and each consumer's treating the water to the level specifically needed for their requirements are the main advantages of this system. Obviously, for purposes such as drinking water, process water for industry, cooling water, irrigation water, or water for flushing toilets, the water quality is not the same. Small and locally used microfiltration units are quickly becoming competitive in price.

Adding pretreatment and some chlorine to the treated water ozonation and UV irradiation in order to maintain chlorine residuals during transport in the distribution network and prevent growth of bacteria originating from biofilms found in the pipe system is needed before discharging disinfected water into the distribution system. As a common practice, a measure of safety is the residual chlorine concentration at the tap (Fig. 6.23). Pre-cholorination is replaced by Ozonation in newer plants. In this chapter, through two simple examples, some of the sizing characteristics of water treatment are presented. The detailed discussion of water treatment is beyond the scope of this book and can be found in related textbooks such as Metcalf and Eddy (2003).

Four major common unit operations in surface and groundwater are discussed in the following sections.

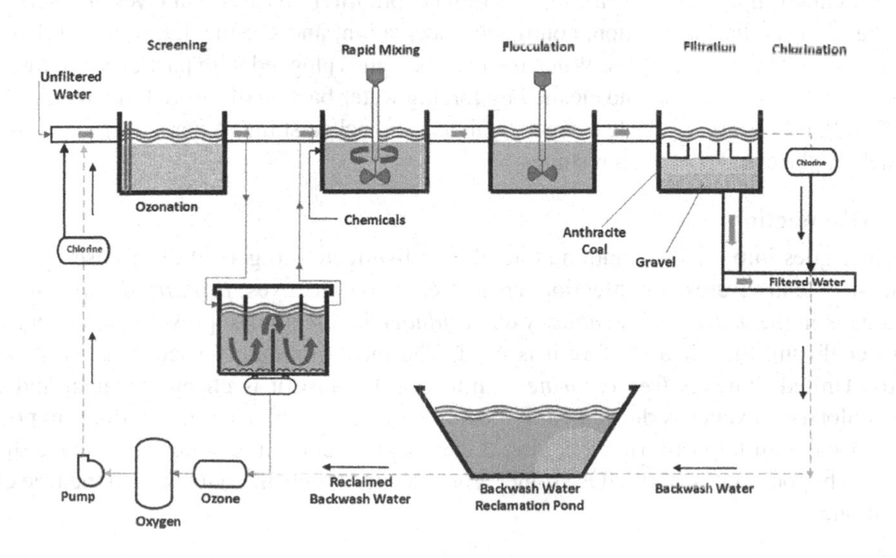

FIGURE 6.23 Water treatment processes including backwash.

6.3.2.1 Coagulation/Flocculation

Most colloids and nonsettleable particles of interest in water treatment remain suspended in solution because they have a net negative surface charge that causes the particles to repel each other. The goal of coagulation is to alter the particle surfaces in such a way as to permit them to adhere to each other. Thus, they can grow to a size that will allow removal by sedimentation or filtration. Coagulation is considered to be a *chemical* treatment process that destabilizes particles (makes them "sticky"), as opposed to a *physical* treatment operation such as flocculation, sedimentation, or filtration. Coagulation is adding liquid aluminum sulfate or alum and/or polymer to raw or untreated water.

6.3.2.2 Sedimentation

After the coagulation/flocculation, water goes into sedimentation basins. There, water moves slowly, making the heavy floc particles settle to the bottom. Floc that accumulates on the bottom is known as sludge. This is carried on to drying lagoons. Sedimentation or gravitational settling of particles from water is one of the oldest and simplest forms of water treatment. Simply allowing water to sit quietly into a reservoir, decanting the water, and then using the undisturbed surface water often considerably improves the water's quality. Sedimentation can remove particles that are contaminants themselves or may harbor other contaminants, such as pathogens or adsorbed metals. Although particles have very irregular shapes, their size may be described by an equivalent diameter that is determined by comparing them with spheres having the same settling velocity.

A sedimentation basin or clarifier is a large circular or rectangular tank designed to hold the water for a long enough time to allow most of the suspended solids to settle out. The longer the detention time, the bigger and more expensive the tank must be, but correspondingly, the better will be the tank's performance. The equivalent diameter is the *hydrodynamic diameter* when we speak of particles settling in water and *aerodynamic diameter* for particles settling in air.

6.3.2.3 Filtration

Filtration is one of the most widely used and effective means of removing small particles from water. This also includes pathogens, which are essentially small particles. In filtration, water passes through a filter, which is made to take away particles from the water. Such filters are composed of gravel and sand or sometimes crushed anthracite. Filtration gathers together impurities that float on water and boosts the effectiveness of disinfection. For drinking water filtration, the most common technique is called *rapid depth filtration*. The rapid depth filter consists of a layer or layers of carefully sieved filter media. Adsorption, continued flocculation, and sedimentation in the pore spaces are important removal mechanisms. When the filter becomes clogged with particles, the filter is shut down for a short period of time and cleaned by forcing water backward through the media (Masters and Ela, 2008). Filters are regularly cleaned by means of backwashing. After a backwash, the media settles back in place and operation resumes.

6.3.2.4 Disinfection

Before water goes into the distribution system, it is disinfected to get rid of disease-causing bacteria, parasites, and viruses. Disinfection has to meet two objectives: *primary disinfection,* to kill any pathogens in the water, and *secondary* (or *residual*) *disinfection* to prevent pathogen regrowth in the water during the period before it is used. The most commonly used method of disinfection in the United States is *free chlorine* disinfection because it is cheap, reliable, and easy to use. Free chlorine in water is developed by dosing with either chlorine gas, sodium hypochlorite (NaOCl), or calcium hypochlorite. The dosed chemical reacts in the water to produce dissolved chlorine gas, hypochlorous acid (HOCl), and hypochlorite, which all contribute to the free chlorine concentration.

Example 6.3

Assume a water treatment plant for a city of 100,000 people with maximum hourly and daily rates of water demand equal to 6.87×10^6 liters per hour (Lph) and 99.0×10^6 liters per day (Lpd), respectively. The detention time for coagulation/flocculation (A) is 25 minutes and the tank is 3.7 m deep. The detention time for the sedimentation tank (B) is 2 hours and the tank is 5.0 m deep. The filtration rate (C) is 110 liters per minute (Lpm)/m^2. Select the appropriate dimensions for these three units. Three parallel sets of tanks are used to provide flexibility in operation and each set includes four filters. Consider that the widths for all the treatment units are equal to 18 m.

Solution:

The required processing rate is the maximum daily water demand. Each tank will handle one-third of this flow or 33×10^6 Lpd (22,916 Lpm). Accordingly, the required capacity of the coagulation/flocculation tank is

$$25\,\text{min}\times 22{,}916\,\text{Lpm} = 572.9\times10^3\,\text{L} = 572.9\,\text{m}^3.$$

So, length A of the coagulation/flocculation tank is 572.9/(18×3.7)=8.6 m.

The required capacity of the sedimentation tank is

$$120\,\text{min}\times 22{,}916\,\text{Lpm} = 2749.9\times10^3\,\text{L} = 2749.9\,\text{m}^3.$$

So, length *B* of the sedimentation tank is 2749.9/(18×5)=30.6 m.

Each filter will handle one-twelfth of the total flow or 5,729 Lpm. Thus, the required area of each filter is 5,729/110=52.1 m^2.

And the length *C* of each filter box is 52.1/4.5=11.6 m.

Example 6.4

Calculate (a) the number of kilograms of chlorine needed per day and (b) the capacity of the contact tank in the water treatment plant for the city of 100,000 people given in the previous example. The chlorine demand is 1 mg/L. It should be noted that at least 1.2 mg of chlorine must be added to every liter to overcome the chlorine demand of 1 mg/L and produce a free available chlorine concentration of 0.2 mg/L.

Solution:

(a) Since the treatment plant must be capable of operating at the maximum daily flow rate, the amount of chlorine needed can be calculated as

$$\frac{\text{kg of chlorine}}{\text{day}} = \text{maximum daily flow rate}\times\frac{1.2\text{ mg of chlorine}}{\text{L}}\times\frac{\text{kg}}{1\times10^6\text{mg}}$$

$$= 99.0\times10^6\text{Lpd}\times1.2\text{ mg/L}\times\frac{1}{10^6}\text{kg / mg} = 118.8\text{ kg/day}.$$

(b) If we assume a minimum contact time of 40 minutes, then

$$\text{Required capacity of the contact tank} = \text{flow rate}\times\text{contact time}$$

$$= 99.0\times10^6(\text{Lpd})\times(\text{day/1,440 minutes})\times40(\text{minutes})$$

$$= 2.75\times10^6\text{L} = 2{,}750\text{ m}^3.$$

6.4 WATER DISTRIBUTION SYSTEM

Providing the desired quantity of water to the appropriate place at a suitable time with acceptable quality is the basic function of water distribution systems. Distribution piping, storage tanks, and pumping stations are three major components of urban water distribution systems. These components can be further divided into subcomponents. For example, structural, electrical, piping, and pumping units are subcomponents of the pumping station component. The pumping unit can be divided into a pump, a driver, controls, power broadcasting, and piping and valves. The definition of components depends on the level of detail of the required analysis and the available data. A hierarchy of building blocks is used to construct the urban water distribution system. The relationship between components and subcomponents is summarized in Figure 6.24.

Pipes, valves, pumps, drivers, power transmission units, controls, and storage tanks are seven sub-subcomponents that can be readily identified for analysis. The pumping unit subcomponent is composed of pipes, valves, a pump, a driver, power transmission, and control sub-subcomponents. The reliability of the urban water distribution systems is elevated by three subcomponents: pumping units, pipe links, and storage tanks. Distribution piping is either branched as demonstrated in Figure 6.25 or looped as presented in Figure 6.26, or is a combination of branched and looped.

6.4.1 System's Components

The most widely used elements in the network are pipe sections, and these contain fittings and other appurtenances, such as valves, storage facilities, and pumps. Pipes are the largest capital investment in a distribution system; they are manufactured in different sizes and are composed of different materials, such as asbestos cement, steel, cast or ductile iron, reinforced or prestressed concrete, polyethylene, polyvinyl chloride, and fiberglass. The American Water Works Association (AWWA) publishes standards for pipe construction, installation, and performance. It is called the C-series standards (continually updated).

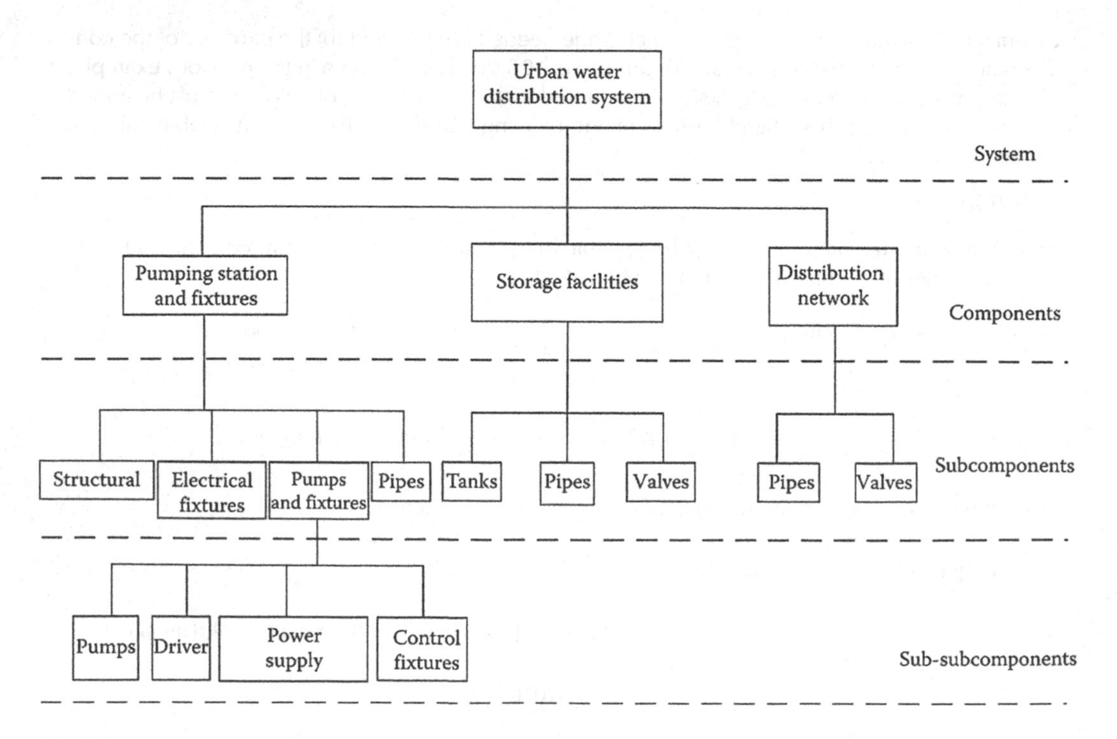

FIGURE 6.24 Different levels of hierarchies in a water distribution system.

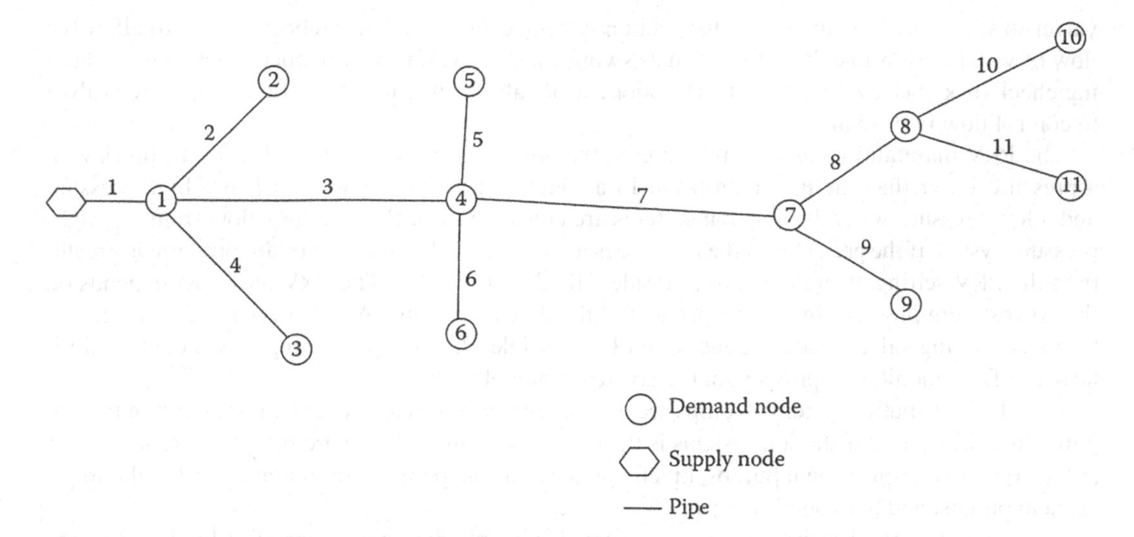

FIGURE 6.25 A typical branch distribution system.

Each end of a pipe is called a node. Junction nodes and fixed grade nodes are two kinds of nodes in a water distribution network. Junction nodes, where the inflow or the outflow is known to have lumped demand, may vary with time. Nodes to which a reservoir is added are referred to as fixed grade nodes. These nodes can take the form of tanks or large constant pressure mains.

The flow or pressure in water distribution systems is regulated by control valves. If conditions exist for flow reversal, the valve will close and no flow will pass. The pressure-reducing or relief (pressure-regulating) valve (PRV) is the most common sort of control valve that is placed at pressure zone boundaries to dissipate pressure. There are many other types of valves, including isolation

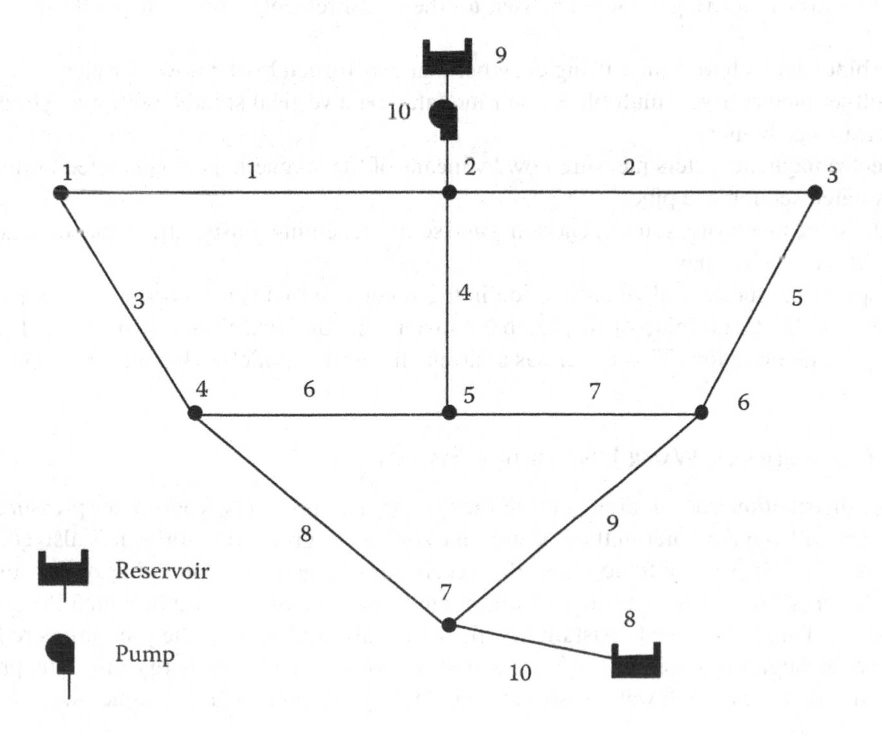

FIGURE 6.26 A typical looped water distribution network.

valves to shut down a segment of a distribution system; direction control (check) valves to allow the flow of water in only one direction, such as swing check valves, rubber flapper check valves, slanting check disk check valves, and double-door check valves; and air release/vacuum breaker valves to control flow in the main.

The PRV maintains a constant pressure on the downstream side of the valve for all flows with a pressure lower than the upstream head. In a water distribution network, when a high-pressure and a low-pressure water distribution systems are connected, the PRV permits flow from the high-pressure system if the pressure on the low side is not excessive. If the downstream pressure is greater than the PRV setting, then the pressure inside will close the valve. The PRV head loss depends on the downstream pressure and is independent of the flow in the pipe. Another type of check valve, a horizontal swing valve, operates under a similar principle. Pressure-sustaining valves operate similarly to PRV's monitoring pressure at the upstream side of the valve.

In water distribution systems, pumps are used to enhance the energy. The most common type of pump used in water distribution systems is the centrifugal pump. There are many different types of pumps (positive displacement pumps, kinetic pumps, turbine pumps, horizontal centrifugal pumps, vertical pumps, and horizontal pumps).

In an urban water distribution system, to provide supply during outages of individual components, to equalize pump discharge near an efficient operating point in spite of varying demands, to provide water for firefighting, and to dampen out hydraulic transients, water storage in the system is needed (Walski, 1996). Distribution is closely associated with the water tank. Tanks are usually used for water storage in a water distribution network and are made of steel. The water tank is used to supply water to meet the requirements during high system demands or during emergency conditions when pumps cannot adequately satisfy the pressure requirements at the demand nodes. The water storage tanks can be at ground level or at a certain elevation above the ground. If, at all times, a minimum quantity of water is kept in the tank, unexpectedly high demands cannot be met during critical conditions. The higher the pump discharge, the lower the pump head. Thus, during a period of peak demand, the amount of available head is low.

A wide array of metering devices are used for the measurement of water mains flow:

- Turbine meters have a measuring chamber that gets turned by the flow of water.
- Multijet meters have a multiblade rotor mounted on a vertical spindle within a cylindrical measuring chamber.
- Electromagnetic meters measure flow by means of the magnetic field generated around an insulated section of a pipe.
- Ultrasonic meters use sound-generating and sound-receiving sensors (transducers) attached to the sides of the pipe.
- Proportional meters utilize abstraction in the water line to divert a portion of water into a loop that holds a turbine or displacement meter. The diverted flow is proportional to the flow in the main line. This meter has a turbine meter in parallel with a multijet meter.

6.4.2 Hydraulics of Water Distribution Systems

In a water distribution system, almost all of the principal components work under pressure. When the pipes are full and the water in the system is moving under gravitation force, it is also considered as pressure flow. The sewage force mains that receive discharge from a pumping station also carry flows under pressure. There is a range of minimum to maximum pressure in which these components operate. Due to frictional resistance by the pipe walls and fittings, the pressure is reduced as water flows through the system. This is measured in terms of the head/energy loss. The principles of hydraulics that are used in water distribution modeling are described in Chapter 4.

6.4.3 Water Supply System Challenges

A number of key challenges required for the management of urban water systems are common to all towns and cities. They include environmental, social, and economic dimensions, but many of the underlying causes are interrelated and overlapping.

Other major challenges suggested by Mays (1996) are as follows:

- Inadequate water flows from excessive and inefficient water use.
- Infection of surface waters and groundwater from uncontrolled or deficiently directed stormwater drainage and wastewater disposal.
- Consumers and ratepayers have increasing expectations about the provision and quality of water services, but there is often a negative reaction to large rate increases or increased charges to fund the required infrastructure development.
- Lack of awareness and understanding of the value of urban water systems and the costs of improving water supplies, wastewater, and stormwater management.
- Poor recreational and bathing water quality, and poor information disclosure.
- Lack of investment and deferred maintenance, in part through incomplete pricing and inadequate financial contributions from new urban developments.
- Institutional and regulatory barriers to improved management.
- Potential risk of infrastructure failure.

6.5 URBAN WATER DEMAND MANAGEMENT

UWDM is a key strategy for achieving the Millennium Development Goals (MDGs) for supplying potable water and sanitation for all the people in an urban area. In a national arena, IWRM strategies and water efficiency plans should be developed before implementing UWDM strategies. A balanced set of objectives of UWDM for the management and allocation of water resources are efficiency, equity, and sustainability. Successful implementation of UWDM as a component of IWRM plays a significant role in the reduction of poverty in society through more efficient and equitable use of the available water resources and by facilitating water service through municipal water supply agencies in a region. UWDM is normally referred to as one of the most useful tools in achieving IWRM (Goldblatt et al., 2000).

UWDM covers a wide range of technical, economic, educational, capacity-building, and policy measures that are useful for municipal planners, water supply agencies, and consumers. UWDM is particularly a major challenge in developing countries. Few important attempts are made by international agencies to allocate water to competing end users and sectors in an equitable fashion.

Important health advantages accrue from adequate access to clean safe water that could lead to reduction of waterborne diseases. Lack of access to safe drinking water supplies and sanitation services is recognized as a poverty indicator, but existing data are inadequate as far as equity issues are concerned. It is important to note that UWDM does not always promote water usage reduction.

6.5.1 Basic Definitions of Water Use

Water use can be classified into two basic categories: consumptive and nonconsumptive uses.

Consumptive use occurs when water is an end in itself. Domestic, agricultural, industrial, and mining water uses are consumptive. Hydropower, transportation, and recreation are the main nonconsumptive water uses. They are also referred to as instream uses that serve the so-called environmental demand as well, in which a use is made of a water body without withdrawing water from it except the water used on-site for maintaining swamps and wetlands for wildlife habitats and ditching and ponding.

In an overall water accounting system, the following water uses are identified by the United States Geological Survey (USGS), (Solley et al., 1993):

1. Water withdrawal for offstream purposes
2. Water deliveries at the point of use or quantities released after use
3. Consumptive use
4. Conveyance loss
5. Reclaimed wastewater
6. Return flow
7. Instream flow

Water use can also be classified as offstream use and instream use. Different water demands resulting from offstream uses can be classified as follows (Karamouz et al., 2003a):

Domestic or municipal water uses include residential (apartments and houses), commercial (stores and businesses), institutional (hospitals and schools), industrial, and other water uses such as firefighting, swimming pools, lawning, and gardening.

These uses require withdrawal of water from surface or groundwater resources systems. Part of the water withdrawn may return to the system perhaps in a different location and time, with a different quality and period of time. The percentage of return flow is an important factor in evaluating the water use efficiency. This should also be considered in water resources management schemes. Table 6.6 shows a more detailed classification of water demands. Water use is also classified as municipal, agricultural, industrial, environmental, and infrastructure (public work). See Chapter 5 of Karamouz et al. (2010) for details.

6.5.2 Water Supply Quantity Standards in Urban Areas

Two terminologies that are commonly used in the texts related to water quantity and quality management are criteria and standards. McKee (1960) has differentiated between standards and criteria as follows.

The term "standard" applies to any definite rule, principle, or measure established by authority. The fact that it has been established by authority makes a standard rather rigid, official, or quasi-legal. This does not mean that the standard is fair, equitable, or based on sound scientific knowledge. A criterion designates a mean by which anything is tried, in forming a correct judgment concerning it. Unlike a standard, it carries no connotation of authority other than fairness and equity nor does it imply an ideal condition. Where scientific data are being accumulated to serve as yardsticks of water quality, without considering legal authority, the term "criterion" is most applicable.

Delivering required fire flows in different parts of the municipality with usage at the maximum daily rate should be considered in the designing of the layout of supply mains, arteries, and secondary distribution feeders. The water quantity standards should be given for the greatest effect that a break, joint division, or other kinds of main failure could have on the supply of water to a system. The minimum pressure in the water distribution main should be 690 kpa for a residential service connection. The maximum allowable pressure is 1,030 kpa (Walski et al., 2001).

The recommended water pressure in a distribution system is about 450–520 kpa, which is considered to be adequate to compensate for local fluctuations in consumption. This level of pressure can provide for ordinary consumption in buildings up to ten stories in height, as well as sufficient supply for automatic sprinkler systems for fire protection in buildings of four or five stories.

Storage is frequently used to equalize pumping rates into the distribution system as well as provide water for firefighting. In determining the fire flow from storage, it is necessary to calculate the rate of delivery during the specified period. Although the amount of storage may be more than adequate, the flow to a hydrant cannot surpass the carrying capacity of the mains, and the residual pressure at the point of use cannot be <140 kpa (Walski et al., 2001). The linear distance between hydrants along streets in residential districts is normally 180m with a maximum of 240m and in high-value districts is normally 90m with a maximum of 150m (Walski et al., 2001).

6.5.3 Water Demand Forecasting

Regression analysis is the most commonly used method for water use prediction. The basis for how the independent variables of the regression model should be selected is the available data about different factors affecting the water use and their relative importance in increasing or decreasing water uses. One of the most important factors in estimating water use in an urban area is the population of each subarea. Some of the variables are population, price, income, air temperatures, and precipitation (Baumann et al., 1998).

To forecast water demands, time series analysis has also been used. For this purpose, the time series of municipal water use and related variables are used to model the historical pattern of variations in water demand. More introductions to water demand forecasting and its different methods are presented in Chapter 5 of Karamouz et al. (2010).

6.5.4 Water Quality Modeling in a Water Distribution Network

Prior to the passage of the US SDWA of 1974, the focus of most water utilities was on treating water, even though it has long been recognized that water quality can deteriorate in the distribution system. However, after the SDWA was amended in 1986, a number of rules and regulations were amended which had a direct impact on water quality in distribution systems. More recently, there has been an increased focus on distribution systems and their importance in maintaining water quality in drinking water distribution systems. There has also been general agreement that the most vulnerable part of a water supply system is the distribution network. To decrease what was considered an unreasonably high risk of waterborne illness, the US Environmental Protection Agency (EPA) promulgated the Total Coliform Rule (TCR) and Surface Water Treatment Rule (SWTR) in 1989 (WHO, 2017).

Water quality models and their potential for use in tracking and predicting water quality movement and changes in water quality are examined.

6.5.4.1 Water Quality Standards

Water quality is evaluated by the physical, chemical, and biological characteristics of water and its intended uses. Water to be used for public water supplies must be drinkable and without any pollution. The presence of any unknown substance (organic, inorganic, radiological, or biological) that tends to degrade the water quality or impairs the usefulness of the water is defined as pollution.

The basis for defining water quality requirements and standards in urban areas is the drinking water security. Pollution control and health authorities give more attention to the protection of surface and groundwater resources. The water pollution control agencies in different countries use classification of the surface water quality standards in urban areas as stream standard, effluent standards, or a combination of them. These standards are associated with particular numerical limits for designated beneficial uses of water resources.

6.5.4.2 Water Quality Model Development

Modeling the movement of a contaminant or a chemical substance such as chlorine within the distribution systems, as it moves through the system from various points of entry (e.g., treatment plants) to water users, is based on three principles:

- Conservation of mass within differential lengths of the pipe.
- Complete and instantaneous mixing of the water entering pipe junctions.
- Appropriate kinetic expressions for the growth or decay of the substance as it flows through pipes and storage facilities.

Advection (movement in the direction of flow) and dispersion (movement in the transverse direction due to concentration difference) are the two important mechanisms for transportation of a substance. The basic equation describing advection–dispersion transport is based on the principle of conservation of mass and Fick's law of diffusion. For a nonconservative substance, the principle of mass conservation within a differential section of a pipe, that is, the control volume, can be stated as

$$|\text{Rate of change of mass in control volume}| = |\text{Rate of change of mass due to advection}|$$
$$+ |\text{Rate of change of mass due to dispersion}| + |\text{Transformation reaction rate}|$$

Considering the first-order reaction of the considered substance with other substances, conservation of mass is given by

$$\delta = -\frac{\partial C(x,t)}{\partial t} = -V\frac{\partial C(x,t)}{\partial x} + E\frac{\partial^2 C(x,t)}{\partial^2 x} - K_R\left[C(x,t)\right]. \tag{6.22}$$

where

$C(x, t)$ is the substance concentration (g/m^3) at position x and time t,
V is the flow velocity (m/s),
E is the coefficient of longitudinal dispersion (m^2/s), and
$K_R\ [C(x, t)]$ is a reaction rate expression in which the (−) sign reflects the decrease in concentration due to decay rate.

Dispersion of common substances such as chlorine is negligible in water distribution systems, and therefore, Equation (6.22) could be reduced to

$$\frac{\partial C(x,t)}{\partial t} = -V\frac{\partial C(x,t)}{\partial x} + -K_R\left[C(x,t)\right]. \tag{6.23}$$

In order to solve Equation 6.23 it is important to know C at $x=0$ for all times (a boundary condition) and the reaction rate expression, $K_R\ [C(x, t)]$.

Equation (6.24) represents the concentration of material leaving the junction and entering a pipe.

$$C_{ij} = \frac{\sum_k Q_{ki}C_{kj}}{\sum_k Q_{ki}}, \tag{6.24}$$

where

C_{ij} is the concentration at the start of the link connecting node i to j, in mg/L,
C_{kj} is the concentration at the end of a link, in mg/L, and
Q_{kj} is the flow from k to i.

Storage tanks can be modeled as completely mixed, variable volume reactors in which the change in volume and concentration over time is

$$\frac{dV_s}{dt} = \sum_k Q_{ks} - \sum_i Q_{sj}, \tag{6.25}$$

$$\frac{dV_sC_s}{dt} = \sum_k Q_{ks}Q_{ks} - \sum_i Q_{sj}C_s + k_{ij}(C_s), \tag{6.26}$$

where

C_s is the concentration for the tanks, in mg/L,
dt is the change in time, in s,
Q_{ks} is the flow from node k to s, in m^3/s,
Q_{sj} is the flow from node s to j, in m^3/s,
dV_s is the change in volume of the tank at nodes, in m^3,
V_s is the volume of the tank at nodes, in m^3,
C_{ks} is the concentration of the contaminant at the end of links, in mg/m^3, and
k_{ij} is the decay coefficient between nodes i and j in s^{-1}.

There are currently several models available for modeling both the hydraulics and water quality in the drinking water distribution system. However, most of this discussion will be focused on a US EPA developed hydraulic/contaminant propagation model called EPANET (Rossman et al., 1994), which is based on mass transfer concepts (transfer of a substance through another on a molecular scale). Islam (1995) developed a model called QUALNET, which predicts the temporal and spatial distribution of chlorine in a pipe network under slowly varying unsteady flow conditions.

6.5.4.3 Chlorine Decay

The chlorine decay in the pipe has two mechanisms. The first mechanism, which is considered as bulk decay, is the reaction of chlorine with other substances available in water. The second mechanism is the reaction of chlorine with substances of the pipe wall and is known as wall decay. The wall decay in distribution networks may be predominant where significant corrosion is present.

6.5.4.3.1 Bulk Decay

Bulk decay is mostly assumed to follow the first-order kinetics, which is formulated as

$$\frac{dC}{dt} = -k_b C, \tag{6.27}$$

or

$$C_t = C_0 e^{-k_b t}, \tag{6.28}$$

where

C_t is the concentration after time t,
C_0 is the initial chlorine concentration, and
k_b is the coefficient of bulk decay.

The bulk decay rate is measured by observing the chlorine concentration, at specified time intervals, from glass bottles that are previously filled with sample water. Then the coefficient of bulk decay is determined by the least-squares curve-fitting method. The bulk decay is a function of initial chlorine concentration as well as the temperature and total organic content (TOC) of water (Powell et al., 2000).

Example 6.5

In the branched network of Figure 6.27, there is a source node and three demand nodes (nodes l, 2, and 3). The pipe characteristics are given in Table 6.3. Assuming the fixed chlorine concentration at source node 1 is equal to 0.8 mg/L, obtain the steady-state concentration at different nodes. Assume that the overall decay rate constant is 6.417×10^{-6}s^{-1} for all pipes.

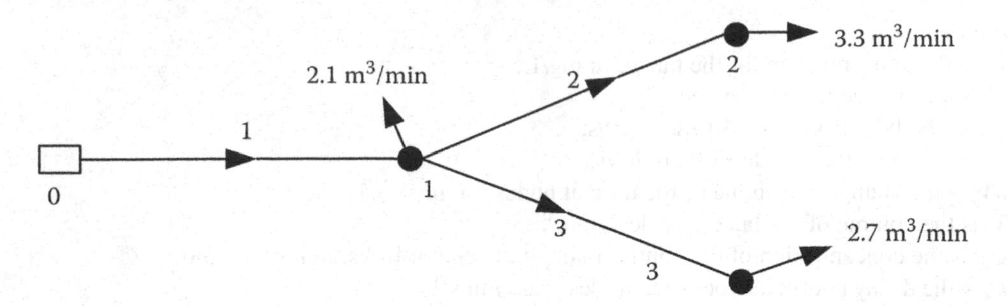

FIGURE 6.27 Branched network of Example 6.5.

TABLE 6.3
Pipe Details for the Branched Network of Example 6.5

Pipe Number	Length, L (m)	Diameter, D (m)	Discharge, Q (m³/min)	Area, A (m²)	Velocity, V (m/min)	Travel Time, t (minutes)
(1)	(2)	(3)	(4)	(5)	(6)	(7)
1	500	0.350	8.1	0.0962	84.20	5.938
2	350	0.200	3.3	0.0314	105.10	3.330
3	400	0.150	2.7	0.0177	152.54	2.622

Solution:

The pipe area, flow velocity, and pipe travel time are calculated for all the pipes as tabulated in columns 5–7 of Table 6.3. The chlorine concentration at the upstream of pipe l, Cj_u, is the same as that for the source node, that is, 0.8 mg/L. Since the travel time for chlorine in pipe 1 is 5.938 minutes, the chlorine concentration would remain constant downstream of pipe l, C_{1d}, after 5.938 minutes. Thus, chlorine concentration at the downstream end of pipe 1 is given by

$$C_{1d} = C_{1u}e^{-6.417\times10^{-6}\times5.938\times60} = 0.80.9977 = 0.7982 \text{ mg/L}$$

Considering mixing at node 1, the chlorine concentration at node 1 (C_1) and upstream of pipes 2 (C_{2u}) and 3 (C_{3u}) can be obtained. Since there is only one supply pipe and no addition of chlorine at node 1, the chlorine concentration at the upstream ends of pipes 2 and 3 would be the same as that at the downstream end of pipe 1. Thus, $C_1 = C_{2u} = C_{3u} = 0.7982$ mg/L.

Now, travel times in pipes 2 and 3 are 3.330 and 2.622 minutes, respectively. Therefore, the steady-state chlorine concentration at the downstream ends of pipe 2 (C_{2d}) and pipe 3 (C_{3d}) observed after 3.330 and 2.622 minutes, respectively, would be

$$C_{2d} = C_{2u}e^{-6.417\times10^{-6}\times3.330\times60} = 0.7982\times0.9987 = 0.7972 \text{ mg/L}$$

and

$$C_{3d} = C_{3u}e^{-6.417\times10^{-6}\times2.622\times60} = 0.7982\times0.9990 = 0.7974 \text{ mg/L}$$

Thus, the steady-state chlorine concentration at nodes 2 (C_2) and 3 (C_3) would be 0.7972 and 0.7974 mg/L, respectively. Total time to reach the steady-state chlorine concentration condition at nodes 2 and 3 would be 9.268 and 8.560 minutes, respectively.

6.5.5 Water Demand and Price Elasticity

In economic aspects, demand is considered to be a general concept that shows the willingness of users to purchase goods, services, or raw materials for the production of new goods. Commonly, it is considered that for a single consumer or group of consumers, there is an inverse relation between the water demanded and the price (cost) per unit. In other words, by increasing the water price, the quantity of demanded water decreases. It should be noted that the water requirement and demand are completely different. The water requirement is not affected by the price, but the water demand is a function of different factors and price is one of them.

The responsiveness of consumers' purchases to price variations is considered as demand elasticity. But the most common concept in water demand analysis is price elasticity, which is the percentage change in water demand in response to 1% change in price. Young (1996) defines the price elasticity of water demand as a measure of the willingness of consumers to give up water use in the face of rising prices, or conversely, the tendency to use more as price falls.

The price elasticity of water demand, η_p, is formulated as follows (Mays, 1996):

$$\eta_p = \frac{\Delta d}{\bar{d}} \div \frac{\Delta p}{\bar{p}}, \tag{6.29}$$

where

$\bar{d}$ is the average water demand,
$\bar{p}$ is the average price,
Δd is the demand variation, and
Δp is the price variation.

If a continuous demand function is considered, Equation (6.29) will be changed (Mays, 1996)

$$\eta_p = \frac{dd}{d} \div \frac{dp}{p}. \tag{6.30}$$

Example 6.6

Interpret the meaning of the following: The price elasticity of water demand is equal to −0.8.

Solution:

A price elasticity of −0.8 means that with a 1.0% increase in the price of water, the water demand will decrease by 0.8% if all other factors affecting the water demand remain constant. This will be valid only if the price–quantities data pairs are close to each other or a smooth demand function fitted to the known data.

The following typical equation is used to estimate the water demand based on price elasticity.

$$Q = cP^E \tag{6.31}$$

where

Q is the quantity of demand for water,
P is the price of water,
c is a constant, and
E is the elasticity of demand, which normally ranges between −1 and 0.

Example 6.7

Consider the following demand function for urban water demand.

$$\ln Q = -1.5 - 0.2P + 0.4\ln(I) - 0.2\ln(F) + 0.4\ln(H),$$

where
Q is the annual water demand (m^3),
P is the average water price (\$/m^3),
I is the average yearly household income (\$),
F is the yearly precipitation (mm), and
H is the average number of residents per meter. Determine the price elasticity of demand calculated at the mean price, \$3.42/m^3.

Solution:

The demand function can be rewritten as

$$Q = e^{-1.50-0.2P} I^{0.4} F^{-0.2} H^{0.4},$$

and the derivative of Q with respect to price is

$$\frac{dQ}{dP} = e^{-1.5-0.2P} I^{0.4} F^{-0.2} H^{0.4}(-0.2) = 0.2Q.$$

Price elasticity of demand is calculated as

$$\varepsilon = \frac{P}{Q}\frac{dQ}{dP} = \frac{P}{Q}(-0.2Q) = -0.2(3.42) = -0.68.$$

The price elasticity of −0.68 indicates that if the price of water increases by 1.0%, the water demand will decrease about 0.68%.

Equation (6.31) is difficult to apply for the water sector as a whole, but for certain subsectors (urban water use, industrial water use, and irrigation), it may serve the purpose of analyzing the effects of tariff changes. The problem with the equation is that E is not a constant. It depends on the price, the water use, and it varies over time. So it is an equation with limited applicability. When we approach the more essential needs of the user, the primary uses of water have a special characteristic in that the elasticity becomes rigid (inelastic; E close to 0). People need water, whatever the price. For the most essential use of water (drinking), few alternative sources of water are available. For sectors such as industry and agriculture, demand for water is generally more elastic (E closer to −1), which is more in agreement with the general economic theory. This is because alternatives for water use exist in these sectors (e.g., introducing water-saving production technologies and shifting to less water-demanding products/crops). For basic needs, however, demand is relatively inelastic or rigid. In urban water supply, elasticity is therefore generally close to 0, unless additional (nonfinancial) measures are taken.

6.6 HYDRAULIC SIMULATION OF WATER NETWORKS

The most commonly used simulation models for water network modeling include EPANET and QUALNET. The mentioned models are discussed in the following sections.

6.6.1 EPANET

EPANET was developed as a software package by the US EPA, Rossman (1999). It is a public domain, water distribution system modeling. This model is extended period simulation of hydraulic and water quality behavior within pressurized pipe networks and is designed to be "a research tool that improves our understanding of the movement and fate of drinking-water constituents within distribution systems."

EPANET for hydraulic simulation of water networks uses the Hazen–Williams formula, the Darcy–Weisbach formula, or the Chezy–Manning formula for calculating the head loss in pipes, pumps, valves, and minor loss. It is assumed that water usage rates, external water supply rates, and source concentrations at nodes remain constant over a fixed period of time, although these quantities can change from one period to another. The default interval period is 1 hour but can be set to any desired value. Various consumption or water usage patterns can be assigned to individual nodes or groups of nodes. For details of the software and modeling as well as the water quality module, see Chapter 12.

6.7 ASSESSING THE ENVIRONMENTAL PERFORMANCE OF URBAN WATER INFRASTRUCTURE

SD index (SDI) should be used in the international, regional, and local scales for assessing the performance of urban water infrastructure. An agency or an organization can turn to different methods to get information, which allows an understanding and improvement of its environmental performance. Examples of tools used for SDI assessment in urban infrastructure are life cycle assessment (LCA) and environmental performance evaluation [by using special performance indicators (PIs)]. While LCA is a technique for assessing the environmental aspects associated with a product and/or a service system, environmental PIs (EPIs) focus on those areas that are under direct control by the agency/organization. Upstream and downstream activities are seldom included. The system boundaries for LCA and EPI are therefore not usually comparable.

The relative importance of environmental indicators depends on local and regional factors. Percentages or dimensionless quotas are fairly easy to relate to, whereas an indicator such as drinking water quality is more difficult, at least for the public, and should therefore be related to a water quality standard. If water is scarce, then the specific water use and leakage are highly relevant in order to save water. Where water is in abundance, other indicators are more important. An indicator should be clearly understandable, without long explanations. The efficiency indicators have been ranked as having moderate importance, since these are not directly related to the objectives of the water sector but still apply to the definition of a sustainable urban water system.

Ease of understanding of indicators should be determined by the users, and this aspect still needs to be investigated.

The use of monetary indicators is a common procedure for companies wishing to monitor their performance or utility efficiency in comparison with others. The objective of the ISO standard is to provide guidelines on the choice, monitoring, and control of EPIs.

Another concept that has been used is the term "ecoefficiency," that is, producing more value with less environmental impact (Schaltegger, 1996). The prefix "eco" refers to both ecological and economic efficiency. Keffer et al. (1999) presented a set of defined general ecoefficiency indicators, as well as specifying guidelines for the selection of sector-specific indicators.

6.8 LIFE CYCLE ASSESSMENT

LCA is a method that aims to analyze and evaluate the environmental impacts of projects, products, or services. The whole chain of activities required for the production of a certain product or service is taken into consideration. Both emissions of potentially harmful substances from these activities and their consumption of natural resources are analyzed.

LCA can briefly be described as a three-step procedure (ISO 14040, 1997). These steps include

- Goal and scope definition, where the purpose of the study is presented and the system boundaries are defined. The functional unit, that is, the basis for comparison in the study, is also defined in this step.

- Life cycle inventory, the phase in which all information on the emissions and the resource consumption of the activities in the system under study is collected from various sources.

Life cycle impact assessment (LCIA) is where the environmental consequences of the inventory data are assessed. Often some kind of sensitivity analysis or discussion of uncertainties is added. LCIA is often described as an indicator system (Owens, 1999). For developing SDI using a Life Cycle Approach, refer to Chapter 8.

6.9 SUSTAINABLE DEVELOPMENT OF URBAN WATER INFRASTRUCTURES

Several attempts have been made to define the SD of urban water infrastructures including water resources and the wastewater system. The following definition has been proposed by American Society of Civil Engineers (ASCE, 1998):

> A sustainable urban water infrastructure should provide required services over a long time perspective while protecting human health and the environment, with a minimum use of scarce resources.

SD of urban water management would be started from an analysis of required services and to identify the needs satisfied by the water system (Larsen and Gujer, 1997). The services or the needs that the urban system should meet include

1. Reliable supply of safe water to all residents for drinking, hygiene, and household purposes
2. Safe transport and treatment of wastewater
3. Drainage of urban areas
4. Recovery of resources for reuse or recycling

Several publications present criteria for the SD of an urban water system, whose summary of the literature is divided into technical, environmental, economic, and social (including health) criteria.

- *Technical performance* can be considered in terms of two aspects, effectiveness and efficiency. Effectiveness is the extent to which the objectives are achieved, such as how well the organization fulfills the needs identified above. Efficiency is the extent to which resources are utilized optimally to fulfill these needs. Suggested indicators of effectiveness are coverage of service, drinking water quality, and percent of wastewater that enters the treatment plant and percentage removal of selected compounds.
- *Reliability, flexibility, and adaptability*: Reliable systems are those that can provide their service even when unexpected events occur such as an electricity delivery stop or a sudden temperature drop. Systems may fail but must be capable of recovering without undue effort or cost. Seasonal variations in loading and climate should be considered. The potential to change (flexibility) and the ability to change (adaptability) need to be promoted (Jeffrey et al., 1997).

  A basic concept of the SD of technology is the separation and selection of subsequent technologies. Separation of waste streams creates flexibility and provides a possibility for a combination of technologies. Municipal wastewater treatment plants (WWTPs) are constructed to treat the wastewater from households and from industries with similar components but not stormwater. In older urban areas, combined sewers are still a problem, causing sewer overflows and contamination of sludge.
- *Durability* refers to the long life of the technical system. The lives of the components water systems are long, 20–40 years, and even longer. A balance between endurance and flexibility is required since the components having long lifetime or high cost might decrease the flexibility.
- *Scale and degree of centralization* need to be considered. A local water supply or wastewater treatment might be advantageous where natural conditions are appropriate and population density is low. In large cities where space is limited but the economic situation and

potential for sophisticated technology are good, large-scale systems may be more appropriate and contribution to other technical systems such as the energy system relevant.

- *Environmental protection*: Environmentally acceptable systems are those that fulfill the objectives of prevention of contamination and a sustainable use of resources. Quantity and quality of water supply and receiving water should be restored and maintained. Emissions of nutrients and oxygen-demanding substances should be as low as is required to maintain the quality and diversity of ecosystems. Harmful substance, such as cleansing agents, chlorine and its compounds, and substances from traffic, should be avoided in order to protect aquatic life and meet quality directives for sewage sludge and prevent negative effects on agriculture and soil. Withdrawal, raw water quality, and protection are indicators that can be used for the assessment of water sources. Emissions to water from point sources and nonpoint sources are important indicators as well as emissions to air of some components.
- *Recycling and reuse of resources* are especially important for an essential and limited resource, such as phosphorus, but should also include other plant nutrients, water, organic compounds, energy, and chemicals. A high degree of recycling or reuse also ensures lower emissions to water and air. Quality of sludge and recycling of nutrients are indicators that can be used for this purpose.
- *Noise, odor, and traffic* are also important factors that need to be considered, for example, for the collection of sewage and transportation of sludge.
- *Cost-effectiveness affordability*: Economic dimensions can be looked upon from different perspectives: household, organization, and society. Households require an affordable service, an organization requires a cost-effective system, and society requires a stable but flexible system. The price of water is a key determinant of both the economic efficiency and the environmental effectiveness of water services.
- *Personnel requirements*: The time required for the operation and maintenance of the urban water system should be considered as well as the level of personnel skills. There is an increasing need, catalyzed by the ideas of SD, for civil engineers to embrace development in microbiology, chemistry, ecology, and IT. A diversity of humans in terms of education and gender are required to ensure a flexible and adaptable organization.
- *Demographics*: Population, population growth and density, specific water use, and coverage of population supplied are examples of traditional indicators that are used to assess future and existing needs.
- *Social dimensions* include the users as well as the staff. All humans have a need for healthy water supply and access to sanitation. Hygiene is a major consideration, and the urban water system should be safe for those who use and operate the system or take care of the final products. It should be convenient as well as socially and culturally acceptable. Some individuals are willing to take a large responsibility and spend considerable time and effort to contribute to a reduction of the environmental impact. Others are not willing or may not be able to do so. The systems should be able to function well under both circumstances. Freedom of choice is therefore important.
- *Awareness and promotion of sustainable behavior*: User behavior affects the function of an urban water system. The direct use of water and chemicals is the most obvious factor, but lifestyle also affects resource use and emissions. For example, if more people would eat a less protein-based diet, less nitrogen would be emitted through the wastewater system. Nonpoint source pollution depends to a high degree on traffic travel habits since pollution from car exhausts reaches rivers and lakes through contaminated stormwater.

6.9.1 Selection of Technologies

The development of the present system towards a more sustainable system may occur by improvements to existing systems and by developing new technology. A number of strategies can be identified, which are described in the following section.

6.9.1.1 Further Development of Large-Scale Centralized Systems

Large-scale, centralized water supply can become more sustainable by increasing the efficiency of treatment processes and decreasing the use of harmful chemicals by optimization of dosages or the use of alternative disinfection processes such as ozonation, nanofiltration, or biological processes. Water use can be decreased by the installation of water-saving appliances, repair of leakage, and by saving, recycling, and reuse. In regions suffering from by water shortage, reuse is an alternative to distant water reservoirs. Wastewater can be reclaimed for nonpotable purposes, such as toilet flushing and irrigation, and stormwater can be used to recharge groundwater and provide water areas in cities. When wastewater is used for irrigation, water and nutrients are recovered.

The centralized, large-scale wastewater system can be further developed towards higher effectiveness and a higher degree of removal through mechanical, biological, and chemical unit processes. Sewage sludge can be treated to extract and separate phosphorus, precipitation chemicals, and heavy metals. Heat may be recovered from the wastewater. Aerobic methods can be substituted by anaerobic methods to save electricity and produce biogas. Organic waste can be added to further increase biogas production. These strategies are by no means new but seem to be the most common way to improve the environmental performance of urban water systems in large cities.

6.9.1.2 Separation for Recycling and Reuse

The separation of flows to improve the opportunities for recycling and reuse has been emphasized by several investigators (Niemczynowicz, 1996; Butler and Parkinson, 1997) and has found to be feasible in both large-scale and small-scale systems. Separation of urine is a new technology that, apart from enabling the recycling of nutrients to agriculture, also has implications for treatment performance in wastewater systems. Human urine is the urban waste flow that contains the greater proportion of nutrients.

Separation of blackwater is also possible, enabling new technologies such as aerobic liquid composting or anaerobic digestion to function well, since blackwater is not diluted with gray water. The possibility of treating different waste streams makes the system flexible but is not yet a generally accepted approach. Whereas solid household waste is source separated into plastics, glass, metals and so on, the separation of liquid waste seems to be more problematic. The technical and economic difficulties and people's willingness to accept that idea should be tested. But before a new technology is introduced, it is important to ensure that it does not lead to new human or environmental health problems.

By appropriate treatment of recycled wastewater and gray water, it can be used as a source of water to offset certain parts of current or future potable water demands. Water reuse can be treated as an additional source of water to the supply system–quantitative value and to reutilize a certain portion of water resources that is otherwise classified as disposable treated wastewater.

The summary of the allowed uses of recycled water corresponding with the degree of treatment is presented in Table 6.4.

TABLE 6.4
Recycled Water Criteria

Treatment Level	Allowed Uses of Recycled Water
Undisinfected secondary	*Surface Irrigation*:
Disinfected secondary	Vineyards—no contact with the edible portion of crop
	Orchards—no contact with the edible portion of crop
	Pasture for animals—not producing milk for human consumption
	Seed crops—not for human consumption
	Ornamental nursery stock
	Sod farms and Christmas trees
	Fodder and fiber crops

(*Continued*)

TABLE 6.4 (*Continued*)
Recycled Water Criteria

Treatment Level	Allowed Uses of Recycled Water
	Others: Flushing sanitary sewers
	Irrigation: Cemeteries Freeway landscaping Restricted access golf courses Ornamental nursery stock Sod farms Pasture for livestock producing milk for human consumption Nonedible vegetation where access is controlled—cannot be used for school yards, playgrounds, and parks Impoundment Landscape impoundments not utilizing decorative fountains
	Cooling: Industrial/commercial cooling that does not use cooling tower. Evaporative condensers or spraying
	Others: Industrial boilers Nonstructural firefighting, backfill Soil compaction Mixing concrete Dust control Cleaning roads and sidewalks Industrial processes where it does not come in contact with workers
Disinfected secondary	*Irrigation*: Food crops—edible portion above the ground and not contacted by recycled water Food corps Parks and playgrounds School yards Residential landscaping Unrestricted access golf courses
	Impoundments: Nonrestricted recreational
	Cooling: Cooling towers Evaporative condensers Spraying or mist cooling
	Others: Flushing toilets/urinals Industrial processes Structural fire fighting Decorative fountains Commercial laundries Commercial car washes—where public is excluded from the process Consolidation of backfill around potable water pipelines Artificial snowmaking for commercial outdoor use

Source: Unofficial California Code of Regulations (CCR) Title 22.

Water reuse can be classified into direct reuse and indirect reuse:

- *Indirect reuse*: Water is taken from a river, lake, or underground aquifer that contains sewage. The practice of discharging sewage to surface waters and its withdrawal for reuse allows the processes of natural purification to occur.
- *Direct reuse*: Planned and deliberate use of treated waste water for some beneficial purpose. Direct reuse of reclaimed waters is practiced for several applications without dilution by natural water resources.

Water can be reused for the following purposes:

- *Agricultural irrigation*: The original form of sewage treatment was sewage farming, or the disposal of sewage on farmlands. Most sewage farms were replaced 70–90 years ago by biological treatment plants, which discharge to the nearest watercourse. Reused water can also be used for urban irrigation. Irrigation with raw or partially treated sewage can conserve water and fertilize crops economically by capturing nutrients that would normally be wasted. This irrigation method is also an effective way to prevent contamination of nearby waterways with disease organisms such as coliform and other bacteria that sewage contains. The most serious drawback of using sewage for irrigation is its role in transmitting infectious diseases to agricultural workers and the general public.
- *Urban irrigation*: Urban irrigation includes parks, golf courses, and landscape medians. Limited exposure risk is presented by areas such as golf courses, cemeteries, and highway medians, where public access is restricted and where water is applied only during night hours without airborne drift or surface runoff into public areas. Other urban uses include toilet flushing, fire protection, and construction.
- *Groundwater recharge*: Recharging underground aquifers with treated sewage is one of the most generally accepted forms of water reuse. Water reuse by groundwater recharge is widely believed to entail a lesser degree of risk than other means of reuse in which the recycling connection is more direct. The degree of treatment depends on the type of application to the soil, soil formation and chemistry, depth to groundwater table, dilution available, and residence time to the point of first extraction.
- *Recreation*: Wastewater is sometime used to fill lakes as part of a recreational system. Biological treatment in a series of lagoons usually provides the quality required for recreational purposes.
- *Industrial use*: The quantity of water used in many manufacturing processes and power generation is very large. The treated wastewater is an appropriate source of water for such industries, especially for regions facing water shortages. The water necessary for cooling purposes does not have to be very pure and in some industries can be supplied from treated wastewater.

6.9.1.3 Natural Treatment Systems

Local, small-scale systems often use the same type of treatment as large-scale systems. Several studies have shown that conventional small-scale systems are often less efficient in terms of energy; they need relatively more maintenance and are therefore more expensive than large-scale systems. A positive aspect of small-scale wastewater systems is the better quality of the recycled product and proximity to suitable land for the treatment and disposal of nutrients, which minimizes transport costs.

The possibility of using natural wastewater treatment systems is usually larger for smaller communities where the population density is low. Natural (or ecological) wastewater treatment systems refer to constructed wetlands, reed beds, or aquaculture, all of which use microbiological and macrophyte communities to assimilate pollutants. The usual purpose of wetlands and reed beds is to remove nutrients, not to recycle them. Aquaculture utilizes a series of ponds of algae, crop plants, and fish to recover nutrients into biomass. The energy requirements of constructed wetlands are low, but if a significant reuse of nutrients is included (aquaculture), the energy requirements would increase significantly (Brix, 1998). Therefore, it is necessary to recognize the importance of recycling of nutrients, not just removal, also in natural systems. Recent research in this direction has been reported (Farahbakhshazad and Morrison, 2000).

6.9.1.4 Combining Treatment Systems

The distinction between conventional and natural technologies is not clear and combinations of these technologies are common, such as combining chemical pretreatment to remove phosphorus and biochemical oxygen demand (BOD) and wetlands to remove nitrogen and remaining BOD. Wetlands are often used as a polishing step after conventional treatment. A combination of these technologies may be one possible way to achieve several of the standard requirements. Whereas conventional systems are reliable and effective for the removal of certain compounds, natural systems may contribute positively to other aspects such as biodiversity or landscape esthetics. Urine separation in combination with conventional treatment facilities also provides an environmentally improved option.

6.9.1.5 Changing Public Perspectives

Over and above the choice of technical system, the behavior of the users and the operators will affect the functionality of urban water infrastructure. Changes of social and institutional aspects are challenging because they involve changing the way individuals (within and outside the organization) think and act. Users of the systems as well as the personnel who design and operate the urban water may need to be informed of the function of systems, the value of water, and how to use the system properly.

6.10 LEAKAGE MANAGEMENT

Throughout the world, a large amount of water is lost through urban water supply systems due to leakage in distribution networks and home appliances or through illegal connections. These water losses are referred to as unaccounted-for water, a measure that is often used to quantify the efficiency of a water supply system. Unaccounted-for water is now widely called nonrevenue water and can be defined as the difference between the amount of water supplied from the water works, as measured through its meters, and the total amount of accounted-for water. Accounted-for water includes water consumption as recorded by customers' meters, water stored in service reservoirs, and authorized free use such as for flushing and sterilization of mains and routine cleaning of service reservoirs. In many cities, unaccounted-for water is estimated to be as high as 50% of water supplied.

The following items can be considered as control measures for reducing the amount of unaccounted-for water:

- *Leakage detection*: A specific program is required to detect the leakage. It can include visual inspection for leaks along transmission and distribution pipelines and also leak detection tests at night for distribution mains. Nighttime is selected because the pressure is usually higher in the system, making it is easier to detect the leaks and also minimizing inconvenience to customers. To locate the leaks, mechanical, electronic, and computerized acoustic instruments such as stethoscopes, geophones, electronic leak detectors, and leak noise correlators can be used. The time frequency of testing is important. It depends on the percentage of unaccounted-for water and share of leakage and the size of the water distribution systems.

- *Leakage control*: Better quality pipes and fittings can be used for leakage control. More durable and corrosion-resistant pipe materials such as ductile iron pipes internally lined with cement mortar, copper, and stainless steel can be used for this purpose. When designing the water distribution network, minimizing the number of joints is important to reduce minor leaks. Teflon packing for repairing valve glands and leaks due to wear and tear and dezincification-resistant brass fittings can also reduce the minor leakage in the water distribution network.
- *Full and accurate metering policy*: Meters with less accuracy such as Venturi/Dall tubes can be replaced with more accurate meters such as electromagnetic meters. A specific program for maintaining and replacing meters regularly should also be considered. This can be done through bulk changing programs.
- *Proper accounting of water used*: In most water distribution networks in urban areas, significant quantities of water are used in the commissioning and filling of new mains, for connection and service reservoirs, for cleaning and flushing the distribution system during maintenance, and for firefighting. Water used for such purposes should be accurately reported to ensure proper accounting.
- *Strict legislation on illegal draw off*: The illegal draw off should be detected, and anyone who is responsible for it should be prosecuted.
- *Education and training*: Public education and training are important issues in achieving the water demand management goals. Various methods have been tried in developed and developing countries such as:
    - Training programs in the mass media
    - Billboards along streets and on public transport vehicles

Key to establish a strategy for management water losses is to gain a better understanding of the reasons for losses and the factors which influence its components. Significant advances have been made by some water utilities in the understanding and modeling of water loss components and in defining the economic level of leakage. Real losses cannot be eliminated completely. The minimum achievable annual volume of real losses is known as unavoidable annual real losses (UARL). UARL can be assessed using a formula as follows:

$$\text{UARL}\left(\text{L/day}\right) = \left(18\ L_m + 0.8\ N_c + 25\ L_p\right) \times P \tag{6.32}$$

where

N_c is the number of service connections,
L_m is the length of mains (km),
L_p is the length of private pipes between the street property boundary and customer meters (km), and
P is the average operating pressure (m).

This formula is the most reliable predictor with real losses for systems with more than 5,000 service connections, connection density (N_c/L_m) more than 20, and average pressure more than 25 m.

In the recent years, International Water Association (IWA) Task Forces have developed and tested the usage of Infrastructure Leakage Index (ILI) as water losses PI. ILI accommodates the fact that real losses will always exist, even in the very best and well managed distribution system.

The international PI defines as "real losses by volume" which is volume of real losses as % of effluent from water treatment plant (WTP), can give the most rational technical basis for comparisons water losses between utilities, which can be used by the operators to measure their attempt in water losses reduction.

The ILI, which in the first few years was known to only a few insiders, is now widely accepted and used by practitioners around the world, as it best describes the efficiency of the real loss management of water utilities. It is a measurement of how well a distribution network is managed (maintained, repaired, and rehabilitated) for the control of real losses, at the current operating pressure. ILI is the ratio of Current Annual Real Losses (CARL) to UARL, or

$$\text{ILI} = \text{CARL/UARL} \tag{6.33}$$

Being a ratio, the ILI has no units and thus facilitates comparisons between countries that use different measurement units (metric, US, British). ILI of less than 1.5 is considered excellent and a value of greater than 3.5 is unacceptable according to the international standards (Liemberger, 2005).

6.10.1 Acceptable Pressure Range

It is approximately between 30m and 90m. The applicability of the ILI in areas with abnormally high pressures or unusually low pressures has been questioned by various experts. A proposed adjustment (Lambert, 2003) to the ILI calculation is to incorporate a variable designed to adjust the ILI value in circumstances where the average pressure is outside the normally accepted range. The ILI is still, however, considered to be a useful indicator in its current form for identifying the areas with the highest leakage.

Leakage management practitioners are well aware that real losses will always exist, even in the very best systems. The volume of UARL is the lowest technically achievable annual real losses for a well maintained and well managed system. The difference between CARL (large rectangle) and UARL (small rectangle) is the potentially recoverable real losses. UARL is a useful concept as it can be used to predict, with reasonable reliability, the lowest technical annual real losses for any combination of mains length, number of connections, customer meter location at current operating pressures, assuming that the system is in good condition with high standards for management of real losses and there are no financial or economic constraints.

In the context of what is presented in Figures 6.28 and 6.29 as Economic Optimum Leakage Reduction Strategy, ideal targets include:

- *Economic*: Optimum between cost and benefit
- *Practical*: In terms of data needs and implementation
- Sustainable in the long term and flexible in the short term
- Consistent with water resources plan
- Be understandable, transparent, simple, and consistent
- Sound understanding of mechanics
- Sensitive to political considerations
- Comparisons between organizations

6.10.2 Economic Leakage Index

The extent of utilities investment in water supply and Non-Revenue Water (NRW) controlling activities must be considered in an overall economic analysis to optimize monetary investment

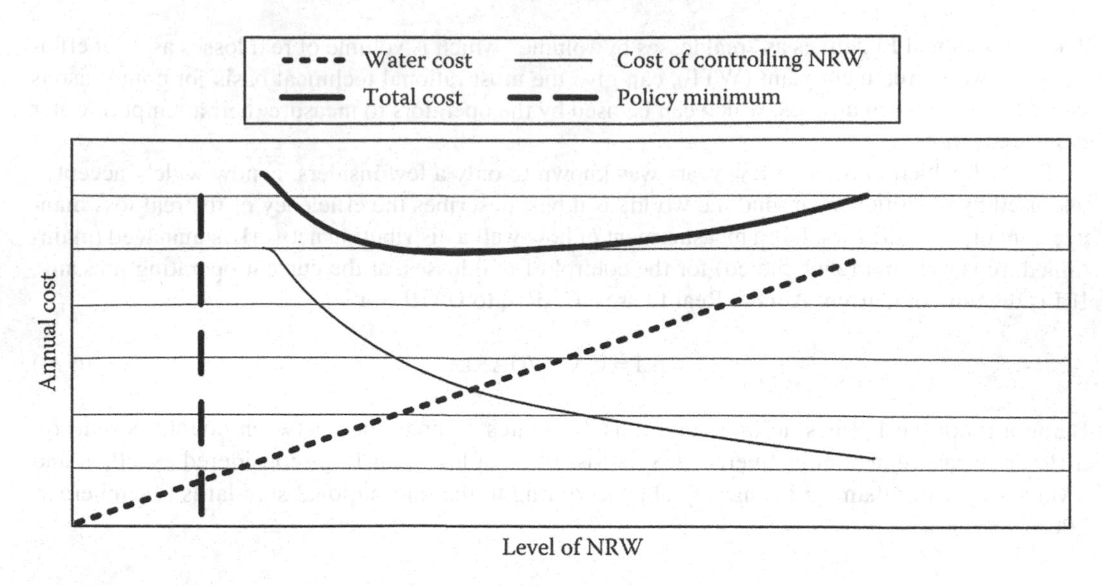

FIGURE 6.28 Trade-off between water cost and NRW cost and the optimum level of `.

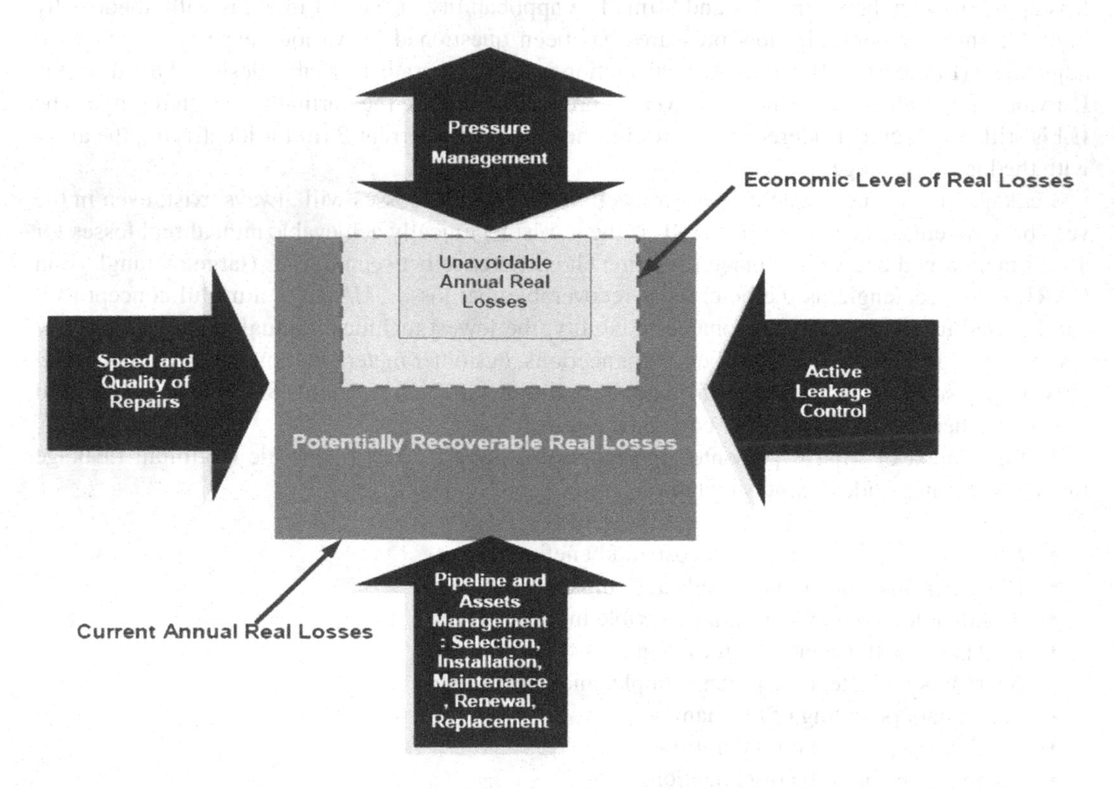

FIGURE 6.29 Economic level of real losses—leakage reduction strategy.

in resources and activities for NRW loss reduction, as compared to the cost of water saved arising from these programs (Figures 6.28 and 6.30). The minimum economically achievable ILI is called economic leakage index (ELI). Figure 6.29 shows the relationship between CARL and UARL.

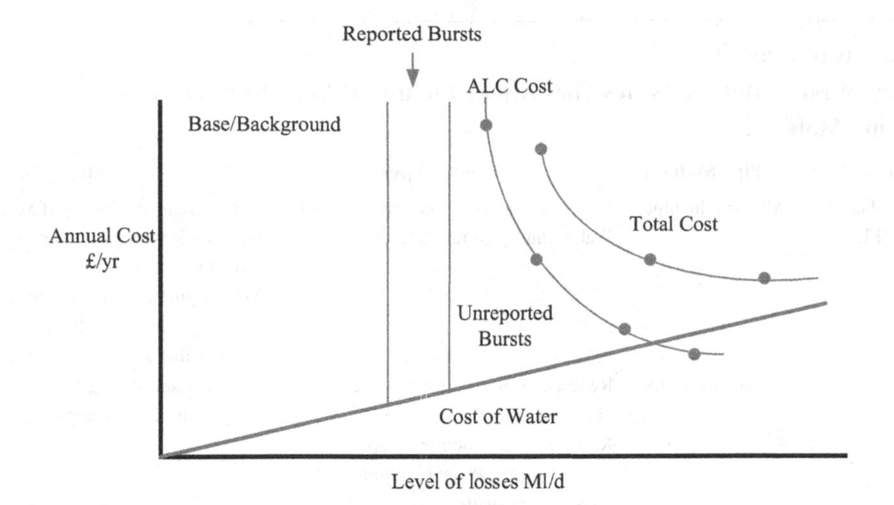

FIGURE 6.30 Economic optimum leakage reduction strategy—ALC: actual leakage control.

6.11 NONDESTRUCTIVE TESTING (NDT)

In this section, the improvement of water mains' structural integrity monitoring (SIM) capability as an approach for reducing high-risk drinking water mains breaks and inefficient maintenance scheduling is explored. Structural integrity of water networks refers to the soundness of the pipe walls and joints for conveying water to its intended locations and preventing egress of water, loss of pressure, and entry of contaminants. SIM is the systematic detection, location, and quantification of pipe wall and pipe joint damage and deterioration (e.g., wall thinning, cracking, bending, crushing, misalignment, or joint separation) of installed drinking water mains. Determination of the present condition and the deterioration rate of the pipe will be made possible by an effective SIM (Royer, 2005). Improving SIM technology for underground pipelines in spite of its difficulty is possible. For example, Table 6.5 lists six nondestructive evaluation (NDE) technologies (i.e., acoustic

TABLE 6.5
Summary of NDE Method Issues That Affect Technique Selection for Various Water Pipe Materials

Inspection Method	Pipe Material	Defect Types	Notes
Acoustic emission	Pretensioned or prestressed concrete pipe	Breaks in reinforcing steel	Pipe not removed from service
		Slippage of broken reinforcement Concrete cracking	Hydrophones left in place for several days to weeks
Electromagnetic	All-metallic pipe	Cracks	Commercial, off-the-shelf availability Detect environmental conditions that are likely to weaken the pipe Does not directly inspect the pipe Totally noninvasive
Impact echo	Concrete pipe containing steel	Delamination and cracks at various concrete/mortar/steel interfaces	Requires dewatering and human access to the interior of the pipe Can be done externally if exterior access is available

(Continued)

TABLE 6.5 (*Continued*)
Summary of NDE Method Issues That Affect Technique Selection for Various Water Pipe Materials

Inspection Method	Pipe Material	Defect Types	Notes
Remote field eddy current (RFEC)	All-metallic pipe	Changes in metal mass graphitization Wall thinning gouges large cracks	Commercial, off-the-shelf availability Pig travels through the pipe via water hydrants May require cleaning before inspection Pig may dislodge material from the pipe wall, requiring flushing
Seismic	All-concrete pipe	Reductions in concrete modulus because of aging Reductions in concrete compression as a result of breakage or slippage of reinforcing steel	Requires dewatering and human access to the interior of the pipe
Ultrasonic	All-metallic pipe	Detection of wall thinning	Not commercially available for water pipes Does not require dewatering of pipes Developed for inspecting oil or gas pipelines systems are long, inflexible, and expensive

Source: Dingus, M., Haven, J., and Austin, R. 2002. *Nondestructive, Noninvasive Assessment of Underground Pipelines.* AWWA Research Foundation, Denver, CO, 116 pp. With permission.

emission, electromagnetic, impact echo, remote field eddy current (RFEC), seismic, and ultrasonic) and describes the pipe materials and defect types to which they are applicable.

More recent technologies include mobile nonintrusive inspection (MNII) systems. Its detector is outside the pipe and can examine a substantial length of pipe from a single location, preferably without excavation, offer the promise of substantially reducing the time, cost, and disruption involved in pinpointing the pipe that should receive detailed inspection. Samples include (a) lamb wave devices that transmit and receive ultrasonic waves for moderate distances (e.g., 30 m) along the pipe wall, which enables 60 m of the pipe to be inspected from one location; (b) electric field monitoring devices, which temporarily electrify the pipe, then detect electric field changes at the ground surface that are indicative of pipe wall thinning, and indicate problem locations for detailed investigation; and (c) aerial or satellite systems for remote monitoring of surface conditions symbol of pipe deterioration or failure.

6.12 WATER SUPPLY INFRASTRUCTURE OF SELECTED CITIES

Water supply systems in selected cities are presented as follows:

6.12.1 Case 1: Philadelphia, USA

Philadelphia is located in the Delaware River Watershed, which begins in New York State and extends 330 miles south to the mouth of the Delaware Bay. The Schuylkill River is part of the Delaware River Watershed (Figure 6.31).

Philadelphia is the largest city in the United States and Commonwealth of Pennsylvania, with a 2018 census-estimated population of 1.5 million in 369.59 km^2 area. The Philadelphia Water Department (PWD) is one of the oldest water departments in the US and it has been providing water

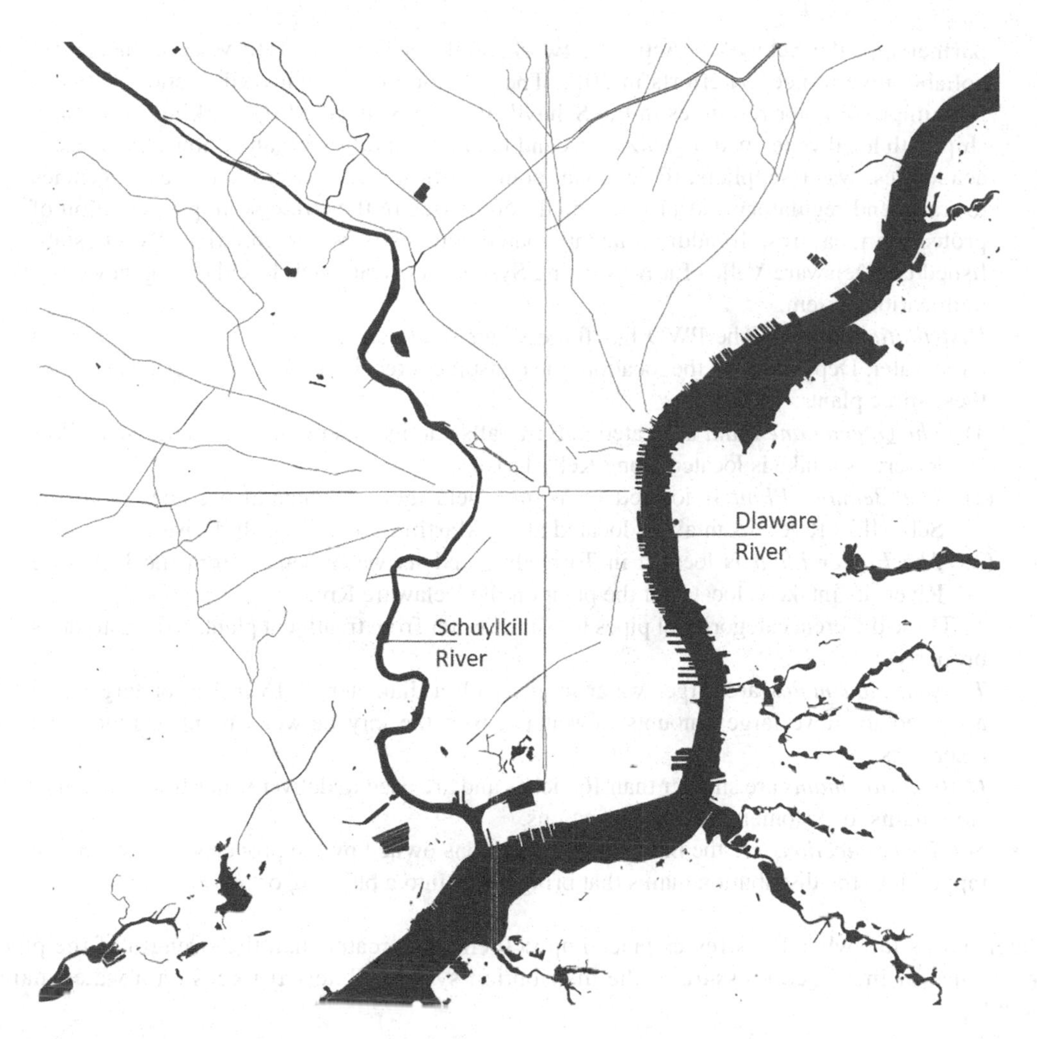

FIGURE 6.31 Water supply sources of Philadelphia from Delaware River.

to citizens since 1801. The department treats 0.929 MCM of water on daily average for its customers and treats 1.78 MCM of wastewater every day.

- *Supply system*: PWD uses proven treatment practices and participates in groundbreaking research to provide drinking water that consistently exceeds EPA standards. There are a number of environmental, social, and developmental trends materializing in Philadelphia's watersheds that raise concerns and pose challenges to the protection of the city's drinking water sources, the Schuylkill and Delaware rivers. Each river contributes one-half of the city's overall supply, and approximately 0.929 MCM of high-quality drinking water is produced on a daily basis. Philadelphia does not use groundwater.

  The Schuylkill and Delaware River Source Water Assessments and Protection Plans provide a comprehensive framework for a watershed-wide effort to protect the quality and quantity of Philadelphia's water supplies. The comprehensive research and analyses completed as part of the Source Water Assessment and Protection Plans in the early 2000s identified the need for a regional partnership in the Schuylkill River Watershed to address priority contaminants through stakeholder workgroups. The resulting watershed partnership, the Schuylkill Action Network, will celebrate its 15th year of successful collaborative protection efforts in 2018. The mission of the Schuylkill Action Network is to improve water resources in the Schuylkill River Watershed by working in partnership with local watershed organizations and land conservation organizations, businesses, academics, water suppliers, recreational communities, local governments, and agencies to transcend regulatory and jurisdictional boundaries in the strategic implementation of protection measures. To address unanticipated sources of water pollution, PWD established the Delaware Valley Early Warning System, a private web-based emergency communication system.

- *Distribution system*: The PWD has three water treatment plants that process untreated river water. Depending on the location, the consumers receive drinking water from one of these three plants:
  (1) *The Queen Lane Plant* is located in East Falls and its water comes from the Schuylkill River; its intake is located along Kelly Drive.
  (2) *The Belmont Plant* is located in Wynne field and its water also comes from the Schuylkill River; its intake is located along Martin Luther King, Jr. Drive.
  (3) *The Baxter Plant* is located in Torresdale and its water comes from the Delaware River; its intake is located at the plant on the Delaware River.

  Three different categories of pipes transport water from treatment plants to a customer's property:
- *Transmission mains* are larger water mains with a diameter of 16 inches or larger and are used to move large amounts of water across the city between pump stations and reservoirs.
- *Distribution mains* are smaller than 16 inches and are used to deliver water from transmission mains to customer service connections.
- *Service connections* are the individual connections owned by the property owner that are tapped into the distribution mains that bring water into a building or house.

Water mains fail when the stresses placed upon them are greater than the strength of the pipe or because of increased pressure in the distribution system. External forces on a water main include:

- Traffic loading
- Temperature changes
- Underground work with direct or indirect impact to the pipe

Ongoing programs to prevent fails and breaks:

- Main replacement
- Leak detection
- Corrosion testing

Corrosion control program, as mandated by Federal law and optimized over the past two decades, minimizes the release of lead from service lines, pipes, fixtures, and solder by creating a coating designed to keep lead from leaching into the water.

The rate of real loss in the City of Philadelphia is calculated and reported to be 23% of all the water supplied due to old water mains. Some of the water mains were first installed in the 1800s. Although most of the old wooden mains are substituted, some of them still have valves that date back to when lead was used to manufacture them. Pipe inspection and water sampling programs have been conducted on a regular basis to speed up the replacement of dangerous components of the distribution system.

6.12.2 Case 2: Los Angeles

The City of Los Angeles is the largest city in California. With an estimated population of nearly 4 million people, it is the country's second most populous city (after New York City) and the third most populous city in North America (after Mexico City and New York City). LASAN (LA sanitation) serves over 4 million residential and industrial customers in the city. Additionally, LASAN also provides conveyance and treatment services for an estimated 600,000 residences outside of the city from its 29 contract agencies.

- *Water sources*: Water resources in this case study include groundwater resources (account for about 30% of LA's water) and Owens River, Northern California, and the Colorado River.
  - *Owens River*: The Owens River, Mono Lake Basin, and reservoirs in the Sierra Nevada Mountains provide 430 MGD of water to LA via the Los Angeles Aqueduct. This represents only about one-third of LA's water supply.
  - *Northern California and the Colorado River*: Accounting for another one-third, the Sierras are the state's largest source of water. When rain falls in the winter at higher elevations, it becomes snow. When that snow melts in the spring time, the melted water becomes runoff and flows into aqueducts and groundwater.
  - *Colorado River*: It can deliver 1 BGD (billion gallons per day) to cities in southern California. In order to conserve the Sierras snowpack, more water is imported from the Colorado River. About half on LA's water flows from the Colorado River via the Colorado River Aqueduct.

The governance system that oversees the distribution, management, and conservation of potable water in Los Angeles County is complex and opaque. Presently, nearly 100 public and private entities are involved in the management of potable water supplies in the region—a system born out of a history of fragmented water rights and governance regimes following rapid urban expansion. These suppliers include cities, special districts, investor-owned utilities (IOUs), and Municipal Water Districts. Each of these has different access to water from different sources.

The urban water management system in LA County is divided into three types of water suppliers: (a) contractors, who receive annual allocations of imported water from the State Water Project and Colorado River authorities; (b) wholesalers, who purchase and resell water from contractors or other wholesalers; and (c) retailers, who sell water directly to residential, commercial, industrial, and agricultural customers.

Researchers at UCLA have developed a comprehensive database of these different suppliers depicting their connections and functions, as well as their scale, role, size, and management structures. Figure 6.32a shows the territories of the region's suppliers.

The region's two contractors (Figure 6.32—top part) are the Metropolitan Water District (MWD), which supplies imported water to 26 member agencies across six counties, and the San Gabriel Valley Municipal Water District, which has four member agencies in the area. In most cases, these contractors sell imported water to wholesalers in the region, who then resell this water to retailers. In certain areas, however, contractors may sell water directly to retailers—certain municipalities purchase water directly from the MWD—or water may move through two different wholesalers before reaching a retailer. There is also variation in water sources, as some retailers pump groundwater or use recycled water for nonpotable demands. This system is further complicated by unclear service area boundaries for suppliers.

Specifically, our research finds that the boundaries for a number of mutual water companies are unclear. Their reporting of boundaries involves low-resolution (sometimes hand-drawn) maps that do not allow for detailed understandings of boundaries. Additionally, in Los Angeles County (Figure 6.32—bottom part), no agency is tasked with overseeing all of these suppliers. The system

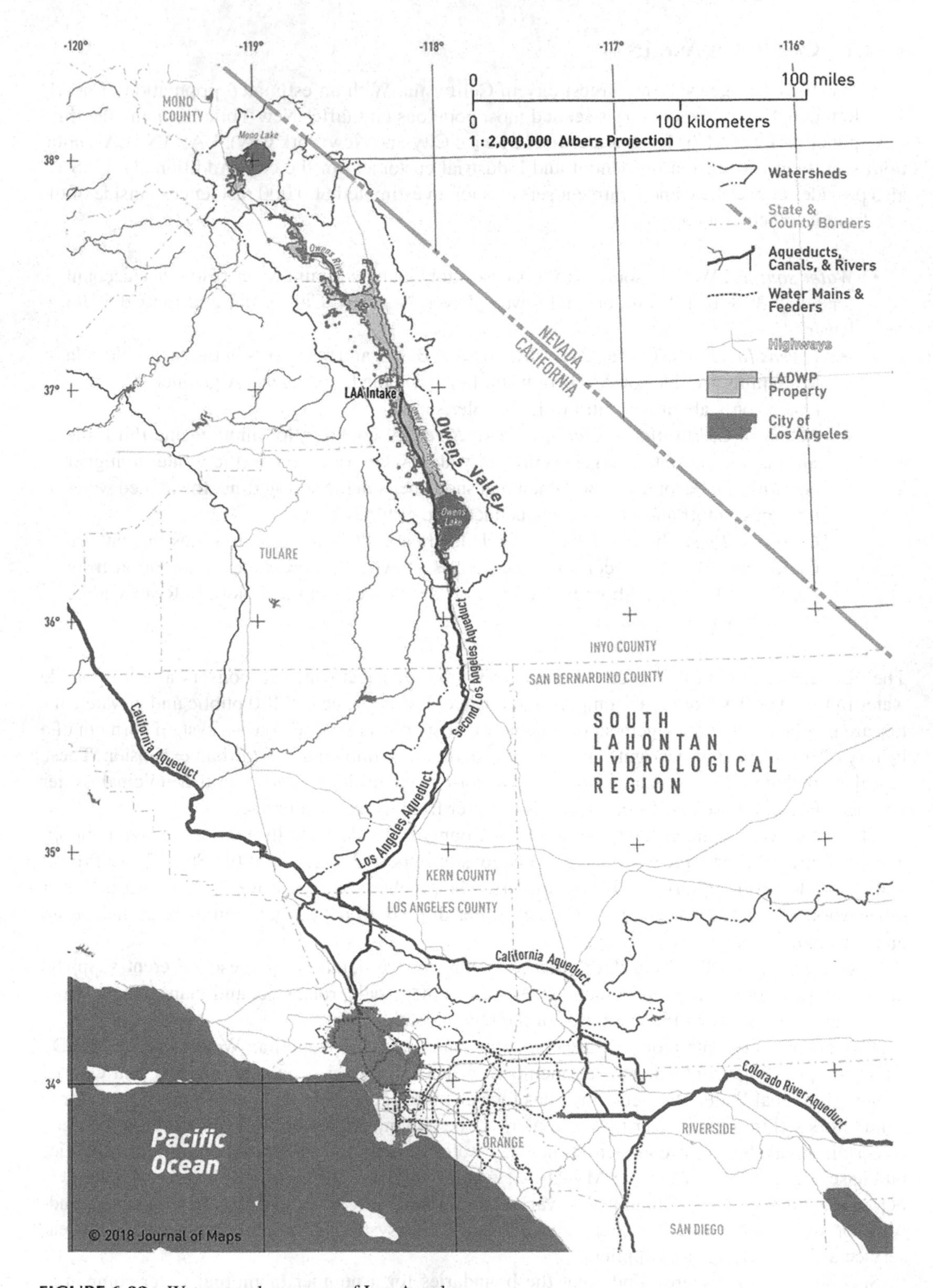

FIGURE 6.32 Water supply sources of Los Angeles.

is comprised of many different types of agencies with varying organizational structures and no systematic reporting or regulatory framework. This results in a lack of supervision, transparency, and accountability in the system, which can lead to uneven and inadequate understandings of water quality, distribution, and service provision according to some analysts.

6.12.3 Case 3: Copenhagen, Denmark

Copenhagen is the capital of Denmark and has 570,000 inhabitants. The average annual rainfall is about 600mm (Figure 6.33).Copenhagen was voted the European Green Capital of 2014 and won the Most Livable City Award for the second time in the same year.

- *Water supply plan*: Copenhagen water supply is mainly based on abstraction of groundwater from the chalk aquifer. The groundwater quality from these deep aquifers is in general high. It takes 40–60 years for the water to penetrate down to the aquifer.

  In the water supply plan, goals are set for a percentage of water analyses which may be above any of the guideline values. Both in the former plan as well as the present, maximum accepted level of exceedance of the guidelines is 2%.

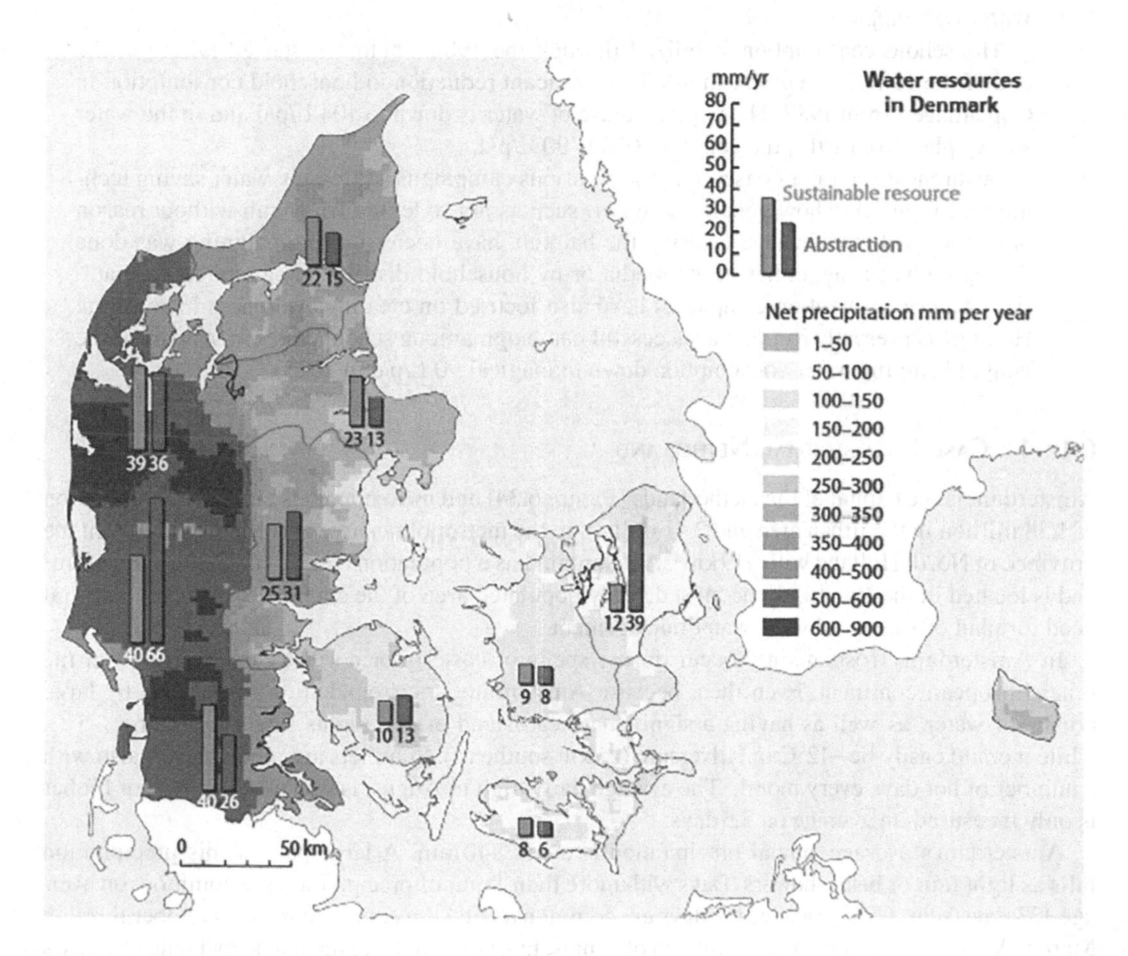

FIGURE 6.33 Water resources in Denmark.

Before 2017 household consumption shall be reduced to 100 L/p/d.

This is to be achieved through further development of water-saving technologies and water conservation campaigns.

To secure the supply, HOFOR company in Copenhagen distributes city heat, and water. The company provides waste water solutions drain operation services, as well as maintains a transmission network. It shall continue to have abstraction permits and production facilities which allow production of at least 25% more than is annually consumed. HOFOR shall continue to maintain the distribution network. The unaccounted consumption shall be as low as possible and not exceed 10%. Reuse of water and use of low-quality groundwater has constituted about 4% of the water consumption. In Groundwater management, which falls under the responsibility of the government and the municipalities, the threat to the use of pure groundwater in regard to both quantity and quality has been realized since the 1970–1980s and an awareness of the importance of protection of the groundwater has been provided to the public. Denmark is classified into three types of groundwater-abstraction areas: particularly valuable, valuable, and less valuable water abstraction areas. The abstraction of groundwater took place for years without any significant problems. The groundwater could be used for drinking after a simple treatment of aeration and filtration before being distributed to consumers.

- *Water consumption*

  Household consumption stabilized through the 1980s, before it started falling at the end of the 1980s. The graph shows the significant reduction in household consumption in Copenhagen from 1987. The per capita use of water is down to 104 L/p/d and in the water supply plan from 2012, the goal for 2025 is 90 L/p/d.

  Awareness on water conservation and various campaigns addressing water saving techniques in regard to how people use water, such as not to let tap water run without reason or to take a shower instead of using the bathtub, have been run. Campaigning was done through advertising, either in the media or by household distributed campaigning material. Water conservation campaigns have also focused on creating awareness in children. HOFOR is presently running a successful campaign among school children, as part of the goal of bringing water consumption down to magical 90 L/p/d in 2025.

6.12.4 Case 4: Amsterdam, Netherland

Amsterdam is the capital of the Netherlands (Figure 6.34) and most populous city with a population of 1.38 million in the urban area and 2.41 million in the metropolitan area and also found within the province of North Holland with 219 km^2. Amsterdam has a population density of 4,439 persons/km^2 and is located in the Randstad, the most densely populated area of the country. So there is essential need for plan of water and wastewater management.

In Amsterdam, frosts mainly occur during spells of easterly or northeasterly winds from the inner European continent. Even then, because Amsterdam is surrounded on three sides by large bodies of water, as well as having a significant heat-island effect, nights rarely fall below −5°C, while it could easily be −12°C in Hilversum (25 km southeast). Summers are moderately warm with a number of hot days every month. The average daily high in August is 22.1°C, and 30°C or higher is only measured on average on 2.5 days.

Amsterdam's average annual precipitation is about 840 mm. A large part of this precipitation falls as light rain or brief showers. Days with more than 1 mm of precipitation are common, on average 133 days/year. Cloudy and damp days are common during the cooler months of October through March. Amsterdam has more than 100 km of canals, most of which are navigable by boat. The city's three main canals are the Prinsengracht, Herengracht, and Keizersgracht.

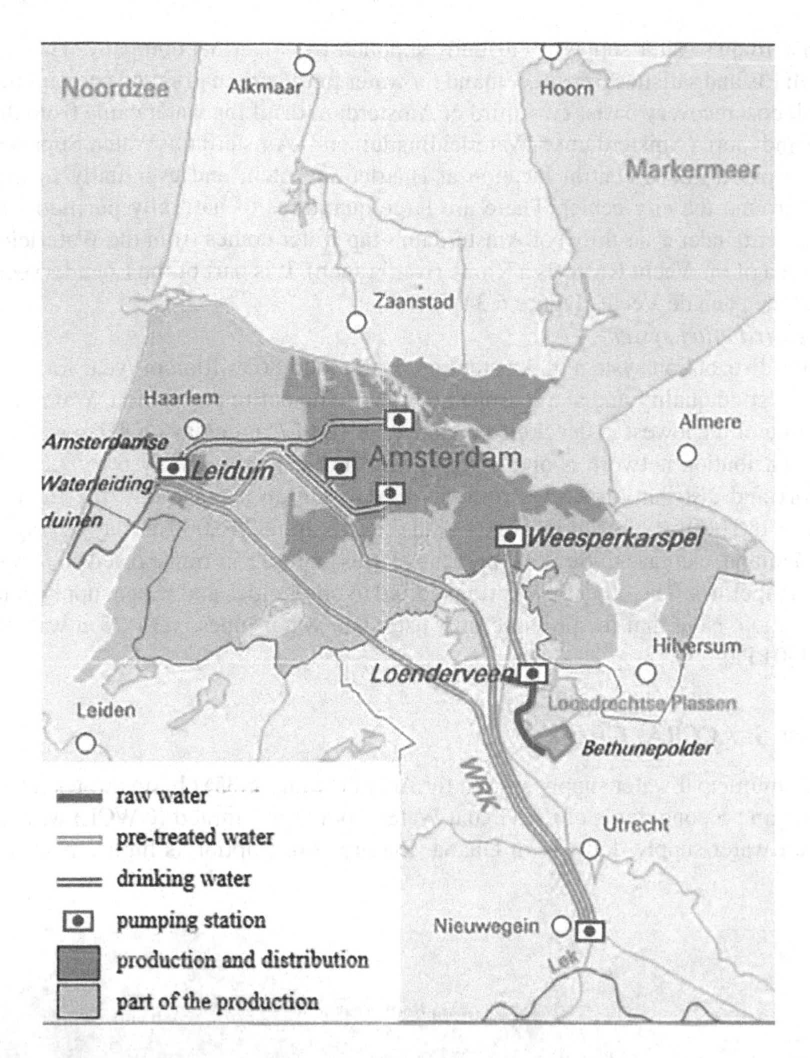

FIGURE 6.34 Amsterdam water supply system.

- *Water supply system:* Water delivery through a pipe network began in Amsterdam in 1853. In the past, drinking water came from wells and springs. These wells and sources were a source of infectious diseases, such as typhus and cholera, especially in the cities. During the seventeenth century, the old chain wells were often replaced by wells with a pump, which made it easier and faster to pump up the water, but hygiene remained poor. There was no bacterial knowledge, so drinking water wells were not treated hygienically. In the same period, water pipes were fed with untreated river water. At the end of the seventeenth century, water towers were also increasingly used, in which the water was stored. From there the water went to the houses of the wealthy bourgeoisie, the ordinary people continued to rely on wells. Around 1900, the water supply via pipes took off and more and more water supply companies were established. Only the province of North Holland opted for a provincial water supply company. Between 1920 and 1930, provincial water supply regulations were drawn up in all provinces to guarantee the quality of drinking water. The National Institute for Drinking Water Supply (Rijksinstituutvoor de Drinkwatervoorziening—RID) was established in 1913.

Amsterdam's water supply is currently supplied by Water net company. This company has high PIs and satisfies the full demand for water in the urban grid and surrounding areas on a full cost recovery basis. Two-third of Amsterdam drinking water came from the dunes near Zandvoort (Amsterdamse Waterleidingduinen—Amsterdam Water Supply Dunes) and was piped into a central location at Haarlemmerplein and eventually to individual homes around the city center. There are large quantities of naturally purified water and also the remainder (one-third) of Amsterdam's tap water comes from the Waterleidingplas in the area of the Vecht River (is a Rhine river branch). It is part of the Loenderveenseplas, near Loenen aan de Vecht (Figure 6.34).

- *Water distribution system*

  Water distribution system in Amsterdam is provided 86 million m^3/year for 1.2 million people in good quality and at a reasonable price to the entire population. Water consumption is one of the lowest in developed countries at 128 L/capita/day and also water leakage in the distribution network is one of the lowest in the world at only 6%. The volume of transport and distribution varies from day to day. On an average day, the Leiduin plant produces 180,000 m^3 of drinking water (This can go up to 240,000 m^3/day during summer when demand increases). Seventy percent of this volume is transported to Amsterdam through pipelines. The other 30% is transported to other cities and water supply companies in the region. Note that the pressure in Amsterdam water pipes is 30–35 m water column (300–350 kPa).

6.12.5 Case 5: ACCRA, Ghana

Currently, the municipal water supply system for Accra (Figure 6.35) obtains water from two main sources: Weija and Kpong dams of the Ghana Water Company Limited (GWCL) which is responsible for urban water supply delivery in Ghana. Energy consumption is higher from Kpong than

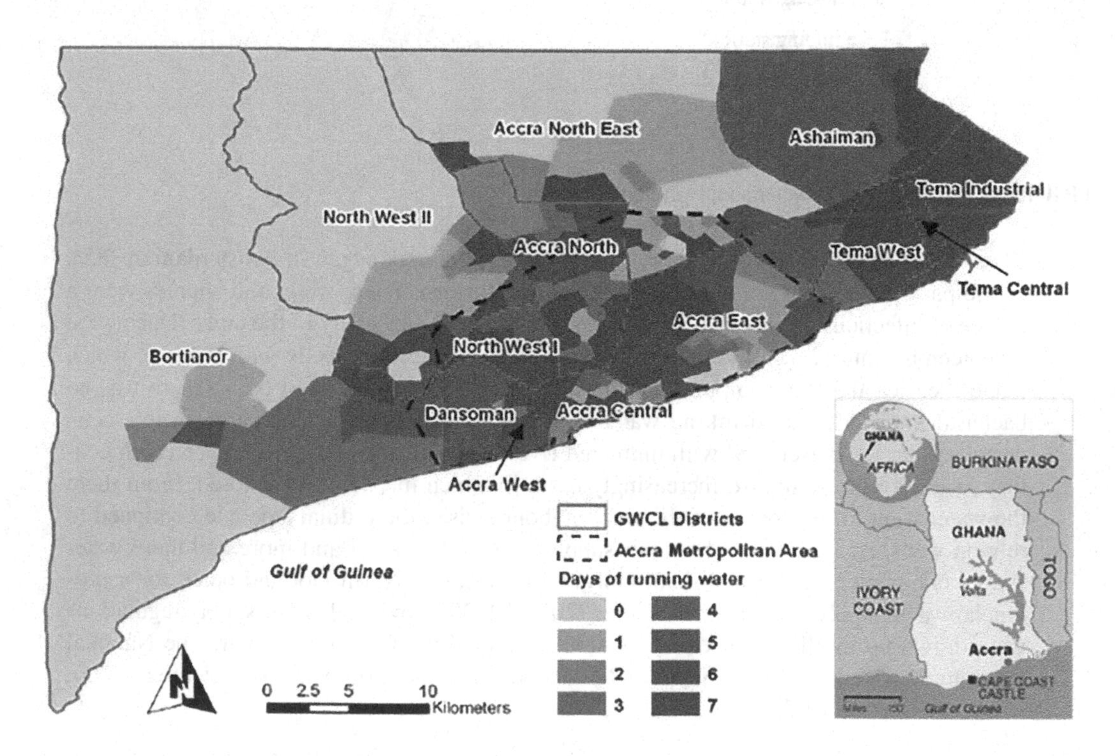

FIGURE 6.35 ACCRA, Ghana location, and days of running water.

from Weija, due to the long pumping distance to Accra whereas distribution from Weija is based on gravity. The total energy consumption for water production and distribution in 2004 was 101,900 MWh. Currently, informal sector providers are seen to play a very important role in the water supply system in Accra. More than 50% of the population does not have house or yard connections. They get their water supply from secondary sources such as water tank operators, motorized cart operators (power tiller tankers), tanker supplied vendors, GWCL direct supply vendors (neighbor sellers), sachet water seller producers, and street water sellers. It has become a norm to see people carrying buckets, pans, jerry cans, etc., looking for water every day. This issue of inadequate water supply affects productivity as people spend long hours to get a bucket of water. The main objectives of the GWCL by the establishment Act 310 of 1965 are to provide, distribute, and conserve water for domestic, public, and industrial purposes; the operations have not been self-sustaining; and it has relied on GoG to still subsidize its operation and maintenance costs and to bear full responsibility for capital investments.

Only about 10% of Accra's waste water is collected for some form of treatment. The remaining wastewater is discharged untreated to open drains, wetlands, and natural channels and finally to the sea. Wastes from houses and public toilets are also discharged directly into the sea at a place known as "lavender hill." Only 77.5% of homes have toilets. Only 30% have flush toilets. The average person in Accra has to share toilets with ten or more persons in public latrines. Waste removal is for the wealthy because they can afford it. Only 60% of the population has regular waste collection. As of June 17, 2012, all three refuse dump sites were closed down. Because of this, open sewers and rains are full of trash. Most of the pipes are in polluted gutters. Broken or vandalized ones are open to germs (Figure 6.35).

6.12.6 Case 6: Stockholm, Sweden

Stockholm is the capital and most populous City of Sweden and the most populous City of Scandinavia and approximately 1.5 million live in the urban area (Figure 6.36). It is located on Sweden's east coast, where the freshwater Lake Malaren (Sweden's third-largest lake) flows out into the Baltic Sea. The central parts of the city consist of 14 islands that are continuous with the Stockholm archipelago. The geographical city center is situated on the water, in Riddarfjärdenbay. Over 30% of the city area is made up of waterways and another 30% is made up of parks and green spaces. The average annual temperature is 7.6°C, and the average rainfall is 531 mm a year.

- *Supply system*: There is no need for groundwater resources, and in modern times, the city gets its water from Lake Malaren (main source) purified by plants at Norsborg and Lovön, together producing 360,000 m^3/day and distributes it. To maintain a high quality of drinking water, measures are continuously taken to protect the source of water supply. Lake Bornsjön, the reserve water source, is today a restricted area, and the work to protect the lake has been so successful that its water today matches drinking water quality without treatment. Water consumption has decreased constantly 2%/year in accordance with the general trend of energy and water conserving attitudes and equipment. At the same time, the population is growing 1%/year in the same area.
- *Water distribution system*: The public water pipe network in Stockholm is 2,200 km long. The average time spent by the water in the network is theoretically 20 hours. There were 364 leaks in water network in 1998, total amount was 10 mm^3. It is estimated that almost one-third of the leaks into sewage pipes involve drinking water. Everyone shall have access to secure drinking water of high quality and in sufficient quantity, through the development of services and products that ensure compliance with stricter demands with regard to the protection of raw water sources and water treatment in a changing climate, and through greater awareness in society of the long-term value of water.

FIGURE 6.36 Location of Stockholm city.

The quality of raw water is strongly affected by climate change in combination with an increasingly polluted environment. Increased rainfall leads to increased runoff and transport to raw water sources of both microbiological and chemical contaminants, as well as organic material from forest and agricultural lands. There are also greater risks of overflows from the wastewater system and of flooding in urban and industrial areas, which also involve risks for raw water sources.

The drinking water sector is one of the most attractive labor markets in Sweden and attracts highly skilled labor from within Sweden and internationally. Risk analyses are conducted in all Swedish municipalities in a structured way that entails the implementation of risk-based measures to prevent outbreaks of waterborne diseases. There is knowledge and technological solutions for safe and effective treatment of raw water with elevated concentrations of natural organic matter. There is knowledge and technical solutions for the characterization of organic matter that facilitate the selection of appropriate water treatment processes.

6.13 CONCLUDING REMARKS

Urban water supply infrastructures include supply, delivery, treatment, and distribution by water mains to its end users. Urban infrastructure systems are critically important for maintaining public health as well as for contributing to the quality of life. As for water quality, it can change significantly as water moves through a water system. Water quality models should be developed to understand the governing factors and to track and monitor water quality changes in drinking water networks.

More attention has been given to the issue of water system's vulnerability, and it is apparent that the most vulnerable portion of a water utility is the network itself. In order to use water quality/hydraulic models correctly, there are two features that must be understood. These are the characterization of system's demands and the proper calibration of these models using data including the tracer tests. A good understanding of both of these aspects could substantially improve the use of models for simulating the system's operating strategies. In the context of managing dependability, groundwater resources are considered as reliable resource for supplying the demands without much delivery and water transfer requirements in an urban system.

In most urban areas in developed regions, government regulations require designers and operators of urban water infrastructures to meet the requirements and standards. Pressures must be adequate for fire protection, water quality must be of high standards to protect public health, and urban drainage of wastewater and stormwater must meet effluent and receiving water body quality standards.

In this chapter, an introduction to different components of urban water supply infrastructures has been presented and their interactions among UWC have also been discussed. Basic concepts on water storage and supply facilities are considered. Reservoir operation, supply, and demand management are discussed. Some basic planning issues of the urban water supply infrastructures have also been discussed in this chapter. SIM of water mains in order to reduce their repair, rehabilitation, and replacement costs/issues and increase the reliability of supplying the demands is discussed. This requires monitoring the use of various methods for detecting leaks and predicting the impacts of alternative urban water systems on the life cycle including operation, maintenance, and repair policies of these systems. The relationship of head/pressure, discharge, and leakage as well as a brief head-driven modeling are presented for water distribution systems. Finally, some examples of water supply systems in selected major cities are briefly described.

Economic and financial analysis plays an important role in infrastructure development. Some examples are given. Environmental performance is investigated in the context of SD. The interactions of systems' components are not often well integrated in the design, construction, rehabilitation, and maintenance of these systems. Its principle and deciding factors/parameters are not well understood and agreed upon by the analysts and decision makers, especially in the developing countries. Inclusion of many elements of water supply system in this chapter could help the potential reader to better comprehend with both analysis and design as well as operational issues of this vital lifeline.

Water quality is regarded in the context of water treatment plants and methods, as well as modeling water distribution systems. The water sector faces serious challenges in this regard. The failure to meet basic human needs for water; difficulties in meeting the financial requirements for maintaining, extending, and upgrading both new and ageing water systems; new regulatory requirements for water quality; increasing water scarcity; competition for limited capital; and global climate change will continue to affect the development of the water sector. These are the main challenges of current urban infrastructure development in addition to the size of capital investment needed to maintain and expand structures to meet the growing demand for urban water. As many existing water infrastructures hit the century-old mark, developed and developing regions such as Organization for Economic Co-operation and Development (OECD) countries are facing immediate needs to replace and upgrade infrastructure. They have to respond to costly new water quality regulations and ensure the security of water supplies in response to any threats and climate change, pollution, and growing populations. In New York City, the aging of water infrastructure combined with complexity of the urban settings have made replacement and rehabilitation projects so expensive that utilities and agencies have to compromise in issues such as CSO discharges not being able to separate combined sewers. For all water systems, there is a growing focus on asset management (see Chapter 8) and the best ways to finance and prioritize improvements in operation and maintenance of systems.

PROBLEMS

1. Calculate the water consumption (average daily rate, maximum daily rate, maximum hourly rate, and flow rate) for a town of 10,000 people in semiarid area with limited water supply resources. As it is an old urban area, the buildings do not have good resistance against fire. Almost more than 90% of the individual buildings in this city have an area less than 200 m^2 and the largest building complex is about 1,000 m^2. The only industry in the town, a wool and textile mill with a production of 100 tons/month, has its own water supply. What will be the design rate of water distribution system in the study area? If there are data missing, make your own assumptions and give your reasons for them.
2. A small community with a population of 1,000 has a trucked water supply system that provides water from a lake (3 km from a village). There are 200 houses, one hotel, one hospital, one school, one nursing station, and two general stores in the community and the largest area is about 2,000 m^2. The total road system in the town is 2 km long. Each house is equipped with a water storage tank of 100 L capacity, with bigger tanks in the other establishments. The average water consumption is 40 Lpcd. It is quite common for winter storms to prevent the trucks from traveling to the lake for up to 3 days. Based on this information, determine (a) the size of the storage reservoir in the village and (b) the number of trucks required, if each truck has a 4,000 L tank. The trucks also serve the purpose of providing fire protection. Make any assumption you feel are necessary to complete the assignment, giving reasons for each assumption.
3. Why are distribution pipes not sized according to maximum hourly demand plus fire flow instead of maximum daily demand plus fire flow?
4. Chlorination is the usual method for disinfecting water.
   a. Name the two parameters that control the extent of disinfection.
   b. Why is it necessary to guard against an overdose of chlorine?
   c. Assume that disinfection with chlorine follows a first-order reaction (the disinfection rate has a direct relation to its concentration). In a chlorinated water sample containing 1.0 mg/L chlorine, the initial concentration of viable bacteria is 100,000/mL. At the end of a 5-minute contact time, the number of viable bacteria has decreased to 10/mL. What effect would a contact time of 10 minutes have on the bacteria count?
5. Water having a temperature of 15°C is flowing through a 150 mm ductile iron main at a rate of 19 L/s. Is the flow laminar, turbulent, or transitional?
6. Assuming that there are no head losses through the Venturi meter demonstrated in Figure 6.37, what is the pressure reading in the throat section of the Venturi? Assume that the discharge through the meter is 0.6 m^3/s.
7. Does conservation of energy apply to the system represented in Figure 6.38? Based on the conservation of energy, the summation of head losses in a loop should be equal to zero. Data describing the physical characteristics of each pipe are presented in Table 6.6 and neglect the minor losses in this loop.
8. Find the pump head needed to deliver water from the lower reservoir to the upper reservoir in Figure 6.39 at a rate of 0.3 m^3/s. The suction pipe length, diameter, and roughness coefficient are 20 m, 200 mm, and 130, respectively. The discharge pipe length and diameter are 200 m and 300 mm and the Hazen–Williams roughness coefficient is 110. Minor head loss coefficients are given in Figure 6.39.
9. Manually find the discharge through each pipeline and the pressure at each junction node of the rural water system shown in Figure 6.40. Physical data for this system are given in Table 6.7.
10. In the network depicted in Figure 6.41, determine the discharge in each pipe. Assume that $f=0.015$.

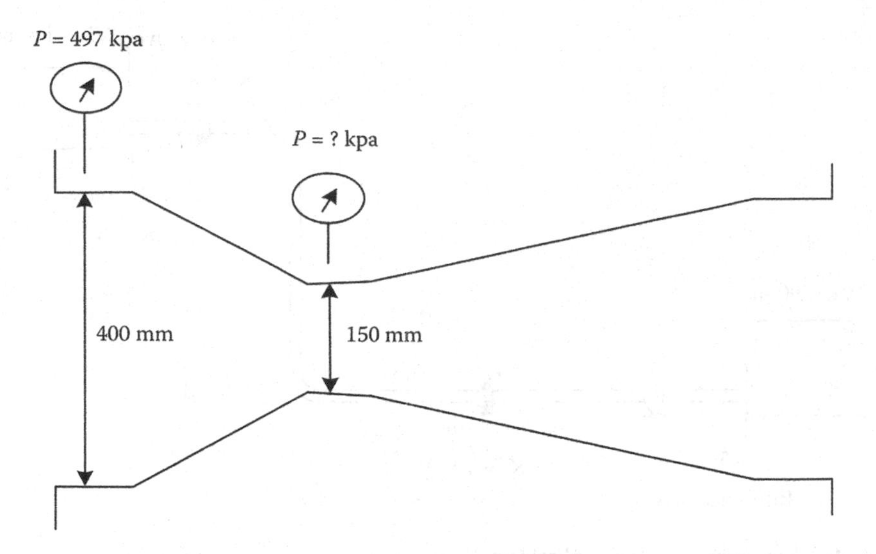

FIGURE 6.37 Venturi of Problem 6.

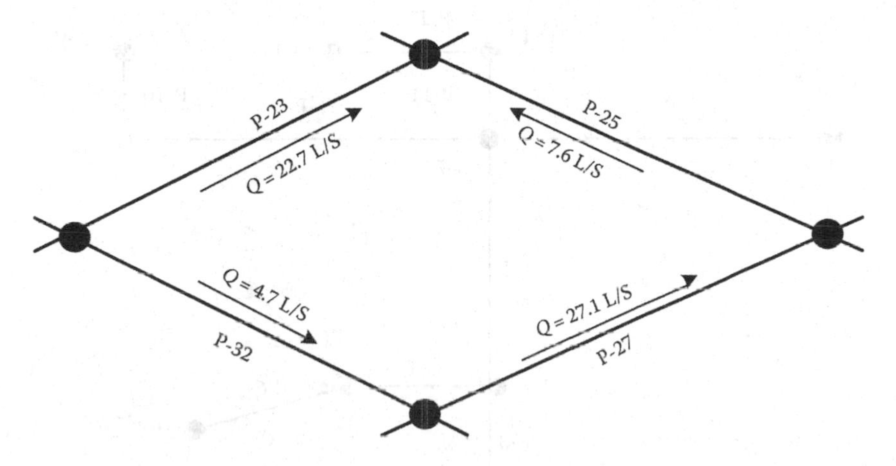

FIGURE 6.38 Pipeline of Problem 7.

TABLE 6.6
Characteristics of the Network of Problem 7

Pipe Label	Length (m)	Diameter (mm)	Hazen–Williams C-Factor
P-23	381.0	305	120
P-25	228.6	203	115
P-27	342.9	254	120
P-32	253.0	152	105

11. A fire hydrant is supplied through three welded steel pipelines (f=0.012) arranged in series (Table 6.8). The total drop in pressure due to friction in the pipeline is limited to 35 m. What is the discharge through the hydrant?
12. The characteristics of the pipe system illustrated in Figure 6.42 are presented in Table 6.9. (a) Determine the diameter of the equivalent pipe of this system with 600 m length and the

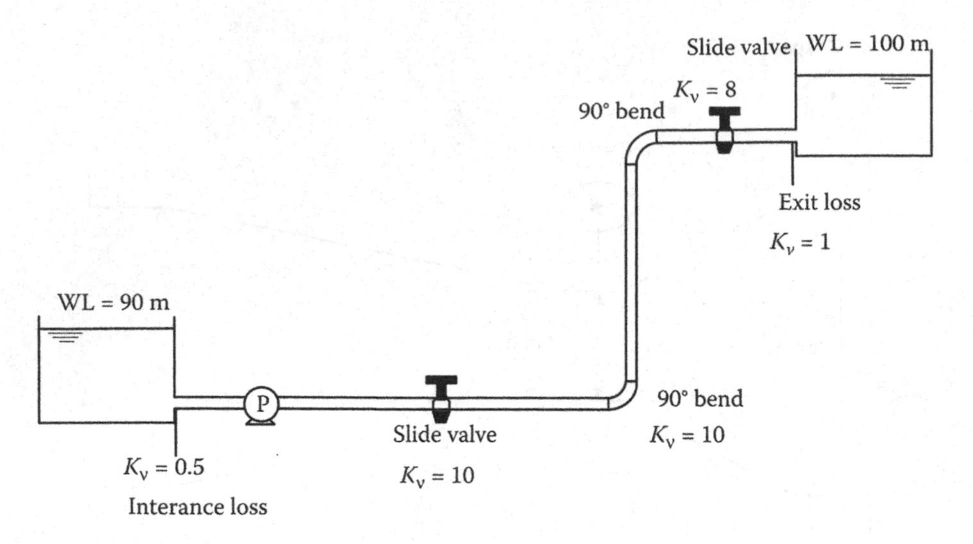

FIGURE 6.39 The reservoir system of Problem 8.

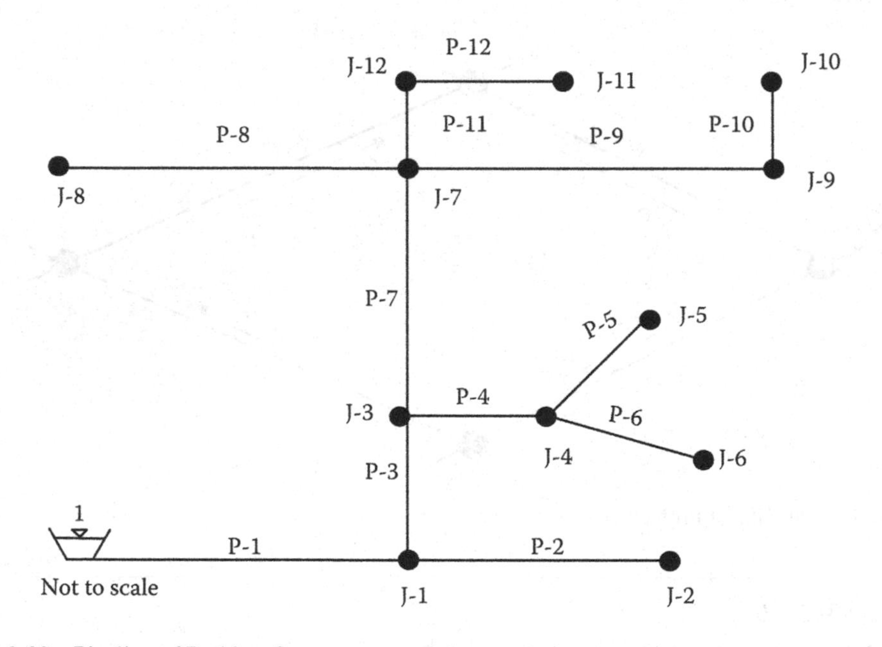

FIGURE 6.40 Pipeline of Problem 9.

TABLE 6.7
Physical Data of the Pipeline System of Problem 9

Pipe Label	Length (m)	Diameter (mm)	Hazen-Williams Roughness Coefficient (C)	Node Label	Elevation (m)	Demand (L/s)
P-1	152.4	254	120	R-1	320.0	N/A
P-2	365.8	152	120	J-1	262.1	2.5
P-3	1,280.2	254	120	J-2	263.7	0.9
P-4	182.9	152	110	J-3	265.2	1.9
P-5	76.2	102	110	J-4	266.7	1.6

(Continued)

TABLE 6.7 (*Continued*)
Physical Data of the Pipeline System of Problem 9

Pipe Label	Length (m)	Diameter (mm)	Hazen-Williams Roughness Coefficient (C)	Node Label	Elevation (m)	Demand (L/s)
P-6	152.5	102	100	J-5	268.2	0.3
P-7	1,585.0	203	120	J-6	269.7	0.8
P-8	1,371.6	102	100	J-7	268.2	4.7
P-9	1,676.4	76	90	J-8	259.1	1.6
P-10	914.4	152	75	J-9	262.1	0
P-11	173.7	152	120	J-10	262.1	1.1
P-12	167.6	102	80	J-11	259.1	0.9
–	–	–	–	J-12	257.6	0.6

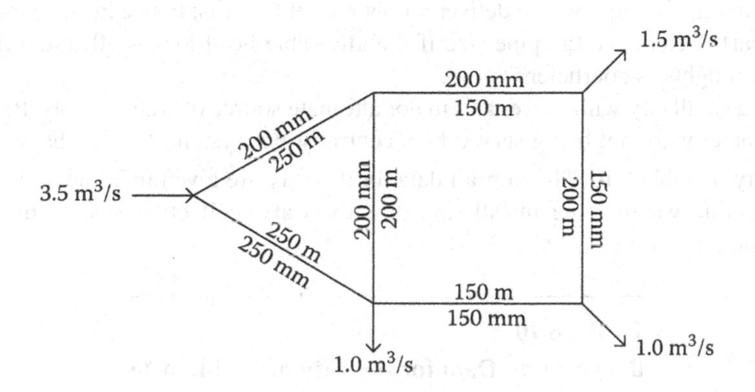

FIGURE 6.41 Layout of the network of Problem 10.

TABLE 6.8
Characteristics of Fire Hydrant System of Problem 11

Pipe	Diameter (mm)	Length (m)
1	150	1,500
2	200	6,000
3	300	48,000

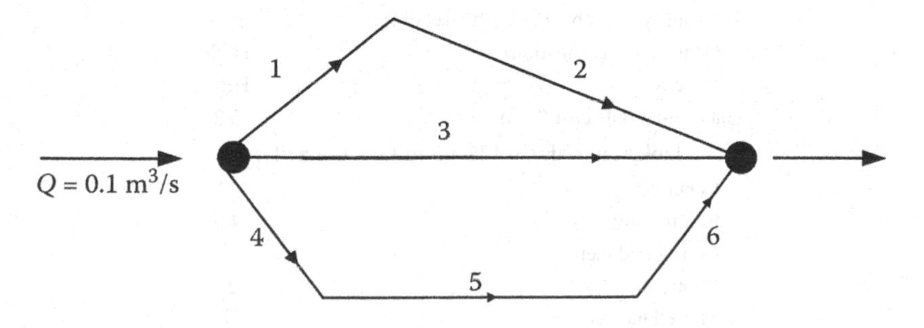

FIGURE 6.42 The layout of network of Problem 12.

TABLE 6.9
Characteristics of Pipe System of Problem 12

Pipe No.	Length (m)	Diameter (mm)	Hazen–Williams Roughness Coefficient
1	300	250	120
2	400	200	110
3	600	200	100
4	200	200	90
5	300	250	110
6	250	200	120

Hazen–Williams roughness coefficient is equal to 130. (b) Determine the Hazen–Williams roughness coefficient of an equivalent pipe of this system with the length and diameter equal to 600 m and 300 mm, respectively.

13. A cast iron pipe is employed to deliver a flow rate of 0.1 m^3/s, between two points that are 800 m apart. Determine the pipe size if the allowable head loss is 10 m and the Hazen–Williams roughness coefficient is 130.
14. Consider a small city with a river as a major alternate source of water supply. People on the edge of the city are not being served by a central water system. See the background data for the city in Table 6.10. The rainfall data for this city are given in Table 6.11.
    A. Determine whether the installation of a rainwater collection system in this city is feasible.

TABLE 6.10
Background Data for the City of Problem 14

Typical Demand Values (L/day/person)	
Drinking and cooking	4.7
Hygiene	7.9
Washing	5.8
Population in the design area (persons)	20,000
Average household size	5
Average roof area (m^2)	40
Roof Material (% of Each Kind in the Region)	
Pervious (thatched, etc.)	5
Impervious (metal, clay, etc.)	80
Nonresidential (%)	15
Other Water Sources	
Distance (km)	2.0
Quantity available (L/day/household)	900
Quality (contamination)	High
Minerals	High
Gutter materials cost ($/m)	0.8
Tank Materials Cost ($/$m^3$ of Tank Capacity)	
Concrete	12
Reinforcing	4.5
Corrugated metal	35
Plastic	28
Miscellaneous	7

TABLE 6.11
Rainfall (mm) for the City of Problem 14

Jan	Feb	Mar	Apr	May	Jun	Jul	Aug	Sep	Oct	Nov	Dec	Yearly Totals
21.4	68.2	27.5	42.8	71.0	1.9	0.0	0.0	13.0	0.4	25.4	21.9	293.5
37.0	47.6	57.0	31.8	41.1	12.2	0.0	3.0	33.0	14.1	20.0	49.2	346.0
40.8	93.0	42.8	25.9	6.8	5.0	0.0	0.0	5.3	16.9	42.6	25.6	304.7
16.8	18.9	39.0	68.5	84.8	10.1	66.4	1.2	15.3	24.5	17.1	19.8	382.4
41.1	46.3	28.0	3.5	27.9	16.4	13.0	0.1	12.3	26.2	26.9	29.7	271.4
24.8	16.6	18.8	17.7	14.6	0.2	0.0	3.0	4.2	0.7	15.1	27.3	143.0
16.8	20.3	43.1	48.0	36.0	1.7	2.0	0.0	2.6	24.5	47.9	30.5	273.4
26.8	25.2	18.0	16.2	3.9	3.0	0.0	7.3	0.0	14.2	26.8	24.9	166.3
16.6	15.1	21.4	49.2	27.9	34.4	10.7	26.0	7.9	11.1	35.4	55.2	310.9
5.5	22.3	97.3	20.8	32.9	0.0	40.2	2.7	2.2	27.6	18.2	13.8	283.5

B. Prepare a list of support services needed for users/owners including what must be done, who is responsible, means to supply service, and benefits.

There are no social conditions that would prevent from the building of the rainfall storage tanks. However, special loans for owners must be arranged because they are not served by the central water system and the people who live in edge of city have low income.

15. Estimate the storage tank size of the city described in Problem 1, using the curve mass method.
16. Historical data for water use, population, price, and precipitation in a city for a 10-year time horizon are given in Table 6.12. Formulate a multiple regression model for estimating the water use data and comment on selecting the independent variables.
17. The monthly data of water price and consumption of a city is given in Table 6.13. Determine the price elasticity. Discuss the effectiveness of applying pricing strategies for decreasing water consumption.
18. Consider a city with about 1 million habitants. The daily water consumption per capital is about 180L, the average temperature is 28°C, and the weather humidity is <35% in most days. The capital daily income is about $60. There are limited surface water resources near

TABLE 6.12
Historical Data for Water Use, Population, Price, and Precipitation for a 10-Year Time Horizon in the City of Problem 16

Year	Population	Price (Monetary Unit/m³)	Annual Precipitation (mm)	Water Use (m³/year)
2000	21,603	0.25	1015	2,540,513
2001	22,004	0.27	810	2,754,901
2002	23,017	0.29	838	2,722,911
2003	23,701	0.30	965	2,832,270
2004	24,430	0.31	838	2,904,727
2005	25,186	0.32	787	3,004,690
2006	26,001	0.33	720	3,309,927
2007	26,825	0.35	695	3,382,633
2008	27,781	0.31	820	3,375,392
2009	28,576	0.34	800	3,406,259

TABLE 6.13
Price and Water Consumption Data of Problem 17

Consumption (MCM)	260	320	185	200	280	310	290	250	210
Price ($/cm)	13	10	20	17	12	9	12	13	18

this city. There is a big aquifer below the city, but its mineral concentration is high, and in some places, it is contaminated by septic tanks. Discuss water demand strategies that could be effective in this city.

19. The linear demand model derived by Hanke and de Marg (1982) for Malmo, Sweden, is

$$Q = 64.7 + 0.00017(Inc) + 4.76(Ad) + 3.92(Ch) - 0.406(R) + 29.03(Age) - 6.42(P),$$

where

Q is the quantity of metered water used per house per semiannual period (m^3).
Inc is the real gross income per house per annum (in Swedish crowns; actual values reported per annum and interpolated values used for mid-year periods).
Ad is the number of adults per house per semiannual period.
Ch is the number of children per house per semiannual period.
R is the rainfall per semiannual period (mm).
Age is the age of the house, with the exception that it is a dummy variable with a value of 1 for houses built in 1968 and 1969, and a value of 0 for houses built between 1936 and 1946.
P is the real price in Swedish crowns/m^3 of water per semiannual period (includes all water and sewer commodity charges that are a function of water use).

Using the average values of $P = 1.5$ and $Q = 81.4$ for the Malmo data, determine the elasticity of demand and interpret the result.

20. Determine the elasticity of demand for the water demand model in Problem 19 using $P = 1.5$ and $Q = 75.2$; $P = 3.0$ and $Q = 75.2$; $P = 2.25$ and $Q = 45$; and $P = 2.25$ and $Q = 100$.
21. Consider a city with current population of 1 million (year 2009). We are planning to develop the water supply facilities of this city during the next 20 years. The water consumption per capita for this city is presently 320 L/day. The share of different consumption components relative to water consumption per capita is given in Table 6.14. Available water supply resources of this city include two surface reservoirs with an average water allocation capacity 100 and 40 MCM/year. The population growth rate of this city has been determined as 2.3%/year. How would the state of water resources availability be in the year 2029 regarding to water consumption?

TABLE 6.14
The Share of Different Consumption Components Relative to Water Use Per Capita for the City of Problem 21

Share of Use (%)	Water Use Per Capita	Users
56	180	Residential
9	30	Public
8	25	Industrial commercial
9	30	Public green space
17	55	Losses
100	320	Sum

22. In order to fill the water supply and demand gap of the city of Problem 21 for the year 2029, the following demand management strategies are recommended:
    a. Water withdrawn from a groundwater source up to 20 MCM/year.
    b. Reduction of network losses from 17% of total water supply to 15% by implementing a pressure management program, improving main breaks management and leakage management.
    c. Reduction of landscaping water demand up to one-third of the present figure by utilizing plants resistant to water shortage and irrigation with recycled water.
    d. Reduction of water demand in residential, public, industrial, and commercial sectors through enforcing legislation and public education up to 10%.

    Discuss how the application of these strategies will be useful in mitigating water shortages of this city in 2029.
23. Consider a small city with a population of 100,000. The water consumption per capita in this city is about 250 L/day. The available water supply system of this city includes a surface reservoir with an annual water supply capacity of 15 MCM. The population growth rate has been determined equal to 2% and the migration rate is about 0.5%. In the current rate of water consumption per capita will increase 1% each year. There is also an aquifer near the city with a water supply capacity of 5 MCM/year. The treatment plant of this city has limitation on the TDS intake of 300 mg/L. The TDS of surface water and groundwater is 250 and 400 mg/L, respectively. The cost of water supply from surface water is about 0.1$/L, where the cost of water withdrawn from groundwater is three times the surface water. The water supply manager of the city has planned some demand management studies with estimated reduction in water consumption and cost as given in Table 6.15. Develop an optimization model for water supply development and demand management schemes in this city for the next 50 years. The objective of the optimization model is to minimize the water deficit and water supply cost. The rate of return on investments is considered equal to 5%. Make any other assumption needed to solve this problem.
24. A survey on a water distribution network with 25% loss shows that 18% of the losses are due to physical losses of the network and the remaining is not a real loss. Fifty percent of physical losses are apparent and reported by people to the water supply agency. The survey result shows that 25% of nonapparent losses could be avoided by low-cost rapid surveys on the system. Determine the efficiency of this water supply system through estimation of different leakage indices.
25. Solve Example 6.4 by using dynamic programming.

    A dam supplies water to a city and downstream agricultural lands. The average monthly domestic and irrigation demands and the monthly reservoir release in a dry year are as shown in Table 6.16. The prices of water for domestic and irrigation uses are $90,000 and $10,000 per million cubic meters, respectively.
    - Formulate the problem for optimizing the water allocation from this reservoir.
    - Solve the problem using linear programming.

TABLE 6.15
Considered Water Demand Management Strategies in the City of Problem 23

Demand Management Strategy	Maximum Reduction in Consumption	Cost ($) Per Each Liter Reduction in Consumption
Training programs	2%	0.02
Leakage reduction	8%	0.5
Using of water consumption reduction strategies	1.5%	0.6
Use of recycled water for landscaping/parks irrigation	6%	0.2

TABLE 6.16
The Average Monthly Domestic and Irrigation Demands and the Monthly Reservoir Release of Problem 35 (Numbers are in Million Cubic Meters)

	Jan	Feb	Mar	Apr	May	Jun	Jul	Aug	Sep	Oct	Nov	Dec
Domestic	8	8.5	8.5	9	9	9.5	9.5	10	9.5	9	8.5	8.5
Irrigation	0	0	20	50	55	70	70	65	20	10	0	0
Release	25	27	40	48	60	42	25	17	14	18	20	25

26. In Problem 23, the shortages should be met by supplying from groundwater resources. The cost of meeting the demand by supplying from groundwater is a function of the volume of water that should be extracted. The cost of water extraction for domestic demands is estimated as $C_{\text{Dom}} = 1000 \cdot x_{\text{Dom}}^2$, where x_{dom} is the volume of water extracted for domestic uses. The cost of water extraction for irrigation purposes is estimated as $C_{\text{irr}} = 250 \cdot x_{\text{irr}}^2$, where x_{irr} is the volume of water extracted for irrigation.
    - Formulate the problem for finding the optimal monthly volumes of water allocation.
    - Solve the problem using linear programming.
27. In Problem 23, the city discharges its wastewater to the river upstream of the agricultural lands. The monthly wastewater discharge rate is about 20% of the water use of the city in each month. In order to keep the quality of the river flow in an acceptable range for irrigation, the pollution load of the city wastewater should be reduced by primary treatment, which has an additional cost equal to $12,000 per million cubic meters.
    - Formulate the problem for optimizing the water allocation from this river.
    - Solve the problem using linear programming.

REFERENCES

ASCE Task Committee on Sustainability Criteria. (1998). *Sustainability Criteria for Water Resource Systems.* ASCE, Reston, VA.

Baumann, D.D., Boland, J.J., and Hanemann, W.M. (1998). *Urban Water Demand Management and Planning.* McGraw-Hill, New York.

Bhardwaj, V. (2001). *Reservoirs, Towers, and Tanks.* Tech Brief, Fall. National Drinking Water Clearinghouse, Morgantown, WV.

Brix, H. (1998). How 'Green' are constructed wetland treatment systems? *Proceedings of 6th International Conference on Wetlands Systems for Water Pollution Control*, Aguas de Sao Pedro, Brazil, September 27–October 2.

Butler, D. and Parkinson, J. (1997). Towards sustainable drainage. *Water Science and Technology*, 35(9), 53–63.

Falkland, A. (ed.) (1991). *Hydrology and Water Resources of Small Islands: A Practice Guide.* United Nations Educational, Scientific, and Cultural Organization (UNESCO), Paris.

Farahbakhshazad, N. and Morrison, G.M. (2000). A constructed vertical macrophyte system for the return of nutrients to agriculture. *Environmental Technology*, 21, 217–224.

Goldblatt, M., Ndamba, J., and van der Merwe, B. (2000). *Water Demand Management: Towards Developing Effective Strategies for Southern Africa.* IUCN, Gland.

Gupta, R.S. (1989). *Hydrology and Hydraulic Systems.* Prentice Hall, Englewood Cliffs, NJ.

Hanke, S.H. and de Marg, L. (1982). Residential water demand: A pooled time series, cross section study of Mamlo. *Water Resources Bulletin*, 18(4), 621–625.

Islam, M.R. (1995). Modeling of chlorine concentration in unsteady flow in pipe networks. Ph.D. thesis, Washington State University.

Jeffrey, P., Seaton, R., Parsons, S., and Stephenson, T. (1997). Evaluation methods for the design of adaptive water supply systems in urban environments. *Water Science and Technology*, 35(9), 45–51.

Karamouz, M., Ahmadi, A., and Akhbari, M. (2020). *Groundwater Hydrology: Engineering, Planning, and Management* (p. 750) 2nd Ed., CRC Press, Boca Raton, FL.

Karamouz, M., Goharian, E., and Nazif, S. (2012). Development of a reliability based dynamic model of urban water supply system: A case study. *World Environmental and Water Resources Congress 2012: Crossing Boundaries*, 20–24 May 2102, Albuquerque, NM, pp. 2067–2078.

Karamouz, M., Moridi, A., and Nazif, S. (2010). *Urban Water Engineering*. CRC Press, Boca Rotan, FL.

Karamouz, M., Szidarovszky, F, and Zahraie, B. (2003a).*Water Resources Systems Analysis* (600 pp.). Lewis Publisher, Boca Raton, FL.

Karamouz, M., Zahraie, B., and Khodatalab, N. (2003b). Reservoir operation optimization: A nonstructural solution for control of seepage from Lar Reservoir in Iran. *IWRA – Water International*, 28(1), 19–26.

Keffer, C., Shimp, R., and Lehni, M. (1999). Eco-efficiency indicators & reporting. Report on the *Status of the Project's Work in Progress and Guideline for Pilot Application*, WBCSD Working Group on Eco-efficiency Metrics and Reporting, WBCSD, Geneva.

Lambert, A. (2003). Assessing non-revenue water and its components: A practical approach. *Water*, 21(2), 50–51.

Larsen, T.A. and Gujer, W. (1997). The concept of sustainable urban water management. *Water Science and Technology*, 35(9), 3–10.

Liemberger, R. (2005). Real losses and apparent losses and the new W392 Guidelines from Germany. Paper presented at the *International Water Association Specialist Workshop*, Radisson Resort, Gold Coast, Queensland Australia, 24 February.

Masters, G.M. and Ela, W.P. (2008). *Introduction to Environmental Engineering and Science* (No. 60457). Prentice Hall, Englewood Cliffs, NJ.

Mays, L.W. (1996). *Water Resources Handbook*. McGraw-Hill, New York.

McKee, R.H. (1960). A new American paper pulp process, part I—Pulping by the full hydrotropic process. *Paper Industry*, 42(4), 255–257, 266.

Metcalf and Eddy. (2003).*Wastewater Engineering, Treatment and Reuse*. (4th ed.). McGraw-Hill, New York.

Milburn, A. (1996). Aglobal freshwater convention—The best means towards sustainable freshwater management. *Proceedings of the Stockholm Water Symposium*, Stockholm, Sweden, 4–9 August, pp. 9–11.

Niemczynowicz, J. (1996). Mismanagement of water resources. *Vatten*, 52, 299–304.

Owens, J.W. (1999). Why life cycle impact assessment is now described as an indicator system. *International Journal of LCA*, 4(2), 81–86.

Powell, J.C., Hallam, N.B., West, J.R., Forster, C.F., and Simms, J. (2000). Factors which control bulk chlorine decay. *Journal of Water Resources Planning and Management*, 126(1), 13–20.

Rossman, L.A. (1999). The EPANET programmer's toolkit for analysis of water distribution systems. Proceeding of ASCE WRPM conference, Tempe, AZ, https://doi.org/10.1061/40430(1999)39

Royer, M.D. (2005). White paper on improvement of structural integrity monitoring for drinking water mains. EPA/600/R-05/038, Cincinnati, OH, 45268.

Safavi, H.R., Golmohammadi, M.H., and Sandoval-Solis, S. (2015). Expert knowledge based modeling for integrated water resources planning and management in the Zayandehrud River Basin. *Journal of Hydrology*, 528, 773–789.

Schaltegger, S. (1996). *Corporate Environmental Accounting*. Wiley, Chichester.

Solley, W.B., Pierce, R.R., and Perlman, H.A. (1993). *Estimated Use of Water in the United States*, U.S. Geological Survey Circular 1081. U.S. Government Printing Office, Washington, D.C.

U.S. Bureau of Reclamation. (1981). *Groundwater Manual*. U.S. Government Printing Office, Denver, CO.

Værbak, M., Ma, Z., Christensen, K., Demazeau, Y., and Jørgensen, B.N. (2019). Agent-based modelling of demand-side flexibility adoption in reservoir pumping. *2019 IEEE Sciences and Humanities International Research Conference (SHIRCON)*, November, IEEE, Lima, Peru, pp. 1–4.

Walski, T.M. (1996). Water distribution. In: L.W. Mays (ed.) *Water Resources Handbook*. Ch. 18, pp. 1–41, McGraw-Hill, New York.

Walski, T.M., Chase, D.V., Savic, D.A., Grayman, W.M., Bechwith, S., and Koelle, E. (2001). *Advanced Water Distribution Modeling and Management*. Heastad Methods, Watertown, CT.

Westfall, D.E. (1996). *Water Conveyance Tunnels. Tunnel Engineering Handbook* (pp. 298–310). Springer, New York. doi: 10.1007/978-1-4613-0449-4_15.

Willis, R. and Yeh, W.W.-G. (1987). *Groundwater Systems Planning and Management*. Prentice Hall, Englewood Cliffs, NJ.

WHO (2017). Guidelines for drinking-water quality, 4th edition, incorporating the 1st addendum. World Health Organization, Geneva, Switzerland, p. 631.

World Resources Institute. (2011). *World Resources 2010–2011*. A Joint Publication by the World Resources Institute, the United Nations Environment Program and the World Resource Institute, London.

7 Wastewater Infrastructure

7.1 INTRODUCTION

It has been estimated that it will cost $16 billion a year to meet the regional Millennium Development Goals (MDGs) for water supply and sanitation to cover all the unserved population. This is a significant number but not an impossible figure. The Millennium Development Task Force defines basic sustainable sanitation for securing access to safe, hygienic, and convenient facilities and services for waste water and sludge disposal. At the same time, it ensures a clean and healthy living environment both at home and in the neighborhood. This will allow low-cost options that can be adopted while governments build their full sewerage collection and treatment system.

Improving poor wastewater management among some other factors stopping the degraded urban water, including water contamination and supplying low-quality water supply are in forefront of the so-called development objectives of many society. This has led to the advocacy of the sustainable urban water system discussed in Chapter 6 with the objective related to this chapter that can be stated as: Collection and treatment of wastewater in order to protect the inhabitants against diseases as well as the environment from harmful impacts.

The urban areas' populations are expected to increase dramatically, especially in Africa, Asia, and Latin America. The population of urban areas in Africa is expected to rise more than two times over the next 25 years. The Caribbean and Latin America's urban population is expected to increase by almost 50% over the same period (WHO and UNICEF, 2000). Consequently, to meet the fast-growing necessities, urban services will face great challenges over the coming decades. Wastewater management with the provision of urban wastewater treatment (WWT) services is the main concern of this chapter.

In the remaining part of this chapter, different issues and challenges in urban wastewater management are presented. WWT, an essential factor in urban wastewater planning, is discussed and followed by a detailed description of each part of the treatment plant. Both quantitative and qualitative aspects of WWT are addressed in this chapter. Then the concepts of wastewater planning are introduced. Afterward, the case studies related to this chapter as well as some selected cities wastewater infrastructures are described. Many other case studies in other chapters are also related to wastewater treatment plants' (WWTPs') vulnerability to coastal flooding. Many WWT facilities are located near water bodies and are prone to system failure due to flooding with considerable consequences for the communities in their sewershed. Some planning and standard issues related to water utilities in general and wastewater in particular are covered in the later part of this chapter.

7.2 THE IMPORTANCE OF WASTEWATER SYSTEMS

According to Ujang and Henze (2006), only 5% of global wastewater, mainly in developed countries, is appropriately treated. As a result, most of the world's population is still struggling with waterborne diseases resulting from the lack of treatment facilities or substandard design and operation. Therefore, the quality of water resources has been rapidly degraded, particularly in poor developing countries. At the beginning of the twenty-first century, about 1.1 billion people lived without access to clean water, 2.4 billion without appropriate sanitation, and 4 billion without sound wastewater disposal.

In recent years, the "conventional sanitation" approach has received widespread criticism, and the so-called sustainable sanitation has been proposed. Sustainable sanitation is flexible and can be applied in any community—poor or rich, urban or rural, water-rich or water-poor countries—and

DOI: 10.1201/9781003241744-7

requires lower investment costs than conventional sanitation approaches. The implementation of sustainable sanitation is much easier in developing countries where water infrastructures are still developing. Ujang and Henze (2006) stated that there are no available public facilities in some developing countries. So it is an excellent opportunity to start a new infrastructure with a new insight to today's challenges.

7.3 WASTEWATER MANAGEMENT

The specific technologies that can be tailor-made to a specific urban and rural setting vary from place to place. Lohani (2005) stated that in dispersed, low-income rural areas, the appropriate technology might be a simple pit latrine, whereas, in a congested urban area with reliable water service, it may be a low-cost sewerage system.

There are some advantages to centralized schemes, but decentralized WWT schemes are more sustainable. The latter is more suitable for achieving sustainability when compared with centralized technology. It makes very small, and community-scale technologies are more easily achievable. Its application in remote and rural areas needs improved technology with integration with other technology sectors (energy and food production) to improve overall sustainability.

There is a high potential in using on-site, small, and community-scale technology in developing countries to achieve sustainability outcomes. It is likely to succeed if the technology is modified to be of lower cost and adaptable to local climates based on the same principles and engineering (Ho, 2004). Technology and technology choices are the key issues, even though technology does not provide the full answer. Even though the classical wastewater management concept has been applied for many years in populated areas of developing countries, there are some disadvantages in utilizing small-scale systems (Ho, 2004).

Decentralized wastewater management systems, with the wastewater treated close to where it is generated, are being considered by various institutions, including the World Bank, as an alternative to the traditional centralized system. However, the degree of technological sophistication that should be applied is in question. The development and application of high-tech on-site treatment plants designed and fabricated by modern industrial techniques can be a solution. When mass-produced, the costs of manufacturing such package plants can be kept at a relatively low level. The plant should produce a hygienically safe effluent that can subsequently be utilized for toilet flushing, washing clothes, cleaning floors, or watering lawns (Wilderer, 2004).

A WWT facility must be able to function under all conditions. Failure to operate can lead to raw or partially treated sewage being discharged into rivers, lakes, and other bodies of water or backing up into streets, houses, and industries. The threat that storms and other hurricanes pose to WWT works is a direct environmental threat to communities and the public. As a result, most wastewater plants have precautions and strategies to remain in service even under extreme conditions. Nonetheless, as storms grow more frequent and more powerful, further improved organization and mitigation measures are desired. Wastewater facilities should prepare for flooding, power outages, equipment damage and failures, and much more. Historically, wastewater plants have been located near the waters to which they discharge. This often makes the facilities exposed to flooding, storm surge, and climate change issues (NEIWPCC, 2016).

7.4 WASTEWATER TREATMENT

Municipal wastewater is typically over 99.9% water. The remaining portion's characteristics vary somewhat from city to city, with the variation depending primarily on inputs from industrial facilities that mix with the somewhat predictable composition of residential flows. Given the almost limitless combinations of chemicals found in wastewater, it is too difficult to list them individually. Instead, they are often described by a few general categories, as shown in Table 7.1. In this table, a distinction is made between total dissolved solids (TDS) and suspended solids (SS). The sum of

TABLE 7.1
Composition of Untreated Domestic Wastewater

Constituent	Abbreviation	Concentration (mg/L)
5-day biochemical oxygen demand	BOD_5	100–350
Chemical oxygen demand	COD	250–1,000
Total dissolved solids	TDS	250–1,000
Suspended solids	SS	100–400
Total Kjeldahl nitrogen	TKN	20–80
Total phosphorous (as P)	TP	5–20

the two is total solids (TSs). The SS portion is, by definition, the portion of TS that can be removed by a membrane filter (having a pore size of about). The remainder (TDS) that cannot be filtered includes dissolved solids, colloidal solids, and very small suspended particles. WWTPs are usually designated as providing primary, secondary, or advanced treatment, depending on the purification degree. These steps include four types of treatment, physical, biological, chemical, and membrane processes.

- Physical processes for mechanical preparation such as aeration, sedimentation, or thermal influence. This also includes the use of screens, filters, and sieves
- Biological processes such as anaerobic WWT, biochemical oxidation, or sludge digestion
- Chemical processes such as neutralization, disinfection, flocculation, and precipitation
- Membrane processes such as filtration, osmosis, and nanofiltration

In primary treatment plants, physical processes, such as screening, skimming, and sedimentation, are used to remove pollutants that are settleable, floatable, or too large to pass through simple screening devices. About 35% of the biochemical oxygen demand (BOD) and 60% of the SSs can be removed by primary treatment. While the most visibly objectionable substances are removed in primary treatment, some degree of safety is generally provided by disinfection after primary treatment. The effluent still has enough BOD to cause oxygen depletion problems and enough nutrients, such as nitrogen and phosphorus, to accelerate eutrophication. The Clean Water Act (CWA), in essence, requires at least secondary treatment for all publicly owned treatment works (POTWs) by stipulating that such facilities provide at least 85% BOD removal (with possible case-by-case variances that allow lower percentages for marine discharges). This translates into an effluent requirement of 30 mg/L for both 5-day BOD and SSs (monthly average). In secondary treatment plants, the physical processes that make up primary treatment are augmented with processes that involve the microbial oxidation of wastes. Such biological treatment mimics nature by utilizing microorganisms to oxidize the organics.

The advantage is that the oxidation can be done under controlled conditions in the treatment plant itself, rather than in the receiving body of water. When properly designed and operated, secondary treatment plants remove about 90% of the BOD and 90% of the SSs. Although the main purpose of primary treatment (in addition to disinfecting the wastes) is to remove objectionable solids, and the principal goal of secondary treatment is to remove most of the BOD, neither is effective at removing nutrients, dissolved material (metals, salts), or biologically resistant (refractory) substances. For example, typically, no more than half of the nitrogen and one-third of the phosphorus is removed during secondary treatment. This means the effluent can still be a major contributor to eutrophication problems. In circumstances where either the raw sewage has particular pollutants of concern or the receiving body of water is especially sensitive, so-called advanced treatment (previously called tertiary treatment) may be required. In many parts of the United States, advanced nutrient removal is now required in POTWs. Advanced treatment processes are varied and specialized, depending

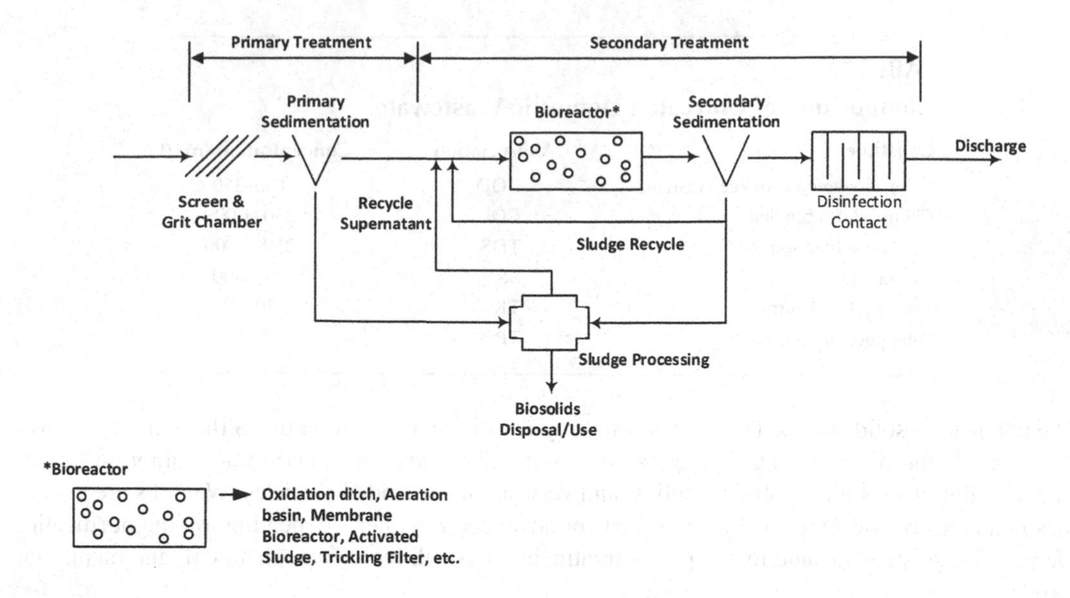

FIGURE 7.1 Schematic of a typical wastewater treatment facility providing primary and secondary treatment (Source: Masters, G.M. and Ela, W.P., *Introduction to Environmental Engineering and Science* (No. 60457), Prentice Hall, Englewood Cliffs, NJ, 2008.)

on the pollutants' nature that must be removed. In most circumstances, advanced treatment follows primary and secondary treatment, although in some cases, especially in the treatment of industrial waste, it may completely replace those conventional processes. An example flow diagram for a WWT plant that provides primary and secondary treatment is illustrated in Figure 7.1. A major distinction of the WWT compared with water treatment presented in Chapter 6 is the Bioreactor.

7.4.1 Primary Treatment

Primary treatment begins with screening, as shown in Figure 7.2. Screening removes large floating objects such as rags, sticks, and whatever else is present that might otherwise damage the pumps or pipes. Screens vary but typically consist of parallel steel bars spaced anywhere from 2 to 7 cm apart, perhaps followed by a wire mesh screen with smaller openings. After screening, the wastewater passes into a grit chamber, where it is held for up to a few minutes. The detention time (the tank volume

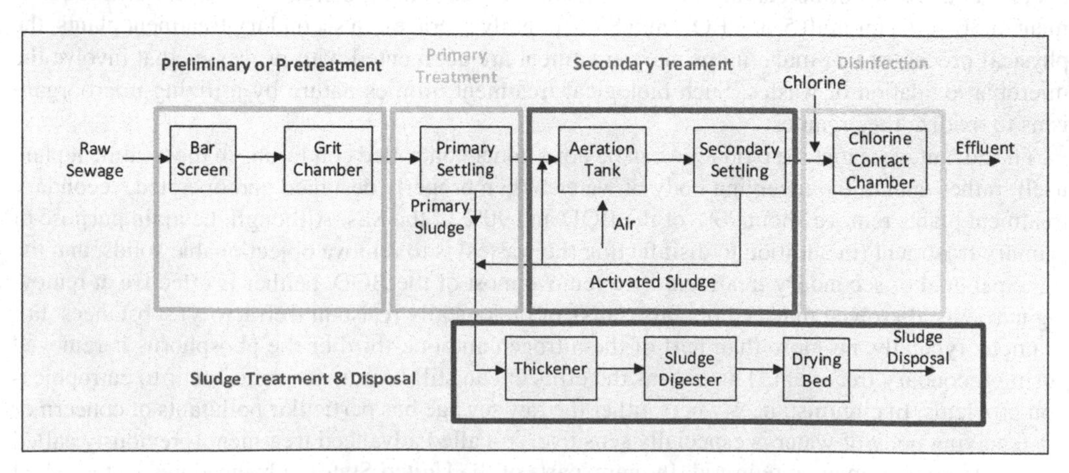

FIGURE 7.2 Components of a municipal wastewater system flow diagram.

divided by the flow rate) is chosen to be long enough to allow sand, grit, and other heavy material to settle out. However, it is too short to allow lighter, organic materials to settle. Detention times are typically 20–30 s (seconds). From the grit chamber, the sewage passes to a primary settling tank (also known as a sedimentation basin or a primary clarifier), where the flow's speed is reduced sufficiently to allow most of the SSs to settle out by gravity. Detention times of approximately 1.5–3 hours are typical, resulting in the removal of 50%–65% of the SSs and 25%–40% of the BOD (Masters and Ela, 2008).

Primary settling tanks are either round or rectangular. The solids that settle, called primary sludge or raw sludge, are removed for further processing, as is the grease and scum that floats to the top of the tank. If this is just a primary treatment plant, the effluent at this point is disinfected (typically by chlorine or UV light) to destroy bacteria and help control odors. Then it is released. By adding various coagulants and ballast (additives such as sand that make the settling particles more dense) and specialized settling equipment within the clarifier, wastewater particles can be made to settle more rapidly and the same particle removal efficiency can be achieved at a higher overflow rate. The range of overflow rates for high-rate clarification is compared for conventional clarification.

7.4.2 Secondary (Biological) Treatment

The objective of secondary treatment is to remove more BOD and SSs than simple sedimentation. There are two approaches, both of which take advantage of the ability of microorganisms to convert organic wastes into stabilized, low-energy compounds. In the first approach, suspended growth treatment, the microorganisms are suspended in and move with the water. In contrast, the microorganisms in attached growth treatment processes are fixed on a stationary surface, and the water flows past the microorganisms.

The WWT objectives are:

a. Transforming the materials available in the wastewater into secure end products that are able to be safely disposed of into domestic water devoid of any negative environmental effects;
b. Protecting public health;
c. Ensuring that wastewaters are efficiently handled on a trustworthy basis without annoyance or offense;
d. Recycling and recovering the valuable components available in wastewaters;
e. Affording feasible treatment processes and disposal techniques;
f. Complying with the legislations, acts and legal standards, and approval conditions of discharge and disposal.

Example 7.1

A town of 30,000 send 0.5 m³ of wastewater per person per day to the WWTP. A conventional circular primary clarifier would be designed to have an average detention time of 2.0 hours and an average overflow rate of 40 m/day, whereas a high-rate clarifier would have an average overflow rate of 1,500 m/day.

a. What would be the dimensions of the conventional clarifier?
b. How much area would be saved by using a high-rate clarifier?

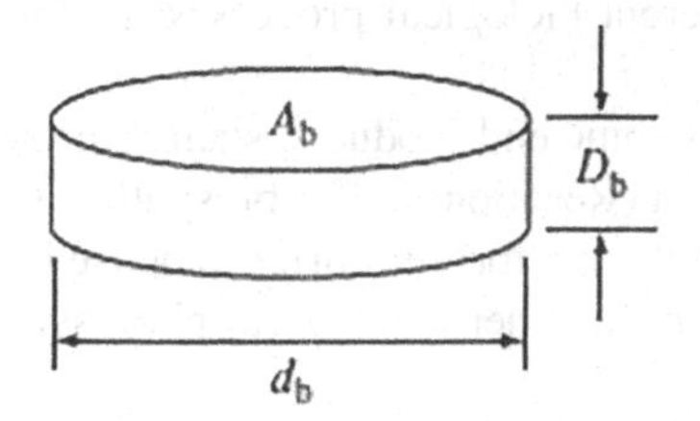

Solution:

At 0.5 m³/person/day, the daily flow for 30,000 people would be 15,000 m³/day. The overflow rate, ν_0 of 40 m/day, then the clarifier surface area is:

$$A_b = \frac{Q}{\upsilon_0} = \frac{15{,}000 \text{ m}^3/\text{day}}{40 \text{ m}^3/(\text{m}^2)/\text{day}} = 375 \text{ m}^2$$

The tank diameter is therefore:

$$d_b = \sqrt{\frac{4\,A_b}{\pi}} = \sqrt{\frac{4\times 375 \text{ m}^2}{\pi}} = 22 \text{ m}$$

The detention time, θ, is the ration of volume to flow rate, and the volume is area times depth, so the clarifier depth is

$$D_b = \frac{Q\theta}{A_b} = \frac{15{,}000 \text{ m}^3/\text{day}\times 2.0 \text{ hours}}{375 \text{ m}^2 \times 24 \text{ hours/day}} = 3.3 \text{ m}$$

As it turns out, this depth is about the minimum that would be considered good design practice.
a. Repeating for the high-rate clarifier case,

$$A_b = \frac{Q}{\upsilon_0} = \frac{15{,}000 \text{ m}^3/\text{day}}{1{,}500 \text{ m}^3/(\text{m}^2)/\text{day}} = 10 \text{ m}^2$$

Thus, the high-rate clarifier would use about 3% of the land area as the conventional clarifier. In reality, the area savings would not be as large as this, because some additional area would be needed for the coagulant/ballast dosing equipment and for a possible flocculation tank.

7.4.2.1 Biological Treatment Processes

The secondary treatment can be defined as the "treatment of wastewater by a process involving biological treatment with a secondary sedimentation." In other words, secondary treatment is a biological process. The settled wastewater is introduced into a specially designed bioreactor where under aerobic or anaerobic conditions, the organic matter is utilized by microorganisms such as bacteria (aerobically or anaerobically), algae, and fungi (aerobically). The bioreactor affords appropriate bioenvironmental conditions for the microorganisms to reproduce and use the dissolved organic matter as energy for themselves. Provided that oxygen and food, in the form of settled wastewater, are supplied to the microorganisms, the biological oxidation process of dissolved organic matter will be maintained. The biological process is mainly carried out by bacteria that form the basic trophic level (the level of an organism is the position it occupies in a food chain) of the food chain inside the bioreactor. The bioconversion of dissolved organic matter into thick bacterial biomass can fundamentally purify the wastewater. Subsequently, it is crucial to separate the microbial biomass from the treated wastewater through sedimentation. This secondary sedimentation is similar to primary sedimentation except that the sludge contains bacterial cells rather than fecal solids. The biological removal of organic matter from settled wastewater is conducted by microorganisms, mainly heterotrophic bacteria and occasionally fungi. The microorganisms can decompose the organic matter through two different biological processes: biological oxidation and biosynthesis (Gray, 2005).

The biological oxidation forms some end products, such as minerals, that remain in the solution and are discharged with the effluent (Reaction 1). The biosynthesis transforms the colloidal and dissolved organic matter into new cells that form in turn the dense biomass that can be then removed by sedimentation (Reaction 2). On the other hand, algal photosynthesis plays an important role in some cases.

Reaction 1—Oxidation:

$$\text{COHNS (Organic matter)} + O_2 + \text{Bacteria} \rightarrow CO_2 + NH_3 \text{ (+Energy)} + \text{Other end products}$$

Reaction 2—Biosynthesis:

$$\text{COHNS(Organic matter)} + O_2 + \text{Bacteria} \rightarrow C_5H_7NO_2$$

The following terms are the most used in biological treatment processes (Lin, 2007):

a. DO: Dissolved oxygen (mg/L)
b. BOD: Biochemical oxygen demand (mg/L)
c. BOD_5: BOD (mg/L), incubation at 15°C for 5 days
d. COD: Chemical oxygen demand (mg/L)
e. CBOD: Carbonaceous BOD (mg/L)
f. NBOD: Nitrogenous (mg/L)
g. SOD: Sediment oxygen demand (mg/L)
h. TBOD: Total BOD (mg/L)

Example 7.2

An aeration basin is 90 ft long, 25 ft wide, and 13 ft deep. If it receives a flow of 0.24 MGD (million gallons per day) at a BOD concentration of 225 mg/L, what is the basin's loading in pounds of BOD/day/1,000 ft^3?

Solution:

$$\text{Loading} = \frac{\text{lbs per day of BOD to aeration tank}}{\text{Volume of aeration tank}}$$

Lbs/day of BOD to aeration tank = Concentration × flow (in MGD) × 8.34 lb/gal
= 225 mg/L × 0.24 MGD × 8.34 lb/gal
= 450.36 lbs/day

Volume of tank = L × W × H = 90 × 25 × 13 ft = 29,250 ft^3
(Note: problem asks for units per 1,000 ft^3. Divide by 1,000 to get 29.25/1,000 ft^3.)

$$\text{Organic loading} = \frac{450.36 \text{ lbs/day}}{29.25\ 1000\ ft^3}$$

7.4.2.1.1 Principles of Biological Treatment

The principles of biological treatment of wastewater consisted of the following (Russell, 2006):

1. The biological systems are susceptible to extreme variations in hydraulic loads. Diurnal variations of greater than 250% are problematic because they will create biomass loss in the clarifiers.
2. The growth rate of microorganisms is highly dependent on temperature. A 10°C reduction in wastewater temperature dramatically decreases the biological reaction rates to half.
3. BOD is efficiently treated in the range of 60–500 mg/L. Wastewaters over 500 mg/L BODs have been treated successfully if sufficient dilution is applied in the treatment process or if an anaerobic process is implemented as a pretreatment process.
4. The biological treatment is effective in removing up to 95% of the BOD. Large tanks are required in order to eliminate the entire BOD, which is not economically feasible.
5. The biological treatment systems are unable to handle "shock loads" efficiently. Equalization is necessary if the variation in the strength of the wastewater is more than 150% or if that wastewater at its peak concentration is more than 1,000 mg/L BOD.

6. The carbon:nitrogen:phosphorus (C:N:P) ratio of wastewater is usually in the range from 100:20:1 to 100:5:1 for most biological processes.
7. If the C:N:P ratio of the wastewater is strong in an element in comparison to the other elements, then poor treatment will result. This is especially true if the wastewater is very strong in carbon. The wastewater should also be neither very weak nor very strong in an element; although very weak is acceptable, it is difficult to treat.
8. Oils and solids cannot be handled in a biological treatment system because they negatively affect the treatment process. These wastes should be pretreated to remove solids and oils.
9. Toxic and biological-resistant materials require special consideration and may require pretreatment before being introduced into a biological treatment system.
10. Although the capacity of the wastewater to utilize oxygen is unlimited, the capacity of any aeration system is limited in terms of oxygen transfer.

7.4.2.2 Aerobic Treatment

Aeration has been used to remove trace volatile organic compounds (VOCs) in water. It has also been employed to transfer a substance, such as oxygen, from the air or a gas phase into the water in a process called "gas adsorption" or "oxidation," that is, to oxidize iron and/or manganese. Aeration also provides the escape of dissolved gases, such as CO_2 and H_2S. Air stripping has also been utilized effectively to remove NH_3 from wastewater and to remove volatile tastes and other such substances in water.

7.4.2.2.1 Oxidation Ponds

Oxidation ponds (Figure 7.3) are aerobic systems where the oxygen required by the heterotrophic bacteria (a heterotroph is an organism that cannot fix carbon and uses organic carbon for growth) is provided not only by transfer from the atmosphere but also by photosynthetic algae. The algae are restricted to the euphotic zone (sunlight zone), which is often only a few centimeters deep. Ponds are constructed to a depth of between 1.2 and 1.8 m to ensure maximum penetration of sunlight and appear dark green in color due to dense algal bloom.

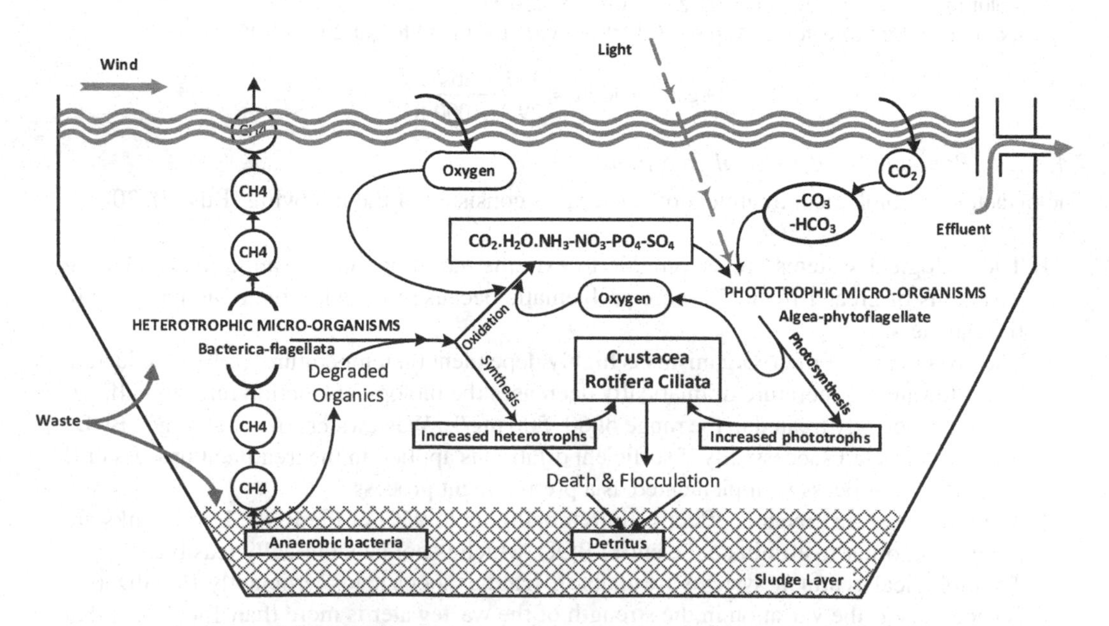

FIGURE 7.3 Aerobic system/oxidation pond.

In oxidation ponds, the algae use the inorganic compounds (N, P, CO_2) released by aerobic bacteria for growth using sunlight for energy. They release oxygen into the solution that in turn is utilized by the bacteria, completing the symbiotic cycle. There are two distinct zones in facultative ponds: the upper aerobic zone where bacterial (facultative) activity occurs and a lower anaerobic zone where solids settle out of suspension to form a sludge that is degraded anaerobically (Gray, 2005).

7.4.2.2.2 Aeration Lagoons

Aeration lagoons are profound (3–4 m) compared to oxidation ponds, where aerators provide oxygen but not by the photosynthetic activity of algae as in the oxidation ponds. The aerators keep the microbial biomass suspended and provide sufficient dissolved oxygen (DO) that allows maximal aerobic activity. On the other hand, bubble aeration is commonly used where the bubbles are generated by compressed air pumped through plastic tubing laid through the base of the lagoon. Predominately bacterial biomass develops, and, although there is neither sedimentation nor sludge return, this procedure counts on adequate mixed liquor formed in the tank/lagoon. Therefore, the aeration lagoons are suitable for strong but degradable wastewater such as wastewaters of food industries. The hydraulic retention time (HRT) ranges from 3 to 8 days based on the influent's treatment level, strength, and temperature. Generally, HRT of about 5 days at 20°C achieves 85% removal of BOD in household wastewater. However, if the temperature falls by 10°C, then the BOD removal will decrease to 65% (Gray, 2005).

7.4.2.3 Anaerobic Treatment

The anaerobic treatments are implemented to treat wastewaters rich in biodegradable organic matter (BOD > 500 mg/L) and for further treatment of sedimentation sludges. Strong organic wastewaters containing large amounts of biodegradable materials are discharged mainly by agricultural and food processing industries. These wastewaters are challenging to be treated aerobically due to the troubles and expenses of the fulfillment of the elevated oxygen demand to preserve the aerobic conditions (Gray, 2005). In contrast, anaerobic degradation occurs in the absence of oxygen. Although the anaerobic treatment is time-consuming, it has a multitude of advantages in treating strong organic wastewaters. These advantages include elevated levels of purification, aptitude to handle high organic loads, generating small amounts of sludges that are usually very stable, and production of methane (inert combustible gas) as an end product. Anaerobic digestion is a complex multistep process in terms of chemistry and microbiology. Organic materials are degraded into essential constituents, finally to methane gas under the absence of an electron acceptor such as oxygen. The basic metabolic pathway of anaerobic digestion is shown in Figure 7.4. To achieve this pathway, the presence of a very different and closely dependent microbial population is required (Ersahin et al., 2011).

Example 7.3

An aeration basin is 80 ft long, 20 ft wide, and 12 ft deep. If the flow rate to it is 0.43 MGD at a CBOD concentration of 150 mg/L and the mixed liquor volatile suspended solid (MLVSS) concentration is 1,350 mg/L, what is the food-to-microorganism ratio (*F/M*)?

Solution:

$$F/M = \frac{\text{lbs of food to aeration tank}}{\text{lbs of solids under aeration}}$$

lbs of food to aeration tank = concentration (mg/L) × flow (in MGD) × 8.34 lbs/gal
= 150 mg/L × 0.43 MGD × 8.34 lbs/gal
= 537.93 lbs/day

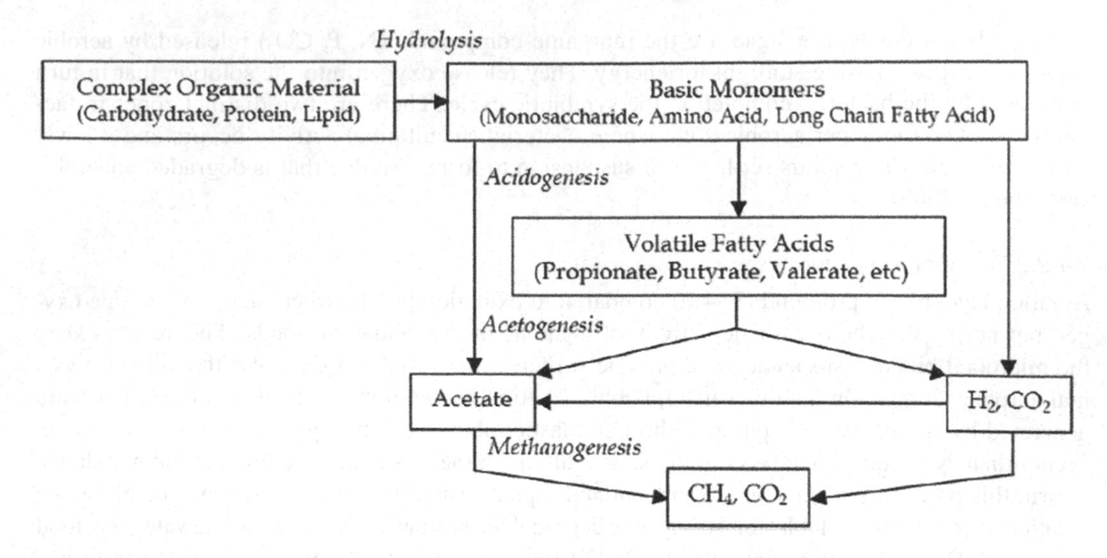

FIGURE 7.4 Steps of the anaerobic digestion process. (From Ersahin et al., *Anaerobic Treatment of Industrial Effluents: An Overview of Applications*, InTech, Rijeka, 2011.)

$$\text{lbs of solids under aeration} = \text{MLVSS (mg/L)} \times \text{tank volume (MG)} \times 8.34\ \text{lb/gal}$$

$$= \overbrace{1{,}350\,\text{mg/L}}^{\text{MLVSS}} \times \overbrace{80 \times 20 \times 12\,\text{ft}}^{\text{Tank Volume}} \times \overbrace{\frac{7.48\,\text{gal}}{1\,\text{ft}^3} \times \frac{1\,\text{MG}}{1{,}000{,}000\,\text{gal}}}^{\text{Conversion factors}} \times 8.34\,\text{lbs/gal}$$

$$= 1616.97\,\text{lbs}$$

$$\frac{F}{M} = \frac{537.93\,\text{lbs/day}}{1616.97\,\text{lbs}} = \frac{0.33}{\text{day}}$$

7.4.2.4 Activated Sludge

The activated sludge process is based on a mixture of thick bacterial populations suspended in the wastewater under aerobic conditions. With unlimited nutrients and oxygen, high rates of bacterial growth and respiration can be attained, which result in the consumption of the available organic matter to either oxidized end products (e.g., CO_2, NO^{3-}, SO_4^{2-}, and PO_4^{3-}) or biosynthesis of new microorganisms. The activated sludge process is based on five interdependent elements: bioreactor, activated sludge, aeration and mixing system, sedimentation tank, and returned sludge (Gray, 2005). The biological process using activated sludge is a commonly used method for treating wastewater, where the running costs are inexpensive (Figure 7.5). However, a considerable quantity of surplus sludge is produced in WWTPs which is an enormous burden in both economic and environmental aspects. The excess sludge contains a lot of moisture and is not easy to treat. The byproducts of WWTPs are dewatered, dried, and finally burnt into ashes. Some are used in farm lands as compost fertilizer (Kabir et al., 2011). However, it is suggested that the dried byproducts of WWTPs are fed into the pyrolysis process rather than the burning process. The sludge volume index (SVI) is an estimation that specifies the tendency of aerated solids, that is, activated sludge solids, to become dense or concentrated through the thickening process. SVI can be computed as follows: (a) allowing a mixed liquor sample from the aeration tank to sediment in 30 minutes; (b) determining the concentration of the SSs for a sample of the same mixed liquor; (c) SVI is then computed as the ratio of the measured wet volume (mL/L) of the settled sludge to the dry weight concentration of mixed liquor suspended solids (MLSS) in g/L (Source: *Water Program*, Dept. of Civil Engineering, Sacramento State Univ., Sacramento, Fl, https://www.owp.csus.edu/glossary/activated-sludge.php.)

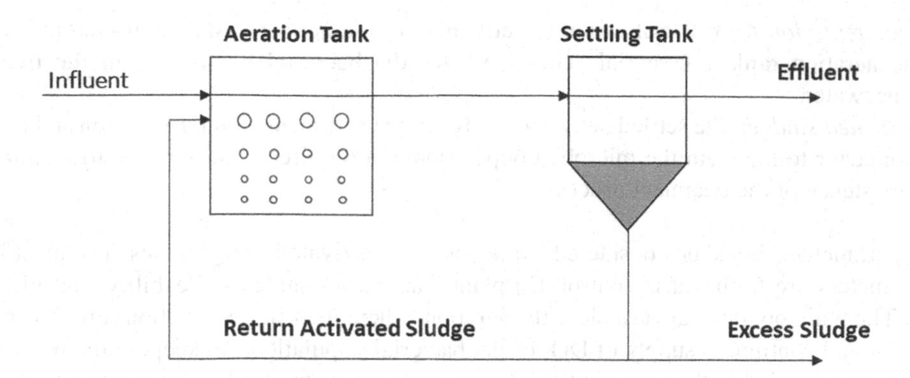

FIGURE 7.5 A schematic of an activated sludge system.

TABLE 7.2
Conventional Activated Sludge

Specification	Value	Unit
BOD sludge loading	0.4	mg/L
BOD volume loading	0.2	mg/L
MLSS	200	mg/L
COD of influent	300	mg/L
Amount of influent	4.48	L/d
Aeration rate	3.00	L/min

During the treatment of wastewater in aeration tanks through the activated sludge process (Table 7.2) there are SSs, where the concentration of the SSs is termed as MLSS, which is measured in milligrams per liter (mg/L). Mixed liquor is a mixture of raw wastewater and activated sludge in an aeration tank. MLSS consists mainly of microorganisms and nonbiodegradable SSs. MLSS is the effective and active portion of the activated sludge process that ensures that there is an adequate quantity of viable biomass available to degrade the supplied quantity of organic pollutants at any time. This is termed as *F/M* ratio or the food-to-mass ratio. If this ratio is kept at a suitable level, then the biomass will be able to consume high quantities of the food, which reduces the loss of residual food in the discharge. In other words, the more the biomass consumes food, the lower the BOD will be in the treated effluent. It is important that MLSS eliminates BOD in order to purify the wastewater for further usage and hygiene. Raw sewage is introduced into the WWT process with a concentration of several hundred mg/L of BOD. The concentration of BOD in wastewater is reduced to less than 2 mg/L after being treated with MLSS and other treatment methods, which is considered to be safe water to use (Kabir et al., 2011).

The main components of all activated sludge systems are:

1. *The bioreactor*: It can be a lagoon, tank, or ditch. The main characteristic of a bioreactor is that it contains sufficiently aerated and mixed contents. The bioreactor is also known as the aeration tank.
2. *Activated sludge*: The bacterial biomass inside the bioreactor consists mainly of bacteria and other flora and microfauna. The sludge is a flocculent suspension of these microorganisms and is usually termed as the MLSS that range between 2,000 and 5,000 mg/L.
3. *Aeration and mixing system*: The aeration and mixing of the activated sludge and the raw influent are necessary. While these processes can be accomplished separately, they are usually conducted using a single system of either surface aeration or diffused air.

4. *Sedimentation tank*: Clarification or settlement of the activated sludge discharged from the aeration tank is essential. This separates the bacterial biomass from the treated wastewater.
5. *Returned sludge*: The settled activated sludge in the sedimentation tank is returned to the bioreactor to maintain the microbial population at a required concentration to guarantee persistence of the treatment process.

Several parameters should be considered while operating activated sludge plants. The most important parameters are (a) biomass control, (b) plant loading, (c) sludge settleability, and (d) sludge activity. The main operational variable is the aeration, where its primary functions are: (a) ensuring a sufficient and continuous supply of DO for the bacterial population, (b) keeping the bacteria and the biomass suspended, and (c) mixing the influent wastewater with the biomass and removing from the solution the excessive CO_2 resulting from oxidation of organic matter.

Example 7.4

Given the following data about an activated sludge plant, calculate solids retention time (SRT) at which it is operating:

Aeration basin dimensions: 110×40× 13 ft
Influent flow rate: 1.47 MGD
Final clarifier dimensions: 50×25×13 ft
MLVSS concentration: 1,500 mg/L
Waste sludge concentration: 7,750 mg/L
Waste sludge flow rate: 20,000 GPD
Effluent TSS: 17 mg/L

Solution:

$$\text{SRT} = \frac{\text{A. Total lbs MLSS in secondary system}}{\text{B. lbs activated sludge wasted per day} + \text{C. lbs TSS lost in effluent per day}}$$

A. Total lbs MLSS (aeration & clarifier) = concentration (mg/L) × total volume (MG) × 8.34 lbs/gal

Total volume = Volume aeration tank + Volume clarifier (Note: includes conversion factors)

$$= 110 \times 40 \times 13\ \text{ft} \times \frac{7.48\ \text{gal}}{1\ \text{ft}^3} \times \frac{1\ \text{MG}}{1{,}000{,}000\ \text{gal}}$$

$$+\ 50 \times 25 \times 13\ \text{ft} \times \frac{7.48\ \text{gal}}{1\ \text{ft}^3} \times \frac{1\ \text{MG}}{1{,}000{,}000\ \text{gal}}$$

$$= 0.549\ \text{MG}$$

$$\text{Total lbs MLSS (aeration \& clarifier)} = 1{,}500\ \text{mg/L} \times 0.549\ \text{MG} \times 8.34\ \text{lbs/gal}$$

$$= 6867.99\ \text{lbs}$$

$$\text{B. lbs activated sludge wasted per day} = \text{concentration (mg/L)} \times \text{flow (MGD)} \times 8.34\ \text{lbs/gal}$$

$$= 7{,}750\ \text{mg/L} \times 0.020\ \text{MGD} \times 8.34\ \text{lbs/gal}$$

$$= 1292.70\ \text{lbs/day}$$

$$\text{C. lbs TSS lost in effluent per day} = \text{concentration (mg/L)} \times \text{flow (MGD)} \times 8.34\ \text{lbs/gal}$$

$$= 17\ \text{mg/L} \times 1.47\ \text{MGD} \times 8.34\ \text{lbs/gal}$$

$$= 208.42\ \text{lbs/day}$$

$$\text{SRT} = \frac{6867.99 \text{ lbs}}{1292.70 \text{ lbs/day} + 208.42 \text{ lbs/day}} = 4.58 \text{ days}$$

For the conventional biological treatment reactor, the model most commonly used is the completely mixed reactor, followed by a secondary sedimentation tank equipped with a recycle line to return microorganisms to the reactor, as shown in Figure 7.5. For this case, the mass balance equations must be modified to introduce the increased volumetric flow rate from the recycling line and introduce the biomass from the recycling line. Equations (7.1) and (7.2) are therefore modified as follows for the substrate uptake and biomass:

$$V\left[\frac{dS}{dt}\right] = QS_{\text{in}} + Q_rS - (Q + Q_r)S + VR_S \tag{7.1}$$

And

$$V\left[\frac{dx}{dt}\right] = QS_{\text{in}} + Q_rS - (Q + Q_r)S + VR_S \tag{7.2}$$

where

Q_r and X_r are the volumetric flow and the concentration of microorganisms in the recycle line, respectively.

Here, a new variable is defined as recycle ratio, R, equal to Q_r/Q. When the mass balance is then divided by Q, θ is substituted for V/Q, a Monod model (relate microbial growth rates in an aqueous environment to the concentration of a limiting nutrient) is substituted for the rate term, and R for Q_r/Q, the above equations become:

$$\theta\left[\frac{dS}{dt}\right] = S_{\text{in}} - S - \frac{K_0SX\theta}{K_M + S} \tag{7.3}$$

$$\theta\left[\frac{dS}{dt}\right] = X_{\text{in}} + RX_r - (1 + R)X + \frac{\mu_{\max}SX\theta}{K_M + S} - k_dX\theta \tag{7.4}$$

where

μ_m is the maximum specific growth rate (time^{-1}),
S is the substrate concentration (typically, mg BOD_5/L), and
K_s is the half velocity constant. K_s is defined as the substrate concentration when the growth rate constant is half of its maximum value (μ_m).

Example 7.5

Determine the theoretical hydraulic detention time and volume of a completely mixed reactor with recycle to be used in an activated sludge treatment plant operating at steady state if the following conditions and constants for the wastewater have been determined:

$$X_{\text{in}} \text{ and } K_d \approx 0$$

$$X = 2{,}000\frac{\text{mg}}{\text{L}}$$

$$K_0 = 0.08 \text{ hour} - 1$$

$$K_m = 75\frac{\text{mg}}{\text{L}}$$

$$\text{Efficiency} = 92\%$$

$$BOD_{in} = 300 \frac{mg}{L}$$

$$Q + Q_r = 0.1 \text{ m}^3/\text{s}$$

Solution:

Using Equation (7.3) at steady state and a known concentration of microorganisms,

$$0 = S_{in} - S - \frac{K_0 S X \theta}{K_M + S}$$

$$0 = \frac{300 \text{ mg}}{\text{L}} - \frac{24 \text{ mg}}{\text{L}} - \frac{(0.08 \text{ hour} - 1)\left(\frac{24 \text{ mg}}{\text{L}}\right)\left(\frac{200 \text{ mg}}{\text{L}}\right)\theta}{\left(\frac{75 \text{ mg}}{\text{L}}\right) + \left(\frac{24 \text{ mg}}{\text{L}}\right)}$$

And, rearranging yields the following result:

$$\theta = \frac{\left(\frac{276 \text{ mg}}{\text{L}}\right)\left(\frac{99 \text{ mg}}{\text{L}}\right)}{\left(\frac{3{,}840 \text{ mg}^2}{\text{L}^2}\right)(\text{hour} - 1)} = 7.1 \text{ hour}$$

With Equation (7.5), the tank volume can be calculated,

$$V = Q\theta$$

$$= \left(\frac{0.1 \text{ m}^3}{\text{s}}\right)(7.1 \text{ hour})\left(\frac{3{,}600 \text{ s}}{\text{hour}}\right) = 2{,}560 \text{ m}^3 \quad (7.5)$$

Example 7.6

Calculate the required aeration tank volume for an activated sludge treatment plant on the bases of the empirical design factors. The anticipated waste volumetric flow rate is 0.044 m³/s and the BOD concentration is 250 mg/L. Assume that the concentration of microorganisms in the reactor is kept at 2,000 mg/L.

Solution:

For 6-hour detention time:

$$V = Q\theta$$

$$\left(\frac{0.044 \text{ m}^3}{\text{s}}\right)(6 \text{ hour})\left(\frac{3{,}600 \text{ s}}{\text{hour}}\right) = 950 \text{ m}^3$$

On the Basis of Volumetric Loading:

For 50 kg BOD/100 m³ of aeration capacity per day:

$$\text{BOD load in rate} = Q \times \text{BOD}$$

$$= \left(\frac{0.044 \text{ m}^3}{\text{s}}\right)\left(\frac{0.250 \text{ kg}}{\text{m}^3}\right)\left(\frac{86{,}400 \text{ s}}{\text{day}}\right) = 950 \text{ kg/day}$$

$$V = \left(\frac{950 \text{ kg}}{\text{day}}\right)\left(\frac{100 \text{ m}^3 - \text{day}}{50 \text{ kg}}\right) = 1{,}900 \text{ m}^3$$

ON THE BASIS OF FOOD/MICROORGANISM:

For a food: microorganism ratio is 0.4 kg BOD/kg microorganisms per day and

BOD load is 950 kg/day (as above) and with 2,000 mg/L of microorganisms, the volume can be calculated as

$$V = \frac{\left(\dfrac{950 \text{ kg BOD}}{\text{day}}\right)}{\left(\dfrac{2 \text{ kg microorganisms}}{\text{m}^3}\right)\left(\dfrac{0.4\text{kg BOD}}{\text{kg microorganisms} - \text{day}}\right)} = 1{,}188 \text{ m}^3$$

Since the largest design volume resulted from the volumetric load in design basis calculation, 1,900 m^3 will be the stipulated aeration tank volume.

7.4.2.4.1 Bacterial Kinetics

The bacterial kinetics can be shown in Figures 7.6 and 7.7. The microbial growth curve that shows bacterial density and specific growth rate at the different growth phases is shown in the first figure. In the second, the microbial growth curves that compare the total biomass and the variable biomass are shown.

The rate of consumption of BOD is the primary rate of interest in secondary treatment. The organic matter that microorganisms consume is generically called a substrate, and it is typically measured as mg/L of BOD. Kinetics is also important in terms of the mass of organisms, which may increase or decrease with time, depending on the growth conditions and availability of substrate. Rather than quantifying the number concentration of a very diverse community of microbes in a bioreactor, they are usually estimated using a surrogate measure such as mg/L of volatile suspended solids (VSS) or mg/L volatile solids (VS). The term volatile arises from the analytic technique used,

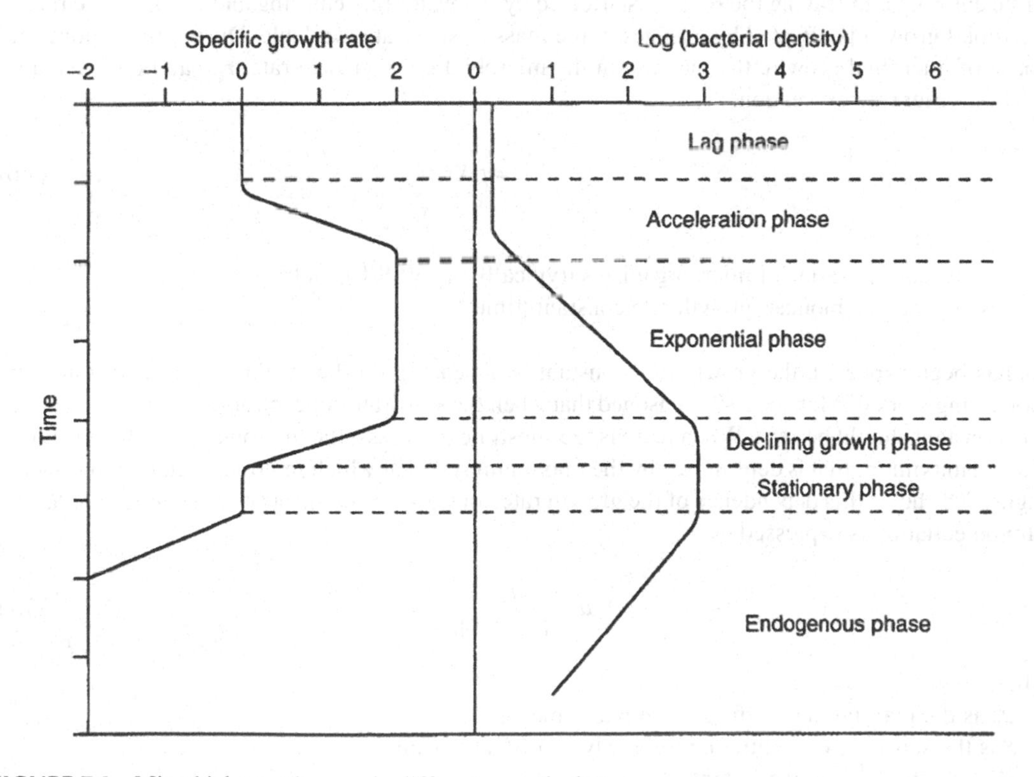

FIGURE 7.6 Microbial growth curve in different growth phases.

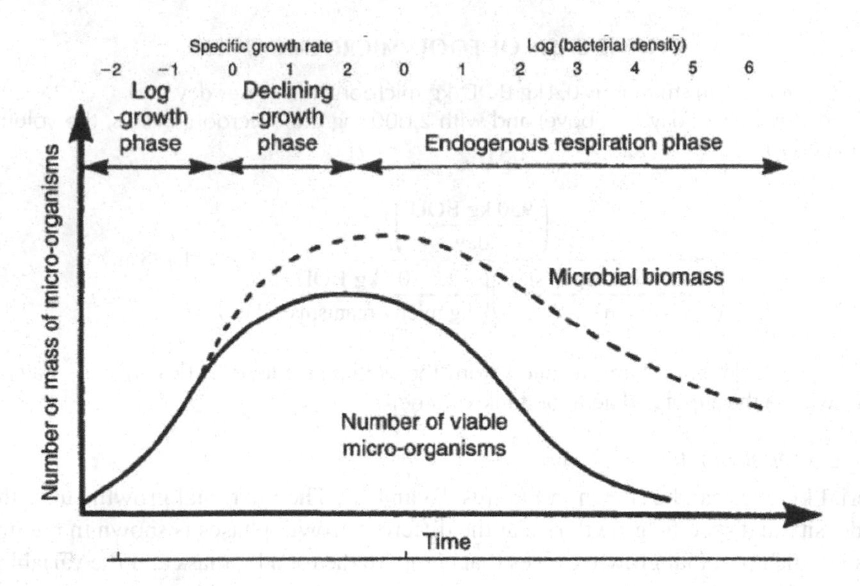

FIGURE 7.7 Microbial growth curves—biomass variations.

which measures the difference in mass of solids in a known volume of water between that weighed value after drying at 105°C and after burning off at 500°C. Although VSS and VS contain more than just the viable mass of microbes, they are simple to measure and have been found to provide a reasonable first approximation of the microorganism concentration. Thus, the kinetic analysis of a bioreactor involves, at a minimum, three mass balances, which quantify the flow and changes in water, substrate, and microbe mass. However, these mass balances are interrelated. The rate of substrate entering and leaving the reactor is affected by the water rate entering and leaving. The rate of microbial growth is affected by changes in the mass of substrate available. During the exponential phase of microbial growth, the increase in the microbial mass growth rate, r_g, can be modeled by the first-order rate expression.

$$r_g = \frac{dX}{dt} = \mu X \tag{7.6}$$

where

X is the concentration of microorganisms (typically mg VSS/L), and
μ is the specific biomass growth rate constant (time^{-1}).

As has been expected, the growth rate constant is dependent on the availability of substrate. The pioneering work of Monod (1949) reasoned that when the substrate concentration is zero, the growth rate constant should be zero. When there is the substrate in excess, the microbial growth rate should reach a maximum that is determined by the maximum speed at which microbes can be reproduced. Figure 7.8 shows this dependence of the growth rate constant on the substrate concentration, S. The Monod equation is expressed as:

$$\mu = \frac{\mu_m S}{(K_s + S)} \tag{7.7}$$

where

μ_m is the maximum specific growth rate (time^{-1}),
S is the substrate concentration (typically, mg BOD_5/L), and
K_s is the half velocity constant.

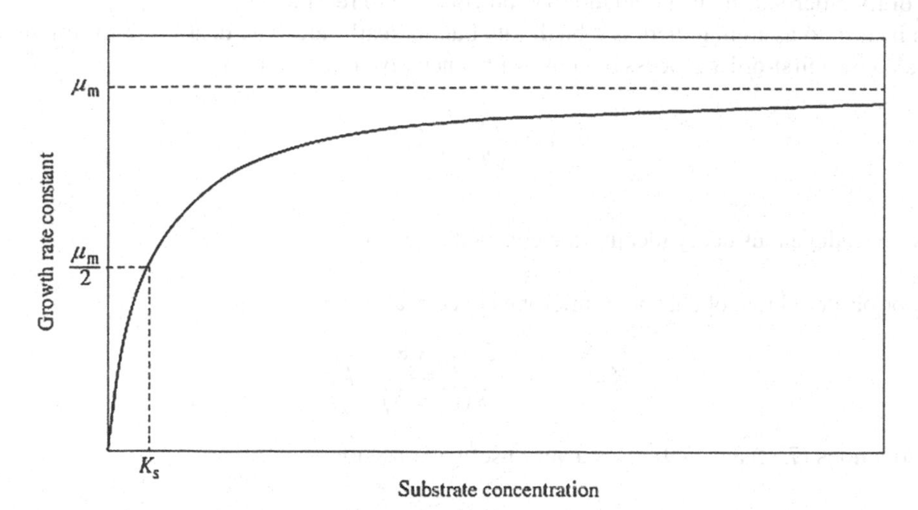

FIGURE 7.8 The Monod equation empirically describes the dependence of the biomass growth rate constant on the substrate concentration.

K_s is defined as the substrate concentration when the growth rate constant is half of its maximum value (μ_m). Based on Equations (7.6) and (7.7), the microbial mass growth rate can be expressed:

$$r_g = \frac{\mu_m XS}{(K_s + S)} \tag{7.8}$$

The microbial growth rate depends on the substrate concentration, but the amount of microbial mass generated should be proportional to the amount of substrate consumed. In other words, for a particular substrate and set of growth conditions, a certain proportion of the substrate mass consumed should be converted into a new microbial cell mass for predictable purposes. The proportionality constant is the yield coefficient, Y, and has the units of mg VSS/mg BOD5. Microbes derive energy for new cell synthesis and normal metabolism by catalyzing oxidation–reduction reactions. The fraction of the substrate that becomes microbial mass depends on the particular biochemical transformation that takes place and is the subject of the study of bioenergetics. For instance, if glucose (a simple sugar substrate) is microbially oxidized by reducing oxygen (O_2) rather than by reducing carbon dioxide (CO_2), about ten times more cell mass is synthesized per gram of glucose consumed. The yield coefficient relates the rate of substrate consumption under particular conditions to the rate of microbe growth:

$$r_{su} = \frac{dS}{dt} = \frac{-r_g}{Y} \tag{7.9}$$

The maximum specific growth rate, μ_m, is related by the yield coefficient to the maximum specific substrate utilization constant, k:

$$k = \frac{\mu_m}{Y} \tag{7.10}$$

Combining the previous three equations, the substrate utilization rate, r_{su}, is expressed as

$$r_{su} = \frac{dS}{dt} = \frac{-\mu_m XS}{Y(K_s + S)} = \frac{-KXS}{(K_s + S)} \tag{7.11}$$

We have made one simplification that cannot be overlooked. In Equations (7.6) and (7.8), we assumed that all of the microbes in the population would continue to reproduce.

This is akin to assuming there is a birth rate but no death rate. The death rate for microbes, r_d, should also be a first-order process but now with a negative reaction term.

$$r_d = \frac{dX}{dt} = -k_d X \tag{7.12}$$

where

k_d is the endogenous decay (death) rate constant (time^{-1}).

The net or observed rate of change of microbe concentration, r'_g, is

$$r'_g = r_g + r_d = \frac{-\mu_m XS}{Y(K_s + S)} - k_d X \tag{7.13}$$

Using Equations (7.12) and (7.9) gives a very useful expression:

$$r'_g = -Yr_{su} - k_d X \tag{7.14}$$

Equation (7.9) tells us the reaction rate in a mass balance equation on microbe mass is the difference between the rate they are generated, Equation (7.8), and the rate they die, Equation (7.12). Similarly, Equation (7.14) states that the reaction rate for microbial growth is proportional to the rate that substrate is consumed r_{su} minus the rate microbes die r_d. Table 7.1 provides ranges and typical values for the microbial kinetic parameters that have been measured in WWT bioreactors.

Example 7.7

The shallow pond depicted in Figure 7.9 stays well mixed due to the wind and the steady flow-through of a small creek.

If the microbes in the pond consume the inflowing biodegradable organic matter according to typical kinetics, determine:

a. The BOD_5 leaving the pond
b. The biodegradable organic matter removal efficiency of the pond
c. The concentration of VSS leaving the pond

Solution:

The pond can be modeled as a steady-state Continuously Stirred-Tank Reactor (CSTR) because it is well mixed and, we will assume the streamflow has been at a steady flow rate and composition for an extended period.

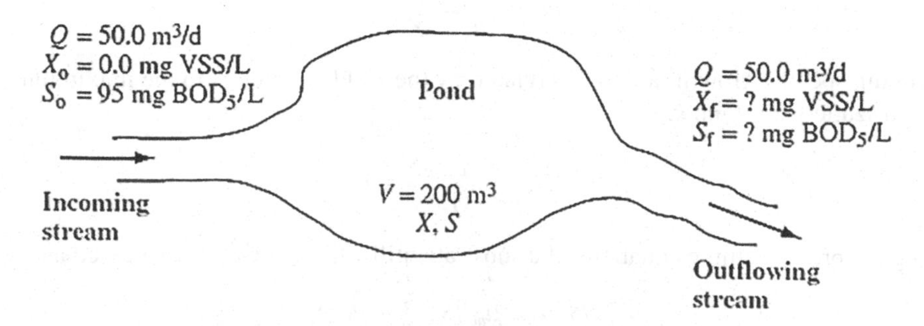

FIGURE 7.9 A pond in which microbial activity consumes organic matter.

a. First, set up a mass balance on the microbial mass using the pond as the control volume. From the variables in Figure 7.9, we can write

$$0 = -QX_0 + QX_f - Vr_g' \tag{7.15}$$

Substituting in Equation (7.8) and recognizing that $X_0=0$, $X_f=X$, and $S_f=S$ for SCTR leads to

$$0 = -QX + V\frac{\mu_m XS}{(K_s+S)} - Vk_d X$$

Because the HRT of the pond is $\theta = V/Q$, Equation (7.15) simplifies to

$$1 = \theta\frac{\mu_m S}{(K_s+S)} - \theta k_d$$

This can be solved algebraically for the substrate concentration to give

$$S = \frac{k_s(1+\theta k_d)}{(\theta\mu_m - 1 - \theta k_d)} \tag{7.16}$$

It is important to note that the effluent substrate concentration, Equation (7.16), is not a function of the influent substrate concentration. Interestingly, it was derived from a microbial mass balance rather than a substrate mass balance.

Using the default values in Table 7.1 and $\theta = \dfrac{200\ \text{m}^3}{50\ \text{m}^3/\text{day}} = 0.4$ day, the outflowing creek's BOD_5 concentration is

$$S = \frac{60\dfrac{\text{mg BOD}_5}{\text{L}}\left(1+4\ \text{day}\times 0.06\ \text{day}^{-1}\right)}{\left(4\ \text{day}\times 3\ \text{day}^{-1} - 1 - 4\ \text{day}\times 0.06\ \text{day}^{-1}\right)} = 6.9\frac{\text{mg BOD}_5}{\text{L}}$$

b. The efficiency of a bioreactor such as the pond is given by

$$\text{Eff} = \frac{(S_0 - S_f)}{S_0}\times 100\% \tag{7.17}$$

$$\text{Eff} = \frac{\left(95 - 6.9\dfrac{\text{mg BOD}_5}{\text{L}}\right)}{95\dfrac{\text{mg BOD}_5}{\text{L}}}\times 100\% = 93\%$$

c. To calculate the microbe concentration (as VSS) in the outflowing stream, we construct a steady-state substrate mass balance for the pond as

$$0 = -QS_0 - QS_f - Vr_{su} \tag{7.18}$$

Substituting in Equation (7.11) and solving for the microbe mass concentration, X, yields

$$X = \frac{(K_s+S)(S_0-S)}{\theta kS} \tag{7.19}$$

Again, using Table 7.1 default values,

$$X = \frac{\left(60+6.9\dfrac{\text{mg BOD}_5}{\text{L}}\right)\left(95-6.9\dfrac{\text{mg BOD}_5}{\text{L}}\right)}{\left(4d\times 5\dfrac{\text{mg BOD}_5}{\text{mg VSS}\cdot\text{day}}\times 6.9\dfrac{\text{mg BOD}_5}{\text{L}}\right)} = 43\frac{\text{mg VSS}}{\text{L}}$$

It is worth remembering that substantially the outflowing creek is underestimated because much of the microbial mass generated could be used as a substrate by other predatory microbes and, in fact, cause a significant oxygen demand. The removal efficiency calculation is based only on the decrease in the original quantity of biodegradable organic matter entering the pond and does not account for the organic matter synthesized in the pond.

7.4.2.5 Suspended Growth

The mathematical expressions just described allowing us to quantify the rate that microbes grow and consume BOD. These can now be applied to understanding and designing the various biological reactors employed in secondary WWT. Most modern plants employ some form of suspended growth bioreactor in which the microbes are freely suspended in the water as it is treated. The two most widely used suspended growth processes are activated sludge (membrane bioreactors, widely used in densely populated urban areas) and aerated lagoons.

7.4.3 Advanced Treatment

Advanced/tertiary WWT is an additional treatment that follows the primary and secondary treatment process. It removes stubborn contaminants that secondary treatment was not able to clean up. More than one tertiary treatment process may be used at any treatment plant. If disinfection is practiced, it is always the final process. Tertiary treatment technologies can be extensions of conventional secondary biological treatment to stabilize oxygen-demanding substances in the wastewater further or remove nitrogen and phosphorus. It may also involve physical–chemical separation techniques such as carbon adsorption, precipitation, membranes for advanced filtration, and reverse osmosis. The tertiary treatment goal is to remove nonbiodegradable toxic organic pollutants, disable disease-causing organisms, viruses, and other synthetic pollutants. Activated carbon filters remove them. Phosphate is removed by precipitation as calcium phosphate, and nitrogen is removed by volatilization as ammonia.

7.4.4 Technologies for Developing Region

Many cities in developing regions are facing surface water and groundwater pollution problems. This decline of water resources needs to be controlled through effective and feasible concepts of urban water management. The Dublin Principles, Agenda21, Vision21, and the MDG provide the basis for developing innovative, holistic, and sustainable approaches. Highly efficient technologies are available; the inclusion of those into a systematic approach is essential for the sustainable management of nutrient flows and other pollutants into and out of cities. There are three steps based on cleaner production principles:

- Minimizing wastewater generation by drastically reducing water consumption and waste generation.
- The treatment and optimal reuse of nutrients and water at the smallest possible level, such as at the on-plot and community levels. Treatment technologies recommended reusing side products for having the best use of them.
- Enhancing the self-purification capacity of receiving water bodies (lakes, rivers, etc.) through the intervention.

The success of this three-step strategic approach requires systematic implementation by providing specific solutions to given situations. This, in turn, requires appropriate planning and legal and institutional responses.

A sanitation system that provides Ecological Sanitation (EcoSan) is a cycle—a sustainable, closed-loop system—which closes the gap between sanitation and agriculture. The EcoSan approach is resource-minded and represents a holistic concept towards ecologically and economically sound sanitation. The underlying aim is to close (local) nutrient and water cycles with as little expenditure on material and energy as possible to contribute to sustainable development. EcoSan is a systemic approach and an attitude; single technologies are the only means to an end and may range from near-natural WWT techniques to compost toilets, simple household installations to complex, mainly decentralized systems. They are picked from the whole range of available conventional, modern, and traditional technical options, combining them to EcoSan systems (Langergraber and Muellegger, 2005).

Whether the classical system of urban water supply and sanitation is appropriate to satisfy the developing world's needs and whether this system meets the general criteria of sustainability is questionable. Decentralized water and wastewater management should be seriously taken into account as an alternative.

Source separation of specific fractions of domestic and industrial wastewater, the different treatment of these fractions, and recovery of water and raw materials, including fertilizer and energy, are the main characteristics of modern high-tech on-site treatment/reuse systems. Mass production of the system's key components could reduce the costs of the treatment units to a reasonable level.

On-site units can be installed independently of the development stage of the urban sewer system. In conjunction with building new housing complexes, a stepwise improvement of the hygienic sanitation in urban and preurban areas could be achieved. Remote control of the satellite systems using modern telecommunication methods would allow reliable operation and comfort for the users. However, intensive research is required to develop this system and bring it to a standard allowing efficient application worldwide (Wilderer, 2004).

The most appropriate technical solution to wastewater collection is simplified sewerage small-diameter sewers laid in a block at relatively flat gradients. In many situations, in developing countries, more appropriate technologies will have to be used. These include waste stabilization ponds, up-flow anaerobic sludge blanket (UASB) reactors, wastewater storage, and treatment reservoirs (or some combination of these). High-quality effluents can be produced that are especially suitable for crop irrigation and fish culture (Mara, 2001).

In developing regions where the climate is warm most of the time or even in subtropical areas, low temperatures do not persist for long periods. Using anaerobic technology is applicable and less expensive, even for treating low-strength industrial wastewaters and domestic sewage. Emphasis should be given to domestic sewage treatment and to the use of compact systems in which sequential batch reactors (SBRs) or dissolved-air flotation (DAF) systems are applied for the posttreatment of anaerobic reactor effluents.

Experiments on the bench and pilot plants have indicated that these systems can achieve high performance in removing organic matter and nutrients during domestic sewage treatment at ambient temperatures (Foresti, 2001). Conventional treatment technology such as the activated sludge process has both relatively high capital and operational costs. An alternative to this dilemma is to develop and apply the technologies with lower costs or better performance. Four technologies, including natural purification systems, highly efficient anaerobic processes, advanced biofilm reactors, and membrane bioreactors, are promising (Qian, 2000).

The UASB or the anaerobic baffled reactor (ABR) as an anaerobic pretreatment system and the reed bed stabilization pond supporting media as a posttreatment system may also be feasible. The pilot- and full-scale treatment plants' results reveal that the anaerobic treatment is indeed a desirable option for municipal wastewater pretreatment at temperatures exceeding 20°C in tropical and subtropical regions. The UASB system has been commonly employed as an anaerobic pretreatment system.

The ABR provides another potential for anaerobic pretreatment. The effluents from the anaerobic treatment system should be posttreated to meet discharge standards. Another cost-effective system, the stabilization pond packed with attached-growth media, is also a potential posttreatment system (Yu et al., 1997).

7.4.5 Wetlands as a Solution

In developing regions, the effective use of natural and constructed wetlands for water quality enhancement is precious and offers many ecological benefits. Constructed wetlands could be potentially valuable, low-cost, appropriate specialized treatment systems for domestic wastewater in rural areas. They can be integrated into agricultural and fish production systems where they are useable and recycled for optimal efficiency.

However, currently, constructed wetlands are rarely installed. The reasons for this are discussed, drawing attention to the limitations of aid programs from donor countries and the need for in-house research, training, and development. A cost-effective, environmentally friendly municipal WWT system using constructed wetlands was developed. The wetland was monitored for 12 months to understand the removal efficiencies of various alternatives. Mathematical models to predict the behavior of the system were also developed. The results demonstrate that the system performs well and can be adopted in upcoming cities in developing countries where sufficient land is cheaply available (Jayakumar and Dandigi, 2003).

A low-tech variation of the system using wetlands for WWT for developing countries consisting of converting existing drains into several kilometers of gravel and redlined channels resulted in the conclusion that it can treat wastewater to high-enough standards for its safe reuse in agriculture.

7.5 SATELLITE WASTEWATER MANAGEMENT

Wastewater management systems are under pressure due to changing economic and environmental conditions. For example, severe water shortages are being experienced in several metropolitan areas due to the combined effects of climate change, continued population growth, and water supplies' overdrafting. Drought conditions have resulted in rationing of remaining water supplies, conflicts between environmental and agricultural uses of water, increases in water value, and restrictions on irrigation and domestic use in urban locations. Many locations worldwide are expected to experience more frequent and extreme drought conditions in the near future. More effective utilization must be made of existing water supplies, especially in urban areas.

Other examples of challenges faced by wastewater management systems include the recovery and utilization of energy and removing nutrients. In terms of energy, wastewater systems are often identified as large consumers of power and targeted for efficiency upgrades (even though the power usage is relatively low on a per capita basis compared to other uses). Effluent water quality requirements have further increased the power demand as more membrane and high-intensity UV systems come online. While anaerobic digestion has been used for sludge management, there is a considerable amount of energy present in wastewater currently not utilized and instead places a demand on the aeration system during aerobic treatment. Hybrid wastewater systems using satellite facilities in urban areas are being considered to overcome these challenges.

7.5.1 Satellite Wastewater Treatment Systems

Satellite WWT systems, located generally in the upper portions of the wastewater collection system, usually lack solids processing facilities. Solids generated from these facilities are returned to the collection system for processing at a centralized treatment plant. Apart from the obvious utility for water reuse, the satellite treatment systems may also be used to reduce wastewater flows to the centralized facilities or remove or decrease discharges to impacted receiving water bodies (Gikas and Tchobanoglous, 2009). Three types of satellite systems demonstrated on Figure 7.10 are described below (for more details, please visit Gikas and Tchobanoglous, 2009).

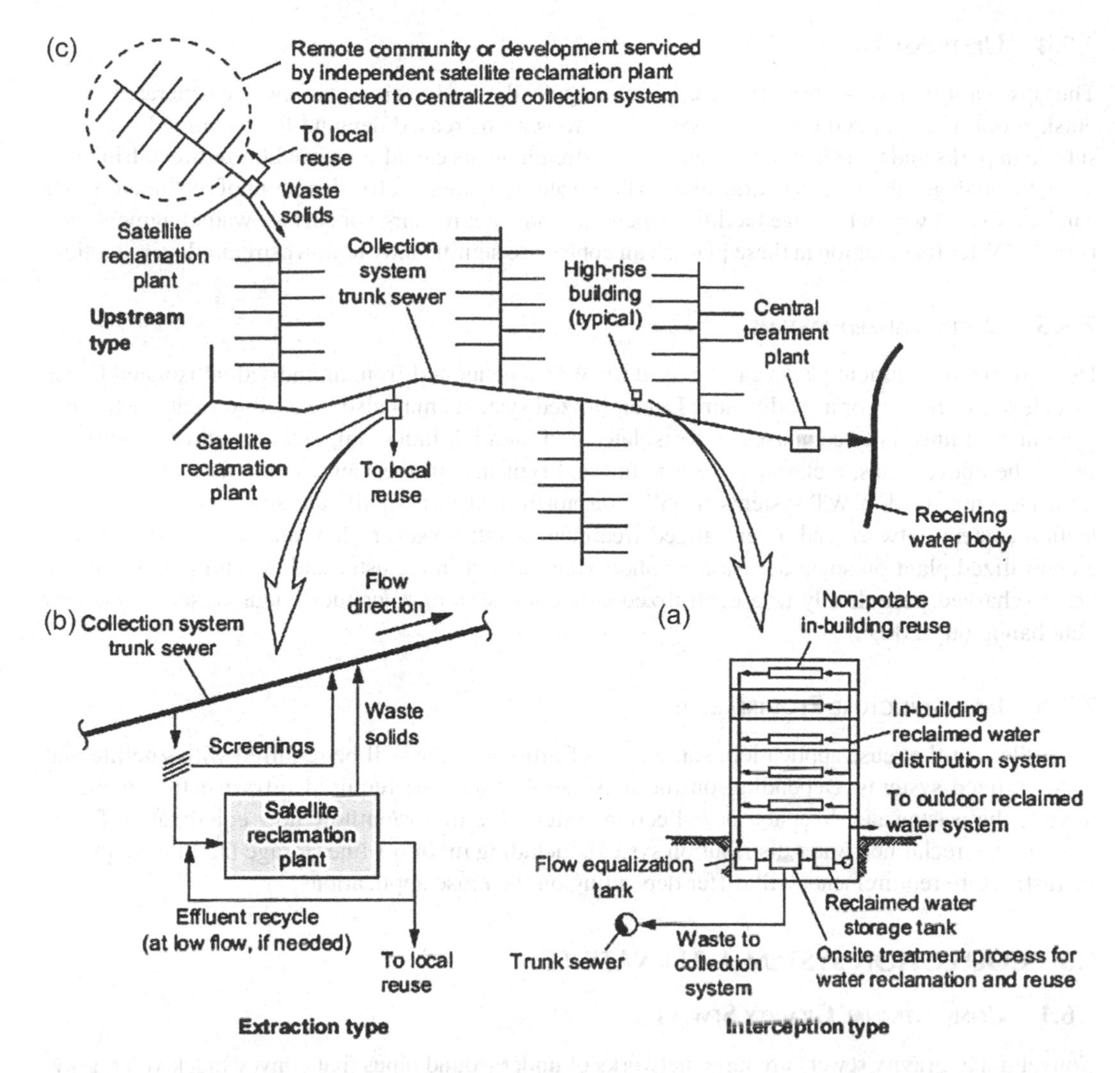

FIGURE 7.10 Schematic illustration of three types of satellite water reclamation and reuse systems: (a) interception type, (b) extraction type, and (c) upstream type. (From Gikas, P. and Tchobanoglous, G., *J. Environ. Manag.*, 90, 144–152, 2009.)

7.5.2 Interception Type

In the interception systems (Figure 7.10a), wastewater is captured before reaching the collection system, diverted to a satellite system for treatment, and reused locally for applications such as toilet and urinal flushing, localized landscaping, including water features, and cooling water in high-rise commercial or residential buildings.

7.5.3 Extraction Type

In the extraction type of satellite systems (Figure 7.10b), wastewater is literally extracted from the collection system en route to the central treatment plant; typical applications of extraction type satellite systems include a park or green-belt irrigation, water reuse in high-rise commercial or residential buildings, and cooling tower applications. Also, indirect drinkable reuse through groundwater recharge has become an alternative option in many places.

7.5.4 Upstream Type

The upstream type of systems (Figure 7.10c) is typically used to treat wastewater generated at the outskirts of a centralized collection system. There is an increased demand for reclaimed water for suburban parks and meridian strip irrigation. Upstream plants can also be used for indirect drinkable reuse through groundwater recharge and surface water irrigation. A fixed amount of reclaimed water can be recycled without storage facilities where groundwater recharge or surface water augmentation is used. Water reclamation in these places can contribute significantly to downstream flow reduction.

7.5.5 Decentralized Systems

Decentralized treatment plants can be used for WWT generated from an individual isolated house to a cluster of houses or a subdivision. Decentralized systems may also be used to treat wastewater generated at university campuses or by isolated commercial, industrial, and agricultural facilities. In all the above cases, reclaimed water is utilized typically in the vicinity of wastewater generation. Decentralized WWT systems usually are not linked to a significant sewer wastewater collection system network and a centralized treatment plant; however, they may be connected with a centralized plant on some occasions. Solids accumulated in a cluster and decentralized systems are discharged periodically to a centralized collection system. (For more details, see Gikas and Tchobanoglous, 2009.)

7.5.6 Infrastructure Requirements

Regardless of the reuse application, some form of infrastructure will be required with satellite and decentralized systems. Depending on the reuse application, the required infrastructure facilities include the wastewater diversion or collection system, the inflow/outflow flow equalization facilities, and the reclaimed water distribution system, including in- or off-line storage facilities required. Infrastructure requirements will differ depending on the reuse applications.

7.6 COLLECTION SYSTEM ALTERNATIVES

7.6.1 Conventional Gravity Sewers

Conventional gravity sewers are large networks of underground pipes that convey blackwater, graywater, and, in many cases, stormwater from individual households to a (semi-) centralized treatment facility, using gravity (and pumps when necessary).

7.6.2 Septic Tank Effluent Gravity (STEG)

Septic tank effluent gravity (STEG) is a type of effluent sewer system used for the gravity collection and conveyance of septic tank effluent from a group of homes to the downstream main sewer system or treatment plant. The network of pipes follows the profile of the ground, using small diameter pipes installed with variable or inflective gradients and clean-outs in place of conventional manholes. Each house has an individual septic tank and effluent filter ahead of the connection to the STEG collection sewers.

7.6.3 Septic Tank Effluent Pumps (STEP)

STEG is a type of pressure sewer system used to collect and convey septic tank effluent from a group of homes to the downstream central sewer system or treatment plant. Each connecting property typically has an individual STEP unit (tank and pump) or a STEG unit (gravity flow) if sufficient elevation is available.

7.6.4 Pressure Sewers with Grinder Pumps

Pressure sewer systems using grinder pumps are the popular choice for contractors and developers looking to avoid the troubles of poor soil conditions, failing septic systems, and environmental damage. Because they can be customized to site-specific needs and expanded upon unit by unit, pressure sewer systems using grinder pumps can be connected to as comprehensive a system as needed yet maintain the reliability of its initial design. A pressure sewer system uses submersible grinder pumps that grind sewage particles into a slurry and moves them along. The system uses smaller diameter pipes that are less expensive and easier to install. Plus, the pipes follow the terrain and move directionally to skirt obstructions.

7.6.5 Vacuum Sewers

Vacuum sewerage systems consist of a vacuum station, where the vacuum is generated, the vacuum pipeline system, collection chambers with collection tanks, and interface valve units. In contrast to conventional gravity sewerage systems with intermediate pumping stations, the permanent pressure within the vacuum system is maintained below atmospheric pressure. Moreover, vacuum technology reduces water consumption considerably, enabling flexible installations regardless of topography and water availability. In addition, it allows for the use of alternative wastewater handling (blackwater and graywater separation).

7.7 WASTEWATER PACKAGE PLANTS

Package plants are premanufactured treatment facilities used to treat wastewater in small communities or on individual properties. According to manufacturers, package plants can be designed to treat flows as low as 0.002 MGD or as high as 0.5 MGD, although they more commonly treat flows between 0.01 and 0.25 MGD (Metcalf and Eddy, 1991). They are designed and engineered specifically for temporarily or permanently outside the reach of municipal waste disposal systems. Prebuilt at the factory and shipped to the project site as a self-contained unit requiring only minimal field assembly. An example of a package system is shown in Figure 7.11.

The most common types of package plants are extended aeration plants, sequencing batch reactors, oxidation ditches, contact stabilization plants, rotating biological contactors, and physical/chemical processes (Metcalf and Eddy, 1991). This fact sheet focuses on the first three, all of which are biological aeration processes.

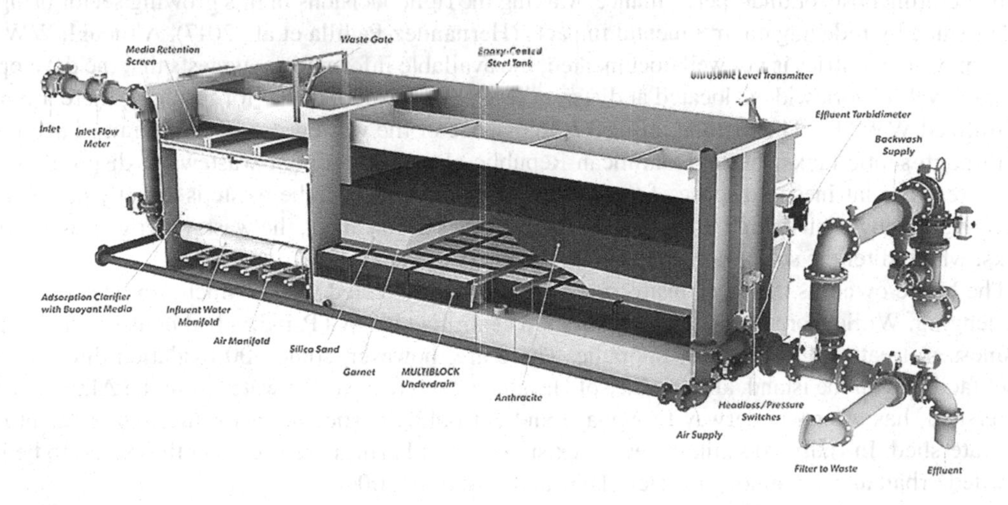

FIGURE 7.11 Example of a complete package plant. Source: https://www.westech-inc.com/hubfs/image/product/trident/lightbox-trident.jpg

Advantages of a package plant include:

1. Service areas not serviced by municipal sewer relatively cheaply
2. Make land available for development where water and roads exist
3. Plants are easy to operate. Many are manned for 2–3 hours/day
4. Superior to septic tank systems in treatment and cost

Disadvantages of package plant include:

1. Promotes leapfrog development
2. Often poorly operated and maintained
3. Allows development in undesirable areas (from the view of planners)

7.8 EXAMPLES OF WASTEWATER TREATMENT DEVELOPMENT

7.8.1 Caribbean Wastewater Treatment

With approximately one-third of the world's water resources and 24,400 m^3 of water per capita per year (WB, 2014), Latin America and the Caribbean (LAC) is a region with high freshwater resource availability (ECLAC/UNW-DPAC, 2012). However, only 20% of municipal wastewater is treated (Balcazar, 2008). Moreover, unplanned urban growth has prioritized water services and sewerage, resulting in an imbalance between available water resources and water quality protection, so treatment of wastewater and solid waste disposal has lagged. Just under half of the region's homes are not connected to a sewer system, and only 17% of homes are connected to adequate collection and treatment systems. A total of 85% of wastewater discharged into the Caribbean Sea is entirely untreated. In the past two decades, 80% of living coral in the Caribbean has been lost, with wastewater considered one of the main culprits (www.fluencecorp.com). Also, according to a survey by Noyola et al. (2012), the majority (38%) of WWTPs in LAC are stabilization ponds (based on the number of facilities). However, activated sludge (mostly conventional and extended aeration) treats the higher accumulated sewage flow that enters the LAC treatment facilities (58%).

Given the infrastructure lag in LAC and the investments required to build new facilities, the determination of treatment systems for sustainable wastewater management is highly relevant and opportune. Furthermore, in most LAC countries, many treatment facilities are abandoned due to high operating costs or underperformance. Making the right decisions in this growing sector brings added value by reducing environmental impacts (Hernández-Padilla et al., 2017). Although WWT in Caribbean countries is not well documented, the available information suggests that the development of WWT worldwide is located at different levels. In Havana, Cuba, for instance, there are no centralized WWTPs. The method utilized is to dispose of the wastewater of communities primarily through septic tanks. In the Dominican Republic also, there is poor wastewater disposal. The rural areas use latrines to dispose of the wastewater. In poor areas, the waste is directly deposited in rivers, thereby contributing to disease. In the country's urban areas, the wastewater goes to septic tanks, which filter the solid wastes and let the water go to the ground.

The house owner is then responsible for emptying the collected waste, which can be costly and challenging. While there is an effort to plan for a centralized WWTP, individual house owners and businesses' treatment is currently sporadic. There are, however, some 300 oxidation ditch treatment facilities on the island, and the City of Havana, which receives its water from the Almendares Watershed, has efforts underway to manage and get public participation for the improvement of the watershed. In Haiti, no sanitary sewers exist. Some pit latrines are used, but these tend to be in crowded urban areas or among the rich (Diaz and Barkdoll, 2006).

There are some septic tanks as well for waste collection. Some institutions have their own septic tanks for their office buildings, but maintenance and disposal are somewhat lacking. In Jamaica,

WWT and disposal is lacking as well. Even in the few existing treatment plants, the operation is not diligent, causing effluent standards to remain unmeet. Again, poverty is a prime cause of substandard treatment. While bringing in some income, a high tourism load also burdens the treatment infrastructure (Diaz and Barkdoll, 2006).

Data from 2,734 WWTP on the distribution of different type of treatment technologies used in the six LAC countries, Figure 7.12, show that the three more adopted technologies in LAC are stabilization ponds (1,106 facilities, 38% of the sample) followed by activated sludge (760, 26%) and the upflow anaerobic sludge blanket reactor, known as UASB (493, 17%). These three technologies correspond to 2,359 facilities (80% of the sample). The first two types turn out to be the more representative in terms of flow treated (Figure 7.13), with a clear contribution of activated sludge (104.1 m^3/s, 58%) followed by stabilization ponds (27.1 m^3/s, 15%). With this typology, the enhanced primary treatment appears in the third place with an accumulated flow of 16.1 m^3/s, 9%) followed by

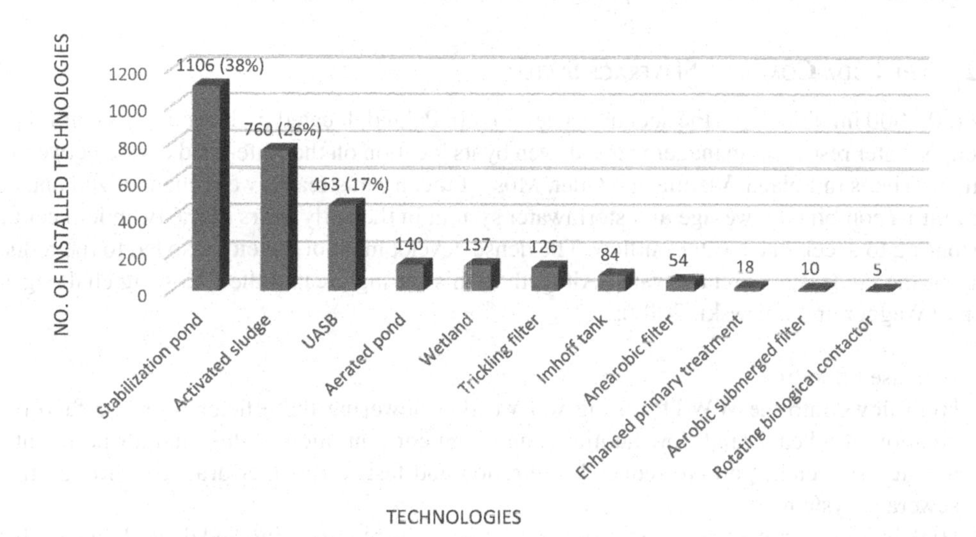

FIGURE 7.12 Distribution of different type of treatment technologies in LAC.

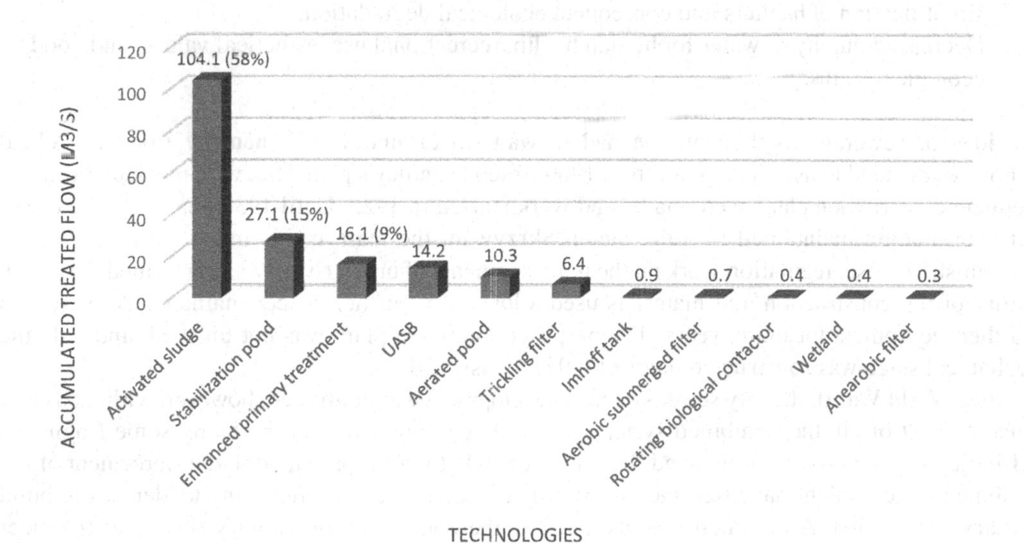

FIGURE 7.13 Accumulated flow treated per each type of technology applied in LAC.

the UASB reactor (14.2 m^3/s, 8%) corresponding altogether to 90% (81% not considering enhanced primary treatment) of the total flow treated by the facilities (Noyola et al., 2012).

Development of wastewater infrastructure in the region is affected by the following (www.gefcrew.org):

- Low priority is given to the development of the wastewater sector
- Capacity constraints of many utilities and other service providers
- A lack of sufficient and stable long-term funding for utilities
- Inadequate and poorly enforced policies and laws
- Poor communication and collaboration among involved agencies
- Limited awareness, knowledge, and understanding of alternative and appropriate treatment technologies
- Limitations on the technical capacity for environmental management

7.8.2 The Lodz Combined Sewerage System

Lodz (800,000 inhabitants) is the second biggest city in Poland, located in the country's central part. The city's water resources management is driven by its location on the watershed divide between the two major rivers in Poland: Vistula and Oder. Most of the urban streams were channelized and converted into a combined sewerage and stormwater system in the early years of the twentieth century, contributing to accelerated water outflow. The dense development of the city also led to the reduced ability of the landscape to retain water. Altogether, this setting created the following challenges to the city (Wagner and Zalewski, 2009):

1. Increased flooding,
2. High flows into the WWTP during wet weather, lowering the efficiency of the facility, creation of a heat island, low humidity, and high concentration of dust and air pollutants during dry weather, due to reduced infiltration and fast stormwater drainage through the sewerage system,
3. High hydraulic stress to aquatic ecosystems, due to overengineering and degradation of the water cycle,
4. Reduction of aquatic ecosystem capacity for water retention and self-purification, due to simplification of habitats and consequent ecological degradation,
5. Decreased quality of water for human health, recreational use, esthetical values, and good ecological status.

The idea of sewerage implementation had to wait till around 1920 when the Lodz in Poland authorities decided to use an old plan from 1909 called "Lindley's plan." So, with much enthusiasm, adequate construction plans were made, and work started in 1925. Until 1939, thanks to the excellent coordination maintained by Eng. Stefan Skrzywan, the main trunk and several side sewers were finished. Also, regulation work on the main segments of urban rivers was performed. The high quality of the construction and materials used allowed for further proper maintenance of sewers and their equipment for many years. The proposed treatment plant was not finished, and only the mechanical stage was constructed before 1939 (Zawilski, 2004).

After World War II, the city sewer system's development was continued, however, without a clear concept. First of all, the combined system was no longer applied; and therefore, some fragments of Lindley's system were not finished. It is not clear what reason prevailed, but enforcement of the combined system might have stemmed from difficult economic conditions and tendencies to build sanitary sewers first. A common practice became the connection of sanitary sewers to the older combined sewerage. The combined system was not promoted apart from some provisional regulations of combined sewer overflow (CSO) weir levels.

Recently, a more modern WWTP has been included in the combined system. The total system's capacity is still sufficient; however, urban floods are to be observed more and more frequently in some areas. It occurs due to an increasing number of sealed surface areas and some bottleneck segments overloaded by storm sewers' improperly organized connections. Since 1999, modern monitoring of CSOs has been available, and thus, information about the spill frequency has been collected. As it turned out, most monitored CSOs are activated too frequently (up to 30–40 spill events per year), causing severe pollution of urban rivers and especially of the primary receiver city area. Therefore, a plan for the combined system renovation was elaborated in 2003 (Zawilski, 2004).

7.8.2.1 Upgrading the Old Sewerage System

The decision taken in the past appeared to be essential for the future modernization of the sewerage systems, which have to meet new European water law regulations. The main problems that appeared during the preparation of the modernization plan are summarized in Table 7.3.

Some technical and nontechnical problems often implement modernization plans difficult or impossible. For Lodz, the introduction of storage tanks close to each CSO had been established. A storage facility and a separator of screening have been proposed. The rest of the free land was utilized for both facilities (Zawilski, 2004).

There are no such promising circumstances in the case of other investment proposals. In general, optimal solutions from the theoretical viewpoint turn out to be impossible to be applied in local conditions. Therefore, quasi-optimal solutions have to be implemented instead. More optimization should be continued following an assessment of the enlarged storage capacity and using a possible introduction of RTC (Zawilski, 2004).

TABLE 7.3
Problems Detected in the Process of Upgrading the Old Lodz Sewerage System

The Idea for Sewerage Upgrading	Problems	Remarks
Storage tanks for reducing CSOs	No space inside the urban infrastructure	Lack of free space or complicated property relationships; collisions with the existing pipelines and planned roads
Storage tanks for reducing CSOs	Deep sewerage causes deep storage tanks and emptying pumping stations	An alternative solution with shallow tanks requires high pumping capacity
Evaluating and improving the CSO hydraulics	Nontypical construction of CSO chambers	Some CSOs have very long weirs and can exchange water with nearby riverbeds
Detention of stormwater inside sewers	Unfavorable level conditions for the idea to be applied	RTC is necessary to be implemented
Separation of screenings from overflowing sewage	No space on overflow sewers; difficult construction conditions in river valleys	An alternative solution with facilities inside CSO chambers is even more difficult for implementation due to the shape of overflow weirs
Disconnection of runoff from the combined system	No space for new storm sewers in streets; unfavorable level conditions for outlets into urban rivers	Also, existing house connections were built into the combined system
Retention of stormwater through infiltration into the ground	Variable and not fully recognized local geological conditions	No distinct favorable effect in densely built-up areas

Source: Zawilski, M. (2004). Modernization problems of the sewerage infrastructure in a large Polish city. In J. Marsalek et al. (eds.) Enhancing Urban Environment by Environmental Upgrading and Restoration (Vol. 43, pp. 151–162). NATO Science Series, IV. Earth and Environmental Sciences. Springer, Dordrecht.

7.9 CASE STUDIES

7.9.1 Case Study 1: Reliability Assessment of Wastewater Treatment Plants Under Coastal Flooding

New York City has a massive wastewater collection system that treats nearly 1.3 billion gallons of wastewater, which helps restore and maintain water quality in New York Harbor (Bloomberg, 2013). Because of the plants' vicinity to the waterfront, the facility's potential for damage is high, mainly due to storm surge and electricity shut down. The occurrence of Superstorm Sandy in October 2012, with a recorded 14 ft surge at the Battery, caused estimated damage of $77 million to NYC wastewater infrastructure. Moreover, approximately 560 million gallons of untreated sewage mixed with stormwater and seawater were released into local waterways, equivalent to approximately half a day's worth of normal WWT (Kenward et al., 2013). In the study by Karamouz and Olyaei (2019), Coney Island, Newton Creek, and Red Hook WWTPs located in Brooklyn are chosen as a case study (Figure 7.14). In Table 7.4, the information related to these plants is presented. All of these plants have an activated sludge system.

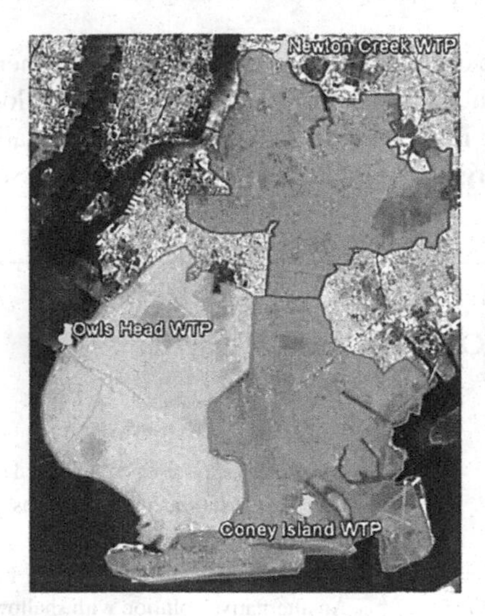

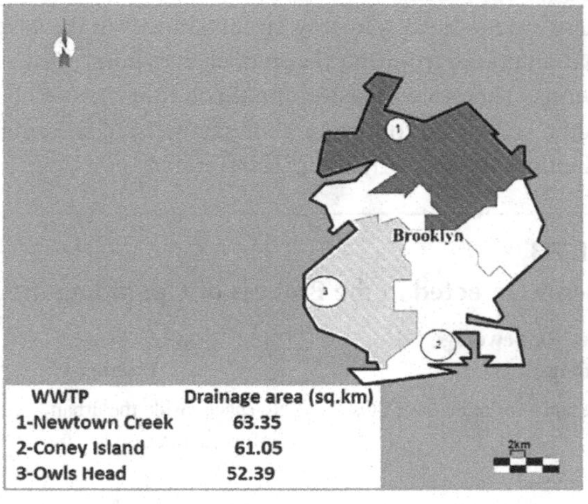

FIGURE 7.14 Location of three wastewater treatment plants in Brooklyn. (From NYC-DEP, *NYC Wastewater Resiliency Plan, Climate Risk Assessment and Adaptation Study*, Department of Environmental Protection, New York, 2013.)

TABLE 7.4
The Related Information of the Wastewater Treatment Plants

WWTP	Coney Island	Newton Creek	Owls Head
Latitude (deg, min, s)	40°35′23.13″N	40°43′58.87″N	40°38′33.71″N
Longitude (deg, min, s)	73°55′59.4″W	73°56′46.18″W	74°2′0.28″W
Plants capacity (cms)	4.82	13.58	5.26
Population under service	596,326	1,068,012	758,007
Plants type	Activated sludge	Activated sludge	Activated sludge
Critical 100-year flood elevation +30 inches of sea-level rise (m)	4.65	4.05	4.35
Superstorm Sandy flood elevation (m)	3.03	3	4.05
Superstorm Sandy flood damage	Major (%)	Minor (%)	Major (%)

The main steps in assessing the WWTP reliability with load–resistance (quantitative method) and fault tree analysis (qualitative method) are shown. In the load resistance method, first, the inundation depth and influent discharge are obtained at the plants' location. Next, the unit operation reduced removal rates are estimated, and the concentrations of effluents are obtained. Finally, the dependency between load and resistance is examined by testing various copula functions and choosing the best one through the ordinary list square (OLS) method. The next task is to develop the diagram based on the fault tree analysis. This objective is achieved by identifying various causes of plant failure in a time of coastal flooding. Figure 7.15 shows the proposed methodology.

- *Influent concentration*: The inflow to the plant system typically consists of two parts: (1) the domestic wastewater which discharges from households and (2) the runoff during a storm. Due to the lack of data for the exact time of food, the influent concentration is estimated based on the proportion of domestic wastewater and runoff considering various land use types in each plant's watershed area. The volume of resulting runoff for considered coastal flooding is also obtained from gridded surface subsurface hydrologic analysis

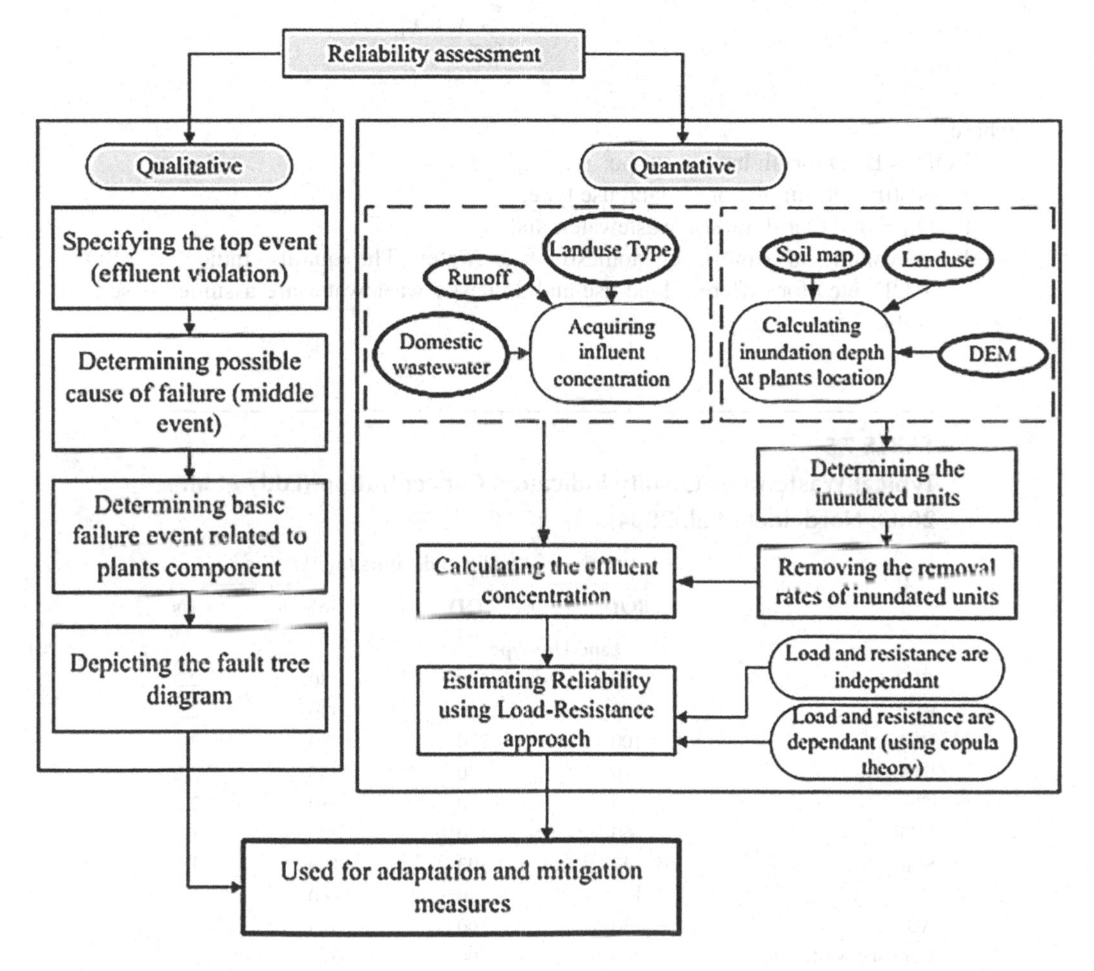

FIGURE 7.15 Proposed methodology for quantitative and qualitative assessment of reliability for a typical wastewater treatment plant under coastal flooding. (From Karamouz, M. and Olyaei, M.A. (2019). A quantitative and qualitative framework for reliability assessment of waste water treatment plants under coastal flooding. *International Journal of Environmental Research*, 13(1), 21–33.)

(GSSHA) software. For estimating the produced daily domestic wastewater in each plant's service area, the following equation is used:

$$\text{Daily domestic wastewater} = W \times C \times P \tag{7.20}$$

where

W is daily water consumption,
C is the coefficient of water conversion to wastewater (which is assumed to be 0.8), and
P is number of residents served.

Finally, based on Equation (7.20), the influent concentration is estimated. The equation is presented for BOD, but it can be used for other influent quality indicators (i.e., COD, TSS, TN) as well:

$$\text{BOD (combined)} = \frac{\sum_{i=1}^{n} \text{BOD}_i \times V_i + \text{BOD}_D \times V_D}{\sum_{i=1}^{n} V_i + V_D} \tag{7.21}$$

where

BOD_i = BOD for ith land use type;
V_i = estimated runoff for ith land use type;
BOD_D = BOD for domestic wastewater, and
V_D = volume of runoff for domestic wastewater. The quality indicators (BOD, COD, etc.) for different land use and domestic wastewater are assumed based on Table 7.5.

TABLE 7.5
Typical Wastewater Quality Indicators Concentration (Eddy et al. 2003; Nordeidet et al. 2004)

	Quality Indicators (mg/L)			
	BOD	COD	TSS	TN
	Land Use Type			
CI	70	140	380	2.2
R	100	200	480	2.1
HW	100	200	480	2.1
IND	70	140	380	2.2
CP	70	140	380	2.2
MFR	60	120	220	1.7
SFR	47	93	160	1.6
G	110	220	280	2
AG	30	100	580	2.2
Domestic wastewater	380	800	370	30

CI, city; R road; HW, highways; IND, industrial; CP, commercial/public; MFR, multifamily residential; SFR single family residential; G, green; AG, agriculture.

- *Removal rate of unit operations*: The performance of a WWTP depends on the proper functioning of different unit operations. In Figure 7.16, the typical unit operation in a WWTP is shown. In this study, the considered unit operations are screening, gritting, primary settling, aeration, and final settling. Each of the units mentioned above operations has a different removal rate in the effluent quality indicators for treatment. According to Eddy et al. (2003), the following removal rate values are proposed (Table 7.6). These removal rates belong to each treatment unit from its influent to the effluent.
- *Case study concluding remarks*: In coastal regions, the growing occurrence of natural hazards has made the critical infrastructures, including WWTPs, prone to destruction and malfunction in their performance. In this study, from a quality perspective, the reliability of three WWTPs in New York City is obtained in the time of coastal flooding. For this purpose, first, the inundation depth at the plants' location is calculated by GSSHA software, and the removal rate of different plants' unit operations (screening, gritting, primary settling, aeration, and final settling) is estimated by comparing the influent discharge with the

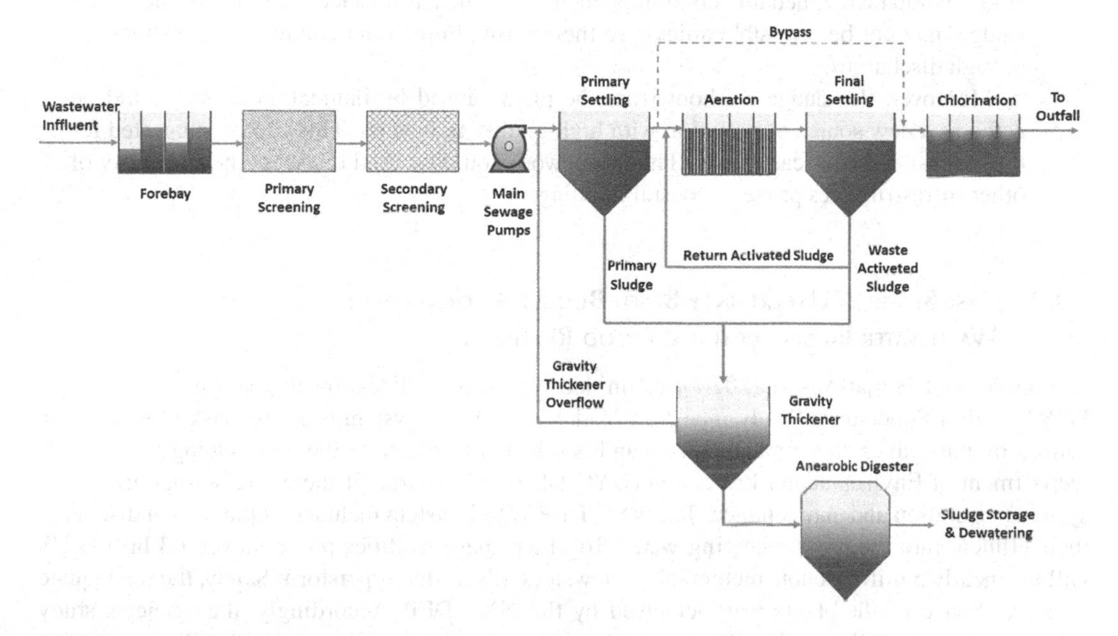

FIGURE 7.16 Typical wastewater treatment plant process (*NYC-DEP. NYC Wastewater Resiliency Plan, Climate Risk Assessment and Adaptation Study.* Department of Environmental Protection, New York, (2013)).

TABLE 7.6
Removal Rate of WWTP's Unit Operations (Eddy et al. 2003; Qasim 1998)

Unit Operation	Screen Chamber (%)	Grit Chamber (%)	Primary Settling (%)	Aeration Tank (%)	Secondary Settling (%)	Chlorination Tank (%)
			Parameters			
BOD	2	3	30–40	45–55	50–60	0
COD	3	6	30–40	45–55	50–60	0
TSS	2	8	60–70	20	80–90	0
TN	0	3	20–30	30	10	0

allowable discharge. Next, the effluent concentrations of various quality indicators (such as BOD, COD, TSS, and TN) are acquired considering the performance of different unit operations of the plant in each period of food occurrence. Finally, the reliability of the plant is calculated using a load-resistance approach.

It should be noted that reliability is estimated and compared, assuming both independence and dependency between load and resistance. The result indicates that the reliability of different effluent quality indicators is varied in different plants, but in general, the reliability of biological indicators (mostly COD) is higher than others. Considering the fact that in time of food, effluent usually becomes diluted, it seems logical. Moreover, SSs and nutrients typically increase during flooding because of washed-out sediments caused by runoff. Furthermore, the result shows a strong correlation between load and resistance; therefore, the joint probability analysis has been considered. Finally, the fault tree diagram indicates the possible cause of quality failure in plants. This diagram accompanied by the qualitative reliability approach could be utilized to design and assess mitigation and adaptation measures' effectiveness. However, some assumptions and considerations in this study should be rectified for actual implementation, such as the fact that units like activated sludge may not be reversible quickly to their optimal operating conditions after flooding or high discharges.

Moreover, the sludge washout from the plant caused by inundation depths could be acted as a new source of pollution with high-quality indicators. This can be suggested for the extension of this case study. This framework could be used to assess the reliability of other infrastructures prone to coastal flooding.

7.9.2 Case Study 2: Uncertainty Based Budget Allocation of Wastewater Infrastructures' Flood Resiliency

Karamouz and Hojjat-Ansari (2020b) examined the budget allocation to the New York City's WWTPs after Superstorm Sandy. The New York City's WWT system is at high risk of inundation caused by hurricanes and tropical storms and is selected to illustrate the methodology. The city's Department of Environmental Protection (NYC-DEP) is in charge of the entire sewage treatment system's operation and maintenance. The NYC-DEP WWT system includes 14 plants that discharge their effluent into the city's receiving water. Together, these facilities process over 1.3 billion US gallons (nearly 5 million cubic meters) of wastewater daily. After Superstorm Sandy, the inadequate flood resilience of the plants was perceived by the NYC-DEP. Accordingly, the agency's study acknowledged that all 14 WWTPs are at risk with a consequence of over $900 million against a 100-year flood. In this respect, the recommended budget to improve these facilities' flood resilience was estimated at $187 million (NYC-DEP, 2013). This budget mainly includes the cost of elevating critical equipment, floodproofing, sandbagging, and constructing barriers.

Figure 7.17 illustrates the plants' names, locations, capacities, and their service area boundaries on the NYC map. According to this figure, some of these treatment plants are directly exposed to surges in the Atlantic Ocean, such as Coney Island and Rockaway. Some of them are exposed after passing through Long Island Sound, such as Hunts Point, and therefore, part of the storm surge is dissipated. Others are exposed by the propagation of surges in the rivers, such as the North River plant. These site-specific exposures to hazards are considered in assigning values to the subcriteria in Table 7.7 and Table 7.8. Table 7.9 shows the proposed coefficients of financial constraints for these plants based on expert advice. Given the considerable budget required for increasing the plants' resilience against coastal flooding, efforts to optimally distribute this budget among them are of great importance. In the next section, the results of such efforts are presented.

Resilience assessment is generally classified into one of two categories: attribute-based (qualitative) and performance-based (quantitative) assessments (Hosseini et al., 2016; Vugrin et al., 2017).

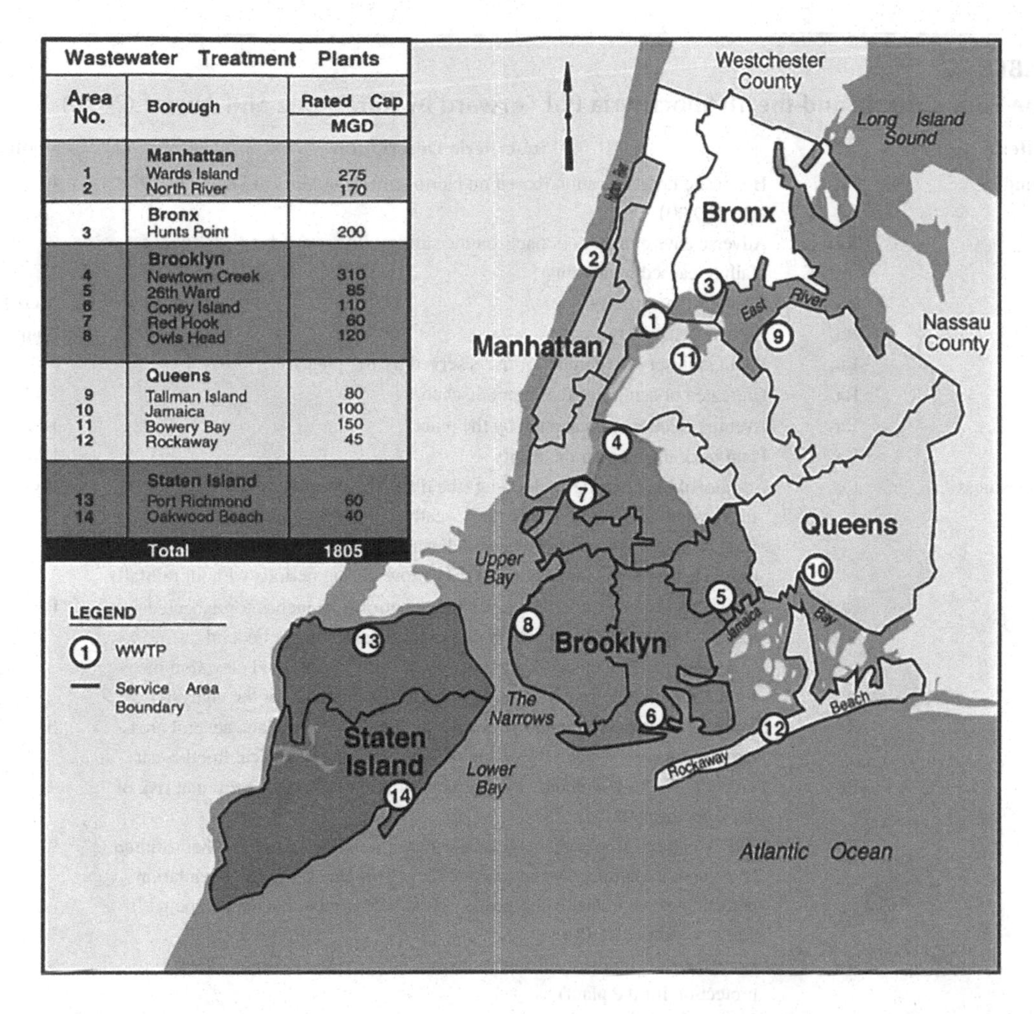

Area No.	Borough	Rated Cap MGD
	Manhattan	
1	Wards Island	275
2	North River	170
	Bronx	
3	Hunts Point	200
	Brooklyn	
4	Newtown Creek	310
5	26th Ward	85
6	Coney Island	110
7	Red Hook	60
8	Owls Head	120
	Queens	
9	Tallman Island	80
10	Jamaica	100
11	Bowery Bay	150
12	Rockaway	45
	Staten Island	
13	Port Richmond	60
14	Oakwood Beach	40
	Total	**1805**

FIGURE 7.17 NYC WWTPs service areas. (Adapted from Wilson, C. et al., *Proc. Water Environ. Fed.*, 2016, 5878–5884, 2016).

Performance-based measuring of resilience requires the system's simulation during hazards. With the increasing number of systems and their complexity, simulation has become more and more difficult. In particular, consideration of systems' interdependencies increases the simulation time burdens. However, it provides a detailed description of systems' performance during hazards. In comparison, in evaluating systems' resilience based on their attributes, there is no need to simulate systems. This approach is a low-cost technique to compare the systems' resilience with a minimum degree of complexity in resource allocation problems. Thus, this case study extends the work of Karamouz et al. (2016, 2018) at using an attribute-based approach to assess the resilience of WWTP systems through four criteria including rapidity, robustness, resourcefulness, and redundancy (four R's) and subcriteria that include different data types, such as hydrological, environmental, economic, and technical data. Also, given the importance of energy for the operation of sewage treatment systems and the possible failure of electrical infrastructures during floods, the relevant subcriteria are identified and incorporated into the framework. The proposed framework is illustrated in Figure 7.18. According to this figure, the framework consisted of four steps: a. Identifying the attributes of systems that are effective in flood resilience. b. Comparing the flood resilience of systems using MCDM. c. Investing in systems and examining their flood resilience improvement. d. Allocating a fixed budget among systems considering uncertainties.

TABLE 7.7
The Four Criteria and the 18 Subcriteria Put Forward by Karamouz and Ansari (2020)

Criteria	ID	Subcriteria Description	Unit
Rapidity	Ra_1	Hurricane flood elevation (based on North American Vertical Datum of 1988 (NAVD88))	ft
	Ra_2	Adverse environmental impacts on the surrounding area (due to treatment failure caused by flooding)	–
	Ra_3	Plant design capacity	MGD[a]
	Ra_4	Poststress recovery	hour
	Ra_5	Population served (number of users served by the plant)	#
	Ra_6	Untreated or semitreated effluent discharge	MG
	Ra_7[b]	Average electricity consumed by the plant	KWH[c]
	Ra_8[b]	Plant back-up generator capacity	KWH
Robustness	Ro_1	Additional load in time of flooding (the difference between WWTP capacity for the total maximum wet and dry weather flow. Maximum wet weather flow is the maximum flow received during any 24-hour period. Maximum dry weather flow is the maximum daily flow during periods without rainfall)	MGD
	Ro_2	Critical flood elevation (100-year flood elevation +30 inches for expected sea level rise by the 2050s, which is determined based on the Federal Emergency Management Agency's new advisory base flood elevation maps for a 100-year flood event, was selected as the baseline for the analysis)	ft[d]
	Ro_3	Maximum inundation depth (due to the flat terrain of the plant, several areas may be flooded by up to this value of water during the critical flood event)	ft
	Ro_4	Percent of not-at-risk equipment (percent of plant items that are not at risk of damage during flood)	%
	Ro_5	DMR violations (the percentage of discharge monitoring reports that resulted in effluent violations. During minimal levels of stress, the DMR violation percentages are indicative of how well each treatment plant can cope with daily operational stresses)	%
	Ro_6	Damage cost from the most severe historical hurricane (without flood protection for the plant)	$
	Ro_7[b]	100-year flood inundation depth at unit substation	ft
	Ro_8[b]	Experiencing power loss during storms	–
Resourcefulness	Rs	Number of plant technical staff	#
	Rs	Availability of dewatering facilities (facilities to drain sludge to decrease 90% of its liquid volume)	–
	Rs	Total risk avoided for every single dollar spent over 50 years	$
Redundancy	Rd_1	Existence of underground tunnel systems	–
	Rd_2	Availability of WWTPs in the neighboring areas (distance from the closest WWTP)	km
	Rd_3	On-site storage (volume of lakes in the WWTP's zone)	ft^3
	Rd_4	Number of back-up generators	#

[a] Million gallons per day (US liquid gallon = 3.78 L).
[b] Subcriteria representing water-energy interdependencies.
[c] Kilowatt hour.
[d] Foot = 0.3048 m.

TABLE 7.8
Values of Subcriteria for WWTP Facilities

WWTP	Bowery Bay	Hunts Point	Tallman Island	Wards Island	Newtown Creek	North River	Oakwood Beach	Port Richmond	Red Hook	26th Ward	Coney Island	Jamaica	Owis Head	Rockaway
							Subcriteria							
Ra_1	11.6	10.2	10.1	10.7	10.0	9.7	13.1	12.1	11.7	12.6	10.1	0.0	13.5	11.4
Ra_2	0	1	0	0	0	0	0	0	0	1	1	1	0	1
Ra_3	150	200	80	275	310	170	39.9	60	60	85	110	100	120	45
Ra_4	0	30	3	0	13	14	167	17	0	30	112	0	16	180
Ra_5	848	685	411	1,062	1,068	589	245	198	192	283	596	728	758	90
Ra_6	0	153.7	7.5	1.7	142.9	8.2	118.7	15.0	0.0	44.5	35.6	0	38.1	118.5
Ra_7	5,786	6,767	1,019	11,388	8,942	6,475	2,576	2,600	1,907	4,116	3,936	3,836	2,645	2,600
Ra_8	3,500	12,000	4,700	14,224	4,000	13,740	3,500	2,000	8,000	9,000	6,733	3,500	275	2,000
Ro_1	150	200	40	275	390	170	80	60	60	85	110	100	120	45
Ro_2	15.5	17.5	15.5	17.5	13.5	12.5	16.5	14.5	14.5	13.5	15.5	13.5	14.5	14.5
Ro_3	5	7	7	6	4	6	5	4	6	5	3	0	4	7
Ro_4	64	45	66	98	92	66	85	55	72	78	73	99	71	62
Ro_5	0	0.2	1.6	1.4	0.9	0.3	2	1.9	0	1.1	2.1	0.9	1.4	1.9
Ro_6	112.6	201.4	45.2	8.7	28.8	94.1	21.0	54.8	67.4	82.4	84.9	1.7	48.4	49.3
Ro_7	0.9	0.0	0.0	0.0	0.0	0.3	2.4	0.0	0.0	0.4	2.5	0.0	0.0	2.9
Ro_8	0	0	0	0	0	1	1	0	0	0	1	0	0	1
Rs_1	81	108	71	118	88	109	59	46	55	93	69	66	68	41
Rs_2	1	1	1	1	0	0	1	0	1	1	0	0	0	0
Rs_3	1.7	10.1	3.0	27.3	1.0	26.0	8.3	5.8	1.3	9.7	22.6	2.2	13.8	13.1
Rd_1	1	0	1	0	1	0	1	1	1	1	1	0	1	1
Rd_2	6.7	3.9	9.3	3.9	4.6	5.0	9.8	9.8	4.6	10.3	9.0	6.6	7.9	9.0
Rd_3	6.4	0	2.2	14.4	0	16.3	1.5	9.2	4.3	0	0	1.6	2.0	53.5
Rd_4	2	6	3	5	2	5	3	1	4	3	5	2	4	2

TABLE 7.9
The Proposed Coefficients' Values of Financial Constraints

Resilience Intervention	Maximum Percentage of Total Budget	Value (%)
Increasing treatment capacity	α	35
Waterproofing, sandbagging, and elevating critical equipment	β	60
Increasing technical workers	γ	10
Increasing on-site storage	λ	35

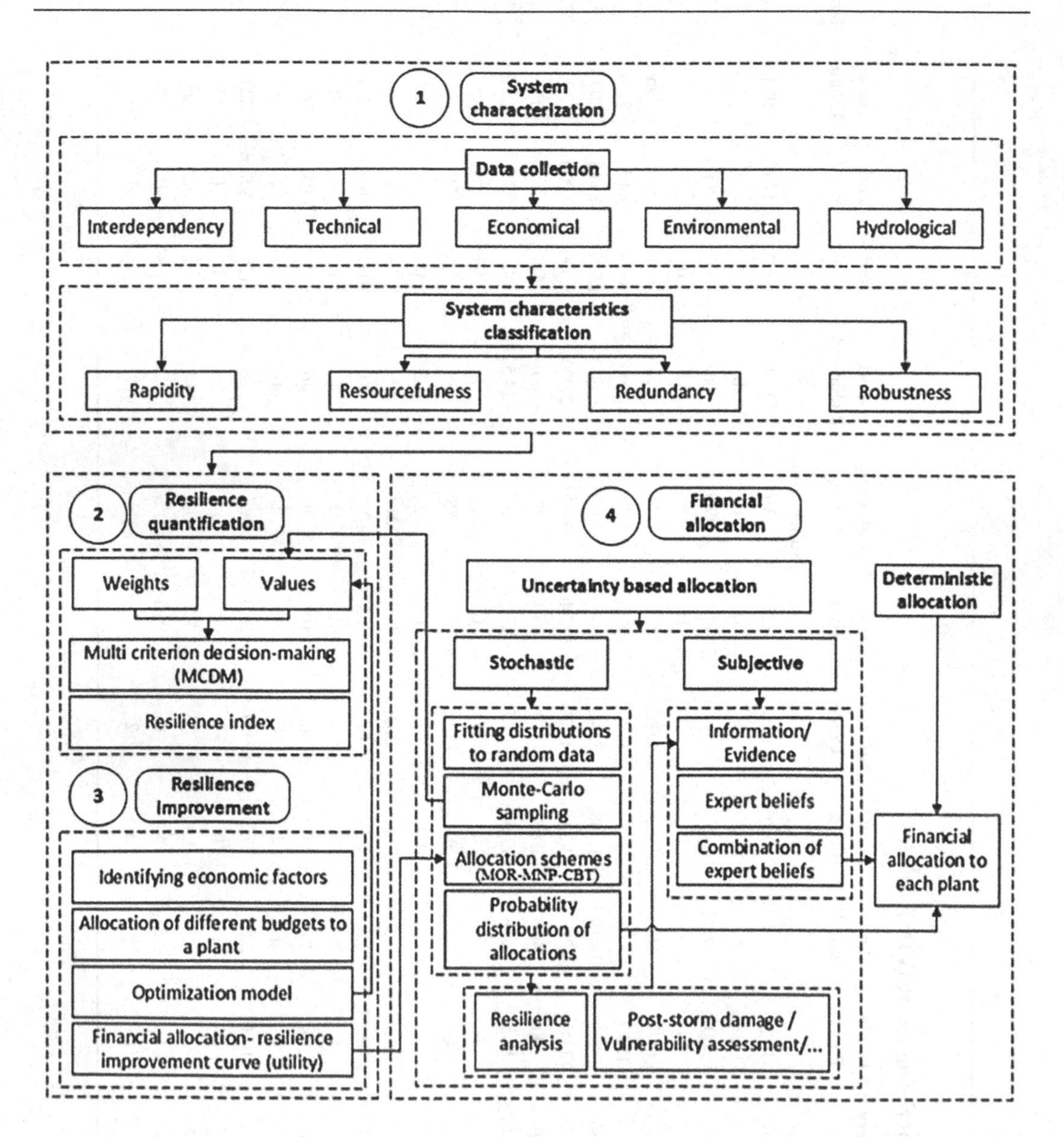

FIGURE 7.18 Proposed frameworks for allocating resources to improve WWTPs flood resilience. (From Karamouz, M. and Hojjat-Ansari, A. (2020b). Uncertainty based budget allocation of wastewater infrastructures' flood resiliency considering interdependencies. *Journal of Hydroinformatics*. doi:10.2166/hydro.2020.145.)

WWTPs are vulnerable infrastructures against coastal storms. The failure of these facilities and the interruption of their recovery will result in irreparable environmental consequences. Improving the flood resilience of these systems is essential and requires considerable funds. However, due to budget limitations, it is inevitable to optimize the allocation of financial resources. The optimal allocation among these systems needs an assessment of their performance indicators such as resiliency. For such an assessment, these systems' characteristics in terms of hydrological, environmental, economic, technical, and operational aspects have been taken into account. Also, given the potential power outages during floods and the urgent need of these systems for power, the dependence of these facilities on energy has been included in assessing their flood resilience. These systems' flood resilience attributes are categorized in terms of rapidity, robustness, resourcefulness, and redundancy (four R's). The allocation of resources among an agency's facilities is typically performed based on nontechnical negotiations and default allocation routine. This study has provided a robust metric based on expert views for reinforcing the resource allocation mechanism of water/wastewater agencies. Using multiple expert views, the relative importance of four R's has been determined. Since more weights were assigned to rapidity and robustness, the importance of these criteria is more dominating to evaluate these facilities' resilience. Assessing NYC's WWTPs' resilience, with regard to interdependencies, shows better agreement with NYC-DEP reports after Superstorm Sandy. In other words, the flood resilience of the plants such as Coney Island that were seriously damaged in Superstorm Sandy has been far less than the plants that were not exposed to major damage.

Taking into account different resilience interventions and related costs, the resilience improvement of each plant was determined for a wide range of allocations and is assumed as their utilities of the allocations. Assuming that these utilities are deterministic, the allocation of resources has been implemented based on maximizing overall resilience (MOR), Nash product (MNP), and consumer behavior theory (CBT). Allocation based on CBT confirms the result of MOR. The Rockaway and Wards Island plants have the highest and the lowest resilience improvement, respectively, in both MOR and MNP. Based on MOR, these two plants have also received the highest and the lowest allocations, respectively. The Wards Island plant has received an insignificant allocation in MOR. However, in MNP, the highest and the lowest allocations are not attributed to these plants. In contrast to MOR, Wards Island has received a significant allocation in MNP. In other words, the resilience improvement of Wards Island with minimal improvement increases in MNP, which demonstrates that this allocation scheme tends to increase the resilience improvement of the system with minimal improvement and to distribute funds more uniformly.

Due to the uncertainties in flood characteristics and some of these facilities' resilience attributes, they were considered in the utility of each plant. Accordingly, allocations were made among the plants using MOR and MNP. Based on the standard deviation and the 95% confidence interval of allocations, it was demonstrated that the level of uncertainty is much lower in MNP compared to MOR. Therefore, the level of uncertainty is influenced by how resources are distributed. Considering the different views of experts in the allocation process, each plant's subjective uncertainties have been examined by the theory of evidence. Accordingly, the degree of inconsistency between experts' beliefs was measured, and the belief, plausibility, and uncertainty were estimated for each proposition of allocation. The results showed that the inconsistency is high in the cases of Bowery Bay, Hunts Point, and Jamaica plants and is far less for the other plants. For plants where the degree of inconsistency is high on the allocation intervals, the Dumpster rule of combination cannot be applied.

The uncertainty intervals of these plants cannot be determined. Subsequently, for other plants, according to the degree of belief in each proposition, the type of allocation has been determined. The results show that the medium type of allocation could be considered for most of the plants. This study demonstrates the significant value of resilience-based funds considering interdependencies and uncertainties related to different views of entities and analyses. The methodology of the paper is applicable to other geographic settings in coastal areas. One limitation of this framework is the sensitivity of its results to model input subjective values. Karamouz et al. (2018) performed a sensitivity analysis on the input of the model which indicates that plants with high investment potential are more sensitive to the model's subjective inputs. Global sensitivity analysis on the dependence of the results on the model inputs is suggested as it was outside this study's scope.

7.9.3 Case Study 3: Margin of Safety-Based Flood Reliability Evaluation of Wastewater Treatment Plants

Karamouz et al. (2020c) and Karamouz and Farzaneh (2020a) worked on a two-part paper to set up a framework for reliability assessment of WWTPs. The Hunts Point plant, shown in Figure 7.19, was selected as the focus of this study. This 67-year-old treatment plant (built in 1952), located on the banks of the river alongside Barrette Point Park, treats wastewater from more than half a million residents of the northeast side of the Bronx. The Bronx is the northernmost point of New York City boroughs, located south of Westchester County and north of Manhattan and Queens. This treatment facility drains approximately 68 km^2 of urbanized land in the eastern section of the Bronx. With a design capacity of 200 million gallons per day (MGD), this facility has one of the lowest altitudes among the New York City's WWTPs, which increase the potential of damage due to flood events. The focus of this study was to better understand the impacts of flood hazard mitigation strategies on reducing the risk of flooding through the implementation of the two types of BMPs. Figure 7.19 demonstrates the New York City's sewer sheds. In this study, since the Hunts Point plant has a combined sewer system, the sewershed of this plant should be considered that includes Bronx River. The spatial characteristics of the basin were identified with the utilization of a DEM of 10 m resolution (USDA, 2018). Furthermore, the physical characteristics of different land use in the catchment are identified using the soil map of the region, Manning's roughness coefficient for natural channels, soil hydraulic conductivity, and porosity parameters.

Figure 7.20 outlines the proposed framework for estimating the marginal and joint probability of climatic and hydrological extremes and quantifying the flood reliability with a load-resistance approach. In the first step, the water level and rainfall time series along with the physical characteristics of the watershed as the input parameters to the hydrologic model are obtained. The second step carries out a trend detection in observational extremes with regard to time. Univariate and bivariate flood frequency analyses are performed to estimate the extreme values of surge and rainfall. In the third step, a probabilistic-based reliability measure is described to evaluate the WWTP performance. A companion part 2 paper (Karamouz et al., 2020c) presented the proposed framework validation, where the utilization of coastal protection strategies in terms of best management practices (BMPs) leads to significant improvement of flood reliability in Hunts Point WWTP located in New York City.

The pertaining concepts in this study consisted of attributes of joint nonstationary statistical and frequency analysis of rainfall and surge; flood remediation design values; load and resistance interplays; flood modeling and mitigation strategies; and reliability quantification and BMPs performance. The four steps, presented in Figure 7.20, outline how these attributes are interplaying and where the hydrologic distributed modeling could simulate a real-world problem. A hydrologic and hydraulic analysis is performed using GSSHA, which is explained in detail in the paper in Part 2.

In this study, the hydrograph of surge and hyetograph of rainfall were developed. Surge data was limited, so the closest station of Kings/Willets point was selected. Two stations were selected for rainfall, and the LaGuardia station was found more suitable with a much higher correlation with the surge station. Results of data analysis showed the surge data to be nonstationary and rainfall data to be stationary. To obtain the extreme rainfall values and surge in the 100-year flood, the GEV

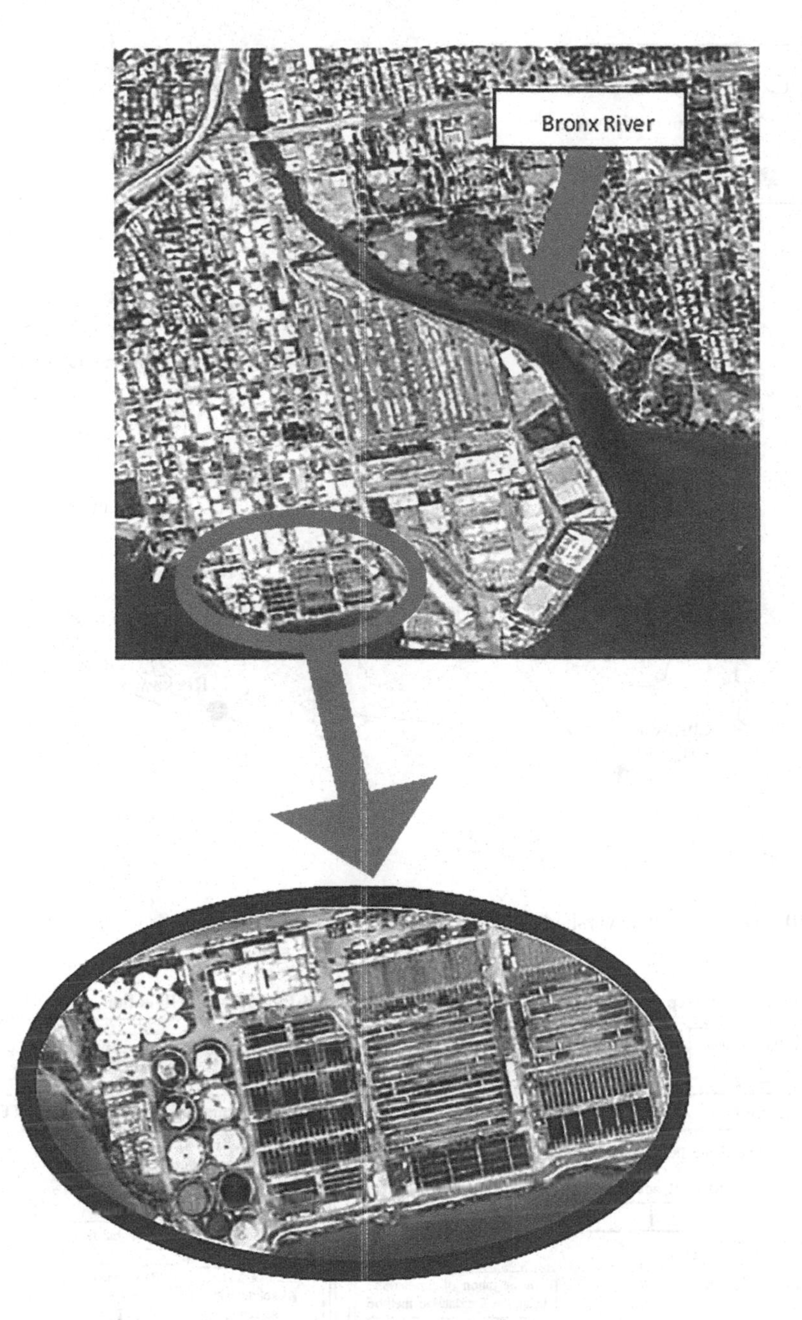

FIGURE 7.19 Location of Hunts Point WWTP and the Bronx River.

distribution was applied. In order to estimate the GEV parameters, different parameter estimation methods are used. The maximum likelihood method (MLE) for stationary analysis and differential evaluation—Markov chain Bayesianbased for nonstationary analysis were used.

Also, to deal with the substantial damages, the BMPs as hazard mitigation strategies were applied in this study. Based on the proposed framework, the reliability index under the deterministic and probabilistic assumptions was quantified to assess the performance of Hunts Point plant in Bronx, one of the five boroughs of New York City. The identified criteria, which are effective on the plant reliability, categorized into load and resistance and can be used to quantify this index.

FIGURE 7.20 Hunts Point sewershed as a part of New York City sewersheds.

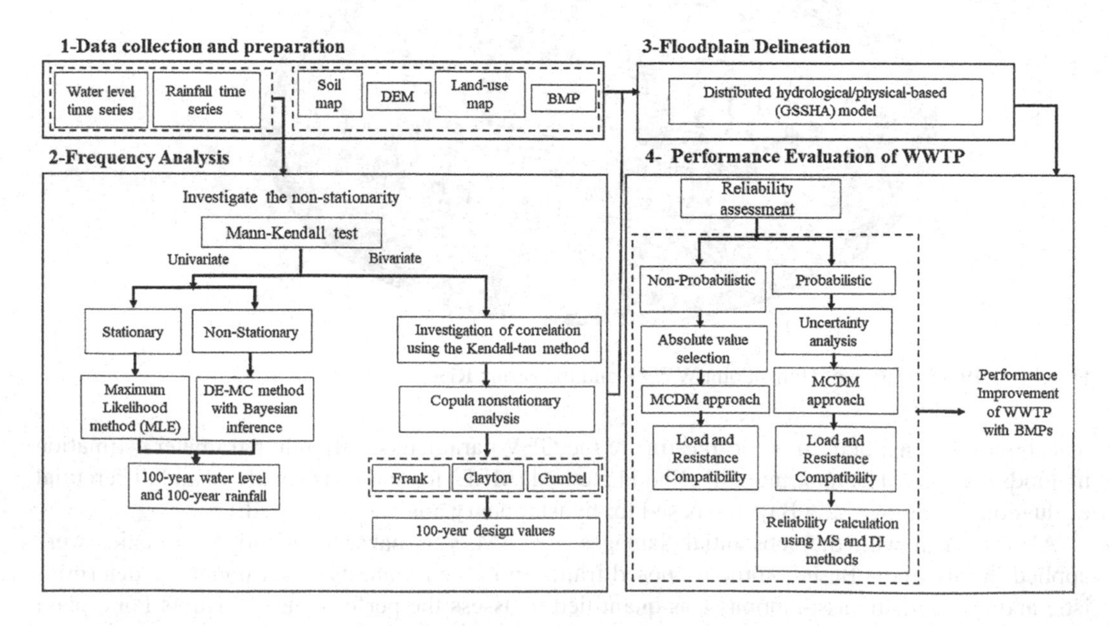

FIGURE 7.21 Proposed frameworks for flood reliability of coastal infrastructure/WWTP (Karamouz, M. and Farzaneh, H.. Margin of safety based flood reliability evaluation of wastewater treatment plants: Part 2 – Quantification of reliability attributes. *Water Resources Management*, 34, 2043–2059, (2020a)).

Using the load-resistance concept based on MCDM and statistical methods of Margin of Safety (MOS) and Direct Integration (DI), the plant's uncertainty-based reliability was estimated. A GIS-based integrated hydrologic model named GSSHA is applied to determine inundation depth at the Hunts Point plant. A series of trade-off curves of rainfall and surge are developed for each time window. In copula nonstationary 100-year flood analysis, the design value for the year 2017 was determined according to the trade-off curves. For the design of BMPs in coastal areas, extreme surge values and the corresponding rainfall were selected for 2017, the last year of record.

The selected BMPs were a combination of a levee and two constructed wetlands for considering the combined mitigation strategies of resist, drainage, and storage effects. The results show how the reliability could be quantified and improved utilizing the concept of MOS and load and resistance attributes. They also showed how the proposed methodologies and strategies could positively affect coastal flood preparedness planning.

7.10 WASTEWATER COLLECTION AND TREATMENT OF SELECTED CITIES

7.10.1 Amsterdam, Netherlands

7.10.1.1 Characteristics of the System

Amsterdam is one of the first cities in the Netherlands where separated sewer systems were built and used since 1930.

Thirty percent of the city's paved area is connected to traditional combined systems, compared to 70% connected to separate systems that directly discharge stormwater into the surface water.

There are more than 300 collection areas, each containing a sewer system and a pumping station. More than 99% of the wastewater is pumped under pressure to a WWTP. The collection systems and pumping stations transporting the wastewater to the treatment plant are mainly switched serially with more than 4,000 km lengths.

7.10.1.2 Improvement and Future Plans

By the year 2000, Amsterdam had three WWTPs: one in the east of Amsterdam (Zeeburger Island), one in the south (Amstel industrial area), and WWTP west port (West Port Area).

By 2005, the WWTPs in the east and the south could no longer comply with the new effluent discharge and environmental requirements. Amsterdam's municipality decided to translate the total treatment capacity of the current WWTPs into one new centralized WWTP in Amsterdam West Port Area that named WWTP Amsterdam West that is the second-largest WWTP in the Netherlands. The process scheme of WWTP Amsterdam west is shown in Figure 7.22.

A network of pressure mains has been constructed to transport the wastewater from the former WWTP sites in Amsterdam South and East to the new WWTP in Amsterdam West. Due to the large distances, the pressure mains are equipped with booster pumping stations.

7.10.2 Stockholm, Sweden

7.10.2.1 Characteristics of the System

Stockholm is the capital and most populous City of Sweden and the most populous Scandinavia city, and approximately 1.5 million live in the urban area. It is located on Sweden's east coast, where freshwater Lake Malaren (Sweden's third-largest lake) flows out into the Baltic Sea, Figure 7.23.

The wastewater is treated with mechanical, chemical, and biological methods and then passes through the sand filtration process that filters out the remaining small particles. The whole purification process takes 24 hours, and the treated water is discharged into the Baltic Sea. Treated wastewater is discharged into Lake Salts Jon, which is part of the archipelago and the Baltic Sea. Wastewater is led away mainly with the help of gravity via the sewers to the treatment plants. A large number of pumping stations and storage facilities supplement the pipe system.

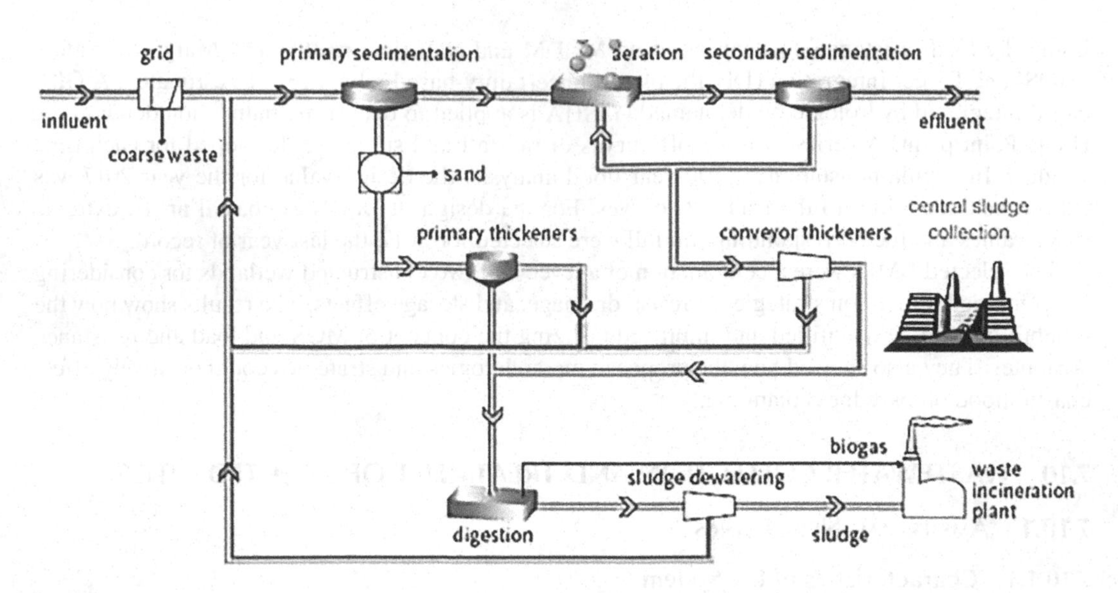

FIGURE 7.22 Process schemes of WWTP Amsterdam West.

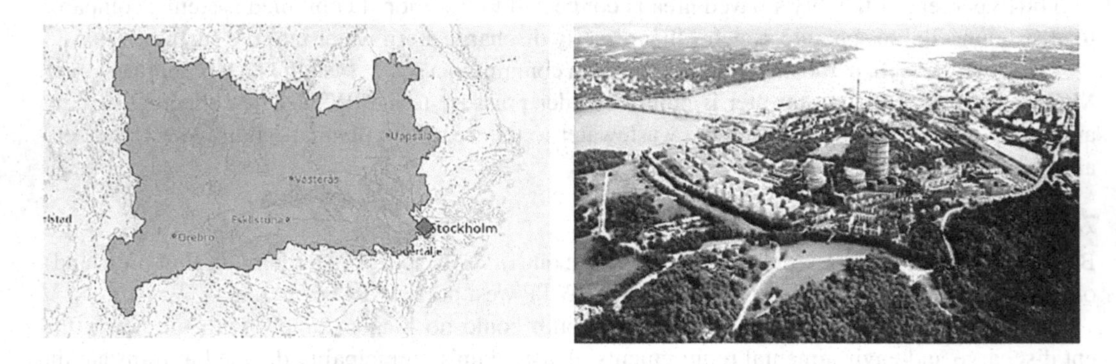

FIGURE 7.23 Location of selected site: Stockholm, Sweden.

7.10.2.2 Improvement and Future Plans

The vision is that the wastewater system can reduce discharges and emissions, despite a changing climate. The objectives for good ecological and chemical status in oceans, lakes, and rivers can be achieved, and that groundwater is not adversely impacted. Simultaneously, the systems contribute to better management of resources such as nutrients and high-value energy like electricity and biogas. WWTPs contribute to a reduced climate impact by reducing greenhouse gas emissions, more efficient energy usage, and the production and efficient use of biogas.

The goals for the year 2050 are:

- Collection networks and WWTPs are operated in an integrated fashion, and the discharge of untreated wastewater via overflows does not occur.
- The wastewater system and treatment plants have become production facilities for recycling water, energy, and nutrients.
- Swedish wastewater systems are an important link to a sustainable cycle of nutrients. In particular, phosphorous, biogas, and other commodities from wastewater systems are used in society, such as nutrient-rich byproducts.

- Swedish companies offer innovative wastewater solutions adapted to local conditions. Dependent on the situation, the systems can function individually, in combination with other systems, and also, in the long term, integrated into a whole system.

7.10.3 Philadelphia, USA

7.10.3.1 Characteristics of the System

Philadelphia is located in the Delaware River Watershed, which begins in New York State and extends 330 miles south to the Delaware Bay's mouth. The Schuylkill River is part of the Delaware River Watershed.

The Water Department's primary mission for wastewater collection and treatment is to plan for, operate, and maintain both the infrastructure and the organization necessary to sustain and enhance the region's watersheds and quality of life by managing wastewater and stormwater effectively.

In areas with combined sewers, a single pipe carries both stormwater from streets, houses, and businesses as well as wastewater from houses and businesses to a WWTP. Many of the older and dense sections of the city are served by a combined sewer system.

In areas with separate sewers, one pipe carries stormwater to the city's streams while another carries wastewater to a WWTP.

The Philadelphia sewer system has nearly 3,000 miles of combined and separate sewers. Wastewater travels along some part of that system to one of three water pollution control plants, Figures 7.24 and 7.25: Southwest, Southeast, and Northeast Water Pollution Control Plants. A combined average of 1.78 MCM of wastewater is cleaned and discharged into the Delaware and Schuylkill Rivers every day in these facilities.

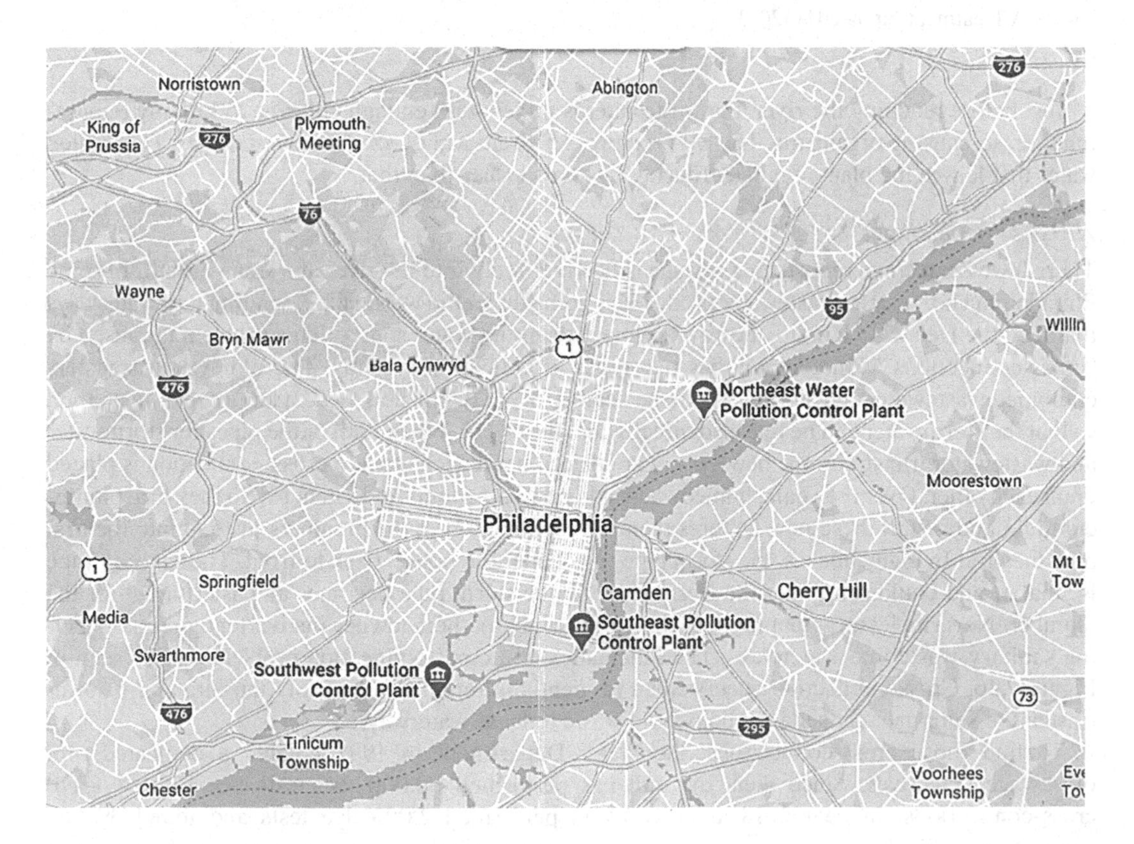

FIGURE 7.24 Location of selected site: Philadelphia.

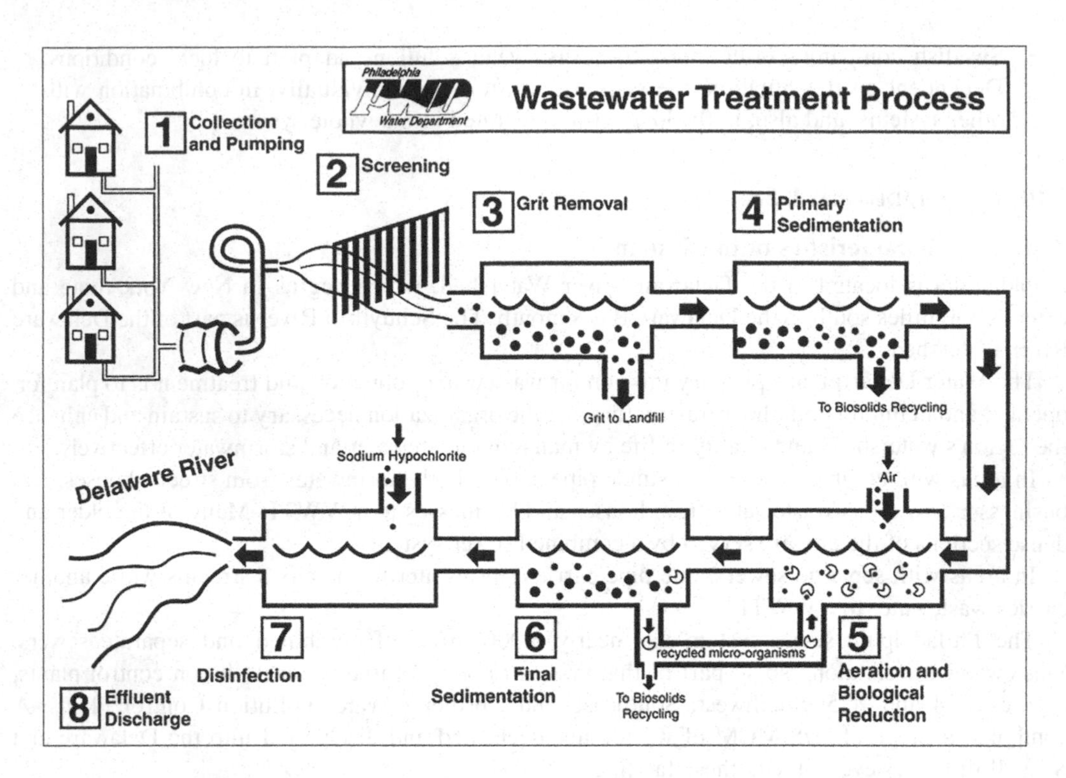

FIGURE 7.25 Wastewater treatment process in Philadelphia. (From http://phillyh2o.org/canvas/mod07/DR_WWTreatment_updatedJan2013.jpg.)

Just before reaching the river, the wastewater is mixed with enough chlorine to kill any remaining disease-causing organisms. The EPA requires 85% removal of SSs from wastewater. The treated water that leaves the plant—called effluent—is even cleaner than that.

7.10.3.2 Challenges

There have been some reports that sewage is flowing freely into the river at a stone outfall. In some Philadelphia homes, the wastewater produced is going down the wrong pipe, sending it to waterways that feed the Delaware River—the city's primary drinking water source. This problem is caused by "cross-connections." Philadelphia has two types of collection systems. The old system transfers combined wastewater and stormwater to the nearest WWTP. The new system has a separate method so that the stormwater can be released directly into the rivers. When both of the pipes for stormwater and wastewater transfer in the new system are placed upon each other, a cross-connection is placed upon each other. Over the years, homeowners add new bathrooms and kitchens. So they hire plumbers to fix and install these new features themselves. Sometimes, the plumbers just tap into the first pipe they hit. In many houses, the storm pipe runs under the house, back to front. The sanitary pipe is smaller and could run under the storm pipe. So the storm pipe becomes an easier target for the plumber. Nevertheless, by tapping into the wrong pipe, the homeowner inadvertently starts injecting sanitary waste into waterways. The schematics of a cross-connection problem can be seen in Figure 7.26. Cross-connection occurs when residential plumbing mistakenly tap into storm sewer (instead of sanitary) that send the household waste to local waterway.

To find those cross-connections, the Water Department in 1995 created a dedicated unit of sewage detectives. Since then, the unit has performed 60,000 dye tests and located 1,440 cross-connections. In year 2018 alone, workers performed 2,093 dye tests and found 39 bad connections.

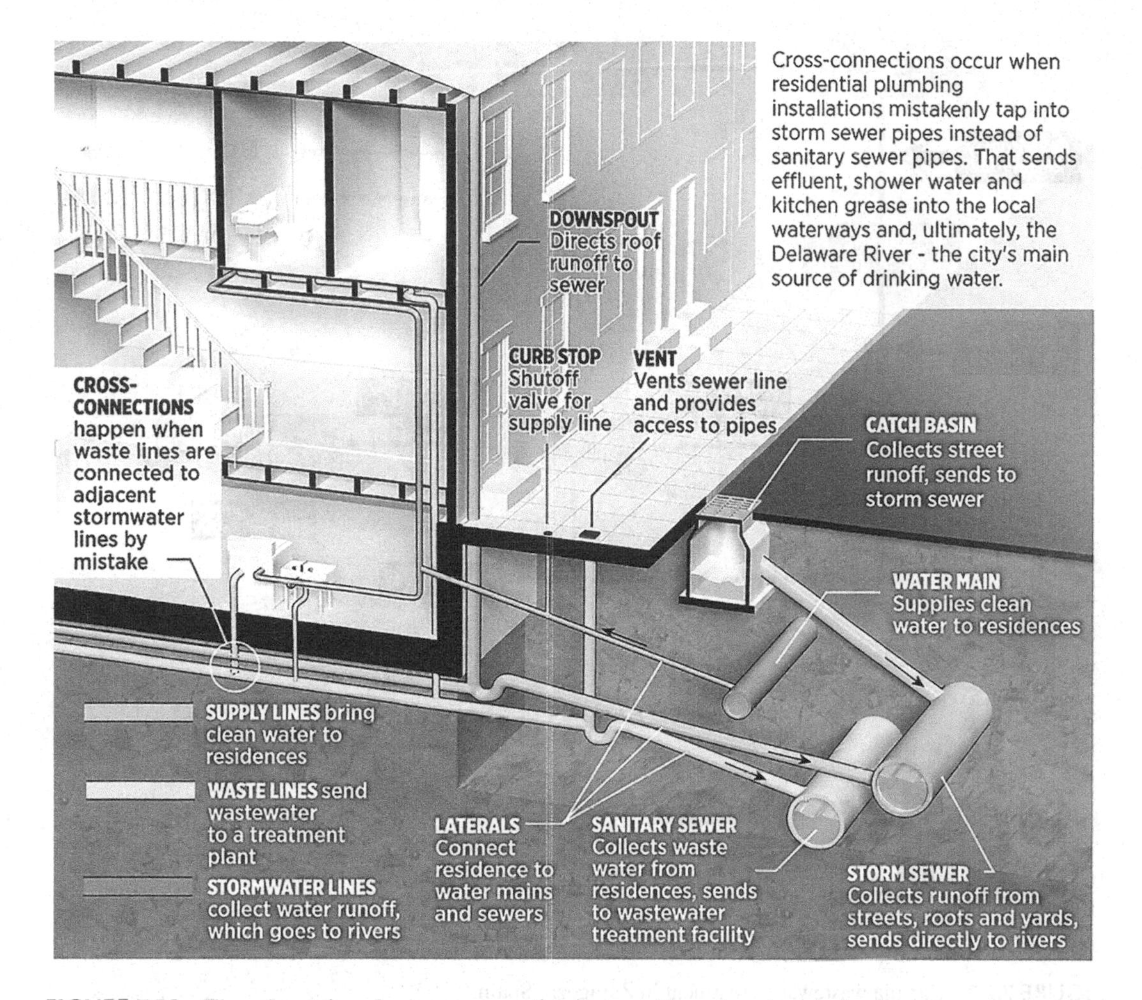
Cross-connections occur when residential plumbing installations mistakenly tap into storm sewer pipes instead of sanitary sewer pipes. That sends effluent, shower water and kitchen grease into the local waterways and, ultimately, the Delaware River - the city's main source of drinking water.
DOWNSPOUT Directs roof runoff to sewer
CURB STOP Shutoff valve for supply line
VENT Vents sewer line and provides access to pipes
CROSS-CONNECTIONS happen when waste lines are connected to adjacent stormwater lines by mistake
CATCH BASIN Collects street runoff, sends to storm sewer
WATER MAIN Supplies clean water to residences
SUPPLY LINES bring clean water to residences
WASTE LINES send wastewater to a treatment plant
STORMWATER LINES collect water runoff, which goes to rivers
LATERALS Connect residence to water mains and sewers
SANITARY SEWER Collects waste water from residences, sends to wastewater treatment facility
STORM SEWER Collects runoff from streets, roofs and yards, sends directly to rivers

FIGURE 7.26 The schematics of a cross-connection issues with wastewater tapping into storm sewer.

7.10.4 Zaragoza, Spain

7.10.4.1 Characteristics of the System

Wastewater has diverged to two different WWTPss, Cartuja and Almozara, built in 1983 and 1989, respectively. Cartuja, the larger plant, can serve a population of 1,200,000 equivalents with a treatment capacity of 259,200 m^3/day, Figure 7.27. Almozara serves a population of 100,000 equivalents with a discharge capacity of 34,560 m^3/day, Figure 7.28. The receiving body for the treated waters is the Ebro River.

Industrial discharges are also diverted to the treatment plants. Industries are subjected to ISO standard monitoring programs to measure pollutant parameters prior to their discharge to the treatment plants or directly to the receiving water body.

7.10.5 Paris, France

7.10.5.1 Characteristics of the System

Nearly 2,600 km long, the Paris sewers constitute all of the underground conduits intended to collect and evacuate waste water. Eugène Belgrand undertook from 1854 the vast site of sanitation from which the current sewer network was developed. Belgrand's innovation is to have designed a network that collects and transports both rainwater and wastewater.

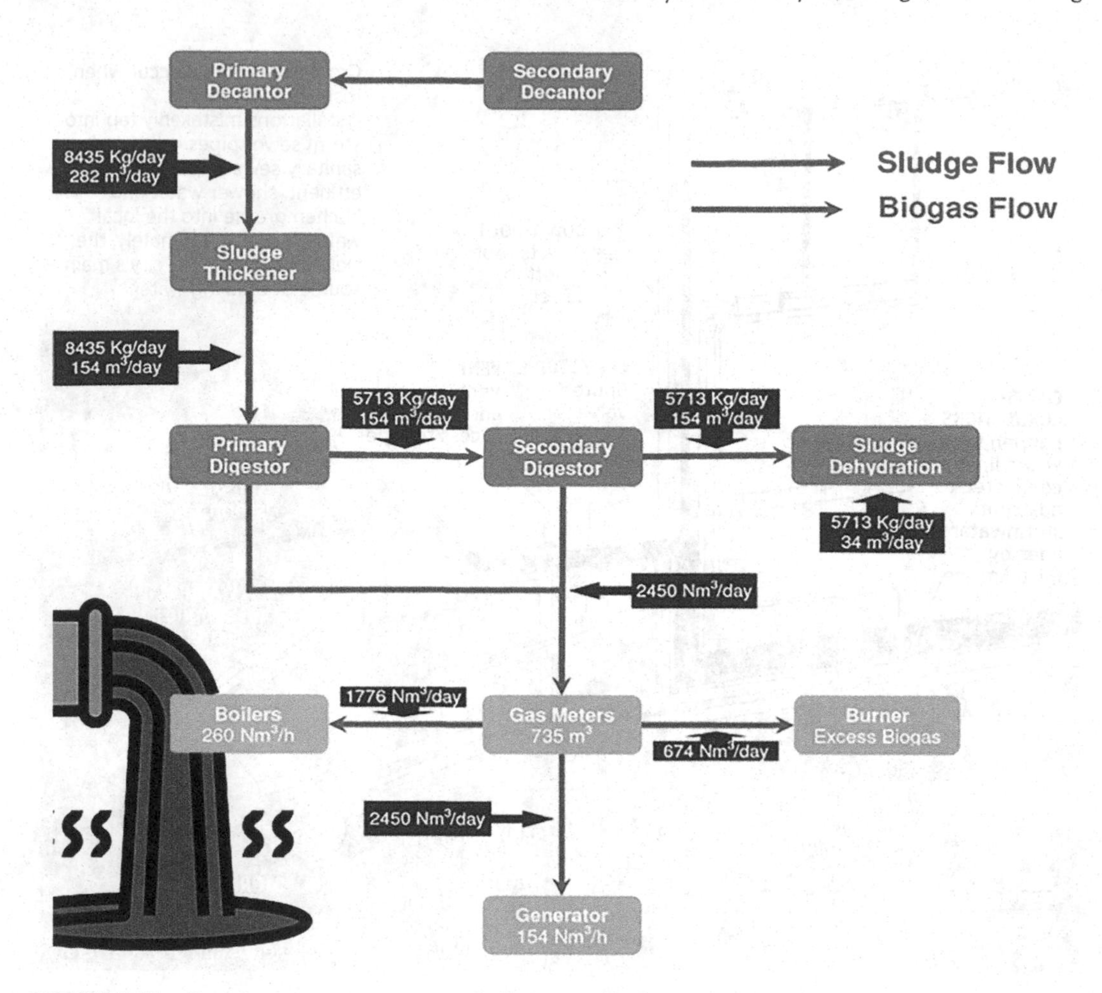

FIGURE 7.27 Cartuja wastewater treatment in Zaragoza, Spain.

Two public bodies carry out the sanitation. One of them, the Section Assainissement de la Ville de Paris (SAP), is part of the Paris technical services and is responsible for collecting all the wastewater, Figure 7.29. The other operates at the level of Greater Paris (le Syndicat Interdépartemental pour l'Assainissement de l'Agglomération Parisienne, SIAAP) and treats all wastewater before it flows back into the rivers.

The wastewater collection is done by two different entities, first SAP and second SIAAP. In what follows first, we will discuss SAP's mission and its struggles, focusing on SIAAP.

In the following section, the five areas of management of SAP are discussed.

1. *Optimize the flow conditions of water*: Wastewater carries waste, part of which settles in the sewers. Thus, it is 6,000 m^3 of sand extracted annually from the network. The flushing tank, a permanent cleaning device generally located at the head of each elementary sewer, automatically fills up before releasing the retained water. The flow thus created cleans up a portion of the sewer. The water used is nonpotable water. The collectors are priests by boat or rail valves, depending on their size. Each section of the sewer system is inspected twice a year.
2. *Maintain and modernize the heritage*: Cracks can create leaks that pollute soil, destabilize the ground, and cause eventual collapse. An effective rehabilitation program has been

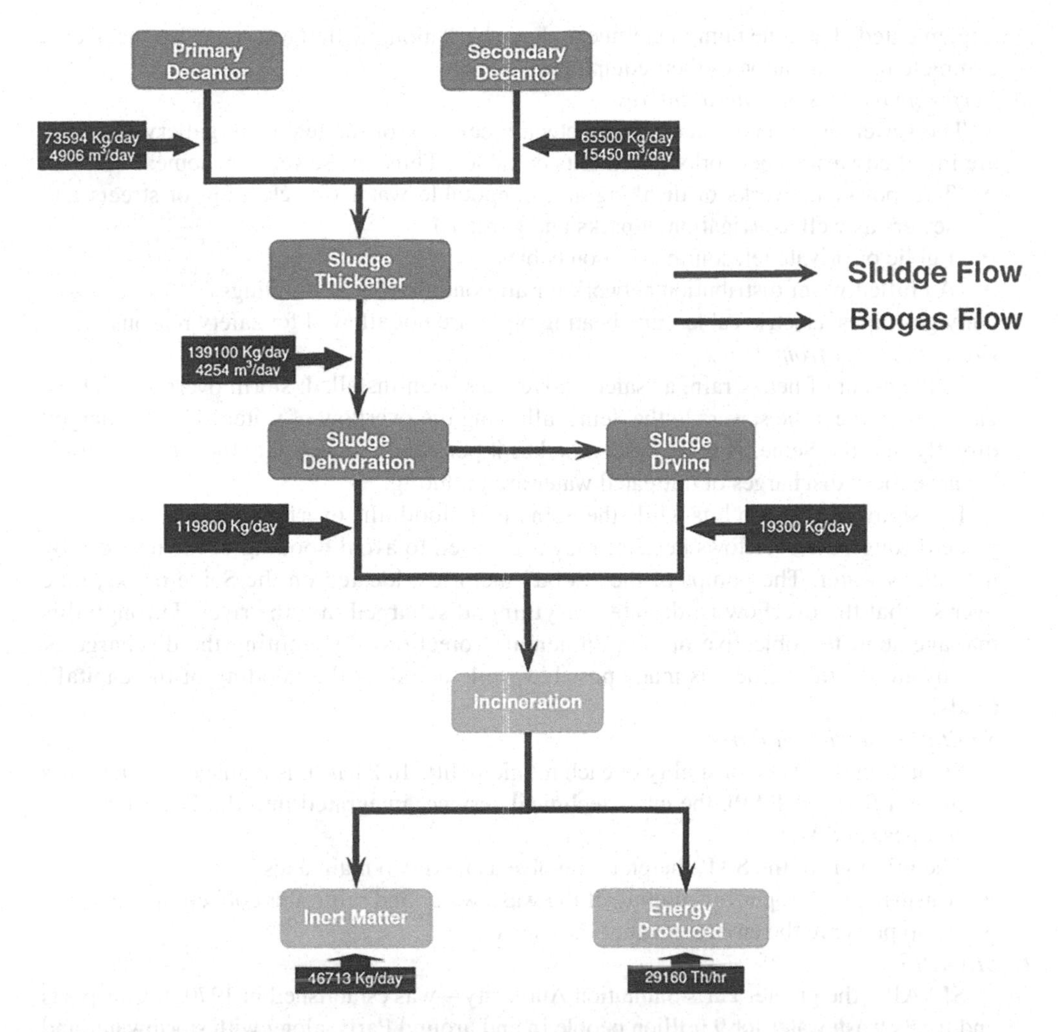

FIGURE 7.28 Almozara wastewater treatment plant in Zaragoza, Spain.

FIGURE 7.29 Location of WWTP: Paris, France.

implemented. The nine pumping effluent plants built along with the sewers also received a complete modernization of their equipment.

3. *Perform missions of general interest*

   The sewers also serve, since the nineteenth century, of the technical gallery in which are installed various networks of conduits or cables. Thus, the sewers are home:

   - Two power networks of drinking and nonpotable water (for cleaning of streets and sewers as well as irrigation of parks and gardens)
   - Public or private telecommunication cables
   - A chilled water distribution network for air conditioning in buildings

   Only gas pipes, electric cables, and heating pipes are not allowed for safety reasons.

4. *Protect the city from floods*

   In the event of heavy rain, a "safety valve" has been installed: storm overflows. These galleries connect the sewers to the Seine, allowing the overflow of water to be discharged directly into the Seine. A valve system makes it possible to reject only the bare essentials because these discharges of untreated water are polluting.

   If a significant storm hits while the Seine is in flood, the overflow can no longer take place through the overflows because they are closed to avoid flooding of the network by the river's water. The pumps of the "flood" factories, located on the Seine banks, take over so that the overflow is, despite everything, discharged into the river. Through this management, the objective of environmental protection—by limiting the discharge of stormwater in the Seine—is made possible while avoiding the flooding of the capital's roads.

5. *Sanitation section of Paris*

   Sanitation is the responsibility of each municipality. In Paris, it is managed by the Paris Sanitation Section (SAP), the city's technical service, integrated into the Department of Cleanliness and Water.

   The missions of the SAP, therefore, revolve around two main axes:

   - Ensuring the proper functioning of the wastewater and rainwater collection network
   - Help preserve the environment of Parisians

- *SIAAP13*

  SIAAP—the greater Paris Sanitation Authority—was established in 1970. It transports and treats wastewater for 9 million people in and around Paris, along with stormwater and industrial water, contributing to a thriving natural environment in the Seine and Marne rivers. SIAAP covers four French districts: Paris, Hauts-de-Seine, Seine-Saint-Denis, and Val-de-Marne, along with 180 municipalities in the Val d'Oise, Essonne, Yvelines, and Seine-et-Marne departments, which have signed an agreement with the SIAAP.

  The protection of biodiversity and preservation of natural resources includes an increased consideration of ecosystems and restrained use of sanitation policy. The SIAAP's action to improve the quality of the Seine and the Marne is its first contribution to improving biodiversity and, more specifically, to the development of aquatic life, notably fish. In a complementary way, the preservation of biodiversity also includes protecting and developing natural land heritage.

  The SIAAP plants, based on the banks of the river, participate in the regional scheme of ecological coherence. These sites house diverse habitats such as spawning grounds, riparian woodland, wetlands, grasslands, and forest habitats, which are refuge areas in the urban area's heart. They constitute living environments essential to the development of species. It is necessary to successfully manage actions aiming to both reestablish ecological continuities and effectively manage industrial heritage.

7.10.6 Copenhagen, Denmark

7.10.6.1 Characteristics of the System

In the City of Copenhagen, wastewater from households and enterprises is managed in closed pipes underground. Most of the city has a joint system where stormwater from roofs and roads, and household wastewater is discharged for treatment in central treatment plants. Separate sewer systems only exist in the part of the city which is close to the ports. The choice of sewer system type has a historical basis. This also applies to the service offered by the city to its citizens and enterprises. The main principle is for wastewater from ground floors to be drained off to sewers through gravitation. The whole city, except for the nature area of Vestamager, has sewers.

The sewers are designed so that statistically, overflow ground level only occurs once every 10 years in joint sewer system areas. In contrast, separate stormwater pipes are designed, so that overflow at ground level only takes place once every 5 years. The sewer systems are furnished with pumping stations and reservoirs for the storage of wastewater during rainfalls. All wastewater is discharged for treatment at the two treatment plants of the Lynettefællesskabet community: Lynetten and Damhusåen. Increasing urban development with accompanying expansion of the sewer system and increased rainfall intensity has increased pressure on central treatment plants. To minimize this, pressure draining off of stormwater in new urban development areas and major renovation work is to be carried out according to the Sustainable Urban Drainage Systems (SUDS) principles. The new Ørestad district founded in 1996 and today is an area of 150 ha and was established with a three-stringed system. In the three-stringed system, household wastewater is discharged to a central treatment plant, and roof water is discharged to recreational canals, whereas road water is treated locally before being discharged to the recreational canals.

7.10.6.2 Infrastructures Sustainability

The Lynetten and Damhusåen treatment plants treat wastewater for organic substances, phosphorus, and nitrogen through mechanical, biological, and chemical processes. The treated wastewater is discharged to Øresund Strait through 1.5 and 1.2 km long outfall sewers furnished with diffusers that ensure that the treated wastewater is well mixed with seawater.

In the past decade, the city has put in great efforts to renovate Copenhagen's sewer systems. As a result, the sewer system is in good condition, although parts of the Copenhagen sewers are very old. Today the sewers only require maintenance at the same rate as degradation. This means that there is no lacking behind of maintenance. No-dig methods are used to the extent possible, taking into account the economy and traffic nuisance.

7.10.6.3 Improvement and Future Plans

Today very few properties in the City of Copenhagen do not have sewers. These are mainly houseboats and garden associations where overnight stays only occur in a limited part of the year. The plan is for housing associations and houseboats to install sewers over 12 years in order to prevent illegal and inappropriate discharges/seepages of wastewater. The City of Copenhagen has adopted the Copenhagen Climate Adaptation Plan, which describes the risk of harm to the city based on a changing climate and provides solutions to address climate change. Several of the solutions provided are about the management of storm and wastewater. These are the two main principles of climate change adaptation of the city:

Climate proofing of sewers by bypassing stormwater is subsequently managed locally either through seepage to the groundwater or by discharge to the recipient.

Systems for the management of rainfall during cloudbursts must be established to protect the city against significant losses due to flooding.

7.11 STANDARDS AND PLANNING CONSIDERATIONS

The primary standards are intended to improve the service to the public, manage waste water utilities, and drinking water services that indirectly affect wastewater services. The planning consideration include legislations, cost factors, linkage to federal requirements, placement of facilities, and maintenance issues. The legislations are limited to US-EPA landmark Acts that are also discussed in Chapter 1. Cost factors and maintenance issues are discussed in Chapters 8 and 9. Sizing and placement of facilities are refer to in different sections of this book. The softer side of planning considerations are only briefly outlined here.

- *Important legislation*:
  - Federal Water Pollution Control Act (1972)
  - Clean Water Act (1977)
  - Water Quality Act (1987)
  - Resource Conservation and Recovery Act (1976)
- *Cost Factor—Sewer System Size and Type*:
  - Building a system of appropriate capacity
  - Federal monies and federal requirements
  - Balancing cost and treatment levels
- *Placement of treatment facilities*:
  - Ideally at the lowest point in the area
  - Away from residential/commercial areas
  - Good transportation access
  - Soil considerations
- *Importance of maintenance*: EPA study showed that while most needs are for new facilities, retrofitting and repairing older facilities are very expensive; therefore, maintenance is essential.

7.11.1 STANDARDS ON WATER AND WASTEWATER SERVICES

Standards of delivering service is different from other often talked about health driven standards of water quality. Water utility activities provide drinking water supply, wastewater, and stormwater services (appropriate to the utility's service area) following the requirements established by the responsible bodies and relevant authorities and the water utility's corporate objectives to a level of service agreed with the registered users. Service delivery can require activities that are not intended to impact users directly but are essential for providing the service (e.g., management of assets). Some activities are essential to maintain service efficiency (e.g., water loss management). Furthermore, some are intended to support users in achieving their efficiency improvements on delivery of the service (e.g., water efficiency management).

The need for early detection and rapid response to abnormal events is encouraging some water utilities to establish innovative means of detecting and classifying causes of substandard/abnormal events. This, allied with recent technological advances, permits more advanced means of online monitoring of water service provision. Water service provision, therefore, requires a diverse range of activities. Not all of these are readily seen by the service users, but they are essential for the effective and efficient operation related to water utility's responsibilities.

The primary standards are under International Organization for Standardization (ISO) and are intended to improve the service to the public. They are listed as ISO 24510, 24511, and 24512 as follows:

- ISO 24510: Guidelines for the assessment and for the improvement of the service to users (service-oriented standard)

- ISO 24511: Guidelines for the management of wastewater utilities and for the assessment of wastewater services (management oriented)
- ISO 24512: Guidelines for the management of drinking water utilities and for the assessment of drinking water services (management oriented)

The water management system is subject to quality, environmental, security, and social attributes, and issues where the first two are under comprehensive guidelines covered by ISO 9001 and 9004 for quality. ISO 14001 and 14004 for the environment. They are shown in Figure 7.30. ISO 14001 of 2015 is the principal management system standard that specifies the requirements for the formulation and maintenance of an Environmental Management System (EMS). ISO 14004 of 2016 provides guidance for an organization on establishing, implementing, maintaining, and improving a robust, credible, and reliable EMS. ISO 9001 of 2015 specifies requirements for a quality management system (QMS). Organizations and utilities use the standard to demonstrate the ability to consistently provide products and services that meet customer and regulatory requirements. ISO 9004 of 2018 is the guideline for enhancing an organization's ability to achieve sustained success. Most of the water utilities are striving to get ISO certifications for compliance with these quid lines. Figure 7.30 shows the further need for policies to ensure sustainability as an intersection of economic, social, and environmental issues a water utility is governed by.

The core of all activities and plans is concentrated on social responsibility (SR). Classification of ISO are accompanied with a particular Technical Committee (TC) within that organization such as ISO/TC 224 with seven core issues that are developed in the draft ISO 26000. SR with main attributes (related to human rights, labor practices, organizational governance fair pricing. environmental, consumer, and social) are presented in Figure 7.31. This applies to the management of any utility systems but more specifically to water and wastewater. ISO/TC 224 aims to increase relevant parties' awareness of water utilities and, where separate, stormwater service providers' duties fulfill

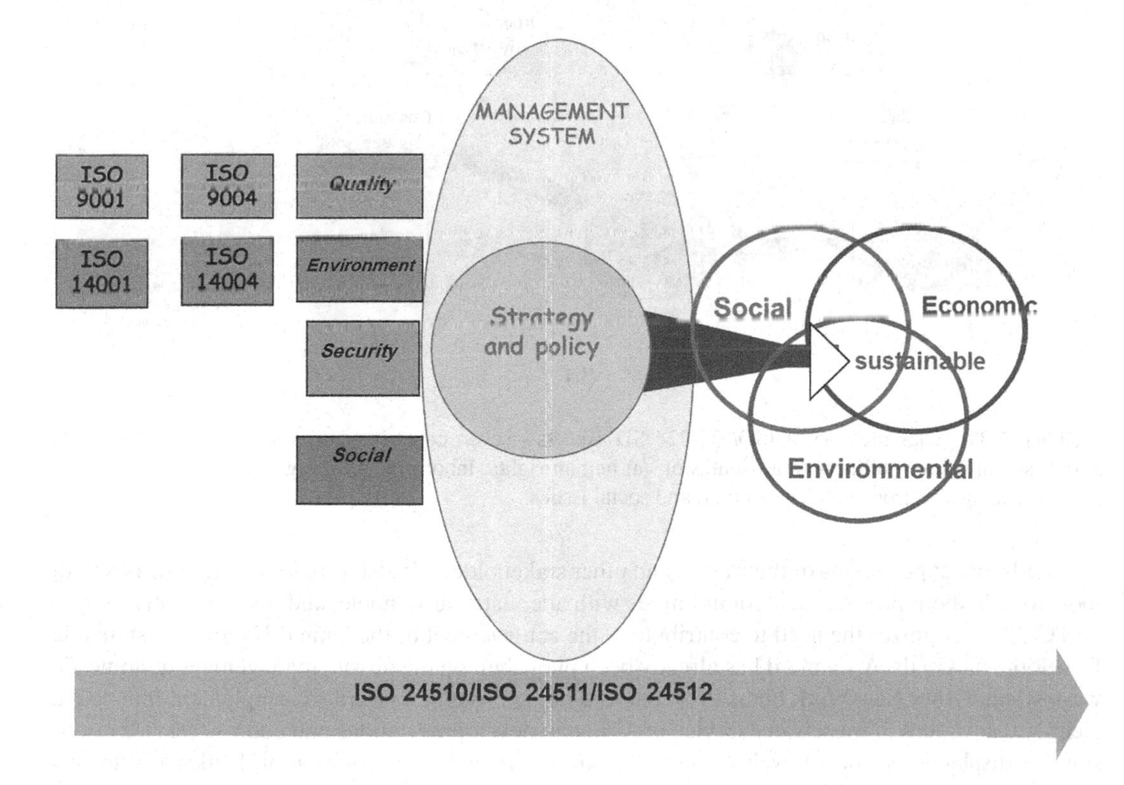

FIGURE 7.30 Standards in a perspective for sustainable development of water services.

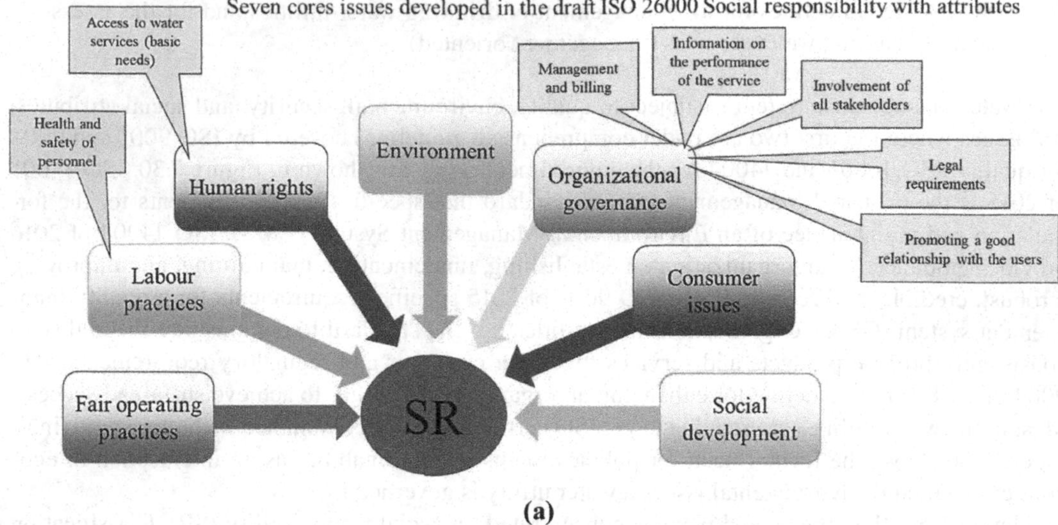

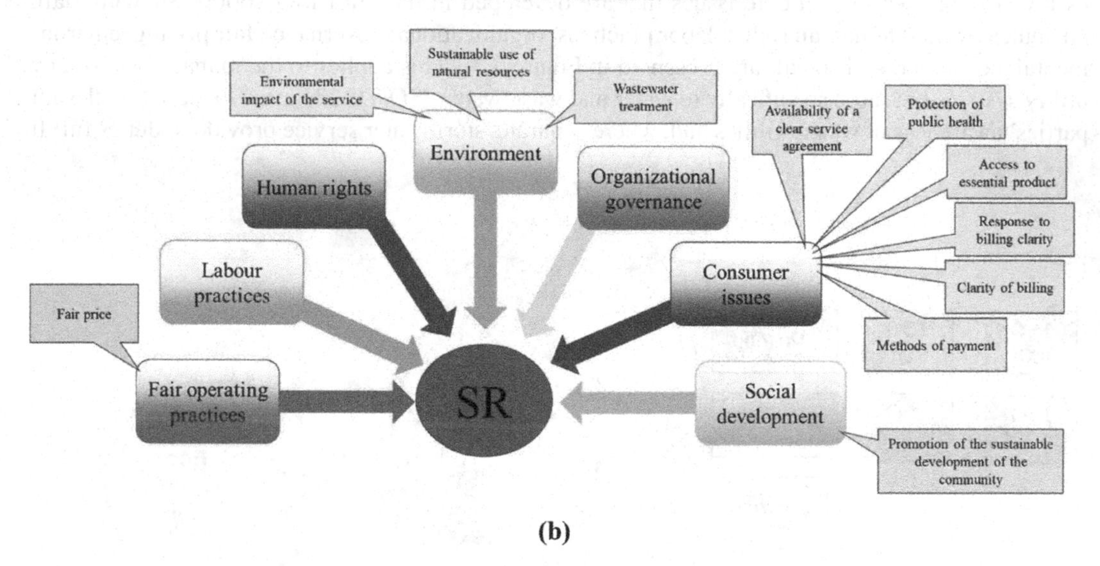

FIGURE 7.31 Classification of ISO/TC 224 SD aspects—seven cores issues developed in the draft ISO 26000 social responsibility with attributes of: (a) human rights, labor practices, organizational governance; (b) fair pricing. Environmental, consumer, and social issues.

the needs and expectations of their users and other stakeholders. It also provides water utilities with tools to help them provide their communities with adequate, sustainable, and resilient services.

TC 224 recognizes the need to contribute to the achievement of the United Nations' Sustainable Development Goals. A standard has already been published on the on-site management of domestic wastewater. Other New Work Items Proposals are being brought forward to complement innovative technical advances in wastewater service provision in developing nations and address service provision for displaced persons. Levels of service examples include the provision of drinking water via users' connections; public standpipes or mobile tankers; connection to a public wastewater system;

the conveyance, treatment, and safe disposal of the wastewater collected; and collection, storage, and safe disposal of stormwater runoff.

7.12 CONCLUDING REMARKS

Wastewater infrastructure are less developed compared to other water infrastructure around the world. Many regions in developing and less developed countries have serious deficiencies or do not have a collection and standard wastewater facilities. Even in developed countries, many large cities such as New York City have combined sewers. Often the collected sewage surpasses the capacity of the treatment facilities and CSO outfalls discharge untreated waste to natural water bodies after any heavy rain.

In this chapter, the principles of waste water infrastructure, its component, collection of sewage, combined sewer issues, planning, and standard issues were described. Case studies presented here show how wastewaters are subject to failure and how a unified resiliency base structure could be used for performance evaluation and resource allocation. Examples of some major cities infrastructure are also included to see how different management and operation styles are practiced in real-world situations. In the standards section, different ISOs related to water in general and waste water in particular are discussed with emphasis on SR. The economic and asset management issues are discussed in Chapter 8.

The main challenge of wastewater systems around the globe are in design, construction, operation, maintenance, funding, and standards of services. Their vulnerability to natural hazard and the public reliance on them have been threatened with more frequent floods. Therefore, as these infrastructures are mostly situated near natural water, the CSO impacts are getting more severe. As these facilities are aging in many cities, the issue of the needed funds to rehabilitate, replace collection pipes, and reconstruction are in for front of challenges many municipalities are facing.

PROBLEMS

1. (a) What is the main constituent of concern in WWT?
   (b) For each of the following unit operations in a WWT train, briefly describe how it removes some of the constituents you identified in part a:
   - Grit chamber
   - Primary sedimentation basin
   - Biological reactor
   - Secondary clarifier
   - Digestor
2. Briefly describe how primary WWT differs from secondary WWT.
3. A rectangular primary clarifier for a domestic wastewater plant is to be designed to settle 2,000 m^3/day with an overflow rate of 32 m^3/m^2day. The tank is to be 2.4 m deep and 4.0 m wide. How long should it be, and what detention time would it have?
4. A final settling tank for a 2-million-gallon-per-day (2 MGD) activated sludge treatment plant has an average overflow rate of 800 g/day ft^2. The tank needs to have a minimum detention time of 2.0 hours and allow proper settling, and it must be at least 11 ft deep. If the tank is circular, what should its diameter and depth be?
5. A perfectly mixed aeration pond with no recycling (return line) serves as the small community's biological reactor. The pond receives 30 m^3/day of influent with a BOD_5 of 350 mg/L that must be reduced to 20 mg/L before discharge. It has been found that the kinetic constants for the system are K_s = 100 mg/L BOD_5, K_d=0.1 day^{-1}, μm = 1.6 day^{-1}, and Y is 0.6 mg VSS/mg BOD_5,
   (a) What must the hydraulic detention time be in the aeration pond?
   (b) What mass of microbes will be produced in the pond each day?

6. Determine the theoretical hydraulic detention time and volume of a completely mixed reactor with recycle to be used in an activated sludge treatment plant operating at steady state if the following conditions and constants for the wastewater have been determined:
   X_{in} and $K_d \approx 0$
   $X = 3{,}000$ mg/L
   $K_0 = 0.075$ hr-1
   $K_m = 68$ mg/L
   Efficiency = 85%
   $BOD_{in} = 420$ mg/L
   $Q + Qr = 0.25 m^3/sec$
7. Calculate the required aeration tank volume for an activated sludge treatment plant on the bases of the empirical design factors. The anticipated waste volumetric flow rate is 0.068 m^3/sec and the BOD concentration is 350mg/L. Assume that the concentration of microorganisms in the reactor is kept at 5,000 mg/L.
8. In case study 1
   a. What is the proposed methodology for quantitative and qualitative assessment of reliability for a typical WWTP under coastal flooding?
   b. What is the main issue and how can it be related to the wastewater infrastructure system?
   c. What parts do the inflow to the plant system typically?
9. What is the volume in cubic feet of a rectangular tank that is 10 ft by 30 ft by 16 ft, and how many gallons can fit in it?
10. What is the volume of a tank in gallons if it is 12 ft deep and has a diameter of 30 ft?
11. How many hours will it take to fill each tank above if the flow entering them is 1.3 MGD?
12. Your superintendent wants to know how efficient your primary clarifier was at removing solids during a major rainstorm a few days earlier. The lab tech tells you that the average 24-hour composite TSS of the sewage entering the primary settling tank on the day in question was 228 ppm, and the average 24-hour composite TSS of the effluent that same day was 87 ppm. What do you tell the superintendent was the approximate percent removal?
13. What is the mixed SSs concentration given the following?

$$\text{Initial weight of filter disk} = 0.45\,\text{gms}$$

$$\text{Volume of filtered sample} = 60\,\text{mLs}$$

$$\text{Weight of filter disk and filtered residue} = 0.775\,\text{gms}$$

14. A WWT facility has three primary clarifiers available for use. They are all circular clarifiers with a radius of 40 ft and a depth of 8 ft. The design engineer wants you to maintain a primary clarification detention time of approximately 3.5 hours. How many tanks will you need to use if the plant flow rate is approximately 2 MGD?
15. What is the daily food to microorganism ratio given the following?

$$\text{Aeration tank } 28' \times 120' \times 15'$$

$$\text{Raw sewage flow} = 7{,}500{,}000$$

$$\text{Primary influent BOD} = 115\,\text{mg/L}$$

$$\text{MLVSS} = 4{,}700\ \text{mg/L}$$

16. Laboratory tests indicate that the volatile content of a raw sludge was 77%, and after digestion, the content is 41%. Calculate the percent reduction of volatile matter.
17. A facility feeds chlorine to the contact tank at a rate of about 280 gallons every day—the plant flow averages about 2.9 MGD. However, due to filamentous bacteria, you must also chlorinate RAS, using about 20 gallons/day for that purpose. Also, the facility superintendent has about 15 gallons/day of chlorine fed to the headworks for odor control in the morning. How much chlorine would be used for disinfection during a 6-hour shift?
18. How many pounds of solids are under aeration in an aeration tank that is 30′ by 70′ by 15′ and if the MLSS is 3,200 mg/L?
19. If an activated sludge aeration basin receives a flow of 0.69 MGD at a BOD concentration of 175 mg/L, how many pounds of BOD enter the aeration basin each day?
20. If the aeration basin from question 11 is 130 ft long, 35 ft wide, and 15 ft deep, what is its organic loading in pounds of BOD/day/1,000 ft^3?
21. An activated sludge aeration basin is 90 ft long, 25 ft wide, and 15 ft deep. The flow rate to it is 0.75 MGD at a CBOD concentration of 210 mg/L. If the MLVSS concentration is 2,500 mg/L, what is the *F/M* ratio?
22. An activated sludge aeration basin is 75 ft long, 25 ft wide, and 13 ft deep. The flow rate to it is 0.68 MGD at a CBOD concentration of 194 mg/L. If the MLVSS concentration is 2,114 mg/L, what is the *F/M* ratio?
23. Given the following data about an activated sludge plant, calculate the SRT at which it is operating:
    Influent flow = 9,020,000 GPD
    Aeration basin dimensions: 160 × 50 × 18 ft
    Final clarifier dimensions: 94 ft diameter; 12 ft deep MLVSS
    Concentration: 2,800 mg/L
    Waste sludge concentration: 5440 mg/L
    Waste sludge flow rate: 91,000 GPD
    Effluent TSS concentration: 10 mg/L
24. Calculate the SRT for the following activated sludge plant.
    Influent flow rate: 1097 GPM
    Aeration basin dimensions: 120 × 35 × 14 ft
    Final clarifier dimensions: 55 ft in diameter and 12 ft deep
    MLVSS concentration: 1,550 mg/L
    Waste sludge concentration: 7,500 mg/L
    Effluent TSS concentration: 24 mg/L
    Waste sludge flow rate: 15 GPM

REFERENCES

Balcazar, C. (2008). Water and sanitation for the urban marginal areas of Latin America (in Spanish). *Proceedings; Water and Sanitation Program*, World Bank, Washington, DC. http://documents.worldbank.org/curated/en/2008/07/ 16699647/agua-y-saneamiento-para-las-zonas-marginales-urbanas-deamerica-latina (Accessed June 2008).

Bloomberg, M. (2013). A stronger, more resilient New York. City of New York, Chapter 1: Sandy and its impacts and Chapter 6: Water and wastewater, PlaNYC report. Available at: http://www.nyc.gov/html/planyc/html/resiliency/resiliency.shtml.

Diaz, J. and B. Barkdoll (2006). Comparison of wastewater treatment in developed and developing countries. *Proceedings of World Environmental and Water Resources Congress*, Ohama, NE.

ECLAC/UNW-DPAC. (2012). Water and a green economy in Latin America and the Caribbean (LAC). *Economic Commission for Latin America and the Caribbean and UN-Water Decade Programme on Advocacy and Communication.* www.un. org/waterforlifedecade/green_economy_2011/interviews.shtml (Accessed June 2014).

Eddy, M., Burton, F.L., Stensel, H.D., and Tchobanoglous, G (2003). *Wastewater Engineering: Treatment and Reuse*. McGraw Hill, New York.

Ersahin, M.E., Ozgun, H., Dereli, R.K., and Ozturk, I. (2011). Anaerobic treatment of industrial effluents: An overview of applications. In F. Einschlag (ed.) *Waste Water – Treatment and Reutilization* (pp. 3–28). InTech, Rijeka. ISBN 978-978-953-307-249-4.

Foresti, E. (2001). Perspectives on anaerobic treatment in developing countries. *Water Science and Technology*, 44(8), 141–148.

Gikas, P. and George Tchobanoglous, G. (2009). The role of satellite and decentralized strategies in water resources management. *Journal of Environmental Management*, 90(1), 144–152. doi:10.1016/j.jenvman.2007.08.016.

Gray, N.F. (2005). *Water Technology: An Introduction for Environmental Scientists and Engineers* (2nd ed.). Elsevier Science & Technology Books, Amsterdam. ISBN 10-0750666331.

Hernández-Padilla, F., Margni, M., Noyola, A., Guereca-Hernandez, L., and Bulle, C. (2017). Assessing wastewater treatment in Latin America and the Caribbean: Enhancing life cycle assessment interpretation by regionalization and impact assessment sensibility. *Journal of Cleaner Production*, 142, 2140–2153.

Ho, G. (2004). Small water and wastewater systems: Pathways to sustainable development? *Water Science and Technology* 48(11–12), 7–14.

Hosseini, S., Barker, K. & Ramirez-Marquez, J.E. (2016). A review of definitions and measures of system resilience. *Reliability Engineering and System Safety*, 145, 47–61.

https://www.fluencecorp.com/the-caribbeans-wastewater-problem/.

https://www.gefcrew.org/index.php/wastewater-management-in-the-wider-caribbean-region-wcr.

Jayakumar, K.V. and Dandigi, M.N. (2003). A cost effective environmentally friendly treatment of municipal wastewater using constructed wetlands for developing countries. *World Water and Environmental Resources Congress*, 2003, 3521–3531.

Kabir, M., Suzuki, M., and Yoshimura, N. (2011). Excess sludge reduction in waste water treatment plants. In F. Einschlag (ed.). *Waste Water – Treatment and Reutilization* (pp. 133–150). InTech, Rijeka. ISBN 978-978-953-307-249-4.

Karamouz, M. and Farzaneh, H. (2020a). Margin of safety based flood reliability evaluation of wastewater treatment plants: Part 2 – Quantification of reliability attributes. *Water Resources Management*, 34, 2043–2059.

Karamouz, M. and Hojjat-Ansari, A. (2020b). Uncertainty based budget allocation of wastewater infrastructures' flood resiliency considering interdependencies. *Journal of Hydroinformatics*. doi:10.2166/hydro.2020.145.

Karamouz, M. and Olyaei, M.A. (2019). A quantitative and qualitative framework for reliability assessment of waste water treatment plants under coastal flooding. *International Journal of Environmental Research*, 13(1), 21–33.

Karamouz, M., Farzaneh, H., and Dolatshahi, M. (2020c). Margin of safety based flood reliability evaluation of wastewater treatment plants: Part 1 – Basic concepts and statistical settings, *Water Resources Management*, 34(2), 579–594.

Karamouz, M., Rasoulnia, E., Olyaei, M.A., and Zahmatkesh, Z. (2018). Prioritizing investments in improving flood resilience and reliability of wastewater treatment infrastructure. *Journal of Infrastructure Systems*, 24(4), 04018021.

Karamouz, M., Rasoulnia, E., Zahmatkesh, Z., Olyaei, M.A., and Baghvand, A. (2016). Uncertainty-based flood resiliency evaluation of wastewater treatment plants. *Journal of Hydroinformatics*, 18(6), 990–1006.

Kenward, A., Yawitz, D., and Raja, U. (2013). *Sewage Overflows from Hurricane Sandy*. Climate Central, Princeton, NJ.

Langergraber, G. and Muellegger, E. (2005). Ecological sanitation – A way to solve global sanitation problems? *Environment International*, 31(3), 433–444.

Lin, S.D. (2007). *Water and Wastewater Calculations Manual* (2nd ed.). McGraw-Hill Companies, Inc., New York. ISBN 0-07-154266-3.

Lohani, B.N. (2005). Advancing sanitation and wastewater management agenda in Asia and Pacific Region. *Sanitation and Wastewater Management – The Way Forward Workshop*, 19–20 September, Manila, Philippines.

Mara, D. (2001). Appropriate wastewater collection, treatment and reuse in developing countries. *Proceedings of the Institution of Civil Engineers: Municipal Engineer*, 145(4), 299–303.

Masters, G.M. and Ela, W.P. (2008). *Introduction to Environmental Engineering and Science* (No. 60457). Prentice Hall, Englewood Cliffs, NJ.

Metcalf and Eddy, Inc. (1991). *Wastewater Engineering: Treatment, Disposal, and Reuse* (3rd ed.). McGraw-Hill, Inc., Singapore.

Monod, J. (1949). The growth of bacterial cultures. *Annual Review of Microbiology*, 3, 371–394.

NEIWPCC. (2016). Preparing for extreme weather at wastewater utilities: Strategies and tips. https://www.neiwpcc.org/neiwpcc_docs/9-20-2016%20NEIWPCC%20Extreme%20Weather%20Guide%20for%20web.pdf.

Nordeidet, B., Nordeide, T, Åstebøl, S.O., and Hvitved-Jacobsen, T. (2004). Prioritising and planning of urban stormwater treatment in the Alna watercourse in Oslo. *The Science of the Total Environment*, 334, 231–238.

Noyola, A., Padilla-Rivera, A., Morgan-Sagastume, J.M., Güereca, P., and HernándezPadilla, F. (2012). Typology of wastewater treatment technologies in Latin America. *Clean–Soil, Air, Water*, 40, 926e932. doi:10.1002/clen.201100707.

NYC-DEP. (2013). *NYC Wastewater Resiliency Plan, Climate Risk Assessment and Adaptation Study*. Department of Environmental Protection, New York.

Qasim, S.R. (1998). *Wastewater Treatment Plants: Planning, Design, and Operation*. CRC Press, Boca Raton, FL.

Qian, Y. (2000). Appropriate technologies for municipal wastewater treatment in China. *Journal of Environmental Science and Health, Part A: Toxic/Hazardous Substances and Environmental Engineering*, 35(10), 1749–1760.

Russell, D.L. (2006). *Practical Wastewater Treatment*. John Wiley & Sons, Inc., Hoboken, NJ. ISBN-13: 978-0-471-78044-1.

Ujang, Z. and Henze, M. (2006). *Municipal Wastewater Management in Developing Countries – Principles and Engineering* (p. 352). IWA, London.

USDA. (2018). Geospatial data gateway: Order data for NY. https://datagateway.nrcs.usda.gov/gdgorder.aspx

Vugrin, E., Castillo, A., and Silva-Monroy, C. (2017). *Resilience Metrics for the Electric Power System: A Performance-Based Approach*. Report: SAND2017-1493. Sandia National Laboratories, Albuquerque, NM.

Wagner, I. and Zalewski, M. (2009). Ecohydrology as a basis for the sustainable city strategic planning: Focus on Lodz, Poland. *Reviews in Environmental Science and Bio/Technology*, 8(3) 209–217.

WB. (2014). *Water in Latin America and the Caribbean*. World Bank. http://web.worldbank.org/ (Accessed August 2014).

WHO and UNICEF. (2000). *Global Water Supply and Sanitation Assessment 2000 Report*. WHO, Geneva.

Wilderer, P.A. (2004). Applying sustainable water management concepts in rural and urban areas: Some thoughts about reasons, means and needs. *Water Science and Technology*, 49(7), 8–16.

Wilson, C., Mahoney, K., Ayotte, F., and Young, P. (2016). Bergen basin bending weirs reduce CSOs. *Proceedings of the Water Environment Federation*, 2016(12), 5878–5884.

Yu, H., Tay, J.-H., and Wilson, F. (1997). Sustainable municipal wastewater treatment process for tropical and subtropical regions in developing countries. *Water Science and Technology*, 35(9), 191–198.

Zawilski, M. (2004). Modernization problems of the sewerage infrastructure in a large Polish city. In J. Marsalek et al. (eds.) *Enhancing Urban Environment by Environmental Upgrading and Restoration* (Vol. 43, pp. 151–162). NATO Science Series, IV. Earth and Environmental Sciences. Springer, Dordrecht.

8 Urban Water Economics—Asset Management

8.1 INTRODUCTION

The engineering economy is concerned with the economic aspects of engineering; it involves the system's evaluation of costs and benefits of technical projects. Environmental economics is applying the principles of economics to the study of how environmental resources are developed and managed.

The entire process of planning, design, construction, operation, and maintenance of water systems entails many important and complex decisions. Besides technological and environmental considerations, economic principles play a significant role in making these decisions. The principles of engineering economics guide the decision makers in selecting the best planning and/or operational decisions.

Considering the expansion of water infrastructures and the cost associated with maintaining them as well as aging and natural hazard impacts such as flooding have changed the way these structures should be economical and financially managed. Asset management (AM) is now the medium for economic analysis and for financial decisions. It could also provide a transparent framework to accept the risk of investments and accounting structure.

AM of urban water infrastructures is the set of processes that water utilities need to have in place in order to ensure that infrastructure performance corresponds to service targets over time. To make sure, risks are adequately managed, and that the corresponding costs, in a lifetime cost perspective, are as low as possible. Lack of sound economic, regulatory frameworks and enforcement setup, as well as poor AM practices, particularly underpriced water services, are common problems throughout the developing regions (Hukka and Katko, 2015). As the primary decision makers in allocating funds to various infrastructures, urban systems' governing bodies need plans to prioritize limited resource allocation. Therefore, this allocation should be in line with the current structure or reformed infrastructure asset management (IAM).

AM is increasingly becoming a key topic in compliance with performance requirements in water supply and wastewater systems. According to Alegre et al. (2013), in AM, effective decision-making requires a comprehensive approach that ensures the desired performance at an acceptable risk level, considering the costs of building, operating, maintaining, and disposing of capital assets over their life cycles.

This chapter discusses the principles of engineering economics and their application to urban water resources planning and operation. Major objectives of economic analysis for water resources development projects are also discussed, and various methods for incorporating the money–time relationship are presented. AM principles are also discussed with emphasis on AM drivers, Sustainable Service Delivery, risk and consequences, and guidelines to select tools and practices that could help municipalities.

8.2 URBAN WATER SYSTEMS ECONOMICS—BASICS

Economics includes analytical tools that are used to determine the allocations of scarce resources by balancing the competing objectives. Different important and complex decisions should be made in the process of planning, design, construction, operation, and maintenance of urban water resources systems. Economic considerations besides the technological and environmental concerns play an important role in the decision-making about urban water resources planning and management.

DOI: 10.1201/9781003241744-8

The economic analysis of urban water infrastructure development projects has two sides. They create value, and they also encounter costs. The preferences of individuals for goods and services are considered in the value side of the analysis. The value of a good or service is related to the willingness to pay. The costs of economic activities are classified into fixed and variable costs. The range of operation or activity level does not affect the fixed costs such as general management and administrative salaries and taxes on facilities, but variable costs are determined based on the quantity of output or other measures of activity levels.

Other economic basics needed for AM are the balance sheet and financial statement simply by implementing local and state tax structure including depreciation, right offs, and credits. It is important to place AM in the core of all economic assessments and financial decisions and see how performance measures such as resiliency of the system include elements of AM that any resource allocation based on those measures could positively stay in line and strengthen the financial viability of the infrastructure/system.

The rest of this section is concentrated on a simple example (accounting of costs and benefits for important economic decisions among alternatives), cost, benefit, value, and other attributes of cash flow and the governing interest base relations.

Example 8.1

There are two choices, A and B, for the trenchless replacement of mains in an urban water distribution system (WDS). A \$20,000 per week should be paid to the traffic control authorities for traffic jams during the project. The other costs of the two trenchless replacement methods are summarized in Table 8.1. Compare the two methods in terms of their fixed, variable, and total costs when 20 km of mains should be replaced. For the selected method, how many meters of the mains should be replaced before starting to make a profit, if there is a payment of \$30/m of mains replacement?

Solution:

As is depicted in Table 8.1, the plant setup is a fixed cost, whereas the payments for traffic jams and rent are variable costs.

As can be seen from Table 8.2, although choice B has higher fixed costs, it has less total cost than choice A. The project will make profit at the point where total revenue equals total cost of the kilometers of mains replaced. For choice B, the variable cost per meter of replacement is calculated as follows:

$$\frac{\$6,000+\$120,000}{20,000}=\$6.3$$

Total cost = total revenue when the project starts to make a profit. Let X be the meters of the mains that should be replaced before starting to make a profit, we will therefore have:

$$\$250,000+\$6.3X=\$30X \quad \text{and} \quad X \text{ is determined as } 10,549\,\text{m}.$$

TABLE 8.1
Costs of Choices A and B for Trenchless Replacement of Example 8.1

Cost	Method A	Method B
Time taken to complete the project	10 weeks	6 weeks
Weekly rental of the site	\$5,000	\$1,000
Cost to set up and remove equipment	\$150,000	\$250,000
Payment for traffic jam	\$20,000/week	\$20,000/week

TABLE 8.2
Analysis of Costs of the Two Choices for Mains Replacement in Example 8.1

Cost	Fixed	Variable	Choice A	Choice B
Rent		✓	$50,000	$6,000
Plant setup	✓		$150,000	$250,000
Traffic jam payment		✓	10 ($20,000) = 200,000	6 ($20,000) = 120,000
Summation			$400,000	$376,000

Therefore, if choice B is selected, the project will begin to make a profit after replacing about 10.5 km of the mains.

Costs are classified as private costs and social costs. The private costs are directly experienced by the decision makers, but all costs of an action are considered in the social costs without considering who is affected by them (Field, 1997). The difference between social and private costs is called external costs or usually environmental costs. Since the firms or agencies do not normally consider these costs in decision-making, they are called "external." However, it should be noted that these costs may be real costs for some members of society. Open access resources such as reservoirs in urban areas are the main source of external or environmental costs.

For example, consider three cities around a lake that use the water of the lake and discharge the sewage back into the lake. Each city should pay about $30,000/year for water treatment. Suppose that a new tourism complex wants to start operating near the lake. Because of the untreated sewages discharged to the lake by this tourism complex, each of these cities should spend $10,000 more per year for water treatment. In this case, the external cost is $30,000 ($10,000×3).

Similarly, external benefits are defined. An external benefit is a benefit that is experienced by some body outside the decision about consuming resources or using services. For example, consider that a private power company constructed a dam for producing hydropower. The benefits experienced by people downstream, such as mitigating both floods and low flows, are classified as external benefits.

Many values of an action cannot be measured in commensurable monetary units as is desired by economists. For example, the effects on human beings physically (through loss of health or life), emotionally (through loss of national prestige or personal integrity), and psychologically (through environmental changes) cannot be measured by monetary units. These sorts of values are called intangible or irreducible (James and Lee, 1971).

In most engineering activities, the capital investment is committed for a long period; therefore, the effects of time should be considered in economic analysis. This is because of the different values of the same amount of money spent or received at different times. The future value of the present amount of money will be larger than the existing amount due to opportunities that are available to invest the money in various enterprises to produce a return over a period. The interest rate is the rate at which the value of money increases from the present to the future. On the contrary, the discount rate is the amount by which the value of money is discounted from the future to the present.

A summary of the relations between general cash flow elements using discrete compounding interest factors is presented in Table 8.3 using the following notations:

i: Interest rate
N: Number of periods (years)
P: Present value of money
F: Future value of money
A: End of the cash flow period in a uniform series that continues for a specific number of periods
G: End-of-period uniform gradient cash flows

More details of time-dependent interest rates and interest formulae for continuous compounding are given in DeGarmo et al. (1997) and Au and Au (1983).

TABLE 8.3
Relations between General Cash Flow Elements Using Discrete Compounding Interest Factors

To Find	Given	Factor by Which to Multiply "Given"	Factor Name	Factor Function Symbol
F	P	$(1+i)^N$	Single payment compound amount	$\left(\frac{F}{P}, i\%, N\right)$
P	F	$\frac{1}{(1+i)^N}$	Single payment present worth	$\left(\frac{P}{F}, i\%, N\right)$
F	A	$\frac{(1+i)^N-1}{i}$	Uniform series compound amount	$\left(\frac{F}{A}, i\%, N\right)$
P	A	$\frac{(1+i)^N-1}{i(1+i)^N}$	Uniform series present worth	$\left(\frac{P}{A}, i\%, N\right)$
A	F	$\frac{i}{(1+i)^N-1}$	Sinking fund	$\left(\frac{A}{F}, i\%, N\right)$
A	P	$\frac{i(1+i)^N}{(1+i)^N-1}$	Capital recovery	$\left(\frac{A}{P}, i\%, N\right)$
P	G	$\frac{(1+i)^N-1-Ni}{i^2(1+i)^N}$	Discount gradient	$\left(\frac{P}{G}, i\%, N\right)$
A	G	$\frac{(1+i)^N-1-Ni}{i(1+i)^N-i}$	Uniform series gradient	$\left(\frac{A}{G}, i\%, N\right)$

Source: DeGarmo, E.P. et al., *Engineering Economy.* Prentice Hall, Upper Saddle River, NJ, 1997.

Example 8.2

A wastewater collection network project in a city produces benefits, as expressed in Figure 8.1: the \$100,000 profit in year 1 is increased in 10 years on a uniform gradient to \$1,000,000. Then it reaches \$1,375,000 in year 25 with a uniform gradient of \$25,000/year and it remains constant at \$1,375,000 each year until year 40, the end of the project life span. Assume an interest rate of 4%. What is the present worth (PW) of this project?

Solution:

First, the present value of the project benefits in years 1–10 is calculated as follows:

$$10{,}000\left(\frac{P}{G}, 4\%, 10\right) + 100{,}000\left(\frac{P}{A}, 4\%, 10\right) = 100{,}000 \times 41.9919 = \$4{,}199{,}190.$$

The same procedure is repeated in years 11–25:

$$1{,}025{,}000\left(\frac{P}{A}, 4\%, 15\right)\left(\frac{P}{F}, 4\%, 10\right) + 25{,}000\left(\frac{P}{G}, 4\%, 15\right)\left(\frac{P}{F}, 4\%, 10\right) = \$8{,}876{,}672$$

The present value of benefits in years 26–40:

$$1{,}375{,}000\left(\frac{P}{A}, 4\%, 15\right)\left(\frac{P}{F}, 4\%, 25\right) = 1{,}375{,}000 \times 11.11838 \times 0.37512 = \$5{,}734{,}749.$$

The total PW is equal to \$18,810,611, the summation of the above three values.

values.

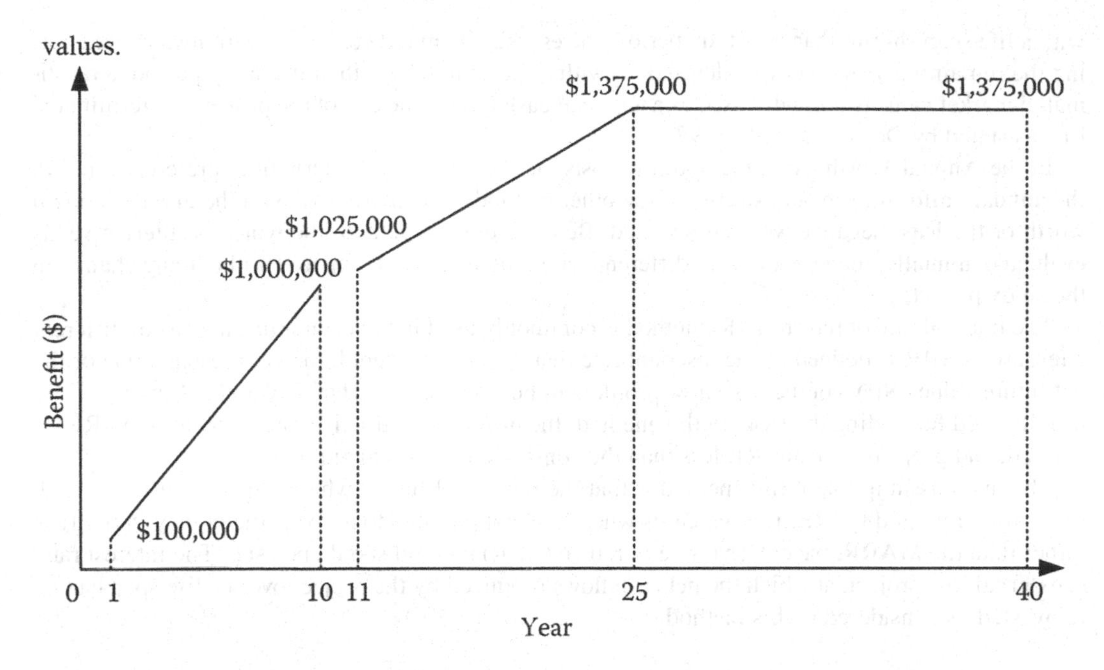

FIGURE 8.1 Cash flow diagram for Example 8.2.

8.2.1 Economic Analysis of Multiple Alternatives

Through economic analysis, it is determined whether a capital investment and the cost associated with the project within the project lifetime can be recovered by revenues or not. It should also be determined how attractive the savings are regarding the involved risks and the potential alternative uses. The five most common methods, including the Present Worth, Future Worth, Annual Worth, Internal Rate of Return, and External Rate of Return methods, are briefly explained in this section. The life span of different projects should be incorporated into economic analysis through the repeatability assumption.

In the PW method, the alternative with the maximum PW of the discounted sum of benefits minus costs over the project life span, namely Equation (8.1), is selected.

$$PW = \sum_{t=1}^{N} (B_t - C_t)\left(\frac{P}{F}, i\%, t\right), \tag{8.1}$$

where

C_t is the cost of the alternative at year t,
B_t is the profit of the alternative at year t,
N is the study period, and
i is the interest rate.

In the case of cost alternatives, the one with the least negative equivalent worth is selected. Cost alternatives are all negative cash flows except for a possible positive cash flow element from the disposal of assets at the end of the project's useful life (DeGarmo et al., 1997).

In the Future Worth method, all benefits and costs of alternatives are converted into their future worth, and then the alternative with the greatest future worth or the least negative future worth is selected. Two choices are available for selecting the study period, including the repeatability assumption and the coterminated assumption. Based on the latter assumption, for the alternatives with a life span shorter than the study period, the estimated annual cost of the activities is used during the remaining years. For the alternatives with a life span longer than the study period, a reestimated market value is normally used as a terminal cash flow at the end of the project's coterminated life, as stated by DeGarmo et al. (1997).

In the Annual Worth (AW) method, the costs and benefits of all alternatives are considered in the annual uniform form, and similar to the other methods, the alternative with the greatest annual worth or the least negative worth is selected. Because in this method, the worth of alternatives is evaluated annually, the projects with different economic lives are compared without any change in the study period.

The internal rate of return (IRR) method is commonly used in the economic analysis of different alternatives. IRR is defined as the discount rate that, when considered, the net present value or the net future value (NFV) of the cash flow profile will be equal to zero. The PW and AW methods are usually used for finding the IRR. In this method, the minimum attractive rate of return (MARR) is defined, and projects with an IRR less than the considered threshold are rejected.

The main assumption of this method is that the recovered funds, which are not consumed each time, are reinvested in IRR. In such cases where it is not possible to reinvest the money in the IRR rather than the MARR, the external rate of return (ERR) method should be used. The interest rate, *e*, external to a project, at which the net cash flows produced by the project over its life span can be reinvested, is considered in this method.

Example 8.3

Two different types of pumps (A and B) can be used in a pumping station of an urban WDS. The costs and benefits of each of the pumps are displayed in Table 8.4. Compare the two pumps economically using the MARR = 10% and considering that the time period of the study is equal to 12 years.

Solution:

Since the life spans of the pumps are less than the study period, the repeatability assumption is used. At first the PW of the benefits and costs of the pumps are calculated.

PW of costs of *A* = \$20,000+(\$20,000−\$500) (*P/F*, 10%, 5)−\$500 (P/F, %%, 10) = \$31,916 PW of benefits of *A* = \$10,000 (*P/A*, 10%, 10) = \$61,440

PW of costs of *B* = \$70,000−\$1,000 (*P/F*, 10%, 10) = \$69,614

PW of benefits of *B* = \$15,000 (*P/A*, 10%, 10) = \$92,170

For selecting the economically best alternative, the graphical method is employed in this example. The location of projects on a graph showing the present value of benefits versus the present value of costs is determined as points A and B and then the slope of the line connecting the points is estimated and compared with a 45° slope line where the points on it have equal present values of costs and benefits (Figure 8.2). When the slope of the line connecting two projects is less (more) than 45°, the project with less (more) initial cost would be selected. As shown in Figure 8.2, pump A, which has less initial cost, should be selected.

TABLE 8.4
Data Used in Example 8.3

Economic Characteristics	Pump A	Pump B
Initial investment	\$20,000	\$70,000
Annual benefits	\$10,000	\$15,000
Salvage value	\$500	\$1,000
Useful life	5	10

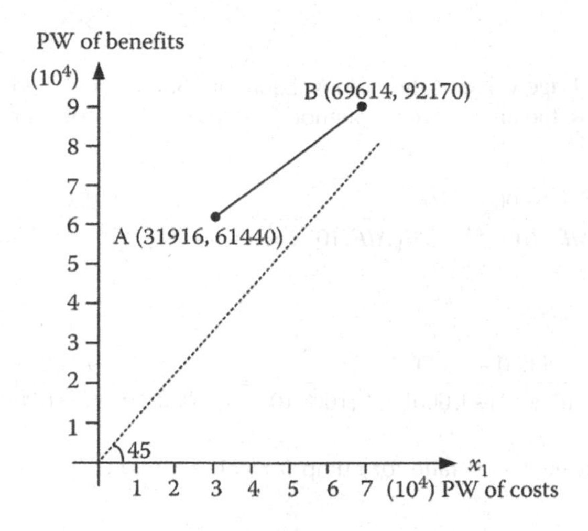

FIGURE 8.2 Comparing the two pumps in Example 8.3 by the rate of return method.

8.2.2 Economic Evaluation of Projects Using Benefit-Cost Ratio Method

The benefit–cost ratio method, which considers the ratio of benefits to costs, has been widely used in the economic analysis of water resources projects. The ratio is estimated based on the equivalent worth of discounted benefits and costs in the form of annual worth, PW, and/or future worth to consider the time value of money. Equations (8.2) and (8.3) are basic formulations of the benefit–cost ratio method.

$$\frac{B}{C}=\frac{B}{I+C}, \tag{8.2}$$

$$\frac{B}{C}=\frac{B-C}{I}, \tag{8.3}$$

where

B is the net equivalent benefits,
C is the net equivalent annual cost, including operation and maintenance costs, and
I is the initial investment.

If the benefit–cost ratio of a project is less than or equal to 1, it is not considered. Through the following equation, the associated salvage value of the investment is considered in the benefit–cost method:

$$\frac{B}{C}=\frac{B-C}{I-S}, \tag{8.4}$$

where

S is the salvage value of investment.

Example 8.4

Assume that the annual costs of operation and maintenance of pumps A and B are equal to $4,000 and $6,000, respectively. Compare the two pumps from an economic aspect, using the benefit–cost ratio method.

Solution:

For considering the salvage value of the pumps, Equation (8.4) is employed. Because of different life spans of the pumps, the annual worth method is employed for more simplicity as follows:

Pump A:

$$\frac{B}{C} = \frac{10{,}000 - 4{,}000}{20{,}000(A/P,\ 10\%, 5) - 500(A/F, 10\%, 5)} = \frac{6{,}000}{20{,}000 \times 0.263 - 500 \times 0.163} = 1.15.$$

Pump B:

$$\frac{B}{C} = \frac{15{,}000 - 6{,}000}{70{,}000(A/P,\ 10\%, 10) - 1{,}000(A/F, 10\%, 10)} = \frac{9{,}000}{70{,}000 \times 0.162 - 1{,}000 \times 0.0627} = 0.79.$$

As can be seen, the benefit–cost ratio for pump B is <1. Therefore, it is rejected and pump A is selected.

8.2.3 Economic Models

Economic models relate to the demands of the consumer on the available resources for demand supply to satisfy the consumers. The demand and production functions for goods and services, and supply functions for resources, are included in this method. Production functions model the production of useful products by utilizing resources and certain technologies. This trade-off between economic activities for goods and services production and the environmental quality is called the production possibility curve (Figure 8.3). Depending on the technological capabilities and limitations in a specific time, each society chooses its locations on this curve. This is called social choice, which depends on the value that people in that society place on the environment (E_i) and the economic productions (c_i) (Field, 1997).

In resource-supply function, the quantity of resources made available per unit time via the price per unit of resources is considered. In water resources development projects, the development of well fields, interbasin water transfer projects, and other related projects should be considered in an accurate estimation of water price. In order to make a correct decision about the development alternatives and water allocation schemes, the distribution of water resources development costs should be considered. In water scarcity situations, more water might be allocated to the users, which could pay high prices for water, such as industries. However, the political and legal issues that are important in developing water allocation schemes affect this process (Viessman and Welty, 1985).

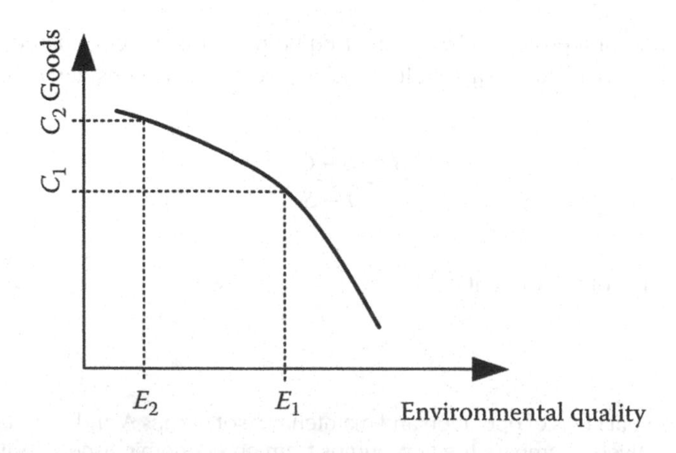

FIGURE 8.3 Production possibility curve. (Adapted from Karamouz, M., Szidarovszky, F., and Zahraie, B., *Water Resources Systems Analysis*. Lewis Publishers, CRC Press, Boca Raton, FL, 2003.)

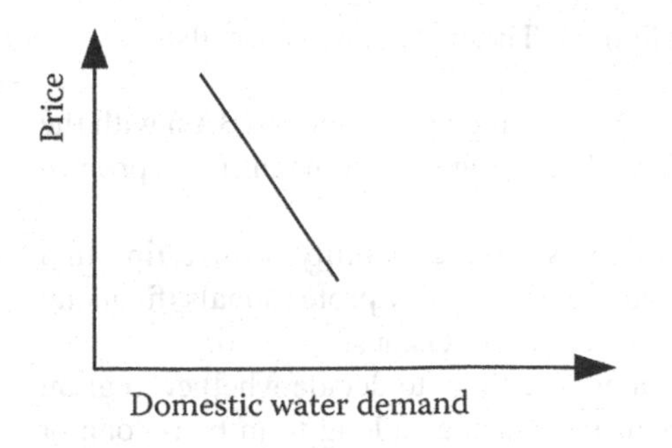

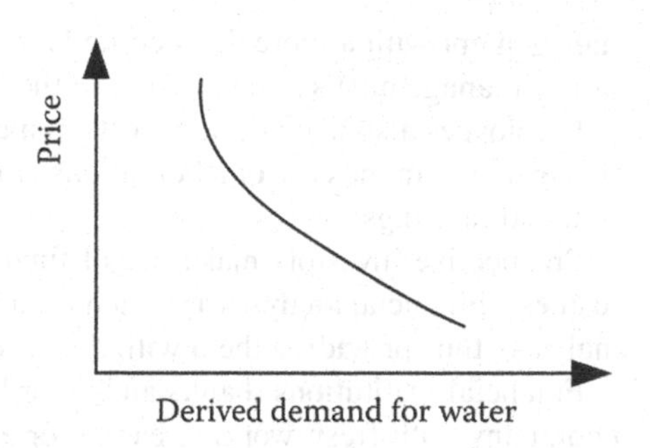

FIGURE 8.4 Examples of demand function. (Adapted from Karamouz, M., Szidarovszky, F., and Zahraie, B., *Water Resources Systems Analysis*. Lewis Publishers, CRC Press, Boca Raton, FL, 2003.)

Demand functions show the demand of goods and services at any particular price. Figure 8.4 shows two typical demand functions. As illustrated in Figure 8.4, the price–quantity relation for domestic water use is very steep. This means that price increase does not affect the water demand significantly, because people would like to supply their needs for cooking, washing, and maintaining health standards. But the demand for products derived from water is significantly sensitive to price changes.

The demand functions that are considered in the engineering and planning process consider the variety of factors that affect the demand and are more complex. For example, the domestic demand functions are formulated as follows (Mays, 1996):

$$Q_W = Q_W(P_W, P_a, P, Y, Z), \tag{8.5}$$

where

Q_W is the consumers' water demands in a specific period of time,
P_W is the price of water,
P_a is the price of an alternative water resource,
P is an average price index representing all goods and services for water supply,
Y is the consumer income, and
Z is a vector representing other factors such as climate and consumer preferences (Mays, 1996).

In the following sections, AM, performance measures, and developing AM plans for water, stormwater, and wastewater utilities are discussed.

8.2.4 Financial Statement

Financial statements are written records that convey the business activities and the financial performance of a company. Financial statements are often audited by government agencies, accountants, firms, etc., to ensure accuracy and for tax, financing, or investing purposes. Financial statements include balance sheets, income statements, and cash flow statements.

The objective of financial statements is often stated as accumulating information about the financial position, performance, and changes in the financial position of a system or an organization that is useful to a wide range of users or in making economic decisions. Financial statements should be understandable, relevant, reliable, and comparable. Reported assets, liabilities, equity, income, and expenses are directly related to an organization's financial position.

Financial statements are intended to be understandable by readers who have a reasonable knowledge of business and economic activities, accounting for those who are willing to study the information diligently. Financial statements may be used by users for different purposes:

Owners and managers require financial statements to make important business decisions that affect its continued operations. Financial analysis is then performed on these statements to provide management with a more detailed understanding of the figures. These statements are also used as part of management's annual report to the stockholders.

Employees also need these reports in making collective bargaining agreements (CBAs) with the management, in the case of labor unions or for individuals in discussing their compensation, promotion, and rankings.

Prospective investors make use of financial statements to assess the viability of investing in a business. Financial analyses are often used by investors and are prepared by professionals (financial analysts), thus providing them with the basis for making investment decisions.

Financial institutions (banks and other lending companies) use them to decide whether to grant a company with fresh working capital or extend debt securities (such as a long-term bank loan or debentures) to finance expansion and other significant expenditures.

8.2.4.1 Balance Sheet

A balance sheet is like a snapshot; it captures the financial position of a company at a particular point in time. The other two statements (income statement and cash flow statement) are for a period of time. As you study the assets, liabilities, and stockholders' equity contained in a balance sheet, you will understand why this financial statement provides information about the solvency (the possession of assets in excess of liabilities) of the business on a specific date.

The balance sheet, sometimes called the statement of financial position, lists the company's assets, liabilities, and stockholders' equity (including dollar amounts) as of a specific moment in time. That specific moment is the close of business on the date of the balance sheet.

Assets of the type which we are considering here are physical items such as plant, machinery, buildings, vehicles, pipes and wires, and associated information and technical control and software systems that are used to serve a business or organizational function. At the outset, and given that AM links closely with financial management, it is important to recognize the accounting definition of assets, and in particular, the split between fixed and current assets.

A *fixed asset* (also called a noncurrent asset) is a physical item that has value over a period exceeding 1 year, for example, land, buildings, plant, and machinery. When fixed assets are acquired, their cost cannot be counted as an expense for tax purposes in the year of acquisition. When we buy or sell fixed assets, we are regarded as having swapped one asset—money—for another asset, a machine, for example. Only the depreciation of the machine in any given year is considered to be an expense in the year. This is important as far as tax treatment is concerned.

Faster moving assets such as cash; accounts receivable; inventory (materials, work in process, finished goods, consumables) are referred to as current assets. Slow moving spares which are normally held for longer than a year should be regarded as fixed assets.

The balance sheet relationship is expressed as; Assets = Liabilities + Equity.

Assets represent things of value that a company owns and has in its possession or something that will be received and can be measured objectively. They are also called the resources of the business; some examples of assets include receivables, equipment, property, and inventory. Assets have value because a business can use or exchange them for producing the services or products of the business.

Liabilities are the debts owed by a business to others—creditors, suppliers, tax authorities, employees, etc. They are obligations that must be paid under certain conditions and time frames. A business incurs many of its liabilities by purchasing items on credit to fund the business operations.

Equity represents retained earnings and funds contributed by a company's owners or shareholders (capital), who accept the uncertainty that comes with ownership risk in exchange for what they hope will be a good return on their investment.

Investors, creditors, and regulatory agencies generally focus their analysis of financial statements on the company as a whole. Since they cannot request special-purpose reports, external users must rely on the general purpose financial statements that companies publish. These statements include the balance sheet, an income statement, a statement of stockholders 'equity, a statement of cash flows, and the explanatory notes that accompany the financial statements.

TABLE 8.5
A Typical Balance Sheet for The Company ABC ending on December 31, 2019 Comparing to The Year Before (USD $ millions).

Assets	2018	2019
Current assets:		
Cash	201,069	224,550
Temporary Investments	10,000	15,000
Accounts Receivable	7,117	7,539
Prepaid expenses	5,998	5,682
Inventory	10,531	11,342
Total current assets	234,715	264,112
Intangible Assets		
Property & Equipment	38,602	37,521
Goodwill	3,870	3,850
Total Assets	**277,187**	**305,483**
Liabilities		
Current liabilities:		
Accounts Payable	5,265	5,671
Accrued expenses	1,865	1,899
Unearned revenue	1,952	1,724
Total current liabilities	9,082	9,294
Long-term debt	36,051	35,909
Total Liabilities	**45,133**	**45,203**
Shareholder's Equity		
Equity Capital	170,000	170,000
Retained Earnings	62,053	90,280
Shareholder's Equity	**232,053**	**260,280**
Total Liabilities & Shareholder's Equity	**277,187**	**305,483**

Users of financial statements need to pay particular attention to the explanatory notes, or the financial review, provided by management in annual reports. This integral part of the annual report provides insight into the scope of the business; the results of operations, liquidity, and capital resources; new accounting standards; and geographic area data. An example of a typical balance sheet is presented in Table 8.5. As it is shown, this company, ABC, has a balance between total assets and the liabilities and stockholders' equity.

8.2.4.2 Financial Analysis

AM decisions involve the application of a combination of technical and financial knowledge. Asset managers play a key role in ensuring that the physical facts and the financial and cost data which are used in making AM decisions are sufficiently correct to enable sound decisions to be made. For this reason, they need to be familiar with the language and methods of accounting and financial analysis.

Fixed assets such as buildings, infrastructure, and plants are items with long lives extending over many years or even decades. Financial analysis related to their acquisition, through life support and disposal, needs to take account of the time value of money. In fact, the financial analysis is the simple part of any given set of facts. The hard part is deciding what factors should be taken into account and estimating the costs, revenues, and risks involved (Hasting, 2010).

8.3 ASSET MANAGEMENT

Sound AM needs to be planned and prioritized and helps to ensure that investments are made to minimize future repair and rehabilitation costs at the right time and to maintain municipal and public assets for their best performance. A Guide for Municipal Asset Management Plans with the sense of working and "Building Together" is presented.

AM levels of service can be complicated and could be defined differently based on the views of stakeholders. Their input is critical for outcomes to be successful. See AECOM's Level of Service Workshop for the City of Cambridge, Ontario, for more details (AECOM, 2013). Even though AM is economically based and its principal applies to all assets, water-related facilities characteristics and the type of vital service they provide are somehow unique and could make AM more challenging. The recent attention to building resilience into water infrastructure and how they are supported financially at the time of disasters also has room for further investigations.

The governmental agencies such as NYC-DEP, which are the primary decision makers in allocating resources to various water infrastructures for improving resiliency, need to see how these infrastructures manage their assets to prioritize limited resource allocation. Therefore, any funding should be in the context of both physical and financial/AM performances (Karamouz and Movahhed, 2021).

The AM paradigm is a continuous process that guides the acquisition, use, and disposal of infrastructure assets. AM is a structured program to deliver the customers' service levels while minimizing asset ownership's whole-life costs and minimizing the whole-life costs of asset ownership. The following program, levels, and ownership issues should be considered:

1. *A structured program*:
    - AM is highly structured
    - Asset decisions should be made in repeatable and supportable ways, based on useful data!
2. *Specified service levels*:
    - AM is not for delivering the optimum serviceability from a utility; it is to deliver a specified level of service (LOS)
    - These levels are selected based on service level/cost trade-offs
    - Ideally, service levels are set on service level/cost trade-offs over specified levels of service.
3. *Minimizing the costs of asset ownership*:
    - All asset decisions (acquire, maintain, refurbish, replace) are made on a life cycle basis
    - Each decision minimizes the present value of all future ownership costs
    - Costs must include economic, environmental, and social costs over a life cycle
    - Decisions must consider risk costs

8.3.1 Attributes of Asset Management

AM as both a concept and a practice is about more than merely maintaining and caring for physical assets, and it is essential for water infrastructure managers to understand this. An efficient AM program should have specific attributes in order to deliver satisfying results. These attributes are listed as follows:

- *Holistic*: Looking at the whole picture, that is, the combined implications of managing all aspects.
- *Systematic*: A methodical approach promoting consistent, repeatable, and auditable decisions and actions.
- *Systemic*: Considering the assets in their asset system context and optimizing the asset system value.

- *Risk-based*: Focusing on resources and expenditure and setting priorities appropriate to the identified risks and the associated cost/benefits.
- *Optimal*: Establishing the best value compromise between competing factors, such as performance, cost, and risk, associated with the assets over their life cycles.
- *Sustainable*: Considering the long-term consequences of short-term activities to ensure that adequate provision is made for future requirements and obligations (such as economic or environmental sustainability, system performance, societal responsibility, and other long-term objectives).
- *Integrated*: Recognizing that interdependencies and combined effects are vital to success. This requires a combination of the above attributes, coordinated to deliver a joined-up approach and net value.

AM elements are best described in the inner loop of value, risk taking, growth, sound decisions, stakeholders' confidence, and performance as shown in Figure 8.5. These elements have to be improved and effectively managed.

8.3.2 Asset Management Drivers

Today with the increasing cost and difficulties in the construction of new projects of water and wastewater infrastructure in major cities, there is less tendency to expand those systems and rather to implement demand management and sources reduction waste water projects and strategies. Therefore, the current assets are more important to manage and sometimes look impossible to replace. A system view of maintaining and protecting assets as it affects the resiliency of the system will have a profound impact on the risk of failure and withstanding the consequences of extreme events, including disasters. The following key drivers and factors should be considered:

- *External forces:*
  - Regularity compliance
  - Growth and demand
  - Public and elected officials

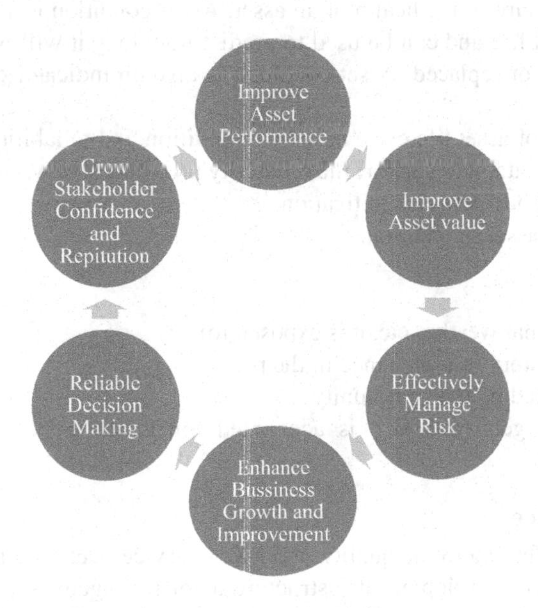

FIGURE 8.5 Benefits of asset management. (From "ISO/CD 55001.2 Asset management – Management systems – Requirements." ISO/TC 251/WG2, ISO. (2012b).)

- *Asset age and condition:*
    - Aging and deteriorating infrastructure
    - Justification for capital and O&M needs
- *Service levels:*
    - Demand for improved reliability
    - Prevention of catastrophic asset
    - Response to resiliency and climate change
- *Cost efficiency:*
    - Do "More with Less" through optimized decisions
    - Move towards a "Businesslike" culture

8.3.4 The Objectives in Asset Management

The objectives for managing infrastructure assets could be diverse depending upon the overall condition of the asset and the risks associated with normal or extreme conditions. Cost plays a dominating factor, but the return if the service and coverage are extended should be optimized. We could face a multiobjective situation. In either case, the physical and financial constraints determine our chance of minimizing the cost as an overall key component.

Often the objective of an AM program is to minimize the total costs (consisted of construction and installation costs, operating costs, maintenance costs, rehabilitation costs, replacement costs, and decommission and salvage costs) of owning and operating assets, while continuously delivering the service levels customers desire at an acceptable level of risk.

8.3.3 Asset Management Steps

In the following sections, the different stages that AM makes a turning point are discussed as suggested in the literature, including current and foreseeable conditions, performance measures; risk; and service delivery. It could include others steps if many external factors are involved.

8.3.3.1 Status and Condition

Asset condition is a measure of the health of an asset. Asset condition is a key parameter in determining remaining useful life and can be used to predict how long it will be before an asset needs to be repaired, renewed, or replaced. Asset condition is also an indicator of how well it is able to perform its function.

Evaluating the status of asset life cycle based on condition and reliability data, with a scientific approach and integrated data, will support the reliability program review, contributing to the reduction of the AM costs and business risk mitigation.

Factors that effect an asset's condition:

- Its age
- Its environment (what weather etc. it is exposed to)
- Its maintenance history (maintenance in the past)
- How well it is treated by the community
- How much use it gets (this also is accounted for in the Service Rating often as a priority 1, 2, 3).

8.3.3.2 Level of Service

LOS is a term in AM referring to the quality of a given service. Defining and measuring levels of service is a key activity in developing infrastructure asset management plans (AMPs). Levels of service may be tied to the physical performance of assets or be defined via customer expectation and satisfaction.

Levels of service also can be seen as technical or strategic. Technical LOS reflects the service provider's perspective, while strategic LOS represents the customer or user's perspective. For instance, in the case of sewer infrastructure, a municipality (as the service provider) may measure the number of microcracks in a pipe or sewer and model its expected lifetime to ensure the quality of the service, but the user's main concern is the availability and reliability of the sewer system, not necessarily the technical aspects of the physical infrastructure.

- *Redundancy*: It is important to determine how many replacement parts or other assets can help to function after a failure. The chance of an asset failing can be derived based on the remaining useful life and redundant elements. Redundancy is the duplication of critical components or functions of a system with the intention of increasing the reliability of the system, usually in the form of a backup or fail-safe to improve actual system performance. Redundancy in AM is the assets that are available for replacement in different parts of the system to ensure its continued operation (Karamouz and Movahhed, 2021). A system could have different levels of redundancy, as expressed in Table 8.6.

8.3.3.3 Risk Management

Risk assessment is a combination of two aspects: first, the probability of a flood event occurring and, second, the consequences that flood will have (Gouldby and Samuels, 2005). An integrated measure of risk is the product of these two components, which can be derived from the risk matrix (Figure 8.6), based on the system's characteristics. As can be seen in Table 8.7, a system could

TABLE 8.6
Levels of Redundancy

General Redundancy	Description
0%	Asset cannot be out of service without loss of function of the system, no backup asset available
50%	Alternative asset(s) provide half of the asset's functioning capability
100%	Alternative asset(s)—all of the it
200%	Alternative asset(s)—double the capability

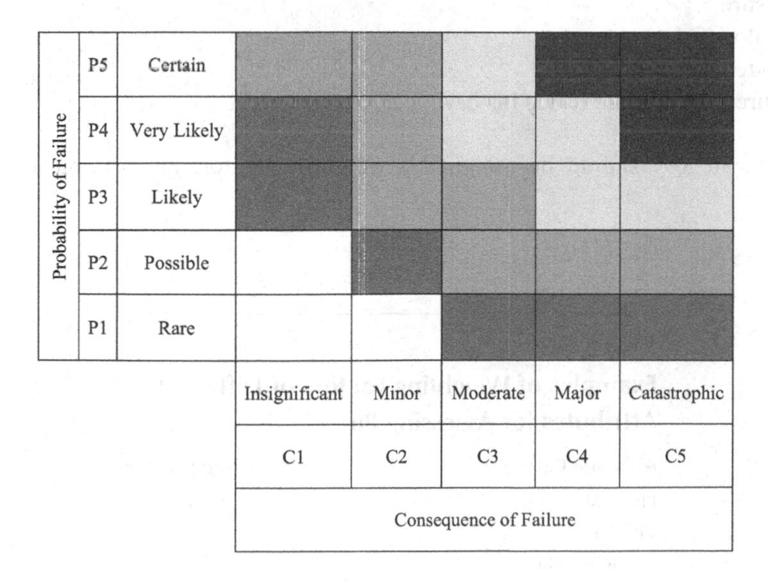

FIGURE 8.6 Determining risk by the probability and consequence of failures.

TABLE 8.7
Different Categories of Risk and Measures Taken for Each One.

Catastrophic	Immediate action to prevent impact to LoS, safety, and environment
Extreme	Gearing up for immediate action
High	Monitoring regime, response plan in place
Moderate	Management responsibility specified
Low	Manage using routine procedures

belong to various categories based on its risk as measures taken for each of these categories are different.

$$\text{Risk} = \text{PoF} \times \text{CoF} \tag{8.6}$$

8.3.3.3.1 Probability of Failure (PoF)

Probability of Failure (PoF) for each system and subsystem depends on different characteristics and attributes of said systems. As an example, some of these attributes for a typical WDS are listed below:

- *Physical attributes*:
  - Age
  - Size
  - Material/linings/coating
- Condition attributes such as pipe conditions and joint condition
- *Environmental attributes*
  - Soil condition
  - Groundwater table location
- *Operational/performance data*
  - Pressure
  - Fire flow/C-factor
  - Maintenance records
  - Failures (breaks and leaks) history

Each attribute category should be assigned a weighting factor; an example can be seen in Table 8.8.

TABLE 8.8
Examples of Weighting Factors of Different Attributes for Assessing PoF

Attribute Category	Weighting Factor
Physical	0.15
Condition	0.35
Environmental	0.25
Operation and performance	0.25

8.3.3.3.2 Consequence of Failure (CoF)

Items to consider when calculating Consequence of Failure (CoF) in water infrastructure:

- Spill, flood, odor
- Water or effluent quality
- Regulatory compliance
- Loss of service to customers
- Equipment and safety
- Economic impact

Quantifying the failure consequences can be done based on Table 8.9.

Generally, consequences include:

- *Social consequences*: number of critical customers affected, roadway impacts, public perception of failure, etc.
- *Economic consequences*: repair difficulty, pipe diameter, replacement cost, compensation for damage to surrounding infrastructure, etc.
- *Environmental consequences*: sensitivity of the failure discharge area, fines that are the result of failure-induced damage, etc.

Effective decision-making for managing assets is all about managing risk.

8.3.3.4 Life Cycle Cost Analysis

All asset decisions should be made based on determining the asset's life cycle cost (LCC). Life cycle costs (LCCs) may include:

- Construction and installation costs
- Operating costs
- Maintenance costs
- Rehabilitation costs
- Replacement costs
- Decommission and salvage costs

As depicted in Figure 8.7, both operational and maintenance costs increase as the assets age, but operations cost increase more rapidly after a certain point in the asset's life span.

As the infrastructure ages, O&M costs increase. So it is important to make plans for optimized O&M based on AM before the cost reduction opportunities in the life span phases of the infrastructure diminish. This is available in different phases of LCC analysis as mentioned in the next case study for the water distribution system depicted in Figure 8.8.

TABLE 8.9
Examples of Different Descriptions of CoF Based on Its Value

CoF	Value
Insignificant	2
Minor	4
Moderate	6
Major	8
Catastrophic	10

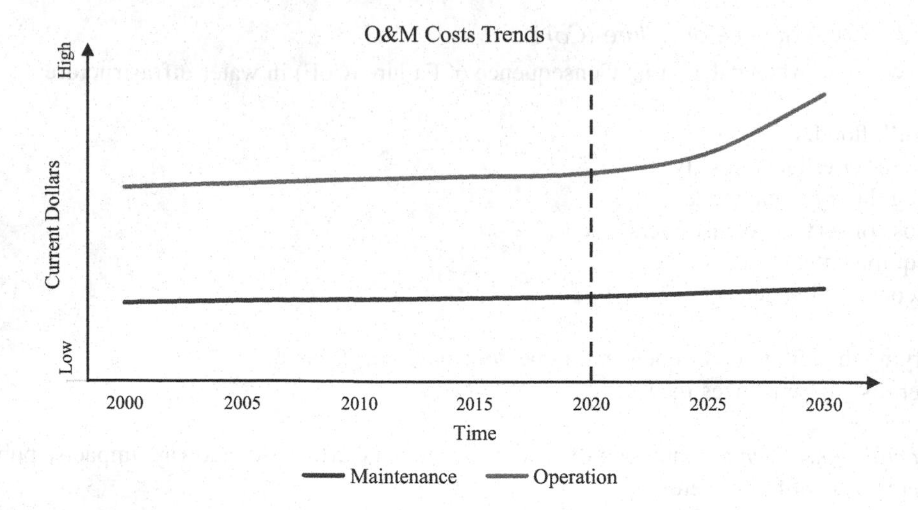

FIGURE 8.7 O&M cost variations over time.

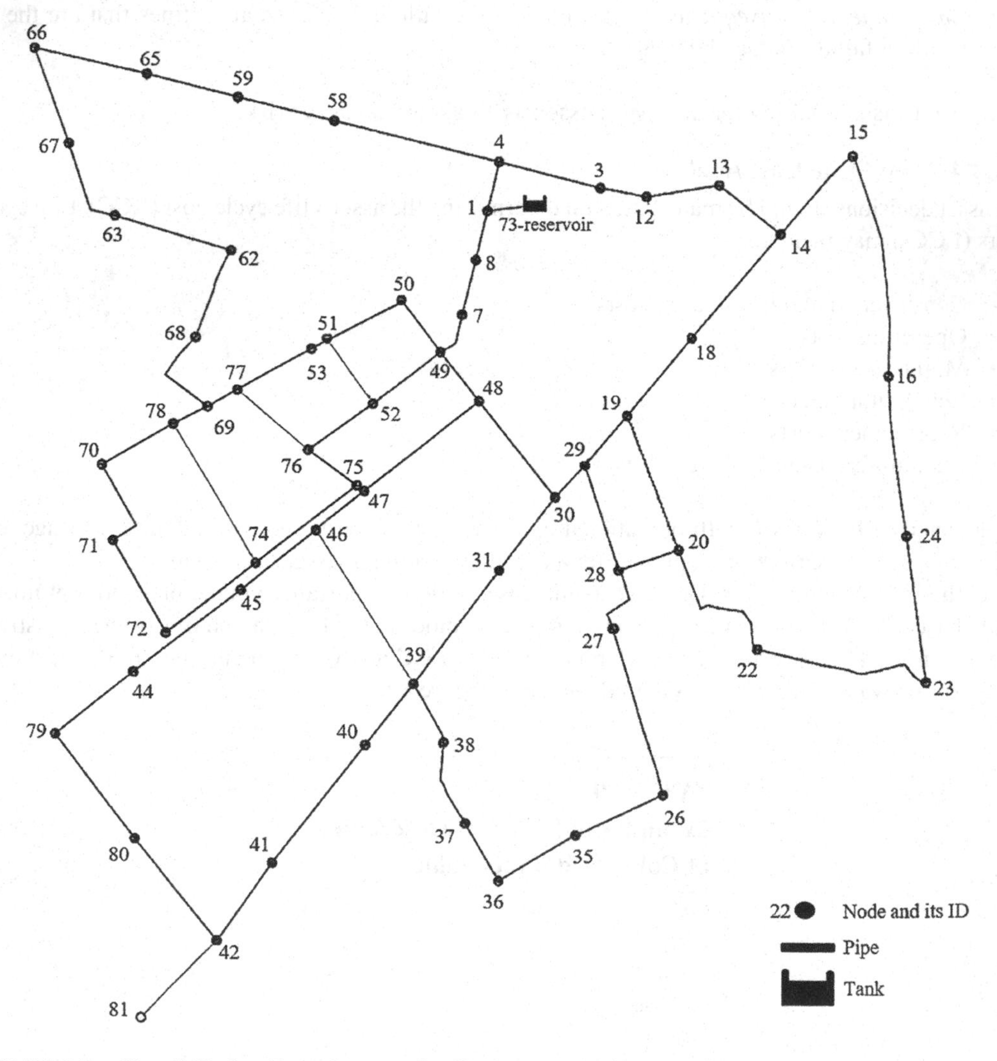

FIGURE 8.8 City of Chahar-Dangeh water distribution network in Iran.

8.3.3.5 Case Study 1: Reliability-Based Assessment of Life Cycle Cost of Urban Water Distribution Infrastructures

WDSs, as urban infrastructure, have an indisputable role in the productivity of a society and the quality of life of its residents. It is necessary to develop policies for evaluation of the urban infrastructure performance and determining the proper time of maintenance and rehabilitation activities to increase system productivity and expand the system's efficient performance time. In the study done by Karamouz et al. (2017), an algorithm is developed to evaluate the efficiency of urban water infrastructure performance, broken down into costs and benefits, and to determine the optimal time of water distribution network (WDN) rehabilitation using a probabilistic approach.

Life cycle modeling is an important decision-making tool because initial costs and agency costs are not illustrative of total LCCs. There has been a paradigm shift over recent decades in the application of maintenance strategies from breakdown maintenance to more sophisticated strategies like condition monitoring and AM. Preventive maintenance (PM) can help minimize the probability of losses due to accidents and unscheduled failure of process units. The growing interest in reliability- and risk-based PM and process safety management (PSM) is driven by the need to develop strategies that lead to an optimum safety versus cost balance (Ghosh and Sandip, 2009).

In this study, an algorithm for LCC assessment from an economic viewpoint has been developed based on system performance reliability index. The WDN operation condition is incorporated in cost–benefit analysis to consider the intangible costs. Therefore, the remaining operation life of WDN is estimated. The main phases of the methodology include model development, WDN evaluation, LCC assessment, and setting strategies.

- *Phase 1 (model development)*: This phase includes the first three stages of the proposed algorithm. At first, the basic data and information of WDN are collected. Then the remaining lifetime of the WDN is estimated. Then, the collected data are used in the development of the hydraulic simulation model of WDN.
- *Phase 2 (WDN evaluation)*: In this phase, WDN failure scenarios are generated, and the hydraulic performance of the WDN is simulated. By using the developed reliability indicator, the WDN performance is quantified. The reliability of WDN is quantified based on the pressure at demand nodes because it shows the amount of delivered water to the demand node in comparison with the requested water.
- *Phase 3 (LCC assessment)*: The WDN operation revenue and costs during its life cycle are estimated in this phase based on generated pipe failure scenarios. In this study, the NFV method is used, which is suitable for the comparison of alternatives. The amount of reduction in WDN revenue is directly related to the decrease in its reliability, which is included in the network total performance reliability (NTPR) evaluation.
- *Phase 4 (setting strategies)*: In this phase, which includes the two last stages of the proposed algorithm, the network's actual remaining life duration is identified based on the results of Phases 2 and 3 with emphasis on WDN reliability. This is the basis for making a decision on how to deal with the network. Phase 4 is a management step in comparison with other technical and engineering steps of the proposed algorithm. It emphasizes that in addition to the engineering aspects, social and governmental concerns should be considered in decision-making.

The case study for this paper is the Chahar-dangeh WDN (Figure 8.8), a small city located in the central part of Iran. The design life of this WDN is 25 years, of which approximately 5 years had passed at the time of the study. During these 20 years, WDN performance is evaluated every 2 years. In real WDN cases, the operation lives of pipelines can be much more than 20 years using periodic maintenance and operation programs.

After estimating the WDN reliability under failure scenarios, the WDN's revenue and cost are calculated to determine its remaining lifetime. When the WDN losses become more than the acceptable value (a marginal loss is considered for determining the WDN operation period), it can be stated that the WDN lifespan is terminated. Based on the socioeconomic data of the studied region (given in the case study section), the base year network gross revenue is estimated as $212,000. To provide an estimation of WDN maintenance costs, the reconstruction cost of the current network is estimated to be approximately $20 million.

It is assumed that the yearly maintenance cost is approximately 3% of the reconstruction cost, which would be $600,000. To estimate the WDN replacement costs, first, the failed pipes in each time step of each scenario, as well as the replacement lengths, are determined.

Considering the enhancement of network losses shown in Figure 8.9, it could be stated that in year 14, network losses increase more than 50% of rehabilitation costs, and in the remaining 6 years, network losses increase twice as much. Also, in the first 14 years, the annual (loss or reconstruction) growth rate is approximately 4%, and in the remaining 6 years, this proportion reached 13%. According to this figure, the remaining network lifespan is approximately 14–16 years, which is less than the initial expectation of 20 years in the WDN design.

The results of this study could be used efficiently in large WDNs to determine the rehabilitation time and avoid the high costs of WDN failure. Considering the actual value of water in this algorithm could help to find more realistic results in the WDN performance evaluation. If the data for the real value of water are available, the B/C ratio will be much higher (more than 1), and therefore, replacement costs are more justifiable. This is more applicable in developed countries when WDNs are managed based on market-driven governance models, but in most of the developing countries, the water price is subsidized, and the results of this study are more applicable.

The proposed methodology in this study can be extended to include all components of the WDN to provide a more realistic scheme of WDN for rehabilitation planning. Furthermore, consideration of variations in demand and the hydraulics of the system through a day (extended period simulation), water quality-based issues, options other than pipe replacement for WDN maintenance and repair, the broken dependency among different system components, and difference in different pipes failure probability could be considered in the future development of this algorithm. Simulation of WDN based on a pressure-driven approach could also be helpful.

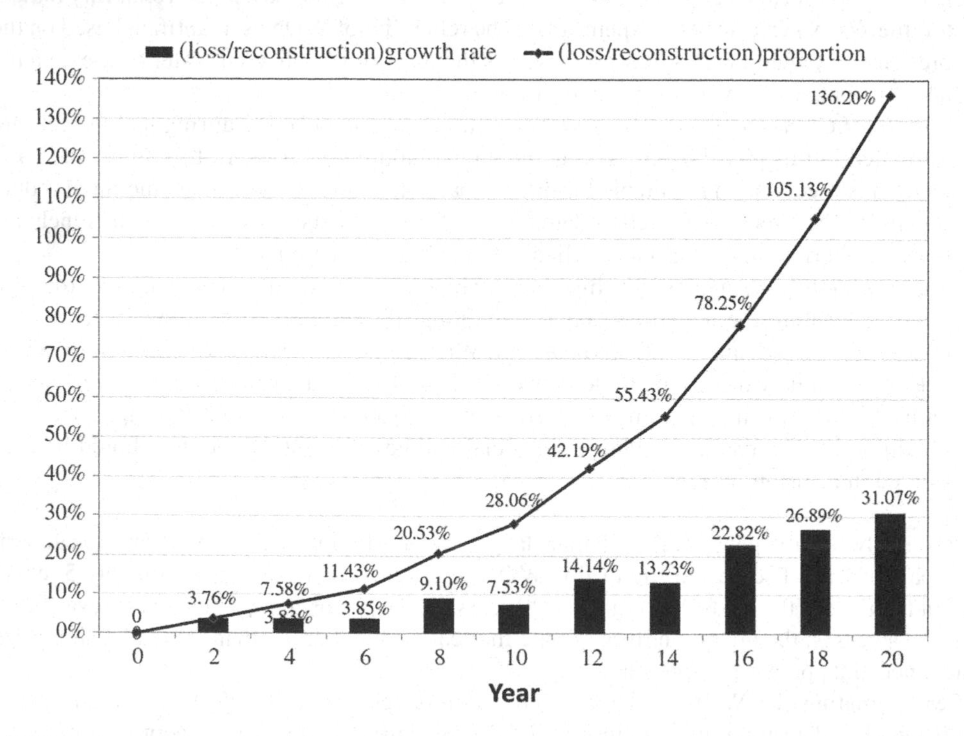

FIGURE 8.9 Loss and reconstruction growth rate during the life cycle of the Chahar-Dangeh WDN.

8.3.4 Sustainable Service Delivery

A successful AM cycle with Sustainable Service Delivery has an outer loop of plan, implement, and access. The AM plan is bounded by the policies and pertaining strategies in order to be integrated with the long-term financial plan. Current practices should be implemented and reported. Assessment should include both evaluating the practices and the state of current assets. The inner loop includes communication, engagement of people, and internal review of asset information and finances. See Figure 8.10 for more details of cycles in these integrations. For Sustainable Service Delivery, the LOS, that the customer desired should be considered are:

- Understanding customer expectations
- Determining LOS
- Establishing performance indicators
- Measuring performance
- Balancing the trade-offs between levels of service and future investments
- Communicating with the customers

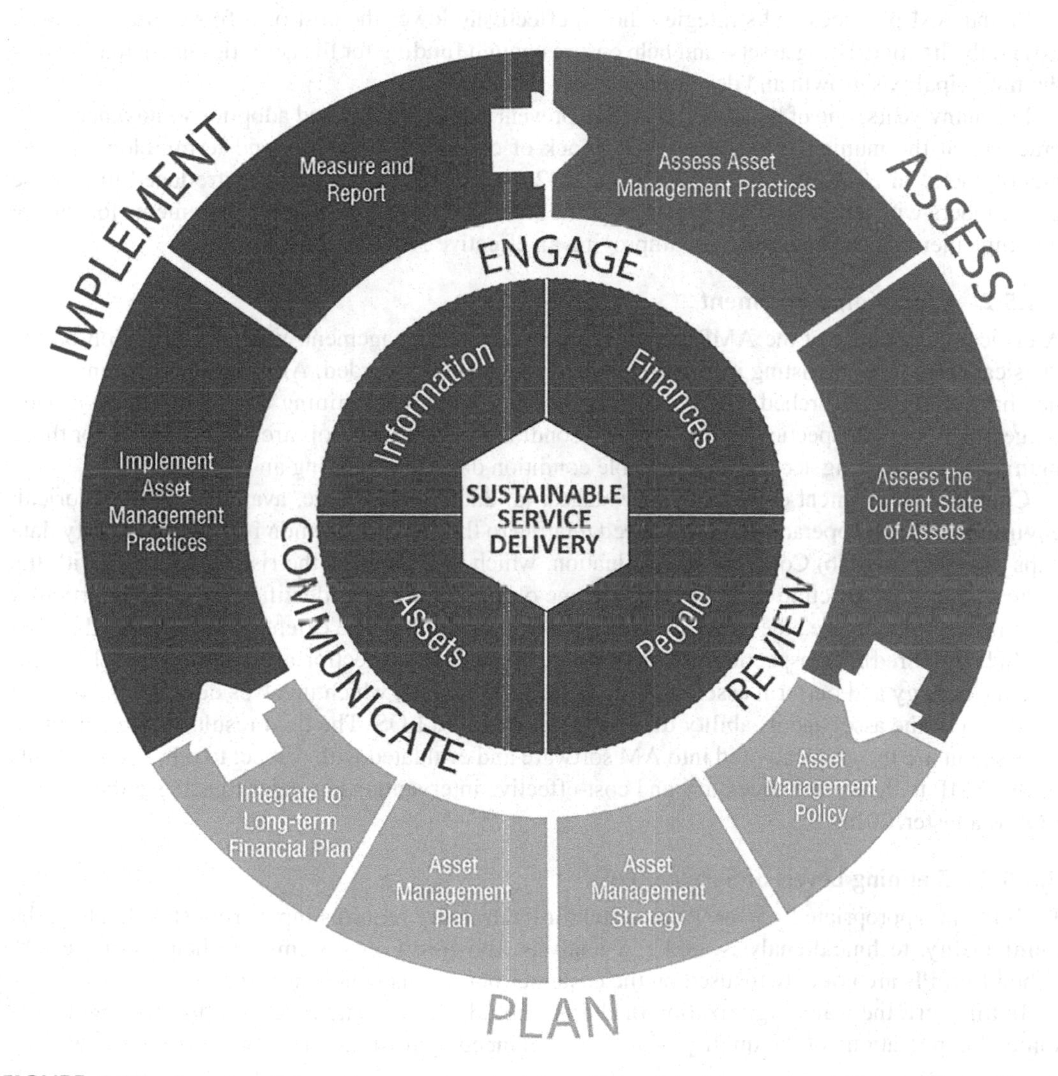

FIGURE 8.10 A successful asset management cycle with sustainable service delivery. (From "Asset Management for Sustainable Service Delivery: A BC Framework (2019)." https://www.assetmanagementbc.ca/asset-management-for-sustainable-service-delivery-a-bc-framework/.)

Defining an appropriate LOS for successful planning of AM for water system infrastructure requires input from stakeholders, the municipality, technical analysis, and risk analysis. Existing LOS systems can help ensure established LOS performance indicators (PIs) are not only focused on the customer but are also relevant and attainable.

8.3.5 Select AM Tools and Practices for Municipalities

In Canada, the following guidelines provide an introduction to select tools and practices that could help municipalities to shift towards an integrated approach to asset managing of water, wastewater, and stormwater infrastructure assets. For more details, refer to a white paper prepared for the Canadian Water Network research project (Harvey, 2015).

8.3.5.1 Asset Management Strategic

Risk-based municipal AM practices are important during an era of infrastructure replacement. AMPs can be developed to document how groups of assets will be managed to provide a satisfactory service level in a sustainable and environmentally responsible manner.

Sound AM practices and strategies should effectively lower the cost of infrastructure renewal, extend the life of existing assets, and help ensure enough funding for the activities needed to sustain the municipality's growth and development over time.

For many years, one of the major obstacles preventing the widespread adoption of advanced AM practices at the municipal level has been a lack of consistent standards and terminology among practitioners. In 2014, the ISO 55000 (ISO, 2012a) series of standards was introduced to provide practitioners with a standardized overview of the AM. It provides the major requirements for setting up, implementing, maintaining, and improving an effective AM system.

8.3.5.2 Condition Assessment

A critical component of the AMP structure involves risk management, where information on the physical condition of existing infrastructure is collected and recorded. Although many municipalities have traditionally relied solely on expert opinion when determining asset conditions, a wide range of advanced inspection technology and condition assessment tools are now available for those municipalities seeking accurate and reliable condition data for providing an LOS target.

Condition assessment consists of four stages; (a) data; at this stage, available data (historical, environmental, and operational) is reviewed relating to the asset in question in order to identify data gaps and priorities. (b) Consequence evaluation, which is evaluating the risks associated with the failure of an infrastructure asset. A key outcome of this stage is the identification of any particular gaps in information (i.e., the provision of a general indication of the extent of asset inspection that might be required). (c) Inspection, the appropriate methods of inspection are evaluated to select suitable technology and perform asset inspection. (d) LOS targets, which involves defining the current condition of the asset and its ability to achieve the required LOS. The final results of the condition assessment are then typically fed into AM software and evaluated with respect to other components of the AMP to determine necessary and cost-effective interventions in order to achieve the desired LOS (Sangster, 2010).

8.3.5.3 Defining Levels of Service (LoS)

Defining an appropriate LoS for water system infrastructure requires input from stakeholders, the municipality, technical analysis, and risk analysis. Existing LoS systems can help ensure established LoS PIs are not only focused on the customer but are also relevant and attainable.

In this part, the water organization in charge should seek to match the LoS provided with customers' expectations of the quality of service, balanced against the price the customer is willing to pay. Customer levels of service include the overall quality, functionality and safety, and capacity of the service, including the receptivity and rapidity (when a systems component fails) that are provided.

Many municipalities consult with consulting engineering firms when developing a so-called "triple-bottom-line approach for water, wastewater, and stormwater" LoS that considers financial, environmental, and social/community/organizational perspectives. By doing so, the resulting levels of service as stated by FCM (2002) are to reflect social and economic goals of the community and may include any of the following parameters: safety, customer satisfaction, quality, quantity, capacity, reliability, responsiveness, environmental acceptability, cost, and availability. Defining LoS for water, wastewater, and stormwater systems can be complex—as trade-offs occur based on the opinion and priorities of the various stakeholders.

8.3.5.4 Software Trends

The ability for a municipality or utility to effectively manage its water and wastewater infrastructure assets will largely be controlled by its ability to organize large amounts of data. A wide variety of technologies that support AM are now available to municipalities and utilities.

AM systems "can range from complex integrated Enterprise Asset Management (EAM) suites to mixed environments of best of breed software, bespoke applications and spreadsheet-based analytics" (IAM, 2015). In general, the software tools adopted by a municipality for managing infrastructure assets should consist of:

1. *The Computerized Maintenance Management Systems (CMMS)*: a type of management software that supports operations and maintenance programs. Use of a CMMS is currently the most widely adopted AM practice by water utilities in North America (McGraw Hill Construction, 2013).
2. A system for managing asset inspection and monitoring where the collected data can be analyzed to enable identification of appropriate solutions for rehabilitation that will ensure continued provision of the desired LOS.

8.3.5.5 Conclusion for Municipalities

A series of barriers to AM still exist in Canada. Embracing standardization of AM practice and state-of-the-art software tools is a necessity for sustainable growth and development.

The 2013 Water Infrastructure Asset Management survey of 451 respondents from the United States and Canada (population range of 3,300 to more than 500,000) indicated the biggest obstacle preventing widespread implementation of advanced AM practices is organizational resistance to change (McGraw Hill Construction, 2013). Such organizational resistance is closely tied to the "silo effect"—where overlapping activities related to managing water, wastewater, stormwater, and watershed have typically been managed by different departments that operate in isolation from one another.

Existing silos need to be broken down to achieve sustainable AM. Municipal water managers, watershed groups, conservation authorities, local businesses, and governments need to start coordinating so that decisions are considered in the full context of actions from various other groups and levels of government.

AMPs structured around industry standards for infrastructure management are an invaluable tool for achieving sustainability at the municipal level and overcoming problems associated with silos. Many of the goals and initiatives defined in these plans will only be achieved if municipalities seek out and adopt leading-edge software tools that can break down communication barriers. This white paper has explored a select set of practices that can assist with communication and ensure AM operations are integrated across all departments

8.4 PERFORMANCE MEASURES

The existing International Water Association (IWA) library contains 70 criterion groups precisely defined performance measures for water supply under seven objectives (45 criterion groups are listed here that have links to other PIs) as follows:

1. Adequacy of the service (three criterion groups)
2. Meeting user's needs and expectations (five criterion groups) such as criterion group 4: continuity of the service; active leakage control repairs; the assessment period; period of time adopted for the assessment of the data; and the total transmission and distribution mains length (service lines not included.
3. Promotion of sustainable development (nine criterion groups)
4. Protection of the environment (two criterion groups)
5. Provision of the service under normal and emergency situations (seven criterion groups)
6. Public health and safety (eight criterion groups)
7. Sustainability of the undertaking (11 criterion groups)

The IWA library also contains 76 PIs for wastewater and stormwater grouped under seven objectives in 59 criterion groups:

1. Meeting user's needs and expectations (six criterion groups)
2. Occupational health protection and safety (two criterion groups)
3. Promotion of sustainable development of the community (15 criterion groups)
4. Protection of the environment (nine criterion groups)
5. Provision of the service under normal and emergency situations (seven criterion groups)
6. Public health and safety (nine criterion groups)
7. Sustainability of the undertaking (11 criterion group)

In Canada, many municipalities evaluate system performance through membership in the National Water and Wastewater Benchmarking Initiative (NWWBI). Since its inception in 1997, the benchmarking initiative has included 50 water utilities, 53 wastewater utilities, and 28 stormwater management programs. The NWWBI now represents 43 of Canada's leading municipalities and regional districts (representing over 60% of the Canadian population). The NWWBI uses a standardized utility management model as shown in Figure 8.11 to establish a framework for the selection and definition of performance measures for a set of seven generic goals that are common to all water and wastewater utilities. Currently, the NWWBI has benchmarked approximately 50 performance measures for each of the water and wastewater treatment, water distribution, and wastewater collection utilities and approximately 15 performance measures for stormwater and drainage utilities. The measures typically consist of a numerator that expresses the level of goal attainment and a denominator that serves as a normalization factor to enable comparisons among different utilities. Municipalities seeking a practical approach to implementing effective management practices for a range of infrastructure assets can refer to the International Infrastructure Management Manual (IIMM) developed by the Institute of Public Works Engineering Australasia.

The model goes through an annual review after an interactive data collection and workshop level analysis (RedZone Robotics, Solo System – The World's Only Self Operating Crawler, 2014).

8.5 DEVELOPING ASSET MANAGEMENT PLANS FOR WATER AND SEWER UTILITIES

US Environmental Protection Agency (USEPA) has defined AM as a process for maintaining a desired level of customer service at the best appropriate cost. Therefore, the core planning issue is

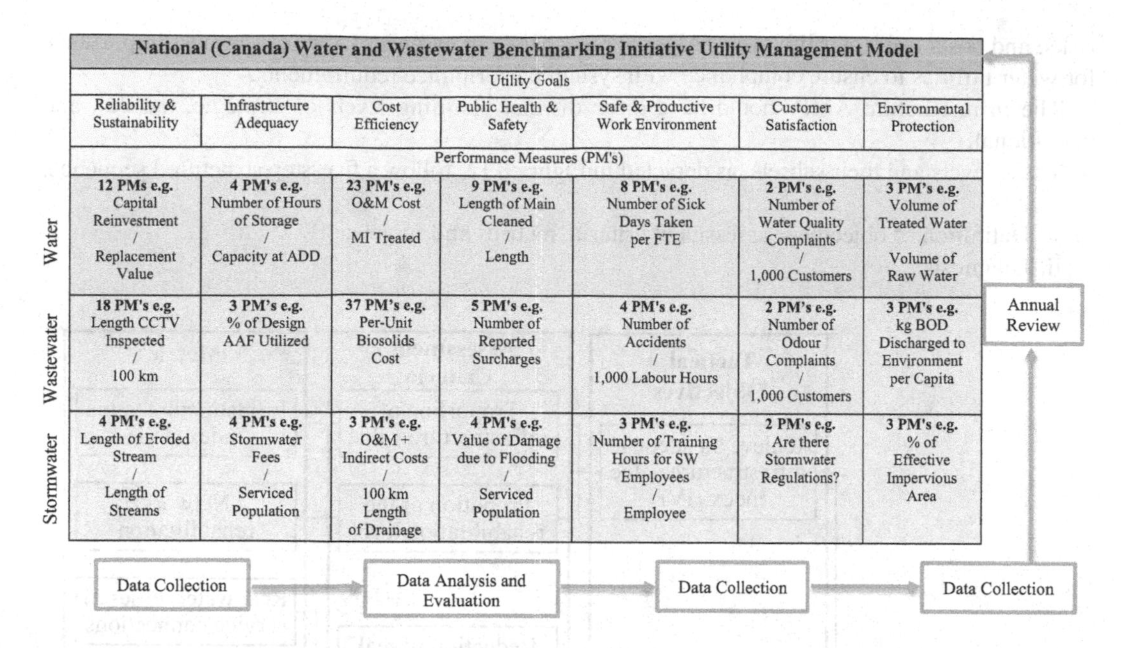

National (Canada) Water and Wastewater Benchmarking Initiative Utility Management Model
Utility Goals
Reliability & Sustainability
Infrastructure Adequacy
Cost Efficiency
Public Health & Safety
Safe & Productive Work Environment
Customer Satisfaction
Environmental Protection
Performance Measures (PM's)
Water
12 PMs e.g. Capital Reinvestment / Replacement Value
4 PM's e.g. Number of Hours of Storage / Capacity at ADD
23 PM's e.g. O&M Cost / Ml Treated
9 PM's e.g. Length of Main Cleaned / Length
8 PM's e.g. Number of Sick Days Taken per FTE
2 PM's e.g. Number of Water Quality Complaints / 1,000 Customers
3 PM's e.g. Volume of Treated Water / Volume of Raw Water
Wastewater
18 PM's e.g. Length CCTV Inspected / 100 km
3 PM's e.g. % of Design AAF Utilized
37 PM's e.g. Per-Unit Biosolids Cost
5 PM's e.g. Number of Reported Surcharges
4 PM's e.g. Number of Accidents / 1,000 Labour Hours
2 PM's e.g. Number of Odour Complaints / 1,000 Customers
3 PM's e.g. kg BOD Discharged to Environment per Capita
Stormwater
4 PM's e.g. Length of Eroded Stream / Length of Streams
4 PM's e.g. Stormwater Fees / Serviced Population
3 PM's e.g. O&M + Indirect Costs / 100 km Length of Drainage
4 PM's e.g. Value of Damage due to Flooding / Serviced Population
3 PM's e.g. Number of Training Hours for SW Employees / Employee
2 PM's e.g. Are there Stormwater Regulations?
3 PM's e.g. % of Effective Impervious Area
Annual Review
Data Collection
Data Analysis and Evaluation
Data Collection
Data Collection

FIGURE 8.11 The National (Canada) utility management model with performance measures for different goals. (From Harvey, R. *An Introduction to Asset Management Tools for Municipal Water, Wastewater and Stormwater Systems*. The Canadian Water Network, Waterloo, ON, (2015).)

customer service/satisfaction. Water utilities are major infrastructures, and planning should realize the shear size and the physical and operational attributes of them and interdependencies with other infrastructures. Table 8.10 shows some of the challenges faced by water systems and the benefit of a sound AM in meeting those challenges.

8.5.1 Water Infrastructure Asset Management

IAM is an integrated and multidisciplinary approach that uses systematic, coordinated activities and practices allowing water utilities to optimally manage their assets and associated performance,

TABLE 8.10
Challenges in Water Systems Management and Their Equivalent Solutions by Utilizing Asset Management

Challenges Faced by Water Systems	Benefit of Asset Management
• Determining the best (or optimal) time to rehabilitate/repair/replace aging assets. • Increasing demand for services. • Overcoming resistance to rate increases. • Diminishing resources. • Rising service expectations of customers. • Increasingly stringent regulatory requirements. • Responding to emergencies as a result of asset failures. • Protecting assets.	• Prolonging asset life and aiding in rehabilitate/repair/replacement decisions through efficient and focused operations and maintenance. • Meeting consumer demands with a focus on system sustainability. • Setting rates based on sound operational and financial planning. • Budgeting focused on activities critical to sustained performance. • Meeting service expectations and regulatory requirements. • Improving response to emergencies. • Improving security and safety of assets

risks, and costs over their life cycle (Alegre et al., 2012). In essence, IAM is of utmost importance for water utilities to ensure compliance with system performance requirements.

The infrastructure AMP should have three distinct planning levels of strategic, tactical, and operational.

These levels and their subsets, as depicted in Figure 8.12, follow a five-step structured sequence:

i. Definition of objectives, assessment criteria, metrics and targets;
ii. Diagnosis;

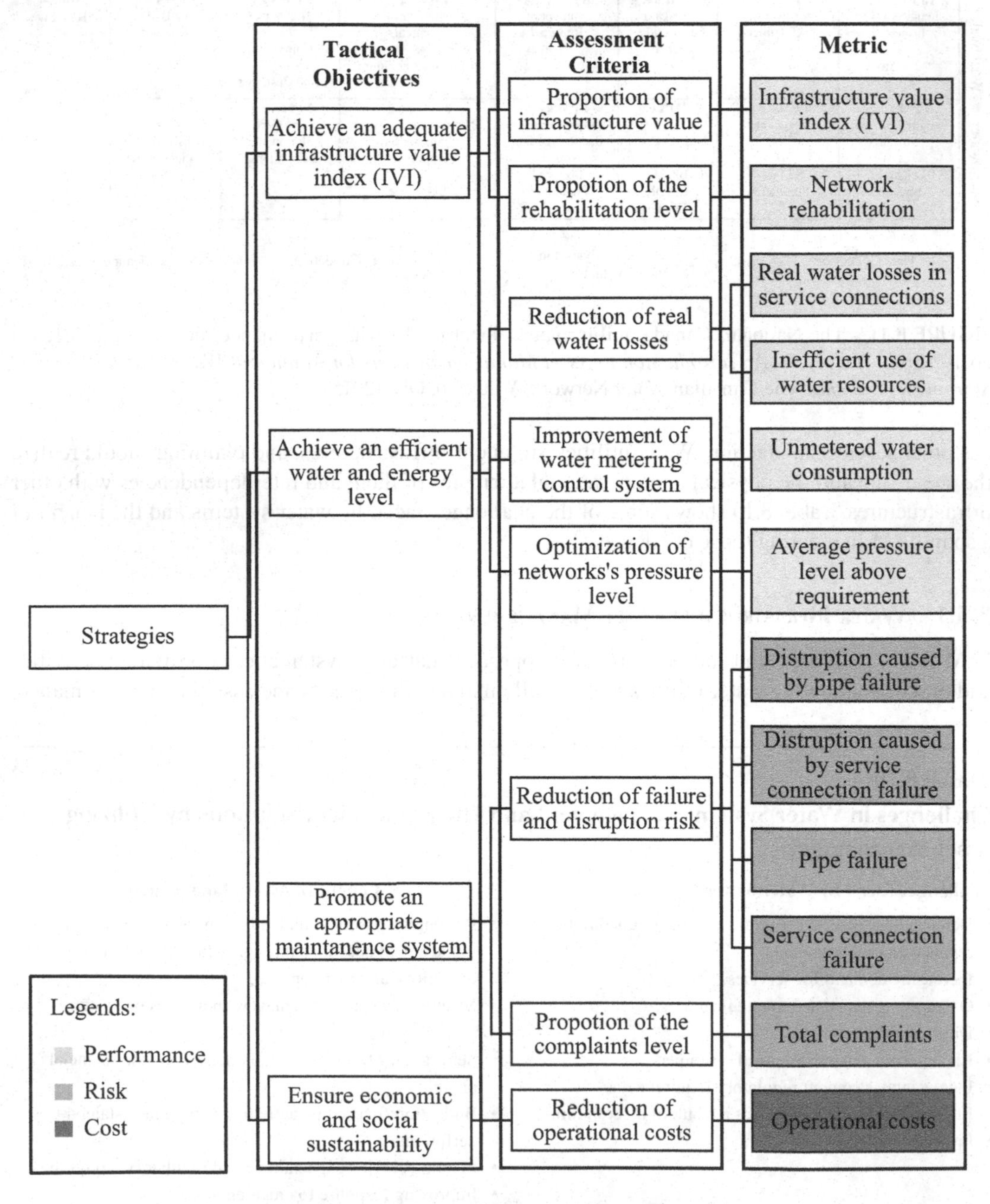

FIGURE 8.12 The three distinct planning levels of infrastructure asset management. (Modified from U.S. EPA, Asset management: A best practices guide. EPA 816/F-08/014, 2008.)

iii. Plan elaboration;
iv. Plan implementation;
v. Plan monitoring and review.

As for tactical objectives, infrastructure value, efficient use of water and energy, high state of maintenance, and economic and social sustainability are emphasized. For assessment criteria, measuring the attainment of objectives including metering, loss reduction, network pressure optimization, reduction of failure risk, complaints, and operation cost. Six metrics are related to performance and 5 related to risk reduction.

Infrastructure value index (IVI), as the ratio between the current (fair) value of an infrastructure and the replacement cost on modern equivalent asset basis, is of interest. It is a powerful modelling tool for long term linear and vertical asset.

8.5.1.1 Stages in Water System's Asset Management

Four horizons/stages could be defined as continuous planning over the infrastructure life cycle; rebuilt or new construction/installations; funding; and information technology.

As a priority planning horizon and on many levels, water system AM is a cornerstone of sustainability and resiliency, supporting the insights, planning, and day-to-day discipline needed to optimize these both operationally and financially over a long-term design period. For water utilities seeking to manage sustainable systems, the challenges are all too familiar and usually start with budget struggles. The conflict between immediate needs and long-term imperatives can quickly derail AM planning before it starts.

As a stage for the action to rebuild or construct, aging infrastructure competes with new construction for investment funds, making for some "apples and oranges" debates, like how to balance the need to renew old systems and also install new installations (such as new tunnels to manage combined sewer overflow (CSO)). Reactive maintenance and repair is the most expensive and inefficient approach that a utility can undertake.

Funding and cash flow management to make sure AM-related policies can be carried out on a timely fashion is a never-ending task. It isn't really planning if you have no way to know how the funding could be provided.

Information technology can provide some answers, but datasets may be incomplete or out of date. Leading approaches and practices of AM can help bring the whole picture into focus, with benefits across operations, ultimately enhancing customer acceptance and stakeholder support.

8.5.2 Asset Management for Water Supply Infrastructures (Dams and Reservoirs)

Dams and reservoirs provide storage of water, including flood water, which can then be supplied for households and irrigation, as well as generation of power, thus reducing fossil fuel depletion and the negative environmental effects of fossil fuel burning. They often emerge as the priority in strategic planning with respect to water and energy. However, similar to the other infrastructure projects, there are also adverse environmental and social impacts that must be minimized or mitigated.

The world has around 55,000 large dams, most of them registered by the International Commission on Large Dams (ICOLD). About half of them are used solely for supplying water for irrigation purposes, and roughly one-third of them are used for multipurpose. No reliable data exist on the total number of "small" dams, that is, those not meeting the ICOLD criteria. A very indicative figure is 800,000, almost all of them used for irrigation and water supply.

In the Asian and Pacific regions, most dams and reservoirs have been built since 1950. Construction peaked in the 1970s, when hundreds of large dams were put into service each year. At that time, Japan, a flood-prone country with 40% of its population and 60% of economic assets located in vulnerable river plains, invested some 2 trillion Yen (about 20 milliard US$) in hydraulic infrastructure, mostly dams and embankments. With these facilities, annual flood losses that, before the 1950s, could reach 20% of GDP were reduced to less than 1% of GDP. Even with its large stock of about 2,800 large dams and reservoirs, Japan still spends about US$9 billion of public funds annually on expanding and maintaining hydraulic infrastructure, and another 100 large dams are under construction (White, 2005).

A major indicator of water resources development is the ratio between the available storage reservoir capacity and the volume of the annual renewable water resources. The ratio in the Asian and Pacific region is 0.10, less than the global average of 0.14 and far behind North America (0.33) and Europe (0.16) (Asian Development Bank, 2001).

Example 8.5

Three water supply projects are implemented in a region. Water demand of a big city, a small city, and a village will be supplied through these projects. Level of income and distribution of the costs and benefits for these projects are presented in Table 8.11. Implementation of these projects reduces the tension on other resources that are expressed as beneficial in Table 8.11. Determine the distribution of costs and benefits and find which projects are progressive and regressive.

Solution:

The percent of income for the village, the small city, and the big city is shown in Table 8.12.

As can be seen in Table 8.12, the first program is regressive because the profits allocated to the groups with higher level of income are higher. The second program is progressive because it has allocated less profit to the people with higher income. In the third program, the difference between benefits and costs is 1% of the income for the village, the small city, and the big city. Therefore, the distribution of costs and benefits has not been very different in this program.

8.5.3 Infrastructure for the Water Distribution System

WDSs include the entire infrastructure from the treatment plant outlet to the household tap. Fire-flow requirements usually control the design aspect of a WDS. The actual layout of water mains, arteries, and secondary distribution feeders should be designed to deliver the required fire flows in all buildings and structures in the municipality, above and beyond the maximum daily usage rate. In developing the WDS layout, the effect of a break, joint separation, or other main failure that could

TABLE 8.11
Level of Income and Distribution of Costs and Benefits Related to Example 8.5

Benefits and Costs	Village	Small City	Big City
Average Income ($)	**1,000**	**20,000**	**70,000**
Water Supply Program 1			
Benefit to other resources	30	1,400	7,700
Costs of project implementation	20	800	4,200
Water Supply Program 2			
Benefit to other resources	140	2,200	7,700
Costs of project implementation	40	1,000	4,200
Water Supply Program 3			
Benefit to other resources	30	300	840
Costs of project implementation	20	100	140

TABLE 8.12
Distribution of Costs and Benefits in the Three Water Supply Projects of Example 8.5

Percentage of Income	Village	Small City	Big City
Income	**1,000**	**20,000**	**70,000**
Water Supply Program 1			
Benefit to other resources	3	7	11
Costs of project implementation	2	4	6
Difference between benefit and costs	1	3	5
Water Supply Program 2			
Benefit to other resources	14	11	11
Costs of project implementation	4	5	6
Difference between benefit and costs	10	6	5
Water Supply Program 3			
Benefit to other resources	3	1.5	1.2
Costs of project implementation	2	0.5	0.2
Difference between benefit and costs	1	1	1

occur during the WDS operation should be considered. Also, most of the water-quantity standards related to the WDS should be observed on the basis of general practices in this field.

In evaluating a WDS, pumps should be considered at their effective capacities when discharging at standard operating pressures. The pumping capacity, in conjunction with storage, should be sufficient to maintain the maximum daily usage rate plus the maximum required firefighting flow.

Storage tanks are frequently used to equalize pumping rates in the distribution system as well as provide water for firefighting. In determining the fire flow from storage, it is necessary to calculate the rate of delivery during the specified period (firefighting). Even though the volume stored may be large, the flow to a hydrant cannot exceed the carrying capacity of the mains, and the residual pressure at the point of use should not be less than what is needed (e.g., 140 kPa), which varies with different standards.

Depending on specific practices, some standards for urban WDNs are noted here for demonstrating the scale and range of figures. As an example, the recommended water pressure in a distribution system is typically 450–520 kPa, which is considered adequate for buildings up to ten stories high as well as for automatic sprinkler systems for fire protection in buildings of four to five stories. For a residential service connection, the minimum pressure in the water distribution main should be 280 kPa; pressure in excess of 700 kPa is not desirable, and the maximum allowable pressure is 1,030 kPa. Fire hydrants are installed at a spacing of 90–240 m and in locations required for firefighting.

Although a gravity system delivering water without the use of pumps is desirable from a fire protection standpoint, the reliability of well-designed and properly safeguarded pumping systems can be developed to such a high degree that no distinction is made between the reliability of gravity-fed and pump-fed systems. Electric power should be provided to all pumping stations and treatment facilities by two separate lines from different sources (Statewide Urban Design and Specifications, 2007). In the following, an introduction to the status of water supply infrastructures in the world based on international reports is presented.

The provision of safe drinking water and improved sanitation has a high priority in the MDG and the Johannesburg Plan of Implementation as a primary means of eradicating poverty. For Asia and the Pacific, meeting the MDG targets, by 2015, the proportion of people without adequate and sustainable access to safe drinking water and improved sanitation presents a particularly formidable challenge. Two-thirds of the world's population do not have access to safe water and more than three quarters of the world's population are not served by adequate sanitation. The drastic increase projected in population over the next century will place enormous pressure on urban water infrastructure. The construction, extension, and rehabilitation of water supply and sanitation facilities, especially those serving the poor nations, will need to be accelerated (Report of the World Summit on Sustainable Development, 2002).

The Asian and Pacific region's lack of sufficient infrastructure denies a large part of its population access to safe water and decent sanitation. In 2002, every sixth person in the region, or an estimated 691 million people, did not have access to safe, sustainable water supplies and almost half the population did not have access to decent sanitation. A huge number of people have gained access to water and sanitation services as a result of the expansion in infrastructure since 1990. However, due to population growth, the absolute number of people without access to such services remained almost the same. Indeed, coverage of the region's urban population actually decreased due to population growth rate that surpassed the rate of development of urban water supply and sanitation (Asian Development Bank, 2005).

8.5.3.1 Water Treatment and Water Mains

Water mains are provided solely for the purpose of supplying potable water and fire protection. "Water mains" refers here to the following categories, but not service lines:

1. A raw water transmission main transports raw water from the source to the treatment plant.
2. A treated water transmission main conveys water from the plant to the storage or directly to an arterial main.
3. Arterial mains transport treated water to the distribution mains from the treatment plant or storage.
4. The distribution mains transport water to the service lines, which convey the water to the end user.

For a given system, the transmission mains are typically larger in diameter, straighter, and have fewer connections than the distribution mains (Smith et al., 2000).

8.5.3.2 Design and Construction of the Water Main System

The important design requirements of water main systems are that they supply each user with sufficient volume of water for a particular designated use plus required fire flows at adequate pressure and that the system maintains the quality of the potable water delivered by the treatment plant. It is important in the design of water main systems to address the maintenance considerations, constantly. The performance of a water main system for health and fire-flow purposes depends on the jurisdiction's ability to maintain the system at an affordable cost. Certain planning considerations related to a new system development or system expansion require the designer to consider factors such as future growth, cost, and system layout. For system layout, all major demand areas should be serviced by an arterial-loop system. High-demand areas are served by distribution mains tied to an arterial-loop system to form a grid without dead-end mains. Areas where adequate water supply must be maintained at all times for health and fire control purposes should be tied to two arterial mains where possible. Minor distribution lines or mains that make up the secondary grid system are a major portion of the grid since they supply the fire hydrants and domestic and commercial consumers. In the following paragraphs, the considerations that should be given to the optimal design of water mains are presented.

In the present context, to optimize a system means to maximize its net benefits to all water users. In order to measure net benefits, one has to evaluate the benefits and costs associated with the system. The benefits of a WDS are vast since modern urban centers cannot exist without them; thus, the fundamental need for a WDS has been taken for granted. Consequently, the benefits are measured by the performance and quality of service that the system provides. The performance of a WDS may be measured by the degree to which the following objectives are accomplished:

1. To provide safe drinking water.
2. To provide water that is acceptable to the consumer in terms of esthetics, odor, and taste.
3. To have an acceptable level of reliability.
4. To be capable of providing emergency flows, for example, for firefighting at an acceptable pressure.
5. To provide the demand for water at an acceptable residual pressure during an acceptable portion of the time which are often subject to regulations and may vary with locale. For example, provide demand flow at a minimum 30-m pressure head for 99% of the time. This is often subject to each region's regulations.
6. To be economically efficient.

The cost of a WDS comprises all direct, indirect, and social costs that are associated with

1. Capital investment in system design, installation, and upgrading
2. System operation—energy cost, materials, labor, monitoring, and so on
3. System maintenance—inspection, breakage repair, rehabilitation, and so on

Thus, a WDS has multiple objectives, of which all but one (economic efficiency) are either difficult to quantify (e.g., reliability, water quality, and esthetics) or difficult to evaluate in monetary terms (e.g., LOS) or both. The situation is somewhat simpler concerning the cost aspects, where direct and indirect costs can be evaluated with relative ease, and social costs can (with some effort) be assessed, albeit with less certainty.

There are various methods to handle multiple objectives and criteria, which are generally non-commensurable and are expressed in different units. However, some of these techniques involve the explicit (or implicit) assignment of monetary values to all objectives, for example, the linear scoring method and goal programming, which may introduce biases, and trying to assign a monetary value to reliability or to water quality. On the other hand, multiobjective evaluation techniques are suitable for only a moderate number of alternatives (e.g., the surrogate worth trade-off method, and utility matrices), which makes them unsuitable for handling the vast number of rehabilitation measures and scheduling alternatives involved in a typical WDS. An alternative approach is to formulate the problem as a "traditional" optimization problem in which the optimization criterion is minimum cost, while all the components that cannot be assigned monetary values are taken into consideration as constraints. A general formulation for the full scope of the problem can then be expressed as follows:

Minimize: {capital investment+operation costs+maintenance costs+rehabilitation costs} Subject to

1. Physical/hydraulic constraints such as mass conservation, continuity equations, and so on
2. Network topology constraints such as the layout of water sources, streets, tanks, and so on
3. Supply pressure head boundaries such as minimum and maximum residual pressure heads
4. Minimal level of reliability constraints
5. Minimal level of water quality constraints
6. Available equipment constraints such as treatment plants, pumps, pipes, appurtenances, and so on
7. Available supply water constraints such as quality, quantity, and source location

All costs and constraints are considered for the entire life of the system. (The "life of a system" is a somewhat misleading idea, which is further discussed in Section 9.5.) It appears at present that any attempt to handle the full scope of the problem would be overly ambitious in light of the knowledge and computational tools available currently.

Example 8.6

Consider two mutually exclusive alternatives for a water main distribution project, given in Table 8.13. Assume that a study period of 15 years is applicable. Using the annual worth and PW methods, determine the preferred project.

Solution:

Using the annual worth method, it can be written that

$$\text{AW of project A} = -\$200{,}000 \times \left(\frac{A}{P}, i\%, 15\right) - \$200{,}000$$

$$\times \left(\frac{P}{F}, i\%, 10\right) \times \left(\frac{A}{P}, i\%, 15\right) + \$100{,}000 - \$44{,}000,$$

$$= -\$200{,}000 \times \left(\frac{i \times (1+i)^{15}}{(1+i)^{15} - 1} - \frac{1}{(1+i)^{10}} \times \frac{i \cdot (1+i)^{15}}{(1+i)^{15} - 1}\right) + \$56{,}000.$$

$$\text{AW of project B} = -\$300{,}000 \times \left(\frac{A}{P}, i\%, 15\right) + \$140{,}000 - \$86{,}000,$$

$$= -\$300{,}000 \times \left(\frac{i \times (1+i)^{15}}{(1+i)^{15} - 1}\right) + \$54{,}000.$$

Figure 8.13 shows the estimated annual worth of projects A and B using arrange of values for the interest rate (i). As can be seen in this figure, the annual worth of project A has always been greater than that of project B regardless of the interest rate. Also, it can be seen that the annual worth of projects A and B becomes negative for the interest rates greater than 30% and 16%, respectively.

Using the PW method, it can be written that

$$\text{PW of project A} = -\$200{,}000 - \$200{,}000 \times \left(\frac{P}{F}, i\%, 10\right) + \$100{,}000$$

$$\times \left(\frac{P}{A}, i\%, 15\right) - \$44{,}000 \times \left(\frac{P}{A}, i\%, 15\right),$$

$$= -\$200{,}000 \times \left(1 + \frac{1}{(1+i)^{10}}\right) + \$56{,}000 \times \left(\frac{i \times (1+i)^{15}}{(1+i)^{15} - 1}\right).$$

TABLE 8.13
Alternatives for a Water Distribution Project (Example 8.6)

	Project A	Project B
Capital investment ($)	200,000	300,000
Annual revenues ($)	100,000	140,000
Annual expenses ($)	44,000	86,000
Useful life (years)	10	15

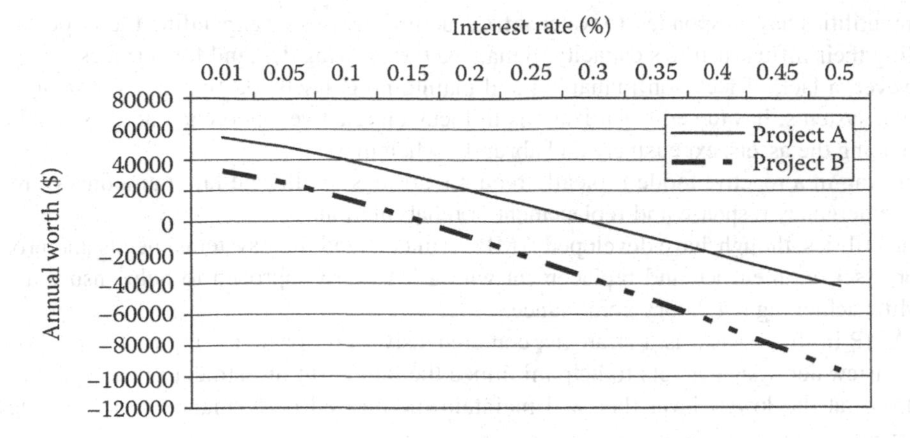

FIGURE 8.13 Variations of the annual worth of the projects in Example 8.6 with respect to interest rate.

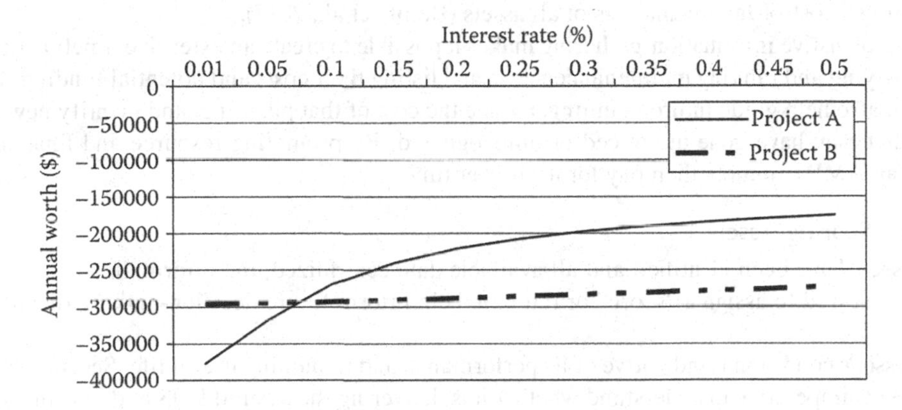

FIGURE 8.14 Variations of the annual worth of the projects in Problem 4 with respect to interest rate.

$$\text{PW of project B} = -\$300{,}000 + \$140{,}000 \times \left(\frac{P}{A}, i\%, 15\right) - \$86{,}000 \times \left(\frac{P}{A}, i\%, 15\right),$$

$$= -\$300{,}000 + \$54{,}000 \times \left(\frac{i \times (1+i)^{15}}{(1+1)^{15} - 1}\right).$$

Figure 8.14 shows the estimated PW of projects A and B using a range of values for the interest rate (i). As can be seen in this figure, for the interest rates greater than 7%, the PW of project A has been greater than that of project B. So, project A is the recommended project.

8.5.4 Asset Management Programs for Stormwater and Wastewater Systems

The 2012 EPA Clean Watersheds Needs Survey conducted to assess the capital investment needed nationwide for publicly owned stormwater and wastewater collection and treatment facilities to meet Clean Water Act water quality goals concluded that about $271 billion in stormwater and wastewater infrastructure capital investment is needed for the nation's approximately 15,000 publicly owned treatment works.

The American Society of Civil Engineers estimated in 2013 that the nation needs to invest about $298 billion of capital in stormwater and wastewater infrastructure over the next 20 years; pipe represents the largest capital need, accounting for three-quarters of total needs.

Many utilities have responded to water infrastructure stresses by expanding their operations and upgrading their infrastructure's capacity to manage the growing demand for services.

However, a lack of focus on managing and maintaining assets, particularly for sanitary sewer collection systems, has forced organizations to focus on reactive emergency actions rehabilitating and replacing the assets, expensively and abruptly, when they fail.

Operating in a reactive mode typically requires utilities to allocate large amounts of resources towards emergency response and replacement or rehabilitation.

Some utilities, though, have developed AMPs to understand their systems' needs and proactively plan for asset maintenance and replacement with a least-cost approach to help ensure a targeted LOS while achieving regulatory compliance.

An AMP in this context is a strategic, comprehensive tool for managing a utility's stormwater and/or wastewater system assets to help minimize the long-term investment in each asset, keeping expenditure at the lowest level that will maintain the desired performance and meet regulatory requirements.

AMPs prioritize the most necessary projects by cataloging assets, identifying performance objectives, completing a life cycle analysis, identifying appropriate maintenance schedules, and conducting a cost-of-failure analysis of all assets (Bonitz et al., 2015).

This exhaustive information-gathering makes it possible to create an extensive timeline for assets by identifying and ranking maintenance needs and listing their costs and potential funding sources.

It can also help guide future planning, reduce the cost of that planning, and identify new system needs that may have gone unnoticed or unrecognized. By promoting resource and financial efficiency, an AMP can more than pay for itself over time.

8.5.4.1 Scoring Assets

Once assets have been identified and all available data centralized, the information can be quantitatively assessed to assign a "score" or rating to help inform future decision-making regarding the assets.

An asset's condition is indicative of its performance and remaining useful life. Scoring an asset's condition is imperative to understand whether it is delivering the desired LOS and how much attention it may need to continue to deliver at the desired level.

Determining remaining useful life is critical to ensuring that an asset continues to meet its performance objectives before unforeseen failure. The remaining useful life values should be refined by regular inspection and performance evaluations as an asset ages. It is important that utilities understand and account for the stressors that the asset was (and will continue to be) subject to when scoring its condition and remaining useful life.

The CoF score indicates the potential for disruption in service and the magnitude of its effect. It informs utilities of which projects should have priority, and most importantly, it illustrates why they should have priority. Calculating an asset's CoF takes into account the social, environmental, and financial consequences of failure, including to community health and safety.

8.5.4.2 Costs of Wastewater Infrastructures

The general form of capital cost estimating equations for conveyance systems, pump stations, storage facilities, water treatment, and WWTPs is

$$C = aX^{b}, \tag{8.7}$$

where

C is the PW of the infrastructure construction cost and

X is the infrastructure size. Parameter C is the total cost, including capital cost plus operation, maintenance, and replacement cost.

The two parameters, a and b, are determined by fitting a power function to the available data. The traditional way to estimate a and b was by plotting the data on log–log paper and finding the parameters of the resulting straight line from the approximation of the data in log–log space. Now, it is simple to find a and b from a least-squares regression that is built into any spreadsheets.

As expressed in Equation (8.7), the economies of scale factor are represented by the exponent b. If b is less than 1.0, then unit costs decrease as size increases. All of these equations illustrated the economies of scale for the output measures of either flow or volume. Pipe flow exhibits very strong economies of scale with $b < 0.5$. The economy of scale factor for treatment plants is about 0.7. A generic economy of scale factor that has been used for years is $b = 0.6$ (Peters and Timmerhaus, 1980).

8.5.4.3 Cost of Stormwater Infrastructures

The cost of treating stormwater varies widely depending on the local runoff patterns and the nature of the treatment. Cost estimates for combined sewer systems are presented in US EPA (1993) for swirl concentrators, screens, sedimentation, and disinfection. Typically, treatment will be combined with storage in order to moderate peak flows and allow bleeding water from storage to the treatment plant. This treatment storage approach can be evaluated using continuous simulation and optimization to find the optimal mix of storage and treatment (Nix and Heaney, 1988). Ambiguities in such an analysis include the important fact that treatment occurs in storage and storage occurs during treatment for some controls (e.g., sedimentation systems).

Owing to the amount of lawn to be watered and the necessity for irrigation water, the cost per dwelling unit (DU) for urban water supply will be mainly variable. In more arid parts of the United States, most of the water entering cities is used for meadow watering. The major factor affecting the variability in wastewater treatment costs is the amount of wastewater. The required lengths of pipe for water supply and wastewater systems can be approximated based on DU and the ratios of the off-site pipe lengths to the on-site pipe lengths. Piping lengths per DU increase if central systems are used because of the longer collection system distances.

Like any other production process, the wastewater treatment sector must optimize the use of the resources necessary to carry out the service (such as energy, personnel, reagents, maintenance). Achieving these criteria implies an efficient management of WWTPs. Not only should WWTP be efficient in terms of contaminant removal, but also in the use of the resources used to achieve this purpose (Ostrom and Wilhelmsen, 2012).

The costs of stormwater systems per DU vary widely as a function of the impervious area per DU and the precipitation in the area. The required stormwater pipe length per DU is almost equal to sanitary sewer lengths for higher-density areas in wetter climates. At the other extreme, very little use is made of storm sewers in arid areas, and runoff is routed down the streets to local outlets. Also, there is a trade-off between the pipe size and the amount of storage provided. Consequently, it is difficult to generalize the expected total cost of stormwater systems (Hernández-Chover at al., 2020).

The term efficiency can have different interpretations depending on the context. In economics, it usually refers to the use of the resources used in a production process. Therefore, the lower the use of resources and the greater the product are, the more efficient the process will be. There exist different methodologies to evaluate the efficiency, which can be divided into two large groups: parametric and nonparametric methods. The first requires establishing the production frontier before the evaluation of the behavior of the units studied. On the contrary, nonparametric models do not require defining the function of the production frontier, being more flexible in terms of the variables used.

Example 8.7

The three alternatives described in Table 8.14 are available for wastewater treatment of an urban area for the next 25 years of their life span. Using 5% discount rate, compare the three projects with the PW method.

TABLE 8.14
Three Alternatives for Supplying a Community Water Supply

Costs	Timing	Project A ($)	Project B ($)	Project C ($)
Initial investment	Year 0	2,000,000	1,000,000	1,500,000
	Year 10	0	1,000,000	1,200,000
	Year 20	0	1,000,000	0
Operation and maintenance costs	Years 1–10	7,000	4,000	6,000
	Years 10–20	8,000	7,000	8,000
	Years 20–25	9,000	9,000	9,000

Solution:

$$\text{PW of costs of project A} = -\$2{,}000{,}000 - \$7{,}000\left(\frac{P}{A},5\%,10\right) - \$8{,}000\left(\frac{P}{A},5\%,10\right)$$

$$\times\left(\frac{P}{F},5\%,10\right) - \$9{,}000\left(\frac{P}{A},5\%,5\right)\times\left(\frac{P}{F},5\%,20\right),$$

$$= -\$2{,}000{,}000 - \$7{,}000\times 7.7216 - \$8{,}000\times 7.7216\times 0.6139 - \$9{,}000$$

$$\times 4.3294\times 0.3769$$

$$= -\$2{,}106{,}390.$$

$$\text{PW of project B} = -\$1{,}000{,}000\times\left(1+\left(\frac{P}{F},5\%,10\right)+\left(\frac{P}{F},5\%,20\right)\right) - \$4{,}000\left(\frac{P}{A},5\%,10\right)$$

$$-\$7{,}000\left(\frac{P}{A},5\%,10\right)\times\left(\frac{P}{F},5\%,10\right) - \$9000\left(\frac{P}{A},5\%,5\right)\times\left(\frac{P}{F},5\%,20\right),$$

$$= -\$1{,}000{,}000\times(1+0.6139+0.3769) - \$7{,}000\times 7.7216 - \$8{,}000$$

$$\times 7.7216\times 0.6139 - \$9{,}000\times 4.3294\times 0.3769$$

$$= -\$1{,}884{,}141.$$

$$\text{PW of project C} = -\$1{,}000{,}000\times\left(1+\left(\frac{P}{F},5\%,10\right)+\left(\frac{P}{F},5\%,20\right)\right) - \$4{,}000\left(\frac{P}{A},5\%,10\right)$$

$$-\$7{,}000\left(\frac{P}{A},5\%,10\right)\times\left(\frac{P}{F},5\%,10\right) - \$9{,}000\left(\frac{P}{A},5\%,5\right)\times\left(\frac{P}{F},5\%,20\right)$$

$$= -\$1{,}500{,}000\times(1+200{,}000\times 0.6139) - \$6{,}000\times 7.7216 - \$8{,}000$$

$$\times 7.7216\times 0.6139 - \$9{,}000\times 4.3294\times 0.3769$$

$$= -\$2{,}335{,}618.$$

The PW of project B is less than that of the others; therefore, it is selected.

8.5.4.4 Wastewater Program Funding

The operation and maintenance of a wastewater utility is typically funded solely by a system's revenue. Some utilities also see a "cash drain" from service revenues collected, which are diverted to fund other "general fund" obligations. Service rates set by the utility will determine the system's

TABLE 8.15
Typical Funding Sources for Wastewater Programs (New Mexico Environmental Finance Center, 2006)

System Revenues	System Reserve Funds	System-Generated Replacement Funds	Nonsystem Revenue
• User fees	• Emergency reserves	• Bonds	• State grants
• Connection fees	• Capital improvement reserves	• Taxes	• State loans
• Stand-by fees	• Debt reserves		• Federal grants
• Late fees			• Federal loans
• Penalties			• State or federal loan/grant combinations
• Reconnect charges			
• Developer impact fees			

revenue, which must also be used to replenish any reserves used for emergency activities. A well-thought-out rate structure will account for system needs in both the current and future years.

Some wastewater utilities have secured and appropriated funding by developing multiannual repair and rehabilitation (R&R) plans for their most necessary projects and concurrent capital improvement plans (CIPs). Funding sources include a variety of user fees (monthly service charges, participation charges from developers, etc.; as well as debt financing options (e.g., bonds)). Table 8.15 mentions some of these typical funding sources.

8.5.4.5 Case Study 2: Asset Management-Based Flood Resiliency of Water Infrastructures

Due to the current worldwide urbanization trend, the population size depending on the near-continuous performance of urban civil infrastructure systems has increased in recent years. Water infrastructures, such as WWTPs, have significant roles in urban systems' functioning. Their performance is directly related to how resilient they are to face and mitigate hazard/flood impacts. Enabling resilience in infrastructure operation and management requires the inclusion of the system's asset base component and preparedness for emergency response and recovery processes. One of the many methods for quantifying resiliency is in the context of four criteria known as 4R's (Robustness, Rapidity, Resourcefulness, and Redundancy). The scope of the study by Karamouz and Movahhed (2021) considered AM as one of the main attributes of robustness and resourcefulness in flood resiliency of water infrastructures. The proposed framework has the potential to be tested on 14 WWTPs in the coastal City of New York. The resilience improvement of these plants could be estimated and compared with different budgets.

The New York City Department of Environmental Protection (NYC-DEP) owns and operates one of the most extensive wastewater collection and treatment systems in the world, with many waterfront facilities that are vulnerable to flooding, as was evident during Hurricane Sandy when several facilities suffered extensive damage. The city's NYC-DEP is in charge of the entire sewage treatment system's operation and maintenance. The NYC-DEP wastewater treatment system includes 14 plants that discharge their effluent into the city's receiving waters. Together, these facilities process over 1.3 billion US gallons (nearly 5 million cubic meters) of wastewater daily (NYC-DEP, 2013).

After superstorm Sandy, the plants' insufficient flood resilience was perceived by NYC-DEP (2013). Studies acknowledge that all 14 wastewater treatment plants and 60% of pumping stations (58 out of 96) are at risk against a 100-year flood event and require additional protection (Balci and Cohn, 2014). In October 2017, the NYC-DEP announced that work is underway on approximately $400 million in resiliency upgrades at the city's critical wastewater collection and treatment facilities to protect them from rising sea levels and storm surge events.

In the proposed methodology, using existing software such as GSSHA and ArcGIS, flood inundation maps could be modeled. The GSSHA model employs the diffusive wave approximation of the Saint-Venant equation to calculate the overland flow from each grid cell into two orthogonal directions and prepares spatial and mapping analysis with raster formats that represent data in different grids of DEM with different land use and soil types.

Then, the resiliency index is quantified of the resilience index for each WWTP. For an accurate estimation of resiliency, a wide range of factors, including hydrological, social, economic, physical, and technical concerns, should be identified. Furthermore, to consider the AM planning's contribution in improving resiliency, each infrastructure's economic status indicated by their financial statements and their level of serviceability for an acceptable level of risk should be included in the study's scope.

For each criterion and subcriterion, a weight representing its overall importance concerning resilience should be assigned. For quantifying resilience, different approaches based on MCDM can be used. PROMETHEE is an efficient decision-making method for a limited set of alternative measures that have been selected and ranked from among the criteria and are often in conflict with each other (Behzadian et al., 2010). PROMETHEE uses geometrical analysis to compare each subcriterion's values for a WWTP with the corresponding values for other plants to combine subjective (weights) and objective (values) data.

By utilizing decision-making techniques such as the analytical hierarchy process (AHP), the weights assigned by the experts for each factor could then be combined so the relative importance of each subcriteria can be determined. As illustrated in Table 8.16, the weights of four R's are not the same since rapidity and robustness play a higher role in determining the systems' resiliency. The proposed parametric approach could then combine values and weights to obtain the resilience index for WWTPs (shown in Figure 8.15).

Rather than aiming to be analytically precise, this method provides relative and indicative measures for quantifying resiliency, aiming to identify those assets that are at risk from flooding and are more threatened in case of failure. These identified assets will be targeted for resiliency improvement via resource allocation.

Here, the optimal improvement of the system's resiliency for a given limited budget is of interest. In doing so, economic factors, which include items that potentially can be improved by investment, will be identified among the subcriteria. For this purpose, subcriteria related to financial/investment characteristics should be identified and targeted. Allocation constraints should be determined for each economic subcriteria (ESC) in an optimization problem. After identifying ESCs and their allocation constraints, optimization will be performed to maximize each WWTP's resilience improvement for a total budget.

In this step, two methods could be investigated to allocate financial resources among the WWTPs. The first method is to maximize overall resilience improvement, and the allocations to the plants are geared towards this goal. In the second method, considering the systems' utilities of the received budgets, the Nash bargaining solution should be used, and the financial allocation will be performed based on maximizing the Nash product.

The initial results show that by including the AMPs in the resiliency framework, compared to previous studies, the resource allocation program for each WWTPs could be done more effectively.

8.5.4.6 Stormwater Program Funding

Stormwater managers often face additional challenges securing funds for program implementation. Though many communities fund stormwater management through property taxes paid into their general funds, stormwater management improvements are often a low priority.

There are alternatives to general funds, though, including service fees based on property type or impervious area, special assessment districts or regional funding mechanisms, system development charges, grants, and low-interest loans. Implementing a stormwater utility fee or user rate structure can be a large barrier for stormwater utilities, especially for municipalities/counties that operate stormwater programs but are not stormwater utilities per se.

TABLE 8.16
Subcriteria Defined for Quantifying WWTPs' Flood Resiliency

Resiliency Criteria	ID	Subcriteria Description	Unit	Weight
Rapidity	Ra_1	Hurricane flood elevation (based on North American Vertical Datum of 1988 (NAVD88))	ft	0.083
	Ra_2	Plant design capacity (a function of the plant users)	MGD.	0.083
	Ra_3	Population served (number of residents that are served by the plant)	#	0.083
	Ra_4	Poststress recovery (refers to any disaster management plan after the flood disaster)	hour	0.074
	Ra_5^*	Adverse environmental impacts on the surrounding area due to treatment failure (level of serviceability)	–	0.019
Robustness	Ro_1	Additional load in time of flooding (the difference between WWTP capacity for the total maximum wet and dry weather flow[a])	MGD	0.046
	Ro_2	Critical flood elevation (100-year flood elevation +30 inches for expected sea-level rise by the 2050s[b])	ft	0.083
	Ro_3	Maximum inundation depth (due to the flat terrain of the plant, several areas may be flooded by up to this value of water during the critical flood event)	ft	0.028
	Ro_4	DMR violations (the percentage of discharge monitoring reports that resulted in effluent violations[c])	%	0.089
	Ro_5	Damage cost from the most severe historical hurricane (without flood protection for the plant)	$	0.088
	Ro_6^*	Assets condition based on their last maintenance check-up	month	
	Ro_7^*	Percentage of critical assets[d] that are not at risk of failure during the time of flooding	%	0.037
Resourcefulness	Rs_1	Availability of dewatering facilities (facilities to drain sludge to decrease 90% of its liquid volume)	–	0.074
	Rs_2^*	Number of plant technical staff (asset operators)	#	0.037
	Rs_3^*	Total risk avoided for every single dollar spent on improving operation and maintenance over 50 years	$	0.028
	Rs_4^*	Percentage increase in total asset's value of each WWTP's financial statement derived from trend analysis	%	
Redundancy	Rd_1	Existence of underground tunnel systems	–	0.065
	Rd_2	Availability of WWTPs in the neighboring areas (distance from the closest WWTP)	km	0.046
	Rd_3	On-site storage (volume of lakes in the WWTP's zone)	ft^3	0.037

[a] Maximum wet weather flow is the maximum flow received during any 24 hours. Maximum dry weather flow is the maximum daily flow during periods without rainfall.

[b] This is determined based on the Federal Emergency Management Agency's new advisory base flood elevation maps for a 100-year flood event, was selected as the baseline for the analysis.

[c] During minimal levels of stress, the Discharge Monitoring Report (DMR) violation percentages indicate how well each treatment plant can cope with daily operational stresses.

[d] These are assets with a low probability of failure but would cause severe consequences to the system performance in case of failure.

* Subcriteria considering asset management.

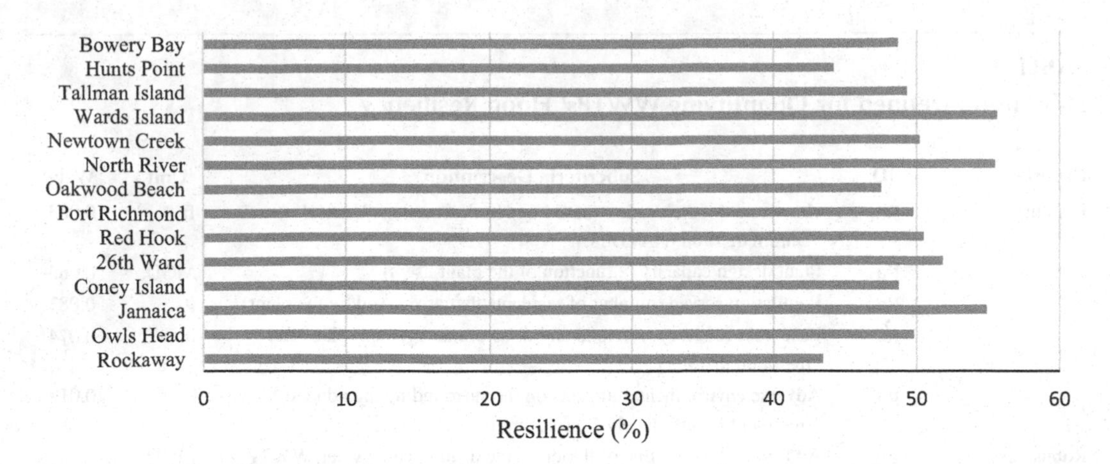

FIGURE 8.15 Estimated resiliency values for 14 WWTPs of NYC utilizing PROMETHEE (Karamouz, M., Rasoulnia, E., Zahmatkesh, Z., Olyaei, M.A., and Baghvand. A. Uncertainty-based flood resiliency evaluation of wastewater treatment plants. *Journal of Hydroinformatics*, 18(6), 990–1006, (2016)).

It is important to inform the public of the inadequacies/deficiencies of the community's current stormwater management program and present the benefits experienced by other communities that use stormwater fees. When an AMP can provide transparent financial and environmental benefits for such funding mechanisms, customers will be more likely to support its implementation.

8.5.5 Tools for Inspecting Water and Wastewater Linear Assets

In this section, examples of state-of-the-art solutions for pipeline condition assessment for proactive AM are presented (Paulson, 2001).

- *Visual Inspection of Internal Condition*:
  - *PureRobotics™ pipeline inspection*: a modular robotic system that can be configured to inspect any pipe with a diameter of at least 300 mm. Capable of providing a variety of high-quality data including closed circuit television (CCTV), profiling SONAR, and laser profiling.
- *Electromagnetic Inspection:*
  - *PipeDiver® platform*: electromagnetic inspection technology for large diameter prestressed concrete cylinder pipe (PCCP), lined cylinder pipe (LCP), and bar-wrapped pipe (BWP).
- *In-Line Leak Detection Systems:*
  - *SmartBall free-swimming acoustic sensor*: used to identify leaks and gas pockets in large diameter pressurized pipe (all material types) with a diameter greater than 150 mm. Consisting of a sensor and data-recording device mounted within a foam ball, the technology is capable of detecting leaks less than 0.026 L/h.
  - *Sahara tethered acoustic sensor*: consisting of a sensor attached to the end of a cable tether, the device allows for extended listening for defects at a particular location along the pipe length and can be removed when unexpected flow conditions are encountered.
  - *SmartBall pipe wall assessment (PWA) tool*: used to identify wall stress in metallic pipelines. Capable of performing long inspections without disruption of regular pipeline service, this free-swimming technology can precisely indicate the position and severity of the damage.
  - *Near real-time risk assessment of large diameter PCCP mains*: proven to be the most reliable acoustic fiber optic monitoring technology for monitoring acoustic

activity associated with the failure of prestressing wires in water and wastewater PCCP pipelines with a diameter larger than 600 mm.
- *Integrated nonrevenue water and asset management software*: developed to allow municipalities to manage their water and wastewater system data.
- Although existing technologies have traditionally focused on the inspection of large diameter pipes, they are expanding their available suite of technologies to include options for assessing the condition of smaller diameter distribution mains.

8.5.5.1 Underground Infrastructure

This section is described through the activities of the Centre for Advancement of Trenchless Technologies (CATT), which was founded in 1994 in Waterloo, Ontario, to help municipalities address critical issues facing underground infrastructure installation, assessment, repair, renewal, and management, which is located at the University of Waterloo.

CATT is a grouping of university, municipal, industrial, business, and government agencies with a mandate to "address infrastructure needs faced by both public sector end-users and industry by providing a common forum for research, education and technology transfer in the area of trenchless technologies" (CATT, 2014).

It frequently hosts technical workshops, short courses, and conferences to provide various end users with an opportunity to both contribute and learn about advanced underground infrastructure management practices. CATT has been working with Canadian municipalities to develop an integrated water and wastewater-specific decision support tool based on system dynamics. Recent publications in this field include:

System Dynamics Model for Financially Sustainable Management of Municipal Water Main Networks (Rehan et al., 2013): provides the first known causal loop diagrams for financially sustainable water systems. These diagrams illustrate feedback loops involving the physical condition of assets, consumer behavior, and finances. The developed system dynamics model captures cost drivers and revenue sources in a system and can be used to achieve a variety of short- and long-term objectives and develop defensible policies.

Financially Sustainable Management Strategies for Urban Wastewater Collection Infrastructure—Implementation of a System Dynamics Model (Rehan et al., 2014): describes how a system dynamics model can be used by utility managers to ensure financial sustainability while maintaining customer expectations for service performance. Demonstrates the interrelationships that exist between asset condition, total life cycle costs, user fees, etc., for a case study area of a medium-sized municipality in Ontario.

http://cattevents.ca/ CATT is heavily involved in research in such areas as buried infrastructure AM, condition assessment of water and wastewater infrastructure, etc. One recent project includes the Plastics Pipe Institute's Pipeline Analysis and Calculation Environment (PPI-PACE): a free online tool developed to help industry professionals complete design calculations for high-density polyethylene (PE) pipe used in pressurized water distribution and transmission systems (PPI-PACE, 2014).

8.5.5.2 Check-Up Program for Small Systems (CUPSS)

Check-Up Program for Small Systems (CUPSS) is a legacy software application that was developed more than 10 years ago to assist drinking water and wastewater utilities in implementing an AMP.

The US EPA developed CUPSS as a user-friendly desktop application to promote the integration of AM activities into utility practices.

CUPSS leads users through a series of modules to collect information on the utility's assets, operation and maintenance activities, and financial status to produce a prioritized asset inventory, financial reports, and a customized AMP.

CUPSS is a free software that makes managing assets easier. It is a desktop system, so no Internet connection is needed, and the software requirements are minimal. CUPSS was designed and developed with input and suggestions from a diverse stakeholder group.

CUPSS is designed to help you establish a successful AM program.

Three important components of AM are a comprehensive list of current assets, including information on their condition and useful life; an understanding of the daily/monthly/yearly tasks to maximize the useful life of the assets; and a clear organization system for financial records, which helps identify trends and determine the full cost of doing business.

Successful AM programs are characterized by a commitment to allocate people, time, and other resources to implement the program:

- Focus on making cost-effective asset decisions
- Provide a sustainable LOS for the community

8.5.5.3 Benefits of Using CUPSS

CUPSS is designed to help small water and wastewater utilities support budget discussions with solid facts and numbers, boost the efficiency of the utility, save your staff time, and improve customer service by ensuring continual service at competitive prices.

It helps to determine the current state of the assets in your utility, the LOS you are aiming to uphold, which of your assets are critically important, what the minimum LCC is, and what your long-term funding strategy is.

CUPSS also helps to prepare work orders and an AMP. An AMP provides valuable information that you can use when you make management decisions about your utility.

CUPSS provides a snapshot of the utility that allows someone with little knowledge to jump in and understand the state of the utility. The generated work orders can be used to help keep operation and maintenance, as well as compliance tasks, on schedule.

The User's Guide is designed to help you work with CUPSS:

https://www.epa.gov/sites/production/files/2015-10/documents/cupssusersguide.pdf

If the CUPSS application is downloaded from its website, a copy of the installation is saved on the computer. In order to navigate that file and open (double-click or right-click and select "Open") the cupss_install.exe file. This launches the installation wizard.

8.6 FINANCING METHODS FOR INFRASTRUCTURE DEVELOPMENT

An essential ingredient of developing and maintaining viable urban water organizations is funding. Integrated urban water management offered the promise of improved economic efficiency and other benefits from the merging of multiple purposes and stakeholders. However, the benefits from integrated watershed management exacerbate problems of financing these more complex organizations because ways must be found to assess a "fair share" of the cost of this operation to each stakeholder (Heaney et al. 1997). Nelson (1995) provides a current overview of utility financing in the water, wastewater, and stormwater areas.

The main financing methods for urban infrastructures are (Debo and Reese, 1995)

1. Tax-funded systems
2. Service charge-funded systems
3. Exactions and impact fee-funded systems
4. Special assessment districts

8.6.1 Tax-Funded System

Usually, the public works department of a city (or similar organizations related to urban infrastructure development) is charged with maintaining and improving urban water infrastructure systems. Projects are funded through the budget of the department, whose source is mainly property tax revenue. However, if property taxes are used, then the infrastructure systems must compete for funds directly with public safety, schools, and other popular programs.

8.6.2 Service Charge-Funded System

In the service charge-funded system, an algorithm that divides the budget for the urban water infrastructure systems by some weighting of the demand for service should be used. This new funding method is being implemented because it has the advantage of separating the funding necessities according to the function on a user pays basis.

8.6.3 Exactions and Impact Fee-Funded Systems

System development charges have emerged as the way to compute the charges to be levied against new developments. This system charges the developer or builder an up-front fee that represents his equity buy-in to the urban water infrastructure systems. Usually, this fee is calculated as a gauge of the depreciated value of the system, plus system-wide funding needs minus the existing users' share. The fee must be reasonable to avoid court challenges. Nelson (1995) defines the rational nexus test of reasonableness of system development charges. These test requirements are as follows:

- Establishment of a connection between new development and new or expanded facilities required to accommodate such development. This establishes the rational basis of public policy.
- Identification of the cost of those new or expanded facilities needed to accommodate new development.
- Appropriate apportionment of that cost to new development in relation to benefits it reasonably receives.

An important feature of this method is the ownership, or equity issue, of existing users. Usually, the existing users are grouped into one class for ease of calculation; however, in actuality, different groups joined at different points in time. At the time of joining, some contractual agreement (written or unwritten) is initiated. A key weakness of the impact fee system is keeping track of these agreements over time and space when setting impact fees is extremely difficult. Because of this added database need and the wide variation in cost allocation techniques for apportioning costs, there can be wide fluctuations in impact fee computations. These shortcomings can be overcome, however, with better accounting and tracking of information.

8.6.4 Special Assessment Districts

This system funds the needs within a designated geographical area by dividing the funds, usually equally, among the parcels within the area. The calculation methods are inherently simple and, usually, the benefits and costs are roughly equally distributed.

In Example 8.8, the optimal planning for capacity building of urban water infrastructure due to finance limitations is presented.

Example 8.8

The variation of annual wastewater generation of a city over a 30-year planning time horizon is tabulated in Table 8.17. Assume that the wastewater generation of this city in the year 2003 was about 90 MCM. Three projects have been proposed to treat wastewater projected generation until 2040. As presented in Table 8.18, three projects are studied for wastewater treatment of this city. The capital investments presented in Table 8.18 are based on prices of the year 2003. Considering the cost of each project and the 3% rate of return, find the optimal sequence of implementing projects using linear integer programming.

TABLE 8.17
Variation of Annual Amount of Wastewater of a City (Example 8.8)

Year	2010	2020	2025	2035	2040
Wastewater (MCM)	100	120	130	145	170

TABLE 8.18
Description of Proposed Projects (Example 8.8)

Project	Capital Investment (10^6 \$)	Capacity (MCM)
Treatment plant A	5	35
Treatment plant B	8	20
Treatment plant C	7	35

Solution:

The formulation of the optimization model based on the proposed methodology for capacity building based on the previous chapter is as follows:

$$\text{Minimize} \quad C = \sum_{i=1}^{3} \sum_{j=1}^{5} X_{i,j} \cdot \text{Cost}_{i,j}.$$

Subject to

$$\text{Cap}_I \times X_{I,J} \geq D_J.$$

$$X_{ij} = \begin{cases} 1 & \text{if project } i \text{ is constructed in time period } j, \\ 0 & \text{otherwise,} \end{cases}$$

where

$\text{Cost}_{i,j}$ is the construction cost of project i at the jth time period,
$X_{i,j}$ is the decision variable for construction of project i at the jth time period,
Cap is the capacity of project i as mentioned in the third column of Table 8.18, and
D_j is the water demand of the city at time period j (Table 8.19).

TABLE 8.19
Present Value of the Initial Investment in the Projects

Year	Project A	Project B	Project C
2003	5	8	7
2010	4.065	6.5	5.69
2020	3.02	4.84	4.23
2025	2.61	4.175	3.65
2035	1.94	3.11	2.72
2040	1.68	2.68	2.34

In order to estimate the present value of investment in year t of the planning time horizon, the following equation can be used:

$$P = \frac{F}{(1+i)^t}.$$

In order to estimate the present value of investment in year t of the planning time horizon, the Single payment present worth equation (Table 8.3) can be used. Table 8.19 shows the present value of the initial investment in the projects for the years in which water demand has increased. Then a linear programming routine is written to determine the best timing for projects construction. According to the results of linear programming that is formulated in LINGO software (Figure 8.16), the optimal projects implementation sequence is: build project A in 2003, project C in 2020, and project B in 2040.

8.7 ASSESSING THE ENVIRONMENTAL PERFORMANCE OF URBAN WATER INFRASTRUCTURE

Sustainable development index (SDI) should be used in the international, regional, and local scales for assessing the performance of urban water infrastructure. At a corporate level, the concept of SD is often limited to an improvement of environmental performance. An agency or an organization can turn to different methods to get information, which allows an understanding and improvement of its environmental performance. Examples of tools used for SDI assessment in urban infrastructure are LCA and environmental performance evaluation [by using special PIs]. While LCA is a technique for assessing the environmental aspects associated with a product and/or a service system, environmental performance indicators (EPIs) focus on those areas that are under direct control by the agency/organization. Upstream and downstream activities are seldom included. The system boundaries for LCA and EPI are therefore not usually comparable. This section briefly describes how these two methods can be used to assess the environmental performance in general and an urban water system in particular and how this affects decisions on which upstream and downstream processes to include in the assessment of the system.

Within the urban water sector, PIs are being developed and used in several countries. A major objective of water utilities is to improve the quality of the service while keeping down the costs. Most of the PIs have been developed as a tool for monitoring these objectives. Environmental aspects have also been recognized as important recently. An international workshop was held in 1997 by the International Water Services Association in order to find a common reference set of PIs for water supply (IWSA, 1997).

The relative importance of environmental indicators depends on local and regional factors. If water is scarce, then the specific water use and leakage are highly relevant in order to save water. Where water is in abundance, other indicators are more important. The efficiency indicators have been ranked as having moderate importance, since these are not directly related to the objectives of the water sector but still apply to the definition of a sustainable urban water system.

An indicator should be clearly understandable, without long explanations. Percentages or dimensionless quotas are fairly easy to relate to, whereas an indicator such as drinking water quality is more difficult, at least for the public, and should therefore be related to a water quality standard. Ease of understanding of indicators should be determined by the users, and this aspect still needs to be investigated.

The use of monetary indicators is a common procedure for companies wishing to monitor their performance or utility efficiency in comparison with others. In recent years, the use of EPIs has become increasingly popular within industry to describe agency pressure on the environment. In order to harmonize efforts, an international standard for environmental performance evaluation has been proposed and will be included in an ISO standard (ISO 14031, 1998). The objective of the ISO standard is to provide guidelines on the choice, monitoring, and control of EPIs. EPIs are

Lingo 18.0 - Lingo Model - Lingo1

ile Edit Solver Window Help

Lingo Model - Lingo1

```
MODEL:
sets:
j1/1..5/:j, demand;
i1/1..3/:i, capacity;
obs1(i1,j1):x, Cost;
k1n..2/:k•,
h2/1..4/:h;
ENDSETS
!************************************;
! initial values;
!************************************;
!************************************;
!Required data
!cost, capacity and demand
!************************************;
DATA:
cost= 4.05 3.02 2.61 1.941.68
6.5 4844.175 3.11 268
5.69 493 3.65 2.72 234,
capacity= 35 20 35
demand=100 120 30 1450 170
ENDDATA
!************************************;
! Objecüve Randon;
!************************************;
MIN = @SUM(j1(i):
@sum(i1(i):X(i,j))*cost(i,j)););
!************************************;
! Constraints;
!************************************;
! Only one project at each duration;
@for (i1(i).@sum(j1(j).X(i,j)-1);
! Each project is only implemented one time ;
@for (j1(j).@sum(i1(i).X(i,j)<=-1);
!Binary variable definition;
@for(j1(i):
@for(i1(i):@bin(x(i,j))););
! Demand allocation increasing during ihe planning horizon;
@sum(i1(i):x(i,5)*cepacity (i)))+
@sum(i1(i):x(i,4)*cepacity (i)))+
@sum(i1(i):x(i,3)*cepacity (i)))+
@sum(i1(i):x(i,2)*cepacity (i)))+
@sum(i1(i):x(i,1)*cepacity (i)))-10)-20)-10)
-15)>=25;
@sum(i1(i):x(i,4)*cepacity (i)))+
@sum(i1(i):x(i,3)*cepacity (i)))+
@sum(i1(i):x(i,2)*cepacity (i)))+
@sum(i1(i):x(i,1)*cepacity (i)))-10)-20)-10)
>=15;
@sum(i1(i):x(i,3)*cepacity (i)))+
@sum(i1(i):x(i,2)*cepacity (i)))+
@sum(i1(i):x(i,1)*cepacity (i)))-10)-20)>=10;
@sum(i1(i):x(i,2)*cepacity (i)))+
@sum(i1(i):x(i,1)*cepacity (i)))-10)>=20;
@sum(i1(i):x(i,1)*cepacity (i)))=10;
```

NUM OVR Ln 57, Col 1 2:18 pm

FIGURE 8.16 Formulation of Example 8.8 in LINGO software.

categorized into three types: operational PIs, environmental condition indicators, and management PIs. In the ISO standard, it is emphasized that the indicators should be selected based on their ability to measure performance against the environmental targets set by the company. It is also stated that a life cycle approach should, if possible, be used to select indicators.

Another concept that has been used is the term "ecoefficiency," that is, producing more value with less environmental impact (Schaltegger, 1996). The prefix "eco" refers to both ecological and economic efficiency. Keffer et al. (1999) presented a set of defined general eco-efficiency indicators, as well as specifying guidelines for the selection of sector-specific indicators.

8.8 CRITICAL INFRASTRUCTURE INTERDEPENDENCIES

The critical infrastructure needs have changed significantly both in the wake of disasters such as the terrorist attacks of 9/11, Hurricane Katrina, and Superstorm Sandy as a result of emerging technologies. It is important to understand critical infrastructure's interdependencies, high-tech opportunities, and the need for long-term investments. Other key aspects are interoperability, real-time monitoring, intelligent networks, and effective modeling and simulation. Underlying all of these is the necessity of communication and education among the various stakeholders—utilities, federal and local governments, businesses, communities, and, of course, engineers. Thinking about critical infrastructure through this subset of lifelines helps clarify common features and provides an effective framework for understanding interdependencies among the different systems (Karamouz and Budinger, 2014).

Critical infrastructure is shaped and characterized by three specific features. First, much of it, especially in cities, is located underground, where it is removed from direct observation unless uncovered, and its state of repair and proximity to other structures is often unknown. Figure 2.16 in Chapter 2 shows complexities of special interdependencies of underground infrastructure in New York City and other major cities that are challenging and costly to touch because of numerous underground structures that compete for limited space for water, sewer, electricity, and natural gas.

Urban congestion increases risk due to the close proximity of many pipelines, cables, and supporting facilities. Damage to one facility, such as a cast-iron water main, can cascade rapidly into damage in surrounding facilities, such as electric and telecommunication cables and gas mains, with system-wide consequences.

Recent flood events, such as Hurricane Katrina and Superstorm Sandy in the United States, have demonstrated the role of interdependencies in cascading failures (Leavitt and Kiefer, 2006; Comes and Van de Walle, 2014; Sharkey et al., 2016). Interdependency is a linkage connecting two infrastructures in a way that the state of one infrastructure influences the state of the others (Rinaldi et al., 2001).

A new paradigm emerged after Hurricane Katrina, centered on the concept of resilience, and much has been written and discussed about this concept. In current parlance, the resilience of an organization or community is an overarching attribute that reflects its degree of preparedness and ability to respond to and recover from sudden shocks. The term has become the scaffolding on which to build a community or organization that is well prepared and responsive to a wide range of demands, including natural hazards and human threats.

Hurricane Sandy's impacts on many infrastructure systems and their recovery were extensive, exacerbated by interdependencies given the density of the region is affected and the unusual conditions that led to record levels of storm surge in coastal areas. These impacts occurred not only because of the direct hit by floodwaters but also from electric power outages. Not only did the infrastructure damage affect electric power, but electric power outages also affected transportation and wastewater treatment and distribution systems (Zimmerman et al., 2017).

Interoperability is another key feature of modern infrastructure. Electric power, natural gas, water, oil, telecommunications, and transportation are all interdependent; as illustrated in Figure 8.17. Electric power is essential for the reliable operation of virtually all other infrastructure systems.

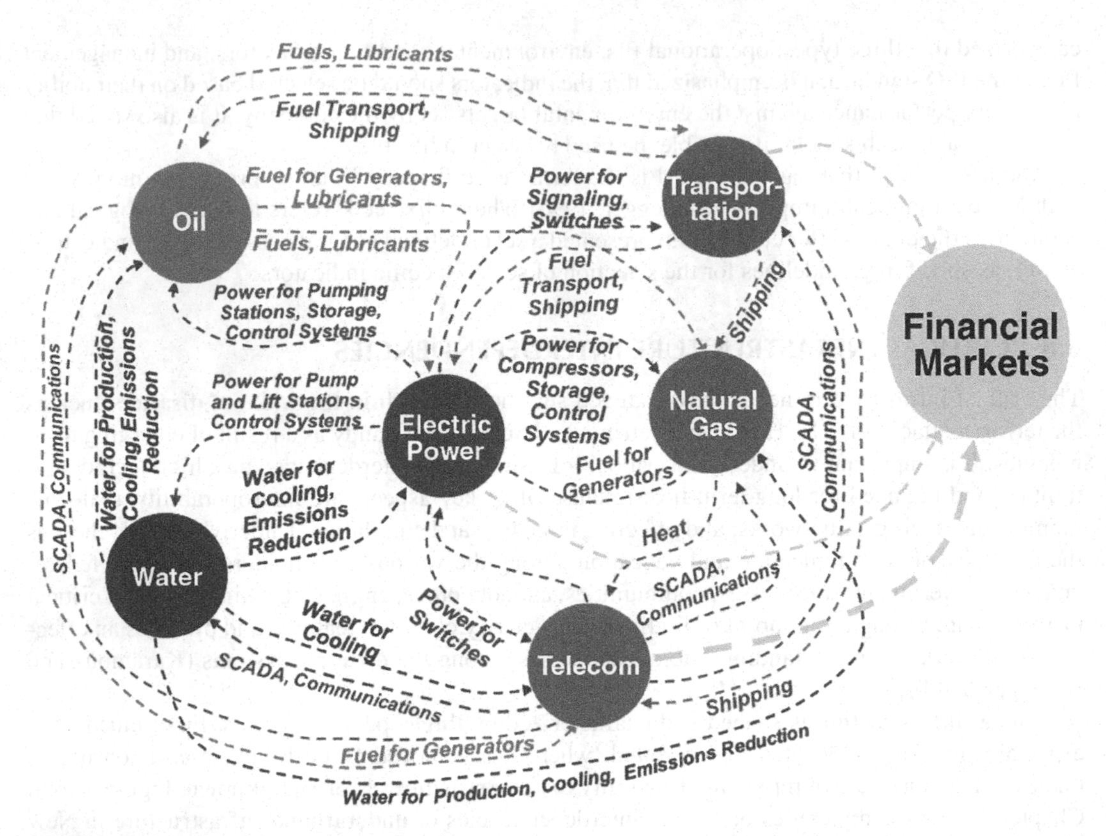

FIGURE 8.17 Infrastructure interoperability. (After Peerenboom, J., Fisher, R., and Whitfield, R., Recovering from disruptions of critical independent infrastructures. Presented at the *CRIS/DRM/IIT/NSF Workshop on Mitigating the Vulnerability of Critical Infrastructures to Catastrophic Failure*, September, Washington, DC, 2001.)

During extreme events, electric power is, in effect, the gateway for local damage to escalate or cascade into other systems.

Rinaldi et al. (2001) identify four principal classes of interdependency: physical, cyber, geographical, and logical. The physical class of dependency refers to assets whose state is dependent on the output of another. For example, a sludge treatment facility may be physically dependent on the material produced by a wastewater treatment plant. Cyber dependencies exist if the functioning of an asset depends on information infrastructure. Such a dependency particularly applies to automatically controlled assets, such as drinking water supplies controlled by the water grid system or flood warning systems triggered by water level gauges. Geographical dependencies exist when a local event affects assets that together form a critical resource. An example of efforts to minimize risk from geographical dependencies was seen after the hurricanes Rita and Katrina in 2005 when the US Energy Policy Act facilitated the construction of new natural gas import terminals in geographically diverse ports (Streips and Simpson, 2007). Logical dependencies cover a range of alternative dependencies, principally relating to human decisions and resources. For example, the availability of staff, pumps, and flood defense equipment may depend on the resources being used elsewhere.

As well as classifying the type of dependency, Rinaldi et al. (2001) referred to the strength of dependency as a 'tight' or 'loose' coupling. This describes the speed with which an asset is affected by a failed dependency: the more immediate the impact, the tighter the dependency. For example, a water treatment plant without a backup generator might have a 'tight' dependency on the electrical supply, but if there is a generator on-site that can keep the site running in case of electricity cut, the dependency would instead be 'loose.'

Like other service providers, water utilities rely on infrastructure networks. As well as their own assets, this includes infrastructure that, for example, provides supplies of power and chemicals and the communication and transport networks. Hence, the scope of evaluating flood risk to water and wastewater service provision should ideally extend beyond individual water utility assets to the broader infrastructure network (Emanuelsson et al., 2014).

Modeling the damage and recovery of critical infrastructure systems by considering different levels of interdependencies is essential for community resilience planning against catastrophic events. However, infrastructure damage and recovery are not deterministic processes due to their inherent uncertainties. It is important to reflect these uncertainties and report the variations in the infrastructure performance estimation to better guide the strategic disaster risk management (He and Cha, 2019). In general, the existing approaches to model the performance of interdependent infrastructure systems under disruptive events can be grouped into six types: empirical-based, agent-based, system dynamics-based, economic-based, network-based, and others (Ouyang, 2014).

8.8.1 Restoration of Interdependent Assets

Restoration efforts, by their nature, involve scheduling resources to activities that restore or repair damaged components in an infrastructure, install new (temporary) components within an infrastructure, or produce some level of functionality within the infrastructure. The term resource is broadly defined in the sense that it could model work crews, machines (e.g., pumps or generators), or individual personnel. Much like how the operations of infrastructures depend on other infrastructures, the restoration efforts of an infrastructure are impacted by the restoration efforts of other infrastructures (Sharkey et al., 2016).

8.9 CONCLUDING REMARKS

In this chapter the principles of engineering economics and their application to urban water resources planning and operation are discussed. Major objectives of economic analysis for water resources development projects and various methods for incorporating the cash flow relationship are presented. AM principles described emphases on AM drivers, sustainable service delivery, risk and consequences, and guidelines to select tools and practices that could help municipalities to manage their infrastructure assets.

Due to the increase in the expansions of water infrastructures for urban communities and the adverse environmental impacts of these projects, IAM of urban water systems will be increasingly critical in the coming decades. This paradigm ensures the performance of water infrastructures and their efficiency in supplying the desired service while minimizing the costs of O&M and capacity expansion projects.

So, regardless of their size, complexity, and level of maturity or development, water utilities need to implement structured IAM approaches that may ensure the sustainable management—Infrastructure Asset Management of Urban Water Systems meant of their systems. There are some key recommendations to be taken into account for successfully implementing an IAM program:

- An efficient AM program should have specific attributes in order to deliver satisfying results.
- A system view of maintaining and protecting assets as it affects the resiliency of the system will have a profound impact on the risk of failure and withstanding the consequences of extreme events, including disasters.
- IAM is not implemented overnight. It is an incremental, step-by-step process that must be kept as simple as possible with a long-term view. The structured approach recommended in this chapter aims at identifying intervention priorities.

- Defining and measuring levels of service is a key activity in developing infrastructure AMPs. Levels of service may be tied to the physical performance of assets or be defined via customer expectation and satisfaction.
- Reliable data are the foundation of successful IAM. Before investing in new data collection, it is vital to get the most out of the existing data through proficient recycling, quality control, analysis, and interpretation.
- Effective decision-making for managing assets is all about managing risk.
- As the infrastructure ages, O&M costs increase. So it is important to make plans for optimized O&M based on AM before the cost reduction opportunities in the life span phases of the infrastructure diminish. This is available in different phases of LCC analysis.
- Condition assessment is a critical component of the AM process as it provides the data necessary to evaluate the risk of failure of individual assets. In this way, it helps utilities make informed decisions and is one of the most effective strategies for allocating funds for the repair, maintenance, and replacement of existing water and wastewater assets.

Throughout this chapter, it is demonstrated how AM can facilitate infrastructure performance corresponds to service targets over time. It helps to make sure risks are adequately managed and that the corresponding costs the life cycle are minimized. Lack of sound economic, regulatory frameworks and enforcement setup, as well as poor AM practices, particularly underpriced water services, are common problems throughout the developing regions. The urban systems' governing bodies need plans to prioritize limited resource allocation. Therefore, this allocation should be in line with the current structure or reformed infrastructure AM.

In AM, effective decision-making requires a comprehensive approach that ensures the desired performance at an acceptable risk level, considering the costs of building, operating, maintaining, and disposing of capital assets over their life cycles. Sustainable management of the system's resources should respond to the growing need for financial stability and sound cash flow and investment strategies in capacity expansion and resource generation.

PROBLEMS

1. Consider the two mutually exclusive alternatives A and B for a water diversion project as described in Table 8.20. Assume that the study period is 15 years and repeatability is applicable. Using the annual worth method, which project do you recommend?
2. A river supplies water to an industrial complex and agricultural lands located downstream of the complex. The average monthly industrial and irrigation demands and the monthly river flows in a dry year are given in Table 8.21 (numbers are in million cubic meters). The prices of water for industrial and irrigation uses are $80,000 and $9,000 per million cubic meters.
   i. Formulate the problem for optimizing the water allocation from this river.
   ii. Solve the problem using linear programming.

TABLE 8.20
Characteristics of Considered Projects in Problem 3

	Project A	Project B
Capital investment	$160,000	$240,000
Annual revenue	$80,000	$112,000
Annual expense	$35,200	$68,800
Useful life (years)	8	12

TABLE 8.21
Average Monthly Industrial and Irrigation Demands and the Monthly River Flows in a Dry Year of Problem 4

	Jan	Feb	Mar	Apr	May	Jun	Jul	Aug	Sep	Oct	Nov	Dec
Industrial demand	8	8.5	8.5	9	9	9.5	9.5	10	9.5	9	8.5	8.5
Irrigation demand	0	0	20	50	55	70	70	65	20	10	0	0
Flow	22.5	24.3	36	43.2	54	37.8	22.5	15.3	12.6	16.2	18	22.5

3. The variation of annual water demand of a city over a 30-year planning time horizon is exhibited in Table 8.22. The current demand of this city is about 90 million cubic meters. Three projects are studied in order to supply the demands of this city. Considering the cost of each project given in Table 8.23 and 3% rate of return, find the optimal sequence of implementing projects using linear programming.
4. Assume that you want to model the water demand of a city in order to forecast the municipal demands in the coming years. The historical data of water use, population, price, and precipitation in a 10-year time horizon are presented in Table 8.24. Formulate a multiple regression model for estimating the water use data and comment on selecting the independent variables.
5. Derive cash flow statement from the given balance sheet (Table 8.25) and income statement of a firm (Table 8.26). Then, complete the balance sheet and check it. Analyze the financial position of the firm. Write your assumptions if there are any. Present the details of your computations. (Hint: Amortization expenses equals $5m. Assume all dividends were cash dividends.)
6. A water pump currently has a resale value of $25,000 and is estimated to have a resale value of $18,000 in 1 year's time. The operating and maintenance cost of the pump for the coming year is $3,500, assumed to be incurred at the end of the year. The life cycle equivalent annual cost of a new pump is estimated at $10,000/year. Should the old pump be replaced now?
   a. Assuming zero interest; and
   b. Assuming interest at 10% p.a.

TABLE 8.22
Variations of Annual Water Demand of the Considered City in Problem 5

Year	2008	2016	2025	2035	2040
Water demand (MCM)	110	125	135	150	170

TABLE 8.23
Cost of Considered Projects for Water Supply of the Considered City in Problem 5

Project	Capital Investment (10^6 $)	Capacity (MCM)
Dam A	4	30
Interbasin water transfer tunnel B	7	17
Dam C	6	30

TABLE 8.24
Historical Data of Water Use, Population, Price, and Precipitation in a 10-Year Time Horizon of Problem 6

Year	Population	Price (Monetary Unit/m³)	Annual Precipitation (mm)	Water Use (m³/year)
1991	21,603	0.25	812	2,032,410
1992	22,004	0.27	648	2,203,921
1993	23,017	0.29	670.4	2,178,329
1994	23,701	0.3	772	2,265,816
1995	24,430	0.31	670.4	2,323,782
1996	25,186	0.32	629.6	2,403,752
1997	26,001	0.33	576	2,647,942
1998	26,825	0.35	556	2,706,106
1999	27,781	0.31	656	2,700,314
2000	28,576	0.34	640	2,725,007

TABLE 8.25
The Balance Sheet of an Example Company

	Assets	2017	2018
Current assets	Cash	34.6+P	?
	Accounts receivable	74.8	82.4
	Inventory	132.6	145.7
	Total current assets	?	?
Fixed (long-term) assets	Property and equipment	212.7	259.3
	Goodwill & Intangible assets	34.2	42.5
	Total fixed assets	?	?
Other assets	Deferred income tax	12.6	34.6
	Other	123.9	145.3
	Total other assets	?	?
	Total assets	?	?
	Liabilities and owner's equity		
Current liabilities	Accounts payable	120.5	130.5
Current liabilities	Notes payable	23.5	34.2
	Current portion of long-term debt	49.6	68.4
	Total current liabilities	193.6	233.1
Long-term liabilities	Long-term debt	214	214
Long-term liabilities	Deferred income tax	43.5	52.3
	Total long-term liabilities	?	?
	Total liabilities	?	?
	Owner's equity		
	Preferred stock	12.4	13.5
	Common stock per value	5.6	7.4
	Additional paid-in-capital	146.4	163.5
	Treasury stock	112.3	123.2
	Retained earnings	122.2	129.9
	Total owner's equity	?	?
	Total liabilities and owner's equity	?	?

TABLE 8.26
Income Statement the Company

Income Statement	
Net revenue	1340.3
Costs of goods sold	423.4
Selling, general, and administrative	123.4
Research and development	56.3
Depreciation and amortization	20
Other operating expenses	12.3
Net interest expenses	45.6
Pretax income	650.3
Income tax expense	227.6
Net income	422.7
Dividends	31.4+P

TABLE 8.27
Available Data of the Current Year for the Company

Equipment Type	Routine Working Hours	Nonroutine Working Hours	Spares and Consumables
Water mains	30,000	10,000	195,000
Wastewater mains	25,000	12,000	185,000
Pumps	28,000	15,000	205,000

7. A company maintains a large network of water and wastewater equipment throughout a city. You are creating a maintenance budget for the coming year. Table 8.27 shows the data available for the current year.

   The direct working cost rate is $30/hour with a multiplier for on-costs and overheads of 2.75. For next year, an increase in activity of 24% has been forecast for all equipment types. A contingency allowance of 15% is applied to the basic budget. What should the total maintenance budget be for the coming year?

REFERENCES

AECOM. (2013). *National Water and Wastewater Benchmarking Initiative – Public Report* (p. 87). AECOM. http://www.nationalbenchmarking.ca/docs/Public_Report_2013.pdf.

Alegre, H., Coelho, S.T., Almeida, M.D.C., Cardoso, M.A., and Covas, D., 2012. An integrated approach for infrastructure asset management of urban water systems. *Water Asset Management International*, 8(2), 10–14.

Alegre, H., Coelho, S.T., Covas, D., Almeida, M.D.C., and Cardoso, M.A. (2013). A utility-tailored methodology for integrated asset management of urban water infrastructure. *Water Science and Technology: Water Supply*, 13(6), 1444–1451.

Asian Development Bank. (2001). *Water for All: The Water Policy of the Asian Development Bank*. Asian Development Bank, Manila.

Asian Development Bank. (2005). *Asia Water Watch 2015: Are Countries in Asia on Track to Meet Target 10 of the Millennium Development Goals? (Summary)*. Asian Development Bank, Manila.

Au, T. and Au, T.P. (1983). *Engineering Economics for Capital Investment Analysis*. Allyn and Bacon, Boston, MA, p. 540.

Balci, P. and Cohn, A. (2014). NYC wastewater resiliency plan: Climate risk assessment and adaptation. *ICSI 2014: Creating Infrastructure for a Sustainable World*, New York City, pp. 246–256.

Behzadian, M., Kazemzadeh, R.B., Albadvi, A., and Aghdasi, M. (2010). PROMETHEE: A comprehensive literature review on methodologies and applications. *European Journal of Operational Research*, 200(1), 198–215.

Bonitz, P., Desai, J., Henderson, B., Lam, B., Powell, J., and Wong, D. (2015). *Asset Management: Balancing Program Constraints*. WE&T (September). http://www.waterenvironmenttechnology-digital.com/waterenvironmenttechnology/september_2015?sub_id=2tgOKY5RCk1Q&pg=80#pg80 (Accessed 27 January 2016).

CATT. (2014). Introduction to CATT. [cited 2014 October 15]. Available from: http://cattevents.ca/introduction/.

Comes, T. and Van de Walle, B.A. (2014). Measuring disaster resilience: The impact of hurricane sandy on critical infrastructure systems. *ISCRAM*, 11(May), pp. 195–204.

Debo, T.N. and Reese, A.J. (1995). *Municipal Storm Water Management*. CRC Press/Lewis Publishers, Boca Raton, FL.

DeGarmo, E.P., Sullivan, W.G., Bontadelli, J.A., and Wicks, E.M. (1997). *Engineering Economy* (647 pp.). Prentice-Hall, Upper Saddle River, NJ.

Emanuelsson, M.A.E., Mcintyre, N., Hunt, C.F., Mawle, R., Kitson, J. and Voulvoulis, N. (2014). Flood risk assessment for infrastructure networks. *Journal of Flood Risk Management*, 7(1), 31–41.

FCM. (2002). *Developing Levels of Service – A Best Practice by the National Guide to Sustainable Municipal Infrastructure*. Federation of Canadian Municipalites and the National Research Council. https://www.fcm.ca/Documents/reports/Infraguide/Developing_Levels_of_ Service_EN.pdf.

Field, B.C. (1997). *Environmental Economics: An Introduction*. McGraw-Hill, New York.

Ghosh, D. and Sandip, R. (2009). Maintenance optimization using probabilistic cost benefit analysis. *Journal of Loss Prevention in the Process Industries*, 22(4), 403–407.

Gouldby, B. and Samuels, P. (2005). *Language of Risk, Project Definitions*, FLOODsite Project Report T32-04-01. FLOODsite Consortium, Oxfordshire.

Harvey, R. (2015). *An Introduction to Asset Management Tools for Municipal Water, Wastewater and Stormwater Systems*. The Canadian Water Network, Waterloo, ON.

He, X. and Cha, E.J. (2019). Modeling the post-disaster recovery of interdependent civil infrastructure network. *13th International Conference on Applications of Statistics and Probability in Civil Engineering, ICASP 2019*, Seoul, Korea.

Heaney, J.P. (1997). Cost allocation in water resources. In C. Revelle and A.E. McGarity (eds.) *Design and Operation of Civil and Environmental Engineering Systems,* Chapter 13 (pp. 567–614). Wiley, New York.

Hernández-Chover, V., Castellet-Viciano, L. and Hernández-Sancho, F. (2020). Preventive maintenance versus cost of repairs in asset management: An efficiency analysis in wastewater treatment plants. *Process Safety and Environmental Protection*, 141, 215–221.

Hukka, J.J. and Katko, T.S. (2015). Resilient asset management and governance for deteriorating water services infrastructure. *Procedia Economics and Finance*, 21, 112–119.

IAM. (2015). *Asset Management – An Anatomy – Version 3* (p. 72). The Institute of Asset Management. https://www.theiam.org/asset-management-an-anatomy/.

ISO. (2012a). ISO/CD 55000.2 Asset management – Overview, principles and terminology, ISO/TC 251/WG 1.

ISO. (2012b). ISO/CD 55001.2 Asset management – Management systems – Requirements, ISO/TC 251/WG 2.

ISO. (1998). ISO/DIS 14031 Environmental Performance Evaluation.

IWSA. (1997). Performance indicators for transmission and distribution systems. *Workshop Postprint*, Lisbon, May 5–7.

James, L.D. and Lee, R.R. (1971). *Economics of Water Resources Planning* (640 pp.). McGraw-Hill, New York.

Karamouz M. and Movahhed, M. (2021). Asset management based flood resiliency of water infrastructure. *Proceedings of Virtual World Environmental and Water Resources Congress 2021*, Milwaukee, WS, June 7–11.

Karamouz, M. and Budinger, T.F. (eds.) (2014). *Livable Cities of the Future: Proceedings of a Symposium Honoring the Legacy of George Bugliarello*. National Academies Press, Washington, DC.

Karamouz, M., Rasoulnia, E., Zahmatkesh, Z., Olyaei, M.A., and Baghvand. A. (2016). Uncertainty-based flood resiliency evaluation of wastewater treatment plants. *Journal of Hydroinformatics*, 18(6), 990–1006.

Karamouz, M., Szidarovszky, F., and Zahraie, B. (2003). *Water Resources Systems Analysis* (590 pp.). Lewis Publishers, Boca Raton, FL.

Karamouz, M., Yaseri, K., and Nazif, S. (2017). Reliability-based assessment of lifecycle cost of urban water distribution infrastructures. *Journal of Infrastructure Systems*, 23(2), 04016030.

Keffer, C., Shimp, R., and Lehni, M. (1999). *Eco-Efficiency Indicators & Reporting. Report on the Status of the Project's Work in Progress and Guideline for Pilot Application, WBCSD Working Group on Eco-efficiency Metrics and Reporting*. WBCSD, Geneva.

Leavitt, W.M. and Kiefer, J.J. (2006). Infrastructure interdependency and the creation of a normal disaster: The case of Hurricane Katrina and the City of New Orleans. *Public Works Management & Policy*, 10(4), 306–314.

Mays, L.W. (1996). *Water Resources Handbook*. McGraw-Hill, New York.

McGraw Hill Construction. (2013). *Water Infrastructure Asset Management – Adopting Best Practices to Enable Better Investments*. McGraw Hill Construction and CH2M HILL. http://ch2m.com/corporate/markets/water/assets/water-infrastructure-asset-management-SMR-2013.pdf.

Nelson, A.C. (1995). *System Development Charges for Water, Wastewater, and Storm Water Facilities*. CRC Press/Lewis Publishers, Boca Raton, FL.

Nix, S.J. and Heaney, J.P. (1988). Optimization of storage-release strategies. *Water Resources Research*, 24(11), 1831–1838.

NYCDEP (New York City Department of Environmental Protection). (2013). *NYC Wastewater Resiliency Plan, Climate Risk Assessment and Adaptation Study. Wastewater Treatment Plants*. NYC DEP, New York.

Ostrom, L.T. and Wilhelmsen, C.A. (2012). Ecological risk assessment. In *Risk Assessment* (pp. 26–55). John Wiley & Sons, Inc. doi.10.1002/9781118309629.ch4.

Ouyang, M. (2014). Review on modeling and simulation of interdependent critical infrastructure systems. *Reliability Engineering & System Safety*, 121, 43–60.

Paulson, P.O. (2001). Continuous monitoring of reinforcements in structures. U.S. Patent 6,170,334. Pure Technologies Ltd.

Peerenboom, J., Fisher, R., and Whitfield, R. (2001). Recovering from disruptions of critical independent infrastructures. Presented at the CRIS/DRM/IIT/NSF *Workshop on Mitigating the Vulnerability of Critical Infrastructures to Catastrophic Failure*, Washington, DC, September.

Peters, M. and Timmerhaus, K. (1980). *Plant Design for Chemical Engineers*. McGraw-Hill, New York.

PPI-PACE (Pipeline Analysis and Calculation Environment). (2014). [cited 2014 November 15]; Available from: http://ppipace.com/.

RedZone Robotics, Solo System – The World's Only Self Operating Crawler. (2014). RedZone Robotics. http://www.redzone.com/wpcontent/uploads/2012/12/Solo.pdf.

Rehan, R., Knight, M.A., Unger, A.J., and Haas, C.T. (2013). Development of a system dynamics model for financially sustainable management of municipal watermain networks. *Water Research*, 47(20), 7184–7205.

Rehan, R., Unger, A.J., Knight, M.A., and Haas, C.T. (2014). Financially sustainable management strategies for urban wastewater collection infrastructure – Implementation of a system dynamics model. *Tunnelling and Underground Space Technology*, 39, 102–115.

Report of the World Summit on Sustainable Development. (2002). Johannesburg, South Africa, August 26–September 4 (United Nations Publication, Sales No. E.03.II.A.1 and Corrigendum), Chapter 1, Resolution 2, Annex.

Rinaldi, S.M., Peerenboom, J.P., and Kelly, T.K. (2001). Identifying, understanding, and analyzing critical infrastructure interdependencies. *IEEE Control Systems Magazine*, 21(6), 11–25.

Sangster, T. (2010). Pipeline condition assessment – The essential engineering step in the asset management process. *Trenchless Technology Conference*, Abu Dhabi.

Schaltegger, S. (1996). *Corporate Environmental Accounting*. Wiley, Chichester.

Sharkey, T.C., Nurre, S.G., Nguyen, H., Chow, J.H., Mitchell, J.E., and Wallace, W.A. (2016). Identification and classification of restoration interdependencies in the wake of Hurricane Sandy. *Journal of Infrastructure Systems*, 22(1), 04015007.

Streips, K. and Simpson, D.M. (2007). *Critical infrastructure failure in a natural disaster: Initial notes comparing Kobe and Katrina*. University of Louisville, Louisville.

Smith, L.A., Fields, K.A., Chen, A.S.C., and Tafuri, A.N. (2000). *Options for Leak and Break Detection and Repair for Drinking Water Systems* (163 pp.). Battelle Press, Columbus, OH.

Statewide Urban Design and Specifications. (2007). *Urban Design Standards Manual*. U.S. Environmental Protection Agency, Washington, DC.

U.S. Environmental Protection Agency (U.S. EPA). (1993). *Combined Sewer Control Manual*. EPA/625/R-93-007. U.S. Environmental Protection Agency, Cincinnati, OH.

U.S. Environmental Protection Agency (U.S. EPA). (2008). *Asset Management: A Best Practices Guide*. EPA 816/F-08/014. U.S. Environmental Protection Agency, Cincinnati, OH.

Viessman, W. and Welty, C. (1985). *Water Management: Technology and Institutions*. Harper and Row Publishers, New York.

White, W. R. 2005. World water storage in man-made reservoirs (FR/R0012). Review of Current Knowledge Series, April (Bucks, United Kingdom, Foundation for Water Research. Available online at www.fwr.org.

Zimmerman, R., Zhu, Q., de Leon, F., and Guo, Z. (2017). Conceptual modeling framework to integrate resilient and interdependent infrastructure in extreme weather. *Journal of Infrastructure Systems*, 23(4), 04017034.

9 Urban Water Systems Analysis and Conflict Resolution

9.1 INTRODUCTION

This chapter is the focus and concentration of one of the main elements of this book, namely, water systems analysis. Needless to say, the design and planning aspects of the materials covered in this book are affected by the way the related systems and infrastructures are analyzed. For considering integration and sustainability in stated resources management, it is necessary to think over the social, economic, and environmental impacts of decisions. This integration in planning and management, especially in urban areas, needs a systematic approach, considering all the interactions among the elements of the system and with the outside world. Simulation of urban water dynamics will give the collective impacts of all possible water-related urban processes on issues such as human health, environmental protection, quality of receiving waters, and urban water demand. Individual processes are then planned and managed in a way that the collective impact, with due consideration of the interaction among processes, is improved as much as possible.

Engineering and management aspects of urban water require utilizing technical tools and techniques that are classified here as simulation, optimization, and economic techniques. Engineering applications primarily use analysis and design models. The overlay between analysis and planning tools in urban water is simulation models, which can be used to test the performance of urban water systems. A simulation technique with an eye on the mathematics of growth that has received a great deal of attention is system dynamics. It utilizes object-oriented programming to simulate, in a physical–mathematical fashion, the behavior of an urban system under different design and operation alternatives.

In the remaining of this chapter, data preparation and processing techniques are discussed, followed by multicriteria decision-making (MCDM). Also, data-driven neural networks and fuzzy inference are introduced. Then mathematics of growth as a basis for systems dynamics is presented. Conventional and evolutionary optimization techniques are introduced. Finally, the conflict resolution in the context of Nash bargaining theory, game theory and agent base modeling are described. There are certain inserts in this chapter from Karamouz et al. (2003) text book that are applied to urban planning and management.

9.1.1 System Representation and Domains

The systems approach is especially useful when a project becomes so large that it cannot be considered as a unit, necessitating its decomposition. However, systems analysis is not an approach that can be used as a routine exercise and without thinking. Usually, the greatest effort of the analyst is to reduce the system to a manageable representation without losing its essential features and relationships. The analyst may overlook important relationships because they may lack access to all necessary data, and usually, time is not sufficient in an actual planning environment to develop the ideal model and test it to its fullest extent or subject it to the scrutiny of several experts. Figure 9.1 represents a system that shows the features of its environment in detail. These incorporate the substances such as intended inputs (resources), unintended inputs (natural or man-made threats and opportunities), default inputs (the goals and objectives of the system), and the outputs of the system that have a stake in the successful operations of the system. Different features incorporate the constraints to the

DOI: 10.1201/9781003241744-9

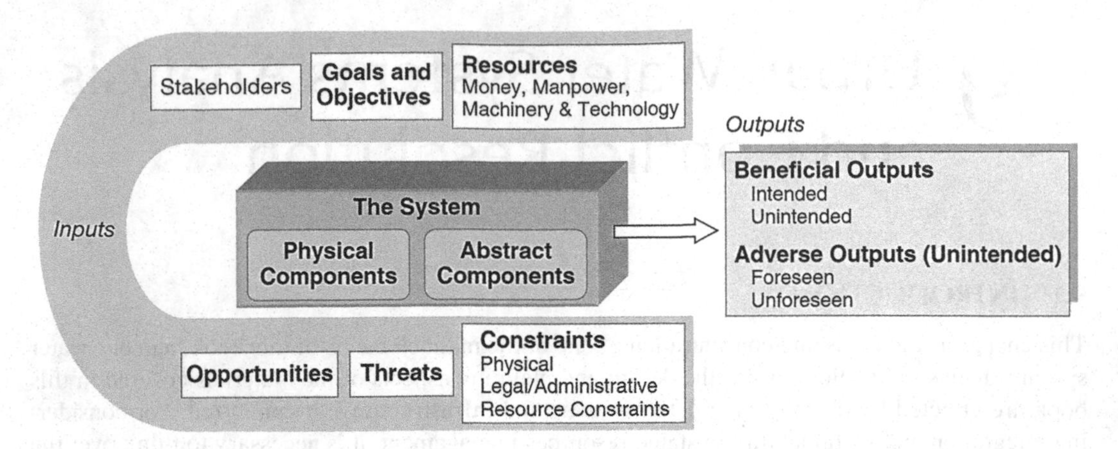

FIGURE 9.1 A more detailed representation of a system showing features of its environment. (Labi, S. *Introduction to civil engineering systems: A systems perspective to the development of civil engineering facilities*, (2014). https://www.wiley.com/en-us/Introduction+to+Civil+Engineering+Systems.)

system's existence or its operations (including physical space constraints, restrictions on the use of resources, quality of outcomes), which could be intended or unintended according to the problem's context. See Labi (2014) for more details.

A physical water system is a collection of various elements, which interact in a logical manner and are designed in response to various social needs in the development and improvement of existing water resources for the benefit of human use. Simonovic (2000) described water resources systems analysis as an approach by which the components of such a system and their interactions are described employing mathematical or logical functions. In an urban setting, the components of a water system are continuously affected by other components and considerable human interactions, which often place more burden on the functionality of them.

Systems inquiry is based on knowledge. It could be categorized into four domains as shown in Figure 9.2: philosophy (knowledge and understanding, the state of being and what is right of and for systems), theory (a set of interrelated principles and concepts that are applicable to all systems), methodology (a set of strategies, techniques, tools, models, and methods for realizing philosophy and theory), and application (the use and their interactions) (Banathy, 1997). Philosophy and theory are associated with knowledge, while method and application are action domains.

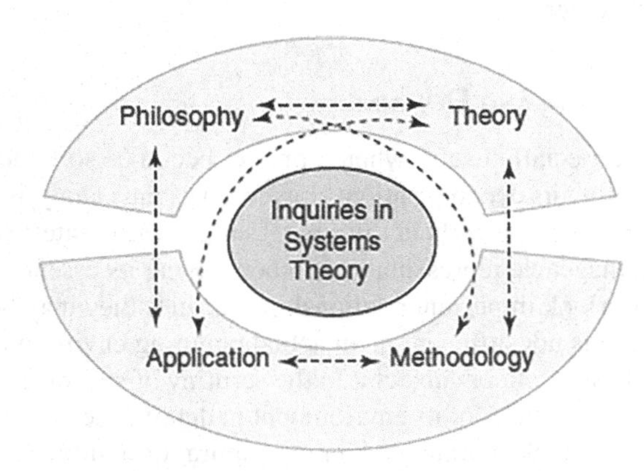

FIGURE 9.2 Domains of inquiry in systems theory. (Labi, S. *Introduction to civil engineering systems: A systems perspective to the development of civil engineering facilities*, (2014). https://www.wiley.com/en-us/Introduction+to+Civil+Engineering+Systems.)

9.1.2 Water Systems Analysis

In general, systems analysis is the study of all the interactions of the components. Very often systems analysis is concerned with finding that combination of components generates an optimum solution, which consists of the best possible combination of elements for satisfying the desired objective. Thus, it involves defining and evaluating numerous water development and management alternatives. This can be done in a very detailed manner representing various possible compromises among conflicting groups, values, and management objectives.

The dynamics of urban water supply and demand components and how an optimum scheme can be developed for urban water management is the first challenge. The second challenge is to operate and effectively maintain the urban water infrastructure. Urban water systems are designed in different zones with known elements of dependency of each zone to others and the components within each zone. Therefore, the analysis can be done in separate zones.

Since models are abstractions of reality, they do not usually describe all the features that are encompassed by a real-world situation. A prerequisite for water resources systems analysis is the description of the system in terms of model components, which permits solutions to be obtained at reasonable costs and within a prescribed time frame. Therefore, the model builder should not attempt to examine the reality and accuracy of individual components but only to meet the overall accuracy requirements for that system. Sensitivity analysis should be performed since urban water models are subject to high levels of uncertainty.

A prerequisite for a systems analysis is that all the system elements can be modeled either analytically or conceptually. It is important to distinguish the difference between a system and a model. A model is the mathematical and/or physical representation of the system and of the relations between the elements of the system. It is an abstraction of the real world. In any particular application, the quality of the model and thus of systems analysis depends on how well the model builder perceives the actual relationship and how well they can describe their functional form. Many advances in making analytical tools such as simulation including system dynamics, optimization, and conflict resolution models have been made that are discussed in this chapter. Conceptual models in the form of laboratory experiments and/or small-scale building prototypes are used to model the real applications physically and in layers that can be focused piece by piece or as a whole in harmony. This later type of systems analysis is not covered in this chapter but partially mentioned in the case studies in Chapters 11–14. Network analysis, pressure, and leakage management are discussed in Chapter 6. Economic and asset management issues are discussed in Chapter 8.

9.2 DATA PREPARATION TECHNIQUES

Input to a system includes different types of data, some are ready to be used and some represent spatial or temporal characteristics. For detailed data preparation for civil engineering systems in general and water infrastructure, refer to Labi (2014). For the remainder of this section, we concentrate on the spatial variability and regionalizing data that is of prominent importance in urban water studies. Criteria and subcriteria are often defined as the information needed to quantify the performance of the system. Their developments are discussed next in the context of MCDM and multiobjective optimization. Finally, parameter imprecision is described in the context of fuzzy logic and fuzzy inference systems (FISs).

9.2.1 Regionalizing Hydrologic Data

Many activities in applied and engineering hydrology involve measurements of one or more quantities at given spatial locations to regionalize the measured quantities at unsampled locations. Often, the ungagged (no samples taken) locations are on a regular grid, and the regionalization is used to produce surface plots or contour maps.

A popular method of regionalization is kriging. This method requires the complete specification (the form and parameter values) of the spatial dependence that characterizes the spatial process. For this purpose, models for spatial dependence are expressed in terms of the distance between any two locations in the spatial domain of interest. These models take the form of a covariance or semivariance function. Spatial prediction, using the kriging method, involves two steps:

1. Model the semivariance or covariance of the spatial process. These measures are typically not known in advance. This step involves computing an empirical estimate and determining both the mathematical form and the values of any parameters for a theoretical form of the dependence model.
2. This dependence model is used to solve the kriging system at a specified set of spatial points, resulting in predicted values and associated standard errors.

9.2.1.1 Theoretical Semivariogram Models

Consider a stochastic spatial process represented by the stationary spatial random field $\{z(x), x \in R^2\}$. The variogram procedure computes the empirical (also known as sample or experimental) semivariance of $z(x)$. Regionalization $z(x)$ at ungagged locations by techniques such as kriging requires a theoretical semivariogram (model) or covariance. A suitable theoretical model must be determined based on what is fitted to the sample data. In one of the simplest models for the definition of a semivariogram, the mean, m, is constant and the two-point covariance function depends only on the distance between two points as follows:

$$E[z(x)] = m \tag{9.1}$$

and

$$E\left[(z(x)-m)(z(x')-m)\right] = \gamma(h), \tag{9.2}$$

where

$$h = \|x - x'\| = \sqrt{(x_1 - x_1')^2 + (x_2 - x_2')^2} \tag{9.3}$$

where h is the distance between sampling location x and x', $x = [x_1, x_2]$, and $x' = [x_1', x_2']$. Equations 9.1 and 9.2 comprise the stationary model; a random function $z(x)$ satisfying these conditions is called stationary. This model is isotropic because it uses only the length and not the orientation of the linear segment that connects the two points. The covariance values at $h = 0$ are known as the variance or the sill of the stationary function. The semivariogram is defined as

$$\gamma(h) = \frac{1}{2} E\left[(z(x) - z(x'))^2\right]. \tag{9.4}$$

Figure 9.3 displays a theoretical semivariogram of a spherical model and points out the semivariogram characteristics. In this figure, $\gamma(h)$ is semivariogram, a_0 is range, c_0 is sill, h is distance, c_n is nugget effect, and σ_0^2 is partial sill. Finally, for selecting the best variogram model, cross-validation is used and is defined in Section 4.10.4. Theoretical semivariogram models such as the Gaussian, exponential, spherical, sine hole effect, and power model are described in the following:

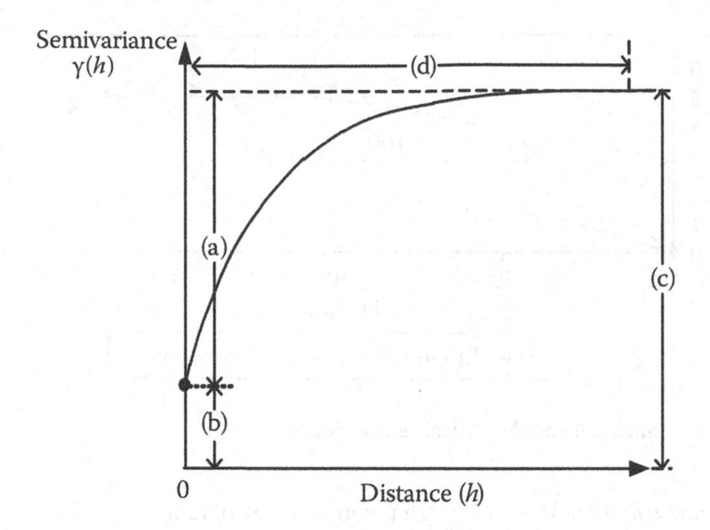

FIGURE 9.3 Theoretical semivariogram of spherical type and its characteristics [(a) partial sill, $\left(\sigma_0^2\right)$, (b) nugget effect (c_n), (c) sill (c_0), and (d) range (a_0).]

a. *The Gaussian Semivariogram Model*: The form of the Gaussian model is

$$\gamma(h) = c_0\left[1-\exp\left(-\frac{h^2}{a_0^2}\right)\right], \tag{9.5}$$

where a_0 is range, c_0 is sill, h is distance.

b. *The Exponential Model*: The form of the exponential model is

$$\gamma(h) = c_0\left[1-\exp\left(-\frac{h}{a_0}\right)\right], \tag{9.6}$$

where $\gamma(h)$ is semivariogram, a_0 is range, c_0 is sill, and h is distance.

c. *The Spherical Model*: The form of the spherical model is

$$\gamma(h) = \begin{cases} \left(\frac{3}{2}\frac{h}{a_0} - \frac{1}{2}\frac{h^3}{a_0^3}\right) & \text{for} \quad (h \le a_0) \\ c_0 & \text{for} \quad (h > a_0) \end{cases} \tag{9.7}$$

d. *The Sine Hole Effect Model*: The form of the sine hole effect is

$$\gamma(h) = c_0\left[1-\frac{\sin\left(\pi h/a_0\right)}{\pi h/a_0}\right], \tag{9.8}$$

e *The Power Model*: The form of the power model is

$$\gamma(h) = c_0 h^{a_0}, \tag{9.9}$$

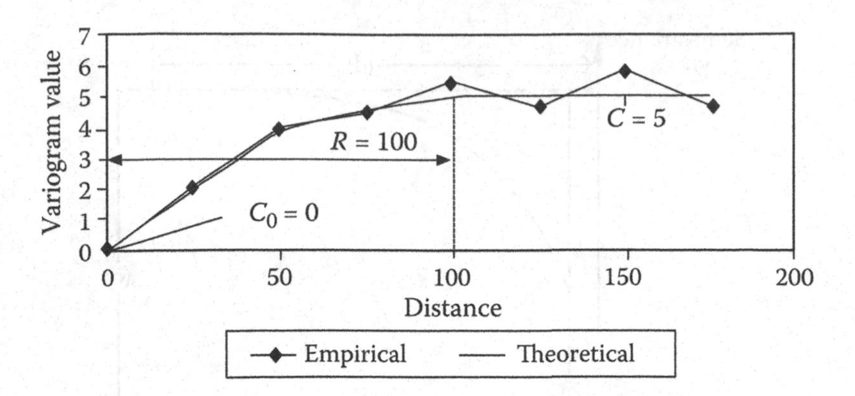

FIGURE 9.4 Pair of empirical and theoretical semivariogram.

f. *Matern Semivariogram Model*: Matern theoretical variogram is

$$\gamma(h) = c_n + c\left[1 - \frac{1}{2^{(v-1)}\Gamma(v)}\left(\frac{h}{a}\right)^a K_v\left(\frac{h}{a}\right)\right] \tag{9.10}$$

where $h > 0$ is the separation distance between two points, Γ is the gamma function, K_v is the modified Bessel function of the second kind of order ν, $v > 0$ is the smoothness parameter, $a > 0$ is range, c is the partial sill, and c_n is the nugget.

An example of empirical and theoretical variograms and the illustration of the above parameters are shown in Figure 9.4.

9.2.1.2 Kriging System

Kriging involves applying the general methodology known as best linear unbiased estimation to intrinsic functions. Given n measurements of z at the location with spatial coordinate $x_1, x_2, x_3, \ldots, x_n$, it estimates the value of z at point x_0. An estimator is simply a procedure or equation that uses data to find a representative value, or estimate, of the unknown quantity.

$$\hat{z}_0 = \sum_{i=1}^{n} \lambda_i z(x_i). \tag{9.11}$$

Thus, the problem is reduced to selecting a set of coefficients $\lambda_1, \lambda_2, \ldots, \lambda_n$. The difference between the estimates $\hat{z}_0$ and the actual value $z(x_0)$ is the estimation error:

$$\hat{z}_0 - z(x_0) = \sum_{i=1}^{n} \lambda_i z(x_i) - z(x_0). \tag{9.12}$$

The coefficient is selected by considering the following specifications:

- *Unbiasedness:* On average, the estimation error must be zero. That is,

$$E\left[\hat{z}_0 - z(x_0)\right] = \sum_{i=1}^{n} \lambda_i m - m = \left(\sum_{i=1}^{n} \lambda_i - 1\right) m = 0. \tag{9.13}$$

However, the numerical value of the mean, m, is not specified. For the estimator to be unbiased for any value of the mean, it is required that

$$\sum_{j=1}^{n} \lambda_i = 1. \tag{9.14}$$

Imposing the unbiasedness constraint eliminates the unknown parameter m.

- *Minimum variance*: The mean square estimation error must be minimized.

$$E\left[\hat{z}_0 - z(x_0)^2\right] = -\sum_{i=1}^{n}\sum_{j=1}^{n} \lambda_i \lambda_j \gamma\left(\|x_i - x_j\|\right) + 2\sum_{i=1}^{n} \lambda_i \gamma(x_i - x_0). \tag{9.15}$$

Thus, the problem of the best (minimum mean square error (MSE)) unbiased estimation of the λ coefficients may be reduced to the constrained optimization problem:

Select the values of $\lambda_1, \lambda_2, \ldots, \lambda_n$ that minimize Equation 9.15 while satisfying Equation 9.14. Equation 9.14 is called the objective function because we try to minimize it; Equation 9.13 is called a constraint because it restricts the values that might be assigned to the coefficients. A constrained optimization problem is formulated with a quadratic objective function and a linear constraint. This problem can be solved easily using Lagrange multipliers, a standard optimization method; the necessary conditions for the minimization are given by the linear kriging system of $n + 1$ equation with $n + 1$ unknowns as follows:

$$-\sum_{i=1}^{n} \lambda_i \gamma\left(\|x_i - x_j\|\right) + \nu = -\gamma\left(\|x_i - x_0\|\right), \quad i = 1,2,\ldots,n$$

$$\sum_{i=1}^{n} \lambda_j = 1, \tag{9.16}$$

where ν is a Lagrange multiplier. It is common practice to write the kriging system in matrix notation. Let X be the vector of the unknowns as defined below:

$$X = \begin{bmatrix} \lambda_1 \\ \lambda_2 \\ \vdots \\ \lambda_n \\ \nu \end{bmatrix}. \tag{9.17}$$

b, the matrix of coefficients, and **A**, the right-hand-side vector, are defined as follows:

$$b = \begin{bmatrix} -\gamma\left(\|x_1 - x_0\|\right) \\ -\gamma\left(\|x_2 - x_0\|\right) \\ \vdots \\ -\gamma\left(\|x_n - x_0\|\right) \\ 1 \end{bmatrix}. \tag{9.18}$$

$$\mathbf{A}=\begin{bmatrix} 0 & -\gamma(\|x_1-x_2\|) & \cdots & -\gamma(\|x_1-x_n\|) & 1 \\ -\gamma(\|x_2-x_1\|) & 0 & \cdots & -\gamma(\|x_2-x_n\|) & 1 \\ \vdots & \vdots & \cdots & \vdots & \vdots \\ -\gamma(\|x_n-x_1\|) & -\gamma(\|x_n-x_2\|) & \cdots & 0 & 1 \\ 1 & 1 & \cdots & 1 & 0 \end{bmatrix} \quad (9.19)$$

Denote by A_{ij} the element of **A** at the ith row and the jth column, and denote by x_i and b_i the element at the ith row of x and b, respectively. Notice that **A** is symmetric, that is, $A_{ij} = A_{ij}$.

The kriging system can be written as

$$\sum_{j=1}^{n+1} A_{ij}x_j = b_i \quad \text{for} \quad i = 1,2,\ldots,n+1. \quad (9.20)$$

The matrix of kriging equations is defined as follows:

$$Ax = b. \quad (9.21)$$

Solving this system, we obtain $\lambda_1, \lambda_2, \ldots, \lambda_n$, and υ. In this manner, the linear estimator of Equation 9.10 is fully specified. Furthermore, the accuracy of the estimate can be quantified through the mean square estimation error. The mean square estimation error may be obtained by substituting in Equation 9.17 the values of $\lambda_1, \lambda_2, \ldots, \lambda_n$ obtained from the solution of the kriging system. Estimation of the variance in an unknown point at x_0 is defined as

$$\sigma_0^2 = E\left[\left(\hat{z}_0 - z(x_0)\right)^2\right] = -\nu + \sum_{i=1}^{n} \lambda_i \gamma(\|x_i - x_0\|). \quad (9.22)$$

9.2.1.3 Fitting Variogram

Variogram (semivariogram) selection is an iterative process that usually starts with an examination of the experimental variogram. How the experimental variogram can be obtained from data has been discussed. Then, a variogram fits the model by selecting one of the equations from the list of basic models of the previous section and adjusting its parameters to reproduce the experimental variogram as closely as possible.

Example 9.1

Consider 70 measurements that were synthetically generated from a one-dimensional random process (see Figure 9.5). Infer the variogram and test the intrinsic model.

Solution:

We always start by plotting and calculating basic statistics for the data (exploratory analysis). For one-dimensional data, the task of exploratory analysis is simplified because a plot of the observations versus location (see Figure 9.6) pretty much conveys a visual impression of the data. In this case, we infer that the data are reasonably continuous but without a well-defined slope. The data clearly indicate that much of the variability is at a scale comparable to the maximum separation distance so that the function does not appear stationary. These observations are essential in selecting the variogram.

Next, we plot the experimental variogram. For a preliminary estimate, we subdivide the separation distances into ten equal intervals. One might proceed to draw a straight line from the origin through the experimental variogram, as shown in Figure 9.6. Our preliminary estimate is that the variogram is linear with a slope of 0.40.

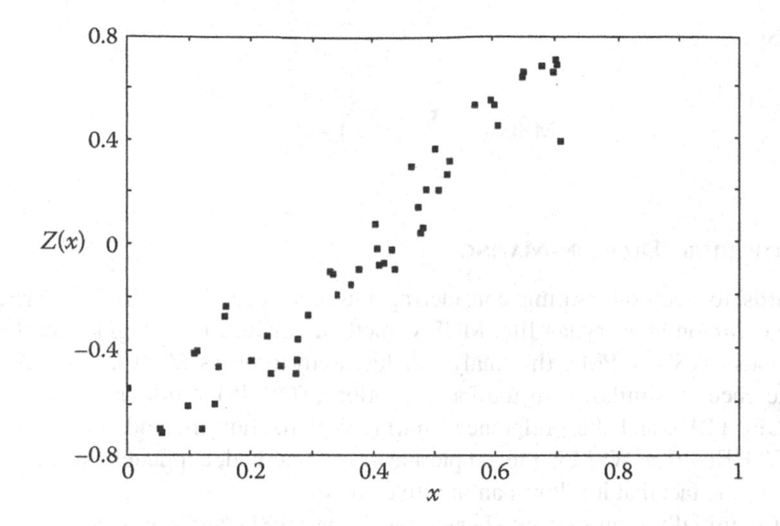

FIGURE 9.5 Plot of observations versus locations for Example 9.1.

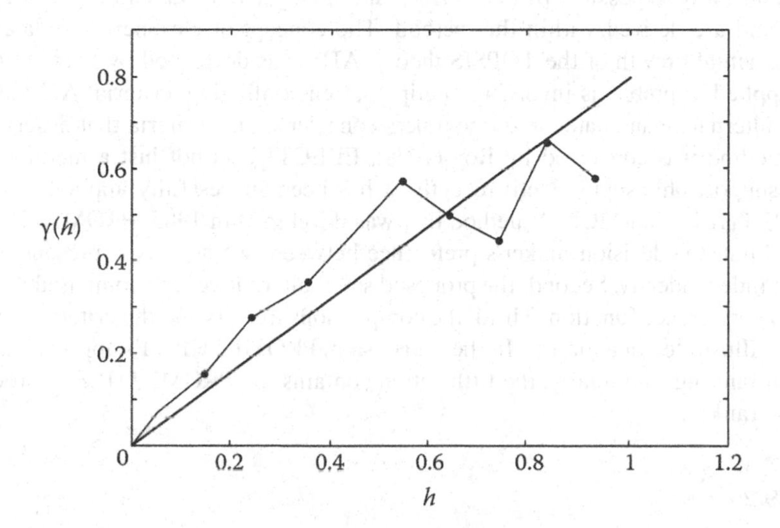

FIGURE 9.6 Experimental variogram and preliminary fit

9.2.1.4 Cross-Validation

A semivariogram model type must be chosen prior to the generation of the kriging model. Cross-validation can be used to assess the impact of many choices on the semivariogram models. It is used to compare estimated and true values using the information available in the sample dataset. An error is calculated as the estimated value minus the true value. Considering kriging system equations and the cross-validation method, the error is estimated by the following equations:

1. The mean error (ME):

$$\text{ME} = \frac{1}{N}\sum_{i=1}^{N}\{z(x_i) - \hat{z}(x_i)\}, \tag{9.23}$$

where $z(x_i)$ is sample data, $\hat{z}(x_i)$ is estimated data, and N is the number of data.

2. The MSE:

$$\text{MSE} = \frac{1}{N}\sum_{i=1}^{N}\left\{z(x_i) - \hat{z}(x_i)\right\}^2, \tag{9.24}$$

9.2.2 Multicriteria Decision-Making

MCDM regards to decision-making considering multiple, usually conflicting, criteria. MCDM problems are common in everyday life. MCDM methods include the weighted sum and weighted product methods (WSM/WPM), the analytical hierarchy process (AHP), the technique for the order of preference by similarity to the ideal solution (TOPSIS), elimination et choix traduisant la realité (ELECTRE), and the preference ranking organization method for enrichment evaluation (PROMETHEE). The WSM is the simplest available method, applicable to single-dimensional problems due to the fact that it follows an intuitive process.

TOPSIS was initially proposed by Hwang and Yoon (1981), and the idea behind it lies in the optimal alternative being as close in the distance as possible from an ideal solution and at the same time as far away as possible from a corresponding negative ideal solution. Both solutions are hypothetical and are derived within the method. The concept of closeness was later established and led to the actual growth of the TOPSIS theory. AHP was developed by Saaty (1980), and it is extensively applied in problems involving multiple, often conflicting, criteria. AHP aims to define the optimum alternative and categorize the others considering the criteria that describe them. The ELECTRE method was conceived by Roy (1978). ELECTRE is not just a method but a different decision support philosophy. Until recently, it has been successfully applied in many diverse fields. PROMETHEE is an MCDM method that was developed in 1985. PROMETHEE is applied in five steps. First, the decision maker's preference between two actions is presented by a preference function independently. Second, the proposed set of alternatives are compared with each other concerning the preference function. Third, the comparisons' results and the criterion's value of each alternative are illustrated in a matrix. In the fourth step, PROMETHEE I's approach is used to sort out the partial ranking, and finally, the fifth action contains the PROMETHEE II process to finish the alternative rankings.

Example 9.2

Consider a one-dimensional function with variogram $\gamma(h) = 1 + h$, for $h > 0$, and three measurements, at location $x_1 = 0$, $x_2 = 1$, and $x_3 = 3$. Estimate the value of the function in the neighborhood of the measurement points, at location $x_0 = 2$.

Solution:

The kriging system of equations, with weights λ_1, λ_2, and λ_3, is

$$-2\lambda_2 - 4\lambda_3 + \nu = -3$$

$$-2\lambda_1 - 3\lambda_3 + \nu = -2$$

$$-4\lambda_1 - 3\lambda_2 + \nu = -2$$

$$\lambda_1 + \lambda_2 + \lambda_3 = 1$$

or, in matrix notation,

$$\begin{bmatrix} 0 & -2 & -4 & 1 \\ -2 & 0 & -3 & 1 \\ -4 & -3 & 0 & 1 \\ 1 & 1 & 1 & 0 \end{bmatrix} \begin{bmatrix} \lambda_1 \\ \lambda_2 \\ \lambda_3 \\ v \end{bmatrix} = \begin{bmatrix} -3 \\ -2 \\ -2 \\ 1 \end{bmatrix}.$$

The mean square estimation error is

$$-\upsilon + 3\lambda_1 + 2\lambda_2 + 2\lambda_3.$$

Solving the system, we obtain $\lambda_1 = 0.1304$, $\lambda_2 = 0.3913$, $\lambda_3 = 0.4783$, and $\upsilon = -0.304$. The mean square error (MSE) is 2.43.

For $x_0 = 0$, that is, a position coinciding with an observation point, the kriging system is

$$\begin{bmatrix} 0 & -2 & -4 & 1 \\ -2 & 0 & -3 & 1 \\ -4 & -3 & 0 & 1 \\ 1 & 1 & 1 & 0 \end{bmatrix} \begin{bmatrix} \lambda_1 \\ \lambda_2 \\ \lambda_3 \\ v \end{bmatrix} = \begin{bmatrix} -1 \\ -2 \\ -4 \\ 1 \end{bmatrix}.$$

By inspection, one can verify that the only possible solution is $\lambda_1 = 0.7391$, $\lambda_2 = 0.2174$, $\lambda_3 = 0.0435$, and $\upsilon = -0.3913$. The MSE = 1.7391.

9.2.2.1 Deterministic MCDM

The term "deterministic" is related to a specific entity. Deterministic models are used to describe one out of many possible results in a reference problem. On the other hand, "stochastic" comes from the Greek "to aim" and refers to a "random" outcome. Several potential outcomes, characterized by their probabilities or likelihood, are best represented through stochastic modeling. Consequently, probabilistic processes denote the set of random variables that are related to a varying factor. Such processes consist of a state space representing the potential values, where the random variables may be related to each other. Deterministic methods are mainly used to describe simple, natural phenomena on the basis of physical laws and are not fit for purpose for large and complicated applications. Consequently, real world behavior is better reflected by employing methods relevant to stochastic simulations. The deterministic MCDM is discussed in a number of case studies in Chapters 8 and 13.

9.2.2.2 Probabilistic MCDM

A probabilistic method (e.g., Monte Carlo simulation [MCS] method) is an alternative to deal with parameter uncertainties in hydrologic modeling. The MCS allows us to perform a comprehensive probabilistic analysis to assess the uncertainty in hydrologic variables. Probabilistic methods can increase the confidence of the decision maker in the final results and analysis and can be more appropriate for cases where the heterogeneity of important factors is critical as the uncertainty of the considered system increases. In general, it is not feasible to obtain an analytical expression for stochastic problems, which would require more computational time and resources to deliver a satisfactory solution. Karamouz et al. (2016) carried out an uncertainty analysis on the resilience measurement of coastal wastewater treatment plants through probability distribution generation for inputs to the resilience evaluation model. Karamouz and Farzaneh (2020) quantified the reliability using MCS, which is explained in the next section and in Chapter 12.

9.2.3 Fuzzy Sets and Parameter Imprecision

Two randomization approaches or fuzzification are used to evaluate the parameter uncertainty in water resources and hydrological modeling. In this section, the fundamentals of fuzzy sets and fuzzy decisions will be introduced.

Let X be a set of certain objects. A fuzzy set A in X is a set of ordered pairs as follows:

$$A=\{[x,\mu_A(x)]\}, \qquad x\in X, \tag{9.25}$$

where μ_A: $X \longmapsto [0, 1]$ is called the membership function, and $\mu_A(x)$ is the grade of membership of x in A. In classical set theory, $\mu_A(x)$ equals 1 or 0 because x either belongs to A or does not.

The basic concepts of fuzzy sets are as follows:

i. A fuzzy set A is *empty* if $\mu_A(x) = 0$ for all $x \in X$.
ii. A fuzzy set A is called *normal* if

$$\sup_x \mu_A(x)=1. \tag{9.26}$$

For normalizing a nonempty fuzzy set, the normalized membership function is computed as follows:

$$\mu_A(x)=\frac{\mu_A(x)}{\sup_x \mu_A(x)}. \tag{9.27}$$

iii. The *support* of a fuzzy set A is defined as

$$S(A)=\{x|x\in X,\mu_A(x)>0\} \tag{9.28}$$

iv. The fuzzy sets A and B are equal if, for all $x \in X$,

$$\mu_A(x)=\mu_B(x). \tag{9.29}$$

v. A fuzzy set A is a *subset* of $B(A\subseteq B)$ if, for all $x \in X$,

$$\mu_A(x)\le\mu_B(x). \tag{9.30}$$

vi. A' is the *complement* of A if, for all $x \in X$,

$$\mu_{A'}(x)=1-\mu_A(x). \tag{9.31}$$

vii. The *intersection* of fuzzy sets A and B is defined as

$$\mu_{A\cap B}(x)=\min\{\mu_A(x);\mu_B(x)\} \quad (\text{for all } x\in X). \tag{9.32}$$

Note that (v) and (vii) imply that $A \subseteq B$ if and only if $A \cap B = A$.

viii. The *union* of fuzzy sets *A* and *B* is given as

$$\mu_{A \cup B}(x) = \max\{\mu_A(x); \mu_B(x)\} \quad (\text{for all } x \in X). \tag{9.33}$$

ix. The *algebraic product* of fuzzy sets *A* and *B* is denoted by *AB* and is defined by the relation

$$\mu_{AB}(x) = \mu_A(x)\mu_B(x) \quad (\text{for all } x \in X). \tag{9.34}$$

x. The *algebraic sum* of fuzzy sets *A* and *B* is denoted by *A* + *B* and has the membership function

$$\mu_{A+B}(x) = \mu_A(x) + \mu_B(x) - \mu_A(x)\mu_B(x) \quad (\text{for all } x \in X). \tag{9.35}$$

xi. A fuzzy set *A* is *convex,* if for all $x, y \in X$ and $\lambda \in [0, 1]$,

$$\mu_A[\lambda x + (1-\lambda)y] \geq \min\{\mu_A(x); \mu_A(y)\}. \tag{9.36}$$

If *A* and *B* are convex, then it can be demonstrated that $A \cap B$ is also convex.

xii. A fuzzy set *A* is *concave* if *A′* is convex. In the case where *A* and *B* are concave, the $A \cup B$ will also be concave.

xiii. Consider $f: X \longmapsto Y$, a mapping from set *X* to set *Y*, and *A* a fuzzy set in *X*. The fuzzy set *B* *induced* by mapping *f* is defined in *Y* with the following membership function:

$$\mu_B(y) = \sup_{x \in f^{-1}(y)} \mu_A(x), \tag{9.37}$$

where $f^{-1}(y) = \{x | x \in X, f(x) = y\}$.

Example 9.3

Consider fuzzy set *A* defined in $X = [-2, 2]$ as follows:

$$\mu_A(x) = \begin{cases} \frac{2+x}{2} & \text{if } -2 \leq x \leq 0, \\ \frac{2-x}{2} & \text{if } 0 < x \leq 2, \\ 0 & \text{otherwise.} \end{cases}$$

Determine the membership function of the fuzzy set induced by the function $f(x) = x^2$.

Solution:

Since $f([-2, 2]) = [0, 4]$, $\mu_B(y) = 0$ if $y \notin [0, 4]$. If $y \in [0, 4]$, then $f^{-1}(y) = \{\sqrt{y}, -\sqrt{y}\}$, and since $\mu_A(\sqrt{y}) = (2-\sqrt{y})/2 = \mu_A(-\sqrt{y})$,

$$\mu_B(y) = \sup_{x \in f^{-1}(y)} \mu_A(x) = \frac{2-\sqrt{y}}{2}.$$

Both $\mu_A(x)$ and $\mu_B(y)$ are illustrated in Figure 9.7.

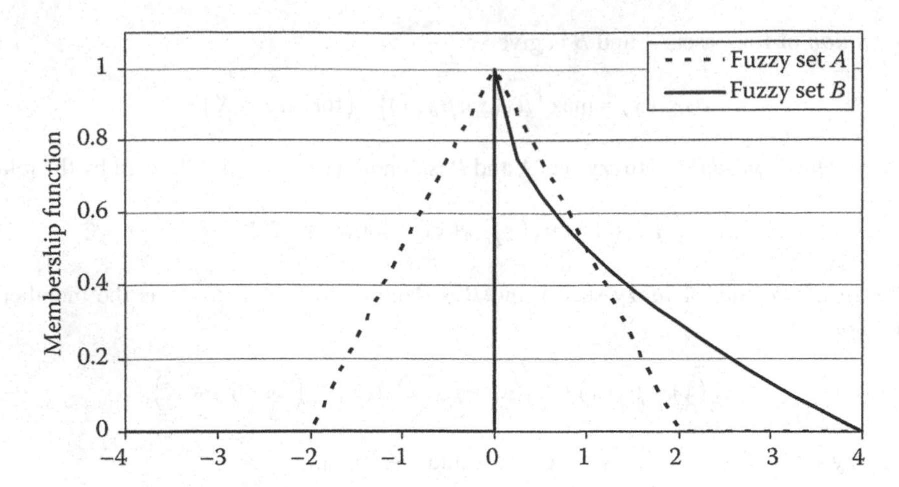

FIGURE 9.7 Illustration of $\mu_A(x)$ and $\mu_B(y)$.

Uncertain model parameters can be considered as fuzzy sets, and the result on any series of operations leads to a fuzzy set. Fuzzy sets can also express uncertainty in constraints and in the objective functions. See Karamouz et al. (2003) for more details.

9.2.4 Fuzzy Inference System

Fuzzy logics are a powerful tool for solving problems using very simple mathematical concepts. It can be stated that fuzzy logic does not solve new problems, but it offers new methods for solving everyday problems. A FIS uses fuzzy logic for providing a relationship between the input and output spaces. This system follows the reasoning process of human language using fuzzy logic for building fuzzy IF–THEN rules. FISs are used in decision-making problems. An example of these rules is indicated as follows:

If rainfall depth is high and the reservoir is full, the flood damages will be serious. The simplest form of a fuzzy "if–then" rule is as follows:

$$\text{If } x \text{ is } A, \text{then } y \text{ is } B,$$

where A and B are fuzzy values defined by fuzzy sets in the universes of discourse X and Y, respectively. x and y are the input and output variables, respectively. This has different meanings in the antecedent and consequent parts of the rule. The antecedent is an interpretation that returns a value between 0 and 1, and the consequent assigns a fuzzy set B to the variable y.

FIS is built based on the knowledge of experts, or in other words, it relies on the know-how of the ones who understand the system. However, FIS is a very flexible system. An FIS can be easily modified by adding or deleting rules without the need for creating a new FIS. Furthermore, imprecise data can be used in FIS (but it does not work with uncertainty) because of the use of elements in a fuzzy set (membership values). An important advantage of FIS is that it can be used together with other classic methods.

In general, as shown in Figure 9.8, an FIS includes four main modules:

- *Fuzzification module*: In this module, the crisp system inputs are transformed into fuzzy sets by applying a fuzzification function.
- *Knowledge base*: Based on the available data of inputs and outputs, some IF–THEN rules are developed by experts and stored in this module.

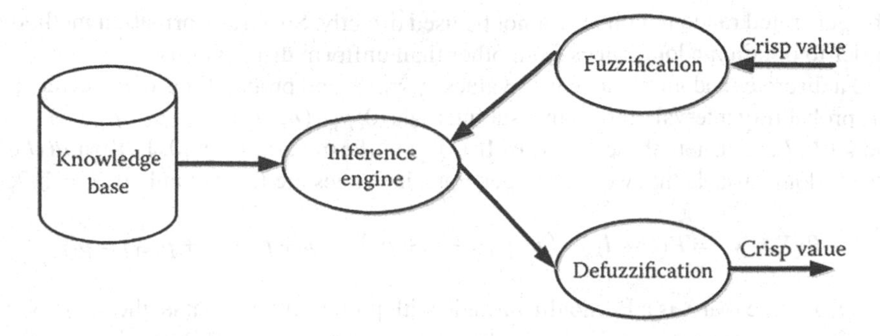

FIGURE 9.8 Four main modules of FIS.

- *Inference engine*: The fuzzy output of the system is determined by making fuzzy inference on the inputs and IF–THEN rules.
- *Defuzzification module*: Finally, the fuzzy output of the system obtained by the inference engine is changed into a crisp value.

Fuzzy inference methods include two groups of direct and indirect methods. Direct methods are the most commonly used, but indirect methods are more complex.

9.3 SIMULATION TECHNIQUES

The mathematical formulation of urban water systems is the basis of the simulation techniques. These techniques are valuable methods for analyzing the urban water systems due to their mathematical simplicity and versatility. Simulation models are used to evaluate system performance under a given set of inputs and operating conditions. They do not identify optimal decisions compared to optimization models but can merely be used to simulate a large number of scenarios. Using the simulation models, a very detailed and realistic representation of the complex physical, economic, and social characteristics of a water resources system can be considered in the form of mathematical formulations. The concepts inherent in the simulation approach are easier to comprehend and communicate than other modeling concepts.

Furthermore, the simulation methods can solve urban water planning models with highly nonlinear relationships and constraints that constrained optimization procedures cannot handle. The process of using a model to study the operation and subsequent performance of a real system is of prominent importance in urban areas. The simulation models may deal with steady-state or transient conditions. For example, to study the operation of an urban water system over a relatively long period of time during which no major alterations in the system occur, we need to perform a steady-state analysis. The simulation models are categorized into two groups, namely, deterministic and stochastic. If the system is subject to random input events or generates them internally, the model is called stochastic. The model is deterministic if no random components are involved.

In this section, stochastic simulation in the context of fitting and assessing the behavior of the system's characteristics is discussed first. Then in deterministic simulation, artificial neural network (ANN), and fuzzy set theory applications are presented. The principle of mathematics growth for exploring the basis of system dynamics is also described. The probabilistic Neural Network (PNN) and Radial Basis Function (RBF) are discussed in chapter 12.

9.3.1 Probabilistic Distribution of the System's Characteristics

To assess the effects of random elements on the sequence of events, probabilistic distributions can be used. The basis for the probabilistic distribution of urban water systems is generating independent random numbers with a uniform distribution. Due to the existence of different distributions in nature, the generated random number cannot be used directly. So a transformation method is necessary in order to obtain random values from other than uniform distributions.

Assume a discrete random variable with values $x_1, x_2, \ldots$ and probabilities of occurrence as p_1, p_2. Divide the probability interval of [0, 1] into subintervals $(0, p_1)$, (p_1, p_1+p_2), $(p_1+p_2, p_1+p_2+p_3)$, and so on, and let $I_1, I_2, \ldots$ denote these intervals. If U is a uniform variable in [0, 1], then $P(U \in I_k) = p_k$. If x_k is the random variable that we wish to generate, its values are the x_k numbers, $k = 1, 2, \ldots$, and

$$P(X = x_k) = P(U \in I_k) = (p_1 + p_2 + \ldots + p_k) - (p_1 + p_2 + \ldots + p_{k-1}) = p_k. \tag{9.38}$$

For example, assume that X is a Bernoulli variable with probability p. That is, the value of X is either 1 or 0. The value of 1 can be considered as a decision about doing something or success in maintaining service in supplying the demand, whereas the value of 0 can be considered as doing nothing or failing to provide adequate service. In that case, $I_1 = (0, p)$, $I_2 = (p, 1)$. So, from a uniform random number, X is 1 if $U < p$, and X is zero otherwise. The evolution of the Bernoulli variable to the normal distribution, which is widely used in urban water studies, is discussed by Karamouz et al. (2003).

Continuous random variables are generated by using their cumulative distribution functions. Let X be a random variable with distribution function $F(x)$ and let U be uniform in [0, 1]. Let u be a random value of U, then the solution of the equation $F(x) = u$ for x gives a random value of X. In order to see this, consider the distribution function of X at any real value t:

$$P(X \le t) = P(F(X) \le F(t)) = P(U \le F(t)) = F(t), \tag{9.39}$$

since U is uniform in [0, 1]. Here we assume that F is strictly increasing, and therefore, equation $F(x) = u$ has a unique solution for all $0 < u < 1$. The events $U = 0$ and $U = 1$ occur with zero probability.

A random variable is called *uniform* in a finite interval $[a, b]$ if its distribution function is

$$F(x) = \begin{cases} 0 & \text{if } x \le a, \\ \dfrac{x-a}{b-a} & \text{if } a < x \le b, \\ 1 & \text{if } x > b. \end{cases} \tag{9.40}$$

By differentiation,

$$f(x) = \begin{cases} \dfrac{1}{b-a} & \text{if } a < x < b, \\ 0 & \text{otherwise.} \end{cases} \tag{9.41}$$

Assume that $\alpha, \beta \in [a, b]$, then

$$P(\alpha \le X \le \beta) = F(\beta) - F(\alpha) = \frac{\beta - a}{b-a} - \frac{\alpha - a}{b-a} = \frac{\beta - \alpha}{b-a} \tag{9.42}$$

showing that this probability depends only on the length $\beta - a$ of the interval $[\alpha, \beta]$ and is independent of the location of this interval. That is why this variable is called *uniform*.

A continuous random variable is called *exponential* if

$$F(x) = \begin{cases} 0 & \text{if } x \le 0, \\ 1 - e^{-\lambda x} & \text{if } x > 0, \end{cases} \tag{9.43}$$

where $\lambda > 0$ is a given parameter. It can be easily seen that

$$f(x) = \begin{cases} \lambda_e^{-\lambda_x} & \text{if } \quad x > 0, \\ 0 & \text{otherwise.} \end{cases} \tag{9.44}$$

The exponential distribution has the forgetfulness property, which means the following: Let $0 < x < x + y$, then

$$P(X > x + y | X > y) = P(X > x), \tag{9.45}$$

showing that if X represents the lifetime of any entity, then at any age y, surviving an additional x time period does not depend on its age. Because of the forgetfulness property, exponential variables are seldom used in lifetime modeling. For such purposes, the gamma or the Weibull distribution is used, which is explained in Karamouz et al. (2010).

In statistical methods, distributions arising from the normal variable are used mostly. The *standard normal* variable is given by the density function

$$\phi(x) = \frac{1}{\sqrt{2\pi}} e^{-x^2/2} (-\infty < x < \infty). \tag{9.46}$$

The distribution function denoted by $\phi(x)$ is tabulated and cannot be given by an analytic form. A general *normal* variable is obtained as follows. Let μ and $\sigma > 0$ be two given parameters, and Z a standard normal variable. Then $X = \sigma Z + \mu$ follows a normal distribution with parameters μ and σ. It is easy to see that, in general,

$$F(x) = \phi\left(\frac{x - \mu}{\sigma}\right) \tag{9.47}$$

and

$$f(x) = \frac{1}{\sigma\sqrt{2\pi}} e^{-\left((x-\mu)^2\right)/2\sigma^2}. \tag{9.48}$$

In the case of normal distribution, the general method cannot be used easily since the standard normal distribution function ϕ is only tabulated. Therefore, the solution of the equation

$$\phi\left(\frac{x - \mu}{\sigma}\right) = u$$

requires the use of a function table, which can be included in the general software. An easier approach is offered by the central limit theorem, which implies that if $u_1, u_2, \ldots, u_n$ are independent uniform numbers in [0, 1], then for large values of n,

$$Z = \frac{\sum_{k=1}^{n} u_k - (n/2)}{\sqrt{n/12}} \tag{9.49}$$

is a standard normal value, so if μ and σ^2 are given, then

$$X = \sigma Z + \mu \tag{9.50}$$

follows the normal distribution with mean μ and variance σ^2.

TABLE 9.1
Expectations and Variances of the Most Popular Distribution Types

Distribution	$E(x)$	$Var(x)$
Bernoulli	p	pq
Binomial	Np	npq
Geometric	$\frac{1}{p}$	$\frac{q}{p^2}$
Negative binomial	$\frac{r}{p}$	$\frac{rq}{p^2}$
Hypergeometric	$n\frac{s}{N}$	$n\frac{s}{N}\left(1-\frac{s}{N}\right)\left[1-\frac{n-1}{N-1}\right]$
Poisson	Λ	λ
Uniform	$\frac{a+b}{2}$	$\frac{(b-a)^2}{12}$
Exponential	$\frac{1}{\lambda}$	$\frac{1}{\lambda^2}$
Gamma	$\frac{\alpha}{\lambda}$	$\frac{\alpha}{\lambda^2}$
Weibull	$\lambda^{-1/\beta}\Gamma\left(1+\frac{1}{\beta}\right)$	$\lambda^{-2/\beta}\Gamma\left(1+\frac{2}{\beta}\right)-\mu^2$
Normal	μ	σ^2
Chi-square	n	$2n$
T	0 (if $n > 1$)	$\frac{n}{n-2}(\text{if } n > 2)$
F	$\frac{m}{m-2}(\text{if } m > 2)$	$\frac{m^2(2m+2n-4)}{m(m-2)^2(m-4)}(\text{if } m > 4)$

Source: Karamouz, M., et al., *Water Resources Systems Analysis.* Lewis Publishers, CRC Press Company, Boca Raton, FL, 2003.

A comprehensive summary of random number generators and simulation methods is presented by Rubinstein (1981). In Table 9.1, the expectations and variances of the most popular distribution types are given.

9.3.2 Stochastic Processes

Assume that a certain event occurs at random time point $0 \le t_1 < t_2 < \ldots$. These events constitute a *stochastic process*. For instance, the times when earthquakes, floods, rainfalls, and so on occur define stochastic processes. The process can be completely defined mathematically, if we know the distribution functions of t_1, $t_2 - t_1$, $t_3 - t_2$. A very important characteristic of the stochastic process is the number of events $N(t)$ that occur in the time interval $[0, t]$. The *Poisson process,* which is defined as follows, is the most frequently used stochastic process as expressed by Karamouz et al. (2003):

1. The number of events is zero at $t = 0$. This shows that the process starts at $t = 0$.
2. The number of events occurring in mutually exclusive time intervals is independent, called "the independent increment assumption." For example, for time intervals of $0 < t_1 < t_2 < t_3 < t_4$, this condition states that the number of events in interval $[t_1, t_2]$

[which is $N(t_2) - N(t_1)$] is independent of the number of events in interval $[t_3, t_4]$, which is $N(t_4) - N(t_3)$.
3. The distribution of $N(t + s) - N(t)$ depends on only s and is independent of t. This means that the distribution of the number of events occurring in any given time interval depends only on the length of the interval and not on its location. This condition is known as the stationary increment assumption.
4. In a small interval of length Δt, the probability that one event occurs is a multiplier of Δt.
5. In a small interval of length Δt, the probability that at least two events take place is approximately zero.

Conditions (4) and (5) can be expressed in the mathematical format as follows:

$$\lim_{\Delta t \to 0} \frac{P(N(\Delta t) = 1)}{\Delta t} = \lambda \tag{9.51}$$

where $\lambda > 0$ is a constant:

$$\lim_{\Delta t \to 0} \frac{P(N(\Delta t) \geq 2)}{\Delta t} = 0. \tag{9.52}$$

Equation 9.53 denotes a Poisson distribution with parameter λ_t for $N(t)$ based on the above mentioned assumption:

$$P(N(t) = k) = \frac{(\lambda t)^k}{k!} e^{-\lambda t}. \tag{9.53}$$

The distribution functions of different time intervals ($X_1 = t_1, X_2 = t_2 - t_1, X_3 = t_3 - t_2, \ldots$), which can be denoted by $F_1, F_2, F_3, \ldots$, are shown in Equations 9.54 and 9.55.

$$F_1(t) = P(t_1 < t) = 1 - P(t_1 \geq t) = 1 - P(N(t) = 0) = 1 - e^{-\lambda t}. \tag{9.54}$$

Therefore, X_1 is exponentially distributed with expectation $1/\lambda$.

For all $k \geq 2$ and $s, t \geq 0$, the distribution function of X_k is the same as that of X_1. All variables $X_1, X_2, X_3, \ldots$ are independent of each other, and they are exponential with the same parameter X as illustrated in Equation 9.55.

$$\begin{aligned} P(X_k > t | X_{k-1} = s) &= P(0 \text{ event occurs in} (s, s+t) | X_{k-1} = s) = P|(0 \text{ event occurs in} (s, s+t)) \\ &= P(0 \text{ event in} (0, t)) = P(N(t) = 0) = e^{-\lambda t}. \end{aligned} \tag{9.55}$$

The distribution functions of the above three distributions cannot be presented in simple equations; therefore, they are tabulated, and so their particular values can be looked up from these tables. For more details and tables, see Milton and Arnold (1995) or Ross (1987).

9.3.3 Artificial Neural Networks "Data-Driven Modeling"

ANNs are robust and easy–to-use global nonlinear function approximates (StatSoft, 2002). ANNs provide a convenient means of either (1) simulating a system in the cases that there is a large dataset but no known mathematical model exists or (2) simplifying an excessively complex model. These models are used in different urban water resource management fields, such as simulating urban

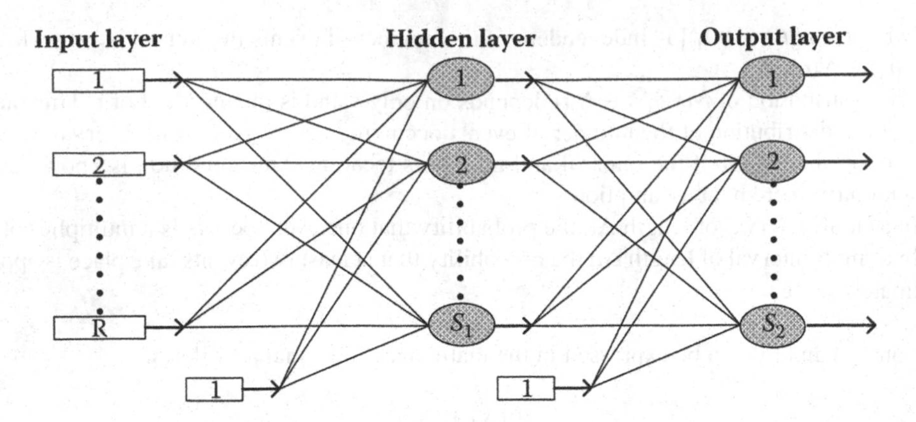

FIGURE 9.9 Simple neural network.

water demand based on the hydroclimatic and socioeconomic parameters and urban runoff based on the rainfall and basin characteristics.

The architectures of ANNs are adapted from biological neural networks, recognizing patterns, and learning from their interaction with the environment. It is believed that the powerful functionality of a biological neural system is attributed to the parallel distributed processing nature of a network of cells known as neurons. An ANN emulates this structure by distributing the computation to small and simple processing units, called artificial neurons, which could receive a number of inputs from either the original data or other neurons. Each connection to a given neuron has a particular strength or weight that can have positive or negative values. Neurons with similar characteristics are arranged into a layer. There are basically three types of layers. The first layer connecting the input variables is called the input layer. The last layer connecting the output variables is called the output layer. Layers between the input and output layers are called hidden layers, which could be more than one.

A simple neural network is shown in Figure 9.9, where we have the input variables in the nodes of the input layer and the output variables in the nodes of the output layer. The hidden layers in the middle could give a considerable level of flexibility in model formulation.

Let z_j denote the artificial variables corresponding to the jth neuron of the hidden layer. In the case of linear neural networks, this variable is calculated as follows:

$$z_j = \sum_{i=1}^{m} w_{ij} x_i, \tag{9.56}$$

where m is the number of input variables. The coefficients w_{ij} are constants, which are determined by the training procedure to be explained later. The relations between the output and hidden variables are also linear:

$$y_k = \sum_{j=1}^{l} \overline{w}_{jk} z_j, \tag{9.57}$$

where I is the number of hidden variables. The coefficients $\overline{w}_{jk}$ are also estimated by the training process. In the case of nonlinear neural network relations (transfer functions), Equations 9.55 and 9.56 have certain nonlinear elements; the training process also determines the parameters of the nonlinear transfer functions. Similar relations to Equations 9.55 and 9.56 could be assumed between the variables of the consecutive hidden layers.

There are many transfer functions used in developing neural networks, but the three most popular ones are introduced here. The hard-limit transfer function is used to create neurons that make classification decisions (Demuth and Beale, 2002). As shown in Figure 9.10, this function limits the neuron's output to either 0, if the net input argument $n < 0$, or 1, if $n \geq 0$.

The linear transfer function is shown in Figure 9.11. Neurons of this type are employed in the output layer as a linear approximator. The sigmoid transfer function depicted in Figure 9.12 takes the input, which restricts the output to the range 0–1 for any input value between plus and minus infinity. This transfer function is commonly used in backpropagation networks because it is differentiable (Demuth and Beale, 2002).

ANNs can be trained using a backpropagation algorithm (BPA), also called the generalized delta rule, during supervised learning. The BPA utilizes the chain rule of differentiation to calculate the gradient of the error function efficiently. In this algorithm, network weights are moved along the

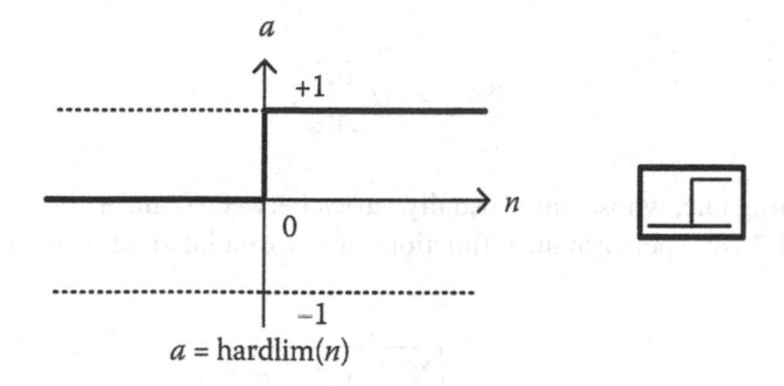

FIGURE 9.10 Hard-limit transfer function.

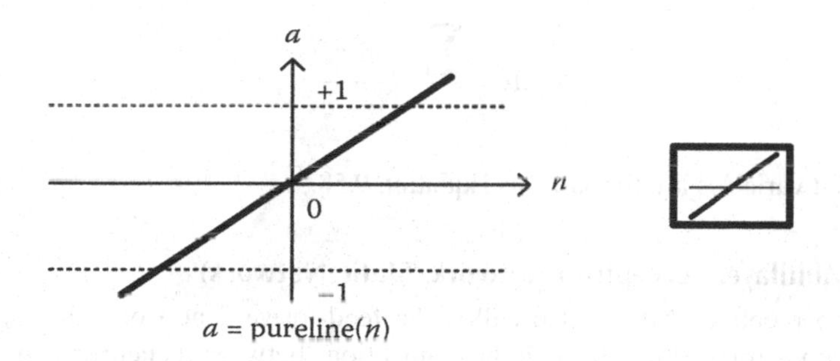

FIGURE 9.11 Linear transfer function.

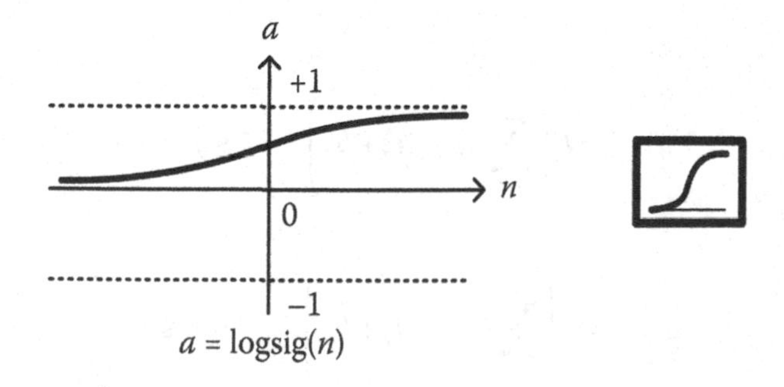

FIGURE 9.12 Log-sigmoid transfer function.

negative gradient of the performance function. The term "back propagation" refers to how the gradient is computed for nonlinear multilayer networks. Several performance functions may be selected to evaluate network performance, such as root mean squared error (RMSE), mean absolute error (MAE), and sum squared error (SSE). Assume that the following equation describes SSE performance function:

$$\text{SSE} = \sum_{k=1}^{S_2} \left(a_k^2 - T_k\right)^2, \tag{9.58}$$

where a_k^2 is the output of the kth neuron of the output layer and T_k is its corresponding target. S_2 is the number of neurons in the output layer. The weights in the training process are updated through each iteration (which is usually called an epoch) in the steepest descent direction as follows:

$$\Delta w_{kj} = -\alpha \frac{\partial E}{\partial w_{kj}}, \tag{9.59}$$

where α is the learning rate, whose value usually varies between 0 and 1.

The RMSE and MAE performance functions are formulated in Equations 9.60 and 9.61, respectively:

$$\text{RMSE} = \sqrt{\frac{\sum_{k=1}^{S_2} \left(a_k^2 - T_k\right)^2}{S_2}}, \tag{9.60}$$

$$\text{MAE} = \frac{\sum_{k=1}^{S_2} \left|a_k^2 - T_k\right|}{S_2}, \tag{9.61}$$

where the applied variables are the same as Equation 9.58.

9.3.3.1 The Multilayer Perceptron Network (Static Network)

The multilayer perceptron (MLP), also called the feed-forward network, is a static network in which information is transmitted through the connections between its neurons only in the forward direction and the network has no feedback, which covers its initial and past states. MLP implements mappings from the input pattern space to the output space. MLP can be trained with the standard BPA. A typical three-layer MLP, which, mathematically, can be expressed in the following formulations:

$$a_j^1(t) = F\left(\sum_{i=1}^{R} w_{j,i}^1 p_i(t) + b_j^1\right), \quad 1 \le j \le S_1, \tag{9.62}$$

$$a_k^2(t) = G\left(\sum_{j=1}^{S_1} w_{k,j}^2 a_j^1(t) + b_k^2\right), \quad 1 \le k \le S_2, \tag{9.63}$$

where t denotes a discrete time, R is the number of input signals, and S_1 and S_2 are the numbers of the hidden and output neurons, respectively. w^1 and w^2 are the weight matrices of the hidden and output layers, b^1 and b^2 are the bias vectors of the hidden and output layers, p is the input matrix, and a^1 and a^2 are the output vectors of the hidden and output layers. F and G are the activation functions of the hidden and output layers, respectively (Karamouz et al., 2007).

Example 9.4

The studies in a city have shown that water consumption is a function of three climatic variables, including precipitation, weather humidity, and average temperature. The monthly data of water consumption, monthly rainfall, monthly average humidity, and temperature of this city are given in Tables 9.2–9.5. The studies have also demonstrated that the relation of water consumption with these parameters is complicated and cannot be formulated in regular form. Develop an MLP

TABLE 9.2
Data of Monthly Water Consumption (MCM) of Example 9.4

Year	1992	1993	1994	1995	1996	1997	1998	1999	2000	2001	2002	2003
January	46.3	50.9	52.6	53.8	57.3	63.0	63.8	64.8	66.9	68.0	66.4	66.9
February	48.2	52.5	53.3	54.7	58.5	64.1	65.2	65.8	67.8	68.6	68.6	67.2
March	50.1	54.0	54.8	56.4	59.4	66.2	67.4	66.8	69.4	70.2	69.2	69.1
April	50.3	55.9	56.9	59.2	61.3	65.6	68.6	68.1	71.0	69.1	66.8	69.9
May	55.2	61.3	63.3	65.0	68.7	72.5	75.8	75.8	76.7	73.6	72.0	73.1
June	63.1	68.8	70.7	71.9	77.9	78.1	84.1	81.5	80.5	76.3	80.2	79.8
July	69.1	71.8	75.4	78.5	82.2	82.0	88.4	83.7	84.1	77.5	83.6	86.0
August	67.6	71.6	74.4	78.3	82.3	80.6	85.5	83.9	81.6	76.4	82.7	84.3
September	61.7	66.9	66.7	71.8	78.3	74.9	80.0	80.4	76.9	72.2	78.8	79.1
October	56.4	57.9	57.8	61.8	70.0	69.5	71.8	72.9	69.5	66.2	73.5	73.1
November	53.4	53.7	54.7	58.2	65.8	65.0	67.9	68.7	67.5	64.2	69.3	70.4
December	51.2	52.6	54.0	57.2	63.9	63.2	66.3	67.2	67.7	65.0	67.2	68.7

TABLE 9.3
Data of Monthly Precipitation (mm) of Example 9.4

Year	1992	1993	1994	1995	1996	1997	1998	1999	2000	2001	2002	2003
January	36	36.1	31.1	7.3	57.2	22.7	34.9	52.2	48.6	35	29.2	36.2
February	24.5	40.1	0.3	30.1	27.9	0	8	64.5	4	16.2	22.6	6.1
March	39.5	12.7	43.2	22	25.5	85.7	27.1	31.7	3	7.4	4.9	13.8
April	1.5	70	3.6	16.5	52.3	9.9	1.2	1.8	0.6	0	8	8.8
May	6	2.6	9.8	0.3	9.6	5	3	0.4	0	0	0	0
June	0	0	0	0	0	0.4	0	0	0	0	0	0
July	0	0	0	0	0	0	0	0	0	0	0	0
August	0	0	0	0	0	0	0	0	0	0	0	0
September	0	0	0	0	0	0	0	0	0	0	0	0
October	0	11.8	23.6	0	1.5	8.5	0	0.1	1.9	2.4	0	0
November	17	4.1	89.3	0.4	1.5	56.2	0.2	11.2	13.8	6.9	0.3	3
December	29.3	0.2	26.1	70.4	1.2	47.8	14.5	66.6	83.6	123.4	49.6	47.3

TABLE 9.4
Data of Monthly Average Temperature of Example 9.4

Year	1992	1993	1994	1995	1996	1997	1998	1999	2000	2001	2002	2003
January	10.1	10.7	14.8	13.2	13.3	13.5	11.7	14.1	12.3	12.2	12.1	13.5
February	13.3	13.8	15	15.5	16.3	12.7	14.9	15.9	14.3	15.2	15.1	16.2
March	16.3	18.7	19.7	19.6	19.5	16.6	18.6	19	19.2	21.3	21	19.3
April	24.9	24.2	26.9	24.6	24.5	23.8	26	26.9	29	27.4	25	26.4
May	31.2	29.8	31.2	31.5	32.9	31.7	32.4	33.3	33.4	32.7	32.7	32.6
June	36.4	34.5	34.9	35.6	35.5	36.3	37.4	37.3	36.2	35.8	35.6	36.7
July	37	36.5	36	36.5	37.9	37.1	37.5	37.7	38.8	37.3	38	37.3
August	37.3	36	35.5	36.6	37.3	35.5	38.5	38.1	38.4	37.6	37.1	36.9
September	32.6	32.2	32.9	31.9	33.1	33	34.1	33.8	32.6	33.3	33.5	33.2
October	25.3	27.1	27.5	26	26	27.9	26.9	28.7	26.6	27.6	28.6	28.4
November	19	18.5	20.8	18.9	19.7	19.9	21.5	19.4	18	18.5	19.1	19.1
December	12.9	15	11.3	12.4	16.8	13.7	16.7	13.4	13.8	16.2	14.2	14.4

TABLE 9.5
Data of Monthly Average Humidity (%) of Example 9.4

Year	1992	1993	1994	1995	1996	1997	1998	1999	2000	2001	2002	2003
January	64	77	68	74	77	64	73	71	69	77	69	67
February	55	65	50	66	68	39	61	69	50	59	62	51
March	52	58	50	55	60	55	57	55	34	48	45	45
April	37	54	42	45	51	49	42	35	30	30	43	40
May	30	40	30	32	38	33	31	25	23	20	23	26
June	23	26	22	24	28	26	27	21	19	16	22	23
July	24	28	27	25	29	21	28	26	25	21	20	21
August	28	30	26	24	29	23	33	30	27	35	20	33
September	35	32	35	32	27	25	33	32	31	32	28	25
October	39	43	47	39	43	47	38	38	37	43	42	46
November	62	52	67	47	51	66	52	53	58	43	50	55
December	75	61	71	68	64	77	54	73	82	78	65	74

model to simulate the monthly water consumption of this city as a function of monthly precipitation, humidity, and temperature.

Solution:

The input data of the MLP model should be completely random. For this purpose, the trend of city water consumption is estimated as a line with the following equation:

$$y = 0.1331x + 58.542.$$

The values estimated by the above equation are subtracted from the observed data. The remaining values are random and are used as input of the MLP model (Figure 9.13). The other data are also checked to determine whether there is any trend. Then the randomized consumption data and climatic data are standardized by subtracting the average of the data and then dividing by the standard division of data. This will help us to get better results from the MLP model, because the transfer functions are more sensitive to variations between 0 and 1.

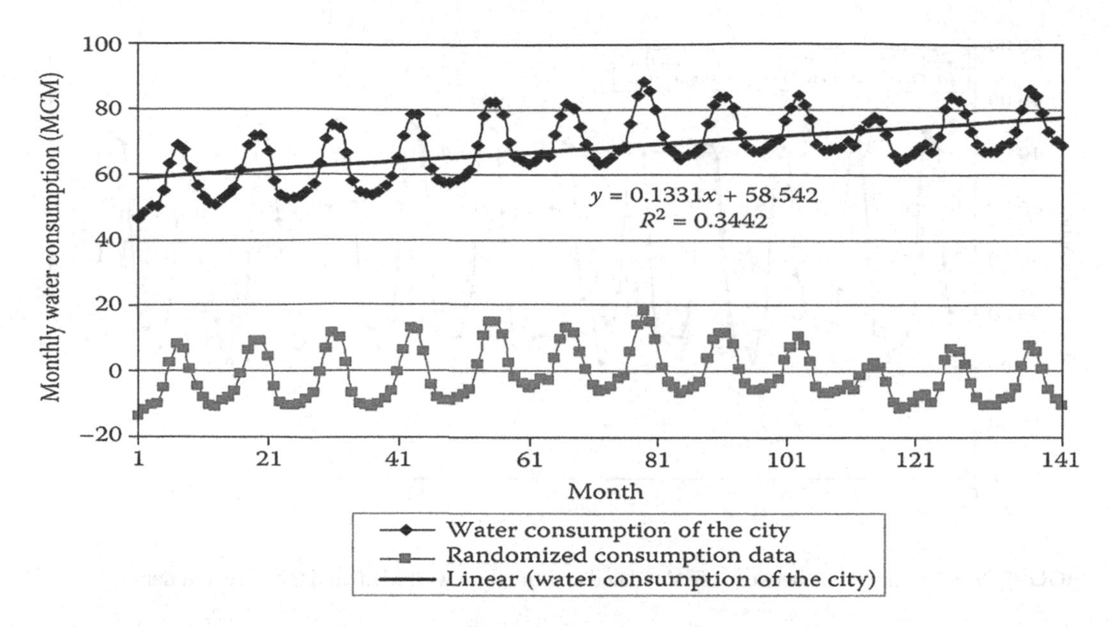

FIGURE 9.13 Randomization of monthly water consumption values of the city.

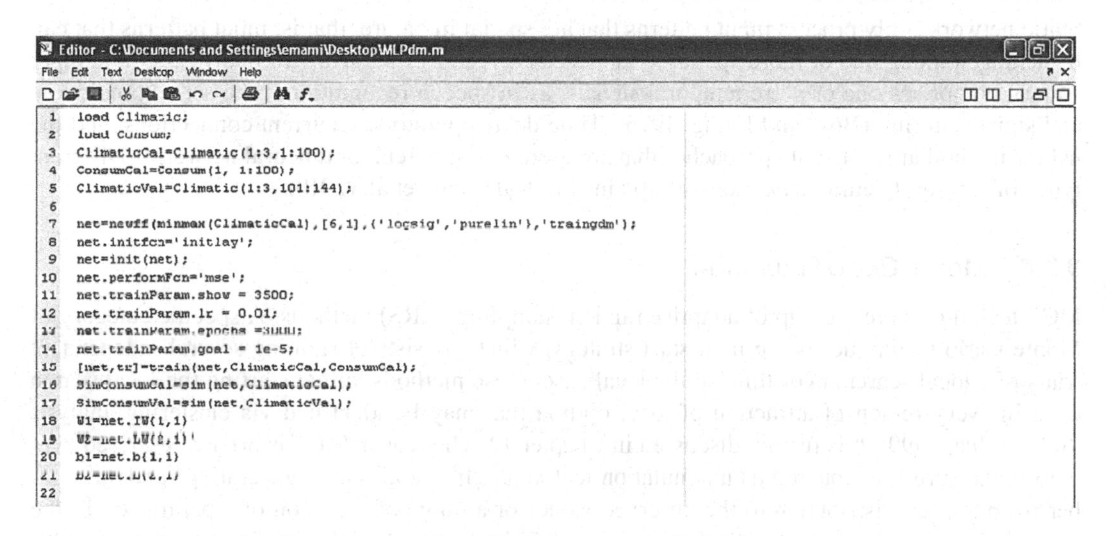

FIGURE 9.14 Structure of the developed MLP model for simulation of monthly water consumption in the city of Example 9.4.

For simulating monthly water consumption, a three-layer neural network with six neurons in the hidden layer and log-sig and linear transit functions in the hidden and output layers is employed as shown in Figure 9.14. One hundred series of data are used for model calibration (ClimaticCal and ConsumCal), and the remaining data are considered for model validation (ClimaticVal and ConsumVal). The model is trained in 3,500 steps (epochs), and its goal is to achieve an error of <0.00001.

The simulated and recorded monthly consumption data are compared in Figure 9.15. As shown in this figure, about 30% of the data have been simulated with <5% error, and 60% of the data have been simulated with <10% error.

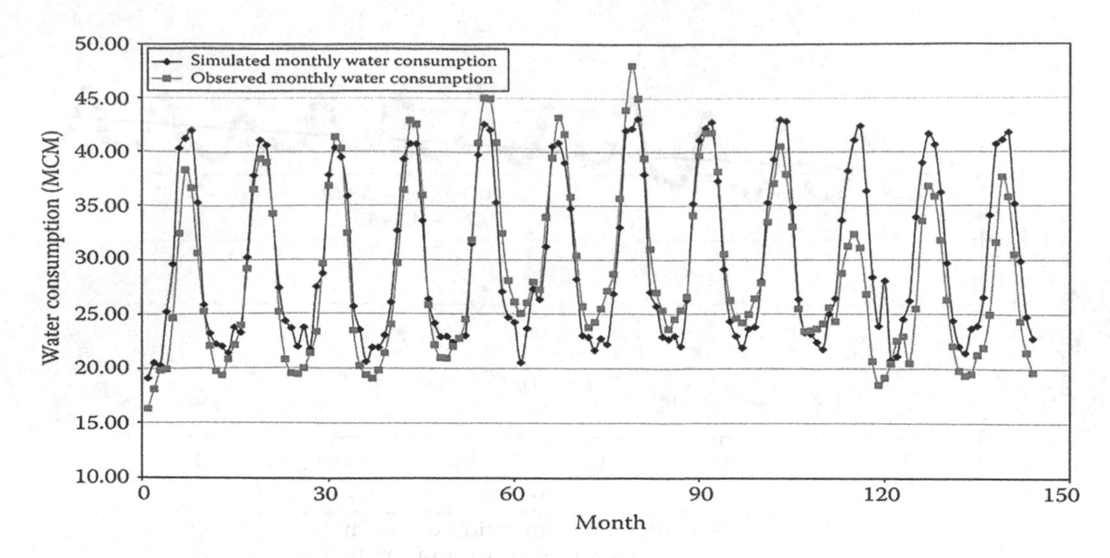

FIGURE 9.15 Comparison between the simulated consumption with MLP and the observed data.

9.3.3.2 Temporal Neural Networks

Static networks only process input patterns that are spatial in nature, that is, input patterns that can be arranged along one or more spatial axes such as a vector or an array. In many tasks, the input pattern comprises one or more temporal signals, as in speech recognition, time series prediction, and signal filtering (Bose and Liang, 1996). Time delay operators, recurrent connections, and the hybrid method are different approaches that are used to design temporal neural networks. Different types of temporal neural networks are explained in Karamouz et al. (2010).

9.3.4 Monte Carlo Simulation

MCS techniques are a group of adaptive random sampling (ARS) methods. A specific category of Monte Carlo techniques is the multistart strategy, which consists of running several independent trials of a local search algorithm. In the ideal case, these methods aim at starting the local search once in every region of attraction of local optima that may be identified via clustering analysis (Solomatine, 1999). It is further discussed in Chapter 12. The reason MCS is briefly described here is to make sure it is counted as a simulation technique. It is mainly for generating random number from a fitted distribution to the observed values or a range of variation of a parameter. In the next section, an extension of MCS for spatial variability as Sequential Gaussian Simulation (SGS) method is presented as it uses kriging method that is explained in detail in this chapter.

9.3.4.1 Sequential Gaussian Simulation

Sequential Gaussian Simulation (SGS) is one of the most common geostatistical approaches for the prediction and uncertainty of simulated data, which considers spatial variability of observations by the kriging method (Goovaerts, 1997). The values simulated by SGS rely on the numerical magnitude of observations in addition to their spatial configuration. Repetition of the simulations at each location thus provides a quantitative and visual measurement of spatial uncertainty. While the derivation of the kriging equations (simple, ordinary, and universal) does not depend on any distributional assumptions, the SGS algorithm requires normally distributed data. Suppose the data distributions do not follow a normal distribution; in that case, it is required to transform the input data to a normal distribution as the discrepancy might affect the SGS algorithm, which is highly dependent on the best fitted distribution of the random function.

The process to generate random fields in the SGS method includes the following steps:

1. Transform the original data to normal-score data with zero mean and unit variance.
2. Assign transformed data into simulation grid.
3. Generate a random path through the grid cells.
4. Visit a cell in the random path and use kriging to estimate a mean and standard deviation at that cell based on surrounding data and variogram. Then, construct the conditional cumulative distribution function using the mean and standard deviation of the variable. Observations at control points and all previously simulated values are used for conditioning the distribution function while maintaining the appropriate covariance structure (Bohling, 2007).
5. Select at random a value from the conditional cumulative distribution function and assign the cell value to that number.
6. Include the newly simulated value as part of the conditioning data.
7. Repeat steps 4 to 6 until all grid cells are taken into account (Wechsler, 2007).
8. Back-transform the realization into the original space.

Implementing the mentioned steps through the SGS algorithm would lead to generating a realization map. Repetition of this process and the consequent realization maps can then be incorporated for uncertainty investigations and producing probability maps. These realization maps can be the generated maps of SGS itself or could be the resulting maps of a model with the SGS generated map as its input. The adequate number of realizations is another question that should be addressed in this context. Ultimately, when the realization maps are prepared, different probability maps could be produced. For example, k% exceedance (or nonexceedance) maps could be generated by finding the relevant quantile in the realization values at all grid cells. These probability maps could be reported and compared with the deterministic approach maps and be a basis for subsequent risk and reliability investigations and prepare a better insight of the situation for decision-makers.

See Karamouz and Fereshtehpour (2019) and Chapter 14 for more details.

9.3.5 Mathematics of Growth

A sense of the future is an essential component in the mix of factors that should be influencing the environmental decisions we make today. Only a few factors may need to be considered in some circumstances, and the time horizon may be relatively short. Accurate predictions of the future cannot be expected, especially when the required time horizon may be extremely long. However, often simple estimates can be made that are robust enough that the insight they provide is most certainly valid. One can say, with considerable certainty, that world population growth at today's rates cannot continue for another 100 years. Also, simple mathematical models can be used to develop very useful "what if" scenarios: If population growth continues at a specific rate, and if energy demand is proportional to economic activity, and so forth, then the following would occur.

Growth is an increase in some quantity over time. Growth is not development until it is to become more mature. Growth by itself does not sustain communities, in fact, it can destroy them. Growth is by itself neither good nor bad. It depends on what is growing and when. Nothing grows forever. Uncontrolled growth could be cancer. The dynamics of uncontrolled growth eventually results in an overshoot and then a collapse. S-shaped growth is a commonly observed mode that reaches an equilibrium after an exponential growth. To understand the structure of underlying S-shaped growth, it is helpful to use the ecological concept of carrying capacity. The carrying capacity of any habitat/system is the number of organisms/elements of a particular type. It can support and is determined by the resources available and required in the environment and the resource for the population. As the carrying capacity is approached, the adequacy of the required resources diminishes, and the fractional net increase rate is forced to decline. The state of the system continues to grow, but at a slower rate, until resources are just scarce enough to halt growth. In general, a population may depend on many resources, creating a negative loop that might limit growth. The most binding constraint determines which of the negative loops will be most influential as the system's state grows.

Any actual quantity undergoing exponential growth can be interpreted as a population drawing on the resources in its environment. The carrying capacity concept is subtle and complex. While it is appropriate to consider the carrying capacity of an environment to be constant in some situations, in general, the carrying capacity of an environment is intimately intertwined with the evolution and dynamics of the species it supports. Similarly, all businesses and organizations grow in the context of a market, society, and physical environment that imposes limits to their growth.

Despite the dynamic character of the carrying capacity, there is, at any moment, a limit to the size of the population (the current carrying capacity), which, if exceeded, causes the population to fall. Further, the carrying capacity itself cannot grow forever. A system generates S-shaped growth only if two critical conditions are met. First, the negative loops must not include any significant time delays (if they did, the system would overshoot and oscillate around the carrying capacity).

The two main components of development are resources and infrastructures (Figure 9.16). Underdeveloped countries have slow growth, developing countries have fast growth, and developed countries also have slow growth because of the less tendency to develop. A good example could be the state of development in a different part of the world. Underdeveloped countries face scarcity of (1) natural resources and/or (2) skilled human resources and inadequacy of (3) infrastructure. The developing regions, usually have (1) & (2) and (3) so developing. The developed countries usually have reached their carrying capacity, their natural resources are getting limited, and technological and economic development advances cannot keep up with some developing countries' competition.

9.3.5.1 Exponential Growth

Exponential growth is quite common, which could occur in any situation where the increase in some quantity is proportional to the amount currently present. Exponential growth is a particular case of a first-order rate process. Suppose something grows by a fixed percentage each year. This can be represented mathematically as follows:

$$N_t = N_0(1+r)t \tag{9.64}$$

where N_0 is the initial amount, N_t is the amount after t years, and r is growth rate (fraction per year)

For most events of interest in the environment, it is usually assumed that the growth curve is a smooth, continuous function without the annual jumps that Equation 9.65 is based on. One way to state the condition that leads to exponential growth is that the quantity grows in proportion to itself, that is, the rate of change of the quantity N is proportional to N. The proportionality constant r is called the rate of growth and has units of (1/time).

$$\frac{dN}{dt} = rN \tag{9.65}$$

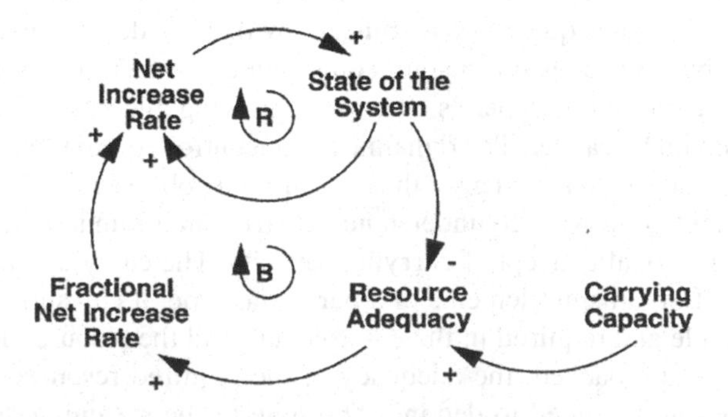

FIGURE 9.16 S-shaped structure and characteristic time path.

The solution is the exponential relation:

$$N = N_0 e^{rt} \tag{9.66}$$

A quantity that is growing exponentially requires a fixed amount of time to double in size, regardless of the starting point. That means it takes the same amount of time to grow from N_0 to $2N_0$, and so on. The doubling time (t_d) of a quantity that grows at a fixed exponential rate r is easily derived. From Equation 9.66, the doubling time can be found by setting $N = 2N_0$ at $t = t_d$ and after taking logarithm from both sides:

$$t_d = \frac{Ln2}{r} \tag{9.67}$$

The equation is independent of N_0. Therefore, doubling the quantity does not depend on how much you start with.

9.3.5.2 Logistic Growth

Population projections are often mathematically modeled with a logistic or S-shaped (sigmoidal) growth curve like the one shown in Figure 9.16. Such a curve has great intuitive appeal. It suggests an early exponential growth phase while conditions for growth are optimal, followed by slower and slower growth as the population nears the carrying capacity of its environment. Biologists have successfully used logistic curves to model populations of many organisms, including protozoa, yeast cells, water fleas, fruit flies, pond snails, worker ants, and sheep (Southwick, 1976).

Mathematically, the logistic curve is derived from the following differential equation:

$$\frac{dN}{dt} = rN\left(1 - \frac{N}{K}\right) \tag{9.68}$$

where N is the population size, K is called the carrying capacity of the environment, and r is the exponential growth rate constant that would apply if the population size is far below the carrying capacity. Strictly speaking, r is the growth rate constant when $N = 0$; the population is zero. However, population growth can be modeled with a constant rate r without significant error when the population is much less than carrying capacity. As N increases, the growth rate slows down, and eventually, as N approaches K, the growth stops altogether, and the population stabilizes at a level equal to the carrying capacity. The factor $\left(1 - N/K\right)$ is often called environmental resistance. As the population grows, the resistance to further population growth continuously increases. The solution to Equation 9.68 is:

$$N = \frac{K}{1 + e^{-r\left(t - t^*\right)}} \tag{9.69}$$

Note that t^* corresponds to the time at which the population is half of the carrying capacity, $K/2$. t^* could be solved by substituting $t = 0$ as the initial condition in Equation 9.69. Therefore, the integrated expression of Equation 9.68 could be written as

$$N(t) = \frac{KN_0}{N_0 + (K - N_0)e^{-rt}} \tag{9.70}$$

where N_0 is the population at time $t = 0$.

To find r from the growth rate measured at R_0 (instantaneous rate constant at $t = 0$). If we characterize growth at $t = 0$ as exponential, then:

$$\left.\frac{dN}{dt}\right|_{t=0} = R_0 N_0 \tag{9.71}$$

But from Equation 9.68:

$$\left.\frac{dN}{dt}\right|_{t=0} = rN_0\left(1-\frac{N_0}{K}\right) \tag{9.72}$$

So r could be obtained as

$$r = \frac{R_0}{1-\dfrac{N_0}{K}} \tag{9.73}$$

The logistic curve can also be used to introduce another helpful concept in population biology called the maximum sustainable yield of an ecosystem. The maximum sustainable yield is the maximum rate that individuals can be harvested (removed) without reducing the population size. Therefore, the maximum sustainable yield will correspond to some populations less than the carrying capacity. In fact, because the yield is the same as $\frac{dN}{dt}$, the maximum yield will correspond to the point on the logistic curve where the slope is a maximum (its inflection point). Setting the derivative of the slope equal to zero, we can find that point. The slope of the logistic curve is given as

$$\text{Yield} = \text{Slope} = \frac{dN}{dt} = rN\left(1-\frac{N}{K}\right) \tag{9.74}$$

Setting the derivative of the slope equal to zero gives:

$$\frac{d}{dt}\left(\frac{dN}{dt}\right) = r\frac{dN}{dt} - \frac{r}{K}\left(2N\frac{dN}{dt}\right) = 0$$

Letting N^* be the population at the maximum yield point gives

$$1-2\frac{N^*}{K} = 0$$

So that for maximum sustainable yield:

$$N^* = \frac{K}{2} \tag{9.75}$$

That is, if population growth is logistic, then the maximum sustainable yield will be obtained when the population is half the carrying capacity. The yield at that point can be determined as

$$\text{Maximum yield} = \left(\frac{dN}{dt}\right)_{\max} = r\frac{K}{2}\left(1-\frac{K/2}{K}\right) = \frac{rK}{4} \tag{9.76}$$

Using Equation 9.73, the maximum yield in terms of the current growth rate R_0 and current size N_0 is as follows:

$$\left(\frac{dN}{dt}\right)_{\max} = \left(\frac{R_0}{1-\dfrac{N_0}{K}}\right)\frac{K}{4} = \frac{R_0\left(K^2\right)}{4(K-N_0)} \tag{9.77}$$

9.3.5.3 Limits to Growth

Population and industrial growth are inherently exponential, and that exponential growth takes one to any existing limit quickly, whatever its magnitude is. Global society will most likely adjust to limits by overshoot and collapse and not by S-shaped growth. However, sustainable development is possible if essential changes are made. Politics and the market are inherently unsuited to adopt constructive policies that can lead to sustainable development. The main components of limits to growth are population, affluence, and economic aspects.

9.3.5.3.1 The Demographic Transition

The theory of demographic transition attempts to explain the observed negative correlation between income level and population growth rate. Figure 9.17 shows the theory of demographic transition.

The four stages mentioned in Figure 9.17 are described as follows:

Stage 1: Low-income economy with high birth and death rates

Stage 2: With rising real incomes, nutrition and improved public health measures, leading to a falling death rate and rapid population growth

Stage 3: Due to some or all of the increasing costs of child treating reduced benefits of large family size, increasing opportunity costs of home employment improved economic and social status of women, the birth rate falls, and the rate of population growth declines

Stage 4: High-income economy with equal and low birth and death rates and constant population size

9.3.5.4 Environmental Limits

Limits to growth have a main category including environmental and social limits. The limits to growth reported the results of a study in which a computer model of the world system, World3, was used to simulate its future. It represented the world economy as a single economy and included interconnections between that economy and its environment. It incorporated a limit to:

- Thc amount of land available for agriculture;
- The amount of agricultural output producible per unit of land in use;
- The amounts of nonrenewable resources available for extraction;
- The ability of the environment to assimilate wastes arising in production and consumption, which limit falls as the level of pollution increases.

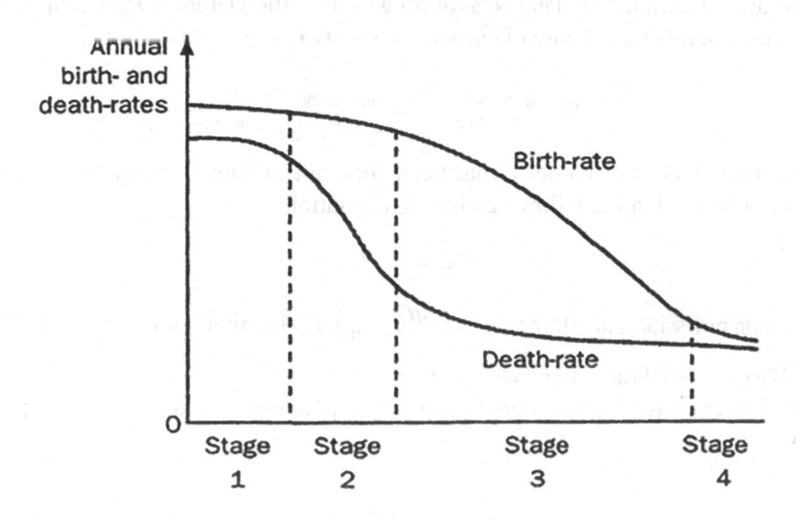

FIGURE 9.17 The theory of demographic transition.

The behavior of the economic system was represented as a continuation of past trends in key variables, subject to those trends being influenced by the relationships between the variables represented in the model. These relationships were represented in terms of positive and negative feedback effects. The behavior over time in the model of each of these variables depends on that of others and affects that of others.

9.3.5.4.1 The Kaya Identity

Predicting future carbon dioxide concentrations depends on numerous assumptions about population growth, economic factors, energy technology, and the carbon cycle itself. One way to build simple models of environmental problems is to start with the notion that impacts are driven by population, affluence, and technology, which is sometimes referred to as the Impact population affluence technology (IPAT) identity (Ehrlich and Holdren, 1971).

$$\text{Environmental impact} = (\text{Population}) \times (\text{affluence}) \times (\text{technology}) \tag{9.78}$$

The Kaya identity is an identity stating that the total emission level of the greenhouse gas carbon dioxide based on IPAT identity can be expressed as

$$C = \text{Population} \times \frac{\text{GDP}}{\text{Person}} \times \frac{\text{Primary energy}}{\text{GDP}} \times \frac{\text{Carbon}}{\text{Primary energy}} \tag{9.79}$$

where

C: carbon emission (GtC/year); GtC: Giga tonnes of Carbon

$\frac{\text{GDP}}{\text{Person}}$: Gross domestic product per capita $\left(\$/\text{person-year}\right)$

$\frac{\text{Primary energy}}{\text{GDP}}$: Primary energy intensity $\left(EJ/\$\right)$

$\frac{\text{Carbon}}{\text{Primary energy}}$: Carbon intensity $\left(GtC/EJ\right)$

Equation 9.79 expresses the carbon emission rate as the product of four terms: population, GDP, carbon intensity, and energy intensity. If each of the factors in a product can be expressed as a growing quantity (or decreasing) exponentially, then the overall rate of growth is merely the sum of the growth rates of each factor. That is, assuming each of the factors in Equation 9.79 is growing exponentially, the overall growth rate of carbon emissions r is given by

$$r = r_{\text{Population}} \times r_{\frac{\text{GDP}}{\text{Person}}} \times r_{\frac{\text{Primary energy}}{\text{GDP}}} \times r_{\frac{\text{Carbon}}{\text{Primary energy}}} \tag{9.80}$$

By adding the individual growth rates as has been done in Equation 9.80, an overall growth rate is found, which can be used in the following emission equation:

$$C = C_0 e^{rt} \tag{9.81}$$

where C is carbon emission rate after t years $\left(GtC/\text{year}\right)$, C_0 is initial emission rate $\left(GtC/\text{year}\right)$, and r is the overall exponential rate of growth (year^{-1}).

The cumulative emission from a quantity growing exponentially at a rate r, over a period of time T, is given by

$$C_{\text{tot}} = \frac{C_0}{r}\left(e^{rt} - 1\right) \tag{9.82}$$

TABLE 9.6
The 1990–2020 Average Annual Growth Rates (%/year) Used in the IPCC IS92a Scenario for Energy-Related Carbon Emissions

Region	Population	$\frac{\text{GDP}}{\text{Person}}$	$r_{\frac{\text{Primary energy}}{\text{GDP}}}$	$\frac{\text{Carbon}}{\text{Primary energy}}$
China and centrally planned Asia	1.03	3.91	−1.73	−0.32
Eastern Europe and ex-USSR	0.43	1.49	−0.66	−0.24
Africa	2.63	1.25	0.26	−0.21
United States	0.57	2.33	−1.81	−0.26
World	1.40	1.53	−0.97	−0.24

Example 9.5

Emissions from fossil-fuel combustion in 2010 are estimated to be 7.6 GtC/year. In the same year, atmospheric CO_2 concentration is estimated to be 390 ppm. Assume the atmospheric fraction remains constant at 0.38.

a. Assuming the energy growth rates shown in Table 9.6, estimate the energy-related carbon emission rate in 2050.
b. Estimate the cumulative energy-related carbon added to the atmosphere between 2010 and 2050.

Solution:

a. The overall growth rate in energy-related carbon emissions is just the sum of the individual growth rates:

$$r = 1.40\% + 1.53\% - 0.97\% - 0.24\% = 1.72\% = 0.0172\,/\,\text{year}$$

With 40 years of growth at 1.72% per year, the emission rate in 2050 would be

$$C_{2050} = C_{2010}e^{rt} = 7.6e^{0.0172\times 40} = 15.1\,GtC/\text{year}$$

b. Over those 40 years, the cumulative energy emissions would be

$$C_{\text{tot}} = \frac{C_0}{r}\left(e^{rt} - 1\right) = \frac{7.6}{0.0172}\left(e^{0.0172\times 40} - 1\right) = 437\ GtC$$

9.3.5.5 Social Limits to Growth

The process of economic growth has increasingly been unable to yield to expected social satisfaction in the general level of material affluence for biological needs of life-sustaining food, shelter, and clothing. The book, *Social limits to growth*, was published by Hirsch (1977), 5 years after *The Limits to Growth* (Meadows et al. 1972) book was published. As the average level of consumption rises, an increasing portion of consumption takes on social and individual aspects, so that "the satisfaction that individuals derive from goods and services depends on increasing measures on their own consumption and the consumption of others."

Once basic material needs are satisfied, further economic growth is associated with increasing income being spent on such positional goods. Consequently, growth in developed economies is a much less socially desirable objective than economists have usually thought. It does not deliver the increased personal satisfaction that was expected. Traditional, down-to-earth conceptions of social welfare may be misleading in such circumstances, as utilities are interdependent. Social limits to growth are not currently a problem in developing countries. Many people now live in conditions such that basic material needs are not satisfied. It is particularly true for people living in the poor nations of the world, but not restricted to them. Even in the wealthiest countries, income and wealth inequalities are such that many people live in material and social deprivation conditions.

A break between individual and social opportunities may occur for many reasons; excessive pollution and congestion are the most commonly recognized results. A neglected general condition that produces this break is a competition by people for a place rather than competition for performance. Society advancement is possible only by moving to a higher place among others by improving one's performance in relation to others. Social interaction of this kind is present, individual action is no longer for fulfilling individual choice: the preferred outcome may be attainable only through collective action. The limits have always been there at some point, but they have not until recent times become obvious. That is the product, essentially, of past achievements in material growth not subject to social limits. Increasing growth without caring about sustainable development may lead to overshoot and collapse. To overshoot means to go too far, to grow so large and fast that limits are exceeded. When an overshoot occurs, it imposes stresses that begin to slow and halt growth.

9.3.6 System Dynamics

Understanding the system and its boundaries, identifying the key variables, representing the physical processes or variables through mathematical relationships, mapping the structure of the model, and simulating the model for understanding its behavior are some of the major steps that are carried out in the development of a system dynamics simulation model. A system dynamics approach relies on understanding the complex interrelationships among different elements within a system. It is interesting to note that the central building blocks of system dynamics are well suited for modeling physical systems and representing nonphysical models developed using a systems thinking paradigm.

System dynamics was introduced in the 1960s by investigators at the Massachusetts Institute of Technology. It was initially rooted in the management and engineering sciences but has gradually developed in other fields. In system dynamics, a system is defined as a collection of elements that continuously interact over time as a unified body. The underlying pattern of interactions among the elements of a system is called the structure of the system. The way in which the elements or variables vary over time is referred to as the system's behavior. An excellent example of water resources engineering of a system is a storage tank. The structure of a tank is defined by the interactions among inflow, storage, outflow, and other site-specific issues such as release values. The structure of a storage tank includes the importance of variables influencing the system. The term "dynamics" refers to change over time. If something is dynamic, it is constantly changing in response to the stimuli influencing it. In the storage tank example, the behavior is described by the dynamics of the storage tank filling and draining, which is influenced by inflow, outflow, losses, and other factors affecting the system characteristics. In an urban setting, storage tanks (Figure 9.18) and the stock of

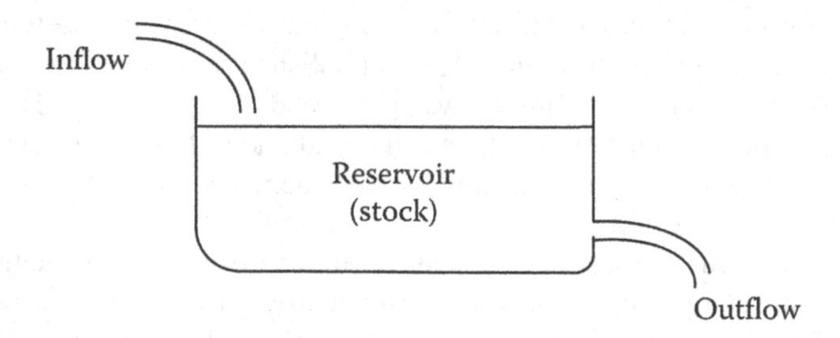

FIGURE 9.18 Schematic of an urban storage tank.

available water as it is affected by input and output present a unique challenge for applying system dynamics in urban water.

By defining the structure of a storage tank, it is possible to use system dynamics to analyze the behavior over time and how structural changes in one part of a system might affect the system's behavior as a whole.

Perturbing a system allows one to test how the system will respond under varying sets of conditions. The visual effects of system dynamics mapping help the decision makers to see the impact of different scenarios clearly. In the case of a storage tank, the impact of storage fluctuations due to supply system shortages or transfer system failure causing the elimination of a particular user could be analyzed.

9.3.6.1 Modeling Dynamics of a System

There are many tools used for implementing the object-oriented modeling approach. Computer software tools such as STELLA developed by High-Performance Systems Inc., VENSIM developed by Ventana System Inc., and POWERSIM developed by Powersim Software AS help execute these processes.

The power and simplicity of the system dynamics that lie in object-oriented programming are quite different from those used in functional algorithmic languages. The power of object-oriented simulation is the ease of constructing "what-if" scenarios and tackling large-scale, messy, real-world problems.

The system dynamics simulation tool contains objects such as stocks, flows, converters, and connectors. Stocks manifest as accumulations. They represent "conditions" within a system, such as water storage in a tank or a reservoir. Flow, representing "actions," represents variables whose values are measured as flow rates in and out of reservoirs (surface and underground tanks, treatment facilities, and distribution systems are examples of this type of object). A converter is used to transform inputs in the form of algebraic relationships and graphs into outputs such as mass balance equations, SCS, and rational method expressions, relating rainfall to runoff in urban catchments. A connector, representing the relationship between other objects, conveys information from one variable to the others. These diagrams also contain "clouds," which represent the boundaries of the system.

The steps in modeling the dynamics of a system are discussed below. The modeling process is *not* linear, and model developers usually find themselves cycling through the steps several times as they work on the model. It is sometimes necessary to cycle back through the entire sequence, as the model is refined based on the challenges that are made to its structure.

9.3.6.1.1 Define the Issue/System

For defining the system, the following steps should be followed:

- Clearly state the purpose of the model.
- Develop a base behavior pattern.
- Develop a system diagram.

In *Step 1* of the modeling process, additional tools will help make the purpose more operational. The first of these is a reference behavior pattern (RBP), a graph over time for at least variables that best capture the dynamic phenomenon and help focus efforts on the behavioral dimensions. In order to focus on the structural dimension, a system diagram composed of the essential actors or sectors could be utilized.

In *Step 2* of the modeling process, the hypotheses responsible for generating the behavior pattern(s) identified in Step 1 are represented. A "dynamic organizing principle" is a stock/flow-based, or feedback loop-based, framework that will reside at the model's core. When the modeler has laid out some stocks and flows, the next step in mapping a hypothesis is to characterize the flows.

Generic flow templates could capture the nature of each flow as it works in reality and achieves an operational specification. Once the flows associated with hypotheses have been characterized, the next step in mapping is to close the loops.

9.3.6.1.2 Test of Hypotheses

For testing the hypotheses responsible for generating the system dynamics, the following steps are taken:

- Mechanical mistake tests
- Robustness tests
- Reference behavior tests

These tests are designed to make the modeler aware of the limitations of the model's utility and to increase the modeler's confidence. In order to ensure learning from each test, before each simulation, sketch out the best guess as to how the sector/model will behave. Then, the behavior should be explained, and after the simulation is complete, either work to resolve any discrepancies in actual versus guessed behavior or resolve it in modeler rationale. If the test results are not acceptable, do not hesitate to cycle back through steps 1 and 2 of the modeling process.

Hypothesis testing always involves two statements:

H_0: The null hypothesis
H_1: The alternative hypothesis (also called HA)

The test involves comparing the value calculated from the given sample data and a threshold value calculated using the given level of significance. The number of tails can categorize hypothesis tests as one-tailed and two-tailed tests. One-tailed tests are also categorized as upper tailed or lower tailed. If the test is upper tailed, then the hypothesis statement is as follows:

$$H_0 : \mu \leq a$$
$$H_0 : \mu > a$$

If the test is lower tailed, then the hypothesis statement is as:

$$H_0 : \mu \geq a$$
$$H_0 : \mu < a$$

If the test is two-tailed, then the hypothesis statement is as follows:

$$H_0 : \mu = a$$
$$H_0 : \mu \neq a$$

Errors in hypothesis testing may arise because of the difference that may exist between a sample and its parent population. These errors may occur due to the inability to achieve perfect randomness, accessibility problems that could preclude inclusion of all diverse elements in a sample, sample sizes, the distribution type utilized, and the equipment or malfunction and human errors. Two types of error could occur:

- *Type 1 error*: A type 1 error, which is also called an error of commission, is the rejection of the null hypothesis when it is actually true. The probability of a hypothesis test's type 1 error is its size. The size of the hypothesis test has a value equal to α (the level of significance).

- *Type 2 error*: A type 2 error, which is also called an error of omission, is the failure to reject the null hypothesis when it is actually false. The symbol β represents the probability of a type 2 error (i.e., not rejecting the null hypothesis when it should be rejected). The power of a hypothesis test is equal to $1 - \beta$.

9.3.6.1.3 Design and Test Policies

For designing and testing the considered policies, the following steps should be done. In the first step, some effective policies/strategies have been identified (policy tests). Next, the sensitivity of their effectiveness to both the behavioral assumptions included in the model and the assumptions about the external environment is determined. Testing the former sensitivity is known as "sensitivity analysis." Testing the latter sensitivity is known as "scenario analysis."

9.3.6.2 Time Paths of a Dynamic System

The first step in modeling the system dynamics is to model the dynamic behavior of systems accurately. Dynamic behavior is considered as the behavior of systems over time, it is also called a time path. The modeler should identify the patterns of system behavior using the most effective and important variables of the system. Then the appropriate relationships and functions are defined to mimic the observed behavior of the system as the model calibration process and then the model can be employed for testing policies aimed at altering a system's behavior in desired ways.

The most fundamental modes of behavior are exponential growth, goal-seeking, and oscillation. Each of these is generated by a simple feedback structure: growth arises from positive feedback, goal-seeking arises from negative feedback, and oscillation arises from negative feedback with time delays in the loop. Other common modes of behavior, including S-shaped growth, S-shaped growth with overshoot and oscillation, and overshoot and collapse, arise from nonlinear interactions of the fundamental feedback structures (Sterman, 2000). Time paths are grouped into five distinct families, which will be introduced briefly in the following sections. The readers are suggested to refer to Karamouz et al. (2010) for more detailed information. Other, more complicated, observable time paths can be modeled as combinations of these five groups.

9.3.6.2.1 Linear Family

Linear growth is the first identifiable family, which is actually quite rare. Linear growth requires that there be no feedback from the state of the system to the net increase rate because the net increase remains constant even as the state of the system changes. The linear family of time paths includes equilibrium, linear growth, and linear decline.

9.3.6.2.2 Exponential Family

The second distinct family of time paths is the exponential family raised from positive (self-reinforcing) feedback. For example, population growth and water demand growth are presented as exponential time paths. Positive feedback need not always generate growth. It can also create self-reinforcing decline, as when a drop in stock prices erodes investor confidence, leading to more selling, lower prices, and still lower confidence.

9.3.6.2.3 Goal-Seeking Family

The goal-seeking family is the third distinct family of time paths. All living systems (and many nonliving systems) exhibit goal-seeking behavior. The difference between goal-seeking and exponential decay is that the exponential decay time path seeks a goal of zero, whereas the goal-seeking time path can seek a nonzero goal. Positive feedback loops generate growth, amplify deviations, and reinforce change. Negative loops seek balance, equilibrium, and stasis. Negative feedback loops act to bring the state of the system in line with a goal or desired state. The state of the system is compared to the goal. If there is a discrepancy between the desired and actual state, corrective action is initiated to bring the state of the system back in line with the goal. For example, the water supply growth in an urban area follows a goal-seeking time path that will eventually converge to a maximum urban water supply potential. Sometimes the desired state of the system and corrective action are explicit and under the control of a decision maker. For example, by transferring water from adjacent watersheds to an urban area, the goal of the water resources development time path can be moved upward, affecting water demand and population growth.

9.3.6.2.4 Oscillation Family

Oscillation is one of the most common dynamic behaviors in the world and is characterized by many distinct patterns. Like goal-seeking behavior, oscillations are caused by negative feedback loops. The state of the system is compared to its goal, and corrective actions are taken to eliminate any discrepancies. In an oscillatory system, the system's state constantly overshoots its goal or equilibrium state, reverses, and then undershoots, and so on.

9.3.6.2.5 S-Shaped Family

No real quantity can grow (or decline) forever: eventually, one or more constraints halt the growth. A commonly observed mode of behavior in dynamic systems is S-shaped growth; growth is a combination of two time paths: exponential growth and goal-seeking behavior. The shape of the curve resembles a stretched-out "S." More precisely, in the case of S-shaped growth, exponential growth gives way to goal-seeking behavior as the system approaches its limit or carrying capacity.

Sometimes, a system can overshoot its carrying capacity. If this occurs and the system's carrying capacity is not completely destroyed, the system tends to oscillate around its carrying capacity. On the other hand, if the system overshoots and its carrying capacity is damaged, the system will eventually collapse. This is referred to as an "overshoot and collapse" system response. As with a "normal" S-shaped pattern, a reverse S-shaped pattern is a combination of two time paths: exponential decay and a self-reinforcing spiral of decline. This behavior can be observed when the urban water demand exceeds the water supply capacity.

Example 9.6

Consider a city with limited resources and without any possibility for water transfer from adjacent basins. The quantity of surface and groundwater resources and their supply potential for domestic usage deteriorate because of the urban development and water quality problems caused by wastewater, solid wastes, and industrial activities. The studies carried out on variations of population, water resources, and their supply potential for domestic usage have resulted in the following three relationships:

$$P_{t+1} = P_t + 0.9\left(\frac{W_t}{P_t} - 0.5\right) \times 10^9,$$

$$W_t = \frac{R_t P_t}{2 \times 10^8},$$

$$R_t = 10^8 + 0.1P_t - 0.01P_t^{1.143},$$

where P_t is the city population at year t, W_t is the water supply potential for domestic usage (MCM) at year t, and R_t is the available water resources (MCM) at year t. Assuming that the city population in the first year is 0.9×10^6, determine the maximum possible population of the city.

Solution:

For determining the maximum possible population of the city, the three equations are solved based on the population factor. At first, R_t (water resources relation) is replaced at W_t (water resources productivity) formulation:

$$W_t = \frac{\left(10^8 + 0.1P_t - 0.01P_t^{1.143}\right)P_t}{2\times 10^8}.$$

Finally, the obtained relation is substituted into the population relation:

$$P_{t+1} = P_t + 0.9\left(\frac{\left(10^8 + 0.1P_t - 0.01P_t^{1.143}\right)P_t / 2\times 10^8}{P_t} - 0.5\right)\times 10^9.$$

The population of each year is more than the previous year so that the expression in brackets should be more than zero (Figure 9.19):

$$0.9\left(\frac{\left(10^8 + 0.1P_t - 0.01P_t^{1.143}\right)P_t / 2\times 10^8}{P_t} - 0.5\right) > 0 \Rightarrow \frac{\left(10^8 + 0.1P_t - 0.01P_t^{1.143}\right)}{2\times 10^8} > 0.5,$$

$$10^8 + 0.1P_t - 0.01P_t^{1.143} > 10^8 \Rightarrow 0.1P_t - 0.01P_t^{1.143} > 0 \Rightarrow P_t > 0.1P_t^{1.143}$$

$$10 > P_t^{0.143} \Rightarrow P_t < 10^7.$$

Therefore, the population cannot exceed 10^7 persons because of the limitations of water resources. Once the population reaches this value, it will fluctuate around it. The variations of P_t, R_t, and W_t are shown in Figures 9.19–9.21, respectively. In Figure 9.20, the graph is rescaled in order to have

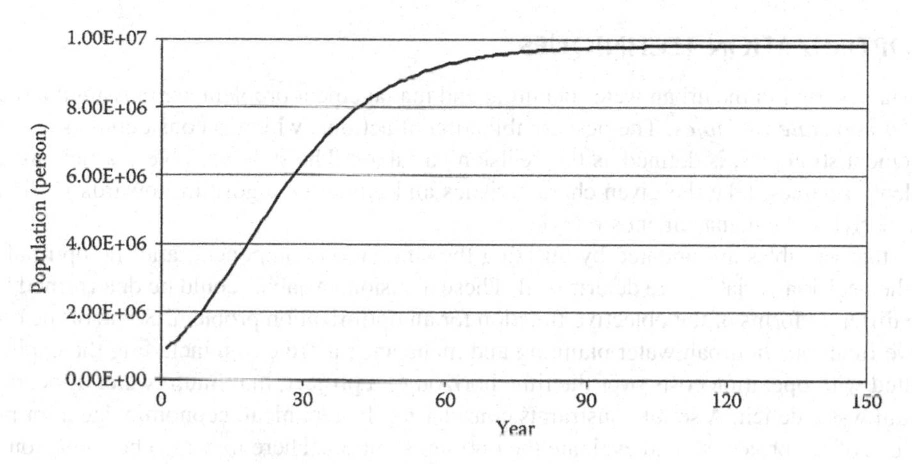

FIGURE 9.19 Variations of city population over time.

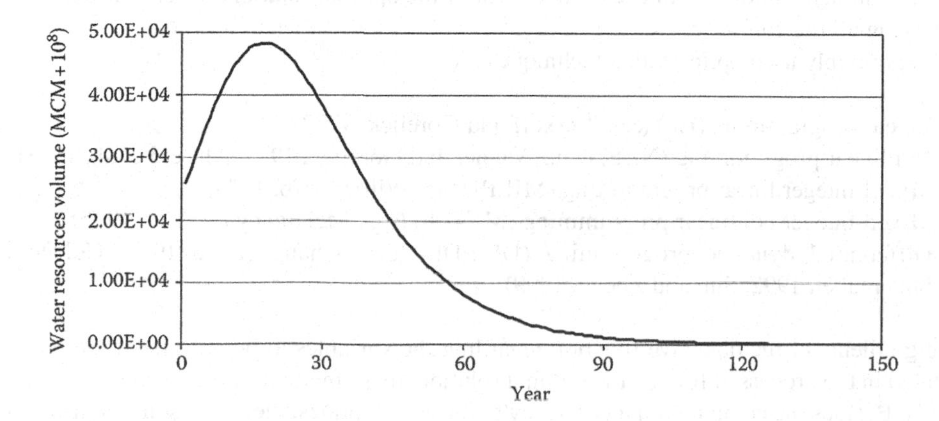

FIGURE 9.20 Variations of available water resources over time.

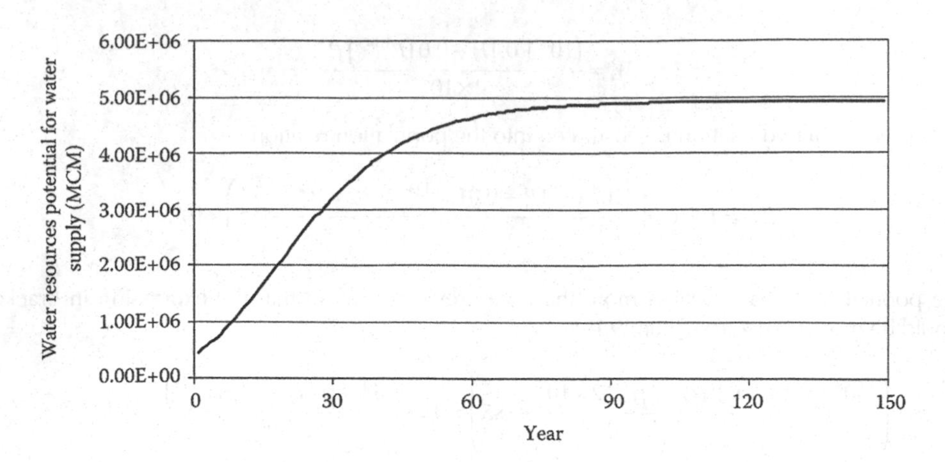

FIGURE 9.21 Variations of water supply potential for domestic usage over time.

a better fit when it is drawn. The vertical axis shows the water volume in excess of 10^8 MCM. Therefore, when the graph converges to zero, it is actually holding a volume of 10^8 MCM. As is shown in these figures, after about 130 years, the curves converge to straight lines, reaching their limits.

9.4 OPTIMIZATION TECHNIQUES

The variables used in the urban water planning and management problem are classified as *decision variables* and *state variables*. The best combination of actions, which is considered as the optimal management strategies, is defined as the decision variables. The state variables, which are usually dependent variables, take the given characteristics and guide the algorithm towards feasible solutions in deriving the management strategies.

The state variables are updated by utilizing the simulation component, and the optimal values for all the decision variables are determined. These decision variables could be determined by optimizing different forms of the objective function for an optimization problem. Some of the common objective functions in urban water planning and management are costs, including the capital costs associated with operation costs over the time horizon of a project; maximum water allocation; and minimum water deficit. A set of constraints considering the technical, economic, legal, or political limitations of the project should evaluate the optimal solution. There may also be some constraints in the form of either equalities or inequalities on decision and state variables.

Optimization techniques could be used to identify the optimal solution, which is feasible in terms of satisfying all the constraints.

The commonly used optimization techniques are

1. Linear programming (LP) (e.g., Lefkoff and Gorelick, 1987)
2. Nonlinear programming (NLP) (e.g., Wagner and Gorelick, 1987; Ahlfeld et al., 1988)
3. Mixed-integer linear programming (MILP) (e.g., Willis, 1976, 1979)
4. Mixed-integer nonlinear programming (MINLP) (e.g., McKinney and Lin, 1995)
5. Differential dynamic programming (DIFFDP) (e.g., Chang et al., 1992; Culver and Shoemaker, 1992; Sun and Zheng, 1999)

As the gradients of the objective function regarding the variables to be optimized are repetitively calculated in the process of finding the optimal solution, these methods are considered as "gradient" methods. Besides the computational efficiency of these techniques, they have some significant limitations. First, when the objective function is highly complex and nonlinear with multiple optimal points, a gradient method may be trapped in one of the local optima and cannot reach the globally optimal solution. Second, sometimes, numerical difficulties in gradient methods result in instability and convergence problems.

More recently, evolutionary optimization methods such as simulated annealing (SA), genetic algorithms (GA), and Tabu search (TS), which are based on heuristic search techniques, have been introduced. These methods are based on natural systems, such as biological evolution in the case of GA, to identify the optimal solution. The evolutionary methods generally require intensive computational efforts; however, they are increasingly being used nowadays because they are able to identify the global optimum, are efficient in handling discrete decision variables, and can be easily applied in different simulation models.

9.4.1 Linear Method

LP is the simplest single criterion optimization method applicable for continuous variables with linear objective function and constraints. Because of computational efficiency, LP has been implemented in a wide range of practical optimization examples.

Consider an optimization problem with n decision variables, $x_1, x_2, \ldots, x_n$, and linear objective functions and constraints. The objective function is formulated as follows:

$$f(x) = c_1 x_1 + c_2 x_2 + \ldots + c_n x_n, \tag{9.83}$$

where $c_1, c_2, \ldots, c_n$ are real numbers.

It is assumed that all decision variables are nonnegative. If any of the decision variables have only negative values, it is reformed to a new nonnegative variable as follows:

$$x_i^- = -x_i. \tag{9.84}$$

If a decision variable can be both positive and negative, then two new nonnegative variables are defined as follows:

$$x_i^+ = \begin{cases} x_i & \text{if } x_i \geq 0, \\ 0 & \text{otherwise,} \end{cases} \tag{9.85}$$

and

$$x_i^- = \begin{cases} x_i & \text{if } x_i < 0, \\ 0 & \text{otherwise,} \end{cases} \tag{9.86}$$

and

$$x_i = x_i^+ - x_i^-. \tag{9.87}$$

All of the constraints should be rewritten as ≤-type conditions. Therefore, constraints with ≥ inequalities are multiplied by (–1) and equality constraints are considered as two inequality constraints as follows:

$$a_{i1}x_1 + a_{i2}x_2 + \ldots + a_{in}x_n = b_i \rightarrow \begin{cases} a_{i1}x_1 + a_{i2}x_2 + \ldots + a_{in}x_n \leq b_i \\ a_{i1}x_1 + a_{i2}x_2 + \ldots + a_{in}x_n \geq b_{i,} \end{cases} \tag{9.88}$$

where the second relation is again multiplied by (–1) to be as ≤-type conditions. The LP problem with the above characteristics can be rewritten into its primal form as follows:

$$
\begin{aligned}
\text{Maximize subject to } & c_1x_1 + c_2x_2 + \ldots + c_nx_n, \\
& x_1, x_2, \ldots, x_n \geq 0, \\
& a_{11}x_1 + a_{12}x_2 + \ldots + a_{1n}x_n \leq b_1, \\
& a_{21}x_1 + a_{22}x_2 + \ldots + a_{2n}x_n \leq b_2, \\
& \vdots \\
& a_{m1}x_1 + a_{m2}x_2 + \ldots + a_{mn}x_n \leq b_m.
\end{aligned}
\tag{9.89}
$$

In the above formulation, maximization is only considered because minimization problems can be reformed to maximization problems by multiplying the objective function by (–1). In the very simple case of the two decision variables problem, Equation 9.89 can be solved by the graphical approach.

According to Rardin (1997), engineers often use deterministic models because they are more tractable (easier to analyze) than their stochastic counterparts and often produce solutions that are valid enough to be useful for the problem at hand. Stochastic models consider random input events, which could be explained as

$$Y = ax + b + \varepsilon \tag{9.90}$$

where ε is error term random number, in which a probability distribution could be utilized to generate random numbers in the error term. These random numbers are multiplied by σ_ε. For example, the error is normally distributed. In this case, the random numbers are generated $N(0, 1)$ multiplied by σ_ε, so we have $N(0, \sigma_\varepsilon^2)$.

Example 9.7

There are three methods for wastewater treatment in an urban area. The methods remove 1.2, 2.5, and 4 g/m^3 amounts of the pollutants, respectively. The third technology is the best, but because of some limitations in allocating enough area, it cannot be applied to more than 40% of the wastewater being treated. The costs of applying the treatment methods are 6, 4, and 3 dollars/m^3. If the discharge of wastewater that should be treated in a day is about 1,200 m^3, and at least 1.677 g/m^3 pollutant has to be removed, determine the optimal combination of proposed methodologies in order to treat the wastewater of the urban areas.

Solution:

A simple LP formulation can be given to model this optimization problem. Let x_1, x_2 be the amount of wastewater (in m^3) for which technologies 1 and 2 are used. Then 1,200 – x_1 – x_2 m^3 is the amount where technology variant 3 is applied. Clearly,

$$x_1, x_2 \geq 0,$$

$$x_1 + x_2 \leq 1{,}200.$$

The condition that the third technology variant cannot be used in more than 40% of the treated wastewater requires that

$$1{,}200 - x_1 - x_2 \leq 480,$$

that is,

$$x_1 + x_2 \geq 720.$$

The total removed pollutant amount in g is

$$1.2x_1 + 2.5x_2 + 4(1,200 - x_1 - x_2).$$

which has to be at least 1.677 g/m³ of the total 1,200 m³ being treated, so

$$1.2x_1 + 2.5x_2 + 4(1,200 - x_1 - x_2) \geq 2,000.$$

This inequality can be rewritten as

$$2.8x_1 + 1.5x_2 \leq 2,800.$$

The total cost is given as

$$6x_1 + 4x_2 + 3(1,200 - x_1 - x_2) = 3x_1 + x_2 + 3,600.$$

Since 3,600 is a constant term, it is sufficient to minimize $3x_1 + x_2$ or maximize $-3x_1 - x_2$. Constraint (10–12) is a ≥-type relation, which is multiplied by (–1) in order to reduce it to a <-type constraint. After this modification is done, we have the following primal-form problem:

$$\begin{aligned} \text{Maximize} \quad & f = -3x_1 - x_2, \\ \text{Subject to} \quad & x_1, x_2 \geq 0, \\ & x_1 + x_2 \leq 1,200, \\ & -x_1 - x_2 \leq -480, \\ & 2.8x_1 + 1.5x_2 \leq 2,800. \end{aligned}$$

Each constraint can be represented in the two-dimensional space as a half-plain, the boundary of which is determined by the straight line representing the constraint as an equality. For instance, the equality $x_1 + x_2 = 1,200$ is a straight line with the x_1-intercept 1,200 and also with the x_2-intercept 1,200. Since the origin, the (0, 0) point satisfies the inequality, the half-plain that contains the origin represents the points that satisfy this inequality constraint. The intercept of the half-plains of all constraints gives the feasible decision space, as it is demonstrated in Figure 9.22. Note that this set has six vertices: (480, 0), (1,000, 0), (769.2, 430.8), (0, 1,200), and (0, 480), and the set is the convex hull, the smallest convex set containing the vertices.

From the theory of LP, we know that if the optimal solution is unique, then it is a vertex, and in the case of multiple optimal solutions, there is always a vertex among the optimal solutions. Therefore, it is sufficient to find the best vertex, which gives the highest objective function value. Simple calculation displays that

$$f(480,0) = -1,440,$$

$$f(1,000,0) = -3,000,$$

$$f(769.2,430.8) = -2738.4,$$

$$f(0,1,200 = -1,200),$$

$$f(0.480 = -480).$$

Since −480 is the largest value, the optimal decision is: $x_1 = 0$, $x_2 = 480$, and $1,200 - x_1 - x_2 = 720$. The minimal cost is therefore $3x_1 + x_2 + 3,600 = \$4,080$.

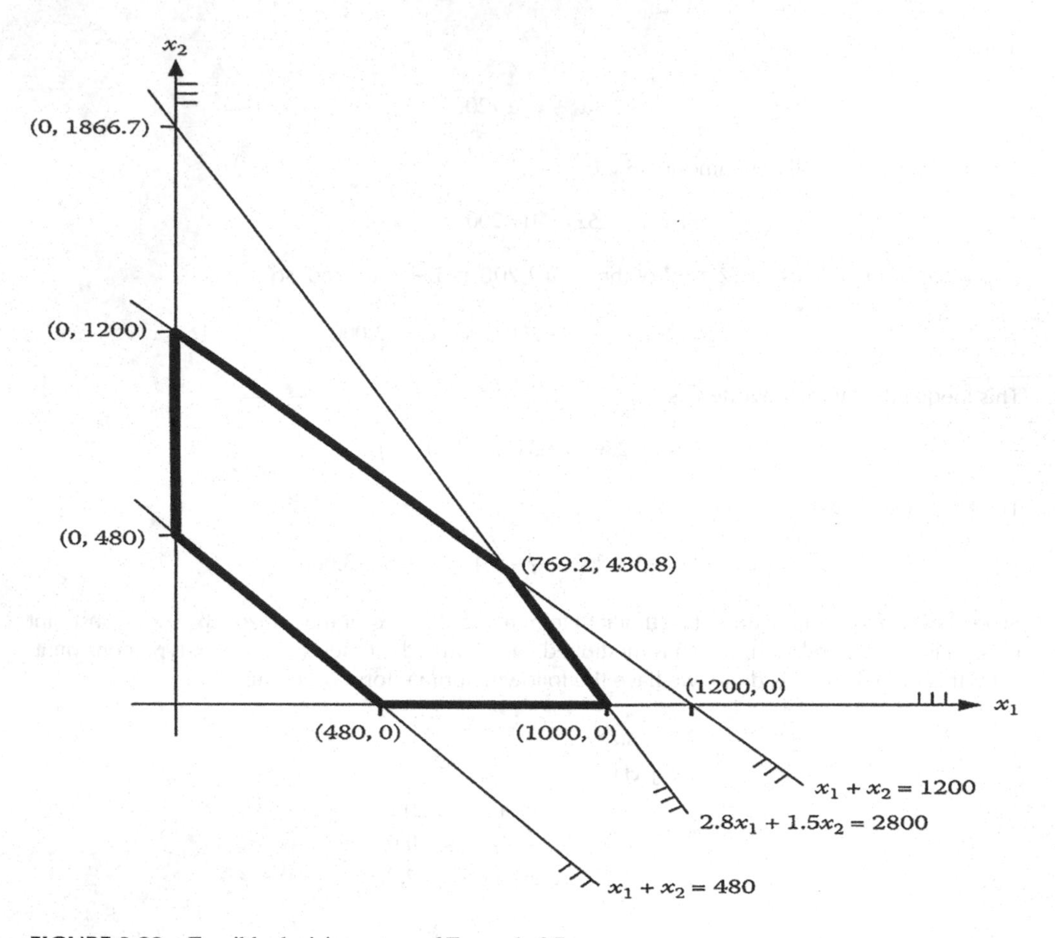

FIGURE 9.22 Feasible decision space of Example 9.7.

9.4.1.1 Simplex Method

In the case of more than two decision variables, so that the graphical representation of the problem is impossible, the *simplex method* is employed. This method systematically searches the feasible decision space to find the optimal solution. The main idea of the simplex method can be summarized as follows: Any vertex of the feasible decision space corresponds to a basis of the column space of the coefficient matrix of the linear constraints. The variables associated with the columns of the basis are called "basic variables." It can be revealed that moving from one vertex to any of its neighbors is equivalent to exchanging one of the basic variables with a nonbasic variable. In the simplex method, the objective function increases in each step, which the following conditions can ensure. Let a_{ij} be the (i, j) element of the coefficient matrix, b_i the ith right-hand-side number, and c_j the coefficient of x_j in the objective function. Then x_i is replaced by x_j if $c_j > 0$, $a_{ij} > 0$, and

$$\frac{b_i}{a_{ij}} = \min_l \left\{ \frac{b_l}{a_{lj}} \right\}. \tag{9.91}$$

Artificial variables called surplus are introduced to all ≥- and =-type constraints to find an initial basic solution. For ≥-type constraints, the slack variables play the same role. In the first phase of the simplex method, all artificial variables are removed from the basis by successive exchanges. In the second phase of the simplex method, the initial feasible basic solution is successively improved until reaching the optimum. Lingo is one of the professional software packages that implement the simplex method.

The simplex method can be summarized in a four-step procedure to get a sense of its functionality:

- Add slack and subtract surplus variables – m equations and n unknowns. Place $n - m$ variables in NB set (all equal to zero). Solve m equations and m unknowns—place them in B set. Find a feasible solution (if all answers are positive. Otherwise rearrange).
- Move an X_j from NB that improves the objective function to B
- Move X_k from B to NB that goes to zero faster than other B elements as the moved X_j becomes positive
- Termination point—no X_j in NB can improve the current objective function

Example 9.8

Based on the population growth analysis, a city's wastewater is estimated to increase over time, as depicted in Figure 9.23. Four phases for treatment plant expansion capacity are studied for the treatment of the future wastewater of this city in a 20-year planning time horizon. The capacity and initial investment of these projects are represented in Table 9.7. The interest rate is considered to be 5%. Find the optimal sequence of implementing projects.

Solution:

The present value of the initial investment of the projects is presented in Table 9.8. In order to find the best timing for the construction of different phases, minimization of the present value of the cost is considered as the objective function of an LP model as follows:

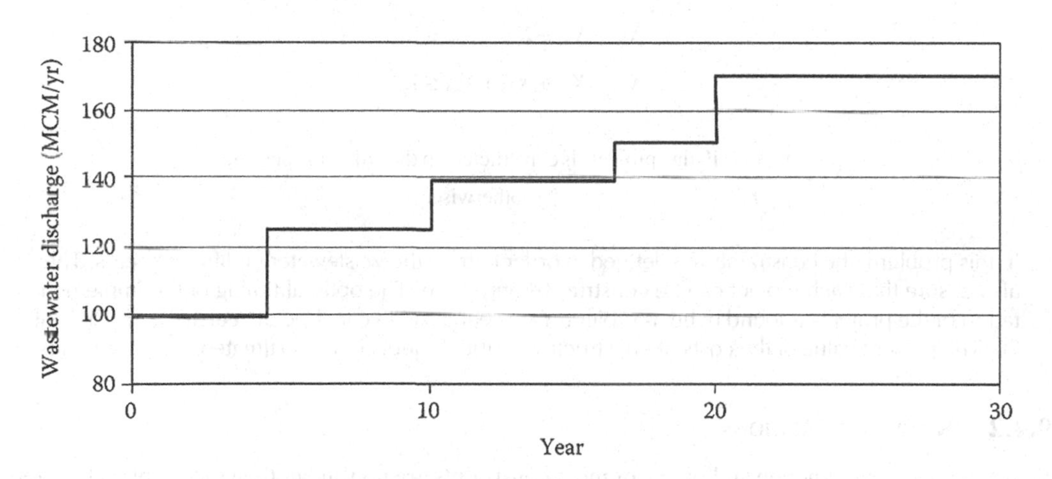

FIGURE 9.23 Four-phase treatment plant capacity expansion.

TABLE 9.7
Initial Investment and Capacity of Different Phases

Treatment Plant Expansion Projects	Initial Investment	Capacity
A	3	15
B	2	20
C	2.5	10
D	5	25

TABLE 9.8
Present Value of Initial Investment for Different Phases

Project	$t = 5$ years	$t = 10$ years	$t = 16$ years	$t = 20$ years
A	2.32	1.8	1.32	1.08
B	1.55	1.2	0.88	0.72
C	1.93	1.5	1.1	0.9
D	3.87	2.99	2.2	1.79

$$\begin{aligned}
\text{Minimize}\, C = {} & 2.32X_{11} + 1.8X_{12} + 1.32X_{13} + 1.08X_{14} \\
& + 1.55X_{21} + 1.2X_{22} + 0.88X_{23} + 0.72X_{24} \\
& + 1.93X_{31} + 1.5X_{32} + 1.32X_{33} + 0.9X_{34} \\
& + 3.87X_{41} + 2.99X_{42} + 2.2X_{43} + 1.79X_{44},
\end{aligned}$$

$$\begin{aligned}
\text{Subject to}\quad & 15X_{11} + 20X_{21} + 10X_{31} + 25X_{41} \geq 30, \\
& 15X_{12} + 20X_{22} + 10X_{32} + 25X_{42} \geq 20, \\
& 15X_{12} + 20X_{22} + 10X_{32} + 25X_{42} \geq 20, \\
& 15X_{14} + 20X_{24} + 10X_{34} + 25X_{44} \geq 10, \\
& X_{11} + X_{12} + X_{13} + X_{14} \leq 1, \\
& X_{31} + X_{32} + X_{33} + X_{34} \leq 1, \\
& X_{41} + X_{42} + X_{43} + X_{44} \leq 1,
\end{aligned}$$

$$X_{ij} = \begin{cases} 1 & \text{if the project } i \text{ is constructed in the } j\text{th time period,} \\ 0 & \text{otherwise.} \end{cases}$$

In this problem, the constraints are defined in order to treat the wastewater in different years. They also ensure that each project can be constructed only once. The optimal timing of the implementation of the projects is found to be as projects *D* at year 5, *A* at year 10, *C* at year 16, and *B* at year 20. The present value of the costs of construction of the projects is also estimated to be 7.49 units.

9.4.2 Nonlinear Methods

When the objective function and/or one or more constraints are nonlinear, then the problem becomes NLP. Because of the nonlinear nature of practical optimization problems, NLP has been widely applied in solving practical problems. In the problems with discrete decision variables, MILP or MINLP must be used. In the case of only two variables, they can be solved by the graphical approach, but it should be considered that in some cases, the feasible decision space does not have vertices or the optimal solution is not a vertex. In such cases, the curves of the objective function with different values have to be compared to find the optimal solution.

Example 9.9

For illustrative purposes, consider the previous example; however, assume that instead of minimizing the linear function $3x_1 + x_2$, a nonlinear function

$$f(\underline{x}) = x_1^2 + x_2^2$$

is minimized.

Solution:

In order to find the feasible solution that minimizes this objective function, we first illustrate the solutions, which give certain special objective function values. The points with zero objective function values satisfy the equation

$$f(\underline{x}) = x_1^2 + x_2^2 = 0$$

when the only solution is $x_1 = x_2 = 0$, that is, the origin. The points with the objective function value of 10,000 satisfy the equation

$$f(\underline{x}) = x_1^2 + x_2^2 = 10{,}000.$$

This is a circle with the origin being its center and having the radius 100. This circle has no feasible point, so 10,000 is a very small value for the objective function. By increasing the value of $f(\underline{x})$, we always get a similar circle with the same center but with an increasing radius. By increasing the circle's radius, we see that the smallest feasible radius occurs when the circle and the line $x_1 + x_2 = 480$ just touch each other. The tangent point is (220, 220), which is the optimal solution to the problem. In more complicated and higher dimensional cases, computer software has to be used. There are several professional packages available to perform this task, so they are usually used in solving practical problems.

Linearization is the most popular nonlinear optimization method. In this method, all nonlinear constraints or objective functions are linearized. The resulting LP problem can be solved by the simplex method. Let $g(x_1, \ldots, x_n)$ be the left-hand side of a constraint or the objective function and $(x_1^0, \ldots, x_n^0)$ be a feasible solution. g is linearly approximated by the linear Taylor's polynomial as follows:

$$g(x_1, \ldots, x_n) \approx g(x_1^0, \ldots, x_n^0) + \sum_{r=1}^{n} \frac{\partial g}{\partial x_r}(x_1^0, \ldots, x_n^0)(x_r - x_r^0). \tag{9.92}$$

The optimal solution of the resulting LP problem is $(x_1^1, \ldots, x_n^1)$. The original problem is linearized again around this solution, and the new LP problem is solved. This iterative procedure is followed until the optimal solution in two iterations does not change. It is known from the theory of NLP that the solutions of the successive LP problems converge to the solution of the original nonlinear optimization problem. Further information about NLP can be found at Hadley (1964).

9.4.3 Dynamic Programming

In water resources planning, the system's dynamism is a major concern that has to be taken into account in modeling and optimization. Dynamic programming (DP) is the most popular tool in optimizing dynamic systems. Consider the following optimization problem for supplying water for an urban area from a multireservoir system:

$$\min Z = \sum_{t=1}^{T} \text{Loss}\left(\sum_{s=1}^{X} R_{s,t}\right), \tag{9.93}$$

where T is the time horizon, X is the total number of reservoirs in the system, and Loss is the cost of operation based on the ratio of supplied water ($R_{s,t}$) from reservoir s in month t to the monthly urban water demand. The continuity or a mass balance of the contents of the reservoir considering regulated and unregulated release (seepage) from the beginning of the month to the next is also included in the model:

$$S_{s,t+1} - S_{s,t} + R_{s,t} = I_{s,t}, \tag{9.94}$$

$$R_{s,t} = R^{R}_{s,t} + R^{U}_{s,t}, \tag{9.95}$$

where $R_{s,t}$ is the release from reservoir s in month t and I_t is the inflow volume to the reservoir in month t. $R^{R}_{s,t}$ and $R^{U}_{s,t}$ are regulated and unregulated releases from reservoirs, respectively. Additional constraints on maximum and minimum allowable release and storage during any season for all sites can be stated as

$$R^{\max}_{s,t} \geq R_{s,t} \geq R^{\min}_{s,t} \qquad t = 1,\ldots,T \quad s = 1,\ldots,X, \tag{9.96}$$

$$S^{\max}_{s,t} \geq S_{s,t} \geq S^{\min}_{s,t} \qquad t = 1,\ldots,T \quad s = 1,\ldots,X, \tag{9.97}$$

$$\left|S_{s,t+1} - S_{s,t}\right| \leq SC_s \qquad t = 1,\ldots,T \quad s = 1,\ldots,X, \tag{9.98}$$

where SC_s is the maximum allowable change in the storage of reservoir s within each month considering dam stability and safety conditions. Because of many decision variables and constraints, finding the direct solution is computationally difficult. This unique optimization problem is changed through the DP procedure to solve many smaller-sized problems in the form of a recursive function.

$$f_{t+1}\left(S_{1,t+1},\ldots,S_{X,t+1}\right) = \min\left(\text{Loss}\left(\sum_{s=1}^{X} R_{st}\right) + f_t\left(S_{1,t},\ldots,S_{X,t}\right)\right),$$

$$\begin{gathered} S_{1,t} \in \Omega_{1,t} \\ \ldots \\ \ldots \\ S_{X,t} \in \Omega_{X,t}. \end{gathered} \tag{9.99}$$

The initial conditions are

$$f_1\left(S_{1,1},\ldots,S_{X,1}\right) = 0, \quad S_{1,1} \in \Omega_{1,1},\ldots,S_{X,1} \in \Omega_{X,1}, \tag{9.100}$$

where $f_t(S_{1,t},\ldots, S_{X,t})$ is the total minimum loss of operation from the beginning of month 1 to the beginning of month t, when the storage volume at the beginning of month t is $S_{1,t}$ at site 1, $S_{R,t}$ at site 2,…, $S_{X,t}$ at site X. $\Omega_{s,t}$ is the set of discrete storage volumes that will be considered for the beginning of month t at site s.

The DP sequences are time- or space-dependent. So the "stages" of the problem are time if, for instance, the operation of the water distribution system in a daily time scale is considered or space if, for example, the design of a storm or sewer collection system is considered.

In most applications, the optimization Equation 9.95 cannot be solved analytically, and recursive functions are determined only numerically. Discrete dynamic programming (DDP) is used in such cases, where the state and decision spaces are discretized. The main problem in using this method is known as the "curse of dimensionality." Because of this, the number of states that should be considered in the evaluation of different possible situations drastically increases with the number of discretizations. Therefore, DDP is used only up to four or five decision and state variables.

Different successive approximation algorithms such as DIFFDP, discrete differential dynamic programming (DDDP), and state incremental dynamic programming (IDP) are developed to overcome this problem. In applying any of these methods, the user should provide an initial estimation of the optimal policy. This estimation is improved in each step until convergence occurs. However, as is usually the case in nonconvex optimization, the limit might only be a local optimum, or the approximating sequence is divergent.

Example 9.10

Assume a reservoir that supplies water for an urban area. The monthly water demand of this urban area (D_t) is about 10 million cubic meters. The total capacity of the reservoir is 30 million cubic meters. Let S_t (reservoir storage at the beginning of month t) take the discrete values 0, 10, 20, and 30. The cost of operation (Loss) can be estimated as a function of the difference between release (R_t) and water demand as follows:

$$\text{Loss}_t = \begin{cases} 0, & R_t \le D_t, \\ (D_t - R_t)^2, & R_t > D_t. \end{cases}$$

(a) Formulate a forward-moving deterministic DP model for finding the optimal release in the next 3 months.
(b) Formulate a backward-moving deterministic DP model for finding the optimal release in the next 3 months.
(c) Solve the DP model developed in part (a), assuming that the inflows to the reservoir in the next 3 months (t = 1, 2, 3) are forecasted to be 10, 50, and 20, respectively. The reservoir storage in the current month is 20 million cubic meters.

Solution:

(a) The forward deterministic DP model is formulated as follows:

$$f_{t+1}(S_{t+1}) = \min\left[\text{Loss}(R_t) + f_t(S_t)\right], \quad t = 1,2,3$$

$$S_t \in \Omega_t,$$

$$S_t \le S\max,$$

$$f_1(S_1) = 0,$$

$$\text{Loss}(R_t) = \begin{cases} 0, & R_t = 10, \\ (10 - R_t)^2, & R_t \ne 10, \end{cases}$$

$$R_t = S_t + I_t - S_{t+1},$$

$$S_{\max} = 30 \text{ MCM},$$

where Ω_t is the set of discrete storage volumes (states) that will be considered at the beginning of month t at site s. S_{t+1} is the storage at the beginning of month $t + 1$.

TABLE 9.9
DP Model Solution

Month	Inflow	S_t	S_{t+1}	R_t	$Loss_t$	R_t^*
1	10	20	0	30	400	10
			10	20	100	
			20	10	0	
			30	0	100	
2	50	20	0	70	3,600	40
			10	60	2,500	
			20	50	1,600	
			30	40	900	
3	20	30	0	50	1,600	20
			10	40	900	
			20	30	400	
			30	20	100	

Note: R_t^* is the optimal policy for each month.

(b) The backward deterministic DP model is formulated as follows:

$$f_t(S_t) = \min\left[\text{Loss}(R_t) + f_{t+1}(S_{t+1})\right], \quad 0 \quad t = 1,2,3.$$

The constraints are as mentioned above.

(c) Loss is calculated using the forward formulation for 3 months. The results are shown in Table 9.9.

9.4.3.1 Stochastic DP

Stochastic Dynamic Programming (SDP) is one of the most powerful and commonly used techniques to aid decision-making in reservoir operation. SDP incorporates a discrete, lag-one Markov process as a streamflow descriptor. The optimal operating policy in SDP is derived using the Bellman's backward recursive relationship (Bellman, 1957). The convergence is determined by two criteria (Nandalal and Bogardi, 2007): stabilization of the incremental change in the optimal value according to the Bellman recursive formula and stabilization of the operating policy. The objective is usually to maximize the total benefit, which consists of current benefits from operations at present and the discounted value coming from future use of stored water within the given planning/operating horizon. SDP is well established in the long-term planning of multireservoir systems (Yeh, 1985). The inflows, electricity demands, and market prices are examples of stochastic variables that may be considered in the reservoir operations planning problem.

Unlike other mathematical programming techniques, such as linear and nonlinear programming, very few general purpose DP solvers are available. An example of software available for solving DP and SDP problems is the CSUDP model, which is generalized DP software developed at the Colorado State University (USA). This software can handle "multidimensional problems, stochastic problems, and certain classes of Markov decision processes" (Labadie, 2004). The general SDP procedure, considering the inflow as the only stochastic variable, is illustrated in Figure 9.24.

9.4.3.2 Markov Chains

A Markov chain is a stochastic and sequential method that involves using a finite number of potential condition states, each of which reflects a certain level of physical deterioration of the system and hence is inherently discrete. Markov chains predict the probability that the system will be in

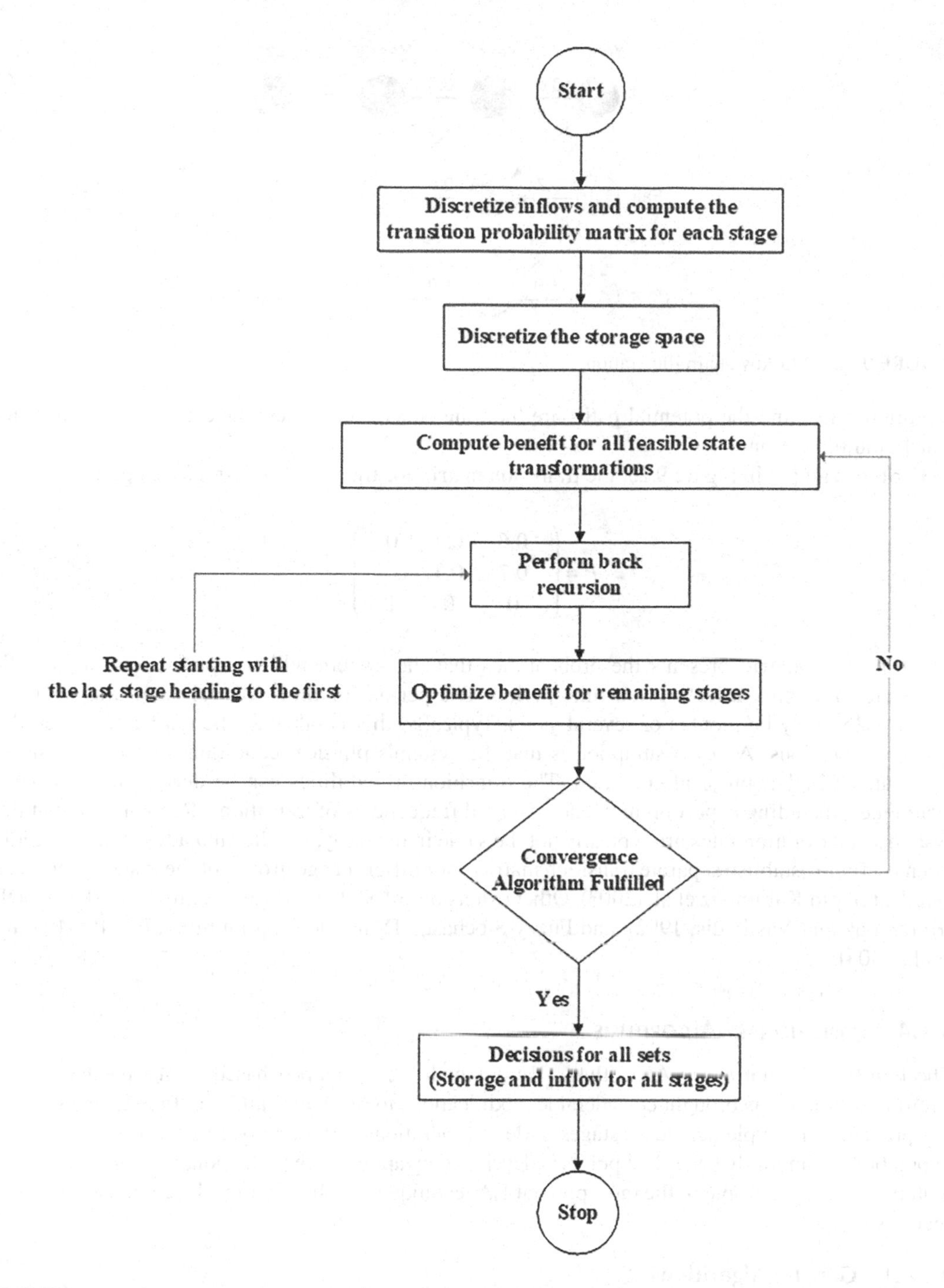

FIGURE 9.24 SDP procedure (Nandalal, K. D. W., and Bogardi, J. J. *Dynamic programming based operation of reservoirs: Applicability and limits*. Cambridge University Press, (2007)).

a certain condition state after a period of discrete time blocks, often years. They are described as memoryless because they assume that the probability that the system transitions from one state to another is based only on the state at a given year and not on the states of past years; an assumption that is convenient from the computational perspective but may not always adequately represent real-world conditions. Often represented as a node-to-node graph (Figure 9.25) where the nodes are the

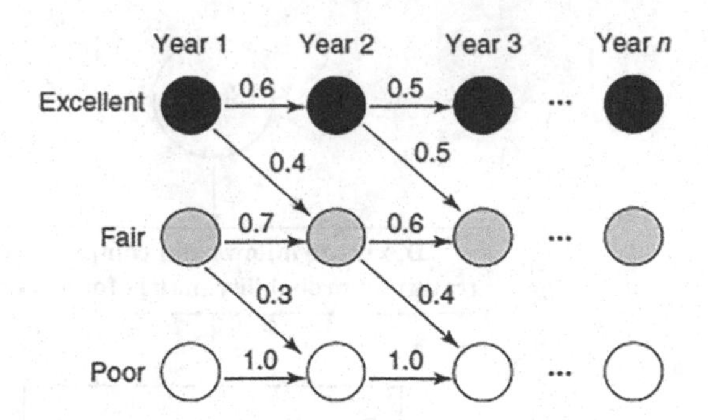

FIGURE 9.25 Markov chain illustration.

condition states and the potential paths are the transitions, Markov chains can be analyzed using matrix multiplication.

For the example in Figure 9.25, the transition matrix for the year 1 to 2 transition period

$$P = \begin{pmatrix} 0.6 & 0.4 & 0 \\ 0.7 & 0.3 & 0 \\ 0 & 0 & 1 \end{pmatrix}$$

The transition matrix presents the probabilities that the system will transition from one condition state to another after a given time period. The period, or time interval (shown as years in Figure 9.25), may be months or several years. Typically, this is taken as the interval between the system inspections. A key assumption is that the system's physical condition does not transition to a state of higher physical condition. The transition probabilities can be derived using several techniques, including expert opinion and observed frequencies of transitions. Recognizing that the system's deterioration rates are typically not the same in the early, middle, and advanced years, it is often useful to establish separate transition matrices for different age groups of the system. For more details, refer to Karamouz et al. (2003). Other extensions of SDP are Bayesian Stochastic DP, BSDP (Karamouz and Vasiliadis, 1992), and Fuzzy Stochastic Dynamic Programming, FSDP (Mousavi et al., 2004).

9.4.4 Evolutionary Algorithms

The family of Evolutionary Algorithms (EAs), which are developed based on the mechanism of natural evolution, has brought considerable modifications to random searching. In EAs, the searching procedure is implemented in stages called "generations." In each one of these techniques, a population of randomly generated points is developed by applying the selection, recombination, and mutation operators. Some of the most popular EA techniques are briefly introduced in the following sections.

9.4.4.1 Genetic Algorithms

GAs are stochastic search methods that imitate the process of natural selection and the mechanism of population genetics. They were first introduced by Holland (1975) and later further developed and popularized by Goldberg (1989). GAs are used in many different application areas ranging from function optimization to solving large combinatorial optimization (CO) problems. The general framework of GAs is described in Figure 9.26. As is illustrated in Figure 9.26, the GA begins with the creation of a set of solutions referred to as a population of individuals. Each individual in a population consists of a set of parameter values that completely describes a solution. A solution is encoded in a string called a chromosome, which consists of genes that can take a number of values. Initially, the collection of solutions (population) is generated randomly. At each iteration (also

```
BEGIN /* genetic algorithm*/
        Generate initial population;
        Compute fitness of each individual;
        WHILE NOT finished DO LOOP
            BEGIN
                Select individuals from old generations for mating;
                Create offspring by applying recombination and/or
mutation
                to the selected individuals ;
                        Compute fitness of the new individuals;
                Kill old individuals to make room for new chromosomes
and                 insert offspring in the new generation;
                IF Population has converged THEN finishes := TRUE ;
            END
END
```

FIGURE 9.26 General framework of GA code.

called generation), a new generation of solutions is formed by applying genetic operators, including selection, crossover, and mutation, analogous to the ones from natural evolution. Each solution is evaluated using an objective function (called a fitness function), and this process is repeated until some form of convergence in fitness is achieved. The objective of the optimization process is to minimize or maximize fitness.

The first step in developing a new generation is selection. Through the selection process, a population of chromosomes is selected via some stochastic selection process. This process aims to propagate the better or fitter chromosomes. The simplest procedure to select chromosomes is known as tournament selection. New chromosomes are created in a process known as crossover. Crossover may be performed using different methods, including single-point, double-point, and uniform methods. The crossover is performed on each selected pair with a specified probability, referred to as crossover probability, which is typically between 0.5 and 0.7. After a pair is selected for crossover, a random number between 0 and 1 is generated and compared with the crossover probability. The crossover will be performed only if the produced random number is less than the crossover probability. In the crossover process, the genetic material of the selected chromosomes is exchanged.

The last step in the production of a new generation is mutation. Some diversity is added to the GA population through the mutation process to prevent from converging to a local minimum. The mutation probability controls the probability of selection of any gene for mutation. Carroll (1996) suggests the following formulation for estimating the mutation probability, p_{mutate}.

$$p_{\text{mutate}} = \frac{1}{n_{\text{popsiz}}}, \tag{9.101}$$

where n_{popsiz} is the size of the population. Similar to the crossover, a random number between 0 and 1 is generated. Only when this number is smaller than the mutation probability, the mutation is performed.

There are several schemes for selection, such as roulette wheel, rank selection, tournament, and elitism selection. The roulette wheel is the original selection mechanism for GAs. Implementation of this approach requires that all the fitness values are positive numbers. Suppose the total fitness function in a population of size N is equal to the circumference of the roulette wheel. In that case, each chromosome in the population will have a corresponding sector bounded by an arc equal to its fitness. When spinning the wheel, the probability that an individual will be selected is proportional to its fitness. In order to obtain N individuals in the mating pool, the wheel is spun N times. The same individual can be chosen several times, which is very likely to happen when it has a large fitness value. Rank selection also works with negative fitness values and is mostly used when the individuals in the population have very close fitness values (this usually happens at the end of the run). This leads to each individual having an almost equal share of the pie (like in the case of proportionate fitness selection). Hence, no matter how fit relative to each other, each individual has approximately the same probability of getting selected as a parent; this leads to a loss in the selection pressure towards fitter individuals, making the GA make poor parent selections in such situations. Tournament selection is a method of choosing the individual from the set of individuals. The winner of each tournament is selected to perform crossover. Often to get better parameters, strategies with partial reproduction are used. One of them is elitism, in which a small portion of the best individuals from the last generation is carried over (without any changes) to the next one.

The GA raises a couple of essential features again. First, it is a stochastic algorithm; randomness has an essential role in the GA. Both selection and reproduction need random procedures. The second crucial point is that the GA always considers a population of solutions. Keeping in memory more than a single solution at each iteration offers many advantages. The algorithm can recombine different solutions to get better answers, and so it can use the benefits of assortment. The robustness of the algorithm should also be mentioned as something essential for the algorithm's success. Robustness refers to the ability to perform consistently well on a broad range of problem types. There is no particular requirement on the problem before using GAs, so it can be applied to resolve any problem. All those are of a GA extension called VLGA (varying chromosome length GA, for more details see Kerachian and Karamouz (2006). For an overview of GA applications in water resources see Nicklow et al. (2010).

Example 9.11

Find the maximum point of the function $f(x) = 2x - x^2$ in the [0, 2] interval using the GA.

Solution:

First, the decision space is discretized, and each is encoded. For illustration purposes, we have four-bit representations. The population is provided in Table 9.10. The initial population is selected next. We have chosen six chromosomes; they are listed in Table 9.11, where their actual values and the corresponding objective function values are also presented.

TABLE 9.10
Population of Chromosomes

Binary Code	Value	Binary Code	Value
0000	0	0110	0.75
0001	0.125	1010	1.25
0010	0.25	1100	1.5
0100	0.5	0111	0.875
1000	1	1011	1.375
0011	0.375	1101	1.625
0101	0.625	1110	1.75
1001	1.125	1111	1.875

The best objective function value is obtained at 1001, but there are two chromosomes with the second-best objective value: 1010, 0110. Randomly selecting the first, we have the chromosome pairs 1001 and 1010. A simple crossover procedure is performed by interchanging the two middle bits to obtain 1000 and 1011. The new population is presented in Table 9.12.

The best objective function value is obtained at 1000, and there is a chromosome with the second-best objective value: 1001. We have the chromosome pair 1000 and 1001. By interchanging their middle bits, we get 1001 and 1000, which are similar to those of their parents. The resulting new population is illustrated in Table 9.13.

Selecting the pair 1000 and 1001, we cannot use the same crossover procedure as before since they have the same middle bits. Since the two chromosomes differ in only one bit, interchanging any substring of them will result in identical chromosomes. Therefore, the mutation is used by reversing its second bit to have 1100. The resulting modified population is depicted in Table 9.14.

It is interesting to examine how the best objective function is evolving from population to population. Starting from the initial table with the value 0.98, it increased to 1, and then it remained

TABLE 9.11
Initial Population

Chromosomes	Value	Objective Function
1010	1.25	0.9375
1100	1.5	0.75
1001	1.125	0.984375
1111	1.875	0.234375
0110	0.75	0.9375
0011	0.375	0.609375

TABLE 9.12
First Modified Population

Chromosomes	Value	Objective Function
1000	1	1
1011	1.375	0.859375
1111	1.875	0.234375
1110	1.75	0.4375
0010	0.25	0.4375
1001	1.125	0.984375

TABLE 9.13
Second Modified Population

Chromosomes	Value	Objective Function
1001	1.125	0.984375
1000	1	1
1010	1.25	0.9375
1011	1.375	0.859375
1110	1.75	0.4375
0010	0.25	0.4375

TABLE 9.14
Third Modified Population

Chromosomes	Value	Objective Function
1001	1.125	0.984375
1000	1	1
1100	1.5	0.75
1001	1.125	0.984375
1000	1	1
1010	1.25	0.9375

the same. Since this value is the global optimum, it will not increase in further steps. In practical applications, we stop the procedure if no, or very small, improvement occurs after a certain (user-specified) number of iterations. More details of GA and possible applications can be found in Goldberg (1989).

In two ways, the genetic algorithm can be used with MATLAB®:

- Calling the genetic algorithm function GA at the command line.
- Using the Genetic Algorithm Tool, a graphical interface to the GA.

To use the Genetic Algorithm and Direct Search Toolbox, you must first write an M-file that computes the function you want to optimize. The M-file should accept a vector, whose length is the number of independent variables for the objective function, and return a scalar. The following steps show how to write an M-file for the function you want to optimize.

1. Select New from the MATLAB File menu.
2. Select M-File. This opens a new M-file in the editor.
3. In the M-file, enter the following two lines of code:
4. function z = my fun(x)
5. Save the M-file in a directory on the MATLAB path.

Example 9.12

Find the best allocations of water to the three water-consuming firms using a genetic algorithm. The maximum allocation to any single user cannot exceed 5, and the sum of all allocations cannot exceed the value of Q, say 6. Only integer solutions are to be considered. The objective is to find each allocation's values that maximize the total benefits, $B(X)$. The benefit function of each water-consuming firm is obtained as below:

$$B_1 = 6X_1 - X_1^2$$

$$B_2 = 7X_2 - 1.5X_2^2$$

$$B_1 = 8X_3 - 0.5X_3^2$$

Solution:

In order to find the best allocation for the water-consuming firms using genetic algorithm in MATLAB, the fitness function, and the GA code could be written as Figure 9.27:

The optimum allocation values are calculated as $X_1 = X_2 = 1$ and $X_3 = 4$ in which the total benefit is 34.5.

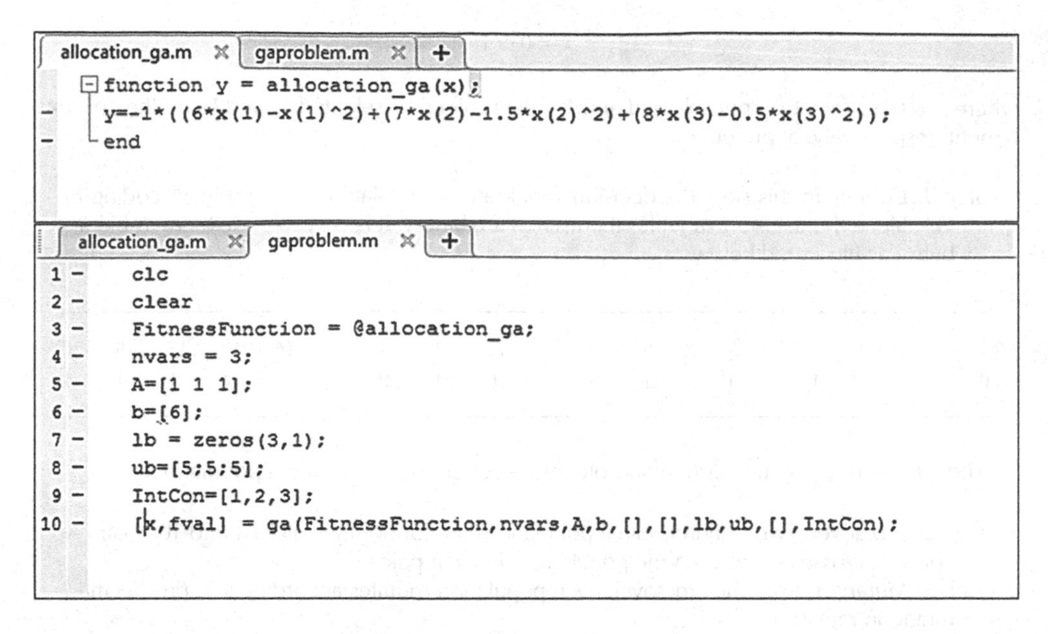

allocation_ga.m
gaproblem.m
function y = allocation_ga(x);
y=-1*((6*x(1)-x(1)^2)+(7*x(2)-1.5*x(2)^2)+(8*x(3)-0.5*x(3)^2));
end
allocation_ga.m
gaproblem.m
1 clc
2 clear
3 FitnessFunction = @allocation_ga;
4 nvars = 3;
5 A=[1 1 1];
6 b=[6];
7 lb = zeros(3,1);
8 ub=[5;5;5];
9 IntCon=[1,2,3];
10 [x,fval] = ga(FitnessFunction,nvars,A,b,[],[],lb,ub,[],IntCon);

FIGURE 9.27 MATLAB code for Example 9.12.

Example 9.13

Consider the following water resource projects with the respective costs and benefits indicated. The total budget is 150 units. Determine the optimal portfolio of projects using the GA **technique.**

Project ID	Cost	Benefit	Project ID	Cost	Benefit
1	10	94	11	18	63
2	18	94	12	19	50
3	10	56	13	14	97
4	20	65	14	12	65
5	18	93	15	10	52
6	10	89	16	16	49
7	13	95	17	13	94
8	16	69	18	12	81
9	13	64	19	18	62
10	19	69	20	11	81

Solution:

This is a knapsack problem that could be formulated as follows:

$$\text{Max } Z = \sum_{j=1}^{20} x_i b_i$$

Subjected to:

$$\sum_{j=1}^{20} x_i c_i \le 150$$

$$x_i = 0 \text{ or } 1$$

where $x_i = 0$ if project i is not selected and 1 if the project i is selected; c_i and b_i are the cost and benefit, respectively, of project i.

Step 1: Coding. In this step, the decision variables are translated into the binary coding in GA. Since the decision variables themselves are binary, it is easy to code them as binary bytes, as illustrated below.

1	2	3	4	5	6	7	8	9	10	11	12	13	14	15	16	17	18	19	20
0	0	0	0	1	1	1	1	1	0	1	0	1	0	0	0	0	1	1	1

Then, randomly generate 200 initial solutions. Each solution is called a population.

Step 2: Crossover. Two hundred initial populations are randomly matched into 100 pairs—conduct crossover between the populations in each pair.

Step 3: Mutation. After the crossover, each population mutates according to a prespecified mutation rate (0.05).

Step 4: Fitness evaluation. The total benefit of each population is calculated as the fitness of that population.

Step 5: Selection. Use tournament selection to select 100 populations based on their fitness values.

Step 6: Repeat steps 2 to 5.

Step 7: If the number of iterations reaches the specified maximum number of iteration (in this case, say, 3,000) or if all the populations in the previous iteration are the same as the offspring, stop iteration and go to step 8; otherwise, go to step 6.

Step 8: Output the population with the maximum benefit, that population is the final solution.

The result for the genetic algorithm is:

Project ID	1	2	3	4	5	6	7	8	9	10	11	12	13	14	15	16	17	18	19	20
Solution	1	1	1	0	0	1	1	0	1	0	0	0	1	1	1	0	1	1	0	1

where "1" means include the project in the portfolio of projects to be implemented, and "0" means otherwise. It can be shown that this portfolio yields a maximum total benefit of 962 units.

9.4.4.2 Simulation Annealing

SA is developed based on the analogy between the physical annealing process of solids and optimization problems. In SA, one can simulate the behavior of a system of particles in thermal equilibrium using a stochastic relaxation technique developed by Metropolis et al. (1953). This stochastic relaxation technique helps the SA procedure escape from the local minimum. SA starts at a feasible solution, q_0 (a real-valued vector representing all decision variables), and objective function $J_0 = J(q_0)$. A new solution q_1 is randomly selected from the neighbors of the initial solution, and the objective function $J_1 = J(q_1)$ is evaluated. Suppose the new solution has a smaller objective function value $J_1 < J_0$ (in minimization problems), the new solution is definitely better than the old one. Therefore, it is accepted, and the search moves to q_1 and continues from there. On the other hand, if the new solution is not better than the current one, $J_1 < J_0$, the new solution may be accepted, depending on the acceptance probability defined as

$$\Pr(\text{accept}) = \exp\left\{\frac{-(J_1 - J_0)}{T}\right\}, \quad (9.102)$$

where T is a positive number that will be discussed later.

For a given value of T, the acceptance probability is high when the difference between the objective function values is small. To decide whether or not to move to the new solution q_1 or stay with the old solution q_0, SA generates a random number U between 0 and 1. If $U <$ Pr(accept), SA moves to q_1 and resumes the search. Otherwise, SA stays with q_0 and selects another neighboring solution. At this step, one iteration has been completed.

The control parameter T plays an important role in reaching the optimal solution through SA because the acceptance probability is strongly influenced by choice of T. When the T is high, the acceptance probability calculated is close to 1, and the new solution will very likely be accepted, even if its objective function value is considerably worse than that of the old solution. On the other hand, if T is low, approaching zero, then the acceptance probability is also low, approaching zero. This effectively precludes the selection of the new solution even if its objective function value is only slightly worse than that of the old solution. Therefore, no new solution for which $J_1 > J_0$ is likely to be selected. Instead, only "downhill" points ($J_1 < J_0$) are accepted; therefore, a low value of T leads to a descending search.

Successful performance of the SA method requires that T be assigned a high initial value and that it be reduced gradually throughout the sequence of calculations by the reduction factor λ. The number of iterations under each constant T should be sufficiently large. However, a higher initial T and a larger number of iterations mean longer simulation time and may severely limit computational efficiency. SA stops either when it has completed all the iterations or when the termination criterion (a specific value of the objective function) is reached.

As a rule of thumb, to balance the computational efficiency and quality of the SA method, the initial T should be chosen so that the acceptance ratio is somewhere between 70% and 80%. It is recommended that a number of simulation runs should be executed first to estimate the acceptance ratio. Second, the number of iterations under a constant T is about 10 times the size of the decision variables. In the literature, the amount of 0.9 is suggested to be a reasonable number for λ as the reduction factor (Egles, 1990).

SA has many advantages, including easy implementation, not requiring much computer memory and coding, and a guarantee for identifying an optimal solution if an appropriate schedule is selected. The last point becomes much more important when the solution space is large, and the objective function has several local minima or changes dramatically with small changes in the parameter values. Examples of the SA application in the hydrological literature include Dougherty and Marryott (1991) and Wang and Zheng (1998).

9.4.4.3 Ant Colony

The ant colony algorithm is an algorithm for finding optimal paths based on the behavior of ants searching for food. At first, the ants wander randomly. When an ant finds a food source, it walks back to the colony, leaving "markers" (pheromones) that show the path has food. Species like ants called leaf cutter is found to leave some chemical called pyrazine, one of the ingredients in pheromone, on their track. These substances are secreted by ants and said to be left on the ground, which ants use as the communication tool and helps to maintain the path between them. In computer science and operations research, the ant colony optimization (ACO) algorithm is a probabilistic technique for solving computational problems which can be reduced to finding good paths through graphs. The main underlying idea, loosely inspired by the behavior of real ants, is that of a parallel search over several constructive computational threads based on local problem data and on a dynamic memory structure containing information on the quality of the previously obtained result. The collective behavior emerging from the interaction of the different search threads has proved effective in solving CO problems.

A CO problem is a problem defined over a set $C = c_1, \ldots, c_n$ of basic components. A subset S of components represents a solution of the problem; $F \subseteq 2^C$ is the subset of feasible solutions; thus, a solution S is feasible if and only if $S \in F$. A cost function z is defined over the solution domain, z: $2C \rightarrow R$, the objective being to find a minimum cost feasible solution S^*, that is, to find S^*: $S^* \in F$ and $z(S^*) \leq z(S)$, $\forall\ S \in F$.

Given this, the function of an ACO algorithm can be summarized as follows. A set of computational concurrent and asynchronous agents (a colony of ants) moves through states of the problem corresponding to partial solutions to the problem. They move by applying a stochastic local decision policy based on two parameters, called trails and attractiveness. By moving, each ant incrementally constructs a solution to the problem. When an ant completes a solution, or during the construction phase, the ant evaluates the solution and modifies the trail value on the components used in its solution. This pheromone information will direct the search of future ants.

Furthermore, an ACO algorithm includes two more mechanisms: trail evaporation and, optionally, daemon actions. Trail evaporation decreases all trail values overtime to avoid an unlimited accumulation of trails over some component. Daemon's actions can be used to implement centralized actions that single ants cannot perform, such as invoking a local optimization procedure or the update of global information to decide whether to bias the search process from a nonlocal perspective.

More specifically, an ant is a simple computational agent, which iteratively constructs a solution for the instance to solve. Partial problem solutions are seen as states. At the core of the ACO algorithm lies a loop, where at each iteration, each ant moves (performs a step) from a state to another one ψ, corresponding to a more complete partial solution. At each step σ, I each ant k computes a set $A_k\sigma(i)$ of feasible expansions to its current state and moves to one of these in probability. The probability distribution is specified as follows. For ant k, the probability $p_{i\psi}^k$ of moving from the state, i to state ψ depends on the combination of two values:

- The attractiveness $\eta_{i\psi}$ of the move, as computed by some heuristic indicating the a priori desirability of that move;
- The trail level $\tau_{\iota\psi}$ of the move, indicating how proficient it has been in the past to make that particular move: it represents therefore an a posteriori indication of the desirability of that move.

Trails are updated usually when all ants have completed their solution, increasing or decreasing the level of trails corresponding to moves that were part of "good" or "bad" solutions, respectively.

The general framework just presented has been specified in different ways by the authors working on the ACO approach.

9.4.4.4 Tabu Search

TS is a ground search algorithm for optimizing complex nonlinear tasks. TS is a neighborhood search descent method that avoids the "local minimum traps" by accepting worse (or even infeasible solutions) and constraining the current solution neighborhood by the solutions' "search history." The search history is stored in the form of a tabu (forbidden) list.

The purpose of this list is to prevent the transfer to previously examined locations. This can expand the search area of the method. In this method, first, a random answer is generated. All of its neighbors are identified, and the value of the objective function is obtained. It is then moved from the first point to the best point, and the previous point profile is stored in the tabu list. Stages one and two are performed again. Now a transfer point is selected with the best pretarget function among all the points and is not in the tabu list.

Example 9.14

Calculate the following optimization answer using the Tabu method.

$$\text{Maximize} \quad Z = 3X_1 + 5X_2$$

$$\text{Subjected to:}$$

$$X_1 \le 4$$

$$2X_2 \le 12$$

$$3X_1 + 2X_2 \le 18$$

$$X_1, X_2 \ge 0$$

Solution:

First, we randomly select a point as the primary answer (A_{22}). The set of neighboring points A_{22} is the points A_{11}, A_{12}, A_{13}, A_{23}, A_{33}, A_{32}, A_{31}, and A_{21}, and the value of the objective function for these points is 0, 6, 12, 22, 26, 20, and 10, respectively. A_{33} point that is not infeasible will be penalized for not entering the answer set

$A_{11}(0,0)$	$A_{12}(2,0)$	$A_{13}(4,0)$
$A_{21}(0,2)$	$A_{22}(2,2)$	$A_{23}(4,2)$
$A_{31}(0,4)$	$A_{32}(2,4)$	$A_{33}(4,4)$
$A_{41}(0,6)$	$A_{42}(2,4)$	$A_{43}(4,6)$

In this step, the taboo list is empty. Among all the neighboring points, point A_{32} is chosen as the basis for the next step. Finally, point A_{22} will be added to the taboo list, and the tabu list will be as follows.

$$\text{TL} = \{A_{22}\}$$

The numeric value 26 was stored as the best answer up to this point. The neighbors of A_{32}, are A_{21}, A_{23}, A_{42}, A_{41}, A_{31}, points. The objective function values in these points are 10, 22, 36, 30, and 20, respectively. Among these points, point A_{42} is selected as the base point of the next step due to having the highest value of the objective function. At the end of this step, point A_{32} is added to the tabu list, and the list is as follows:

$$\text{TL} = \{A_{32}, A_{22}\}$$

The value of the variable is the best answer which is 36. The neighborhood complex of point A_{42}, are points A_{41}, and A_{31}. The amount of target waste in these neighboring areas is 30 and 20, respectively. Among these points, point A_{41} is chosen as the basis for the next step, as it has the highest value of the objective function. At the end of the point, A_{42}, it will be added to the tabu list, and the list will be as follows:

$$\text{TL} = \{A_{42}, A_{32}, A_{22}\}$$

Because the value of the target function at point A_{41} is smaller than the value stored as the best answer, the best answer does not change. By repeating the above steps, the tabu list will increase, but the best answer will not change. Therefore, it can be expected that the optimal general answer to the problem has been calculated.

The main advantage of this method is the vast search area and the lack of reevaluation of the previously examined points.

9.4.5 Multiobjective Optimization

In MCDM problems, several criteria are contributing to one or more objectives. However, if explicitly more than one objective function is considered, it should be called multiobjective optimization. Consider an optimization problem with I objectives $f_1,\ldots,f_I$, as criteria with a set X illustrates which decisions can be made. In the presence of multiple criteria, we should represent the set of all possible outcomes, H, as follows:

$$H = \{(f_1(x),\ldots,f_I(x)) | x \in X\}. \tag{9.103}$$

This set is called the criteria space, which shows what can be achieved by selecting different alternatives. Any point of $\underline{f} \in H$ is called dominated if there is another point $\underline{f}^1 \in H$ such that $\underline{f}^1 \le \underline{f}$ and if there is strict inequality in at least one component. Similarly, a point $\underline{f} \in H$ is called nondominated if there is not another point $\underline{f}^1 \in H$. In the case of single-objective optimization problems, optimal solutions are selected, and if multiple optimal solutions exist, it does not matter which one is selected since they give the same objective function value. In MCDM, nondominated points will serve as the solutions to the problem. In multiobjective problems, the different nondominated solutions give different objective function values, so selecting the optimal solution is of paramount since they lead to different outcomes. For a set of nondominated solutions, additional preference information is needed from the decision makers to decide on trade-offs between the objectives.

In these cases, the multicriteria problem could be reformed to a single-criterion problem using weighting. In this method, the weight of different criteria is determined based on the decision maker's judgment and criteria priorities. Weighting is the most popular MCDM method, which could be described as follows:

$$\begin{cases} \text{Maximize} & F(x) = [f_1(x), f_2(x),\ldots,f_I(x)] \\ \text{Subject to} & x \in X \end{cases}$$
$$\Rightarrow \begin{cases} \text{Maximize} & F(x) = \sum_{k=1}^{I} W_k f_k(x) \\ \text{Subject to} & x \in X, \end{cases} \tag{9.104}$$

where W_k is the weight of the kth objective function that is a nonzero positive real number. In the weighting method, all the criteria are reformed to maximization by multiplying the minimizing criteria by −1. Some methods for MCDM were discussed before such as TOPSIS (a technique for order preference by similarity to ideal solution) and AHP. In multiobjective optimization, the linear or nonlinear optimization techniques are used.

Example 9.15

Assume that a combination of three alternative technologies removes two pollutants. They remove 3, 2, and 1 g/m³, respectively, of the first kind of pollutant and 2, 1, and 3 g/m³ of the second kind. If 1,000 m³ of each pollutant has to be treated in a day, determine the optimal combination of methodologies.

Solution:

The total removed amounts of the two pollutants are

$$3x_1 + 2x_2 + 1(1{,}000 - x_1 - x_2) = 2x_1 + x_2 + 1{,}000$$

and

$$2x_1 + x_2 + 3(1{,}000 - x_1 - x_2) = -x_1 - 2x_2 + 3{,}000.$$

If no condition is given on the usage of the third technology, there is no constraint on the minimum amount to be removed, and no cost is considered, then this problem can be mathematically formulated as

$$\begin{aligned} &\text{Maximize} \quad f_1(\underline{x}) = 2x_1 + x_2, \quad f_2(\underline{x}) = -x_1 - 2x_2, \\ &\text{Subject to} \quad x_1, x_2 \geq 0, \\ &\qquad\qquad\quad x_1 + x_2 \leq 1{,}000. \end{aligned}$$

The feasible decision space is exhibited in Figure 9.28.

The criteria space can be determined as follows:

$$f_1 = 2x_1 + x_2 \quad \text{and} \quad f_2 = -x_1 - 2x_2.$$

Therefore,

$$x_1 = \frac{2f_1 + f_2}{3} \quad \text{and} \quad x_2 = \frac{-f_1 - 2f_2}{3}.$$

The constraints $x_1, x_2 \geq 0$ can be rewritten as

$$\frac{2f_1 + f_2}{3} \geq 0 \quad \text{and} \quad \frac{-f_1 - 2f_2}{3} \geq 0,$$

that is,

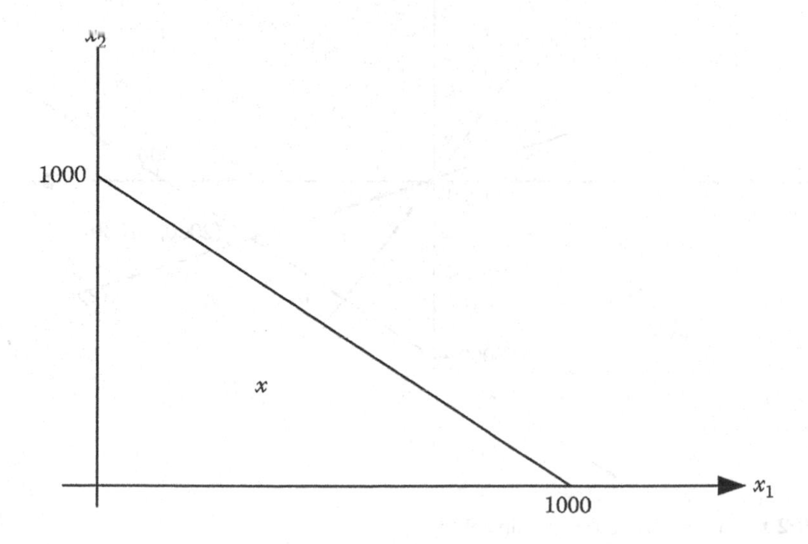

FIGURE 9.28 Feasible decision space for Example 9.15.

$$2f_1 + f_2 \geq 0 \quad \text{and} \quad f_1 + 2f_2 \leq 0.$$

The third constraint $x_1 + x_2 \leq 1{,}000$ now has the form

$$\frac{2f_1 + f_2}{3} + \frac{-f_1 - 2f_2}{3} \leq 1{,}000,$$

that is, $f_1 - f_2 \leq 3{,}000$.

Hence, set H is characterized by the following inequalities:

$$2f_1 + f_2 \geq 0,$$

$$f_1 + 2f_2 \leq 0,$$

$$f_1 - f_2 \leq 3{,}000,$$

and is illustrated in Figure 9.29. Note that H is the triangle with vertices (0, 0), (1,000, −2,000), and (2,000, −1,000). The point (0, 0) is nondominated, but it is not better than the point (1,000, −2,000), which is dominated, since 1,000 > 0 gives a higher value in the first objective function. The points on the linear segment connecting the vertices (0, 0) and (2,000, −1,000) cannot be dominated by any point of H, since none of the objectives can be improved without worsening the other. This, however, does not hold for the other points of H, where both objectives can be improved simultaneously. Therefore, all points between (0, 0) and (2,000, −1,000) can be accepted as the best solutions.

We can say that (0, 0) is better than (2,000, −1,000), if a 1,000 unit loss in the second objective is not compensated by a 2,000 unit gain in the first objective. These kinds of decisions cannot be made based on only the concepts being discussed above. Depending on the additional preference and/or trade-off information obtained from the decision makers, different solution concepts and methods are available.

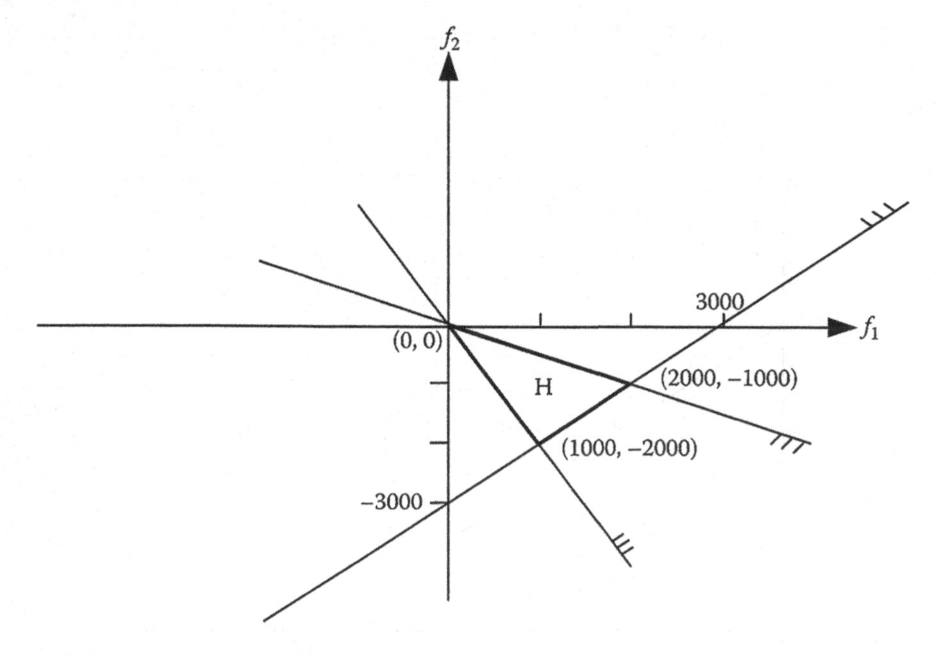

FIGURE 9.29 Criteria space for Example 9.15.

9.5 CONFLICT RESOLUTION

Water management comprises of sharing water and resolving conflicts among users and stakeholders. A stakeholder is either directly affected by the decision or has the power to influence or block the decision. White et al. (1992) examined three sources of provoking conflicts. The first is human intervention in the environment when one or more of the stakeholders see the activity as a disturbance to physical, biological, and social processes. The second source is disagreement over the management of the water supply at one location as it affects its use elsewhere. The third source is climatic variability independent of any human activity, which places more stress on the water resources. Urban areas are more subject to climate change due to added heat flux and the fate of contaminants in the urban area.

The use of water always involves an interaction between human users and soil, water, and air resources. The key indicators of the water conflicts are related to a number of issues, including water quantity, water quality, management of multiple uses, and level of national development. The political differences, geopolitical settings, and hydropolitical issues that are at stake, and institutional control of water resources are the governing factors in water conflicts (Wolf, 1998).

9.5.1 Conflict Resolution Process

Many disciplines have approached the conflict resolution process, such as law, economics, engineering, political economy, geography, and systems theory. An excellent source of selected disciplinary approaches is available in Wolf (2002).

This section differentiates between traditional versus system approaches to water conflicts. Traditional conflict resolution approaches such as the judicial systems, state legislatures, commissions, and similar governmental systems provide resolutions in which one party gains at the expense of the other. This is referred to as the "zero-sum" or "distributive" solution. A negotiation process referred to as alternative dispute resolution (ADR) is involved in water and environmental conflict resolution. ADR seeks a "mutually acceptable settlement." ADR generally moves parties to what is referred to as "positive-sum" or "integrative" solutions (Bingham et al., 1994). Negotiation, collaboration, and consensus building are the key issues that facilitate ADR.

A systems approach to conflict resolution among stakeholders helped by exploring the underlying structural causes of conflict. It can transform problems into opportunities for all parties involved. Cobble and Huffman (1999) have explored the systems approach to conflict resolution in management science.

Some elements of the systemic approach have also been present in the work of Simonovic and Bender (1996), which proposes collaboration and a collaborative process with the active involvement of stakeholders who agree to work together to identify problems and develop mutually acceptable solutions. Consensus building processes constitute a form of collaboration that explicitly includes the goal of reaching a consensus agreement on water conflicts. Wolf (2000) initiated an indigenous approach to water conflict reduction. Such methods include (a) prioritizing different demand sectors, (b) protecting downstream and minority rights, and (c) practicing forgiveness and compromise.

Four steps can be implemented in this process:

Step 1: Create the space and the intention among stakeholders to address a conflict. This can be achieved by encouraging participants to explore the source of the conflict. Stakeholders should identify critical issues, actions, and the thinking that led to the conflict situation.

Step 2: Build a shared understanding of the conflict through inquiry and the creation of a systems map. Causal diagrams like the one in Figure 9.30 may be of help. It is important to look for places of disagreement in the diagram.

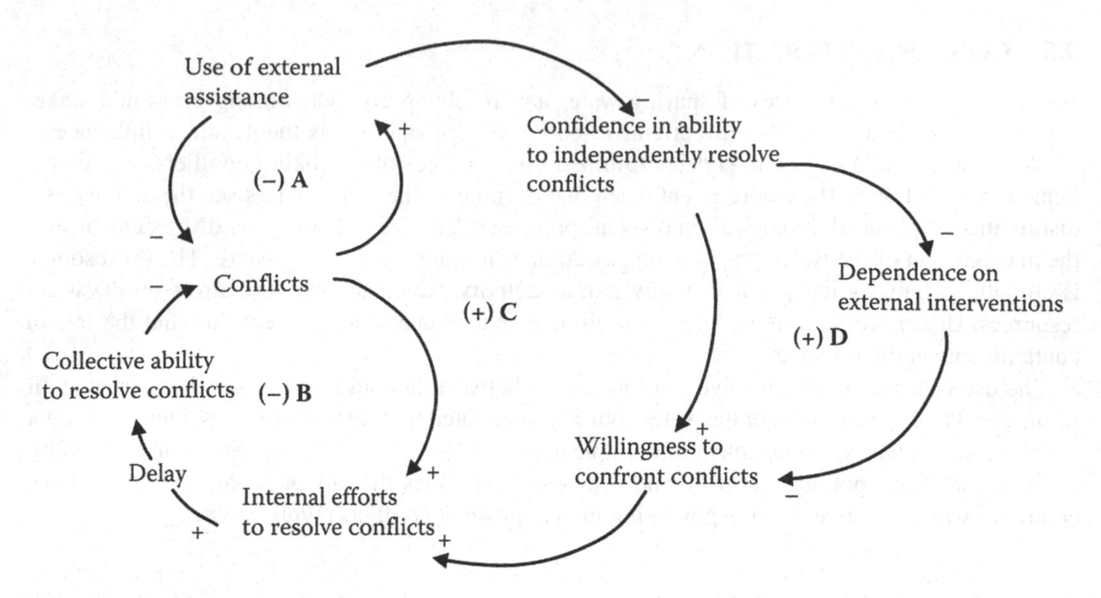

FIGURE 9.30 Advantages of a systematic approach to conflict resolution. (From Cobble, K. and Huffman, D., *Syst. Thinker*, 10(2), 8–16, 1999. With permission.)

Step 3: Build dialogues so that participants can directly address sources of conflict and understand their own role in it. During this process, modeling "hot spots" should be considered rather than general problems.

Step 4: Create an action plan to develop and implement alternative solutions—new ways to work and interact. Participants are expected to reach agreements for trying out new solutions and behaviors.

9.5.2 A System Approach to Conflict Resolution

A systematic approach plays three major roles in water conflict resolution. First, scientific investigation determines the relationship among the various components. Second, it helps us to describe the characteristics of the various components, including the physical systems, the ecosystems, affected social groups, and organizations with their preferences and modes of action. Third, it offers a means of estimating the significance of impacts in terms of physical quantities and perceived impacts by the people and organizations affected.

A causal loop diagram (CLD) is a qualitative SD technique for mapping the problem structure (i.e., variables, relationships, feedback loops, delays). It is one of the most commonly used diagramming tools in system dynamics (Lane, 2000). According to Sweeney and Sterman (2000), a CLD can be used for mapping and communicating the problem structure and eliciting individual and team mental models. Expanding the developed CLDs could help people to think beyond immediate outcomes and linear relationships to consider the effects of delays and feedback loops. Developing the CLDs is essential for sharing knowledge to build a shared understanding of the problem's emergence (Karamouz and Zare, 2021).

By providing a user-friendly interface, the software Vensim® has been widely adopted for simulation applications by system dynamics. Primarily by building various simulations from stock and flow or causal loops, they made it possible to visualize how complex systems work dynamically (Wen et al., 2016).

A systematic approach's advantages are identified in Figure 9.30 using the systems language of causal diagrams. Negative (balancing) feedback loop A shows that relying on outside assistance, like hiring an outside mediator, for example, to respond to conflict may serve parties involved in the short term. However, over the long term, it reduces the stakeholders' confidence in their own ability to resolve problems and willingness to confront conflicting situations, as shown by the positive feedback C. Another unintended consequence is a rise in the stakeholders' dependence on external intervention, further decreasing their comfort with handling conflicting situations as indicated by the positive feedback D. The proposed approach offers a solution through building the stakeholders' conflict resolution skills (negative feedback B).

9.5.3 Conflict Resolution Models

As mentioned earlier, the presence of multiple decision makers makes the decision-making process much more difficult since they usually have different priorities and goals. Therefore, reaching a compromise between the conflicting objectives is sometimes very difficult.

There are several alternative ways to solve conflict situations. One might consider the problem as a multiobjective optimization problem with the objectives of the different decision makers. Conflict situations can also be modeled as social choice problems in which the rankings of the decision makers are taken into account in the final decision. A third way of resolving conflicts was offered by Nash, who considered a particular set of conditions the solution has to satisfy and proved that precisely one solution satisfies his "fairness" requirements. In this section, the Nash bargaining solution will be outlined (Karamouz et al., 2003).

Assume that there are I decision makers. Let X be the decision space and f_i: $X \mapsto R$ be the objective function of decision maker i. The criteria space is defined as

$$H=\left\{\underline{u}\,\middle|\,\underline{u}\in R^I,\underline{u}=(u_i),u_i=f_i(x)\text{ with some }x\in X\right\}. \tag{9.105}$$

It is also assumed that in the case when the decision makers are unable to reach an agreement, all decision makers will get low objective function values. Let d_i denote this value for decision maker i, and let $\underline{d} = (d_1, d_2, \ldots, d_I)$. Therefore, the conflict is completely defined by the pair $(H,\underline{d})$, where H shows the set of all possible outcomes and $\underline{d}$ shows the outcomes if no agreement is reached. Therefore, any solution to the conflict depends on both H and $\underline{d}$. Let the solution be therefore denoted as a function of H and $\underline{d}$: $\varphi(H,\underline{d})$. It is assumed that the solution function satisfies the following conditions:

1. The solution has to be feasible: $\varphi(H,\underline{d})\in H$.
2. The solution has to provide at least the disagreement outcome to all decision makers: $\varphi(H,\underline{d})\geq \underline{d}$
3. The solution has to be nondominated. That is, there is no $f\in H$ such that $\underline{f}\neq\varphi(H,\underline{d})$ and $\underline{f}\geq\varphi(H,\underline{d})$.
4. The solution must not depend on unfavorable alternatives. That is if $H_1\subset H$ is a subset of H such that $\varphi(H,\underline{d})\in H_1$, then $\varphi(H,\underline{d})=\varphi(H_1,\underline{d})$.
5. Increasing linear transformation should not alter the solution. Let T be a linear transformation on H such that $T(\underline{f})=(\alpha_1 f_1+\beta_1,\ldots,\alpha_I f_I+\beta_I)$ with $\alpha_i > 0$ for all i, then $\varphi(T(H),T(\underline{d}))=T(\varphi(H,\underline{d}))$.
6. If two decision makers have equal positions in the conflict definition, they must get equal objective values at the solution. Decision makers i and j have an equal position if $d_i = d_j$ and any vector $\underline{f}=(f_1,\ldots,f_I)\in H$ if and only if $(\bar{f}_1,\ldots,\bar{f}_I)\in H$ with $\bar{f}_1=f_j,\bar{f}_j=f_i,\bar{f}_l$ for $l\neq i, j$. Then we require that $\varphi_i(H,\underline{d})=\varphi_j(H,\underline{d})$.

Before describing the method to find the solution satisfying these properties, some remarks are in order. The feasibility condition requires that the decision makers cannot get more than the amount is available. No decision maker would agree to an outcome that is worse than the amount they would get anyhow without the agreement. This property is given in condition 2. The requirement that the solution is nondominated shows that there is no better possibility available for all. The fourth requirement says that if certain possibilities become infeasible, but the solution remains feasible, then the solution must not change. If any one of the decision makers changes the unit of their objective, then a linear transformation is performed on H. The fifth property requires that the solution must remain the same. The last requirement shows a certain kind of fairness, stating that if two decision makers have the same outcome possibilities and the same disagreement outcome, there is no reason to distinguish them in the final solution.

If H is convex, closed, and bounded, and there is at least one $\underline{f} \in H$ such that $\underline{f} > \underline{d}$, then there is a unique solution $\underline{f}^* = \varphi(H, \underline{d})$, which can be obtained as the unique solution of the following optimization problem:

$$\begin{aligned} &\text{Maximize}: (f_1 - d_1)(f_2 - d_2),\ldots,(f_I - d_I), \\ &\text{Subject to}: f_i \geq d_i (i = 1,2,\ldots,I), \\ &\underline{f} = (f_1,\ldots,f_I) \in H. \end{aligned} \tag{9.106}$$

The objective function is called the *Nash product.* Note that this method can be considered as a special distance-based method when the geometric distance is maximized from the disagreement point $\underline{d}$.

In many practical cases, conflict resolution is made when the decision makers have different powers. In such cases, Equation 9.106 is modified as

$$\begin{aligned} &\text{Maximize}: (f_1 - d_1)^{c_1} (f_2 - d_2)^{c_2},\ldots(f_I - d_I)^{c_I}, \\ &\text{Subject to}: f_i \geq d_i (i = 1,2,\ldots,I), \\ &(f_1,\ldots,f_I) \in H. \end{aligned} \tag{9.107}$$

Here c_1, c_2, …, c_I shows the relative powers of the decision makers. The solution to this problem is usually called the nonsymmetric Nash bargaining solution. The general format of the utility function for different water users/stakeholders is considered in Figure 9.31. In this figure, the utility function of different sectors varies between 0 and 1 when the allocated water to each sector is in the range of $b - c$.

Conflict resolution is a special section in game theory. See Forgo et al. (1999) for more details.

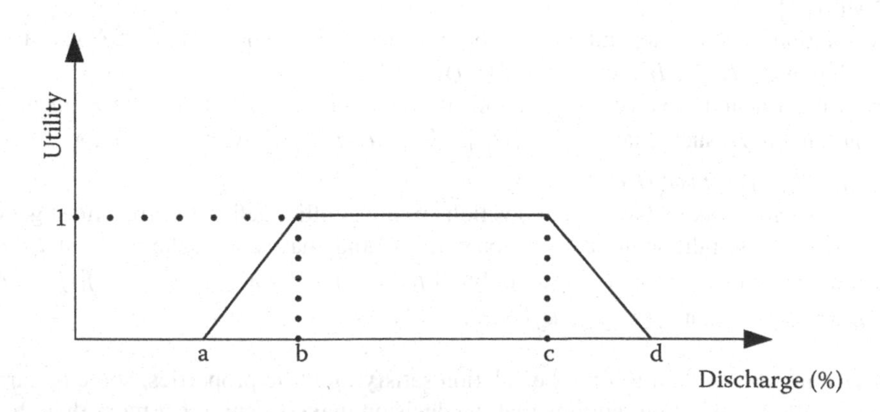

FIGURE 9.31 Utility function of the allocated water.

Example 9.16

Consider a multipurpose reservoir upstream of an urban area with conflicting objectives. Discuss the general considerations in resolving conflicts over a multipurpose reservoir operation.

Solution:

The objectives of the operation of the reservoir have been categorized as follows:

1. *Downstream water supply:* There are several users downstream of the reservoir with urban, agricultural, and environmental demands. The major conflicts in reservoir operation occur mostly in dry seasons when there is not enough water and the reservoir is not capable of supplying the demands.
2. *Flood damage control:* The reservoir is also intended for flood control, which is necessary to protect cities, infrastructures, and irrigation projects from floods downstream of the reservoir. This objective conflicts with water conservation strategies for storing water in high-flow seasons.
3. *Environmental water demand*: Supplying downstream river instream requirements are in conflict with reservoir water conservation and supplying other downstream demands.
4. *Power generation*: Power generation is another objective that has a conflict with the other objectives. This is because the trends in supplying demands and power generation do not follow the same pattern. Therefore, supplying one of these demands in a specific period of time might conflict with supplying the others.

First, the utility functions are defined based on the priorities and favorable ranges of water supply for the agencies associated with each demand for resolving conflicts. Different stakeholders and water users and their objectives are described in the following example.

Each of these sectors has its own set of priorities for allocating water to different demands within their own line of operation and responsibilities considering the relative priority of each water use. For example, household demands may have the highest priority for the city department of water supply.

Example 9.17

In Example 9.16, consider the following data that show the demand and the utility function for different sectors that are obtained from compromising sessions. If the relative weights of agriculture, domestic water use, industrial water use, environmental protection, and power generation are considered as 0.135, 0.33, 0.2, 0.2, and 0.135, respectively, find the optimal solution and the most appropriate water allocation scheme for a year with an average flow of 15 MCM and an initial water storage of 25 MCM.

The results of the compromising sessions are as follows:

- *Environmental sector:* The environmental water quantity and quality in the downstream river is the main concern of this sector. Assume that the discharge of 8 MCM per month is needed for the river ecosystem; the environmental utility function for river flow (f_{env}) is formulated as

$$f_{env}(Q_{env}) = \begin{cases} 1 & \text{if } Q_{env} \geq 8 \text{ MCM}, \\ 1 - 0.33(8 - Q_{env}) & \text{if } 5 \leq Q_{env} < 8 \text{ MCM}, \\ 0 & \text{if } Q_{env} < 5 \text{ MCM}, \end{cases} \tag{9.108}$$

  where Q_{env} is the instream flow downstream.
- *Agricultural sector*: The main objective of this sector is water supply with acceptable quality to fulfill agricultural demands and reduce the return flow removal cost. The utility of this sector related to the water supply is based on the water supply reliability. The considered agricultural water demand in this example is about 8 MCM. Therefore, the utility function for agricultural demand is assumed to be

$$f_{\text{agr}}\left(Q_{\text{agr}}\right)=\begin{cases}1 & \text{if } Q_{\text{agr}}>8,\\ 1-0.17\left(8-Q_{\text{agr}}\right) & \text{if } 2<Q_{\text{agr}}\leq 8,\\ 0 & \text{if } 0<Q_{\text{agr}}\,2,\end{cases} \quad (9.109)$$

where f_{agr} is the utility function related to the water supply reliability and Q_{agr} is the water allocated to agricultural water

- *Industrial sector*: The main objective of this sector is water supply to fulfill industrial demands. The utility function of the decision makers in this sector for assessing the reliability of the industrial water supply is as follows:
- *Industrial sector*: The main objective of this sector is water supply to fulfill industrial demands. The utility function of the decision makers in this sector for assessing the reliability of the industrial water supply is as follows:

$$f_{\text{ind}}\left(Q_{\text{ind}}\right)=\begin{cases}1 & \text{if } Q_{\text{ind}}>8,\\ 1-0.18\left(8-Q_{\text{ind}}\right) & \text{if } 3<Q_{\text{ind}}\leq 8,\\ 0 & \text{if } 0<Q_{\text{ind}}\leq 3,\end{cases} \quad (9.110)$$

where f_{ind} is the utility function related to the water allocated to the industrial complex (Q_{ind}). *Water and wastewater sector*: The main objective of these companies is water supply with acceptable quality to domestic demands and wastewater collection and disposal. The utility function of the decision makers in this sector for water allocated to domestic water users is assumed to be

$$f_d=\begin{cases}1 & \text{if } Q_d>10,\\ 1-0.14\left(10-S_d\right) & \text{if } 3<Q_d<10,\\ 0 & \text{if } 0<Q_d\leq 3,\end{cases} \quad (9.111)$$

where f_d is the utility function related to domestic water supply reliability and Q_d is the percentage of the supplied domestic water demand.

- *Water supply and energy production sector*: The main objectives of this sector are electrical power generation and water storage for future demands. The reservoir storage utility is developed considering the minimum and maximum allowable water storage and water level over the hydropower intake in each month. The utility function of the decision makers in this sector for reservoir water storage is

$$f_{s,m}\left(S_{m+1}\right)=\begin{cases}0 & \text{if } S_{t+1}\leq 10 \text{ million cubic meters},\\ 0.0066\times\left(S_{t+1}-30\right) & \text{if } 10<S_{t+1}<25 \text{ million cubic meters},\\ 1 & \text{if } 25\leq S_{t+1}\leq 30 \text{ million cubic meters},\\ 0 & \text{if } S_{t+1}>30 \text{ million cubic meters},\end{cases} \quad (9.112)$$

where $f_{s,m}$ is the utility function related to the reservoir storage, and $S_m + 1$ is the reservoir storage at the end of month m.

Solution:

The nonsymmetric Nash solution of the example is the unique optimal solution of the following problem:

$$\text{Maximize}: Z=\prod_{i=1}^{5}\left(f_i-d_i\right)^{w_i},$$

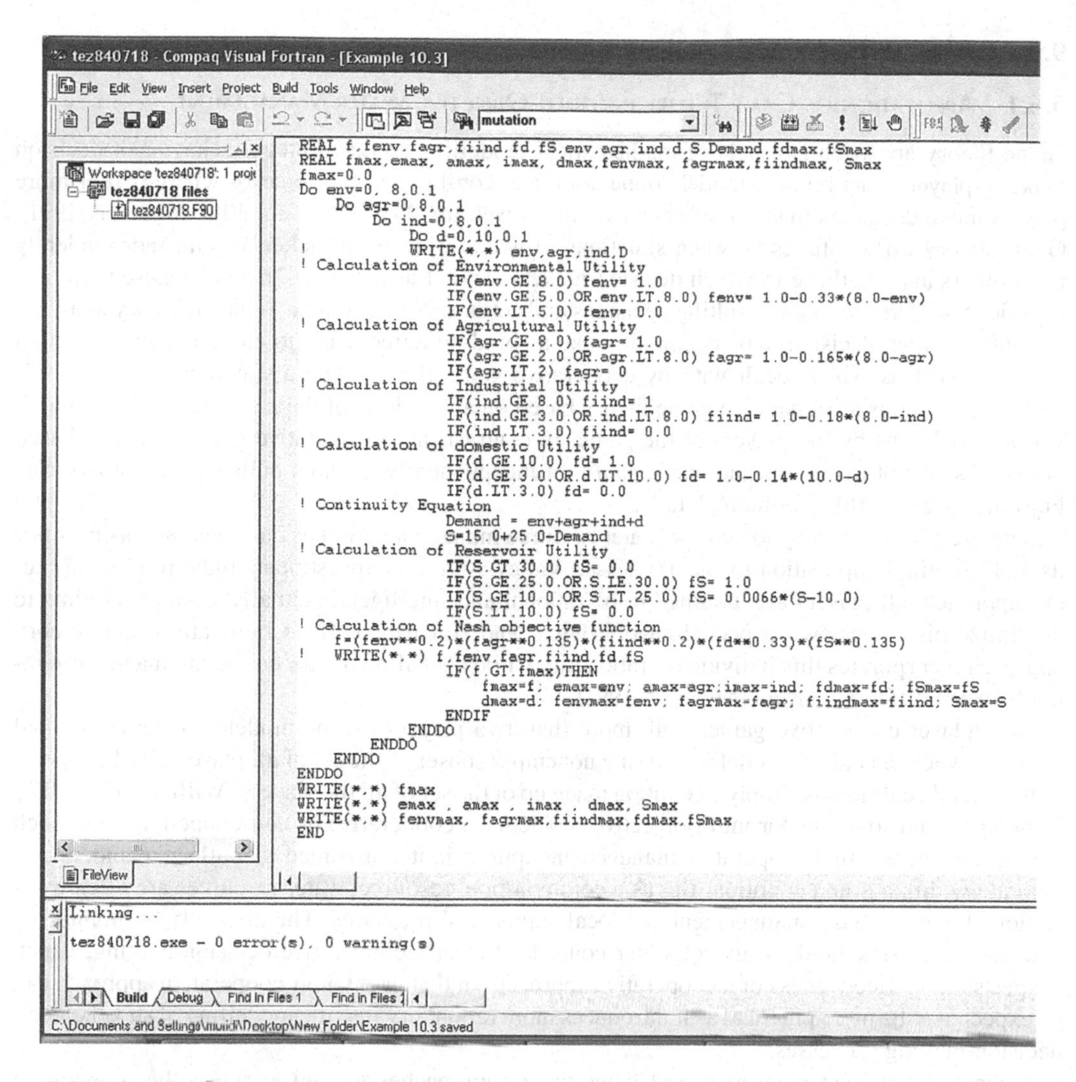

tez840718 - Compaq Visual Fortran - [Example 10.3]
File Edit View Insert Project Build Tools Window Help
mutation
Workspace 'tez840718': 1 proje
tez840718 files
tez840718.F90
REAL f,fenv,fagr,fiind,fd,fS,env,agr,ind,d,S,Demand,fdmax,fSmax
REAL fmax,emax, amax, imax, dmax,fenvmax, fagrmax,fiindmax, Smax
fmax=0.0
Do env=0, 8,0.1
Do agr=0,8,0.1
Do ind=0,8,0.1
Do d=0,10,0.1
! WRITE(*,*) env,agr,ind,D
! Calculation of Environmental Utility
IF(env.GE.8.0) fenv= 1.0
IF(env.GE.5.0.OR.env.LT.8.0) fenv= 1.0-0.33*(8.0-env)
IF(env.LT.5.0) fenv= 0.0
! Calculation of Agricultural Utility
IF(agr.GE.8.0) fagr= 1.0
IF(agr.GE.2.0.OR.agr.LT.8.0) fagr= 1.0-0.165*(8.0-agr)
IF(agr.LT.2) fagr= 0
! Calculation of Industrial Utility
IF(ind.GE.8.0) fiind= 1
IF(ind.GE.3.0.OR.ind.LT.8.0) fiind= 1.0-0.18*(8.0-ind)
IF(ind.LT.3.0) fiind= 0.0
! Calculation of domestic Utility
IF(d.GE.10.0) fd= 1.0
IF(d.GE.3.0.OR.d.LT.10.0) fd= 1.0-0.14*(10.0-d)
IF(d.LT.3.0) fd= 0.0
! Continuity Equation
Demand = env+agr+ind+d
S=15.0+25.0-Demand
! Calculation of Reservoir Utility
IF(S.GT.30.0) fS= 0.0
IF(S.GE.25.0.OR.S.LE.30.0) fS= 1.0
IF(S.GE.10.0.OR.S.LT.25.0) fS= 0.0066*(S-10.0)
IF(S.LT.10.0) fS= 0.0
! Calculation of Nash objective function
f=(fenv**0.2)*(fagr**0.135)*(fiind**0.2)*(fd**0.33)*(fS**0.135)
! WRITE(*,*) f,fenv,fagr,fiind,fd,fS
IF(f.GT.fmax)THEN
fmax=f; emax=env; amax=agr;imax=ind; fdmax=fd; fSmax=fS
dmax=d; fenvmax=fenv; fagrmax=fagr; fiindmax=fiind; Smax=S
ENDIF
ENDDO
ENDDO
ENDDO
ENDDO
WRITE(*,*) fmax
WRITE(*,*) emax , amax , imax , dmax, Smax
WRITE(*,*) fenvmax, fagrmax,fiindmax,fdmax,fSmax
END
FileView
Linking...
tez840718.exe - 0 error(s), 0 warning(s)
Build Debug Find in Files 1 Find in Files 2
C:\Documents and Settings\muidi\Desktop\New Folder\Example 10.3 saved

FIGURE 9.32 Source code of the program for solving Example 9.17.

$$\text{Subject to: } S_i + Q_i - R_i = S_{i+1},$$

where w_i is the relative weight, f_i is the utility function of each sector, d_i is the disagreement point, S_i is the reservoir storage at each stage, I_i is the inflow to the reservoir at each stage, and R_i is the reservoir release at each stage which is equal to the summation of water allocation to each community at each stage.

In order to find the optimal solution of the Nash product in this example, a Fortran code is generated for searching all feasible environments, which is shown in Figure 9.32. After running the shown program, the optimal solution of the problem is as follows:

The water allocated for environmental flow: 7.5 MCM.
The water allocated to the agricultural demand: 4.4 MCM.
The water allocated to the industrial demand: 6.1 MCM.
The water allocated to the domestic demand: 9 MCM.

The maximum value of the Nash objective function based on the above optimum values is about 0.445.

9.6 GAME THEORY AND AGENT BASED MODELLING

9.6.1 Application of Game Theory in Multi-Objective Water Management

Game theory approach is a field of mathematical modelling of the strategic behavior of decision makers (players) that aims to model cooperation and conflict and in scenarios where two or more players make decisions that will affect each other's welfare (William et al., 2017; Myerson, 1991). Game theory differentiates between situations in which a decision-maker acts (a) independently from others and (b) those in which decision makers can act as a group. This is because there is a significant difference in the resulting decisions and actions. Non-cooperative game theory analyzes case (a) in which decision-makers cannot make a "binding agreement" to enforce some action on one another. Case (b) is dealt with by cooperative game theory. The distinction follows Nash's statement which is a cooperative game theory based on an analysis of the interrelationships among various coalitions by the players of the game. In contrast, non-cooperative game theory is based on the absence of coalitions and participants act independently, without collaboration others. See Fujiwara-Greve, (2005) for more details.

The use of game theory to address water resource management issues has been increasing since its 1942 seminal application to the Tennessee Valley Authority investment study. In a game theory approach, all players are assumed to be rational and intelligent. Actually, each agent aims to maximize his or her own expected payoff (Dinar and Hogarth, 2015). Cooperative game theory approach incorporates this individual's rationality while also allowing for cooperation and bargaining between different players (Myerson, 1991).

Multiplayer cooperative games with more than two players can be modeled using coalitional analysis, where a coalition is defined as any nonempty subset of the set of all players. Full cooperation (a grand coalition) is simply a coalition made up of the set of all the players (William et al. 2017). In order to find solutions for multi-objective problems, a cooperative or non-cooperative approach can be considered. In a cooperative management approach, it is assumed that all stakeholders collaborate within a grand coalition. But as a common-pool resource, water resources are often overexploited with growing management and local authorities' pressures. The difficulty of monitoring and managing stakeholders' use of water could lead to non-cooperative decision-making, which suggests the ineffectiveness of a cooperative approach. In that event, non-cooperative approach can be expected to be more practical as it introduces more rationality into the modelling of stakeholders' decision-making processes.

Cooperative management as one of game theory approaches is used to assess the cooperation among involved beneficiaries and allocate fairly and efficiently among them. The long-term benefits of cooperation are considered as long-term common-pool resource planning and management horizon. Shapley value, as it is explained below, based one of the cooperative solution methods is applied to fairly share the benefits among the beneficiaries. Moreover, the core condition (Individual and group rationality, and efficiency conditions) and Loehman stability index is used to assess the stability and acceptability of the shares (Esteban and Dinar, 2013; Liu et al., 2020; Mahmoodzadeh and Karamouz, 2021). The Shapley value is based on the weighted average of their contributions to all possible coalitions (Shapley 1953) and is defined as follows (Equation 9.113):

$$\varphi_p(v_c)=\sum_{p\in|s_c|\in N_c}\frac{\left[(|s_c|-1)!\times(N_c-|s_c|)!\right]}{N_c!}\times\left[v(s_c)-v(s_c\setminus p)\right];\; v(s_c)=\sum_{p\in|s_c|}\varphi_p(v_c) \quad (9.113)$$

where v_c is the value function of players, $\phi_p(v_c)$ is the Shapley value for the player p (such as farmers), $N_c=\{1, 2, \ldots, n\}$ is the total number of beneficiaries, s_c is a feasible coalition in the game, $|s_c|$ is the number of members of coalition s_c, $v(s_c)$ is the total acquirable benefits by the members of coalition s_c and $v(s_c \setminus p)$ is the total acquirable benefits of the coalition s without contributing the player p, $v(s_c)$ is the validity of the Shapley value. Based on Chalkiadakis et al. (2011) and Ghadimi and Ketabchi (2019), a fair and efficient allocation needed in a demand management problem to satisfy the core conditions which are stated as follows:

The individual rationality (Equation 9.114): for each player, the cooperative solution should be preferred to the individual case (i.e. non-cooperation).

$$v(p) \le u_p \tag{9.114}$$

where u_p is the allocated value under cooperation, $v(p)$ is the value of the non-cooperative coalition of player p.

The group rationality (Equation 9.115): the allocated share of full (grand) coalition should be preferred to any allocation in any possible partial coalition or sub-coalition

$$v(s_c) \le \sum_{p \in N_c} u_p \tag{9.115}$$

The efficiency (Equation 9.116): the total acquirable benefits under the grand coalition, $v(N)$, should be fully allocated to the members of that coalition.

$$v(N) = \sum_{p \in N_c} u_p v(N) \tag{9.116}$$

The other selected method for measuring the stability of a cooperative game is the Loehman power/stability index (L_p) (Loehman et al., 1979). This index compares the obtained benefits to a beneficiary with the obtained benefits in the coalition which is described as follows:

$$L_p = \frac{u_p^* - v(p)}{\sum_{p \in N_c} (u_p^* - v(p))}, \quad \sum_{p \in N_c} L_p = 1 \tag{9.117}$$

where u_p^* is the allocated share to player p. This index demonstrates stability and acceptability of the allocation solution which is defined as:

$$M_L = \frac{\sigma_L}{\bar{L}}, \quad 0 \le M_L \le 1 \quad M_L = \frac{\sigma_L}{\bar{L}}, \tag{9.118}$$

where M_L is the coefficient of variation (L is used in lieu of L_p to prevent double subscripting) which is calculated over all players in a given allocation solution. The greater the value of M_L, the larger the instability of the allocation solution. Further information about types of stability definitions is provided by Mahmoodzadeh and Karamouz (2021).

In the application of non-cooperative institutions, Nazari and Ahmadi, (2019) displayed two challenging issues in groundwater beneficiaries: (i) the need to reconcile the conflicting interests of many beneficiaries and (ii) the existence of externalities which may affect groundwater level. Given the nature of groundwater as a common-pool resource, assume that there are two groups of beneficiaries with different objectives and preferences, namely, the government and farmers. The optimization model for such scenario consists of two objectives. The objective function for the farmers is to maximize their total profit during the modelling period, Z_1 (Equation 9.119), and the objective function of the government sector is to minimize the mean change in water storage (here represented by groundwater level drawdown, Z_2) as shown in Equation 9.120. Monthly allocation of water to each crop and its corresponding cultivated area (CA) during the planning period are decision variables in this model.

$$\text{Maximize } Z_1 = \sum_{d=1}^{D} (B_d - C_d) \tag{9.119}$$

$$\text{Minimize } Z_2 = \frac{\sum_{d=1}^{D} \Delta h_d}{D} \tag{9.120}$$

where, D is the number of zones in the study area, B_d is the profit of crop cultivation during planning horizon in zone d (US Dollar), C_d is the cost of cultivation in zone d (US Dollar). Δh_d is the groundwater level drawdown in zone d during planning horizon.

Farmers' net benefit is the difference between their revenue and costs in each zone. The revenue is calculated using Equation 9.121. Their revenue is proportional to crop yields, cultivated area, and the sale price of products as expressed in Equation 9.122.

$$B_d = \sum_{t=1}^{T}\sum_{p=1}^{P}\left(Ya_{dtp} \times T_{tp} \times A_{dp}\right) \tag{9.121}$$

$$Ya_{dtp} = Ym_p \times \prod_{m=1}^{M}\left(1 - Ky_{pm}\left(1 - \left(\frac{Vg_{dtpm}}{\text{Demand}_{dtpm}}\right)\right)\right) \tag{9.122}$$

where, T is the total number of years in the plan horizon, M is the total number of months in one year, Ya_{dtp} is the actual crop production for crop p in year t and zone d (kg/ha), T_{tp} is the price of crop p in year t (US Dollar/kg), A_{dp} is the production area of crop p in zone d (ha), Y_{mp} is the maximum yield of crop p in month m (kg/ha), Ky_{pm} shows the sensitivity coefficient for crop p in month m (dimensionless), Vg_{dtpm} is the groundwater allocation to crop p in zone d, month m, year t (m^3/ha), Demand_{dtpm} is the water demand of crop p in zone d, year t, month m (m^3/ha). The model also considers salinity effects of water on crop yield. Higher water salinity could reduce the yield of products.

Cultivation costs, C_d, include fixed cultivation costs and the pumping cost, calculated by Equation 9.123:

$$W_{P_d} = \frac{9.8 \times 10^6}{3600 \times \eta} \times \sum_{t=1}^{T}\sum_{m=1}^{M}(G_{dtm} \times \Delta h_{dtm} \times \text{Pr}_t) \tag{9.123}$$

$$C_d = W_{P_d} + \sum_{t=1}^{T}\sum_{p=1}^{P} FC_{tp} \times A_{dp} \tag{9.124}$$

where, W_{P_d} is the pumping cost for groundwater extraction (US Dollar), FC_{tp} is the cultivation cost of crop p in year t (US Dollar/ha), η is the pumping efficiency (%), G_{dtm} is the volume of groundwater extraction from wells in zone d, month m, year t (m^3), Δh_{dtm} is the groundwater level drawdown in zone d, year t, month m (m), Pr_t is the price of electricity needed for groundwater pumping (US Dollars/kwh).

The constraints of the optimization model include cultivated area limitations expressed as follows:

$$\sum_{p=1}^{P} A_{dp} \le MaxA_d \tag{9.125}$$

where, $MaxA_d$ is the total cultivated area in zone d (ha).

Finally, the cultivated area must be in allowable cultivated area ranges in each zone as follows:

$$A_{dp,\min} \leq A_{dp} \leq A_{dp,\max} \tag{9.126}$$

where, $A_{dp,\min}$ represents the minimum cultivated area for product p in zone d (ha) and $A_{dp,\max}$ is the maximum cultivated area for this product in the same zone. The Pareto solutions are obtained using multi objective optimization techniques. In the work presented here, an interactive relationship is established by considering conflicts, diverging objectives, and non-cooperative behaviors of beneficiaries that leads to maintain groundwater resources and increase the users' benefits.

9.6.1.1 Non-Cooperative Stability Definitions

There are several types of stability definitions in non-cooperative games, namely Nash, General Meta-Rationality (GMR), Symmetric Meta-Rationality (SMR), and Sequential Stability (SEQ). Types of stability can be categorized based on three criteria. (a) The foresight that indicates the number of examined movements in a model, (b) Willingness to dis-improvement which represents a player's willingness to choose a strategy that hurts their own payoff in order to hurt others, and (c) The decision-maker's awareness about other individuals' priorities (Fang et al., 1989; Selbirak, 1994).

The Nash theory is the most widely used theory for determining a stable response in non-cooperative games. In fact, this theory models the behavior of risk-averse decision-makers. Assuming the players' complete rationality and lack of foresight in this theory, the strategy is determined by each decision-maker to maximize their payoff.

In fact, the Nash solution is created when the decision-maker i cannot make any improvement in the outcome by changing the strategy; and if the answer j is the Nash response for all players, Nash solution will be the solution of the entire game. The basic problem with the Nash solution concept is the player's lack of knowledge about opponents' selected strategies; hence, the Nash solution usually fails to identify the exact output of the game. Therefore, other theories have been developed in non-cooperative games in order to reinforce the Nash theory and apply opponents' reactions to decision-making. See Section 9.5 for more details on Nash theory.

GMR is another stable definition in a non-cooperative game. According to decision-makers, a state is the GMR stable if any unilateral move, which improves the payoff, is limited by other opponents. As another stable definition, SMR provides a more limited definition than GMR. For each decision-maker, SMR refers to a definition in which, the strategy change from x to z is limited by the opponent, but no y choice exists with a higher payoff than x for the first player. SEQ is defined like the GMR, but players only do moves that improve their payoff and they are not intended to reduce the opponent payoff. In fact, players are conservative and also risk taker. Further information about types of stability definitions is provided by Madani and Hipel (2011).

9.6.1.1.1 A Cooperative Water Allocation Example

This example is used from Wang et al. (2003). The problem stated here is to model as a 5-year plan by the cooperative water allocation model with an annual time step. Suppose there are three stakeholders, Irrigation Water Association (IWA), City 1 and City 2, along a river as shown in the node-link flow network in Figure 9.33. The IWA has two crop areas located upstream. The return flow coefficients from irrigation is 20% and for cities are 90%. The minimum demands from the Crop 1, Crop 2, City 1, and City 2 are 40, 50, 20 and 25 million m^3/year, respectively, while the maximum demands are 100, 120, 40 and 50 million m^3/year, respectively, as given in Table 9.15. $Q(k_1, k, t)$ is the flow (m^3/s) from node $k1$ to k. $C_p(k_1, k, t)$ and $Z(k_1, k, t)$ are the concentration and the load of salinity in (mg/l) and (Kg) respectively between node k_1 and k, all at every time step t.

The statistical functions of net benefits and salinity in return flows given in Table 9.16 are designed according to experience by Booker and Young (1994). Now the water allocation plan is developed for a hypothetical drought period with a series of annual upstream flows $Q(1, 2, t)$ and corresponding average salinity concentrations $C_p(1, 2, t)$ shown in Table 9.17.

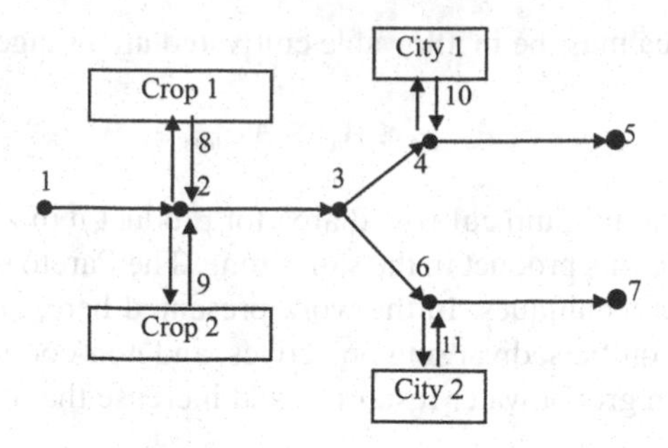

FIGURE 9.33 Flow network and water uses.

For a general node k, the water and pollutant balance equations for time step t can be written as:

$$S(k,t)-S(k,t-1)=\sum_{(k_1,k)\in A} Q(k_1,k,t)-\sum_{(k_1,k)\in A} Q_l(k_1,k,t)+Q_0(k,t)-Q_c(k,t)$$

$$-\sum_{(k,k_2)\in A} Q(k,k_2,t), \forall k \in V \tag{9.127}$$

where t is the index of time steps (step length is Δt), $t \in T=\{1, 2, \ldots, \tau\}$ (τ is the largest index of time step); $S(k, t)$ is the storage variable for node $k \in$ {aquifers and reservoirs} during time step t; Q_0 (k, t) is inflow adjustment of node k during step t for the recharge from the local catchment's rainfall drainage or small tributaries; $Q(k_1, k, t)$ is the flow from node k_1 *to* k (or k to k_2 in the last term of Eq. 9.127); Q_l (k_1, k, t) is the conveyance losses because of evaporation, leakage and seepage of the flow from node k_1 to k; Q_c (k, t) is the water consumed at node k because of economic activities and evaporation.

Inflow adjustment and evaporation losses are not considered in the following analysis that is presented in three parts:

1. *Initial Water Rights and Optimal Economic Allocation*: The problem stated above is modeled as a 5-year plan by the cooperative water allocation model with an annual time step. Water flows listed in Table 9.16 are allocated using a modified dual-priority riparian water rights allocation method (see Wang et al., 2003), in which initial water rights allocation is solved. In each of the five years, water is firstly allocated to meet all the minimum annual demands of the four uses from upstream to downstream, and then the remaining water is allocated to the maximum demands of these riparian uses along the river. The water flows at node 2 are allocated proportionally to the annual demands of both crops, and the water flows at node 3 are allocated proportionally to the annual demands of both cities, respectively. The results show that the minimum water demands of all uses and the maximum demands of both crops are met in each time step, but the maximum demands of cities cannot be satisfied during drought years. The salinity concentrations are calculated according to the water and pollutant balance equations after the water flows are allocated. Then, the net benefits for individual stakeholders are calculated by the benefit functions given in Table 9.16.
2. *Estimation of Coalition Payoffs*: Table 9.17 summarizes the overall net benefits during the 5-year period for all independent stakeholders and all possible coalitions. Since irrigation has lower marginal net benefits than the cities, water is transferred from IWA to City 1 and City 2 when they form coalitions. For example, at the last points of abstracting water according to their water rights during the drought year 2, Crop 1 and Crop 2 have very low

TABLE 9.15
Net Benefit and Return Flow Salinity Functions

Stakeholder	Water Use	Net Benefit Function NE_{jt} (\$10³)	Return Flow Salinity $Z(j, k, t)$ (10⁶kg)	(10⁶m³/ year) *Minimum demand*	(10⁶m³/ year) *Maximum demand*	*Return flow ratio*
1. IWA	Crop 1	$-1{,}000+60Q(2,8,t)-0.2Q(2,8,t)^2$	$0.3Q(2,8,t)-0.0008Q(2,8,t)^2$	40	100	0.2
	Crop 2	$-1{,}100+60Q(2,9,t)-0.2Q(2,9,t)^2$	$0.3Q(2,9,t)-0.0008Q(2,9,t)^2$	50	120	0.2
2. City 1	City 1	$700Q(4,10,t)-0.3Q(4,10,t)^2-0.25Q(4,10,t)\times\max(C_p(4,10,t)-400, 0)$	$2.5Q(4,10,t)-0.0008Q(4,10,t)^2$	20	40	0.9
3. City 2	City 2	$680Q(6,11,t)-0.3Q(6,11,t)^2-0.25Q(6,11,t)\times\max(C_p(6,11,t)-400, 0)$	$2.5Q(6,11,t)-0.0008Q(6,11,t)^2$	25	50	0.9

Units of flow and salinity are *MCM* and *mg/L*, respectively.

TABLE 9.16
Total Upstream Inflows and Initial Water Rights Allocations

Time Step	Unit	Year 1	Year 2	Year 3	Year 4	Year 5
Q(1,2,t)	(10^6m^3)	280	260	240	240	260
C_p (1,2,t)	(mg/L)	400	410	420	430	410
Q (2,8,t)	(10^6m^3)	100	100	100	100	100
C_p (2,8,t)	(mg/L)	400	410	420	430	410
Q (2,9,t)	(10^6m^3)	120	120	120	120	120
C_p (2,9,t)	(mg/L)	400	410	420	430	410
Q (4,10,t)	(10^6m^3)	40	37.33	28.44	28.44	37.33
C_p (4,10,t)	(mg/L)	677.69	748.57	857.5	860.63	748.57
Q (6,11,t)	(10^6m^3)	50	46.67	35.56	35.56	46.67
C_p (6,11,t)	(mg/L)	677.69	748.57	857.5	860.63	748.57
Q (4,5,t)	(10^6m^3)	42.22	33.6	25.6	25.6	33.6
C_p (4,5,t)	(mg/L)	2,437.98	2,744.59	2,752.49	2,752.49	2,744.59
Q (6,7,t)	(10^6m^3)	52.78	42	32	32	42
C_p (6,7,t)	(mg/L)	2,430.4	2,736.3	2,746.17	2,746.17	2,736.3
NB1,t	($\$10^3$)	6,220	6,220	6,220	6,220	6,220
NB2,t	($\$10^3$)	24,743.08	22,461.87	16,415.05	16,392.83	22,461.87
NB3,t	($\$10^3$)	29,778.85	27,013.33	19,731.85	19,704.07	27,013.33
Total Net Benefit	($\$10^3$)	60,741.92	55,695.2	42,366.9	42,316.9	55,695.2

TABLE 9.17
Overall Net Benefits in the 5-year Planning Period ($\$10^3$)

Net Benefit (103$)	Coalition {1,2}	Coalition {1,3}	Coalition {2,3}	Grand coalition {1,2,3}
NB1	27,415.48	25,943.71	31,100	17,966.29
NB2	125,063.89	108,725.59	117,547.33	130,543.92
NB3	129,661.24	152,222.51	108,675.39	157,429.9
Total	282,140.61	286,891.82	257,322.72	305,940.11

marginal net benefits given as 0.02 and 0.01 $\$/m^3$, respectively, while City 1 and City 2 have higher marginal net benefits calculated as 0.59 and 0.56 $\$/m^3$, respectively. Therefore, water is transferred from crops to cities when they form coalitions. In the grand coalition situation, Crop 1, Crop 2, City 1 and City 2 have marginal net benefits of 0.04, 0.03, 0.64 and 0.62 $\$/m^3$, respectively. The reason that cities' marginal net benefits increase as the amount of the water received increases is that the water quality improves when Crop 1 and Crop 2 use less water. For the same reason, although only additional 2.67×10^6 and $3.33\times10^6 m^3$ of water are received by City 1 and City 2 to obtain the maximum grand coalition net benefit, the amount of water received by Crop 1 and Crop 2 is reduced by 59.42×10^6 and $34.78\times10^6 m^3$, respectively. This implies that the hydrology-based cooperative water allocation model can be applied to allocate water flows as well as pollutant trading.

3. *Reallocation of the Grand Coalition Net Benefit*: In Figure 9.34 the triangle slope plane shows the set of all possible nonnegative allocations of the total net benefit of the grand coalition among competing stakeholders. For each point in the triangle, the perpendicular distances from three edges indicate allocated benefit to each stakeholder. The distance

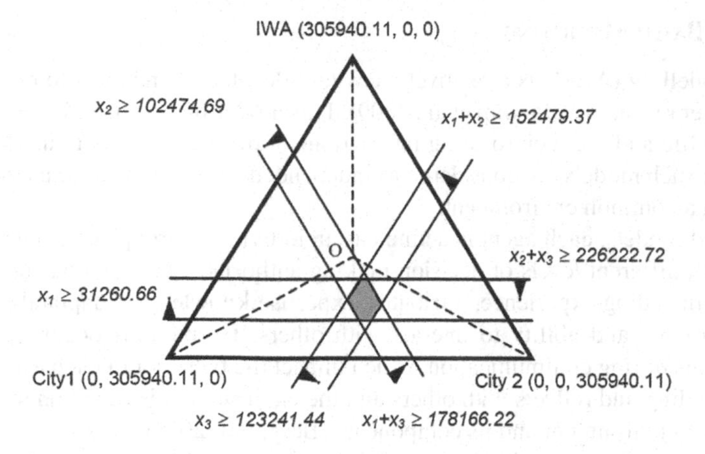

FIGURE 9.34 The core of a cooperative water allocation game.

from the lower edge gives allocation to IWA, the perpendicular distances from upper-left and upper-right edges provide allocations to Cities 1 and 2, respectively. Only the shaded area, the core, is the subset of allocations satisfying individual and group rationality. Note that the triangle can be used to explain the allocation only in three-player cooperative games. For games with more than three players, the core cannot be drawn.

By solving with Shapley value concept, we get the overall 5-year and subsequent annual schedules of equitable and efficient allocation of net benefits in this cooperative water allocation project as shown in Table 9.18. These Pareto optimal schedules provide the alternatives needed for further negotiation or reaching a final decision.

TABLE 9.18
Equitable Allocations of the Net Benefits of the Grand Coalition ($10³)

Period	Stakeholder	Shapley Value
Overall 5 years	IWA	54,480.93
	City 1	114,116.19
	City 2	137,342.99
Year 1	IWA	11,019.68
	City 1	23,081.92
	City 2	27,779.93
Year 2	IWA	10,924.21
	City 1	22,881.94
	City 2	27,539.25
Year 3	IWA	10,823.4
	City 1	22,670.78
	City 2	27,285.11
Year 4	IWA	10,789.56
	City 1	22,599.9
	City 2	27,199.81
Year 5	IWA	10,924.07
	City 1	22,881.65
	City 2	27,538.9

9.6.2 Agent Based Modelling

Agent-based modelling (ABM) is a relatively new tool developed and used to model complex systems such as water resources (Bandini et al., 2009). This modelling tool is utilized for the simulation of human social life and the reciprocating interactions of different agents/units (Macy and Willer, 2002). Agents in such models are considered as independent institutions or units that cooperate and work together in a common environment.

In agent-based models, each agent is defined as an individual, group, stakeholder, institution, or organization with different levels of decision-making authority. The agent has unique characteristics and powers including experience, attributes, expertise knowledge, adaptability, independence, memory, updatability, and ability to interact with others. Interactions occur as a result of relations among agents during communication, which impact the behavior of each agent involved. Each agent's understanding and reflects with others and the environment is based on the feedback that it receives from that environment and its components (Berglund, 2015).

Figure 9.35 shows the general influence of the surrounding environment and the agents on each other in a hypothetical system with three key agents: state or regulatory agencies, environmental sector, and agricultural sector that diverts water. In water resources systems, because environmental conditions determine water availabilities and limitations of the system, the perceptions of agents are strongly influenced by the environment. The environmental sector's perception is only influenced by the environment. Both agricultural water users (diversions) and environmental sector influence the perception of the State agent by informing it about their concerns and water demands, and justifying the importance of their goals. In addition, the State affects the perception of the agricultural water users by informing them about the new policies, regulations, educational plans, assigned penalties, incentives, etc.

Kelly et al. (2013) reviewed five common modelling approaches for integrated environmental assessment and management including systems dynamics, Bayesian networks, coupled component models, agent-based models and knowledge-based models. They summarized that agent-based models simulate the dynamics of individuals or groups of humans or animals. These agents usually have their own goals, use the environment to achieve these goals, share resources, and communicate with each other. They react to changes based on the "perceived" changes in the environment according to pre-defined rules. They concluded that agent-based modelling helps develop a framework of techniques promoting thinking about system structures and their interactions. There are different simulation methods for this type of modelling, ranging from simple cellular automata to detailed, to disaggregated system dynamics models.

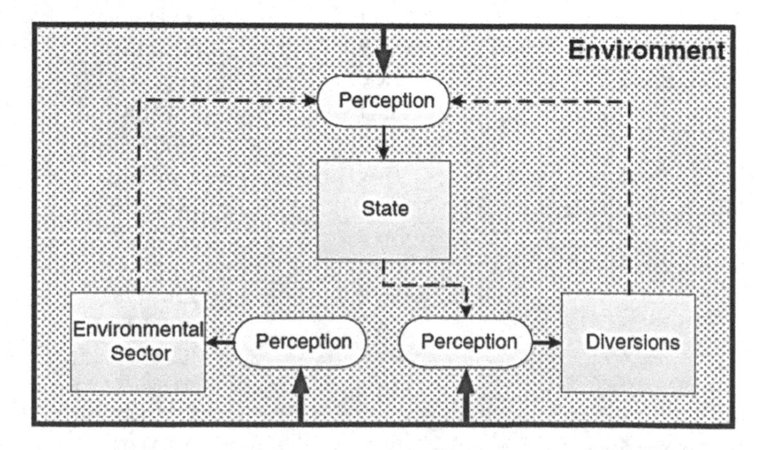

FIGURE 9.35 Influence of the environment and the agents' interactions (Akhbari, M., & Grigg, N. S. A framework for an agent-based model to manage water resources conflicts. *Water resources management*, 27(11), 4039–4052. 2013.)

9.6.2.1 Agent Based Modelling for Water Management

A typical agent-based model has three major elements: (a) Key agents, their characteristics and distinct behaviors; (b) The environment within which, agents interact with each other and the surrounding environment; and (c) Agents' relationships and methods of interactions with each other and with the environment, which in this context can be defined as the aquifer or basin of interest.

ABMs can be defined as simulating the actions and interactions of autonomous agents where the agents interact with each other and the environment according to rules of behavior. Agents may have the ability to make independent decisions and behave in self-selected and varying ways as determined by their behavior rules, their internal state, and the environment. Agents' interactions may include sending messages, exchanging resources, or cooperating to achieve goals. It is an "approach to modelling systems composed of autonomous, interacting agents" (Macal and North 2010), and a modelling paradigm as stated by Galán et al. (2009) for entities with the defining characteristic, within the target system, to be modeled. The interactions between them are explicitly and individually represented in the model in contrast to other models where some entities are represented via average properties.

Agent-based models are well suited to model those situations that are according to Bandini et al. (2009) are characterized by the presence of a number of autonomous entities whose behaviors (actions and interactions) determine (in a non-trivial way) the evolution of the overall system. The agents' rules of behavior and the interactions between the agents may combine to produce emergent behavior, i.e., complex behavioral effects not explicitly encoded in the agents' behavior rules.

Generally, in the ABM, different tools such as theorems and rules are utilized for quantifying the agent's behaviors. These tools comprise a set of governing equations that characterizes the system state resulting from agents' autonomy and interactions among the agents (the choice of water use, the water infrastructural development, etc.) and between the agent and the environment (Zhao et al. 2013). An important issue is that the variables and equations used in ABM are usually static, which inhibits the ability of capturing feedback mechanisms. For example, in Rasoulkhani et al. (2018), the ABM can be created based on a number of theoretical elements including the theories of Innovation Diffusion, Peer Effect, and Affordability based on the demographic and building characteristics, external factors, and social interactions. As an example, in water conservation, the theory of Innovation Diffusion can be used to capture the coupled effect of income level, education, ownership status, infrastructure age, water pricing regimes, rebate availability, technology cost, and social networks concurrently.

Another important issue is that the ABM is a case-specific method, however, usually not every demographic characteristic of an individual can have could be accounted for, such as religious identity, race, sexual orientation, or even number of children in the household. That is not to say that all of these demographics would have had an impact on the utility value and household's adoption state, but it could have fostered more inclusive results. These characteristics were not considered due to a lack of information.

Figure 9.36 presents a conceptual framework for a shared water resource to define different levels of decision-making and interactions in an agent-based model (Castilla-Rho et al., 2015). The first level contains the government agencies, environmental sector and regulators which are institutional agents. In the second level, includes water user associations, water management districts, or water markets. The individual agents or stakeholders such as farmers, domestic users, and drinking water wells are in the third level.

9.6.2.2 Agents and Their Characteristics

Examples of agents normally involved in water decisions include: agricultural, industrial, urban/domestic, environmental, and regulatory sectors. They can be characterized based on their attributes, behavioral rules, memory, decision-making sophistication (the amount of information an agent requires to make decisions), and resources/flows (Macal and North 2006). In water resources

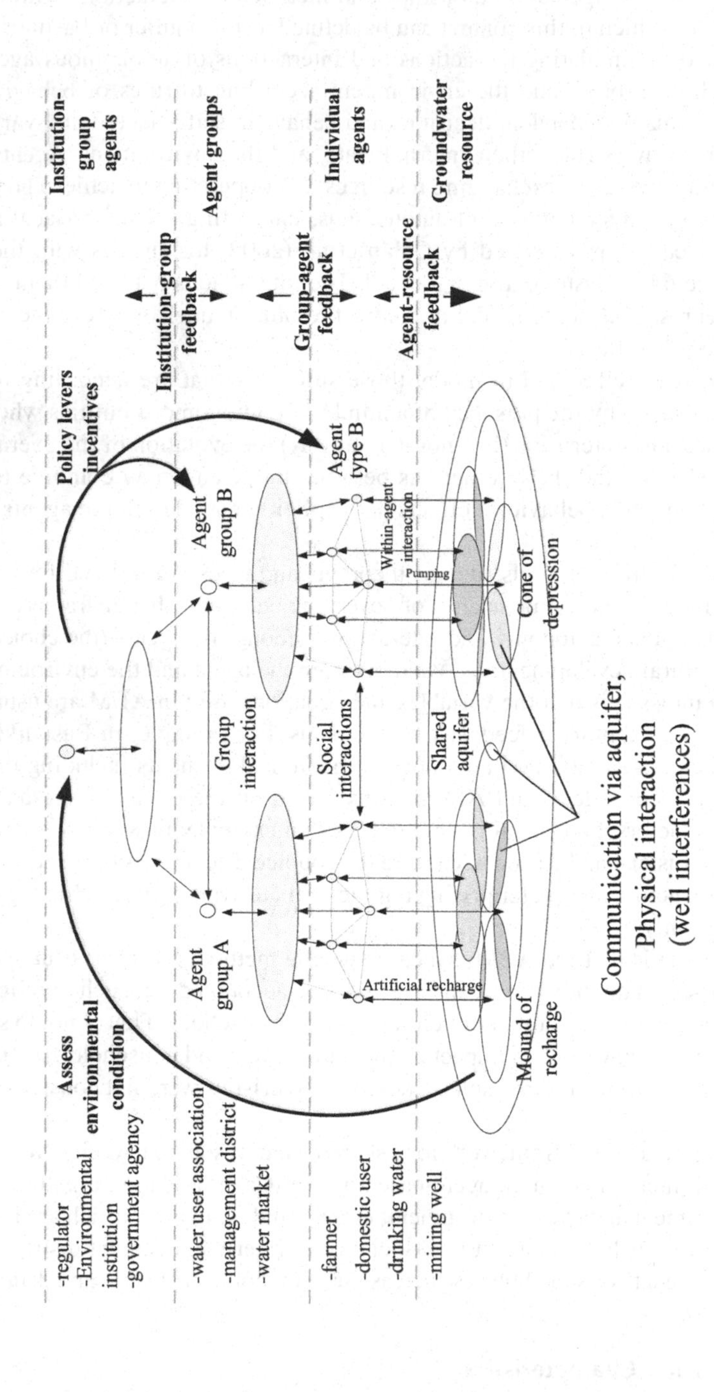

FIGURE 9.36 A conceptual framework for a shared water resource to define different levels of decision-making and interactions in an agent-based model. (Castilla-Rho, J. C., Mariethoz, G., Rojas, R., Andersen, M. S., & Kelly, B. F. An agent-based platform for simulating complex human–aquifer interactions in managed groundwater systems. *Environmental Modelling & Software*, 73, 305–323. 2015.)

management, the world of perception about water also influences agents' characteristics and their interactions with the others.

Wolf (2008) defined four worlds of perception: physical, emotional, knowing/intellectual, and spiritual. In a conflict resolution process, consideration and understanding the agents' perceptions become meaningful because different sides of a conflict have different perceptions about water, e.g. one side may have intellectual perception and the other side emotional perception. Agents with physical or intellectual perceptions are more likely to relax their demand through negotiations, incentives, etc. However, the agents with spiritual or emotional perceptions are expected to be stricter and harder to change their water demand.

Characterizing agents is case-specific; however, Figure 9.37 suggests an overall framework to characterize some types of agents typically involved in water resources systems. All agents except regulators demand water. The decision-making sophistication level is relatively simple for agricultural and industrial agents as there is a simple relationship between their water demands and potential benefits through production opportunities.

See also Case study 3 in Chapter 7 of Karamouz et al. (2020).

9.7 CASE STUDY

In this section, the application of some of the topics discussed in this chapter is presented in a case study by Karamouz and Farzaneh (2020).

9.7.1 Reliability Evaluation of Wastewater Treatment Plants Using MCDM Approach and Margin of Safety Method

The performance of WWTPs is investigated through reliability calculation by the load-resistance concept. A reliable plant is defined when the applied load is lower than its carrying resistance capacity. The failure of infrastructure occurs when the external load during flood exceeds the infrastructure's resistance, implying that the system is no longer functional. Using the MCDM approach, performance-related criteria and subcriteria are determined to assess a reliability/performance index.

The resistance and load criteria could be defined as the potential resistance capacity of the plant and the intensity of the design flood. On the other hand, the load criteria are categorized into two main classifications: triggers and demands. Triggers (physical load) include the vulnerability criterion such as the longitudinal extent of the infrastructure adjacent to the coastal line), hydrologic criteria such as average inundation depth and critical flood elevation, and the quality criterion (biological treatment parameters). The demands load includes additional load during the flood and the population being served.

The resistance criteria are categorized into structural and nonstructural subcriteria. The structural criteria of resistance show the intrinsic capability of a WWTP to perform three stages of treatment on five biological parameters of BOD, COD, TSS, TN, and TP. In the aeration tank, four kinetic coefficients are defined, including the maximum rate of substrate utilization per unit mass of microorganism (k), endogenous decay coefficient (k_s), substrate concentration at one-half the maximum growth rate (k_d), and yield coefficient over a finite period of log growth (Y). The nonstructural criteria of resistance are divided into human resources (number of staff) and resiliency (the distance between the Hunts Point plant, the nearest WWTP, and the coastal line). The load and resistance criteria are shown in Figure 9.38a and b.

Note: k is the maximum substrate rate; K_d: delay coefficient; k_s: substrate concentration; and Y: yield coefficient.

Karamouz and Farzaneh (2020) weighted the load and resistance subcriteria using the AHP, a technique in MCDM method. The weight and values associated with each subcriterion are applied to an MCDM approach named PROMETHEE to calculate the reliability of the WWTP (Karamouz et al., 2016).

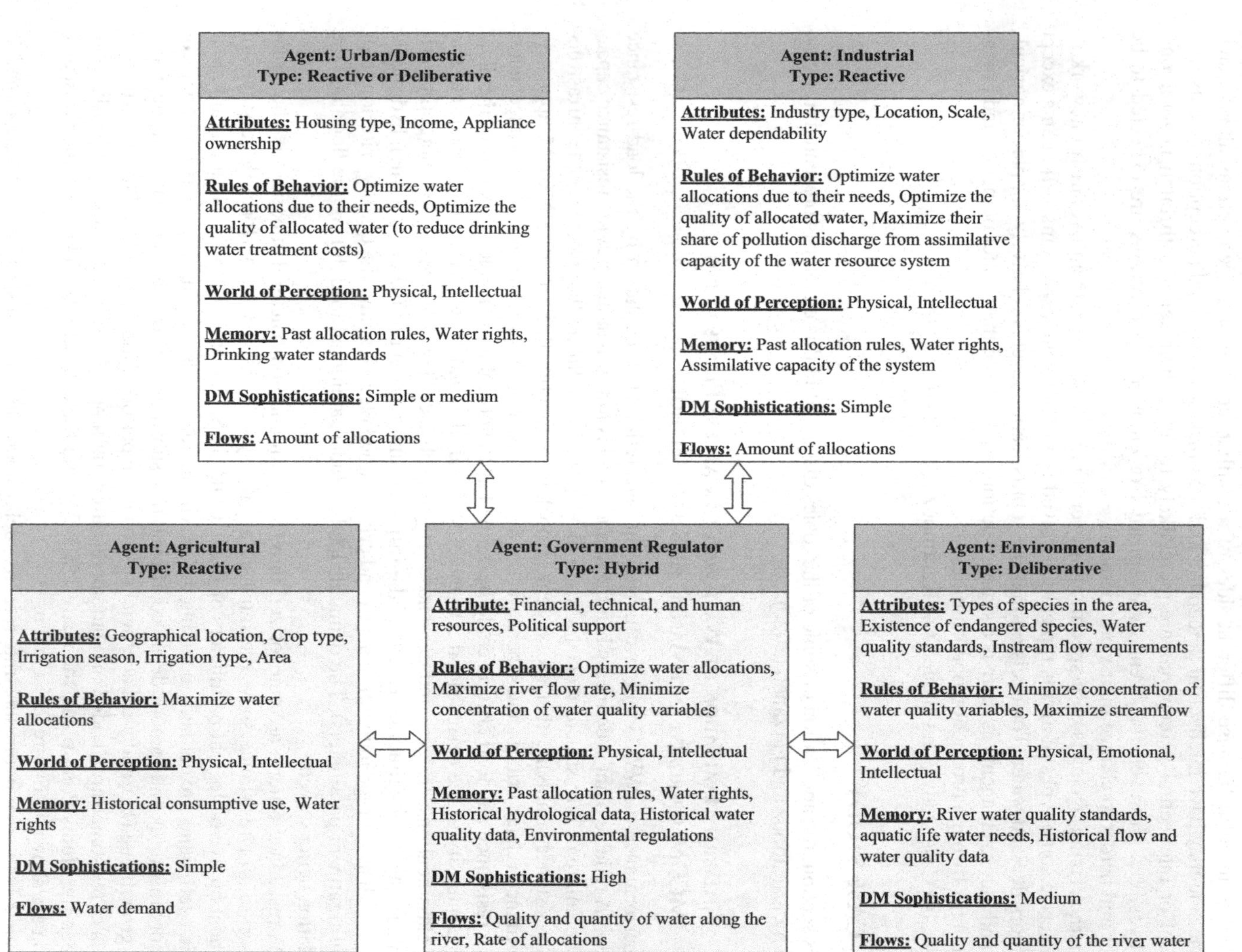

FIGURE 9.37 Proposed characteristics for some agents typically involved in water resources systems (arrows show interactions). (From Akhbari and Grigg, 2015.)

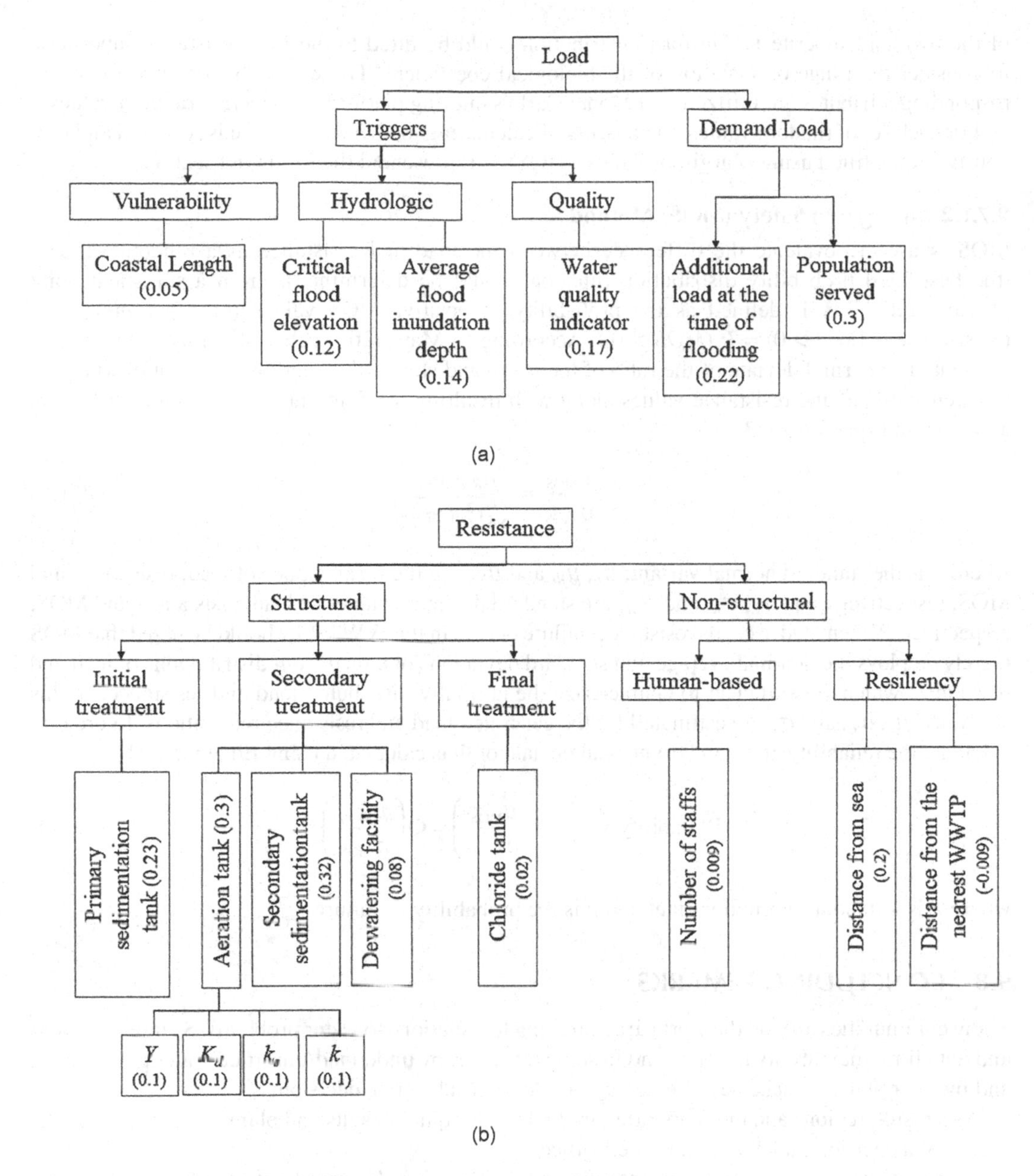

FIGURE 9.38 Hierarchical categories of (a) load and (b) resistance and the subcriteria corresponding to each category.

9.7.1.1 Probabilistic Load and Resistance Reliability

The stochastic nature of load and resistance factors could be evaluated in a nonprobabilistic MCDM approach to quantify the reliability. Two approaches could be used to take into account the uncertainty of the subcriteria assigned to the MCDM approach. In the first approach, the load subcriteria with random nature are investigated. In the second approach, random nature is also utilized for resistance. In this study, additional load during the flood and critical flood elevation parameters are incorporated in the load category. For resistance, the potential sources of uncertainty could be found in the biological kinetic coefficient of k, k_d, k_s, and Y, which fluctuate within specific ranges (Qasim, 1998). The best probability distribution functions could be fitted to incorporate the uncertainty of the two load subcriteria. Uniform distribution could be fitted to the four resistance subcriteria to consider the range of variation of the biological coefficient. These distributions and their corresponding attributes are utilized in a Monte Carlo sampling platform to generate random values.

For each set of random values, the process of calculating load and resistance is repeated and reliability is quantified using Margin of Safety (MOS) as represented the following sections.

9.7.1.2 Margin of Safety (MOS) Method

MOS is used to evaluate the difference between the load and resistance as a random variable (the best-fitted probability distribution function should be determined). From a statistical point of view, reliability is defined as the probability of positive MOS values (i.e., *Reliability* = *P* (Resistance – Load > 0) = *P* (*MOS* > 0)). According to Mays (2010), the reliability index can be calculated as normal deviate of the ratio of the mean and standard deviation of the MOS when the independent load and resistance values along with resulting MOS maintain a normal distribution, as shown in Equation 9.113.

$$Z = \frac{\mu_{\text{MOS}}}{\sigma_{\text{MOS}}} = \frac{\mu_R - \mu_L}{\sqrt{\sigma_L^2 + \sigma_R^2}} \tag{9.113}$$

where Z is the standard normal variant; μ_L, μ_R, and μ_{MS} are the mean values of load, resistance, and MOS, respectively; and σ_L, σ_R, and σ_{MS} are standard deviation values of load, resistance, and MOS, respectively. When load exceeds resistance, failure occurs in the WWTP. It should be noted that MOS merely employs the nominal average and standard deviations obtained from distributions of load and resistance, which are sufficient to characterize the normally distributed load and resistance. In this study, μ_L, μ_R, σ_L, and σ_R are estimated for the generated load and resistance realizations. In order to calculate the reliability, the standard normal deviate of Φ is calculated using Equation 9.114.

$$\text{Reliability} = p\left(Z \le \frac{\mu_{\text{MOS}}}{\sigma_{\text{MOS}}}\right) = \Phi\left(\frac{\mu_{\text{MOS}}}{\sigma_{\text{MOS}}}\right) \tag{9.114}$$

where Z is a standard normal variant and p is the probability of failure.

9.8 CONCLUDING REMARKS

Many communities around the world are searching for solutions to water problems. Scarce resources and data limit their ability to solve, and in some instances by undefined/unsettled water governance, and by their planning schemes affected by politics and subjective decisions.

As a result, regions and municipalities are looking for quick results and planning schemes for the period of a decision maker's term in their office.

In this chapter, the system representation and domains are discussed with the essence of water system analysis. Data preparation and processing techniques are discussed followed by MCDM. Then, data-driven neural networks and fuzzy inference are introduced. Furthermore, mathematics of growth as a basis for systems dynamics is presented. Conventional and evolutionary optimization techniques are also introduced. Finally, the conflict resolution and Nash bargaining theory, game theory and agent base modelling are described that can be utilized to bring consensus among the stakeholder and decision makers.

It is demonstrated that traditionally the first challenge is the dynamics of urban water supply and demand components and how an optimum scheme can be developed for urban water management. The second challenge is to operate and effectively maintain the urban water infrastructure. The new challenge in planning, more than the development of models and toolboxes, is geared towards making transparent data and algorithms that are adaptable to regional and local needs and can be used to bring different decision makers and stakeholders together and create consensus.

A shared vision planning is needed to make the selected techniques and allocation schemes useable, adaptable, and expandable. Participatory planning and decision-making is the key to sound water management. System dynamics and conflict resolution techniques can be used to formulate the problems with the intension of bringing stakeholders into the decision-making process.

PROBLEMS

1. In an unconfined aquifer system, the following agencies are affected by the decisions made to discharge water from the aquifer to fulfill water demands:

   Agency 1: Department of Water Supply
   Agency 2: Department of Agriculture
   Agency 3: Industries
   Agency 4: Department of Environmental Protection

   Department of Water Supply has a twofold role, namely, to allocate water to different purposes and to control the groundwater table variations. The decision makers in different agencies are asked to set their utility functions as given in Table 9.19.

   The analyst should set the weights shown in Table 9.20 on the role and authority of different agencies in the political climate of that region:

   The initial surface, volume, and TDS concentration of the water content of the aquifer are 600 km^2, 1,440 million m^3, and 1,250 mg/L, respectively. The net underground inflow is 1,000 million cubic meters/year, with a TDS concentration of 1,250 mg/L. Assume that 60% of allocated water returns to the aquifer as the return flow and that the average TDS concentration of the return flow is 2,000 mg/L and the average storage coefficient of the aquifer is 0.06. Find the most appropriate water allocation scheme for this year using the NBT.

TABLE 9.19
Utility Function Parameters for Different Sectors

Sector	*a*	*b*	c	*d*
Department of Agriculture	50	90	200	500
Department of Domestic Water	20	80	100	150
Department of Water Supply	80	95	100	150
Industries	60	95	100	150
Department of Environmental Protection	40	60	120	132

TABLE 9.20
Relative Weights or Relative Authority of Agencies

Agencies	Relative Weight (Case 1)	Relative Weight (Case 2)
Department of Agriculture	0.133	0.17
Department of Domestic Water	0.33	0.2
Department of Water Supply	0.2	0.23
Industries	0.133	0.1
Department of Environmental protection	0.2	0.3

2. Determine the monthly water allocation to domestic, industrial, agricultural, and recreation demands in a river system shown in Figure 9.39 The average monthly river discharges upstream of the system, in a 2-year time horizon, are presented in Table 9.21.

   The return flow of domestic and industrial sectors is assumed to be 20% of allocated water, and the initial volume of the lake is 30 million cubic meters. The utility functions of different sectors are presented in Table 9.21 and Figure 9.40. The values of utilities have been normalized between 0 and 1, and the higher utility shows the higher priority of a decision maker or a sector. The shape of utilities is considered to be trapezoidal, and the array (*a*, *b*, *c*, and *d*) in Table 9.21 shows the values of water allocated to the agricultural sector, corresponding to utilities of 0, 1, 1, and 0, respectively.
3. In Figure 9.41, a typical dynamic model of urban water management has been shown. Using your own city data, create this model with system dynamic simulation software. Explain the relations you have considered between different parts of the urban water system. Discuss the future of water if water demand could be decreased by 5%.
4. Considering the urban water dynamic model in Figure 9.41, what are the main parts of urban water system? How are different parts of the urban water supply affected by each other?

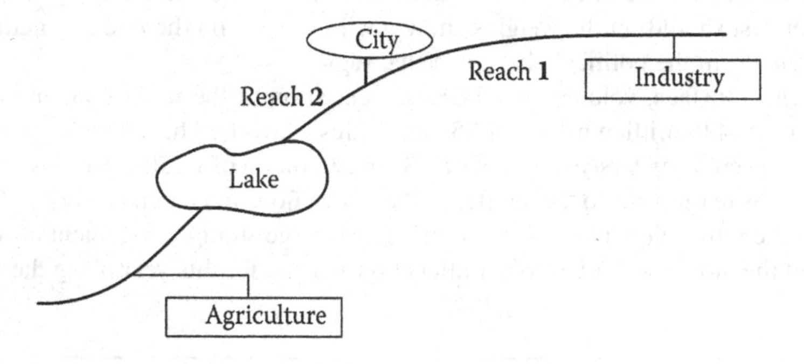

FIGURE 9.39 Components of the river system.

TABLE 9.21
Monthly River Discharge and Utility Functions of Agricultural Sector

Agriculture Demand (MCM/Month)	River Discharge Upstream of the System (MCM)	Month
(0, 10, 15, 25)[a]	31	1
(0, 8, 12, 20)	31	2
(0, 10, 15, 25)	31	3
(0, 10, 15, 25)	40	4
(0, 10, 15, 25)	45	5
(0, 5, 10, 15)	60	6
(5, 10, 25, 50)	75	7
(15, 30, 60, 90)	100	8
(20, 35, 55, 90)	90	9
(25, 35, 60, 90)	51	10
(30, 40, 60, 90)	31	11
(25, 35, 60, 10)	31	12

[a] The entries in the parentheses are a, b, c, and d in the utility function, respectively.

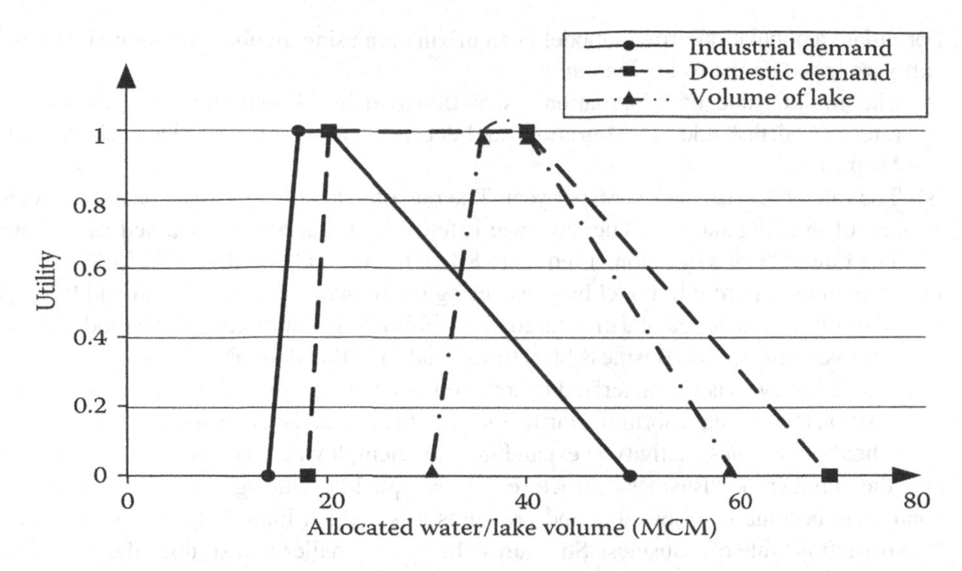

FIGURE 9.40 Utility function of different sectors for water allocated to different sectors and volume of the lake.

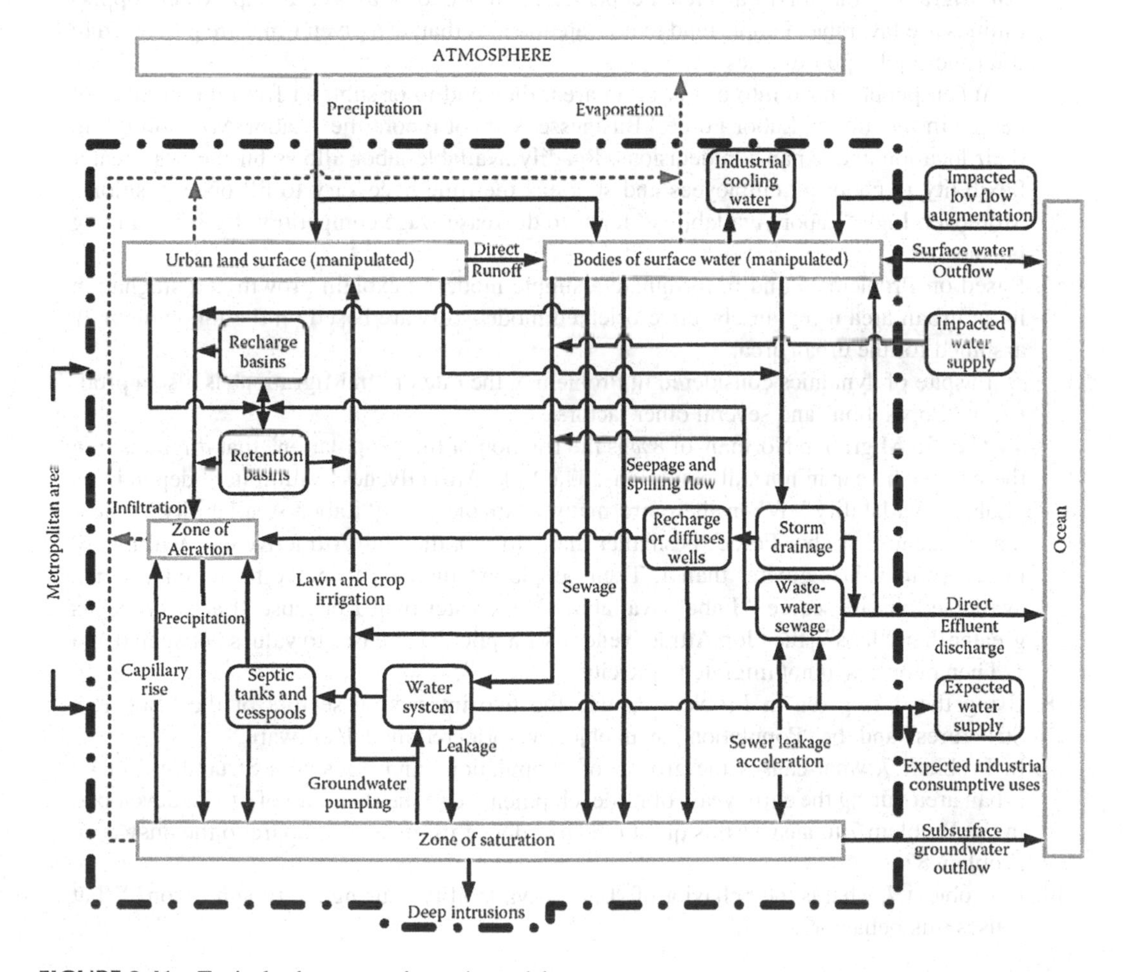

FIGURE 9.41 Typical urban water dynamic model.

5. Formulate a population growth model in an urban area using an objective-oriented model/software based on the following data:
   - The initial value of "Population" is 50,000 people. "Population" increases at the rates of "Births" and "In Migration" and decreases at the rates of "Deaths" and "Out Migration."
   - The rate of "Births" is ~1.5% per year. The rate of "Deaths" is also a constant percentage of the "Population." The "Average Lifetime" of a person is assumed as 67 years. The rate of "Out Migration" is equal to 8% of the initial "Population."
6. Develop an urban growth model by considering the dynamics of population and business growth within a fixed area of a city based on the following assumptions. The model should contain two sectors: the "Business Structures" and the "Population" sectors.

   A city's life cycle is characterized by a period of economic growth followed by a period of transition towards equilibrium. During the growth period, the city appears to be economically healthy: business activity is expanding and unemployment is low. The "Population" and the number of "Business Structures" grow quickly. During the transition period, conditions become less desirable and pressures arise which impede further growth. The "Construction" rate of "Business Structures" becomes smaller and so does the rate of "In Migration" into the city. In equilibrium, when the "Population" and the number of "Business Structures" stop growing, the city suffers from problems such as high unemployment.

   The availability of "Jobs" and promise of higher incomes are the prime motivations for "In Migration" into an urban area. People tend to move to areas where employment opportunities are favorable. People tend to migrate to cities that, at a given time, are perceived to offer more job opportunities.

   When people move into or out of an area, they add to or subtract from the number of people in the area's "Labor Force." Businesses cannot ignore the "Labor Availability" in their location and expansion decisions. Readily available labor allows businesses greater flexibility in choosing employees and shortens the time necessary to fill open positions. Moreover, high "Labor Availability" tends to decrease wage competition for labor among businesses.
7. Based on Problems 5 and 6, formulate a simple model to explain growth and stagnation in an urban area using an objective-oriented model/software based on the following data assumed for the urban area:

   In spite of dynamics considered in Problem 5, the rate of "In Migration" is also a product of "Population" and several other factors.

   The "In Migration Normal" of 8% is the fraction of the "Population" that migrates into the city each year in normal conditions. The "Job Attractiveness Multiplier" depends on "Labor Availability." When there are many available jobs ("Labor Availability" is less than 1 because "Labor Force" is smaller than "Jobs"), the "Job Attractiveness Multiplier" increases to values greater than 1. Then people are inclined to move to the city. When available jobs are scarce ("Labor Availability" is greater than 1 because "Labor Force" is greater than "Jobs"), the "Job Attractiveness Multiplier" decreases to values between 0 and 1. Then people will not migrate to the city.
8. Using the data given in Problem 7, map the two interacting sectors of the "Business Structures" and the "Population" in an objective-oriented model/software.
9. In Problem 7, what causes the growth of "Population" and "Business Structures" of an urban area during the early years of its development? Use the structure of model developed in the Problem 7 to answer this question. How does this answer compare to the answer in Problem 8?
10. In Problem 7, what is the behavior of "Labor Availability" during the time horizon? What causes this behavior?

11. In Problem 7, how does the finite "Land Area" limit the population growth? Will all the available "Land Area" be occupied in equilibrium condition (20 km^2)? Use the model to show the effects of annexing surrounding land to increase the "Land Area" to 30 km^2. Initialize the model in equilibrium. Use the STEP function to increase the "Land Area" after 10 years to 40 km^2. How does the behavior change? Will all the "Land Area" be occupied in equilibrium condition?
12. In Problem 7, does the assumption of a fixed "Land Area" invalidate the model? Most cities can and have expanded from their original areas. How does such expansion influence the results given by the model? For example, what would be the likely consequences of expanding the "Land Area" within which the city is allowed to grow? Simulate the model with a sequence of such expansions, say every 20 years.
13. Consider the following disaggregation of carbon emissions:

$$\text{Carbon emission}\left(\text{kg}\frac{\text{C}}{\text{year}}\right) = \text{Population} \times \frac{\text{Energy}\left(\frac{\text{kJ}}{\text{year}}\right)}{\text{Person}} \times \frac{\text{Carbon}(\text{kgC})}{\text{Energy}(\text{kJ})}$$

Using the following estimates for the United States and assuming that growth rates remain constant,

	Population	**Energy/Person**	**Carbon/Energy**
1,990 amounts	250×10^6	320×10^6	15×10^{-6}
Growth, r (%/year)	0.6	0.5	−0.3

a. Find the carbon emission rate in 2030.
b. Find the carbon emitted in those 40 years.

14- Consider the following problem with two objective functions:

Maximize	$f_1 = x_1 + x_2,$	$f_2 = x_1 - x_2,$
Subject to	$x_1, x_2 \geq 0,$	
	$x_1 + 2x_2 \leq 4,$	
	$2x_1 - x_2 \geq 2,$	

i. Represent graphically the decision space.
ii. Display graphically the criteria space.

15. Consider the following linear programming problem:

Maximize	$x_1 + x_2,$
Subject to	$x_1, x_2 \geq 0,$
	$2x_1 + x_2 \leq x,$
	$x_1 + 2x_2 \geq 3,$

where ξ is randomly distributed with $\mu = 3$ and $\sigma^2 = 1$. Give the chance constraint formulation of the problem, and find the optimal solution.

Consider a reservoir that supplies water to a city. The monthly water demand of this complex is 20 million cubic meters. The total capacity of the reservoir is 40 million cubic meters. Let *St* (reservoir storage at the beginning of month *t*) take the discrete values 0, 10, 20, 30, and 40. The cost of operation (Loss) can be estimated as a function of the difference between release (R_t) and water demand as follows:

$$\text{Loss}_t = \begin{cases} 0, & R_t \geq 10. \\ (20 - R_t)^2, & R_t > 10. \end{cases}$$

i. Formulate a forward-moving deterministic DP model for finding the optimal release in the next 3 months.
ii. Formulate a backward-moving deterministic DP model for finding the optimal release in the next 3 months.
iii. Solve the DP model developed in part (a), assuming that the inflows to the reservoir in the next 3 months ($t = 1, 2, 3$) are forecasted to be 10, 50, and 20, respectively. The reservoir storage in the current month is 20 million cubic meters.

16. Find the minimum point of the function $f(x) = 2x^2 - 3x + 2$ in the $[-2, 2]$ interval using the GA.
17. Fuzzy sets A and B are defined in $X = [-\infty, \infty]$, by the membership functions as follows:

$$\mu_A(x) = \begin{cases} 2x & \text{if } 0 \leq x \leq 5 \\ 15 - x & \text{if } 5 < x \leq 10 \\ 0 & \text{otherwise.} \end{cases}$$

$$\mu_B(x) = \begin{cases} 9 - x & \text{if } 1 \leq x \leq 5 \\ 4 & \text{if } 5 \leq x \leq 15 \\ x - 11 & \text{if } 5 \leq x \leq 20 \\ 0 & \text{otherwise.} \end{cases}$$

Determine the membership functions of $A \cap B$, $A \cap B$, AB, and $A + B$ fuzzy sets.

18. Solve Example 8.2 with the changes assumed in Problem 8.

REFERENCES

Ahlfeld, D.P., Mulvey, J.M., Pinder, G.F., and Wood, E.F. (1988). Contaminated groundwater remediation design using simulation, optimization and sensitivity theory. 1. Model development. *Water Resources Research*, 24(5), 431–441.

Akhbari, M., and Grigg, N.S. (2013). A framework for an agent-based model to manage water resources conflicts. *Water Resources Management*, 27(11), 4039–4052.

Akhbari, M., and Grigg, N.S. (2015). *Managing water resources conflicts: modelling behavior in a decision tool.* Water Resources Management, 29(14), 5201–5216.

Banathy, B.H. (1997). *A Taste of Systemics.* International Society for the Systems Sciences, Pocklington.

Bandini, S., Manzoni, S., and Vizzari, G. (2009). Agent based modeling and simulation: an informatics perspective. *Journal of Artificial Societies and Social Simulation*, 12(4), 4.

Berglund, E.Z. (2015). Using agent-based modeling for water resources planning and management. *Journal of Water Resources Planning and Management*, 141(11), 04015025.

Bellman, R. (1957). A Markovian decision process. *Journal of Mathematics and Mechanics*, 6(5), 679–684.

Bingham, G., Wolf, A., and Wohlgenant, T. (1994). *Resolving Water Disputes: Conflict and Cooperation in the U.S., Asia, and the Near East*. U.S. Agency for International Development, Washington DC.

Bohling, G.C. (2005). Introduction to geostatistics and variogram analysis. *Kansas Geological Survey*, 1, 1–20.

Bohling, G.C. (2007). Introduction to geostatistics. Kansas Geological Survey Open File Report (2007–26), p. 50.

Booker, J.F. and Young, R.A. (1994). Modeling intrastate and interstate markets for Colorado River water resources. *Journal of Environmental Economics and Management*, 26, 66–87.

Bose, N.K. and Liang, P. (1996). *Neural Network Fundamentals with Graphs, Algorithms, and Application*. McGraw-Hill, New York.

Carroll, D.L. (1996). Genetic algorithms and optimizing chemical oxygen-iodine lasers. In H.B. Wilson, R.C. Batra, C.W. Bert, A.M.J. Davis, R.A. Schapery, D.S. Stewart, and F.F. Swinson (eds.) *Developments in Theoretical and Applied Mechanics* (Vol. XVIII, pp. 411–424). School of Engineering, The University of Alabama, Birmingham.

Castilla-Rho, J.C., Mariethoz, G., Rojas, R., Andersen, M.S., and Kelly, B.F. (2015). An agent-based platform for simulating complex human–aquifer interactions in managed groundwater systems. *Environmental Modelling & Software*, 73, 305–323.

Chalkiadakis, G., Elkind, E., and Wooldridge, M. (2011). Computational aspects of cooperative game theory. *Synthesis Lectures on Artificial Intelligence and Machine Learning* 5(6), 1–168.

Chang, L.C., Shoemaker, C.A., and Liu, P.L.-F. (1992). Optimal time-varying pumping rates for groundwater remediation: Application of a constrained optimal control theory. *Water Resources Research*, 28(12), 3157–3174.

Cobble, K. and Huffman, D. (1999). Learning from everyday conflict. *The Systems Thinker*, 10(2), 8–16.

Culver, T.B. and Shoemaker, C.A. (1992). Dynamic optimal control for groundwater remediation with flexible management periods. *Water Resources Research*, 28(3), 629–641.

Demuth, H. and Beale, M. (2002). *Neural Network Toolboox for MATLAB*. User's Guide. Available at http://www.mathworks.com/.

Dinar, A., and Hogarth, M. (2015). Game theory and water resources critical review of its contributions, progress and remaining challenges. *Foundations and Trends® in Microeconomics*, 11(1–2), 1–139.

Dougherty, D.E. and Marryott, R.A. (1991). Optimal groundwater management. 1. Simulated annealing. *Water Resources Research*, 27(10), 2493–2508.

Egles, R.W. (1990). Simulated annealing: A tool for operational research. *European Journal of Operational Research*, 46, 271–281.

Ehrlich, P. R., and Holdren, J. P. (1971). Impact of population growth. Science, 171(3977), 1212–1217.

Esteban, E. and Dinar, A. (2013). Cooperative management of groundwater resources in the presence of environmental externalities. *Environmental and Resource Economics*, 54(3), 443–469.

Fang, L., Hipel, K.W., and Kilgour, D.M. (1989). Conflict models in graph form: Solution concepts and their interrelationships. *European Journal of Operational Research*, 41(1), 86–100.

Forgo, F., Szep, J., and Szidarovszky, F. (1999). *Introduction to the Theory of Games*. KluwerAcademic Publishers, Dordrecht, the Netherlands.

Galán, J.M., López-Paredes, A., and Del Olmo, R. (2009). An agent-based model for domestic water management in Valladolid metropolitan area. *Water Resources Research*, 45(5).

Ghadimi, S., and Ketabchi, H. (2019). Possibility of cooperative management in groundwater resources using an evolutionary hydro-economic simulation-optimization model. *Journal of Hydrology*, 578, 124094.

Goldberg, D.E. (1989). *Genetic Algorithms in Search, Optimization, and Machine Learning*. Addison-Wesley, Reading, MA.

Goovaerts, P. (1997). *Geostatistics for Natural Resources Evaluation*. Oxford University Press on Demand, Oxford.

Hadley, A. (1964). *Nonlinear and Dynamic Programming*. Addison-Wesley, Reading, MA.

Hirsch, F. (1977). *Social Limits to Growth*. Harvard University Press, Cambridge, MA.

Holland, J. (1975). *Adaptation in Natural and Artificial Systems*. The University of Michigan Press, Ann Arbour, MI.

Hwang, C.L. and Yoon, K. (1981). *Multiple Attribute Decision Making: Methods and Applications*. Springer-Verlag, New York.

Karamouz, M. and Farzaneh, H. (2020). Margin of safety based flood reliability evaluation of wastewater treatment plants: Part 2-quantification of reliability attributes. *Water Resources Management: An International Journal, Published for the European Water Resources Association (EWRA)*, 34(6), 2043–2059.

Karamouz, M. and Fereshtehpour, M. (2019). Modeling DEM errors in coastal flood inundation and damages: A spatial nonstationary approach. *Water Resources Research*, 55(8), 6606–6624.

Karamouz, M., Moridi, A., and Nazif, S. (2010). *Urban Water Engineering and Management.* CRC Press, Boca Raton, FL.

Karamouz, M. and Vasiliadis, H.V. (1992), Bayesian stochastic optimization of reservoir operation using uncertain forecasts. *Water Resources Research*, 28(5), 1221–1232.

Karamouz, M. and Zare, M.R. (2021). Carbon footprint of water use in industrial expansion. *World Environmental and Water Resources Congress 2021*, American Society of Civil Engineers, Atlanta, GA.

Karamouz, M., Szidarovszky, F., and Zahraie, B. (2003). *Water Resources Systems Analysis.* Lewis Publishers, CRC Press, Boca Raton, FL.

Karamouz, M., Razavi, S., and Araghinejad, Sh. (2007). Long-lead seasonal rainfall forecasting using time-delay recurrent neural networks: A case study. *Journal of Hydrological Processes*, 22(2), 229–238.

Karamouz, M., E. Rasoulnia, Z. Zahmatkesh, M.A. Olyaei, and A. Baghvand (2016). Uncertainty-based flood resiliency evaluation of wastewater treatment plants. *Journal of Hydroinformatics*, 18(6), 990–1006. doi: 10.2166/hydro.2016.084.

Kelly, R.A., Jakeman, A.J., Barreteau, O., Borsuk, M.E., ElSawah, S., Hamilton, S.H., ... and Voinov, A.A. (2013). Selecting among five common modelling approaches for integrated environmental assessment and management. *Environmental Modelling & Software*, 47, 159–181.

Kerachian, R., and Karamouz, M. (2006). Optimal reservoir operation considering the water quality issues: A stochastic conflict resolution approach. *Water Resources Research*, 42(12).

Labadie, J.W. (2004). Optimal operation of multireservoir systems: State-of-the-art review. *Journal of Water Resources Planning and Management*, 130(2), 93–111.

Labi, S. (2014). Introduction to civil engineering systems: A systems perspective to the development of civil engineering facilities. https://www.wiley.com/en-us/Introduction+to+Civil+Engineering+Systems.

Lane, D.C. (2000). Should system dynamics be described as a 'hard' or 'deterministic' systems approach? *Systems Research and Behavioral Science*, 17(1), 3–22. doi: 10.1002/(sici)1099-1743(200001/02)17:1<3::aid-sres344>3.0.co;2-7.

Lefkoff, L.J. and Gorelick, S.M. (1987). AQMAN: Linear and quadratic programming matrix generator using two-dimensional groundwater flow simulation for aquifer management modeling. U.S. Geological Survey Water Resources Investigations Report, 87-4061.

Liu, D., Ji, X., Tang, J., and Li, H. (2020). A fuzzy cooperative game theoretic approach for multinational water resource spatiotemporal allocation. *European Journal of Operational Research*, 282(3), 1025–1037.

Loehman, E., Orlando, J., Tschirhart, J., and Whinston, A. (1979). Cost allocation for a regional wastewater treatment system. *Water Resources Research*, 15 (2), 193–202.

Macal, C.M., and North, M.J. (2006, December). Tutorial on agent-based modeling and simulation part 2: How to model with agents. In *Proceedings of the 2006 Winter simulation conference* (pp. 73–83). IEEE.

Macal, C. and North, M. (2010). Tutorial on agent-based modelling and simulation. *Journal of Simulation 4*, 151–162. https://doi.org/10.1057/jos.2010.3

Macy, M.W., and Willer, R. (2002). From factors to actors: Computational sociology and agent-based modeling. *Annual Review of Sociology*, 28(1), 143–166.

Madani, K., and Hipel, K.W. (2011). Non-cooperative stability definitions for strategic analysis of generic water resources conflicts. *Water Resources Management*, 25(8), 1949–1977.

Mahmoodzadeh, D. and Karamouz, M. (2021). A hydro-economic simulation-optimization framework to assess the cooperative game theory in coastal groundwater management. *Journal of Water Resources Planning and Management* (Tentatively Accepted)

Mays, L. W. (2010). *Water Resources Engineering.* John Wiley & Sons, New York, NY.

Meadows, D.H., Meadows, D.L., Randers, J., and Behrens, W.W. (1972). *The Limits to Growth*, 102, 27. New York.

McKinney, D.C. and Lin, M.D. (1995). Approximate mixed-integer nonlinear programming methods for optimal aquifer remediation design. *Water Resources Research*, 31(3), 731–740.

Metropolis, N., Rosenbluth, A., Rosenbluth, M., Teller, A., and Teller, E. (1953). Equation of state calculations by fast computing machines. *Journal of Chemical Physics*, 21, 1087–1092.

Milton, J.S. and Arnold, J.C. (1995). *Introduction to Probability and Statistics* (3rd ed.). McGraw-Hill, New York.

Mousavi, S. J., Karamouz, M., and Menhadj, M.B. (2004). Fuzzy-state stochastic dynamic programming for reservoir operation. *Journal of Water Resources Planning and Management*, 130(6), 460–470.

Myerson, R. (1991), *Game Theory: Analysis of Conflict*, 1st edition, Harvard University Press, Cambridge, MA.

Nandalal, K. D. W., and Bogardi, J. J. (2007). Dynamic programming based operation of reservoirs: Applicability and limits. Cambridge University Press. Cambridge, UK.

Nazari, S., and Ahmadi, A. (2019). Non-cooperative stability assessments of groundwater resources management based on the tradeoff between the economy and the environment. *Journal of Hydrology*, 578, 124075.

Nicklow, J., Reed, S., Savic, D., Dessalegne, T., Harrell, L., Chan-Hilton, A, Karamouz,M., Minsker, B., Ostfeld, A., Singh, A., and Zechman, E. (2010). State of the art for genetic algorithms and beyond in water resources planning and management. *Journal of Water Resources Planning and Management*, 136(4), 412–432.

Qasim, S.R. (1998). *Wastewater Treatment Plants: Planning, Design, and Operation* (2nd ed.). CRC Press, Boca Raton, FL.

Rardin, R. (1997). Optimization in Operation Research, 1st edition, Prentice Hall, New Jersey.

Rasoulkhani, K., Logasa, B., Presa Reyes, M., and Mostafavi, A. (2018). Understanding fundamental phenomena affecting the water conservation technology adoption of residential consumers using agent-based modeling. *Water*, 10(8), 993.

Ross, Sh.M. (1987). *Introduction to Probability and Statistics for Engineers and Scientists*. Wiley, New York.

Roy, B. (1978). ELECTRE III: Un algorithme de classements fondé sur une représentation floue des préférences en présence de critères multiples.

Rubinstein, R. Y. (1981). *Simulation and Monte Carlo Method*. John & Wiley & Sons, New York.

Saaty, T.L. (1980). *The Analytic Hierarchy Process: Planning, Priority Setting, Resource Allocation (Decision Making Series)*. McGraw-Hill, New York.

Selbirak, T. (1994). Some concepts of non-myopic equilibria in games with finite strategy sets and their properties. *Annals of Operations Research*, 51(2), 73–82.

Shapley, L.S. (1953). A value for n-person games. *Contributions to the Theory Games*, 2(28), 307–317.

Simonovic, S.P. (2000). Tools for water management: One view of the future. *Water International*, 25(1), 76–88.

Simonovic, S.P. and Bender, M.J. (1996). Collaborative planning-support system: An approach for determining evaluation criteria. *Journal of Hydrology*, 177, 237–251.

Solomatine, D.P. (1999). Two strategies of adaptive cluster covering with descent and their comparison to other algorithms. *Journal of Global Optimization*, 14(1), 55–78.

Southwick, C. H. (1976). Ecology and the Quality of Our Environment (No. 304.2 S68).

StatSoft. (2002). Electronic statistics textbook. Available at http://www.statsoftinc.com/textbook/stathome.html.

Sterman, J. (2000). Business Dynamics. McGraw-Hill, Inc., New York.

Sun, M. and Zheng, C. (1999). Long-term groundwater management by a MODFLOW based dynamic groundwater optimization tool. *Journal of American Water Resources Association*, 35(1), 99–111.

Sweeney, L.B. and Sterman, J.D. (2000). Bathtub dynamics: Initial results of a systems thinking inventory. *System Dynamics Review: The Journal of the System Dynamics Society*, 16(4), 249–286.

Wagner, B.J. and Gorelick, S.M. (1987). Optimal groundwater quality management under parameter uncertainty. *Water Resources Research*, 23(7), 1162–1174.

Wang, M. and Zheng, C. (1998). Application of genetic algorithms and simulated annealing in groundwater management: Formulation and comparison. *Journal of American Water Resources Association*, 34(3), 519–530.

Wang, L.Z., Fang, L., and Hipel, K.W. (2003). Water resources allocation: a cooperative game theoretic approach. *Journal of Environmental Informatics*, 2(2), 11–22.

Wechsler, S.P. (2007). Uncertainties associated with digital elevation models for hydrologic applications: A review.

Wen, L., Bai, L., and Zhang, E. (2016). System dynamic modeling and scenario simulation on Beijing industrial carbon emissions. *Environmental Engineering Research*, 21(4), 355–364. doi: 10.4491/eer.2016.049.

White, I.D., Mottershead, D.N., and Harrison, S.J. (1992). *Environmental Systems: An Introductory Text* (2nd ed.). Chapman & Hall, New York.

William, R., Garg, J., Stillwell, A.S. (2017). A game theory analysis of green infrastructure stormwater management policies. *Water Resources Research*, 53(9), 8003–8019.

Willis, R.L. (1976). Optimal groundwater quality management, well injection of waste waters. *Water Resources Research*, 12, 47–53.

Willis, R.L. (1979). A planning model for the management of groundwater quality. *Water Resources Research*, 15, 1305–1313.

Wolf, A. (1998). Conflict and cooperation along international waterways. *Water Policy*, 1(2), 251–265.

Wolf, A. (2000). Indigenous approaches to water conflict negotiations and implications for international waters. *International Negotiation*, 5(2), 357–373.

Wolf, A. (2002). *Conflict Prevention and Resolution in Water Systems*. Edward Elgar, Cheltenham, UK.

Wolf, M. (2008). *Complex Adaptive Systems: An Introduction to Computational Models of Social Life: Miller, JH, & Page, SE (2007)*. Princeton, NJ: Princeton University Press (284 pp., hb= 65.00,ISBN13:978-0-691-13096-5.sb= 24.95, ISBN 13: 978-0-691-12702-6).

Yeh, W.W.G. (1985). Reservoir management and operations models: A state-of-the-art review. *Water Resources Research*, 21(12), 1797–1818.

Zhao, J., Cai, X., and Wang, Z. (2013). Comparing administered and market-based water allocation systems through a consistent agent-based modeling framework. *Journal of Environmental Management*, 123, 120–130.

10 Risk and Reliability

10.1 INTRODUCTION

The risk of failure in extreme events such as failure of flood-related structures (culverts and bridges) is related to the return period T or exceedance probability P of the design flood. The risk of failure is typically defined as the probability of at least one flood equal to or greater than the design flood in an n-year period. Besides this classical definition, the risk of extreme event is much higher when hazard such as flooding hits an urban area with sever vulnerabilities due to socioeconomic and environmental condition. Water-related risk issues are discussed in a number of topics and case studies in other chapter including the application of Best Management Practices in urban and coastal area. In this chapter, the focus is on the principle of risk analysis as well as reliability of affected areas and structures.

In order to understand risk, it is essential to be familiar with different components that create risk. Risk is the occurrence of an extreme event such as earthquake, flood, drought, hurricane, and storm, resulting from natural forces, with or without human influences. Although hazard events are a significant condition, it is only one component in the realization of risk. The next components in risk characterization are that somebody or something has vulnerability to a hazard. The last component in risk is the system exposure to the hazard or hazard probability of a system. Hence, the risk is the probability of a loss that depends on three aspects: hazard, vulnerability, and exposure. Furthermore, the risk is defined as a function of hazard, exposure, and vulnerability.

Figure 10.1 shows the risk as to the overlay and intersection of these components. While exposure refers, for example, to floods, only to the question of whether assets or people are physically in the floodwaters' path or not, vulnerability is defined as the conditions determined by social, economic, environmental, and physical factors, which increase the susceptibility of a community to the impact of hazards (Glossary of the UN International Strategy for Disaster Reduction, 2004). The risk increases or decreases respectively when any of these three risk factors decreases or increases (Crichton, 1999).

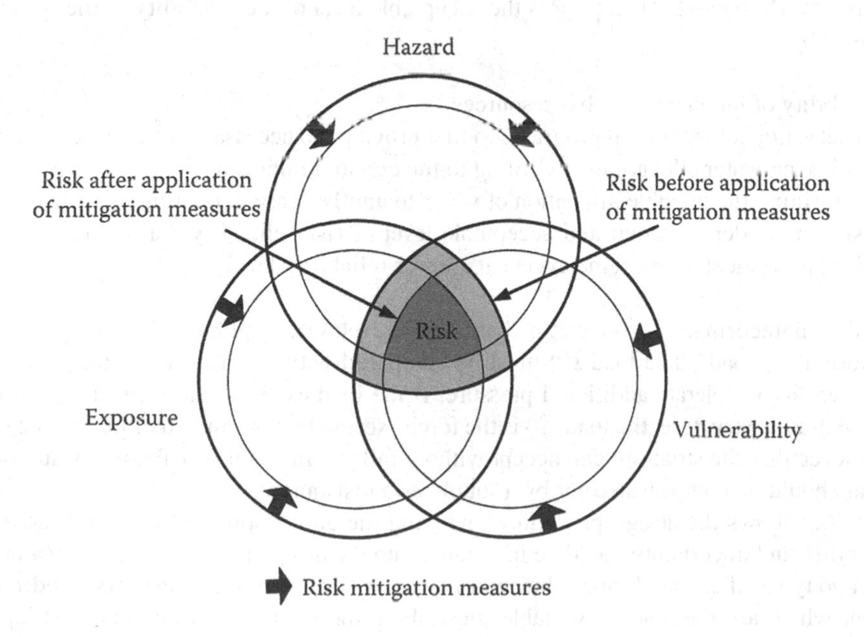

FIGURE 10.1 Risk components and lower risk with mitigation of hazard, exposure, and vulnerability space.

DOI: 10.1201/9781003241744-10

By changing the system and its surrounding environmental characteristics, the risk components and risk are changed. For example, through applying mitigation practices, system vulnerability and exposure would decrease (as shown by arrows in Figure 10.1), resulting in a decrease in system's risk. Besides this, by decreasing the possible hazards, system risk would decrease significantly. The hazards can be changed by changing the system's surrounding environment. For example, by implementing the suitable zonings in coastal urban area and determining the flood plains and evacuation zones, flood hazard, exposure, and vulnerability could be reduced. Perhaps the reason for many growing risks related to urban stormwater management challenges is the lack of appropriate land use planning and excessive human intervention that could exuberate to disasters.

Climate change can highly affect system exposure to hazards (mostly natural hazards), which should be considered in future assessment of the system's situation. Aside from this, social changes may result in changes in the system's adaptive capacity. These changes may be the result of better infrastructures and warning systems. These changes should be considered in an integrated fashion in risk components' evaluation and management.

The topics discussed in this chapter need to cover the objective of placing design and analysis of infrastructure into a reliability framework. To do that, the concept of probability; basic statistical analysis; common probability distributions; extreme value (flood) and frequency analysis—design values; and basic concept of risk and uncertainty are discussed. Then reliability in the context of serial and parallel components as well as load and resistant concept are described. The performance indicator with emphasis on vulnerability and resiliency are presented including a case study. Finally, the basis of uncertainty analysis is covered with a summary of how error and uncertainties have been quantified in the case studies of this book. The entropy theory including trans-information (measure of redundancy in information) is also discussed in the context of water resources issues to measure the information content of random variables and models, evaluate information transfer between hydrological processes, evaluate data acquisition systems, and design monitoring networks. The risk attributes that are related to disaster management are discussed in Chapter 11.

10.2 DESIGN BY RELIABILITY

What level of reliability are we expecting for an urban water system? How this reliability, defined as *Prob* (Supply > Demand) $\geq P$ (where P is the acceptable level of the reliability of the system), could be increased by

a. Possibility of tapping into other resources
b. Conservation and rationing programs to first provide the necessary water needs of a region
c. Prioritizing water allocations according to the demand importance
d. Determining the possible allocation of water to another case at the time of disaster
e. Designing under a normal and acceptable level of risk reliability but making provisions and plans to meet the extreme events at a lower reliability level

In hydraulic engineering, a design can be based on hydrological and biological inputs, encompassing the notion of "load." The load $s(t)$ must be compared with the resistance $r(t)$, which could be a system's ability to tolerate additional pressure. If the load exceeds that resistance, a failure will occur. In hydraulic structure, the load $r(t)$ is the force exerted by flow and other loads. The resistance $r(t)$ is the force that the structure can accept without failing. In water distribution, water demand is a load that should be compensated for by a supply as a resistance.

Figure 10.2 shows the design procedure involving the above notion of load and resistance and how reliability and uncertainty could be integrated into the design process. In urban water systems, there are two types of essential input: the natural input such as rainfall, temperature, sediment yield, and so on, which are the random variables describing the natural environment, and inputs coming from human activities, such as water demand, sewage effluents, pollutants, and so on, which are random numbers describing the interaction of man with his environment. The hydrological,

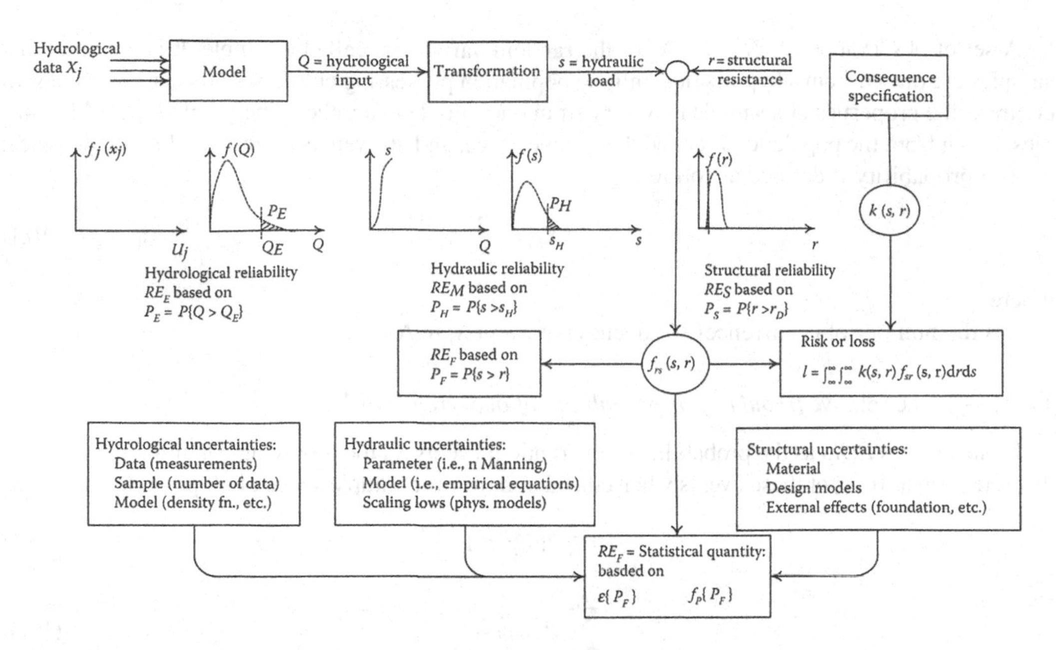

FIGURE 10.2 Generalized concept of risk and reliability analysis for structures. (From Plate, E.J. and Duckstain, L., *Water Resour. Bull.*, 24(2), 235–245, 1988.)

hydraulic, and structural uncertainties are identified and coupled with a loss function to determine the expected damage resulting from a disaster/failure.

The notations used and further deliberation of Figure 10.2 are as follows:

The hydrologic variable $Q(t)$ is usually not the variable that is to be compared with the resistance in determining the probability of failure. It is transformed through hydraulic calculations into a design variable. A typical design variable could be a dimension of a canal for drainage which has to be designed. When we construct many structures for the design, the variability of the resistance leads to an unexpected value of $\varepsilon\{P_F\}$ which yields to an average failure rate. The concept of calculating the probability of failure P_F, in spite of the variability of both load and resistance, is a single value of for the set of experiments. If all the assumptions leading to the calculation of P_F were correct, then such a value could be trusted. In actual cases, however, this true value cannot be found because of uncertainties, and P_F, and thus also RE (reliability), is a random variable with pdf fp (P_F).

10.3 PROBABILISTIC TREATMENT OF HYDROLOGIC DATA

Hydrologic variables and events such as rainfall, runoff, floods, and drought are usually investigated by analyzing their records of observations. Many characteristics of these processes seem to vary in a way not amenable to deterministic analysis. In other words, deterministic relationships presented so far do not seem to be applicable for analyzing these characteristics. For hydrologic analysis, the annual peak discharge is considered to be a random variable. Methods of probability and statistics are employed for the analysis of random variables.

In all hydrologic analysis fields, we face uncertainty arising from hydrologic phenomena that cannot be predicted accurately. Any prediction is uncertain, and in the mathematical modeling of hydrological processes, this uncertainty has to be taken into account.

A random variable X is a variable described by a probability distribution. The distribution specifies the chance that observation x of the variable will fall in a specified range of X. For example, suppose X is annual precipitation at a specified location. In that case, X's probability distribution specifies the chance that the observed annual precipitation in a given year will lie in a defined range, such as less than 40 in, and so on.

A set of observations X_1, X_2, …, X_n of the random variable is called a sample. It is assumed that samples are drawn from a hypothetical infinite population possessing constant statistical properties. In contrast, the properties of a sample may vary from one sample to another. The set of all possible samples drawn from the population is called the sample space, and an event is a subset of the sample space.

The probability is defined as follows:

$$P(X_i) = \frac{n}{N} \tag{10.1}$$

where

n is the number of occurrences (frequency) of event X_i in N trials.

Thus, $\frac{n}{N}$ is the relative *frequency or probability of occurrence* of X_i.

If one defines $P(X_i)$ as the probability of the random events X_i, the following condition holds on the discrete probabilities of these events when considered over the sample space of all possible outcomes:

$$0 \le P(X_i) \le 1 \tag{10.2}$$

$$\sum_{i=1}^{N} P(X_i) = 1. \tag{10.3}$$

10.3.1 Discrete and Continuous Random Variables

Its probability distribution may describe the behavior of a random variable. Every possible outcome of an experiment is assigned a numerical value according to a discrete probability mass function or a continuous probability density function (PDF). In hydrology, discrete random variables are most commonly used to describe the number of occurrences that satisfy a certain criterion, such as the number of floods that exceed a specified value or the number of storms at a given location. For discrete probabilities:

$$P(a \le x \le b) = \sum_{a \le x_i \le b} P(x_i). \tag{10.4}$$

The cumulative distribution function (CDF) is defined as

$$F(x) = P(X \le x) = \sum_{x_i \le x} P(x_i) \tag{10.5}$$

Continuous random variables are usually used to represent hydrologic phenomena such as flow, rainfall, volume, depth, and time. Values are not restricted to integers, although continuous variables might be commonly rounded to integers. For a continuous random variable, the area under PDF $f(x)$ represents probability.

$$P(x_1 \le x \le x_2) = \int_{x_1}^{x_2} f(x)dx, \tag{10.6}$$

where the entire area under the PDF equals 1.0. The continuous CDF is defined similarly to its discrete counterpart:

$$P(x_1 < x \le x_2) = F(x_2) - F(x_1). \tag{10.7}$$

10.3.2 Moments of Distribution

A PDF is functional from whose moments are related to its parameters. Thus, if moments can be found, then the parameters of the distribution can be estimated. The moments themselves are also indicative of the shape of the distribution. For a discrete distribution, the M_h moment about the origin can be defined as follows:

1. For a discrete distribution

$$\mu_N' = \sum_{i=-\infty}^{\infty} x_i^N P(x_i). \tag{10.8}$$

2. For a continuous distribution

$$\mu_N' = \int_{-\infty}^{\infty} x^N f(x)dx. \tag{10.9}$$

The first moment, μ, is the mean value and is defined as follows:

1. For a continuous distribution

$$E(x) = \mu = \sum_{-\infty}^{\infty} x_i P(x_i) \tag{10.10}$$

Example 10.1

The probability mass function of floods is shown in Figure 10.3. Estimate the mean number of floods in a 10-year period where $f(x_0) = f(x_{10}) = 0.0010$.

Solution:

Equation (10.10) can be used to compute the mean for a discrete random variable of floods in a 10-year period:

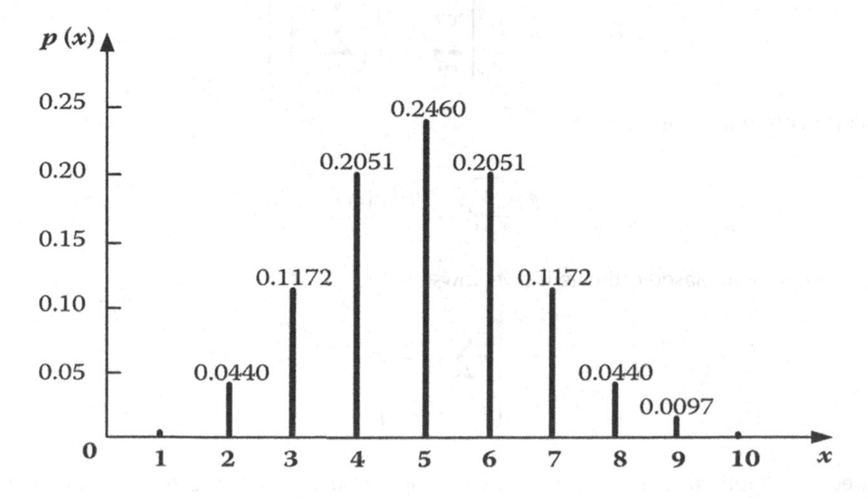

FIGURE 10.3 Probability mass function for a discrete random variable.

$$E(x) = \mu = 0(0.0010) + 1(0.0097) + 2(0.0440) + 3(0.1072) + 4(0.2051) + 5(0.2460) + 6(0.2051)$$
$$+ 7(0.1072) + 8(0.0440) + 9(0.0097) + 10(0.0010) = 5.$$

Thus, the mean number of floods in 10 years is 5.

The central moment about the mean may be defined as follows:

1. For a discrete distribution

$$\mu_N = \sum_{-\infty}^{\infty} (x_i - \mu)^N P(x_i). \tag{10.11}$$

2. For a continuous distribution

$$\mu_N = \int_{-\infty}^{\infty} (x - \mu)^N f(x)\,dx. \tag{10.12}$$

Furthermore, the first central moment is zero. The second central moment is called the variance and is computed as follows:

1. For a discrete distribution

$$\text{Var}(x) = \sigma^2 = \mu_2 = E\left[(x - \mu)^2\right] = \sum_{-\infty}^{\infty} (x - \mu)^2 P(x_i) \tag{10.13}$$

2. For a continuous distribution

$$\text{Var}(x) = \sigma^2 = \mu_2 = E\left[(x - \mu)^2\right] = \int_{-\infty}^{\infty} (x_i - \mu)^2 f(x)\,dx. \tag{10.14}$$

An equivalent measure is the standard deviation, which is simply the square root of the variance. Higher moments are subject to bias in their estimates. An unbiased estimate is one for which the expected value of the estimate equals the population value. For the variance, an unbiased estimate is

$$\sigma^2 = S^2 = \frac{1}{n-1} \sum_{i=1}^{n} (x_i - \mu)^2 \tag{10.15}$$

or

$$\sigma^2 = S^2 = \frac{1}{n-1} \left[\sum_{i=1}^{n} x_i^2 - \frac{1}{n} \left(\sum_{i=1}^{n} x_i \right)^2 \right]. \tag{10.16}$$

2. For a continuous distribution

$$g = \sum_{i=1}^{n} (x - \mu)^3 f(x_i) \tag{10.17}$$

An approximately unbiased estimate is as follows:

$$g = \frac{n \sum_{i=1}^{n} (x - \bar{x})^3}{(n-1)(n-2)S^3}. \tag{10.18}$$

The three conditions of the skew are shown in Figure 10.4. A skew is zero for the symmetric distribution and the skew is positive or negative when the distribution is nonsymmetrical. If the

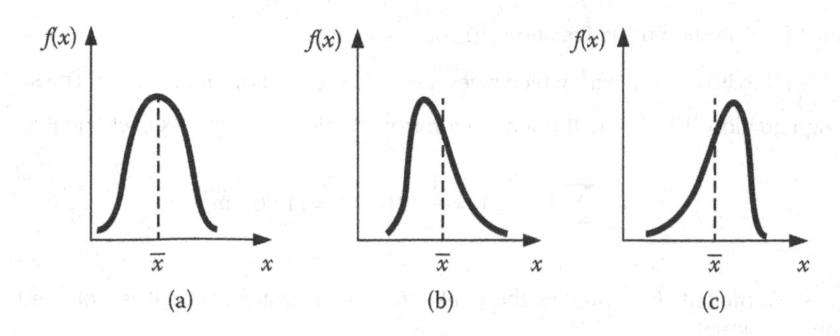

FIGURE 10.4 Comparison of the skew of distributions: (a) g = 0; (b) g > 0; and (c) g < 0.

more extreme tail of the distribution is to the right, the skew is positive and it is negative when the more extreme tail is to the left of the mean.

Example 10.2

Estimate the mean, variance, standard deviation, and skew of 18-year precipitation data on a small basin given in Table 10.1.

Solution:

Values of $(x-\bar{x}),(x-\bar{x})^2,(x-\bar{x})^3$ are estimated for precipitation data each year; the summation of the annual data is given in the last row. The mean, variance, standard deviation, and skew are estimated as follows:

$$\bar{x}=\frac{1}{n}\sum x_i=\frac{1}{18}(68.12)=3.78\ \text{cm.} \tag{10.19}$$

TABLE 10.1
Computations of Moments of June Precipitation

Year	Precipitation x (cm)	$(x-\bar{x})$	$(x-\bar{x})^2$	$(x-\bar{x})^3$
1954	4.27	0.486	0.236	0.1145
1955	2.92	−0.864	0.747	−0.6460
1956	2.00	−1.784	3.184	−5.6821
1957	3.34	−0.441	0.198	−0.0878
1958	2.32	−1.464	2.145	−3.1406
1959	2.89	−0.894	0.800	−0.7156
1960	2.79	−0.994	0.989	−0.9834
1961	5.52	1.736	3.012	5.2278
1962	4.09	0.306	0.093	0.0285
1963	9.86	6.076	36.912	224.2632
1964	3.90	0.116	0.013	0.0015
1965	6.85	3.066	9.398	28.8090
1966	2.33	−1.454	2.115	−3.0767
1967	4.16	0.376	0.141	0.0530
1968	2.10	−1.684	2.837	−4.7794
1969	0.85	−2.934	8.611	−25.2684
1970	5.00	1.216	1.478	1.7961
1971	2.93	−0.854	0.730	−0.6238
Sum	68.12	0.008	73.639	215.2898

The variance is computed using Equation (10.16).

$S^2 = \frac{1}{18-1}(73.639) = 4.332\ \text{cm}^2$, which gives a standard deviation of 2.081 cm. The skew is estimated using Equation (10.17) and the summation from Table 10.1, with $f(x_i)$ set equal to $1/n$:

$$g = \frac{1}{n}\sum (x - x_i)^3 = \frac{1}{18}(215.28) = 11.96\ \text{cm}^3. \tag{10.20}$$

The kurtosis coefficient, *k*, expresses the peakedness of a distribution. It is obtained from the fourth central moment:

$$m_4 = \frac{1}{n}\sum_{i=1}^{N}(x_i - \mu)^4 \tag{10.21}$$

$$k = \frac{m_4}{m_2^2} = \frac{m_4}{\left(S^2\right)^2} \tag{10.22}$$

Its significance relates mainly to the normal distribution, for which $k = 3$. Distributions that are taller than normal have $k > 3$; flatter ones have $k < 3$.

10.3.3 Flood Probability Analysis

Flood frequency analysis of the US Water Resources Council (USWRC, 1981) proposed a method in order to analyze the flood frequency; the procedure that is described here is based on this method. The probability distribution of runoff is highly skewed because the extreme flood events in flow time series are very far from average flows. The sensitivity of the skewness coefficient is very high and it is related to the size of the sample; therefore, it is difficult to get a precise estimation from small samples. Hence, it is recommended that generalized estimation of the skewness coefficient is used when short records of the skew approximation are needed. Weighted averages between skewness coefficients give a generalized skew. Increasing record length usually leads to a more reliable skew. The weighted skew, G_w, is achieved as follows:

$$G_w = WG_s + (1 - W)G_m, \tag{10.23}$$

where

W is weight,

G_s is the skew coefficient computed using the sample data, and

G_m is a map skewness coefficient, the values of which for the United States are found in Figure 10.5.

A weighting procedure is a function of the variance of the derived sample skew and the variance of the derived map skewness. In this procedure, the uncertainty of deriving skewness coefficients from both sample data and regional or map values is considered to find a generalized skew that reduces the uncertainty to the minimum based upon information known.

Assessment of the sample skew coefficient and the map skewness coefficient is assumed to be independent with the same mean and respective variances. Assuming independency of G_s and G_m, the variance (mean square error) of weighted skew, $V(G_w)$, can be expressed as

$$V(G_w) = W^2 \cdot V(G_s) + (1 - W)^2 \cdot V(G_m), \tag{10.24}$$

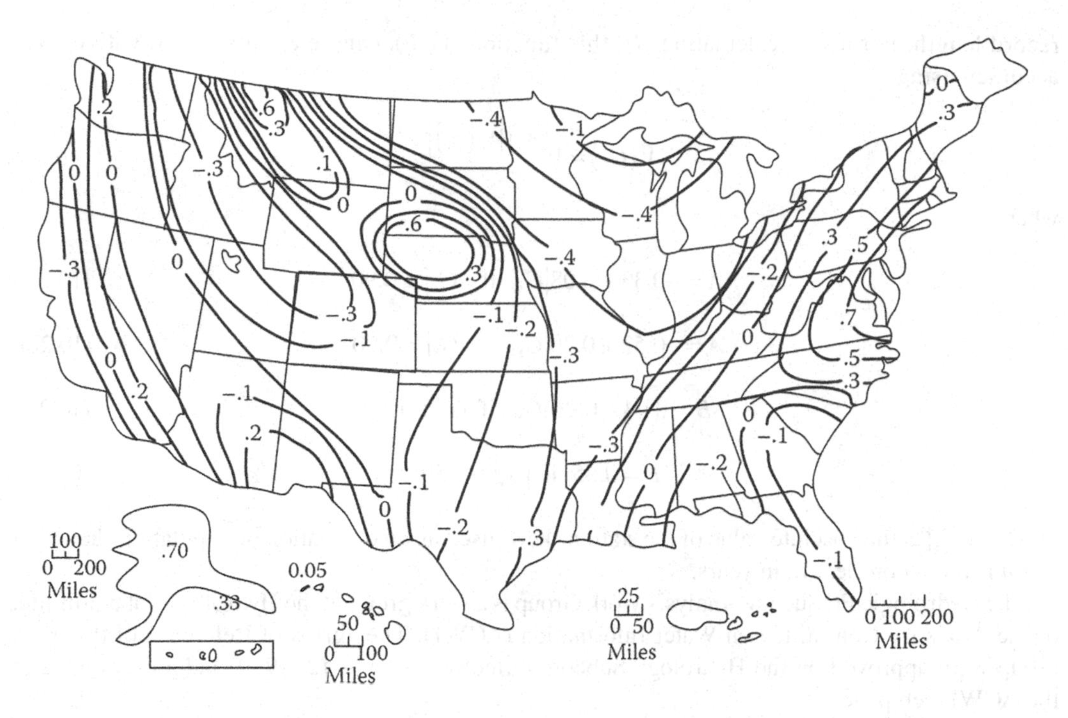

FIGURE 10.5 Generalized skew coefficients of annual maximum streamflow. (From USWRC, *Estimating Peak Flow Frequencies for Natural Ungauged Watersheds—A Proposed Nationwide Test.* Hydrology Subcommittee, US Water Resources Council, Washington, DC, 1981. With permission.)

where

$V(G_s)$ is the variance of the sample skew and

$V(G_m)$ is the variance of the map skewness.

The skew weight that minimizes the variance of the weighted skew can be determined by differentiating Equation (10.24) with respect to W and solving $d[V(G_w)]/dW = 0$ for W to obtain

$$W = \frac{V(G_m)}{V(G_s)+V(G_m)}. \tag{10.25}$$

The values of $V(G_m)$ and $V(G_s)+V(G_m)$, which are estimated from the map skewness coefficient as the square of the standard deviation of station values of skew coefficients about the isolines of the skew map, are required in order to determine W using Equation (10.25). In the USWRC (1981) skew map, the value of $V(G_m)$ is 0.3025. Otherwise, $V(G_m)$ could be obtained from a regression study relating the skew to physiographical and meteorological characteristics of the basins and determining $V(G_m)$ as the square of the standard error of the regression equation (Tung and Mays, 1981).

The weighted skew, G_w, can be determined by substituting Equation (10.25) into Equation (10.23), resulting in

$$G_w = \frac{V(G_m)\cdot G_s + V(G_s)\cdot G_m}{V(G_s)+V(G_m)}. \tag{10.26}$$

The results of Monte Carlo experiments by Wallis (1974) give the variance (mean square error) of the station skews for log-Pearson Type III (see next section) random variables. The results of the study showed that $V(G_m)$ of the logarithmic station skew is a function of population skew and record length. For use in calculating G_w, this function, $V(G_s)$, can be approximated with enough accuracy using

$$V(G_s) = 10^{A-B\left[\log\left(\frac{n}{10}\right)\right]}, \tag{10.27}$$

where,

$$A = -0.33 + 0.08|G_s| \text{ if } |G_s| \leq 0.90 \tag{10.28a}$$

$$A = -0.52 + 0.30|G_s| \text{ if } |G_s| > 0.90 \tag{10.28b}$$

$$B = 0.94 - 0.26|G_s| \text{ if } |G_s| \leq 1.50 \tag{10.28c}$$

$$B = 0.55 \text{ if } |G_s| > 1.50, \tag{10.28d}$$

in which $|G_s|$ is the absolute value of the station skew (used as an estimation of population skewness) and n is the record length in years.

The Hydrologic Frequency Analysis Work Group is a work group of the Hydrology Subcommittee of the Advisory Committee on Water Information (ACWI). The Terms of Reference of this work group were approved by the Hydrology Subcommittee on October 12, 1999, and are available on the ACWI web page.

10.4 COMMON PROBABILISTIC MODELS

Many discrete and continuous PDFs are used in hydrology. However, this subsection focuses on only a few of the most common. For discrete analysis, there may be interest in both of the CDFs. However, for continuous analysis, the value of the PDF itself is rarely of interest. Instead, only the CDF for the continuous random variable needs to be evaluated. These distributions will be seen as the various distributions are presented.

10.4.1 The Binomial Distribution

It is common to examine a sequence of independent events for which the outcome of each can be either a success or a failure; for example, either the Γ-year flood occurs or it does not. The number of possible ways of choosing x events out of n possible events is given by the binomial coefficient: where n is trials and x is occurrence. Thus, the desired probability is the product of the probability of any one sequence, and the number of ways in which such a sequence can accrue is as follows:

$$P(x) = \binom{n}{x} P^x (1-P)^{n-x}. \tag{10.29}$$

The CDF is defined as follows:

$$F(x) = \sum_{i=0}^{x} \binom{n}{i} p^i (1-p)^{n-i}. \tag{10.30}$$

10.4.2 Normal Distribution

The normal distribution is a well-known probability distribution. Two parameters are involved in a normal distribution: the mean and the variance. A normal random variable having a mean μ and variance σ^2 is herein denoted as $X \cong N(\mu, \sigma^2)$ with a PDF given as follows:

$$f(x) = \frac{1}{\sqrt{2\pi}} \exp\left[-\frac{1}{2}\left(\frac{x-\mu}{\sigma}\right)^2\right]. \tag{10.31}$$

A normal distribution is bell-shaped and symmetric concerning $x = \mu$. Therefore, the skew coefficient for a normal random variable is zero. A random variable Y that is a linear function of a normal random variable X is also normal. That is, if $X \cong N(\mu, \sigma^2)$ and $Y = aX + b$, then $Y \cong N(\alpha\mu + b, \alpha^2\sigma^2)$. An extension of this is that the sum of normal random variables (independent or dependent) is also a normal random variable.

Probability computations for normal random variables are made by first transforming to the standardized variant as follows:

$$Z = (x - \mu)/\sigma \tag{10.32}$$

in which Z has a zero mean and unit variance. Since Z is a linear function of the random variable *X*, Z is also normally distributed. The cumulative probability of normal distribution uses Tables 10.A1 (a) and (b) in this chapter's Appendix. The PDF of Z, ϕ, called the standard normal distribution, can be expressed as follows:

$$\phi(z) = \frac{1}{\sqrt{2\pi}} \exp\left(-\frac{z^2}{2}\right) \quad \text{For} -\infty < z < \infty. \tag{10.33}$$

Computations of probability for $X \approx N(\mu, \sigma^2)$ can be performed as follows:

$$p(X \le x) = p\left[\frac{X-\mu}{\sigma} \le \frac{x-\mu}{\sigma}\right] = p(Z \le z). \tag{10.34}$$

10.4.3 The Exponential Distribution

Consider a process of random arrivals such that the arrivals (events) are independent. The process is stationary, and it is not possible to have more than one arrival at an instant in time. If the random variable t represents the interarrival time (the time between events), it is found to be exponentially distributed with one parameter, λ, PDF

$$f(t) = \lambda e^{-\lambda t}, \quad t \ge 0 \tag{10.35}$$

The mean of the distribution is

$$E(t) = \frac{1}{\lambda}, \tag{10.36}$$

and the variance is

$$\text{Var}(t) = \frac{1}{\lambda^2} \tag{10.37}$$

The CDF is evaluated as follows:

$$F(t) = \int_0^t \lambda e^{-\lambda t} = 1 - e^{-\lambda t} \tag{10.38}$$

10.4.4 The Gamma Distribution

This distribution receives extensive use in hydrology simply because of its shape and its well-known mathematical properties. The frequency factor K is a function of the skewness, C_s, and return period (or CDF), and values are given in Table 10.A2 of this chapter's Appendix. Thus, the T-year flood, is computed as:

$$Q_T = \bar{Q} + K(C_s, T)\ S_Q \tag{10.39}$$

The two-parameter gamma distribution corresponds to setting the left boundary to zero. So the C_s is considered as twice as much as CV (coefficient of variation).

Example 10.3

What is the magnitude of the 100-year flood for a river $\left(\bar{Q} = 4.14 \text{ cubic meter per second (cms)}, S_Q = 3.31 \text{ cms and } C_s = 1.981 \text{ cm}\right)$ using the gamma-3 and gamma-2 distribution?

Solution:

For gamma-3, $C_s = 1.981$. Linear interpolation in the table in this chapter's Appendix. yields to a K *value of* 3.595. Thus,

$$Q_{100} = 4.14 + 3.595 \times 3.31 = 16.04 \text{ cms.}$$

For gamma-2 *g* or $C_s = 2CV = 1.277$. Linear interpolation from the same table a K *value of* 3.197. Thus,

$$Q_{100} = 4.14 \text{ cms} + 3.197 \times 3.31 \text{ cms} = 14.72 \text{ cms.}$$

10.4.5 The Log Pearson Type 3 Distribution

The three-parameter gamma distribution is applied to the logs of the random variables and hydrology because it has been recommended to apply to flood flow. The shape of the LP3 is quite flexible due to its three parameters.

Its use is entirely analogous to the lognormal; however, the moments of the transformed and untransformed variables will not be related here. Instead, the data are transformed by taking logarithms, and the gamma-3 distribution is applied precisely as in the preceding section.

10.5 RETURN PERIOD OR RECURRENCE INTERVAL

Return period is defined as the average number of trials required to the first occurrence of an event $D \geq D_0$ or an event that is greater than or equal to a particularly critical event or design event D_0 (Bras, 1990). The preceding definition assumes that an event $D \geq D_0$ occurred in the past. A finite time τ has elapsed since then, and the interest is in the residual or remaining waiting time N for the next occurrence of $D \geq D_0$. For example, such a critical event, D_0, may be flood (flood of a given return period T or *T*-year flood) or drought (a drought of a given return period). Definition of variables involved in estimation of return period and risk of failure is shown in Figure 10.6. In this figure, Y_t is a hydrological process, Y_0 is a threshold value, e is an event representing a continuous sequence in which $Y_t < Y_0$. Thus, the events $e_1, e_2, \ldots, e_n$ occur at $t_1, t_2, \ldots, t_n$, respectively. In addition, the events e can be described by a certain characteristic of interest D and the resulting sequence $D_1, D_2, \ldots, D_n$ occurring at times $t_1\ t_2, \ldots, t_n$. Furthermore, the sequence $D_1, D_2, \ldots, D_n$ may be censored using a critical value D_0. Figure 10.6 shows an example related to droughts, where the duration of a drought is considered to be the property of interest, and critical drought duration on censoring level D_0 is used to distinguish the common droughts from the critical drought. In the case of annual floods, the events $e_1, e_2, \ldots$ are simply the sequence $Y_1, Y_2, \ldots$ and $D_0 = Y_0$ is the design flood. D shows the hydrological events characteristics and D_0 corresponds to the characteristics in the design event.

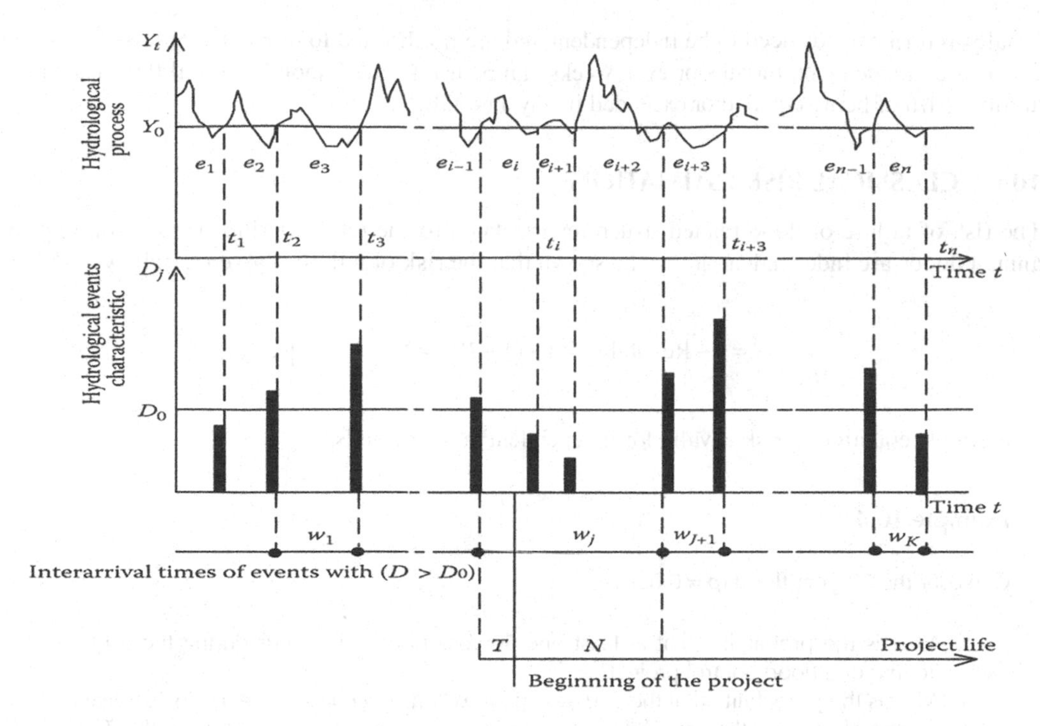

FIGURE 10.6 Definition of variables involved in estimation of return period and risk of failure.

The return period is defined as follows:

$$T = \frac{1}{P}, \tag{10.40}$$

where

T is the return period, and
P is the probability of an event.

An annual maximum event has a return period (or recurrence interval) of T years if its magnitude is equaled or exceeded once, on average, every T years. *T*'s reciprocal is the exceedance probability of the event, that is, the probability that the event is equaled or exceeded in any 1 year. Thus, the 50-year flood has a probability of 0.02, or 2%, of being equaled or exceeded in any single year. The concept of a return period implies independent events. It is usually found by analyzing the series of maximum annual data. The largest event in 1 year is assumed to be independent of the largest event in any other year. However, it is also possible to apply such an analysis to the n largest independent events from an n-year period, regardless of the year in which they occur. In this case, if the second-largest event in 1 year was more significant than the largest event in another year, it could be included in the frequency analysis. This section of n largest (independent) values is called the series of annual exceedance, as opposed to an annual maximum series. Series of annual exceedances and annual maximum are used in hydrology, with little difference at high return periods. There are likely to be more problems of ensuring independence when using annual exceedances. However, for low return periods, annual exceedances give a more realistic lower return period for the same magnitude than do annual maxima. The relationship between return period based on annual exceedances T_e and annual maxima T_m is as follows (Chow et al. 2013).

$$T_e = \frac{1}{\ln T_m - \ln(T_{m-1})}. \tag{10.41}$$

Finally, return periods need to be independent and are not limited to units of years. As long as the events are independent, months or even weeks can be used. The 6-month rainfall thus has a probability of 1/6 of being equaled or exceeded in any 1 month.

10.6 CLASSICAL RISK ESTIMATION

The risk of failure of flood related structures is related to the return period. Thus, assuming that annual floods are independent, it may be shown that the risk of failure is given as follows:

$$R = 1 - \text{Reliability} = 1 - (1 - P)^n = 1 - \left(1 - \frac{1}{T}\right)^n. \tag{10.42}$$

This simple equation of risk is valid for independent annual floods.

Example 10.4

Consider the 50-year flood ($p = 0.02$).

a. What is the probability that at least one 50-year flood will occur during the 30-year lifetime of a flood control project?
b. What is the probability that the 100-year flood will not occur in 10 years? In 100 years?
c. In general, what is the probability of having no floods more significant than the T-year flood during a sequence of T_{year}?

Solution:

a. This is just the risk of failure discussed above, and the distribution of the number of failures is $B(30, 0.02)$. Thus, from Equation (10.42),

$$\begin{aligned} \text{Risk} &= 1 - (1 - 0.02)^{30} \\ &= 1 - 0.98^{38} \\ &= 1 - 0.545 \\ &= 0.455. \end{aligned}$$

b. For $n = 10$, $P(x = 0) = 1 - (1 - p)^{10} = 0.99^{10} = 0.92$.
   For $n = 100$, $P(x = 0) = 1 - (1 - p)^{100} = 0.99^{100} = 0.37$.

c. $P(x = 0) = 1 - \left(1 - \frac{1}{T}\right)^T$.

Example 10.5

A cofferdam has been built to protect homes in a floodplain until a major channel project can be completed. The cofferdam was built for a 20-year flood event. The channel project will require 3 years to complete. Hence, the process is B(3, 0.05). What are the probabilities that

a. The cofferdam will not be overtopped during the 3 years (the reliability)?
b. The cofferdam will be overtopped in any 1 year?
c. The cofferdam will be overtopped precisely once in 3 years?
d. The cofferdam will be overtopped at least once in 3 years (the risk)?
e. The cofferdam will be overtopped only in the third year?

Solution:

a. $\text{Reliability} = \left(1 - \frac{1}{20}\right)^3 = 0.95^3 = 0.86.$

b. $\text{Prob} = \frac{1}{T} = 0.05.$

c. $P(x=1) = \binom{3}{1} p^1 (1-p)^2 = 3 \times 0.05 \times 0.95^2 = 0.135.$

d. Risk = 1 − Reliability = 0.14.

e. Prob = (1 − p) (1 − p) p = 0.903 × 0.05 = 0.045.

Example 10.6

Determine the design return period, if the acceptable hydrologic risk is 0.15 (or 15%) during the 25-year lifetime of a culvert.

Solution:

Equation (10.42) is employed again for the determination of the return period of the design rainfall event as follows:

$$0.15 = 1 - \left(1 - \frac{1}{T_r}\right)^{25}.$$

T_r has been calculated to be 154 years where, in practice, a return period of 150 years is considered to determine culvert size.

Risk is defined differently in engineering applications as follows:

a. *Risk versus probability*: Some descriptions of risk focus only on the event occurring probability but other explanations include both the probability and the event's consequences. For example, the probability of a severe earthquake may be minimal. However, the effects are so disastrous that it would be classified as a high-risk event.
b. *Risk versus threat*: In some disciplines, there is a difference between a threat and risk.
   A threat can be any event with low probability and enormous negative consequences. However, a risk is defined as a higher probability event, where there is enough information to assess consequences.
c. *All outcomes versus adverse outcomes*: Definitions of risk mostly focus on the downside scenarios, while all consequences are considered. Risk is defined as the product of the probability of an event occurring and the corresponding damage it causes.

10.7 RELIABILITY

In a broader context, reliability can be explained in this section. The system output, X_t, as a random variable, is classified into two groups of failure and success outputs. The system reliability is the probability that X_t belongs to the success outputs group (Hashimoto et al., 1982), which can be formulated as follows:

$$\alpha = \text{Prob}\left[X_t \in S\right] \forall t, \tag{10.43}$$

where
S is the set of all satisfactory outputs.

Whether natural or human-made, all systems may fail for various reasons, including structural inadequacies, natural causes exceeding the design parameters of the system (e.g., droughts and floods), and human causes such as population growth that raises the system's demands above the supply capacity. Thus, reliability is conceptually related to the probability of system failure and the rate and consequences of failure. It can be measured in several different but related ways, depending on the particular situation's needs and relevance.

While reliability terms are frequently used in water resources planning and management, there are some difficulties in practical cases. The main difficulties are as follows:

- Agencies or institutions do not formally define reliability. There is no existing and widely accepted framework of reliability measurement applicable in all aspects of influence and applications.
- Reliability is considered with different water management entities' operating and planning processes with substantial variation.
- Reliability is often considered in qualitative rather than quantitative terms.
- Reliability at key points within the sphere of influence ultimately depends on the complexity of interactions of a web of rights, regulations, laws, and rules that affect all institutions that manage water in a municipality. It will be necessary to understand these interactions in order to develop meaningful definitions of reliability, and based on these definitions, related indicators could be defined.

In reliability theory, the Weibull distribution is assumed to a model lifetime (the time between two consecutive system failures) most appropriately. The cumulative probability function of the system failures could be formulated as follows:

$$F(t) = 1 - e^{-\omega t^{\beta}} \, (t > 0). \tag{10.44}$$

where

$\alpha, \beta > 0$ are parameters to be estimated.

A definition of reliability is the probability that no failure will occur within the planning horizon (Karamouz et al., 2003). Therefore, using Equation (10.44), reliability could be formulated as follows:

$$R(t) = 1 - F(t) = e^{-\alpha t^{\beta}}. \tag{10.45}$$

The density function of the time of failure occurrences is obtained by simple differentiation of Equation (10.45) as follows:

$$f(t) = \alpha \beta t^{\beta - 1} e^{-\alpha t^{\beta}} \, (t > 0). \tag{10.46}$$

In the particular case of $\beta = 1$, $f(t) = \alpha e^{-\alpha t}$, the density function of the exponential distribution. The exponential distribution is only seldom used in reliability studies since it has the so-called forgetfulness property. If X is an exponential variable indicating the time when the system fails, then for all t and $\tau > 0$:

$$P(X > t + \tau | X > \tau) = P(X > t). \tag{10.47}$$

This relation shows that the working condition's probability in any time length t is independent of how long the system was working before. Equation (10.47) can be shown as

$$P(X > t + \tau \mid X > \tau) = \frac{P(X > t + \tau)}{P(X > \tau)} = \frac{R(t + \tau)}{R(\tau)} = \frac{e^{-\alpha(t+\tau)}}{e^{-\alpha \tau}} = e^{-\alpha t} = R(t) = P(X > t). \tag{10.48}$$

In most practical cases, the breakdown probabilities are increasing in time as the system becomes older; thus, the exponential variable is inappropriate in such cases.

The hazard rate is given as follows:

$$\rho(t)=\lim_{\Delta t\to 0}\frac{P\left(t\le X\le t+\Delta t|t\le X\right)}{\Delta t}. \tag{10.49}$$

It shows how often failures will occur after time period t as follows:

$$\rho(t)=\lim_{\Delta t\to 0}\frac{P(t\le X\le t+\Delta t)}{P(t\le X)\Delta t}\lim_{\Delta t\to 0}\frac{F(t+\Delta t)}{\Delta t}.\frac{1}{R(t)}=\frac{f(t)}{R(t)}. \tag{10.50}$$

Therefore, the hazard rate can be computed as the ratio of the density function of time between failure occurrences and the reliability function (Karamouz et al., 2003).

Example 10.7

Assume the time between drought occurrences in a watershed follows the Weibull distribution. Formulate the reliability and hazard rate of this watershed in dealing with droughts.

Solution:

The watershed reliability is calculated as follows:

$$R(t)=1-F(t)=e^{-\alpha t^{\beta}},$$

Moreover, from Equation (10.50), the hazard rate is calculated as follows:

$$\rho(t)=\frac{\alpha\beta t^{\beta-1}e^{-\alpha t^{\beta}}}{e^{-\alpha t^{\beta}}}=\alpha\beta t^{\beta-1},$$

which is an increasing polynomial of t, showing that failure will occur more frequently for larger values of t. In the particular case of an exponential distribution, $\beta = 1$, so $\rho(t) = \alpha$ being a constant.

It is an essential problem in reliability engineering to reconstruct $F(t)$ or the reliability function if the hazard rate is given (Karamouz et al., 2003). Notice first that

$$\rho(t)=\frac{f(t)}{1-F(t)}=\frac{(F(t))'}{1-F(t)}=-\frac{(1-F(t))'}{1-F(t)}.$$

By integration of both sides in the interval $[0, t]$, it will achieve

$$\int_0^t\rho(\tau)d\tau=\left[-\ln\left(1-F(\tau)\right)\right]_{\tau=0}^{t}=-\ln\left(1-F(t)\right)+\ln\left(1-F(0)\right).$$

Since $F(0) = 0$, the second term equals zero; hence,

$$\ln\left(1-F(t)\right)=-\int_0^t\rho(\tau)d\tau,$$

implying that

$$1-F(t)=\exp\left(-\int_0^t\rho(\tau)d\tau\right).$$

and, finally,

$$F(t)=1-\exp\left(-\int_0^t \rho(\tau)d\tau\right). \tag{10.51}$$

Example 10.8

Construct the CDF of failure occurrences in two given situations: (a) The hazard rate is constant, $\rho(i) = \alpha$; and (b) $\rho(t) = \alpha\beta t^{\beta-1}$.

Solution:

Using Equation (10.51) for $\rho(t) = \alpha$, it is obtained that

$$F(t)=1-\exp\left(-\int_0^t \rho(\tau)d\tau\right)=1-\exp\left(-\int_0^t \alpha\, d\tau\right)=1-e^{-\alpha t}.$$

Similarly, when $\rho(t) = \alpha\beta t^{\beta-1}$, and then showing that the distribution is Weibull,

$$F(t)=1-\exp\left(-\int_0^t \alpha\beta\tau^{\beta-1}\, d\tau\right)=1-\exp\left(-\left[\alpha\tau^{\beta}\right]_0^t\right)=1-e^{-\alpha t^{\beta}}.$$

In contrast with $F(t)$, which is increasing, $F(0) = 0$ and $\lim_{t\to\infty} F(t)=1$, $R(t)$ is decreasing, $R(0) = 1$, and $\lim_{t\to\infty} R(t)=0$. In the case of $\beta > 2$ and Weibull distribution, $\rho(0) = 0$, $\lim_{t\to\infty} \rho(t)$ and $\rho(t)$ is a strictly increasing and strictly convex function. If $\beta = 2$, then $\rho(t)$ is linear, and if $1 < \beta < 2$, then $\rho(t)$ is strictly increasing and strictly concave. If $\beta = 1$, then $\rho(t)$ is a constant. In the case of $\beta < 1$, the hazard rate is decreasing in t, in which case defective systems tend to fail early. Thus, the hazard rate decreases for a well-made system. Notice that

$$\rho'(t)=\frac{f'(t)(1-F(t))+f(t)^2}{(1-F(t))^2}$$

which is positive if and only if the numerator is positive:

$$f'(t)>-\frac{f(t)^2}{1-F(t)}=-f(t)\,\rho(t). \tag{10.52}$$

If the inequity of Equation (10.52) is satisfied, $\rho(t)$ is locally increasing; otherwise, $\rho(t)$ decreases.

The above discussions about reliability are applicable to a component of a system. The reliability of a system with several components depends on what kind of configuration of the components is the entire system. Systems could be arranged in "series," "parallel," or in combinations. The failure of any component in a series system results in system failure, but in a parallel system, only when all components fail simultaneously does failure occur.

A typical series combination is shown in Figure 10.7. If $R_i(t)$ denotes the reliability function of component i ($i = 1, 2, \ldots, n$), then the reliability function of the system could be estimated as follows:

$$R(t)=P(X>t)=P\big((X_1>t)\cap(X_2>t)\cap\ldots\cap(X_n>t)\big), \tag{10.53}$$

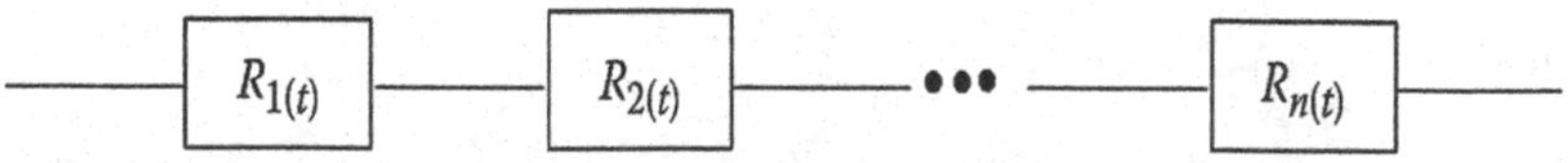

$R_{1(t)}$ — $R_{2(t)}$ — ••• — $R_{n(t)}$

FIGURE 10.7 Series combination of components.

where

X is when the system fails, and

$X_1, \ldots, X_n$ are the same for the components.

It can be assumed that if the failures of the different system components occur independently of each other, then

$$R(t) = P(X_1 > t) = P(X_2 > t) \ldots P(X_n > t) = \prod_{i=1}^{n} R_i(t). \tag{10.54}$$

Note that the inclusion of a new component into the system results in a smaller reliability function. It is multiplied by the new factor $R_{n+1}(t)$, which is below 1.

In a parallel system, as shown in Figure 10.8, the system fails, if all components fail; hence, the system failure probability is calculated as follows:

$$P(X \leq t) = P\big((X_1 \leq t) \cap (X_2 \leq t) \cap \ldots \cap (X_n \leq t)\big), \tag{10.55}$$

If the components fail independently of each other, then

$$R(t) = 1 - F(t) = 1 - P(X \leq t) = 1 - \prod_{i=1}^{n} P(X_i \leq t) = 1 - \prod_{i=1}^{n} F_i(t) = 1 - \prod_{i=1}^{n} (1 - R_i(t)), \tag{10.56}$$

where F and F_i are the cumulative distribution until the system's first failure and component i, respectively, in this case, including a new component in the system will increase the system reliability since the second term is multiplied by $1 - R_{n+1}(t)$, which is less than 1. In practical cases, commonly, a mixture of series and parallel connections between systems components is used. In these cases, Equations (10.53) and (10.55) should be combined appropriately as shown in the following example.

Example 10.9

The system illustrated in Figure 10.9 includes five components, where components 2, 3, and 5 are parallel. Calculate the system reliability at $t = 0.1$ assuming that $R_1(t) = R_4(t) = e^{-2t}$ and $R_2^i(t) = R_3^j(t) = R_5^i(t) = e^{-t}\ (1 \leq i\ 2,\ 1 \leq j \leq 3)$.

Solution:

At first, the reliability of parallel components is calculated using Equation (10.54)

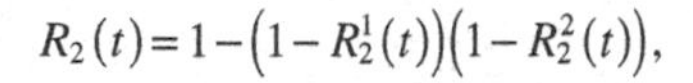

$$R_2(t) = 1 - \big(1 - R_2^1(t)\big)\big(1 - R_2^2(t)\big),$$

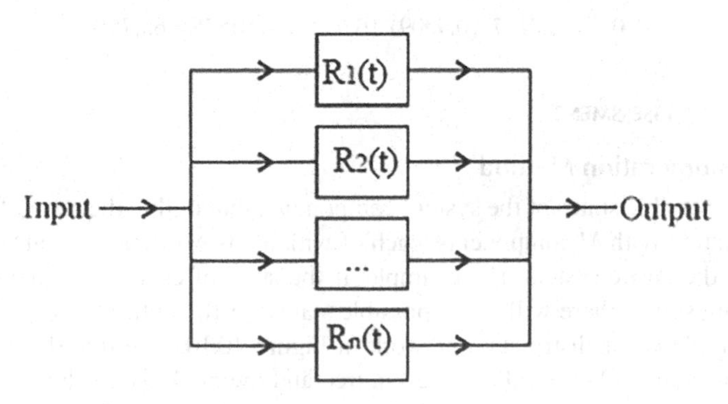

FIGURE 10.8 Parallel combinations of system components.

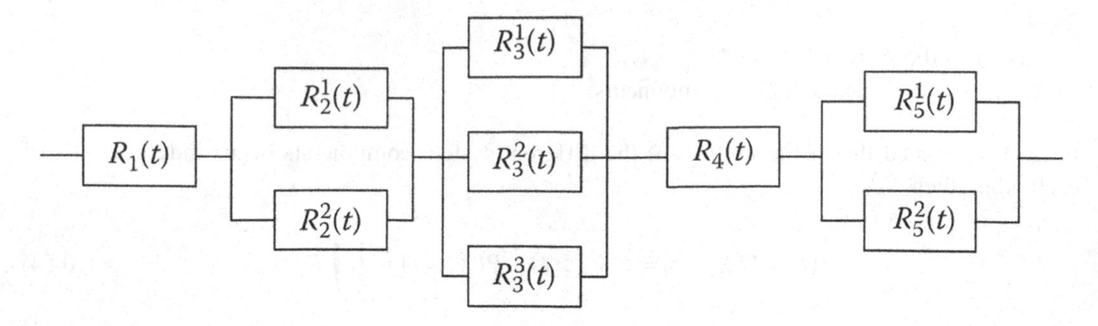

FIGURE 10.9 Combined connections in a system.

$$R_3(t) = 1 - \prod_{i=1}^{3}\left(1 - R_2^i(t)\right),$$

$$R_5(t) = 1 - \left(1 - R_5^1(t)\right)\left(1 - R_5^2(t)\right).$$

The reliability function of the system using Equation (10.56) is estimated as

$$R(t) = R_1(t)1 - \left[1 - \prod_{i=1}^{2}\left(1 - R_2^i(t)\right)\right]\cdot\left[1 - \prod_{i=1}^{3}\left(1 - R_3^i(t)\right)\right]\cdot R_4(t)\cdot\left[1 - \prod_{i=1}^{n}\left(1 - R_5^i(t)\right)\right].$$

At $t = 0.1$,

$$R_1(0.1) = R_4(0.1) = e^{-0.2} = 0.8187,$$

$$R_2^i(0.1) = R_3^j(0.1) = R_5^i(0.1) = e^{-0.1} = 0.9048.$$

Then,

$$R_2(0.1) = R_5(0.1) = 1 - (1 - 0.0952)^2 = 0.9909,$$

$$R_3(0.1) = 1 - 0.0952^3 = 0.9991.$$

Hence,

$$R(0.1) = 0.8187^2(0.9909)^2(0.9991) = 0.6575 = 65.75\%.$$

10.7.1 Reliability Assessment

10.7.1.1 State Enumeration Method

In this method, all possible states of the system components that define the state of the entire system are listed. For a system with M components, each of which has N operating states, there will be N^M possible states for the entire system. For example, if the state of each component is classified into failed and operating states, there will be 2^M possible states for the entire system.

Consider the simple water distribution network in Figure 10.10. The tree diagram for a five-pipe system as shown in Figure 10.11 is called an event tree and the analysis involving the construction of an event tree is referred to as an event-tree analysis. An event tree simulates the topology of a system as well as the sequential or chronological operation of the system, which is highly important (Mays, 2004).

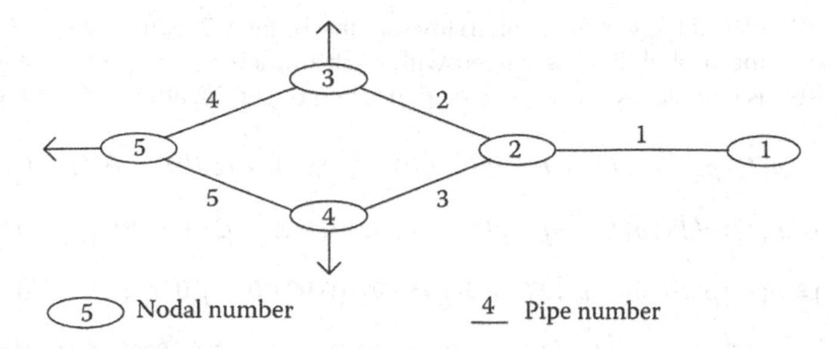

FIGURE 10.10 Example water distribution network.

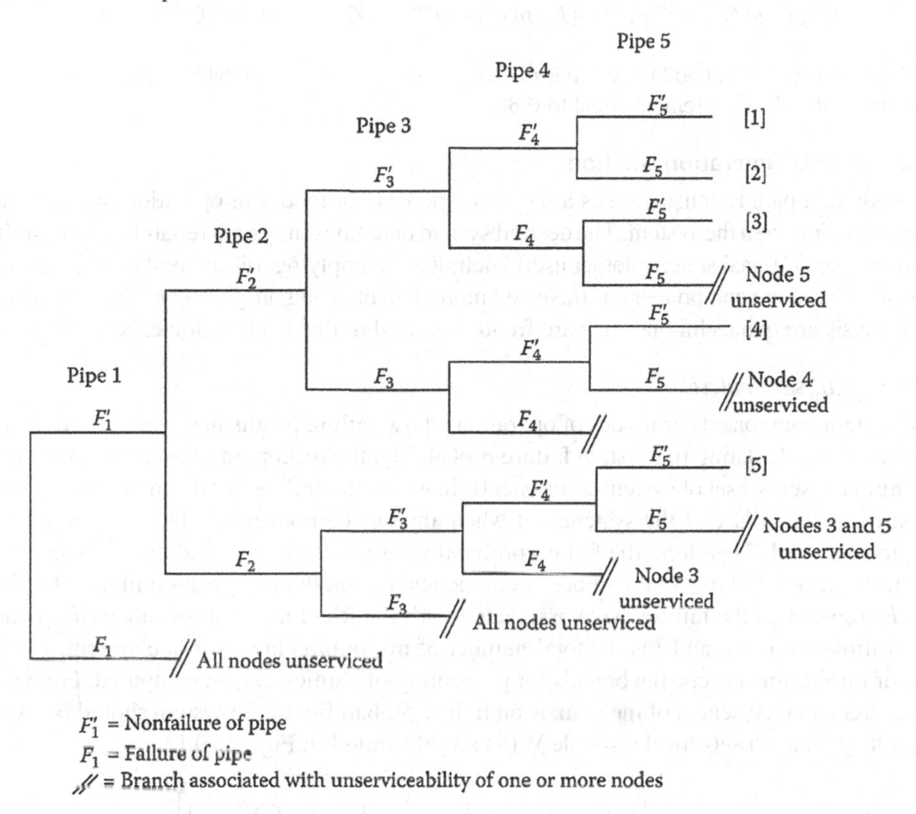

FIGURE 10.11 Event tree for the reliability of example water distribution network.

Example 10.10

Calculate the system reliability of the water distribution network shown in Figure 10.7 considering that node 1 is the source node and nodes 3, 4, and 5 are demand nodes. All of the pipes have the same failure probability equal to 8%. System reliability is defined as the ability of a system to supply water to all of the demand nodes. The pipes' performances are independent of each other.

Solution:

The event tree of the subjected water distribution system (WDS) is shown in Figure 10.11. The complete event tree of this WDS has $2^5 = 32$ branches because of two possible states for each of five pipes. However, the event tree is summarized based on the role of each pipe component in the network connectivity. For example, if pipe 1 fails, all demand nodes regardless of the state of other pipes cannot receive water, and this situation causes a system failure. So, branches in the event tree beyond this point are not shown. As for Figure 10.11, there are only five branches corresponding to system success.

Considering that $p(B_i)$ indicates the probability that the branch B_i of the event tree provides full service to all users, the probability associated with each branch resulting in satisfactory delivery of water to all users is calculated, owing to independence of serviceability of individual pipes, as

$$p(B_1) = p(F_1')p(F_2')p(F_3')p(F_4')p(F_5') = (0.92)(0.92)(0.92)(0.92)(0.92) = 0.659,$$

$$p(B_2) = p(F_1')p(F_2')p(F_3')p(F_4')p(F_5') = (0.92)(0.92)(0.92)(0.92)(0.08) = 0.057,$$

$$p(B_3) = p(F_1')p(F_2')p(F_3')p(F_4')p(F_5') = (0.92)(0.92)(0.92)(0.08)(0.92) = 0.057,$$

$$p(B_4) = p(F_1')p(F_2')p(F_3')p(F_4')p(F_5') = (0.92)(0.92)(0.92)(0.92)(0.08) = 0.057,$$

$$p(B_5) = p(F_1')p(F_2')p(F_3')p(F_4')p(F_5') = (0.92)(0.92)(0.92)(0.92)(0.08) = 0.057,$$

Therefore, the system reliability, which is the sum of all the above probabilities associated with the operating state of the system, is equal to 0.887.

10.7.1.2 Path Enumeration Method

In this method, a path is considered as a set of components or modes of operation, which result in a specified outcome from the system. The desired system outcomes in system reliability analysis include failed state or operational state. Another useful definition in applying this method is "minimum path," which is a path that no component has traversed more than once in going along it. Tie-set analysis and cut-set analysis are two techniques that are frequently used in this methodological category.

10.7.1.2.1 Cut-set analysis

A set of system components or modes of operation whose failure results in system failure is referred as a cut set. For calculating the system failure probability, the minimum cut-set concept is utilized. A minimum cut set is a set of system components, in which the failure of all components at the same time results in the failure of the system, but when any one component of the set does not fail, the system does not fail. Therefore, the failure probability of a system is formulated as where C_i is the h minimum cut set; J_i is the total number of components or modes of operation in the ith minimum cut set; F_{ij} represents the failure event associated with the jth component or mode of operation in the ith minimum cut set; and I is the total number of minimum cut sets in the system. For a large number of minimum cut sets, the bounds for probability of a union can be computed. For achieving adequate precision, closeness of the bounds on failure probability of the system should be examined (Mays, 2004). The cut sets for the simple WDS are illustrated in Figure 10.12.

$$pf,sys = p\left[\bigcup_{i=1}^{I} C_i\right] = p\left[\bigcup_{i=1}^{I} C_i\right] = p\left[\left(\bigcup_{i=1}^{I}\bigcap_{j=1}^{J_i} F_{ij}\right)\right], \quad (10.57)$$

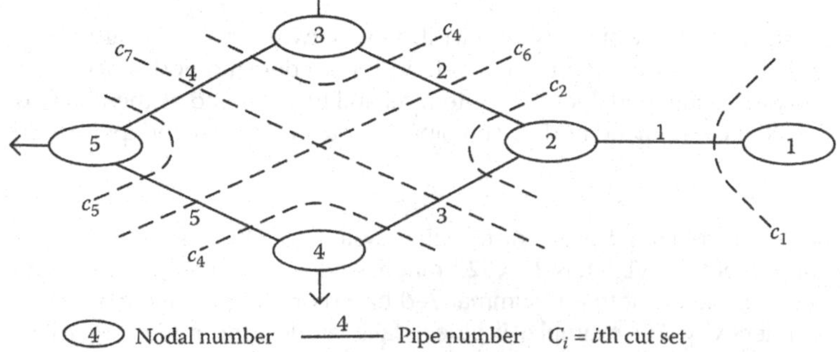

FIGURE 10.12 Minimum cut sets for the water distribution network of Figure 10.7.

Example 10.11

Evaluate the reliability of the water distribution network of Example 10.10 using the minimum cut-set method.

Solution:

Based on the definition of the minimum cut set, seven minimum cut sets are characterized in the example water distribution network as follows (Figure 10.12):

$$C_1: F_1 \quad C_2: F_2 \cap F_3 \quad C_3: F_2 \cap F_4 \quad C_4: F_3 \cap F_5 \quad C_5: F_4 \cap F_5 \quad C_6: F_2 \cap F_5 \quad C_7: F_4 \cap F_3,$$

where
F_k is the failure state of pipe k.

The system failure probability, $pf,_{sys}$, is equal to the probability of occurrence of the union of the cut sets and the system reliability, $p_{s,\,sys}$, is obtained by subtracting it from 1 as follows.

$$p_{s,sys} = 1 - p\left[\bigcup_{i=1}^{7} C_i\right].$$

Based on the probability theory, the above equation is reformed as

$$p_{s,sys} = 1 - p\left[\bigcup_{i=1}^{7} C_i\right] = p\left[\bigcup_{i=1}^{7} C_i'\right].$$

Since all cut sets behave independently, all their complements also behave independently. The probability of the intersection of a number of independent events is

$$p_{s,sys} = p\left[\bigcup_{i=1}^{7} C_i'\right] = \prod_{i=1}^{7} p(C_i')$$

and based on the definition of minimum cut sets, the probability of each of them is calculated as

$$p(C_1') = 0.92, \quad p(C_1') = p(C_3') = \ldots = p(C_7') = 0.92 \times 0.92 = 0.85.$$

Thus, the total system reliability of the example water distribution network is 0.34.

10.7.1.2.2 Tie-set analysis

A tie set is a minimal path of the system in which system components or modes of operation are arranged in series. Therefore, failure of any component or mode in a tie set results in tie set failure. Since the tie sets are connected in parallel, the system serviceability continues if any of its tie sets act successfully. In this method, the system reliability is formulated as

$$p_{s,sys} = p\left[\bigcup_{i=1}^{I} T_i\right] = p\left[\bigcup_{i=1}^{I}\left(\bigcap_{j=1}^{J_i} F_{ij}'\right)\right] \tag{10.58}$$

where
T_i is the ith tie set;
F_{ij}' refers to the nonfailure state of the jth component in the ith tie set;
J_i is the number of components or modes of operation in the ith tie set; and
I is the number of tie sets in the system.

In the cases with a large number of tie sets, bounds for system reliability could be computed.

The main shortcoming of this method is that it is not directly related to failure states. The minimum tie sets of WDS are depicted in Figure 10.13.

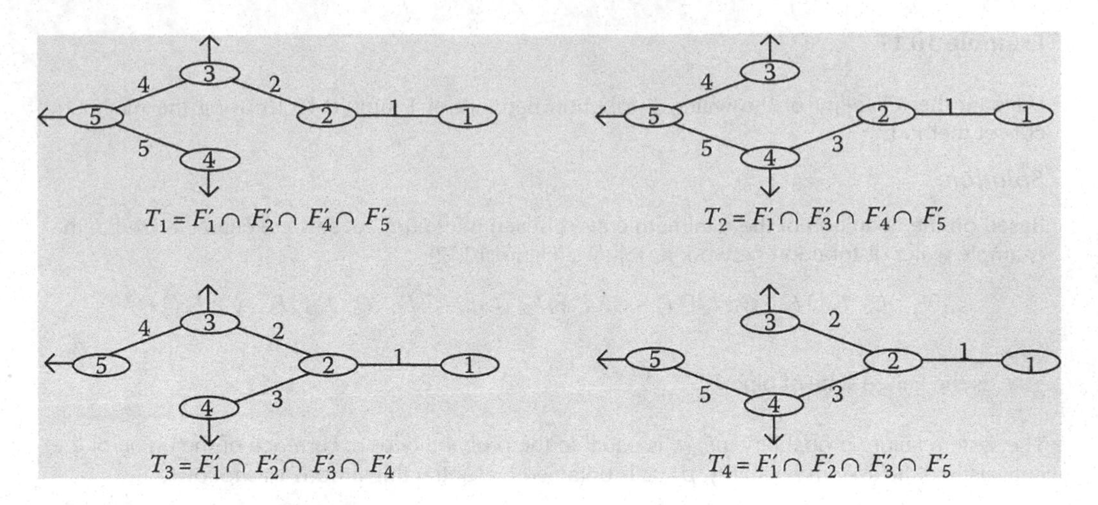

FIGURE 10.13 Minimum tie sets of the illustrated water distribution network.

Example 10.12

Calculate the reliability of the water distribution network of Example 10.10 using the tie-set method.

Solution:

The four distinguished minimum tie sets of the system can be expressed as

$$T_1 : F_1' \cap F_2' \cap F_4' \cap F_5';$$

$$T_2 : F_1' \cap F_3' \cap F_4' \cap F_5';$$

$$T_3 : F_1' \cap F_2' \cap F_3' \cap F_4';$$

$$T_4 : F_1' \cap F_2' \cap F_3' \cap F_5'.$$

The system reliability is equal to the union of these paths and is formulated as follows based on the probability theory:

$$p_{s,sys} = p(T_1 \cup T_2 \cup T_3 \cup T_4)$$

$$= [p(T_1) + p(T_2) + p(T_3) + p(T_4)]$$

$$-[p(T_1,T_2) + p(T_1,T_3) + p(T_1,T_4) + p(T_2,T_3) + p(T_2,T_4) + p(T_3,T_4)]$$

$$+[p(T_1,T_2,T_3) + p(T_1,T_2,T_4) + p(T_1,T_3,T_4) + p(T_2,T_3,T_4)]$$

$$-p(T_1,T_2,T_3,T_4).$$

Because of independency of pipes, the joint probability of tie-set occurrence is equal to the product of the probabilities of individual events. This results in

$$p(T_1) = p(F_1')p(F_2')p(F_4')p(F_5') = (0.92)^4 = 0.7164,$$

$$P(T_2) = P(T_3) = P(T_4) = 0.7164.$$

It should be noted that in this example, the union of two and more tie sets is equal to the intersections of the nonfailure state of all five pipes. As an example,

$$T_1 \cup T_2 = \left(F_1' \cap F_2' \cap F_4' \cap F_5'\right) \cup \left(F_1' \cap F_3' \cap F_4' \cap F_5'\right) = \left(F_1' \cap F_2' \cap F_3' \cap F_4' \cap F_5'\right).$$

Therefore, the system reliability could be summarized as

$$p_{s,sys} = \left[p(T_1) + p(T_2) + p(T_3) + p(T_4)\right] - 6p\left(F_1' \cap F_2' \cap F_3' \cap F_4' \cap F_5'\right)$$

$$+4p\left(F_1' \cap F_2' \cap F_3' \cap F_4' \cap F_5'\right) - p\left(F_1' \cap F_2' \cap F_3' \cap F_4' \cap F_5'\right)$$

$$= 4(0.7164) - 3(0.92)^5 = 0.8884.$$

In summary, the path enumeration method involves the following steps (Henley and Gandhi, 1975):

1. Distinguish all minimum paths.
2. Calculate the required unions of the minimum paths.
3. Give each path union a reliability expression in terms of module reliability.
4. Compute the total system reliability based on module reliabilities.

10.7.2 Reliability Analysis—Load-Resistance Concept

External loads combined with uncertainties in analysis, design, build, and operation lead to system failure. A system fails when the external loads, L (e.g., generated runoff over the watershed), exceed the system resistance, R_e (e.g., carrying capacity of the river). Reliability (*R*) of any component or the entire system is equal to its safety probability:

$$R = p\left[L \le R_e\right]. \tag{10.59}$$

In hydrology and hydraulics, load and resistance are functions of some random variables:

$$R_e = h\left(X_{R_e}\right) \tag{10.60}$$

$$L = g(X_L). \tag{10.61}$$

Thus, reliability is a function of random variables:

$$R = p\left[g(X_L) \le h\left(X_{R_e}\right)\right]. \tag{10.62}$$

Reliability variations with time are not considered in the above equation. X_{Re} and X_L are stationary random variables, and the resulting model is called the static reliability model. The word "static," from the reliability computation point of view, represents the worst single stress, or load, applied. The loading applied to many hydrologic systems is a random variable. Also, the number of times a loading is imposed is random. The static reliability model is used to evaluate system performance in a particular situation (the most critical loading).

For reliability analysis, a performance function is proposed by Mays (2001) as follows:

$$W(X) = W\left(X_L, X_{R_e}\right). \tag{10.63}$$

The system reliability is defined based on this function as follows:

$$R = p\left[W\left(X_L X_{R_e}\right) \ge 0\right] = p\left[W(X) \ge 0\right]. \tag{10.64}$$

$W(X) = 0$ is called the failure surface or limit. $W(X) > 0$ is the safety region and $W(X) < 0$ is the failure region. The performance function $W(X)$ can be defined in the following forms:

$$W_1(X) = R_e - L = h(X_{R_e}) - g(X_L) \tag{10.65}$$

$$W_2(X) = \left(\frac{R_e}{L}\right) - 1 = \left[h(X_{R_e})/g(X_L)\right] - 1 \tag{10.66}$$

Another reliability index is the β reliability index, which is equal to the inverse of the coefficient of variation of the performance function, $W(X)$, and is calculated as follows:

$$\beta = \frac{\mu_w}{\sigma_w}. \tag{10.67}$$

σ_κ and μ_κ are mean and standard deviations of the performance function, respectively. By assuming an appropriate PDF for random performance functions, $W(X)$, the reliability is estimated as

$$R = 1 - F_W(0) = 1 - F_{W'}(-\beta), \tag{10.68}$$

where

F_W is the CDF of variable W and

W' is the standardized variable as follows:

$$W' = \frac{W - \mu_W}{\sigma_W}. \tag{10.69}$$

10.7.3 Direct Integration Method

Following the reliability definition, the reliability is expressed by Mays (2001) as follows:

$$R = \int_0^\infty fR_e(r)\left[\int_0^r f_L(l)dl\right]dr = \int_0^\infty f_{R_e}(r)F_L(l)dr, \tag{10.70}$$

where

$f_R()$ and $f_L()$ are the PDFs of resistance and load functions, respectively.

If load and resistance are independent, then reliability is formulated as

$$R = E_{R_e}\left[F_L(r)\right] = 1 - E_L\left[F_{R_e}(l)\right], \tag{10.71}$$

$E_r[F_L(r)]$ is the expected value of CDF of load in the probable resistance limitation. The reliability computations require the knowledge of the probability distributions of loading and resistance. A schematic diagram of the reliability analysis by Equation (10.71) is shown in Figure 10.14.

To illustrate the computation procedure involved, the exponential distribution has been considered for loading L and the resistance R_e:

$$f_L(l) = \lambda_L e^{-\lambda_L l}, \quad l \geq 0 \tag{10.72}$$

$$f_{R_e}(r) = \lambda_{R_e} e^{-\lambda_{R_e} r}, \quad r \geq 0 \tag{10.73}$$

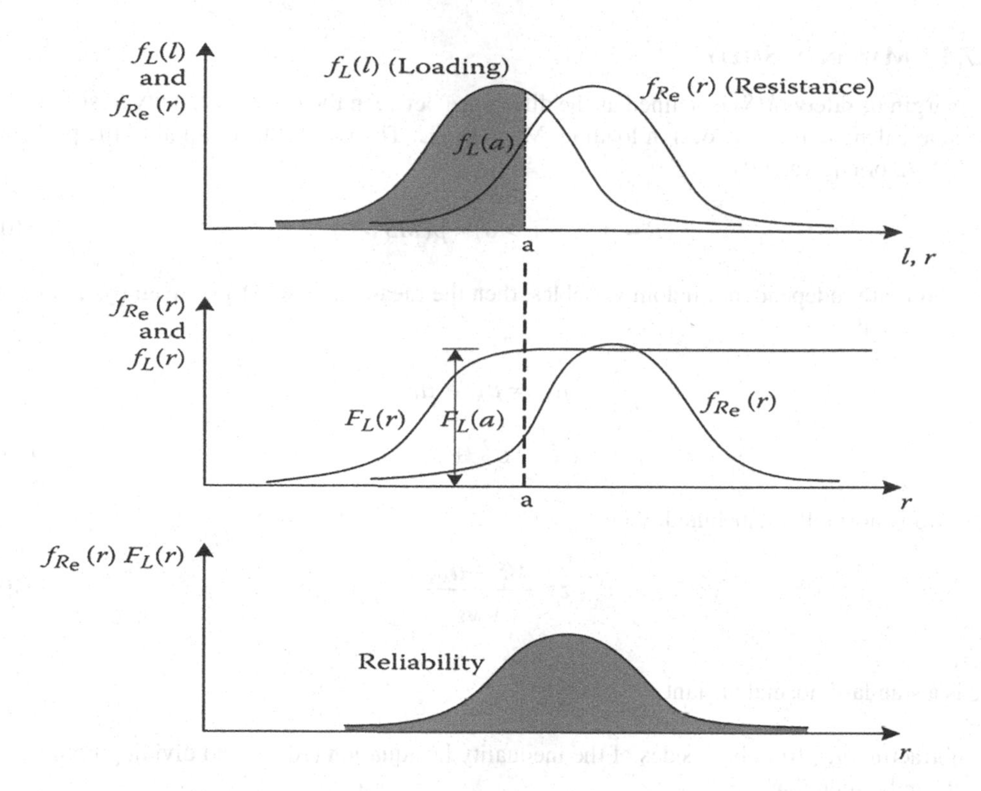

FIGURE 10.14 Graphical illustration of the steps involved in reliability computation (Mays, L.W. *Water Resources Engineering*. John Wiley and Sons Inc., New York, (2001)).

The static reliability can then be derived as

$$R = \int_0^\infty \lambda_{R_e} e^{-\lambda_{R_e} r} \left[\int_0^r \lambda_L e^{-\lambda_L l} dl \right] dr = \int_0^\infty \lambda_{R_e} e^{-\lambda_{R_e} r} \left[1 - e^{-\lambda_L r} \right] dr = \frac{\lambda_L}{\lambda_{R_e} + \lambda_L}. \tag{10.74}$$

For some particular combinations of load and resistance distributions, the static reliability can be derived analytically in closed form. Kapur and Lamberson (1977) considered the loading L and resistance R_e lognormally distributed and computed reliability as

$$R = \int_0^\infty \phi(z) dz = \Phi(z), \tag{10.75}$$

where

$\phi(z)$ and $\Phi(z)$ are the PDF and the CDF, respectively, for the standard normal variant z:

$$z = \frac{\mu_{\ln R_e} - \mu_{\ln L}}{\sqrt{\sigma_{\ln R_e}^2 - \sigma_{\ln L}^2}}. \tag{10.76}$$

The CDF $\Phi(z)$ values for the standard normal variant are given in normal distribution tables.

By considering exponential distribution for the load and the normal distribution for the resistance, the reliability was expressed by Kapur and Lamberson (1977) as

$$R = 1 - \Phi\left(\frac{\mu_{R_e}}{\sigma_{R_e}} \right) - \exp\left[-\frac{1}{2} \left(2\mu_{R_e} \lambda_L - \lambda_L^2 \sigma_{R_e}^2 \right) \right] \times \left[1 - \Phi\left(-\frac{\mu_{R_e} - \lambda_L \sigma_{R_e}^2}{\sigma_{R_e}} \right) \right]. \tag{10.77}$$

10.7.4 Margin of Safety

The margin of safety (*MS*) is defined as the difference between the project capacity (resistance) and the value calculated for the design loading, MS = R_e−L. The reliability is equal to the probability that $R_e > L$, or equivalently,

$$R = p(R_e - L > 0) = p(MS > 0). \tag{10.78}$$

If R_e and L are independent random variables, then the mean value of MS is given by, and its variance is given by

$$\mu_{MS} = \mu_{R_e} - \mu_L \tag{10.79}$$

$$\sigma_{MS}^2 = \sigma_{R_e}^2 + \sigma_L^2. \tag{10.80}$$

If the *MS* is normally distributed, then

$$z = \frac{MS - \mu_{MS}}{\sigma_{MS}}, \tag{10.81}$$

where

z is a standard normal variant.

By subtracting μ_{MS} from both sides of the inequality in Equation (10.78) and dividing both sides by σ_{MS}, it can be seen that

$$R = p\left(z \le \frac{\mu_{MS}}{\sigma_{MS}}\right) = \Phi\left(\frac{\mu_{MS}}{\sigma_{MS}}\right). \tag{10.82}$$

The key assumption of this analysis is that it considers that the *MS* is normally distributed but does not specify the distributions of loading and capacity. Ang (1973) indicated that, provided $R > 0.001$, R is not greatly influenced by the choice of distribution for R_e and L and the assumption of a normal distribution for MS is satisfactory. For lower risk than this (e.g., R = 0.00001), the shape of the tails of the distributions for R_e and L becomes critical, in which case accurate assessment of the distribution of *MS* of direct integration procedure should be used to evaluate the risk or probability of failure.

10.7.5 Factor of Safety

The factor of safety *(FS)* is given by the ratio R_e/L, and the reliability can be specified by $P\,(FS > 1)$. Several factors of safety measures and their usefulness in hydraulic engineering are discussed by Yen (1978) and Mays (2001). By taking the logarithm of both sides of this inequality, it yields to

$$R = p(FS > 1) = p(\ln(FS) > 0) = P\left(\ln\left(\left(\frac{R_e}{L}\right) > 0\right)\right) \tag{10.83}$$

$$R = p\left(z \le \frac{\mu_{\ln FS}}{\sigma_{\ln FS}}\right) = \Phi\left(\frac{\mu_{\ln FS}}{\sigma_{\ln FS}}\right), \tag{10.84}$$

where $z = \dfrac{\ln(FS) - \mu_{\ln FS}}{\sigma_{\ln FS}}$; moreover, Φ is the standard cumulative normal probability distribution. It is assumed that the *FS* is normally distributed.

10.8 WATER SUPPLY RELIABILITY INDICATORS AND METRICS

To use the reliability concept in design, it must be quantified based on several indicators. No single indicator can adequately measure and communicate all the critical dimensions of reliability. Multiple indicators are needed, many of which will be strongly interrelated due to the water supply system's network characteristics. Some indicators will be more useful for monitoring supply reliability for specific service types (e.g., residential/municipal versus environmental) or for different functions (e.g., real-time operations versus policy development). The following paragraphs discuss a variety of features that could distinguish reliability indicators.

Some indicators will require greater breadth, reflecting the effects of many factors. Simultaneously, some will be more narrowly focused on isolating a particular factor or reflecting a specific system scale. Operation functions are mainly concerned with having indicators that measure reliability in terms of the factors under their control and broader indicators dominated by actual inflow levels. Some of the dimensions of breadth or focus are as follows:

- *Type of service*: Clearer definitions of the services currently provided to different end users (e.g., consumptive, environmental, and inflow) would facilitate the development of indicators and metrics. In other industries, service is often described by specifying requirements related to the purpose of use, quantity and temporal/geographical distribution of deliveries, and quality and delivery curtailment provisions.
- *Temporal scale and distribution*: For example, hourly, monthly, or annual.
- *Spatial (or geographical) scale and distribution*: The scale refers to, for example, storage data that are aggregated to the combined project level or are disaggregated at the individual reservoir level.
- *Climate normalization*: For the available supply indicators, both metrics that are normalized for climatic variability and those that are not will be valuable. When normalized, metrics will reflect how efficiently the existing storage and transportation system is being used. When not normalized, the system's reliability, including the biggest failure in supply availability or inflow, is tracked.
- *Condition and outcome indicators/metrics*: Factors that affect reliability can be translated into indicators of reliability, for which metrics can be crafted. Indicators are statements on what is expected to change if the program shows progress towards the objective or if problems begin to arise. Indicators may identify outcomes or conditions and features that are believed to influence outcomes. Indicators of the condition can provide insight into the mechanics of reliability and may be components of outcome indicators.
- *Physical and financial indicators/metrics*: Reliability indicators that reflect physical phenomena such as the supply delivered or allocated need to be paired with financial indicators that describe the costs associated with reliability as the level of those physically based indicators change. It is essential to monitor cost to develop policy positions on the desired level of reliability and answer questions about the cost of different possible reliability levels.
- *Forecast versus actual*: Some outcome indicators could compare actual to forecasted or expected/desired service dimensions. Some metrics might utilize forecast data, while others may reflect what has happened. For example, one measure of reliability at the project level may be the initial monthly allocation to contractors (developed in the first quarter of the water year) relative to the final allocation produced in the third quarter of the water year (summer). For some customers (e.g., agricultural), the closer these values are, the more reliable project supplies are. Furthermore, for indicators/metrics that embody forecast values, the dependency of the outcome on the forecast's nature must be carefully considered. For example, for metrics that involve the allocation as a measure of supply, the degree of conservativeness in the runoff forecast will influence the metric reliability values. In other words, the less risky (more conservative) the projects are in the inflow forecasts that are the basis of allocations, the more likely they will be to meet that allocation.

- *Deterministic and probabilistic indicators*: Both deterministic and probabilistic metrics (discrete probabilities or probability distributions) will be useful.
- *Quantitative and qualitative indicators/metrics*: Some metrics will be quantitative while others will be qualitative (e.g., those related to institutional failures).

10.8.1 Risk Analysis Methods and Tools

Risk is a characteristic of a situation in which several outcomes are possible; the particular one that will occur is uncertain. At least one of the possibilities is undesirable. Risk analysis is for extreme or low probability events (e.g., floods and hurricanes) and any situation in which there is a range of possibilities, such as natural events that have profound economic, environmental, and safety implications. Statistical measures can provide important insight into the variability associated with these processes and their consequences. Incorporating factors describing uncertainty and variability into the decision-making process comes under the general term of risk analysis. Risk analysis provides an important decision-making tool.

An event tree (decision tree) is a tool that provides a structure for risk quantification. The necessary steps in decision tree analysis include the following:

- Specification of the decision context
- Development of a decision model (decision tree), including
- Management options
- The consequences of each option
- Likelihood and desirability of each outcome

The decision tree structure develops sequentially from base to terminal ends based on the components of nodes, branches, and outcomes. There are three types of nodes, including decision (choice) nodes, chance (probability) nodes, and terminal nodes. For example, in decision-making on flood planning, decision nodes can apply a specific flood management practice; chance nodes are the corresponding probability of flooding and not flooding before and after the application of flood management practices. The terminal nodes will indicate corresponding flood damage in different situations.

The branches indicate different choices if they are extended from a decision node with different outcomes. Each of the branches emanating from a chance node has associated probabilities. Each of the terminal ends had associated utilities or values. The branches representing choices must be mutually exclusive, meaning that only one branch can be chosen at a time. In the case of chance nodes, the branches have to correspond to mutually exclusive and collectively exhaustive outcomes. This means only one of them can happen, and the different branches include all possible outcomes. Chance events placed in front of decision nodes indicate the chance of that decision to happen.

When consequences are assigned to the ultimate nodes of the tree, the tree can be rolled back to determine, using the rules of probability and decision rules at decision nodes, the best choice (i.e., the decision that maximizes or minimizes the desired value) based on the various event probabilities. More sophisticated decision tree software allows for the more complex specification of probability distributions. Through the use of Monte Carlo simulation techniques, it can display probability distributions for desired decision variables.

10.8.2 Event Tree of Risk Assessment

An event tree is a tool that provides a structure for risk quantification. Event trees provide a structured approach for risk quantification. In this figure, assume that the initiating event is the placement of hazardous materials at a site with two liners. Systems 1 and 2 represent the primary (top) liner and the secondary (bottom) liner, respectively. If System 1 fails and System 2 works, there is no damage. But, if both systems fail, there is the possible release of the hazardous materials to the subsurface environment. Figure 10.15 shows how the probabilities of each branch of the event tree can be calculated.

Figure 10.16 shows an actual event tree provided by Lee (2011) for a risk assessment of liquefied natural gas (LNG) spills in a coastal harbor as a typical treat to water and soil resources. Decision trees can also be developed as a part of a risk quantification process.

In risk evaluation, socially acceptable risks are determined. One approach for risk evaluation is to compare the quantified risk with various possible existing risks in everyday life. This evaluation approach is called the "revealed societal preferences" approach. Risk evaluation is a controversial and challenging task in risk assessment.

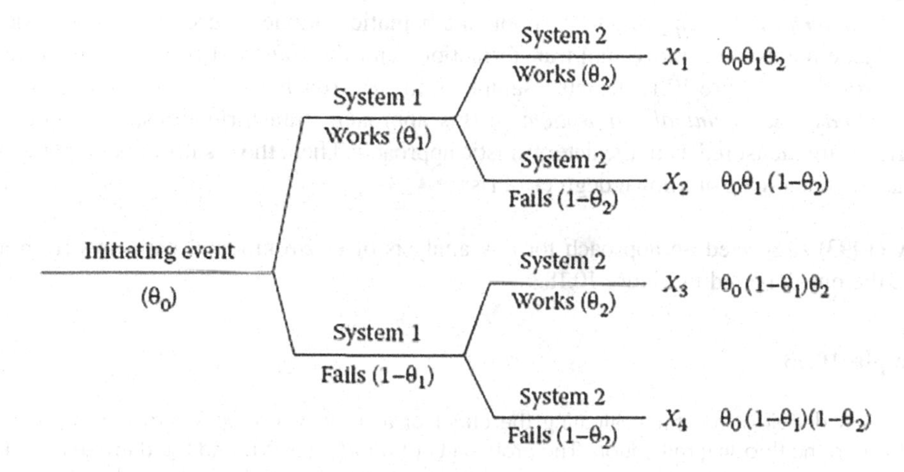

FIGURE 10.15 Risk quantification using an event tree. (From Lee, W., *J. Portfolio Manage.*, 37(4), 11–28, 2011.)

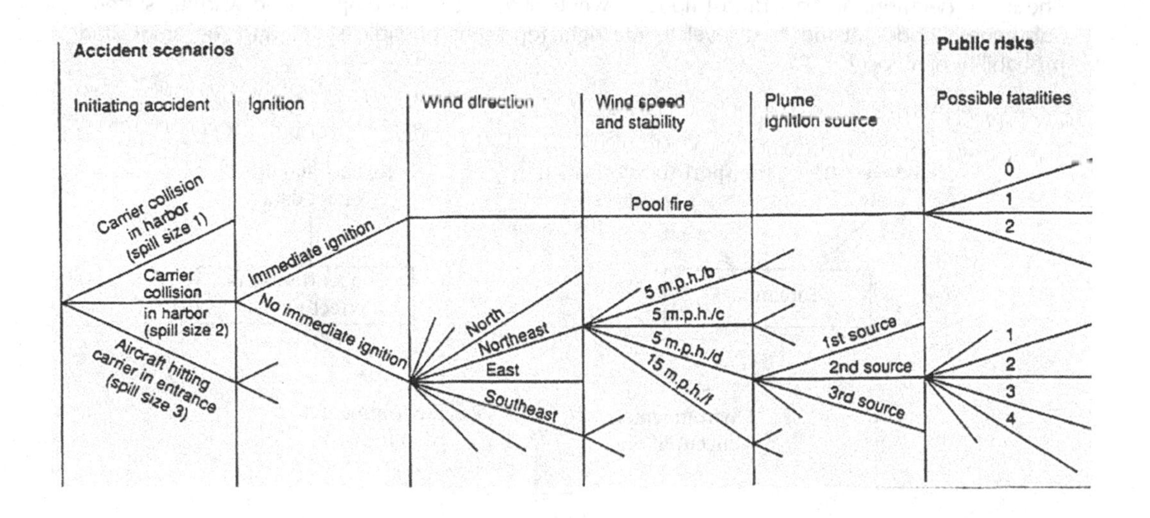

FIGURE 10.16 Event tree analysis of LNG spills as a threat *t*. (From Lee, W., *J. Portfolio Manage.*, 37(4), 11–28, 2011.)

10.8.3 Environmental Risk Analysis

There are four significant approaches suggested by Conway (1993) for environmental risk analysis. These approaches are

1. *The stochastic/statistical approach*: In this approach, large amounts of data are obtained under various conditions. Then, the correlations between the input of a particular material and its observed concentrations and effects are determined in various environmental compartments.
2. *The model ecosystem approach*: In this approach, which is also called the "microcosm" approach, a physical model of a given environmental situation is constructed. A chemical is then applied to the model. The fate and effects of the chemical are observed through the model.
3. *The "deterministic" approach*: A simple mathematical model is used in this approach to describe the rates of individual transformations and the transport of the chemical in the environment. Figure 10.17 depicts a sample of this approach.
4. *The "baseline chemical" approach*: In this approach, transformations, transports, and effects are measured as in the deterministic approach. Then, the results are compared with data on chemicals of known degrees of risk.

Conway (1993) suggested an approach for risk analysis of environmental pathways from a waste site, like the one depicted in Figure 10.18.

Example 10.13

Develop a decision tree for evaluating the effect of a flood warning system development for $20,000 on the flooding mitigation. The probability of flooding is 0.1, and the damage associated with flooding in the absence of the warning system is $50,000. However, it is reduced to $5,000 with the flood warning system.

Solution:

The initial decision, at the leftmost node, is whether or not to develop a flood warning system. Subsequent nodes at the next level to the right represent possible states and the associated probability of a flood.

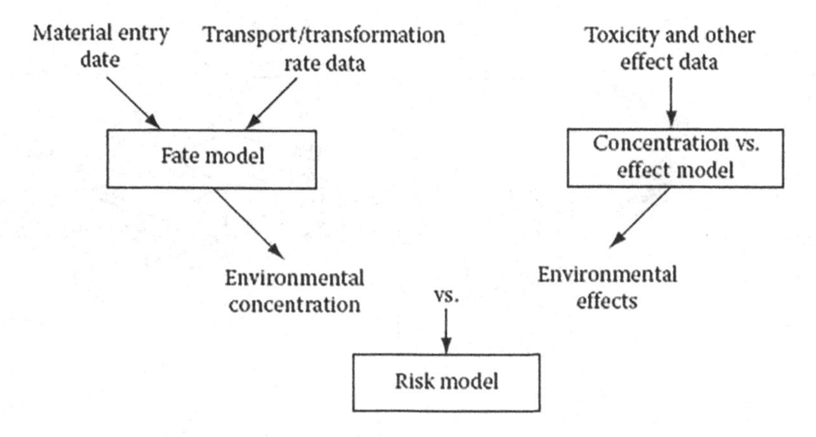

FIGURE 10.17 Sample of the deterministic approach to environmental risk analysis. (From Conway, R.A. ed., *Environmental Risk Analysis for Chemicals*. Van Nostrand Reinhold, New York, 1993.)

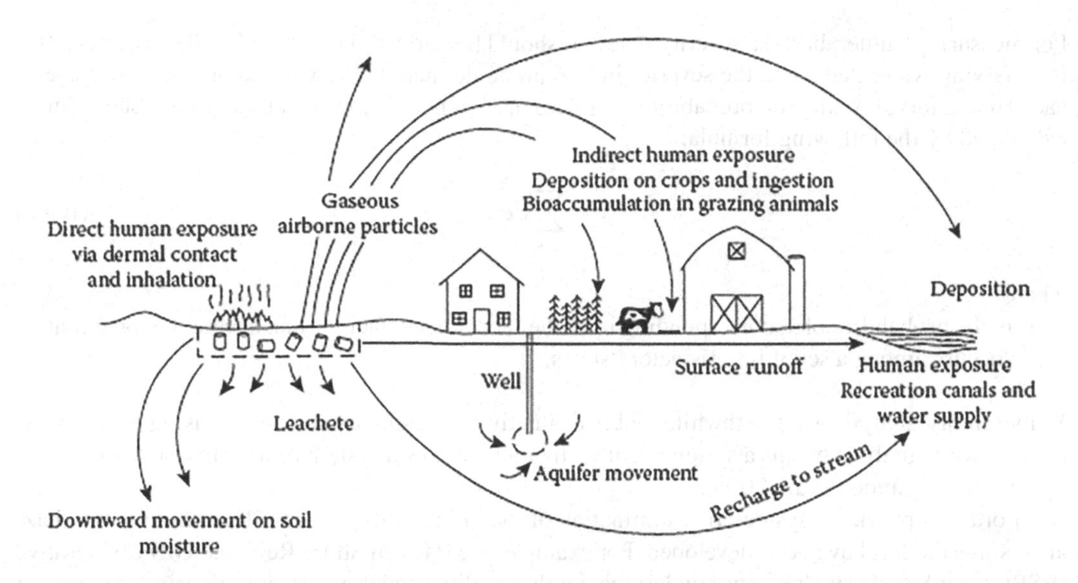

FIGURE 10.18 Environmental pathways from a waste site. (From Schweitzer, G.E., *Borrowed Earth, Borrowed Time: Healing America's Chemical Wounds.* Springer Nature, New York, 2013.)

If the flood warning system is developed, for the case of a flood happening, the expected cost would be 20,000 (for flood warning development) plus 5000×0.1 (probability of this situation), which is equal to 20,500. If the flood warning system is not developed, the expected cost would be 50,000×0.1 = 5,000. Since the maximum cost is less in the second case, the decision tree analysis suggests that the preferred choice is not to build the flood warning system. While this simple decision tree is based on expected value analysis, more complex decision tree technologies provide more complete risk profiles.

10.9 VULNERABILITY

The most critical risk factor is vulnerability, which determines whether or not exposure to a hazard constitutes a risk that may result in a disaster. There are three types of vulnerability: physical vulnerability of people and infrastructure, unfavorable economic and organizational conditions, and motivations and attitudes (Associated Programme on Flood Management [APFM], 2008).

10.9.1 Vulnerability Estimation

The vulnerability can be defined as a failed character state's distance from the historical time series' desired condition. There are three ways for vulnerability estimation. Some researchers consider the maximum distance from the desired situation. Other groups consider the expected value of system failure cases, and the last group considers the probability of failure. Among these different ideas, the second idea is the most common.

Similarly, the techniques used to measure vulnerability have also varied according to the objective and the discipline trying to assess it. The Third Assessment Report of the Intergovernmental Panel on Climate Change (IPCC) defines vulnerability due to climate change as the extent to which a natural or social system is susceptible to sustaining climate change damage. It is defined as a function of the system's sensitivity to change in climate (hazard), its adaptive capacity, and the degree of exposure of the system to climatic hazards (McCarthy et al., 2001). Quantitatively characterizing vulnerability will allow us to assess security risks. The vulnerability can also be considered as a function of hazard, exposure, and adaptive capacity.

$$\text{vulnerability} = f\left(\text{hazard, exposure, adaptive capacity}\right). \tag{10.85}$$

For measuring vulnerability, a severity index, s_j, should be defined. For example, when the objective is satisfying water demands, the severity index can be defined as the volume of water shortage in each time interval. Using the probability of failure occurrence, the vulnerability of a system can be calculated by the following formula:

$$V = \sum_{j \in F} s_j e_j, \tag{10.86}$$

where

e_j is the probability of x_j corresponding to s_j, the most unsatisfactory and severe outcome that occurs among a set of unsatisfactory states.

Vulnerability analysis is a worthwhile tool in evaluating the entire event chain causing water disaster. Recovery analysis means a systematic investigation on how a system returns to a state of normal operation (Karamouz et al., 2003).

In order to provide a systematic examination of the vulnerability of a built environment to hazards, some models have been developed. For example, the Eastern Shore Regional GIS Cooperative (ESRGC) at Salisbury University undertook a vulnerability modeling effort to riverine and coastal flooding. Using the HAZUS-MH (Hazards US Multi-Hazard) vulnerability analysis modeling software (which is described in the next section) of the Federal Emergency Management Agency (FEMA), the ESRGC sought to generate maps and tables of Maryland's potential for loss related to buildings from flooding on a county-by-county basis. This potential for loss, or the degree of vulnerability, was measured using four different factors: the amount of county land area susceptible to a 100-year flood, the amount of square footage of buildings potentially damaged, the number of buildings potentially damaged, and the amount of direct economic losses related to buildings. These four-loss measures help give a complete picture of the very complex issue of vulnerability to floods.

To perform a vulnerability assessment of hydrologic systems, it is essential to review indices and indicators of vulnerability developed by previous authors. Some of these indices have been developed as indicators of general human welfare, economic well-being, or development status, while others specifically address the vulnerability. A framework for vulnerability assessment should include indicators that could cover all components of a water supply system. The vulnerability process is divided into five main stages (Mohamed et al., 2009):

Stage 1: There are many approaches to indicator development. However, the cause–effect approach is the most widely used (Bossel 1999; Meadows 1998). The Driving force–Pressure–State–Impact–Response (DPSIR) approach developed by Karageorgis et al. (2005) can be used for indicator development and classification. DPSIR provides a cause–effect relationship and can be used as an analytical framework for assessing water issues.

Generally in DPSIR, the "Drivers" term includes all the driving forces acting on regional water resource systems. It is basically comprised of human needs such as water and food. However, increase in human needs due to population or economic growth will cause pressures on the water resources of the area. Thus, "pressures" will include all human activities undertaken due to increase in needs, such as installation of treatment plants or desalination industry. These human activities will apply pressures on the environment and available resources which will lead to changes in "state" of the environment. These changes in state of environmental compartments will cause certain impacts on the environment, human welfare and ecosystem services.

Under the "impacts" term of DPSIR, the humans are considered as the center of the ecosystem and any loss of functions or services to humans categorized under impacts due to marine water pollution will affect the businesses. At the end, the "response" term includes the suggested management options that can help to deal with identified causes and effects of the problem. These responses can be suggested against any part of the chain from drivers to impacts or there can be a wholesome approach.

Thus, the model was used in this research to identify different indicators and to depict the general chain of effects that shows how the drivers (D) (e.g., population growth) are putting pressures (P) on water resources in, which are resulting in changes in state of environment (S) for example, brine discharge. The impacts (I) to human welfare are then identified and several management actions and responses (R) are suggested to achieve water resource sustainability and security.

This approach allows a comprehensive assessment of the issues by examining the relevant driving forces and pressures on the system, their consequent state, and impacts, and the responses are undertaken as shown in Figure 10.19.

Stage 2: The indicators are categorized according to the system components that they correspond to.

Stage 3: A screening process for the prepared list of indicators is carried out regarding the predefined system objectives and the system's spatial and temporal scales. Two general approaches are considered in the procedures for indicator selection. One approach is based on a theoretical understanding of relationships, and the other one is based on statistical relationships. However, conceptual understanding plays a vital role in both approaches (Adger et al. 2004). Avoiding an overlap between the indicators is of primary importance at this stage. Suppose two indicators measure similar or overlapping aspects of the water supply system in the screening process. In that case, only one could be used in the vulnerability assessment.

Stage 4: Indicator evaluation is carried out based on data availability.

Stage 5: The evaluated indicators are presented using a graphical display method for each cluster of indicators.

The presented indicators were used to analyze water supply systems' status and outline possible vulnerability mitigation strategies.

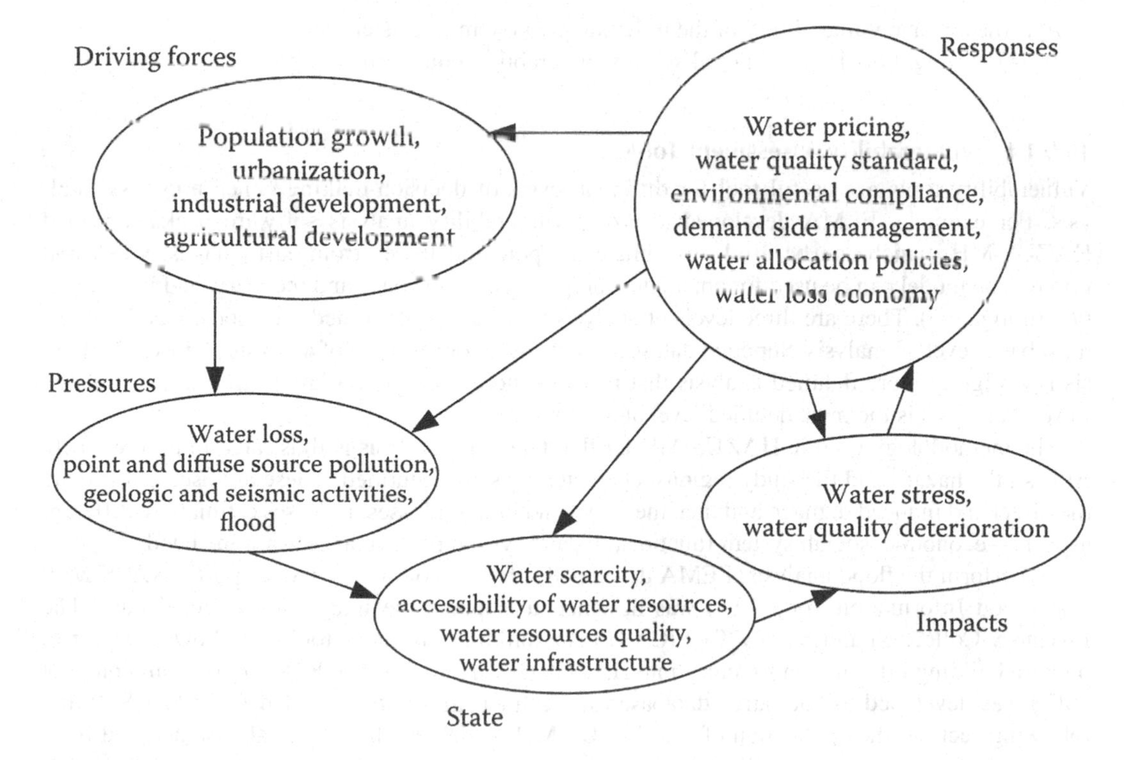

FIGURE 10.19 DPSIR framework to water vulnerability assessment.

Calculations are confined to simple mathematical operations needed to arrive at the required form of the indicator value. There are two significant steps in developing indices: (a) calculating values for vulnerability indicators and (b) the subsequent aggregation of these subindices into an overall index. An aggregation scheme is usually used to summarize all the indicators into a few water system vulnerability indices. On the other hand, aggregation causes a loss of information.

Displaying indicators in diagrams require absolute consistency in the displayed value; it is impossible to display two indicators of different units or different range on the same diagram.

Thus, before displaying and interpreting the indicators' results, it is essential to alter the indicators' values. The details of data manipulation and alteration can be found in Hamouda (2006). These alterations are done to ease the interpretation of the graphical display; the main alterations include the following:

1. Some indicators are reversed. The trend is that of an increase in the indicator value leading to greater vulnerability.
2. All indicators are standardized as the ratio of their respective thresholds.
3. After standardizing the values, a cutoff value (here 4) is used for values that are very high as they can cause graphical distortion.

Furthermore, for aggregation of the results of diagrams, in order for them to be better employed in the decision-making process, the readiness index (RI) is calculated as follows:

$$RI = \frac{W_i A_i}{\sum_{i=1}^{n} W_i}. \tag{10.87}$$

where

A_i is the average vulnerability of the water supply system to cluster i and
W_i is the weight of cluster i in total system vulnerability condition.

10.9.1.1 Vulnerability Assessment Tools

Vulnerability tools are useful tool for different levels of decision-making concerning risk analysis. For example, FEMA developed a hazard vulnerability analysis software package named HAZUS-MH, which can be used to estimate the potential losses from earthquakes, wind, and floods. The model can be used for emergency preparedness, response and recovery, and loss reduction (mitigation). There are three levels of analysis that can be performed for floods. Level 1 is the most basic level of analysis. Supplied datasets can be used for this type of analysis. A Level 2 analysis is a slightly more detailed analysis that requires more accurate building information. Finally, Level 3 analysis is the most detailed level of analysis.

The methodology used in HAZUS-MH for flood damage analysis is illustrated in Figure 10.20. At first, the hazard and the study regions' characteristics are identified. These are used to estimate the direct and induced damage and then the direct and indirect losses. In losses estimation, different aspects—economic, social, system functionality, and system performance—are included.

To perform the flood analysis, FEMA developed software components to support HAZUS-MH. The Flood Information Tool (FIT) was designed to support the integration of local data. The Inventory Collection and Survey Tool (CAST) is a building inventory tool that allows the user to prepare building information for entry into HAZUS (Figure 10.21). The Building Information Tool (BIT) was developed to take large databases and extract information needed for HAZUS. In the following section, the application of the HAZUS-MH model in a hypothetical case adopted from http://www.fema.gov/hazus/ is illustrated.

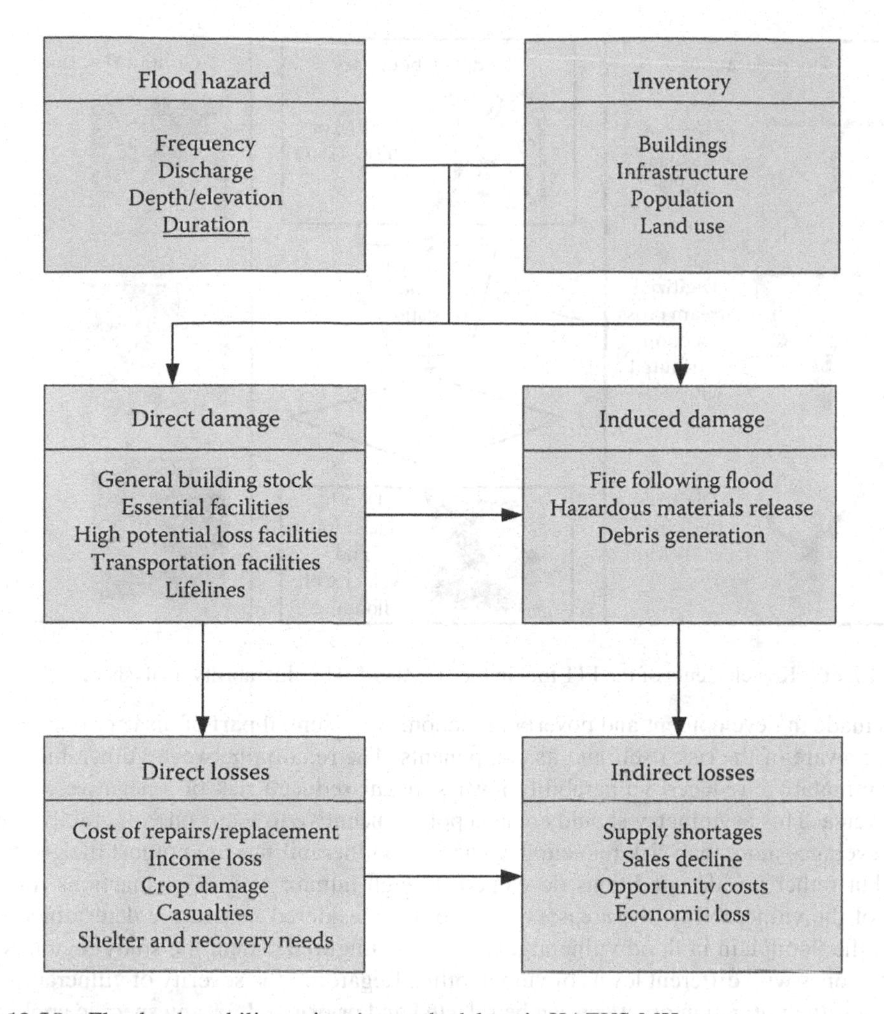

FIGURE 10.20 Flood vulnerability estimation methodology in HAZUS-MH.

10.9.2 Risk Reduction through Reducing Vulnerability

Reducing the vulnerability of communities to hazards is a crucial aspect of risk management. Poverty and lack of resources increase vulnerability, weaken coping strategies, and delay the recovery process. Poor people everywhere, especially in urban areas, are the most at risk. Local authorities in developing countries are usually ill-equipped to provide sufficient infrastructure and services in urban areas. As a result, most of the world's poor live in densely populated squatter settlements on the periphery of cities, which lack the basics of life, leaving many inhabitants caught in a spiral of increasing vulnerability.

Demand for commercial and residential land in cities has led to unsuitable terrain prone to natural hazards. Therefore, many informal settlements are located in dangerous or unsuitable areas, such as floodplains, unstable slopes, or reclaimed land. Moreover, these cities are often unable to manage rapid population growth; poorly planned urbanization and increasing numbers of inadequately constructed and badly maintained buildings further increase the level of vulnerabilities in cities. Ironically, most of today's largest cities are in areas where earthquakes, floods, landslides, and other disasters are likely to happen. Therefore, reducing the risk of societies to hazards by reducing vulnerability is critical in risk management.

Risk reduction is an approach to identifying, assessing, and reducing the risks of disaster. This concept aims to reduce socioeconomic vulnerabilities to disasters and deal with the environmental and other hazards that trigger them. Reducing disaster risks is essential if we are to consolidate the

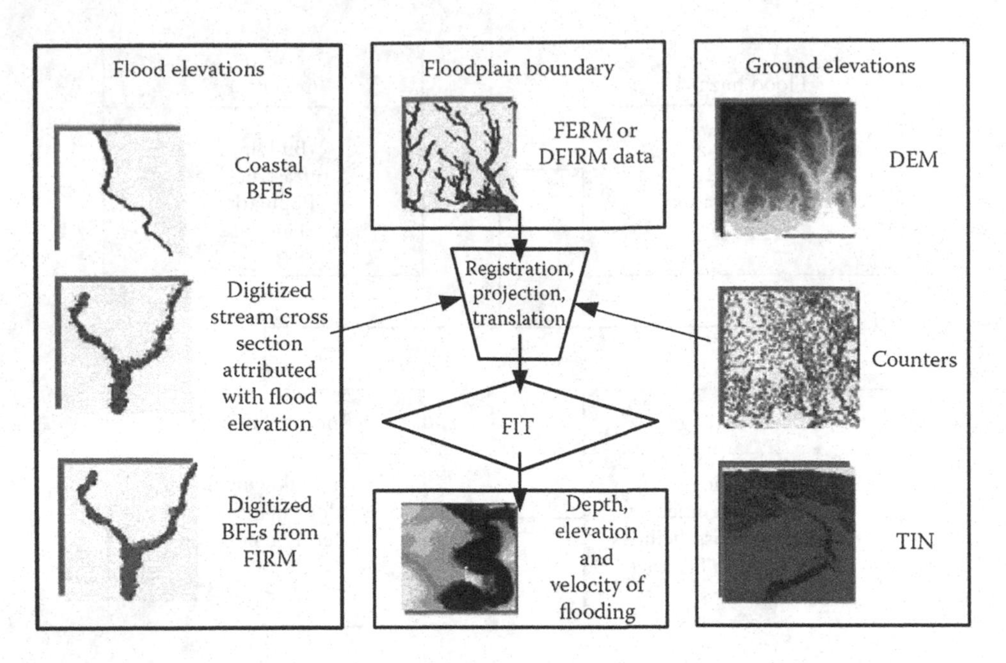

FIGURE 10.21 Key elements of the FIT tool in the HAZUS-MH vulnerability analysis model.

progress made in development and poverty reduction. An essential part of disaster risk reduction is becoming aware of the risk itself and its components. The relation between vulnerability and risk is not commutative: reduced vulnerability always means reduced risk of a negative outcome, but not vice versa. This asymmetry should create a policy incentive to focus on vulnerability reduction since it leverages more than risk reduction. Generally, vulnerabilities are not just the given circumstances but rather unsafe conditions developed through human actions or inactions. In a spatial analysis of the vulnerability, the areas exposed to the considered hazard are determined first. For example, the floodplain in flood vulnerability is first configured. Then, the study region is divided into some zones with different levels of vulnerability. Regarding the severity of vulnerability in different zones, the mitigation practices can be selected and prioritized for any specific application. In flood planning, the floodplain is combined with insurance zones to determine the evacuation zones with different levels of flood hazards. It is further discussed in Chapter 13.

10.10 RESILIENCY

Over the past decades, there has been a shift away from structural and large-scale defense against hazards towards integrated risk management. The modern risk management concept acknowledges that hazards such as floods cannot be stopped from occurring and emphasizes reducing vulnerability, exposure, and sensibility of risk-prone communities. The resilience concept provides a practical framework for this end.

Resiliency describes how quickly a system recovers from failure once a failure has occurred. In other words, resiliency is a measure of the duration of an unsatisfactory condition. It is perhaps the most critical indicator of crisis recovery and how successful the disaster has been managed. By this description, the inverse of average break time is considered as the resiliency index in time t (Res_t), which can be formulated as follows:

$$\text{Res}_t = \frac{k}{\sum_{t=1}^{k} T_t} \times 100, \tag{10.88}$$

where

k is the number of system breaks in the considered time, and

T_t is the duration of break t.

Resilience is a dynamic quality within a community that is used to develop and strengthen over time. A community can increase the capacity of its organizations, resources, processes, and people to respond to influence change.

However, resiliency definition and implications have changed drastically after some recent water disasters in the US such as superstorm Sandy in 2012. It is explained in the context of four criteria known as four Rs (Robustness, Rapidity, Resourcefulness, and Redundancy). Rapidity is in line with the classical definition of resiliency as the system's ability to achieve the expected level of performance in the shortest possible time. Other three Rs represent a wide spectrum of other characteristics that a system should attain to withstand the normal and extreme conditions in an increasing interdependent world. A detailed explanation of resiliency characteristics and quantification are presented in the Section 11.8.3.

10.11 SUSTAINABILITY INDEX

10.11.1 Case Study 1: Uncertainty Analysis of the Water Supply and Demand Indicators

In Karamouz et al. (2021), an index called PI-Plus (P for Paydari as it is an indigenous term for sustainability) is extended, and its interdependencies with another index called Sustainability Group Index (SGI) are taken into account. A water balance platform is used that utilizes hydrologic cycle attributes and could be considered as a systems approach to sustainability evaluation. A hypothetical case study is selected that has the components of a real-world problem. A time series of water balance data is synthesized by utilizing basic climatic and hydrologic data from a real watershed. This could be considered as a contribution and a practical engineering solution to generate water balance data in the developing regions. PI-Plus consists of five indicators representing estimation of gross annual available water (using Physical Input–Output Table approach), economic efficiency of delivered water, system's performance in the context of reliability of supply, and maintaining aquifer storage and river instream flow requirements. Furthermore, a framework for considering a new fifth indicator for social aspects is developed by employing Nash bargaining theory.

Uncertainty assessment is an essential part of any modeling procedure especially in environmental and water resources studies. Use of uncertainty analysis can assist in identifying the gaps and checking the robustness of the composite indicator, which further enhances the transparency and credibility of the indices. Without this step, prediction and assessment of the results can become misleading (Nasseri et al., 2014; Juwana et al., 2016; Ahmadi et al., 2019).

Indicators of PI-Plus are calculated based on the water balance data described in the case study. Even though the data from these reports is considered as reliable, they are subjected to uncertainties associated with many field and calculation errors.

In order to implement uncertainty analysis, first, random-natured variables are selected. These variables include: precipitation, infiltration, groundwater inflow, discharge from spring, inflow to reservoir, seepage/drainage of groundwater to and from surface water; economic elements of crop yield, water price, and water allocated sectors; performance elements of aquifer recharge, industrial, domestic, agricultural, and environmental demands; and groundwater discharge.

Second, by finding the distribution of each variable from the 23 years of available data, random numbers are generated with the same distribution of observed data. Subsequently, 100 random numbers for each variable are generated to estimate their corresponding indicators. The observed data were considered as about 30% of data utilized for validation, by the use of indicators' value and finding the best distribution function for each indicator, the PDFs (Figure 10.22) and CDFs

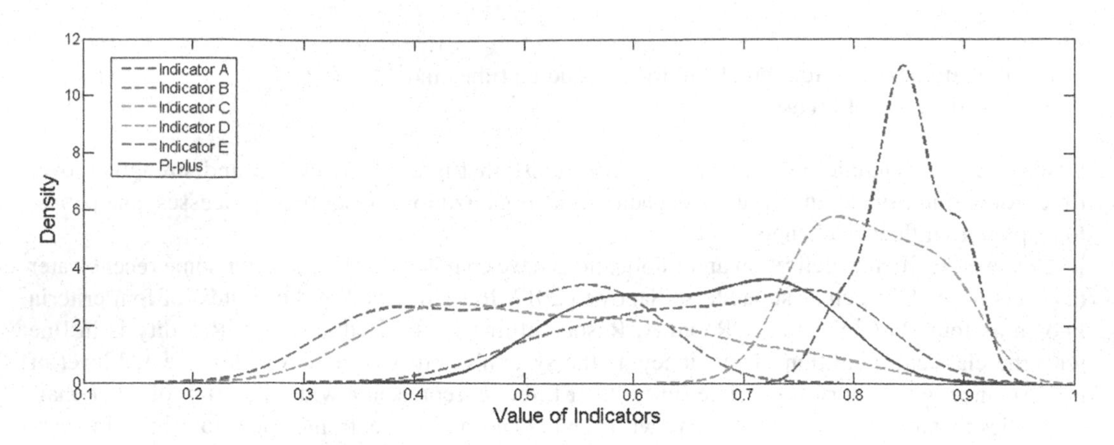

FIGURE 10.22 Probability density functions (PDFs) of the five main indicators and PI-Plus.

are plotted. One of the important aspects of this study is to determine the margin of error in calculating indicators and estimating the extent at which the indicators can be improved. By determining the confidence level of 5% and 10%, the 95%, 90% confidence interval for each indicator, and subsequently PI-Plus is estimated. This range also presents the potential room for improvement of each indicator.

To have a complementary view of the range of indicators and their state in each year, CDFs are plotted in Figure 10.23.

Considering uncertainties and the positioning of 23 water years on the CDF plots, one can see how the values for different year move between indicators and make a comparative analysis of indicators and the timing of low and high values and how they rebound. PI-Plus can be utilized as a useful tool in determining water for planning utilizing CDF graphs in Figure 10.23. With the rather consistent and normal characteristics of PI-Plus presented, if a range of value close to the current value is selected, the next year estimates cannot be too different, and by back tracing the probability associated with that value, a range of values of A–E from their CDFs (Figure 10.23) could be found. This has the potential as a predictive tool to see a range of variations for many attributes of water balance including water for planning figures for the following year. In this chapter's Appendix, an alternative way of estimating water for planning for the next year is presented. Following the platform provided in this study, for an actual implementation, one could provide many needed information for decision makers to reassess the state of water cycle in their region.

In order to perform the uncertainty analysis, the range of variability of each indicator can be investigated (Figure 10.22). Indicator B and C have the highest variability. As for B, it is mainly due to sudden drop in year 1997 (during the final stage of dam construction) and then significant drop in years 2010–2012. For C, its variability is due to sudden changes between years 2004 and 2008. High range of variability of indicators B and C can be attributed to the volatile value of different demands and prices as well of dam construction that affect the variability of these two indicators. Indicator A has the lowest variability and maintains it relatively high value compared to others during the study duration. The others two have moderate variability. As stated, indicator D does not have a wide range of variation since there is limited reliance on the aquifer in the hypothetical case study. Indicator E is skewed but relative display a well-rounded distribution with a linearly decreasing trend. It is interesting to see PI-Plus distribution resembles normal distribution and it falls between the other distributions that display a wide range of elements in the indicators with varying volatility involved. It is symmetrical and it is approximately within $\pm 3\sigma$ ($\sigma = 0.1$) of its mean (0.67). This adds to the validity of this index that collectively normalizes the effects so it could be a consistent and reliable measure to be used in decision-making process.

10.12 UNCERTAINTY ANALYSIS

Hydrology is a highly uncertain science. The main reason for this uncertainty is that many hydrological processes' intrinsic dynamics are still not known. Moreover, the geometry of hydrological control volumes (river beds, subsurface preferential flow paths, etc.) and most of the related initial and boundary conditions and biogeochemical processes cannot be observed in detail and, consequently, cannot be mathematically represented. Finally, hydrologists are typically working under conditions of data scarcity, limiting the efficiency of an inductive approach for tackling the above problems. Uncertainties are caused by a lack of perfect understanding of hydrologic phenomena

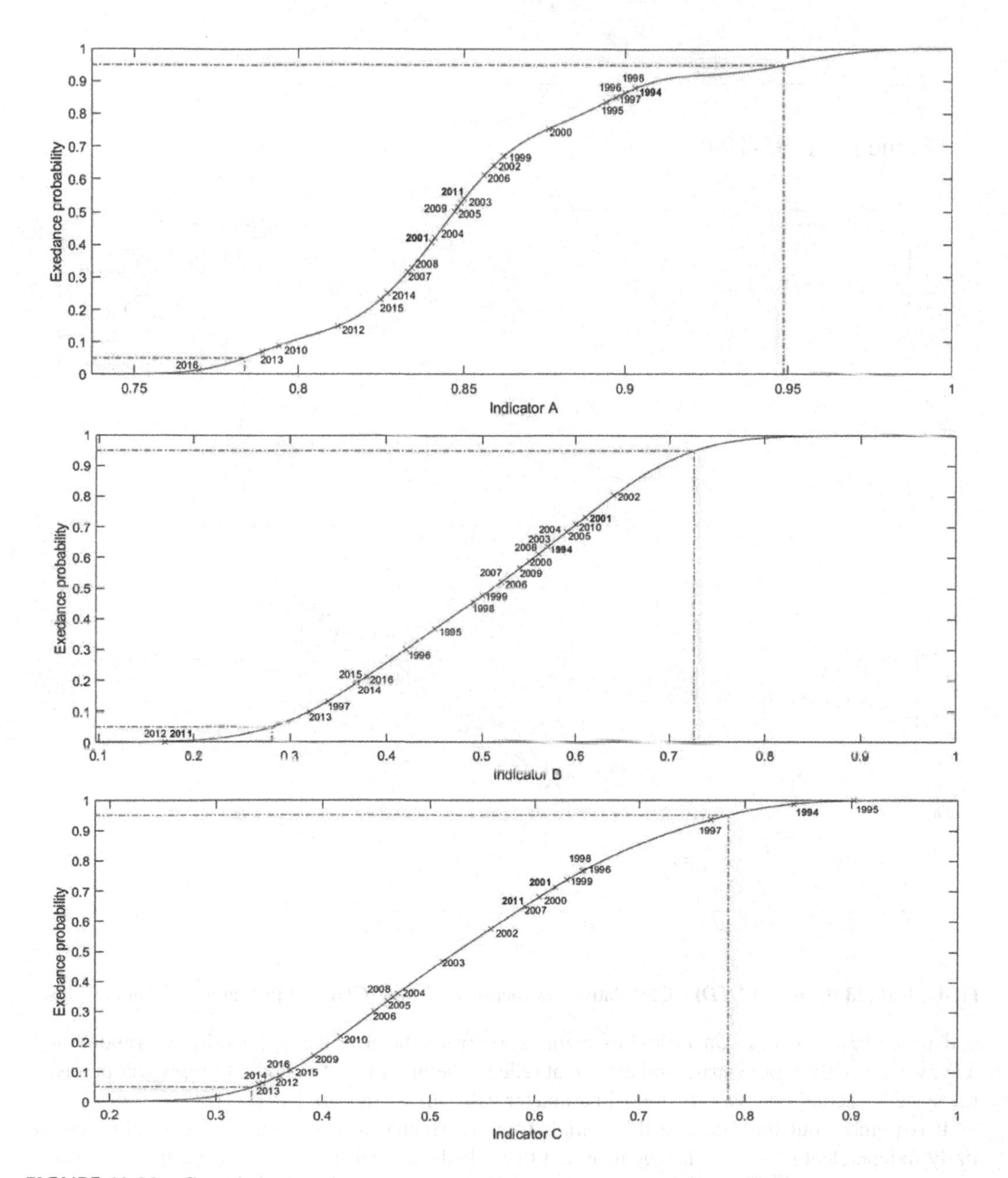

FIGURE 10.23 Cumulative distribution functions (CDFs) of indicators A-E and PI-Plus.

(Continued)

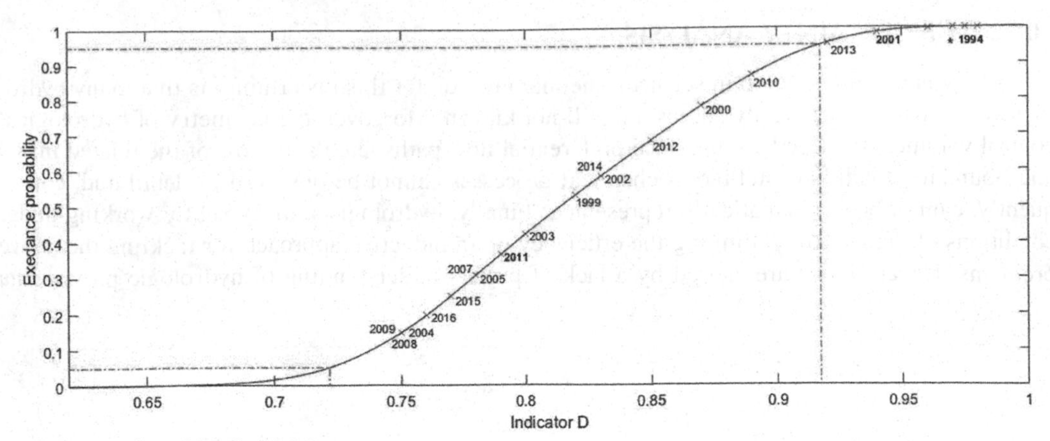

* Same for 1995-1998.

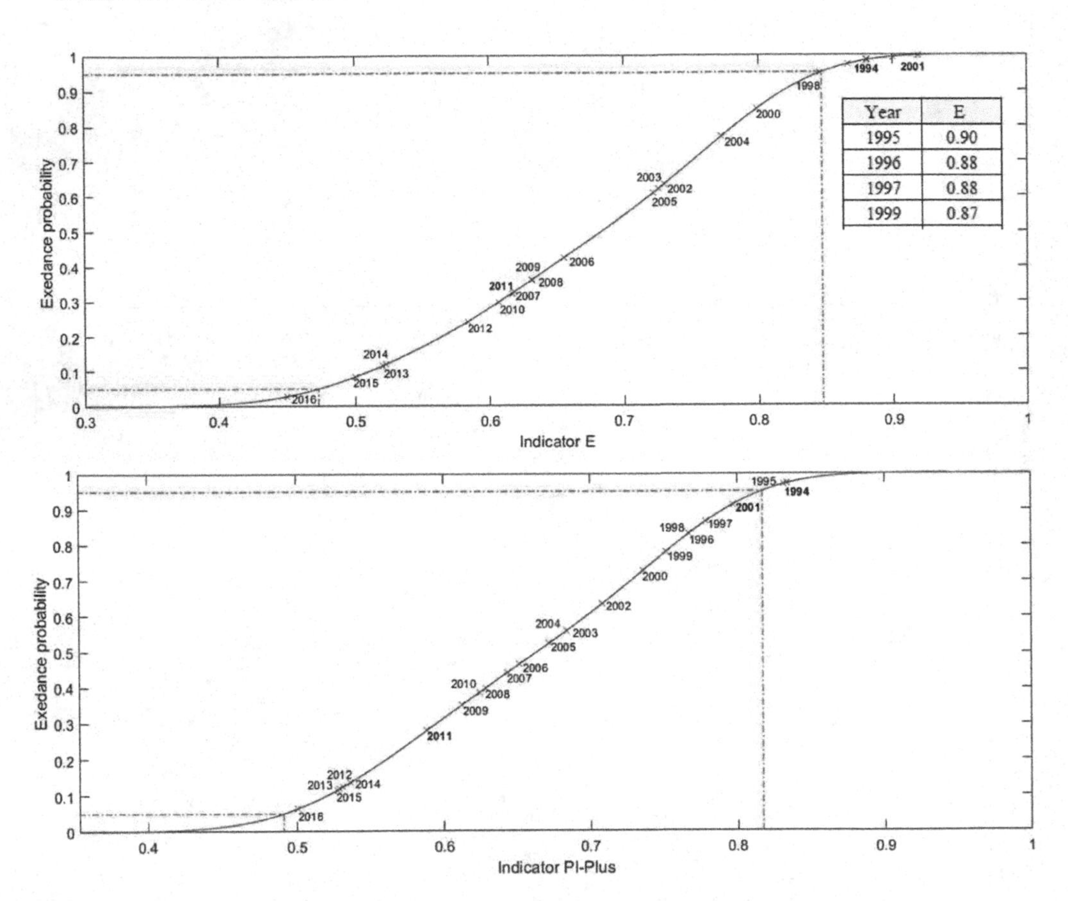

FIGURE 10.23 (*CONTINUED*) Cumulative distribution functions (CDFs) of indicators A-E and PI-Plus.

and processes involved. Uncertainties could arise from the following: (a) inherent randomness (e.g., weather), (b) model structural error that reflects the inability of a model to represent precisely the system's actual behavior, (c) model parameter value error, and (d) data error.

It is pointed out that, despite this initial stimulus, stochastic hydrology is a discipline essentially independent of water management and that a hydrological uncertainty need not necessarily be significant for a water management decision and may carry a different degree of importance in different contexts.

Hydrology design and analysis deal with water in various parts of a hydro system and its effects on environmental, ecological, and socioeconomic settings. Due to the too complex nature of the physical, chemical, biological, and socioeconomic processes involved, major efforts have been devoted by different investigations to better understand the processes.

In general, uncertainty due to the inherent randomness of physical processes cannot be eliminated. On the other hand, uncertainties such as those associated with lack of complete knowledge about the parameters, data, process, and models could be reduced through research, data collection, and careful design and manufacturing. In hydrology engineering, uncertainties can be divided into four basic categories: hydraulic, structural, hydrologic, and economical. More specifically, in hydrology engineering analyses and designs, uncertainties could arise from different sources, including intrinsic or natural uncertainties, parameter uncertainties, model uncertainties, operational uncertainties, and data uncertainties.

Natural uncertainty is associated with the inherent randomness of natural processes such as floods and precipitation events. The occurrence of hydrological events often shows variations in space and in time. A model is only an abstraction of the reality, which generally involves certain simplifications; model uncertainty reflects the inability of a model or design technique to represent precisely the system's proper physical behavior. Parameter uncertainties result from the inability to quantify model parameters and inputs accurately. Parameter uncertainty could also be caused by a change in hydraulic structures' operational conditions, inherent variability of inputs and parameters in time and space, and lack of sufficient data.

Data uncertainties include (a) measurement errors, (b) nonhomogeneity and inconsistency of data, (c) transcription and data handling errors, and (d) inadequate representation of data sample due to space and time limitations. Operational uncertainties include deterioration, manufacturing, construction process, human factors, and maintenance impact. This type of uncertainty is mainly dependent on the craft and quality control during manufacturing and construction.

Another measure of a quantity's uncertainty is expressing it in terms of a reliability domain such as the confidence interval. A confidence interval is a numerical interval that would capture the quantity subject to uncertainty with specified probabilistic confidence. Nevertheless, the use of confidence intervals has a few drawbacks: (a) the parameter population may not be normally distributed as assumed in the conventional procedures, and this problem is fundamental when the sample size is small; and (b) there are no means available to directly combine the confidence intervals of individual contributing random components to give the overall confidence interval of the system. A useful alternative to quantify the level of uncertainty is to use the statistical moments associated with a quantity subject to uncertainty. In particular, the variance and standard deviation, which measure a stochastic variable's dispersion, are commonly used. By increasing the standard deviation of data, the corresponding uncertainty increases due to a higher range of data variations.

Figure 10.24 shows the design procedure involving the above discussion of load and resistance and how reliability and uncertainty could be integrated into the design process. In urban water systems, there are two types of necessary input: natural input, such as rainfall, temperature, and

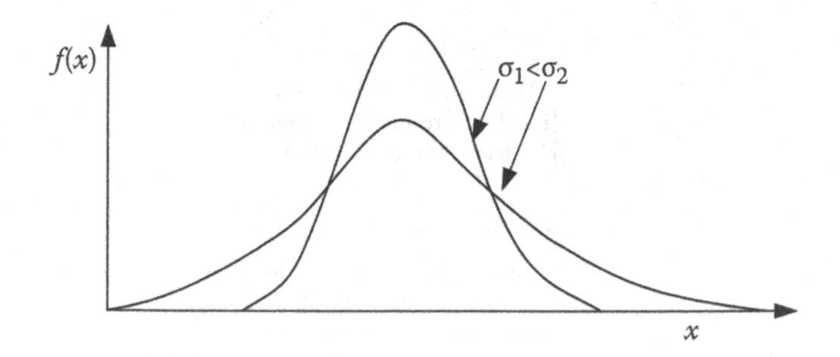

FIGURE 10.24 Relationship between standard deviation and uncertainty.

sediment yield, which are random variables describing the natural environment; and inputs coming from human activities, such as water demand, sewage effluents, and pollutants, which are random numbers describing the interaction of people with the environment. The hydrologic, hydraulic, and structural uncertainties are identified and coupled with a loss function to determine the expected damage resulting from a disaster/failure.

Figure 10.25 shows the framework of the proposed algorithm in dealing with risk and uncertainty. Probability distribution functions are developed for hydraulic and/or structural behavior of the system as well as quality of supplied water considering different values for uncertain variables and parameters and simulation of system performance under these values. The identified uncertainties by these PDFs are incorporated in the base simulation model and then alternative models are developed to evaluate uncertainties that are affecting the system performance.

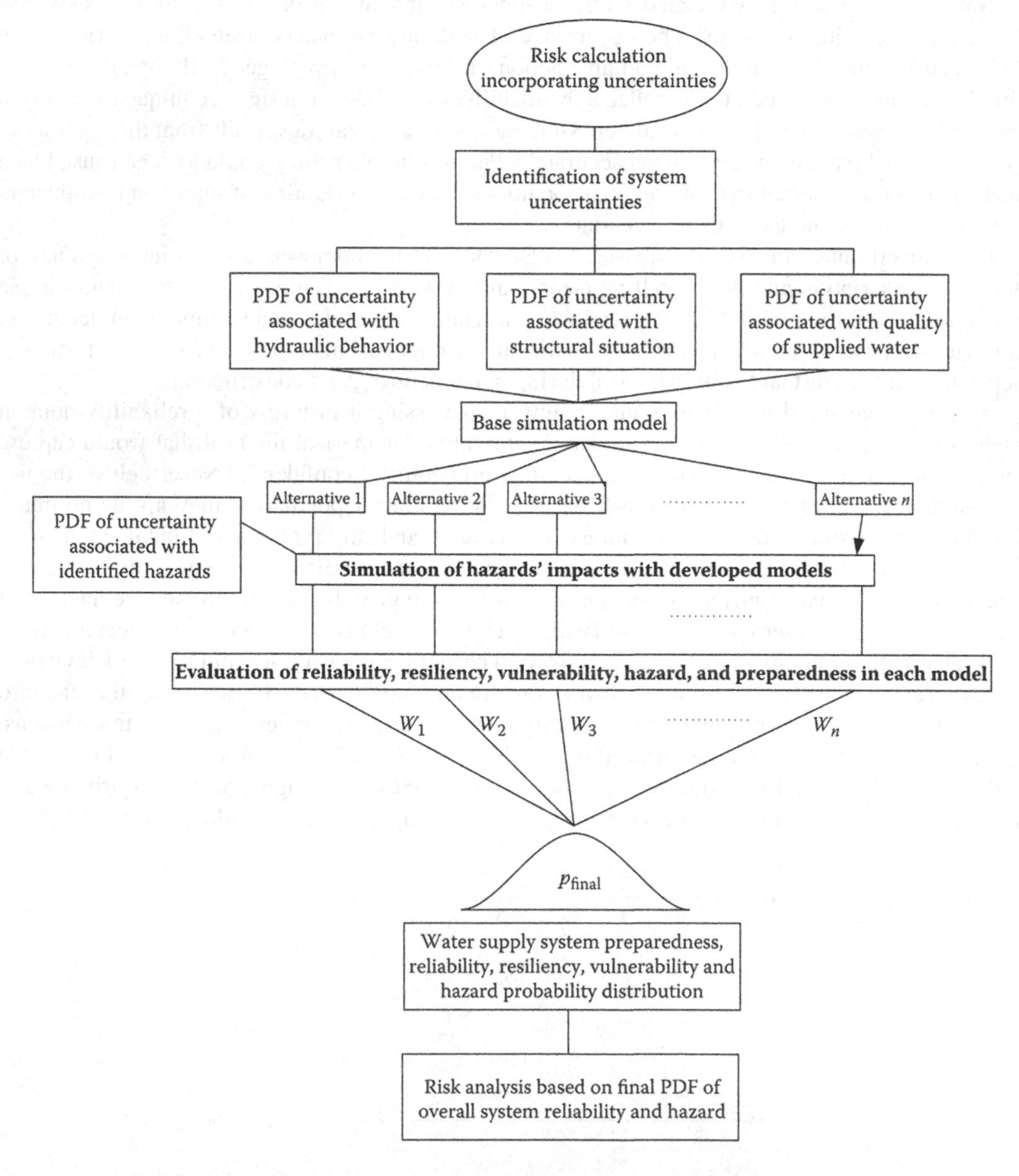

FIGURE 10.25 Generalized concept of risk and reliability analysis for water supply structure.

10.12.1 Implications of Uncertainty

In hydrologic design and modeling, the design quantity and system output are functions of several system parameters, not all of which can be quantified with absolute accuracy. The task of uncertainty analysis is to quantify the system outputs' uncertainty features as a function of model characteristics and the stochastic variables involved. It provides a formal system framework to quantify the uncertainty associated with the system's output. Furthermore, it offers the designer useful insights regarding each stochastic variable's contribution to the overall uncertainty of the system outputs. Such knowledge is essential to identify the "important" parameters to which more attention should be given to assess their values better and, accordingly, to reduce the overall uncertainty of the system outputs.

10.12.2 Uncertainty of Hydrological Forecasting

An important issue in modeling hydrological processes and forecasting is the results' confidence for decision-making in real-time applications. In other words, each model is an approximation to reality; therefore, a decision maker needs to know how reliable the results from a model are and how they affect the final decision. There are different levels of uncertainty in almost all circumstances and at all times, which is a common experience in everyday life.

All-natural events in the hydrologic cycle are inherently uncertain. The primary source of uncertainty is related to data and information. As an example, there is an uncertainty of 20% or more inflow measurements. Therefore, the models that are developed using these data will have an uncertainty of 20%, at least. Uncertainty arises from system complexity, ignorance, imprecision, different randomness classes, knowledge unclarity, and so forth.

Many different factors are affecting the final quality of hydrological forecasts. According to Krzysztofowicz (2001), these sources of uncertainty could be divided into three groups based on their origin:

1. Operational uncertainty is caused by an unpredictable event during the forecasting process, such as dam breaks and ice jams. Hydrologists operating the model significantly impact the final hydrological forecast (human impact also belongs to the group of operational uncertainty). Unfortunately, operational uncertainty could not be quantified in advance and, therefore, remains unexpressed in deterministic and probabilistic approaches.
2. Hydrological uncertainty is represented by the uncertainty of hydrological tools and techniques. The uncertainty of model parameters' calibration and the uncertainty of rating curves and measurement could be mentioned, to name a few.
3. Input uncertainty is represented by other uncertainty of input data. The most important member of that group is the Quantitative Precipitation Forecast (QPF) uncertainty. Experience shows that QPF is the most sensitive factor affecting the result of hydrological forecast in headwater areas, including the Czech Republic area.

Different sources of uncertainty in hydrological simulation can be classified: uncertainty of model, the uncertainty of input, the uncertainty of parameter, and operational and natural uncertainty, such as unforeseen causes and system components, erroneous and missing data, human errors, and mistakes.

Although input uncertainties have been treated (through climate change studies) within water resources management models, and although a full range of uncertainties has been included in hydrological models, studies investigate and address end-to-end impacts of hydrological modeling uncertainties from a water supply point of view are lacking.

10.12.3 Measures of Uncertainty

In statistical analysis, the uncertainty measurement is based on measures of the distribution that describes the uncertainty. The most ordinary estimate of uncertainty is variance or coefficient of variation (ratio of standard deviation to mean). The variance of a calculated parameter describes how the parameter calculation would be different in repeated sampling. Other procedures of uncertainty are based on distribution quantiles such as lower and upper quantiles. The likelihood method can also be used to estimate the uncertainty when parameter values are assigned. The likelihood is associated with the data and how likely the observed data are, given the estimated parameter(s) (Ramsey and Schafer, 1997). The likelihood is usually used to evaluate the uncertainty of models.

Several expressions have been used to describe the level of uncertainty of a parameter, a function, a model, or a system. In general, the uncertainty associated with all three results from the combined effect of the uncertainties of the contributing parameters. The most complete and ideal description of uncertainty is the PDF of the quantity that is subject to uncertainty. However, in most practical problems, such a probability function cannot be precisely derived or obtained.

In a number of case studies in this book especially in Chapters 12–14, uncertainty has been quantified also in terms of error analysis. The theme of Uncertain Flood Inundation was a topic of, in a 2021 ASCE-EWRI virtual Congress panel, discussion entitled "State-of-the-Art Hydro climate Modeling and Water Management under Deep Uncertainties." Here a summary of uncertainty elements of four case studies from different chapter of this book was presented in that topic

- Uncertain Soil Moisture Estimation by Machine Learning, Geostatistical Methods and Satellite Data
- Flood Inundation Maps, Machine Learning (Kalman Filter) and SMAP Soil Moisture
- Flood Inundation Probability Map and DEM Error Estimation
- Load and Resistance Concept and Probabilistic Multi-criteria Decision Making

10.12.3.1 Uncertain Soil Moisture (SM) Estimation

The primary objective of this study is to develop a platform for estimating soil moisture (SM) from satellite in 1-km cells without the need for in situ SM data.

Location: State of Utah, with 158 monitoring stations from Soil Climate Analysis Network (SCAN).

Step 1: SM estimation by adjusting SMAP (Soil Moisture Active Passive) downscaled SM using ancillary real-time data. Combining machine learning (ANN) and a geostatistical method (kriging) to estimate from in situ data, the 1-km estimations of SM (D-SMAP).

Steps 2 and 3: Estimating SMAP SM in any parts of the study area with a finer resolution (from 36 to 1 km) by coregionalization of the estimated SM from ANN-kriging and 36-km SMAP data.

Step 4: Fitting a relationship between in situ SM with the ancillary real-time date (including Antecedent Moisture Condition, AMC_p, and normalized difference vegetation index, NDVI), the downscaled satellite SM, the resulting adjusted SM (AD-SMAP) values were adjusted and presented by adding a stochastic term.

Error and uncertainty attribute of this step/study includes considering an error term (ε) as the stochastic component by adding a random error to the following expression:

$$\text{AD-SMAP} = f\left(\text{D-SMAP, Altitude, Slope, NDVI, } AMC_p\text{, Air Temp}\right) + \varepsilon$$

Generating the error randomly from the distribution fitted to the residuals.

The source of uncertainty is in error in correlating downscaled SMAP data and ancillary data such as AMC in soil estimation has been realized. See Sections 2.4 and 14.2.4 in Chapters 2 and 14 for details of SMAP data and how to download it.

10.12.3.2 Flood Inundation Maps, Machine Learning (Kalman Filter), and SMAP Soil Moisture

This study aims to develop a framework to study the changes in soil moisture using an inundation model using precise antecedent soil moisture derived by the data assimilation model.

Location: Central and western parts of Mississippi State in The Vicinity of Mississippi River. The atmospheric forcing data, SM, and temperature were collected.

Step 1: Assimilating (Ensemble Kalman Filter) the observational variability of the SMAP soil moisture data by Land Surface VIC (Variable Infiltration Capacity) model to improve initial SM conditions.

Step 2: RVIC (Routing) model with 30 m DEM (subject to error and uncertainty) was used in order to estimate inundation depth and volume.

The final product (hazard map) of this study can be applied in assessment and management of extreme events such as flash floods. The improvement in estimation (reduction of error and uncertainty) of flooding characteristics allows better assessment of damage and enhances our preparedness planning efforts.

The improvement in the initial soil moisture estimation reduces error and uncertainty in flood inundation. Furthermore, modeling with 30 m DEM is subject to error and uncertainty that is addressed in case study 2 in Chapter 14.

10.12.3.3 Inundation Probability Map

The objective is to generate flood probability map for a 100-year design value.

Location: The sewershed of Coney Island wastewater treatment facility in Brooklyn, NY.

The rainfall data recorded in JFK airport, the storm surge data at the Battery Park Station, and Delft3D (Coupled Flow and Wave Modules) model is used to estimate flood water level, in an ungauged location in southern Brooklyn.

- There are many uncertainties in calculating water accumulation volume in storm event. Digital Elevation Model (DEM) is an essential input among the needed data in assessing flood impacts.
- Developed 500 DEM realizations by comparing a 1-m Light Detection and Ranging (LiDAR) and 10-m National Elevation Dataset (NED) DEM at 500 Ground Control Points (GCPs) locations using Sequential Gaussian Simulation (SGS) to estimate and quantify the effect of DEM error uncertainties on the assessment of flood inundation.
- The 2D hydrodynamic flood model, LISFLOOD-FP, has been utilized to generate 500 successive estimates of accumulated water in the sewershed and finally a probability map is prepared.

Sources of uncertainty are DEM error and through 500 realization of flood inundation maps, it has been realized. See case study 1 in Chapter 12.

10.12.3.4 Load and Resistance Concept and Probabilistic Multicriteria Decision-Making

The primary purpose is to improve the infrastructure performance and mitigate the flood damage through BMP application considering the uncertainties associated with the criteria with random nature.

Location: The Hunt's Point WWTP as one of the largest 14 New York City's wastewater treatment plants located in East Bronx, NY—experienced two major hurricanes of Irene 2011 (with landfalls wind of 100 km/h) and Sandy 2012 (with 160 km/h wind).

Floodplains corresponding to the flood scenarios are developed using the gridded surface-subsurface hydrological analysis (GSSHA) model developed by US Corps of Engineers (USCOE) to estimate the performance of BMPs. Time series of Kings Point annual water level and LaGuardia Airport rainfall data were obtained for the period of 1946–2017 for joint and nonstationary analysis.

- A DEM with 10×10 m resolution, land use, soil hydraulic conductivity, and soil porosity is used to define the catchment characteristics in the distributed hydrologic model.
- To quantify load and resistance, the probabilistic multicriteria decision-making (MCDM) method is developed (including subcriteria with random nature).
- To statistically quantify the value of reliability, *MS* (difference between load and resistance) and direct integration were used considering the uncertain range of factors and variables.

Uncertainties in evaluating load and resistant of flooding were captured by utilizing a probabilistic MCDM and through fitting a probability distribution to certain subcriteria with random nature for flooding of WWTPs. See the case study in Chapter 9.

The topic of the next section is Entropy Theory. Its concept is the central role of information theory, sometimes referred to as the measure of uncertainty. The entropy of a random variable is defined in terms of its probability distribution. It can be shown to be a good measure of randomness or uncertainty. This part mainly deals with its characterizations and properties. Properties of discrete finite random variables are studied. The study is extended to random vectors with finite and infinite values. The idea of entropy series is explained. Finally, the continuous case generally referred to as differential entropy with different probability distributions and power inequality is studied.

10.13 ENTROPY THEORY

The entropy (or information) theory, developed by Shannon (1948), has recently been applied in many different fields. The entropy theory has also been applied in hydrology and water resources to measure the information content of random variables and models, evaluate information transfer between hydrological processes, evaluate data acquisition systems, and design water quality monitoring networks.

There are four necessary information measures based on entropy: marginal, joint, and conditional entropies and transformation. Shannon and Weaver (1949) were the first to define the marginal entropy, $H(x)$, of a discrete random variable x as

$$H(x) = -\sum_{i=1}^{n} p(x_i)\log_b p(x_i), \tag{10.89}$$

where

n represents the number of elementary events with probabilities $p(x_i)$ ($i = 1, \ldots, N$) and
b is the logarithm base e.

Example 10.14

Let x_i be the uncertain variable of rainfall depth of event i with the uncertain distribution of inundation probability as follows:

$$p(x_i) = \begin{cases} 10\% & i=1 \\ 30\% & i=2 \\ 50\% & i=3 \\ 70\% & i=4 \\ 100\% & i=5 \end{cases}$$

Using the entropy definition for a discrete random variable, and assuming $b = 10$ as the logarithm base, calculate the entropy of x.

Solution:

It follows from the definition of entropy that

$$H(x) = -\sum_{i=1}^{n} p(x_i)\log_b p(x_i) = -\sum_{i=1}^{5} p(x_i)\log_b p(x_i)$$

$$= -(0.1\times -1 + 0.3\times -0.52288 + 0.5\times -0.30103 + 0.7\times -0.1549 + 1\times 0)$$

$$\Rightarrow \quad H(x) = 0.51581.$$

The total entropy of two independent random variables, x and y are equal to the sum of their marginal entropies.

$$H(x,y) = H(x) + H(y), \tag{10.90}$$

where

x and y are stochastically dependent, their joint entropy is less than the total entropy.

Conditional entropy of x given y represents the uncertainty remaining in x when y is known, and vice versa.

$$H(x|y) = H(x,y) - H(y). \tag{10.91}$$

Transinformation is another entropy measure that measures the redundant or mutual information between x and y. It is described as the difference between the total entropy and the dependent x and y.

$$T(x,y) = H(x) + H(y) - H(x,y) \tag{10.92a}$$

or

$$T(x,y) = H(x) - H(x|y) = H(y) - H(y|x). \tag{10.92b}$$

The above expression can be extended to the multivariate case with M variables (Harmancioglu et al., 1999). The total entropy of independent variables X_m ($m = 1, \ldots, M$) equals

$$H(X_1, X_2, \ldots, X_m) = \sum_{m=1}^{M} H(X_m). \tag{10.93}$$

If the variables are dependent, their joint entropy can be expressed as

$$H(X_1, X_2, \ldots, X_m) = H(X_1) + \sum_{m=2}^{M} H(X_m | X_1, X_2, \ldots, X_{m-1}). \tag{10.94}$$

It is sufficient to compute the joint entropy of the variables to estimate the conditional entropies of Equation (10.94) as the latter can be obtained as the difference between two joint entropies; for example,

$$H(X_m | X_1, X_2, \ldots, X_{m-1}) = H(X_1, X_2, \ldots, X_m) - H(X_1, X_2, \ldots, X_{m-1}). \tag{10.95}$$

Finally, when the multivariate normal distribution is assumed for $f(X_1, X_2,\ldots,X_M)$, the joint entropy of X, with X being the vector of M variables, can be expressed as

$$H(x_1,x_1,\ldots,x_M)=(M/2)\ln 2\Pi+(1/2)\ln|C|+M/2-M\ln(\Delta x), \quad (10.96)$$

where
- $|C|$ is the determinant of the covariance matrix C, and
- Δx is the class interval size assumed to be the same for all M variables.

The design of water quality monitoring networks is still a challenging issue. There are difficulties in selecting temporal and spatial sampling frequencies, the variables to be monitored, the sampling duration, and sampling objectives (Harmancioglu et al., 1999). The entropy theory can be used in the optimal design of water quality monitoring systems. By adding new stations and gathering new information, the uncertainty (entropy) in the water quality is reduced. In other words, transinformation can show the redundant information obtained from a monitoring system, which is due to spatial and temporal correlation among the values of water quality variables. Therefore, this index can be effectively used for the monitoring stations' optimal location and for determining the sampling frequencies. Refer to Karamouz et al. (2009) for more information. In their study, an entropy-based approach is presented for design of an online water quality monitoring network for the Karoon River, which is the largest and the most important river in Iran. In the proposed algorithm of design, the number and location of sampling sites and sampling frequencies are determined by minimizing the redundant information, which is quantified using the entropy theory. A water quality simulation model is also used to generate the time series of the concentration of water quality variables at some potential sites along the river. As several water quality variables are usually considered in the design of water quality monitoring networks, the pair-wise comparison is used to combine the spatial and temporal frequencies calculated for each water quality variable. After selecting the sampling frequencies, different components of a comprehensive monitoring system such as data acquisition, transmission, and processing are designed for the study area, and technical characteristics of the online and off-line monitoring equipment are presented. Finally, the assessment for the human resources needs, as well as training and quality assurance programs are presented considering the existing resources in the study area.

10.14 PROBABILITY THEORY—BAYES' THEOREM

The probability theory can be used in two ways for uncertainty evaluation: subjective or Bayesian view and frequents view. When using Bayes' theorem, a rule is provided for updating the probability of H based on additional information E and background information I:

$$P(H|E,I)=\frac{P(H|I)P(E|H,I)}{P(E|I)}, \quad (10.97)$$

where
- $P(H|E,I)$ indicates the posterior probability, which shows the probability of H after incorporating the effect of information E in the context I.
- $P(H|I)$ is the posterior probability of H just based on background information I. In other words, prior probability is the belief in H before considering the additional information of E.
- The term $P(E|H,I)$ is the likelihood indicating the probability of E when the hypothesis H and background information I are true.
- $P(E|I)$ is the prior probability of E, which is used as a normalizing or scaling constant.

In frequency analysis, the situation that an event can be repeated without any limitation in identical conditions is considered; however, the event consequence is random. See more details of Bayes' theorem in the pattern recognition section, 14.5.3, of Chapter 14.

Example 10.15

In order to drill a well, the water depth needs to be known. In preliminary designs, water depth is divided into four categories: less than 5 m, between 5 and 10 m, between 10 and 15 m, and more than 15 m. The following assumptions are known based on the experiences of the hydrology of the area:

The probability of depth to be less than 5 m = Pr $[B_1]$ = 0.6.
The probability of depth to be between 5 and 10 m = Pr$[B_2]$ = 0.2.
The probability of depth to be between 10 and 15 m = Pr $[B_3]$ = 0.15.
The probability of depth to be more than 15 m = Pr $[B_4]$ = 0.05.

The depth h is measured and tabulated in Table 10.2 (the depth includes measurement error).

The h value is obtained at 7 m from the measurements. Considering the error measurements and other assumptions, calculate the probability of h to be in each category.

Solution:

In this example, B probabilities are the prior probabilities of the water depth in the well. Sampling aims to update and correct the information by knowing that sampling results include measurement error (sample likelihood). Using Bayes' theorem, posterior probability can be obtained from the following equation:

$$\Pr\left[B_i \middle| \text{Sample no. 2}\right] = \frac{\Pr\left[\text{Sample no. 2} \middle| B_i\right]\Pr\left[B_i\right]}{\sum_{i=1}^{4}\Pr\left[\text{Sample no. 1} \middle| B_i\right]\Pr\left[B_i\right]}.$$

Therefore,

$$\sum_{i=1}^{4}\Pr\left[\{5 < h \le 10\} \middle| B_i\right]\Pr\left[B_i\right] = 0.07\times0.6 + 0.88\times0.20 + 0.10\times0.15 + 0.06\times0.05 = 0.236$$

$$\Pr\left[B_1 \middle| \text{Sample no. 2}\right] = \frac{0.07\times0.6}{0.236} = 0.178$$

$$\Pr\left[B_2 \middle| \text{Sample no. 2}\right] = \frac{0.88\times0.2}{0.236} = 0.746$$

$$\Pr\left[B_3 \middle| \text{Sample no. 2}\right] = \frac{0.10\times0.15}{0.236} = 0.063$$

$$\Pr\left[B_4 \middle| \text{Sample no. 2}\right] = \frac{0.06\times0.05}{0.236} = 0.13.$$

TABLE 10.2
Probability of Correctness of Measurements

		$h \le 5$	$5 < h \le 10$	$10 < h \le 15$	$h > 15$
$i = 1$	$h \le 5$	0.9	0.05	0.03	0.02
$i = 2$	$5 < h \le 10$	0.07	0.88	0.06	0.06
$i = 3$	$10 < h \le 15$	0.03	0.05	0.85	0.12
$i = 4$	$h > 15$	0.0	0.02	0.06	0.8

10.15 CONCLUDING REMARKS

Water-related risk is the occurrence of an extreme event such as flood, drought, hurricane, and storm that are combined with human interventions. Although hazard events are the basis for realization of risk, it should be characterized in a way that somebody or something has exposure and vulnerability to a hazard. Hence, the risk is the probability of a loss that depends on three processes of hazard, vulnerability, and exposure.

The topic discussed in this chapter includes the concept of probability, basic statistics, common probabilistic models; and extreme value (flood) frequency analysis—design values and basic concept of risk are discussed. Then reliability in the context of serial and parallel components as well as load and resistant concept are described. The performance indicator with emphasis on vulnerability and resiliency is presented including a case study. Finally, the basis of uncertainty analysis is covered with a summary of how error and uncertainties have been quantified in the case studies of this book. The entropy theory including trans-information (measure of redundancy in information) is also discussed in the context of water resources issues to measure the information content of random variables and models, evaluate information transfer between hydrological processes, evaluate data acquisition systems, and design monitoring networks.

The materials covered in this chapter allow a realistic assessment of how to characterize and manage risk. The adaptation and mitigation issues are further discussed in other chapters' case studies.

Land cover could affect the runoff characteristics of an urban watershed and the volume and timing of runoff and maximum flood flow rate would change significantly. Consequently, land cover and use are among the primary problems in assessing increased risk in the urban watersheds. The design values are significantly higher as more frequent hazard event (such as flood) happening, and there are more stringent and complex spatial distribution of vulnerable socioeconomic and environmental conditions in urban areas. The total systems of hazard, human and institutional, and built environment discussed in Chapter 1 are getting more complex with growing population and increasing interdependencies in large cities that make us more susceptible to water disasters that should be prepared for.

PROBLEMS

1. Calculate the sample mean, sample standard deviation, and sample coefficient of skewness of the annual precipitation data given in Table 10.3.
2. Consider the 100-year flood ($P = 0.01$).
   a. What is the probability that at least one 100-year flood will occur during the 50-year lifetime of a flood control project?
   b. What is the probability that the 200-year flood will not occur in 50 years? In 100 years?

TABLE 10.3
Annual Precipitation Data for Problem

Year	Precipitation (cm)
1970	44.9
1971	41.7
1972	41.6
1973	61.2
1974	49.7
1975	41.2
1976	36.3
1977	46.4
1978	52.1
1979	31

c. In general, what is the probability of having no floods more significant than the T-year flood during a sequence of T years?

3. A cofferdam has been built to protect homes in a floodplain until a major channel project can be completed. The cofferdam was built for the 40-year flood event. The channel project will require 5 years to complete. What are the probabilities that
   a. The cofferdam will not be overtopped during the 5 years (the reliability)?
   b. The cofferdam will be overtopped in any 1 year?
   c. The cofferdam will be overtopped exactly one in 5 years?
   d. The cofferdam will be overtopped at least once in 5 years (the risk)?
   e. The cofferdam will be overtopped only in the third year?
4. During a year, about 150 independent storm events occur in two cities, and their average duration is 6.4 hours. Ignoring seasonal variations, in a year of 8,760 hours,
   a. Estimate the average time interval between two storms.
   b. What is the probability that at least 5 days will elapse between storms?
   c. What is the probability that the separation between two storms will be precisely 15 hours?
   d. What is the probability that the separation between two storms will be less than or equal to 15 hours?
5. Determine the hydrologic risk of a roadway culvert with 25 years of expected service life designed to carry a 50-year storm.
6. Construct the CDF of failure occurrences in two hazard rate situations: (a) $\rho(t) = t\alpha$; and (b) $\rho(t) = e^{t-1}$.
7. Calculate the system's reliability shown in Figure 10.26 at $t = 0.5$ using reliability functions of system components.

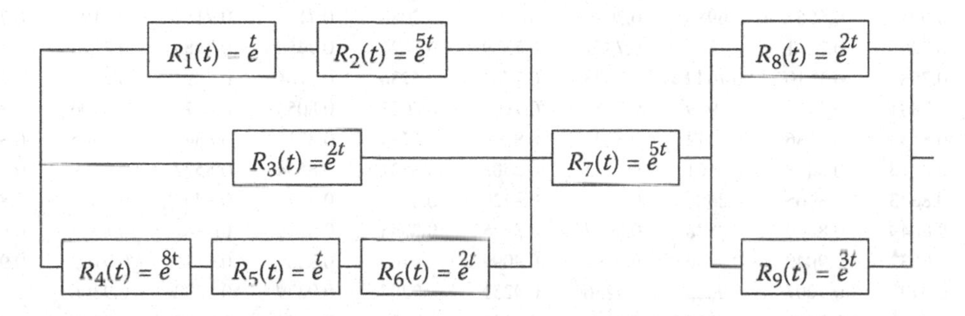

FIGURE 10.26 System in Problem 7.

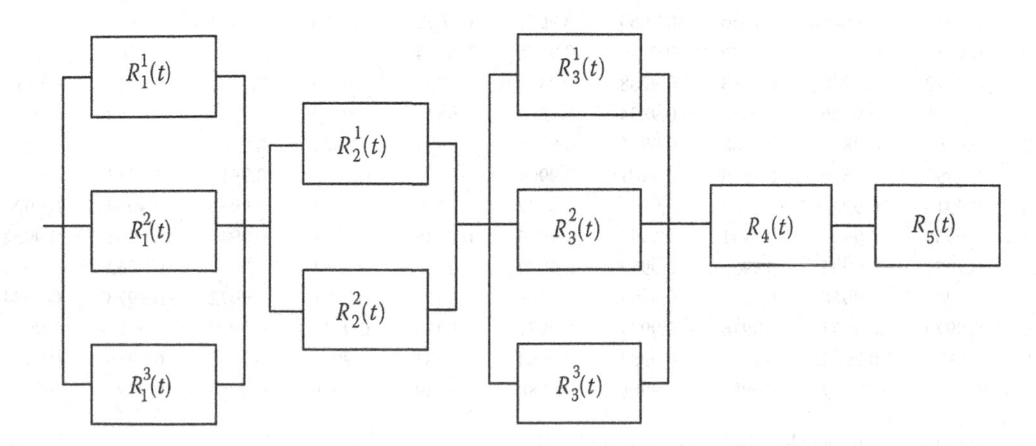

FIGURE 10.27 System in Problem 8.

8. Calculate the reliability of the system shown in Figure 10.27 at $t = 0.1$, assuming that $R_4(t) = R_5(t) = e^{-3t}$ and $R_2^i(t) = R_2^j(t) = R_3^i(t) = e^{0.8t}$ $(1 \le i \le 3, 1 \le j \le 2)$.
9. Determine the hydrologic risk of a roadway culvert flooded with 50 years of expected service life designed to carry a 100-year storm.
10. Explain the stated case study in terms of probability distributions.

APPENDIX

Cumulative probability of normal distribution probability contents from −∞ to +∞ is presented in Tables 10.A1(a) and (b). Frequency factors *K* for Gamma and Log-Pearson type III distributions are tabulated in Table 10.A2.

TABLE 10.A1(A)
Cumulative Probability of Normal Distribution Probability Content from −∞ to *Z*

*Z**	0	0.01	0.02	0.03	0.04	0.05	0.06	0.07	0.08	0.09
0	0.5	0.504	0.508	0.512	0.516	0.5199	0.5239	0.5279	0.5319	0.5359
0.1	0.5398	0.5438	0.5478	0.5517	0.5557	0.5596	0.5636	0.5675	0.5714	0.5753
0.2	0.5793	0.5832	0.5871	0.591	0.5948	0.5987	0.6026	0.6064	0.6103	0.6141
0.3	0.6179	0.6217	0.6255	0.6293	0.6331	0.6368	0.6406	0.6443	0.648	0.6517
0.4	0.6554	0.6591	0.6628	0.6664	0.67	0.6736	0.6772	0.6808	0.6844	0.6879
0.5	0.6915	0.695	0.6985	0.7019	0.7054	0.7088	0.7123	0.7157	0.719	0.7224
0.6	0.7257	0.7291	0.7324	0.7357	0.7389	0.7422	0.7454	0.7486	0.7517	0.7549
0.7	0.758	0.7610	0.7642	0.7673	0.7704	0.7734	0.7764	0.7794	0.7823	0.7852
0.8	0.7881	0.791	0.7939	0.7967	0.7995	0.8023	0.8051	0.8078	0.8106	.8133
0.9	0.8159	0.8186	0.8212	0.8238	0.8264	0.8289	0.8315	0.834	0.8365	0.8389
1	0.8413	0.8438	0.8461	0.8485	0.8508	0.8531	0.8554	0.8577	0.8599	0.8621
1.1	0.8643	0.8665	0.8686	0.8708	0.8729	0.8749	0.877	0.879	0.881	0.883
1.2	0.8849	0.8869	0.8888	0.8907	0.8925	0.8944	0.8962	0.898	0.8997	0.9015
1.3	0.9032	0.9049	0.9066	0.9082	0.9099	0.9105	0.9131	0.9147	0.9162	0.9177
1.4	0.9192	0.9207	0.9222	0.9236	0.9251	0.9265	0.9279	0.9292	0.9306	0.9319
1.5	0.9332	0.9345	0.9357	0.937	0.9382	0.9394	0.9406	0.9418	0.9429	0.9441
1.6	0.9452	0.9463	0.9474	0.9484	0.9495	0.9505	0.9515	0.9525	0.9535	0.9545
1.7	0.9554	0.9564	0.9573	0.9582	0.9591	0.9599	0.9608	0.9616	0.9625	0.9633
1.8	0.9641	0.9649	0.9656	0.9664	0.9671	0.9678	0.9686	0.9693	0.9699	0.9706
1.9	0.9713	0.9719	0.9726	0.9732	0.9738	0.9744	0.975	0.9756	0.9761	0.9767
2	0.9772	0.9778	0.9783	0.9788	0.9793	0.9798	0.9803	0.9808	0.9812	0.9817
2.1	0.9821	0.9826	0.983	0.9834	0.9838	0.9842	0.9846	0.985	0.9854	0.9857
2.2	0.9861	0.9864	0.9868	0.9871	0.9875	0.9878	0.9881	0.9884	0.9887	0.989
2.3	0.9893	0.9896	0.9898	0.9901	0.9904	0.9906	0.9909	0.9910	0.9913	0.9916
2.4	0.9918	0.992	0.9922	0.9925	0.9927	0.9929	0.9931	0.9932	0.9934	0.9936
2.5	0.9938	0.994	0.9941	0.9943	0.9945	0.9946	0.9948	0.9949	0.9951	0.9952
2.6	0.9953	0.9955	0.9956	0.9957	0.9959	0.996	0.9961	0.9962	0.9963	0.9964
2.7	0.9965	0.9966	0.9967	0.9968	0.9969	0.997	0.9971	0.9972	0.9973	0.9974
2.8	0.9974	0.9975	0.9976	0.9977	0.9977	0.9978	0.9979	0.9979	0.998	0.9981
2.9	0.9981	0.9982	0.9982	0.9983	0.9984	0.9984	0.9985	0.9985	0.9986	0.9986
3	0.9987	0.9987	0.9987	0.9988	0.9988	0.9989	0.9989	0.9989	0.999	0.999

* *Z* in this table is considered as it is shown in Figure 10.A1.

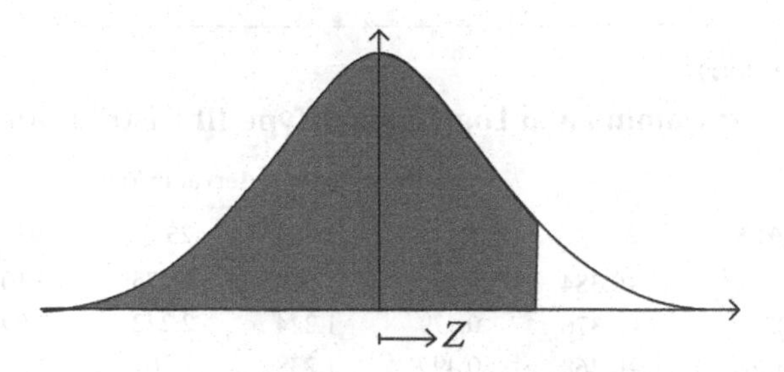

FIGURE 10.A1 The estimation of Z in Table 10.A1(a).

TABLE 10.A1(B)
Cumulative Probability of Normal Distribution Probability Content from Z to $+\infty$

Z*	Cumulative Probability	Z	Cumulative Probability	Z	Cumulative Probability	Z	Cumulative Probability
2	0.02275	3	0.00135	4	3.17E-05	5	2.867E-7
2.1	0.01786	3.1	0.000968	4.1	2.07E-05	5.5	1.899E-8
2.2	0.0139	3.2	0.000687	4.2	1.34E-05	6	9.866E-10
2.3	0.01072	3.3	0.000483	4.3	8.54E-06	6.5	4.016E-10
2.4	0.0082	3.4	0.000337	4.4	5.41E-06	7	1.28E-12
2.5	0.00621	3.5	0.000233	4.5	3.4E-06	7.5	3.191E-14
2.6	0.004661	3.6	0.000159	4.6	2.10E-06	8	6.221E-16
2.7	0.003467	3.7	0.000108	4.7	1.3E-06	8.5	9.48E-18
2.8	0.002555	3.8	7.24E-05	4.8	7.933E-7	9	1.129E-19
2.9	0.001866	3.9	4.81E-05	4.9	4.792E-7	9.5	1.049E-21

* Z in this table is considered as it is shown in Figure 10.A2.

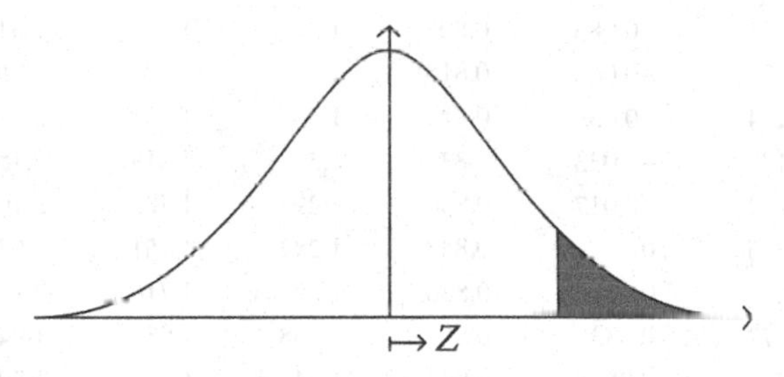

FIGURE 10.A2 The estimation of Z in Table 10.A1(b).

TABLE 10.A2
Frequency Factors K for Gamma and Log-Pearson Type III Distributions

Weighted	Recurrence Interval in Years							
	1.0101	2	5	10	25	50	100	200
Skew Coefficient	Percent Chance (≥) = 1 − F							
C_s	99	50	20	10	4	2	1	0.5
3	−0.667	−0.396	0.42	1.18	2.278	3.152	4.051	4.97
2.9	−0.69	−0.39	0.44	1.195	2.277	3.134	4.013	4.904

(Continued)

TABLE 10.A2 (*Continued*)
Frequency Factors *K* for Gamma and Log-Pearson Type III Distributions

	Recurrence Interval in Years							
Weighted	**1.0101**	**2**	**5**	**10**	**25**	**50**	**100**	**200**
2.8	–0.714	–0.384	0.46	1.21	2.275	3.104	3.973	4.847
2.7	–0.74	–0.376	0.479	1.224	2.272	3.093	3.932	4.783
2.6	–0.769	–0.368	0.499	1.238	2.267	3.071	3.889	4.718
2.5	–0.799	–0.36	0.518	1.25	2.262	3.048	3.845	4.652
2.4	–0.832	–0.351	0.537	1.262	2.256	3.023	3.8	4.584
2.3	–0.867	–0.341	0.555	1.274	2.248	2.997	3.753	4.515
2.2	–0.905	–0.33	0.574	1.284	2.24	2.97	3.705	4.444
2.1	–0.946	–0.319	0.592	1.294	2.23	2.942	3.656	4.372
2	–0.99	–0.307	0.609	1.302	2.219	2.912	3.605	4.298
1.9	–1.037	–0.294	0.627	1.31	2.207	2.881	3.553	4.223
1.8	–1.087	–0.282	0.643	1.318	2.193	2.848	3.499	4.147
1.7	–1.14	–0.268	0.66	1.324	2.179	2.815	3.444	4.069
1.6	–1.197	–0.254	0.675	1.329	2.163	2.78	3.388	3.99
1.5	–1.256	–0.24	0.69	1.333	2.146	2.743	3.33	3.91
1.4	–1.318	–0.225	0.705	1.337	2.128	2.706	3.271	3.828
1.3	–1.383	–0.21	0.719	1.339	2.108	2.666	3.210	3.745
1.2	–1.449	–0.195	0.732	1.34	2.087	2.626	3.149	3.661
1.1	–1.518	–0.18	0.745	1.341	2.066	2.585	3.087	3.575
1	–1.588	–0.164	0.758	1.34	2.043	2.542	3.022	3.489
0.9	–1.66	–0.148	0.769	1.339	2.018	2.498	2.957	3.401
0.8	–1.733	–0.132	0.78	1.336	1.993	2.453	2.891	3.312
0.7	–1.806	–0.106	0.79	1.333	1.967	2.407	2.824	3.223
0.6	–1.88	–0.099	0.8	1.328	1.939	2.359	2.755	3.132
0.5	–1.955	–0.083	0.808	1.323	1.91	2.310	2.686	3.041
0.4	–2.029	–0.066	0.816	1.317	1.88	2.261	2.615	2.949
0.3	–2.104	–0.05	0.824	1.309	1.849	2.210	2.544	2.856
0.2	–2.178	–0.033	0.83	1.301	1.818	2.159	2.472	2.763
0.1	–2.252	–0.017	0.836	1.292	1.785	2.107	2.4	2.67
0	–2.326	0	0.842	1.282	1.751	2.054	2.326	2.576
–0.1	–2.4	0.017	0.846	1.27	1.716	2	2.252	2.482
–0.2	–2.472	0.033	0.85	1.258	1.68	1.945	2.178	2.388
–0.3	–2.544	0.05	0.853	1.245	1.643	1.89	2.104	2.294
–0.4	–2.615	0.066	0.855	1.231	1.606	1.834	2.029	2.201
–0.5	–2.686	0.083	0.856	1.216	1.567	1.777	1.955	2.108
–0.6	–2.755	0.099	0.857	1.2	1.528	1.72	1.88	2.016
–0.7	–2.824	0.106	0.857	1.183	1.488	1.663	1.806	1.926
–0.8	–2.891	0.132	0.856	1.166	1.448	1.606	1.733	1.837
–0.9	–2.957	0.148	0.854	1.147	1.407	1.549	1.66	1.749
–1	–3.022	0.164	0.852	1.128	1.366	1.492	1.588	1.664
–1.1	–3.087	0.18	0.848	1.107	1.324	1.435	1.518	1.581
–1.2	–3.149	0.195	0.844	1.086	1.282	1.379	1.449	1.501
–1.3	–3.210	0.21	0.838	1.064	1.24	1.324	1.383	1.424
–1.4	–3.271	0.225	0.832	1.041	1.198	1.27	1.318	1.351
–1.5	–3.33	0.24	0.825	1.018	1.157	1.217	1.256	1.282

(Continued)

TABLE 10.A2 (*Continued*)
Frequency Factors *K* for Gamma and Log-Pearson Type III Distributions

	Recurrence Interval in Years							
Weighted	1.0101	2	5	10	25	50	100	200
−1.6	−3.88	0.254	0.817	0.994	1.106	1.166	1.197	1.216
−1.7	−3.444	0.268	0.808	0.97	1.075	1.106	1.14	1.155
−1.8	−3.499	0.282	0.799	0.945	1.035	1.069	1.087	1.097
−1.9	−3.553	0.294	0.788	0.92	0.996	1.023	1.037	1.044
−2	−3.605	0.307	0.777	0.895	0.959	0.98	0.99	0.995
−2.1	−3.656	0.319	0.765	0.869	0.923	0.939	0.946	0.949
−2.2	−3.705	0.33	0.752	0.844	0.888	0.9	0.905	0.907
−2.3	−3.753	0.341	0.739	0.819	0.855	0.864	0.867	0.869
−2.4	−3.8	0.351	0.725	0.795	0.823	0.83	0.832	0.833
−2.5	−3.845	0.36	0.710	0.710	0.793	0.798	0.799	0.8
−2.6	−3.899	0.368	0.696	0.747	0.764	0.768	0.769	0.769
−2.7	−3.932	0.376	0.681	0.724	0.738	0.74	0.74	0.741
−2.8	−3.973	0.384	0.666	0.702	0.712	0.714	0.714	0.714
−2.9	−4.013	0.39	0.651	0.681	0.683	0.689	0.69	0.69
−3	−4.051	0.396	0.636	0.66	0.666	0.666	0.667	0.667

REFERENCES

Adger, W.N., Brooks, N., Bentham, G., Agnew, M., and Eriksen, S. (2004). New indicators for vulnerability and adaptive capacity. In T. Centre (ed.) *Technical Report 7* (pp. 1–128). Tyndall Centre, Norwich.

Ahmadi, A., Nasseri, M., and Solomatine, D. P. (2019). Parametric uncertainty assessment of hydrological models: Coupling UNEEC-P and a fuzzy general regression neural network. *Hydrological Sciences Journal*, 64(9), 1080–1094. doi: 10.1080/02626667.2019.1610565.

Ang, A.H.S. (1973). Structural risk analysis and reliability based design. *Journal of the Structural Engineering Division. ASCE*, 99(ST9), 1891–1910.

APFM. (2008). Urban flood risk management. A tool for integrated flood management version 1. http://www.apfm.info/pdf/ifm_tools/Tools_Urban_Flood_Risk_Management.pdf.

Bossel, H. (1999). *Indicators for Sustainable Development: Theory, Method, Applications. A Report to the Balaton Group.* International Institute for Sustainable Development, Winnipeg. http://www.iisd.org/pdf/balatonreport.pdf.

Bras, R.L. (1990). *Hydrology: An Introduction to Hydrologic Science.* Addison-Wesley, Reading, MA.

Chow, V.T., Maidment, D.R., and Mays, L.W. (2013). *Applied Hydrology.* 2nd edition. Tata McGraw-Hill Education, New York, NY.

Conway, R.A. (ed.) (1993). *Environmental Risk Analysis for Chemicals* (558 pp.). Van Nostrand Reinhold, Miami Beach, FL.

Crichton, D. (1999). The risk triangle. In J. Ingleton (ed.) *Natural Disaster Management* (pp. 102–103). Tudor Rose, London.

Glossary of the UN International Strategy for Disaster Reduction. (2004). http://www.unisdr.org/eng/library/lib-terminology-eng%20home.htm.

Hamouda, M.A.A. (2006). Vulnerability assessment of water resources systems in the Eastern Nile Basin to environmental factors. Master's thesis Cairo University, Institute of African Research and Studies, Department of Natural Resource.

Harmancioglu, N.B., Fistikoglu, O., Ozkul, S.D., Singh, V.P., and Alpaslan, M.N. (1999). *Water Quality Monitoring Network Design.* Kluwer, Boston, MA.

Hashimoto, T., Stedinger, J.R., and Loucks, D.P. (1982). Reliability, resiliency, and vulnerability criteria for water resources performance evaluation. *Water Resources Research*, 18(1), 14–20.

Henley, E.J. and Gandhi, S.L. (1975). Process reliability analysis. AIChE Journal, 21(4), 677–686.

Jorgensen, S.E. and Bendoricchio, G. (2001). *Fundamentals of Ecological Modeling*. Gulf Professional Publishing, Amsterdam.

Juwana, I., Muttil, N., and Perera, B.J.C. (2016). Uncertainty and sensitivity analysis of West Java water sustainability index—A case study on Citarum catchment in Indonesia. *Ecological Indicators* 61, 170–178.

Kapur, K.C. and Lamberson, L.R. (1977). *Reliability in Engineering Design*. Wiley, New York.

Karageorgis, A.P., Skourtos, M.S., Kapsimalis, V., Kontogianni, A.D., Skoulikidis, N.Th., Pagou, K., Nikolaidis, N.P., Drakopoulou, P., Zanou, B., Karamanos, H., Levkov, Z., and Anagnostou, Ch. (2005). An integrated approach to watershed management within the DPSIR framework: Axios River catchment and Thermaikos Gulf. *Regional Environmental Change*, 5, 138–160.

Karamouz, M., Nokhandan, A.K., Kerachian, R., and Maksimovic, C. (2009). Design of on-line river water quality monitoring systems using the entropy theory: A case study. *Journal of Environmental Modeling and Assessment*, 155, 63–81.

Karamouz, M., Rahimi, R., and Ebrahimi, E. (2021). Uncertain water balance-based sustainability index of supply and demand. *Journal of Water Resources Planning and Management*, *147*(5), 04021015.

Karamouz, M., Szidarovszky, F., and Zahraie, B. (2003). *Water Resources System Analysis*. Lewis Publishers, CRC Press, Boca Raton, FL.

Krzysztofowicz, R. (2001). Integrator of uncertainties for probabilistic river stage forecasting: Precipitation dependent model. *Journal of Hydrology*, 249, 69–85.

Lee, W. (2011). Risk-based asset allocation: A new answer to an old question?. *The Journal of Portfolio Management*, 37(4), 11–28.

Mays, L.W. (2001). *Water Resources Engineering*. John Wiley and Sons Inc., New York.

Mays, L.W. (2004). Water *Supply Systems Security*. McGraw-Hill Education, Arizona, pp. 1–475.

McCarthy, J.J., Canziani, O.F., Leary, N.A., Dokken, D.J. and White, K.S. (2001). *Climate Change 2001: Impacts, Adaptation and Vulnerability*. Cambridge University Press, Cambridge, MA.

Meadows, D. (1998). *Indicators and Information Systems for Sustainable Development. A Report to the Balaton Group*. The Sustainability Institute, Hartland Four Corners, VT, Bilthoven.

Mohamed, M.O.S., Neukermans, G., Kairo, J.F., Dahdouh-Guebas, F., and Koedam, N. (2009). Mangrove forests in a peri-urban setting: The case of Mombasa (Kenya). *Wetland Ecology and Management*, 17, 243–255.

Nasseri, M., Ansari, A., and Zahraie, B., (2014). Uncertainty assessment of hydrological models with fuzzy extension principle: Evaluation of a new arithmetic operator. Water Resources Research, 50(2), 1095–1111.

Plate, E.J. and Duckstein, L. (1988). Reliability-based design concepts in hydraulic engineering. *Water Resources Bulletin*, 24(2), 235–245.

Ramsey, F.L. and Schafer, D.W. (1997). *The Statistical Sleuth: A Course in the Methods of Statistics*. Duxbury Press, Belmont, CA.

Schweitzer, G.E. (2013). *Borrowed Earth, Borrowed Time: Healing America's Chemical Wounds* (298 pp.). Springer Nature, New York.

Shannon, C.E. (1948). A mathematical theory of communication, part I. *Bell Systems Technical Journal*, 27, 379–423.

Shannon, C.E. and Weaver, W. (1949). *A Mathematical Model of Communication*. University of Illinois Press, Urbana, IL.

Tung, Y.K. and Mays, L.W. (1981). Risk models for flood levee design. Water Resources Research, 17(4), 833–841.

Wallis, K.F. (1974). Seasonal adjustment and relations between variables. Journal of the American Statistical Association, 69(345), 18–31.

Yen, B.C. (1978). Safety factor in hydrologic and hydraulic design. *Proceedings of the International Symposium on Risk Reliability in Water Resources*, University of Waterloo, Waterloo, Ontario, Canada.

11 Urban Water Disaster Management

11.1 INTRODUCTION

Water is a vital sustenance for life, and there is no substitute for this essential resource. Therefore, water shortages and pollution, flood disasters, aging water infrastructures as well as the increasing demand for water, and other problems associated with water resources are becoming more devastating and the need for disaster planning and management is growing. The rapid osculation of these problems has the gravest effects on developing countries. Therefore, water disaster of a higher magnitude can affect everyone with rapid expansion of water sectors and inadequate institutional and infrastructural setups. The developed societies are also becoming more vulnerable due to their high dependence on water and the concerns for water security.

Natural disasters cause suffering to people around the world. This can be particularly devastating for those living in developing countries, where the social infrastructure is not fully in place. In recent years, more frequent floods, droughts, and hurricanes (due to climate change and variability) have caused disasters of the records. Furthermore, the world is getting more hostile and political and economic motives have caused water security issues and threats to water supply systems. Hurricane Katrina smashed the entire neighborhoods in New Orleans, Louisiana, in the United States and killed at least 185 people. This event shows that even developed countries are subject to water disaster that has not been experienced before.

In this chapter, the nature of a water disaster and the factors contributing to the formation and extent of changes caused by a disaster are discussed. The notion of water hazard (including water scarcity) as a "load" and our ability to withstand it as a "resistance" is discussed in the context of reliability and risk-based design. The elements of uncertainty and how risk management can be coupled with disaster management (DM) are presented. This chapter is divided into 11 sections with the following focus areas: First, an introduction to urban water disaster management (UWDM) is presented. Then the planning process for UWDM is presented, followed by situation analysis, disaster indices, risk, and uncertainties. Finally, guidelines for UWDM and preparedness planning are presented.

11.2 SOURCES AND KINDS OF DISASTERS

Many regional problems are the direct result of the regional climates, geography, and hydrology. Others are as a result of human activities and mismanagement. Traditionally, water disaster has been attributed to the quantity of water that causes droughts and floods. A drought is a shortage of water beyond a level that a water supply/distribution system can deliver to meet the basic demands. Floods are the result of excess runoff, which can increase depending on various factors such as intensity of rainfall, snow melt, soil type, soil moisture conditions, and land use and land cover.

In addition to the quantity of water, its quality might also be the cause of a water disaster. In many developed regions, issues such as widespread contamination of water mains and water supply reservoirs by accidental or intentional acts have been the focus of many activities.

DOI: 10.1201/9781003241744-11

11.2.1 Drought

When we think of water disasters, we tend to think of cyclones, hurricanes, storms, and floods. Less attention is given to droughts, which is also classified as a disaster or even the most long-term disaster with intense social and environmental predicaments.

What is a drought? It is viewed as a sustained and regionally extensive occurrence of below average or any low-flow thresholds on natural water availability (Figure 11.1). The more precise definitions for specific areas of concern that are most commonly used are

- Meteorological or climatic drought is a period of well-below-average or normal rainfall that spans from a few months to several years.
- Agricultural drought is a period when soil moisture is inadequate to meet the demands for crops to initiate and sustain plant growth.
- Hydrological drought is a period of below-average or normal stream flow where rivers dry up completely and remain dry for a very long period or where there is significant depletion of water in aquifers.
- Socioeconomic drought refers to the situation that occurs when physical water shortage begins to affect people due to inability of the transfer and distribution systems to perform.

Among different types of droughts, hydrological drought has more severe effects on urban areas because of the high rate of water use in these areas and significant social and economic issues related to water shortages in urban areas.

Developing a national or regional drought policy and an emergency response program is essential for reducing societal vulnerability and disasters and hence reducing the water disaster impacts. Resiliency in the context of drought management has received attention (Karamouz et al., 2004).

Strategies for drought management focus on reducing the impact of droughts by regional analysis to determine the most vulnerable part of a region. Figure 11.2 shows the vulnerable areas of Lorestan Province for a drought with 10-year return period. The area with dark shadows requires more attention if the region is facing a 10-year drought. Drought planning and management must take into account not only the risk of potential economic and social (psychological) damages resulting from

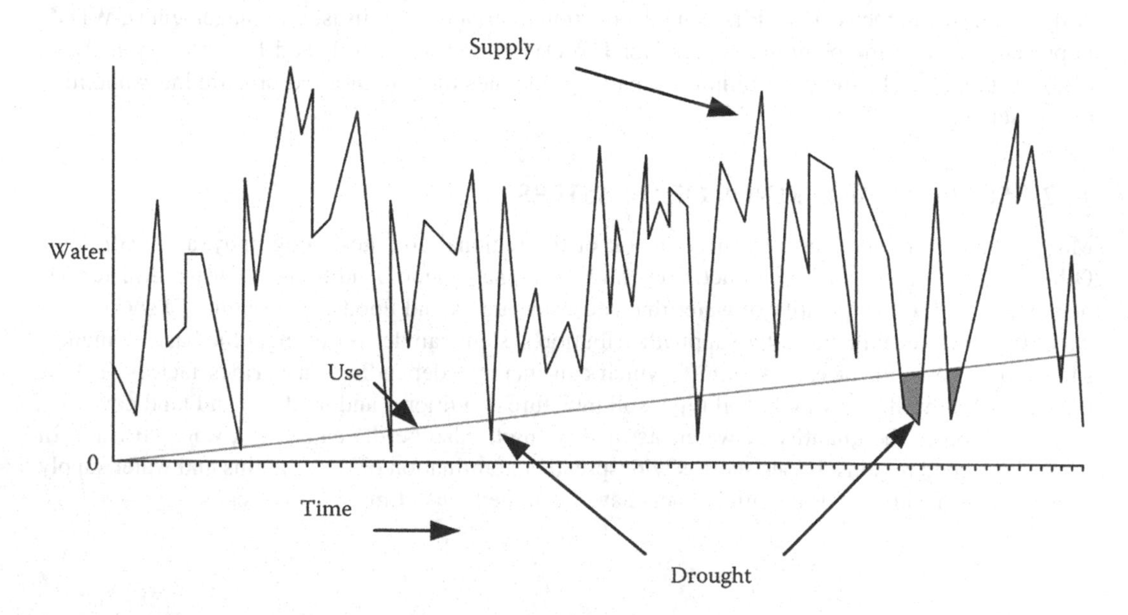

FIGURE 11.1 Graphical definition of drought.

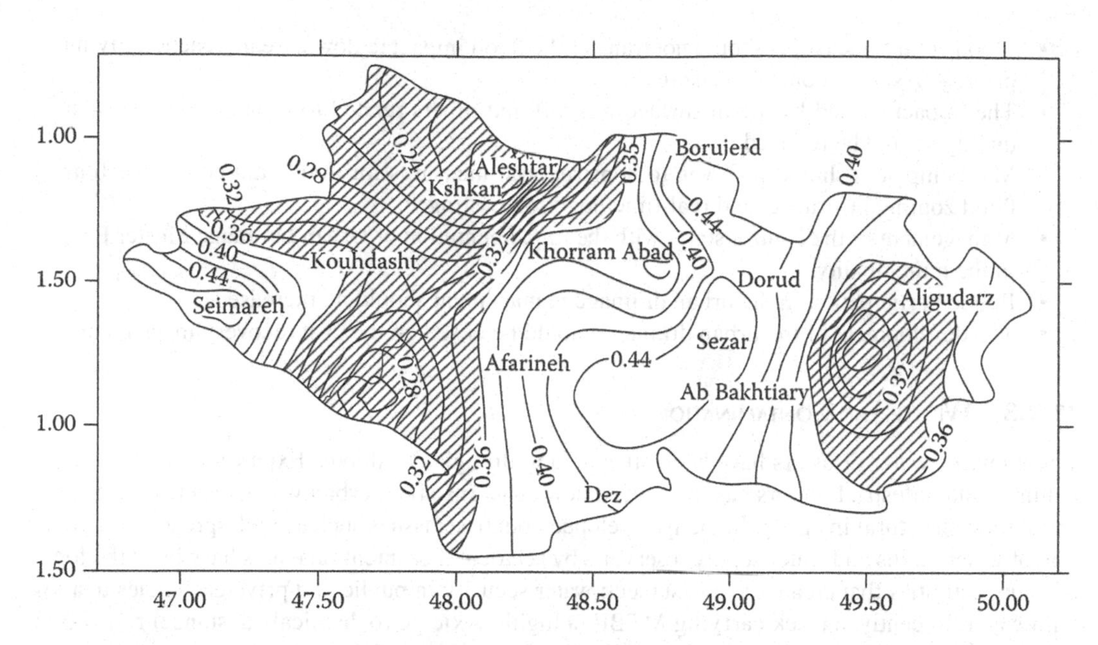

FIGURE 11.2 Vulnerable areas of Lorestan Province in Iran in a drought with 10-year return period. The darkened areas require more attention if the region is facing a 10-year drought.

droughts but also the ecological and economic costs and benefits of exercising alternative options such as aquifer recharge and capacity (physical) building (filling the reservoir) schemes and helps us into determining how these options can reduce potential damages and provides a way of facing future droughts (Karamouz, 2002).

11.2.2 Floods

Floods are the result of excess runoff, which can increase depending on various factors such as intensity of rainfall, snow melt, soil type, soil moisture conditions, and land use and land cover. Runoff from rural and urban areas is generally a response of excess water after the processes of infiltration and evapotranspiration have taken place. Obviously, urban regions have more impervious surfaces and the surface runoffs are more intense. So they are more prone to flooding (Herath, 2008). According to the Federal Emergency Management Agency (FEMA) of the United States, floods are the second most common and widespread of all natural disasters (drought is first, globally) (FEMA, 2008). More explanations are given in Chapter 13.

Flooding is caused primarily by hydrometeorological mechanisms, acting either by a single factor or by a combination of different factors. Most of the urban areas are subject to some kind of flooding after spring rains, heavy thunderstorms, or winter snow thaws. Coastal flooding is a major challenge as the frequency of extreme events and major storms has been increased and devastated major cities such as New York City. In Hurricane Irene, the lack of understandings of vulnerabilities and in Superstorm Sandy, the inadequacy of infrastructure integration was realized.

11.2.2.1 Principles of Urban Flood Control Management

The reliable experience in flood control of many countries has now led to some main principles in urban drainage and flood control management as follows:

- Flood control evaluation should be done in the whole basin and not only in specific flow sections.
- Urban drainage control scenarios should take into account future city developments.

- Flood control measures should not transfer the flood impact to downstream reaches, giving priority to source control measures.
- The impact caused by urban surface washoff and others related to urban drainage water quality should be reduced.
- More emphasis should be given to nonstructural measures for floodplain control such as flood zoning, insurance, and real-time flood forecasting.
- Management of the control starts with the implementation of Urban Drainage Master Plan in the municipality.
- Public participation in the urban drainage management should be increased.
- The development of the urban drainage should be based on the cost recovery investments.

11.2.3 Widespread Contamination

Traditionally, water disasters have been attributed to droughts or floods. Expansion of urban, agricultural, and industrial sectors has introduced new concerns about urban water systems' safety and about their structural integrity. In many developed countries, issues such as widespread contamination of water mains and water supply reservoirs by accidental or intentional acts have been the focus of many activities that created a new subject; water security in public and private agencies dealing with water. Recently, a truck carrying MTBE (a highly toxic petrochemical substance) rolled over near a water supply reservoir of a major city in the western part of Iran and contaminated about 100 MCM of water. Karamouz et al. (2017b) investigated the widespread contamination by biological (coliform bacteria) and chemical (arsenic) contamination in a water distribution system using the EPANET.MS software environment.

Pollution is one of the major contributors to the urban water problems. It affects rivers, lakes, and groundwater, and much of it comes from poor sanitation and uncontrolled solid waste disposal. The main sources of ground water pollution include

- The municipal system (urban runoff, sewer leakage, wastewater effluent, septic tanks, sewage sludge, urban runoff, landfill, and latrines)
- Industries (processes, water treatment, industrial effluent, leaking tanks, and pipelines)
- Agriculture (return flows, fertilizers, pesticides, herbalists, and animal wastes)
- Mining (solid wastes and liquid wastes)

Industrial effluent is another major source of pollution, and urban stormwater runoff affects the quality of receiving waters (Akan and Houghtalen, 2003).

11.2.4 System Failure

System failures are uncertain events due to a combination of factors. The failure can lead to supply interruption, physical damage, and/or unacceptable water quality. The reaction to a failure is repair, replacement, and/or some other types of protective actions. In order to quantify the uncertainty of failure, there is a wide range of failure models that can be used. The output of such models is the distribution function and/or probabilities of outcomes (e.g., failures). The state of the network can be assumed either through some linear models (sometimes nonlinear) or through monitoring conditions. However, due to high costs, this is primarily considered only if the consequences of failure are very high.

11.2.5 Earthquakes

Somehow an urban water disaster (UWD) is a consequence of a more destructive disaster such as earthquakes or tsunamis. All the water systems are subject to seismic activities and are highly vulnerable due to nonelastic design of components. Lessons should be drawn from past earthquakes' consequences on water systems and simulate in to regions with a similar setup. It helps in the planning for earthquakes and tsunami-type events, which result in flooding and/or water supply and distribution collapses.

11.3 WHAT IS UWDM?

UWDM is a systematic process for the reduction of water disaster impacts on social, economic, and environmental systems. Here are a few reasons why many decision makers argue that an UWDM is necessary to implement.

- Water disasters are increasingly under pressure from population growth, economic activity, and intensifying competition for the water among users.
- Water withdrawals have increased more than twice as fast as population growth, and currently one-third of the world's population live in countries that experience medium to high water stress. The higher the water stress, the higher the vulnerability to disaster-scale water shortages.
- Pollution is further enhancing water scarcity by reducing water usability.
- Shortcomings in the management of water, a focus on developing new sources rather than managing existing ones better, and top-down sector approaches to water management result in uncoordinated development and management of the disaster.
- More and more development means greater impacts on the environment.
- Current concerns about climate variability and climate change require improved management of water disasters to cope with more intense floods and droughts.

11.3.1 Policy, Legal, and Institutional Framework

Attitudes are changing as officials are becoming more aware of the need to manage disasters efficiently. Officials also see that the construction of new infrastructure has to take into account environmental and social impacts and the fundamental need for systems to be economically viable for maintenance purposes against disaster. However, they may still be inhibited by the political implications of such a change. The process of revising water policy is therefore a key step, requiring extensive consultation and demanding political commitment.

For many reasons, the governments of developing countries consider water disaster planning and management to be a central government responsibility. However, public and voluntary participation is the key issue in DM. This view is consistent with the international consensus that promotes the concept of government as a facilitator and regulator, rather than an implementer of projects even at the time of disaster. The challenge is to reach mutual agreement about the level at which, in the same specific instance, government responsibility should cease or be partnered by autonomous water services management bodies and/or community-based organizations. The role of government and national guards should be converted from legislations into laws in order to allow them to take over basic operation and management of facilities and equipment at the time of disasters.

In order to bring UWDM into effect, institutional arrangements are needed to enable

- The functioning of a consortium of stakeholders and water users to get involved in decision-making, with representation of all sections of society in planning for DM
- Water disasters management based on levels of disaster and how widespread it is
- Organizational structures at provincial and local levels to enable decision-making at the lowest appropriate level
- Government to coordinate the national management of water disasters across water use sectors

11.4 SOCIETAL RESPONSIBILITIES

Protecting people's lives and property from disasters and keeping social assets safe are the basic issues for the development of cities and their activities. Thus, an important problem is to make the cities more resistant to different types of disasters. In order to create a city for the twenty-first century where inhabitants can feel secure, all parties need to come together to promote disaster-resistant urban development through the mutual integration of livability. Water-related disasters are of significant importance because they include every facet of our urban life with high social and economic drawbacks.

The core of all activities and plan is concentrated on social responsibility (SR). As discussed in Chapter 7, classification of service activities of sustainable development aspects relating to drinking water supply, wastewater, and stormwater systems (ISO/TC 224) is developed in the draft ISO 26000. It has seven core issues of human rights, labor practices, organizational governance, fair pricing, environmental, consumer, and social. SR with main attributes related to the core issues should be considered as they are more sensitive and critical at the time of disasters. These are applicable to the management of any utility systems but more specifically to water and wastewater. ISO/TC 224 aims to increase relevant parties' awareness of water utilities and where separate stormwater service providers' duties in fulfilling the needs and expectations of their users and other stakeholders. It also provides water utilities with tools to help them provide their communities with adequate, sustainable, and resilient services.

Urban developers are instructed to take safety measures, regulations, and providing guidelines on the urban planning and structure of buildings, including those on the design of the building and the minimum open spaces for passage of water mains, tributaries, pumping stations, and service connections as well as shafts and conduits for distributing water through neighborhoods and buildings.

11.5 PLANNING PROCESS FOR UWDM

11.5.1 Taking a Strategic Approach

The recognition that water disaster in most parts is a result of failure of water management systems leads to long-term planning. The impact is more obvious at the time of water disasters. What we expect to achieve is a UWDM plan, endorsed and implemented by government to overcome the additional stress imposed on systems that normally fail to meet normal levels of water demand and services in many municipalities.

Being strategic means seeking the solutions that attack the causes of the water problems rather than their symptoms. It takes a long-term view. Understanding the underlying forces that cause water-related problems helps in building up a shared water vision and commitment to make that vision come true. In that sense, a strategy sets the long-term framework for incremental action that moves towards a basis for disaster prevention and formatting of the UWDM principles.

11.5.2 Scope of the Strategy Decisions

A strategy seeks to meet certain goals through specific actions and investments. In a strategy, the possible disasters and options to reach the goals of minimizing their unsuitable effects have to be assessed and a plan is devised for managing the probable disasters with expected consequences.

The scopes include policy, vision, goals, targets, institutional roles, and action plans.

A policy (or a problem statement) is often the starting point being a statement of intent. The essential difference is to translate the policies into a strategy.

The vision is water security across the board. It involves educating the public about disasters and how ready the system is to face disasters and what is expected from them. Calm, cooperative, and careful handling of disasters brings people together rather than separate them with anguish, sense of not belonging, and fear of isolation and disparity at the time of water disaster.

The strategic goals describe how the vision might be achieved. Each goal should cover a given issue (problems or certain opportunities may arise from a disaster), address the main changes required to make the transition to minimize unsuitable effects in a way that is broad enough to encompass all aspects of the issue, and ensure all relevant stakeholders.

Targets for each goal describe specific and measurable activities, accomplishments, or thresholds to be achieved within a given duration after a disaster's occurring. These form the core of any action plan and serve to focus disasters and guide the selection of options for action. Reaching such goals will often require a legal and institutional reform supported by specific management skills and instruments.

Institutional roles cover the roles, partnerships, and systems required to implement the strategy. This may include linkage between the UWDM plan and other strategic plans and between plans at different spatial levels: national, subnational, local, or for different sectors or geographical regions. It would identify which institutions are responsible for which parts of the strategy action plan. It might also signal to a rationale for streamlining institutions (especially where responsibilities overlap or conflict).

The action plan is developed from the outcome of the strategy. There is an inseparable link that refers the action plan back to strategy as further assessment and adjustment takes place.

11.5.3 UWDM as a Component of a Comprehensive DM

In a long-term strategic plan, a UWDM is a component of a comprehensive plan for lifeline disasters, which turns into a comprehensive regional and national plan for DM. Activities and guidelines of UWDM should be synchronized with the plans provided to compensate for the impacts of large-scale disasters such as an earthquake.

A disaster-resistant city (DRC) with targeted districts covers areas that are close to earthquake centers and seismic activities and their outskirts in the adjacent cities. The plan for DRC should aim at completion over a medium time horizon (say 10 years), but in districts in urgent need of development, completion within a shorter time period is required.

The minimum size of a disaster-prevention base should be determined so that adequate protection can be provided at the time of disaster. Disaster-prevention bases should be located in districts where there is much land space that can be purchased, such as factory sites and large factories that can be moved, as well as public lands.

Swift and systematic urban restoration following a major earthquake is an extremely important issue that will decide the future course of the city and the lives of the people who live in the area. Therefore, it is necessary to study plans and procedures to rebuild the city before another major disaster strikes. A guideline for the procedures and the planning of the administrative process must follow to ensure the swift and systematic revival of urban areas. Urban revival simulation training should be carried out with a carefully designed manual as a guide, with the aim of strengthening the coordination and deploying new methods (Tokyo Metropolitan Area, 2006).

11.5.4 Planning Cycle

Planning is a logical process, which is most effective when viewed as a continuous cycle. This cycle entails a vision and work plan for UWRM surrounded by situation analysis, strategy choice through implementation, and evaluation process (CIDA, 2005).

UWDM planning requires a team to organize and coordinate efforts and facilitate a series of actions before, during, and after the water disasters. An important starting point for government commitment is an understanding of UWDM principles and ensuring the allocation of water at least at a minimal level at the time of disaster.

Commitment from the public is necessary as they are the ones who strongly influence handling the disaster, through voluntary participation, and change in water use habits. Thus, planning requires recognizing and mobilizing relevant interest groups, stakeholders, volunteers, and the decision makers, despite their multiple and often conflicting goals in the use of water at the time of disaster. Politicians are a special group of stakeholders as they are both responsible for approving a plan and are also held accountable for its success or failure. Thus, a UWDM contains:

- Management of the process
- Maintaining political commitment
- Ensuring effective public participation
- Creating awareness of UWDM guidelines to be followed by different players in DM
- Possible involvement of National Guard or Army at the time of disaster

11.6 WATER DISASTER MANAGEMENT STRATEGIES

Possible solutions could be sought at the same time or immediately after defining the problems. Such solutions need to be analyzed, considering the requirements, the advantages and disadvantages involved, and their feasibility. Establishing the goals for the UWDM plan is important at this stage now that the extent of the problem and the hurdles to be faced are known. For each goal, the most appropriate strategy is selected and assessed for feasibility as well as its conformity to the overall goal of DM. The scope for technical and managerial action is considerable given the complexity of the water sector.

A national vision that water bridges between many cultural, social, and economic ties and incentives combined with a national drive for resource management and protection on conversion captures many common grounds about the use and management of water in a country. A vision provides direction to the actions on utilizing water and protecting it with high- or low-flow consequences and, in particular, guides the planning process. The vision may or may not be translated into water policies but would be expected to provide a guideline towards sustainable supply of the demands. The vision is particularly important in facing water disasters. It goes beyond quantitative and qualitative aspects of water and captures the heart of a nation and city and a community faced with water disaster. The question to be answered is whether a vision for DM is tied to with a water vision on integral water resources management. The answer could be found in the complexity of water-related activities of social vulnerability of society on water.

A change of paradigm is also in shifting from supply management to demand and shortfall managements, and it has made a drastic shift in our vision and actions on DM. In order to define the actions needed to reach such a change in vision, it is important to know the existing situation and to consult with stakeholders and various government entities to understand the needs for contingency plans for water disasters.

On the basis of the vision for disaster prevention and management, the situation analysis and the water disasters strategy can be prepared. Several drafts may be required not only to achieve feasible and realistic activities and budgets but also to get politicians and stakeholders to agree on the various trade-offs and the level of risk that can be taken. Approval by government is essential for disaster mobilization and implementation.

These are not dealt with in this training material. To obtain the UWDM plan is a milestone but not the end product. Too often, plans are not implemented and the main reasons are important to know and avoid:

- Lack of political commitment to the process, usually due to the drive coming from external sources or a lack of engagement of key decision makers in initiating the process.

- Unrealistic planning with disaster requirements beyond the reach of local government.
- Unacceptable plans that are rejected by one or more influential groups due to inadequate consultation or unrealistic expectations of compromise. With water disasters, where economic losses are incurring, adequate consultation is vital.

11.6.1 Disaster Management—Governance Perspective

According to UNEP (2005), major observations and lessons learned from the way governments handle DM are as follows:

- *Enhance the communication system*: Due to a failure of the telecommunication system, direct information on damages and intensity of the disaster will not be properly transmitted to other municipalities.
- *Incorporate environment issues in disaster management*: Current DM plans of the municipality do not concretely incorporate environmental issues. In contrast, environmental issues are handled by a separate department that has few links to disaster issues.
- *Relief versus long-term rehabilitation and reconstruction*: In most cases, the emergency committee, after the disaster, focuses on the emergency response and interagency coordination only. Reconstruction efforts are handled by individual departments as part of their everyday duties. Thus, there is a lack of integration of different issues and disciplines, and environment aspects are no exception. Long-term impacts of environment are often overlooked as a result.
- *A proactive link between research and practice*: It is often found that the practitioners at municipality and city governments are not aware of research results conducted in universities and research institutions. Thus, opportunities for information exchange in the form of study groups and/or participatory training programs should be created, where practitioners can interact with experts and resource persons on a continual basis.
- *Training of community leaders*: Similar to the above interaction, opportunities for training of community leaders should also be created and interactions among the practitioners and the community members promoted.
- *Dissemination of experiences to other municipalities*: It is important and relevant to share experiences of good practices with other municipalities and regions. Incentives to the municipality to undertake certain innovative approaches are also important, for example, in the form of awards or prizes. The municipality should take pride in the incentive and should promote its dissemination to other municipalities, thereby motivating its neighbors (UNEP, 2005).
- *Interlinkages between disaster preparedness and environmental management*: The interlinkages between disaster preparedness and environmental management in policies and plans at the local and national levels are the most important point in urban DM. DM plans should incorporate environmental dimensions and should anticipate the impact of disasters on the environment as well as the impact of environmental practices (e.g., forest and river management) on the impacts of disasters.
- *Plans and programs*: While disaster and environmental management should be linked to development planning, broader river basin management should also be considered in disaster programs and should be linked to the overall city management strategy.
- *Dissemination, adaptation, and implementation*: Many of the lessons from the past disasters in the same country are relevant to other cities/areas but are often not disseminated properly. Cooperation of cities and municipalities in this regard is essential for adaptation and implementation of the lessons learned.

- *Training and human resource development*: Proper training and human resource development programs should be undertaken not only for disaster managers but also for high-level decision makers such as city mayors. Training in decision-making systems is essential in this regard to integrate disaster preparedness and response into the larger development and management of cities in general (UNEP, 2005).

11.6.2 Initiation

The initiation of a UWDM planning process may arise from several sources. Internationally, governments have agreed at the global summit on system dynamics to put in place plans for sustainable management and development of water disasters by 2005.

At the national level, many governments are aware of the problems that their own water sector is facing from issues, such as pollution, scarcity, emergencies, competition for use, and have identified action as a priority. The focus on specific water problems or problem areas is also adequate stimulus for the government to act, and even if this results in a more focused plan of action to solve a specific issue, it may lead to the gradual development of a fully integrated approach to water DM. The process for the development of the UWDM plan requires a different process than is usually taken in governmental planning. Key differences include

- *A multisectoral approach*: To manage water in an integrated way means developing linkages and structures for management across sectors. For such strategies to be successful, the main water use sectors should be involved in planning and strategy development from the outset.
- *A dynamic process*: The development of a sustainable management system for water disasters and the integrated approach will be a long process. This will require regular review, adaptation, and possibly reformulation of plans to remain effective.
- *Stakeholder participation*: Because most problems with UWDM are felt at the lowest levels and therefore changes in water management are required down to the individual participation, the strategy development process requires extensive consultation with the stakeholders.

There are several key activities regardless of the underlying reasons; they include obtaining government commitment, raising awareness on principles for management of water disasters (UWDM), and establishing a management team.

11.6.2.1 Political and Governmental Commitment

Political support and commitment are essential for the success of any part of the DM process, and the changes across ministries or legal and institutional structures require the highest level of political commitment (cabinet, head of the state). Some of the reasons for strong political support are as follows:

- Multiple levels will be involved at the time of disaster, and only through the right political process can a coordinated emergency be handled properly.
- Ensure that the water disasters vision and objectives incorporate political goals consistent with other national goals especially for allocation of immense financial resources needed at the time of widespread water disaster.
- Conversely, ensure that the water vision and objectives are reflected in political aspirations.
- Ensure that the policy implications of the strategy are followed and considered throughout the process and not merely at some formal end point (to allow a continuous improvement approach to the work).

- Make decisions on recommended policy plans with legal and institutional supports.
- Ensure that the plan is adopted and followed through.
- Commit government funds (and, if necessary, mobilize donor assistance).

Political commitment needs to be of long term and therefore across political parties so that it is not rejected when a new government takes office. For this reason, a compelling vision is that all parties can aspire to provide a good foundation for action.

11.6.2.2 Policy Implications for Disaster Preparedness

Urban disasters (natural and man-made) can cause large loss of life and have enormous economic and financial costs. The drive for mitigation increases as the effects of disasters, and the costs of failures become more immediate and widely spread.

Institution establishments are needed that can motivate action in advance of disaster and can share the costs and benefits of preventive measures among affected people in a fair manner. Hazard mitigation requires improving knowledge, building constituencies for risk reduction, and strengthening institutions and partnerships across levels of government and the private sector.

Stakeholder analysis essentially involves identification of key players in DM.

Certain important policy developments and actions in countries facing different disasters include (UNEP, 2005)

- Raising awareness about the integration of environment and disaster issues.
  - *Proposed actions: Arrange meetings, seminars, and forums at different levels.*
- Documenting and disseminating information.
  - *Proposed actions: Undertake field surveys after different disaster events to document concrete examples and disseminate it through the Internet, and in training and study programs.*
- Bridging the gaps between knowledge and practice.
  - *Proposed actions: Undertake and facilitate training programs involving experts, resource persons, practitioners, and community leaders through forums.*
- Implementing practical exercises and techniques.
  - *Proposed actions: Undertake small projects at different locations having different socioeconomic contexts. These will be learning experiences, which can be also disseminated to other areas.*
- Developing guidelines and tools on environment and DM.
  - *Proposed actions: Guidelines and tools should be developed based on the field experiences, and incorporating the comments and expertise of practitioners and professionals from different parts of the world.*
- Establishing a continuous monitoring system.
  - *Proposed actions: A network or partnership should be developed, which will continuously monitor different activities, identify best practices, and disseminate through training and capacity-building programs (UNEP, 2005).*

Building knowledge about the physics of hazards may be inadequate or absent, even among residents at risk, yet community awareness of physical hazards is fundamental for mitigation efforts. Comprehensive vulnerability assessments using remote sensing, satellite imagery, and risk and loss estimation modeling can help document and reduce physical, social, and economic vulnerability. Changing physical infrastructure and innovative techniques for retrofitting buildings can improve disaster prevention. Soft nonstructural methods can increase hazard information, create new knowledge, build physical and human resources capacities, and train and raise the awareness of decision makers and the communities at risk.

11.6.2.3 Public Participation

The provision of a foundation and strategy for involving the stakeholders in the various stages of preparing and implementing the UWDM plan is needed, so that the stakeholder engagement strategy runs efficiently through the planning and political processes as well as planning and supplementation processes.

11.6.2.3.1 Benefits of Public Involvement

- It can lead to informed decision-making as stakeholders often possess a wealth of information that can benefit the project.
- Stakeholders are the most affected by lack of or poor water disasters management.
- Consensus at early stages of the project can reduce the likelihood of conflicts at the time of water disaster and public participation in successful handling of the disaster.
- Stakeholder involvement contributes to the transparency of public and private actions, as these actions are monitored by the different stakeholders that are involved.
- The involvement of stakeholders can build trust between the government and civil society, which can possibly lead to long-term collaborative relationships.

11.6.2.3.2 Methods for Public Participation

Stakeholders should be engaged at all critical steps in the process of developing the plan. These stages should be planned and the work plan should identify the timing, the purpose, the target stakeholders, the method, and the expected outcome. The scale and strategy of stakeholder participation must be carefully determined and be coordinated with the emergency response team and perhaps national guards.

Methods may include

- Stakeholder workshops, in which selected stakeholders are invited to discuss water issues
- Representation in the management structure for the planning process
- Local consultations "on the ground"
- Surveys
- Consultations with collaborating organizations (such as NGOs and academic institutions)

Using multiple sources of information not only has the advantage that the information obtained is more likely to be accurate but especially the participatory methods of information gathering can also contribute to creating a sense of local responsibilities of the process and consensus about the actions to be taken. Stakeholder participation techniques range from a low level of involvement to a high level of involvement depending on the state of disaster.

11.6.2.4 Lessons on Community Activities

- *Perception versus action*: There is a difference between perception and the action taken. Although people are quite aware of the risks involved and of risk mitigation measures, it is not reflected in the action that they take.
- *Preparedness for evacuation*: The experiences in different cities also demonstrate the need for the local government and community to work together in organizing evacuation simulation exercises, and designating "disaster evacuation areas" for residents to assemble during a disaster.
- *Self-reliability versus dependence*: A new system of emergency radios that the city had introduced made people more reliant on the system. A balanced approach that introduces new ideas without destroying local existing systems is necessary.

- *Public awareness*: It is important to raise awareness among people and communities about the link between environmental management and disaster preparedness. The link between upstream and downstream issues should not only be reflected in river basin management but should also be incorporated in community awareness raising and implementing projects at the local level (UNEP, 2005).

11.6.3 Steps in Drought Disaster Management

The ability of a water supply system to face a disaster caused by drought can be improved through long-term management such as construction of infrastructures development and rehabilitation, reduction of water distribution and water use losses, dual WDS, and reuse of wastewater.

In short-term water DM, the severity of a hydrological drought can be reduced by supply/demand management through delivery restrictions, augmentation of water availability by a lower quality or by a greater cost than ordinary sources, and making better use of available water resources by optimal-operation-of-reservoir rules.

Common UWDM during a drought consists of the following steps:

1. Monitoring drought indices/forecasting of water resources and demands
2. Consideration of drought management options
3. Establishment of levels of indicators that trigger the various options of a water disaster program
4. Adaptation of a management plan at the levels indicated by the drought indices

11.6.4 Drought Management Case—Georgia, USA

Periods of drought have naturally come and gone throughout history. In recent years, however, some regions of the United States such as Georgia have become more vulnerable to the effects of drought as pressures increase on natural resources. Rain deficits are rising, crops are dying on the vine, and lake levels are plummeting. Meanwhile, concern is growing among water managers and others. Already, 31 states have implemented drought management plans, five of which attack the situation proactively, rather than emphasizing an emergency response. Georgia will soon be among the proactive.

Georgia's plan links indicators, which characterize stages of drought severity, with responses. For instance, when an indicator shows that a drought is developing, a response could be to curtail nonessential water uses. Then during a drought, the plan also provides guidance, such as when and how to make water use restrictions more stringent, or when and how to implement water use surcharges to manage demand. The idea is that it is more effective—and less painful—to prepare for drought, and take timely actions, than to wait until a full-blown disaster has developed.

An example of such strategies at an individual level is an alternative for chemical applications to lawns, which is one of the major sources of nonpoint pollution. This alternative, called xeriscaping, uses more native vegetation to reduce the need for water, pesticides, and fertilizers. Xeriscaped lawns also cost less to maintain while lowering public health risks. It is important to think about the impacts of each individual decision.

Businesses and industries can make a difference by lessening their dependence on water. At the local government level, it is hoped that officials will understand the need for drought planning, particularly as population growth occurs, because drought occurs when demands exceed supplies. Demand for water has skyrocketed in Georgia, while new supplies are limited.

At the state level, officials are beginning to take a risk management approach to drought management, which is a slow process to change the historic way of doing things.

11.6.5 Flood Management Case—Northern California, USA

An unusual series of storms from January 5 through 26, 1995, caused heavy, prolonged, and, in some cases, unprecedented precipitation across California. This series of storms resulted in widespread minor to record-breaking floods from Santa Barbara to the Oregon border. Several stream-gaging stations used to measure the water levels in streams and rivers recorded the largest peaks in the history of their operation.

11.6.5.1 Flood Characteristics

Before the January storms, rainfall was near normal across northern California. The most intense storms occurred during the week of January 8 and produced an average 33 cm of precipitation over most of northern California. Precipitation amounts of as much as 61 cm were recorded for the week. Maximum 1-day rainfall data of these storms were compared with the theoretical 100-year, 24-hour, precipitation, and rainfall amounts were greatest in Humboldt, Lake, Mendocino, Napa, Sacramento, Shasta, Sonoma, and Trinity Counties.

Flooding was significant throughout northern California from January 9 to 14. The highest peak flows in the region occurred from January 8 to 10 with a peak height of 8 m. Flooding in small basins was unusually rapid because of the high-intensity, short-duration microbursts of rainfall. Small streams rather than large rivers caused most of the damage in the Central Valley. Flooding along large rivers in the Central Valley was controlled by diversions and flood-control reservoirs that had large storage capacities available. Small streams and rivers in the coastal areas also caused widespread flooding. More than 10 years of drought and low stream flows resulted in the accumulation of dense riparian vegetation in most stream channels. Flooding along the Russian River was due, in part, to the accumulation of vegetation and debris that had reduced the capacity of the stream channel. This reduced capacity resulted in higher water levels with less stream flow.

11.6.5.2 Response

In recent years, information compilation has been improved by the use of electronic data collection platforms (DCPs). These platforms use automated earth-satellite telemetry for the immediate transmission of data from remote sites. Cellular telephones and modems also are being used to obtain timely information, but not all stream-gaging stations are equipped with this instrumentation. This on-time information is used for early warning of flood. When on-time information shows increasing of water levels at upstream rain gauges, a warning is given that a flood is about to occur. According to the amount of increase in water levels, suitable decisions are needed to mitigate or decrease impacts of flood in habitable areas downstream. During the recent floods, the USGS was able to rapidly compile and disseminate near-real-time information for many of its gaging stations by using the telephone and computer networks. The USGS has also made data available on the Internet to provide immediate access to flood data and other hydrological data. During January 1995, most of data were made available within hours of a flood peak.

11.7 SITUATION ANALYSIS

Output from the situation analysis is a report elaborating the progress with implementing improved management of water disasters, the outstanding issues, the problems, and some of the solutions. The purpose of this step is to help characterize the present situation and to use the information to predict possible future UWDs for developing a UWDM approach.

For the purposes of a UWDM plan, the situation analysis is assessed against the principles of those embodied in the UWDM approach. Analysis and interpretation made against predefined goals and the national DM vision or policy can be focused and targeted to address the main constraints and causes rather than the symptoms. The analysis report should be shared widely, and this means summarized as appropriate.

An essential aspect in the management of UWDs is the anticipation of change including the changes in the natural system due to geomorphological processes, the changes in the engineered components due to aging, the changes in the demands, and even changes in the supply potential of water, possibly due to changing climate and weather. System dynamics is a new approach that helps managers to see all the impacts of a disaster. System dynamics is the essence of the limit to growth and how understanding that limit could help us to prevent disaster and/or manage it intelligently. We are not elaborating on this, but we are making a reference to the ongoing study on the future water allocation from Karkheh Reservoir in Iran (Karamouz et al., 2004).

11.7.1 Steps in the Development of Situation Analysis

11.7.1.1 Approach

There is a role for specialist expertise in conducting the analysis when high-tech skills are required, large baseline surveys need to be done, or there is particular need for an independent viewpoint. There are several related principles for coordinating the collection of knowledge:

- Multistakeholder groups themselves should design the information gathering, analysis, and research process to ensure ownership of the strategy and its results.
- All the "analysis" tasks are best implemented by bringing together, and supporting, existing centers of technical expertise, learning, and research.
- Since analysis is central to strategy development, it should be commissioned, agreed, and endorsed at the highest level (i.e., by key government ministries or by the planning steering committee). This will increase the chance that analysis will be well focused and timely in relation to the plan's evolution and timetable and that it will be implemented.
- In the same way, analysis needs good coordination. It is logical for the DM team to coordinate the analysis, but it should not undertake all the analyses itself.

11.7.1.2 Objectives

The objectives of the analysis must be clear. The field of UWDM is large and covers a huge number of issues. The purpose of the situation analysis is to study the UWDs in terms of the UWDM principles. Weaknesses, problems, and issues identified in UWDM may arise from the following areas:

- Policies affecting UWDM
- Legislation on preventive measures as well as mobilization of people/equipment during disaster and relief effort after that
- Charter of water DM institutions and commissions
- Practicing UWDM for different types of disaster

An analysis of the present water disasters management situation in the country should therefore identify gaps in the management framework and allow a prioritization of action.

11.7.1.3 Data Collection

Data for the situation analysis comes from a variety of sources. For reasons of efficiency and effectiveness, the planning process should build on and explore earlier knowledge and experience and draw on lessons learned. Part of such useful knowledge is not readily available or well documented. It often exists in an ad hoc form among professionals and practitioners as well as among government and nongovernment staff within disaster management and water-relevant sectors. The political level holds important knowledge on the various processes involved in achieving overall endorsement of the goals of the plan and rallying support to its implementation.

The knowledge to be compiled and made available includes the following areas:

- UWDM experience at the country level, where elements of UWDM frameworks may have been completed in part or in full. National DM policies, management organizations, and water disasters assessment tools are constantly changing in many countries around the world and constitute important requisites for UWDM.
- International UWDM experience, which can mean both experiences collected from several countries or groups of countries and experiences where regional aspects of disasters are dominating.
- Experience from past and present national planning processes within other sectors and in particular those cutting across several sectors. Examples of such processes are development of damage reduction strategies, strategies and plans for sustainable level of water demand, development of water rationing, and conservation strategies.

11.7.2 Urban Disasters Situation Analysis

The situation analysis should examine all types of UWDs with respect to different quantity and quality problems. It should identify the pertinent parameters of the hydrological cycle and urban morphological condition and evaluate the cost of damages caused by different types of disasters. The analysis should pinpoint the major water disasters issues, their severity, and social implications, as well as the risks and hazards of them. For the purposes of UWDM planning, care should be taken not to embrace an approach that is too technical but to emphasize the implementation process and the enabling environment for efficient and effective handling of the disaster.

Socioeconomic aspects are important when looking at the impacts of the water disasters on water users (including environment) and society as a whole. Articulating priority goals can focus more attention on the future management situation

11.8 DISASTER INDICES

Disaster indices provide ways by which we can quantify relative levels of disaster. They can be defined in a number of ways. One way is to express relative levels of disaster as separate or weighted combinations of reliability, resilience, and vulnerability, the measures of various criteria that contribute to human welfare in time and space. These criteria can be economic, environmental, ecological, and social. To do this, one must first identify the overall set of criteria, and then for each one decide which ranges of values are satisfactory and which ranges are not. These decisions are subjective. They are generally based on human judgment or social goals, not scientific theories. In some cases, they may be based on well-defined health and safety standards. Most criteria will not have predefined or published standards or threshold values separating what is considered satisfactory and what is not. For many criteria, the duration as well as the extent of individual and cumulative failures of water resources systems should be determined for each criterion (Karamouz and Moridi, 2005).

11.8.1 Reliability

System performance indices are essential for the evaluation of system performance to identify optimal situations. These indices show the system's ability to work without any problem. System performance can be desired or undesired. Undesired conditions are named as failure. A system's weaknesses in delivering water with the desired pressure and the requested quantity and quality are different aspects of system failure. Mean and standard deviation of the system's output are applicable for the evaluation of system performance but are not sufficient. Figure 11.3 shows the weakness of mean and standard deviation indices for definition of severity and frequency of system failure.

Diagrams of Figures 11.3a and b are symmetrical to the axis x, so their mean and standard deviation are the same. There are two failure events in Figure 11.3b, but there is no failure situation in the other figure. Furthermore, it cannot be identified how an increase or decrease in the system's means can affect the system performance. Because of these weaknesses in using mean and standard deviation indices for the evaluation of probability of failure and system performance, reliability is used.

The random variable of X_t is the system's output classified into two groups of failure and success outputs. The system reliability is the probability that X_t belongs to the success outputs group (Hashimoto et al., 1982). The reliability of a system can be calculated by

$$\alpha = \text{Prob}\left[X_t \in S\right], \quad \forall t, \tag{11.1}$$

where

S is the set of all satisfactory outputs. Based on this definition, reliability is the opposite of risk, in which the probability of system failure could be expressed upon.

Reliability can be measured in several different but related ways, depending on the needs and relevance of a particular situation. For example, it may be useful to express and characterize the expected length of time between successive failures (i.e., time to failure), similar to the notion of a 100-year flood event, an event that is expected to occur on average once in 100 years.

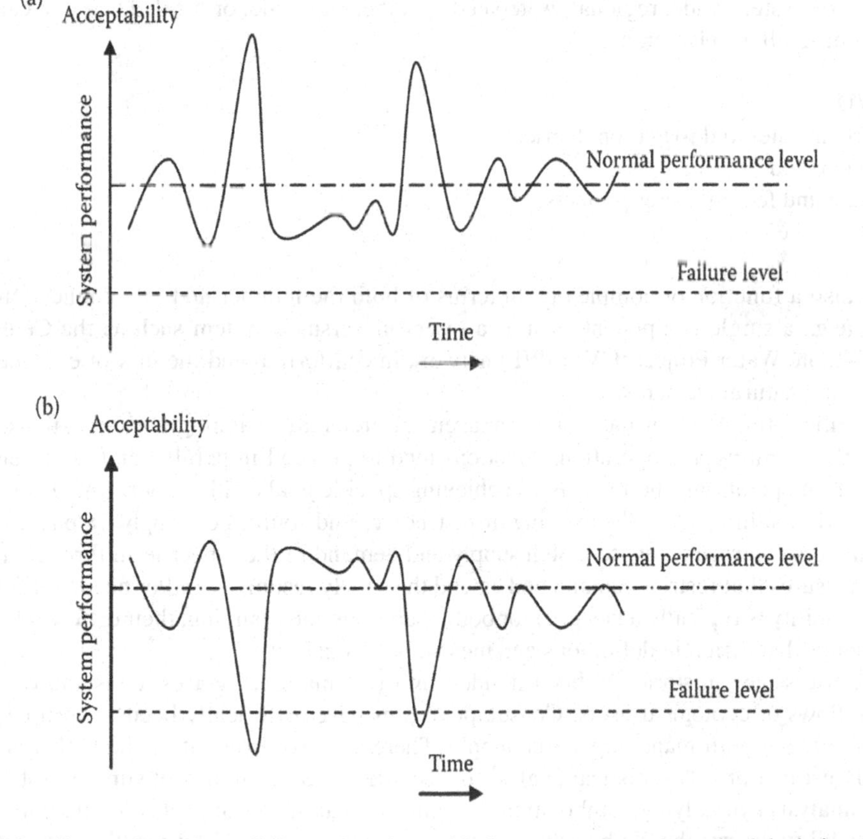

FIGURE 11.3 Comparison of two different functions with the same mean and standard deviation (a and b). (After Hashimoto, T., Stedinger, J.R., and Loucks, D.P., *J. Water Resour. Res.*, 18(1), 14–20, 1982.)

While reliability is frequently used in the planning and operations documents, trade, and water, it is

- Often not formally defined by agencies or institutions
- Concerned with different water management entities' operating and planning processes with substantial variation
- Often considered in qualitative rather than quantitative terms

Thus, there is no existing and widely accepted framework of definitions and measurement approaches that are applicable in all aspects of influence and applications.

Reliability at key points within the sphere of influence ultimately depends on the complexity of interactions of a web of rights, regulations, laws, and rules that affect all institutions that manage water in a municipality. It will be necessary to understand these interactions in order to develop meaningful definitions of water supply reliability and, based on these definitions, related indicators could be defined. These interactions play out across numerous dimensions that include, primarily

- System scale and complexity
- Specific management function (i.e., operations, planning)
- Purpose of the service, that is, the intended use of water

A major determinant of how reliability is defined as the scale or slice of the water stock. Discussions with water professionals and a review of the literature indicate that differences in the system scale (i.e., the boundaries of the system) will have a significant influence on how reliability is defined. Scale can be system-wide, regional, watershed-specific, city-wide, or local. The main categories of scale, from smaller to larger, are

- WDS
- Urban water and irrigation districts
- Watershed
- State and federal water projects
- Statewide

Scale is also a function of complexity, in terms of both the number and interconnections of components (e.g., a single component such as a reservoir versus a system such as the Central Valley Project—State Water Project (CVP/SWP) network in California) and the mix of engineered infrastructure and natural elements.

Two critical functions in the water management arena are planning and operations. In public utilities, the planning and operations functions tend to proceed in parallel and are often not fully integrated. In operations, the focus is on achieving specific goals within short time frames (hourly, weekly, and monthly), given the existing infrastructure and sources of supply. In planning, in contrast, planners determine how to match supply and demand in the longer term through a combination of measures that restrict demand and extend the supply resources and/or needed infrastructure. While reliability is typically a key goal for both operations and planning, their different time frames lead to somewhat different definitions and measures of reliability.

Water users have historically been divided into two main categories: consumptive users and instream flows or ecological users. Consumptive uses have historically been the primary focus of agencies charged with managing water supply. There are two different methods that are used for reliability estimation: (1) statistical analysis of recorded data of failures of similar systems and (2) indirect analysis by studying combination of significant parameter impacts on reliability.

In reliability theory, the Weibull distribution is assumed to model lifetime (the time between two consecutive system failures) most appropriately. See Karamouz et al. (2012) for more details.

11.8.1.1 Reliability Indices

For reliability analysis, a performance function *W(X)* of X_L and X_{Re}, which are system load and resistance, respectively, proposed by Mays (2001) is defined.

$$W(X) = W(X_L, X_{Re}), \tag{11.2}$$

$$R = p\left[W(X_L, X_{Re}) \geq 0\right] = p\left[W(X) \geq 0\right]. \tag{11.3}$$

$W(X)=0$ is called failure surface or limit. $W(X) \geq 0$ is the safety region and $W(X) < 0$ is the failure region. The performance function *W(X)* could be defined in different forms such as

$$W_1(X) = R_e - L = h(X_{Re}) - g(X_L), \tag{11.4}$$

$$W_2(X) = \left(\frac{R_e}{L}\right) - 1 = \left[\frac{h(X_{Re})}{g(X_L)}\right] - 1, \tag{11.5}$$

$$W_3(X) = \ln\left(\frac{R_e}{L}\right) = \ln\left[h(X_{Re})\right] - \ln\left[g(X_L)\right]. \tag{11.6}$$

Another reliability index is the β reliability index, which is equal to the inverse of the coefficient of variation of the performance function, *W (X)*, and is calculated as

$$\beta = \frac{\mu_w}{\sigma_w}, \tag{11.7}$$

where

μ_w and σ_w are mean and standard deviation of the performance function, respectively.

By assuming an appropriate probability density function (PDF) for random performance functions, *W (X)*, using Equation (11.1) the reliability is estimated as

$$R = 1 - F_W(0) = 1 - F_{W'}(-\beta), \tag{11.8}$$

where

F_W is the cumulative distribution function of variable *W*, and
W′ is the standardized variable.

$$W' = \frac{W - \mu_W}{\sigma_W}. \tag{11.9}$$

$$R = p(FS > 1) = p(\ln(FS) > 0) = p\left(\ln\left(\frac{Re}{L}\right) > 0\right),$$

$$R = p\left(z \leq \frac{\mu_{\ln FS}}{\sigma_{\ln FS}}\right) = \Phi\left(\frac{\mu_{\ln FS}}{\sigma_{\ln FS}}\right), \tag{11.10}$$

where

$z = \left(\ln(FS) - \mu_{\ln FS}\right)/\sigma_{\ln FS}$ and

Φ is the standard cumulative normal probability distribution.
It is assumed that the FS is normally distributed.

If the resistance and loading are independent and log normally distributed, then the risk can be expressed as (Mays, 2001).

$$\bar{R} = \Phi\left\{\frac{\ln\left[\left(\bar{Q}_{Re}/\bar{Q}_L\right)\sqrt{\left(1+\Omega_L^2\right)/\left(1+\Omega_{Re}^2\right)}\right]}{\ln\left(\sqrt{\left(1+\Omega_L^2\right)\left(1+\Omega_{Re}^2\right)}\right)}\right\}. \quad (11.11)$$

where
$\bar{Q}_{Re}$, $\bar{Q}_L$ are the resistance and load flow discharges, respectively and
Ω_L, Ω_{Re} are corresponding coefficients of variation for resistance and load discharges.

11.8.1.2 Mean Value First-Order Second Moment (MFOSM) Method

Mean and standard deviation of the performance function, $W(X)$, are estimated using the first-order variance method in uncertainty analysis (a detailed discussion on uncertainty analysis is given in Section 9.3). These values are used for estimation of the β reliability index as

$$\beta = \frac{W(\mu)}{\sqrt{S^T C(X) S}}, \quad (11.12)$$

where
μ is the mean vector,
$C(X)$ is the covariance matrix of random variable X, and
S is the N-dimensional matrix equal to the partial differential of W to X (Mays, 2001).

Studies have shown that in a confidence boundary of <0.99, the reliability value is not sensitive to the distribution function of W and assumption of a normal distribution for W would be sufficient. But for confidence levels that are more restrict than 0.99, such as 0.995, the extreme edges of the distribution function of W will be critical.

The application of this model is easy; however, it must be noted that there are some weaknesses because of the nonlinearity and noninvariability of the performance function in critical situations. Therefore, the application of this method in the following situations is not recommended:

- Estimation of risk and reliability with high precision
- Highly nonlinear performance function
- Existence of random variables with high skewness in performance function

In order to decrease the errors of this method, another method called the advanced first-order second-moment (AFOSM) method has also been developed by Tung et al. (2004).

11.8.1.3 AFOSM Method

The main thrust of the AFOSM method is to reduce the error of the MFOSM method, while keeping the advantages and simplicity of the first-order approximation. The expansion point in the AFOSM method lies on the failure surface defined by $W(X)=0$. Among all the possible values of X that fall on the limit state surface, one is more concerned with the combination of stochastic variables that would yield the lowest reliability or highest risk. The point on the failure surface with the lowest reliability is the one having the shortest distance to the point where the means of stochastic variables are located. This point is called the design point or the most probable failure point. With the mean and standard deviation of the performance function computed at the design point, the AFOSM reliability index can be determined.

Owing to the nature of nonlinear optimization, the algorithm AFOSM does not necessarily converge to the true design point associated with the minimum reliability index. Therefore, different initial trial points are used and the smallest reliability index is selected to measure the reliability (Mays, 2001).

11.8.2 Time-to-Failure Analysis

Any system will fail eventually; it is just a matter of time. Due to the presence of many uncertainties that affect the operation of a physical system, the time when the system fails and its performance is unsatisfactory are intended as random variables. Instead of considering detailed interactions of resistance and loading over time, a system or its components can be treated as a black box or a lumped parameter system and their performances are observed over time. This reduces the reliability analysis to a one-dimensional problem involving time as the only random variable.

11.8.2.1 Failure and Repair Characteristics

For a repairable system or component, its service life can be extended indefinitely if repair work can restore the system as if it was new. Intuitively, the probability of a repairable system available for service is greater than that of a nonrepairable system.

For repairable urban water systems, such as pipe networks, pump stations, and storm runoff drainage structures, failed components within the system can be repaired or replaced so that the system can be put back into service. The time required to have the failed system repaired is uncertain and, consequently, the total time required to restore the system from its failure state to its operational state is a random variable. Like the time to failure (TTF), the random time to repair (TTR) has the repair density function describing the random characteristics of the time required to repair a failed system when failure occurs at time zero. The repair probability is the probability that the failed system can be restored within a given time period and it is sometimes used for measuring the maintainability or repairability. They are usually expressed in the form of resiliency discussed in Section 11.8.3.

11.8.2.2 Availability and Unavailability

The term "availability" is generally used for repairable systems to indicate the probability that the system is in operating condition at any given time *t*. On the other hand, reliability is appropriate for nonrepairable systems indicating the probability that the system has been continuously in its operating state starting from time zero up to time *t*.

Availability can also be interpreted as the percentage of time that the system is in operating condition within a specified time period. On the other hand, unavailability is the percentage of time that the system is not available for the intended use during a time period, given that it is operational at time zero.

11.8.3 Resiliency

Resiliency describes how quickly a system recovers from failure once failure has occurred. In other words, resiliency is basically a measure of the duration of an unsatisfactory condition. Resiliency could be considered as the most important indicator of system performance in disaster recovery and how successful the management of disaster strategies has been. In general form, the resiliency of a system can be calculated as (Karamouz et al., 2003).

$$\beta = \text{Prob}\left[X_{t+1} \in S \middle| X_t \in F\right], \tag{11.13}$$

where

S and F are the set of all satisfactory and unsatisfactory outputs, respectively.

The application of this definition in practical cases has been discussed in Section 11.11.1.

Developing resilience systems and living with disaster are the major challenges of the recent decades. Alternative solutions have been explored and the concept of resilience has been introduced into risk management of all the water planning and management fields. The principle of resilience is from ecology. Holling (1973) used the following definition for resiliency: "a measure of the persistence of systems and of their ability to absorb changes and disturbances and still maintain the same relationships between populations and state variables." Pimm (1991) has defined resiliency as follows: "the speed with which a system disturbed from equilibrium recovers some proportion of its equilibrium." More generally, "resiliency" could be considered as a tendency to stability and as resistance to perturbation. There is no general and overall accepted definition for resiliency, although everybody agrees that it is desirable.

Resilience has also been defined as the ability of a system to persist if exposed to a perturbation by recovering after the response. Using this definition of resiliency, it is the opposite of resistance, which is the ability of a system to persist if disturbed, without showing any reaction at all.

The resiliency concept definition and formulation is specialized regarding the field of application. For example, De Bruijn and Klijn (2001) defined resilience in the context of flood risk management. Strategies for flood risk management in which resilience is used focus on reducing the impact of floods by "living with floods" instead of "fighting floods," as in the traditional strategy. Therefore, resilient flood risk management is the flood risk management that aims at giving room to the floods but with concurrent impact minimization. This also implies that the consequences of floods have to be taken into account and that safety standards must be differentiated on the basis of land use and spatial planning. The area as a whole is more resilient if the less valuable parts are flooded prior to the more valuable parts. A resilient flood risk management strategy also considers measures to reduce the impacts of flooding, such as the design of warning systems and evacuation plans and the application of spatial planning and building regulations. Resilience strategies may also include measures to accelerate the recovery after a disaster, for example, damage compensation and insurances.

Bruneau et al. (2003) proposed one of the many methods for quantifying resiliency which is in the context of four criteria known as 4R's (Robustness, Rapidity, Resourcefulness, and Redundancy). Rapidity is the system's ability to achieve the expected level of performance in the shortest possible time, while robustness is the system's resistance to maintain its function against stressors. On the other hand, resourcefulness is defined as the ability to employ material and human resources to meet the expected level of service, and redundancy is the units that are available for replacement in different parts of the system to ensure its continued operation.

Robustness describes the ability of the WWTP to withstand a certain level of flooding without suffering degradation and maintaining the wastewater treatment function. On the other hand, rapidity describes the system's ability to meet goals and priorities to avoid service disruption promptly. Therefore, together, these two terms can represent the ends of resiliency for a WWTP. If the estimations of the WWTP's robustness and rapidity are both high, then it could be considered resilient. Moreover, redundancy and resourcefulness are means by which the system can become more resilient (Karamouz and Movahhed, 2021). For further explanations on how to apply this method in an MCDM problem see case study 2 in Section 13.11.1.1 in Chapter 13.

11.8.4 Vulnerability

Vulnerability is the distance of failed character state from the desired condition in historical time series. There are three ways of vulnerability estimation. Some researchers consider the maximum distance from the desired situation, other groups consider the expected value of system failure cases, and the last group considers the probability of failure occurrence. Among these different ideas, the second idea is more common. Considering vulnerability as the expected value of system failure, the vulnerability of a system is defined as

$$\text{vulnerability} = \frac{\text{no. of situations with positive } (X_T - X_t)}{\text{no. of failure situtations}},$$

where
X_T is the desired situation and
X_t is the system situation.

Using the probability of failure occurrence, the vulnerability of a system can be calculated by the following formula:

$$\nu = \sum_{j \in F} S_j e_j, \tag{11.14}$$

where
e_j is the probability that x_j corresponds to s_j.

However, in this definition, we usually assume that events and failures are singular and independent, that is, several simultaneous events and cascading effects of failures are unconsidered. For instance, if a water intake (spring, tank, etc.) is out of service, water is delivered from other intakes (redundant capacity, emergency supply, etc.). Hence, the flow regime and subsequently the pressure distribution in the distribution network are changed. These changed pressure conditions can increase the probability of, for example, pipe bursts. An example for simultaneous events in an urban drainage system is the simultaneous interruption of the power supply for an emergency pump at a combined sewer overflow (CSO) and a fluvial flood surcharge of the outlet structure, both caused by a thunderstorm (Sitzenfrei et al., 2011).

Vulnerability analysis is a worthwhile tool in evaluating the entire event chain causing water disaster. Recovery analysis means a systematic investigation of how a system returns to the state of normal operation (Karamouz et al., 2003). With various formulation of vulnerability system and assessment framework, researchers have targeted the vulnerabilities and adaption strategies of different urban infrastructures including the energy and power, water and wastewater, building environment, transportation, and telecommunication systems, while climate change impacts on these systems from multiple causes such as temperature increase, sea level rise, humidity change, and precipitation pattern variation have been widely discussed (Dong et al., 2020). Further discussion on using this index for the evaluation of system performance in practical cases is given in the following section. Vulnerability in the context of risk management is explained in Chapter 10, Section 10.9 and in Section 11.10.1.

11.8.5 Sustainability Index

In order to making sustainability, a better understanding of the variables which affect the sustainability of water supply and demand and the relationship among them must be realized. Therefore, an index is needed for connecting the elements of water balance in order to quantify the state of water sustainability and the extent of water conservation needed for sound water resources management (Karamouz et al., 2006).

In recent years, several studies were done to develop sustainability indices based on identifying indicators of systems performance and/or economic returns. Among them are sustainability index (SI) by Loucks (1997), water poverty index (WPI) by Sullivan (2002), Canadian water sustainability index (CWSI) by the Policy Research Initiative (2007), watershed sustainability index (WSI) by Chaves and Alipaz (2007), and West Java water sustainability index (WJWSI) by Juwana et al. (2010).

Sandoval-Solis et al. (2011) extended SI developed by Loucks (1997) while modifying some factors including structure, scale, and content to make it more adjustable to the demands of each water user, type of use, and basin. They also developed the sustainability group index (SGI) for the Rio Grande basin by assigning weights to the SI of each water user.

Karamouz et al. (2017a) developed a supply and demand SI called the Paydari Index (PI). They also performed an uncertainty analysis to assess the random nature of variables in estimating water balance and quantifying water sustainability. This index includes parameters that are the difference between supply and demand, percentage of the satisfied demand, productivity of water resources, and an indicator for evaluating the reduction of aquifer storage. Finally, these methods are compared and a hybrid index combining the indices is developed.

In the context of developing indices, many investigators barely allude to the social issues of water allocation, ensuring some level of social and stakeholders' satisfaction including the application of Nash bargaining theory in the social welfare optimization and conflict resolution (Brânzei et al., 2017). Karamouz et al. (2021) brought social aspects into SI of supply and demand. An index called PI-Plus (P for Paydari as it is an indigenous term for sustainability) is extended, and its interdependencies with another index called SGI are taken into account: PI and proposed PI-Plus. Different indices have been used to assess the sustainability of water supply–demand within a catchment. The most important elements of water sustainability in a region are its water supply–demand, environmental, and socioeconomic attributes. For further explanations, see the case study in Chapter 10.

11.8.6 Drought Early Warning Systems

In comparison with drought vulnerability mapping, early warning systems are based on longer-term meteorological forecasting. Groundwater drought early warning is usually accomplished at two levels: global and regional levels. The global level is in international meteorological scales and considers climate and climate variation, including long-term changes. At the regional level, which is in government, donor, and NGO scales, watershed/aquifer scale signals of groundwater drought and possible consequences are considered (Verhagen, 2006).

Reliable data and long-term observed and antecedent meteorological and hydrological time series, such as water levels and yields, are required for developing a meaningful groundwater drought early warning system. These data are then compiled and analyzed and user needs are determined. Finally, an early warning system is developed to identify drought management areas.

11.9 UNCERTAINTIES IN URBAN WATER ENGINEERING

In general, uncertainty due to the inherent randomness of physical processes cannot be eliminated. On the other hand, uncertainties such as those associated with lack of complete knowledge about the process, data, models, parameters, and so on could be reduced through data collection, research, and careful design and implementation/manufacturing. In urban water engineering, uncertainties involved can be divided into four basic categories: hydrological, hydraulic, structural, and economic. More specifically, in urban water engineering analyses and designs, uncertainties could arise from the various sources including natural or intrinsic uncertainties, model uncertainties, parameter uncertainties, data uncertainties, and operational uncertainties.

Natural uncertainty is associated with the inherent randomness of natural processes such as the occurrence of precipitation and flood events. The occurrence of hydrological events often displays variations in time and space. Their occurrences and intensities could not be predicted precisely in advance. Because a model is only an abstraction of the reality that generally involves certain simplifications, model uncertainty reflects the inability of a model or design technique to represent precisely the system's true physical behavior. Parameter uncertainties result from the inability to quantify accurately the model inputs and parameters. Parameter uncertainty could also be caused by a change in operational conditions of hydraulic structures, inherent variability of inputs and parameters in time and space, and a lack of sufficient data.

Data uncertainties include (a) measurement errors, (b) inconsistency and nonhomogeneity of data, (c) data handling and transcription errors, and (d) inadequate representation of the data sample due to time and space limitations. Operational uncertainties include those associated with construction, manufacturing, deterioration process, maintenance impact, and human factors. The magnitude of this type of uncertainty is largely dependent on the workmanship and quality control during the construction and manufacturing.

11.9.1 Implications and Analysis of Uncertainty

In urban water engineering design and modeling, the design quantity and system output are functions of several system parameters; not all of them can be quantified with absolute accuracy. The task of uncertainty analysis is to quantify the uncertainty features of the system outputs as a function of model characteristics and the stochastic variables involved. It provides a formal and a system framework to quantify the uncertainty associated with the system's output. Furthermore, it offers the designer useful insights regarding the contribution of each stochastic variable to the overall uncertainty of the system outputs. Such knowledge is essential to identify the "important" parameters to which more attention should be given to have a better assessment of their values and, accordingly, to reduce the overall uncertainty of the system outputs.

11.9.2 Measures of Uncertainty

Several expressions have been used to describe the level of uncertainty of a parameter, a function, a model, or a system. In general, the uncertainty associated with the latter three is a result of the combined effect of the uncertainties of the contributing parameters. The most complete and ideal description of uncertainty is the PDF of the quantity subject to uncertainty. However, in most practical problems, such a probability function cannot be derived or obtained precisely.

Another measure of the uncertainty of a quantity is to express it in terms of a reliability domain such as the confidence interval. A confidence interval is a numerical interval that would capture the quantity subject to uncertainty with a specified probabilistic confidence. Nevertheless, the use of confidence intervals has a few drawbacks: (a) The parameter population may not be normally distributed as assumed in the conventional procedures, and this problem is particularly important when the sample size is small. (b) No means is available to directly combine the confidence intervals of individual contributing random components to give the overall confidence interval of the system. A useful alternative to quantify the level of uncertainty is to use the statistical moments associated with a quantity subject to uncertainty. In particular, the variance and standard deviation that measure the dispersion of a stochastic variable are commonly used.

11.9.3 Analysis of Uncertainties

In the design and analysis of hydrosystems, many quantities of interest are functionally related to a number of variables, some of which are subject to uncertainty. A rather straightforward and useful technique for the approximation of uncertainties is the first-order analysis of uncertainties, sometimes called the delta method.

The use of the first-order analysis of uncertainties is quite popular in many fields of engineering because of its relative ease in application to a wide array of problems. First-order analysis is used to estimate the uncertainty in a deterministic model formulation involving parameters that are uncertain (not known with certainty). More specifically, first-order analysis enables one to estimate the mean and variance of a random variable that is functionally related to several other variables, some of which are random. By using first-order analysis, the combined effect of uncertainty in a model formulation, as well as the use of uncertain parameters, can be assessed.

Consider a random variable y that is a function of k random variables (e.g., the pressure of a specific point in WDS is a function of demands in different points that could be considered as random variables):

$$y = g\left(x_1, x_2, \ldots, x_k\right). \tag{11.15}$$

This can be a deterministic equation such as the rational formula or the Darcy–Weisbach equation or this function can be a complex model that must be solved on a computer. The objective is to treat a deterministic model that has uncertain inputs in order to determine the effect of the uncertain parameters $x_1, x_2, \ldots, x_k$ on the model output y.

Equation (11.15) can be expressed as $y = g(X)$, where $X = x_1, x_2, \ldots, x_k$. Through a Taylor series expansion about k random variables, ignoring the second- and higher-order terms, we obtain

$$y \approx g\left(\bar{X}\right) + \sum_{i=1}^{k}\left[\frac{\partial g}{\partial x_i}\right]_{\bar{X}}\left(x_i - \bar{X}\right). \tag{11.16}$$

The derivations $\left[\partial g/\partial x_i\right]_{\bar{X}}$ are the sensitivity coefficients that represent the rate of change of the function value $g\left(\bar{X}\right)$ at $x = \bar{X}$ (Mays, 2001).

Assuming that the k random variables are independent, the variance of y is approximated as

$$\sigma_y^2 = \mathrm{Var}[y] = \sum a_i^2 \sigma_{x_i}^2. \tag{11.17}$$

And the coefficient of variation, Ω_y, which is often used as the measure of uncertainty, is estimated as

$$\Omega_y = \left[\sum_{t=1}^{k} a_i^2 \left(\frac{\bar{x}_i}{\mu_y}\right)^2 \Omega_{x_i}^2\right]^{1/2}, \tag{11.18}$$

where

$a_i = \left(\partial g/\partial x\right)_{\bar{X}}$.

Refer to Mays and Tung (1992) for a detailed derivation of Equations (11.47) and (11.48).

Example 11.1

Appling the first-order analysis, formulate σ_Q and Ω_Q in Manning's equation $[Q = (0.311/n)S^{1/2}D^{8/3}]$, where diameter D is a deterministic parameter and n and S are considered to be uncertain.

Solution:

Since n and S are uncertain, Manning's equation can be rewritten as

$$Q = Kn^{-1}S^{1/2},$$

where

$K = 0.311\ D^{8/3}$.

The first-order approximation of Q is determined using the above equation, so that

$$Q \approx \bar{Q} + \left[\frac{\partial Q}{\partial n}\right]_{\bar{n},\bar{S}}(n-\bar{n}) + \left[\frac{\partial Q}{\partial S}\right]_{\bar{n},\bar{S}}(S-\bar{S})$$

$$= \bar{Q} + \left[-K\bar{n}^{-2}\bar{S}^{1/2}\right](n-\bar{n}) + \left[0.5K\bar{n}^{-1}\bar{S}^{-1/2}\right](S-\bar{S}),$$

where

$$\bar{Q} = K\bar{n}^{-1}\bar{S}^{1/2}.$$

The variance of the pipe capacity can be computed as

$$\sigma_Q^2 = \left[\frac{\partial Q}{\partial n}\right]^2_{\bar{n},\bar{S}}\sigma_n^2 + \left[\frac{\partial Q}{\partial S}\right]^2_{\bar{n},\bar{S}}\sigma_S^2,$$

$$\sigma_Q = \left\{\left[\frac{\partial Q}{\partial n}\right]^2_{\bar{n},\bar{S}}\sigma_n^2 + \left[\frac{\partial Q}{\partial S}\right]^2_{\bar{n},\bar{S}}\sigma_S^2\right\}^{1/2}.$$

The coefficient of variation of Q is determined as

$$\Omega_Q^2 = \sum_{i=1}^{2}\left[\frac{\partial Q}{\partial x_i}\right]^2\left[\frac{\bar{x}_i}{\bar{Q}}\right]^2\Omega_{x_i}^2,$$

$$\Omega_Q^2 = \left[\frac{\partial Q}{\partial n}\right]^2_{\bar{n},\bar{S}}\left[\frac{\bar{n}}{\bar{Q}}\right]^2\Omega_n^2 + \left[\frac{\partial Q}{\partial n}\right]^2_{\bar{n},\bar{S}}\left[\frac{\bar{S}}{\bar{Q}}\right]^2\Omega_S^2,$$

$$\Omega_Q^2 = \left[\frac{-K\bar{S}^{1/2}}{\bar{n}^2}\right]^2\left[\frac{\bar{n}}{\bar{Q}}\right]^2\Omega_n^2 + \left[\frac{0.5K}{\bar{n}\bar{S}^{1/2}}\right]^2\left[\frac{\bar{S}}{\bar{Q}}\right]^2\Omega_S^2,$$

$$\Omega_Q^2 = \left[\frac{-K\bar{S}^{1/2}}{\bar{Q}}\right]^2\left[\frac{1}{\bar{n}^2}\right]\Omega_n^2 + (0.5)^2\left[\frac{K}{\bar{n}\bar{S}^{1/2}}\right]^2\left[\frac{\bar{S}}{\bar{Q}}\right]^2\Omega_S^2,$$

$$\Omega_Q^2 = \left[\bar{n}^2\right]\left[\frac{1}{\bar{n}^2}\right]\Omega_n^2 + 0.25\left[\frac{1}{\bar{S}}\right]\left[\bar{S}\right]\Omega_S^2,$$

$$\Omega_Q^2 = \Omega_n^2 + 0.25\Omega_S^2,$$

$$\Omega_Q = \left[\Omega_n^2 + 0.25\Omega_S^2\right]^{1/2}.$$

Example 11.2

Determine the mean capacity of a storm sewer pipe, the coefficient of variation of the pipe capacity, and the standard deviation of the pipe capacity using Manning's equation (refer to Example 11.1). The following parameter values can be considered as shown in Table 11.1.

Solution:

Manning's equation for full pipe flow is $Q = (0.311/N)S^{1/2}D^{8/3}$, so for first-order analysis we have

$$\bar{Q} = \frac{0.311}{0.015}(0.001)^{1/2}(1.5)^{8/3} = 1.93 \text{ m}^3/\text{s}.$$

TABLE 11.1
Parameters of the Considered Storm Sewer Pipe in Example 11.2

Parameter	Mean	Coefficient of Variation
N	0.015	0.01
D	1.5 m	0
S	0.001	0.05

Using the results of Example 11.1, it is found that

$$\Omega_Q = \left[(0.01)^2 + 0.25(0.25)^2\right]^{1/2} = 0.027,$$

$$\sigma_Q = \bar{Q}\Omega_Q = 1.93(0.027) = 0.052\left(\text{m}^3/\text{s}\right).$$

11.10 RISK ANALYSIS: COMPOSITE HYDROLOGICAL AND HYDRAULIC RISK

The risk of a hydraulic component, subsystem, or system is defined as the probability of the loading exceeding the resistance, that is, the probability of failure. The relationship between reliability R and risk $\bar{R}$ is

$$R = 1 - \bar{R}. \tag{11.19}$$

Therefore, the same formulations of reliability can be used for the analysis of system risk.

11.10.1 Risk Management and Vulnerability

Systems that are highly exposed, sensitive, and less able to adapt are vulnerable. This is illustrated in Figure 11.4. In order to develop adaptation strategies, at first the systems/elements that are vulnerable to change should be identified. Then the scope should be determined to increase the coping capacity of those systems—their resilience—to decrease the vulnerability.

As illustrated in Figure 11.4, the vulnerability of a system is a function of three instinct factors, namely, exposure, sensitivity, and adaptive capacity. This approach to vulnerability assessment is important because it highlights the key elements that are combined to alleviate the risks and costs that disasters can impose on a system. Taking into account these factors can help us identify the extent of disasters. Developing action plans in each of these areas can help us reduce or deal with those different types of disasters.

There are many changes that may adversely affect the management of urban water systems. These changes may involve risks that can be caused by climatic variation (causing droughts or floods), mechanical failures, poor management, and health risks from pollution and waterborne diseases. The consequences can be considered in terms of financial, environmental, health, cultural and professional ethic impacts, short-term, long-term, and cumulative risks. Managing risk involves understanding the factors that contribute to the cause of that risk.

By understanding the complex interrelationships in the urban system, one can actively work towards reducing the factors that contribute to the consequences of risk. These factors include communications across resources availability and allocation and/or ecosystem variability. In this way, better decision-making processes can be implemented to create improved solutions for UWDs

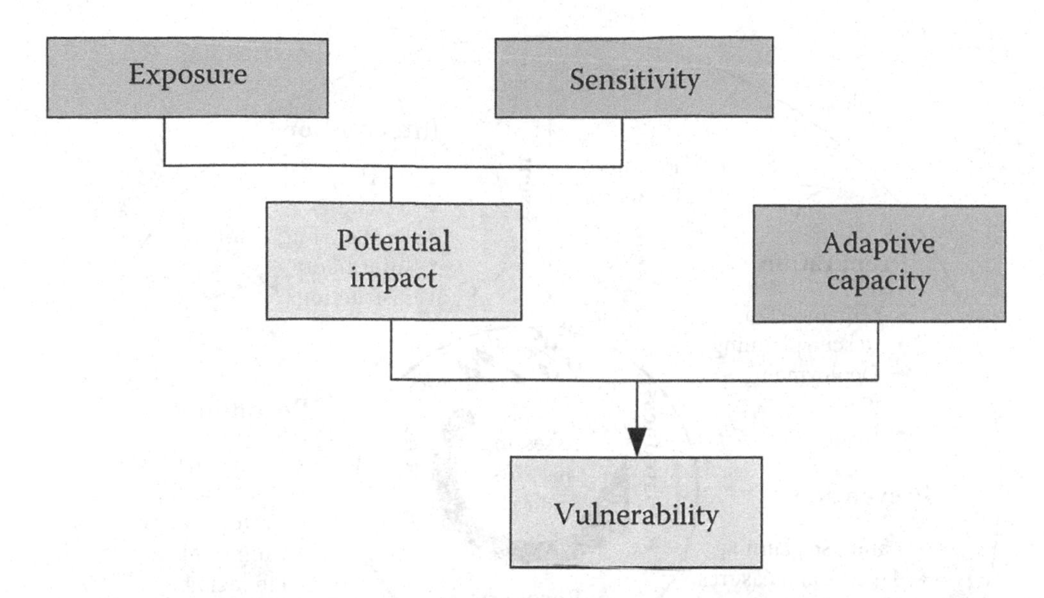

FIGURE 11.4 Vulnerability and its components. (From Allen Consulting Group, *Climate Change Risk and Vulnerability Promoting an Efficient Adaptation Response*. Department of Environment and Heritage Australian Green House Office. Available at: http://www.greenhouse.gov.au /impacts/publications/pubs/risk-vulnerability.pdf, 2005.)

throughout a region. An example of this process might be the comparison between major water schemes and disaggregated household or community schemes.

Once the risk attributes are better understood, the effect of implementing suitable control strategies to address a particular risk can be evaluated. Sound risk management planning does reduce risk but rarely eliminates it. Contingency plans should be developed to effectively manage risk issues. For example, adequate capability to monitor public health and respond to outbreaks of widespread disease contamination should be provided.

In the United States, disaster prevention started with coalitions of scientists, emergency-relief organizations, professional associations, and other civic groups who lobbied governments to fund research and hazard mitigation strategies. This movement received impetus when the FEMA, armed with a federal mandate and incentives, took the lead and promoted local and state initiatives (such as the regional Earthquake Preparedness Projects in California), but still worked through civic and professional partners.

The public needs to decide on acceptable levels of risk, comparing the immediate benefits of expenditures on other social priorities with the delayed benefits of reduced loss of life and asset replacement cost following a potential disaster. These trade-offs can be eased when well-designed incentives change private behavior to help prevent hazards. Examples include reducing insurance premiums on residential property when basic hazard-resistant steps are taken, offering disaster insurance with strict enforcement of building code provisions, or providing tax holidays or grants for mitigation. Poor residents, for whom insurance or fiscally based incentives may not be practical, would benefit from urban planning for slum prevention, enforceable environmental zoning in cities, and resettlement combined with community-based upgrading.

In Australia, there is a new DM authority to coordinate with all aspects of the response, working with NGOs, private sectors, universities, local communities, and external donors. The program includes predisaster preparedness and postdisaster response, reconstruction, and disaster prevention. Incentives are being introduced to build constituencies for disaster prevention by capitalizing on the population's awareness and willingness to change (Allen Consulting Group, 2005).

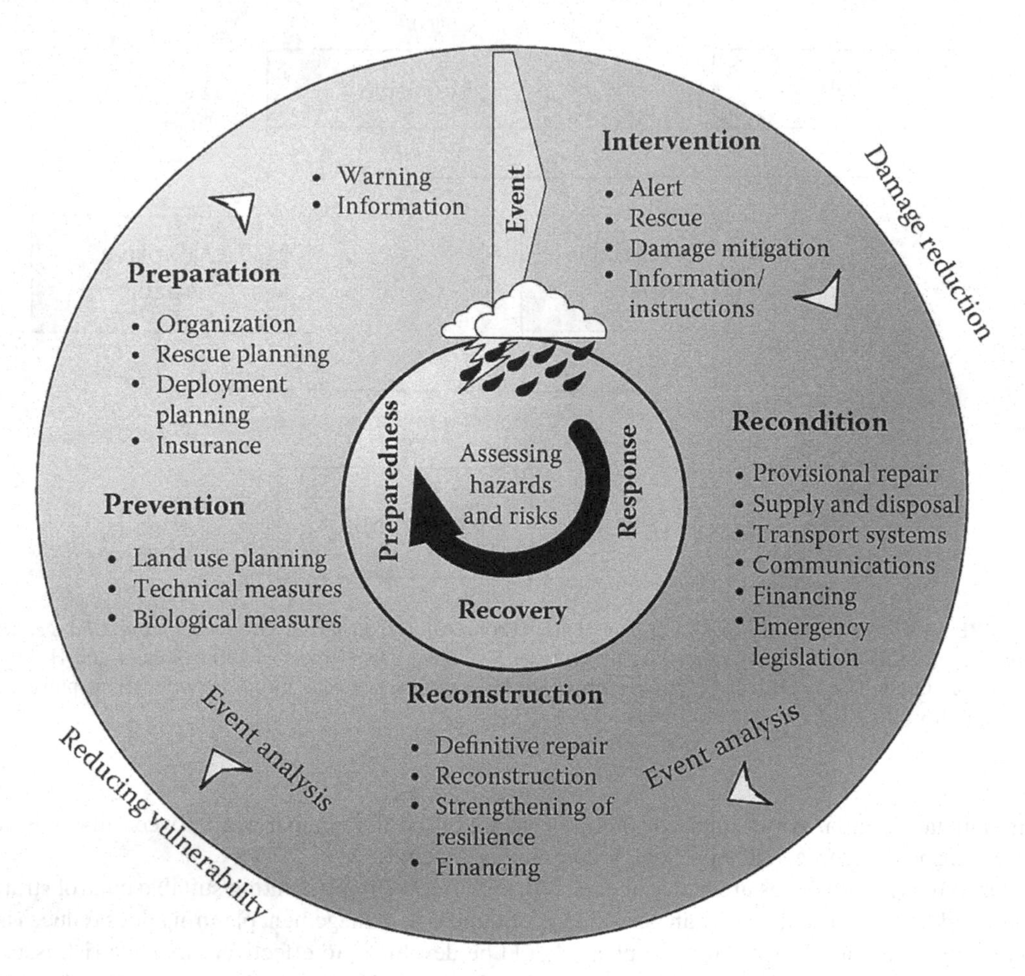

FIGURE 11.5 Risk management cycle.

Risk management has to follow the stages of a risk cycle through recovery, preparedness, and response (Figure 11.5). Preparedness decreases the potential risks of disasters at social and individual levels. This includes risk improvement to an acceptable and affordable level and the preparation of activities to manage residual risks. Response and adaptive measures are needed during and after a hazard occurrence. Preparation/preparedness is necessary when responding to emergencies. The respective measures deal with the mitigation of emergencies and the conditional recondition of essential services and infrastructure.

In most cases, there is a tendency to overlook preparatory and preventive measures. The response is certainly the preferred method of risk management, especially in medium- and low-income countries. Furthermore, decision makers often are willing to invest more in preparedness only after critical disasters. However, preparatory and preventive measures are generally more sustainable and cost-efficient than emergency response measures. Based on the information that suitable preparedness measures are still missing in many regions, the structural and nonstructural measures can be used to one or more of the three factors of risk, either as joint or decreased.

To make possible classification of aims and to decide the possible approaches towards risk reduction, it is necessary to distinguish between measures that address mitigation of exposure, hazard, and vulnerability. Hazard mitigation measures refer to what can be done to mitigate extreme adverse events in quality and quantity. Exposure mitigation addresses primarily spatial planning, which aims to reconcile the spatial demands of watercourses and development to minimize the location of infrastructure and vulnerable people in harm's way. Measures in the field of vulnerability plan to mitigate the tangible and intangible susceptibility of infrastructure and people.

It must be identified that specific measures apply to more than one of the three categories, for example, hazard transfer measures can also be understood as measures of exposure reduction. Therefore, the discussion has to be seen as an exercise in practice rather than making theoretical distinctions. By considering the selection of measures, experience shows that the most promising approach is to adopt a mixed strategy that addresses multiple risk components. Finally, it is essential to know that all suggested measures are subject to local situations. Their suitability and applicability will depend on socioeconomic and political conditions.

11.10.2 Risk-Based Design of Water Resources Systems

Reliability analysis can be applied to the design of various hydraulic structures with or without considering the risk costs that are associated with the failure of hydraulic structures or systems. The risk-based least cost design of hydraulic structures promises to be, potentially, the most significant application of reliability analysis. The risk-based design of water resources engineering structures integrates the procedures of economic, uncertainty, and reliability analyses in the design practice. Engineers using a risk-based design procedure consider trade-offs among various factors such as risk, economics, and other performance measures in hydraulic structure design. When risk-based design is embedded into an optimization framework, the combined procedure is called optimal risk-based design.

Because the cost associated with the failure of a hydraulic structure cannot be predicted from year to year, a practical way to quantify it is to use an expected value on the annual basis. The total annual expected cost is the sum of the annual installation cost and the annual expected damage cost.

In general, as the structural size increases, the annual installation cost increases, while the annual expected damage cost associated with failure decreases. The optimal risk-based design determines the optimal structural size, configuration, and operation such that the annual total expected cost is minimized.

In the optimal risk-based designs of hydraulic structures, the thrust of the exercise is to evaluate annual expected damage cost as a function of the PDFs of loading and resistance, damage function, and the types of uncertainty considered. The conventional risk-based hydraulic design considers only the inherent hydrological uncertainty associated with the random occurrence of loads. It does not consider hydraulic and economic uncertainties. Also, the probability distribution of the load to the water resources system is assumed to be known, which generally not the case in reality is. However, the evaluation of annual expected cost can be made by further incorporating the uncertainties in hydraulic and hydrological models and parameters.

To obtain an accurate estimation of annual expected damage associated with structural failure, it would require the consideration of all uncertainties, if they can be quantified. Otherwise, the annual expected damage would, in most cases, be underestimated, leading to an inaccurate optimal design.

11.10.3 Creating Incentives and Constituencies for Risk Reduction

Indispensable for mitigation strategies are strong disaster prevention proponents and the political will to lead regulatory changes and financial appropriations. With limited resources, developing countries must rely on partnerships of all actors. In the United States, disaster prevention started with coalitions of scientists, emergency-relief organizations, professional associations, and other civic groups who lobbied governments to fund research and hazard mitigation strategies. This movement received impetus when the FEMA, armed with a federal mandate and incentives, took the lead and promoted local and state initiatives (such as the regional Earthquake Preparedness Projects in California) but still worked through civic and professional partners.

The public needs to decide on acceptable levels of risk, comparing the immediate benefits of expenditures on other social priorities with the delayed benefits of reduced loss of life and asset replacement cost following a potential disaster. These trade-offs can be eased when well-designed incentives change private behavior to help prevent hazards. Examples include reducing insurance premiums on residential property when basic hazard-resistant steps are taken, offering disaster insurance with strict enforcement of building code provisions, or providing tax incentives or grants for mitigation. Poor residents, for whom insurance or fiscally based incentives may not be practical, would benefit from urban planning for slum prevention, enforceable environmental zoning in cities, and resettlement combined with community-based upgrading and tenure regularization schemes.

Recent disasters can motivate countries to undertake some of these measures and instill in them longer-term thinking to better enforce risk reduction. Gujarat state in India is trying to establish effective DM institutions following the January 2001 earthquake that killed 15,000 people. The state has a new DM authority to coordinate with all aspects of the response, working with NGOs, private sectors, universities, local communities, and external donors. The program includes predisaster preparedness and postdisaster response, reconstruction, disaster prevention, and risk reduction. Incentives are being introduced to build constituencies for disaster prevention by capitalizing on the population's heightened awareness and willingness to change.

Reducing risks caused by climate change may be more difficult since the risks mount gradually and less visibly but no less urgently. Coastal cities and other population centers (especially small island states) will need to invest in protective barriers and possibly to relocate residences and essential public facilities through managed retreat.

Priorities for such adaptation should be given to build areas and infrastructure that require urgent attention in any case, such as vulnerable informal settlements and outgrown sanitation and drainage systems. Adaptive expenditures will place a significant burden on the public sector, private utility companies, and, indirectly, on the urban economy. Low-income residents living with social and economic stress will need particular assistance.

11.11 SYSTEM PREPAREDNESS

For increasing urban water systems readiness, the following steps can be taken:

- *Raising awareness on the integration of environment and disaster issues*: This is the first step and should be done at different levels, targeting (a) policy makers, to initiate policy dialog at central, prefecture, and city levels; (b) professionals, to raise awareness among managers; and (c) individuals, to raise awareness at community levels.
  - *Proposed actions: Arrange meetings, seminars, and forums at different levels.*
- *Documenting and disseminating examples*: Documentation and its dissemination is one of the best ways to raise awareness among different stakeholders on the need for good cooperation for disaster preparedness.
  - *Proposed actions: Undertake field surveys after different disaster events to document concrete examples and disseminate it through the Internet and in training and study programs.*
- *Bridge gaps between knowledge and practice*: To enhance the understanding of the need to integrate environment management and DM, gaps between knowledge and practice should be reduced. Experts, resource persons, and practitioners should have opportunities to interact and learn from each other's experiences, for example, in preparing disaster contingency plans.
  - *Proposed actions: Undertake and facilitate training programs involving experts, resource persons, practitioners, and community leaders through forums.*
- *Implementing practical examples*: While it is important to raise awareness, provide training, and initiate policy dialog, it is also important to implement environment and DM practices at the local level.

   - *Proposed actions: Undertake small projects at different locations having different socioeconomic contexts. These will be learning experiences, which can also be disseminated to other areas.*
- *Developing guidelines and tools on environment and disaster management*: To provide a standard of environment and DM practices, guidelines and tools will have to be developed.
   - *Proposed actions: Guidelines and tools should be developed based on the field experiences, and incorporating the comments and expertise of practitioners and professionals from different parts of the world.*
- *Continuous monitoring of systems*: It is required to continuously monitor the progress in the practice of environment management and disaster preparedness and to distribute the lessons to develop policies and strategies.
   - *Proposed actions: A network or partnership should be developed that will continuously monitor different activities, identify best practices, and disseminate information through training and capacity building programs.*

11.11.1 Evaluation of WDS Preparedness

The objective of every urban WDS is to deliver enough water of acceptable quality with adequate pressure to different demand points. There are many interruptions and occasional disasters that impact the performance of the WDS; some could be quite devastating. The most common disasters are main breaks that may cause considerable water losses and bring up the system to partial or complete shutdown. Evaluation of the state of the system's preparedness helps managers make better decisions to prevent disasters and respond better to emergencies. Nazif and Karamouz (2009) have quantified the state of the system's preparedness in dealing with disasters using a hybrid index called the system readiness index (SRI). The three indices of a system's performance, namely reliability, resiliency, and vulnerability, are integrated through the demand-weighted average of pressure deficits at the critical nodes of WDS. Critical nodes are those with high demand and/or pressure variations. Available pressures at the demand nodes are the governing factor in assessing WDS performance. This is a new approach in system preparedness evaluation.

The SRI in each week is calculated based on the demand-weighted average of pressure at the critical nodes:

$$\mathrm{SRI}_t = -\sum_{j=1}^{k} \frac{\sum_{i=1}^{m} D_i \mathrm{PD}_{ijt}}{\sum_{i=1}^{m} D_i} \times \frac{T_{jt}}{168}, \quad i \in S\text{: } \{\text{Set of critical nodes}\}, \tag{11.20}$$

where

PD_{ijt} is the pressure deficit from a normal situation at the critical node i in the jth water main-break event at time step t,
D_i is the demand at the ith critical node,
T_{jt} is the duration of the jth water main-break event during time step t (hours),
m is the number of selected critical nodes, and
k is the total number of water main-break events that have happened in a given week.

In the above equation, 168 is the total hours of operation per week and is used to incorporate the weekly fraction of failure in the equation. In this index, by increasing pressure deficit, the system's readiness decreases. The state of different nodes in the system's readiness has been considered in the calculation of demand-weighted average of pressure deficits.

TABLE 11.2
System Preparedness Based on SRI Classification

The State of System Preparedness	SRI Class	Probability of Failure
Sound	1	<0.05
Good	2	0.05–0.35
Fair	3	0.35–0.65
Poor	4	0.65–0.95
Critical	5	>0.95

It should be noted that two groups of nodes are selected as critical nodes. The first group consists of those with higher water dependencies (water demands of top 20 percentile) and the second group consists of nodes with highest head loss (i.e., away from the storage tank(s)). This step plays an important role in achieving reliable results on the state of the system's readiness. It must be noted that the critical nodes should be revised during the simulation procedure if the hydraulic state of the system changes.

The negative sign in the SRI formulation is to show more severity of system failure as the SRI absolute value increases. Five main categories of SRI are defined in this study, namely, sound, good, fair, poor, and critical. These classes are determined based on a probability distribution fitted to the SRI values. The ranges of probabilities on SRI classifications are shown in Table 11.2. It is assumed that if there is more than 95% probability of failure, the system should be in full alert (red zone). The other zones are assumed arbitrarily to cover its range of variation. As the class number of SRI increases, the system could be considered at a lower state of preparedness. According to Table 11.2, the system preparedness is categorized as sound when the SRI has the probability of failure of <5% based on the probability distribution of observed/generated SRI values. As another sample of SRI interpretation, one could say for example if SRI is in the fourth class, the system could be subject to failure with 65%–95% probability of occurrence, with 80% as a representative of that range. In this situation, the critical nodes have high pressure deficits, and there is no water delivery in certain nodes. The characteristics of the other classes are given in Table 11.2.

The following steps are carried out for SRI determination using a probabilistic neural network (PNN):

1. Reliability, resiliency, and vulnerability indices of the system during the main breaks in the weekly time intervals are calculated.
2. Critical nodes in WDS are identified based on the hydraulic analysis of the WDS.
3. Deficit pressure from normal situations in the critical nodes and their demand-weighted average are calculated.
4. SRI values for main-break events in each week using demand-weighted average pressures at the critical nodes are determined using Equation 11.28.
5. A PNN is used to train a system for prediction of SRI, as the new information on the state of the system becomes available. The flowchart of the proposed approach for determining the SRI is shown in Figure 11.6.

The reliability index in time t (Rel_t) (during each week) is estimated as

$$\mathrm{Rel}_t = \left(1 - \frac{\sum_{j=1}^{k} T_{jt}}{168}\right) \times 100, \tag{11.21}$$

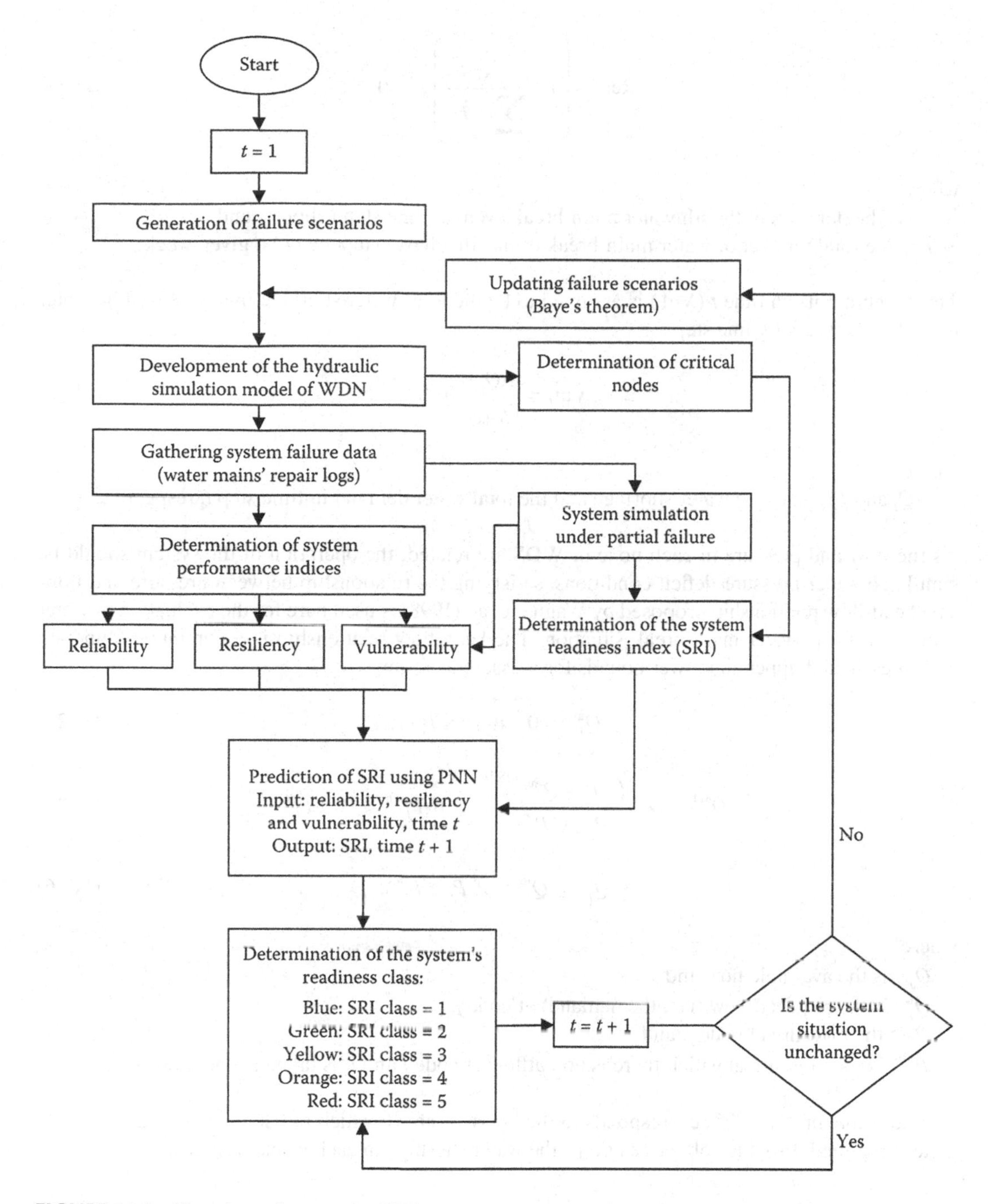

FIGURE 11.6 Flowchart of measuring SRI.

where

T_{jt} is the duration of the jth water main-break event at time step t (hours) and
k is the total number of water main-break events that have happened in a given week.

In the above equation, 168 is the total hours of operation per week.

The inverse of average break time has been considered as the resiliency index in time t (Res_t):

$$\text{Res}_t = \left(1 - \frac{k}{\sum_{j=1}^{k} T_{jt}}\right) \times 100, \tag{11.22}$$

where

T_{jt} is the duration of the jth water main-break event at time step t (hours) and
k is the total number of water main-break events that have happened in a given week.

The vulnerability in time t (Vul_t) is assumed to be the total water shortage that is caused by water main-break events in time step t:

$$\text{Vul}_t = \frac{\Delta Q_t}{Q_{\text{req},t}} \times 100, \tag{11.23}$$

where

ΔQ_t and $Q_{\text{req},t}$ are the water shortage and the total water demand in time step t, respectively.

As the flow and pressure in each node of WDS are related, the operation of the system should be simulated under pressure-deficit conditions, satisfying the relationship between pressure and flow. The head-flow relationship proposed by Wagner et al. (1998) is used here for the evaluation of water shortages in the water main-break situation. The head-flow relationship is a continuous function with preselected upper and lower bounds for pressure as follows:

$$Q_j^{\text{avl}} = 0 \quad \text{if } P_j \leq P_j^{\min}, \tag{11.24}$$

$$Q_j^{\text{avl}} = Q_j^{\text{req}} \left(\frac{P_j - P_j^{\min}}{P_j^{\text{des}} - P_j^{\min}} \right)^{1/n_j} \quad \text{if } P_j^{\min} < P_j < P_j^{\text{des}}, \tag{11.25}$$

$$Q_j^{\text{avl}} = Q_j^{\text{req}} \quad \text{if } P_j \geq P_j^{\text{des}}, \tag{11.26}$$

where

Q_j^{avl} is the available flow and
Q_j^{req} is the required flow (i.e., the demand) at node j,
P_j is the pressure at node j, and
$P_j^{\min}$ is the pressure at which there is no outflow at node j and it is taken as the minimum.

The desirable pressure P_j^{des} corresponds to the pressure above which the demand at node j is completely supplied. From the above relations, the water shortage in each week is calculated as

$$\Delta Q_t = \sum_{j=1}^{m} \sum_{i=1}^{n} Q_{jit}^{\text{req}} - Q_{jit}^{\text{avl}}, \tag{11.27}$$

where

ΔQ_t is the water shortage in time step t.
Q_{jit}^{req} and Q_{jit}^{avl} are the required and the available water at node i in the jth system failure in time step t, respectively.
The n and m denote the number of nodes in the water distribution network and the number of failures in each time step, respectively.

Example 11.3

Evaluate the preparedness of the Tehran (capital of Iran) WDS. Because of the aging water distribution infrastructure of Tehran, the water pipe breaks are a common problem in this city. As there is not any rehabilitation program of aging pipes in this city, this problem is getting more severe over the time. The total length of pipes in the Tehran WDS is about 9,323 km and an informal figure of 700 pipe breaks in a day is quoted. This results in a break rate of about 8/100 km of pipes per day, which causes considerable water losses during a day. Furthermore, a lack of emergency teams in the nearby posts (say within 10-minute response time) has resulted in oscillating adverse factors of disengagements. So evaluation of system preparedness in dealing with this problem and developing strategies for increasing the system readiness are essential for the Tehran WDS.

The considered part of Tehran water distribution supplies water for 120,000 residents in an area of about 450 ha (1 ha = 10,000 m^2). Maximum daily water consumption in this system is about 60,000 m^3. The simulated part of the WDS of Tehran has been shown in Figure 11.7. There are 35 main demand nodes with the total demand of about 0.51 cm and 78 water mains. The characteristics of the network are shown in Table 11.3. The minimum and desirable pressures for this water distribution network are 5 and 20 m, respectively.

Solution:

The selected WDS has been simulated under failure scenarios for 60 weeks. The system performance indices of reliability, resiliency, and vulnerability have been calculated. The results have been shown in Table 11.4 and Figure 11.8.

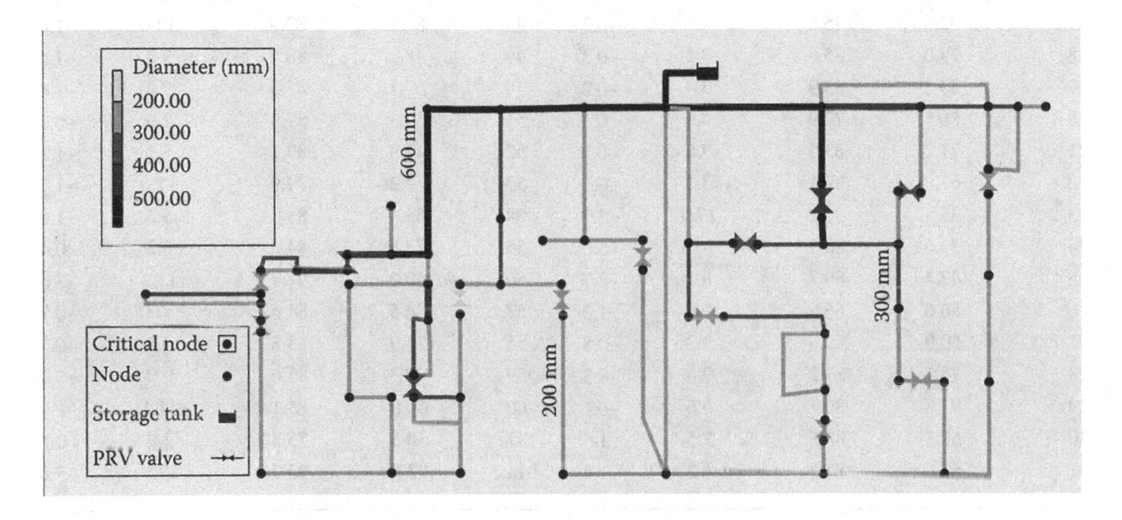

FIGURE 11.7 Schematic of the WDS of Tehran supported by storage tank no. 83.

TABLE 11.3
Characteristics of Water Mains in the Simulated WDS

Diameter (mm)	Number of Pipes	Total Length (m)
200	30	9,458
250	18	4,960
300	11	2,931
350	5	690
400	5	1,585
>400	9	3,197

TABLE 11.4
System Performance Indices

Week Number	Reliability	Resiliency	Vulnerability	SRI	Week Number	Reliability	Resiliency	Vulnerability	SRI
1	58.3	85.7	14.7	−3.7	32	73.7	93.2	5.7	−1.0
2	64.2	85.0	3.7	−0.6	33	56.2	91.9	4.3	−1.0
3	67.4	87.2	4.8	−0.5	34	77.7	86.7	4.3	−0.4
4	79.3	88.5	11.26	−0.3	35	61.8	89.1	3.4	−0.6
5	67.1	87.3	15.1	−1.4	36	73.7	86.4	15.0	−1.3
6	67.7	83.4	8.3	−1.4	37	63.5	86.9	16.5	−1.5
7	68.1	92.5	8.8	−0.5	38	63.2	88.7	5.7	−0.8
8	65.2	88.0	11.9	−0.9	39	60.1	91.0	15.0	−2.9
9	80.5	78.6	8.8	−0.4	40	90.6	87.4	3.0	−0.1
10	81.6	83.8	9.2	−0.3	41	69.1	88.4	8.3	−1.1
11	84.3	88.7	7.2	−0.2	42	57.7	84.5	6.3	−1.0
12	86.1	78.6	3.3	−0.3	43	57.3	84.7	11.2	−2.0
13	62.4	87.3	7.3	−0.7	44	30.3	90.6	5.6	−1.3
14	60.5	84.9	5.5	−0.8	45	56.7	90.4	6.8	−1.3
15	75.7	87.7	3.9	−0.5	46	84.4	88.6	4.4	−0.3
16	64.8	88.2	5.6	−1.1	47	57.2	87.5	3.6	−0.7
17	92.6	75.8	5.1	−0.3	48	86.1	82.9	12.6	−1.0
18	79.0	85.8	3.1	−0.3	49	48.2	88.5	9.3	−1.5
19	85.2	83.9	3.4	−0.2	50	67.1	85.5	9.4	−1.4
20	86.5	77.9	4.1	−0.2	51	80.3	87.9	2.9	−0.3
21	71.2	83.5	7.6	−0.7	52	44.9	89.2	5.1	−1.1
22	67.3	89.1	3.2	−0.5	53	79.2	79.9	11.0	−1.7
23	45.3	90.2	17.9	−3.3	54	65.4	88.0	5.3	−1.0
24	33.6	86.5	16.8	−5.0	55	73.5	84.3	3.2	−0.4
25	47.7	89.7	6.1	−2.7	56	40.0	91.1	11.2	−3.6
26	86.6	86.6	9.6	−1.0	57	84.5	84.6	7.2	−0.7
27	50.9	90.3	3.2	−0.8	58	77.9	83.8	4.1	−0.7
28	76.7	87.2	3.6	−0.5	59	79.3	85.6	9.0	−0.8
29	93.8	80.7	3.6	−0.1	60	60.0	85.1	11.1	−1.4
30	68.8	88.5	7.5	−1.5	**Min**	**30.3**	**75.8**	**2.9**	**−0.1**
31	57.5	84.6	6.2	−1.4	**Max**	**93.8**	**93.2**	**17.9**	**−5.0**

As can be seen, the system reliability is relatively low and varies between 30.3% and 93.8% in a given week. Only in 35% of the time period, the reliability is more than 75%, but in most of the time period (88%), the reliability is more than 50%. Although the low end of reliability figures is acceptable, the high end is not satisfactory. The general low reliability of the system could be because of the lack of well-equipped emergency response groups and the absence of the preventive maintenance program in the study area.

The resiliency of the system varies between 75.8% and 93.2%. There are 42 out of 60 weeks with a resiliency of more than 85%, which shows the system's ability to recover well after any system failure. Implementing some modification and rehabilitation programs as well as extending the current GIS-based monitoring system, the resiliency of the system could be even better than the current situation.

The minimum and maximum values of vulnerability (percent of water shortage) are 2.9% and 17.9%, respectively. In only 13% of the time, the water shortage is above 10% and, almost always, 80% of the demand is supplied, which shows the system's ability to continue its water delivery relatively well. The high values of the vulnerability index are due to some primary water main

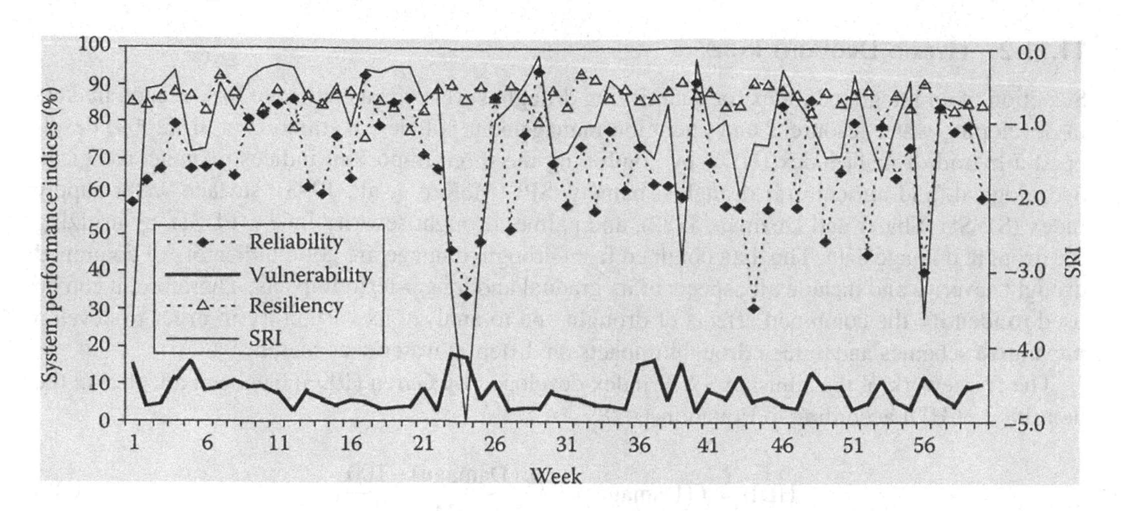

FIGURE 11.8 Variation of system performance indices in comparison with SRI.

breaks that supply water to the relatively large section of the system. This is perhaps an inherent characteristic of WDSs with large water mains without enough redundancy.

The calculated SRI values based on the proposed algorithm are shown in Table 11.4. It should be noted that the SRI values are rescaled to be within 0 and −5 in order to bring it to a manageable range as is done for many indices. This has been done by dividing the SRI values by the minimum calculated SRI value, multiplied by −5. Figure 11.8 shows a close correlation between the system performance indices and SRI. The SRI has relatively high correlations with vulnerability ($R=0.83$) and reliability ($R=0.82$). Figure 11.8 also illustrates that when all of the system performance indices have low values, the value of SRI suddenly decreases and the system preparedness situation goes to the red zone, which is expected. The worst-case scenarios are caused due to the failure of large-size mains (pipes) in the WDSs. In such cases, there is not any water delivery to some of the water demand nodes. Then, the flow to those sections is stopped and diverted to other nodes. These additional flow causes increase in pressure losses and pressure deficits in those nodes. This leads to high values of system vulnerability. Furthermore, to recover large failures, more time is needed and this will adversely affect the reliability and resiliency indices and lower the SRI. The failure of smaller pipes may result in less or zero pressure deficit because of the system redundancy. It also takes less time to recover, so the critical situations are not happening too often and the system preparedness is classified as fair or better. As can be seen in Figure 11.8, only in few weeks during the operation horizon, the SRI takes values <-3.0. During those weeks, the situation could be devastating and could result in extensive structural and economic damages. By prediction of these situations, the expected damages could be reduced by declaring a state of emergency and implementing the corrective measures such as pipe replacement and rehabilitation program.

In order to determine the probability ranges of SRI, the log-normal probability distribution has been fitted to the observed SRI values. The corresponding values of the probability range of SRI classes based on the fitted log-normal distribution are shown in Table 11.5.

TABLE 11.5
Corresponding Values of the Probability Limits of SRI Classes

SRI Class	Probability of Occurrence	SRI Value
1	<0.05	$-0.2<\text{SRI}\leq 0.0$
2	0.05–0.35	$-0.6<\text{SRI}\leq -0.2$
3	0.35–0.65	$-1.1<\text{SRI}\leq -0.6$
4	0.65–0.95	$-3.0<\text{SRI}\leq -1.1$
5	>0.95	$\text{SRI}<-3.0$

11.11.2 Hybrid Drought Index

Selection of an integrated index for quantifying drought severity is a challenge for decision makers in developing water resources and operation management policies. Karamouz et al. (2009) developed a hybrid drought index (HDI) by combining the three important indices of meteorological, hydrological, and agricultural droughts, namely, SPI (McKee et al., 1993), surface water supply index (SWSI) (Shafer and Dezman, 1982), and palmer drought severity index (PDSI), by utilizing the drought damage data. The data obtained from drought damage are good indicators of combined drought severity and include all aspects of its gradual and longer-term impacts. Therefore, it can be used to identify the combined effects of drought and to analyze its variability in order to develop preventive schemes and reduce drought impacts on different water-user sectors.

The framework of the adjusted SWSI index developed by Garen (1993) has been adapted in the definition of HDI according to Equation 11.28:

$$\mathrm{HDI}_t = f(\mathrm{Damage}_t) = \frac{p_t(\mathrm{Damage}) - 100}{24}, \tag{11.28}$$

where

p_t is the cumulative probability of damage in month t.

The characteristics of the damages for the three types of droughts are complex with a significant degree of nonlinearity and uncertainties. Therefore, for analyzing the nonlinear relations between them, the ANN models are applied.

The following steps are carried out for HDI determination and utilization for the drought severity prediction using the ANN:

1. Calculation of the time series of SPI, SWSI, and PDSI as indices for meteorological, hydrological, and agricultural aspects of droughts
2. Analysis and adjustment of historical drought damage data and estimating the monthly damage
3. Finding the most appropriate cumulative probability distribution that fits to the historical monthly damages
4. Determination of the HDI values from the cumulative probability of damages (Equation 11.28)
5. Classification of the HDI values into subcategories where class 1 and the last class denote the normal and extremely severe drought conditions, respectively
6. Estimation of HDI subcategories and consequently the range of drought damage through calculated indices of different drought types by training a system

The flowchart of the proposed approach for determining the HDI is shown in Figure 11.9. In comparison to the SWSI, HDI is utilized only for evaluation of dry spans (unlike SWSI, HDI only has negative values, so it can be used only for drought classification). The HDI values vary within (–4, 0), which makes it comparable with other indices of drought such as SWSI.

According to the HDI, drought starts when the HDI goes below –2 and continues unless it reaches above –1. The reason for selecting –2 as a threshold of drought beginning is that when the HDI is above –2, there is no significant change in the amount of damages, especially for the HDI values > –1. It can also be mentioned that at the beginning of drought spells, observed impacts are not very high because of gradual occurrence of droughts.

There are four main categories of drought severity according to the HDI: negligible, low, moderate, and high severity. This classification of drought is shown in Table 11.6 which is based on the cumulative probability of drought damage. For damages with the cumulative probability and

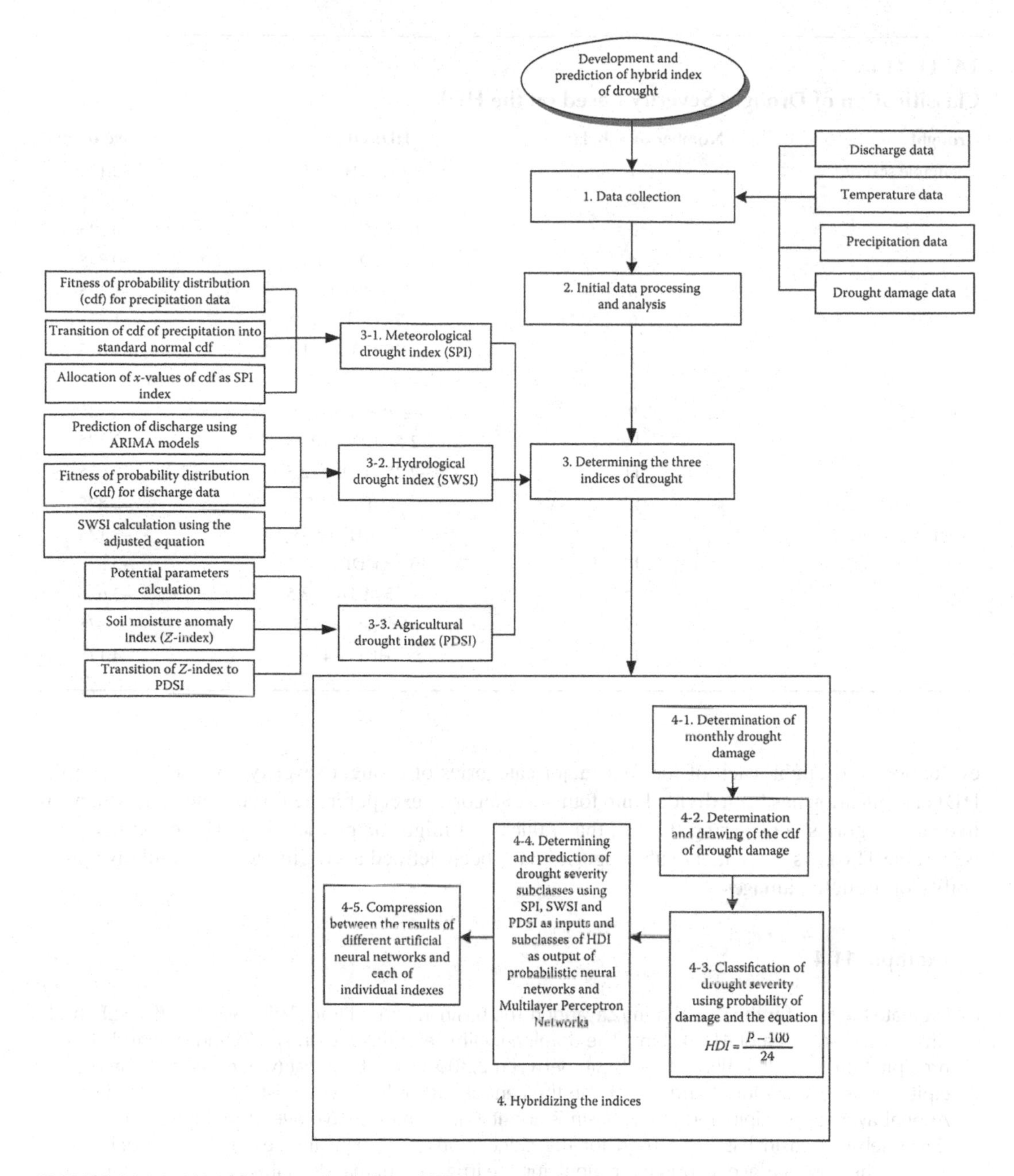

FIGURE 11.9 Algorithm for calculation of HDI.

the HDI value of <25% and −1, respectively, drought severity is considered to be negligible. This is because of the fact that damage exists, but it is not significant enough to be attributed only to drought occurrence. When the value of HDI reaches −2, the damage caused by drought is moderate and it can be assumed that drought span starts. Finally, when the value of HDI reaches −3, different water users, especially agriculture and animal husbandry sectors, suffer from water shortages and it is considered the critical stage of drought.

In regions where generally drought damages are high, small climate changes may cause extensive variations in damages. Therefore, in these regions, in order to evaluate the variations of drought impact, the classification of drought severity in smaller intervals can be helpful. For a more accurate

TABLE 11.6
Classification of Drought Severity Based on the HDI

Drought	Number of Subclasses	HDI Interval	Average of HDI
Negligible severity	1	−0.25<HDI<0	−0.125
	2	−0.5<HDI<−0.25	−0.375
	3	−0.75<HDI<−0.5	−0.625
	4	−1<HDI<−0.75	−0.875
Mild	5	−1.25<HDI<−1	−1.125
	6	−1.5<HDI<−1.25	−1.375
	7	−1.75<HDI<−1.5	−1.625
	8	−2<HDI<−1.75	−1.875
Moderate	9	−2.25<HDI<−2	−2.125
	10	−2.5<HDI<−2.25	−2.375
	11	−2.75<HDI<−2.5	−2.625
	12	−3<HDI<−2.75	−2.875
Extremely severe	13	−3.25<HDI<−3	−3.125
	14	−3.5<HDI<−3.25	−3.375
	15	−3.75<HDI<−3.5	−3.625
	16	−4<HDI<−3.75	−3.875
	17	HDI<−4	−4.125

evaluation of drought, each of the four major categories of drought severity obtained through the HDI classification has been divided into four subcategories except for the fourth category, which has five subcategories considering the fact that values<−4 might be predicted by the neural networks (see Table 11.6). As a result, 17 subcategories have been defined according to the cumulative probability of drought damages.

Example 11.4

Evaluate the HDI for the Gavkhooni/Zayandeh-rud basin in central Iran. This basin has five subbasins with a total area of 41,347 km^2. The dominant climate in this region is arid and semiarid. The precipitation varies throughout the basin between 2,300 mm in the west (where most of the precipitation is in snow form) and 130 mm in the central part of Iran (where Isfahan City is located). Annual average precipitation in this basin is about 1,500 mm. The average of precipitation in the Zayandeh-rud basin has been used for the calculation of drought indices. The Zayandeh-rud River is the main surface resource for supplying the irrigation demands in this basin. As water and energy demands increase in Isfahan, water withdrawals from the river increase and it is important to incorporate climate variability into water resources' decision-making.

The Zayandeh-rud reservoir controls the stream flow with a volume of 1,470 MCM. The location of the Zayandeh-rud reservoir is shown in Figure 11.10. The average annual inflow to the Zayandeh-rud reservoir is about 1,600 MCM, of which an average flow of 600 MCM is transferred from the adjacent Karoon river basin. Drought trends in the basin have been studied between years 1971 and 2004.

The Gavkhooni/Zayandeh-rud basin has special effects on the development and economy of the region through agricultural, industrial, and tourism activities. Statistics show that the amount of precipitation, especially in the high altitudes from October 1999 to April 2000, has decreased ~35%–45% compared to the long-term average and resulted in 250 MCM (million cubic meters) water shortages in the Isfahan region in the year 2000 (Karamouz et al., 2009).

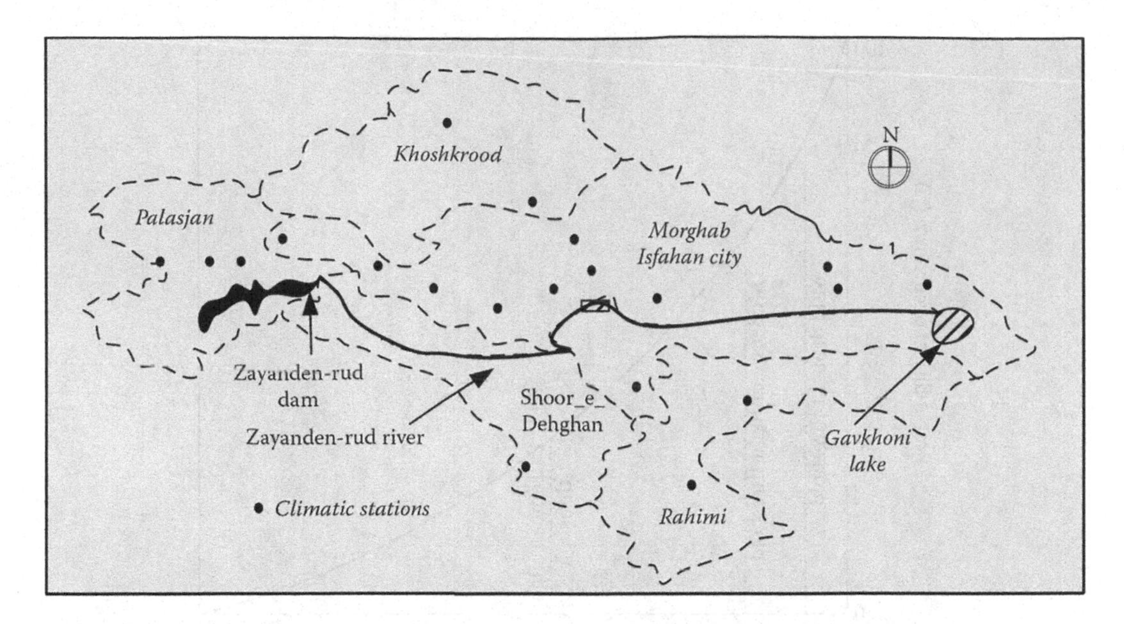

FIGURE 11.10 The Gavkhooni/Zayandeh-rud basin with subbasins and climatic stations.

Solution:

As the first step in the HDI calculations, the monthly time series of SPI, modified SWSI, and PDSI are calculated from 1971 to 2004. The time series of damage is converted to year 2006 values for comparison purposes. The interest rate is considered as 17%. This time series of monthly damages are shown in Figure 11.11. As mentioned in the steps of HDI calculation, the cumulative probability function of damage occurrences is developed using the Weibull probability distribution (Figure 11.12) and used for determining the HDI values (Equation 11.26). Different thresholds of drought intensity according to the HDI, which are obtained from the probability damage diagram, are also shown in Figure 11.12.

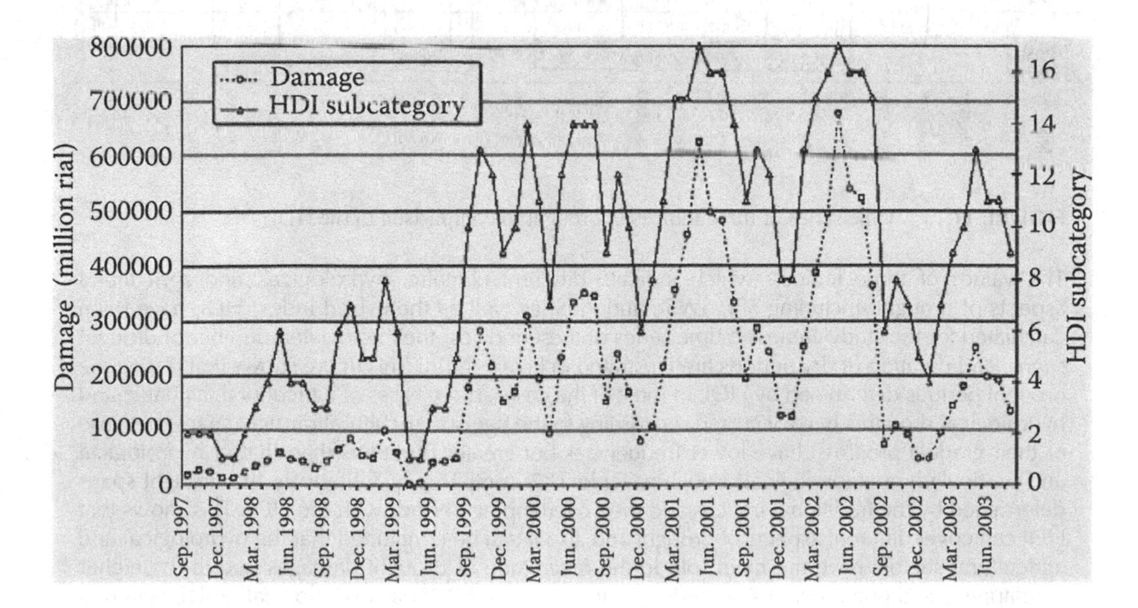

FIGURE 11.11 Monthly time series of damage (converted to values in 2006).

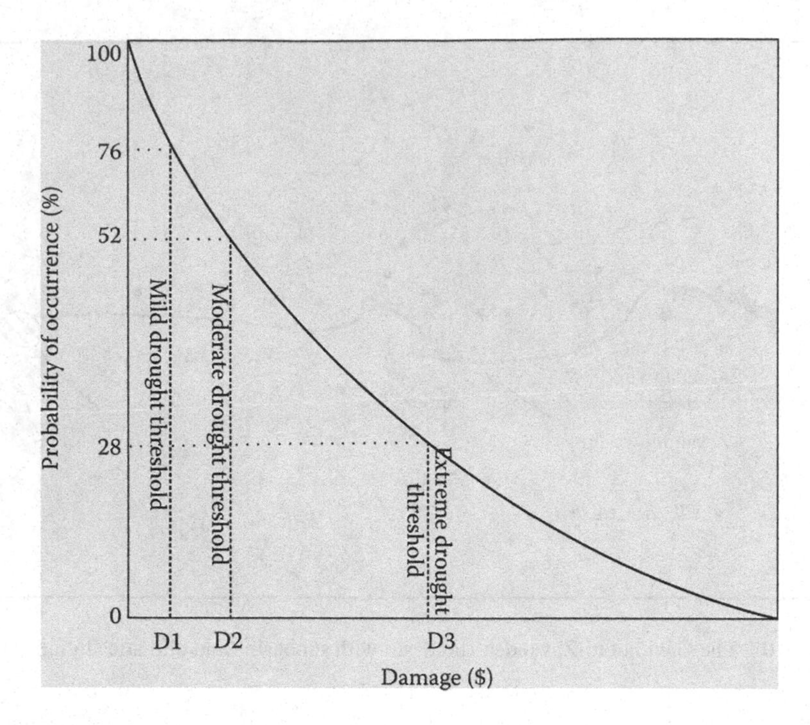

FIGURE 11.12 Diagram of cumulative probability of historical damages and thresholds of different drought severity.

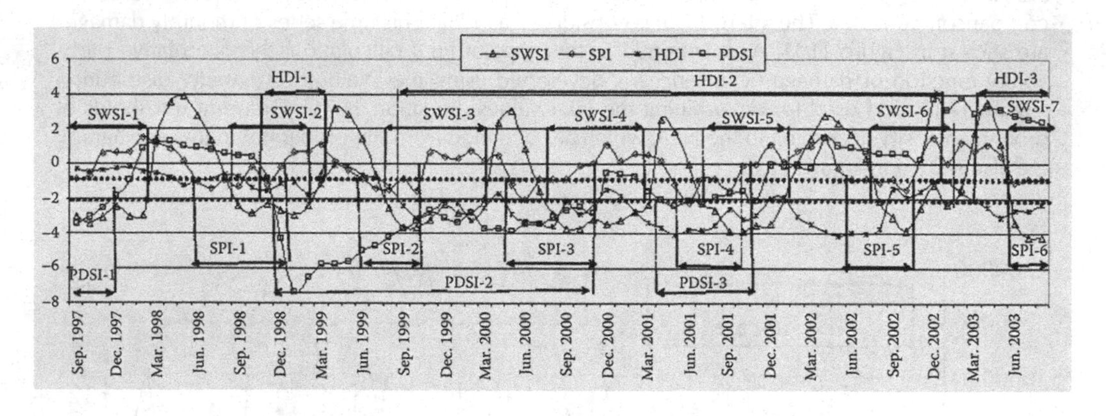

FIGURE 11.13 Time series of three indices of drought in comparison to the HDI.

The values of three indices (which indicate different climatic, hydrological, and agricultural aspects of drought) including SPI, SWSI, and PDSI as well as the hybrid index, HDI, have been calculated for the study area. The time series of these indices, their associated number of drought spans, and duration of dry periods are illustrated in Figure 11.13. This figure shows that during the drought periods determined by HDI, in most of the cases, three types of agricultural, climatic, and hydrological droughts have occurred. According to the figure, agricultural drought spans, because of their gradual progress, have lower frequencies but greater durations than that of hydrological and particularly meteorological droughts. Table 11.7 compares quantitatively the drought spans determined by the traditional indices and their overlapping periods with the HDI. This shows that HDI can cover different aspects of drought and also it can be concluded that the hydrological and agricultural sectors have the main role in the magnitude of drought damages due to the higher overlapping of drought periods according to the SWSI and PDSI as hydrological and agricultural drought indices with those reported according to the HDI.

TABLE 11.7
Durations of Drought Spans (month)

Drought Span's Number	SPI	SWSI	PDSI	HDI
1	7	6	4	5
2	–	6	24	–
3	5	8		40
4	7	8	–	–
5	6	7	8	–
6	5	6	–	–
7	3	3	–	6
Overlapping with HDI	22	34	27	51

11.11.3 Disaster and Scale

If we maintain too broad an interpretation of DM, it becomes difficult to determine the progress made in achieving it. In particular, concern only with the water disaster of a major city could overlook the unique attributes of particular local watershed economies, environments, ecosystems, resource substitution, and human health. On the other hand, not every hectare of land should be relieved. This highlights the need to consider the appropriate spatial scales when applying DM to a specific urban region.

We also need to consider the appropriate temporal scales when dealing with reliability of water disaster of a specific urban region. We cannot prevent disasters from happening; given the variations in natural water supplies and the fact that floods and droughts do occur, it is impossible, or at least very costly, to design and operate urban water systems that will never fail. During periods of failure, the economic benefits derived from such systems may decrease. However, ecological benefits may depend on these events. One of the challenges faced in measuring our success is to identify the appropriate temporal scales in which these measurements should be made.

An example of temporal scale and disaster variability is as follows: If a city of 2–3 million habitants faces water disaster as a drought over a period of 2–5 years, the tension could be reduced by many well-thought water conservation and rationing schemes as well as taping into other resources. If the same city faces a water tunnel collapse or a massive accidental or terrorist act of water contamination, then there will be a disaster with a potential for many losses of life and epidemic illnesses. The second scenario requires a different level of the systems' preparedness at a much shorter time scale (Karamouz and Moridi, 2005).

11.11.4 Disaster and Uncertainty

There are many uncertainties associated with the occurrence, extend, and intensity of disasters. Data and information including forecasting data estimation of which areas are effected such as inundations and water depth thresholds, intensity versus duration for rainfall, and the state of antecedent soil moisture condition are the main sources of uncertainty in hydrology. Figure 11.14 illustrates the design procedure when considering uncertainties in disaster management. It is a simplified version of what was presented in Section 10.2. Despite the variability of both load and resistance, the notion of estimating the probability of failure, *P*, as just described produces a single value of *P* for the set of experiments. Such a value could be trusted if all of the assumptions that led to the calculation of *P* are valid.

Different methodologies can be used to analyze the effects of uncertainty in estimating *P*. The second moment analysis (standard deviation or coefficient of variation), for example, is widely

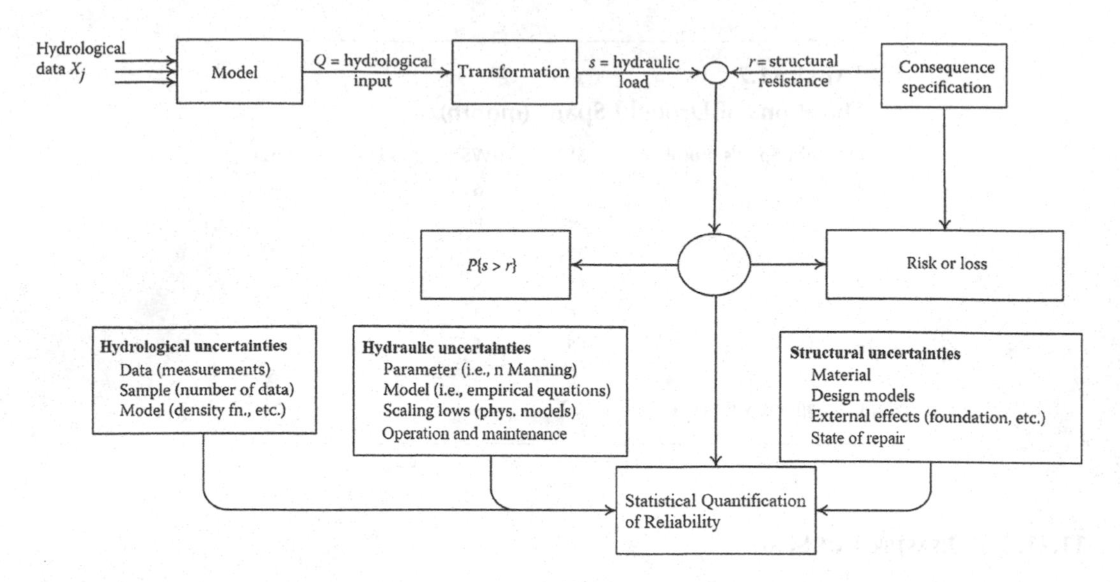

FIGURE 11.14 Generalized concept of risk and reliability analysis for structures. (Modified from Plate, E. J. and Duckstain, L. 1988. *Water Resources Bulletin*, 24(2).)

employed; Bayesian analysis, which requires the use of a loss function, can also be utilized. These methods provide P with ranges and pdfs, but they can only be determined by assigning the value based on the risk taking attitude and engineering judgment of the analyst or decision-maker.

Hydraulics and structural uncertainties are less dynamic than hydrologic uncertainties but could be more crucial in the extent and severity of disasters. The state of repairs and maintenance of hydraulic structures such as levees and the state of operation and positioning of unit operations in a wastewater treatment plant or even electro-mechanical equipment are subject to hydraulics and structural uncertainties that should be considered in disaster preparedness.

In Section 10.12, examples of how uncertainty can be measured in different applications. For models of water inundation that are used for flood disaster management, the uncertainties associated with DEM error, and error in utilizing and correlating satellite data with ancillary data for estimating soil moisture, are quantified. When designing BMPs for flood remediation, the structural integrity of BMPs are often subject to parameter and scaling laws uncertainties. For developing evacuation zones for flood hazards, many of the initiatives in this section and throughout the book in case studies and in development of flood probability maps can be used.

11.11.5 Water Supply Reliability Indicators and Metrics

In order to use reliability concept in design, it must be quantified based on a number of indicators. No single indicator can adequately measure and communicate all the critical dimensions of reliability. Multiple indicators are needed, many of which will be strongly interrelated due to the network characteristics of the water supply system. Some indicators will be more useful for monitoring supply reliability for certain types of service (e.g., residential/municipal versus environmental) or for different functions (e.g., real-time operations versus policy development). The following paragraphs discuss a variety of features that could distinguish reliability indicators.

Some indicators will require greater breadth, reflecting the effects of many factors, while some will be more narrowly focused to isolate a particular factor or reflect a certain system scale. Operations functions are particularly concerned with having indicators that measure reliability in terms of the factors under their control, in addition to broader indicators dominated by actual inflow levels. Some of the dimensions of breadth or focus are as follows:

- *Type of service*: Clearer definitions of the services currently provided to different end users (e.g., consumptive, environmental, and inflow) would facilitate the development of indicators and metrics. In other industries, service is often described or defined by specifying requirements related to the purpose of use, quantity and temporal/geographical distribution of deliveries, and quality and delivery curtailment provisions.
- *Temporal scale and distribution*: For example, hourly, monthly, or annual.
- *Spatial (or geographical) scale and distribution*: The scale refers to, for example, storage data that are aggregated to the combined project level or are disaggregated at the individual reservoir level.
- *Climate normalization*: For the available supply indicators, both metrics that are normalized for climatic variability and those that are not will be valuable. When normalized, metrics will reflect how efficiently the existing storage and transportation system is being used. When not normalized, reliability of the system including the biggest failure in supply availability or inflow is tracked.
- *Condition and outcome indicators/metrics*: Factors that affect reliability can be translated into indicators of reliability, for which metrics can be crafted. Indicators are statements on what is expected to change if the program shows progress towards the objective or if problems begin to arise. Indicators may identify outcomes or conditions and features that are believed to influence outcomes. Indicators of condition can provide insight into the mechanics of reliability and may be components of outcome indicators.
- *Physical and financial indicators/metrics*: Reliability indicators that reflect physical phenomena such as the supply delivered or allocated need to be paired with financial indicators that describe the costs associated with reliability as the level of those physically based indicators change. It is important to monitor cost in order to develop policy positions on the desired level of reliability and to answer questions about the cost of different possible levels of reliability.
- *Forecast versus actual*: Some outcome indicators could compare actual to forecasted or expected/desired service dimensions. Some metrics might utilize forecast data, while others may reflect what has actually happened. For example, one measure of reliability at the project level may be the initial monthly allocation to contractors (developed in the first quarter of the water year) relative to the final allocation that is produced in the third quarter of the water year (summer). For some customers (e.g., agricultural), the closer these values are, the more reliable project supplies are to them. Furthermore, for indicators/metrics that embody forecast values, the dependency of the outcome on the nature of the forecast must be carefully considered. For example, for metrics that involve the allocation as a measure of supply, the degree of conservativeness in the runoff forecast will influence the values of the reliability metric. In other words, the less risky (more conservative) the projects are in the inflow forecasts that are the basis of allocations, the more likely they will be to actually meet that allocation.
- *Deterministic and probabilistic indicators*: Both deterministic metrics and probabilistic metrics (discrete probabilities or probability distributions) will be useful.
- *Quantitative and qualitative indicators/metrics*: Some metrics will be quantitative while others will be qualitative (e.g., those related to institutional failures).

11.11.6 Issues of Concern for the Public

The public plays an important role in the situation analysis being able to reach beyond numbers and statistics to the real impact of the UWDM on society. The aim of including them at this stage is to identify, prioritize, and formulate the problems in a clear way and with a common understanding. It is important to be aware of the conflicts, views, and interest of the public. Negotiation skills and conflict resolution techniques will be a useful skill.

This first step analyzes the major DM actors, their interests and goals, and their interrelationships. It aims to shed light on the social reality and power relationships prevailing in the institutional setting of the DM. Major actors include potential losers. Furthermore, it is important to determine who volunteers to help.

11.12 WATER RESOURCES DISASTER

Groundwater and surface water resources, like any other water resources, are exposed to various disasters and contaminations. However, due the unique specification of these resources, different approaches should be utilized for their DM.

11.12.1 Prevention and Mitigation of Natural and Man-Induced Disasters

There are five phases in a disaster event. These phases include anticipatory, warning, impact, relief, and rehabilitation (Dooge, 2004). Within each phase of disaster prevention and mitigation, the following activities related to public and domestic drinking water supplies can be implemented:

- *Anticipatory phase*: Identification and assessment of the potential risk and vulnerability of existing public and domestic water supply systems as well as the identification, investigation, delineation, and evaluation of groundwater and surface water resources resistant to natural hazards.
- *The warning phase*: The activities of this phase are strongly related to the activities in the previous phase. The main activities of the warning phase are the establishment and operation of early warning monitoring systems, geological monitoring systems, and integrated hydro-climatological monitoring systems. The important elements of the warning phase are the formulation of suitable indicators of disaster risk and vulnerability as well as relevant groundwater and surface water indicators.
- *The impact and relief phases*: These phases focus on rescue efforts during and after disasters as well as on immediate external help These efforts include the distribution of drinking water, which requires identification, development, and setting aside rescue activities for the immediate emergency In many regions affected by natural disasters, water supplies depend on the transport of water by tankers from somewhere outside of the area or on the import of bottled water.
- *The rehabilitation phase*: This is a long-term phase and involves the reconstruction of water supply systems and the water distribution infrastructure, the remediation of polluted water, and the intensive pumping of existing deep wells or developing deeper aquifers of low vulnerability in areas with known properties. Other effective activities are well cleaning, rehabilitation of both well and pumping mechanisms and well dewatering and disinfection, and well deepening (in case of drought). Postevaluation of all phases of the rescue process, the preparation of plans for rehabilitation, including water management plans, and an assessment of emergency costs can help the situation.

11.12.2 Disaster Management Phases

DM includes all activities, programs, and measures that can be attended to before, during, and after a disaster with main purposes of avoiding a disaster, reducing its impacts, and recovering from its losses. There are three phases in DM as shown in Figure 11.15:

1. *Before a disaster (predisaster)*: Initiatives are taken to reduce human and property losses caused by a potential disaster. For example, conducting awareness campaigns, strengthening the existing weak structures, providing DM plans at the household and

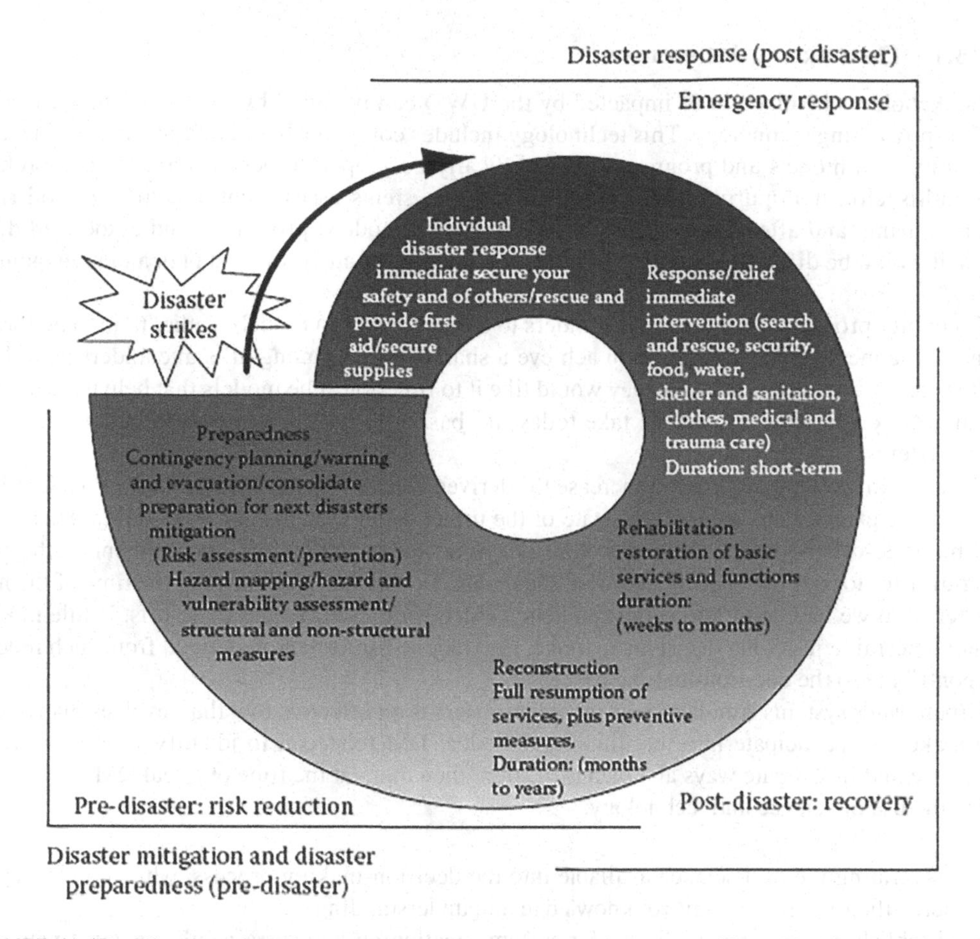

FIGURE 11.15 Disaster mitigation, preparedness, response, and recovery. (From Toshimaru, N., *Are You Prepared? Learning from the Great Hanshin-Awaji Earthquake Disaster, Handbook for Disaster Reduction and Volunteer Activities*. Hyogo University, Kokogawa, Hyogo, 1998.)

community level, etc. These risk reduction measures are called "mitigation and preparedness activities."

2. *During a disaster (disaster occurrence)*: Activities are undertaken to ensure that the needs of victims are met and impacts are minimized. Activities in this stage are called "emergency response activities."
3. *After a disaster (postdisaster)*: Activities are immediately undertaken right after a disaster for the recovery and rehabilitation of affected communities. These activities are termed as "response and recovery activities."

11.13 OTHER ATTRIBUTES OF DISASTER MANAGEMENT

A set of criteria should be selected, and for each criterion, a range of values that are satisfactory should be defined. The duration and the extent of individual and cumulative failures of water resources systems should be determined for each criterion. Other attributes include a) Disaster and Technology: with the fast paste of technology advancement there are many new tools and techniques that can be utilized. b) Disaster and training: The engineers, personnels, and first responders should be further trained to meet the growing needs for protecting and assisting the higher standards of living. c) Institutional Role: Institutions in charge have more responsibilities to coordinate among interdependent infrastructures for public safety and awareness.

11.13.1 Disaster and Technology

All stakeholders involved in or impacted by the UWD can be aided by the use of modern information-processing technology. This technology includes computer-based interactive optimization and simulation models and programs, all specifically developed to perform a more comprehensive multisector, multipurpose, multiobjective water systems assessment at a short period right before, during, and after a water disaster. Without such models, programs, and associated databases, it would be difficult to predict the expected impacts of any proposed plan and management policy.

Computer programs allow the stakeholders to create their own models, rather than to be forced to use someone else. Models can help achieve a shared vision among all stakeholders as to how their system functions, if not how they would like it to function. The models that help us to predict the impacts of possible actions we take today are based on the current conditions of our urban water systems.

What we might do to improve or increase the derived benefits, however measured, of our UWDM is, to a large extent, dependent on the state of the urban water systems that exist today. There may well be trade-offs between what we would like to do today to plan for management of the next water disaster for our own benefit and what the managers and decision maker at the time of the next disaster wish we had done. Modeling can help us identify these possible trade-offs. While models cannot determine just what decisions to make, the trade-off information derived from such models can contribute to the decision-making process.

Urban water systems simulation during the disasters is an effective tool that involves many decision makers to participate in a real-time hypothetical DM rehearsal to identify where the weaknesses are and investigate ways and means to strengthen them at the time of a real DM.

On the use of science and technology:

- Integrating the best science available into the decision-making process, while continuing scientific research to improve knowledge and understanding
- Establishing baseline conditions for system functioning and sustainability against which change can be measured
- Monitoring and evaluating actions to determine if goals and objectives are being achieved

Finally, it should be stated that disaster of any magnitude could be converted to opportunities if we learn from them and utilize the unification of efforts that has been developed through the disaster for better practices and implementation of sound water resources, urban water systems, and demand management as well as disaster prevention.

11.13.2 Disaster and Training

A key to UWDM is the existence of sufficiently well-trained personnel in all of the disciplines needed in the planning, development, and management processes. In regions where such a capacity is needed but does not exist, it should be developed. Training and education are a key input and requirement of DM. While outside experts and aid organizations can provide temporary assistance, each urban region must inevitably depend primarily on its own professionals to provide the know-how and experience required for UWDM.

Capacity building is one of the most essential and important long-term conditions required for DM. Another important factor in DM is that the local people must not only be capable but must also be willing to assume the responsibility for their urban water systems. One of the drawbacks of a centralized dominating government that takes the responsibility for local system design and operation is that the local people become accustomed to looking to the government for help, rather than looking to them. The ideal local urban water systems managers are well-trained persons who know the behavior of that system, have experience with its disasters such as floods and droughts, and know the concerns and customs of the people of the region, a group to which they belong.

11.13.3 Institutional Roles in Disaster Management

The government institutions, agencies, local authorities, the private sector, civil society organizations, and partnerships all constitute an institutional framework that ideally should be geared towards the implementation of the policy and the legal provisions. Whether building existing water management institutions or forming new ones, the challenge will be to make them effective and this requires capacity building. Awareness creation, participation, and consultations should serve to upgrade the skills and understanding of decision makers, water disaster and water resources managers, and professionals in all sectors. The key goals for the institutional framework are as follows:

- Separate water disasters management functions from service delivery functions (water supply, sanitation, and sewerage) and consolidate the government as the main manager of the water disasters—the enabler but not the provider of services. This will avoid conflicts of interest and encourage commercial autonomy.
- Manage surface water disasters within the boundaries of urban, not within administrative boundaries, decentralizing regulatory and service functions to the lowest appropriate level and promoting stakeholder involvement and public participation in planning and management decisions.
- To ensure balance between the extent and complexity of regulatory functions and the skills and humans required to deal with them. A continued capacity building program is required to develop and maintain the appropriate skills to facilitate, regulate, and encourage the private sector potential contributions to financing and delivery of services (water supply, sanitation, and sewerage).

11.14 A PATTERN OF ANALYZING SYSTEM'S PREPAREDNESS

The following assessment steps for system's preparedness against three causes of disasters are:

- *Water Shortage/Drought/Flood*:
  - Identification of the system's components, data gathering, and analysis
  - Assessment of the baseline estimates for the appropriate operation of a system from the hydrological and hydraulic viewpoints
  - Developing an algorithm to convert the indicators of the existing and anticipated situations of an urban water system to the disaster triggers
  - Evaluation of the severity of a disaster by analyzing the following criteria:
    - Reliability
    - Resiliency
    - Vulnerability
  - Assessment of the alternatives for water supply during a disaster according to its scale and severity
  - Assessment of the system's readiness from the hydrological viewpoint
- *Widespread Contamination*:
  - Assessment of the baseline estimates for the appropriate operation of a system from the hydraulic and structural standpoints

- Evaluation of the severity of a disaster by an analysis of the reliability and vulnerability criteria
- Assessment of the system's preparedness from the contamination viewpoint

Guidelines for the Mitigation and Compensatory Activities include: Assessment of the consequences of applying alternate policies during a disaster; managing water distribution during a disaster; identification of the most vulnerable areas and priorities to be considered in compensatory activities; isolation of the most vulnerable areas to prevent the spread of a disaster.

11.14.1 A Monitoring System for the Water Supply and Distribution Networks

The steps describing a monitoring system are as follows:

- Identification of the characteristic variables, proper indicators, and disaster triggers in a system
- Providing necessary facilities and equipment for the system
- Using a telemetric system for either online or offline data/information transformation
- Design and application of a monitoring system on the water pressure and heavy metals concentration in pipelines
- Design and application of a nondestructive monitoring system
- Analyzing the life cycle data of different components of a system

11.14.2 Organization and Institutional Chart of Decision Makers in a Disaster Committee

As an example, Figure 11.16 shows the organization chart of the group of UWDM in the Tehran Metropolitan Water and Wastewater Company, Tehran, Iran. This chart, for a more general water shortage situation, has been shown in Figure 11.17.

The described steps for evaluation of system preparedness are summarized in Figure 11.18. As can be seen from this figure, there are five phases in system preparedness evaluation that finally

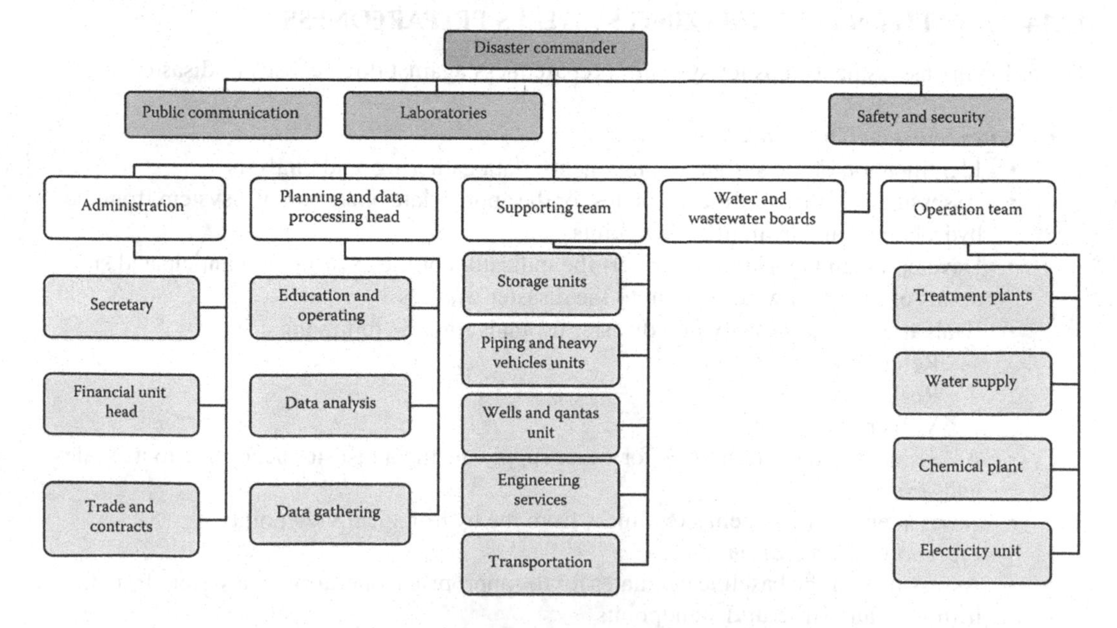

FIGURE 11.16 Network of responsibilities during a UWD for a typical WWTP.

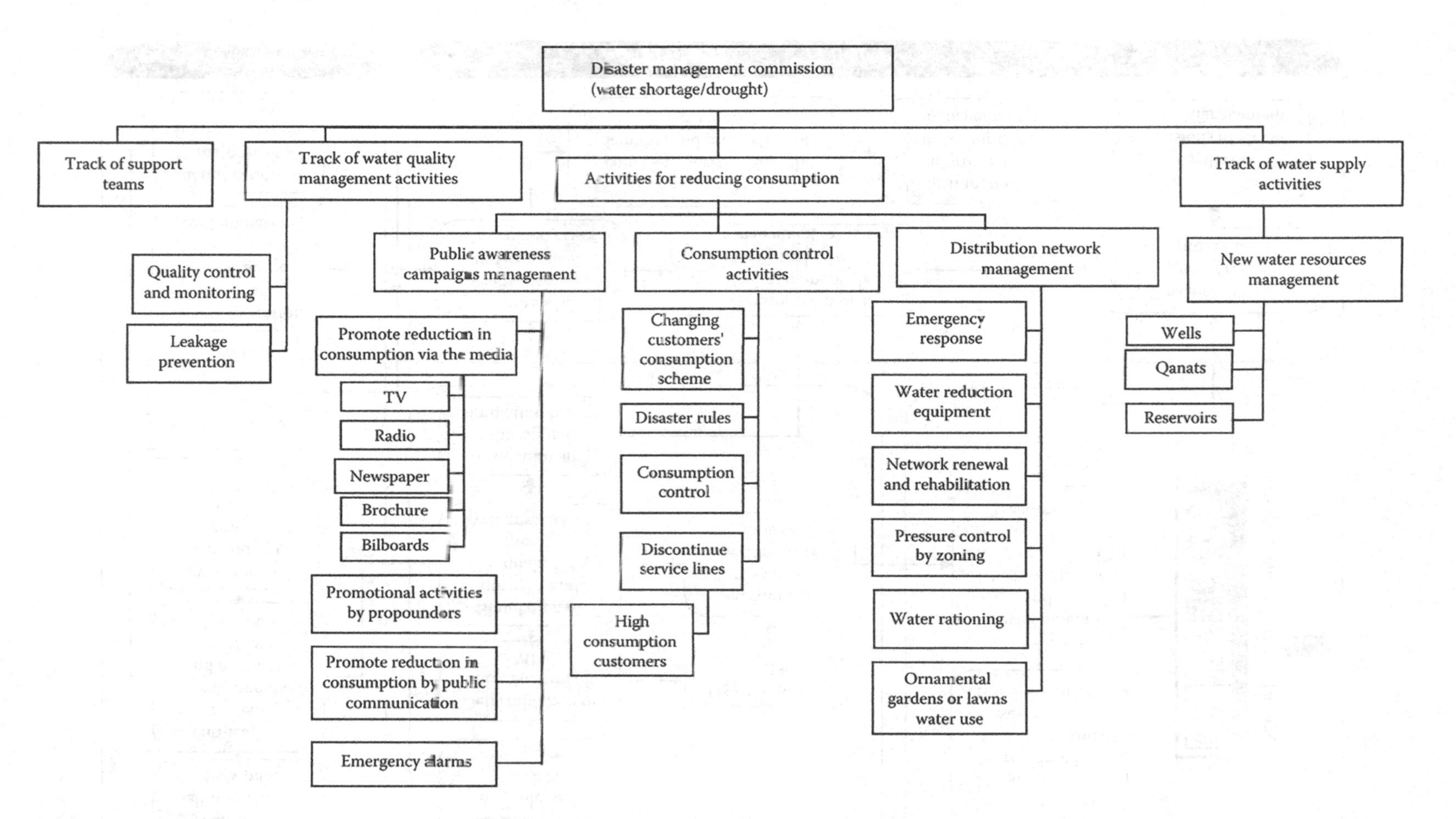

FIGURE 11.17 Organization chart for water shortage disaster management commission.

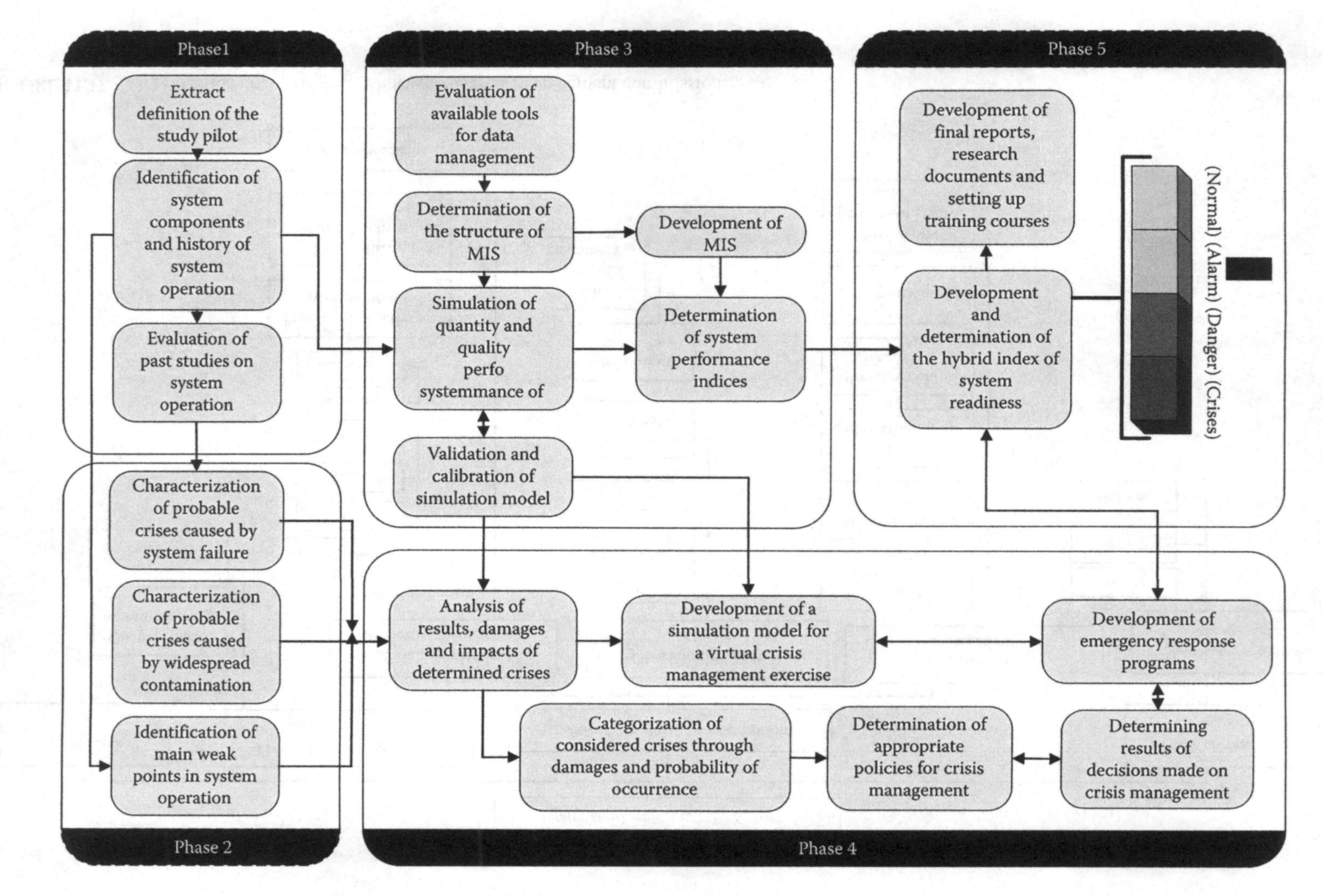

FIGURE 11.18 Diagram of system preparedness evaluation methodology.

result in a hybrid index that determines what is the system's preparedness situation. According to the hybrid index, appropriate decisions are made for increasing system preparedness to deal with disasters. This hybrid index in this methodology is predicted based on system performance indices, which have been described before. This is very useful to avoid disaster occurrence. For this purpose, the system performance indices are calculated based on the available data in the management information system (MIS) and then they are accumulated and result in the hybrid index of system readiness/prepardness.

11.15 CONCLUDING REMARKS

In this chapter, three levels of water system failure to the points of widespread disaster have been discussed. Flood, drought, and widespread contamination are the most concerning issues. Water scarcity in many regions could easily be triggered to become a disaster when these regions are facing drought or any other impact that makes their limited resources scarce or unusable. Flooding could become devastating with more frequent rainfall and storm surges in the coastal areas. It is more critical when the domino effect of interdependency results in failure of multiple facet of life in a city including, water delivery, wastewater treatment, electrical, gas, and communication shut downs. The widespread contamination results in when intentionally or accidentally, a water supply reservoir or a distribution system is contaminated.

The solution to the water disaster is closely linked to improving the water governance of our cities. A paradigm shift is urgently needed in urban water governance. What is needed is a broad-based partnership of public, private, and community sectors. The private sector brings in efficiency gains in water management. Community participation facilitates transparency, equity, and sense of ownership and helps in cost recovery. The government's role is most important in policy setting and as a regulatory agency. The new paradigm must build on the relative strengths of all actors, avoiding overlaps, and redundancies.

Furthermore, in this chapter, many initiatives for UWDM have been presented. Action items at this stage involve raising questions and trying to create consensus among the decision makers and stakeholders. Without a shared vision on UWDM, we are only relying on sacrifices, sporadic and noncoordinated actions, and outside helps to reduce the level of suffering and discomfort caused by a disaster. The accelerated development plans to build, expand, and invest in the water infrastructures prevent us from sitting back and thinking hard about disaster prevention and management and coming with solid contingency plans for different probable disasters to prevent public and water infrastructure facilities.

Finally, it should be stated that disaster of any magnitude could be converted to opportunities if we learn from the past disasters and utilize the unification of efforts that has been resulted from disaster. Experience has shown that the chance of better practices and implementation of sound water resources and demand management as well as disaster prevention is higher when people and societies have experienced or witnessed the suffering of societies/regions. Perhaps what happened in 2005 around the world dealing with water disasters and what happened in New York City with Superstorm Sandy in 2012 have been or could be a strong motive for many governments/municipalities to place action plans for water DM, especially in urban areas on their national/regional agenda.

PROBLEMS

1. Construct the CDF of failure occurrences in two given situations. (1) $\rho(t)=\alpha t$, and (2) $\rho(t)=e^t$.
2. Calculate the system reliability of the water distribution network of Figure 11.19 using the state enumeration method. Node 1 is the source node and other nodes are demand nodes.

All of the pipes have the same failure probability equal to 5%. System reliability is defined as the ability of the system for supplying water to all of the demand nodes. The pipes' performances are independent of each other.

3. Calculate the reliability of the network of Problem 2 using cut-set analysis.
4. Employ the tie-set analysis to calculate the reliability of water network described in Problem 2.
5. Determine the coefficient of variation of the loading and the capacity for a main pipe of a water distribution network with the parameters given in Table 11.8. Consider a uniform distribution for definition of the uncertainty of each parameter.
6. Using the results of Problem 5, determine the risk of the loading exceeding the capacity of the main pipe. Consider that $MS = Q_c - Q_L$ *and MS follow the normal distribution.*
7. Determine the coefficient of variation of the loading and the capacity for a sewer pipe with the parameters given in Table 11.9. The uncertainty of each parameter is defied with a triangular distribution.

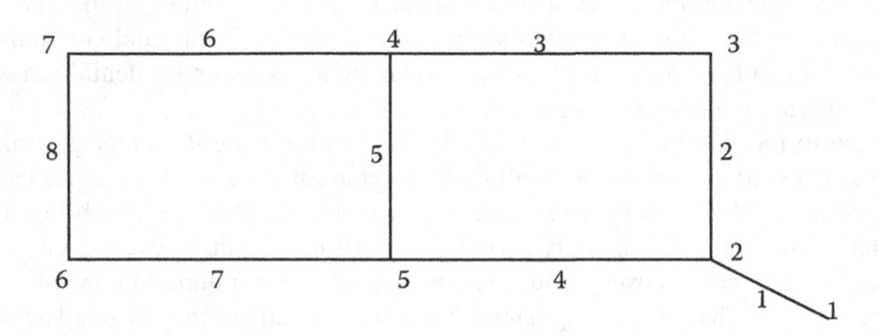

FIGURE 11.19 Layout of water distribution network of Problem 2.

TABLE 11.8
Characteristics of Main Pipe of Problem 5

Parameter	Mode	Range
Water demand	$10\,m^3/s$	$8–12\,m^3/s$
C_{HW}	150	145–155
D	800 mm	600–1000 mm
S_0	0.001 m/m	0.0009–0.0011 m/m

TABLE 11.9
Characteristics of Sewer Pipe of Problem 7

Parameter	Mode	Range
C	0.65	0.6–0.70
I	20 cm/hour	18–22 cm/hour
A	7 ha	6.9–7.1 ha
F	0.002	0.0019–0.0021
D	150 mm	145–155 mm
S_0	0.0005 m/m	0.0004–0.0006 m/m

8. Determine the risk of the loading exceeding the capacity of the sewer pipe in Problem 7. The MS is normally distributed and is calculated as $MS = Q_c/Q_L$.
9. Apply the first-order analysis of uncertainty to the Darcy–Weisbach equation. Consider that S_0, f, and d are uncertain.
10. Consider an 800-mm water transfer tunnel with mean loading equal to 2 m³/s and coefficient of variation equal to 0.20. Calculate the risk of failure of a pipe using safety margin approach when MS is normally distributed. The energy slope of tunnel is 0.0002 with a coefficient of variation of 0.11. The Darcy–Weisbach friction factor is 0.0002 and has a coefficient of variation of 0.15.
11. The monthly supplied water for an urban area in 3 years is presented in the Table 11.10. The monthly mean water demand of the urban area is equal to 100 MCM. Evaluate the reliability, resiliency, and vulnerability of this urban water supply system for supplying more than 90% of urban water demand. How would the system reliability change if the water demand in months June to August is decreased up to15%?
12. The concentration of a toxic chemical in an aquifer is 9.8 mg/L. This aquifer is being used as the source of drinking water. Assume that the average weight of adult women in the area is 60 kg and they drink 1.5 L of water per day, determine the total amount of the chemical that an adult woman may intake after 25 years and the chronic daily intake?
13. Calculate the reliability of the water supply system shown in Figure 11.19 at $t=0.1$ assuming that $R_1(t)=R_4(t)=e^{-2t}$ and $R_2^i(t)=R_3^j(t)=R_5^i(t)=e^{-t}, (1 \le i \le 2; 1 \le j \le 3)$.
14. TDS concentration data of removed water from a well is presented in Table 11.11. The standard value of TDS concentration in this region is considered as 1,400 (mg/L). Calculate the reliability, resiliency, and vulnerability of water quality requirement.
15. The monthly water withdrawals from an aquifer and monthly water demand during 3 years are presented in Table 11.12. Calculate the reliability, resiliency, and vulnerability of this aquifer of water supply

TABLE 11.10
Monthly Supplied Water for Urban Area Described in Problem 11

Month	First Year	Second Year	Third Year
January	101	82	93
February	120	102	110
March	110	110	107
April	85	86	83
May	77	64	55
June	73	76	67
July	84	93	59
August	105	100	72
September	111	119	94
October	106	92	101
November	112	99	121
December	118	108	111

TABLE 11.11
TDS Concentration of Removed Water from a Well (mg/L)

	Year 1	Year 2	Year 3	Year 4
January	1505	1422	1230	1556
February	1215	1498	1409	1545
March	1559	1410	1371	1316
April	1244	1526	1449	1600
May	1532	1210	1574	1486
Jun	1260	1468	1387	1549
July	1480	1300	1206	1317
August	1228	1549	1472	1275
September	1312	1360	1487	1300
October	1574	1319	1406	1539
November	1309	1200	1568	1491
December	1468	1438	1271	1506

TABLE 11.12
Water Release and Water Demand During 3 Years (MCM)

	Water Release		Water Demand	
January	8	8	8	10
February	7	6	5	11
March	7	7	6	12
April	8	9	5	12
May	10	10	7	13
Jun	11	11	9	12
July	10	9	10	11
August	12	10	12	9
September	13	12	11	7
October	10	8	9	8
November	12	10	11	9
December	11	11	10	9

REFERENCES

Akan, A.O. and Houghtalen, R.J., 2003. *Urban Hydrology, Hydraulics, and Stormwater Quality: Engineering Applications and Computer Modeling*. John Wiley & Sons.

Brânzei, S., Gkatzelis, V., and Mehta, R. (2017). Nash social welfare approximation for strategic agents. *Proceedings of the 2017 ACM Conference on Economics and Computation*, pp. 611–628, Association for Computing Machinery, New York.

Bruneau, M., Chang, S.E., Eguchi, R.T., Lee, G.C., O'Rourke, T.D., Reinhorn, A.M., Shinozuka, M., Tierney, K., Wallace, W.A., and Winterfeldt, D.V. (2003). A framework to quantitatively assess and enhance the seismic resilience of communities. *Earthquake Spectra*, 19(4), 733–752.

Canadian International Development Agency (CIDA). (2005). *Integrated Water Resources Management, Training Manual and Operation Guide*. Available at: http://dsp-psd.pwgsc.gc.ca/Collection/BT31-4-27-1998E.pdf.

Chaves, H.M. and Alipaz, S. (2007). An integrated indicator based on basin hydrology, environment, life, and policy: The watershed sustainability index. *Water Resources Management*, 21(5), 883–895. doi:10.1007/s11269-006-9107-2.

De Bruijn, K.M. and Klijn, F. (2001). Resilient flood risk management strategies. In: L. Guifen and L. Wenxue (eds.) *Proceedings of the IAHR Congress* (pp. 450–457). Tsinghua University Press, Beijing, China. ISBN 7-302-04676-X/TV.

Dong, X., Jiang, L., Zeng, S., Guo, R., and Zeng, Y. (2020). Vulnerability of urban water infrastructures to climate change at city level. *Resources, Conservation and Recycling*, 161, 104918.

Dooge, J. (2004). *Ethics of Water Related Disasters*. Series on Water and Ethics, Essay 9. UNESCO, France.

Federal Emergency Management Agency (FEMA). (2008). http://www.fema.gov/library.

Garen, D. (1993). Revised surface water supply index for Western United States. *Journal of Water Resource Planning and Management ASCE*, 119(4), 437–454.

Hashimoto, T., Stedinger, J.R., and Loucks, D.P. (1982). Reliability, resiliency and vulnerability criteria for water resources performance evaluation. *Journal of Water Resources Research*, 18(1), 14–20.

Herath, G. (2008). Soil erosion in developing countries: A socio-economic appraisal. *Journal of Environmental Management*, 68(4), 343–353.

Holling, C.S. (1973). Resilience and stability of ecological systems. *Annual Review of Ecology and Systematic*, 4, 1–24.

Juwana, I., Perera, B.J., and Muttil, N. (2010). A water sustainability index for West Java. Part 1: Developing the conceptual framework. Water Science Technology, 62, 1629–1640.

Karamouz, M. (2002). Decision support system and drought management in Lorestan Province. Final Report to Lorestan Water Board, Iran.

Karamouz, M., Mohammadpour, P., and Mahmoodzadeh, D. (2017a). Assessment of sustainability in water supply-demand considering uncertainties. *Water Resource Management*. 31(12), 3761–3778. doi:10.1007/s11269-017-1703-9.

Karamouz, M. and Moridi, A. (2005). Water crisis management. *Joint Workshop of Tehran-Madrid-Melbourne on Optimal Management of Water Demand and Consumption*, Tehran, Iran.

Karamouz, M. and Movahhed, M. (2021). Asset management based flood resiliency of water infrastructure. *Proceedings of Virtual World Environmental and Water Resources Congress 2021*, June 7–11, Milwaukee, WS.

Karamouz, M., Nazif, S., and Ahmadi, A. (2006). Sustainability indexes in water resources management. *6th Biennial Conference of Community of Environmental Professionals of Iran*, Tehran, Iran.

Karamouz, M., Nazif, S. and Falahi, M. (2012). *Hydrology and Hydroclimatology: Principles and Applications*. CRC Press, London.

Karamouz, M., Rahimi, R., and Ebrahimi, E. (2021). Uncertain water balance-based sustainability index of supply and demand. *Journal of Water Resources Planning and Management*, 147(5), p.04021015.

Karamouz, M., Rasouli, K., and Nazif, S. (2009). Development of a hybrid index for drought prediction: A case study. *ASCE Journal of Hydrologic Engineering*, 14(16), 617–627.

Karamouz, M., Szidarovszky, F., and Zahraie, B. (2003). *Water Resources Systems Analysis*. Lewis Publishers, CRC Publishing, Boca Raton, FL.

Karamouz, M., Torabi, S., and Araghi Nejad, Sh. (2004). Analysis of hydrologic and agricultural droughts in central part of Iran. *ASCE, Journal of Hydrologic Engineering*, 9(5), 402–414.

Karamouz, M., Zanjani, S., and Zahmatkesh, Z. (2017b). Vulnerability assessment of drinking water distribution networks to chemical and biological contaminations: case study. *Journal of Water Resources Planning and Management*, 143(6), 06017003.

Loucks, D.P. (1997). Quantifying trends in system sustainability. *Hydrological Sciences Journal*, 42(4), 513–530. doi:10.1080/02626669709492051.

Mays, L.W. (2001). *Water Resources Engineering*. Wiley, New York.

Mays, L.W. and Tung, Y.K. (1992). *Hydrosystems Engineering and Management*. McGraw-Hill, New York.

McKee, T.B., Doesken, N.J. and Kleist, J. (1993). The relationship of drought frequency and duration to time scales, *8th Conference on Applied Climatology*, Anaheim, CA.

Nazif, S. and Karamouz, M. 2009. An algorithm for assessment of water distribution system's readiness: Planning for disasters. *ASCE Journal of Water Resources Planning and Management*, 135(4), 244–252.

Pimm, S.L. (1991). *The Balance of Nature? Ecological Issues in the Conservation of Species and Communities*. University of Chicago Press, Chicago, IL.

Plate, E.J. and Duckstain, L. (1988). Reliability-based design concepts in hydraulic engineering. *Water Resources Bulletin*, 24(2).

Policy Research Initiative. (2007). *Canadian water sustainability index. Ultimo Accesso*, 30, 2013.

Sandoval-Solis, S., McKinney, D.C., and Loucks, D.P. (2011). Sustainability index for water resources planning and management. *Journal of Water Resources Planning and Management*, 137(5), 381–390.

Shafer, B.A. and Dezman, L.E. (1982). Development of a surface water supply index (SWSI) to assess the severity of drought conditions in snow pack runoff area. *Proceeding of the Western Snow Conference*, pp. 164–175, Colorado State University, Fort Collins, CO.
Sitzenfrei, R., Mair, M., Möderl, M. and Rauch, W., 2011. Cascade vulnerability for risk analysis of water infrastructure. *Water Science and Technology*, 64(9), 1885–1891.
Sullivan, C. (2002). Calculating a water poverty index. *World Development*, 30(7), 1195–1210.
Tokyo Metropolitan Area. (2006). Urban disaster prevention project. Available at http://www.toshiseibi.metro.tokyo.jp/plan/pe-011.htm.
Toshimaru, N. (1998). *Are You Prepared? Learning from the Great Hanshin-Awaji Earthquake Disaster, Handbook for Disaster Reduction and Volunteer Activities*. Hyogo University, Kakogawa.
Tung, Y.K., Mays, L.W., and Yen, B.C. (2004). Risk/Reliability Models for Design. In L.W. Mays (ed.), *Urban Stormwater Management Tools*, Chapter 7, pp. 7.1–7.25. McGraw-Hill, New York, NY.
United Nations Environment Program (UNEP). (2005). Environmental management and disaster preparedness. Available at https://www.unep.or.jp/ietc/wcdr/unep-tokage-report.pdf.
Verhagen, B.T. (2006). Drought: Identification, investigation, planning and risk management. *Proceedings of the International Workshop Groundwater for Emergency Situations*, 29–31 October 2006, Tehran, Iran.
Wagner, J.M., Shamir, U., and Marks, D.H. (1998). Water distribution reliability: Simulation methods. *ASCE Journal of Water Resource Planning and Management*, 114(3), 276–294.

12 Urban Hydrologic and Hydrodynamic Simulation

12.1 INTRODUCTION

In this chapter, many attributes of hydrological and hydraulic behaviors of water cycles are modeled with water infrastructures as the main physical medium especially in the urban watersheds. For examination and synthesizing watershed's occurrences, simulation of hydrologic occasions is required. Hydrologic simulation models are utilized to give a prediction/projection as a better and more precise comprehension of water movement inside the hydrologic cycle both in time and space. An assortment of modeling approaches is utilized for this reason. These models are a representation of theoretical demonstration of a watershed or part/segment of that. Various characteristics are considered for hydrologic simulation. As for hydraulics and hydrodynamic models, different types are explained and three case studies are explained in details.

In hydrologic simulation, three categories of hydrologic models are frequently characterized as:

1. *Data-driven mathematical models (DDM)*: These models are discussed in Chapter 9, and some specific modeling capabilities of artificial neural network (ANN) in the context of probabilistic neural network (PNN) and radial basis function (RBF) are discussed here. In this chapter, the application of numerical and statistical inferences is utilized to build up a connection between model input(s) and output(s). The methodologies utilized in these models are regression analysis, transfer functions, ANNs, fuzzy inference, and system identification. These models can be deterministic or stochastic.
2. *Physical (conceptual based) models (PBM)*: These models endeavor to recreate the actual cycles that occur in reality through the hydrological cycle dependent on recognized physical and experimental connections. Regularly, these models incorporate demonstration of surface/subsurface runoff formation, evapotranspiration, and channel flows/streamflow.
3. System dynamics models are a combination of data-driven and functional relationships that could be considered as a combination of DDM and PBM and are discussed in Chapter 9.

In the initial segment, data-driven mathematical models' applications with accentuation on new information-based models, for example, ANN are discussed. In the subsequent part, various kinds of physical-based hydrologic models of rainfall–runoff including lumped models and semidistributed models like IHACRES (acronym for Identification of unit Hydrographs and Component flows from Rainfall, Evaporation, and Streamflow data), StormNET, and HBV (Hydrologiska Byråns Vattenbalansavdelning) are briefly discussed, and distributed models with applications of HEC-HMS (Hydrologic Modeling System), Gridded Surface-Subsurface Hydrological Analysis (GSSHA), and LISFLOOD as hydrological distributed models are presented. A hydrodynamic model of Delft3D with two modules of FLOW and WAVE and hydraulic-driven models such as EPANET and QAULNET are presented.

12.2 MATHEMATICAL SIMULATION TECHNIQUES

The principle of simulation methods is the numerical demonstration of hydrologic systems' behavior. Because of the numerical simplicity and adaptability of these strategies, they can be handily utilized for investigating the water cycle frameworks. Numerical models, which are called here as data-driven

DOI: 10.1201/9781003241744-12

models (not the technique used for solving partial differential governing equations), are utilized for assessment of a framework execution under a given arrangement of inputs and operating conditions to reproduce and then be used for analyzing different scenarios or for prediction of system's functioning. They can be utilized for simulating countless situations. By utilizing the reproduction models, exceptionally itemized and reasonable demonstration of a complex process/component of the water cycle can be given in terms of properties or variables numerical values. The ideas inherent in the mathematical simulation approach are simpler to comprehend and communicate than physical modeling concepts.

Moreover, simulation strategies can represent the profoundly nonlinear behavior of the water cycle. The reproduction models may cope with steady-state or transient conditions. For instance, to consider the water equilibrium of a metropolitan territory over a generally extensive stretch of time during which no significant adjustments in the framework happen, the steady-state investigation is utilized. The investigation of drainage frameworks that transients through flood propagation lies in the zone of moving from one equilibrium to another. The simulation models are classified into two categories of deterministic and stochastic. If the framework is dependent upon irregular information occasions, or produces them verifiably, the model is called stochastic. The model is deterministic if no irregular components are included and parameters are fixed. In this chapter, stochastic simulation is discussed briefly first as that title appears in Chapters 9–11 representing different behaviors and the random characteristics of parameters and variables. Stochastic approach in terms of random behavior can be added to any conceptual, numerical, and/or data-driven models in the form of sensitivity or uncertainty analysis. Here we only discuss these headings briefly to have a basis for Monte Carlo simulation that is often utilized in modeling and uncertainty analysis. As for fuzzy set theory, fuzzy inference system, and adaptive neuro-fuzzy inference system (ANFIS) and modeling applications, they are also discussed in Chapter 9. ANN applications of pattern recognition and probabilistic features of input data are briefly discussed in this chapter.

12.2.1 Stochastic Simulation

The conduct of stochastic frameworks is naturally nondeterministic and a system's subsequent state is composed of a predictable and random element. The reason for stochastic simulation of the hydrologic framework is the generation of independent random numbers that are uniformly distributed or generated by other fitted distributions. The probability attributes of stochastic simulation are discussed in detail in Chapter 10. The most appropriate probability function for dealing with a specific hydrologic variable is determined by analyzing its historical behavior and using goodness-of-fit tests, some of which were discussed in Chapter 3.

12.2.2 Stochastic Processes

Assume certain events occur at random time points $0 \le t_1 < t_2 < \ldots$. These events constitute a *stochastic process*. The process can be mathematically defined if the distribution functions of t_1, $t_2 - t_1$, $t_3 - t_2$, … are known. A very important characteristic of the stochastic process is the number of events $N(t)$ that occur in the time interval $[0, t]$. The *Poisson process* is the most frequently used stochastic process as expressed by Karamouz et al. (2003) and discussed in Chapter 9.

12.2.3 Markov Processes and Markov Chains

Most of the stochastic hydrologic processes, $X(t)$, follow the Markov process. In this process, the reliance of future estimations of the cycle on past qualities is summed up by the current worth. The Markov chain is an extraordinary sort of Markov measure, wherein the state $X(t)$ can take on just discrete qualities. Nevertheless, in viable investigations of the hydrologic cycle, the continuous stochastic cycles can be approximated by Markov chains. This encourages the improvement of stochastic reproduction models. The fundamental notation and properties of Markov chains are introduced in Chapter 9.

Consider a stationary time series of a continuous random variable that would be simulated through Markov chains. This continuous random variable can be approximated by a discrete random variable Q_y, in year y, which takes on values q_i, with unconditional probabilities p_i where

$$\sum_{i=1}^{n} p_i = 1. \tag{12.1}$$

q_i is the representative of the ith state of the Q variable. In most hydrologic cases, the value of Q_{y+1} is dependent on Q_y, whose Markov chain process can be used to develop transition probabilities. See Chapter 3 for more details.

12.2.4 Monte Carlo Technique/Simulation

Some hydrologic studies depend on various situations of conceivable hydroclimatic factors or, generally, random sample generation methods. For producing assessment focuses in a more methodical manner, adaptive random sampling (ARS) procedures have been created. The fundamental thought of ARS is that the new produced point around the real one should improve the goal work; else, it is dismissed (Rubinstein, 1981). Price (1965) proposed an advanced ARS system, which is the controlled irregular inquiry procedure that presented the idea of an advancing populace of plausible focuses. This idea is the premise of most current worldwide enhancement techniques. At each progression, a simplex is framed from an example of the populace, which is transformed by reflecting one of its vertices through its centroid.

Monte Carlo reenactment strategies are a gathering of ARS techniques. A particular classification of Monte Carlo strategies is the multistart methodology, which comprises of running a few independent trials of a local search algorithm. In an ideal case, these methods aim at starting the local search once in every region of attraction of local optima that may be identified via clustering analysis (Solomatine, 1999). A simple step-by-step procedure for using Monte Carlo simulation to calculate a desired quantity u_k in the e_kth quintile of variable a that has m quintiles is proposed by Fujiwara et al. (1988) as follows:

1. Select the random independent variables to produce a.
2. Generate random numbers and calculate m independent values for a.
3. Label the calculated values of a as Y_1, Y_2, … Y_m and sort them in ascending order $Y_1 \le Y_2 \le \ldots \le Y_m$.
4. Determine Y_i where i is the first $i > m \cdot e_k$. Y_i is the e_kth quintile of a_k.
5. Let $Z_1 = Y_i$ and replicate the above procedure to determine Z_2, …, Z_N.
6. Calculate the mean and variance of sample Z_i as follows:

$$\bar{Z}_N = \frac{\sum_{j=1}^{N} Z_j}{N} \tag{12.2}$$

$$S_N^2 = \frac{\sum_{j=1}^{N} \left(Z_j - \bar{Z}_N\right)^2}{(N-1)}. \tag{12.3}$$

7. Calculate the relative precision (RP) using Equation (12.3).

$$RP = \frac{\delta(N,\alpha)}{\bar{Z}_N}. \tag{12.4}$$

$\delta(N, \alpha)$ is calculated using Equation (12.5).

$$\delta(N,\alpha) = t_{N-1,1-\frac{\alpha}{2}} \cdot \left(S_N^2 / N \right), \tag{12.5}$$

where

$t_{N-1,1-\frac{\alpha}{2}}$ is the $(1 - \alpha/2)$ quintile of t distribution with $N - 1$ degree of freedom.

8. The Monte Carlo simulation terminates when the relative precision becomes less than a given value β $(0 < \beta < 1)$, and the desired quintile u_k is $\overline{Z}_N$; otherwise, N is increased until the desired precision is reached.

Other more practical ways of dealing with Monte Carlo simulation is to fit a distribution to a random variable or the range of parameter variability and then use that distribution to generate several realizations (say 500). These realizations can be utilized in modeling error and quantifying uncertainties. For parameter uncertainty, if we only have a range of variations with a minimum and a maximum, then a uniform distribution can be fitted to generate random numbers. When spatial variability of errors (such as DEM) is considered, Sequential Gaussian Simulation (SGS) as a geostatistical approach is utilized. See case study 1 in this chapter based on the work of Karamouz and Fooladi (2021). SGS is discussed in Chapter 9.

Example 12.1

Use the Monte Carlo technique for a flood control simulation. Use the random number of a uniform distribution. The value of U (random number from uniform distribution) is transformed into a value of x as follows:

$$\begin{cases} x = a + \sqrt{(b-a)(c-a)U} & \text{for } 0 \le U \le \dfrac{b-a}{c-a} \\ x = c - \sqrt{(c-b)(c-a)(1-U)} & \text{for } \dfrac{b-a}{c-a} \le U \le 1 \end{cases},$$

where

x is the value of benefit and
$a(\$) = 10{,}000$,
$b(\$) = 20{,}000$, and
$c(\$) = 50{,}000$.

The value of cost for flood control is obtained by a similar procedure and $a(\$) = 8{,}000$, $b(\$) = 15{,}000$, and $c(\$) = 40{,}000$. Similarly, a value of benefit and a value of cost are obtained for each of the proposed projects. Determine the expected value of B/C for the project.

Solution:

The value of U is drawn from a table of random numbers and gives a value of x as follows for example when U is equal to 0.20:

$$\begin{cases} x = 10{,}000 + \sqrt{(10{,}000)(40{,}000)U} & \text{for } 0 \le U \le 0.25 \\ x = 50{,}000 - \sqrt{(30{,}000)(40{,}000)(1-U)} & \text{for } 0.25 \le U \le 1 \end{cases} \rightarrow \text{if } \; U = 0.2, \text{ then } x(\$) = 18{,}945.$$

The same procedure would be repeated for all benefit values, and these are summarized for total project benefits and the same for value of total project cost. By repeating the same process for 20 times, the outcomes from a probability distribution are obtained and a mean and standard deviation are calculated (Table 12.1). Based on the values in this table, the mean and standard deviation

TABLE 12.1
Loss and Benefit Calculations for Random Numbers Generated Using Monte Carlo Method

Series Number	**1**	**2**	**3**	**4**	**5**	**6**	**7**	**8**	**9**	**10**
U	0.51	0.32	0.14	0.95	0.23	0.75	0.59	0.34	0.52	0.29
B	25,628.5	21,478.3	17,571.9	42,260.3	19,679.6	32,703.2	27,797.5	21,835.8	26,023.2	20,919.9
C	380,100.8	376,712.7	13,566.3	393,680.6	375,249	385,877.2	381,871.7	377,004	380,423	376,256.2
B/C	0.067	0.057	1.286	0.0107	0.052	0.085	0.073	0.058	0.068	0.056
Series Number	**11**	**12**	**13**	**14**	**15**	**16**	**17**	**18**	**19**	**20**
U	0.13	0.24	0.53	0.35	0.12	0.99	0.17	0.76	0.3	0.37
B	17,105.2	19,875.4	26,272.5	22,133	16,907	47,417.4	18,226.8	32,857.4	21,020.9	22,416.5
C	13,317	375,404.2	380,626.6	377,246.7	13,168.7	397,891.4	14,156.4	386,003.1	376,338.6	377,478.1
B/C	1.28	0.053	0.069	0.059	1.28	0.12	1.29	0.085	0.056	0.059

of the B/C values are calculated as 0.31 and 0.5, respectively. The coefficient of variation (the result of division of standard deviation to mean) is also equal to 0.25 for this example. This value reflects too much uncertainty or may be within the acceptable limit of the evaluating agency. A second interpretation is based on confidence limits. For example, for a 95% confidence interval based on a normal distribution of errors (where Z is determined to be equal to 1.96 from a normal probability table), B/C will be in the range of:

$$B/C = 0.31 \pm 1.96 \times 0.5 = 0.31 \pm 0.98.$$

Thus, 0 (because B/C cannot be less than zero) $\leq B/C < 1.19$. Therefore, with 95% confidence, B/C would fall within the range indicated.

12.3 ARTIFICIAL NEURAL NETWORKS

ANNs have a high potential to be utilized as an elective demonstrating instrument in practical hydrology as they have power to display distinctive complicated hydrologic measures, for example, precipitation, precipitation spillover, surface streams, groundwater, and water quality in surface water and groundwater assets. The presentation and advancement of ANNs are very data dependent, and there is no settled procedure for their effective plan and usage.

ANN imitates the structure of biological neural networks by distributing the computation to small and simple processing units, called artificial neurons, or nodes. In ANNs, neurons receive a number of inputs from either the original data or other neurons. Each connection to a given neuron has a particular weight, which could get positive or negative values. A layer can be seen as a group of neurons with similar characteristics, which are connected to other layers or the external environment, which have no interconnections. Transfer functions are partitioned underneath into four classifications including step-like functions, radial and sigmoidal functions, multivariate functions with independent parameters for each dimension, and lastly, the most adaptable transfer functions, for example, universal transfer functions (for more information refer to Karamouz et al., 2012). These and other attributes of ANN including multilayer perceptron network (static network) and general regression neural network are that are well covered in literature and in Chapter 9. Only the temporal and probabilistic ANN are discussed briefly here.

Static networks just process input designs that are spatial in nature, input designs that can be arranged along at least one spatial axis, for example, a vector or a cluster. In numerous errands, the input design contains at least one worldly signal, as in speech recognition, time series prediction, and signal filtering (Bose and Liang, 1996). Time delay operators, recurrent connections, and the hybrid method are various methodologies that are utilized to temporal neural networks.

In temporal neural networks, the feedback from past information can be provided through tapped delay lines (TDLs). TDLs help networks to process dynamically through the flexibility in considering a sequential input. TDLs are a combination of time delay operators in a sequential order. It includes a buffer containing the N most recent inputs generated by a delay unit operator D. Given an input variable $p(t)$, D operating on $p(t)$ yields its past values $p(t-1), p(t-2), \ldots, p(t-N)$, where N is the TDL memory length. Thus, the output of the TDL is an (N+1)-dimensional vector including the input signal at the current time and the previous input signals.

12.3.1 Probabilistic Neural Network

The PNN model, due to their probabilistic nature and simplicity in their application, can be an efficient tool in the classification of common states of variables into specified groups such as classification of system preparedness situation. PNN is based on a Bayesian rule and has been used for functional approximation in time series modeling and in pattern classification. PNN models can generalize the results and can estimate anomalies.

There are four units in a PNN structure, namely, input, pattern, summation, and output layers. An input vector $X = (x_1, \ldots, x_P)^T$ is applied to the n neurons in the input layer and is passed to the pattern layer. In this layer, the neurons are divided into k groups equal to the output classes. The output of the ith neuron in the kth group of pattern layer is calculated using a Gaussian kernel as follows:

$$F_{k,t}(X) = \frac{1}{\left(2\pi\sigma^2\right)^{n/2}} \exp\left(-\frac{\left\|X - X_{k,t}\right\|^2}{2\sigma^2}\right), \quad (12.6)$$

where

$X_{k,t} \in R^P$ is the mean and

σ is the spread of the kernel function.

The summation layer of the network computes the approximation of the conditional class probability functions, $G_k(X)$, through a combination of the previously observed densities as follows:

$$G_k(X) = \sum_{t=1}^{M_k} \omega_{k,t} F_{k,t}(X), \quad k \in \{1,\ldots,K\}, \quad (12.7)$$

where

M_k is the number of pattern neurons of class k and

$\omega_{k,t}$ are positive weights satisfying $\sum_{t=1}^{M_k} \omega_{k,t} = 1$.

The input vector X belongs to the class that has maximum output in the summation unit.

A PNN is a specific form of neural network used to perform Bayesian classification techniques incorporating Parzen univariate estimation.

The Parzen window method is a widely used nonparametric approach to estimate a probability density function p (x) for a specific point p (x) from a sample p (x_n) that doesn't require any knowledge or assumption about the underlying distribution. Each point of the pdf will be estimated. (In other parametric estimation techniques for instance maximum likelihood, a distribution for example normal distribution is chosen, then mean and variance are estimated.)

Then, using PDF of each class, the class probability of a new input data is estimated and Bayes' rule is then employed to allocate the class with highest posterior probability to new input data.

To classify two classes, Bayesian classifiers which can be used are as follows:

$$d(X) = \begin{cases} C_1 & \text{if } l_1 h_1 f_1(X) > l_2 h_2 f_2(X) \\ C_2 & \text{if } l_1 h_1 f_1(X) < l_2 h_2 f_2(X) \end{cases} \quad (12.8)$$

where

X is a p-dimensional random vector,

d (X) is an image of X in a set of classes,

C_i is the ith class,

l_i is the loss associated with misclassifying a vector of the ith class into other class,

h_i is the prior probability of f occurrence in the ith class, and

f_i (x) is the probability distribution function (pdf) for ith class.

The aim of the above equation is to minimize the expected risk in classification. The product of h_i and f_i (x) is a posterior probability from Bayesian theorem which permits the updating of available knowledge h_i with new information f_i (x). The available knowledge h_i could be obtained from a previous sample or the opinion of an expert, and an established mathematical foundation determines f_i (x) to estimate the univariate pdf of a population from its sample. The loss l_i can be calculated or subjectively estimated, but it is usually assigned the same value for all classes. To estimate the multivariate density function, some methods such as parametric and nonparametric methods can be used for estimating the pdf, in PNN we usually use Parzen method.

Example 12.2

In this example, the concentration of nitrate in ppm of the two classes shown in Table 12.2 is related to nitrate from domestic and agricultural sources, respectively. If the concentration of nitrate is 3 ppm, which group does it belong to?

Suppose X_{ij} for class i =1, there are five data points (j = 1–5) are:

Solution:

Using the Gaussian window function with $\sigma = 1$, the Parzen pdf for class 1 and class 2 at x are:

$$y_1(x) = \frac{1}{5}\sum_{i=1}^{5}\frac{1}{\sqrt{2\pi}}\exp\left(-\frac{(x_{1,i}-x)^2}{2}\right)$$

And

$$y_2(x) = \frac{1}{3}\sum_{i=1}^{3}\frac{1}{\sqrt{2\pi}}\exp\left(-\frac{(x_{2,i}-x)^2}{2}\right)$$

respectively. The PNN classifies a new x by comparing the values of y_1 (x) and y_2 (x). If y_1 $(x) > y_2$ (x), then x is assigned to class 1, Otherwise class 2.

For this example, y_1 $(3) = 0.2103$.

Note: The Gaussian window function is a bell-shaped function that is zero-valued outside of some chosen interval, normally symmetric around the middle of the interval, usually near a maximum in the middle.

For Figure 12.1, Parzen window pdf for two classes is:

$$y_2(3) = \frac{1}{3\sqrt{2\pi}}\left\{\exp\left(-\frac{(6-3)^2}{2}\right) + \exp\left(-\frac{(6.5-3)^2}{2}\right) + \exp\left(-\frac{(7-3)^2}{2}\right)\right\}$$

$$= 0.0011 < 0.2103 = y_1(x)$$

So, the sample x = 3 will be classified as class 1 using PNN.

TABLE 12.2
The Concentration of Nitrate in ppm

$X_{1,1}$	2
$X_{1,2}$	2.5
$X_{1,3}$	3
$X_{1,4}$	1
$X_{1,5}$	6
$X_{2,1}$	6
$X_{2,2}$	6.5
$X_{2,3}$	7

12.3.2 Radial Basis Function

In the field of numerical modeling, an RBF neural network is an artificial neural organization that utilizes RBF as actuation capacities. The yield of the organization is a straight blend of RBF of the information sources and neuron boundaries. RBF networks have numerous utilizations, including function approximation, time series prediction, classification, and system control. An RBF is a real-valued function whose worth relies just upon the distance between the info and some fixed point, either the root or some other fixed point called a center. Accumulation of RBFs is regularly used to rough given capacities. This guess cycle can likewise be deciphered as a basic sort of neural organization; this was the setting where they were initially applied to machine learning.

RBF networks are recognized from other neural organizations because of their general estimation and quicker learning speed. An RBF network is a kind of feed-forward neural organization made out of three layers, in particular the input layer, the hidden layer, and the output layer.

Radial functions are a unique class of function. Their trademark is that their reaction diminishes (or increments) monotonically with distance from a central point. An architecture of these sorts of ANNs appeared in Figure 12.2. V is the input, O is the radial function, and w is the weight.

The middle, the distance scale, and the exact state of the radial function are factors of the model all fixed in the event that it is linear. Functions that rely just upon the separation from a middle vector are radially symmetric about that vector, henceforth, the name is radial basis function. In the essential structure, all sources of information are associated with each hidden neuron. The standard is ordinarily taken to be the Euclidean distance and the RBF is usually considered to be Gaussian.

A typical radial function is the Gaussian which in the case of a scalar input:

$$h(x) = \exp\left(-\frac{(x-c)^2}{r^2}\right) \tag{12.9}$$

where

c is the center and
r is the radius.

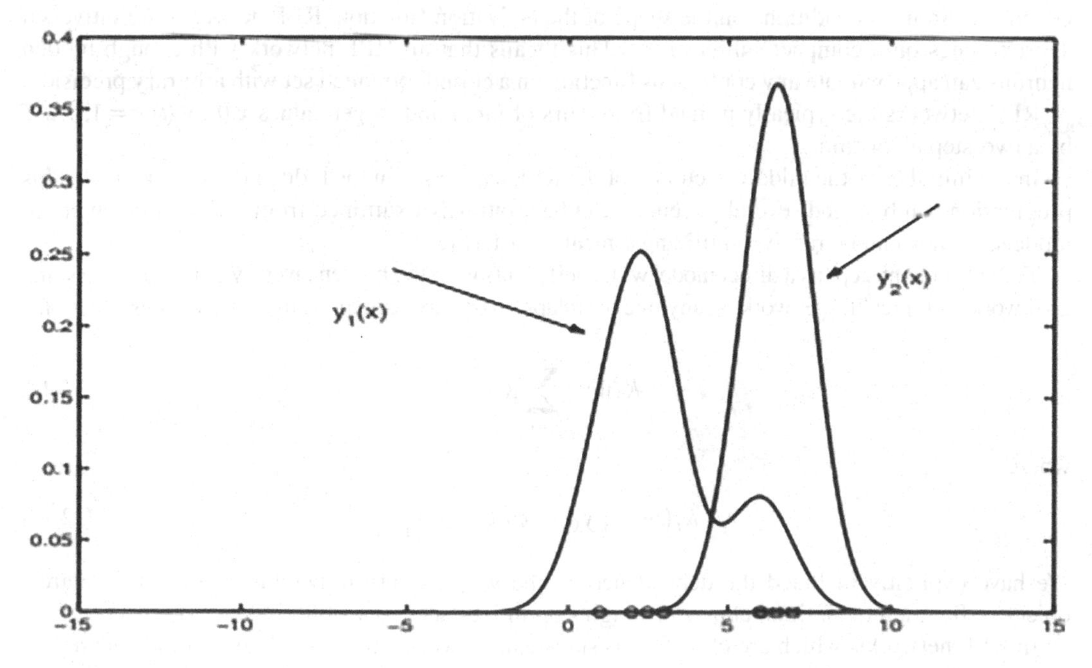

FIGURE 12.1 Two classes of Parzen.

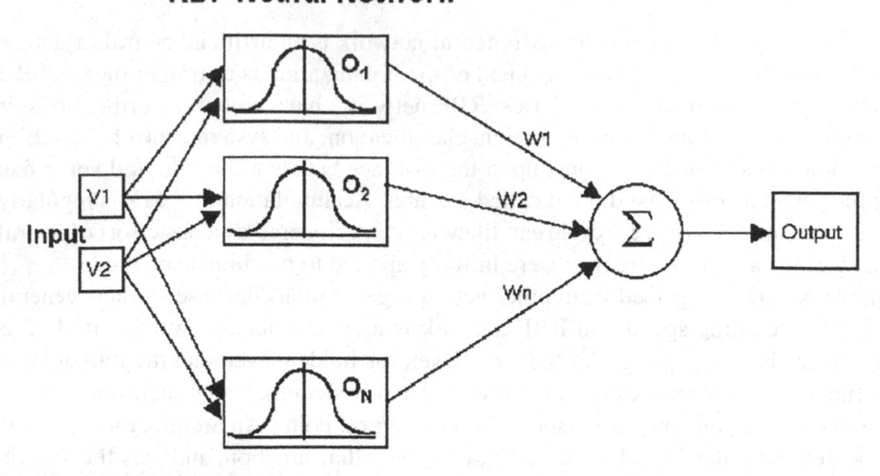

FIGURE 12.2 Architecture of radial basis function artificial neural network (ANN).

RBF networks typically have three layers: an input layer, a hidden layer with a nonlinear RBF activation function, and a linear output layer. The input can be modeled as a vector of real numbers $x \in \mathbb{R}^n$. The output of the network is then a scalar function of the input vector, and is given by:

$$\varphi(x) = \sum_{i=1}^{N} a_i \rho\left(\left\|x - c_i\right\|\right) \tag{12.10}$$

where

N is the number of neurons in the hidden layer,
c_i is the center vector for neuron i, and
a_i is the weight of neuron i in the linear output neuron.

Given certain mild conditions on the shape of the activation function, RBF networks are universal approximates on a compact subset of $\mathbb{R}^n$. This means that an RBF network with enough hidden neurons can approximate any continuous function on a closed, bounded set with arbitrary precision.

RBF networks are typically trained from pairs of input and target values x (t), y (t), $t = 1, \ldots, T$ by a two-step algorithm.

In the initial step, the middle vectors c_i of the RBF capacities in the hidden layer are picked. This progression can be acted severally; centers can be arbitrarily examined from some arrangement of models, or they can be resolved utilizing k-means clustering.

The subsequent step fits a direct model with coefficients w_i to the hidden layer's yields regarding some goal work. A typical target work, at any rate for relapse/work assessment, is the least squares function:

$$K(w) = \sum_{t=1}^{T} K_t(w) \tag{12.11}$$

where

$$K_t(w) = \left[y(t) - \varphi\left(x(t), w\right)\right]^2 \tag{12.12}$$

We have explicitly included the dependence on the weights. Minimization of the least squares' objective function by optimal choice of weights optimizes accuracy of fit.

In RBF networks, which are classified as statistical networks, the same logic is implemented by considering the centroid of the transfer functions (mean of the Gaussian transfer function) of the neurons as the observed historical input vectors (contrary to the MLPs, in which transfer functions of the neurons are independent of the training data). It can be mentioned that the transfer function of the utilized MLP is log-sig, but the transfer function of the RBF network is Gaussian. Therefore, the use of term statistical is due to the Gaussian nature of the transfer function, which is normally distributed, in the hidden layer. In RBF method, the effect of historical dataset is directly related to the distance of input vector from the mean vector of the Gaussian functions. For this purpose, "h" is defined as the spread of the Gaussian transfer function which is similar to three standard deviations in normal distribution. So, it determines the distance that each neuron can affect the input data vector. Therefore, "h" is the influence radius for the new input vector data. Its optimal value is calculated by an iterative method. If the distance of input vector data is less than "h," the output vector is affected by that neuron, and by reducing the distance of the input vector from the mean of Gaussian transfer function of the neuron, the output vector would be larger. In other words, that historical data plays a more important role in determining the final output value. As the distance increases, the effect of that neuron on the output vector decreases and eventually approaches to zero.

Training procedure of the RBFs consists of two main stages. In the first stage, the hidden layers' architecture would be regulated, and in the second stage, weights of the connections (from hidden layer to the output layer) would be determined. An RBF network is shown in Figure 12.3. This is further discussed in case study 1 of Chapter 14.

12.4 OVERLAND FLOW SIMULATION

Physical hydrological models are divided into four groups: lumped, semidistributed, distributed, and hydrodynamics models. Based on these different modeling approaches, different software is developed for the hydrologic analysis and development of the corresponding flood hydrograph for a given storm.

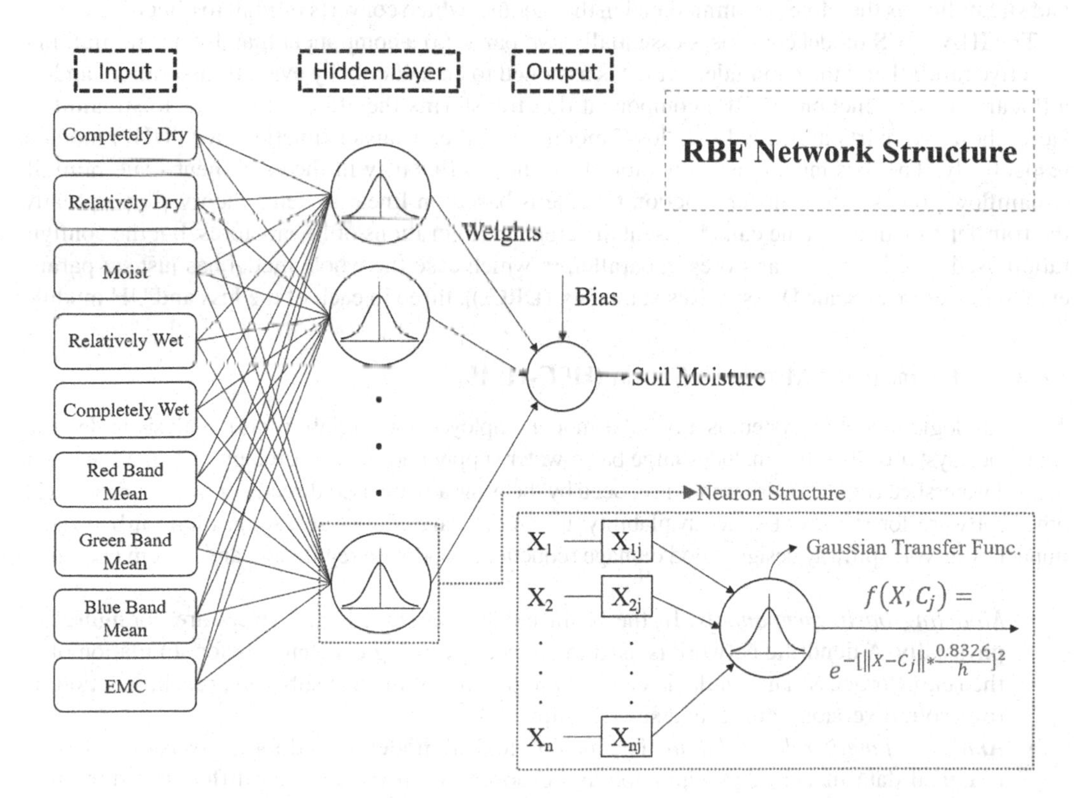

FIGURE 12.3 The architecture of the radial basis function (RBF) model for quantifying the soil moisture classifications, using image processing attributes and an explanatory moisture condition (EMC).

The choice of these models is highly dependent on the objectives of the study, and different considerations are taken into account. Some of these considerations are as follows:

1. Availability and the quality of hydrological data, especially rainfall and runoff data
2. Model structure and regionalization ability of the model
3. Catchment characteristics and flow homogeneity. For example, in arid and semiarid regions, due to flash floods and rapid variations of flow, only some models can be employed successfully for flow simulation.

Therefore, enough attention should be paid to the selection of the most appropriate model based on the characteristics of the study region. In this section, an example of a lumped model named IHACRES and three examples of semidistributed models including HEC-HMS, StormNET, and HBV are described.

12.4.1 IHACRES

IHACRES is a catchment-scale rainfall–runoff model that aims to characterize the dynamic relationship between rainfall and runoff. The only field data required is time series of rainfall and streamflow and a third variable by which evapotranspiration effects can be approximated. The third variable can be air temperature but pan evaporation, or potential evaporation derived from hydrometeorological measurements, can also be used as alternatives if they are available. It is assumed that there is a linear relationship between effective rainfall and streamflow. This allows the application of the well-known unit hydrograph (UH) theory, which concentrates the catchment as a configuration of linear storages acting in series and/or parallel. All of the nonlinearity that is commonly observed between rainfall and streamflow is therefore accommodated in the module, which converts rainfall to effective rainfall.

The IHACRES model comprises essentially two parts: (a) a component that divides rainfall into effective rainfall and the remainder, which is assumed to be lost only by evapotranspiration; and (b) a linear transfer function (or UH) component that transforms the effective rainfall to streamflow. Here, these two parts are called the "loss" module and the "transfer-function" (or "UH") module, respectively. The loss module accounts for all of the nonlinearity in the catchment-scale rainfall-streamflow process; the transfer function module is based on linear systems theory. Conceptually, the transfer function module can represent different configurations of linear stores, but the configuration used here is two linear stores in parallel, in which case the whole model has just six parameters (or catchment-scale Disaster-Resistant City (DRCs)), three in each of the loss and UH module.

12.4.2 Hydrologic Modeling System (HEC-HMS)

The hydrologic modeling system is a popular model employed for rainfall–runoff analysis in dendrite watershed systems. Its range includes large basin water supply and flood hydrology and small urban or natural watershed runoff. Hydrographs produced by the program are used directly or in conjunction with other software for studies of water availability, urban drainage, flow forecasting, future urbanization impact, reservoir spillway design, flood damage reduction, floodplain regulation, and system operation.

- *Modeling basin components*: In the basin model, watersheds and rivers are configured physically. A dendritic network is used to connect hydrological elements for simulation of the runoff process. The available elements in the model include subbasin, reach, junction, reservoir, diversion, source, and sink (Figure 12.4).
- *Analysis of meteorological data*: The meteorological model is used for analysis of meteorological data including precipitation and evaporation. It includes six different historical and synthetic precipitation methods as well as one evaporation method. A tutorial of this model is demonstrated in this chapter's Appendix.

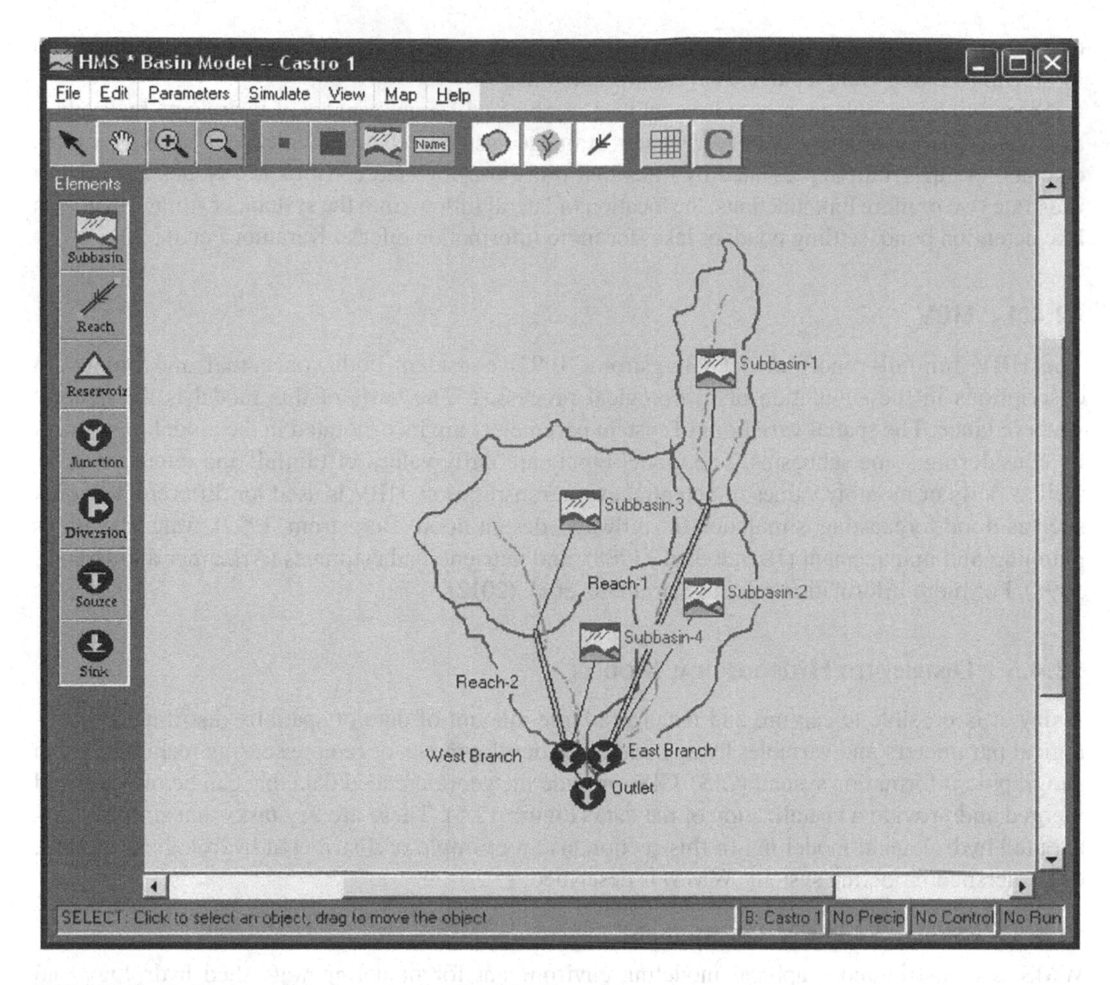

FIGURE 12.4 Basin model screen in the HEC-HMS model.

12.4.2.1 Rainfall–Runoff Simulation

The time span of a simulation is controlled by control specifications. Control specifications include a starting date and time, ending date and time, and computation time step.

12.4.2.2 Parameters Estimation

Most parameters for methods included in subbasin and reach elements can be estimated automatically using the optimization manager. The observed discharge that must be available for at least one element upstream of the observed flow can be estimated. Different objective functions are available to estimate the goodness-of-fit between the computed results and the observed discharge. Two different search methods can be used to find the best fit between the computed results and the observed discharge. Constraints can be imposed to restrict the parameter space of the search method (for more examples refer to Karamouz et al., 2012).

12.4.3 StormNET

StormNET is one of the most advanced, powerful, and comprehensive stormwater modeling packages available for analyzing and designing urban drainage systems, stormwater sewers, and sanitary sewers. The StormNET model can combine hydrology, hydraulics, and water quality issues in a graphical, user-friendly interface. Different graphical symbols are used for representing network elements such as pipes, channels, and detention ponds. A variety of modeling elements are provided in StormNET. See Gong et al. (1996) for more details.

StormNET is a link-node-based model that combines different aspects of hydrology, hydraulics, and water quality in analysis of stormwater drainage systems. Hydraulic elements such as a pipe, channel, or culvert are represented by links that transport flow and constituents. Nodes are used to illustrate two or more link junctions, the location of lateral inflows into the system, or storage elements like detention pond, settling pond, or lake (for more information refer to Karamouz et al., 2012).

12.4.4 HBV

The HBV rainfall–runoff model (Bergstrom, 1992) considers both conceptual and numerical descriptions in the simulation of hydrological processes. The basis of this model is the general water balance. The spatial variations of system parameters are incorporated in the modeling process by considering some subbasins. The model inputs are daily values of rainfall and temperature as well as daily or monthly values of potential evapotranspiration. HBV is used for different purposes such as flood forecasting, simulation of spillways' design flood (Bergstrom, 1992), water resources planning and management (Brandt et al., 1988), and nutrient load estimates (Arheimer and Brandt, 1998). For more information refer to Karamouz et al. (2012).

12.4.5 Distributed Hydrological Models

Today, it is possible to capture and manage a large amount of data of spatially distributed hydrological parameters and variables by application of developments of remote sensing technology and geographic information system (GIS). GISs provide the georeferenced data that can be overlaid and merged and provide a visualization of the data (Figure 12.5). These are key tasks that simplify distributed hydrological modeling. In this section, as an example of distributed hydrological models, the watershed modeling system (WMS) is described.

12.4.5.1 Watershed Modeling System

WMS is a distributed graphical modeling environment for modeling watershed hydrology and hydraulics. WMS is empowered by different tools that facilitate the modeling procedures such as

- Automated basin delineation
- Geometric parameter calculations
- GIS overlay computations such as CN, rainfall depth, and roughness coefficients
- Cross-section extraction from terrain data.
- Automated watershed delineation

WMS is used primarily to set up and run hydrologic models. Though the software has expanded to provide additional hydraulic and hydrologic tools that engineers will find useful, the original focus of the software remains the same. If digital elevation map (DEM) is used, WMS can automatically determine watershed boundaries (Figure 12.6). After determining the watershed boundary, different useful basin data such as area, slope, mean elevation, and maximum flow distance are calculated. WMS locates all flow paths on the whole terrain model, which make it possible to examine flow patterns in different parts of the basin. The longest flow path in each subbasin, which is used for estimation of the time of concentration, is also calculated.

- *Floodplain modeling and mapping*: Using the interpolation algorithms of WMS, flood extents and flooding depth maps can be developed based on DEMs, and water surface data can be obtained from hydraulic analysis. There are fast and easy channel hydraulics

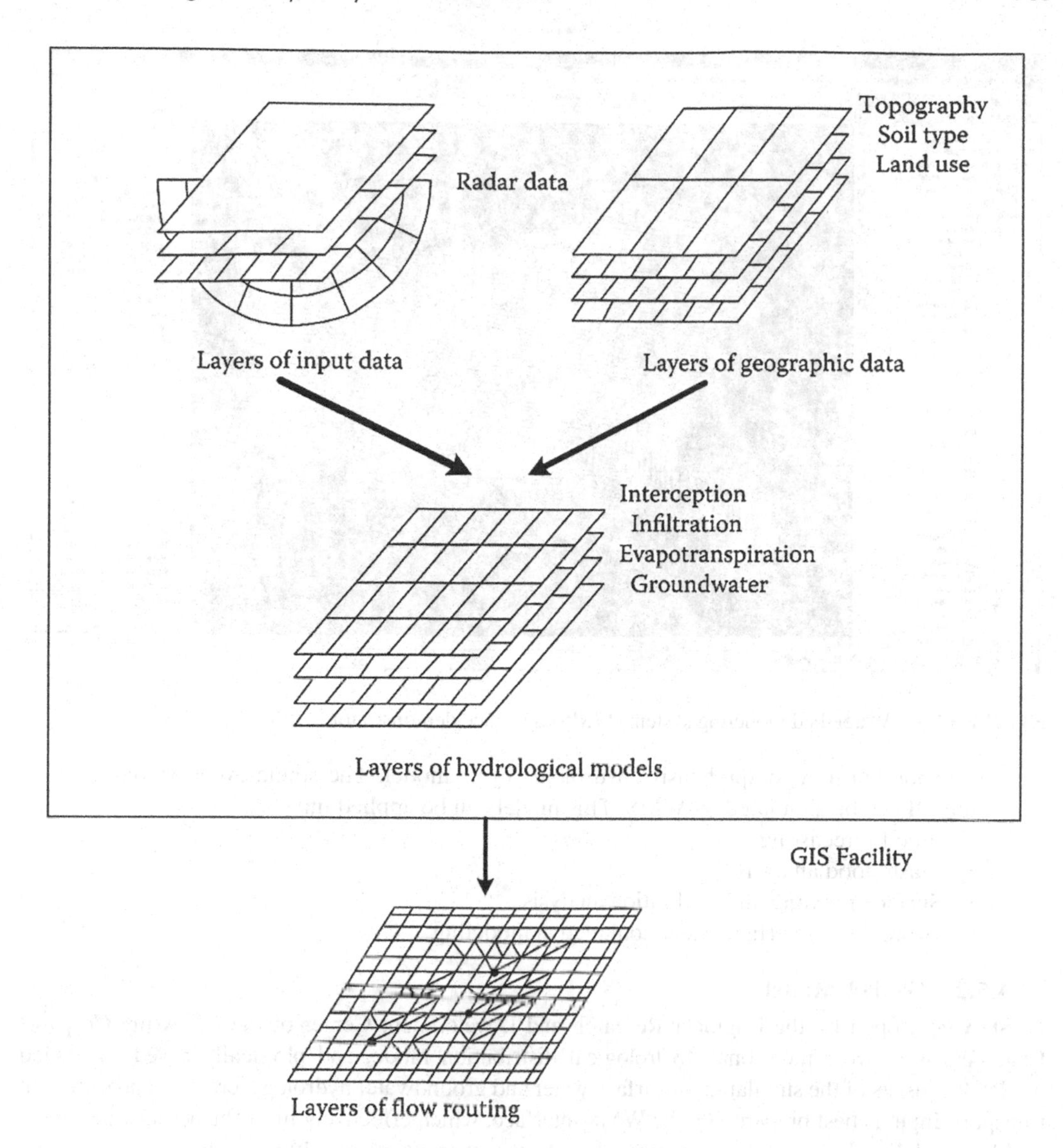

FIGURE 12.5 Schematic of the distributed hydrological modeling system.

analysis tools in WMS to create approximate maps. There is an interface with HEC-RAS (Hydrologic Engineering Center River Analysis System) and flood mapping tools in WMS for detail analysis.

- *Stochastic modeling*: The model parameters' uncertainty can be analyzed using the automated stochastic modeling tools in WMS. These tools automatically change the values of certain parameters (such as CN or roughness) in the model input file and the simulation can be run with different values.
- *Storm drain network modeling*: WMS can simulate the storm drain system. A storm drain network with different components is created in integration with the surface runoff model or any other hydrologic model supported by WMS.
- *2D (Distributed) hydrology*: A 2D finite-difference grid analysis approach is used for surface runoff analysis. Also, 1D channel hydraulics and groundwater interaction is

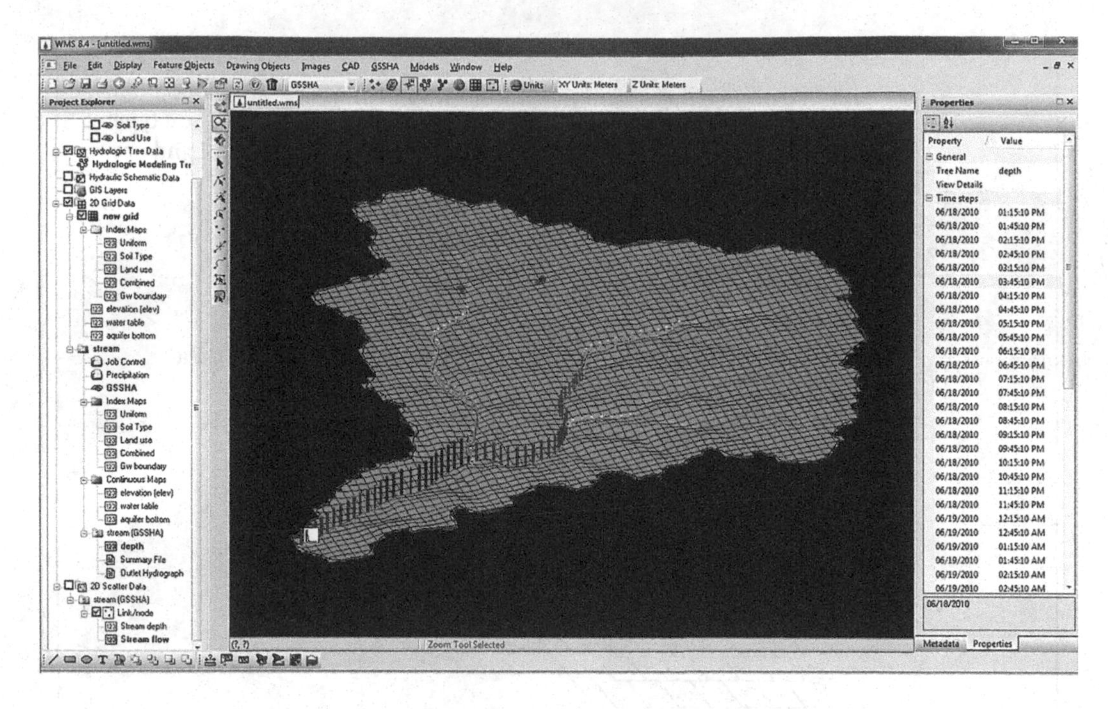

FIGURE 12.6 Watershed modeling system (WMS) interface demonstrator.

incorporated in a comprehensive hydrologic cycle model. The single-event or long-term rainfall can be simulated by WMS. This model can be applied in:

- Flood forecasting
- Flash flood analysis
- Surface ponding and infiltration analysis
- Groundwater/surface water conjunctive modeling

12.4.5.2 GSSHA Model

GSSHA developed by the Engineer Research and Development Center of the US Army Corps of Engineers is a two-dimensional, hydrological distribution mode, and physically based watershed model. It consists of the simulation of surface water and groundwater hydrology, erosion, and sediment transport. Input is best prepared by the WMS interface, which effectively links the model with GIS.

This model enables spatial and temporal analysis in raster format, which attributes altitude, land use, and soil properties data to each cell. GSSHA is one of the modules defined in WMS software. Distribution modeling in GSSHA includes hydraulic and hydrological analyzes. The basis of the flood simulation model is continuity and momentum equations. The Green–Amp equation is also used to estimate the infiltration. Finally, the model uses the hydrograph of the water level entering the basin, the flood zone, and the water height in different cells. GSSHA also has the ability to model flood mitigation strategies such as coastal dams, lakes, and wetlands. The following is a detailed modeling process in GSSHA.

The inputs consist of following factors which can be applied to the model by Hydrologic Modeling Wizard:

- DEM
- Land use
- Soil type
- Rainfall input
- Water level hydrograph
- Evapotranspiration using the Penman–Monteith equation

- Infiltration
- Hydraulic structures
- Overland erosion and sediment transport
- Two-dimensional groundwater

12.4.5.2.1 Starting the program

The Hydrologic Modeling Wizard is used to create and manage projects. It also provides access to components, analysis tools, and other datasets such as land use and soil type data. This part includes the following (Karamouz and Heydari, 2020) demonstrated in Figure 12.7. This figure shows different parts of GSSHA modeling.

The first step is to provide all the data needed to the model. For this, the "Watershed Data" of GSSHA is selected and data of land use, DEM, soil-type data are entered to the model and the boundary information are automatically determined. Most of this data is obtainable from National Geographic Data Center (NGDC) and National Oceanic and Atmospheric Administration (NOAA) sites. Then, the streams paths are determined by the software in Compute Flow Direction section and smoothed by the user in Define and Smooth Streams part. The outlets are also defined by the users. In "Defined precipitation part," method of computation such as uniform and varied form is selected. Moreover, in "Define land use and soil type," infiltration and roughness assigned by the user are applied to the model.

12.4.5.2.2 Digital Elevation Model Data Collection

In order to identify the topographic changes of the study area, the DEM is used, which is both a web service tool in the GSSHA model and directly can be accessed from the USGS site at the address (https://gdg.sc.egov.usda.gov/) and called to the model. Access to different resolutions of the data is provided. As mentioned in the previous section, the model is able to navigate within networks, which is called resampling. In addition to altitude data, soil and land use maps are required to determine runoff and infiltration parameters. This data is available through the site (https://gdg.sc.egov.usda.gov/) to interpret soil and area parameters. For more information on this section refer to Chapter 14

Project Filename

Project Filename
Define Project Bounds
Watershed Data
Download Data (Web Service
Read Data (Catalog)
Compute Flow Directions and
Choose Outlet Locations
Delineate Watershed
Select Model
Define and Smooth Streams
Create 2D Grid
Job Control
Define Land Use and Soil Dat
Hydrologic Computations
Define Precipitation
Clean Up Model
Run Hydrologic Model

FIGURE 12.7 The Hydrologic Modeling sections devised in Gridded Surface-Subsurface Hydrological Analysis (GSSHA) model.

12.4.5.2.3 *Identify Catchment Area*

Using topographic parameters, the catchment area is determined. By receiving the elevation figure model and checking the height of different points, high-altitude points around the basin outlet are identified and the basin boundary is obtained for the flow direction and permeable areas. Waterways should be modified to avoid computational problems such as negative slopes. The smoothing method is used and the redistribution of waterways is done according to the size of the applied network. Thus, elevation data and canals have acceptable overlap and adaptation. These works are applied to the project through Define and Smooth Stream option shown in Figure 12.7.

12.4.5.2.4 *Gridding*

In order to determine the size of the grids, the resolution of the input data is essential. Building a network with a resolution greater than the resolution of the input data increases the execution time of the model and sometimes leads to instability of the courant number within the allowable range. Therefore, according to the steps described in previous section, the appropriate size of gridding that has optimal conditions in terms of accuracy and speed of the model is selected. The different sizes of the network and the increased concentration of networking in one area contain more details. The gridding is applied to the model via Create 2D Grid option.

12.4.5.2.5 *Parameter Estimation*

Utilizing soil type and land use maps by calling them to the model, parameters such as Manning coefficient, Hydraulic Conductivity, Capillary Head, Porosity, Pore Index, Residual Saturation, Field Capacity, and water content (wilting point) are determined. After receiving the data, GSSHA defines a characteristic value for each cell and attributes the parameters related to soil properties to it. This characteristic value is not present in some cells located in the study area, especially near the border, which indicates a lack of information. In these cells, the characteristic value must be adjusted manually according to the values of the adjacent cells. The manual changing is viable through Hydrologic Computation option.

12.4.5.2.6 *Infiltration*

The infiltration rate is a measure of the soil's ability to absorb rainwater or irrigation and is measured in millimeters per hour, which decreases as the soil saturates. When the rate of water flow exceeds the rate of infiltration, runoff flows to the surface and its extent depends on the hydraulic conductivity of the soil near the ground. Permeability is achieved by the interaction of gravity and capillary processes.

While the smaller empty space is more resistant to gravity, it draws water up through the capillary process and acts against gravity. There are different methods in calculating and estimating the volume of water infiltrated into the soil. While the initial water content of the soil is considered constant, equations such as Green–Ampt are used to solve the infiltration flow for an event resulting in flooding (Green and Ampt, 1911). According to Equation (12.13), this is a function of soil suction head, porosity, and hydraulic conductivity and time.

$$\int_0^{F(t)} \frac{F}{F+\psi\Delta\theta} dF = \int_0^t K\,dt \tag{12.13}$$

where

ψ is the soil suction,
θ is the water content,
K is the hydraulic conductivity, and
$F(t)$ is the accumulated depth of infiltration.

After integration, the equation can be used for both the infiltration rate and momentary infiltration rate. The accumulated depth of infiltration is also can be calculated as Equation (12.14).

$$F(t) = Kt + \psi\Delta\theta \ln\left[1 + \frac{F(t)}{\psi\Delta\theta}\right] \tag{12.14}$$

The momentary infiltration rate $f(t)$ is obtained through Equation (12.15).

$$f(t) = K\left[\frac{\psi\Delta\theta}{F(t)} + 1\right] \tag{12.15}$$

Infiltration parameters are given to the model according to soil cover in the study area.

12.4.5.2.7 Roughness Coefficient

Parameters related to the roughness coefficient in accordance with land use are introduced. Each land use has a certain roughness coefficient and according to the existing uses in the area, the roughness coefficient values are called to the model. It should be noted that the roughness coefficient in the area of interaction of the flow with the plant mass needs to be determined. Both infiltration and roughness coefficient are applied in the Hydrologic Computation option.

12.4.5.2.8 Outputs

GSSHA provides inundation maps and also, water level in each cell can be obtained manually. A figure of water inundation map simulated in GSSHA is demonstrated in Figure 12.8. Finally, a table containing all cells' water level is obtained with which total volume of water inundation can be accessed.

FIGURE 12.8 Water inundation map in GSSHA model in a watershed in south coast of Bronx Borough in New York City.

12.4.5.3 LISFLOOD Model

A simplified raster-based 2D model called LISFLOOD-FP (Bates et al., 2013) developed by the University of Bristol is used for DEM uncertainty and error analysis. LISFLOOD-FP is a raster-based flood inundation model (Bates and De Roo, 2000; De Roo et al., 2000; Horritt and Bates, 2001). The main goal of designing LISFLOOD-FP is to produce a simple physical representation of dynamics of flood spreading (Horritt and Bates, 2002). The computational efficiency becomes important when multiple simulations are conducted as part of uncertainty analysis. Using simple flood inundation schemes such as what was used in LISFLOOD-FP is more suitable for multiple simulations. It allows the needed interface with programming codes to make the run times more manageable.

The important parameters, such as surface roughness and infiltration, should also be estimated. A flood plain roughness has been suggested and used in previous studies (Yin et al., 2017). The GSSHA model uses the Green–Ampt method to estimate the area's infiltration, but the LISFLOOD model uses a spatially uniform infiltration rate constant for the entire area. Therefore, the infiltration rate in the LISFLOOD model is calibrated to get an accumulation figure close to what the GSSHA model estimates as it is a more elaborated tool in estimating the infiltration rate and the excess storm than LISFLOOD.

This model is easy to set up. The basic formulations of LISFLOOD-FP (V5.9.6) are described in detail by De Almeida et al. (2012). The 2D continuity and momentum equation is the basis for overflow modeling, which is a function of the difference in inundation level between adjacent cells (Horritt and Bates, 2001; Wu et al. 2018). In this model, the continuity equation is discretized as follows:

$$\frac{dh^{i,j}}{dt} = \frac{Q_x^{i-1,j} + Q_x^{i,j} + Q_y^{i,j-1} + Q_y^{i,j}}{\Delta x \Delta y} \tag{12.16}$$

where

$h^{i,j}$ is inundation level at the cell (i, j),

Δx and Δy denote cell sizes in x and y direction,

n is floodplain roughness coefficient.

The flowrate between adjacent cells, Q_x and Q_y, are defined by a momentum equation as follows:

$$Q_x^{i,j} = \pm \frac{h_{\text{flow}}^{\frac{5}{3}}}{n} \left(\frac{\left| h^{i+1,j} - h^{i,j} \right|}{\Delta x} \right)^{1/2} \Delta y \tag{12.17}$$

where

h_{flow} is the hydraulic head and the difference of maximum inundation level and the highest bed elevation between adjacent cells.

12.4.5.3.1 LISFLOOD Data Preparation

Floodplain topography and flow boundary conditions are the required two types of datasets to run LISFLOOD-FP (Bates et al., 2013; Karamouz and Fereshtehpour, 2019). High-resolution topographic data are increasingly available but still is an important shortage for many urban areas throughout the world. The availability of high-resolution data facilitates developing more accurate tools/models based on resampled low-resolution data. The alternative is to incorporate the error associated with lower-resolution NED DEMs compared to 1-m DEM constructed from LiDAR point cloud and generate multiple realizations of the true state of the elevation.

In terms of flow boundary conditions, different types of boundary conditions such as zero flux, uniform flow, fixed and time-varying free surface elevation, and fixed and time-varying flow into the domain can be applied on the edge of the simulation domain. For coastal flood modeling, both fixed and time-varying free surface elevations are suitable, but the latter is more precise. The fixed free surface elevation equal to a given stillwater elevation is adopted for the edges of the domain (i.e., shorelines; edges of the domain refer to the sides of a rectangular that covers the area of interest.).

The important parameters, such as surface roughness and infiltration, should also be estimated. A flood plain roughness, $n = 0.06$, has been suggested and used in previous studies (Yin et al., 2017). The GSHHA model uses the Green–Ampt method to estimate the area's infiltration, but the LISFLOOD model uses a spatially uniform infiltration rate constant for the entire area. Therefore, the infiltration rate in the LISFLOOD model is calibrated to get an accumulation figure close to what the GSHHA model estimates. It is used in uncertainty analysis of the paper by Karamouz and Fooladi (2021).

12.5 HYDRODYNAMIC (OFFSHORE) MODELING

Hydrodynamic modeling structures the reason for many modeling studies, regardless of whether silt transport, morphology, waves, water quality, and biological changes are being explored. Examination is being done to improve the portrayal of tides, waves, flows, and flood in seaside waters.

In the investigation of the coastal hydrodynamic cycles, modeling (physical and numerical) is frequently utilized to recreate major physical events in the waterfront locale. Physical models (PM) allude to the utilization of research center models at a proper scale (miniature, little, medium, and enormous scope models) for examining the significant cycle, numerical models (NM) allude to the utilization of PC codes (business, open source, home-made programming).

12.5.1 Physical Models

Wind, waves, and swell can be replicated acceptably by a physical model which must be picked subject to different undertaking and site-explicit boundaries like:

1. Model test goals.
2. Proposed structure plan and reasoning behind the proposed structure. These incorporate regular cross areas, definite drawings of the structure, and material-related data, for example, thickness of material and grain size data.
3. Bathymetric subtleties of the encompassing zone (3D) or the wave approach heading (2D) with the goal that a delegate base arrangement can be built. Great-quality bathymetric information will guarantee that wave change is very much recreated in the model.
4. Environmental plan conditions (wave height, period, water level, wave spectra, and wave direction).
5. Structure execution models, for example, passable harm level, greatest wave run-up, and allowable wave transmission at configuration wave and water-level conditions.

The utilization of physical models for understanding the effect of water waves is basic for the plan of marine structures and evaluation of waterfront improvement because of characteristic causes just as human intercession. For physical model tests, wave attributes must be chosen like

1. *The kinds of the waves to be utilized (standard, sporadic, multidirectional, and so on)*: The determination of wave conditions in the model should be viewed as corresponding to the real issue.
2. *The model scale proportions*: The biggest conceivable scale for the accessible trial facility is by and large chosen. Different constraints than actual measurements may assume a significant job in scale proportion determination (wave maker limit, towing speed, etc.).
3. *Representative ocean states*: Wave attributes from the genuine project area will be the reason for determination of delegate ocean states for a model test program. The determination likewise relies upon the application.

4. *The span of the time series*: At times, exceptionally long testing time is needed to acquire enough data to infer configuration esteems.
5. *Free and bound long waves*: Any structure or coastline is profoundly reflective when presented to long waves. While long waves can frequently get away to profound water under common conditions, they are definitely entangled in traditional laboratory tests. Dynamic ingestion frameworks can help the straightforwardness of seaward limits to long waves. Albeit settled by and large, there are serious viable issues in taking care of long waves by the wave generator. In shallow water, the essential stroke of the wave generator is exceptionally enormous and may in this way confine propagation off extraordinary ocean states at an agreeable model scale.
6. *2D/3D waves*: Determination of 2D or 3D wave testing relies upon the issue to be explored; notwithstanding, an overall encounter shows that 2D testing is adequate for a huge scope of issues. The overall absence of good quality directional wave information is likewise utilized as motivation not to utilize 3D testing. In 3D wave basin, the treatment of limit impacts, for example, wave reflection or damping at the limits of the wave bowl requires bigger endeavors than in 2D demonstrating. Checking of 3D wave field inside the bowl can likewise be risky. In 3D basin, the wave conditions are more inhomogeneous in space than in 2D flumes. Thus, wave conditions in the 3D basin require higher goal checking of wave boundaries.

The planning and execution of experiments and the measurement and the analysis of laboratory waves are vital parts of physical modeling.

12.5.2 Numerical Modeling

Floods are simulated in detail for an assortment of reasons, including most, if not all, of the previous segments for genuine shorelines as an element of time. One potential explanation is the expectation of the 50-or 100-year storm flood at a waterfront site for the planned water levels for seaside structures or the foundation of perils and protection rates for beach front networks. Another reason, including real-time modeling, is for danger relief and public well-being. Then again, the hindcasting of a given chronicled storm flood might be completed to decide the nature and degree of the flood or to align and confirm "another" flood model. The last issue is far simpler than the initial two for frequently information concerning storm factors and the wind fields can be gotten. Inferable from the absence of long-term water level records or any records whatsoever, measurable flood data is normally inaccessible. For genuine flood events, frequently very little information is accessible besides from a couple of setup tide measure locales and site-explicit proof, for example, high water levels inside structures, rise of wave harm, and different pointers of storm water level.

For any flood model, a sufficient demonstration of a wind field is vital on the grounds that the spatial degrees of the wind and pressure fields related with a storm are required as input information. Further, the path of the tempest and its right forward speed are vital.

Barotropic tides and floods are by and large simulated utilizing the "shallow-water conditions" since they have long frequencies (many kilometers) contrasted with the water profundity (Goring, 1978).

Nonetheless, tidal models are as yet second rate compared to consonant investigation and forecast for shallow water tides at areas where beach front tide check information are accessible. The tide–flood model expectations mirror this in utilizing the model flood along with symphonious forecasts for tides to give the all-out water level (Flather and Williams, 2004). Figure 12.9 shows a schematic image of a numerical displaying's outcomes.

The mild-slope equation in simulating wave spread from profound water to shallow water is successful, in light of the fact that the vertical profile of the speed is recommended "effectively" as indicated by the direct wave hypothesis. The mild-slope equation can be applied to a wave framework with different wave parts as long as the framework is linear and these segments do not interface with one another. In applying the mild-slope equation to a huge district in waterfront zone, one

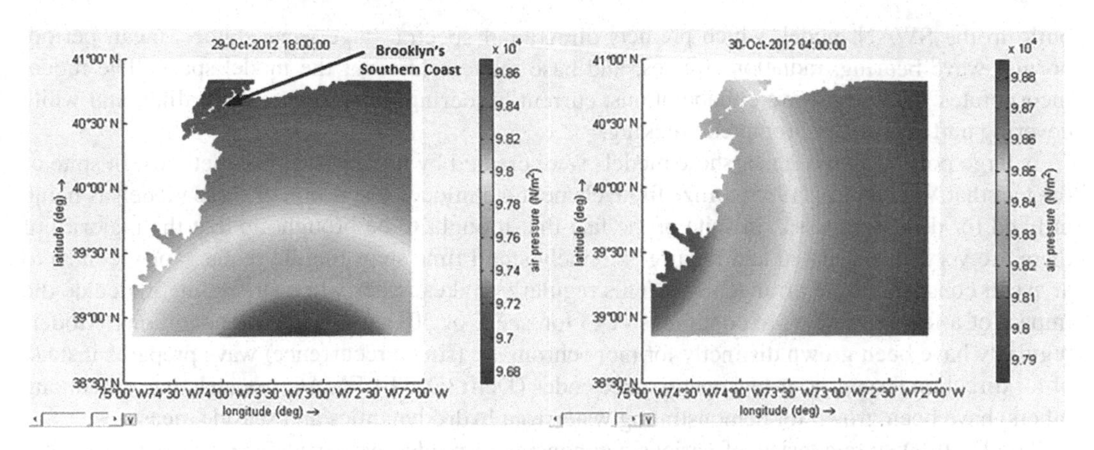

FIGURE 12.9 A schematic of the simulated sea-level pressure variation from the storm front (Karamouz, M. and Mahmoudi, S.. Shallowing of bathymetry (seabed) for flood preparedness: towards designing nearshore BMPs. Journal of Water Resources Planning and Management, (2021)).

experiences the trouble of indicating limit conditions along the shoreline, which are fundamental for settling the elliptic-type mild-slope equation. The mild-slope equation is:

$$\nabla \cdot (c_p \cdot c_g \cdot \nabla \eta) + k^2 \cdot c_p \cdot c_g \cdot \eta = 0 \tag{12.18}$$

where

$\eta(x, y)$ is the complex-valued amplitude of the free-surface elevation $\zeta(x, y)$;
(x, y) is the horizontal position;
ω is the angular frequency of the monochromatic wave motion;
i is the imaginary unit;
∇ is the horizontal gradient operator;
$\nabla \cdot$ is the divergence operator;
k is the wavenumber;
c_p is the phase speed of the waves; and
c_g is the group speed of the waves.

The viable use of wave transformation normally requires the recreation of directional irregular waves. Due to the linear qualities of the mild-slope equation and the parabolic estimation, the guideline of superposition of various wave recurrence segments can be applied. By and large, parabolic models for spectral wave conditions require contributions of the approaching directional arbitrary ocean at the seaward limit. The two-dimensional information spectra are discretized into a limited number of recurrence and course wave segments. Utilizing the parabolic condition, the advancement of the amplitudes of all the wave segments is registered at the same time. In view of the counts for all segments, and accepting a Rayleigh distribution, statistical amounts, for example, the significant wave height can be determined at each grid point. Nonetheless, one should practice alert in expanding the nonlinear Stokes wave hypothesis into the shallow water; extra condition should be fulfilled.

The mild-slope equation might be altered in a straightforward way to oblige events by including an energy dissipation function depicting the pace of progress of wave energy. The energy dissipation function is generally characterized experimentally as per diverse dissipative cycles (Dalrymple et al., 1984).

12.5.2.1 Open-Source Models

These are the ongoing models that regularly incorporate the communications of wave fields with flows and bathymetry and are labeled as open-source models. The contribution of wave energy by the breeze and wave breaking is included in these models. For instance, Holthuijsen et al. (1993)

built up the SWAN model, which predicts directional spectra, huge wave stature, mean period, normal wave bearing, radiation stresses, and base movements over the model space. The model incorporates nonlinear wave collaborations, current hindering, refraction and shoaling, and white covering and profundity prompted breaking.

A large portion of these nearshore models were created by finite difference methods, in spite of the fact that Wu and Liu (1985) utilize finite element techniques. A portion of these models is being utilized for designing work, in spite of the fact that it ought to be brought up that the majority of them are very PC escalated and require extremely small time steps (on the request for seconds) to arrive at consistent state arrangements. This regularly makes issues when attempting to decide the impact of a few days of wave conditions or to foresee 1 or 50 years of seaside conditions. Models regularly have been grown distinctly for monochromatic (single recurrence) wave prepares instead of for directional spectra. A few open-source codes (Delft3D, TELEMAC, opentelemac.org among others) have been grown for demonstrating waterfront hydrodynamics and seaside measures.

The Delft3D suite consists of various components to model the particular physics of the water system, such as the hydrodynamics, morphology and water quality. Delft3D allows you to simulate the interaction of water, sediment, ecology, and water quality in time and space. The suite is mostly used for the modeling of natural environments such as coastal, river, and estuarine areas, but it is equally suitable for more artificial environments such as harbors and locks. Delft3D consists of a number of well-tested and validated programs, which are linked to and integrated with one another. These programs are D-Flow, D-Morphology, D-Waves, D-Water Quality, D-Ecology, and D-Particle Tracking. The Delft3D wave component can be used to simulate the propagation and transformation of random, short-crested, wind-generated waves in coastal waters which may extend to estuaries, tidal inlets, barrier islands with tidal flats, channels and more.

12.5.2.1.1 Delft3D-FLOW

Delft3D-FLOW is a multidimensional (2D or 3D) hydrodynamic (and transport) simulation program that figures nonconsistent stream and transport events that outcome from flowing and meteorological driving on a rectilinear or a curvilinear, limit-fitted framework. Storm tide levels (surge in addition to tide) are recreated utilizing Delft3D-FLOW hydrodynamic model. The model considers different actual properties including water density gradients, tides, flows actuated by wind, and pressure gradients just as Coriolis power. In the reproduction model, the nonlinear shallow water conditions can be settled in two and three measurements. These conditions are accessed by averaging the full Navier–Stokes conditions the vertical way, accepting the vertical speeding up is unimportant contrasted with the flat part. By making this suspicion, the vertical energy condition was streamlined to a hydrostatic pressure condition. Taking the kinematic limit conditions at water surface and bed level into account, the profundity found the middle value of congruity condition was acquired by incorporating the mass equilibrium condition for incompressible fluids over the total depth. The overseeing condition as indicated by Deltares (2011) is:

$$\frac{\partial \zeta}{\partial t}+\frac{1}{\sqrt{G_{xx}}\sqrt{G_{yy}}}\frac{\partial\left((d+\zeta)U\sqrt{G_{yy}}\right)}{\partial x}+\frac{1}{\sqrt{G_{xx}}\sqrt{G_{yy}}}\frac{\partial\left((d+\zeta)V\sqrt{G_{xx}}\right)}{\partial y}=(d+\zeta)Q \quad (12.19)$$

where

x and y are the Cartesian directions.

ζ and d represent the water level above the reference plane and the water depth, respectively.

G_{xx} and G_{yy} are coefficients transforming curvilinear coordinates to rectangular grids of xx and yy.

U and V are the depth-averaged components of velocity in the computational domain, and

Q is the source/sink term, due to the discharge or withdrawal of water, precipitation, and evaporation, which is defined according to Equation (12.20):

$$Q = H\int_{-1}^{0}(q_{\text{in}} - q_{\text{out}})\,d\sigma + P - E_V \tag{12.20}$$

where

H is sum of d and ζ, P, and
E_V are precipitation and evaporation, respectively.
q_{in} and q_{out} are the input and output fluxes.

The model is computationally effective in recreating a wide scope of flood occasions. The Tropical Cyclone tool kit on the Delft Dashboard (DDB), shown in Figure 12.10, encourages the displaying cycle of storm floods (development of storm eye). The low-pressure framework and the breeze speed field brought about by storms are characterized through this tool kit. The figure also shows the global height variation due to sea level besides different toolboxes used in this interface. Bathymetry, tropical cyclone, and tide stations are of the most important ones.

DDB is utilized for water-level simulation including storm surge and wave demonstration. DDB is an independent MATLAB®-based graphical UI pointed towards supporting modelers in setting up new and altering existing models. The interface is combined with Delft3D-FLOW, which permits clients to do progressed flowing investigation and reenact the wind speed and pressure drop brought about by a typhoon. The model domain is chosen as per the examination zone and the storm track. Gauges are utilized to decide the offshore boundary states of the model. The underlying state of the model is considered regarding the underlying water level towards the beginning of the reproduction. The track of the storm's eye is reenacted, and the comparing wind speed and pressure drop are spatially registered as done by Holland and Webster (2007) in various time steps. Next, the offshore boundary condition is changed as the constrain drops to incorporate a precise gauge of wind and storm pressure. This displaying arrangement is planned to create yields that incorporate the fields of storm tides, wind pressure, and speed and the profundity arrived at the midpoint of speeds of flows. The tropical cyclone can be simulated in toolbox=>tropical cyclone.

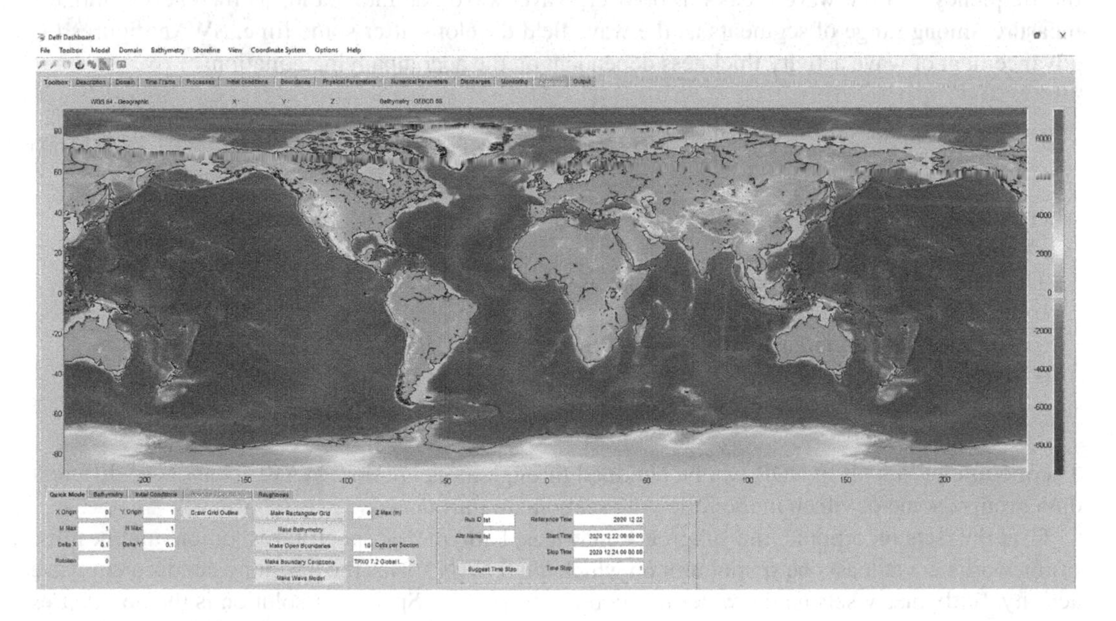

FIGURE 12.10 Delft Dashboard interface.

12.5.2.1.2 Delft3D-Wave

The Delft3D-Wave module consists of a slightly adapted version of the open-source model SWAN and a shell allowing for interaction with Delft3D modules like Delft3D-FLOW and providing additional options for model input like "spiderweb wind fields." SWAN was developed by Delft University of Technology (TUD) with support of the Office of Naval Research (ONR). It is managed and maintained by TUD with funding of the Dutch Ministry of Public Works. TUD releases authorized versions of SWAN in the public domain as open-source code under the GNU GPL license. The SWAN model is applicable in deep, intermediate, and shallow waters, and the spatial model grid may cover any model surface area of up to more than 50 by 50 km.

The added value of Delft3D Wave is the capability of Delft3D hydrodynamics, morphology, and waves to perform a so-called online calculation, in which information is transferred from flow and morphology to wave and back again. This online coupling allows for the simulation of complex water systems in which flow–wave (wave currents interaction as well as wave setup) or flow–wave–morphology (effect of radiation stress on sediment transport and seabed changes) are important.

SWAN is a phase-averaged wave model which is less or not applicable in regions where complex phenomena occur within relatively short distances, for example, near coastal structures or steep sloping beaches, and within harbors. For those areas, phase-resolving models are required to obtain more accurate wave predictions. Examples of these models are Boussinesq-type models and multilayer models.

In practice, Delft3D-Wave (in combination with other modules) is used to transform offshore information such as wind speed statistics to nearshore wave conditions, or more concrete, hydraulic loads on revetments, dune retreat, resulting ship motion, etc. Several models and techniques are required for this transfer that is coupled in a so-called coastal engineering platform.

To reproduce the progression of waves in coastline and oceanic waters, a WAVE module is used in Delft3D. Delft3d-WAVE is a module that engages the coupling of the third era seaward short-wave reproduction (SWAN) with FLOW. SWAN is a medium-stage wave model used to mimic the improvement of self-assertive short waves. The SWAN created by the University of Delft designs the progression of unpredictable short peak waves in waterfront zones with significant, medium, and shallow flows. The SWAN model displays the strategies of wave age through wind, dissipation because of white capping, bed harshness, and wave breaking because of profundity and nonlinear wave-to-wave associations. White capping happens when the wave height is significantly more than the frequency and the wave breaks. Moreover, wave–wave communication is the energy-initiated measure among range of segments as the wave field develops after some time. SWAN figures the advancement of wave activity thickness dependent on the accompanying equation:

$$\frac{\partial N}{\partial t}+\nabla_{x,y}\cdot\left[\left(\vec{c}_g+\vec{U}\right)N\right]+\frac{\partial}{\partial \theta}(c_\theta N)+\frac{\partial}{\partial \sigma}(c_\sigma N)=\frac{S_{\text{tot}}}{\sigma} \tag{12.21}$$

where

(σ,θ) is spectral space (°),

The quantities c_σ and c_θ are the propagation velocities in spectral space (σ,θ),

$\vec{U}$ is ambient current,

σ is the frame of reference moving with current velocity, and

θ is the propagation direction (the direction normal to the wave crest of each spectral component).

The module of WAVE in DDB can be obtained through model option. WAVE consists of different data groups, some of which in the following section are introduced.

Grid datasets incorporate the subgroups matrices, bathymetry, spectral resolution, and nesting. Frameworks are at least one spatial unit on which the SWAN solves the harmony condition of wave activity. Bathymetry sets up the water profundity of the zone. Spectral resolution is the boundaries and resolution of the directional space that SWAN used to perform counts. Nesting is when at least two computational matrices are characterized (for instance, one network with more subtleties and more modest, however, the other one with less subtleties and bigger).

Boundary data group is utilized to decide boundaries, their location, type, and other parameters that portray them which are characterized as follows: At data group boundaries, spectral wave conditions are endorsed at the limit of the principal network and just the primary computational lattice. Any remaining processing organizations (e.g., settled organizations) get their limit data from different organizations. Limit conditions in SWAN can be indicated regarding indispensable (parametric) wave boundaries or they can be perused from an outside document. In the event that the conditions are determined to the uniform alternative and the wave, other worldly qualities are determined to boundary choice, at that point the boundaries are significant wave height, wave period, and the direction of the wave in uniform boundary conditions.

12.5.2.1.3 Flow and Wave Coupling

When FLOW and WAVE are coupled, they become powerfully related in which the FLOW module moves streams, water level, wind, and any geological changes to the WAVE module. At that point, WAVE module ships off the FLOW module at the orbital speed of the wave just as the wave powers.

They can be coupled in three different ways which are as follows: (a) WAVE calculation that utilizes user-defined current characteristics, (b) coupling WAVE with Delft3D-FLOW in offline mode, and (c) web-based coupling (online coupling) of WAVE with Delft3D-FLOW.

With both types of coupling between the FLOW and WAVE modules, the data is exchanged using an arranged correspondence report that contains the latest computational outcomes. To be more explicit, Delft3D-FLOW, as the core of modeling, generates water development and addresses shallow unsteady water conditions in two or three measurements. SWAN spectral wave model is likewise utilized in a fixed computational mode to spread waves from seaward to shore and furthermore to mimic wave-actuated sedimentation.

As a summary of what is discussed in this section, a number of overland models in watershed and urban scales are discussed. There is a tutorial in this chapter's appendix and three case studies that demonstrate the capabilities and real-world applications of these models. Furthermore, a series of offshore and nearshore model in the context of DELFT 3D was presented with detailed explanation of modeling and data preparation. There is also a case study that demonstrates the practical application of these models in the later part of this chapter. In the next section, the modeling attributes of water distribution systems are presented.

12.6 HYDRAULIC-DRIVEN SIMULATION MODELS

There are several different ways to show the motivation behind the simulation study. A model can be exceptionally little and direct to understand the regular reaction of a framework, for example, stability or sensitivity to parameter changes. Or there are models that tend to simulate exceptionally detailed water movement with great deal of nonlinearities to foresee. The hydraulic-driven models are mainly conceptual type, using the basic laws of fluid mechanics and hydraulic engineering. The governing principles are discussed in Chapters 3, 5, and 6. Here the emphasis is placed on models for simulating water distributions including water quality issues.

12.6.1 EPANET

EPANET for hydraulic simulation of water networks uses the Hazen–Williams formula, the Darcy–Weisbach formula, or the Chezy–Manning formula for calculating the head loss in pipes, pumps, valves, and minor loss. It is assumed that water usage rates, external water supply rates, and source concentrations at nodes remain constant over a fixed period of time, although these quantities can change from one period to another. The default period interval is 1 hour but can be set to any desired value. Various consumption or water usage patterns can be assigned to individual nodes or groups of nodes.

EPANET solves a series of equations for each link using the gradient algorithm. Gradient algorithms provide an interactive mechanism for approaching an optimal solution by calculating a series of slopes that lead to better and better solutions. Flow continuity is maintained at all nodes after the first iteration. The method easily handles pumps and valves.

A typical hydraulic-driven model utilized for demonstrating the response of chlorine with contaminants is EPANET. EPANET-Multi-Species Extension (MSX) is likewise utilized for displaying the quality of water in dense wavelength division multiplexing (DWDN) and recreating the response of chlorine with different contamination sources (Shang et al., 2011). This expansion, while holds the capacities of EPANET for pressure-driven displaying of the organization, permits the cycle and demonstration dependent on numerical governing expressions of contamination reaction, spread, and decay.

For water quality simulation, the EPANET model uses the flows from the hydraulic simulation to track the propagation of contaminants through a distribution system. Water quality time steps are chosen to be as large as possible without causing the flow volume of any pipe to exceed its physical volume. Therefore, the water quality time step dt_{wq} source cannot be larger than the shortest time of travel through any pipe in the network, that is,

$$dt_{wq} = \min\left\{\frac{V_{ij}}{q_{ij}}\right\} \quad \text{for } j = 1,\ldots,M+N. \tag{12.22}$$

where
V_j is the volume of pipe i, j and
q_{ij} is the flow rate of pipe i, j.

Pumps and valves are not part of this determination because transport through them is assumed to occur instantaneously. Based on this water quality time step, the number of volume segments in each pipe (n_{ij}) is

$$n_{ij} = \text{INT}\left\{\frac{V_{ij}}{q_{ij}dt_{wq}}\right\}, \tag{12.23}$$

where
INT(x) is the largest integer less than or equal to x.

There is both a default limit 100 for pipe segments or the user can set dt_{wq} to be no smaller than a user-adjustable time tolerance. EPANET models both types of reactions using first-order kinetics. There are three coefficients used by EPANET to describe reactions within a pipe. These are the bulk rate constant k_b and the wall rate constant k_w, which must be determined empirically and supplied as input to the model. The mass transfer coefficient is calculated internally by EPANET.

EPANET can also model the changes in age of water and travel time to a node. The percentage of water reaching any node from any other node can also be calculated. Source tracing is a useful tool for computing the percentage of water from a given source at a given node in the network over time.

There are several different modules in EPANET, two of which are discussed here. For reenactment of biological contamination, two unique methodologies are thought of. In the first approach, EPANET toolbox (EPANET-T) is utilized for mimicking virtual biological pollution dependent on the head driven simulation method (HDSM) in MATLAB programming based on the head-driven simulation method, in which, the measure of chlorine drops down dependent on virtual passageway of pollution. In this technique, the water-driven reproduction is actualized thinking about cooperation among pressure and request. In the subsequent methodology, EPANET-MSX is utilized to mimic the response of a biological contaminant with chlorine, in a MATLAB interface. In this augmentation, notwithstanding the remaining chlorine focus, attributes (i.e., decay rate and concentration) of the organic foreign substance are additionally needed as sources of info.

EPANET-MSX is additionally utilized for the reenactment of chemical contamination (i.e., arsenic) in the organization.

It empowers EPANET to display complex responses between different synthetic and organic species in both the mass stream and at the line divider. This ability has been incorporated into both an independent executable program just as a toolbox library of capacities that developers can use to construct modified applications. EPANET-MSX permits users the adaptability to demonstrate a wide scope of substance responses of premium, including autodisintegration of chloramines to smelling salts, the development of cleansing side-effects, natural regrowth, consolidated response rate constants in multisource frameworks, and mass exchange restricted oxidation pipe divider adsorption responses.

12.6.2 QUALNET

Islam et al. (1995) developed a model called QUALNET, which predicts the spatial and temporal distributions of chlorine residuals in pipe networks under slowly varying unsteady flow conditions. Unlike other available models, which use steady-state or extended period simulation (EPS) of steady flow conditions, QUALNET uses a lumped-system approach to compute unsteady flow conditions and includes dispersion and decay of chlorine during travel in a pipe. The pipe network is first analyzed to determine the initial steady-state conditions. The slowly varying conditions are then computed by numerically integrating the governing equations by an implicit finite difference, a scheme subject to the appropriate boundary conditions. The one-dimensional dispersion equation is used to calculate the concentration of chlorine over time during travel in a pipe, assuming a first-order decay rate.

Numerical techniques are used to solve the dispersion, diffusion, and decay equation. Complete mixing is assumed at the pipe junctions. The model has been verified by comparing the results with those of EPANET for two typical networks. The results are in good agreement at the beginning of the simulation model for unsteady flow; however, chlorine concentrations at different nodes vary when the flow becomes unsteady and when reverse flows occur. The model may be used to analyze the propagation and decay of any other substance for which a first-order reaction rate is valid (Chaudhry et al., 1987). The water quality simulation process used in the previous models is based on a one-dimensional transport model in conjunction with the assumption that complete mixing of material occurs at the junction of pipes.

These models consist of moving the substance concentrations forward in time at the mean flow velocity while undergoing a concentration change based on kinetic assumptions. The simulation proceeds by considering all the changes to the state of the system as the changes occur in chronological order. Based on this approach, the advective movement of substance defines the dynamic simulation model. Most water quality simulation models are interval oriented, which in some cases can lead to solutions that are either prohibitively expensive or contain excessive errors.

12.6.3 Event-Driven Method

WDM is extremely simple in concept and is based on a next-event scheduling approach. In this method, the simulation clock time is advanced to the time of the next event to take place in the system. The simulation scheduled is executed by carrying out all the changes to a system associated with an event, as events occur in chronological order. Since the only factors affecting the concentration at any node are the concentrations and flows at the pipes immediately upstream of the given node, the only information that must be available during the simulation is the different segment concentrations. The technique makes an efficient use of the water quality simulation process.

The advective transport process is dictated by the distribution system demand. The model follows a front-tracking approach and explicitly determines the optimal pipe segmentation scheme with the smallest number of segments necessary to carry out the simulation process. To each pipe, pointers (concentration fronts), whose function is to delineate volumes of water with different concentrations, are dynamically assigned. Particles representing substance injections are processed in chronological order as they encounter the nodes. All concentration fronts are advanced within their respective pipes based on their velocities. As the injected constituent moves through the system, the position of the concentration fronts defines the spatial location behind which constituent concentrations exist at any given time. The concentration at each affected node is then given in the form of a time–concentration histogram.

The primary advantage of this model is that it allows for dynamic water quality modeling that is less sensitive to the structure of the network and to the length of the simulation process. In addition, numerical dispersion of the concentration front profile resolution is nearly eliminated. The method can be readily applied to all types of network configurations and dynamic hydraulic conditions and has been shown to exhibit excellent convergence characteristics.

12.7 CASE STUDIES

12.7.1 Case Study 1: DEM Error Realizations in Hydrologic Modeling

There are many uncertainties in calculating water accumulation volume that should be addressed mainly because of the resolution of spatial data and the way they affect the implemented models.

Karamouz and Fooladi (2021) produced 500 DEM realization by comparing a 1-m LiDAR and 10 by 10 NED DEM using SGS (discussed in Chapter 9) method to estimate the error associated with inundation estimates and account for data uncertainties. The 2D hydrodynamic flood mapping, LISFLOOD-FP, has been utilized to determine the successive estimate of accumulated water in the sewershed and determine the appropriate ranges and a probability distribution that better represents the uncertain nature of these estimates rather than relying on a single deterministic estimate.

12.7.1.1 Methodology

This paper presents a comprehensive framework for uncertainty analysis of spatial data in the R programming environment. R is a programming language and software environment for statistical computing and data analysis. With its multiple libraries and highly advanced graphical display capabilities, this environment significantly increases the efficiency of spatial and statistical processing in the face of large data and is developing faster than other comparable data-intensive environments. In addition to statistical packages, it has many spatial packages that can be used effectively in the coding process. Almost all the GIS-based operations that can be done in ArcMap can be implemented in R with higher flexibility.

In the first step, 500 random points in the region are selected as ground control points (GCPs) and the error values are calculated by comparing the LIDAR values and the original DEM. The error random path variogram is modeled. Namely, the experimental variogram is obtained using observational errors at control points, and then a theoretical model, such as spherical, exponential, Gaussian, and Matern, is fitted to this experimental variogram. In this paper, the Matern model has been used. The SGS algorithm randomly selects one of the points in the region each time. If this point is one of the control points, the error value is maintained at this point. Otherwise, the values of the existing side neighbors, including control points and values that have already been simulated, are used to interpolation kriging at those points. As previously defined, SGS may be referred to as a Gaussian conditional simulation because the error value is maintained throughout the solution path at GCPs. Kriging interpolation produces a mean value and variance for errors (Aerts et al., 2003). Then, based on these values, provides a conditional cumulative distribution function (CCDF) for each computational cell as described by Bivand et al. (2008). By determining a random path, each realization is then generated by a random value from the estimated CCDF. Control dataset and previously simulated values are used as conditional information. This process is repeated until all are simulated.

12.7.1.2 Results

For uncertainty analysis, a 2D hydrodynamic model called LISFLOOD-FP was used along with an SGS model to quantify the effect of DEM error uncertainties on inundation. A 500 DEM realization provided enough water accumulation data to fit a normal probability distribution and showed about 12% underestimation when DEM error is ignored. This makes a big difference when damages are estimated. The flood probability map in Figure 12.11 prepared and showed the significant value of considering DEM error. It shows only 8% in a condensed coastal area of south shore Brooklyn; however, the incremental value of that and the implication of the proposed methodology could be profound in a more open and less congested areas.

12.7.2 Case Study 2: Simulation of Ungagged Coastal Flooding—Nearshore and Inland BMPs

Karamouz and Mahmoudi (2022) have utilized an approach towards alleviating the consequences of flood damages. Coastal areas are of great strategic importance for socioeconomic and environmental reasons. However, these areas are often exposed to natural disasters such as hurricanes and sea level rise. Severe storms account for up to 38% of natural disasters (Kron, 2013) and are the most significant natural hazard in the United States. Given the global population growth in coastal areas and the occurrence of more extreme incidents, with having many of the important infrastructure in those areas, the importance of protecting coastal areas from hurricanes and storms is inevitable. Moreover, climate change in recent years has intensified the frequency and severity of these severe events (Holland and Webster, 2007), all of which together have increased the damage caused by floods in these areas. Coasts as complex geological structures act differently in response to natural disasters such as floods (Balica et al., 2012). Infrastructures in urban areas such as refineries, power plants, and intercity trains are the most vital urban infrastructure that can disrupt urban life in the case of failure. Therefore, in order to protect these infrastructures, measures should be considered, one of which is to reduce the burden on (Hawkes et al., 2008) the infrastructures. Using the best management strategies (BMPs) is one of the ways to deal with flood remediation for infrastructure protection. Current approaches to flood reduction include gray infrastructures, green infrastructures, or a combination of green and gray infrastructures. Gray BMPs refer to rigid engineering

FIGURE 12.11 Flood probability flooding map for a 100-year design value for Coney Island sewershed in Brooklyn. The top view is the base deterministic flood inundation image that is also marked as the dash lines on the map. (Karamouz M. and Mahani F. F. 2021).

structures designed to drain water at sea. Green BMPs, meanwhile, refer to measures such as conservation ponds, detention ponds, and vegetation strips that help reduce runoff and help the natural and ecofriendly aspects of landscape architecture.

For reaching this objective, two different strategies were taken into account. First, to utilize offshore BMPs and shallow the seabed by raising its bathymetry in the form of an apron. An apron is a cube filled with soil (including vegetated by mangrove) or concrete which can possess different sizes.

The idea of an apron goes back to the pioneering work of McCowan (1894). He stated that the wave breaking happens when the velocity of particles and waves are equal the ratio of wave height to water depth is 0.78. Hence, if an apron is placed to the point where the wave breaks, it could push back the wave break for more energy dissipation as the wave has to travel more distance to reach the shoreline.

Further, they utilized inshore BMPs which included a levee and/or a green levee. A green levee consists of a levee having vegetation cover on its crest. Moreover, a combination of the offshore and inshore BMPs was investigated.

12.7.2.1 Area Characteristics

Brooklyn, a county in New York City, and Superstorm Sandy were chosen as the case study and the reference storm, respectively. More than 20 hurricanes have hit Brooklyn since the seventeenth century, some of which are Bob (1991), Ernesto (2006), Irene (2011), and Sandy (2012) as the most important of these storms. Hurricane Irene was one of the most devastating coastal hurricanes to hit much of the eastern United States. Hurricane Irene increased the maximum wind speed to 100 km/hour and the total damage caused by the storm is estimated at $15.6 billion. Superstorm Sandy has created waves with a height of 3.4 m, which has caused floods in various places, power outages, and disruption of the public transportation system. The total damage from Superstorm Sandy is estimated at $71 billion. A schematic figure of Superstorm Sandy's track is shown in Figure 12.12. Hence, selecting Brooklyn as one the frequent areas to experience flood damages and Superstorm Sandy as the most severe storms the area has ever experienced was reasonable.

The data buoy was used to capture wave information in offshore simulation. As shown in Figure 12.12, the closest data buoy to the case study area is the 44,065 data buoy with coordinates 40.36 northward and 73.70 westward.

Figure 12.13 also shows the points where in this study the water level is monitored in offshore simulation. Each of these points is located 500 m from each other, behind the apron on the shoreline. The central point (M2) shown in this figure with the coordinates of longitude 73.96 and latitude 40.56 is considered as a virtual station in an ungagged area in which the hydrograph of the water surface is simulated. The other two points are located 500 m across the left and right edges of the M2. The desired location for the tablecloth is also specified in this figure. Other points are used to observe any lateral changes in the water level.

Furthermore, for extracting the initial water level information to enter the offshore software, the battery station was used, which is the closest station with complete information to the study area and is shown in Figure 12.13.

12.7.2.2 Methodology

In this study, for offshore modeling, Delft3D software was utilized which is able to couple flow and wave components (for more information refer to Section 12.5.2.1). This software has also a toolbox for changing the bathymetry which is called QUICKIN. Thus, the bathymetry obtained from the software can be utilized in this toolbox in order to shallow the seabed bathymetry in desired places. Afterward, different placement and sizes of apron were investigated in order to find the best scenario of bathymetric shallowing which means the apron with the best effect on reducing water level especially at the time of storm's peak hour. Eventually, as outputs, two different water level hydrographs, before and after applying the apron, were obtained.

An example of McCowan's index is illustrated in Figure 12.14. For a slope of 0.007 (similar to Brooklyn nearshore) and a significant wave height of 2.7 m, the breaking depth for the waves is 3.5 m

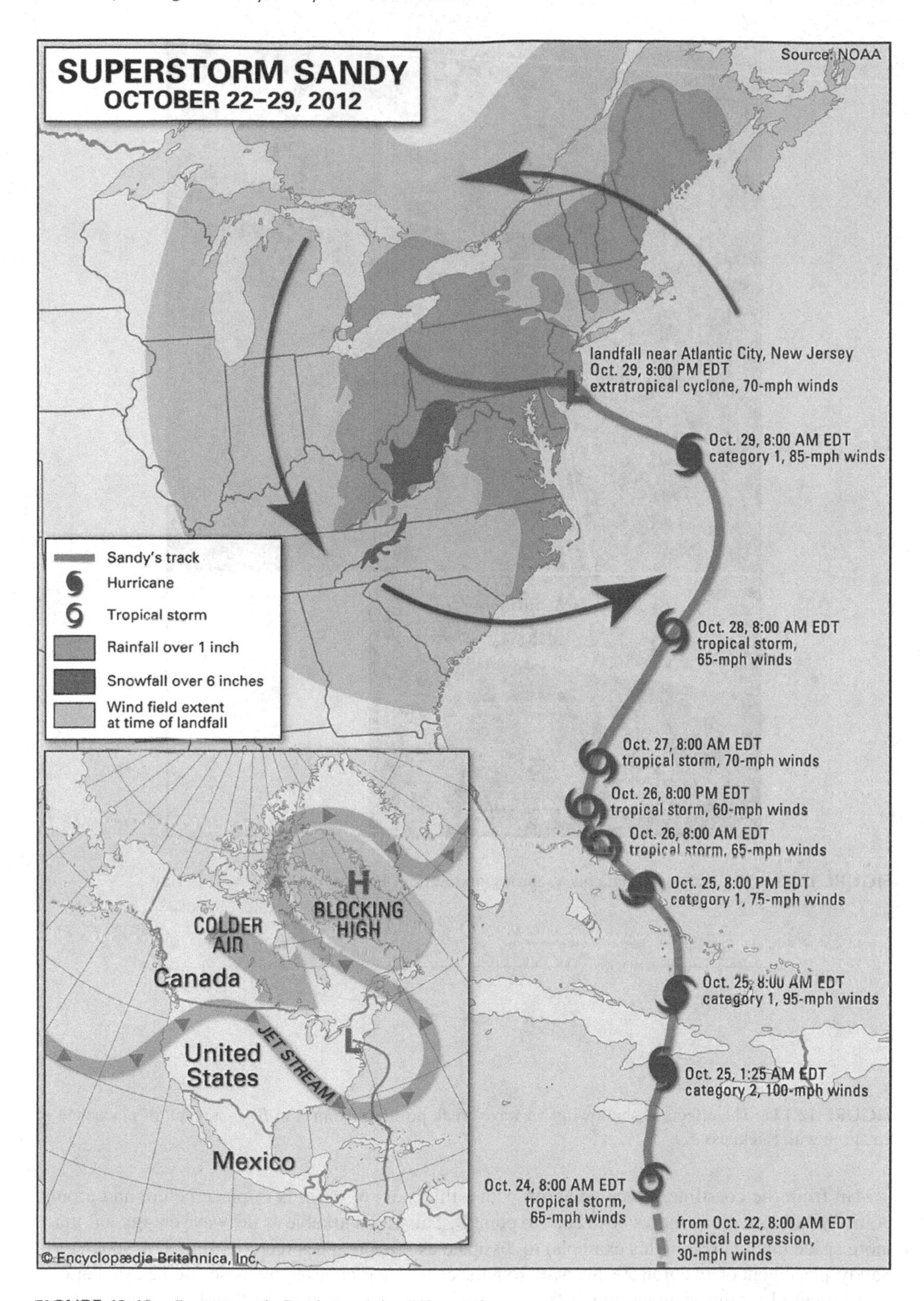

FIGURE 12.12 Superstorm's Sandy track in different time.

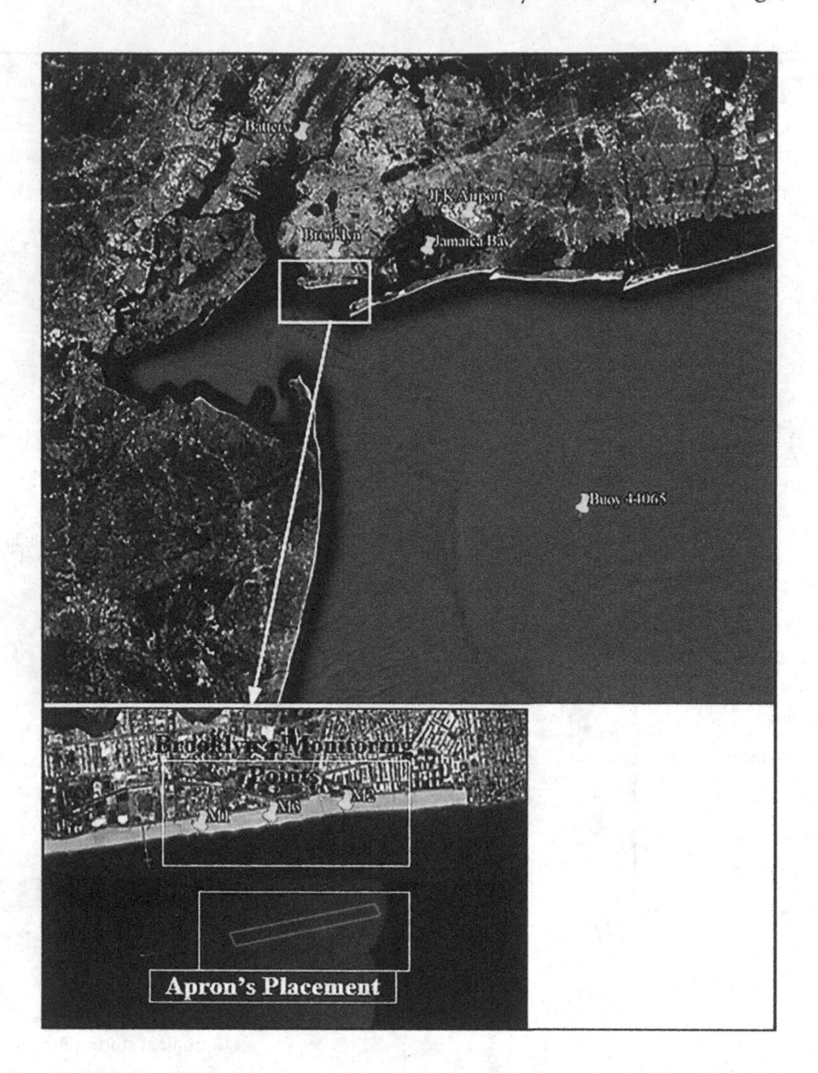

FIGURE 12.13 The location of data buoy, aprons, and Battery Park station in New York Bay.

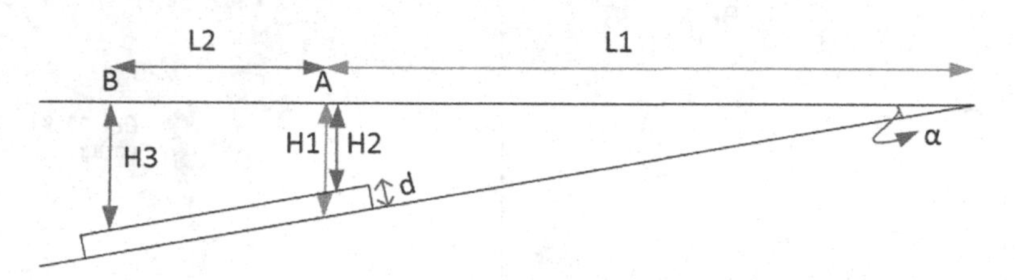

FIGURE 12.14 The effects of shallowing on wave break position (from *A* to *B*) as a result of placement of an apron with thickness *d*.

(494 m from the coastline). Once an apron with a thickness of 0.6 m is in place before and around point A, the wave break moves 86 m back to point B. This is remarkable as the wave energy has much more space (about 20% in this example) to dissipate as it approaches the coastline. For Superstorm Sandy, placement of an apron 500 m away from the coastline is checked by these simple calculations. The data used for this example comes from that storm moving towards the Brooklyn south shore.

Further, GSSHA model, which includes different kinds of BMPs, was used for inshore simulation (for more information see Section 12.4.5.2). As inputs, water level hydrographs obtained from previous paragraph were taken into use and two different models were constructed. Moreover, the levee and the green levee were applied to these two models to see their effects. The hydrograph with an apron after applying the choices of levee and the green levee led us to see the effect of combination of offshore and inshore BMPs.

12.7.2.3 Results

As mentioned before, the Delft3D software's most wanted output for this study is water level hydrograph. In this study, two other components of water level (storm tide and wave components) are simulated as well and the results are shown in Figure 12.15.

Several different scenarios were taken into account and finally the best scenario was found to have 200 m width, 1,000 m length, and 0.6 m thickness with being 500 m away from the shoreline. This scenario reduces water level and wave height as much as 11% and 48%, respectively.

Afterward, water level hydrographs obtained from this section were utilized as input for GSSHA model. After applying the inshore BMPs, the inundation maps before and after inshore BMPs were obtained and are demonstrated in Figure 12.16. The statistical results show that some areas experienced a 100% water level reduction which in case of combination of inshore and offshore BMPs, these areas were more than applying an offshore BMP. Moreover, by applying only the apron, the water volume was reduced from 711,030 to 372,733 m^3.

As can be seen in these figures, the apron alleviates some amount of the flood damages at the sea section, and further in the coastal areas, inland BMPs reduce a huge amount of water inundation. Nevertheless, although usage of green levee reduces the water level, the trade-off between the construction cost and the reduction in damages costs should be evaluated.

12.7.2.4 Concluding Remarks

The case study is the southern coastline of Brooklyn, which has confronted the most extreme waves in NYC at the time of storms that are passing through the east shoreline of United States in the Atlantic Ocean. Superstorm Sandy was modeled to produce the water level hydrograph in the southern shoreline of Brooklyn. Delft3D model was used by coupling its two modules, Delft3D-WAVE

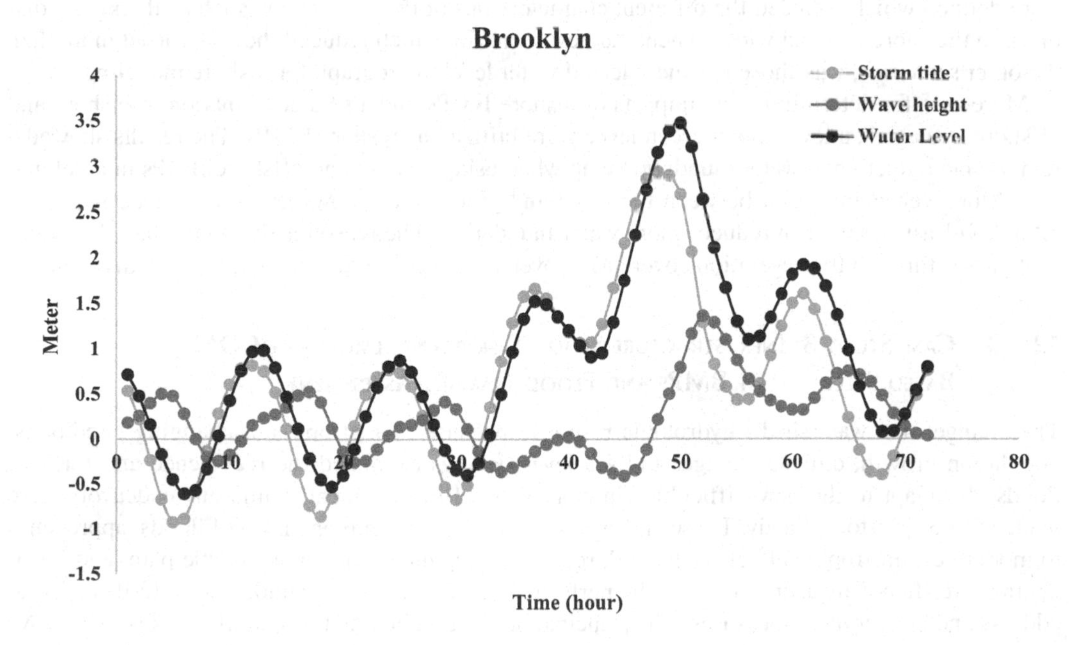

FIGURE 12.15 Different components of simulated Brooklyn's water level during Superstorm Sandy (Karamouz M. and Mahmoudi S. 2022).

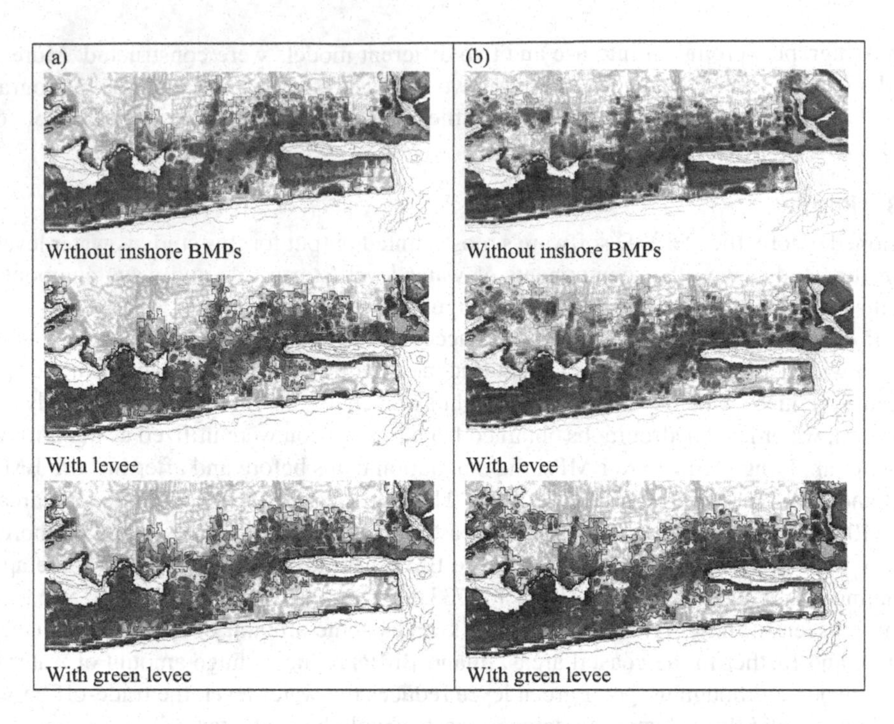

FIGURE 12.16 Water inundation maps. (a) Superstorm Sandy with apron. (b) Superstorm Sandy without apron.

and Delft3D-FLOW, to reproduce the wave height and the storm tide for the Battery Park station and ungagged areas in Brooklyn.

In this study, the impact of bathymetric shallowing on water level decrease during storms is examined as an antecedent for nearshore BMP plan. For achieving this goal, different scenarios were defined which included the different characteristics of the shallowing, such as thickness, distance to the shoreline, and width. Then, the best scenario, which reduced the water level more than the other scenarios, was chosen as the selected water level hydrograph for inshore modeling.

Moreover, for determining the impacts of inshore BMPs and also a combination of inshore and offshore strategies, a levee and a green levee were utilized as inshore BMPs. The results showed a remarkable reduction in water inundation map when using inshore and offshore BMPs in combination. Moreover, comparison between the levee and green levee shows that green levee is a more reliable infrastructure as it reduces more water inundation. The reason is due to the fact that water as it passes through the vegetation cover has to overcome the drag force, so, its energy dissipates.

12.7.3 Case Study 3: Infrastructure Flood Risk Management—MCDM-Based Selection of BMPs and Flood Damage Assessment

The change of a watershed's hydrologic response because of metropolitan changing conditions, population growth, climate change, and sea level rise has expanded the recurrence and force of floods. To adapt to the new difficulties in coastal flood management, numerous endeavors were made after Superstorm Sandy. These endeavors call for better comprehension of floods, approaches to moderate catastrophic effects of hazard, reconstructing endeavors by adaptable plan, and introducing a resilience measure to assess the performance of systems. Karamouz et al. (2019) tried to address and deliver these measures. The principal motivation behind this examination is to improve resilience in systems, especially wastewater treatment plants (WWTPs), which are considered as a vital service to the metropolitans' life lines.

Accordingly, the first step is flood hazard identification by application of flood inundation maps. Then, effective factors for resilience are recognized and partitioned into four main categories of the so-called 4R's. The principal objective of this paper is the development of a framework of combined structural and nonstructural/operational approaches to enhance WWTPs' performance. Figure 12.17 shows a flowchart for the taken steps to reach the overall objective. The main stages of the methodology are described in Steps 1–4 of the flowchart.

In step 1, the 100-year water level has been defined as the study's flood scenario by utilizing observed data from the water level station and frequency analysis. To show how this part of the methodology could be applied, only the flood inundation depth for one WWTP was obtained and tested. For other 13 WWTPs, the depth of the 100-year flood at those facilities was obtained from NYCDEP (2013). Next, frequency analysis is done in order to find flood inundation depth at one of the 14 WWTPs in the study area, the water level time series of coastal flooding is needed. Water levels during a coastal flood consist of astronomical tide and any abnormal rise of water over the predicted astronomical tide, which is called a storm surge. In this study, flood frequency analysis is performed on storm surge data to find the intensity and recurrence of extreme floods and maximum water level of a 100-year coastal flood. Distributed modeling of flood events and their routing over time is an important issue in floodplain delineation. GSSHA is an integrated GIS-based rainfall–runoff/flood routing model that has been utilized in this paper. It prepares spatial and mapping analysis with raster formats that represent data in different grids of DEM with different land use and soil types. To evaluate flood inundation depth obtained from the distributed hydrologic model (GSSHA), HAZUS-MH version 3.1 software was used. This software has an efficient two-dimensional (2D) flood modeling tool. It also consists of damage curves for kinds of buildings and an algorithm for anticipating the direct and indirect flood damages (FEMA, 2012).

Step 2 of the flowchart is described here. To estimate the resilience index, a wide range of subcriteria including hydrological, social, economic, and technical concerns have been identified.

Having a more resilient WWTP at the time that a flood occurs is the prime challenge in this study. Resources should be allocated to improve the performance of its different components in order to reduce a flood's devastating damages. In this manner, the main issues are the amount and priority of financial allocation, investment effectiveness, and the amount of resilience improvement resulted from every dollar spent on different components. In the third step of the flow chart, cooperative games in the context of game theory are utilized to demonstrate joint operations among WWTPs and their interaction process. In step 4 of the flowchart, due to infeasible joint operation in some WWTPs, in regard to the WWTPs' location and whether they are separated by water bodies, adaptive flood hazard mitigation strategies such as BMPs have been used. Proposing the most qualified BMP for the study area is done based on the idea of the competitors of Rebuild by Design competition, mostly the Hunts Point Lifeline's proposal and the Hudson River Project's proposal.

12.7.3.1 Methodology

To improve WWTP system execution and measure the readiness condition of these facilities, the resilience index was used. In like manner, the initial step is flood hazard identification using flood inundation maps. The 100-year water level was characterized as the study's flood scenario by using observations of the water level station and frequency analysis. In this examination, flood frequency analysis is implemented on storm surge information to discover the intensity and frequency of extreme floods and the most extreme water level of a 100-year coastal flood. To complete frequency analysis, Log-Pearson Type III distribution was fitted on storm surge information. Then, the inundation maps were obtained from the gridded surface subsurface hydrologic analysis (GSSHA—https://www.aquaveo.com/software/wms-gssha).

To assess flood inundation depth from the distributed hydrologic model (GSSHA), HAZUS-MH was utilized. In HAZUS, the flood inundation zone and local's DEM are used. Hence, the 100-year still water depth in each point of the inundated zone is determined by reducing the DEM from the 100-year still water.

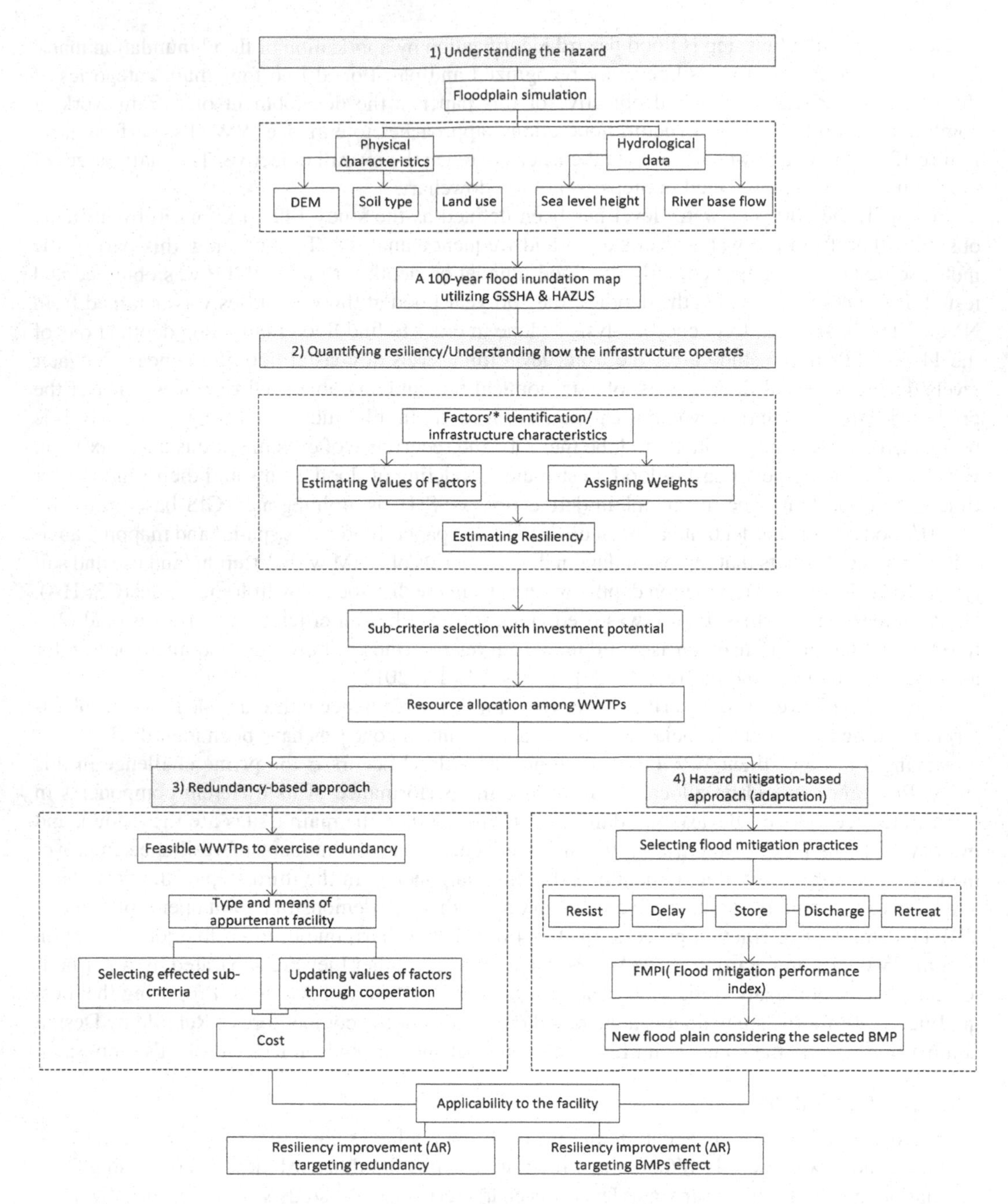

FIGURE 12.17 The study's flow chart. (Adapted from Karamouz, M., Taheri, M., Khalili, P., and Chen, X., *J. Water Resour. Plan. Manage.*, 145(4), 04019004, 2019.)

To appraise the resilience index, a wide scope of subcriteria including hydrological, social, economic, and specialized are considered utilizing analytical hierarchy process (AHP), a technique in multicriteria decision-making (MCDM).

A more resilient WWTP during a flood is the ultimate goal in this approach. Assets ought to be made to improve the presentation of its various parts to alleviate a flood's overwhelming harms. Thus, the principal issues are the sum and need of monetary assignment, venture adequacy, and the measure of flexibility improvement came about because of each dollar spent on various segments.

For this reason, resources are allocated to the financial aspects based subcriteria (including plant capacity, percent of not at-risk equipment, number of plant technical staff, availability of dewatering facilities, and on-site storage).

In conditions where joint activity in some WWTPs is not possible due to the WWTPs' location and isolation by water bodies, application of flood mitigation and adaptation strategies, for example, implementation of BMPs, were suggested. A flood mitigation performance index (FMPI) is suggested and quantified here based on the MCDM approach to evaluate the performance of different categories of BMPs in confronting floods. Proposing the top BMP for the examination zone was done dependent on the possibility of the contenders of Rebuild by Design competition.

12.7.3.2 Results

The frequency analysis results based on the Log-Pearson Type III distribution are presented in Table 12.3.

Figure 12.18 shows the 100-year flood inundation maps utilizing the GSSHA hydrologic model (part a) and HAZUS model (part b)]. The attention is on the lower Bronx territory because of broad industrial exercises and the presence of a Hunts Point WWTP, as appeared in (parts c and d). In utilizing GSSHA for floodplain depiction, notwithstanding the waterfront flooding, riverine examination is incorporated, which is not considered in HAZUS. In many parts of the investigation zone, flood inundation by HAZUS model is overestimated comparing with the in-house model; however, the thing that matters is not excessively critical, except for Bronx Whitestone Bridge. In general, the outcomes show great agreement between the two models. Moreover, some strategic points are selected, as shown on the figures, to obtain water level information during simulations.

To calculate the flood mitigation performance index (FMPI) for each training of retreat, discharge, store, delay, and resist, certain factors have been distinguished, including hazard mitigation degree, capital cost, operation and maintenance costs, esthetics, and level of adaptability. The weight of each factor has been determined by using a pairwise comparison using AHP, as shown in Figure 12.19. Each category is compared based on these five factors and a final FMPI is estimated for them. Based on the results of this study, resist category of BMPs had the highest FMPI value and is considered as the best strategy.

To make this strategy more adaptable, the application of a multipurpose levee (MPL) is proposed. The MPL has an opposition effect as a sea wall and can give proper setting to public exercises including walkways. Besides, application of vegetation improves the common similarity (adaptability) and style of this structure.

Effectiveness of MPL in decreasing the inundation profundity at Hunts Point WWTP has been investigated by demonstrating in GSSHA. This model can simulate different flood mitigation structures, for example, levees (using the wide peaked weir condition). Prior to MPL application, the flood inundation depth at Hunts Point was 2.65 m. This value has been diminished to 0.7 m. Figure 12.20 depicts the flood zones before (Figure 12.20a) and after (Figure 12.20b) development of the MPL.

TABLE 12.3
Results of Design Values for Different Return Periods

Return Period (Years)	Surge (m)
2	1.22
5	1.58
10	1.91
25	2.42
50	2.89
100	3.45
200	4.1

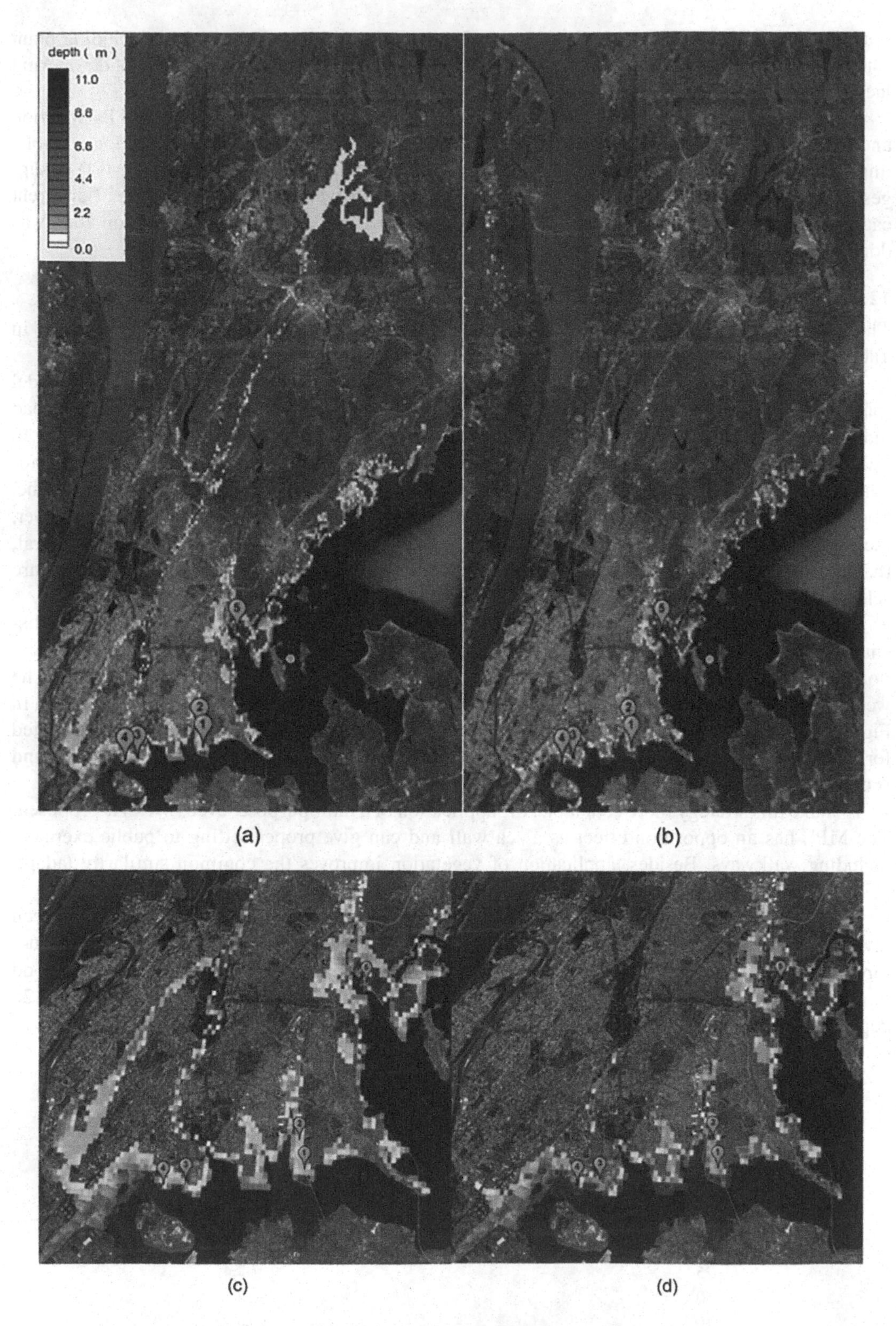

FIGURE 12.18 Different flood 100-year inundation maps with some strategic points for Bronx. (a) GSSHA simulation model. (b) HAZUS, FIS (2013). (c) Close-up of plot (a). (d) Close-up of plot (b).

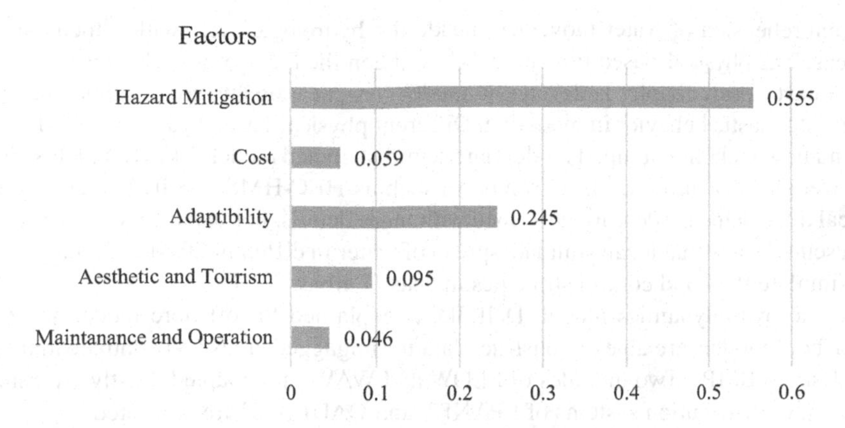

FIGURE 12.19 Weight of factors in qualifying the value of flood mitigation performance index (FMPI).

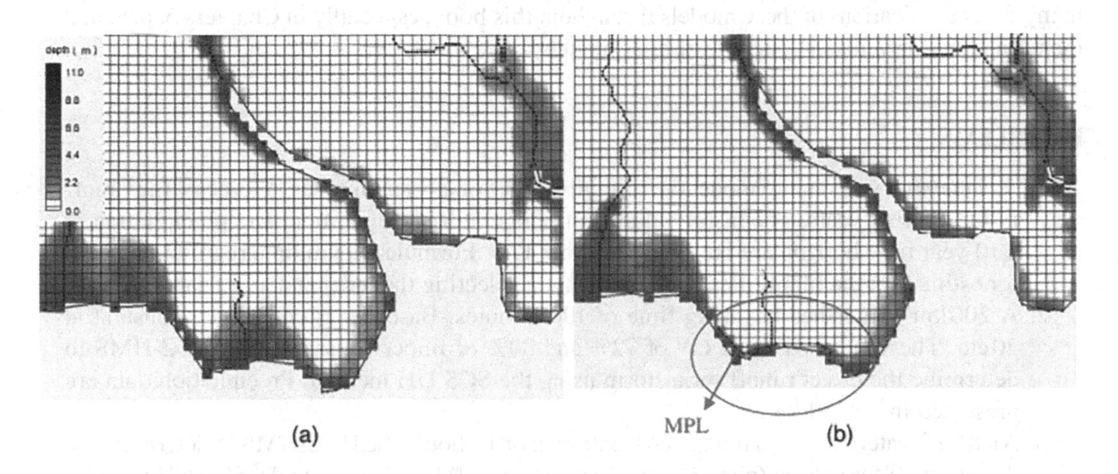

FIGURE 12.20 Flood inundation. (a) Before and (b) after levee construction in Hunts Point wastewater treatment plant (WWTP), south of Bronx.

12.7.3.3 Case Study Concluding Remarks

Developing urbanization with complex interconnected infrastructures in coastal area have raised concerns of flood hazards such as vulnerability of natural and human made systems. Additionally, climate change has increased the intensity and recurrence of extreme events. Study of historical floods on the East Coast of the United States reveals that New York metropolitan area is a flood-vulnerable area. The main goal of this paper was to provide a framework to define resilience index of infrastructures and recognize approaches to improve resiliency of WWTPs specifically during flooding by using soft and hard risk management strategies.

To point this, several steps had been taken, including (a) better comprehension of flood impacts, with two GIS-based models of GSSHA and HAZUS, and systems functioning; (b) developing a resilience index for estimating functionality of the system using an MCDM approach; and (c) improvement of this index by adopting the prosed adaptive strategies.

12.8 SUMMARY AND CONCLUSION

In this chapter, many attributes of hydrological and hydraulic behaviors of water cycles are discussed. For studying and synthesizing watershed's occurrences, simulation of hydrologic events is required. Hydrologic simulation models are used to produce a hydrologic prediction/projection as a better comprehension of water movement inside the hydrologic cycle with a focus in urban and coastal areas. The physical-based type models have been the focus of this chapter.

The first part of this chapter is devoted to model classification and some simple methods incorporating the stochastic behavior in modeling. Different physical-based hydrologic models of coastal and overland flow including lumped models and semidistributed models like IHACRES, StormNET, and HBV were briefly introduced. Also models such as HEC-HMS, GSSHA, and LISFLOOD as hydrological distributed models are described with more details. For instance, GSSHA model applications presented show that it can simulate spread of water in different DEM cells and also it has the ability to simulate the flood control strategies in coastal areas.

Then a true hydrodynamics model, Delft3D, is explained for offshore modeling. It is shown that the offshore models are able to construct data for ungagged areas and simulate and investigate different offshore BMPs. Two modules of FLOW and WAVE are coupled. Lastly, hydraulic-driven models for water distribution systems of EPANET and QAULNET are presented.

The three case studies presented show how these models can help the engineers and the decision makers to better prepare for water-related incidents such as inland and coastal floods. There are many other applications of these models throughout this book, especially in Chapters 5, 6, 9, and 13 including data-driven and system dynamics models.

PROBLEMS

1. Assume that you want to model the rainfall–runoff of a basin in order to forecast the runoff in the coming years. The historical data of temperature, evaporation, and precipitation in a 10-year time horizon are presented in Table 12.4. Formulate a multiple regression model for estimating the runoff data and comment on selecting the independent variables.
2. A 200 km^2 watershed has a lag time of 100 minutes. Baseflow is considered constant at 20 cm. The watershed has a *CN* of 72% and 30% of imperviousness. Use HEC-HMS to determine the direct runoff for a storm using the SCS UH method. Precipitation data are presented in Table 12.5.
3. An 8 km^2 watershed has a time of concentration of 1.0 hour. Use HEC-HMS to determine the direct runoff for a storm (rainfall hyetograph given in Table 12.6) using the SCS UH method.
   Hint:
   a. You must create new basin, Meteorologic, and Control Specifications files either in the same project or in a new project.

TABLE 12.4
Historical Data of Temperature, Evaporation, and Precipitation in a 10-Year Time Horizon in Problem 1

Year	Temperature (°C)	Evaporation (mm)	Annual Precipitation (mm)
1991	6.56	745	812
1992	8.2	967	648
1993	7.42	833	670.4
1994	6.68	712	772
1995	7.3	850	670.4
1996	7.45	897	629.6
1997	7.9	915	576
1998	8.8	942	556
1999	8.64	888	656
2000	8.02	832	640

TABLE 12.5
Precipitation data for Problem 2

Date	2000-01-01, 00:00	2000-01-01, 03:00	2000-01-01, 06:00	2000-01-01, 09:00	2000-01-01, 12:00
Incremental rainfall (cm)	0	0.5	1.25	1.25	0.5

TABLE 12.6
Precipitation Data for Problem 3

Time (min)	Excess Rainfall (cm)
0	0
10	0.6
20	1.5
30	1.2
40	0.1
50	0
60	0

b. You will need only one subbasin element in the basin file with area and SCS hydrograph details. You will only need the transform method. Specify None for loss method (input provided is excess rainfall) and baseflow.
c. The input to the Meteorologic file involves using the Time Series Data Manager from the Components menu. This will add a time series data folder in the watershed explorer and you will expand this folder to specify the rainfall hyetograph. When you edit the Meteorologic model in the component editor, you will use the Specified hyetograph (default) method. You will then click on the specified hyetograph option in the watershed explorer and link the time series data to the Meteorologic model.
d. In the Control Specifications model, use 10-minute time interval and 12-hour simulation time (end time = 12 hours past start time).

4. The parameters of a small undeveloped watershed are listed in Tables 12.7 and 12.8. A UH, and Muskingum routing coefficients are known for subbasin 3, as shown in Figure 12.21. *TC* and *R* values for subbasins 1 and 2 and associated SCS CNs are provided as shown.

TABLE 12.7
Properties of Subbasins in Problem 4

Subbasin Number	*TC* (hours)	*R* (hours)	SCS CN	% Impervious	Area (km^2)
1	2.5	5.5	66	0	2.5
2	2.8	7.5	58	0	2.7
3	_	_	58	00	3.3

TABLE 12.8
UH for Subbasin 3 in Problem 4

Time (hours)	0	1	2	3	4	5	6	7
U (cm)	0	200	400	600	450	300	150	0

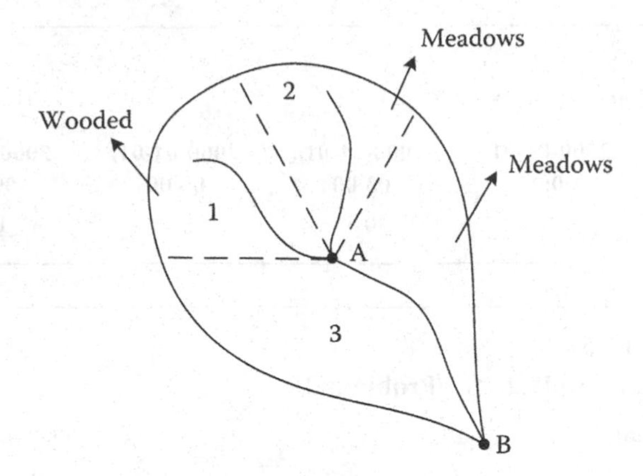

FIGURE 12.21 Watershed in Problem 4.

A 5-hour rainfall hyetograph (in cm/hour) is shown in Figure 12.22 for a storm event that occurred on June 19, 2010. Assume that the rain fell uniformly over the watershed. Use the information given to develop an HEC-HMS input dataset to model this storm. Run the model to determine the predicted outflow at point B (Muskingum coefficient: $X = 0.15$, $K = 3$ hours, area = 3.3 km^2).

5. Develop an MLP model with two hidden layers and three perceptrons with logsig and linear function in each layer, respectively. The input of the MLP model is rainfall, evaporation, and temperature, and the inflow must be simulated. Use the given data in Table 12.9 for model development.
6. Develop a TDL, GRNN, and RNN model for the data of Problem 5 and then compare and discuss the results of different simulations with ANN's models.
7. Calibrate the HBV model for the catchment described in Tables 12.10 and 12.11 for the period of March 1, 2002, to May 1, 2002. Discuss, before running the model, what effects you expect and then make a note of each change in parameter value and its effect on the simulation.
8. Simulate the rainfall–runoff model in Problem 5 with IHACRES and then compare the difference between the next 2 months' predictions with these two models.

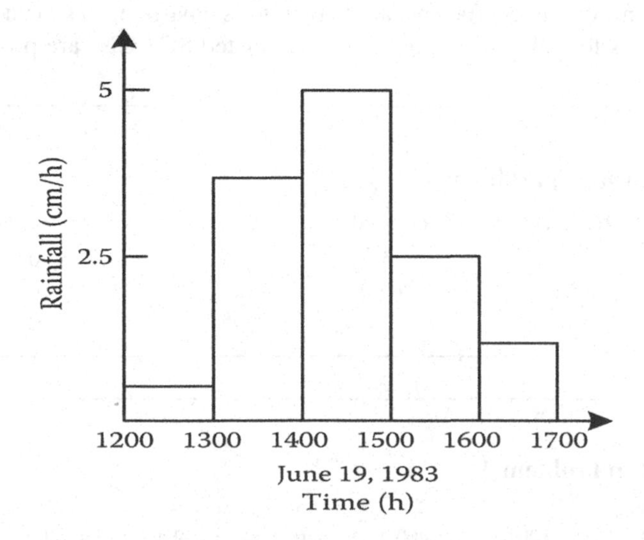

FIGURE 12.22 Rainfall hyetograph in Problem 4.

TABLE 12.9
Observed Rainfall, Evaporation, and Temperature, and Inflow Data for Problem 5

Inflow (m^3/s)	Temperature (°C)	Rainfall (mm)	Evaporation (mm)	Day	Inflow (m^3/s)	Temperature (°C)	Rainfall (mm)	Evaporation (mm)	Day
3.32	14.7	4.5	2.4	1	5.93	6.9	0	0	31
4.31	15.1	0	3.5	2	6.57	4.7	6	0	32
4.33	14.2	0	3.5	3	6.02	5	0	0	33
4.29	14.9	10	2.3	4	6.68	5.6	0	0	34
4.51	8.7	2.5	1.6	5	6.81	8.6	0	0	35
4.53	12.1	3.5	1.7	6	6.55	6.3	0	0	36
5.54	13.3	2.5	2.5	7	6.67	9.4	0	0	37
5.41	13.5	0	3.9	8	7.31	9.3	1.5	0	38
3.23	10.3	0	2.2	9	7.3	6.2	0	0	39
3.21	9.1	0	2.2	10	8.07	3.9	9.5	0	40
3.29	11.2	0	1.7	11	7.49	5	4.3	0	41
3.28	11.6	0	3.1	12	6.83	6.1	0	0	42
3.34	13.7	0	1.3	13	7.5	3.6	5	0	43
3.03	15.7	0	3.1	14	6.82	4.6	11	0	44
3.01	15.7	0	3.5	15	6.45	6.5	0	0	45
3.06	13.2	1.7	2.2	16	6.68	6.6	0	0	46
3.23	11	2.2	2.2	17	6.66	5.8	0.5	0	47
3.25	13.2	0	2.6	18	6.04	7.7	0	0	48
3.19	13.8	0	3.1	19	6.13	8.6	0	0	49
3.45	11.8	0	2.6	20	7	5.3	1	0	50
4.34	9.3	8	2.3	21	6.99	6	2.5	0	51
4.36	8.6	4.3	1.9	22	8.86	5.6	0	0	52
4.51	10.2	0	2.2	23	8.86	6.7	0	0	53
4.59	12.2	0	2.6	24	7.06	7.1	6.5	0	54
4.4	12.4	0	2.2	25	7.07	7.4	1	0	55
4.01	8.6	0	1.3	26	6.02	4.6	0	0	56
11.31	6.8	26	0.2	27	4.86	5.7	0	0	57
9.38	1	23	0	28	8.08	5.8	3	0	58
8.33	0.3	0	0	29	8.12	7.3	0.8	0	59
6.5	4	0	0	30	8.06	5.9	0	0	60

TABLE 12.10
Monthly average of evaporation in Problem 7

Month	1	2	3	4	5	6	7	8	9	10	11	12
Evaporation (mm)	1.44	0	3.33	93.25	195.33	272.92	288.74	280.9	212.02	137.88	55.96	16.22

TABLE 12.11
Observed Rainfall, Temperature, and Inflow in Problem 7

Date	Precipitation (mm)	Temperature (°C)	Q (m³/s)	Date	Precipitation (mm)	Temperature (°C)	Q (m³/s)
2002-03-01	0	6.6	0.68	2002-04-01	0	8.6	1.973
2002-03-02	0	8.3	0.778	2002-04-02	15.5	6.2	2.808
2002-03-03	0	9.3	0.865	2002-04-03	16.2	5.8	3.284
2002-03-04	0	12.1	0.823	2002-04-04	2.5	5.6	2.523
2002-03-05	0	13.1	0.943	2002-04-05	0.5	8.3	2.283
2002-03-06	0	8.8	1.03	2002-04-06	0	10.1	2.369
2002-03-07	0	6.1	1.03	2002-04-07	2	12.3	2.494
2002-03-08	0	7.2	1.038	2002-04-08	0	13.1	2.686
2002-03-09	0	9.6	0.823	2002-04-09	0	13.8	2.798
2002-03-10	0	7.2	0.804	2002-04-10	0	13.2	2.914
2002-03-11	0	6.7	0.827	2002-04-11	5.7	9.8	2.917
2002-03-12	0	8.7	0.815	2002-04-12	13.1	6.2	3.061
2002-03-13	0	11.2	0.889	2002-04-13	18.8	5	3.484
2002-03-14	0	13.7	0.898	2002-04-14	0	7.7	3.183
2002-03-15	0	14	0.88	2002-04-15	0	11.2	3.038
2002-03-16	1.7	11.5	0.91	2002-04-16	2.5	12.8	2.543
2002-03-17	1.5	10.1	1.059	2002-04-17	6	15.5	3.432
2002-03-18	1	8.3	1.63	2002-04-18	5	13.3	4.867
2002-03-19	13.5	9.5	2.816	2002-04-19	35.6	11.5	6.546
2002-03-20	3.5	13.3	1.832	2002-04-20	4	10.5	6.911
2002-03-21	7.5	8.6	1.901	2002-04-21	1	11	5.117
2002-03-22	1.7	6.2	1.985	2002-04-22	2	9.1	3.867
2002-03-23	6.7	6.2	2.283	2002-04-23	0	9.1	3.87
2002-03-24	0	9.7	1.834	2002-04-24	0	10.4	3.753
2002-03-25	0	12.3	1.849	2002-04-25	6	12.6	3.725
2002-03-26	0	15	1.729	2002-04-26	0	11.8	3.789
2002-03-27	0	14.8	2.065	2002-04-27	1.5	11.1	4.052
2002-03-28	5.7	12.7	2.184	2002-04-28	0	16.2	3.826
2002-03-29	7	6.5	2.237	2002-04-29	0.7	15.8	3.945
2002-03-30	0	6.4	1.237	2002-04-30	0	14.2	4.092
2002-03-31	3.5	6.2	1.973	–	–	–	–

9. The transition probabilities of a Markov chain model for streamflow in two different seasons are given in Tables 12.12 and 12.13.
   Calculate the steady-state probabilities of flows in each interval in each season.
10. The data presented in Table 12.14 is the annual maximum series (Q_p) and the percentage of impervious area (I) for an urbanized watershed for the period from 1930 to 1977. Adjust the flood series to an eventual development of 50%. Estimate the effect on the estimated 2-, 10-, 25-, and 100-year floods.

TABLE 12.12
Streamflow of Season 1 in Problem 9

	Streamflow Following Season 2		
Streamflow in Season 1	**0–3 m³/s**	**3–6 m³/s**	**>6 m³/s**
0–10 m³/s	0.25	0.50	0.25
>10 m³/s	0.05	0.55	0.40

TABLE 12.13
Streamflow of Season 2 in Problem 9

	Streamflow Following Season 2	
Streamflow in Season 2	**0–10 m³/s**	**>10 m³/s**
0–3 m³/s	0.70	0.30
3–6 m³/s	0.50	0.50
>6 m³/s	0.40	0.60

TABLE 12.14
Data of Problem 10

Year	I (%)	Q_p (cm)	Year	I (%)	Q_p (cm)	Year	I (%)	Q_p (cm)
1930	21	1,870	1946	35	1,600	1962	45	2,560
1931	21	1,530	1947	37	3,810	1963	45	2,215
1932	22	1,120	1948	39	2,670	1964	45	2,210
1933	22	1,850	1949	41	758	1965	45	3,730
1934	23	4,890	1950	43	1,630	1966	45	3,520
1935	23	2,280	1951	45	1,620	1967	45	3,550
1936	24	1,700	1952	45	3,811	1968	45	3,480
1937	24	2,470	1953	45	3,140	1969	45	3,980
1938	25	5,010	1954	45	2,410	1970	45	3,430
1939	25	2,480	1955	45	1,890	1971	45	4,040
1940	26	1,280	1956	45	4,550	1972	46	2,000
1941	27	2,080	1957	45	3,090	1973	46	4,450
1942	29	2,320	1958	45	4,830	1974	46	4,330
1943	30	4,480	1959	45	3,170	1975	46	6,000
1944	32	1,860	1960	45	1,710	1976	46	1,820
1945	33	2,220	1961	45	1,480	1977	46	1,770

11. Residents in a community at the discharge point of a 614-km² watershed believe that the recent increase in peak discharge rates is due to the deforestation by a logging company that has been occurring in recent years. Analyze the annual maximum discharges (Q_p) and an average forest coverage (FC) for the watershed data given in Table 12.15.
12. Analyze the data of Problem 5 to evaluate whether or not the increase in urbanization has been accompanied by an increase in the annual maximum discharge. Apply the Spearman test with both a 1% and a 5% level of significance. Discuss the results.
13. Consider two watersheds with different capacities for storage, such as sandy soil and clay with potential maximum retention of 7 and 5 cm, respectively, and also initial abstraction

TABLE 12.15
Data of Problem 11

Year	Q_p (cm)	FC (%)	Year	Q_p (cm)	FC (%)	Year	Q_p (cm)	FC (%)
1982	8,000	53	1987	12,200	54	1992	5,800	46
1983	8,800	56	1988	5,700	51	1993	14,300	44
1984	7,400	57	1989	9,400	50	1994	11,600	43
1985	6,700	58	1990	14,200	49	1995	10,400	42
1986	11,100	55	1991	7,600	47	–	–	–

before ponding is 1.5 cm. If a storm of 10 cm during 10 minutes occurs in both watersheds, determine the percentage of effective rainfall in each watershed.

14. A development project on a small upland watershed is shown in Figure 12.23. The developed portion of the area is 0.7 km² in which 21% is impervious area. The developed area is graded so that runoff is collected in grass-lined swales at the front of the lot and drained to a paved swale that flows along the side of the main road. Flow from the paved swales passes through a pipe culvert to the upper end of a stream channel. The upper portion of

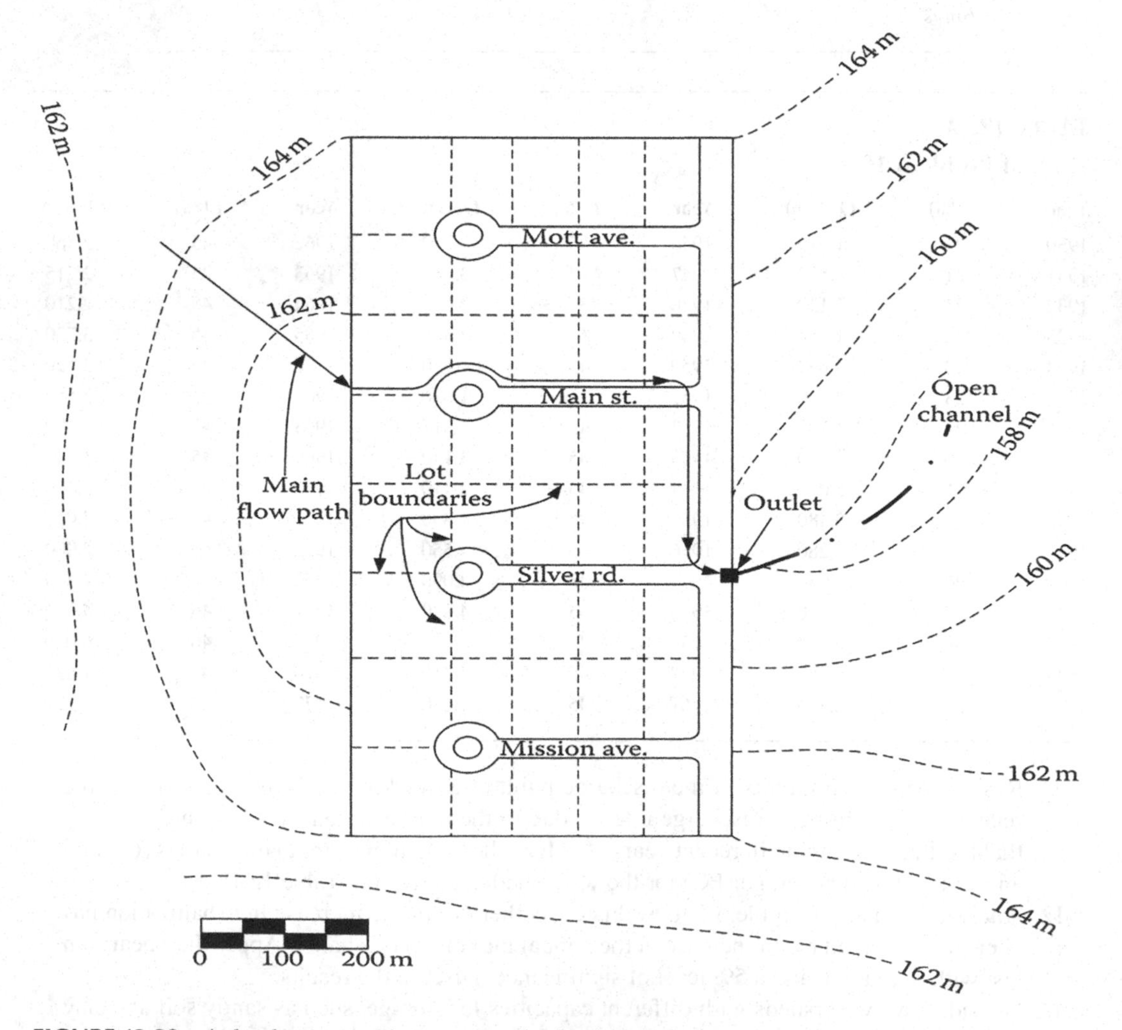

FIGURE 12.23 A development project on a small upland watershed.

the watershed with a maximum height of 164 m is a forest with B-type soil ($CN = 60$) and has an area of 0.3 km^2. Estimate the peak runoff of this area if the design return period of the drainage system is 25 years. Use the IDF curve of Problem 3.

15. Determine the infiltration losses and excess rainfall from the rainfall data given in Table 12.16 on a watershed using the modified Horton equation. The Horton infiltration capacity in the watershed is as follows: $f = 0.2 + 0.4e^{-0.6t}$.
16. A watershed with an area of 230 km^2 is under further development during a 10-year horizon. After finishing the development project, the percentage of imperviousness, watershed storage, and basin development factor will change from 25, 4, and 3 to 45, 6, and 9, respectively. Evaluate the effect of the urbanization on the peak discharge of this watershed. Assume that the average slope of watershed is 0.02 m/m. The IDF curve of Problem 3 could be applied in this watershed. The peak discharges from different return periods before urbanization are given in Table 12.17.
17. The ordinates of a 4-hour UH of a catchment are given in Table 12.18. Derive the flood hydrograph in the catchment for the storm figures presented in Table 12.19.

    The storm loss rate (Φ index) for the catchment is estimated as 0.43 cm/hour. The base flow at the beginning is 10 m^3/s and will increase by 1.5 m^3/s every 8 hours until the end of the direct runoff hydrograph.
18. In case study 1, answer the following questions:
    a. What types of boundary conditions can be applied on the edge of the simulation domain for coastal flood modeling?
    b. How are the probability maps obtained? How are their numbers generated?

TABLE 12.16
Hyetograph of Rainfall of Problem 15

Time (hours)	*i* (cm/hour)	Time (hours)	*i* (cm/hour)
0	1.2	2.00	2.0
0.25	1.2	2.25	1.7
0.50	1.2	2.50	1.7
0.75	1.8	2.75	1.7
1.00	1.8	3.00	0.9
1.25	1.8	3.25	0.9
1.50	2.0	3.50	0.4
1.75	2.0	3.75	0.4
–	–	4.00	0

TABLE 12.17
Estimated Peak Discharges for Watershed of Problem 16 in Different Urbanization Conditions

Return Period (year)	Rural Peaks (m^3/s)
2	32.5
5	49.8
10	67.3
25	86.1
50	104.8
100	150.3
500	210.2

TABLE 12.18
Ordinates of 4-Hour UH of Problem 17

Time (hours)	0	2	4	6	8	10	12	14	16	18	20	22	24	26	28
Q (m³/s)	0	25	50	85	125	160	185	160	110	60	36	25	16	8	0

TABLE 12.19
Hyetograph of Storm of Problem 17

Time from start of storm (h)	0	4	8	12
Accumulated rainfall (cm)	0	3.5	11.0	16.5

c. What is the process of specifying uncertainty in DEM? What is a DEM error?
d. What is the differernce between deterministic and probabilistic maps?
e. State the inputs required for LISflood model.
f. Express the continuity equation in the LISflood model. How is the flow between adjacent cells obtained?

19. In case study 2, answer the following questions:
    a. What are the initial inputs in a hydrodynamic model like Delft3D? How is this information obtained? Data buoys are of those tools utilized to obtained wave characteristics, explain the data buoys' characteristics, what kind of information they provide, and how the information is used in hydrodynamic modeling?
    b. Do research and provide more information about Delft3D environment. The explanation should include information about different modules of this software, how can they be coupled, what outputs does each of them provide, what is their usage in offshore modeling, and what are the offshore BMPs they can model? The answer should especially involve explanation about FLOW and WAVE modules.
    c. How can the Delft3D software be installed? What is the prerequirement software for this package? The information is available at Delft3D website.
    d. While modeling, users come across different terms. Provide information about each of them:
        - Cell resolution (also, explain what is the appropriate cell resolution for offshore modeling. Hint: different resolutions are used for different purposes).
        - Domain
        - Open boundaries
        - Nesting
20. Consider a slope of 0.005 and a significant wave height of 3.4 m. Calculate the depth at which the wave breaks and also, if an apron with 1 m thickness is placed at this point, calculate the new wave breaking point. What should be the suggested minimum width (along the wave movement) of the apron?
21. In case study 3, answer the following questions:
    a. What are the initial inputs for a distributed hydrological model like GSSHA? There are two ways to obtain this information, via the model itself and via the web search. Search and try to obtain that information. Provide information about the data those sites offer.
    b. Provide brief explanations about GSSHA capabilities and different environment in which it operates. Also, bring about a more detailed explanation about its outputs, their usage, and how can it be utilized in research.

c. While modeling, users face different challenges. Provide information about the following items:
   - Computed flow directions and accumulation
   - Defining smooth stream
   - Job control
   - Defining land use and soil data
   - Hydrologic computation

APPENDIX

HEC-HMS Tutorial

The intent of this example is to introduce the structure and some of the functions of the HEC-Hydrologic Modeling System (HEC-HMS), by simulating the runoff hydrographs resulting from a design storm in a watershed.

Consider the watershed illustrated in Figure 12.A1 which contains four major catchments. Precipitation data for a storm is available for three gages in the watershed. The goal is to estimate the effect of proposed future urbanization on the hydrologic response.

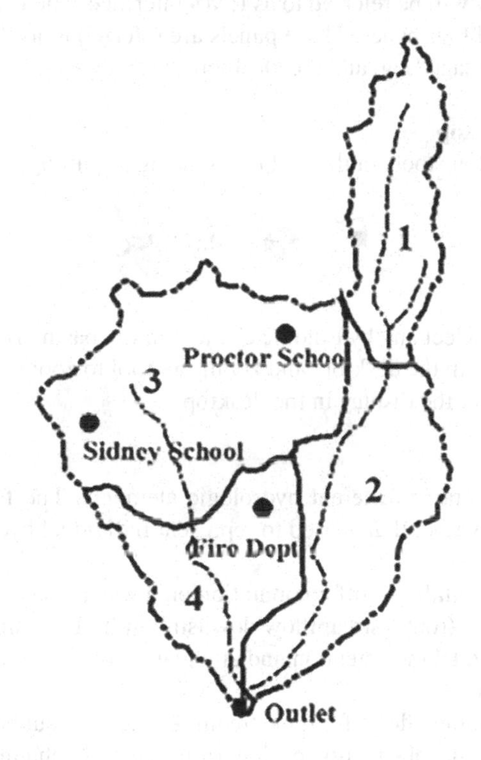

FIGURE 12.A1 Schematic view of watershed.

Getting Started

Start HEC-HMS by clicking on the HEC-HMS icon. After a few seconds, the following should appear:

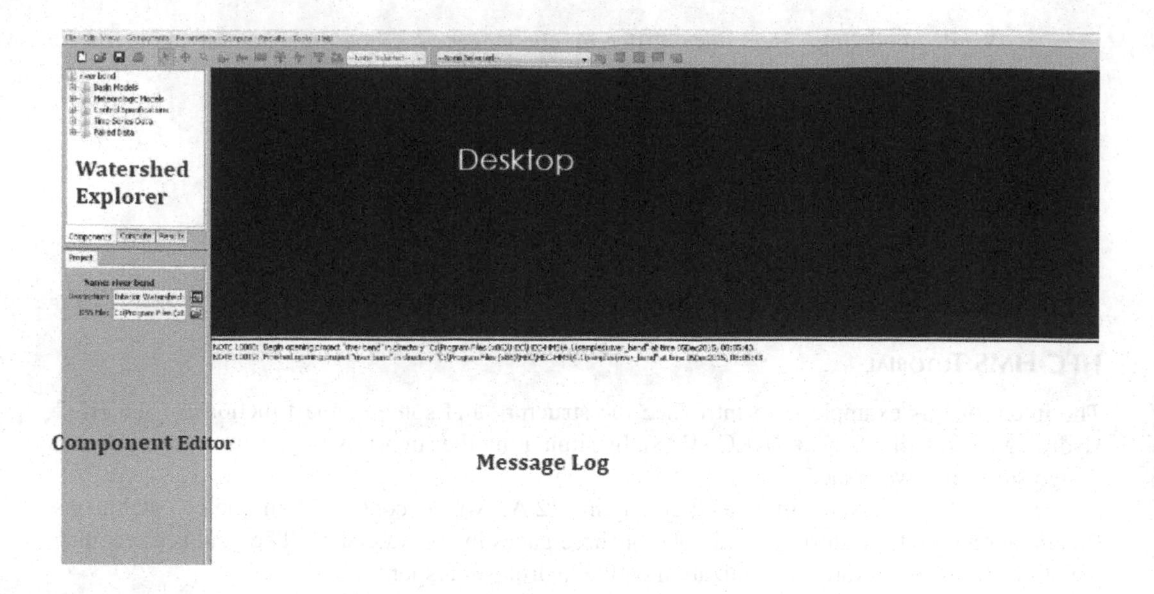

Henceforth, this window will be referred to as HMS interface. The HEC-HMS interface consists of a menu bar, tool bar, and four panes. These panels are referred to as the Watershed Explorer, the Component Editor, the Message Log, and the Desktop.

Navigating the HMS Desktop

You can use the following four tools in the tool bar to navigate through the HMS desktop:

The arrow tool lets you select any hydrologic element in the basin. You can use the zoom-in tool to zoom-in to a smaller area in the desktop, and zoom-out tool to zoom out to see a larger area. The pan tool can be used to move the display in the desktop.

Hydrologic Elements

Our example watershed contains different hydrologic elements. The following description gives brief information on each symbol that is used to represent individual hydrologic element.

- *Subbasin*: Used for rainfall–runoff computation on a watershed.
- *Reach*: Used to convey (route) streamflow downstream in the basin model.
- *Reservoir*: Used to model the detention and attenuation of a hydrograph caused by a reservoir or detention pond.
- *Junction*: Used to combine flows from upstream reaches and subbasins.
- *Diversion*: Used to model abstraction of flow from the main channel.
- *Source*: Used to introduce flow into the basin model (from a stream crossing the boundary of the modeled region). Source has no inflow.
- *Sink*: Used to represent the outlet of the physical watershed. Sink has no outflow.

According to the given information, the model can be developed:

1. Add four subbasin elements. Select the subbasin icon on the tool bar. Place the icons by clicking the left mouse button in the basin map.

2. Add two reach elements. Click first where you want the upstream end of the reach to be located. Click a second time where you want the downstream end of the reach.
3. Add three junction elements.
4. Connect Sub-basin-2 downstream to Junction-1. Place the mouse over the subbasin icon and click the right mouse button. Select the Connect Downstream menu item. A connection link shows the elements are connected.
5. Connect the other element icons using the same procedure used to connect Sub-basin-2 downstream to Junction-1. Move an element by placing the mouse over the icon in the basin model map. The upper and lower ends of a reach element icon can be moved independently.

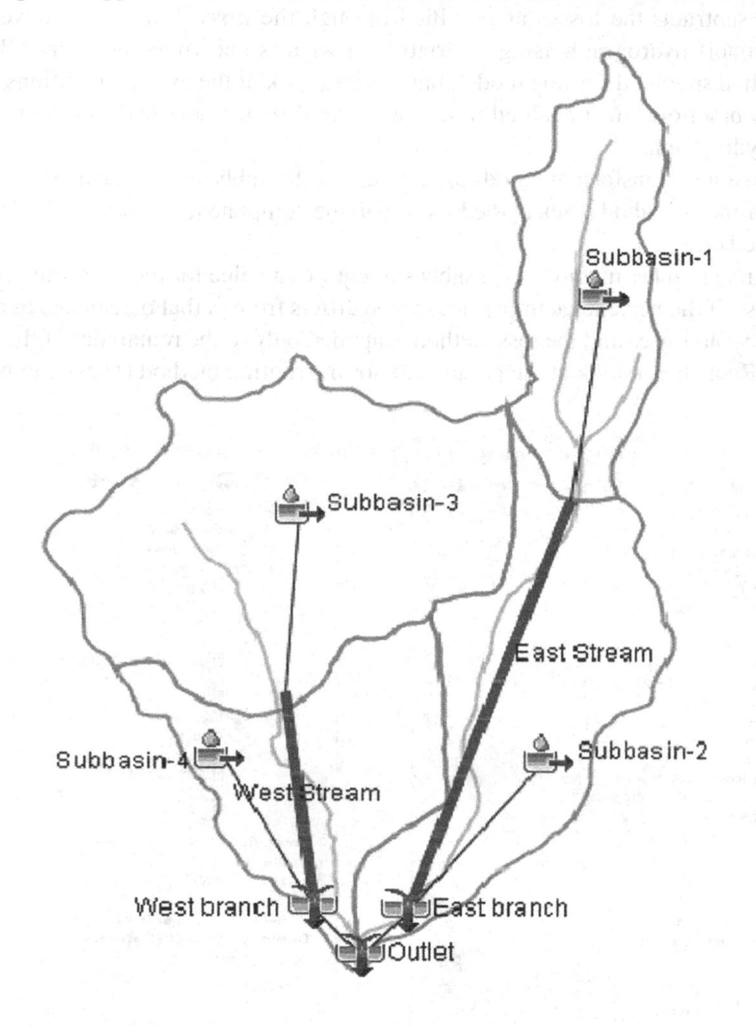

After constructing the model, information should be given to the program. A basin model will be created from the following parameter data using the pertaining assumptions:

- Initial constant loss
- Subbasins' areas
- Snyder UH transform
- Recession base flow methods.
- Theissen polygon weights
- Rainfall data from gages
- Strom rainfall distributed in time using the temporal pattern of incremental precipitation

Remember the subbasin element is used to convert rainfall to runoff. So the information on methods used to compute loss rates, hydrograph transformation and baseflow is required for each subbasin element. The loss method allows you to choose the process which calculates the rainfall losses absorbed by the ground. In this case, we are using the SCS curve number method to compute losses and get excess rainfall from the total rainfall. Some options are Initial and Constant, Soil Moisture Accounting, and Green and Ampt.

The transform method allows you to specify how to convert excess rainfall to direct runoff. Again, click on the drop-down menu to view your options. You may remember these options from class notes. This model employs the SCS technique. The modClark model takes gridded rainfall data, subtracts the losses as specified through the Loss Rates, and converts the excess rainfall to a runoff hydrograph using a variation of what is known as the Clark UH. There is no baseflow method specified for this model, but you can look at the available options. If we specify baseflow, this baseflow will be added to the resulting direct runoff hydrograph to produce total streamflow hydrograph.

Once the loss and transform methods are chosen for the subbasin, the next step is to specify the parameters for these methods. Select the Loss tab in the component editor to look at the parameters for the loss method.

For SCS curve number method, each subbasin requires a value for the curve number and percent imperviousness. If the percentage impervious value differs from 0, that percentage of the land area is assumed to have no losses and the loss method is applied only to the remainder of the drainage area.

Select the Route tab to look at the parameters for the routing method (Muskingum).

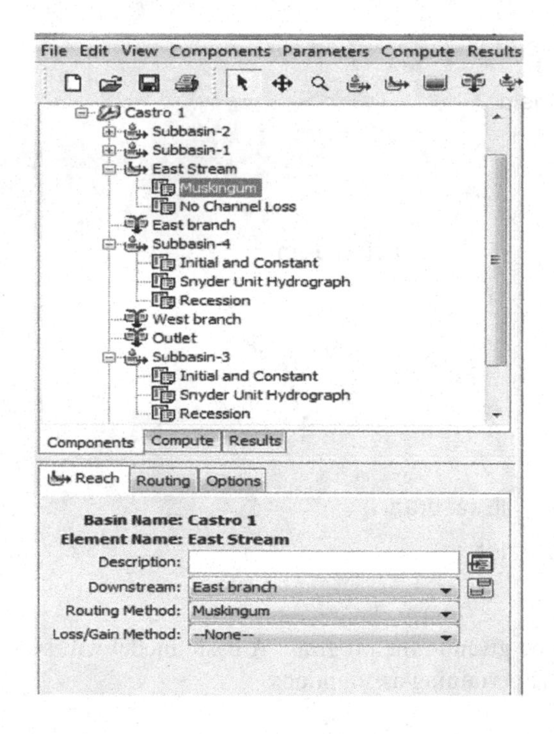

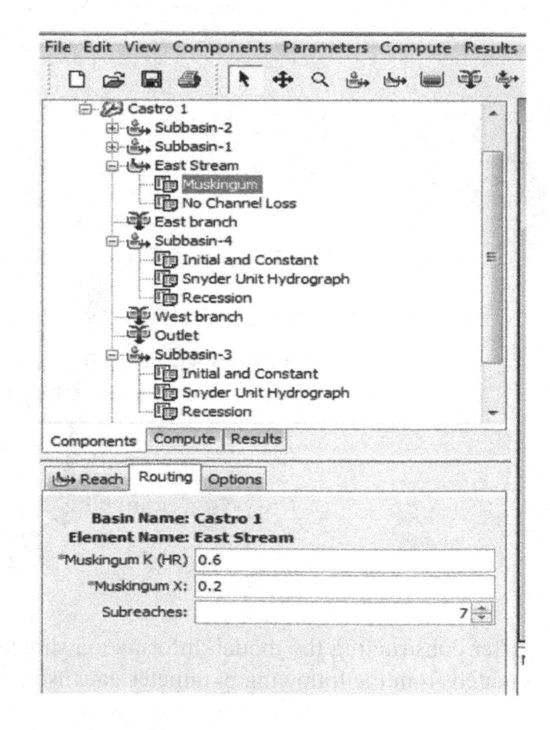

Creating a Meteorologic Model

This is done by specifying a time series of rainfall at a rainfall gauge and associating this gauge with each individual subbasin or all subbasins. If you click on the time series folder in the Watershed Explorer, you will see that there is a gage associated with each subbasin.

Next, in the Watershed Explorer, expand the Time Series folder to provide data for the rain gauge. We will use an observed event in the basin to populate the information for the rain gauge.

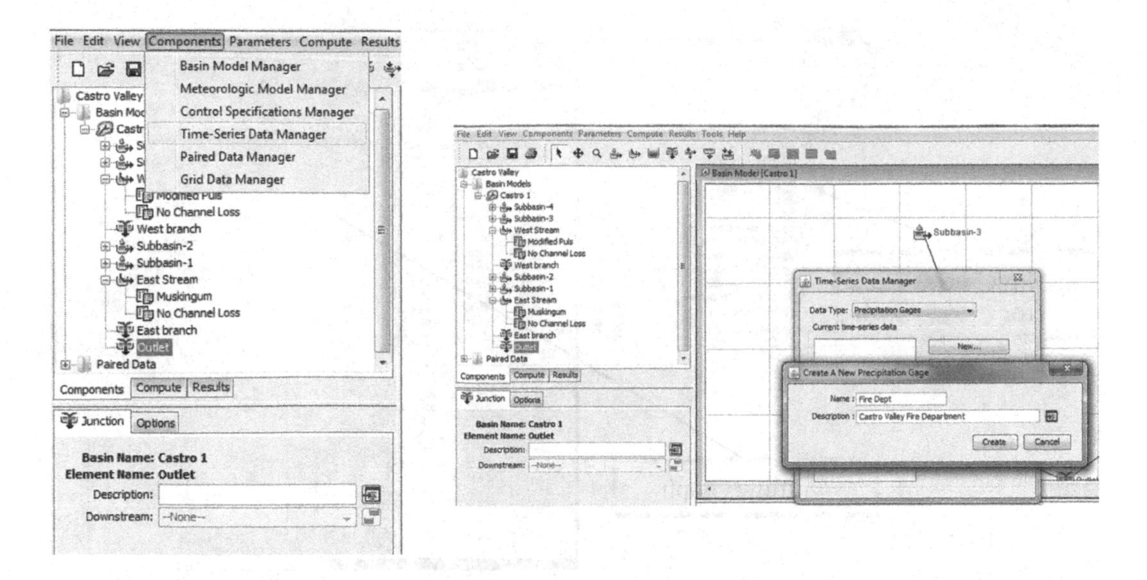

Now for Creating a discharge gage (for the observed hydrograph at the watershed outlet), the same procedure as the precipitation gage is executed again. The model is now complete.

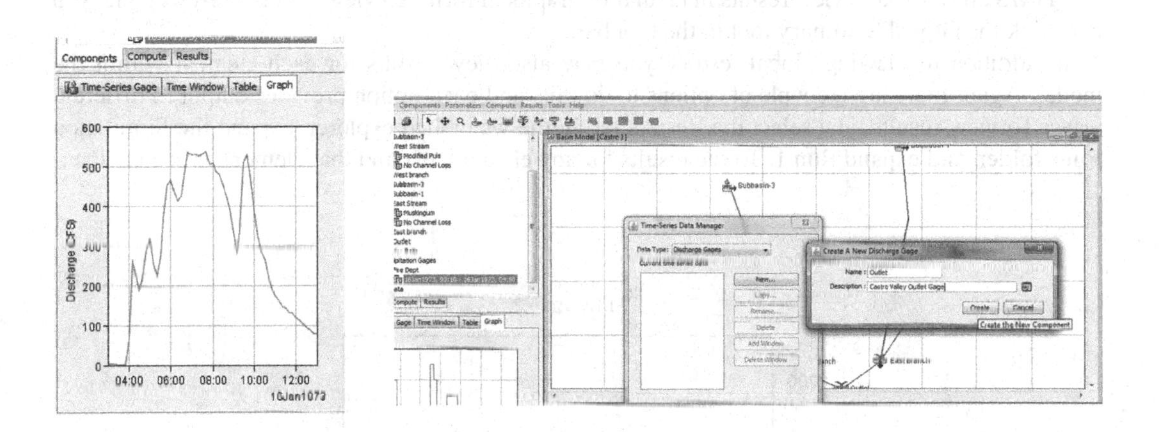

The last step is to run the model. Select Compute → Create Simulation Run. Accept the default name for the run (Run 1), click Next to complete all the steps and finally Click Finish to complete the run. Now to run the model, you can click the compute run tool in the tool bar.

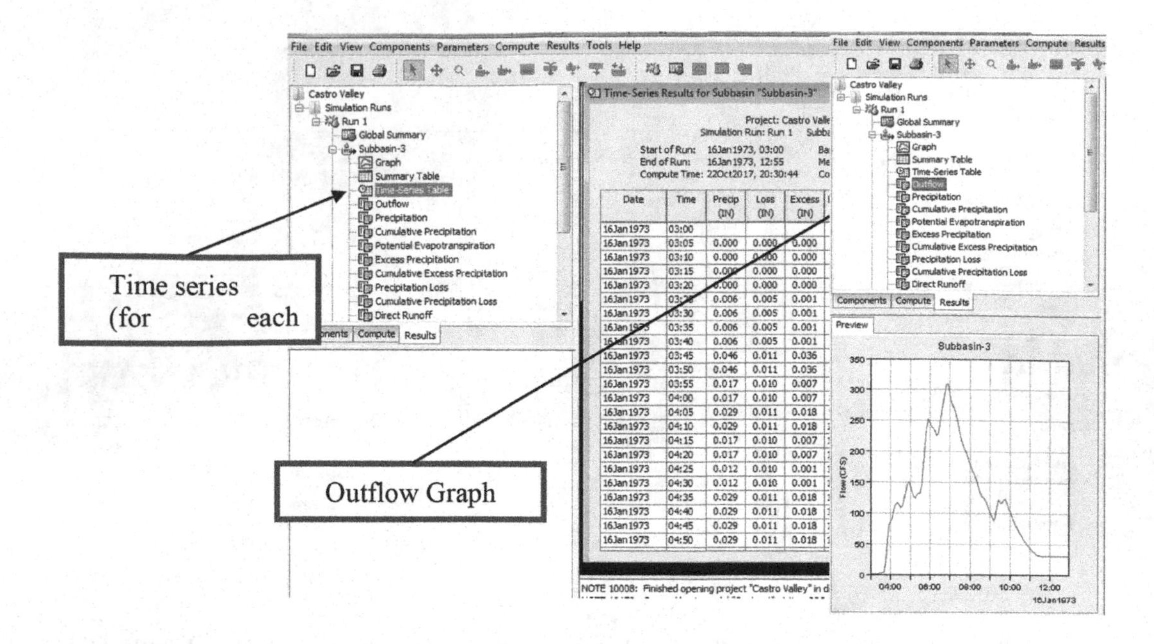

Viewing HMS Results

The HMS allows you to view results in tabular or graphical form. To view a global results table, you can click the Global Summary tool in the tool bar.

In addition to viewing global results, you may also view results for each element within the model. Again there are a couple of options to do this, and each option provides output in different ways. To view results, you select the Results tab in the watershed explorer, expand the Simulation Runs folder, and expand Run 1. To see results for any element, expand that element as seen below:

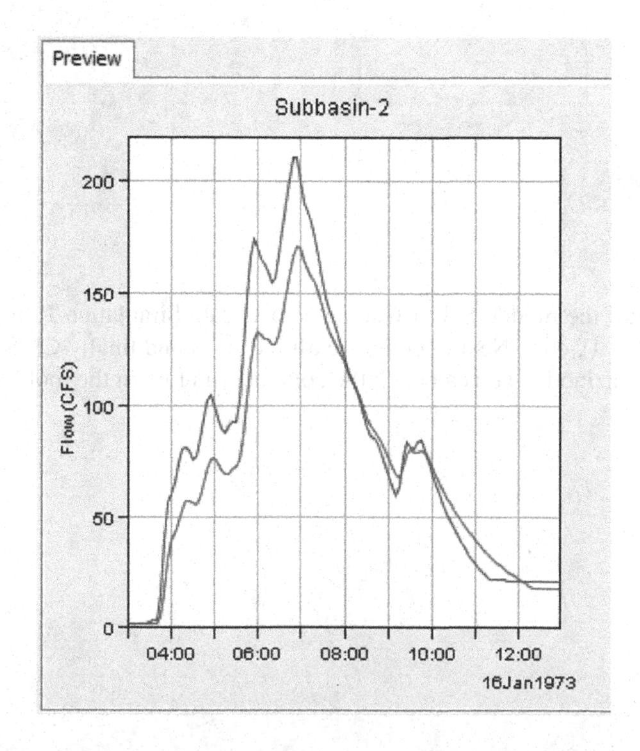

Effect of Urbanization (Defining Scenarios)

Suppose the following changes have been made due to future urbanization:

For Sub-basin 2, change the percent imperviousness from 8% to 17% and the Snyder t_p from 0.28 to 0.19 hours.

Meteorological model and Control Specifications remain same but a modified basin must be created to reflect anticipated changes to the watershed. Run the new model and compare the results by holding down Ctrl Key and clicking on Outflow Hydrograph in Run 1 and Run 2 to display both graph simultaneously:

For more information on the software, visit the official website of the model at www.hec.usace.army.mil.

Assignment

This class project is designed with the aim of modeling a watershed with various hydrological elements in HEC-HMS software. The schematic view of the watershed is depicted in the following figure:

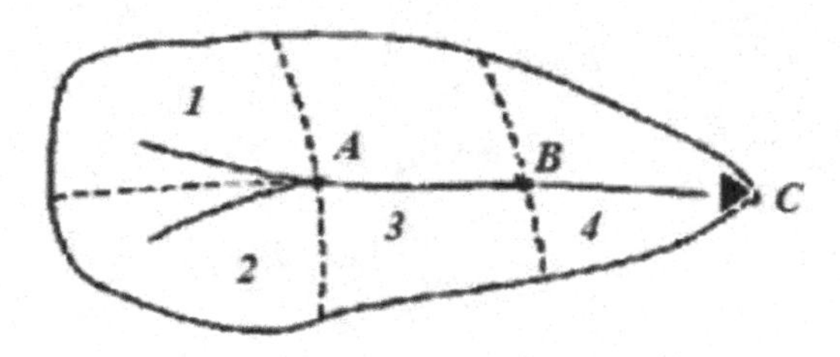

Inputs for this model are listed below:

Show Elements: All Elements

Subbasin	Area (KM2)
Subbasin-2	60
Subbasin-1	51
Subbasin-3	80
Subbasin-4	100

Show Elements: All Elements Sorting: Hydrologic

Subbasin	Initial Loss (MM)	Constant Rate (MM/HR)	Impervious (%)
Subbasin-2	0.5	3	4
Subbasin-1	0.5	3	10

Show Elements: All Elements   Sorting: Hydrologic

Subbasin	Initial Abstraction (MM)	Curve Number	Impervious (%)
Subbasin-3		70	8
Subbasin-4		66	15

Show Elements: All Elements   Sorting: Hydrologic

Standard | Ft Worth | Tulsa

Subbasin	Lag Time (HR)	Peaking Coefficient
Subbasin-2	0.18	0.16
Subbasin-1	0.2	0.16

Show Elements: All Elements   Sorting: Hydrologic

Subbasin	Graph Type	Lag Time (MIN)
Subbasin-3	Standard	12
Subbasin-4	Standard	15

Show Elements: All Elements

Reach	Muskingum K (HR)	Muskingum X	Number of Subreaches
Reach-1	1.8	0.1	1
Reach-2	1.5	0.2	1

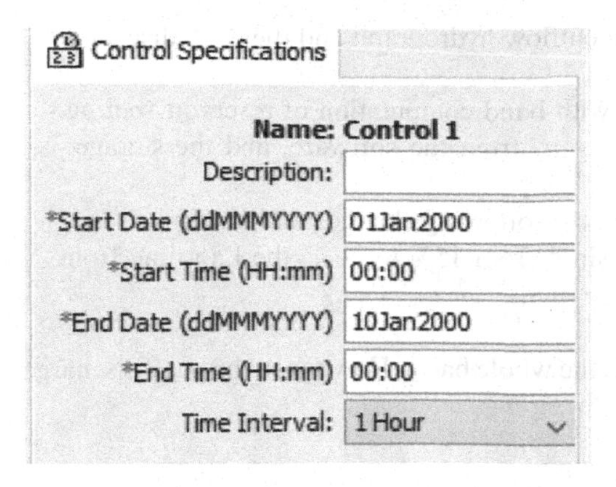

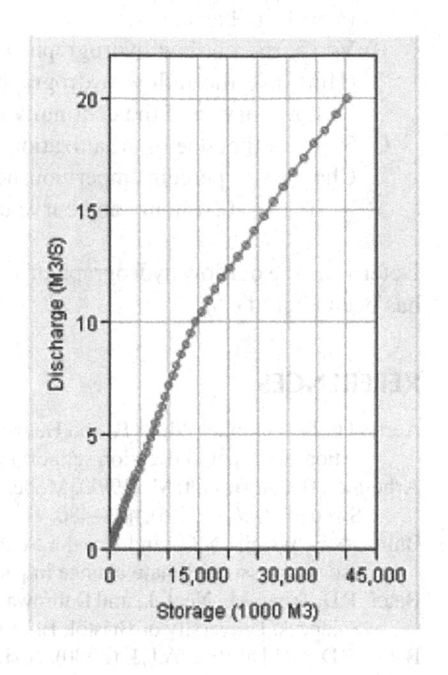

Total Storm Gages

Met Name: Gage Wts

Gage Name	Total Depth (MM)
Gage 1	49
Gage 2	66.6

Element Name: Subbasin-1

Gage Name	Depth Weight	Time Weight
Gage 1	0.2	
Gage 2	0.8	
Gage 3	0	1

Element Name: Subbasin-2

Gage Name	Depth Weight	Time Weight
Gage 1	0.5	
Gage 2	0.3	
Gage 3	0.2	1

Element Name: Subbasin-3

Gage Name	Depth Weight	Time Weight
Gage 2	0.7	
Gage 3	0.3	1

Element Name: Subbasin-4

Gage Name	Depth Weight	Time Weight
Gage 1	0.33	
Gage 2	0.33	
Gage 3	0.33	1

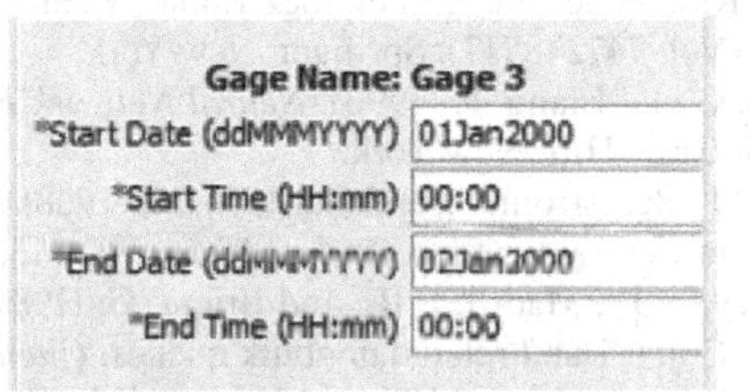

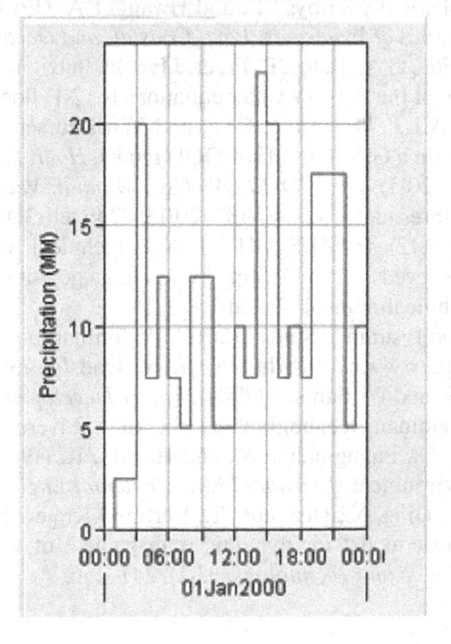

A. Run the HEC-HMS model and determine the outflow hydrograph and the peak discharge in each subbasin.
B. Verify the outflow hydrograph from part (a) with hand computation of reservoir routing. (Hint: use the inflow hydrograph of the reservoir, from the software, and the storage–outflow curve and the continuity equation).
C. Suppose that due to urbanization in the future, the following changes will occur:
   Change the percent imperviousness of subbasin 4 from 15% to 20%, the Lag time from 0.25 to 0.18 hours and curve umber from 66 to 8.

Determine the outflow hydrograph of subbasin 4 and the whole basin. How much the peak discharge has been altered (%)?

REFERENCES

Aerts, J.C.J.H., Goodchild, M.F., and Heuvelink, G.B.M. (2003). Accounting for spatial uncertainty in optimization with spatial decision support systems. *Transactions in GIS*, 7, 211–230.

Arheimer, B. and Brandt, M. (1998). Modelling nitrogen transport and retention in the catchments of southern Sweden. *Ambio*, 27(6), 471–480.

Balica, S.F., Wright, N.G., and Van der Meulen, F. (2012). A flood vulnerability index for coastal cities and its use in assessing climate change impacts. *Natural Hazards*, 64(1), 73–105.

Bates, P.D, Trigg, M., Neal, J., and Dabrowa, A. (2013). *LISFLOOD-FP. User Manual*. School of Geographical Sciences, University of Bristol, Bristol.

Bates, P.D. and De Roo, A.P.J. (2000). A simple raster-based model for flood inundation simulation. *Journal of Hydrology*, 236(1–2), 54–77.

Bergstrom, S. (1992). THE HBV MODEL—its structure and applications. Stockholm, Sweden. Retrieved from http://www.smhi.se/polopoly_fs/1.83589!/Menu/general/extGroup/attachmentColHold/mainColl/file/RH_4.pdf

Bivand, R.S., Pebesma, E.J., Gomez-Rubio, V., and Pebesma, E.J. (2008). *Applied Spatial Data Analysis with R* (Vol. 747248717). Springer, New York.

Bose, N.K. and Liang, P. (1996). *Neural Network Fundamentals with Graphs, Algorithms, and Application*. McGraw-Hill, New York.

Brandt, M., Bergstrom, S., and Gardelin, M. (1988). Modelling the effects of clearcutting on runoff – Examples from Central Sweden. *Ambio*, 17(5), 307–313.

Chaudhry, M.L., Madill, B.R., and Briere, G. (1987). Computational analysis of steady-state probabilities of M/G a, b/1 and related nonbulk queues. *Queueing systems*, 2(2), 93–114.

Dalrymple, R.A., Kirby, J.T., and Hwang, P.A. (1984). Wave diffraction due to areas of energy dissipation. *Journal of Waterway, Port, Coastal, and Ocean Engineering*, 110(1), 67–79.

De Almeida, G.A., Bates, P., Freer, J.E. and Souvignet, M. (2012). Improving the stability of a simple formulation of the shallow water equations for 2-D flood modeling. *Water Resources Research*, 48(5).

De Roo, A.P.J., Wesseling, C.G., and Van Deursen, W.P.A. (2000). Physically based river basin modelling within a GIS: The LISFLOOD model. *Hydrological Processes*, 14(11-12), 1981–1992.

Deltares. (2011). *Delft3D-FLOW User Manual. Version 3.15.14499*. Rotterdam. Available from https://oss.deltares.nl/documents/183920/185723/Delft3D-FLOW_User_Manual.pdf.

FEMA. (2012). HAZUS-MH 2.1 flood technical manual. Dept. of Homeland Security, Washington, DC. Retrieved from https://www.fema.gov/sites/default/files/2020-09/fema_hazus_earthquake-model_technical-manual_2.1.pdf

FIS (Flood Insurance Study). (2013). Preliminary flood insurance. New York City, New York. Retrieved from http://www.nyc.gov/html/sirr/downloads/pdf/360497NY-New-York-City.pdf

Flather, R. and Williams, J. (2004). *Future Development of Operational Storm Surge and Sea Level Prediction*. Proudman Oceanographic Laboratory, Liverpool.

Fujiwara, O., Puangmaha, W., and Hanaki, K. (1988). River basin water quality management in stochastic environment. *Journal of Environmental Engineering, ASCE*, 114(4), 864–877.

Gong, N., Ding, X., Denoeux, T., Bertrand-Krajewski, J.-L., and Clément, M. (1996). StormNet: A connectionist model for dynamic management of wastewater treatment plants during storm events. *Water Science and Technology*, 33(1), 247–256.

Goring, D.G. (1978). *Tsunamis – The propagation of Long Waves onto a Shelf.* Report No. KH-R-38. W. M. Keck Laboratory of Hydraulics and Water Resources, California Institute of Technology, Pasadena, CA.

Green, W.H. and Ampt, G.A. (1911). Studies on soil physics. *The Journal of Agricultural Science*, 4(1), 1–24.

Hawkes, P.J., Gonzalez-Marco, D., Sánchez-Arcilla, A., and Prinos, P. (2008). Best practice for the estimation of extremes: A review. *Journal of Hydraulic Research*, 46(S2), 324–332.

Holland, G.J. and Webster, P.J. (2007). Heightened tropical cyclone activity in the north Atlantic: Natural variability or climate trend?. *Philosophical Transactions of the Royal Society A: Mathematical, Physical and Engineering Sciences*, 365(1860), 2695–2716.

Holthuijsen, L.H., Booij, N., and Ris, R.C. (1993). A spectral wave model for the coastal zone. *Proceedings 2nd International Symposium on Ocean Wave Measurement and Analysis*, New Orleans, LO, 25–28 July 1993, ASCE, New York, pp. 630–641.

Horritt, M.S. and Bates, P.D. (2001). Predicting floodplain inundation: Raster-based modelling versus the finite-element approach. *Hydrological Processes*, 15, 825–842.

Horritt, M.S. and Bates, P.D. (2002). Evaluation of 1D and 2D numerical models for predicting river flood inundation. *Journal of Hydrology*, 268(1–4), 87–99.

Islam, N., Aftabuddin, M., Moriwaki, A., Hattori, Y., and Hori, Y. (1995). Increase in the calcium level following anodal polarization in the rat brain. *Brain Research*, 684(2), 206–208.

Karamouz, M. and Fereshtehpour, M. (2019). Modeling DEM errors in coastal flood inundation and damages: A spatial nonstationary approach. *Water Resources Research*, 55(8), 6606–6624.

Karamouz, M. and Fooladi, F. (2021). DEM uncertainty based coastal flood inundation modeling considering water quality impacts. *Water Resource Management*, 35, 3083–3103. doi:10.1007/s11269-021-02849-9.

Karamouz, M. and Heydari, Z. (2020). Conceptual design framework for coastal flood best management practices. *Journal of Water Resources Planning and Management*, 146(6), 04020041.

Karamouz, M. and Mahmoudi, S. (2022). Shallowing of bathymetry (seabed) for flood preparedness: towards designing nearshore BMPs. *Journal of Water Resources Planning and Management* (accepted).

Karamouz, M., Nazif, S., and Falahi, M. (2012). Hydrology and Hydroclimatology: Principles and Applications. CRC Press, Boca Raton, FL.

Karamouz, M., Szidarovszky, F., and Zahraie, B. (2003). *Water Resources Systems Analysis.* Lewis Publishers, CRC Press, Boca Raton, FL.

Karamouz, M., Taheri, M., Khalili, P., and Chen, X. (2019). Building infrastructure resilience in coastal flood risk management. *Journal of Water Resources Planning and Management*, 145(4), 04019004.

Kron, W. (2013). Coasts: The high-risk areas of the world. *Natural Hazards*, 66(3), 1363–1382.

McCowan, J. (1894). XXXIX. On the highest wave of permanent type. *The London, Edinburgh, and Dublin Philosophical Magazine and Journal of Science*, 38(233), 351–358.

NYCDEP (New York City Department of Environmental Protection). (2013). NYC wastewater resiliency plan, climate risk assessment and adaptation study. Chap. 2 In *Wastewater Treatment Plants.* Dept. of Environmental Protection, New York.

Price, W.L. (1965). A controlled random search procedure for global optimization. *Computer Journal*, 7, 347–370.

Rubinstein, R.Y. (1981) *Simulation and Monte Carlo Method.* Wiley, New York.

Shang, L., Dong, S., and Nienhaus, G.U. (2011). Ultra-small fluorescent metal nanoclusters: Synthesis and biological applications. *Nano Today*, 6(4), 401–418.

Solomatine, D.P. (1999). Two strategies of adaptive cluster covering with descent and their comparison to other algorithms. *Journal of Global Optimization*, 14(1), 55–78.

Wu, C. and Liu, P.L. (1985). Finite element modelling of nonlinear coastal currents. *Journal of Waterway, Port, Coastal and Ocean Engineering*, 111(2), 417–432.

Wu, X., Wang, Z., Guo, S., Lai, C., and Chen, X. (2018). A simplified approach for flood modeling in urban environments. *Hydrology Research*, 49(6), 1804–1816.

Yin, J., Yu, D., Lin, N. and Wilby, R.L. (2017). Evaluating the cascading impacts of sea level rise and coastal flooding on emergency response spatial accessibility in Lower Manhattan, New York City. *Journal of Hydrology*, 555, 648–658.

13 Flood Resiliency of Cities

13.1 INTRODUCTION

Flood is an overflow or overland expansion of water that could come from the sea, lakes, rivers, canals, or sewers, as well as direct rainwater and storm tides (sea surge and tide). Floods are considered to happen when the excess rainfall and storm surge (runoff) escapes the boundaries of the flowing water body and damages infrastructures, a village, a city, or other inhabited areas. Besides the seasonal and intraannual variations in river flow, this results in very high flow periods and high tides. Flooding is caused primarily by hydrometeorological mechanisms, acting either as a single factor or in combination with different characteristics. Whatever the type of flooding, it is described as a volume of water that enters a certain area that cannot be discharged quickly enough.

Floods in coastal areas, which are enhanced by sea-level rise, especially in coastal regions of the United States, Japan, and southeastern Asia, cause severe damage to human societies. Most urban areas are subject to some kinds of flooding after spring rains, heavy thunderstorms, or winter snow thaws because of high percentages of imperviousness. Various structures such as dams, levees, constructed wetlands, and weirs are employed to control the floods and decrease damage. However, floods continue to be the most destructive natural hazards in terms of short-term damage and economic losses to a region. The large damage caused by floods are due to the short warning time before flooding occurs, which makes it difficult to move people and valuables from the flooding area. Often flood plains and evacuation zones are outdated as it was evident at the time of 2012 Superstorm Sandy in New York City. Climate change and climate variability as well as population increase in coastal area, and human interventions have exacerbate flooding intensity and damages.

However, despite of excessive damage, floods also bring benefits to people in developing nations, especially those in arid and semiarid regions, as floods provide water resources for irrigation and development and fertilize the ground for agricultural purposes. These benefits can be utilized through integrated management of flood and the flooding regions. Furthermore, flood-related property losses can be minimized by updating the zoning ordinance and making flood insurance available on reasonable terms and encouraging its purchase by people who need protection, particularly those living in flood prone areas.

For effective flood management, it is necessary first to determine the characteristics of the flood in the study region. This includes determining the severity of floods with different return periods as well as the corresponding hydrographs. The most important information obtained from flood hydrographs is peak flow, timing, and flood duration (maximum and total accumulation). Furthermore, the flood design and the appropriate flood control and conveyance practices are determined for different facilities. In this regard, regional characteristics should be incorporated in design procedure, where the most essential concept is quantifying the flooding risk based on its components. Nonstationarity of data should be tested and the joint probability of rain and storm surges in coastal areas should be considered. Digital elevation information should only be used with meaningful resolution for flood applications (at least 1–30 m) and preferably be resampled with 1 m Light Detection and Ranging (LiDAR).

What constitutes a flood resilient city includes building smart communities, tools, models, and data processing and information management; flood hazard characterization and warning apparatus; inner and other links and interdependency characterization, infrastructure risk, sound asset; and financial management and performance measures. A number of these issues are discussed in

DOI: 10.1201/9781003241744-13

this chapter and elaborated in the case studies. At the end, a gap analysis section discusses the remaining challenges and looking-forward perspective for a livable city of the future.

In this chapter, flood resiliency is not discussed explicitly until Section 13.9. However, what is explained prior to that is needed to set the stage and to bring resiliency as the natural product of flood risk management. First flood types and characteristics are introduced since it is important to understand the problem before any solution can be proposed followed by a detailed description as to how flood analysis must be carried out. Discussion of historical floods and real-world flood problems of water infrastructures applies to one of the most crucial systems, New York City's wastewater treatment plants (WWTPs). As this chapter progresses, flood frequency analysis (including partial) flood routing and urban floods are discussed. Next, factors contributing to flood hazards, evacuation zones are discussed. This leads us to resiliency and flood risk management. The application of resiliency concept and how it can be used as a metric for performance evaluation and resource allocation are explored through a number of case studies.

13.2 SETTING THE STAGE—FLOOD TYPES AND FORMATIONS

With regard to the intensity, characteristics, and causes of flooding, different types of floods are experienced. More details are given in Karamouz et al. (2012).

A. *Flash floods*: Heavy rains in areas with steep slopes may result in a riverbed with little or no water to suddenly overflow with fast flowing water. In these floods, two key factors are rapid formation of runoff and short time of concentration. The timing and the speed of the flash floods are comparable with the resulting flood from a dike break. The flash floods stop as suddenly as they start and develop.
B. *Urban floods*: The specification of urban flooding is in its cause, which is lack of drainage in an urban area. High-intensity rainfall can cause flooding when the city sewage system and draining canals do not have the necessary capacity to drain away the amount of rain that is falling.
C. *River floods*: River (fluvial) floods result from heavy rainfall (over an extended period and an extended area) or snowmelt upstream or tidal fluctuations downstream. In river floods, a major river overflows its banks (exceeding local flow capacities). Sometimes, river floods are the result of failure or bad operation of drainage or flood control practices upstream.
D. *Ponding or pluvial flooding*: Ponding flood is typical in relatively flat areas. In this case, rain is the source of the flood: not water coming from a river, but water on its way to the river. That is why it is also called "pluvial flood."

13.2.1 Inland and Coastal Flooding

In another way, floods can be categorized into two main groups: inland and coastal. In coastal areas of the United States, land-induced floods are often combined with coastal flood storm surges with a higher probability of occurrence as could be quantified by classical flood frequency analysis. Coastal floods occur due to hurricanes or tropical storms.

13.2.1.1 Inland Flooding

Inland flooding occurs when the surface runoff volume exceeds the drainage system capacity. The depth of overflow is dependent on its volume as well as on the topography of the basin. Figure 13.1 shows the changes in basin and the increasing percentage of impervious areas result in increasing surface runoff volume and, due to the limited capacity of the drainage system, increasing overflow.

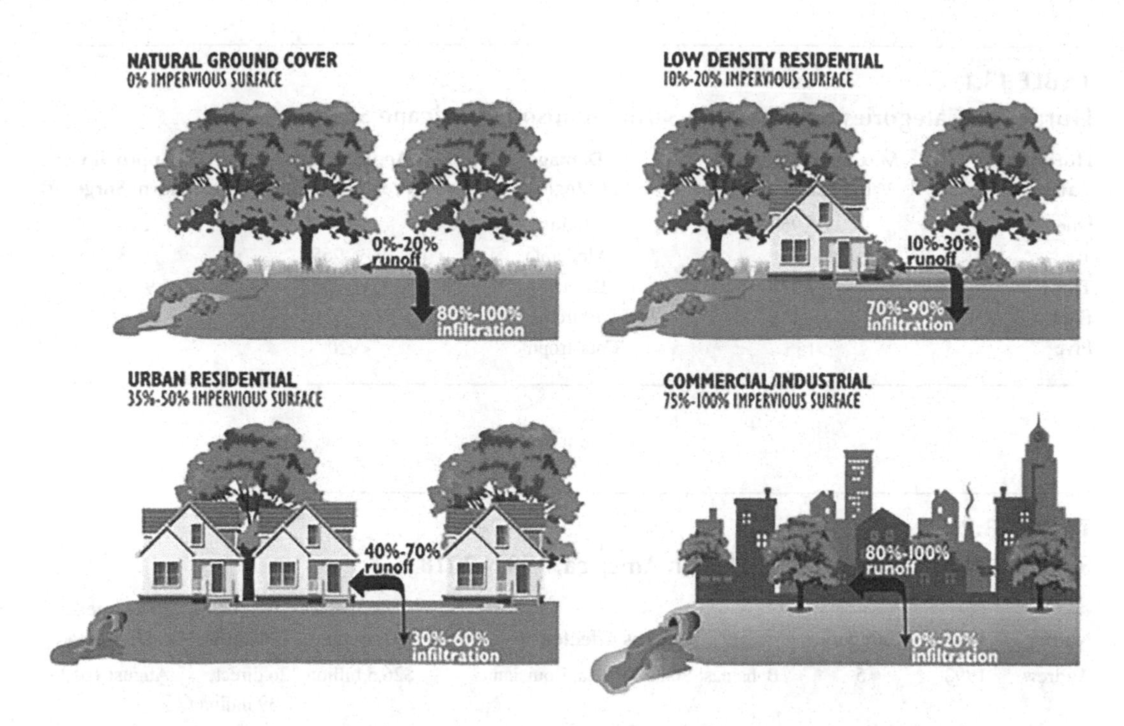

FIGURE 13.1 Influence of urbanization on decreasing rate of infiltration and increasing surface runoff. (From "How to manage stormwater. City of Oberlin." https://www.cityofoberlin.com/city-government/departments/public-works/stormwater-management/how-to-manage-stormwater/.)

Reducing the flooding depth is an essential step in decreasing flood damage. Different best management practices (BMPs) can be employed for this purpose. Providing warnings for flood events as soon as possible also helps to mitigate casualties especially in extreme flood events due to people's preparedness to deal with flooding.

13.2.1.2 Coastal Flooding

In coastal floods, the coast is flooded by the sea due to a severe storm. In such cases, the storm wind pushes the water up and creates high waves. The resulting high tides and storm surges from tropical depressions and cyclones can produce coastal floods in estuaries, tidal flats, and low-lying land near the sea.

Coastal flooding occurs due to the deadly storm surges of hurricanes, tropical storms, or tropical depressions, which greatly affect shore locations, causing landslides. The water carried by storm winds towards the coast is called a "storm surge." Hurricane storm tide is caused by the combination of this surge in movement with normal tides, and the water level may increase 15 ft or more after this point. With the help of official definitions from the National Hurricane Center, according to the Saffir–Simpson hurricane scale, five hurricane categories are defined as given in Table 13.1.

Table 13.2 lists some hurricanes in different categories that have impacted the North American continent since 1992. This section will briefly highlight three storms, including 2005 hurricane Katrina, 2011 hurricane Irene and the Tohoku tsunami, and 2012 Superstorm Sandy.

13.2.1.2.1 Hurricane Katrina (August 2005, LA, USA)

The most damaging storm ever hit the United States was Hurricane Katrina, which set records as the costliest US natural disaster and the third deadliest hurricane to strike this country. Lashing the southeast Louisiana coast with winds up to 140 mph and a 27 ft storm surge, the storm devastated costal Mississippi towns, breached the Lake Pontchartrain levees, and flooded 80% of New Orleans (Figure 13.2).

TABLE 13.1
Hurricane Categories Based on the Saffir–Simpson Hurricane Scale

Hurricane Category	Maximum Sustained Wind Speed (mi/h)	Damage Category	Approximate Pressure (mb)	Approximate Storm Surge (ft)
One	74–95	Minimal	>980	3–5
Two	96–110	Moderate	979–965	6–8
Three	111–130	Extensive	964–945	9–12
Four	131–155	Extreme	944–920	13–18
Five	>155	Catastrophic	<920	>18

TABLE 13.2
Selected Hurricanes Affecting North America, 1992–2018

Name	Year	Category	Areas Affected	Damage	Deaths	Formed–Dissipated
Andrew	1992	5	Bahamas; South Florida, Louisiana	$26.5 billion	26 direct, 39 indirect	August 16–28
Alberto	1994	Tropical storm	Florida Panhandle, Alabama, Georgia	$1 billion	32 direct	June 30–July 10
Opal	1995	4	Guatemala, Yucatán Peninsula, Alabama, Florida Panhandle, Georgia	$3.9 billion	59 direct, 10 indirect	September 27–October 6
Floyd	1999	4	Bahamas, US East Coast from Florida to Maine, Atlantic Canada	$4.5 billion	57 direct, 20–30 indirect	September 7– 19
Allison	2001	Tropical storm	Texas, Louisiana	$5.5 billion	41 direct, 14 indirect	June 4–18
Isabel	2003	5	The Greater Antilles, Bahamas, most western and southern US states, Canada (Ontario)	$3.6 billion	16 direct, 35 indirect	September 6–20
Charley	2004	4	Jamaica, Cayman Islands, Cuba, Florida, South Carolina, North Carolina	$16.3 billion	15 direct, 20 indirect	August 9–15
Frances	2004	4	British Virgin Islands, Puerto Rico, US Virgin Islands, Bahamas, Florida, Georgia, North and South Carolina, Ohio	$12 billion	7 direct, 42 indirect	August 24–September 10
Ivan	2004	5	Venezuela, Jamaica, Grand Cayman, Cuba, Alabama, Florida, Louisiana, Texas	$18 billion	91 direct, 32 indirect	September 2–24
Jeanne	2004	3	US Virgin Islands, Puerto Rico, Dominican Republic, Haiti, Bahamas, Florida	$7 billion	3,035 direct	September 13–28
Dennis	2005	4	Grenada, Haiti, Jamaica, Cuba, Florida, Alabama, Mississippi, Georgia, Tennessee, Ohio	$4 billion	42 direct, 47 indirect	July 4–13
Katrina	2005	5	Bahamas, South Florida, Cuba, Louisiana, Mississippi, Alabama, Florida Panhandle	$108 billion	1,833	August 23–30

(Continued)

TABLE 13.2 (*Continued*)
Selected Hurricanes Affecting North America, 1992–2018

Name	Year	Category	Areas Affected	Damage	Deaths	Formed–Dissipated
Rita	2005	5	Arkansas, South Florida, Cuba, Florida Panhandle, Louisiana, Mississippi, Texas	$12 billion	97–125	September 18–26
Ophelia	2005	1	Northeast Florida, North Carolina, Massachusetts, Atlantic Canada	$70 million	1 direct, 2 indirect	September 6–23
Wilma	2005	5	Jamaica, Haiti, Cuba, Honduras, Nicaragua, Belize, Florida, Bahamas	$29.1 billion	23 direct, 39 indirect	October 15–26
Humberto	2007	1	Southeast Texas, Louisiana	$50 million	1	September 12–14
Gustav	2008	4	Dominican Republic, Haiti, Jamaica, Cuba, Florida, Louisiana, Mississippi, Alabama, Arkansas	$6.61 billion	112 direct, 41 indirect	August 25–September 4
Dolly	2008	2	Guatemala, Yucatan Peninsula, western Cuba, Northern Mexico, South Texas, New Mexico	$1.35 billion	1 direct, 21 indirect	July 20–25
Ike	2008	4	Bahamas, Haiti, Dominican Republic, Cuba, Florida, Mississippi, Louisiana, Texas, Mississippi, Ohio, eastern Canada	$37.6 billion	103 direct, 92 indirect	September 1–14
Irene	2011	3	Bahamas, eastern United States (landfalls in North Carolina, Connecticut, New Jersey, and New York), eastern Canada	$10.1 billion	49 direct, 7 indirect	August 20–28
Sandy	2012	3	Caribbean, Bahamas, United States East Coast, eastern Canada	$68.7 billion	67 direct, 38 indirect	October 22–29
Ingrid	2013	1	Mexico	$1.5 billion	32	September 12–17
Erika	2015	TropicalsStorm	Lesser Antilles, Hispaniola	$511 million	35	August 24–28
Joaquin	2015	4	Bahamas, Bermuda	$200 million	34	September 28–October 8
Matthew	2016	5	Caribbean, Southeastern United States	$15.1 billion	585 direct, 18 indirect	September 28, October 9
Otto	2016	3	Panama, Costa Rica, Nicaragua	$192 million	23	November 20–26
Harvey	2017	4	Texas, Louisiana	$125 billion	68 direct, 35 indirect	August 17–September 1
Irma	2017	5	Caribbean, Southeastern United States	$77.2 billion	11 direct, 115 indirect	August 30–September 12
Maria	2017	5	Lesser Antilles, Puerto Rico	$91.6 billion	3,057	September 16–30
Nate	2017	1	Central America, United States Gulf Coast	$787 million	48	October 4–9
Florence	2018	4	Eastern United States	$24 billion	22 direct, 30 indirect	August 31–September 17
Michael	2018	5	Central American, United States Gulf Coast	$25.1 billion	31 direct, 43 indirect	October 7–11

FIGURE 13.2 Demonstration of the inundation following hurricane Katrina. (Adapted from the Interagency Performance Evaluation Task Force (IPET) Report (2009).)

Katrina began as a tropical depression on August 13, 2005, located 175 miles northeast of Nassau, Bahamas. By the next day, Katrina was anticipated to hit the Florida coast near Miami as a category 1 hurricane.

The hurricane made landfall on August 25 at 6:30 PM and traveled quickly across the state to exit into the Gulf of Mexico considerably weaker than before. However, by early morning of August 26, it became apparent that Katrina would intensify into a serious threat and affect much of the Gulf Coast. The above-average sea surface temperatures and reduced wind shear helped the storm grow from a tropical storm to a category 5 hurricane within 48 hours. By 10 PM of August 27, the Katrina hurricane warning from the National Weather Service indicated a major and ominous storm approaching the City of New Orleans and warned of massive storm surges of 15–20 ft from Morgan City to the Alabama/Florida border. New Orleans was bracing itself for a devastating storm since the levees protecting the city were only built to withstand a category 3 hurricane and had never been fully tested. When Katrina made landfall near Buras, Louisiana, at 6:10 AM Central Time, August 29, the maximum wind speed had reached 125 mph and a central pressure of 920 mph officially made it a category 3 hurricane. Nonetheless, New Orleans suffered massive damage due to levee failure. Katrina then continued inland, weakening as it went, becoming a category 1 by 6 PM that day, a tropical storm around midnight as it passed through Northern Mississippi, and a tropical depression near Clarksville, Tennessee, on August 30.

13.2.1.2.1.1 Background—the Hurricane/Flood Protection System (HPS/FPS) The hurricane/flood protection system installed in the region, illustrated on the map in Figure 13.3, includes ~350 miles of protective strategies and engineering structures. Out of the mentioned 350 miles, 56 are floodwalls. The majority of the installed floodwalls are I-walls with only a small proportion of T- and L-Walls. Figure 13.4 depicts a schematic of the basic configuration of these structures. The elevation of the installed hurricane/flood protection structures is meaningfully below the initially

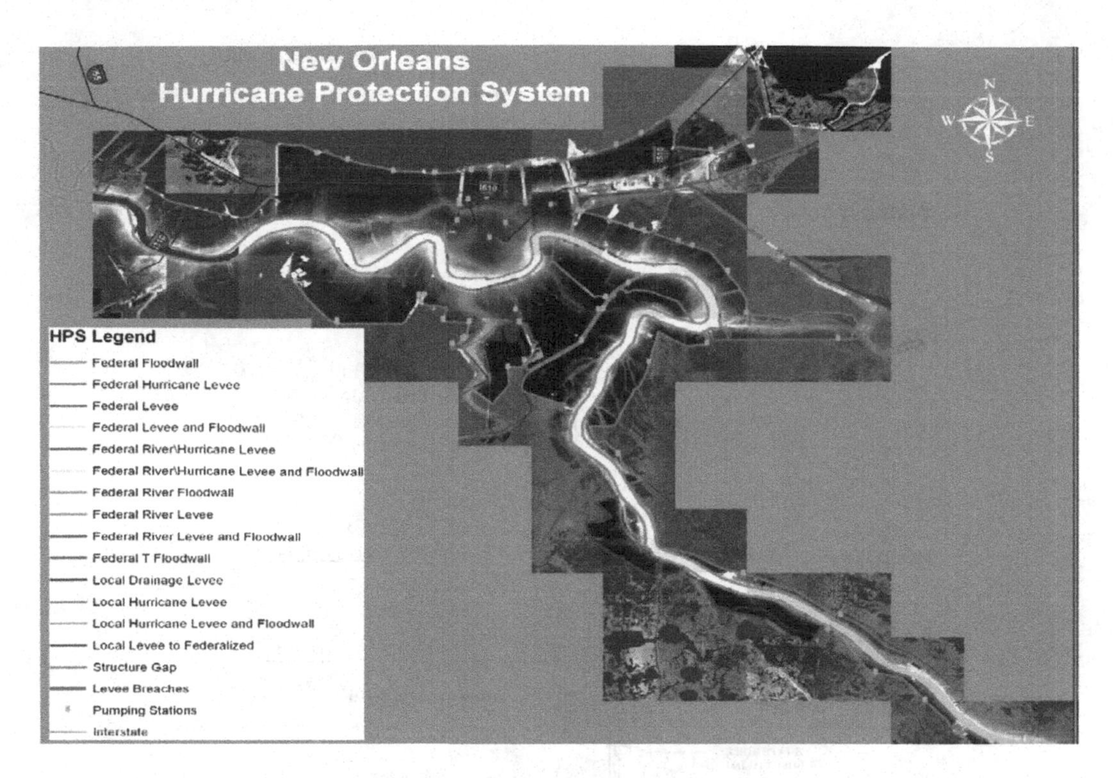

FIGURE 13.3 Outline of the New Orleans and Southeast Louisiana Hurricane Protection System. (Adapted from the IPET Report (2009).)

planned heights due in part to the errors and discrepancies in the initial constructed elevations and the high rate of subsidence in the region.

13.2.1.2.1.2 The Interagency Performance Evaluation Task Force (IPET) After Hurricane Katrina, due to the massive casualties sustained to the region, an intense performance evaluation of the New Orleans and Southeast Louisiana hurricane protection system (HPS) during Hurricane Katrina was conducted to find the shortcomings of the system. The evaluation was carried out by the Interagency Performance Evaluation Task Force (IPET), an outstanding group of engineers and scientists from the government of the United States, the academic and the private sectors. All task force members dedicated themselves solely to their mission shortly after Hurricane Katrina struck through the start of the next hurricane season.

A key objective of the IPET was to understand the behavior of the New Orleans Hurricane Protection Sytem in response to Hurricane Katrina and assist in the application of that knowledge to the reconstitution of a more resilient and capable system. The results of their work are publicly available online. The task forced presented the outcome of their work to the public in a comprehensive report in nine volumes.

The configuration of this report (Figure 13.5) has been planned in a way that provides answers to the five major questions that comprise the IPET mission. The five questions are (US Army Corps of Engineers (USACE), 2009):

- *The system*: What were the pre-Katrina characteristics of the HPS components; how did they compare to the original design intent?
- *The storm*: What was the surge and wave environment created by Katrina and the forces incident on the levees and floodwalls?

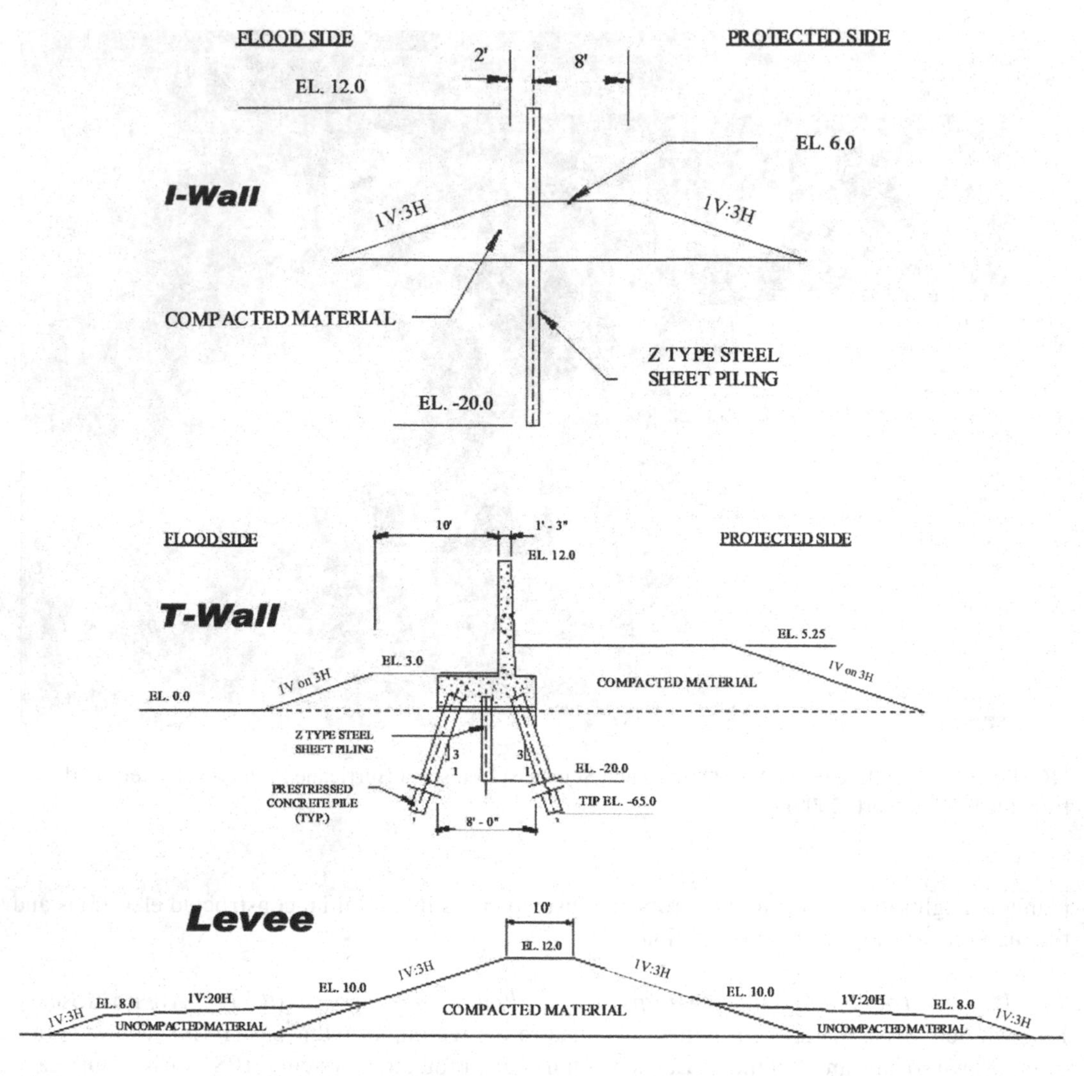

FIGURE 13.4 General schematic of major hurricane protection structures used in New Orleans and Vicinity. (From the IPET Report (2009).)

- *The performance*: How did the levees and floodwalls perform, and what insights can be gained for the effective repair of the system, and what is the residual capacity of the undamaged portions? This also involved understanding the performance of the interior drainage and pump stations and their role in flooding and unwatering of the area.
- *The consequences*: What were the societal-related consequences of the flooding from Katrina to include economic, life and safety, environmental, and historical and cultural losses?
- *The risk and reliability*: What was the risk and reliability of the HPS prior to Katrina, and what will it be following the planned repairs and improvements?

Findings of the IPET have been presented in three tiers. Tiers one and two were introduced in the first volume of their report, whereas the third tier was presented in individual volumes. The first tier is the titled the "Overarching Findings" and it represents a synthesis of the component analyses carried out in the region. The second tier, also presented in Volume 1 as a synopsis of principal findings, is a synthesis of the findings from the component analyses of the HPS. The detailed findings are presented in Volumes II through to VIII.

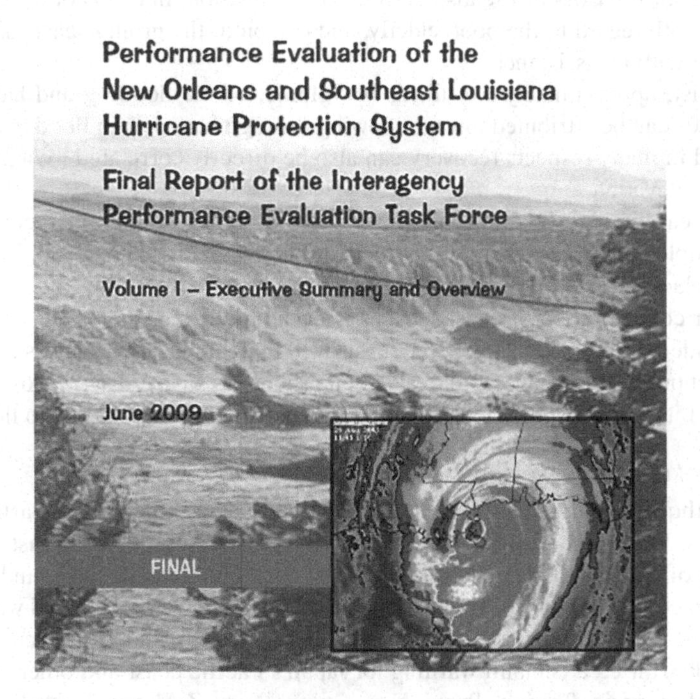

FIGURE 13.5 Cover page of performance evaluation of the New Orleans and Southeast Louisiana Hurricane Protection System. (From "Performance Evaluation of the New Orleans and Southeast Louisiana Hurricane Protection System: Final Report of the Interagency Performance Evaluation Task Force." Interagency Performance Evaluation Task Force (IPET). Volume I (June): Executive Summary and Overview. (2009). Retrieved from https://ipet.wes.army.mil.)

13.2.1.2.1.3 Key Findings The most significant findings of the IPET were as follows:

- The system did not function in a systematic manner; flood protection systems are series systems—if a single levee or floodwall fails, the entire area is put at risk. In this case, the system's performance was compromised by the incompleteness of the system, the inconsistency in levels of protection, and the lack of redundancy.
- The storm exceeded all expectations but also the FPS/HPS did not live up to the expectations as well: sections of the HPS were in a variety of ways overwhelmed by the conditions created by Hurricane Katrina.
- While overtopping the installed structures and extensive flooding from Katrina were unavoidable, a complete system at authorized elevations would have reduced the losses incurred.
- The designs were developed to deal with a specific hazard level, the Standard Project Hurricane as defined in 1965; however, little consideration was given to the system's performance if the design event or system requirements were exceeded.
- Some sites experienced I-wall breaches due to overtopping and scour behind the walls which reduced the stability of the structures. These breaches added to the flooding in Orleans (East Bank) and the Lower Ninth Ward.

- The flooding and the consequences of the flooding were not only pervasive but also concentrated. Consequences of the flooding and the associated losses were greater than any previous disaster in New Orleans and, in themselves, create a formidable barrier to recovery. Loss of life was concentrated by age, with more than 75% of deaths being people over the age of 60. Loss of life also correlated to elevation, in terms of depth of flooding, especially with regard to the poor, elderly, and disabled, the groups least likely to be able to evacuate without assistance.
- The majority, approximately two-thirds by volume, of the flooding and half of the economic losses can be attributed to water flowing through breaches in floodwalls and levees.
- Losses and in many respects recovery can also be directly correlated to depth of flooding and thus to elevation.
- In some areas flooded by Katrina, where water depths were small, recovery has been almost complete. In areas where water depths were greater, little recovery or reinvestment has taken place.

  Another concentration of consequences is in the nature of the losses. Twenty-five percent of residential property values were destroyed by Katrina and this loss represents 78% of all direct property damages. Nonresidential properties suffered a 12% loss in total value or half the rate of residential. Clearly residential areas were more prone to flooding.

13.2.1.2.2 Tohoku Earthquake and Tsunami (March 2011, Japan)

The Tōhoku earthquake and tsunami on March 11, 2011, was a 9-magnitude earthquake followed by tsunami waves. It happened 130 km off Sendai, Miyagi Prefecture, on the east coast of Tōhoku, Japan, at a depth of 24.4 km. This makes it the largest earthquake to hit Japan and the seventh biggest earthquake recorded in history. The earthquake triggered powerful tsunami waves that reached heights of up to 40.5 m in Miyako area and traveled up to 10 km inland in the Sendai area.

The earthquake started a tsunami warning for Japan's Pacific coast and other countries, including Philippines, Indonesia, Taiwan, Papua New Guinea New Zealand, Australia, Russia, Guam, Nauru, and Hawaii Northern Marianas (United States). It was a warning that the wave could be as much as 10 m high. A 0.5 m high wave hit Japan's northern coast. News reported that a 4 m high tsunami hit the Iwate Prefecture in Japan. A wave 2 m high reached California, after traveling across the Pacific Ocean at a speed of 500 km/h.

13.2.1.2.3 Hurricane Irene (August 2011, United States)

Hurricane Irene was a costly hurricane that occurred in the northeastern United States. Irene originated from an Atlantic tropical wave, showing signs of organization east of the Lesser Antilles. The system was labeled as Tropical Storm Irene on August 20, 2011, because of the development of atmospheric convection and a closed center of circulation. After intensifying, while crossing through St. Croix, it made landfalls as a strong tropical storm. However, after a day, it strengthened into a category 1 hurricane. The storm continued along the offshore of Hispaniola and intensified gradually to become a category 3 hurricane. At that time, Irene peaked with wind speeds up to 120 mph (195 km/h). After reaching the Bahamas, the storm slowly leveled off in intensity and then curved northward. Before reaching the Outer Banks of North Carolina on August 27, Irene weakened to a category 1 hurricane. Hurricane Irene devastated the Caribbean and US East Coast.

The storm took the lives of over 40 people as well as causing $6.5 billion in property damages, unleashing major flooding on the citizens of the affected areas, downing numerous trees, infrastructures and power lines, and forcing road closures, major evacuations, and immense rescue efforts. Figure 13.6 shows an example of the damage dealt to critical infrastructures caused by Hurricane Irene. The effects of Hurricane Irene were witnessed from the US Virgin Islands and Puerto Rico all the way to the Canadian Maritime Provinces and even as far as west as the Catskill Mountains of the state of New York. The storm was responsible for the generation of widespread, devastating

FIGURE 13.6 Damage to critical infrastructure caused by Hurricane Irene. (From https://www.nytimes.com/2011/08/31/us/31floods.html.)

flooding in the states of Vermont, New Hampshire, New York, and New Jersey. The hurricane's damaging storm surge was experienced along the coasts of North Carolina and Connecticut and left over 8 million people without power, some for as long as a week, which, in turn, resulted in the closure of several major airports, the suspension of Amtrak train service, and the historic closure of the New York City mass transit system. It was apparent that until then, there was an inadequate understanding of vulnerability as New York City was paralyzed for hours/days. It was a turning point in looking at the domino effect of disasters in coastal urban areas. Figure 13.7 illustrates Hurricane Irene's water-level variations based on the observations from the Battery Park Station.

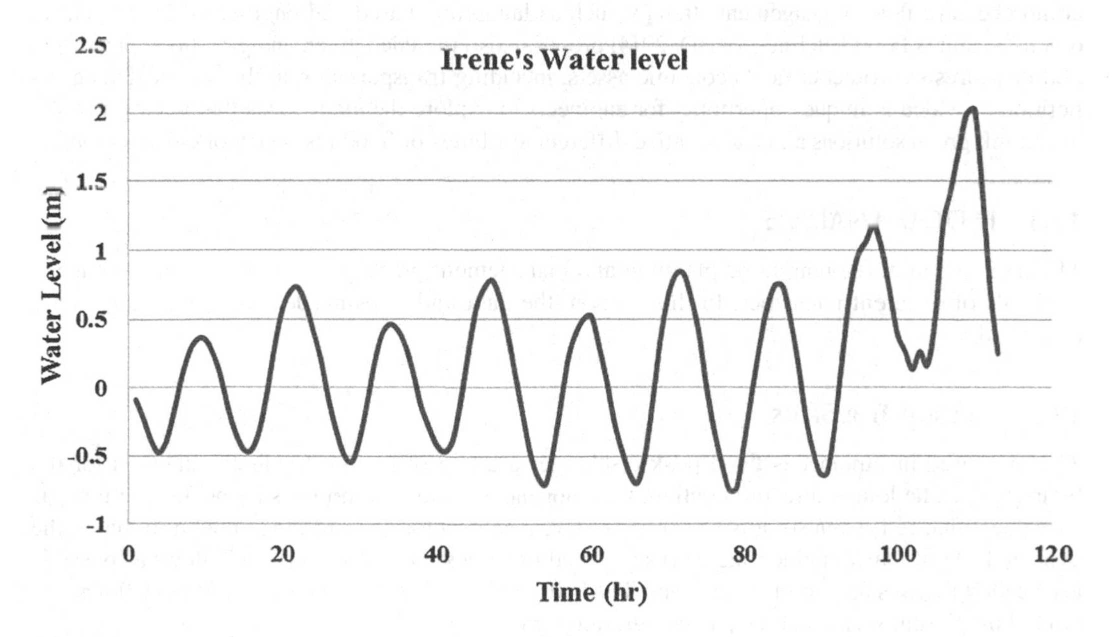

FIGURE 13.7 Hurricane Irene's water level based on the observations from the Battery Park Station, New York—August 24–29, 2011.

13.2.1.2.4 Superstorm Sandy (October 2012, United States)

Superstorm Sandy's occurrence was a real wake-up call. It showed how insufficient the integration of critical infrastructures is. The lower Manhattan as a global strategic financial location was shut down for a week due to malfunctioning of an electrical substation. The enormity of the consequence of this occurrence was so bad and showed that despite of advances in the use of technology and with all available resources, how naïve and vulnerable coastal communities are. Despite the remarkable critics of Katerina recovery and rebuilding efforts through IPET report and other publications, Irene's proof of serious vulnerabilities did not provide enough chance to revisit the way flood preparedness and response should be. However, Sandy was a turning point.

There was an apparent lack of holistic/system-based thinking and an inadequate understanding of the vulnerability of the region that had led to large-scale casualties. Nationwide concerns led to the formation of a comprehensive effort through what was called "Rebuild by Design." In December 2013, President Obama signed an executive order creating the Hurricane Sandy Rebuilding Task Force to ensure that the Federal government continues to provide appropriate resources to support affected state, local, and tribal communities to improve the region's resilience, health, and prosperity by building for the future. The Task Force exists to ensure cabinet-level, Government-wide and region-wide coordination to help communities as they are making decisions about long term rebuilding.

Following the presidential executive order, the US Department of Housing and Urban Development (HUD) initiated a competition with the collaboration of the Netherlands government called Rebuild by Design (RBD). RBD consisted of ten teams, including the finalists BIG U, New Meadowlands, Hudson River Project, and Hunts Point Lifelines, made up of experts, landscape architects, and engineers to generate ideas and conceptual designs for flood risk management. The main objective was to find solutions suitable for flood control infrastructure during both extreme events and normal weather conditions. The BIG U represented the notion of integrating a city park with floodwalls.

The New Meadowlands team suggested an integrated linked system of embankments and wetlands to provide flood protection through the Meadowlands in New Jersey. The Hudson River Project proposed the green and gray infrastructure approach for reducing flood risk and achieving a more comprehensive flood management strategy such as landscape-based and engineered-based coastal defenses. Hunts Point Lifelines' (RBD, 2014) proposal also included green and gray flood protection and measures to protect critical economic assets, including transportation in the region. This competition provided a unique opportunity for engineers to explore the many ideas that were generated to rethink green solutions and truly realize different attributes of flood resiliency of cities/regions.

13.3 FLOOD ANALYSIS

The first step in developing flood planning and management practices is to analyze the frequency of floods of different intensities. In this section, the data and steps needed for flood analysis are described.

13.3.1 Flood Time Series

The measured instantaneous flood peak discharges are one of the most valuable datasets for the hydrologist. The longer a record continues, homogeneous and with no missing peaks, the more is its value enhanced. Even so, it is very rare to have a satisfactory record long enough to match the expected life of many engineering works required to be designed. As many peak flows as possible are needed in assessing flood frequencies. The hydrologist defines two data series of peak flows: the annual maximum series and the partial duration series.

13.3.1.1 Peaks Over Threshold Series

The annual maximum series takes the single maximum peak discharge in each year of record so that the number of data values equals the record length in years. For statistical purposes, it is necessary to ensure that the selected annual peaks are independent of one another. This is sometimes difficult, for example, when an annual maximum flow in January may be related to an annual maximum flow in the previous December. For this reason, it is sometimes advisable to use the water year rather than the calendar year. In hydrology, any 12-month period, usually selected to begin and end during a relatively dry season, used as a basis for processing streamflow and other hydrologic data is called a water year. The definition of the water year depends on the seasonal climatic and flow regimes; for example, the period from October 1 to September 30 is most widely used in the United States. The partial duration series takes all the peaks over a selected level of discharge, a threshold. Hence, the series is often called the "Peaks over Threshold" (POT) series. There are generally more data values for analysis in this series than in the annual series. Still, there is more chance of the peaks being related and the assumption of true independence is less valid.

The annual series and the partial duration series of peak flows result in different probability distributions. Still, for return periods of 10 years and more, the differences are minimal and the annual maximum series is usually analyzed.

13.3.2 Partial Frequency Analysis

The main purpose of partial frequency analysis is to find the design value of extreme events with different return periods when not having a complete set of information. There are several methods for this type of analysis, such as Log Pearson Type III distribution. If log Pearson type III is used, three parameters, and if other types of distributions (symmetric distributions) are used, two parameters are taken into account. The log Pearson type III distribution requires average, standard deviation, and skewness to be fitted. However, other distributions such as normal or log-normal only require average and standard deviation. In this section, the log Pearson distribution is explained, nevertheless, if other distributions are to use, the skewness calculations are eliminated. When intending to carry out calcaulations, first the stationarity or nonstationarity of the data should be examined. If the data are proven to be stationary, stationary analysis is used, otherwise, nonstationary analysis needs to be performed to determine the extreme and design values.

13.3.2.1 Stationary Analysis

In general, when the sample is small, the weighted skew can be used, because in these small samples the accuracy of the calculations is questioned. The total years in which the analysis should be done consist of years the data is available and unavailable. This analysis needs maximum annual extreme events' data in some consecutive years and some years when major historical events had happened. Further, to determine the design values for different return periods. This analysis can be implemented as follows:

1. *Obtaining the design value*: The logarithm of the historically adjusted flood is obtained from the following method:

$$\hat{Y}_{hi} = \overline{Y_h} + K\sigma_h \tag{13.1}$$

Where $\overline{Y_h}$ is the average, σ_h^2 is the standard deviation, and K corresponds to the exceedence percentage. Here, if K is calculated with a 1% exceedence probability, the result would be the 100-year flood stationary water level.

Then, removing the logarithm form, the final number is obtained which is the design value of desired return period.

$$\hat{X}_{hi} = 10^{\hat{Y}_{hi}} \tag{13.2}$$

2. *Weight estimation*:
   1. For each historical flood (Z), a weight of 1 is assigned to each.
   2. For consecutive annual floods (Y), a weight of W is assigned to each of them.

$$W = \frac{H - N_Z}{N_n + L} \tag{13.3}$$

The maximum annual flow for a particular station in consecutive years is N_n, L is the number of zero flows or flows below a measurable base (usually zero), H is the number of years in which the flood should be estimated, and N_z is the number of historical peaks.

3. *Estimation of the moments*: Calculation of mean values, standard deviation, and skewness coefficient are done using Equations 13.4–13.6:

$$\bar{Y}_h = \frac{W\sum_{i=1}^{N_N} Y_i + \sum_{i=1}^{N_Z} Z_i}{H - WL} \tag{13.4}$$

$$\sigma_h^2 = \frac{W\sum_{i=1}^{N_N} \left(Y_i - \bar{Y}_h\right)^2 + \sum_{i=1}^{N_Z} \left(Z_i - \bar{Y}_h\right)^2}{H - WL - 1} \tag{13.5}$$

$$g_h = \frac{H - Wl}{(H - WL - 1)(H - WL - 2)} \times \left(\frac{W\sum_{i=1}^{N_N} (Y_i - \bar{Y}_h)^3 + \sum_{i=1}^{N_Z} (Z_i - \bar{Y}_h)^3}{S_{yh}^3} \right) \tag{13.6}$$

where $\bar{Y}_h$ is the average, σ_h^2 is the variance, g_h is the sample skewness, Y_i refers to data of consecutive years, and Z_i refers to extreme events data.

Weighted skewness should then be calculated using skewness and generalized skewness:

$$g_{hw} = \frac{\text{MSE}_{\bar{g}}(g_h) + \text{MSE}_{g_h}(\bar{g})}{\text{MSE}_{\bar{g}} + \text{MSE}_{g_h}} \tag{13.7}$$

where $\bar{g}$ represents generalized skew and is calculated using the map depicted in Figure 10.5. MSE is the mean square error (variance). See Section 10.3 for more details.

4. *Estimation of skewness coefficient*: Using the skewness map, $\text{MSE}_{\bar{g}}$ is estimated to be equal to 0.302. g_h is the same skewness that was calculated in the previous step. For calculating MSE_{g_h}, which is the standard root of the skew error standard as per Equations 13.8–13.10:

$$A = \begin{cases} -0.33 + 0.08|g_h| \text{ if } |g_h| \le 0.9^0 \\ -0.52 + 0.3|g_h| \quad \text{if} \quad |g_h| > 0.9 \end{cases} \tag{13.8}$$

$$B = \begin{cases} 0.94 - 0.26|g_h| \quad \text{if} \quad |g_h| \le 1.5^0 \\ 0.55 \quad \text{if} \quad |g_h| > 1.5 \end{cases} \tag{13.9}$$

$$\mathrm{MSE}_{gh} = \left[10^{A-B\log_{10}\left(n/10\right)}\right] \tag{13.10}$$

5. *K Estimation*: Then, having the weighted skewness and also the recurrence interval, the K value is obtained to estimate the design value using Table A.1 in the Appendix of this chapter. It should be noted that all of these represented calculations are related to log Pearson type III distribution. If other distributions (symmetric distributions) are required to be used, then the calculations due to the skewness should be removed. By using average, standard deviation, and probability exceedance, the coefficient should be obtained utilizing specific tables related to each distribution.

13.3.2.2 Non-stationary Analysis

In this section, the nonstationary form of calculations is represented. In order to analyzeextreme events, it is better to use nonstationary form of the analysis if nonstationarity was detected in the data. Again, in this section, different distributions (symmetric distributions) can be used with two parameters. But here we aim to present the nonstationary form of log Pearson type III with three parameters. To that end, the average, standard deviation, and weights are obtained using previous equations numbers. However, skewness formula is obtained differently. In this analysis, an explanatory variant is required which is usually considered as time variate.

1. ***Obtaining design values***: Equation 13.11 is used to estimate the probability of nonexceedance of a given return period:

$$\hat{X}_{plw} = \exp\left(\bar{y} + \hat{\beta}\left(t_n - \bar{t}\right) + k_p\sqrt{\hat{\sigma}_y{}^2 - \hat{\beta}^2 \times \hat{\sigma}_w^2}\right) \tag{13.11}$$

   where t_n is the total number of years and Equation 13.8 $\bar{t}$ is the total median of the time period. That demonstrates the design value for an extreme event with a specific return period. If β is considered zero, then the stationary form of this analysis will be obtained.

2. *Finding parameters*: Equations 13.12–13.17 are utilized to find the skewness and the coefficient.

$$\hat{G}_{ylt} = \left(1 + \frac{6}{n}\right) \times \left(\frac{\left(N_n \times W \times \sum_{i=1}^{Nn}\left(y_i - \bar{y}\right)^3\right) + \left(N_z \times \sum_{i=1}^{Nz}\left(Z_i - \bar{y}\right)\right)}{(n-1)(n-2)\hat{\sigma}_y{}^3}\right) \tag{13.12}$$

$$A = \max\left(\frac{2}{\hat{G}_{ylt}} 0.4\right) \tag{13.13}$$

$$B = 1 + 0.0144 \times \max\left(0.\hat{G}_{ylt} - 2.25\right)^2 \tag{13.14}$$

$$F = \hat{G}_{ylt} - 0.063 \times \max\left(0.\hat{G}_{ylt} - 1\right)^{1.85} \tag{13.15}$$

$$H = \left(B - \frac{2}{\hat{G}_{ylt} \times A}\right) \tag{13.16}$$

$$k_p = A \times \left(\max\left(H, 1 - \left(\frac{F}{6}\right)^2 + z_p \times \left(\frac{F}{6}\right)^3 \right) \right) - B \tag{13.17}$$

Here k_p is the three-parameter gamma variable. Equations 13.18 and 13.19 are used:

$$\rho = \frac{\sum_{i=1}^{n} (t_i - \bar{t})(y_i - \bar{y})}{\left[\sum_{i=1}^{n} (t_i - \bar{t})^2 \sum_{i=1}^{n} (y_i - \bar{y})^2\right]^{0.5}} \tag{13.18}$$

$$\beta = \rho \times \frac{\hat{\sigma}_y}{\hat{\sigma}_t} \tag{13.19}$$

where t refers to explanatory variate, which is considered as time in this case.

13.3.2.3 Ungagged Flood Data

Here, the southern part of Brooklyn coastline is chosen as the case study. The data of this area are obtained by using the Delft3D software which is a hydrodynamic model. In this software, FLOW and WAVE are coupled to obtain the water level hydrograph. The Brooklyn's coastline is an ungagged area which does not have water level observed data. Then, utilizing Delft3D leads us to having the water level data in some years. Different years were simulated in this research. First, ten historical hurricanes which affected Brooklyn the most were simulated. Then, some hurricanes in consecutive years were simulated. Finally, their stationarity or nonstationarity should be evaluated. Their characteristics have been shown in Table 1 3.3.

Three of the historical events (Sandy, Irene, and Ernesto) were common in the two tables. Thus, these three events are considered as consecutive year data. Hence, we have seven historical storms and 14 consecutive storms which are utilized to obtain the design values. This leads us to having 21 years of information. The simulated water levels in the case study are shown in Table 13.4.

TABLE 13.3
Names and Characteristics of the Selected Storms

14 Years		10 Historical Hurricanes	
Event Name	**Date of Occurrence**	**Name of Event**	**Date of Occurrence**
Ernesto	31/08/2006 to 04/09/2006	Connie	12/08/1955 to 15/08/1955
Barry	01/06/2007 to 05/06/2007	Donna	11/09/1960 to 14/09/1960
Hannah	05/09/2008 to 08/09/2008	Doria	26/08/1971 to 29/08/1971
Bill	20/08/2009 to 25/08/2009	Belle	08/08/1976 to 10/08/1976
Earl	01/09/2010 to 05/09/2010	Gloria	26/09/1985 to 29/09/1985
Irene	24/08/2011 to 29/08/2011	Bob	16/08/1991 to 19/08/1991
Sandy	26/10/2012 to 31/10/2012	Floyd	15/09/1999 to 17/09/1999
Andrea	06/06/2013 to 08/06/2013	Ernesto	31/08/2006 to 04/09/2006
Arthur	02/07/2014 to 07/07/2014	Irene	24/08/2011 to 29/08/2011
Joaquin	03/10/2015 to 08/10/2015	Sandy	26/10/2012 to 31/10/2012
Hermine	02/09/2016 to 05/09/2016		
Nate	08/10/2017 to 11/10/2017		
Michael	11/10/2018 to 13/10/2018		
Dorian	06/09/2019 to 09/09/2019		

TABLE 13.4
Maximum Water Levels of the Selected Storms

14 Years			10 Historical Hurricanes		
Name	Date	Maximum Water Level (m)	Name	Date	Maximum Water Level (m)
Ernesto	31/08/2006 to 04/09/2006	0.858	Connie	12/08/1955 to 15/08/1955	1.012
Barry	01/06/2007 to 05/06/2007	1.075	Donna	11/09/1960 to 14/09/1960	2.105
Hannah	05/09/2008 to 08/09/2008	0.967	Doria	26/08/1971 to 29/08/1971	0.978
Bill	20/08/2009 to 25/08/2009	1.034	Belle	08/08/1976 to 10/08/1976	1.434
Earl	01/09/2010 to 05/09/2010	1.003	Gloria	26/09/1985 to 29/09/1985	1.798
Irene	24/08/2011 to 29/08/2011	2.011	Bob	16/08/1991 to 19/08/1991	1.254
Beryl	27/05/2012 to 02/06/2012	1.291	Floyd	15/09/1999 to 17/09/1999	1.135
Andrea	06/06/2013 to 08/06/2013	1.032	Ernesto	31/08/2006 to 04/09/2006	0.858
Arthur	02/07/2014 to 07/07/2014	0.697	Irene	24/08/2011 to 29/08/2011	2.011
Joaquin	03/10/2015 to 08/10/2015	1.09	Beryl	27/05/2012 to 02/06/2012	1.291
Hermine	02/09/2016 to 05/09/2016	1.08			
Nate	08/10/2017 to 11/10/2017	1.133			
Michael	11/10/2018 to 13/10/2018	1.414			
Dorian	06/09/2019 to 09/09/2019	0.931			

It should be noted that Superstorm Sandy is replaced by another storm as it was chosen as outlier due to the following explanations: The USWRC method suggests that outliers be identified and adjusted according to their proposed methods. Data points that deviate significantly from the remaining data process are called outliers. Maintaining or eliminating these outliers can significantly impact the amount of statistical parameters calculated from the data, especially for small samples. The methods of dealing with outliers require judgment, which is considered both mathematically and hydrologically. Equation 13.20 is used to determine large outliers:

$$\overline{y}_H = y + K_N S_y \tag{13.20}$$

where $\overline{y}_H$ is a large outlier, y is the mean of the dataset, S_y is the standard deviation, and K_N is obtained from Table A.2 in this chapter's Appendix. Hence, using the previous equation:

$$\overline{y}_H = y + K_N S_y = 0.0433 + 2.408 \times 0.108 = 0.304$$

$$\rightarrow 10^{0.304} = 2.1144 \text{ m}$$

So, Sandy with 3.46 m water level is an outlier and another storm from that year with the peak water level is simulated.

At this point, by using equations explained in the partial frequency analysis section and the simulated data, the design value of different return periods is obtained.

In this section, $N=72$ years, $H=14$ years, $Z=7$ years, and $L=0$ are considered. The weight of each part is obtained (it should be noted that the weight is considered equal to one for seven historical events). The results of these weights can be seen in the fourth column of Table 13.5.

Now, the stationary calculation is demonstrated if the data is shown to be stationary.

After calculating the weight of each event, it is time to calculate the mean, standard deviation, and skewness.

$$\overline{Y}_h = \frac{W\sum_{i=1}^{n} Y_i + \sum_{i=1}^{n} Z_i}{H - WL} = \frac{4.642 \times 0.481 + 0.883}{72} = 0.0437$$

TABLE 13.5
Values of the Weights Corresponding to Each Year

	Year	Y = Water Level (m)	Log (Y) = X	Weight = W
Zi	1960	2.105	0.323	1
	1985	1.798	0.255	1
	1976	1.434	0.157	1
	1991	1.254	0.098	1
	1999	1.135	0.055	1
	1955	1.012	0.005	1
	1971	0.978	−0.01	1
Yi	2011	2.011	0.303	4.071
	2018	1.414	0.15	4.071
	2012	1.291	0.111	4.071
	2017	1.133	0.054	4.071
	2015	1.09	0.037	4.071
	2016	1.08	0.033	4.071
	2007	1.075	0.031	4.071
	2009	1.034	0.015	4.071
	2013	1.032	0.014	4.071
	2010	1.003	0.001	4.071
	2008	0.967	−0.015	4.071
	2019	0.931	−0.031	4.071
	2006	0.858	−0.067	4.071
	2014	0.697	−0.157	4.071

$$S_{yh}^2 = \frac{W\sum_{i=1}^{N}\left(Y_i-\bar{Y}_h\right)^2+\sum_{i=1}^{n}\left(Z_i-\bar{Y}_h\right)^2}{H-WL-1} = \frac{4.642\times 0.148+0.143}{72-1} = 0.011$$

$$S_{yh} = 0.108$$

$$g_h = \frac{H-Wl}{(H-WL-1)(H-WL-2)} \times \left(\frac{W\sum_{i=1}^{N}\left(Y_i-\bar{Y}_h\right)^3+\sum_{i=1}^{n}\left(Z_i-\bar{Y}_h\right)^3}{S_{yh}^3}\right)$$

$$= \frac{72}{71\times 70} \times \left(\frac{4.642\times 0.009+0.032}{0.108^3}\right) = 0.876$$

According to the map, generalized skew in the Brooklyn area (New York) is 0.7. Then MSE_{g_h} can be calculated as :

$$A = \begin{cases} -0.33+0.08|g_h| & \text{if } |g_h| \le 0.9 \\ -0.52+0.3|g_h| & \text{if } |g_h| > 0.9 \end{cases}$$

$$A = -0.33+0.08*0.876 = -0.259$$

$$B = \begin{cases} 0.94-0.26|g_h| & \text{if } |g_h| \le 1.5 \\ 0.55 \; if \; |g_h| > 1.5 \end{cases}$$

TABLE 13.6
Calculation of the Water Level for each Return Period

Probability	K	s×K	Log (Y)	Y (m)
95	−1.423	−0.154	−0.11	0.776
50	−0.116	−0.013	0.031	1.075
20	0.79	0.085	0.129	1.346
10	1.333	0.144	0.188	1.541
4	1.967	0.213	0.256	1.805
2	2.407	0.26	0.304	2.014
1	2.857	0.309	0.353	2.253

$$B = 0.94 - 0.26 \times 0.876 = 0.712$$

$$\mathrm{MSE}_{gh} = \left[10^{A - B\log_{10}\left(n/10\right)}\right]^{0.5} = 0.569$$

Thus, the weighted skewness is as follows:

$$g_{hw} = \frac{\mathrm{MSE}_{\bar{g}}(g_h) + \mathrm{MSE}_{g_h}(\bar{g})}{\mathrm{MSE}_{\bar{g}} + \mathrm{MSE}_{g_h}} = \frac{0.302 \times 0.876 + 0.569 \times 0.7}{0.302 + 0.569} = 0.761$$

Therefore, having this skewness and the desired probability of exceedance, the height of the water level for different return period is obtained and shown in Table 13.6. The 100-year value is 2.253:

From this point forward, the nonstationary calculations are represented if the data prove to be nonstationary.

The average and standard deviations are as before; however, other calculations are different.

$$\bar{t} = \frac{n+1}{2} = \frac{21+1}{2} = 11$$

$$\hat{\sigma}_t = \left[\frac{n(n+1)}{12}\right]^{0.5} = \left[\frac{21 \times 22}{12}\right]^{0.5} = 6.055$$

$$\rho = -0.11$$

$$\beta = \rho \times \frac{\hat{\sigma}_y}{\hat{\sigma}_t} = -0.005$$

Then, the skewness is obtained as much as 0.75, and K_p equals to 3.26. Finally, the design value of 100 year return period is 2.44 m for the nonstationary assumption of water level data for Brooklyn's southern coastline. Afterward, for obtaining the 100-year water level hydrograph, a storm is considered as reference storm. Here we chose Irene as the reference hydrograph (Figure 13.7). Then, the design value of the 100 years obtained from partial frequency analysis was proportionally superimposed on hydrograph of hurricane Irene to obtain the 100-year water level hydrograph (Figure 13.8).

13.3.3 Testing Outliers

The USWRC method suggests that outliers be identified and adjusted according to their recommended methods. Data points that depart significantly from the trend of the remaining data are called outliers. The preservation or removal of these outliers can considerably affect the magnitude

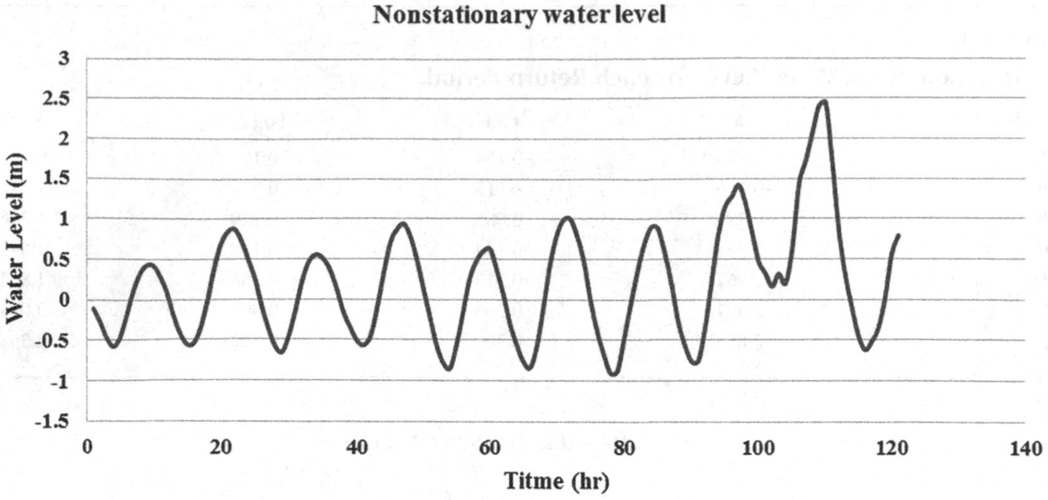

FIGURE 13.8 The 100-year nonstationary water-level hydrograph based on Hurricane Irene template for southern shore of NYC.

of statistical parameters computed from the data, particularly for small samples. Procedures for treating outliers need judgment involving both mathematical and hydrologic considerations. Tests for high outliers are considered first if the station skew is greater than +0.4 according to the USWRC (1981). Otherwise, tests for low outliers should be applied before removing any outliers from the dataset.

Equation 13.21 can be used to identify high outliers:

$$\overline{y}_H = y + K_N S_y \tag{13.21}$$

where $\overline{y}_H$ is the high outlier threshold in log units, K_N is the K value from Table A.2 in this chapter's Appendix for sample size N, and S_y is the standard deviation. High outliers are the peaks in the sample when they are greater than $\overline{y}_H$ in Equation 13.21. Then they should be compared with historic flood data and flood information in close sites. According to the USWRC (1981), if information is available indicating that a high outlier is the maximum over an extended period of time, the outlier is treated as historic flood data. If useful historic information is not available to adjust for high outliers, then the outliers should be retained as part of the record. Equation 13.22 can be used to detect low outliers:

$$\overline{y}_L = y - K_N S_y \tag{13.22}$$

where $\overline{y}_L$ is the low outlier threshold in log units. Flood peaks considered low outliers are removed from the record and a conditional probability adjustment explained in USWRC (1981) is applied. Use of the K values in Table A.2 is equivalent to a one-sided test that detects outliers at the 10% level of significance. The K values are based on a normal distribution for detection of single outliers.

Example 13.1

The mean, standard deviation, and skewness coefficient for a gauge are 3.338, 0.653, and −0.3, respectively. Compute the 25-year and the 100-year peak discharges using the USWRC guideline. The map skewness coefficient for this location is −0.015.

Solution:

In this example, $\overline{y} = 3.338$, $S_y = 0.653$, and $G_s = -0.3$.

Step 1: Compute A, B, and $V(G_s)$ using Equations 13.21, 13.22, and 13.20.

$$A-0.33+0.08|-0.3|=-0.306$$

$$B=0.94-0.26|-0.3|=0.86$$

$$V(G_s)=10^{A-B\left[\log\left(\frac{n}{10}\right)\right]}=10^{-0.306-0.864\log\left(\frac{47}{10}\right)}=0.13.$$

Step 2: Use Equation 13.19 to compute the weighted skewness coefficient using $V(G_m)=0.302$ [as estimated in USWRC (1981)]:

$$G_w=\frac{0.302(-0.3)+0.130(-0.015)}{0.302+0.130}=-0.12.$$

Step 3: Use Table A.1 in this chapter's Appendix (K_T values for Pearson Type III distribution) to obtain the frequency factor equation to determine Q_{25} and Q_{100}:

$$\log Q_{25}=\bar{y}+K(25-0.125)S_y=3.338+1.676(0.653)=4.432 \qquad Q_{25}=27{,}070\text{ cfs}$$

$$\log Q_{100}=3.338+2.171(0.653)=4.432 \qquad Q_{100}=56{,}975\text{ cfs}$$

13.4 FLOOD RECURRENCE INTERVAL

The concept of the return period states that the T-year event/flood, Q_T, is the average chance of exceedance once every T years over a long record. However, it is often required to know the actual probability of exceedance of the T-year flood in a specific period of n years. So if a flood protection is designed based on that return period, the risk of failure could be estimated.

$F(X)$ is the probability of X not being equaled or exceeded in any 1 year. Assuming that annual maxima are independent, the probability of no annual maxima exceeding X in the whole n years is given by $(F(X))^n$. The probability of exceedance of X at least once in n years is then $1-(F(X))^n$, which may be signified by Equation 13.23:

$$P_n(X)=1-(1-C\,P(X))^n. \tag{13.23}$$

This is also called the risk of failure in n years and 1 minus that could be considered reliable for not failing in n years. Thus, for a return period of 100 years and $n=100$ years, the following is obtained:

$$P_n(X)=P(Q_{100}\geq X)=1-\left(\frac{99}{100}\right)^{100}=0.634. \tag{13.24}$$

This means that the chance of the 100-year flood being equaled or exceeded at least once in 100 years is 63.4%; that is, there is roughly a two in three chance that a 100-year event in any given 100-year record will happen (and a one in three chance/reliability that it will not). Rearranging Equation 13.24 gives:

$$n=\frac{\log\left[1-P_n(X)\right]}{\log\left[\frac{T(X)-1}{T(X)}\right]}. \tag{13.25}$$

Thus, for a required probability, $P_n(X)$, and return period, T, the length of record, n years, in which the T-year event of one or more equal or higher magnitude has the probability $P_n(X)$, which can be calculated. Such information is essential to engineers designing flood protection structures.

13.5 FLOOD ROUTING

A flood hydrograph is modified in two ways as the stormwater flows downstream. First, and obviously, the time of the peak rate of flow occurs later at downstream points. This is known as translation. Second, the magnitude of the peak rate of flow is diminished at downstream points, the shape of the hydrograph flattens out, and the volume of floodwater takes longer to pass a lower section. This modification to the hydrograph is called attenuation (Figure 13.9).

The derivation of downstream hydrographs like B in Figure 13.9 from an upstream known flood pattern A is essential for river managers concerned with forecasting floods in the lower parts of a river basin. The design engineer also needs to be able to route flood hydrographs to assess the capacity of reservoir spillways, design flood protection schemes, or evaluate the span and height of bridges or other river structures. In any situation where it is planned to modify the channel of a river, it is necessary to know the likely effect on the shape of the flood hydrograph in addition to that on the peak stage, that is, the whole hydrograph of water passing through a section, not just the peak instantaneous rate.

Flood routing methods may be divided into two main categories, differing in their fundamental approaches to the problem. One category of methods uses the principle of continuity and a relationship between discharge and the temporary storage of excess volumes of water during the flood period. The calculations are relatively simple and reasonably accurate and, from the hydrologist's point of view, generally give satisfactory results. The second category of methods, favored by hydraulic engineers, adopts the more rigorous equations of motion for unsteady flow in open channels. Still, complex calculations, assumptions, and approximations are often necessary, and some of the terms of the dynamic equation must be omitted in certain circumstances to obtain solutions. In coastal flooding, a distributed model should be used, classified as hydrodynamic models such as Gridded Surface Subsurface Hydrologic Analysis (GHHSA) developed by USCOE. The model and the governing equations are discussed later in this chapter.

13.5.1 Storage-Based Routing

Routing techniques can be classified into two major categories: simple hydrologic routing and more complex hydraulic routing. Hydraulic routing is more complex and accurate than hydrologic routing. It is based on the solution of the continuity equation and the momentum equation for unsteady flow in open channels. These differential equations are usually solved by explicit or implicit numerical methods known as the Saint Venant equations.

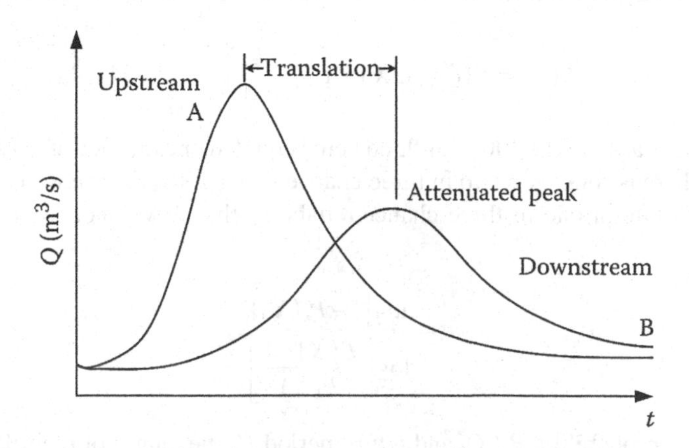

FIGURE 13.9 Flood peak translation and attenuation.

Hydrologic routing involves balancing inflow, outflow, and volume of storage through the continuity equation. A second relationship, the storage–discharge relation, is also required between outflow rate and storage in the system. Applications of hydrologic routing techniques to problems of flood prediction, flood control measures, reservoir design and operation, watershed simulation, and urban design are numerous. Examples of hydrologic river routing and hydrologic reservoir routing techniques can be Muskingum and storage indication methods, respectively. In flood remediation efforts in urban coastal areas through the use of BMPs such as constructed wetlands, storage-based routing is utilized to attenuate the flood peak volume and time.

The storage in the reaches of stream channels is often used as an index of the timing and shape of flood waves at successive points along the path of a river. Among the most prominent users of the technique are the US Army Corps of Engineers who route hypothetical floods through river systems to determine the effectiveness of proposed flood-control projects and strategies, the US Weather Bureau whose forecasts of river stages are based primarily on flood routing and the operators of hydroelectric power systems who schedule their operations according to the projected progress of a flood wave. US Geological Survey is another user of such methods. The primary use of these methods in the Geological Survey is in testing and enhancing the overall consistency of records of discharge during major floods in river basins. The number of direct observations of discharge during such flood periods is generally limited by the short duration of the flood and the inaccessibility of certain stream sites. Through the use of flood-routing techniques, all observations of discharge and other hydrologic events in a river basin may be combined and used to evaluate the discharge hydrograph at a single site.

13.6 URBAN FLOODS

Through the development of new urban areas, infiltration reductions and water retention characteristics of the natural land should be considered. The increasing quantity and the short duration of drainage produce high flood peaks that exceed the conveyed secure capacities of urban streams. The total economic and social damage in urban areas tends to impact the national economies as flood frequencies increase significantly. Recent events even in North America have been devastating in urban areas such as New Orleans.

With uncontrolled urbanization in the floodplains, a sequence of small flood events could be managed, but with higher flood levels, damage increases and the municipalities' administrations have to invest in population relief. The costs of applying structural solutions for flood mitigation are higher, but they are commonly used when damage is greater than development costs or because of intangible social aspects. Nonstructural measures have lower costs but with less chance to be implemented because they are not politically attractive in developing regions (Tucci, 1991).

13.6.1 Urban Flood Control Principles

Certain principles of urban drainage and flood control are proposed by Tucci (1991):

- Promoting an urban drainage master plan
- Involvement of the whole basin in evaluating flood control
- Any city developments should be within urban drainage control plans
- Priority is given to source control and keep the flood control measures away from downstream reaches
- Application of nonstructural measures for flood control such as flood warning, flood zoning, and flood insurance
- Public participation in urban flood management
- Consideration of full recovery investments

In developing countries, a set principle for urban drainage practices has not been established due to fast and unpredictable developments (Dunne, 1986). Furthermore, the local regulations are neglected in urbanization of preurban areas such as unregulated developments and invasion of public areas.

In many developing urban areas, the above principles are not incorporated because of the following:

- Insufficient funds
- Lack of appropriate waste collection and disposal system resulting in decrease of water quality and the capacity of the urban drainage network due to filling
- Absence of a preventive program for risk area occupation
- Lack of adequate knowledge on how to deal with floods
- Lack of institutional organization at a municipal level for urban drainage management

One of the most common problems facing a practicing civil engineer is the estimation of the hydrograph of the rise and fall of a river at any given point on the river during a flood event. The problem is solved by the techniques of flood routing, which is the process of following the behavior of a flood hydrograph upstream or downstream from one point to another point on the river. The challenge is to simulate the water level and inundated areas in ungagged locations in urban distributed areas. Nearshore models such as Delf3D and inland models such as GSSHA and Lisflood are used in coastal urban areas.

13.7 UNDERSTANDING FLOOD HAZARDS

Flooding is the result of a combination of meteorological and hydrological conditions, some of which are indicated in Table 13.7. These conditions are influenced by human factors. These influences are very different but they commonly increase flood hazards by accentuating flood peaks. Thus, flood hazards are considered to be the consequence of natural and man-made factors.

13.7.1 Climate Change and Flooding

Floods have always been a source of major concerns to human societies in different regions. Although there has been significant improvement in understanding floods and technological means

TABLE 13.7
Factors Contributing to Flooding

Meteorological Factors	Hydrological Factors	Human Factors
• Rainfall • Cyclonic storms • Small-scale storms • Temperature • Snowfall and snowmelt	• Soil moisture level • Groundwater level prior to storm • Natural surface infiltration rate • Presence of impervious cover • Channel cross-sectional shape and roughness • Presence or absence of over bank flow, channel network • Synchronization of runoffs from various part of watershed • High tide impeding drainage	• Land-use changes (e.g., surface sealing due to urbanization, deforestation) increase runoff and maybe sedimentation • Occupation of the floodplain obstructing flows • Inefficiency or non-maintenance of infrastructure • Too efficient drainage of upstream increases flood peaks • Climate change affects magnitude and frequency of precipitations and floods • Urban microclimate may enforce precipitation events

Source: APFM, *Flood Management Tools Series,* 2008. With permission.

to deal with them, people still suffer from severe floods. Floods cause damage to different infrastructures such as settlements, roads, and railways and destroy human heritage. Climate change has a potential to increase the frequency of extreme events as well as flooding in many regions of the world. In coastal area, sea level rise exacerbates the intensity and consequences.

Global warming may result in intensifying many subsystems of the global water cycle. This can increase the flood magnitude as well as flood frequency, making weather less predictable and increasing precipitation uncertainty. Some other factors, such as urbanization may increase thunderstorm activity because of heat islands and create local air circulation. Dust particles that circulate act as nuclei for condensation of the moisture in clouds, forming rain droplets. These droplets can eventually develop into large raindrops in a major thunderstorm.

The Intergovernmental Panel on Climate Change, IPCC, has carried out several studies on the potential impacts of climate change on flooding. These studies show 15% potential increase in flood peaks in temperate zones due to increased storm activity and precipitation depth. Currently, it is not possible to predict potential increases in flood peaks due to climate change for specific basins with the level of accuracy needed for design and planning purposes, but it seems that the freeboard on levees and other flood control structures can accommodate the potential changes in extremes due to climate change, if some changes are made in operating procedures of control structures.

13.7.2 Sea Level Rise and Storm Surge

As a result of climate change impacts, sea-level rise is an added factor that could intensify the inverse flood impacts. Sea-level rise increases the risk of coastal flooding, specifically in the case of storm surges. The sea-level projections by the 2080s show that millions of people are in areas with flooding potential due to sea-level rise. The risk is higher in densely populated, low-lying areas with low adaptive capacity.

Coastal communities should consider sea level rise, tsunamis, and ocean storm surge in preparing and dealing with flooding events. As an example, sea-level rise results in decreased river slopes at the upstream of river entries to the ocean. This decreases the secure channel capacity of passing flood flows and increases coastal cities' elevation. Due to the slow rate of sea level rise, the long-term protective works or floodplain management exercises are planned to incorporate the predicted rise. Some studies have indicated that there is potential for increased frequency of storm surges, which result from high winds and increased barometric pressure. Tsunamis can also be devastating natural disasters and must be considered like flooding. Forecasting and emergency responses to these events must be based on acceptable risk and advance planning principles.

13.8 EVACUATION ZONES

To deal with floods and prevent loss of human life, evacuation zones are defined in the coastal regions. Evacuation zones in different coastal regions are determined based on (a) vulnerability of respondents, (b) types of communities, (c) demographic characteristics of the populations, (d) sample size, (e) type of housing, and (f) time between the evacuation and the flood occurrence (Nelson et al., 1989).

The purpose of a regional evacuation study is to provide emergency management officials with realistic data by quantifying the major factors in hurricane evacuation decision making. These data are provided as a framework of information that counties can use to update and revise their hurricane evacuation plans and operational procedures to improve their response to future hurricane threats. It should be noted that hurricane evacuation zones are different from flood insurance risk zones designated by Federal Emergency Management Agency (FEMA) and published in the form of Flood Insurance Rate Maps.

Each zone will be evacuated with the evacuation maps depending on the hurricane's track and projected storm surge. For example, in New York City (NYC), hurricane contingency plans are

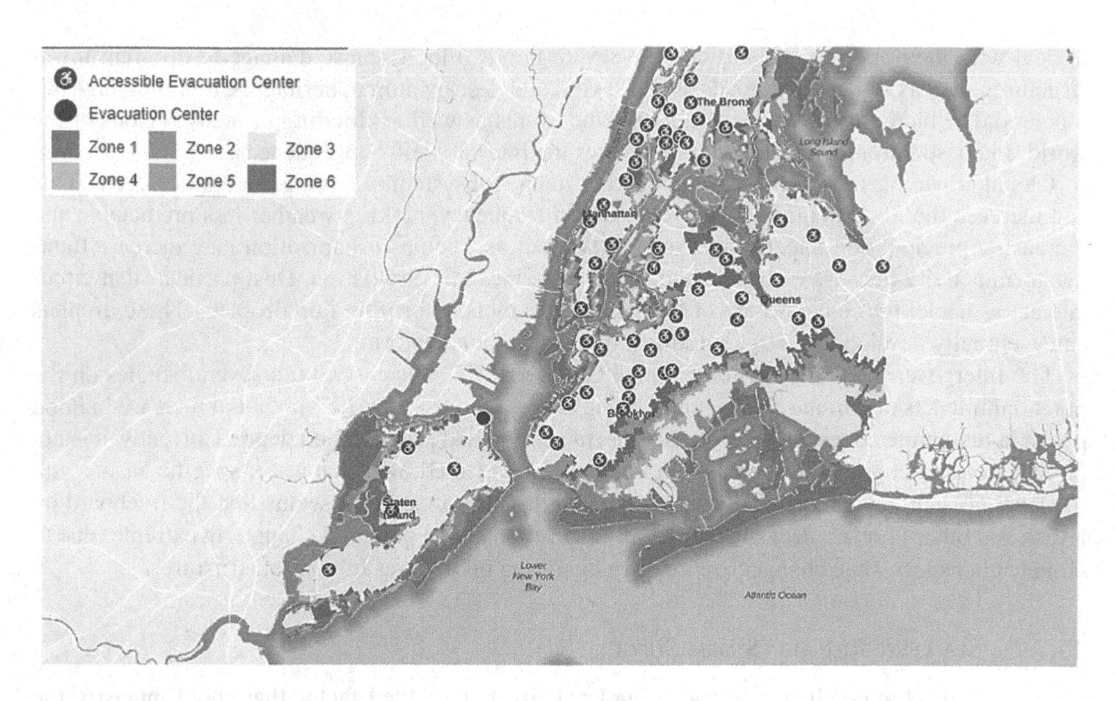

FIGURE 13.10 New York City evacuation zone. (From https://maps.nyc.gov/hurricane/#.)

based on three evacuation zones (Figure 13.10). These zones represent varying threat levels of coastal flooding resulting from storm surge. Manhattan and its surrounding areas have been sliced up into Zones A, B, and C depending on the strength of hurricanes. Residents in Zone A were asked to evacuate a day before hurricane Irene hit NYC. They were facing the highest risk of flooding from a hurricane category that became a storm surge when it reached NYC. Zone A includes all low-lying coastal areas and other areas that could experience storm surge from a hurricane making landfall close to NYC. Residents in Zone B may experience storm surge flooding from a moderate (category 2 and higher) hurricane. Residents in Zone C may experience storm surge flooding from a major (category 3 and 4) hurricane making landfall just south of NYC. A major hurricane is unlikely in NYC, but not impossible. Zone A in this region is determined based on 1% per year probability of flooding plus 26% probability (risk) of failure in 30 years. This is equivalent to another 1% per year due to insurance and home considerations.

Residents in the Zone A region are at the highest risk of flooding from a storm surge. Almost any hurricane or near-hurricane-type event making landfall in NYC could cause these areas to flood, requiring evacuation.

In order to reduce flood impacts, current flood risk management strategies and floodplains and evacuation zones, which are often outdated by a couple of decades, need a major overhaul and should be reassessed to meet the public's expectation of accuracy, responsiveness, and safety and security. The concept of resilience is used in various disciplines and has become multi-interpretable. Resilience theory offers insights into the behavior of complex systems and the importance of system criteria such as system memory and self-organization. Resilience is a measure of how our municipalities and emergency response unit are ready to resume a close-to-normal operation in a disaster area.

The schematic of the overlay of inland and coastal flood impacts is given in Figure 13.11. Based on this figure, the overlay of inland and coastal floods resembles the most severe flood zones. The results of these events should be analyzed considering the conjunctive impacts of these floods in addition to each event individually. A combined probability distribution should be considered to analyze the impact of simultaneous inland and coastal floods on an urban area.

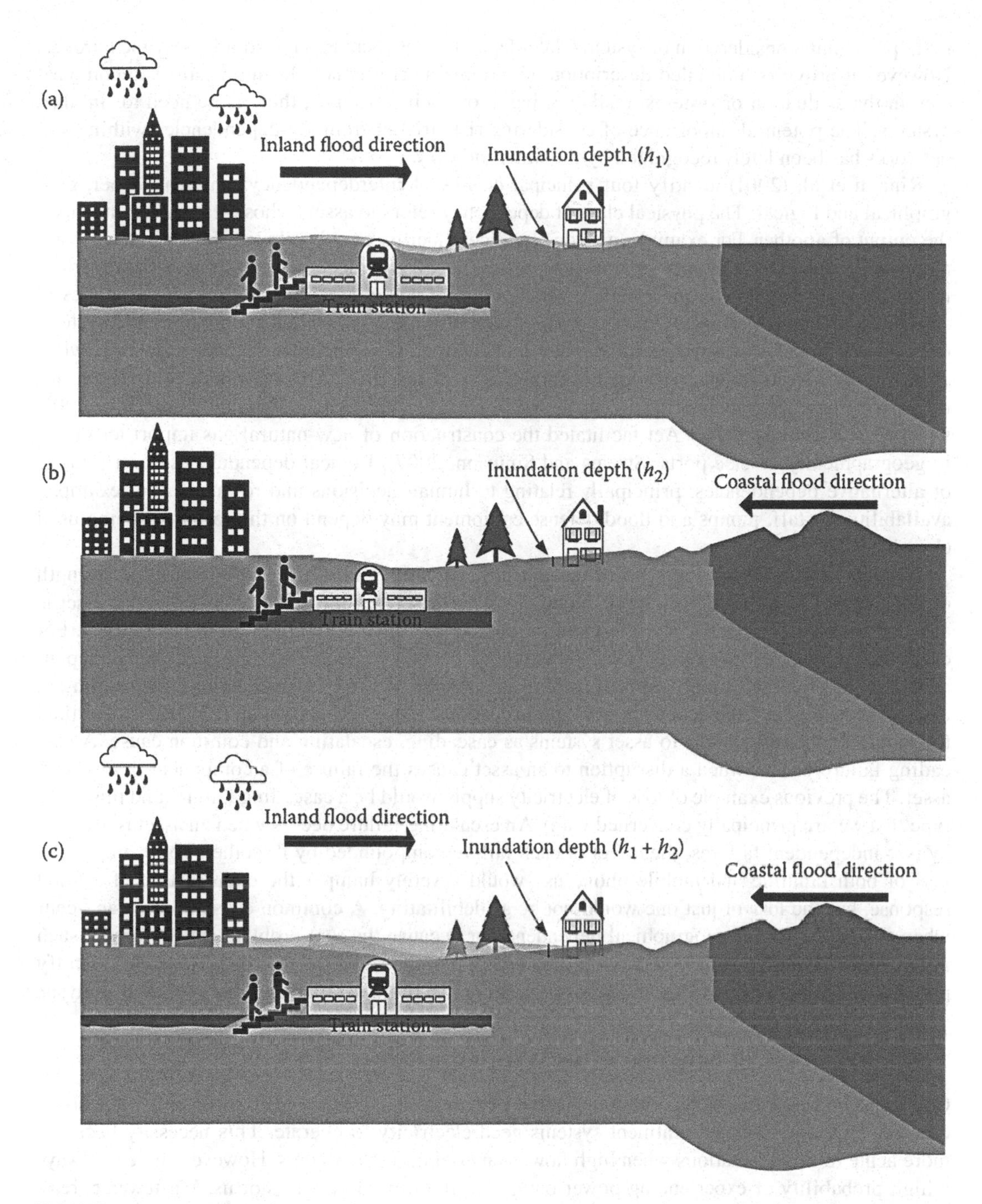

FIGURE 13.11 (a) Inland flood; (b) coastal flood; and (c) the integration of inland and coastal floods' impacts on a region.

13.9 INTERDEPENDENCIES ROLE ON WATER INFRASTRUCTURE PERFORMANCE

Recent flood events such as hurricane Katrina and superstorm Sandy in the US, have demonstrated the role of interdependencies in cascading failures (Leavitt and Kiefer, 2006; Comes and Van de Walle, 2014; Sharkey et al., 2015). Interdependency is a linkage connecting two infrastructures in a way that the state of one infrastructure influences the state of the others (Rinaldi et al., 2001).

In particular, consideration of systems' interdependencies increases the simulation time burdens. However, it provides a detailed description of systems' performance during hazards. In comparison, in the evaluation of systems' resilience based on their attributes, there is no need to simulate systems. The potential importance of considering risk arising from the dependencies within asset networks has been lately recognized by the water industry.

Rinaldi et al. (2001) identify four principal classes of interdependency: physical, cyber, geographical and logical. The physical class of dependency refers to assets whose state is dependent on the output of another. For example, a sludge treatment facility may be physically dependent on the material produced by a sewage treatment works (STW). Cyber dependencies exist if the functioning of an asset depends on information infrastructure. Such a dependency particularly applies to automatically controlled assets, such as drinking water supplies controlled by the water grid system, or flood warning systems triggered by water level gauges. Geographical dependencies exist when a local event affects assets that together form a critical resource. An example of efforts to minimize risk from geographical dependencies was seen after Hurricanes Rita and Katrina in 2005, when the US Energy Policy Act facilitated the construction of new natural gas import terminals in geographically diverse ports (Streips and Simpson, 2007). Logical dependencies cover a range of alternative dependencies, principally relating to human decisions and resources. For example, availability of staff, pumps and flood defense equipment may depend on the resources being used elsewhere.

As well as classifying the type of dependency, Rinaldi et al. (2001) referred to the strength of dependency as a "tight" or "loose" coupling. This describes the speed with which an asset is affected by a failed dependency: the more immediate the impact, the tighter the dependency. For example, a water treatment works (WTW) without a backup generator might have a "tight" dependency on the electrical supply. Still, if there is a generator on site that can keep the site running in case of electricity cut, the dependency would instead be "loose." Rinaldi et al. (2001) also classified the nature of the disruptions to asset systems as cascading, escalating and common cause. A cascading failure occurs when a disruption to an asset causes the failure of a component in a second asset. The previous example of loss of electricity supply would be a cascading failure (and this is the type that we are principally concerned with). An escalating failure occurs when an asset is affected by two independent failures, the effect of each failure compounded by the other. For example, the loss of both landline and mobile phone use would severely hamper the coordination of a flood response, but the loss of just one would not be as debilitating. A common cause failure can occur when two assets have a geographical dependency or because the root problem is widespread, such as two WTW within a supply network failing because of the same storm event. This helps identify potentially critical assets within the network, leading to a more detailed and more focused analysis.

13.9.1 Resiliency of New York City's Wastewater System

One important aspect which is often neglected in examining systems' resilience is interdependencies among them. Sewage treatment systems need electricity to operate. This necessity becomes more acute in flood situations when high flow is entered into the systems. However, there is always a high probability of experiencing power outages at the time of severe storms. Wastewater treatment plants (WWTPs) are usually equipped with backup power to support their performance in emergencies. However, due to the high cost of backup generators, the total electrical power capacity required is not provided. Therefore, the number and capacity of generators in critical situations play a significant role in the rapid recovery.

Furthermore, the safety of substations inside the plants and electrical feeders against floods should also be considered because high inundation depth at the unit substations, main plant switchgear, and transformers may also result in complete shutdown of the plants. Their flood resilience has also been assessed based on historical observation of disruption to electricity connected to WWTPs. This number is set to zero and one, which one indicates disruption during storm. All the above-mentioned factors are specified into five sub-criteria in Table 13.8 and are assumed to represent water-energy interdependencies.

Assessing the resilience of NYC's WWTPs, concerning interdependencies, shows better agreement with NYCDEP reports after Superstorm Sandy. This study demonstrates the significant value of resilience-based funds considering interdependencies and uncertainties related to different views of entities and analyses. See Karamouz and Ansari (2020) for a resiliency based fund allocation metrics.

TABLE 13.8
The Four Criteria and the 23 Subcriteria, Karamouz and Ansari (2020)

Criteria	ID	Subcriteria Description	Unit
Rapidity	Ra_1	Hurricane flood elevation (based on North American Vertical Datum of 1988 (NAVD88))	ft
	Ra_2	Adverse environmental impacts on the surrounding area (due to treatment failure caused by flooding)	–
	Ra_3	Plant design capacity	MGD[a]
	Ra_4	Poststress recovery	Hour
	Ra_5	Population served (number of users served by the plant)	#
	Ra_6	Untreated or semitreated effluent discharge	MG
	Ra_7^b	Average electricity consumed by the plant	KWH[c]
	Ra_8^b	Plant backup generator capacity	KWH
Robustness	Ro_1	Additional load in time of flooding (the difference between WWTP capacity for the total maximum wet and dry weather flow. Maximum wet weather flow is the maximum flow received during any 24-hour period. Maximum dry weather flow is the maximum daily flow during periods without rainfall.)	MGD
	Ro_2	Critical flood elevation (100-year flood elevation +30 inches for expected sea-level rise by the 2050s, which is determined based on the Federal Emergency Management Agency's new advisory base flood elevation maps for a 100-year flood event, was selected as the baseline for the analysis)	ft[d]
	Ro_3	Maximum inundation depth (due to the flat terrain of the plant, several areas may be flooded by up to this value of water during the critical flood event)	ft
	Ro_4	Percent of not-at-risk equipment (percent of plant items that are not at risk of damage during flood)	%
	Ro_5	DMR violations (the percentage of discharge monitoring reports that resulted in effluent violations. During minimal levels of stress, the DMR violation percentages are indicative of how well each treatment plant can cope with daily operational stresses)	%
	Ro_6	Damage cost from the most severe historical hurricane (without flood protection for the plant)	$
	Ro_7^b	100-year flood inundation depth at unit substation	ft
	Ro_8^b	Experiencing power loss during storms	–
Resourcefulness	Rs_1	Number of plant technical staff	#
	Rs_2	Availability of dewatering facilities (facilities to drain sludge to decrease 90% of its liquid volume)	–
	Rs_3	Total risk avoided for every single dollar spent over 50 years	$
Redundancy	Rd_1	Existence of underground tunnel systems	–
	Rd_2	Availability of WWTPs in the neighboring areas (distance from the closest WWTP)	km
	Rd_3	On-site storage (volume of lakes in the WWTP's zone)	ft^3
	Rd4[b]	Number of backup generators	#

[a] Million gallons per day (US liquid gallon = 3.78 L).
[b] Sub-criteria representing water-energy interdependencies.
[c] Kilowatt hour. [d] Foot = 0.3048 m.

13.10 FLOOD DAMAGE

Flood vulnerability (the severity of the damage) depends on the flood type and characteristics such as flow depth, velocity, quality, duration, and sediment load. In rural and undeveloped areas, flood damage is mostly loss of agricultural production, but the damage in the urban region is more complex. Flood damage can be categorized as follows:

- *Direct losses:* These losses result from direct contact of floodwater with buildings, infrastructure, and properties.
- *Indirect losses*: These losses are event consequences but not direct impacts. Examples of indirect losses are transport disruption, business losses that cannot be made up, and family income losses.
- Both direct and indirect losses include two subcategories: tangible and intangible losses.
- *Tangible losses*: These refer to loss of things with a monetary (replacement) value such as buildings, livestock, and infrastructure.
- *Intangible losses*: These refer to loss of things that cannot be bought and sold. Examples of intangible losses are lives and injuries, heritage items, and memorabilia.

Various techniques have been used to calculate direct damage. Grigg and Helweg (1975) used three categories of techniques: aggregate equations, historical damage curves, and empirical depth damage curves. One of the more familiar aggregate equations is that suggested by James (1972):

$$C_D = K_D U M_S h A, \tag{13.26}$$

where C_D is the flood damage cost for a particular flood event, K_D is the flood damage per foot of flood depth per dollars of market value of the structure, U is the fraction of floodplain in urban development, M_S is the market value of the structure inundated in dollars per developed area, h is the average flood depth over the inundated area in feet, and A is the area flooded in acres. Eckstein (1958) presented the historical damage curve method in which the historical damage of floods is plotted against the flood stage.

The US FEMA has developed a program HAZUS for damage estimations. FEMA's HAZUS provides researchers and engineers with standardized data and tools necessary to estimate risk from such as earthquakes, tsunamis, hurricanes, and floods. HAZUS model put together the expertise from many disciplines in order to generate actionable risk information that enhances community resilience. HAZUS program is distributed as a Geographical Information System (GIS)-based desktop application with a comprehensive collection of simplified open-source tools. For more information on damage estimation and damage estimation techniques, refer to Chapter 12.

13.10.1 Stage-Damage Curve

The stage–damage (depth–damage) curve is a relationship between extents of damage corresponding to the floodwater level (Figure 13.12a). It is a graphical representation of the expected losses from a specified flooding depth. These curves are typically used for housing and other structures. The stage or depth is the depth of water inside a building and the damage refers to the expected damage from that depth of water (Emergency Management Australia, 2007).

The use of empirical depth–damage curves requires a property survey of the floodplain and either an individual or aggregated estimate of depth (stage) versus damage curves for structures, roads, crops, utilities, etc. that are in the floodplain. This stage–damage is then related to the relationship for the stage discharge (Figure 13.12b) to derive the damage–discharge relationship (Figure 13.12c), which is then used along with the discharge–frequency relationship to derive the damage–frequency curve, as shown in Figure 13.12e.

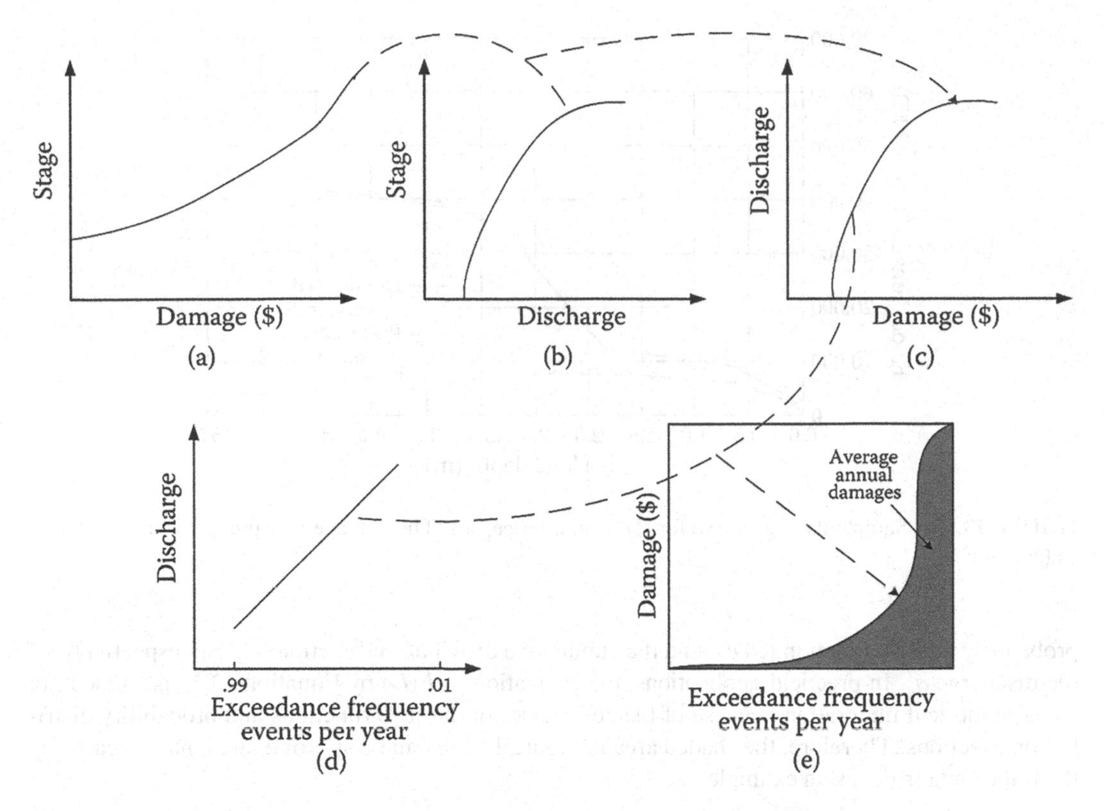

FIGURE 13.12 Development of damage–probability curves of floods in a region (a) damage–stage relationship; (b) discharge–stage relationship; (c) damage–discharge relationship; (d) exceedance frequency–discharge relationship; and (e) exceedance frequency–damage relationship (dashed line shows the process of damage–probability curve development) (Mays, 2001).

There are two ways for developing stage–damage curves. The first way involves using data on building contents and structure repair costs to generate synthetic or artificial estimates of damage curves. In the second way, the information on damage caused by previous floods and corresponding depths is used.

Vulnerability can be represented through vulnerability functions, illustrating how a selected consequence variable depends on a selected threat parameter. It can be necessary to use many such functions to represent a particular vulnerability comprehensively. This level of representation is similar to damage functions used in many vulnerability assessments. For river flood management, a classic example of vulnerability would relate a threat (flood height above the floodplain) to its consequence (damage) (Figure 13.13).

13.10.2 Expected Damage

The annual expected damage cost $E(D)$ is the area under the damage–frequency curve as shown in Figure 13.12e, which can be expressed as

$$E(D)=\int_{q_c}^{\infty}D(q_d)f(q_d)dq_d=\int_{q_c}^{\infty}D(q_d)dF(q_d), \tag{13.27}$$

where q_d is the threshold discharge beyond which damage would occur, $D(q_d)$ is the flood damage for various discharges q_d, which is the damage–discharge relationship, and $f(q_d)$ and $F(q_d)$ are the

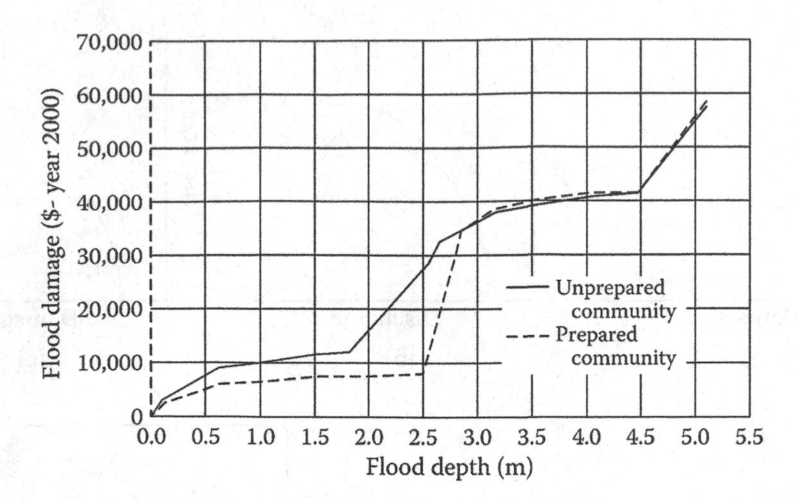

FIGURE 13.13 Sample damage curve for residential properties. The damage is expressed as a function of water depth.

probability density function (PDF) and the cumulative distribution function (CDF), respectively, of the discharge q_d. In practical applications, the evaluation of $E(D)$ by Equation 13.29 is carried out using numerical integration because of the complexity of damage functions and probability distribution functions. Therefore, the shaded area in Figure 13.12e can be approximated, numerically, by the trapezoidal rule, as an example,

$$E(D)=\sum_{j=1}^{n}\int_{q_c}^{\infty}\frac{\left[D(q_j)+D(q_{j+1})\right]}{2}\left[D\left(F_{j+1}-F(q_j)\right)\right] \text{ for } q_c=q_1\leq q_2\leq\ldots\leq q_n<\infty, \quad (13.28)$$

in which q_j is the discredited discharge in interval (q_c, ∞).

Example 13.2

Use the damage–frequency relationships in Table 13.9 for flood control alternatives to rank their merits based on expected flood damage reduction.

Solution:

The economic merit of each flood control alternative can be measured by the annual expected savings in flood damage of each alternative, which can be calculated as the difference between the annual expected damage at the existing condition (without flood control measures) and the annual expected damage with a given flood control measure under consideration. Using Tables 13.10 and 13.11, the damage reduction associated with each flood control measure at different return periods is determined.

From the data in Table 13.10, the average damage reduction for each flood control alternative and incremental probability can be developed as shown in Table 13.11. The best alternative with the highest benefit of annual expected flood reduction is to build a detention basin upstream.

In urban areas with different land use, damage could vary from block to block and from tax zone to the next. FEMA developed a program called HAZUS that has damage stage information for different land use. It was applied to many urban areas to get estimates of damages. Karamouz et al. (2016b) developed an alternative model called Flood Damage Estimator, FDE, that is briefly explained in the case studies of this chapter.

TABLE 13.9
Damage–Frequency Relationships in Example 13.2

Exceedance Probability (%)	Damage[a] (10^6)	Damage[b] (10^6)	Damage[c] (10^6)	Damage[d] (10^6)	Damage[e] (10^6)
20	0	0	0	0	0
10	6	0	0	0	0
7	10	0	0	0	0
5	13	13	2	4	3
2	22	22	10	12	10
1	30	30	20	18	12
0.5	40	40	30	27	21
0.2	50	50	43	40	35
0.1	54	54	47	43	45
0.05	57	57	55	50	56

[a] Existing condition.
[b] Dike system.
[c] Upstream diversion.
[d] Channel modification.
[e] Detention basin.

TABLE 13.10
Damage–Frequency Relationships in Example 13.2

Exceedance Probability (%)	Damage Reduction (10^6)			
	1	2	3	4
20	0	0	0	0
10	6	6	6	6
7	10	10	10	10
5	0	9	7	10
2	0	11	9	12
1	0	10	12	18
0.5	0	10	13	19
0.2	0	7	10	15
0.1	0	5	9	9
0.05	0	2	7	1

Note: 1: Dike system; 2: Upstream diversion; 3: Channelization; 4: Detention basin.

13.10.2.1 Case Study 1: Coastal Flood Damage Estimator: An Alternative to FEMA's HAZUS Platform

Estimation of damages made to urban infrastructure or the possible damages that could be inflicted on them is not something than can be neglected when it comes to the topic of flood resilient cities. One of the primary steps of building a resilient city that could remain resilient in the future is to be able to plan for its flood resistance by being able to estimate damages and think of ways to prevent them from happening. The following paper by Karamouz et al. (2016a) highlights the importance of damage estimation and introduces an effective alternative to FEMA's damage estimator, HAZUS.

TABLE 13.11
Damage–Frequency Relationships in Example 13.2

	Damage Reduction ($\$10^6$)			
Increm. Prob. (ΔF)	**1**	**2**	**3**	**4**
0.1	3	3	3	3
0.03	8	8	8	8
0.02	5	9.5	8.5	10
0.03	0	10	8	11
0.01	0	10.5	10.5	15
0.005	0	10	12.5	18.5
0.003	0	8.5	11.5	17
0.001	0	6	9.5	12
0.0005	0	3.5	8	5
$\Sigma(\Delta D \times \Delta F)$	0.64	1.218	1.251	1.378

Note: 1: Dike system; 2: Upstream diversion; 3: Channelization; 4: Detention basin.

Floods are the most devastating natural hazards in the United States, and these events lead to significant economic consequences. Estimation of damages and losses at different parts of the floodplain can be used to identify the vulnerable areas susceptible to flood. Floodplain maps combined with damage assessment could identify high-flood-risk areas and present a platform to develop guidelines and zoning resolutions for proper land use, building new infrastructures, and placing and retracting residential areas.

A framework is proposed for flood-damage assessment in urban areas which can be seen in Figure 13.14. A comprehensive collection of various sets of data are used (step 1). In order to process and analyze this data, GSSHA, ArcGIS, and HAZUS models have been employed (step 2). Using these models, flood damage is estimated for a coastal region according to a proposed GIS-based approach, FDE, as well as the HAZUS software. Flood depth and area (floodplain delineation) are determined based on the 100-year flood scenario in the region. Depth of water from flood hazard is obtained from two different approaches.

In one approach, a 100-year floodplain based on the information of water elevation provided by FIS in 2007 and 2013 is determined and used. In the other approach, a Gridded Surface-Subsurface Hydrological Analysis (GSSHA) hydrologic model is used to create floodplain from 100-year rainfall hyetograph and sea-level variation hydrograph, to simultaneously consider inland and coastal floods. Results of flood damage simulations by the FDE and HAZUS models are then compared (step 3). In this step, HAZUS uses its default socioeconomic data, whereas FDE uses the data provided in this study. In FDE, all operations to estimate the damage are defined in the GIS ModelBuilder tool to be done automatically.

Two different models are used and compared for estimation of flood damage. Both models are GIS based. The first model, developed in this study, combines the floodplain with the depth-damage functions, different land uses, and DEM of the region to estimate flood damage. This is the FDE model. The results are compared with the HAZUS model, which FEMA develops for estimation of damage from flood hazard. The HAZUS platform has been changed to adapt to a floodplain externally developed in this study, rather than using the FEMA floodplain as a built-in capability of HAZUS. Manhattan, one of the most important economic and strategic hubs in the world, is selected as the case study. A 100-year floodplain based on the information provided by FEMA in FIS 2007 and 2013 studies is utilized in both models to estimate floodwater depth at different locations of the study area. A flood plain developed in-house by a distributed hydrologic model named

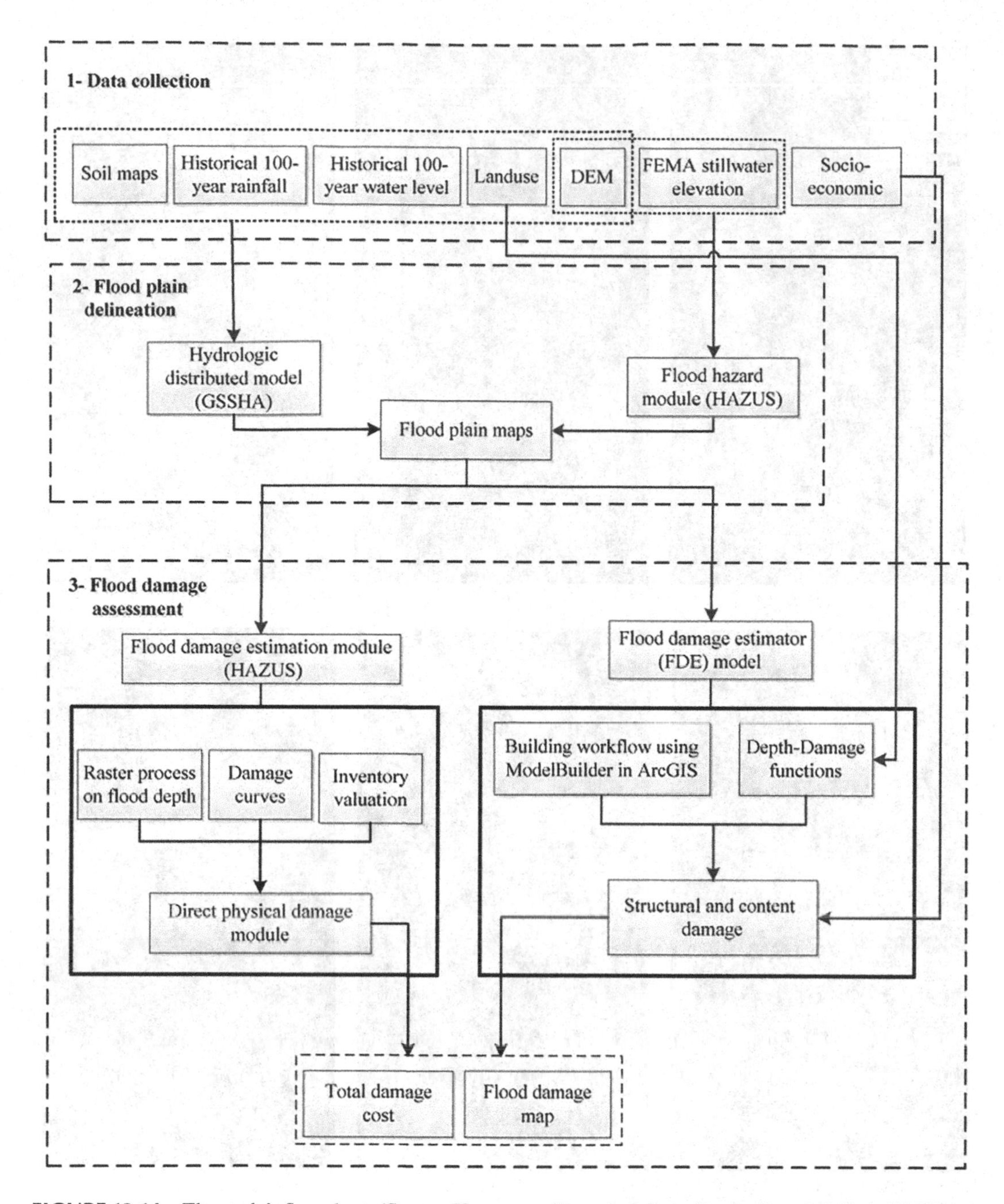

FIGURE 13.14 The study's flow chart. (*Source*: Karamouz, M. et al., *J. Irrig. Drain. Eng.*, 142(6), 1–12, 2016a.)

GSSHA is also used for considering the combined effect of 100-year inland and coastal flooding to determine the flood damage.

3.10.2.1.1 Major Findings

In Figure 13.15, flood risk maps developed for lower Manhattan obtained from the FDE and HAZUS models corresponding with different floodplains and land uses are shown. According to this figure, spatial distribution of damage in the proposed FDE approach has finer resolution for most parts (on some occasions, tax parcels are bigger than census blocks), which helps achieve better damage assessment.

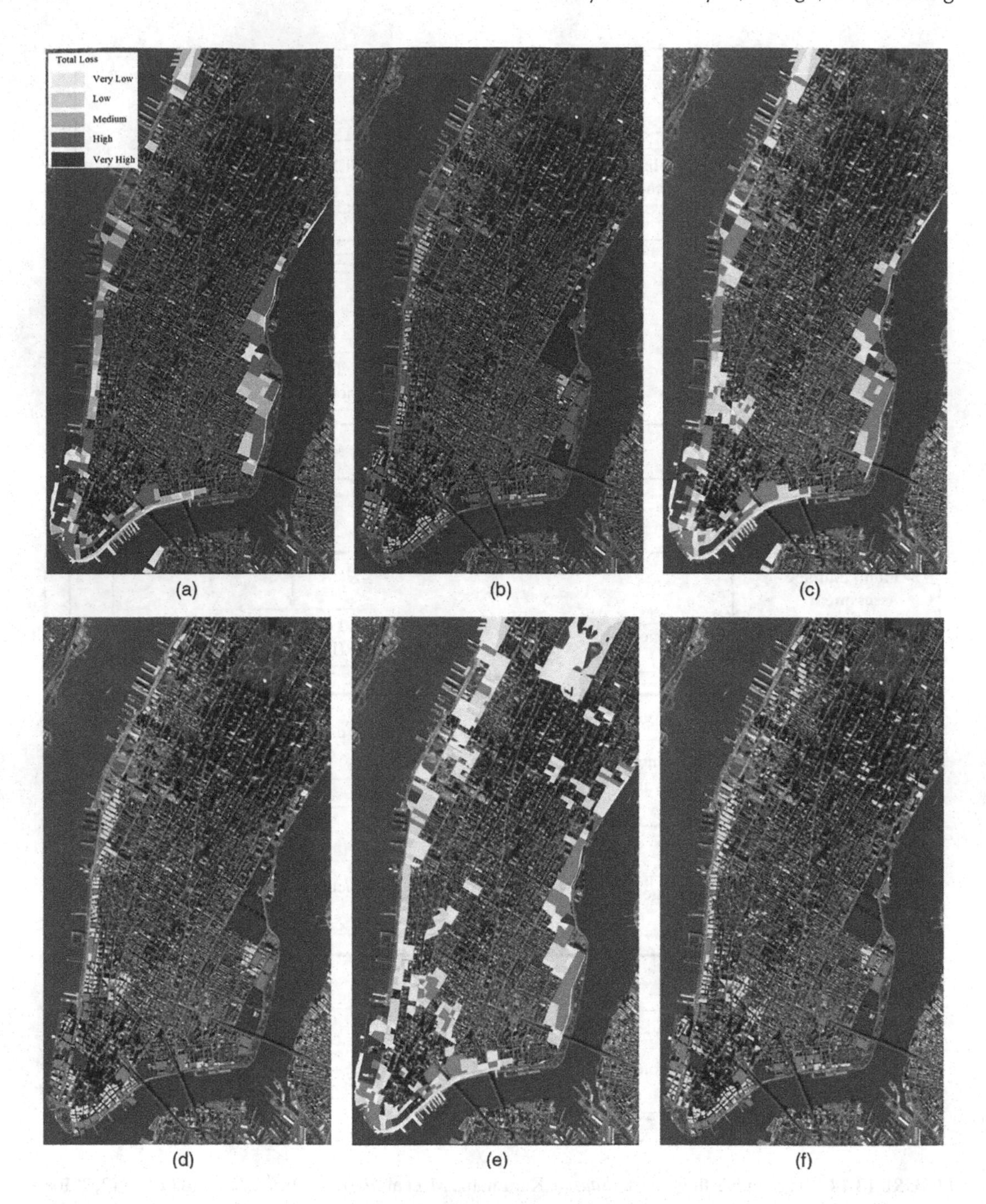

FIGURE 13.15 Risk map for 100-year flood considering structural and content damages using HAZUS and FDE for different floodplains of (a) and (b) FIS 2007; (c) and (d) FIS 2013; (e) and (f) GSSHA. (Adapted from Karamouz, M. et al., *J. Irrig. Drain. Eng.*, 142(6), 1–12, 2016a.)

The results of this study show that:

1. Using the proposed FDE model, compared with the results from HAZUS, for three floodplains of FIS 2007 and 2013 and from the GSSHA model, estimates the damages within a margin of 0.27%, 3.93%, and –5.74%, respectively.

2. In both models, the GSSHA floodplain represents damages between FIS 2007 and FIS 2013 by a margin of 7.5% and −6.21%. The GSSHA floodplain considers inland flooding, and the corresponding estimated damage could be considered more realistic.
3. Therefore, it could be concluded that perhaps FIS 2013 is overestimated.
4. The GIS-based FDE model could be considered an alternative to a more sophisticated HAZUS flood model because it provides better cell resolution (tax parcel versus census block) and presents similar results with a much simpler structure.
5. Using a tax parcel can also provide a better basis for determining flood insurance rates and governmental actions for financial compensation of damaged buildings.

13.11 FLOOD RISK MANAGEMENT

Flood risk management for a system in a narrow sense is the process of managing a flood risk situation. In a wider sense, it includes the planning of a system that will reduce the flood risk.

These two aspects of flood risk management will be considered separately, starting with the management of a system that consists of the processes indicated in Figure 13.16. Risk management for the operation of a flood protection system is the sum of actions for a rational approach to flood disaster mitigation. Its purpose is the control of flood disasters, in the sense of being prepared for a flood, and to minimize its impact.

It includes the process of risk analysis, which provides the basis for long-term management decisions for the existing flood protection system. Continuous improvement of the system requires a reassessment of the existing risks and an evaluation of the hazards depending on the newest

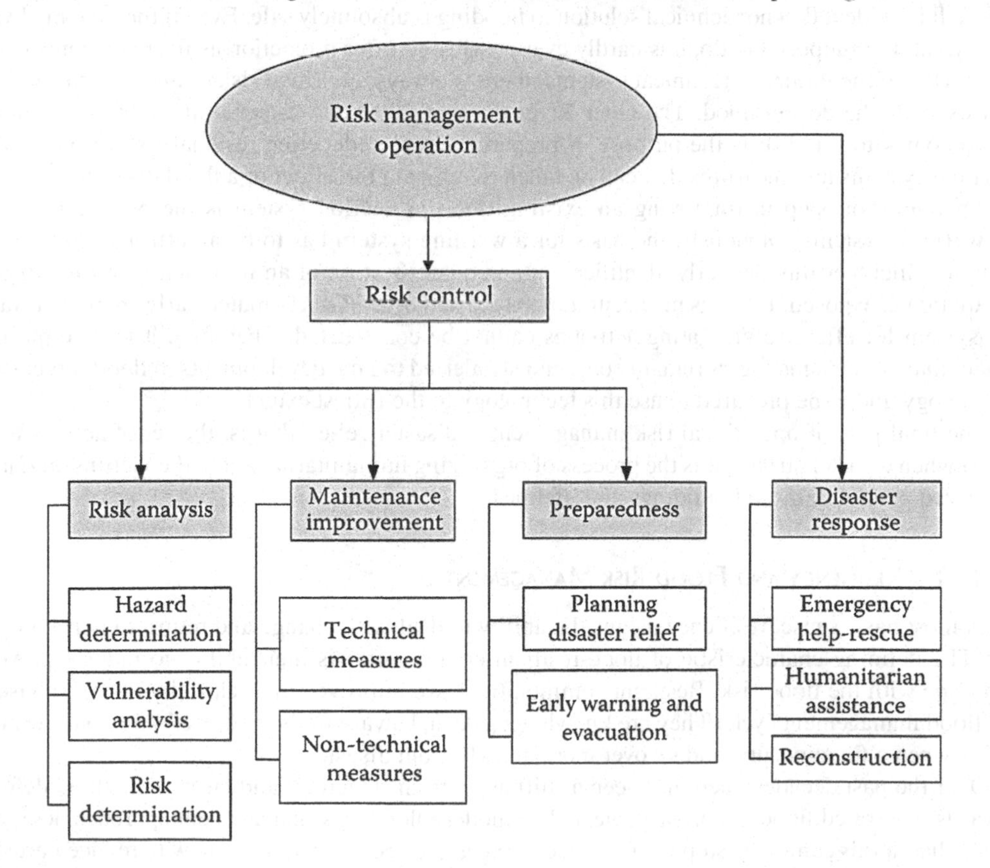

FIGURE 13.16 Stages of operational risk management. (Adapted from Eikenberg, C., *Journalistenhandbuch zum Katastrophenman-agement* (5th ed.), German IDNDR-Committee, Bonn, 1998.)

information available: on new data, on new theoretical developments, or on new boundary conditions, for example, due to change of land use.

The hazards are to be combined with the vulnerability into the risk. The vulnerability of the persons or objects (the elements at risk) in an area, which is inundated if a flood of a certain magnitude occurs, is weighted with the frequency of occurrence of that flood. A good risk analysis process yields hazard or risk maps, which today are drawn by means of GIS-based on extensive surveys of vulnerability combined with topographic maps. Such maps serve to identify weak points of the flood defense system or indicate a need for action, which may lead to a new project. Other weaknesses of the system become evident during extreme floods. For example, the Oder River flood of 1997 in Poland has indicated that weak points contributing to flooding of a city in a floodplain are not only failures of dikes but also seepage through the dikes and penetration of floodwaters through the drainage system, that is, through the sewerage system or water courses inside the city (Kowalczak, 1999). The 1997 Central European flood or the 1997 Oder Flood of Oder and Morava river basins in July 1997 affected Poland, Germany, and Czech Republic, taking lives of about 74 people (in Czech Republic and Poland) and causing material damages estimated at $4.5 billion (3.8 billion euros in Czech Republic and Poland and 330 million euros in Germany).

Risk analysis forms the basis for decisions on maintaining and improving the system, which is the second part of the operation of an existing system. It is a truism that a system requires continuous maintenance to be always functioning as planned, and new concepts of protection may require local improvements of the existing system.

The third part of the management process is the preparedness stage, whose purpose is to provide the necessary decision support system for the case that the existing flood protection system has failed. It is evident that no technical solution to flooding is absolutely safe. Even if the system always does what it is supposed to do, it is hardly ever possible to offer protection against any conceivable flood. Due to the failure of technical systems, there is always a residual risk or due to the rare flood that exceeds the design flood. The Oder River flood of 1997 (Bronstert et al., 1999; Grunewald, 1998) comes to mind. It is the purpose of preparedness to reduce the residual risk through early warning systems and measures that can be taken to mitigate the effect of a flood disaster.

An important step in improving an existing flood protection system is the provision of better warning systems. Obviously, the basis for a warning system has to be an effective forecasting system, which permits the early identification and quantification of an imminent flood to which a population is exposed. If this is not accurately forecasted or at least estimated early enough, a warning system for effective mitigating activities cannot be constructed. Therefore, it is an important aspect that systems managers remain continuously alerted to new developments in flood forecasting technology and to be prepared to use this technology to the fullest extent.

The final part of operational risk management is disaster relief, that is, the set of actions to be taken when disaster strikes. It is the process of organizing humanitarian aid to the victims, and later reconstruction of damaged buildings and lifelines.

13.11.1 Resiliency and Flood Risk Management

In its most basic sense, resilience means the ability to deal with change and return to normal operate. The defining characteristic of flood-resilient communities is their ability to reduce, prevent, and cope with the flood risk. Resilient communities have improved their capacity in each phase of the flood management cycle. They are knowledgeable and aware of the risk, well prepared, respond better when a flood occurs, and recover more quickly from disasters.

Over the past decades, there has been a shift away from structural and large-scale flood defense towards integrated flood risk management. The modern flood risk management approach acknowledges that floods cannot be stopped from occurring and places emphasis on how to reduce hardship and vulnerability of risk-prone communities (e.g., Bharwani et al., 2008; Krywkow et al., 2008; Vis et al., 2003).

The impacts of disasters on the community are significantly influenced by community preparedness. Community resilience is the group's capacity to recover from, withstand, and respond positively to adversity or crisis. Community resilience is often defined as the following three properties:

1. Resistance is the disruption degree that can be accommodated without the community undergoing long-term change. Before undergoing long-term change, a highly resilient community can withstand considerable disruption.
2. Recovery is the community's ability to pull through or bounce back to its predisaster state. A highly resilient community returns to its predisaster state or moves beyond that quicker than a less-resilient community.
3. Learning process is the community's ability to build on learning of a crisis or disaster, gain an improved functioning level, and increase resilience levels. A highly resilient community will adapt to its new conditions and learn from the experience of disaster.

Resiliency could be considered as the most important indicator of system performance in disaster recovery and how successful the management of disaster strategies have been. Communities are vulnerable to natural disasters. In some areas, there is a high risk of floods in the plain and landslide in the hills, and the frequency of such disaster is increasing year by year. Low awareness level in terms of disaster preparedness and management and the lack of efficient mechanisms and capacity to deal with these natural disasters have had severe impacts on the lives of the people, property, and economy at large. Climate change has also added to these challenges, especially for the poor in rural areas.

Therefore, resiliency works to reduce the impacts by helping people meet their household food needs through home-based agriculture, cash crop production, and nonfarm alternative livelihood strategies. It has also helped reduce the loss of lives and assets—from the landslide, floods, and other forms of a natural disaster—through the use of early warning systems, improved physical infrastructure, and local preparedness planning, among other initiatives.

The defining and thus taken-for-granted characteristic of resilient communities is the ability to reduce, prevent, and cope with the risk. Resilient communities have improved their capacity in each phase of the management cycle, as shown in Figure 13.17. There are four steps in disaster management, including mitigation, preparedness, response, and recovery. Mitigation and preparedness are considered before a disaster happens. Response refers to the activities during the disaster, and recovery refers to the activities after a disaster. Preparedness activities help build community resilience to disasters. Preparation helps prevent disasters and lessen the impact of disasters when they do occur. Strategies, tools, and training are provided to assist in the preparation for disasters. In mitigation

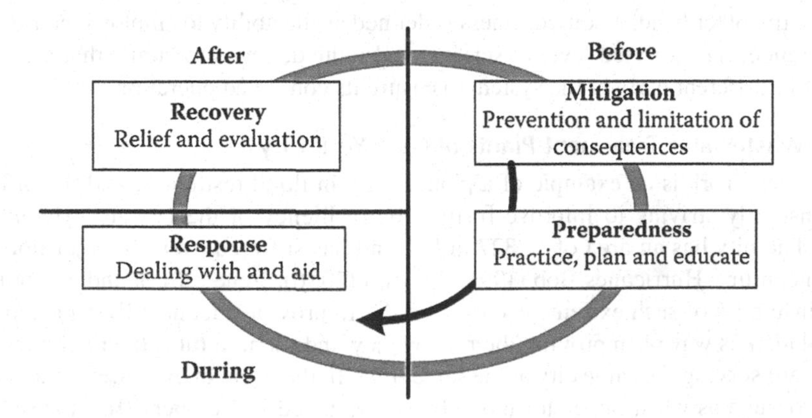

FIGURE 13.17 Resilient communities have an improved capacity in each phase of the disaster management cycle. (From Joosten-ten Brinke, D. et al., *Educ. Res. Rev.*, 3, 51–65, 2008. With permission.)

prevention and limitation of disaster, consequences are considered and preparedness includes practicing, planning, and education of the society to deal with disasters. They are knowledgeable and aware of the risk, are well prepared and respond better when a hazard occurs, and recover more quickly from disasters. Resilience is also promoted through programs that encourage, create, and develop resources and connections drawn on in times of crisis.

Arguably, the overall performance of the risk management system can be improved by adopting a resilient approach. While a resistant approach is directed to maintain the system's structure and functions, the resilient approach enhances the capacity of the system to recover from nonstructural changes in dynamics. The resilience of a system relates to three aspects that determine the reaction of a system to a hazard: (a) the amplitude of the reaction, (b) the gradual increase of reaction with increasing disturbances, and (c) the recovery rate. A system's resilience is more considerable when the amplitudes (i.e., amount of damage) are smaller. Design strategies based on suitable indicators could provide a larger magnitude of reactions. This will lead to a more gradual slope of the damage frequency curve and enhance system resiliency.

Resilience strategies for flood risk management focus on minimizing flood impacts and enhancing recovery compared to resistance strategies that aim to prevent floods entirely. Since resilience strategies are expected to result in improved flood risk management, they deserve careful evaluation. For the evaluation and comparison of strategies for flood risk management, it is necessary to be able to quantify the resilience of the resulting flood risk management systems. Therefore, indicators for quantifying the resilience of flood risk management should be developed and evaluated. To evaluate the behavior of these indicators, they can be applied to a case study such as NYC or hypothetical cases that resemble the challenges we are facing in the eastern coastal area of the United States.

Resilience as a criterion for the operation and design of water supply systems in water management has been quantified by Karamouz et al. (2014) and others. There have been some other studies on adopting resilience to measure the performance of water distribution systems (Zongxue et al., 1998), to characterize reservoir operation rules (e.g., Burn et al., 1991; Moy et al., 1986), or to quantify sustainability of water resource systems (ASCE and UNESCO, 1998). In these applications, resilience is described as how quickly a system recovers from failure and is a measure of the duration of an unsatisfactory condition. A flood risk management system is a complex system and assessing its recovery time for the future is difficult. Because responses and recoveries are not easily measured, indicators for resilience have to be found.

One widely used framework has been proposed by Bruneau et al. (2003) that is based on quantifying system's resilience dimensions including rapidity, robustness, resourcefulness, and redundancy (four R's). Rapidity is the ability of a system to achieve the expected level of performance in the shortest possible time while robustness is the system's resistance to maintain its function against stressors. On the other hand, resourcefulness is defined as the ability to employ material and human resources to meet the expected level of service, and redundancy is the units that are available for replacement in different parts of the system to ensure its continued operation.

13.11.1.1 Wastewater Treatment Plants of New York City

The City of New York is an example of a pioneer city in flood resiliency, and the officials of the city are constantly striving to improve further the resiliency of the city in different aspects of the matter. The city has an area of 11,327 miles^2 and has suffered over 20 major floods after the seventeenth century. Hurricanes Bob (1991), Ernesto (2006), Irene (2011), and Superstorm Sandy (2012) are instances of such extreme events. In order to provide adequate living standards for the resident and also as way of improving the redundancy and then, in turn, the resiliency of the city, 14 WWTPs are serving the same city at the same time. In the event of extreme hurricanes and such other disaster such as when Superstorm Sandy had occurred in October 2012, the redundancy of the WWTP system of NYC was of excellent use. Resiliency is one of the primary features of the livable cities of the future and it is impossible to achieve resiliency without being assured that the

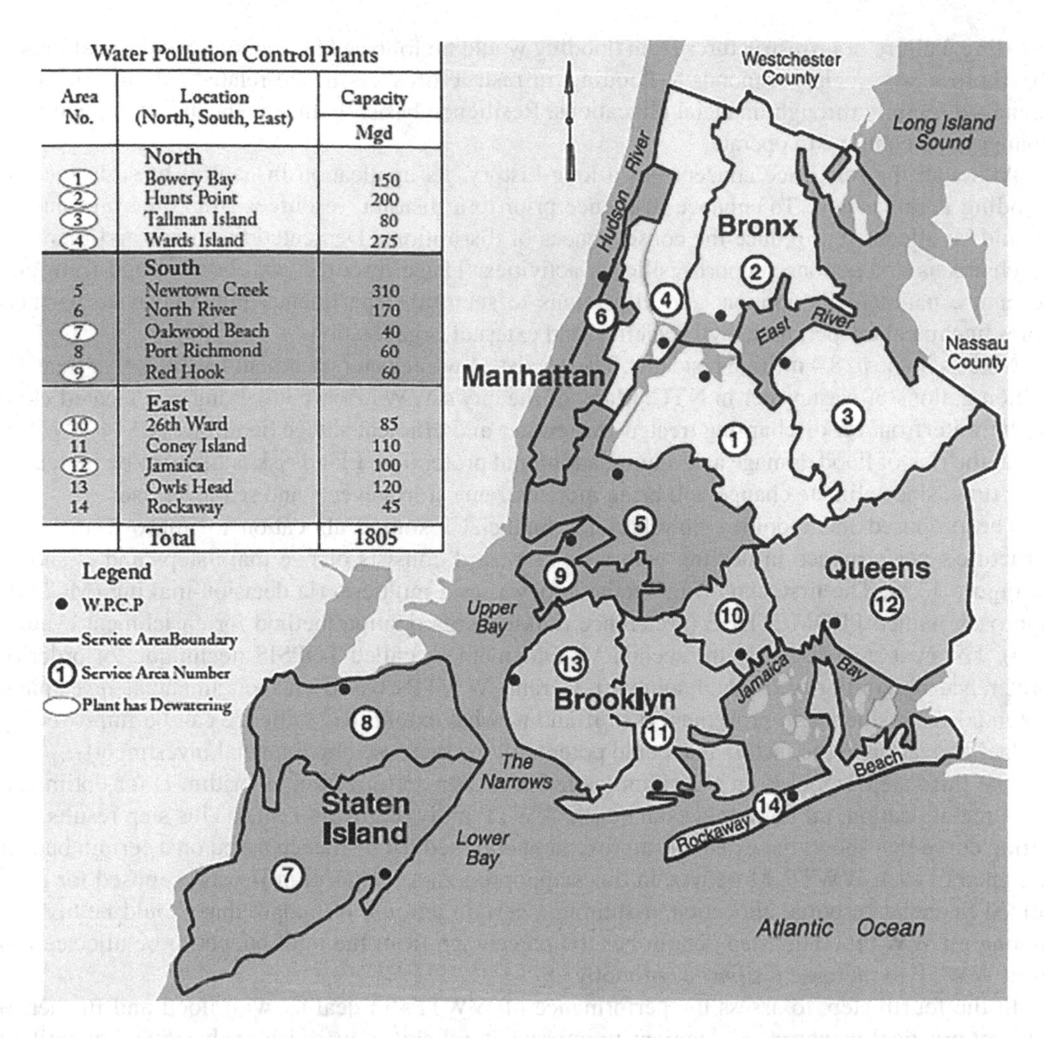

Water Pollution Control Plants

Area No.	Location (North, South, East)	Capacity Mgd
	North	
1	Bowery Bay	150
2	Hunts Point	200
3	Tallman Island	80
4	Wards Island	275
	South	
5	Newtown Creek	310
6	North River	170
7	Oakwood Beach	40
8	Port Richmond	60
9	Red Hook	60
	East	
10	26th Ward	85
11	Coney Island	110
12	Jamaica	100
13	Owls Head	120
14	Rockaway	45
	Total	**1805**

FIGURE 13.18 New York City's boroughs, its wastewater treatment plants, and some of the main characteristics of them. (*Source*: "New York's City's Wastewater Treatment System" by New York City's Department of Environmental Protection.)

system is sufficient in terms of redundancy. A map of the five boroughs of the city, the 14 WWTPs mentioned above and the areas and volume of wastewater they each cover have been illustrated in Figure 13.18.

13.11.1.2 Case Study 2: Prioritizing Investments in Improving Flood Resilience and Reliability of Wastewater Treatment Infrastructure

Monetary resources are not easily granted and almost never abundant. Therefore, managing assets and prioritizing investments are of utmost significance when building resilience in coastal communities. Flood resilient cities are metropolitan societies armed with the ability of applying finite available resources when and where they are most needed. This case study (Karamouz et al., 2018) discusses the importance of the mentioned subject, applied to NYC.

For urban societies, every activity requires promised financial resources. Most of these activities (e.g., infrastructure rehabilitation) are resource-intensive, and with emerging new threats, complexity of technologies, and shifting tangible assets to intangible assets (such as information), their cost continues to increase. Urban activities, especially in coastal areas, are threatened by hazards such as flooding. Failure of infrastructures from flooding would be followed by economic and social losses. To improve how society responds to flooding, infrastructures' resilience-related activities must be managed, mainly through financial allocations. Resilience basically means the ability to deal with changes and continue to operate.

Although the resilience concept has a long history, its application in natural hazards such as flooding is rather new. To enhance resilience prior to a disaster, resources from a certain budget should be allocated to reduce the consequences of disruptions. Dedicated resources and financial mechanisms and actions support resilience activities. These resources can be obtained from city revenues, national distribution, and allocations to sectoral departments, public–private partnerships, technical cooperation, civil societies, and external organizations.

NYC is home to 8.4 million residents. There exist 14 wastewater treatment facilities that treat 1.3 billion gallons of wastewater in NYC. Many of the city's WWTPs are low-lying and located close to the waterfront for discharging treated wastewater and efficient sludge handling. All 14 WWTPs are at the risk of flood damage and require additional protection. Flood risk is likely to be increased over time, since climate change will bring more extreme storm events and sea-level rise.

The proposed methodology flowchart for financial resource allocation to improve an infrastructure's performance in dealing with a flood hazard consists of five main steps and is shown in Figure 13.19. The first step of the methodology uses a multicriteria decision-making (MCDM) approach named PROMETHEE (preference ranking organization method for enrichment evaluation). The next step discusses the second MCDM method, called TOPSIS (technique for order of preference by similarity to ideal solution), to rank WWTPs based on their financial investment potential for resilience improvement (i.e., if and to what extent the resilience can be improved by budget allocation to the factors that could potentially be increased by financial investment).

The third step proposes an algorithm (called hereafter optimization algorithm I) for optimized resource allocation, taking into account each WWTP individually at a time. This step results in a fitting curve that shows the extent of improvement obtained for resilience based on a certain budget assigned to each WWTP. Moreover, in this step, optimization algorithm II was proposed for optimized financial resource allocation, assuming a certain amount of budget that should be divided among all WWTPs. This step determines the percentage from the total budget to be allocated to each WWTP to increase resilience optimally.

In the fourth step, to assess the performance of WWTPs in dealing with flood and further to suggest practical management solutions to improve it, reliability was also analyzed and quantified based on the load-resistance concept. To quantify load and resistance, the PROMETHEE MCDM method was used. For this purpose, factors associated with the load and resistance were identified and grouped into two classes. By comparing load and resistance, the reliability state of the plants was estimated. Then, a statistical method was proposed for probabilistically quantifying reliability using the margin of safety approach. Finally, sensitivity of resilience to changes in the inputs (values of financial subcriteria) was investigated.

13.11.1.2.1 Major Findings

1. Results show that the improvement of the selected factors can increase the resilience; however, the marginal improvement will be reduced significantly. Therefore, allocating a budget up to a certain level is justifiable; after that, it cannot significantly improve the resilience.
2. By comparing the values of load and resistance, it was shown that six WWTPs did not pass the reliability threshold.
3. The Bower Bay plant was found as the WWTP with the highest reliability. Moreover, reliability for the six aforementioned plants was considered negligible. Except for one plant (Bower Bay), the reliability status and resilience had a direct relation; that is, WWTPs with higher values of resilience had better performance based on the reliability. Therefore, it was concluded that, although resilience and reliability metrics measure the performance

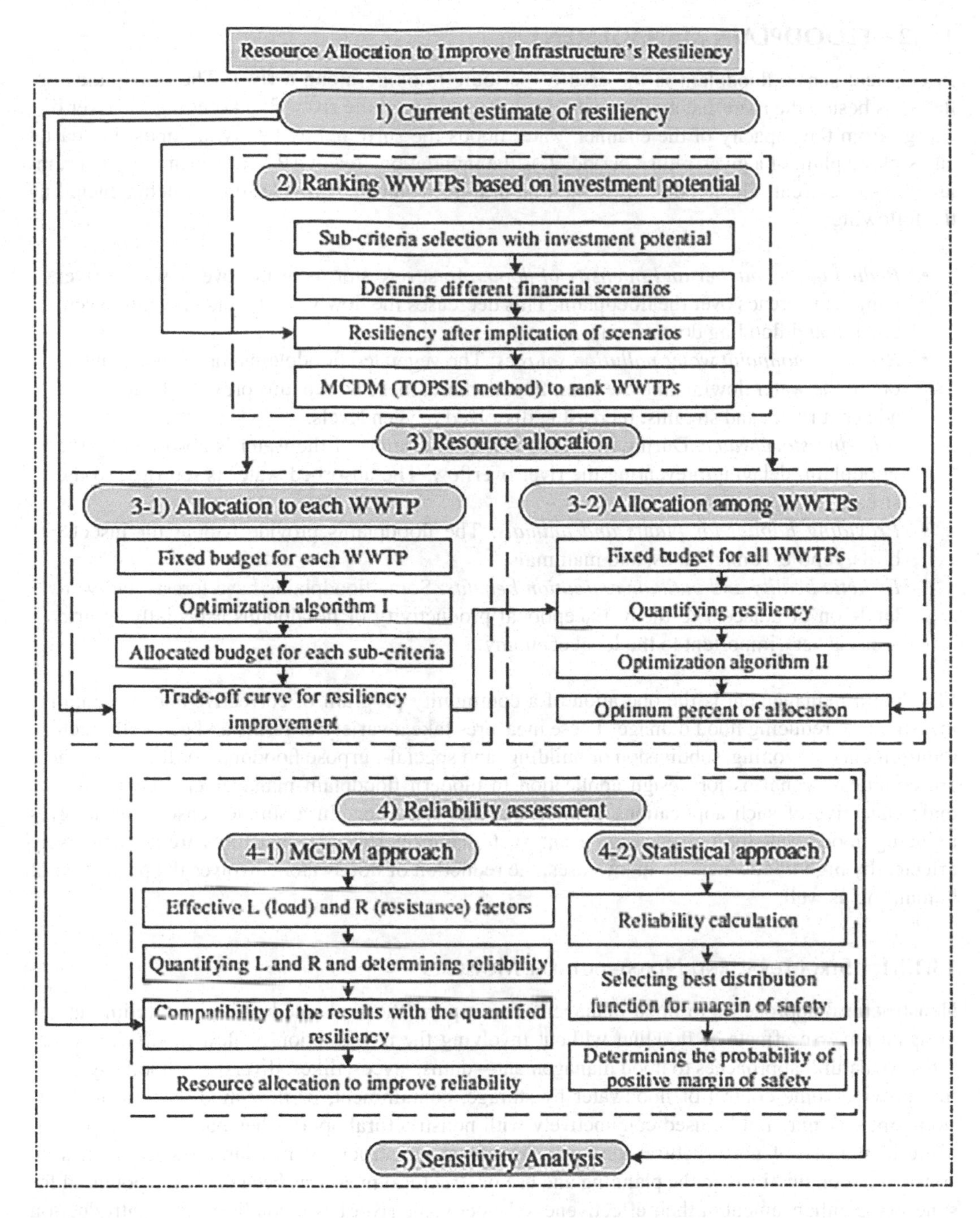

FIGURE 13.19 The study's flow chart. (*Source*: Karamouz, M. et al., *J. Infrastruct. Sys.*, 24(4), 04018021, 2018.)

of the system based on different concepts, their application for assessing the performance results in compatible findings.

4. A variance-based sensitivity analysis was performed on the optimization model inputs (financial subcriteria) and outputs (resilience improvement and the invested budget), which indicated that plants with high investment potential are more sensitive to the changes in the budget.

13.12 FLOODPLAIN MANAGEMENT

In classical sense, floodplain is the area of periodic flooding along a river. The floodplains are flatlands beside the river that are formed due to the actions of the river. By exceeding the river discharge from the capacity of the channel, water floods the surrounding low-lying lands. In coastal cities, flood plain of a given return period, T, is the inundation area of a T year storm or a joint rainfall and surge event. The floodplains in their natural form can also provide some benefits, including the following:

- *Reducing the number and intensity of floods*: In severe storms, water overflows the river bank and spreads over the floodplain. This decreases the flow velocity and prevents severe erosion and flooding downstream.
- *Reducing nonpoint water pollution sources*: The vegetated floodplains filter contaminants out of the water flowing into the river. Furthermore, these floodplains provide shade for the adjacent rivers and streams, increasing dissolved oxygen levels.
- *Filtering stormwater*: During increased water flow, some of the water is absorbed by the floodplain, delaying/preventing the river overflow. The absorbed water is returned to the stream in dry seasons.
- *Providing habitat for plants and animals*: The floodplains provide habitat for insects, birds, reptiles, amphibians, and mammals.
- *Esthetic beauty and outdoor recreation benefits*: Some floodplains have forests and wetlands on or adjacent to them. Exceptional productivity of floodplains especially in arid areas is very important to the local economy.

Floodplain management is the operation of a community program of corrective and preventative measures for reducing flood damage. These measures take a variety of forms and generally include requirements for zoning, subdivision or building, and special-purpose floodplain ordinances. There are countless scenarios for design application in modern floodplain management. Typically, the main objective of such applications is flood damage reduction. In a simple sense, this involves reducing flood levels for a given storm event, such that properties and structures are not negatively affected by high floodwaters. In many cases, the reduction of floodwaters involves the protection of human life as well.

13.12.1 Structural and Nonstructural Measures

Nonstructural approaches to flood management comprise those planned activities to eliminate or mitigate adverse effects of flooding without involving the construction of flow-modifying structures. Structural approaches to flood management—dams, levees, dikes, diversions, floodways, etc., that provide some control of floodwater by storage, containment, or flow modification or diversion—may or may not be used conjunctively with nonstructural approaches but are not a prerequisite to the use of nonstructural measures. However, nonstructural measures should always be considered conjunctively in the planning and use of structural measures because of the potential for synergistic enhancement of their effectiveness. Under some river basin conditions, the introduction of nonstructural methods to limit flood damage may alone be more cost-effective than alternatives involving structural methods.

Nonstructural approaches to flood management fall naturally into two categories:

- Those anticipatory measures that can be assessed, defined, and implemented in the floodplains to reduce the risk to property from identifiable potential floods
- Those planned emergency response measures that are applied when a damaging flood is forecast, imminent, or underway to help mitigate its damaging effects

Accordingly, a distinction is made between planning measures and response measures:

- Planning measures:
  a. Flood forecasting
  b. Control of floodplain development
  c. Flood insurance
  d. Floodproofing
  e. Catchment management
- Response measures:
  a. Flood emergency response planning
  b. Flood fighting
  c. Flood warning
  d. Evacuation
  e. Emergency assistance and relief

13.12.2 BMPs and Flood Control

Flooding is a natural phenomenon that cannot be prevented; therefore, flood control practices should focus on the mitigation of flood damage instead of flood prevention. The government should help people by allocating needed resources to adapt their lifestyle to their natural environment. Some of these indigenous solutions are changing the housing structures and crop patterns for reducing flood damage. For sustainable economic development and reduction of environmental degradation, good governance and appropriate environmental laws, acts, and ordinances are necessary.

A better understanding of the processes that result in increased flood damage can also help in more efficient mitigation of the adverse effects of floods on human lives, environment, and economy. For flood mitigation, both the government and the people should shift their paradigms and adopt nature-based BMPs, land-use planning, and adaptation strategies. The flood BMPs will result in less runoff, more carrying capacity of the drainage system, and increased land elevations concerning sea level or riverbeds. In Chapter 6, BMPs for urban coastal areas are presented.

13.12.2.1 Case Study 3: Integration of Inland and Coastal Storms for Flood Hazard Assessment Using a Distributed Hydrologic Model

The simultaneous occurrence of inland flood with coastal storms could result in a much more devastating flood. Therefore, it is necessary to consider the combined effect of rainfall events and storm surges in flood-resilient cities. This case study (Karamouz et al., 2017) is an example of using a distributed hydrologic model for better and more effective planning and management of coastal cities and making them more resilient.

In recent years, inland and coastal flood modeling has greatly improved through integration of the hydraulic and hydrologic models with geographic information systems (GISs) and digital elevation models (DEMs). Recent destructive flood events in the NYC coastal areas, hurricane Irene, and Superstorm Sandy, have brought ample awareness and concerns for having reliable coastal floodplain delineation tools. This study provided a framework for coastal floodplain delineation with a less data-intensive methodology within a reasonable time and with relatively acceptable accuracy. This study mainly focuses on the lower Manhattan. Manhattan is the most densely populated borough of NYC.

The flowchart of the proposed methodology is illustrated in Figure 13.20. This scheme includes four main steps. First, required input data are collected and their adequacy and accuracy are checked through application of statistical methods. Frequency analysis is performed on the maximum rainfall (representative of inland floods) and extreme water level (EWL) (representative of coastal flood)

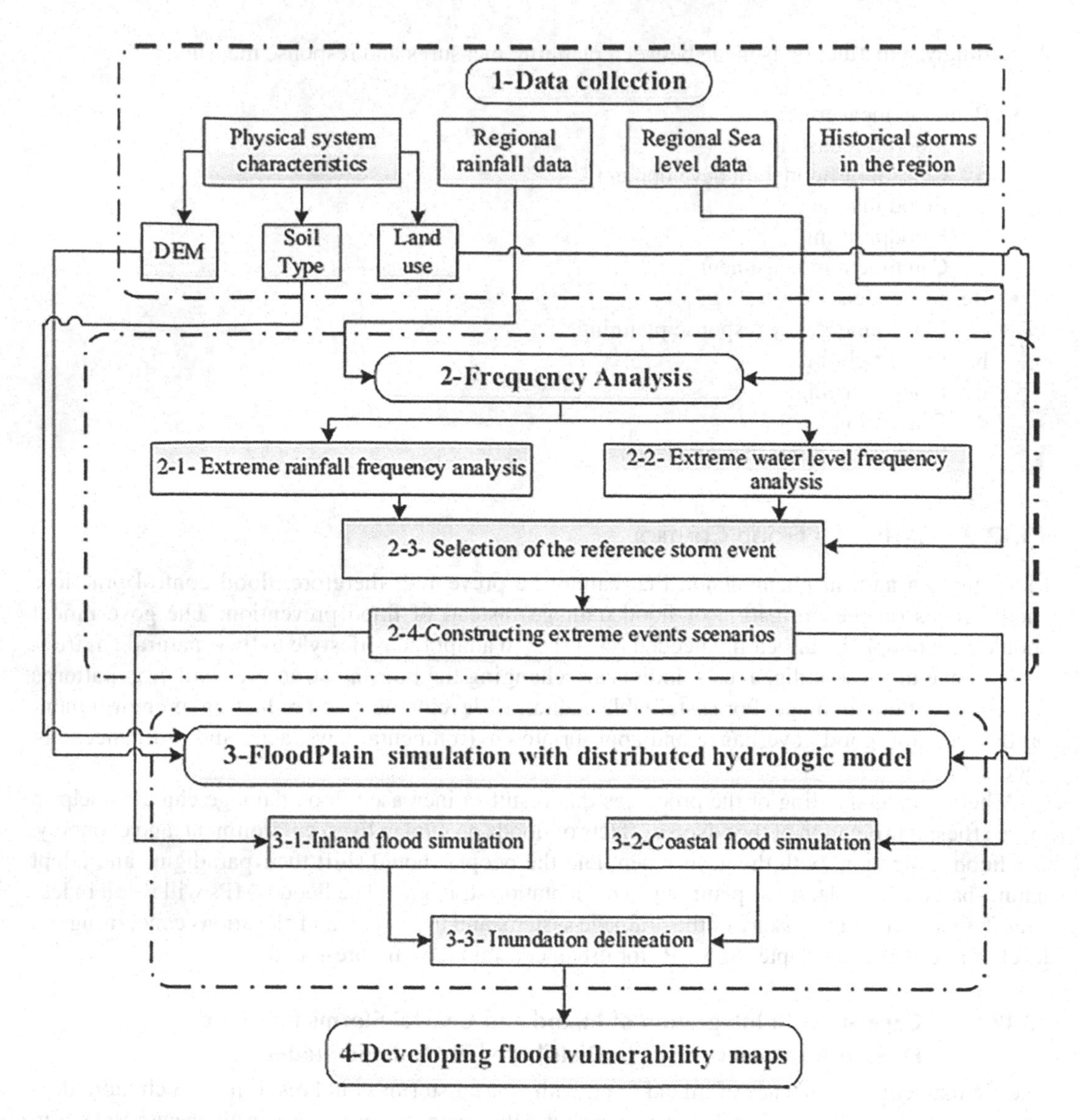

FIGURE 13.20 The study's flow chart. (*Source*: Karamouz, M. et al., *Environ. Earth Sci.*, Springer Berlin Heidelberg, 76(11), 395, 2017.)

data, and rainfall and sea levels corresponding to different return periods (recurrence intervals) are determined. Historical storms are investigated to select a "referenced storm"—in this case, hurricane Irene—considering the characteristics needed as a representative extreme event. Then extreme inland, coastal, and combined flooding scenarios with 100- and 200-year return periods are defined based on the referenced storm. The corresponding floodplains are then delineated using GSSHA model. The simulation results are compared with the FEMA floodplains.

The results show that the joint probability distribution better represents the severity of storms. Based on the joint probability distribution, Sandy is the most severe recorded storm in the study region. It should be noted that Superstorm Sandy has been classified in the previous studies as an event of 500 or more year return period. However, there is a general belief that Sandy is more like a 150- to 200-year event, especially by the agencies involved in flood assessment. The result of joint probability distribution has verified this belief. The 100-year and Superstorm Sandy floods are simulated and compared with the maps released by FEMA. At some strategic points in the region, results show that the mapped flooding area for Sandy is in agreement with that determined by FEMA from standpoints of the extent and depth of the flood. It is also concluded that storm surge is the governing factor in flood inundation for NYC. In comparison with the generated flood maps in this study with those developed by FEMA, some differences were observed that can be attributed to not considering inland flooding in coastal floodplain mapping by FEMA.

This is a main difference between the methodology of this study and FEMA's study. Moreover, the procedure that FEMA follows is time-consuming and perhaps with allocation of considerable resources involved. In contrast, although the procedure outlined in this paper is a compromise, it could be done and updated at a low cost more frequently.

In Figure 13.21, the lower Manhattan flooding area with 100-year return period, developed by FEMA (blue area), is shown and compared with the floodplain simulated from scenario S4 (100-year EWL based on Irene with 100-year rainfall). Total flooded area for lower Manhattan (south of 33rd street) from FEMA and scenario S4 are 4.07 and 4.17 km^2, respectively. Flooding areas developed by FEMA and simulated in this study are quite similar. This shows that the proposed scheme is promising and of significant value in flood inundation vulnerability studies.

FIGURE 13.21 Comparing 100-year floodplain for lower Manhattan developed by FEMA (blue areas) and resulted from the hydrologic model (yellow lines). (From Karamouz, M. et al., *Environ. Earth Sci.*, Springer Berlin Heidelberg, 76(11), 395, 2017.)

13.12.2.1.1 Major Findings

1. Climate change can adversely impact severity of flooding, particularly with its effect on sea-level rise and storm intensity in coastal areas.
2. Design floods which are determined based on the frequency analysis of extreme rainfall and water level data may significantly be different when climate change impacts are taken into account in floodplain delineation studies. Therefore, consideration of uncertainties associated with climate change impacts in determining the extent of floodplain is suggested to extend the current work.

13.12.2.2 Case Study 4: Nonstationary-Based Framework for Performance Enhancement of Coastal Flood Mitigation Strategies

Nonstationarity of data can lead to nonnegligible discrepancies that could in turn lead to catastrophes beyond imagination. Livable cities of the future are flood-resilient cities designed and constantly managed while considering nonstationarity. In this study (Karamouz and Mohammadi, 2020), increasing the reliability of the urban infrastructure has been addressed by a more accurate understanding of flood hazards and coastal protection strategies.

The first step in this study is the hydrologic data analysis. Climate change and human activities have often caused an invalidation of stationarity assumption for simulation and forecasting flood hazards. Also, hydrological events often coincide with two or more natural events. Therefore, for a more accurate analysis, three approaches are introduced to assess the nonstationarity in data. Then, the nonstationary frequency analysis on a univariate distribution function and the joint probability distribution have been applied to define the frequency of flood hazards. First, a tradeoff curve is plotted to contain all results of a nonstationary joint probability analysis that can be used to select the design value curve based on decision makers' preferences. The second step is the selection of coastal protection strategies. Strategies are selected based on their suitability and the characteristics of the protected area. The final step is to utilize a hydrologic distribution model, with the ability to simulate both rainfall and storm surge for obtaining a flood inundation map in order to test the strategies. Although, this model has certain limitations of not including wave dynamics and setup and wind speed. It succeeds in integrating hydrologic and systems analyses, which is the main advantage of the methodology. The results show the significant value of a nonstationary analysis for obtaining design values and selecting comprehensive coastal protection strategies.

In this paper, a methodology is presented to develop a framework for the design of coastal protection strategies. A nonstationary frequency analysis is applied to the storm surge and rainfall data, and 100-year design values are obtained. Then, coastal protection strategies are chosen based on their suitability and utilized in the simulation model. Figure 13.22 shows the flowchart for the steps taken to reach the overall objective of the proposed flood protection strategies.

This paper focuses on developing a framework for obtaining flood inundation maps as a tool for flood hazard mitigation and selecting flood protection strategies based on land use and site characteristics. A hydrologic distributed model, GSSHA, has been utilized for estimating the inundation depth in the east part of the Bronx County in New York City. The model is then used for flood inundation map delineation with four flood scenarios, including a 100-year flood based on separate storm surge and rainfall analysis, a 100-year joint events analysis, a stationary analysis, and a joint analysis envelope. One of the primary contributions of this paper is to present and assess the nonstationarity detection generalized extreme value (GEV) distribution approach by the L-moment method and sequential time windows. Sequential time windows are used to create a relation between time and large-scale climate indices with GEV parameters. The nonstationary GEV method considers the observed trend and incorporates climate scenarios for changing water level elevations and its frequencies of occurrence in the future.

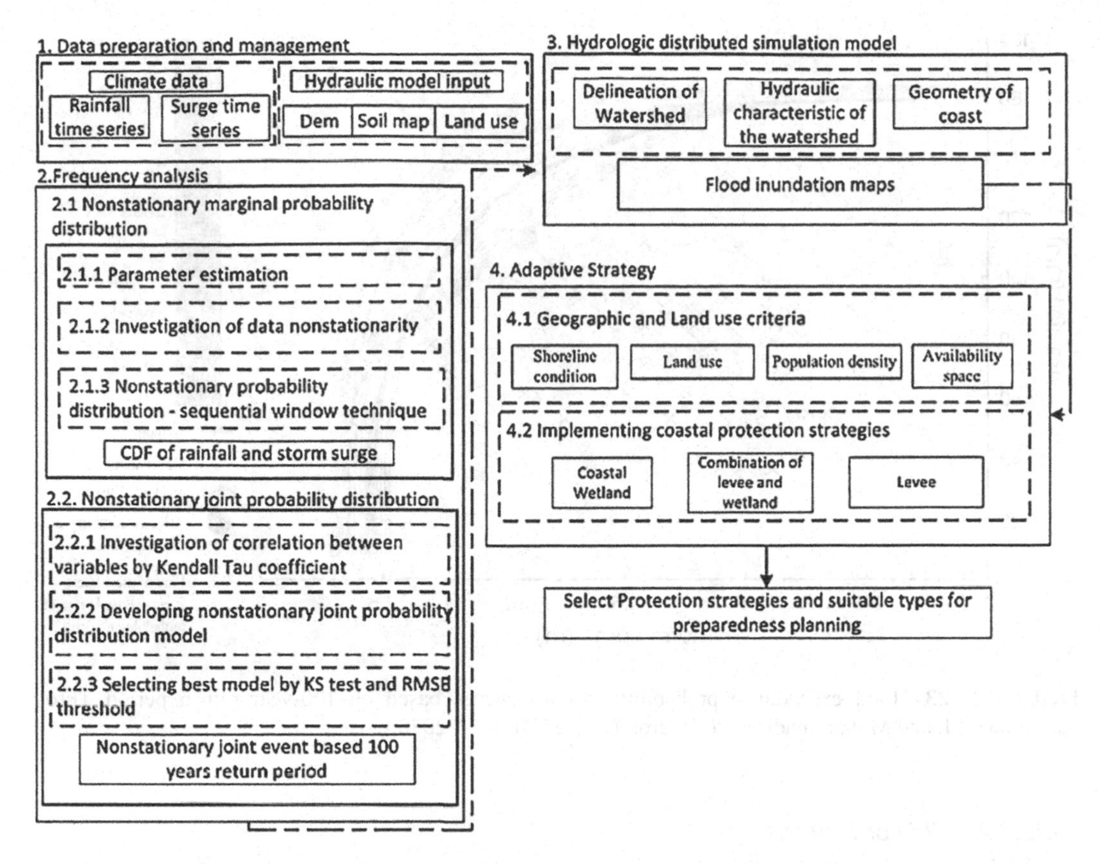

FIGURE 13.22 The study's flow chart. (From Karamouz, M. and Mohammadi, K., *J. Hydrol. Eng.*, 25(6), 1–17, 2020.)

The GEV distribution is utilized to model the marginal distribution and the Archimedean copula functions as a joint distribution function to model the dependency structure in each time window. For developing the nonstationary framework, the acceptable copula functions are selected by the KS test and root mean square error (RMSE) in each time window. Later on, by using the Kendall Tau coefficient as the dependence parameter in each time window, the best copula parameter is estimated. The marginal distribution parameters play an essential role in obtaining a nonstationary 100-year joint flood design value based on the results. In the joint analysis section, the time window of a particular year is a period of 50 years, leading up to that specific year. In this part, the most likely design value is obtained in each time window. Another contribution of this paper is developing a trade-off curve based on joint exceedance probability isolines curves (Figure 13.23). This method can perhaps change the way we have looked at design values in coastal regions. It provides design alternatives for engineers and decision makers. In the last part of the paper, three coastal protection strategies are simulated by a hydrologic distribution model. The coastal protection strategies are selected based on four characteristics of each of the three neighborhoods and their spatial constraints. It should be noted that some coastal protection approaches, although they attenuate the wave height, produce an increase in wave setup and thus could have negative effects on flood risk. This is beyond the scope of this study but should be considered in future developments. The main advantage of the methodology lies in the integration of the hydrologic and systems analyses.

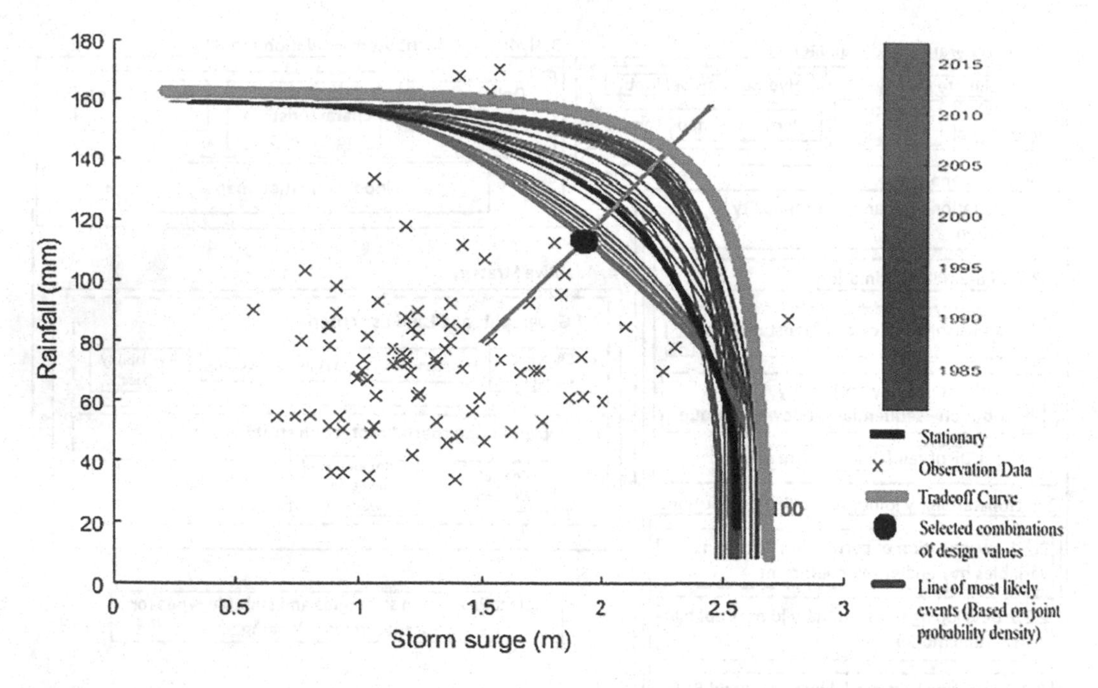

FIGURE 13.23 Joint exceedance probability isolines curves based on 100-year return period. (From Karamouz, M. and Mohammadi, K., J. Hydrol. Eng., 25(6), 1–17, 2020.)

13.12.2.2.1 Major Findings

The main conclusions are presented as follows:

1. The nonstationary approach used in this study provides many possibilities and measures for assessing the future risk of failure by considering certain hydrological and climate indices, such as Sea Surface Temperature (SST) and Southern Oscilation Index (SOI).
2. The results indicate that the model based on the SST index has a better performance than the model based on the time covariate. Furthermore, by a nonstationary joint frequency analysis, design values based on a joint probability curve are conservative and seem more practical with fewer consequences for a given risk of failure. In this way, the priorities of a decision maker can be taken into consideration by weighing more on surge or/and rainfall extreme events.
3. The results also show the significant value of utilizing the proposed methodology in selecting design values for coastal protection strategies that can also be utilized in similar coastal settings.

13.12.2.3 Case Study 5: Conceptual Design Framework for Coastal Flood Best Management Practices

Over the past few decades, the development of municipal societies along flood-prone areas such as waterfronts and riverbanks has increased rapidly, causing flood-induced disasters more often than ever before. Flood defense systems have defaulted to gray infrastructure for many centuries. However, more recent storms such as Superstorm Sandy revealed the shortcomings of gray solutions, and therefore, the notion of adaptability has been directed to the use of green infrastructure. As is becoming more and more obvious. Livable cities of the future will benefit from the numerous advantages of hybrid green–gray flood protection strategies. This study (Karamouz and Heydari, 2020) puts forward a framework for designing nature-based best management practices.

In flood risk management, nature-based strategies have already proven considerable flood-reducing potential while effectively conserving the ecosystem. However, while nature-based approaches are extremely beneficial in preserving the ecological characteristics of the region, they cannot be solely relied on when it comes to rarely occurring events such as extreme floods. BMPs are one of the most commonly used safeguards in which natural elements can be integrated into traditional infrastructure designs. In this paper, different strategies are proposed to highlight the immense effects of combining green and gray approaches on flood mitigation and prevention costs. These strategies focus on flow through vegetation (dense plants), which could be patterned as flow through porous media and were applied to Westchester Creek, New York. Suggested conceptual designs were then simulated in a hydrologic model. Flood inundation maps generated through the modeling process were then compared in five different scenarios including before and after BMPs. The results of this study showed how the combination of levee and vegetation could significantly lower surge effects while proving to be cost-effective. The proposed framework as demonstrated in Figure 13.24 has three main steps. First, hydrological and physical characteristics data including selection of a suitable DEM resolution were collected. The next step was simulating the flood inundation map using GSSHA. Finally, different strategies were suggested corresponding to flood depths and extents to mitigate both. Strategies were modeled and new inundation maps were developed.

New perspectives on coastal flood mitigation strategies in coastal regions, with a focus on nature-based solutions, have been proposed to attenuate flood inundation and extents efficiently. The initiatives outlined here could lead to a safer yet more cost-effective flood mitigation strategy. The 2-day storm surge water surface elevation data during Superstorm Sandy were selected from the observation dataset by the National Oceanic and Atmospheric Administration (NOAA) for Westchester Creek in New York. A distributed hydrologic simulation model was used to test the effectiveness

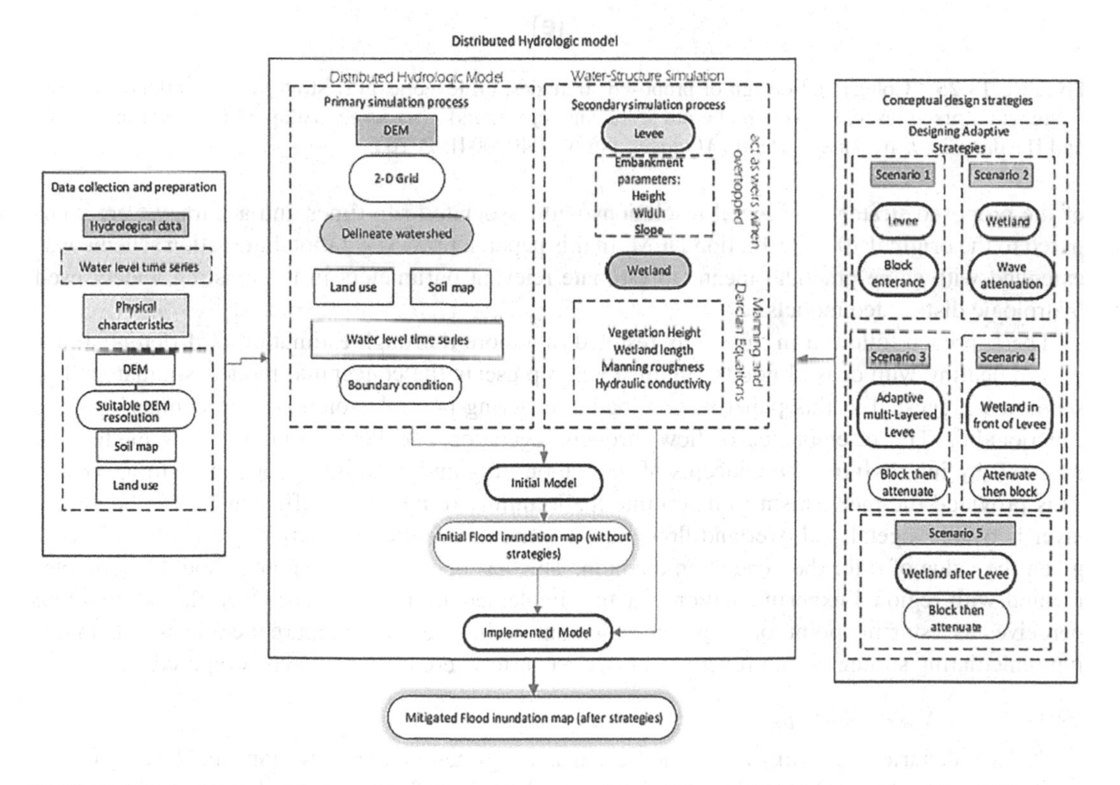

FIGURE 13.24 The study's flow chart. (From Karamouz, M. and Heydari, Z., J. Water Res. Plann. Manage., 146(6), 04020041, 2020.)

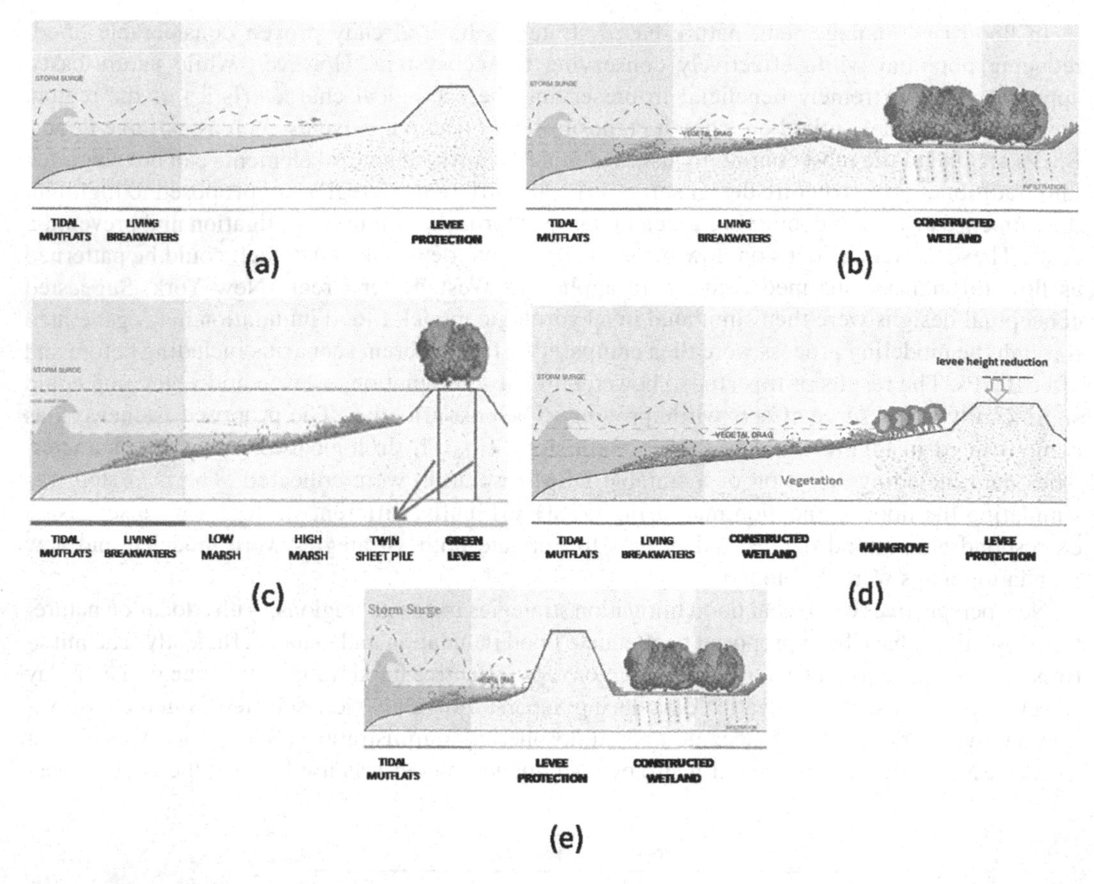

FIGURE 13.25 Conceptual design of proposed strategies: (a) levee as gray strategy; (b) wetland as green strategy; (c) green levee; (d) wetland before levee; and (e) wetland after levee. (Adapted from Karamouz, M. and Heydari, Z., *J. Water Res. Plann. Manage.*, 146(6), 04020041, 2020.)

of the proposed strategies. Several resolutions with associated run times and accuracy were compared to the original 10-m-resolution DEM. In this paper, a flow–vegetation interaction scheme was extended with some practical means to estimate relevant parameters in the existing widely used hydrologic distributed models.

The approach followed in this study resulted in a more thorough estimation of different strategies in dealing with coastal floods. Figure 13.25 represents the conceptual models simulated. The scenarios suggested in this paper were aimed at covering possible solutions and comparing their functionality. The resemblance of flow through vegetation and flow through porous media was explored by intertwining the concepts of vegetation drag and Manning's roughness to approach. This provided the mechanism to determine the Manning roughness coefficient for the boundary layer between vegetal and overland flow. Furthermore, hydraulic conductivity (K) was estimated given the value of n for the vegetation medium. This has been a challenge and should be complemented with regional experimentation if actual implementation is desired. Overall, this could be perceived as a starting point for future investigations in this area. To compare the effects of combining remediating strategies, the functions of five scenarios were evaluated and compared.

13.12.2.3.1 Major Findings

1. Two scenarios, including a green levee and an integrated system of wetland and levee, were chosen as the final strategies for this study. The rate of attenuation showed remarkable decrease in both inundation and extent when combined strategies were applied.

2. An added benefit of the strategies was the overall cost savings of more than 40% for areas with fewer spatial constraints (west bank of the creek) and up to 17% for the combined areas with imperative land use near the shore.

13.12.2.4 Case Study 6: Improvement of Urban Drainage System Performance

In recent years, climate change and its consequences have affected flood characteristics especially in urban areas. In this way, incorporating the climate change effects in urban flooding studies can help achieve more reliable results that will be applied in real-time planning of urban areas through selection of BMPs. Karamouz et al. (2011) proposed an algorithm for selecting the BMPs to improve system performance and reliability in dealing with urban flash floods, considering the anthropogenic changes and climate change effects. The suggested algorithm is applied in the Tehran metropolitan area, as a case study. This case study complements the case study in chapter 6.

In recent years, Tehran, the capital of Iran, has undergone rapid development without considering the diverse impacts on the environment especially the water cycle. This has resulted in a wide range of challenges and obstacles in water- and sanitation-related infrastructures in the area. Lack of a systematic approach to runoff management in Tehran has led to the frequent overflow of channels and some health hazard problems in rainy seasons. This case is a mountainous area with high population density. Therefore, urban floods can cause considerable damage and the evaluation and simulation of the effects of floods in this area are very important.

The drainage system of the study area is composed of natural and man-made channels. These channels carry large amounts of sediments especially during floods, which drastically decrease their safe carrying capacity. These channels are used as a combined facility for surface water and wastewater drainage and snowmelt discharge.

In recent years, some river training projects have been performed on these channels, which have changed the characteristics of the surface water collection system. For evaluation of the effects of these projects on the runoff collection system of Tehran, three scenarios have been considered:

Scenario 1: In this scenario, the surface water collection system of about 10 years ago is simulated.

Scenario 2: In this scenario, the present surface water collection system of Tehran is modeled. The effectiveness of development projects utilized in recent years for improving the performance of the drainage system is evaluated in this scenario.

Scenario 3: Future plans for the improvement and development of the drainage system of the case study have been modeled in this scenario. Here, a channel and two detention ponds are added to improve the performance of the drainage system.

The possible changes in rainfall patterns of the study area under climate change impact are simulated. The effectiveness of present and future planned projects for the improvement of drainage system performance in transmitting urban floods is evaluated under different scenarios. Also, the effect of solid wastes and sediments carried with surface runoff on system performance is considered. In suggestion of BMPs, feasibility and effectiveness are related to the cost and benefit.

In this case, an algorithm is developed for evaluating the performance of the urban drainage system considering development projects and climate change effects. Three scenarios—past, present, and future—for the drainage system have been considered. The proposed algorithm for evaluating climate change impact on the urban water drainage system is illustrated in Figure 13.26. More details are provided in a case study in Chapter 6.

The downscaling model used in this case is Statistical Downscaling Model (SDSM). In this study, StormNET developed by Boss International (2005) is used as a rainfall–runoff model. Information on channel and pipe links, snow packs, subbasins, detention ponds, and flow diversions is needed to develop this model. In order to choose the critical daily rainfall of each season that may result in a flood, the maximum value of each season's rainfall is selected among the rainfall data of each year as the most critical rainfall of that season of that year.

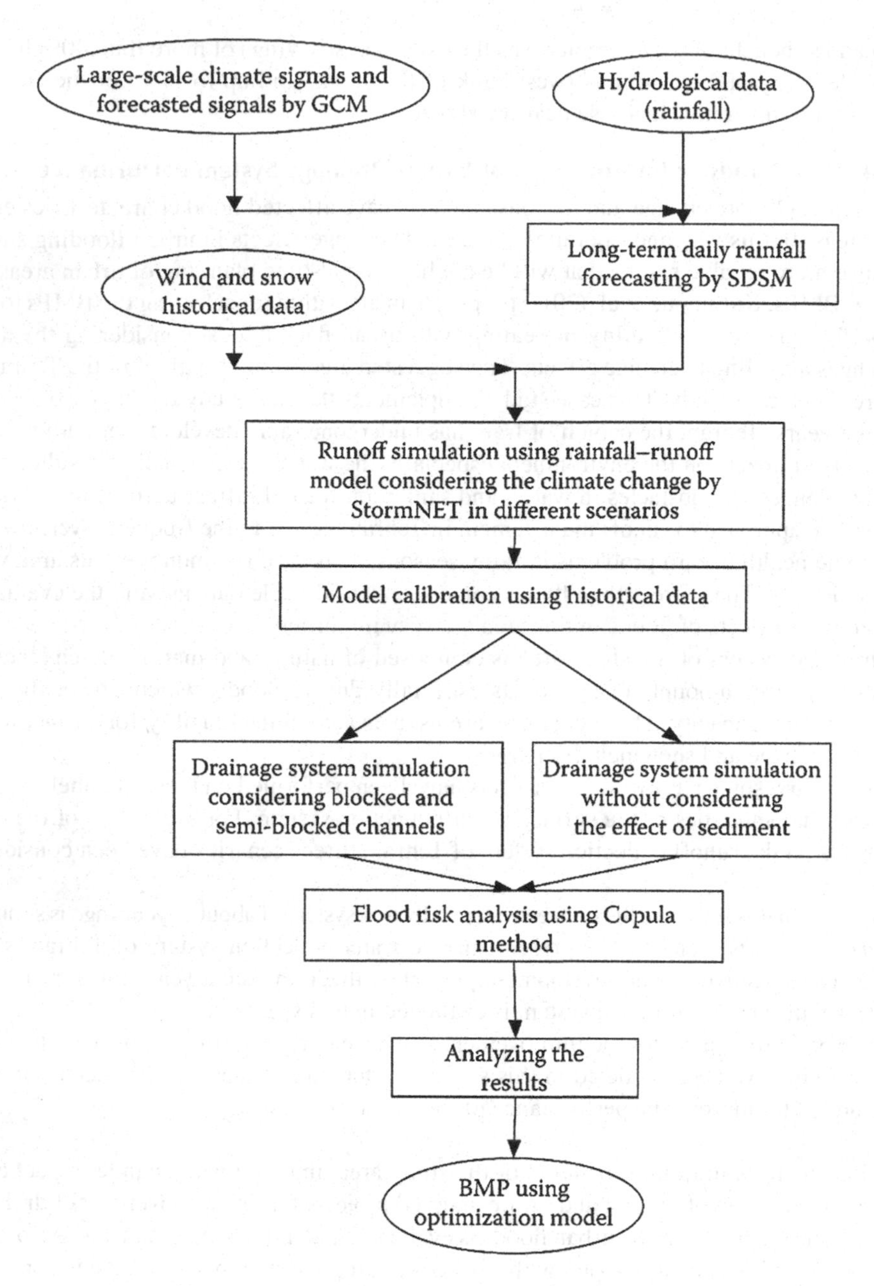

FIGURE 13.26 Proposed algorithm of case study 2 for the improvement of drainage system performance in urban areas.

To provide a realistic configuration of the performance of the urban drainage system, simultaneous simulation of rainfall and sediment load during flood is needed. For this purpose, the Copula theory has been used in this study to calculate the joint return period of rainfall and sediment. Investigation of the sediment data shows that the sediment load in the spring and winter is much more than that in the fall and summer, and sediment load is at a minimum level in the summer. According to the available reports on recent floods, the volume of channels that have blockage problems has decreased based on the sediment data in each season. Then, the model is analyzed for different sediment loads in different seasons. In these models, the 20-, 50-, 100-, and 200-year return period rainfalls are considered.

The volume of surface flood of the entire study area in scenario 3 is a little less than that in scenario 2 (an average of 5%) due to little improvement of the drainage system performance in scenario 3 in dealing with high sediment loads. However, because of the decrease in perviousness in scenario 2, the surface flood volume of the entire study area increased by about 40% compared to the first scenario. Factors such as urbanization, which has destroyed green space and replaced it with impervious structures, and construction regardless of the channels' right of way result in high flood volumes in scenario 3.

The joint return period of rainfall and sediment is considered in this study for flood risk evaluation. The results show that the frequency of flood events will considerably decrease in the future owing to climate change effects and modification of drainage systems.

The selected area includes 15 main subbasins; the most important ones are Velenjak, Sadabad, Darband, Golabdare, Jamshidie, and Kashanak (Figure 13.27). The study area has a short concentration time because of steep slopes, especially upstream. In this kind of situation, enhancing the density of grasslands can help lag the peak time of urban floods. Since the upstream subbasins of the study area are nonresidential, enhancing the green space, which results in increased soil penetration and a decrease of soil erosion, is considered a BMP. Also, due to steep slopes of upstream mountainous subbasins, detention ponds are proposed to decrease flood hazards by increasing the lag time of the flood peak as well as decreasing the sediment load. The land use pattern of the study areas should be cautiously considered when increasing the capacity of the existing channels. Since developing a closed channel system is more reasonable and economical in an urban area, it is also considered as another BMP. The risk of flooding is drastically increased downstream of junctions that receive large channels. In these cases, some parts of channel flows can be diverted to adjacent channels with less flooding risks and damages or extra capacity for safe stormwater transfer. Therefore, the development of appropriate diversion systems is considered as another BMP in developing flood management schemes in the study area.

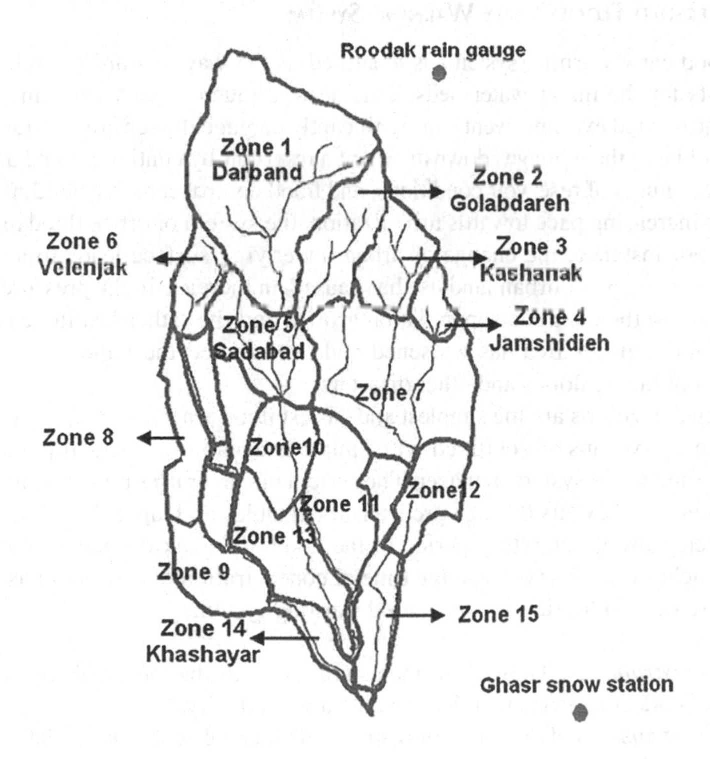

FIGURE 13.27 Overall view of the study area.

The BMPs' performance is evaluated through their employment in scenario 3 of the watershed drainage system considering the climate change effects on rainfall and sediment. An optimization process is followed to determine the optimal BMPs and their location in the drainage system. Since the considered case study here has limited choices for application of BMPs, the optimal composition of BMPs is determined by evaluating all possible situations with regard to the objective function and constraints of the proposed optimization model.

Cities are often vulnerable to floods, and hence, they have to be reinforced through the usage of BMPs. In many cases, some projects will be proposed and implemented by the in-charge authorities without enough initial studies. Such new projects can impair and modify many of the characteristics of the area in terms of the components of the water cycle such as the drainage system. On the other hand, stationary estimations and predictions are no longer accurate and the effects of climate change cannot be overseen. This study demonstrates how the effectiveness of announced projects should be evaluated considering the consequences of climate change and what can be done to improve the overall responsiveness of cities to flash floods.

13.12.2.4.1 *Major Findings*

1. In this optimum BMP, the total cost of flooding in the study area has decreased about 75% in comparison with the condition that no BMP is developed.
2. Using the optimal combination of BMPs in the study area has resulted in a significant decrease in flood volume and flooded area, especially at downstream subbasins.
3. The results show that in the future, the intensity of extreme events will decrease because of climate change effects.
4. The results of the study show that the probability of accruing flash floods with high intensity in short time periods is increasing; also the best way to decrease the risk of floods in urban areas is to pay more attention to waterways and maintain their capacity.

13.12.3 Watershed Flood Early Warning System

The need for flood early warning systems is identified as the way forward in preflooding and post-flooding incidents for the major watersheds. This need is much more serious in low-lying coastal areas since an unexpected extreme event can significantly damage the sensitive infrastructure. A flood warning system reduces the damages downstream of a reservoir by continuous and active monitoring, attending to the readiness of reservoir conditions, and flood control capacity building in the reservoir.

With the ever-increasing pace towards urbanization, the system of urban flood disasters is changing by the day. For instance, the change of urban underlying surface leads to a change of urban runoff; the rapid expansion of urban land use has caused an increase in the pressure on the drainage system of these areas; the continuous population growth and the rather fast increase in the number of properties in the built-up area has worsened and exacerbated the vulnerability of these urban areas in the event of facing floods and other disasters.

Flood warning algorithms are the simplest and easiest flood warning systems handled by a manual operation. These systems are outfitted with simple gauges in the critical points of the system. The data are reported to the system manager. The system manager uses real-time hydrometeorological data to estimate flood events through predetermined tables and algorithms. The predictions can be included in determining the return period of the flood, the peak discharge of the flood and the time the flood reaches the reservoir, and the time of concentration of each sub-basin. Three groups of instruments are needed for developing a flood warning system:

- *Measuring instruments*: These instruments are set up in the meteorologic and hydrologic stations and collect hydrometeorological data automatically.
- *Hardware instruments*: This group of instruments includes computers, data transfer tools, and receiver and amplifier centers.

- *Software instruments*: Databanks, data processors, flood prediction models (including rainfall–runoff and flood routing models), visualization models, and user interfaces are included in this group.

In this chapter, as an example, a general framework for the design of a flood warning system for the Kajoo watershed located in the southeastern part of Iran is presented (Figure 13.28). The main steps of the algorithm of designing such a system for an urban are more or less the same as the one illustrated here in Figure 13.28 for Kajoo river. A main component of this flood warning system is the assessment of the observed cumulative precipitation in the upstream climatic stations as well as the observed flood hydrograph in the upstream hydrometric stations. To forecast floods, control points have to be determined and flood characteristics can be estimated once the precipitation is observed using rainfall–runoff models such as Hydrologic Engineering Center– Hydrologic Modeling System (HEC-HMS). This flood warning algorithm is developed based on the data from weather and hydrometric stations upstream of the watershed. The components of this algorithm are rainfall and runoff thresholds, corresponding time of thresholds in existing and proposed stations, and the state of reservoir management, including the flood control capacity and the river carrying capacity downstream of the reservoir. Floodplains can be determined using the HEC–River Analysis System (RAS) model, and different warning levels can be announced depending on the expected flow progression in the floodplains. When used for urban areas, a few of the inputs to these simulation models may change. For instance, many hydrological characteristics of urban areas are somewhat different from the same attributes of the surrounding areas. Moreover, urbanization noticeably reduces the extent of the drainage areas of a watershed since much of the soil and vegetation is replaced by street covers such as concrete. These variations have to be considered when designing an early flood warning system for a metropolitan area.

This type of system could provide early warning to other reservoir–river systems and could be utilized if the information about upstream weather and hydrometric stations could be gathered in a timely fashion. This information will guide the decision makers and/or operators to lower the water level in the reservoir and/or undertake other flood control alternatives. It will also give an early warning to farmers and the public.

13.12.4 Flood Insurance

Flood insurance is one of the approaches to reduce vulnerability in flood risk assessment. Disaster vulnerability and poverty are mutually reinforcing. Low income, poor housing and public services, and lack of social security and insurance coverage make poor people behave in a way that puts them at greater risk. Since natural disasters significantly affect poor people, specific policies and strategies are required for tackling the link between poverty and disaster vulnerability.

An effective way for lessening damages and increasing resiliency is risk sharing and transfer at national, community, and household levels (UN/ISDR, 2002). An example of this approach is insurance. Insurance can improve the situation of individuals by compensation and spreads the risk of disaster across society.

In the 1960s, nonstructural measures, such as warning systems based on real-time flood forecasting techniques, floodplain zoning to restrict occupancy of the plain to certain uses, local flood-proofing, and flood insurance programs, started to receive more attention in flood risk management. Therefore, flood insurance has often been advocated as a long-term nonstructural measure for building resilience among flood victims. However, the unsatisfactory status of flood insurance in developing countries shows that financial support mechanisms (subsidies, funds, and loans for spreading the financial burden in terms of equity and fairness) should be combined in flood management to incorporate equity with economic effectiveness.

Flood insurance is an effective way for reducing economic vulnerability especially in industrialized societies. Insurance is a classic form of risk-sharing since it distributes financial risk

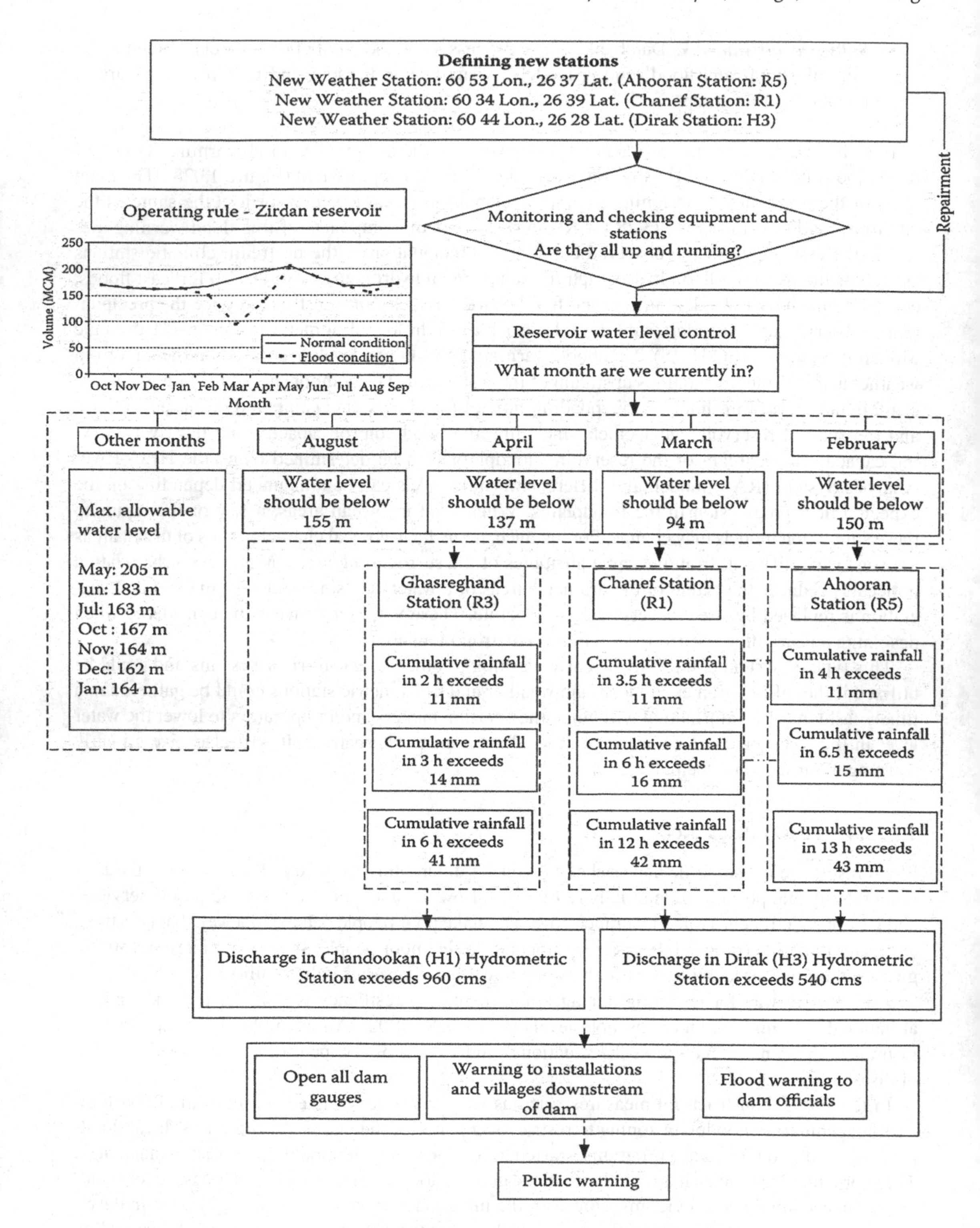

FIGURE 13.28 Flood warning system algorithm.

approximately evenly among all policyholders. However, for the poor people, who are more vulnerable to floods, flood insurance is often economically unfeasible. In such cases, cooperation between the government and insurance companies is necessary in order to encourage the latter to provide affordable flood insurance policies for low-income groups. Risk-adequate insurance premiums are distributed spatially as a function of the size of the risk community to make premiums more affordable by enlarging the number of policyholders (APFM, 2008).

In spite of the various efforts for mitigating flood damage, floods damage properties, and interrupt economic activities. Some of the tangible losses are absorbed by the retained risks (a component of general risk). In a systematic risk management process, some of the tangible risk is transferred through insurance as the last step for sharing the cost of recovery. Insurance has different benefits: it protects capital, enhances solvency, and allows recovery. A well-designed insurance scheme could also encourage risk reduction behavior. For small-scale floods, which are usually predictable, risk reduction methods are more advisable. High-risk floods (low-probability and high-consequence events) easily destroy the insurance market. Therefore, insurance instruments are appropriate for middle levels of risk.

13.13 LIVABLE CITIES OF THE FUTURE

Smart Cities and Livable Cities of the Future are what city planner with engineers, social scientist, and IT experts are starving to design. There are many initiatives and models. Among them, a symposium to honor the legacy of George Bugliarello in late October 2012, that was hosted by the Polytechnic Institute of New York University (NYU-Poly), was remarkable. This event brought more than 200 engineers, civic leaders, educators, and futurists to discuss how George Bugliarello's works and his technical point of view manifest themselves in innovating urban planning for livable cities of tomorrow.

The symposium objectives were to generate innovative ideas for best practices and strategies for sustainable urban development and further enhance New York City's evolution to a real-life laboratory for modern urban innovation. Participants heard the perspectives and experiences of representatives from both private and public service operators and sectors, infrastructure agencies, and academic individuals. Stakeholders in urban and other industries and elected officials examined issues critical to resilient and sustainable cities, such as energy, water supply and treatment, public health, security infrastructure, transportation, telecommunications, and environmental protection.

The presentations and speeches of the symposium were recorded and transcribed into a proceeding of the symposium by Karamouz and Budinger (2014). Some designated parts of the proceedings were selected to be presented here since they would paint a clear picture of the challenges faced and the must-have attributes of the livable cities of the future. Each of the presentations is used here to reflect the view point of an office, an agency, a former Mayor, and an IT expert.

13.13.1 Mayor's Office Point of View

Robert K. Steel the depute Mayor of New York City at the time of the symposium made the keynote speech. He argues that Mayor Bloomberg's administration considers economic development strategy is to be defined by four pillars. The first pillar is a high quality of life characterized by public safety, excellent schools, beautiful parks, clean water, and cultural amenities. The second is a pro-business environment with sensible regulation that ensures safety while encouraging business growth. The third pillar is investment in the future by developing transportation infrastructure, office space, and housing, which are the city's long-term success. The final pillar is innovation and economic transformation.

He stated that Mayor Bloomberg's applied sciences initiative would generate billions of dollars of economic activity and create thousands of new jobs connected with the campuses' construction, operations, and spinoff companies. Cornell NYC Tech, NYU, and Columbia would more than double the number of engineering faculty and graduate students in NYC over the next decade. Taken together, these three new campuses will dramatically transform the city economy's ability to compete successfully in the twenty-first century and beyond.

13.13.2 Infrastructure Renewal: An Agency View Point

13.13.2.1 The Big Picture

According to the President of NY-NJ port Authority, with the ever-growing demand for more secure, affordable, safe, and sustainable metropolitan water and energy resources and supply systems, national and local governments and urban utilities face the challenge of upgrading their infrastructure monitoring and system management capacity. Growing eco risks of climate change impacts and accompanying uncertainties pose conditions, financial, environmental, operational, and societal challenges to metropolitan governments and urban utilities for the strategic and operational deployment of their natural resources and the management of their urban distribution systems.

Sustainable urban economic growth and development over the coming decades will depend on cities' potential to react well when facing these challenges, which in turn will require reinventing regional planning practices, adapting sustainable development strategies, and implementing intelligent control systems and proactive incident detection and mitigation measures. Furthermore, as energy and water utilities face growing uncertainties of eco risks and greater frequency of extreme events, they have a critical need for smart control capabilities for integrated and real-time system management, early incident detection, and preemptive mitigation.

13.13.2.2 The Critical Role of Transportation

Transportation is a significant player in any city. In particular, intelligent livable towns and cities of the future must have smart and sustainable transpotation systems. In NYC, the Port Authority is the central administration in charge of the transportation sector. The Port Authority of New York and New Jersey is a bistate agency that operates and maintains the following transportation infrastructure assets:

- Six bridges and tunnels connecting the two states;
- Five airports that constitute the busiest commercial aviation system in the United States;
- Two bus terminals, including the Port Authority Bus Terminal, the oldest and most active bus facility in the world;
- The Port Authority Trans-Hudson (PATH) rail system, used by more than a quarter of a million commuters every weekday;
- Marine facilities on both sides of the Hudson River that are the largest destination for cargo on the East Coast and third largest port in the country; and
- The 16-acre World Trade Center site.

The Port Authority is responsible for the transportation infrastructure that drives economic development and prosperity in our region. It is committed to accomplishing the dual mission of sustainably meeting the region's transportation needs, now and in the future, while catalyzing jobs and economic development. Without sustainable transportation systems, there will not be livable cities.

13.13.2.3 Sustainable Urban Renewal

Urban infrastructure for water, energy transmission, and transportation requires significant investment to manage its economic and environmental impacts effectively. Investment in *water infrastructure* must be increased to prevent higher costs to businesses and households and protect almost 700,000 jobs, personal incomes, gross domestic product (GDP), and US exports. Investment in *energy transmission infrastructure* can prevent or minimize the impact of future blackouts and brownouts, yielding further protection for 529,000 jobs, personal incomes, GDP, and US exports. Finally, investment in *surface transportation infrastructure* can create millions of jobs, protect existing positions, save nearly 2 billion hours in travel time, save each family $1,060 per year, and add $2,600 in GDP for every person in the United States. These investments can also help improve the overall standard of living. In all these different fields, engineers play a critical role in shaping the world and making the environment sustainable.

To conclude, engineers are at the forefront of the nation's infrastructure upgrades, with a critical opportunity to approach infrastructure repair, enhancement, and construction in new ways that incorporate sustainability and resiliency to make existing and emerging cities and megaregions smarter and greener.

13.13.2.4 Energy as the Core of NYC

There are five critical components to the livable cities of the future: public safety, reliability, affordability, reduced environmental impact, and more innovative and more secure facilities. Cities must also prepare for both a growing demand for power and the effects of increasingly severe weather patterns that threaten the grid. This became dramatically clear with Superstorm Sandy, which hit the New York City region 3 days after this symposium, on October 29, 2012. Utilities, urban planners, climate experts, government leaders, and regulators must all collaborate to determine the best approaches to fortify the city's infrastructure and protect residents and businesses from future threats.

New York City's largest energy provider is taking steps to address challenges and meet needs to ensure delivery of these critical components. For a city to be livable, safety is critically important, as are reliability, risk reduction, and affordability. As urban sites continue to grow, reduced environmental impacts, more intelligent systems, and more secure infrastructure are also paramount to the future of the cities and their inhabitants all over the world.

13.13.2.5 Using Water for Urban Renewal

Needs for clean water in all aspects of urban development and maintenance can be met through by integrating schemes for capturing and collecting precipitation, wastewater sanitation management, and modernization of infrastructure maintenance technologies. New York City has had success in using natural systems to provide clean drinking water and manage storm runoff. The city has saved billions of dollars through integration of diverse methods for controlling water quality, distribution, use, and reuse.

To meet future urban water and sanitation challenges, we have to rethink the way we manage urban water systems. We need a paradigm shift. A more integrated approach may transform threats into opportunities and address the challenges of urban water management in both developed and developing countries. Do we continue to spend money to treat water to drinking water quality only to use it to fight fires or carry wastes to a WWTP? Do we continue to spend money on the traditional (gray) infrastructure? In addition to the integrated approach, do we consider more decentralized approaches where beneficial? Water reuse, energy recovery, the use of local water sources, and waterless toilets all foster decentralization. I predict that we will see many more integrated decentralized approaches to urban renewal efforts in the future, in the cities of both developing and developed countries.

13.13.3 Fighting Climate Change—A Former Mayor View Points

This part is related to climate change, the global issue humanity is facing. Cities around the world are acting to both fight climate change and adapt to it. Like Toronto's Clean Air and Climate Change Action Plan, they have strategic plans, and executing those plans is lowering carbon emissions and building new resilient infrastructures. The role of cities is critical because as of 2008 most residents of the world lived in cities, which is where most jobs are and most carbon emissions occur. By addressing carbon emissions in three sectors—heating and cooling buildings, energy generation, and transportation—emissions can be dramatically reduced, the livability and resilience of cities improved, and new jobs created.

Future strategies should not focus only on economic or environmental sustainability. Social sustainability must also be at the forefront of a city's planning strategies. The City of Toronto calls it prosperity, livability, and opportunity. Environmental strategies can provide many skilled job opportunities. With trade unions, Toronto has developed strategies to help train young people in low-income neighborhoods with the skills needed to obtain jobs. Combining all these effects, the plan will help meet critical environmental goals, create jobs, and support social sustainability. This could be done in all cities around the universe.

13.13.4 Urban Challenges: The Way Forward—An IT Expert View Points

According to an IBM Vice President, there are many challenges in urban communities to become livable cities in the years to come by becoming more intelligent. "Smarter cities" use interconnection, instrumentation, and intelligence to raise awareness and enhance coordinated responsiveness to different activities and events. Interconnected technologies evolve the way the universe literally works, and cities are arguably the best opportunity to begin working towards a more intelligent world.

13.13.4.1 The Challenges

In the universally interconnected business and IT world, natural disasters fueled by anthropogenic interventions, large-scale systems' failure, and dangerous threats emphasize the complexity and vulnerability of urban systems, whether natural or developed for global business and government operations and purposes. Moreover, reported cases of disasters/incidents around the world have been increasing in recent years in both number and severity.

Moreover, the growth of populations, changes to the balance between economic status of individuals and nations, migration, political instability, and environmental complications are leading to constraints on vital resources such as water, food, and energy. This "stress nexus," as shown in Figure 13.29, is limiting economic development in many places around the world.

13.13.4.2 The Path to Success

Collaboration is indeed the way forward and the only way to succeed. Industries and businesses, universities, and governments can cooperate to bring actual economic development to cities and regions. For instance, just north of New York City, IBM was created as a center for nanotechnology

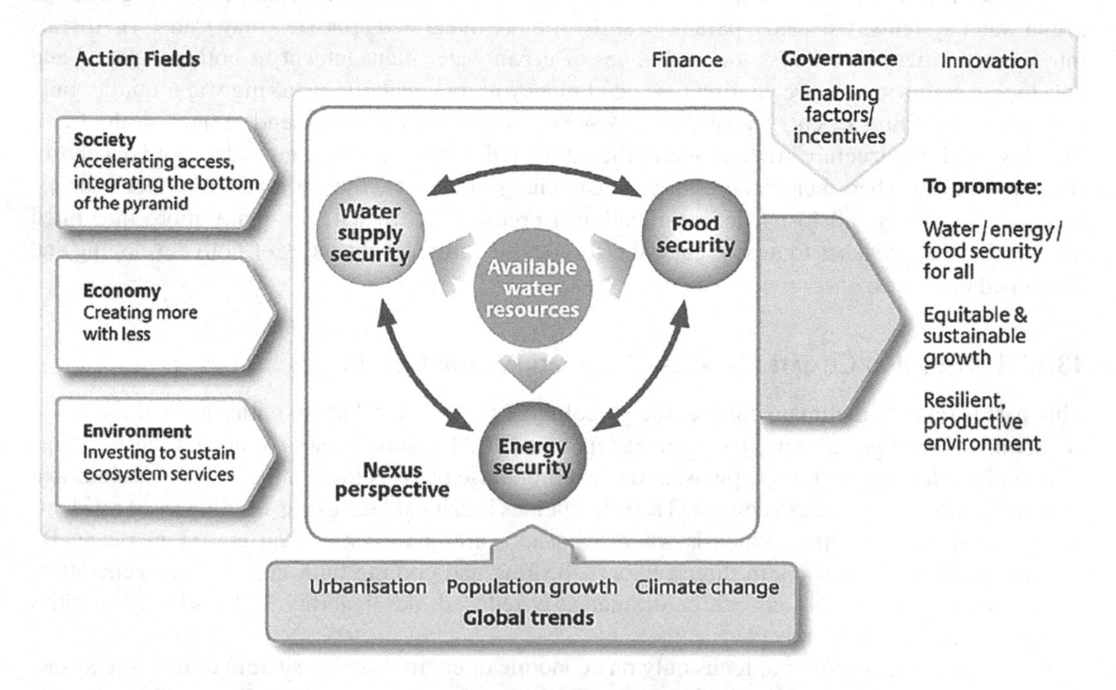

FIGURE 13.29 The stress nexus. (*Source*: Hoff, H. (2011) Understanding the Nexus. Background Paper for the Bonn 2011 Conference: The Water, Energy and Food Security Nexus, Stockholm Environment Institute (SEI), Stockholm, Sweden.)

and microelectronics. The project brought the local university, industry, and government together in order to create benefits to the region's financial state and employment. A presentable outcome was that the government invested in the project, helping to underwrite not only the facility, that some microelectronics companies share, but also the infrastructure needed to support future research for that industry of the region. Before long, this initiative attracted every major company/organization in the industry to send a research team to the facility in Albany.

Gatherings such as this symposium pave the way for future projects and expansions. Bringing scientists and specialists together for projects is one of the best possible practices that can be further explored in future in times of hardship as it has been used in the past in the form of special task forces.

13.14 CONCLUDING REMARKS

This chapter is dedicated to the flood resiliency of cities. Hence, many of the essential concepts of a flood-resilient city such as different types of flood, flood analysis, flood routing, flood damage, and partial frequency analysis are covered in this chapter. In addition, this chapter has employed a case-study-based approach towards explaining various necessary concepts of flood-resilient cities that would be emerging for smart water cities for years to come. The case-study-based format of this chapter is meant to facilitate grasping some of the practical details. An innovative concept, "Livable Cities of the Future," is introduced and developed throughout this chapter, and different aspects of such a city are made by presenting the view points some experts. The topics discussed in this chapter could be used to improve/create cities' long-term management plans, in particular, low-lying coastal areas which are at a significantly higher risk of flooding.

Six case studies are presented. In the first one, the focus is on the process and tools needed for coastal flood damage estimation; in the second one, the idea of prioritizing investments and allocating funds based on a resiliency metrics to increase flood resilience in coastal cities is discussed. The third one points out a subtle fact of combined coastal and inland floods. The fourth case study emphasizes the significance of nonstationary analysis and also discussed BMPs for coastal flood mitigation. The fifth one presents a conceptual design framework for the said BMPs taking into account nature-based solutions, and the sixth and last case study of this chapter seeks the improvement of urban drainage systems due to system expansion and climate change. If all the major finding in this chapter are incorporated into the flood planning agenda of urban communities, we can expect higher adaptive capacity and lower vulnerability to more frequent extreme floods.

Livable cities of the future should be resilient cities to expected threats such as flood. They should be resiliently and economically planned and equipped with smart tools and techniques and should be expanded or contained based on resiliency metrics.

PROBLEMS

1. In case study 1, find answers to the following questions:
   a. What does FEMA stand for? Summarize the responsibilities undertaken by FEMA. What organization would be your own nation's equivalent of FEMA? Elaborate.
   b. What does HAZUS stand for and how and on what basis does in operate? What inputs would be fed to HAZUS? What output do you expect to receive from HAZUS?
   c. Is it feasible to use the FDE model instead of the sophisticated HAZUS in a project of national significance? Provide a complete answer.
2. In case study 2, find answers to the following questions:
   a. What does resiliency mean and what are the main components of it? How would you assess New York City's resiliency in terms of the criteria of resiliency?
   b. What concept was used to quantify reliability? Could the same concept be used to quantify the reliability of other infrastructure? Elaborate with an example.
   c. Which plants were more sensitive to budget changes and how was that discovered?

3. In case study 3, find answers to the following questions:
   a. What do DEM and GIS stand for and how have they contributed to the subject of flood simulation?
   b. Which distribution represents the severity of floods more accurately? Why do you think that is?
   c. What does floodplain delineation mean and what tool/tools are used when conducting it?
4. List some of the implications of the major findings of the first three case studies in building resilient cities of the future.
5. In case study 4, find answers to the following questions:
   a. What does nonstationarity mean and how is it detected in a set of data? What would the presence of nonstationarity in a set of hydrologic data mean?
   b. What are univariate and bivariate distributions and which is better suited for flood occurrence probability studies?
   c. What are flood protection strategies? Mention instances of use of such strategies in flood protection systems of coastal cities. On what basis were these strategies designed in this case study?
6. In case study 5, find answers to the following questions:
   a. Name two other terms that are used to refer to structural and nature-based flood mitigation strategies. Which are more effective? Why?
   b. What does BMP stand for? Can green BMPs be solely relied upon? Elaborate.
   c. What does vegetation resemble when it comes to simulations? What scheme was used to model vegetation in this study? What model was used for the simulation?
   d. Which scenario generated the best results and why do you think that is?
7. In case study 6, find answers to the following questions:
   a. Describe the state of the drainage system of the study area. What is the main flaw of the drainage channels during flooding events?
   b. What is downscaling and when are downscaling methods used? What downscaling method was used in this study? Search for and name two more downscaling methods and provide a brief description/comparison of the two.
   c. What is the best way to decrease the risk of floods in urban areas as stated in this study? What other ways could you recommend? Name and discuss at least two.
8. Estimate the 25-year and the 100-year peak discharges using the data in Table 13.12. The map skew for this location is −0.015.
9. Using the data in Table 13.13, compute the 25-year and the 100-year peak discharges. The map skew for this location is 4.4.
10. Use the data of Table 13.14 to estimate the 10-year and 200-year peak discharge using the log-Pearson Type III distribution.
11. In Problem 10, what is the 100-year flood using the normal distribution?
12. Use the damage–frequency relationships in Table 13.15 for flood control alternatives to rank their merits on the basis of expected flood damage reduction.
13. Determine the optimal design return period from a flood control project with damage and capital costs as given in Table 13.16.
14. The frequency–discharge–storage–damage data for existing conditions at a particular area along a river is given in Table 13.17. The storage–discharge data make up the rating curve, the discharge–frequency data compose the frequency curve, and the storage–damage data make up the storage–damage curve. Plot these three relationships.

TABLE 13.12
Annual Peak Discharge in Problem 8

Year	Annual Peak Discharge (m³/s)	Ranked Annual Peak Discharge (m³/s)	Weibull Plotting Position	y	y^2	y^3
1962	2,930	2190	0.521	3.34044	11.15854	37.27443
1963	574	1960	0.542	3.29226	10.83898	35.68473
1964	159	1800	0.563	3.25527	10.59678	34.49539
1965	1,400	1750	0.583	3.24304	10.51731	34.10805
1966	1,800	1510	0.604	3.17898	10.10591	32.1265
1967	2,990	1420	0.625	3.15229	9.93693	31.32409
1968	4,470	1400	0.646	3.14613	9.89813	31.14082
1969	2,280	1630	0.667	3.13354	9.81907	30.76846
1970	734	1240	0.688	3.09342	9.56925	29.6017
1971	4,710	1080	0.708	3.03342	9.20164	27.91243
1972	1,360	1050	0.729	3.02119	9.12759	27.57618
1973	265	734	0.75	2.8657	8.21224	23.53381
1974	592	592	0.771	2.77232	7.68576	21.30738
1975	1,050	574	0.792	2.75891	7.61158	20.99968
1976	1,960	560	0.813	2.74819	7.55255	20.75584
1977	4,660	465	0.833	2.66745	7.11529	18.97968
1978	297	427	0.854	2.63043	6.91916	18.20037
1979	400	400	0.875	2.60206	6.77072	17.61781
1980	560	297	0.896	2.47276	6.11454	15.11979
1981	38	265	0.917	2.42325	5.87214	14.22966
1982	1,420	159	0.938	2.2014	4.84616	10.66834
1983	427	119	0.958	2.07555	4.30791	8.941278
1984	119	38	0.979	1.57978	2.4957	3.942665

15. If a dike system with a capacity of 15 m³/s were used as one alternative for the situation in Problem 14, develop the damage frequency curve for this alternative.
16. If an upstream permanent diversion that will protect up to a natural flow of 15 m³/s is used for the situation in Problem 14, develop the damage frequency curve for this alternative.
17. If channel modification is used for the situation in Problem 14 to increase the conveyance capacity of the river up to 15 m³/s, develop the damage frequency curve for this alternative.
18. Considering Problems 14 and 15, determine the return period for the dike system that maximizes annual expected benefit.
19. Considering Problems 14 and 17, determine the return period for the diversion capacity with maximum expected annual benefit.
20. Considering Problems 14 and 18, determine the return period associated with the flow capacity that maximizes the expected annual benefit in the channel modification.

TABLE 13.13
Annual Peak Discharge in Problem 9

Year	Annual Peak Discharge (m^3/s)
1962	1.53
1963	2.18
1964	7.89
1965	4.54
1966	3.72
1967	1.45
1968	4.76
1969	2.1
1970	42.15
1971	5.01
1972	1.11
1973	0.39
1974	1.51
1975	1.49
1976	2.53
1977	2.95
1978	2.95
1979	6.22
1980	1.77
1981	1.55
1982	1.3
1983	1.59
1984	5.83

TABLE 13.14
Annual Peak Discharge in Problem 10

Year	Annual Peak Discharge (m^3/s)
1990	10.2
1991	11
1992	3.5
1993	6.7
1994	4.2
1995	5.3
1996	1.8
1997	9.2
1998	4.5
1999	7.2
2000	1.1

TABLE 13.15
Damage–Frequency Relationships in Problem 12

Exceedance Probability (%)	Damage[a] (10^6)	Damage[b] (10^6)	Damage[c] (10^6)	Damage[d] (10^6)	Damage[e] (10^6)
20	0	0	0	0	0
10	6	0	0	0	0
7	10	0	0	0	0
5	13	13	2	4	3
2	22	22	10	12	10
1	30	30	20	18	12
0.5	40	40	30	27	21
0.2	50	50	43	40	35
0.1	54	54	47	43	45
0.05	57	57	55	50	56

a–e Damages sustained corresponding to each exceedance probability if the first-fifth alternative is implemented, respectively.

TABLE 13.16
Return Period, Damage Cost, and Capital Cost in Problem 13

T (Return period)	1	2	5	10	15	20	25	50	100	200
Damage cost ($ $\times 10^3$)	0	40	120	280	354	426	500	600	800	1,000
Capital cost ($/year $\times 10^3$)	0	6	28	46	50	54	58	80	120	160

TABLE 13.17
Frequency–Discharge–Storage–Damage Data in Problem 14

Exceedance Probability	Q (m^3/s)	Stage (m)	Damage (10^6)
20	339.84	103.7	0
10	396.48	120.9	6
7	424.8	129.6	10
5	452.8	138.1	13
2	509.4	155.4	22
1	566.4	172.8	30
0.5	623.04	190.0	40
0.2	708	215.9	50
0.1	736.32	224.6	54
0.05	792.96	241.9	57

APPENDIX

This appendix contains two tables for the *K* values used for design value based on return periods for log Pearson type III and *K* values used for estimation of outliers.

TABLE 13.A1
K for Log Pearson Type III Based on Weighted Skewness Values for Different Recurrence Intervals

	Recurrence Interval in Years							
Weighted	**1.0101**	**2**	**5**	**10**	**25**	**50**	**100**	**200**
Skewness	Percent chance (≥) = 1 − *F*							
Gw	99	50	20	10	4	2	1	0.5
3	−0.667	−0.396	0.42	1.18	2.278	3.152	4.051	4.97
2.8	−0.714	−0.384	0.46	1.21	2.275	3.114	3.973	4.847
2.6	−0.769	−0.368	0.499	1.238	2.267	3.071	3.889	4.718
2.4	−0.832	−0.351	0.537	1.262	2.256	3.023	3.8	4.584
2.2	−0.905	−0.33	0.574	1.284	2.24	2.97	3.705	4.444
2	−0.99	−0.307	0.609	1.302	2.219	2.912	3.605	4.298
1.8	−1.087	−0.282	0.643	1.318	2.193	2.848	3.499	4.147
1.6	−1.197	−0.254	0.675	1.329	2.163	2.78	3.388	3.99
1.4	−1.318	−0.225	0.705	1.337	2.128	2.706	3.271	3.828
1.2	−1.449	−0.195	0.732	1.34	2.087	2.626	3.149	3.661
1	−1.588	−0.164	0.758	1.34	2.043	2.542	3.022	3.489
0.8	−1.733	−0.132	0.78	1.336	1.993	2.453	2.891	3.312
0.6	−1.88	−0.099	0.8	1.328	1.939	2.359	2.755	3.132
0.4	−2.029	−0.066	0.816	1.317	1.88	2.261	2.615	2.949
0.2	−2.178	−0.033	0.83	1.301	1.818	2.159	2.472	2.763
0	−2.326	0	0.842	1.282	1.751	2.054	2.326	2.576
−0.2	−2.472	0.033	0.85	1.258	1.68	1.945	2.178	2.388
−0.4	−2.615	0.066	0.855	1.231	1.606	1.834	2.029	2.201
−0.6	−2.755	0.099	0.857	1.2	1.528	1.72	1.88	2.016
−0.8	−2.891	0.132	0.856	1.166	1.448	1.606	1.733	1.837
−1	−3.022	0.164	0.852	1.128	1.366	1.492	1.588	1.664
−1.2	−3.149	0.195	0.844	1.086	1.282	1.379	1.449	1.501
−1.4	−3.271	0.225	0.832	1.041	1.198	1.27	1.318	1.351
−1.6	−3.88	0.254	0.817	0.994	1.116	1.166	1.197	1.216
−1.8	−3.499	0.282	0.799	0.945	1.035	1.069	1.087	1.097
−2	−3.605	0.307	0.777	0.895	0.959	0.98	0.99	0.995
−2.2	−3.705	0.33	0.752	0.844	0.888	0.9	0.905	0.907
−2.4	−3.8	0.351	0.725	0.795	0.823	0.83	0.832	0.833
−2.6	−3.899	0.368	0.696	0.747	0.764	0.768	0.769	0.769
−2.8	−3.973	0.384	0.666	0.702	0.712	0.714	0.714	0.714
−3	−4.051	0.396	0.636	0.66	0.666	0.666	0.667	0.667

TABLE 13.A2
Outlier Test *K* Values: 10% Significant Level *K* Values

Sample Size	*K* Value	Sample Size	*K* Value	Sample Size	*K* Value	Sample Size	*K* Value
10	2.036	45	2.727	80	2.940	115	3.064
11	2.088	46	2.736	81	2.945	116	3.067
12	2.134	47	2.744	82	2.949	117	3.070
13	2.175	48	2.753	83	2.953	118	3.073
14	2.213	49	2.76	84	2.957	119	3.075
15	2.247	50	2.768	85	2.961	120	3.078
16	2.279	51	2.775	86	2.966	121	3.081
17	2.309	52	2.783	87	2.970	122	3.083
18	2.335	53	2.79	88	2.973	123	3.086
19	2.361	54	2.798	89	2.977	124	3.089
20	2.385	55	2.804	90	2.981	125	3.092
21	2.408	56	2.811	91	2.984	126	3.095
22	2.429	57	2.818	92	2.989	127	3.097
23	2.448	58	2.824	93	2.993	128	3.100
24	2.467	59	2.831	94	2.996	129	3.102
25	2.486	60	2.837	95	3.000	130	3.104
26	2.502	61	2.842	96	3.003	131	3.107
27	2.519	62	2.849	97	3.006	132	3.109
28	2.534	63	2.854	98	3.011	133	3.112
29	2.549	64	2.86	99	3.014	134	3.114
30	2.563	65	2.866	100	3.017	135	3.116
31	2.577	66	2.871	101	3.021	136	3.119
32	2.591	67	2.877	102	3.024	137	3.122
33	2.604	68	2.883	103	3.027	138	3.124
34	2.616	69	2.888	104	3.030	139	3.126
35	2.628	70	2.893	105	3.033	140	3.129
36	2.639	71	2.897	106	3.037	141	3.131
37	2.650	72	2.903	107	3.040	142	3.133
38	2.661	73	2.908	108	3.043	143	3.135
39	2.671	74	2.912	109	30.460	144	3.138
40	2.682	75	2.917	110	3.049	145	3.140
41	2.692	76	2.922	111	3.052	146	3.142
42	2.700	77	2.927	112	3.055	147	3.144
43	2.710	78	2.931	113	3.058	148	1.146
44	2.719	79	2.953	114	3.061	149	3.148

Source: USWRC, Estimating Peak Flow Frequencies for Natural Ungauged Watersheds, A Proposed Nationwide Test, Hydrology Subcommittee, US Water Resources Council, 1981. With permission.

REFERENCES

APFM. (2008). Urban flood risk management. Flood Management Tools Series.

ASCE and UNESCO. (1998). *Sustainability Criteria for Water Resource Systems.* ASCE, Reston, VA.

Bharwani, S., Magnuszewski, P., Sendzimir, J., Stein, C., and Downing, T.E. (2008). Vulnerability, adaptation and resilience. Progress toward incorporating VAR concepts into adaptive water resource management, Report of the NeWater Project, New Approaches to Adaptive Water Management under Uncertainty.

Boss International. (2005). *StormNet and Wastewater Model User's Manual.* Available at https://sourceforge.net/projects/stormdotnet

Bronstert, A., Ghazi, A., Hljadny, J., Kundzevicz, Z.W., and Menzel, L. (1999). *Proceedings of the European Expert Meeting on the Oder Flood*, May 18, Potsdam, Germany, European Commission.

Bruneau, M., Chang, S.E., Eguchi, R.T., Lee, G.C., O'Rourke, T.D., Reinhorn, A.M., et al. (2003). A framework to quantitatively assess and enhance the seismic resilience of communities. *Earthquake Spectra*, 19(4), 733–752. doi:10.1193/1.1623497.

Burn, D.H., Venema, H.D., and Simonovic, S.P. (1991). Risk based performance criteria for real time reservoir operation. *Canadian Journal of Civil Engineering*, 18(1), 36–42.

Comes, T., and Van De Walle, B. (2014). Measuring disaster resilience: The impact of hurricane sandy on critical infrastructure systems. In S. R. Hiltz, M. S. Pfaff, L. Plotnick, and P. C. Shih (eds.), ISCRAM 2014 Conference Proceedings – 11th International Conference on Information Systems for Crisis Response and Management (pp. 195–204). University Park, Pennsylvania.

Dunne, T. (1986). Urban hydrology in the tropics: Problems, solutions, data collection and analysis. In *Urban Climatology and Its Application with Special Regards to Tropical Areas: Proceedings of the Mexico Tech Conference*, November 1984, Mexico, World Climate Programme, WMO.

Eckstein, O. (1958). *Water Resources Development, the Economics of Project Evaluation.* Harvard University Press, Cambridge, MA.

Eikenberg, C. (1998). *Journalistenhandbuch zum Katastrophenman-agement* (5th ed.). German IDNR-Committee, Bonn.

Emergency Management Australia. (2007). *Disaster Loss Assessment Guidelines*, Australian Emergency Manuals Series, Part III, Volume 3, Guide 11. Available at: http://www.ema.gov.au/.

Grigg, N.S. and Helweg, O.J. (1975). State of the art of estimating damage in urban areas. *Water Resources Bulletin,* 11(2), 379–390.

Grunewald, U. (1998). The causes, progression, and consequences of the river oder floods in summer 1997, Including Remarks on the Existence of Risk Potential, German IDNDR Committee for Natural Disaster Reduction, German IDNDR Series No. 10e, Bonn.

Hoff, H. (2011). Understanding the Nexus. Background Paper for the Bonn 2011 Conference: The Water, Energy and Food Security Nexus, Stockholm Environment Institute (SEI), Stockholm, Sweden.

Interagency Performance Evaluation Task Force (IPET). (2009). "Performance Evaluation of the New Orleans and Southeast Louisiana Hurricane Protection System: Final Report of the Interagency Performance Evaluation Task Force." Volume I (June): Executive Summary and Overview. Retrieved from https://ipet.wes.army.mil

James, L.D. (1972). Role of economics in planning floodplain land use. *Journal of the Hydraulics Division, ASCE,* 98(HY6), 981–992.

Joosten-Ten Brinke, D., Sluijsmans, D.M.A., Brand-Gruwel, S., and Jochems, W.M.G. (2008). The quality of procedures to assess and credit prior learning: Implications for design. *Educational Research Review*, 3(1), 51–65. doi:10.1016/j.edurev.2007.08.001.

Karamouz, M. and Heydari, Z. (2020). Conceptual design framework for coastal flood best management practices. *Journal of Water Resources Planning and Management*, 146(6), 04020041.

Karamouz, M. and Mohammadi, K. (2020). Nonstationary based framework for performance enhancement of coastal flood mitigation strategies. *Journal of Hydrologic Engineering*, 25(6), 1–17.

Karamouz, M., Hosseinpour, A., and Nazif, S. (2011). Improvement of urban drainage system performance under climate change impact: A case study. *ASCE Journal of Hydrologic Engineering,* 15(5), 395–412.

Karamouz, M., Fallahi, M., and Nazif, S. (2012). Evaluation of climate change impact on regional flood characteristics. *Iranian Journal of Science and Technology*, 36(2), 225–238.

Karamouz, M., Zahmatkesh, Z., and Nazif, S. (2014). Quantifying resilience to coastal flood events: A case study of New York City. *World Environmental and Water Resources Congress 2014: Water Without Borders – Proceedings of the 2014 World Environmental and Water Resources Congress*, Portland, Oregon, 911–923.

Karamouz, M. and Budinger, T. (2014). *Livable Cities of the Future.* National Academies Press, Washington, DC.

Karamouz, M., Fereshtehpour, M., Ahmadvand, F., and Zahmatkesh, Z. (2016a). Coastal flood damage estimator : An alternative to FEMA' s HAZUS platform. *Journal of Irrigation and Drainage Engineering*, 142(6), 1–12.

Karamouz, M., Rasoulnia, E., Zahmatkesh, Z., Olyaei, M. A., and Baghvand, A. (2016b). Uncertainty-based flood resiliency evaluation of wastewater treatment plants. *Journal of Hydroinformatics*, 18(6), 990–1006.

Karamouz, M., Razmi, A., Nazif, S., and Zahmatkesh, Z. (2017). Integration of inland and coastal storms for flood hazard assessment using a distributed hydrologic model. *Environmental Earth Sciences*, Springer Berlin Heidelberg, 76(11), 395.

Karamouz, M., Rasoulnia, E., Olyaei, M. A., and Zahmatkesh, Z. (2018). Prioritizing investments in improving flood resilience and reliability of wastewater treatment infrastructure. *Journal of Infrastructure Systems*, 24(4), 04018021.

Karamouz, M. and Ansari, A. (2020). Uncertainty based budget allocation of wastewater infrastructures' flood resiliency considering interdependencies. *Journal of Hydroinformatics*, 22(4), 768–792.

Kowalczak, P. (1999). Flood 1997, infrastructure and urban context, Bronstert, A., ed. *Proceedings of the European Expert Meeting on the Oder Flood*, May 18, Potsdam, Germany, European Commission, 99–104.

Krywkow, J., Filatova, T., and van der Veen, A. (2008). Flood risk perceptions in the Dutch province of Zeeland: Does the public still support current policies? *Flood Risk Management: Research and Practice, CRC Press/Balkema Proceedings and Monographs in Engineering, Water and Earth Science*, Taylor & Francis Group, Boca Raton, FL, 1513–1521.

Leavitt, W.M. and Kiefer, J.J. (2006). Infrastructure interdependency and the creation of a normal disaster: The case of Hurricane Katrina and the City of New Orleans. Public Works Management & Policy, 10(4), 306–314. doi:10.1177/1087724X06289055.

Mays, L.W. (2001). *Water Resources Management*. John Wiley and Sons Inc., New York.

Moy, W., Cohon, J.L., and Revelle, C.S. (1986). A programming model for analysis of the reliability, resilience and vulnerability of a water supply reservoir. *Water Resources Research,* 22(4), 489–498.

Nelson, C.E., Crumely, C., Fritzsche, B., and Adcock, B. (1989). Lower southeast Florida hurricane evacuation study. Report prepared for the US Army Corps of Engineers, Jacksonville, FL.

New York City's Department of Environmental Protection. New York City's Wastewater Treatment System.

RBD (Rebuild by Design). (2014). The New Meadowlands. Retrieved from http://www.rebuildbydesign.org/data/files/672.pdf

Rinaldi, S.M., Peerenboom, J., & Kelly, T. (2001). Identifying, understanding, and analyzing critical infrastructure interdependencies. IEEE Control Systems, 21(6), 11–25. doi:10.1109/37.969131.

Sharkey, T.C., Cavdaroglu, B., Nguyen, H., Holman, J., Mitchell, J.E., & Wallace, W.A. (2015). Interdependent network restoration: On the value of information-sharing. European Journal of Operational Research, 244(1), 309–321. doi:10.1016/j.ejor.2014.12.051.

Sklar, A. (1959). Fonctions de répartition à n dimensions et leus marges. *Publications de l'Institut de Statistique de L'Université de Paris,* 8, 229–231.

Tucci, C.E.M. (1991). Flood control and urban drainage management. Available at http://www.cig.ensmp.fr/~iahs/maastricht/s1/TUCCI.htm.

Streips, K. and Simpson, D.M. (2007). *Critical infrastructure failure in a natural disaster: Initial notes comparing Kobe and Katrina.* University of Louisville, Louisville.

UN/ISDR. (2002). Living with Risk: A global review of disaster reduction initiatives. Preliminary version prepared as an Interagency effort coordinated by the ISDR Secretariat, Geneva, Switzerland.

US Army Corps of Engineers. (2009). Performance evaluation of the New Orleans and Southeast Louisiana hurricane protection system: Volume I: Executive summary and overview, US Army Corps of Engineers.

US Water Resources Council. (1981). Estimating peak flow frequencies for natural ungauged watersheds. A proposed nationwide test, Hydrology Subcommittee, US Water Resources Council.

Vis, M., Klijn, F., De Bruijn, K.M., and Van Buuren, M. (2003). Resilience strategies for flood management in the Netherlands. *International Journal of River Basin Management,* 1(1), 33–40.

Zongxue, X., Jinno, K., Kawamura, A., Takesaki, S., and Ito, K. (1998). Performance risk analysis for Fukuoka water supply system. *Water Resources Management,* 12(1), 13–30.

14 Environmental Visualization

14.1 INTRODUCTION

The significance of visualization for information representation lies in its functionality to serve as a storage mechanism, a processing and research instrument, and a communication tool. Specifically, as a tool, visualization technology provides three key functions: (a) it enables large volumes of data to be processed and organized in a simple, easy-to-use visual format; (b) it documents site activities and physical and chemical conditions through time; and (c) it allows information to be readily communicated.

Visualization technology is capable of transferring information into a simple image or animation. Environmental visualization (EV) is one of the hottest areas that many detection and presentation technology have been realized with many more potential for ground -breaking new developments. It can readily document many types of site data, including the ground surface, boring lithology and geologic strata, well construction information, groundwater conditions, and chemical concentrations of soil, water, and vapor. The generated image is typically a compilation of hundreds of pages of information from large and cumbersome reports. As a result, visualization acts as a data management tool that collates, organizes, and displays large volumes of information (Ling and Chen, 2014). It has special application in disaster management and flood inundation that can be utilized by a variety of users with low- to very high- technical capabilities.

The water movement and accumulation and water infrastructures need to be continuously monitored. The materials presented in this chapter include: the use of sensors, pattern recognition tools and techniques, satellite technology and data collections, and development of citizen science applications. By completion of this chapter, we should realize a number of opportunities for better water recourses assessment, protection and conservations, water hazard prevention, and effective visual communication. The advancement of EV and how we should prepare to utilize it is presented.

14.1.1 Sensed Water Infrastructure

Advances in sensing technology are rapidly changing which has yielded to improved monitoring and managing water infrastructure systems. For example, deployable, in situ water quality sensors are now available to collect "real-time" data at second to minute to hour intervals, a significant improvement over traditional methods with larger time intervals (USEPA, 2009a). At present, the information that sensors can collect is limited to a few parameters and they generally require frequent calibration and maintenance in the field. Eventually, networks of these sensors will be widely deployed, producing a wealth of data. Sensors allow us to understand how water management systems like water tanks and green roofs can reduce the loading to storm sewers. Sensors in drinking water distribution systems can track changes to our drinking water quality that could show problems with treatment or even an intentional attack on our water distribution systems. This type of real-time data collecting in our water infrastructure systems will transform the understanding of the systems and improve the ability to manage these systems for improved human and ecosystem health.

The use of real-time sensors for water flow and monitoring water quality parameters has even more improved when coupled with distribution system modeling and sensor placement algorithms (see, e.g., Isovitsch and van Briesen (2008) and Xu and Brambilla (2008)). Sensor networks within distribution systems also are proposed and installed to enhance security (Ailamaki et al., 2003; Krause et al., 2008). As an example of the potential of sensors for improved operation and performance in drinking water distribution systems, online sensors for measuring chlorine residual have

DOI: 10.1201/9781003241744-14

already been widely used in many water distribution networks. Chlorine sensors, usually in conjunction with other sensors, can be used to detect events (accidental or intentional contamination) as well as to monitor chlorine residual during normal operation. Control of chlorine boosters and interpretation of sensor data is currently done by water engineers, but it is expected to be done by using cyber infrastructure to model, evaluate, and control the system autonomously (Harley et al., 2009).

In addition to the use of new technologies to make measurements more rapid in water infrastructure systems, other advanced technics are needed to receive these data in real time and enable extraction of relevant information from the basic data to present it in such a way that it is understandable to human senses. The general problem of pattern discovery in a large number of coevolving groups of data streams is in the focus of extensive study that requires Big Data (Papadimitriou et al., 2005) and direct application to drinking water data streams by investigators (Faloutsos and VanBriesen, 2006; Sun et al., 2006) and by groups within government agencies (USEPA, 2009b). In order to use these real-time data to manage our infrastructure systems in real time, the relevant information have to be organized in ways that enable us to make decisions, to understand to trouble shoot, and to act.

Research and development are in progress of creating water infrastructure sensor technology and networks, as well as the signal and data handling, processing, and analysis systems for improved human decision-making capability for water infrastructure management. Integration of these various components requires water system experts who know what to measure in water and why; sensor network experts who understand how signals and data are measured, collected, stored, transmitted, and processed; and information technology experts who are able to transform streaming data into information useful for design, operation, and maintenance and decision-making in water infrastructure systems (Dzombak et al., 2012).

14.2 ENVIRONMENTAL SENSING

14.2.1 Introduction

With a growing interest in environmental data and the need to consider various environmental factors earlier in the planning processes, it becomes more important to disseminate this type of information to different target groups in a comprehensible fashion. To support easier decision-making, many cities and municipalities are increasingly using digital city models where it is possible to integrate different types of information based on simulation and visualization of future scenarios. Such tools have high potential, but the visual representation of data still needs to be developed.

The increasing development of digital tools to support dialog processes also requires that expertise from different research areas such as environmental research and human–computer interaction (HCI) can be linked. Different types and levels of information should be conveyed in a visually comprehensible manner in both 2D and 3D, depending on what is most relevant to show. Knowledge on how to represent environmental data in 3D media needs to be developed. One step towards this is to translate traditional knowledge from visualizing environmental data in 2D maps to interactive 3D visualizations. In street view, the spatial context can be shown in more detail, while planar view provides an overview. The level of abstraction and information in a visualization should be connected to factors such as who the target audience is, their previous knowledge of the project, and how much information it is suitable to convey in order to maximize understanding.

Moreover, an important design challenge is the esthetic considerations of the visualization in themselves, such as visual style and the use of colors. Esthetic considerations have, according to Hullman et al., been underexplored in many efficiency-motivated studies. If the visualization is appealing, the likelihood is that we will be positively attuned to the content, even though the aspects of visual appearance and information are unrelated. Esthetic considerations are also important for improving the readability of a visualization. One challenge lies in how to combine map material and abstract coloring of environmental data. Traditionally, colors play an important role in cartographic visualization. Borland and Tayloralso point out the importance of appropriate choices of color and exemplify with the misuse of the commonly used rainbow color scale. The common use of the rainbow scale to highlight environmental data is referred to by several authors. Instead of using this scale, Grainger and Dufau (2012) propose the use of form attributes such as shapes or the use of a single-color ranging from low to high intensity. Cultural or natural connotations of specific colors are also important to take into consideration, based on the target audience. When using different visual styles, such as symbolic objects or images in an otherwise visually photorealistic setting, it is important to work with a design language that distinguishes between visual realism and visual nonrealism, for example, by a knowledgeable use of colors (Wästberg et al., 2020).

For many years, there was a mismatch between the scale of collected or estimated data by satellite, general circulation models (GCMs), and what the environmental impact models have required (from 50 km down to <1 m), Figure 14.1. The gap is closing with the development of finer resolution data and digital elevation model (DEM) including 1 m LiDAR information; however, the never-ending challenge is to bring the models run time to something that widely available high- speed processors/computers can handle.

Most of images used in EV are the compilation of layers of information. Figure 14.2 shows how from GCM data of coarse resolution other layers of data and medium for interpolation and processing including downscaling should be used to see and apply in a scale (i.e., watershed) that is suitable for many assessment/engineering and planning applications.

14.2.2 Ubiquitous Environmental Sensor Technologies

New sensor technologies and methodologies for using them are being created more and more quickly (Kanoun and Trankler, 2004). Advances in modern sensor technology across all spatial and performance scales will continue and accelerate. Smaller and more efficient sensors are expected to be adaptable to widespread environmental and water resources monitoring systems and be more

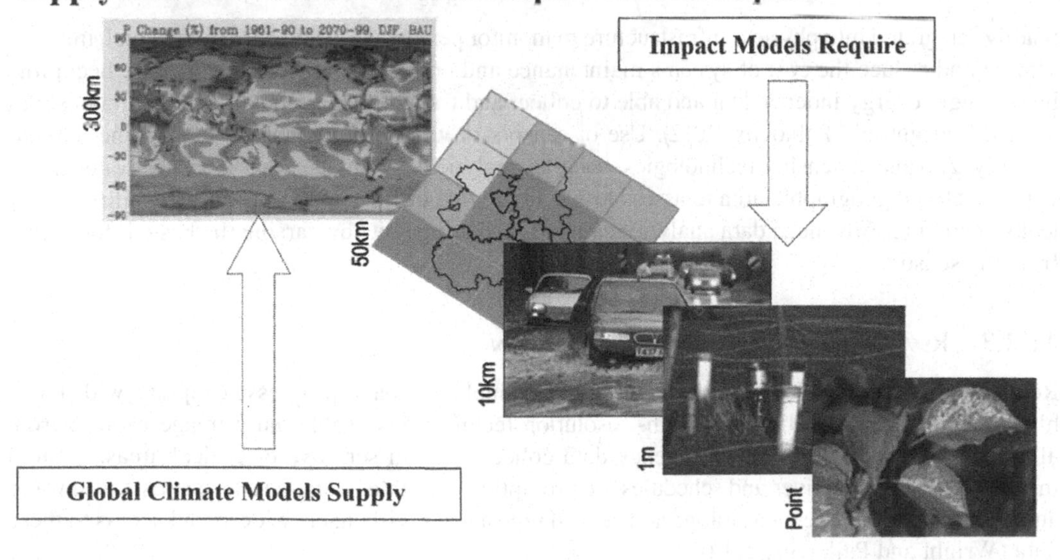
Because there is a mismatch of scales between what climate models can supply and what environmental impact models require
Impact Models Require
300km
50km
10km
1m
Point
Global Climate Models Supply

FIGURE 14.1 A mismatch of scales between what climate models can supply and what environmental impact models require. (*Source*: Canadian Institute for Climate Studies (CICS), https://climatechoices.ca/.)

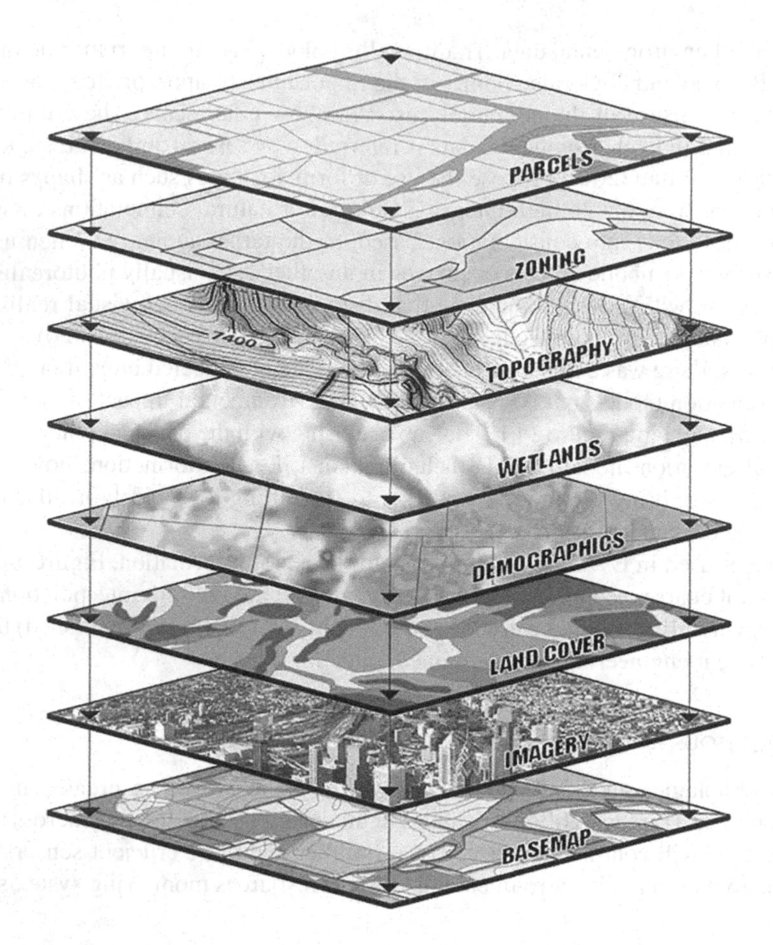

FIGURE 14.2 Layers of information that constitute the environmental visualization images from GCM to watershed scales. (*Source*: Wright, W.R. and Palkovits, R., ChemSusChem, 5(9), 1657–1667, 2012.)

readily integrated into physical infrastructure to monitor performance and condition, which improve service and reduce the cost of systems maintenance and operation. Sensor networks are becoming increasingly energy independent and able to collect and transmit data reliably and continuously as needed (Wright and Palkovits, 2012). Use of sensors in water infrastructure has become popular recently. Advanced sensing technologies, such as analytical techniques that measure water usage within a closed geographic area to assess loss, can provide utilities with real-time visibility on any leaks occurring. Advanced data analytics can add further insights by parsing the basic information from the sensors.

14.2.3 Remote Sensing and Earth Observation

Remote sensing and related earth observation technologies have progressed rapidly with much higher resolution and reliability. High- resolution technologies enable daily image capture from all populated areas with high- frequency data collection from sensitive or critical areas. Custom image capture perspectives and schedules are promptly accessible to engineers, scientists, and water authorities. Large -scale data storage and central repositories will enable widespread access to these data (Wright and Palkovits, 2012).

Remote sensing capabilities and techniques are appropriate for the monitoring of regional-scale precipitation, water budgets, soil moisture, and some measures of water quality. Other possible satellite imagery and remote sensing applications include catchment characterization, water quality monitoring, soil moisture assessment, water extent and level monitoring, irrigation service monitoring, urban and agricultural water demand modeling, evapotranspiration estimation, ground water management, hydrological modeling, and flood mapping/forecasting. Various satellite data products have been leveraged to aid water and sanitation programs. Key examples include (Andres et al., 2018):

The Sentinel-1 Program for water management in low-income countries was part of a large research initiative that explored the performance and opportunities provided by the European satellite Sentinel-1 for water resource management applications. The satellite data was utilized to estimate water volumes retained by small reservoirs to assess the feasibility of small water supply systems (Amitrano et al., 2014).

In particular, LandSat 8 data allow for a calculation of the Normalized Differential Vegetation Index (NDVI). The LandSat 8 NDVI allows an estimation of land surface emissivity (Sobrino et al., 2004) and land cover classification (Weng el al., 2004), as well as surface temperature. These measures can be used by remote sensing experts to allow a planning-level estimation of watershed health across a broad region. Additionally, land use classification can identify the vegetation and population densities of rural versus urban, built environments.

SERVIR, a cooperative effort of NASA and the United States Agency for International Development (USAID), "works in partnership with leading regional organizations world-wide to help developing countries use information provided by Earth observing satellites and geospatial technologies for managing climate risks and land use" (servirglobal.net). With three regional offices, SERVIR has been able to partner with remote sensing experts and national decision-making bodies. Among other activities, SERVIR focuses on monitoring water bodies to observe effects from "human activities, climate change, and other environmental phenomena." It takes advantage of LandSat, ASTER, MODIS, and other satellite assets to monitor water quality and changes. Specifically, SERVIR is developing rainfall and runoff models to study the availability and quality (as sediment or total dissolved solids) of surface water over the next several decades.

Using the Tropic Rainfall Measuring Mission (TRMM) data, the Nile Basin Initiative in partnership with NASA provides flood forecasts and water balance estimates for the Eastern Nile Basin. Similarly, the Surface Hydrology Group at Princeton University developed the Africa Drought Monitor and provides maps of rainfall, temperature, and other hydrologic variables (García-Peñalvo, 2016).

The USAID Famine Early Warning System Network (FEWS NET) monitors rainfall and crop production with satellite assets and combines these data with socioeconomic insights to identify population groups that may be vulnerable to food insecurity (www.fews.net).

NASA's Terra satellite includes two instruments that have been leveraged for watershed monitoring. The Moderate Resolution Imaging Spectroradiometer (MODIS, 1999, 2002) and the Multiangle Imaging Spectroradiometer (MISR) satellite can be used to determine aerosol optical depth, land surface temperature, enhanced vegetation index, and middle infrared reflectance. Some of these data can be used to assess water quality parameters including chlorophyll, cyanobacterial pigments, colored dissolved organic matter, and suspended matter in water bodies on a large scale (www.nasa.gov). In Nigeria, the World Bank recently used geographic information system (GIS) mapping techniques to compare household survey data against MODIS land use estimates to generate spatial distribution estimates of water and sanitation indicators, including service access (World Bank, 2017; Ajisegiri et al., 2019).

In Dhaka, Bangladesh, the World Bank and other partners (University of Massachusetts Boston, and Earth Observation for Sustainable Development/GiSat) used high-resolution imagery in slums to correlate satellite imagery to public service availability (World Bank, 2018).

The Inter-American Development Bank (IDB) developed the Hydro-BID platform to assist countries in Latin America and the Caribbean with water management through the mapping and tracking of over 230,000 water catchment areas. The Hydro-BID platform is leveraged by government agencies and water utilities for regional water management and infrastructure planning (www. hydrobidlac.org). Other water-related satellites are SMOS (Soil Moisture and Ocean Salinity), GRACE (Gravity Recovery and Climate Experiment), and SMAP (Soil Moisture Active Passive). SMAP is explained with some details in the next section.

14.2.4 Soil Moisture Active Passive (SMAP)

SMAP that was launched in Jan uary 31, 2015, is a NASA remote sensing observatory, designed to carry two instruments that will map soil moisture and determine the freeze or thaw state of the same area being mapped. Soil moisture content can be mapped via the radiometer data at a spatial resolution of 36 km. A combination of radar and radiometer measurements would lead to a soil moisture product at a spatial resolution of 9 km. The freeze/thaw mapping could be accomplished using unique properties of the radar system's measurements at a spatial resolution of 3 km. After the failure of SMAP radar in early July 2015, the SMAP radiometer data has been used to produce the freeze/thaw map at a resolution of 36 km. Both the radar and radiometer share a common antenna and feed assembly externally, but their electronics inside SMAP are different.

The radar sent pulses of radio waves down to a spot-on Earth and measured the echo that returned a few microseconds later. The strength and "shape" of the echoes could be interpreted to indicate the moisture level of the soil, even though moderate levels of vegetation. Since the radar actively sent and received radio waves, this was where the "active" in SMAP came from.

The radiometer detects radio waves that are emitted by the ground from the same area. The strength of the emission is an indicator of the brightness temperature of the ground in that location. Since a radiometer passively makes these brightness temperature measurements, this is where the "passive" in SMAP comes from. SMAP is designed to measure soil moisture, every 2–3 days. This permits changes, around the world, to be observed over time scales ranging from major storms to repeated measurements of changes over the seasons.

Everywhere on Earth not covered with water or not frozen (have snow packs), SMAP measures water content in the top layer of soil. It also distinguishes between ground that is frozen or thawed. SMAP also measures the amount of water found between the minerals, rocky material, and organic particles found in soil.

With SMAP, data global maps of soil moisture can be produced. Scientists will use these to help improve our understanding of how water, energy, and carbon fluxes (in its various forms) maintain our climate and environment. Table 14.1 shows SMAP characteristics.

TABLE 14.1
SMAP Characteristics

	Characteristics
Resolution	Radar sensor: 3 km Radiometer sensor: 36 km
Equatorial crossing time	18:00 hours [6:00 AM (descending node) and 6:00 PM (ascending node)]
Revisit	2–3 days
Repeat cycle	8 days
Orbit	Near-polar, sun-synchronous
Orbit duration	98.5 minutes
Altitude	685 km

14.2.4.1 Step-by-Step Procedure to Download SMAP Data for a Specific Date at a Specific Location

1. Go to https://search.earthdata.nasa.gov/search
2. Go to Earth data login and **register** your account
3. Go to search for collections or topics and search **SMAP**
4. Choose the **time period** (2018-01-01 00:00:00 to 23:59:59)
5. Select the **polygon** from the top left browser, spatial (Box) shape, and choose **Utah State** from the map as shown below (Select the entire State) (Figure 14.3)
6. Choose SMAP L3 Radiometer Global Daily 36 km EASE-Grid **Soil Moisture V007** from matching collections and download all
7. Choose **customize** from select a data access method
8. Choose **GeoTIFF** for output file format
9. **Click to enable** the spatial subsetting
10. Choose **soil_moisture** and **soil_moisture_pm** from SPL3SMP in band subsetting
11. Click **done** and download data
12. For this specific date and location, soil moisture was retrieved and emailed a tiff file
13. Open the **ArcMap,** select **Add Data** and choose the **tiff file** – See Table 14.2 for a given date.
14. Select **Identify** from toolbar and read **four-pixel (Cell)** SM values with the coordination showed below (Figure 14.4).

In the next section, pattern recognition principles are discussed for development of the tools needed for EV.

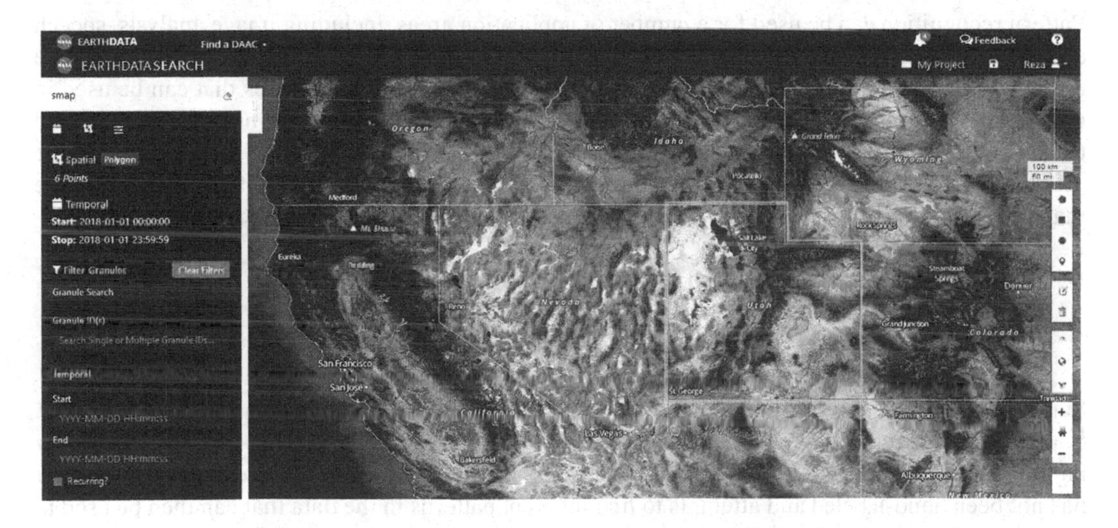

FIGURE 14.3 EARTHDATA application window.

TABLE 14.2
Four Cells of 36-km Soil Moisture Readings

Number	Coordination (Decimal Degrees)	SMAP 36-km Soil Moisture (%)
1	112°W, 40°N	8.65
2	114°W, 39°N	7.08
3	109°W, 38°N	6.69
4	111°W, 37°N	2.00

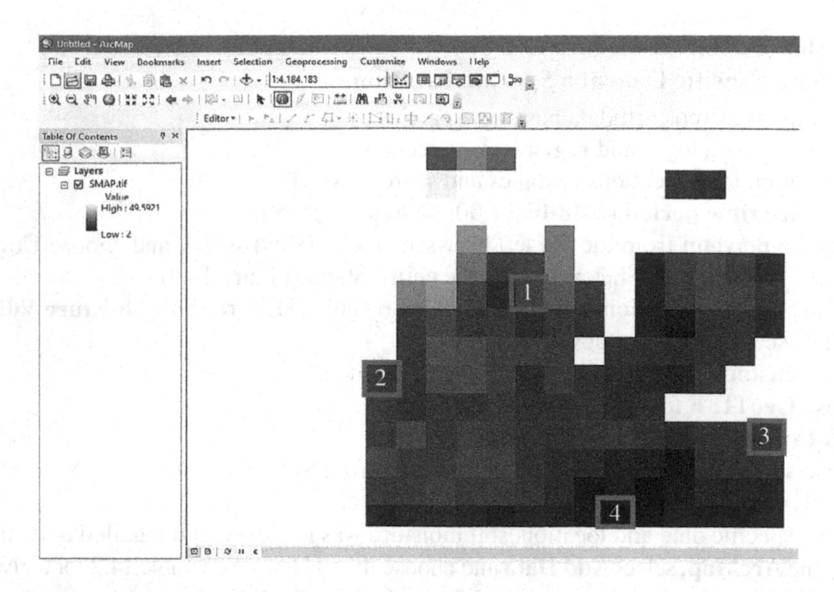

FIGURE 14.4 ArcMap application window and the selected 36-km cells.

14.3 PATTERN RECOGNITION

14.3.1 Introduction

Pattern recognition can be used for a number of application areas, including image analysis, speech and audio recognition, biometrics, bioinformatics, data mining, and information retrieval. In spite of their differences, these areas share, to a large extent, a corpus of techniques that can be used in retrieving available data, information related to data categories, important hidden patterns, and trends. Pattern recognition is the act of utilizing basic/raw data and taking an action based on the "category" of the pattern. A classifier is a device that performs such task. Classification is one of the major applications of pattern recognition. There are three types of classification:

- Supervised classification (when set of training data is available)
- Unsupervised classification (when set of training data is not available) → Clustering
- Semisupervised classification

Supervised classification assumes that a set of training data (the training set) has been provided. Supervised learning/classification consists of a set of instances that have been properly labeled by hand with the correct output. Unsupervised learning, on the other hand, assumes training data that has not been hand-labeled and attempts to find inherent patterns in the data that can then be used to determine the correct output value for new data instances.

Training and Learning in Pattern Recognition: Learning is a phenomenon through which a system gets trained and becomes adaptable to give result in an accurate manner. Learning is the most important phase as how well the system performs on the data provided to the system depends on which algorithms are used on the data. Entire dataset is divided into two categories, one which is used in training the model called training set and the other that is used in testing the model after training, referred as testing set.

- *Training set*: Training set is used to build a model. It consists of the set of images which are used to train the system. Training rules and algorithms used give relevant information on how to associate input data with output decision. The system is trained by applying these algorithms on the dataset, all the relevant information is extracted from the data and results are obtained. Generally, 80% of the data of the dataset is taken for training data.

- *Testing set:* Testing data is used to test the system. It is the set of data which is used to verify whether the system is producing the correct output after being trained or not. Generally, 20% of the data of the dataset is used for testing. Testing data is used to measure the accuracy of the system.

Identifying vegetation, waterways, and man-made structures from remote sensing images involves large amount of data processing for statistical pattern recognition methods. With regard to the last mentioned, classification of remote sensing imagery which are emerging to interpret the cluster centers of an image and to reveal a suitable number of classes to overcome the disadvantage of unsupervised classification. The application of high-resolution remote sensing technology and advanced image processing using advanced algorithms and statistical pattern recognition will significantly improve the ability to monitor and assess water resources (Aloui and Dincer, 2018).

Another way of looking at a pattern is to go through the stages shown in Figure 14.5. This is with the assumption of having sensors as means of collecting data, generating features from the collected data, and go through feature selection explained later in this section. The classifier design depends on how the features are generated and selecting the best for the specific task. It is important to measure the classification error rate.

In a more elaborate fashion, a pattern recognition investigation may consist of several stages enumerated below. Not all stages may be present; some may be merged together so that the distinction between two operations may not be clear, even if both are carried out. There could be some application of specific data processing that may not be regarded as one of the stages listed below. However, the investigation stages are fairly typical.

1. *Formulation of the problem*: having a clear understanding of the purpose of the investigation and planning the remaining stages.
2. *Data collection*: making measurements on appropriate variables and recording details of the data collection procedure (ground truth).
3. *Initial examination of the data*: checking the data, calculating summary statistics, and producing plots in order to have a sense for the structure.
4. *Feature selection or feature extraction*: selecting variables from the measured set that are appropriate for the task. These new variables may be obtained by a linear or nonlinear transformation of the original set (feature extraction). To some extent, the partitioning of the data processing into separate feature extraction and classification processes is synthesized, because a classifier often includes the simulation of a feature extraction stage as part of its design.
5. *Unsupervised pattern classification or clustering*: This may be viewed as exploratory data analysis, and it may provide a successful conclusion to a study. On the other hand, it may be a means of preprocessing the data for a supervised classification procedure.
6. *Discrimination or regression procedures*: To design the classifier using a training set of exemplar patterns.
7. *Assessment of results*: This may involve applying the trained classifier to an independent test set of labeled patterns.
8. *Interpretation.*

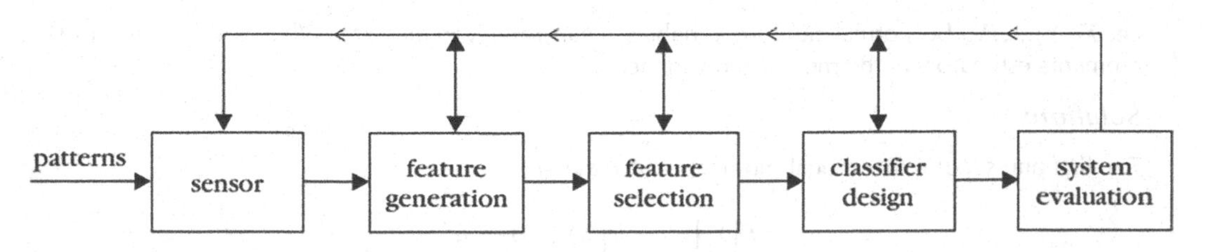

FIGURE 14.5 Typical stage of using patterns for system evaluation.

The above stages are necessarily an iterative process in order to analyze the results and generate new hypotheses that require further data collection. The cycle may be terminated at different stages. The questions that are posed may be answered by an initial examination of the data or it may be discovered that the data cannot answer the initial question and the problem must be reformulated.

An example of a classical model for PR for differentiating the flooded area could be:

1. A feature extraction step extracts feature from raw data (e.g., audio, image, weather data)
2. A classifier received X and assigns it to one of c categories, class1, class 2, …, class c (i.e., labels the raw data)

X are the features of raw data, for a flood susceptibility mapping x_1, x_2, …, x_n can be altitude, slope, aspect, drainage density, soil depth, land use, and soil hydrological groups.

c are the classifications such as c_1 as flood area and c_2 as nonflood area.

This section is organized to provide the basics of the statistical concepts and parameter estimation methods from the view point of pattern recognition as well as feature extraction and selection, supervised and unsupervised classification, image processing, map resolution, and data assimilation (DA).

14.3.2 Parameter Estimation

14.3.2.1 Moments Method

In statistics, the method of moments is a method of estimation of population parameters. It starts by expressing the population moments (i.e., the expected values of powers of the random variable under consideration) as functions of the parameters of interest. Those expressions are then set equal to the sample moments. The number of such equations is the same as the number of parameters to be estimated. Those equations are then solved for the parameters of interest. The solutions are estimates of those parameters. Suppose that the problem is to estimate k unknown parameters $\theta_1,\theta_2,\ldots,\theta_k$ characterizing the distribution $f_W(\omega;\theta)$ of the random variable W. Suppose the first k moments of the true distribution (the "population moments") can be expressed as functions of the θ_s:

$$\mu_k = E\left[W^k\right] = g_k(\theta_1,\theta_2,\ldots,\theta_k) \tag{14.1}$$

Suppose a sample of size n is drawn, resulting in the values $w_1,\ldots,w_n$. For j = 1, …, k, let

$$\hat{\mu}_j = \frac{1}{n}\sum_{i=1}^{n} w_i^j \tag{14.2}$$

be the jth sample moment, an estimate of μ_j. The method of moments estimator for $\theta_1,\theta_2,\ldots,\theta_k$ denoted by $\widehat{\theta}_1,\widehat{\theta}_2,\ldots,\widehat{\theta}_k$ is defined as the solution (if there is one) to the equations:

$$\hat{\mu}_k = g_k\left(\hat{\theta}_1,\hat{\theta}_2,\ldots,\hat{\theta}_k\right) \tag{14.3}$$

Example 14.1

Let $X_1,X_2,\ldots,X_n$ be normal random variables mean μ and variance σ^2. What is the method of moments estimators of the mean μ and variance σ^2?

Solution:

The first and second theoretical moments about the origin are:

$$E[X_i] = \mu \quad E\left[X_i^2\right] = \sigma^2 + \mu^2$$

(Incidentally, in case it's not obvious, that second moment can be derived from manipulating the shortcut formula for the variance.) In this case, we have two parameters for which we are trying to derive method of moments estimators. Therefore, we need two equations here. Equating the first theoretical moment about the origin with the corresponding sample moment, we get:

$$E[X] = \mu = \frac{1}{n}\sum_{i=1}^{n} X_i$$

And, equating the second theoretical moment about the origin with the corresponding sample moment, we get:

$$E[X] = \sigma^2 + \mu^2 = \frac{1}{n}\sum_{i=1}^{n} X_i^2$$

Now, the first equation tells us that the method of moments estimator for the mean μ is the sample mean:

$$\hat{\mu}_{MM} = \frac{1}{n}\sum_{i=1}^{n} X_i = \bar{X}$$

And, substituting the sample mean in for μ in the second equation and solving for σ^2, we get that the method of moments estimator for the variance σ^2 is:

$$\hat{\sigma}^2_{MM} = \frac{1}{n}\sum_{i=1}^{n} X_i^2 - \mu^2 = \frac{1}{n}\sum_{i=1}^{n} X_i^2 - \bar{X}^2$$

which can be rewritten as:

$$\hat{\sigma}^2_{MM} = \frac{1}{n}\sum_{i=1}^{n} \left(X_i - \bar{X}\right)^2$$

For this example, the method of moments estimators is the same as the maximum likelihood (ML) estimators.

In some cases, rather than using the sample moments about the origin, it is easier to use the sample moments about the mean. Doing so provides us with an alternative form of the method of moments.

14.3.2.2 Maximum Likelihood (MLE)

In statistics, maximum likelihood estimation (MLE) is a method of estimating the parameters of a probability distribution by maximizing a likelihood function, so that under the assumed statistical model, the observed data is most probable. The point in the parameter space that maximizes the likelihood function is called the ML estimate. The logic of ML is both intuitive and flexible, and as such, the method has become a dominant means of statistical inference.

If the likelihood function is differentiable, the derivative test for determining maxima can be applied. In some cases, the first-order conditions of the likelihood function can be solved explicitly; for instance, the ordinary least squares estimator maximizes the likelihood of the linear regression model. Under most circumstances, however, numerical methods will be necessary to find the maximum of the likelihood function.

From the vantage point of Bayesian inference, MLE is a special case of maximum a posteriori (MAP) estimation that assumes a uniform prior distribution of the parameters. In frequentist inference, MLE is a special case of an extremum estimator, with the objective function being the likelihood.

Assumptions:

1. The density function $p(x;\theta)$ is known up to a parameter $\theta \epsilon \Theta$. The column vector of parameters $\theta = [\theta_1,\dots,\theta_d] \epsilon \mathbb{R}^d$.
2. A measured data (set of examples) $D = \{x_1,\dots,x_n\}$ independently drawn from the identical distribution $p(x;\ \theta^*)$ is available.
3. The true parameter $\theta^* \epsilon \Theta$ is unknown, but it is fixed.
4. The probability $p(x;\ \theta)$ assumes that n examples D were generated (measured) for a given Θ is called the likelihood function

$$p(D;\theta) = \prod_{i=1}^{n} p(x_i;\theta) \tag{14.4}$$

MLE $\hat{\theta}_{ML}$ = argmax $p(D;\ \theta)$ seeks for the parameter $\hat{\theta}_{ML}$, which best explains the examples D.

Example 14.2

Suppose the weights of randomly selected American female college students are normally distributed with unknown mean μ and standard deviation σ^2. A random sample of ten American female college students yielded the following weights (in pounds):

115 122 130 127 149 160 152 138 149 180

Based on the definitions given above, identify the likelihood function and the ML estimator of μ, the mean weight of all-American female college students. Using the given sample, find a ML estimate of μ as well.

Solution:

The probability density function (pdf) of X_i is:

$$f(x_i;\mu,\sigma^2) = \frac{1}{\sigma\sqrt{2\pi}} \exp\left[-\frac{(x_i-\mu)^2}{2\sigma^2}\right]$$

for $-\infty < x < \infty$. The parameter space is $\Omega = \{(\mu,\sigma): -\infty < \mu < \infty$ and $0 < \sigma < \infty$. Therefore, (you might want to convince yourself that) the likelihood function is:

$$L(\mu,\sigma) = \sigma^{-n}(2\pi)^{-\frac{n}{2}} \exp\left[-\frac{1}{2\sigma^2}\sum_{i=1}^{n}(x_i-\mu)^2\right]$$

for $-\infty < \mu < \infty$ and $0 < \sigma < \infty$. It can be shown upon maximizing the likelihood function with respect to μ, that the ML estimator of μ is:

$$\hat{\mu} = \frac{1}{n}\sum_{i=1}^{n} X_i = \bar{X}$$

Based on the given sample, a ML estimate of μ is:

$$\hat{\mu} = \frac{1}{n}\sum_{i=1}^{n} x_i = \frac{1}{10}(115 + \dots + 180) = 142.2\,\text{pounds}$$

Note that the only difference between the formulas for the ML estimator and the ML estimate is that:

- The estimator is defined using capital letters (to denote that its value is random), and
- The estimate is defined using lowercase letters (to denote that its value is fixed and based on an obtained sample)

14.3.2.3 Maximum Posteriori (MAP)

In Bayesian statistics, an MAP probability estimate is an estimate of an unknown quantity that equals the mode of the posterior distribution. The MAP can be used to obtain a point estimate of an unobserved quantity on the basis of empirical data. It is closely related to the method of ML estimation but employs an augmented optimization objective which incorporates a prior distribution (that quantifies the additional information available through prior knowledge of a related event) over the quantity one wants to estimate. MAP estimation can therefore be seen as a regularization of MLE.

The posterior distribution, $f_{X|Y}(x \mid y)$ (or $P_{X|Y}(x \mid y)$), contains all the knowledge about the unknown quantity X. Therefore, we can use the posterior distribution to find point or interval estimates of X. One way to obtain a point estimate is to choose the value of x that maximizes the posterior PDF (or PMF). This is called the MAP estimation. The MAP estimate of the random variable X, given that we have observed $Y = y$, is given by the value of x that maximizes

$f_{X|Y}(x \mid y)$ if X is a continuous random variable,
$P_{X|Y}(x \mid y)$ if X is a discrete random variable.

The MAP estimate is shown by $\hat{x}_{MAP}$.

To find the MAP estimate, we need to find the value of x that maximizes

$$f_{X|Y}(x \mid y) = \frac{f_{Y|X}(y \mid x) f_X(x)}{f_Y(y)} \tag{14.5}$$

Note that $f_Y(y)$ does not depend on the value of x. Therefore, we can equivalently find the value of x that maximizes

$$f_{Y|X}(y \mid x) f_X(x)$$

This can simplify finding the MAP estimate significantly, because finding $f_Y(y)$ might be complicated. More specifically, finding $f_Y(y)$ usually is done using the law of total probability, which involves integration or summation.

To find the MAP estimate of X given that we have observed $Y = y$, we find the value of x that maximizes

$$f_{Y|X}(y \mid x) f_X(x) \tag{14.6}$$

If either X or Y is discrete, we replace its PDF in the above expression by the corresponding PMF. See Reich and Ghosh (2019) for more details on the matter.

Example 14.3

Let X be a continuous random variable with the following PDF:

$$f_X(x) = \begin{cases} 2x, & 1 \leq x \leq 0 \\ 0, & \text{otherwise} \end{cases}$$

Also, suppose that

$$Y \mid X = x \sim \text{Geometric}(x)$$

Find the MAP estimate of X given $Y = 3$.

Solution:

We know that $Y \mid X = x \sim \text{Geometric}(x)$, so

$$P_{Y|X}(y \mid x) = x(1-x)^{y-1}, \quad \text{for } y = 1, 2, \ldots$$

Therefore,

$$P_{Y|X}(3 \mid x) = x(x-1)^2$$

We need to find the value of $x \in [0,1]$ that maximizes

$$P_{Y|X}(y \mid x) f_X(x) = x(1-x)^2 \cdot 2x = 2x^2(1-x)^2$$

We can find the maximizing value by differentiation. We obtain

$$\frac{d}{dx}\left[x^2(1-x)^2\right] = 2x(1-x)^2 + 2(1-x)x^2 = 0$$

Solving for x (and checking for maximization criteria), we obtain the MAP estimate as

$$\hat{x}_{\text{MAP}} = \frac{1}{2}$$

14.3.2.4 Nonparametric Density Estimation—Parzen Method

In order to introduce a nonparametric estimator for the regression function m, we need to introduce first a nonparametric estimator for the density of the predictor X. This estimator is aimed to estimate f, the density of X, from a sample $X_1, \ldots, X_n$ without assuming any specific form for f. This is, without assuming, for example that the data is normally distributed.

No assumption is needed on the form of the distribution. However, many more training examples are required in comparison with parametric methods.

Suppose there is some data and we want to get a feeling about the underlying probability distribution which produced them, but there is no further information like the type of distribution (t-distribution, Poisson, etc.) or its parameters (mean, variance, etc.). Further, assuming a normal distribution does not seem to be the right solution. This situation is quite common in reality and there are some helpful techniques. *One is known as* kernel density estimation (*also known as* Parzen window density estimation *or* Parzen-Rosenblatt window method).

Parzen window is a nonparametric density estimation technique. Density estimation in Pattern Recognition can be achieved by using the approach of the Parzen windows. Parzen window density estimation technique is a kind of generalization of the histogram technique.

A popular application of the Parzen window technique is to estimate the class-conditional densities (or also often called 'likelihoods') $p(x \mid \omega_i)$ in a supervised pattern classification problem from the training dataset (where $p(x)$ refers to a multidimensional sample that belongs to particular class ω_i) that is a grouping of values by which data is binned for computation of a frequency distribution. See the illustration of Parzen windows in Figure 14.6.

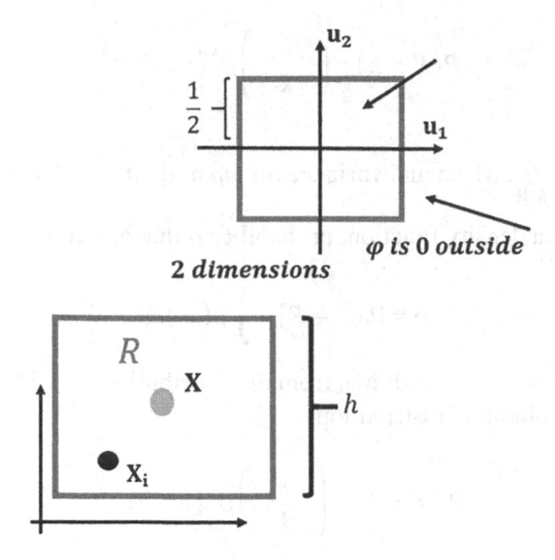

FIGURE 14.6 The illustration of Parzen windows.

This can be linked to design of a Bayes for solving a statistical pattern classification task using Bayes' rule:

$$P(\omega_i \mid x) = \frac{P(x \mid \omega_i) \times P(\omega_i)}{P(x)} \tag{14.7}$$

The likelihood function can be estimated by the use of Parzen method, which is investigated in this section. If the parameters of the class-conditional densities (also called likelihoods) are known, it is pretty easy to design the classifier. However, it becomes much more challenging, if there is no prior knowledge about the underlying parameters that define the model of our data.

If we want to design a classifier for a pattern classification task where the parameters of the underlying sample distribution are not known, we wouldn't need the knowledge about the whole range of the distribution; it would be sufficient to know the probability of the particular point, which we want to classify, in order to estimate that. And here we are going to see how we can estimate this probability from the training sample.

However, the only problem of this approach would be that we would seldom have exact values — if we consider the histogram of the frequencies for an arbitrary training dataset. Therefore, we define a certain region (i.e., the Parzen window's length h) around the particular value to make the estimate.

$$\Pr[x \in R] = \frac{\text{No of samples in } R}{\text{Total no of samples}} \tag{14.8}$$

Thus, the density at a point x inside R can be approximated by:

$$p(x) = \frac{\text{No of samples in } R}{\text{Total no of samples}} \cdot \frac{1}{\text{Volume}(R)} \tag{14.9}$$

Now let's derive this formula more formally.

Let us flip a coin n times (each one is called "trial"). Probability of head ρ, probability of tail is $1 - \rho$. Binomial random variable K counts the number of heads in n trials.

$$P(K=k)=\binom{n}{k}\rho^{k}(1-\rho)^{n-k} \tag{14.10}$$

Where, $\binom{n}{k}=\frac{n!}{k!(n-k)!}$. Mean and variance are $n\rho$ and $n\rho(1-\rho)$, respectively.

From the definition of a density function, probability ρ that a vector x will fall in region R is:

$$\rho=\Pr[x\in R]=\int p(x')dx' \tag{14.11}$$

Suppose we have samples $x_1, x_2, \ldots, x_n$ drawn from the distribution $p(x)$. The probability that k points fall in R is then given by binomial distribution:

$$\Pr[K=k]=\binom{n}{k}\rho^{k}(1-\rho)^{n-k} \tag{14.12}$$

Suppose that k points fall in R, we can use MLE to estimate the value of ρ. The likelihood function is:

$$p(x_1,x_2,\ldots,x_n \mid \rho)=\binom{n}{k}\rho^{k}(1-\rho)^{n-k} \tag{14.13}$$

This likelihood function is maximized at $\rho=\frac{k}{n}$. Thus, the MLE is $\hat{\rho}=\frac{k}{n}$.

Assume that $p(x)$ is continuous and that the region R is so small that $p(x)$ is approximately constant in R.

$$\int p(x')dx'\cong p(x)V \tag{14.14}$$

Recall from Equation 14.11, we have:

$$p(x)=\frac{\frac{k}{n}}{V} \tag{14.15}$$

Where, x is inside some region R, V is volume of R, n is total number of samples inside R, and k is number of samples inside R.

For a particular number n (total number of points), we use volume V of a fixed size and observe how many points k falls into the region. In other words, we use the same volume to make an estimate at different regions.

It is assumed that the region R is a d-dimensional hypercube with side length h; thus, its volume is h^d.

To estimate the density at point x, simply center the region R at x (*point of interest*), count the number of samples in R, and substitute everything in Eq uation 14.21.

We wish to have an analytic expression for our approximate density R. Let us define a window function:

$$\varphi(u)=\begin{cases}1, & |u_j|\le\frac{1}{2},\quad j=1,2,\ldots,d\\ 0 & \text{otherwise}\end{cases} \tag{14.16}$$

where u is $\frac{x - x_i}{h}$ What this function basically does is assigning a value 1 to a sample point if it lies within 1/2 of the edges of the hypercube, and 0 if lies outside (note that the evaluation is done for all dimensions of the sample point.

$$\varphi\left(\frac{x - x_i}{h}\right) = \begin{cases} 1 & \text{if } x_i \text{ is inside the hypercube with width } h \\ & \text{and center at } x \text{ (point of unknown density)} \\ 0 & \text{otherwise} \end{cases}$$

We count the total number of sample points $x_1, x_2, \ldots, x_n$ which are inside the hypercube with side length of h and centered at x by:

$$k = \sum_{i=1}^{n} \varphi\left(\frac{x - x_i}{h}\right) \quad (14.17)$$

Recall Equation 14.17, substituting for $p(x) = \frac{k/n}{V}$; thus, here is the desired analytical expression for the estimate of density $p_\varphi(x)$:

$$p_\varphi(x) = \frac{1}{n}\sum_{i=1}^{n} \frac{1}{h^d}\varphi\left(\frac{x - x_i}{h}\right) \quad (14.18)$$

Let's make sure $p_\varphi(x)$ is in fact a density.

$$p_\varphi(x) \geq 0 \quad \forall x \qquad \text{and} \quad \int p_\varphi(x)dx = 1 \quad (14.19)$$

We have to know which window size we should choose. The window width is a function of the number of training samples:

$$h_n \propto \frac{1}{\sqrt{n}} \quad (14.20)$$

This is because the number of k_n points within a window grows much smaller than the number of training samples, although:

$$\lim_{n \to \infty} k_n = \infty \quad (14.21)$$

We have $k < n$.

This is also one of the biggest drawbacks of the Parzen window technique, since in practice, the number of training data is usually (too) small, which makes the choice of an "optimal" window size difficult (Bishop, 2006).

Example 14.4

Suppose there are seven samples $D = \{2, 3, 4, 8, 10, 11, 12\}$. Assume the window width $h = 3$. Estimate density at $x = 1$ with Parzen method. d is one because only one set of data (dimension) is available (Figure 14.7).

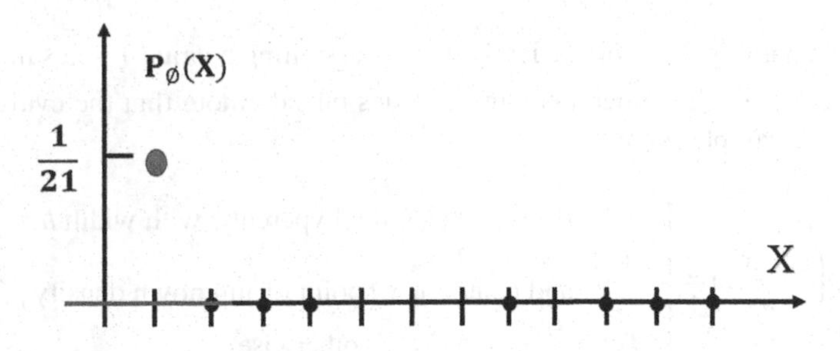

FIGURE 14.7 The Parzen windows example in 1D.

Solution:

$$p_\varphi(1)=\frac{1}{7}\sum_{i=1}^{n}\frac{1}{3^1}\varphi\left(\frac{1-x_i}{3}\right)=\frac{1}{21}\left(\underset{\left|-\frac{1}{3}\right|\le\frac{1}{2}}{\varphi\left(\frac{1-2}{3}\right)}+\underset{\left|-\frac{2}{3}\right|>\frac{1}{2}}{\varphi\left(\frac{1-3}{3}\right)}+\cdots+\underset{\left|-\frac{11}{3}\right|>\frac{1}{2}}{\varphi\left(\frac{1-12}{3}\right)}\right)$$

$$p_\varphi(1)=\frac{1}{21}[1+0+\cdots+0]=\frac{1}{21}$$

To assume the Parzen window as a different distribution (instead of Binomial) such as the following Gaussian function:

$$f(x)=\frac{1}{\sqrt{2\pi}}\exp\left[-\frac{(x_i-x)^2}{2\sigma}\right] \tag{14.22}$$

Then it replaces $\frac{1}{h^d}\varphi\left(\frac{x-x_i}{h}\right)$ as $\frac{k}{V}$ and dividing by n, then the average of $f(x)$ values yield to the density function at a specific point.

Example 14.5

Given a set of five data points $x_1=2,\ x_2=2.5,\ x_3=3,\ x_4=1$ and $x_5=6$, find Parzen pdf estimates at $x=3$, using the following Gaussian function with $\sigma=1$ as window function.

Solution:

$$f(x)=\frac{1}{\sqrt{2\pi}}\exp\left[-\frac{(x_i-x)^2}{2\sigma}\right]$$

for the given x_i, the $f(x)$ values are 0.2420, 0.3521, 0.3989, 0.0540, 0.0044, respectively

$$p(x=3)=\frac{0.2420+0.3521+0.3989+0.0540+0.0044}{5}=0.2103$$

The Parzen window can be graphically illustrated next. Each data point makes an equal contribution to the final pdf denoted by the solid line. Each data point is the mean of a Gaussian dist. and with $\sigma=1$, the distribution shows dot lines are depicted. The solid line distribution is the average of the five dash line distributions shown in Figure 14.8.

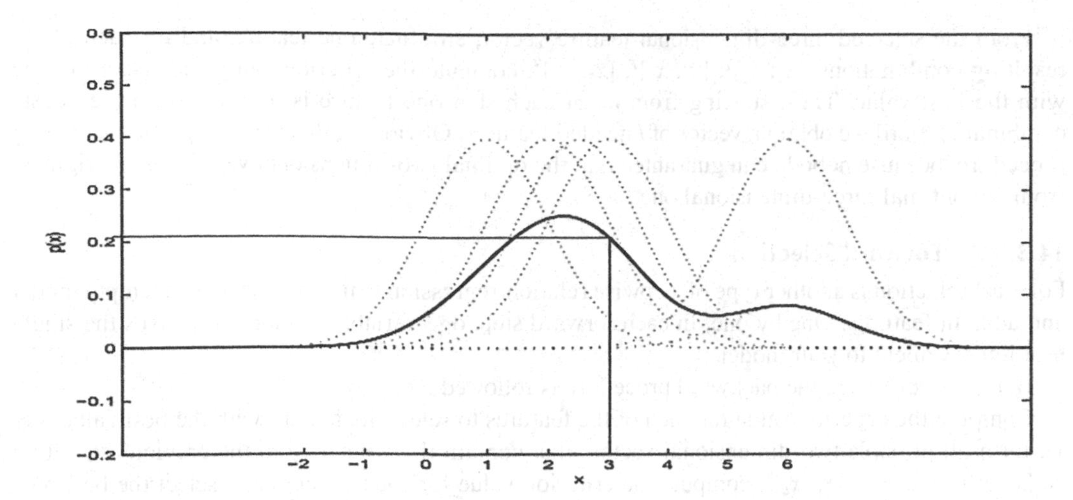

FIGURE 14.8 The dotted lines are the Gaussian functions centered at five data points and the Parzen window pdf function as average of the functions.

14.3.3 Feature Extraction and Selection

Feature extraction is a quite complex concept concerning the translation of basic data into the inputs that a particular machine learning algorithm requires. The model is the engine, but it needs fuel to work. Features must represent the information of the data in a format that will best fit the needs of the algorithm that is going to be used to solve a problem.

While some inherent features can be obtained directly from raw data, we usually need derived features from these inherent features that are actually relevant to target the underlying problem. A poor model fed with meaningful features will surely perform better than an amazing algorithm fed with low-quality features.

Feature extraction fills this requirement to build valuable information from basic data. By reformatting, combining, transforming primary features into new ones until it yields a new set of data that can be utilized by the machine learning models to achieve a set of goals.

Feature selection, for its part, is a clearer task to screen a set of potential features. Feature selection is applied either to prevent redundancy and/or irrelevancy in the features or just to get a limited number of features to prevent over fitting.

Designing a feature extractor is problem specific; the ideal feature extractor would produce the same feature vector x for all patterns in the same class and different feature vectors for patterns in different classes. Furthermore, we should remember different inputs to the feature extractor always produce different feature vectors, but we hope that the within-class variability is small relative to the between-class variability.

14.3.3.1 Backward Elimination

Backward elimination (or backward deletion) is the reverse process. All the independent variables are entered into the feasible set first, and each one is deleted one at a time if they do not contribute to the selected model such as a regression equation. We will demonstrate the method via an example. Let $m = 4$ to be the number of variables, and the original available features are x_1, x_2, x_3, x_4. If two of them need to be selected, then the selection procedure consists of the following steps:

Adopt a class reparability criterion, C, and compute its value for the feature vector $[x_1, x_2, x_3, x_4]^T$, as T stands for transpose. Eliminate one feature and for each of the possible resulting combinations, that is, $[x_1, x_2, x_3]^T$, $[x_1, x_2, x_4]^T$, $[x_1, x_3, x_4]^T$, $[x_2, x_3, x_4]^T$, compute the corresponding criterion value. *Select* the combination with the best value, say $[x_1, x_2, x_3]^T$.

From the selected three-dimensional feature vector, eliminate one feature, and for each of the resulting combinations, $[x_1, x_2]^T$, $[x_1, x_3]^T$, $[x_2, x_3]^T$, compute the criterion value and select the one with the best value. Thus, starting from m, at each step one feature is dropped from the "best" combination until we obtain a vector of l needed features. Obviously, this is a *suboptimal* searching procedure, because nobody can guarantee that the optimal two-dimensional vector has to originate from the optimal three-dimensional one.

14.3.3.2 Forward Selection

Forward selection is another type of stepwise relation/regression which begins with an empty model and adds in features. One by one, in each forward step, one variable is added that gives the single best improvement to your model.

Here, the reverse to the backward procedure is followed:

Compute the criterion value for each of the features to select the feature with the best value, say x_1; form all possible two-dimensional vectors that contain the winner from the previous step, that is, $[x_1, x_2]^T$, $[x_1, x_3]^T$, $[x_1, x_4]^T$, compute the criterion value for each of them and select the best one, say $[x_1, x_3]^T$; and so on.

14.3.4 Discrimination Analysis

Three different types of analysis are discussed here, namely, Fisher Discriminant Analysis (FDA), linear discriminant analysis (LDA), and principal of component analysis (PCA). See Bouveyron and Brunet (2012) for more details.

14.3.4.1 Fisher Discriminant Analysis (FDA)

FDA is a commonly used method for linear dimensionality reduction in supervised classification. FDA aims to find a linear subspace that well separates the classes in which a linear classifier (usually LDA) can be learned. FDA is a popular method which works very well in several cases. However, FDA does have some very well-known limitations. In particular, FDA produces correlated axes and its prediction performances are very sensitive to outliers, unbalanced classes, and label noise. Moreover, FDA has not been defined in a probabilistic framework, and its theoretical justification can be obtained only under the homoscedastic assumption on the distribution of the classes.

The terms Fisher's linear discriminant and LDA are often used interchangeably, although Fisher's original article actually describes a slightly different discriminant, which does not make some of the assumptions of LDA such as normally distributed classes or equal class covariances.

Consider input data set matrix is $X \in R^{n*m}$ by m variables and n observations for each variable, p is the number of classes, and n_j number of observations in j^{th} class. Let x_i represents the transpose i^{th} of row of matrix X. The standard FDA algorithm is briefly formulated as follows:

Step 1: Compute total scatter matrix S_t

$$S_t = \sum_{i=1}^{n} \left(X_i - \bar{X}\right)\left(X_i - \bar{X}\right)^T \tag{14.23}$$

Where $\bar{X}$ represents the total mean vector.

Step 2: Compute within-class scatter matrix S_w

$$S_w = \sum_{j=1}^{p} S_j \tag{14.24}$$

where

$$S_j = \sum_{X_i \in X_j}^{n} (X_i - \overline{X_j})(X_i - \overline{X_j})^T \tag{14.25}$$

where $\overline{X_j}$ is the mean of jth class.

Step 3: The between-class scatter matrix S_b

$$S_b = \sum_{j=1}^{p} n_j (\overline{X}_j - \overline{X})(\overline{X}_j - \overline{X})^T \tag{14.26}$$

$$S_t = S_w + S_b \tag{14.27}$$

where n_j is the number of samples in j^{th} class

Step 4: FDA vectors are obtained by solving the generalized eigenvalue problem solving the following expression.

$$S_b W = \lambda S_w W \tag{14.28}$$

Step 5: Since rank $(S_b)<p$, there exist at most $p - 1$ eigenvectors which are associated with nonzero eigenvalues (Fukunaga, 1990). This implies that FDA can find at most $p - 1$ meaningful features (the residual features created by FDA are arbitrary). This is an essential restriction of FDA in dimensionality reduction that is greatly restrictive in application.

This is a fundamental restriction of FDA in dimensionality reduction and is greatly restrictive in the application. Let k represents the number of non zero eigenvalues, then:

$$W_k = [w_1 \quad w_2 \quad \dots \quad w_k] \tag{14.29}$$

Step 6: FDA transformation vectors are calculated by the following equation:

$$Z_i = W_K^T X_i \tag{14.30}$$

FDA considers the information between various classes and calculates the transformation vectors.

14.3.4.2 Linear Discriminant Analysis

LDA, normal discriminant analysis (NDA), or discriminant function analysis is a generalization of Fisher's linear discriminant, a method used in statistics and other fields, to find a linear combination of features that characterizes or separates two or more classes of objects or events. The resulting combination may be used as a linear classifier, or, more commonly, for dimensionality reduction before later classification. LDA works when the measurements made on independent variables for each observation are continuous quantities. When dealing with categorical independent variables, the equivalent technique is discriminant correspondence analysis. This method can be used to separate the alteration zones. For example, when different data from various zones are available, discriminant analysis can find the pattern within the data and classify it effectively.

In LDA for two classes, a set of observations $\vec{x}$ (also called features, attributes, variables or measurements) is consider for each sample of an object or event with known class y. This set of samples is called the training set. The classification problem is then to find a good predictor for the class y of any sample of the same distribution (not necessarily from the training set) given only an observation $\vec{x}$.

LDA approaches the problem by assuming that the conditional pdfs $p(\vec{x} \mid y=0)$ and $p(\vec{x} \mid y=1)$ are both normally distributed with mean and covariance parameters $(\vec{\mu}_0, \Sigma_0)$ and $(\vec{\mu}_1, \Sigma_2)$, respectively. Under this assumption, the Bayes optimal solution is to predict points as being from the second class if the log of the likelihood ratios is bigger than some threshold T, so that:

$$(\vec{x}-\vec{\mu}_0)^T \Sigma_0^1 (\vec{x}-\vec{\mu}_0) + \ln|\Sigma_0| - (\vec{x}-\vec{\mu}_1)^T \Sigma_1^1 (\vec{x}-\vec{\mu}_1) - \ln|\Sigma_1| > T \quad (14.31)$$

Without any further assumptions, the resulting classifier is referred to as quadratic discriminant analysis (QDA).

LDA instead makes the additional simplifying homoscedasticity assumption (i.e., that the class covariances are identical, so $\Sigma_0 = \Sigma_1 = \Sigma$) and that the covariances have full rank. In this case, several terms cancel:

$$\vec{x}^T \Sigma_0^{-1} \vec{x} = \vec{x}^T \Sigma_1^{-1} \vec{x} \quad (14.32)$$

because Σ_i is Hermitian and the above decision criterion becomes a threshold on the dot product

$$\vec{\omega} \cdot \vec{x} > c \quad (14.33)$$

for some threshold constant c were

$$\vec{\omega} = \Sigma^{-1}(\vec{\mu}_1 - \vec{\mu}_0) \quad (14.34)$$

$$c = \vec{\omega} \cdot \frac{1}{2}(\vec{\mu}_1 + \vec{\mu}_0) \quad (14.35)$$

This means that the criterion of an input $\vec{x}$ being in a class y is purely a function of this linear combination of the known observations.

It is often useful to see this conclusion in geometrical terms: the criterion of an input $\vec{x}$ being in a class y is purely a function of projection of multidimensional space point $\vec{x}$ onto vector $\vec{\omega}$ (thus, we only consider its direction). In other words, the observation belongs to y if corresponding $\vec{x}$ is located on a certain side of a hyperplane perpendicular to $\vec{\omega}$. The location of the plane is defined by the threshold c.

14.3.4.3 Principal of Component Analysis (PCA)

PCA is an ordination-based statistic data exploration tool that converts a number of potentially correlated variables (with some shared attribute, such as points in space or time) into a set of uncorrelated variables that capture the variability in the underlying data. As such, PCA can be used to highlight patterns within multivariable data. PCA is a nonparametric analysis and is independent of any hypothesis about data probability distribution (Abdi and Williams, 2010).

PCA uses orthogonal linear transformation to identify a vector in N-dimensional space that accounts for as much of the total variability in a set of N variables as possible the first principal component (PC) where the total variability within the data is the sum of the variances of the observed variables, when each variable has been transformed so that it has a mean of zero and a variance of one. A second vector (second PC), orthogonal to the first, is then sought that accounts for as much of the remaining variability as possible in the original variables. Each succeeding PC is linearly uncorrelated to the others and accounts for as much of the remaining variability as possible (Jolliffe, 2002). PCA can therefore be used as descriptive, statistical approach to data transformation as a means of overcoming variable incommensurability. The ranking of the PCs in order of their significance (based on how much of the variability in the data they capture) is denoted by the eigenvalues associated with the vector for each PC.

PCA finds the most accurate data representation in a lower dimensional space. It is a simple, nonparametric method of extracting relevant information from confusing datasets. Statistical procedure that uses an orthogonal transformation to convert a set of observations of possibly correlated variables in to a set of values of linearly uncorrelated variable is called principal components. It is a way of identifying patterns in data. Since finding patterns in high- dimensional data is hard to find, PCA is a powerful tool for analyzing data. An important concept of PCA is to reduce the number of variables or reduce dimensionality. It helps in finding eigenvalues and eigenvectors. PCA is a standard tool in multivariate analysis for examining multidimensional data and to reveal patterns between objects that would not be apparent in a univariate analysis.

PCA reduces a correlated dataset (values of variables $\{x_1, x_2, \ldots, x_p\}$) to a dataset containing fewer new variables by axis rotation. The new variables are linear combinations of the original ones and are uncorrelated. The PCs are the new variables (or axes) which summarize several of the original variables. If nonzero correlation exists among the variables of the data set, then it is possible to determine a more compact description of the data, which amounts to finding the dominant modes in the data.

PCA can explain most of the variability of the original dataset in a few new variables (if data are well correlated). Correlation introduces redundancy (if two variables are perfectly correlated, then one of them is redundant because if we know x, we know y). PCA forms a new coordinate system defined by the eigenvectors (new coordinated will have lower dimensions). The data will be mapped into a new space.

PCA exploits this redundancy in multivariate data to pick out patterns and relationships in the variables and reduce the dimensionality of the dataset without significant loss of information.

To better understanding the PCA approach, Figure 14.9 is presented. Based on the 3D cluster of data, there could be three variables (X, Y, Z) or three locations (X, Y, Z). In the first step, it is shown that the variables are corrected and there is more variance in certain direction (Figure 14.9 a). The first pricipal component (PC_1) is the direction along which there is large variance (Figure 14.9 b). This is equivalent to axis rotation and expressing the data in a new coordinate system. The second PC (PC_2) is in the direction uncorrelated to the first component which along which the data shows the largest variation (Figure 14.9 c). PC_1 and PC_2 are uncorrelated or orthogonal (Figure 14.9 d).

The first PC is the linear combination of variables that explains the largest amount of variation and corresponds to the largest eigenvalue. Often, most of the variability observed across the variables (> 80%) can be explained in two to three PCs (Haak and Pagilla, 2020).

PCA can be used for interesting capabilities. It is applicable in different scopes such as:

- *Detection of outliers,* such as, data containing outliers and missing elements (Serneels and Verdonck, 2008)
- *Identification of clusters (grouping, regionalization),* such as, regionalization of precipitation regimes (Fazel et al., 2018), regional flood frequency analysis (Rahman and Rahman, 2020)
- *Reduction of variables (data preprocessing, multivariate modeling),* such as, PCA of turbulent combustion data: Data preprocessing and manifold sensitivity (Parente and Sutherland, 2013).
- *Data compression,* such as, digital image compression (Santo, 2012), Ensemble Assimilation of Hyperspectral Satellite Observations with Data Compression (Lu and Zhang, 2019).
- *Analysis of variability in space and time,* such as, the spatiotemporal variability of the MODIS images (de Almeida et al., 2015).
- *New interpretation of the data (in terms of the main components of variability),* such as, evaluation of natural support capacity of resources (Cao et al., 2020); water stress vulnerabilities (Karamouz and Ebrahimi (2022)).

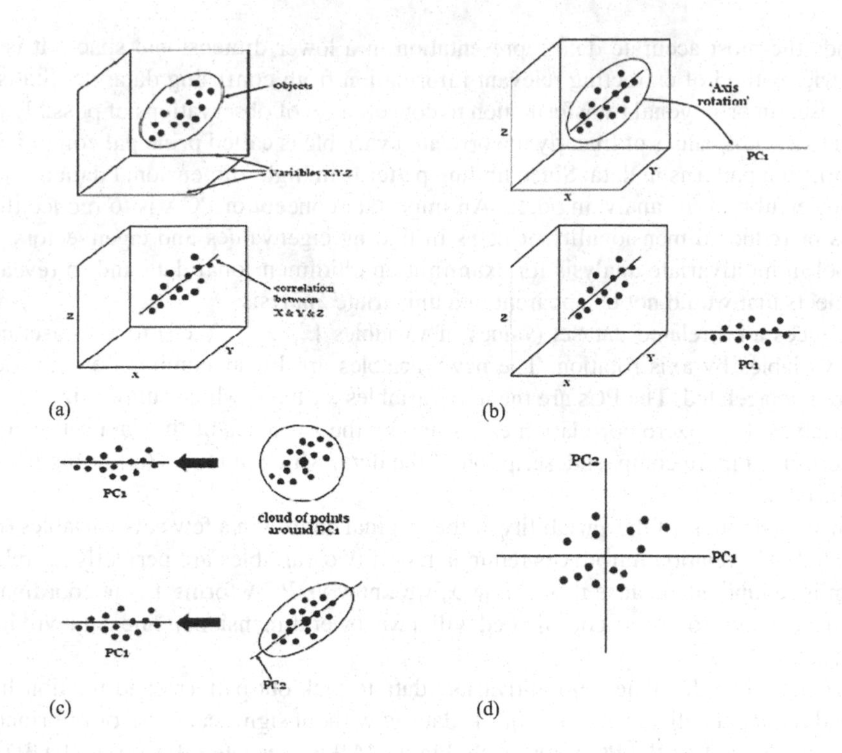

FIGURE 14.9 Process of principal of component analysis: (a) variance of variables in certain direction, (b) PC_1 in the direction along which there is large variance (expressing the data in a new coordinate system), (c) PC_2 in the direction uncorrelated to the PC_1 showing the largest variation, and (d) PC_1 and PC_2 are uncorrelated or orthogonal (Hu, Y.P. and Tsay, R.S. Principal volatility component analysis. *Journal of Business & Economic Statistics*, 32(2), 153–164, (2014)).

- *Forecasting (finding relationships between variables),* such as, improved estimation of electricity demand function (Kheirkhah et al., 2013), prediction of carbon dioxide emissions (Sun and Sun, 2017).

In the context of calculating eigenvalues and eigenvectors, the following paragraph is considered for estimating algebra vectors. The product Ax of a matrix A and a vector x is itself a vector. For a given matrix A, what are the vectors x for which the product Ax is a scalar multiple of x? That is, what nonzero vectors x (eigenvector) satisfy the equation $Ax = \lambda x$ for some scalar λ called eigenvalue?

To answer this question, we first perform some algebraic manipulations upon the equation $Ax = \lambda x$. We note first that, considering the I as the unit matrix, if $I = I_n$ (the $n \times n$), then we can write:

$$\begin{aligned} &Ax = \lambda x \\ &\leftrightarrow Ax - \lambda x = 0 \\ &\leftrightarrow Ax - \lambda I x = 0 \\ &\leftrightarrow (Ax - \lambda I)x = 0. \end{aligned} \tag{14.36}$$

Remember that we are looking for nonzero x that satisfies this last equation. But $A - \lambda I$ is an $n \times n$ matrix and, determinant should be nonzero, this last equation will have exactly one solution, namely $x = 0$. Thus, our question above has the following answer:

The equation $Ax = \lambda_i\, x$ has nonzero solutions for the vector x if and only if the matrix $A - \lambda I$ has zero determinant.

For a given matrix A, there are only a few special values of the scalar λ for which $A - \lambda I$ will have zero determinant, and these special values are called the eigenvalues of the matrix A. Based upon the answer to our question, it seems we must first be able to find the eigenvalues $\lambda_1, \lambda_2, \ldots, \lambda_n$ of A and then solve the individual equations $Ax = \lambda_i x$ for each $i = 1, \ldots, n$.

Example 14.6

Find the eigenvalues and associated eigenvectors of the matrix $A = \begin{bmatrix} -1 & 2 \\ 0 & -1 \end{bmatrix}$.

Solution:

Compute

$$\det(A - \lambda I) = \begin{vmatrix} -1-\lambda & 2 \\ 0 & -1-\lambda \end{vmatrix} = (\lambda+1)^2$$

Setting this equal to zero, we get that $\lambda = -1$ is a (repeated) eigenvalue. To find any associated eigenvectors, we must solve for $x = (x_1, x_2)$ so that $(A+I)\,x = 0$; that is,

$$\begin{bmatrix} 0 & 2 \\ 0 & 0 \end{bmatrix}\begin{bmatrix} x_1 \\ x_2 \end{bmatrix} = \begin{bmatrix} 2x_2 \\ 0 \end{bmatrix} = \begin{bmatrix} 0 \\ 0 \end{bmatrix}$$

$$x_2 = 0$$

Thus, the eigenvectors corresponding to the eigenvalue $\lambda = -1$ are the vectors whose second component is zero.

So the steps to calculate PCA are as follows:

Step 1: Mean of variables
Step 2: Form a matrix with deviations from mean
Step 3: Calculate the covariance matrix

Covariance measures how two features vary with each other. A positive covariance indicates that features increase and decrease together. Whereas, a negative covariance indicates that the two features vary in the opposite directions. For two feature vectors x_j and x_k, the covariance between them σ_{jk} can be calculated using the following equation:

$$\lambda_i = \frac{1}{n-1}\sum_{i=1}^{n}\left(x_j^i - \mu_j\right)\left(x_k^i - \mu_k\right) \tag{14.37}$$

A covariance matrix contains the covariance values between features and has shape $d \times d$. For our dataset, the covariance matrix should, therefore, look like the following:

$$\Sigma = \begin{bmatrix} \sigma_1^2 & \sigma_{12} & \sigma_{13} & \sigma_{14} \\ \sigma_{21} & \sigma_2^2 & \sigma_{23} & \sigma_{24} \\ \sigma_{31} & \sigma_{32} & \sigma_3^2 & \sigma_{34} \\ \sigma_{41} & \sigma_{42} & \sigma_{43} & \sigma_4^2 \end{bmatrix} \tag{14.38}$$

The diagonal entries of the covariance matrix are the variances and the other entries are the covariances. Since the feature columns have been standardized and, therefore, they each have a mean of zero, the covariance matrix Σ can be calculated by the following:

$$\Sigma = \frac{1}{n-1} X'X \tag{14.39}$$

Step 4: Calculate the eigenvalues and eigenvectors of the covariance matrix

The eigenvectors represent the PCs (the directions of maximum variance) of the covariance matrix. The eigenvalues are their corresponding magnitude. The eigenvector that has the largest corresponding eigenvalue represents the direction of maximum variance. An eigenvector v satisfies the following condition:

$$\sum \vartheta = \lambda \tag{14.40}$$

where λ is a scalar and known as the eigenvalue. The manual computation is quite elaborate and could be a post all its own.

Step 5: Choose components and form a feature vector

Now that the eigenpairs have been computed, they now need to be sorted based on the magnitude of their eigenvalues.

Step 6: Derive the new data set

Now that the PCs have been sorted based on the magnitude of their corresponding eigenvalues, it is time to determine how many PCs to select for dimensionality reduction. This can be done by plotting the cumulative sum of the eigenvalues. The cumulative sum is computed as the following:

$$e_i = \lambda_i \Big/ \sum_{i=1}^{p} \lambda_i \tag{14.41}$$

Now that it has been decided how many of the PCs to make up the projection matrix W, the scores Z can be calculated as follows:

$$Z = XW \tag{14.42}$$

So we can use the coefficient of linear regression as:

$$\text{PC}_1 = a_1 X_1 + a_2 X_2 \tag{14.43}$$

Example 14.7

The data set of the yearly rainfall and the yearly runoff of a catchment for 10 years is given. Use the PCA to find the PCs.

Year	1	2	3	4	5	6	7	8	9	10
Rainfall (cm)	105	115	103	94	95	104	120	121	127	79
Runoff (cm)	42	46	26	39	29	33	48	58	45	20

Solution:

Step 1: Mean of rainfall = 108.5 cm and mean of runoff = 38.3 cm

Step 2: Form a matrix with deviations from mean

Original matrix			Matrix with deviation from mean	
105	42		−1.3	3.4
115	46		8.7	7.4
103	26		−3.3	−12.6
94	39		−12.3	0.4
95	29	X =	−11.3	−9.3
104	33		−2.3	−5.6
120	48		13.7	9.4
121	58		14.7	19.4
127	45		20.7	6.4
79	20		−27.3	−18.6

Step 3: Calculate the covariance matrix

$$\text{cov}(X,Y) = s_{X,Y} = \frac{\sum_{i=1}^{n}(X_i - \bar{X})(Y_i - \bar{Y})}{n-1}$$

$$\begin{bmatrix} \text{cov}(X,X) & \text{cov}(X,Y) \\ \text{cov}(Y,X) & \text{cov}(Y,Y) \end{bmatrix} = \begin{bmatrix} 216.67 & 141.35 \\ 141.35 & 133.38 \end{bmatrix}$$

Step 4: Calculate the eigenvalues and eigenvectors of the covariance matrix

$$A = \begin{bmatrix} 216.67 & 141.35 \\ 141.35 & 133.38 \end{bmatrix}$$

Eigenvalues:

$$|A - \lambda I| = 0$$

$$\lambda_1 = 322.4 \quad \text{and} \quad \lambda_2 = 27.7$$

Eigenvectors:

$$(A - \lambda I)X = 0$$

$$X = \begin{bmatrix} 0.801 & -0.599 \\ 0.599 & 0.801 \end{bmatrix}$$

Step 5: Choose components and form a feature vector

The fraction of the total variance accounted for by the *j*th PC is

$$\frac{\lambda_j}{\text{Trace}(S)}$$

where

$$\text{Trace}(s) = \sum \lambda_j$$

In linear algebra, the trace of a square matrix A, denoted tr(A), is defined to be the sum of the eigen values of A.

$$\text{Trace}(S) = 322.4 + 27.7 = 350.1$$

The total variance accounted for by the first PC is

$$\frac{\lambda_j}{\text{Trace}(S)} = \frac{322.3}{350.1} = 0.92$$

that is, 92% of total system variance is represented by the first PC and the remaining 8% is represented by the second component.

Hence, the second PC can be neglected and only the first one considered.

From the two eigenvectors, the feature vector is selected

$$A = \begin{bmatrix} 0.801 \\ 0.599 \end{bmatrix}$$

Step 6: Derive the new data set

$$ZA = X$$

$$\begin{bmatrix} -1.3 & 3.4 \\ 8.7 & 7.4 \\ -3.3 & -12.6 \\ -12.3 & 0.4 \\ -11.3 & -9.3 \\ -2.3 & -5.6 \\ 13.7 & 9.4 \\ 14.7 & 19.4 \\ 20.7 & 6.4 \\ -27.3 & -18.6 \end{bmatrix} \begin{bmatrix} 0.801 \\ 0.599 \end{bmatrix} = \begin{bmatrix} 0.995 \\ 11.39 \\ -10.2 \\ -9.61 \\ -14.8 \\ -5.20 \\ 16.60 \\ 23.39 \\ 20.41 \\ -33.0 \end{bmatrix}$$

Using both the eigenvalues, the new data set is

$$\begin{bmatrix} -1.3 & 3.4 \\ 8.7 & 7.4 \\ -3.3 & -12.6 \\ -12.3 & 0.4 \\ -11.3 & -9.3 \\ -2.3 & -5.6 \\ 13.7 & 9.4 \\ 14.7 & 19.4 \\ 20.7 & 6.4 \\ -27.3 & -18.6 \end{bmatrix} \begin{bmatrix} 0.801 & -0.599 \\ 0.599 & 0.801 \end{bmatrix} = \begin{bmatrix} 0.995 & 3.510 \\ 11.39 & 0.716 \\ -10.2 & -8.11 \\ -9.61 & 7.687 \\ -14.8 & -0.92 \\ -5.20 & -3.11 \\ 16.60 & -0.68 \\ 23.39 & 6.732 \\ 20.41 & -7.27 \\ -33.0 & 14.55 \end{bmatrix}$$

So in this example, the PCA method is applied on the temporal scale for rainfall and runoff, the matrix X is the new projection of the data set, and deriving linear regression here is not feasible. In the context of clustering or demonstrating new interpretation of the data (in terms of the main components of variability), linear equation can be derived by application of PCA method.

14.3.5 Supervised Classification

In supervised classification, the user or image analyst "supervises" the pixel classification process. The user specifies the various pixel values or spectral signatures that should be associated with each class. This is done by selecting representative sample sites of a known cover type called training sites or areas. The computer algorithm then uses the spectral signatures from these training areas to classify the whole image. Ideally, the classes should not overlap or should only minimally overlap with other classes.

To solve a given problem of supervised learning, one has to perform the following steps:

1. *Determine the type of training examples*: Before doing anything else, the user should decide what kind of data is to be used as a training set. In the case of handwriting analysis, for example, this might be a single handwritten character, an entire handwritten word, or an entire line of handwriting.
2. *Gather a training set*: The training set needs to be representative of the real-world use of the function. Thus, a set of input objects is gathered and corresponding outputs are also gathered, either from human experts or from measurements.
3. *Determine the input feature representation of the learned function*: The accuracy of the learned function depends strongly on how the input object is represented. Typically, the input object is transformed into a feature vector, which contains a number of features that are descriptive of the object. The number of features should not be too large, because of the curse of dimensionality but should contain enough information to accurately predict the output.
4. *Determine the structure of the learned function and corresponding learning algorithm*: For example, the engineer may choose to use support vector machines (SVMs) or decision trees.
5. *Complete the design*: Run the learning algorithm on the gathered training set. Some supervised learning algorithms require the user to determine certain control parameters. These parameters may be adjusted by optimizing performance on a subset (called a validation set) of the training set, or via cross-validation.
6. *Evaluate the accuracy of the learned function*: After parameter adjustment and learning, the performance of the resulting function should be measured on a test set that is separate from the training set.

14.3.5.1 Bayes Decision Theory

The approach to be followed builds upon probabilistic arguments stemming from the statistical nature of the generated features. This is due to the statistical variation of the patterns. Adopting this reasoning as our kickoff point, we will design classifiers that classify an unknown pattern in the most probable of the classes. Thus, our task now becomes that of defining what "most probable" means. Given a classification task of M classes, ω_1, ω_2, …, ω_M, and an unknown pattern, which is represented by a feature vector x, we form the M conditional probabilities $P(x_i|x)$, $i=1, 2,\ldots, M$. Sometimes, these are also referred to as a posteriori probability. In words, each of them represents the probability that the unknown pattern belongs to the respective class ω_i, given that the corresponding feature vector takes the value x. Who could then argue that these conditional probabilities are not sensible choices to quantify the term most probable? Indeed, the classifiers to be considered in section compute either the maximum of these M values or, equivalently, the maximum of an appropriately defined function of them. The unknown pattern is then assigned to the class corresponding to this maximum.

The first task we are faced with is the computation of the conditional probabilities. The Bayes rule will once more prove its usefulness! A major effort in this chapter will be devoted to techniques for estimating pdfs, based on the available experimental evidence, that is, the feature vectors corresponding to the patterns of the training set (Theodoridis, 2003).

14.3.5.2 Density Function Estimation

In probability and statistics, density estimation is the construction of an estimate, based on observed data, of an unobservable underlying pdf. The unobservable density function is thought of as the density according to which a large population is distributed; the data are usually thought of as a random sample from that population.

A variety of approaches to density estimation are used, including Parzen windows and a range of data clustering techniques. The most basic form of density estimation is a rescaled histogram.

A discrimination rule may be constructed through explicit estimation of the class-conditional density functions and the use of Bayes' rule. One approach is to assume a simple parametric model for the density functions and to estimate the parameters of the model using an available training set. The Gaussian classifier and its variants are introduced. The more powerful approach of mixture models is then presented.

14.3.5.3 Parametric Density Estimation

A particular form of the density function (e.g., Gaussian) is assumed to be known and only its parameters $\theta \in \Theta$ need to be estimated (e.g., the mean, the covariance).

Sought: $p(x, \theta)$;
Performed: $\{x_1,\ldots,x_n\} \to \hat{\theta}$.

So far, we have learned several nonparametric methods for density estimation. In fact, we can use a simple parametric method for density estimation.

We will start with a simple example by assuming the data is from a Gaussian (normal) distribution. Recall that we observe

$$X_1, X_2, \ldots, X_n \sim P, \tag{14.44}$$

where P is the underlying population CDF and it has a PDF p. If we fit a Gaussian distribution to the data, we need to find the two parameters of Gaussian: the mean μ and the variance σ^2. While there are many approaches for estimating them (e.g., method of moments, or ML method), we use a very simple estimator: the sample mean and sample variance.

Let

$$\hat{\mu}_n = \bar{X}_n = \frac{1}{n}\sum_{i=1}^{n} X_i, \quad \hat{\sigma}_n^2 = S_n^2 = \frac{1}{n-1}\sum_{i=1}^{n}\left(X_i - \bar{X}_n\right)^2 \tag{14.45}$$

be the sample mean and sample variance. Then our density estimator is

$$\hat{p}_n(x) = \frac{1}{\sqrt{2\pi\hat{\sigma}_n^2}} e^{-\frac{1}{2\hat{\sigma}_n^2}(x-\hat{\mu}_n)^2} \tag{14.46}$$

If the true PDF p is close to a Gaussian distribution, then probably the parametric approach is a good one. But if p is very far away from being a Gaussian, this method is going to give us a huge bias.

Example 14.8

For the following random samples, find the likelihood function:
$X_i \sim \text{Binomial}(3,\theta)$, and we have observed $(x_1,x_2,x_3,x_4) = (1,3,2,2)$.

Solution:

When we have a random sample, X_i's are i.i.d., so we can obtain the joint PMF and PDF by multiplying the marginal (individual) PMFs and PDFs.

If $X_i \sim \text{Binomial}(3,\theta)$, then

$$P_{Xi}(x;\theta) = \binom{3}{x}\theta^x(1-\theta)^{3-x}$$

Thus,

$$L(x_1,x_2,x_3,x_4;\theta) = P_{X_1X_2X_3X_4}(x_1,x_2,x_3,x_4;\theta) = P_{X1}(x_1;\theta)P_{X2}(x_2;\theta)P_{X3}(x_3;\theta)P_{X4}(x_4;\theta)$$

$$= \binom{3}{x_1}\binom{3}{x_2}\binom{3}{x_3}\binom{3}{x_3}\theta^{x_1+x_2+x_3+x_4}(1-\theta)^{12-(x_1+x_2+x_3+x_4)}$$

Since we have observed $(x_1,x_2,x_3,x_4)=(1,3,2,2)$, we have

$$L(1,3,2,2;\theta) = \binom{3}{1}\binom{3}{3}\binom{3}{2}\binom{3}{2}\theta^8(1-\theta)^4$$

14.3.5.4 k-Nearest Neighbor Estimation (k-NN)

In statistics, the k-nearest neighbors' (k-NN) algorithm is a nonparametric machine learning method. It is used for classification and regression. In both cases, the input consists of the k closest training examples in feature space. The output depends on whether k-NN is used for classification or regression:

In k-NN classification, the output is a class membership. An object is classified by a plurality vote of its neighbors, with the object being assigned to the class most common among its k-NNs (k is a positive integer, typically small). If $k = 1$, then the object is simply assigned to the class of that single nearest neighbor.

In k-NN regression, the output is the property value for the object. This value is the average of the values of k-NNs.

k-NN is a type of instance-based learning, or lazy learning, where the function is only approximated locally and all computation is deferred until function evaluation. Since this algorithm relies on distance for classification, if the features represent different physical units or come in vastly different scales, then normalizing the training data can improve its accuracy dramatically.

Both for classification and regression, a useful technique can be to assign weights to the contributions of the neighbors, so that the nearer neighbors contribute more to the average than the more distant ones. For example, a common weighting scheme consists in giving each neighbor a weight of $1/d$, where d is the distance to the neighbor.

The neighbors are taken from a set of objects for which the class (for k-NN classification) or the object property value (for k-NN regression) is known. This can be thought of as the training set for the algorithm, though no explicit training step is required.

A peculiarity of the k-NN algorithm is that it is sensitive to the local structure of the data.

Example 14.9

Suppose that a fish packing plant wants to automate the process of sorting incoming fish. We set up a camera, take some sample images, and begin to note some physical differences between the two types of fish and suggest features to explore for use in our classifier. We also notice noise or variations in the images such as variations in lighting, position of the fish on the conveyor, even "static" due to the electronics of the camera itself. Given that there truly are differences between the population of sea bass and that model of salmon, we view them as having different models. The process of feature extraction of fish classification is shown in Figure 14.10. The overarching goal and approach in pattern classification are to hypothesize the class of these models, process the sensed data to eliminate noise (not due to the models), and for any sensed pattern choose the model that corresponds best.

The objects to be classified are first sensed by a transducer (camera), whose signals are preprocessed, then the features extracted and finally the classification emitted (here either "salmon" or "sea bass"). Although the information flow is often chosen to be from the source to the classifier ("bottom-up"), some systems employ "top-down" flow as well, in which earlier levels of processing can be altered based on the tentative or preliminary response in later levels (gray arrows). Yet others combine two or more stages into a unified step, such as simultaneous segmentation and feature extraction.

Suppose somebody at the fish plant tells us that a sea bass is generally longer than a salmon. These, then, give us our tentative models for the fish, namely, sea bass have some typical length (l^*), and this is greater than that for salmon. Then length becomes an obvious feature. Suppose that we take samples and obtain the histograms (Figure 14.11).

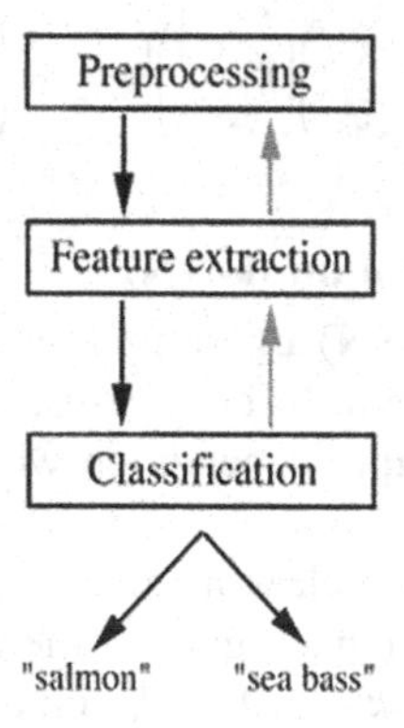

FIGURE 14.10 The process feature extraction of fish classification.

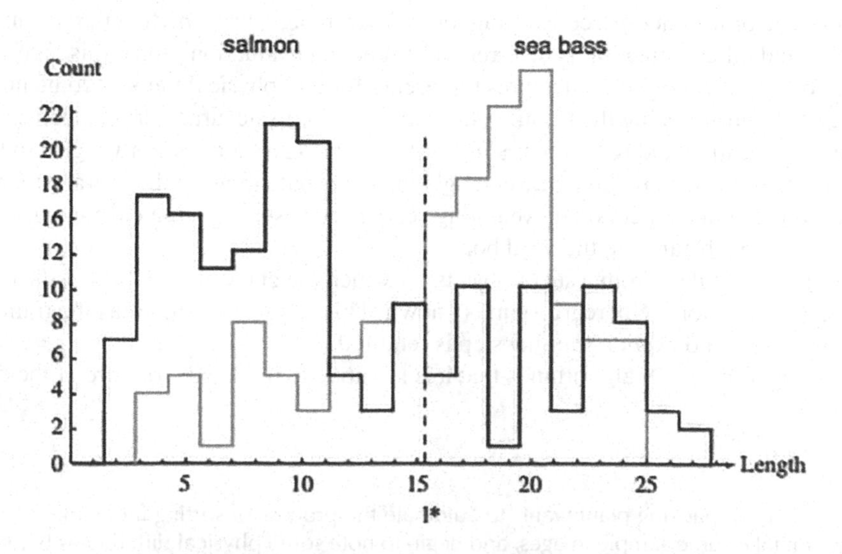

FIGURE 14.11 Histograms for the length feature for the two categories.

Solution:

No single threshold value l^* (decision boundary) will serve to unambiguously discriminate between the two categories by using length alone. There are some errors. The value l^* marked will lead to the smallest number of errors, on average.

We try another feature, the average lightness of the fish scales (x^*). Histograms are much more satisfactory, yet no single threshold value x^* (decision boundary) will serve to unambiguously discriminate between the two categories (Figure 14.12).

No single threshold value x^* (decision boundary) will serve to unambiguously discriminate between the two categories; using lightness alone, the value x^* marked will lead to the smallest errors, on average.

To improve recognition, then, we must resort to the use of more than one feature at a time. The dark line might serve as a decision boundary of our classifier. Overall classification error on the data shown is lower than if we use only one feature as in Figure 14.9, but there will still be some errors.

The two features of lightness and width for sea bass and salmon are shown in Figure 14.13. The dark line might serve as a decision boundary of our classifier.

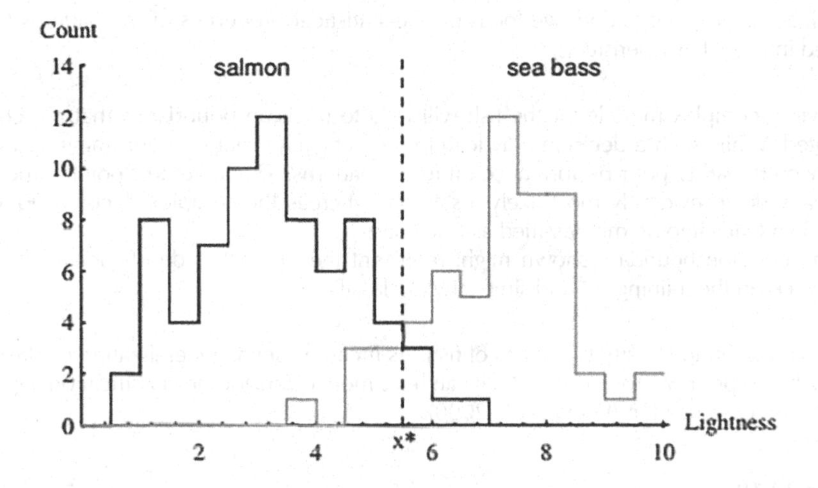

FIGURE 14.12 Histograms for the lightness feature for the two categories.

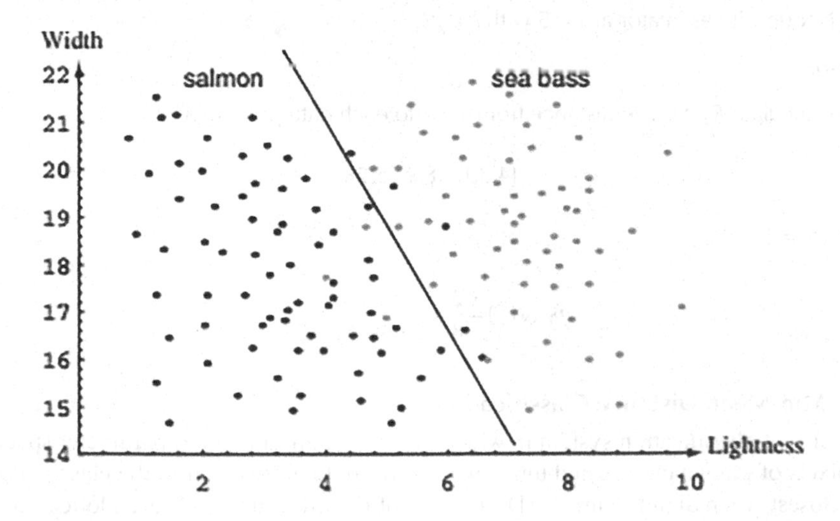

FIGURE 14.13 The two features of lightness and width for sea bass and salmon.

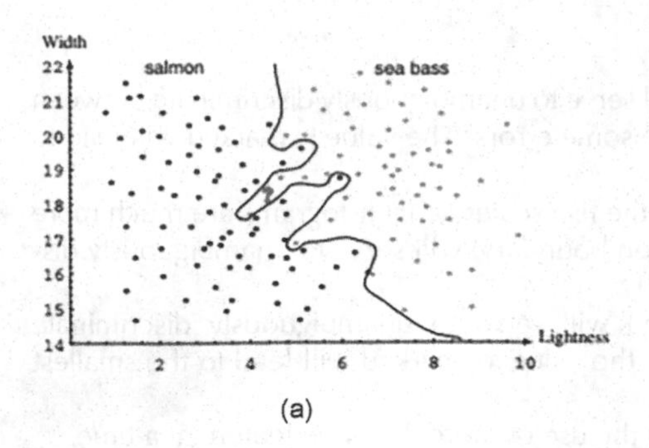

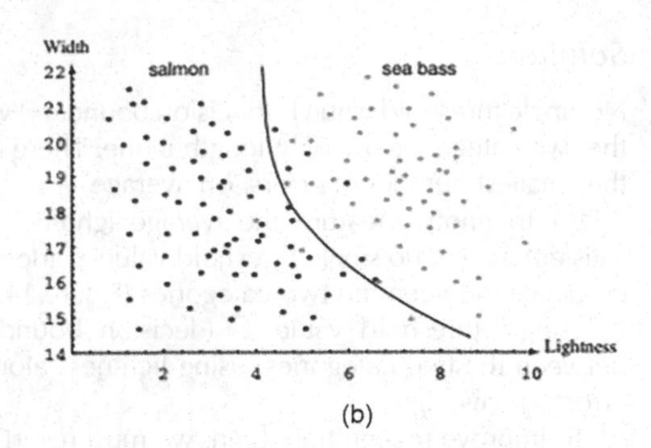

FIGURE 14.14 (a) Overly complex models for the fish—complicated decision boundaries and (b) the decision boundary representing the optimal trade-off.

Since classification is, at base, the task of recovering the model that generated the patterns, different classification techniques are useful depending on the type of candidate models themselves. In statistical pattern recognition, we focus on the statistical properties of the patterns (generally expressed in probability densities),

a. Overly complex models for the fish will lead to decision boundaries that are complicated. While such a decision may lead to perfect classification of our training samples, it would lead to poor performance on future patterns. The novel test point marked red heavy dot is evidently most likely a salmon, whereas the complex decision boundary shown leads it to be misclassified as a sea bass.
b. The decision boundary shown might represent the optimal trade-off between performance on the training set and simplicity of classifier.

There is a trade-off to classify two types of fish. As it can be seen, generalization of classifying of Figure 14.14 a is poor. So, for a new data, it can have more errors for the classification. Figure 14.14 b is the better classification (Duda et al., 2000).

Example 14.10

We consider a simple example in $d = 1$. Assume our data is $X = \{1, 2, 6, 11, 13, 14, 20, 33\}$. What is the k-NN density estimator at $x = 5$ with $k = 2$?

Solution:

First, we calculate $R_2(5)$. The distance from $x = 5$ to each data point in X is

$$\{4,3,1,6,8,9,15,28\}.$$

Thus, $R_2(5) = 3$ and

$$\hat{p}_{\text{k-NN}}(5) = \frac{2}{8} \times \frac{1}{2 \times R_2(5)} = \frac{1}{24}$$

14.3.5.5 Min-Mean Distance Classification

A remote sensing classification system in which the mean point in digital parameter space is calculated for pixels of known classes and unknown pixels are then assigned to the class which is arithmetically closest when digital number (DN) values of the different bands are plotted. See also box classification and maximum-likelihood classification.

14.3.5.6 Support Vector Machines (SVM)

In machine learning, SVMs (also support vector networks) are supervised learning models with associated learning algorithms that analyze data for classification and regression analysis. Developed at AT&T Bell Laboratories by Vapnik with colleagues (Boser et al., 1992 ; Guyon et al., 1993; Vapnik et al., 1997), SVMs are one of the most robust prediction methods, being based on statistical learning frameworks or VC theory proposed by Vapnik and Chervonenkis (1974) and Vapnik (1982, 1995). Given a set of training examples, each marked as belonging to one of two categories, an SVM training algorithm builds a model that assigns new examples to one category or the other, making it a nonprobabilistic binary linear classifier (although methods such as Platt scaling exist to use SVM in a probabilistic classification setting). An SVM maps training examples to points in space so as to maximize the width of the gap between the two categories. New examples are then mapped into that same space and predicted to belong to a category based on which side of the gap they fall.

In addition to performing linear classification, SVMs can efficiently perform a nonlinear classification using what is called the kernel trick, implicitly mapping their inputs into high-dimensional feature spaces.

When data are unlabeled, supervised learning is not possible, and an unsupervised learning approach is required, which attempts to find natural clustering of the data to groups, and then map new data to these formed groups. The support vector clustering algorithm, created by Hava Siegelmann and Vladimir Vapnik, applies the statistics of support vectors, developed in the support vector machines algorithm, to categorize unlabeled data, and is one of the most widely used clustering algorithms in industrial applications.

More formally, a SVM constructs a hyperplane or set of hyperplanes in a high- or infinite-dimensional space, which can be used for classification, regression, or other tasks like outlier's detection. Intuitively, a good separation is achieved by the hyperplane that has the largest distance to the nearest training data point of any class (so-called functional margin), since in general the larger the margin, the lower the generalization error of the classifier.

14.3.6 Unsupervised Classification (Clustering)

A clustering algorithm can be employed to reveal the groups in which feature vectors are clustered in the one-dimensional feature space. The labels of the classes are not determined. Points that correspond to the same ground cover type, such as water, are expected to cluster together and form groups. The unsupervised classification analyst can identify the type of each cluster and a sample of points is associated in each group. Clustering is also widely used in the social sciences in order to study and correlate survey and statistical data and draw useful conclusions, which will then lead to the right actions. In water resource management, unsupervised clustering can be used for land cover classification (Figure 14.15).

14.3.6.1 Sequential Clustering

The use of the sequential clustering algorithm is more complex than the use of other algorithms (such as K-MEANS) due to the extra effort needed to adjust all its parameters. The algorithm begins choosing the first cluster center among all the pattern samples arbitrarily. Then, it processes the remaining of the patterns sequentially. It computes the distance from the actual pattern to its nearest cluster center. If it is smaller than or equal to distance threshold used to form new clusters (threshold), the pattern is assigned to its nearer cluster. If not, a new cluster is formed with the actual pattern.

$$dm(x) = \text{Min}_{i=1}^{c} d(xi, wi) \tag{14.47}$$

$$dm(x) < \text{th} \rightarrow \text{absorb to the cluster}$$

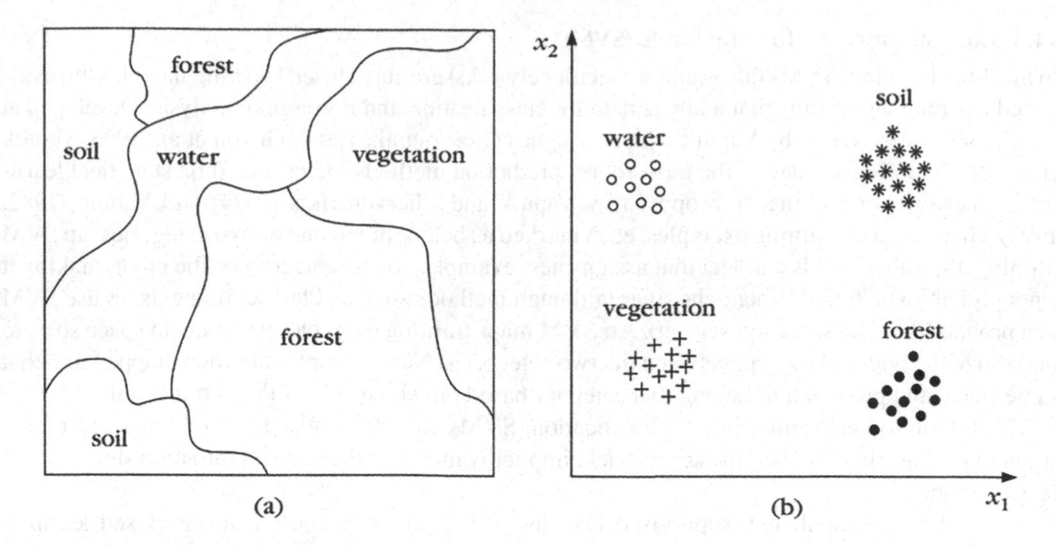

FIGURE 14.15 (a) An illustration of various ground covers and (b) clustering of the respective features for multispectral imagining using two bands.

$$dm(x) > \text{th} \rightarrow \text{assign to a new cluster}$$

Distance threshold = th

W_i is the cluster i

X_i is the new point which has to be assigned to a cluster

Another important factor affecting the result of the clustering algorithm is the choice of the threshold. This value directly affects the number of clusters formed. If it is too small, unnecessary clusters will be created. On the other hand, if it is too large, a smaller than appropriate number of clusters will be created. In both cases, the number of clusters that best fits the data set is missed.

14.3.6.2 Optimization Based Clustering

14.3.6.2.1 K-Means Clustering Algorithm

K-means is a clustering method in which clusters and cluster centers would be determined through an iterative approach. Figure 14.16(a)–(i) demonstrate steps of clustering a two-component dataset into two clusters. (a) and (b): all data points and two random cluster centers ($k = 2$) selected during the first iteration. (c)–(h): three more iterations, which were needed for converge. (i): final position of clusters and cluster centers. See Bishop (2006) for more details.

Euclidean distance of data points with cluster centers is a criterion for assigning points to clusters. Points that are closer to a certain cluster's centroid belong to that cluster. The new centroid (mean) of all data points in each cluster would be calculated during each iteration. By repeating this iterative process, the final position of clusters would be determined. The final position is when the same points are assigned to each cluster.

14.3.6.2.2 Example of K-Means Clustering Application

Digital image processing techniques could be utilized to investigate the correlation between soil color and moisture. In the first step, image of each soil sample should be classified into two classes by utilizing k-means unsupervised classification scheme to distinguish soil image from its background. In the second step, a Red–Green–Blue (RGB) analysis would be implemented on soil cluster. See Figure 14.17 in which the background is removed before RGB analysis.

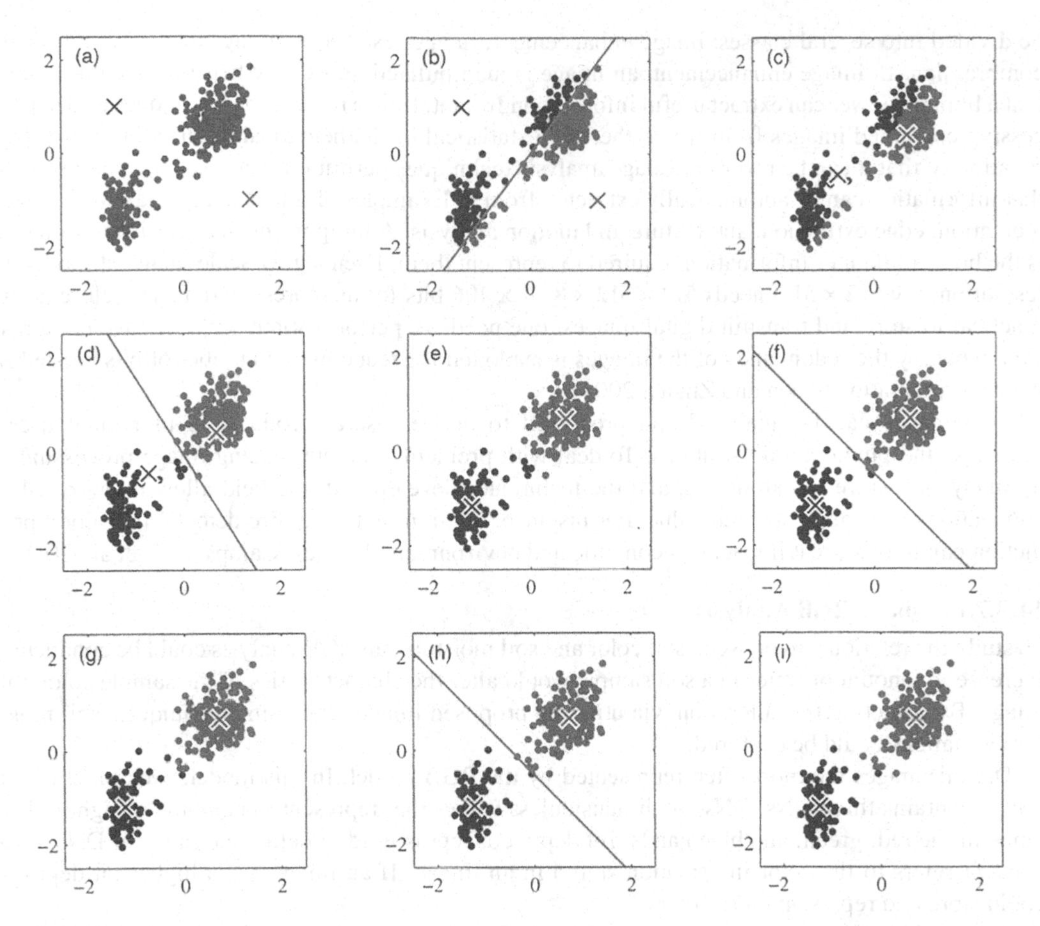

FIGURE 14.16 Steps of clustering a two-component dataset into two clusters by utilizing k-means technique. (a) and (b): all data points and two random cluster centers ($k = 2$) selected during the first iteration. (c–h): three more iterations, which were needed for converge. (i): final position of clusters and cluster centers. (From Bishop, C.M., *Pattern Recognition and Machine Learning*. Springer, Berlin, Germany, 2006.)

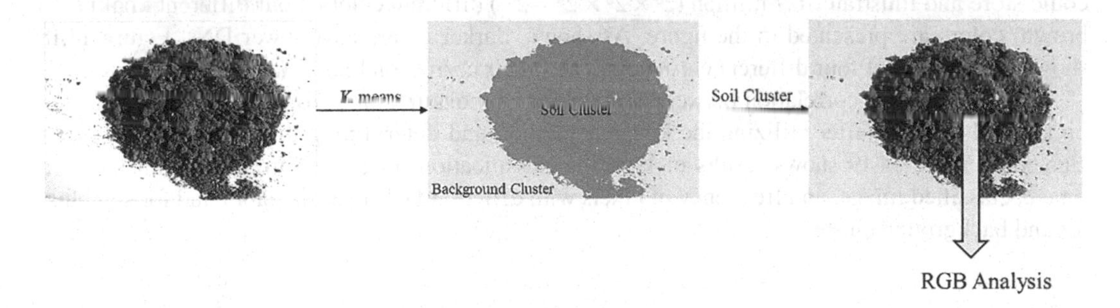

FIGURE 14.17 RGB analysis for soil moisture (Karamouz, M., Ebrahimi, E., & Ghomlaghi, A. Soil moisture data using citizen science technology cross-validated by satellite data. *Journal of Hydroinformatics*, doi: 10.2166/hydro.2021,029, (2021)).

14.3.7 Image Processing

Digital image processing consists of the manipulation of images using digital computers. Its use has been increasing exponentially in the last decades. Its applications range from medicine to entertainment, passing by geological processing and remote sensing. The processing of digital images can be divided into several classes: image enhancement, image restoration, image analysis, and image compression. In image enhancement, an image is manipulated, mostly by heuristic techniques, so that a human viewer can extract useful information from it. Image restoration techniques aim at processing corrupted images from which there is a statistical or mathematical description of the degradation so that it can be reverted. Image analysis techniques permit that an image be processed so that information can be automatically extracted from it. Examples of image analysis are image segmentation, edge extraction, and texture and motion analysis. An important characteristic of images is the huge amount of information required to represent them. Even a gray-scale image of moderate resolution, say 512×512, needs $512 \times 512 \times 8 \approx 2 \times 106$ bits for its representation. Therefore, to be practical to store and transmit digital images, one needs to perform some sort of image compression, whereby the redundancy of the images is exploited for reducing the number of bits needed in their representation (Chen and Zhang, 2004).

The images can be analyzed and processed to deliver useful products to individual users, agencies, and public administrations. To deal with problems, remote sensing image processing is nowadays a mature research area, and the techniques developed in the field allow many real-life applications with great societal value. For instance, urban monitoring, fire detection, or flood prediction can have a great impact on economic and environmental issues (Camps-Valls et al., 2011).

14.3.7.1 Image RGB Analysis

To study the relationship between soil color and soil moisture, an RGB analysis could be conducted. Increase in amount of water in a soil sample would alter the characteristics of the sample color and image. By detecting this alteration, via utilizing proposed image processing techniques, soil moisture variations could be captured.

Digital images are most often represented by the RGB model. In this model, each pixel of the image contains three DNs. DNs are dimensionless values that represent the amount of light reflectance in the red, green, and blue bands. Bit depth concept is used to define the range of DNs. This concept refers to the color information stored in an image. If an image has a higher bit depth, it could store and represent more colors.

For an N-bit digital image, each DN ranges from 0 to $2^N - 1$ (for an 8-bit image, the DN could be as high as 255). Because the light reflectance in "bright" colors is more than "dark" ones, the "bright" pixels have high DNs, while the "dark" pixels have low DNs. In other words, "0" means no reflection (black), and "$2^N - 1$" is maximum reflection in each color band (for white). An 8-bit image could store and illustrate 16.7 million ($2^8 \times 2^8 \times 2^8 = 2^{24}$) different colors. Four different khaki (bro brown) colors are presented in the figure. As shown, darker colors have lower DNs. Figure 14.18 shows an example of four different earth colors and their corresponding DNs.

To investigate the correlation between soil color and moisture, an RGB analysis was conducted on the soil clusters (after utilizing the k-means method and determining the soil and background clusters). Figure 14.19 shows results of k-means classification for a random sample, unclassified image, classified image, and frequency of pixels with different DNs in each color band for soil cluster and background cluster.

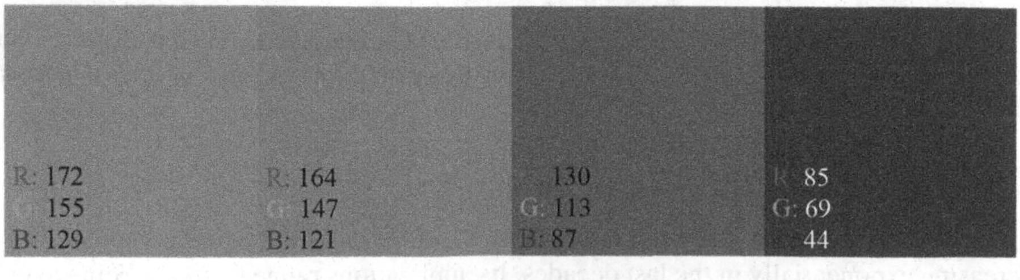

FIGURE 14.18 Four different earth colors and their corresponding digital numbers.

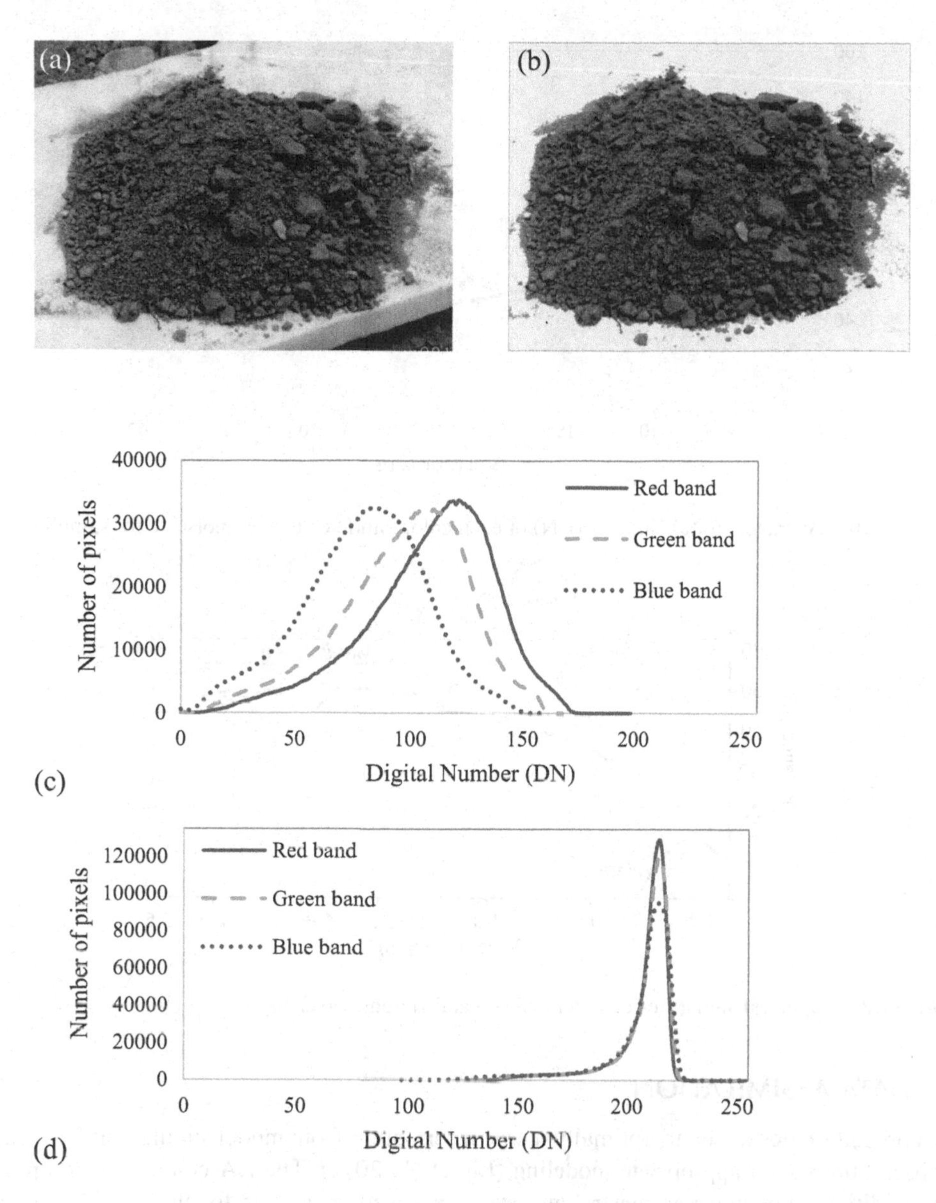

FIGURE 14.19 Results of k-means classification for a random sample (a), unclassified image (b), classified image (c), and (d) frequency of pixels with different DNs in each color band for soil and background clusters.

To study the correlation between soil moisture and color characteristics of images, the average DN of the red, green, and blue bands was calculated for the soil cluster pixels in each image. Then, the average DNs were plotted versus gravimetric soil moisture data (Figure 14.20).

By increase in soil moisture, the average DN declines in all three bands. This decrease is due to the characteristics of water. As shown in the figure, water reflectance (light reflection of the water) is smaller than soil reflectance in visible spectra. By increasing the water content, the reflectance of the sample decreases as well as the average DN. In other words, any increase in amount of water in a soil sample would alter the characteristics of the sample's color and its image.

By detecting this alteration, via utilizing proposed image processing techniques, soil moisture variations could be captured. In other words, soil color could be utilized as a predictor to predict soil moisture (Figure 14.21).

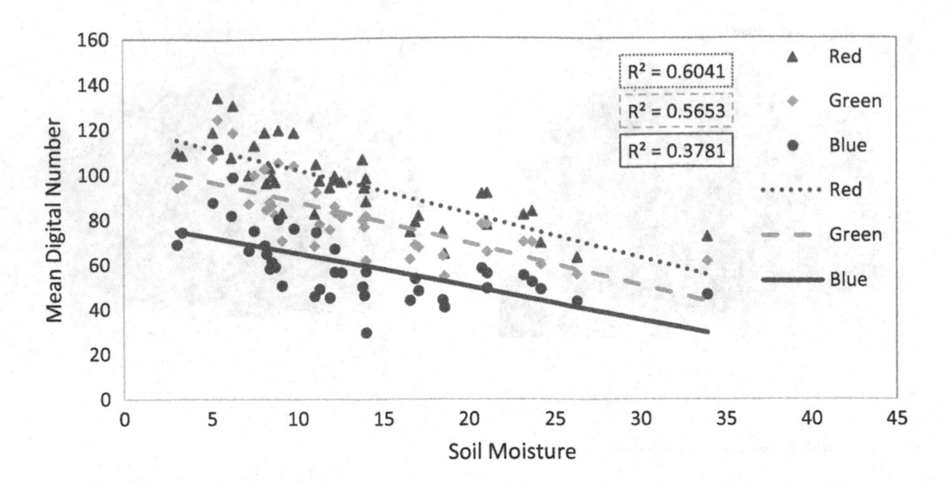

FIGURE 14.20 Average Digital Number (DN) of each color band versus soil moisture for sampling data.

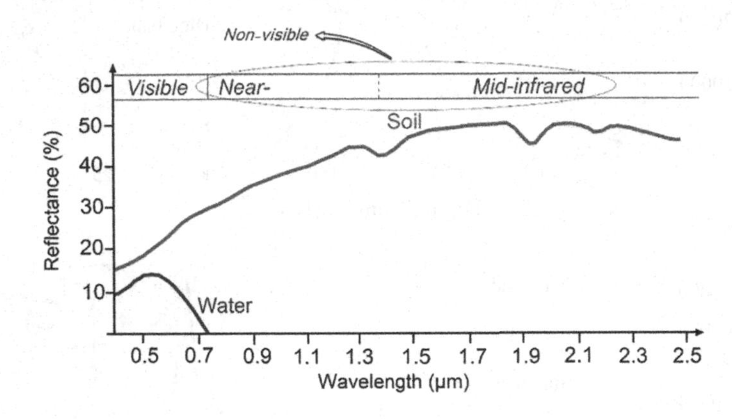

FIGURE 14.21 Spectral signatures of water and soil at different wavelengths.

14.4 DATA ASSIMILATION

DA is a procedure developed to optimally merge information from model simulations and independent observations with appropriate modeling (Liu et al., 2012). The DA could provide optimized initial conditions, updated parameters, and even improved structures for the dynamic model. A DA system aims at improving model state estimates by weighted averaging these estimates with observational data whenever available. To maintain an optimal weight distribution, it simultaneously updates and propagates model state variables and uncertainties. Since land surface models (LSMs) are typically too complex to do this analytically, model forecast uncertainties are often propagated using Monte Carlo simulations, commonly referred to as ensemble Kalman filtering (Evensen, 2003). Ensembles of the forecast states and of the observations are generated by perturbing them with random noise artificially drawn from an a priori assumed probability distribution. Each ensemble member represents an equally probable model or observation trajectory, and the variability of the ensembles is used to diagnose the fore cast uncertainty pre update and postupdate. In this study, we aim to optimize such filtering by estimating the—typically a priori assumed—error distribution parameters as part of the DA system.

DA was first used in the 1950s for numerical weather forecast modeling; however, hydrologists did not pay much attention to it until the 1990s (Evensen, 1994b; McLaughlin, 1995). Proper use of DA may help in handling uncertainties from model inputs, the initialization and propagation of states, the model structures, and even the model parameters (Vrugt et al., 2006; Liu and Gupta, 2007; He et al., 2012). Meanwhile, global DA may improve regional field estimation by achieving more accurate external boundary condition estimations (Robinson and Lermusiaux, 2000), and local DA could also lead to improved global estimations (Clark et al., 2008). The development of remote sensing has promoted the application of DA in hydrological models. The remotely sensed hydrological data that have had, or currently have, the potential to be applied in hydrological models include (Houser et al., 2005):

1. Overland parameters (e.g., topography, land cover, albedo);
2. Forcing inputs (e.g., precipitation, air humidity and temperature);
3. States (e.g., soil moisture, snow cover);
4. Fluxes (e.g., carbon flux).

Many approaches have been applied in hydrological DA (Houser et al., 2005). The most commonly used methods include the variational method, particle filters (PFs), and the Kalman filters (KFs). Although traditionally dominant in numerical weather forecasts, variational methods have not been widely used in hydrological DA.

14.4.1 Kalman Filter (KF)

In DA, Kalman filtering, also known as linear quadratic estimation (LQE), is an algorithm that uses a series of measurements observed over time, containing statistical noise and other inaccuracies, and produces estimates of unknown variables that tend to be more accurate than those based on a single measurement alone, by estimating a joint probability distribution over the variables for each timeframe. The filter is named after Rudolf E. Kálmán, one of the primary developers of its theory.

Furthermore, the KF is a widely applied concept in time series analysis used in fields such as LSMs. The algorithm works in a two-step process. In the prediction step, the KF produces estimates of the current state variables, along with their uncertainties. Once the outcome of the next measurement (including random noise) is observed, these estimates are updated using a weighted average, with more weight being given to estimates with higher certainty. The algorithm is recursive. It can run in real time, using only the present input measurements and the previously calculated state and its uncertainty matrix; no additional past information is required.

Optimality of the KF assumes that the errors are Gaussian. The primary sources are assumed to be independent Gaussian random processes with zero mean; the dynamic systems will be linear. Though regardless of being a Gaussian process, if the process and measurement covariances are known, the KF is the best possible linear estimator in the minimum mean square error sense.

In conclusion, it can be stated that the output of DA is a weighted average of measurements and model prediction with their covariance as weights.

$$\bar{X} = \left(\frac{\sigma_y^2}{\sigma_y^2 + \sigma_x^2}\right) X_t + \left(\frac{\sigma_x^2}{\sigma_y^2 + \sigma_x^2}\right) Y_t \tag{14.48}$$

In this equation, X_t and Y_t are model output and measurement in time step t, σ_x^2 and σ_y^2 are models and measurement covariances, and $\bar{X}$ is assimilation output for each time step. Model and measurement covariances should be updated in each time step.

14.4.2 VIC Model Application with Data Assimilation

The variable infiltration capacity (VIC) model (Liang et al., 1999) is a distributed LSM which solves both water and energy balance equations and, in each time step, the VIC model utilizes the last time step state condition. In the last decade, the VIC has been vastly utilized in a number of simulations.

The grid cell size of the VIC simulations can vary, but the minimum grid size is 1 km. VIC inputs are mainly consisting of atmospheric forcing, soil, and vegetation parameters. Also, each cell can be subdivided into sections representing specific land covers and vegetation types.

The ensemble Kalman filter (EnKF) is a Monte Carlo approach for nonlinear filtering problems (Evensen, 1994a). Although the propagation of state error covariance does not require the linearization of the model, the parallel computation of the ensemble means the EnKF is an expensive method to use. Nevertheless, the EnKF is a widespread approach in hydrological DA.

The EnKF which is introduced by Evensen (1994a) is a sequential DA technique. It has been widely used for state and parameter estimation in hydrological simulations (Cammalleri and Ciraolo, 2012). For instance, Moradkhani et al. (2005) proposed the dual- state parameter estimation of hydrologic models using the EnKF.

The VIC model doesn't represent exchanges of moisture or runoff between neighboring grid cells. Therefore, a routing model should be utilizing in order to model streamflow transportation across the study area. RVIC is a source-to-sink routing model that simulates routing via solving a linearized version of the Saint-Venant equations (USACE, 1995). The linearized Saint-Venant equations are a one-dimensional model describing unsteady flow in terms of two time-invariant parameters, flow velocity, and diffusivity. RVIC uses flow direction rasters (FDRs) which are derived from topographic information files such as digital elevation maps (DEMs) in order to specify the flow path and distance for each source–sink pair. The travel path flow is parameterized as a linear and time-invariant unit impulse response function (IRF) to runoff generated at each grid cells by the VIC. The IRF is often known as a unit hydrograph (UH). See case study 2 for more details.

The state variables of a single forecast ensemble member are represented in the vector x_k^{if}, where k indicates the time step and i is the ensemble member. The superscript f refers to the forecast variables and the variables that are obtained prior to assimilation of the external data. The state variables that are updated in this study are the first- and the second- layer soil moisture estimates in order to enhance the simulations of associated runoff/baseflow. The main motivation therefore is that SMAP provides observations of the top surface SM and, hence, is expected to improve the runoff simulation. State variables that are obtained after the assimilation of the external data are referred to as the analyzed variables and are denoted with the superscript a. The analyzed state vector at time step $k-1$ is propagated to the next time step using:

$$x_k^{if} = f_{k-1}\left(x_{k-1}^{ia}\right) + w_{k-1}^{i}, \tag{14.49}$$

where f_{k-1} is the model and w_{k-1}^{i} is a realization of the forecast error.

The system is represented as:

$$y_k = h_k\left(x_k, v_k\right), \tag{14.50}$$

where h_k is the nonlinear observation operator, mapping the state variables to observation space, and v_k is the observation error. One of main aspects in DA process is the estimation of the uncertainty in the forecasts. In the EnKF, the forecast error covariance is estimated as:

$$P_k^f = \frac{1}{N-1} D_{xk} D_{xk}^T, \tag{14.51}$$

$$D_{xk} = \left[x_k^{1f} - \overline{x}_k^f, \ldots, x_k^{Nf}, x_k^{Nf}\right], \tag{14.52}$$

$$\overline{x}_k^f = \frac{1}{N} \sum_{i=1}^{N} x_k^{if}, \tag{14.53}$$

where N is the number of ensemble members and superscript T indicates the transpose operator. The Kalman Gain (K_k) is then calculated as:

$$K_k = P_k^f H_k^T \left[H_k P_k^f H_k^T + R_k \right]^{-1}, \tag{14.54}$$

where R_k is the observation covariance, and H_k is the Jacobian of the observation operator h_k. For nonlinear observation systems, the process of explicit calculation of H_k is bypassed by the use of the ensemble statistics:

$$P_k^f H_k^T = \frac{1}{N-1} D_{xk} D_{yk}^T, \tag{14.55}$$

$$H_k P_k^f H_k^T = \frac{1}{N-1} D_{yk} D_{yk}^T, \tag{14.56}$$

$$D_{yk} = \left[y_k^{1f} - \bar{y}_k^f, \ldots, y_k^{Nf}, y_k^{Nf} \right], \tag{14.57}$$

$$\bar{y}_k^f = \frac{1}{N} \sum_{i=1}^{N} y_k^{if}, \tag{14.58}$$

Pauwels and De Lannoy (2009) showed that using this approach has better performance in comparison to the explicit calculation of h_k through linearization of the observation system. By using the Kalman Gain, the states of the individual ensemble members are then updated by:

$$x_k^{ia} = x_k^{if} + K_k \left[y_k - h_k\left(x_k^{if}\right) + v_k^i \right], \tag{14.59}$$

where $h_k(x_k^{if})$ is the simulation of the observation for ensemble member i which is written as y_k^{if}. At last, v_k i is a random realization of the observation error.

In this study, the state variable can be written in each grid cell as follows:

$$x_k^{if} = \left[\theta_{1,k}^{if}, \theta_{2,k}^{if}, \ldots, \theta_{m,k}^{if} \right]^T \tag{14.60}$$

where $\theta_{m,k}^{if}$ indicates simulated soil moisture for model grid m and layer 1 inside each SMAP grid, for time step k and ensemble member i ($i \in N$).

The bias in the first order moment (MEAN) was removed by comparing the corresponding pixel-based mean values of the VIC simulations and SMAP observations as:

$$\text{SM}^* = \overline{\text{SM}}_{\text{sim}} + \left(\text{SM}_{\text{obs}} - \overline{\text{SM}}_{\text{obs}} \right) \tag{14.61}$$

where SM^* is the rescaled SMAP soil moisture observation, SM_{obs} is the 2-year mean of the soil moisture observations, and $\overline{\text{SM}}_{\text{sim}}$ is the 2-year mean of the SM simulations in April (46% in study area), which are aggregated to the SMAP grid. In addition, the analysis update was limited at the soil porosity value (Lievens et al., 2015).

VIC model that is shown in Figure 14.22 is a macroscale hydrologic model that solves full water and energy balances, originally developed by Liang et al. (1994) at the University of Washington. VIC is a research model, and in its various forms, it has been applied to most of the major river basins around the world, as well as globally. The VIC model is distributed under the GNU GPL v2.0 license. Some great works including those by Liang et al. (1994) and Hamman et al. (2018) for VIC-5 should be acknowledged.

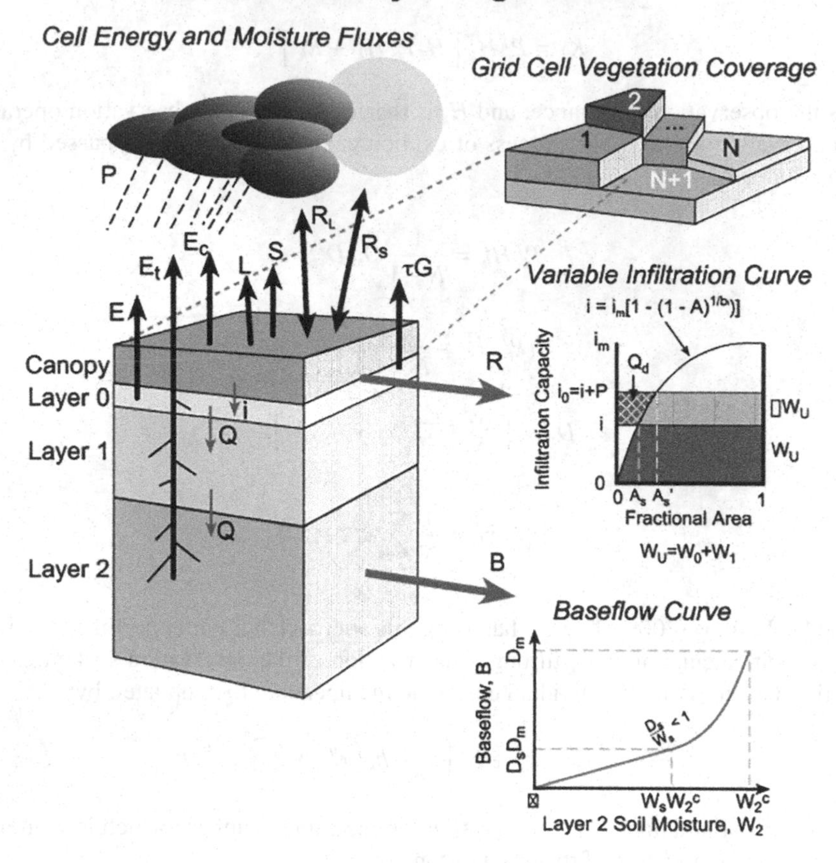

FIGURE 14.22 A schematic of variable infiltration capacity model performance. (From Gao, H., Tang, Q., Shi, X., Zhu, C., Bohn, T., Su, F., and Wood, E. "Water budget record from Variable Infiltration Capacity (VIC) model." (2010).)

Development and maintenance of the current official version of the VIC model is led by the UW Hydro | Computational Hydrology group in the Department of Civil and Environmental Engineering at the University of Washington. Every new application addresses new problems and conditions that the model may not currently be able to handle, and as such, the model is always under development. The VIC model is an open-source development project, and the input data includes precipitation, temperature, vapor pressure, and land use (vegetation cover). The calculations are made in 1 km cell size. In this figure, cell and moisture fluxes are shown among the top canopy and three layers of soil. Canopy layer is classified by grid cell vegetation coverage coupled with the infiltration curve (to estimate SM in layers 0 and 1, $W_0 + W_1$) land use fractional area. Baseflow curve as a function of soil moisture, W_2, at layer 2 is also demonstrated.

14.5 ENVIRONMENTAL VISUALIZATION ATTRIBUTES

14.5.1 Intelligent Visualization and Image Analysis Systems

New and continuously improved intelligent and adaptive systems will be created to perform real-time analyses of images and sensor signals. The information from the analyses of these images will be combined seamlessly with data from distributed sensor networks to provide immediate and reliable measurements of the condition and health of environmental and water resource systems. Autonomous learning agents and adaptive systems will form the structure for intelligent scene-driven data analysis algorithms, providing opportunities for the development of more cost-effective and responsive early warning systems (Wright and Palkovits, 2012).

14.5.2 Water-Related Environmental Visualization

Flood imaging and obtaining flood and flood progression maps have always been fundamental issues in flood studies, especially in coastal areas. Examples of these maps have been designed by FEMA or other research teams for a 100-year return or flood event.

In flood risk analysis, a topographic map of the area is required. Topography plays an important role in determining the correct flooding map. DEM is a raster data set that contains information about the topography of the area and is a prerequisite for hydraulic modeling. DEM accuracy directly affects hydrological and hydraulic modeling. The spatial resolution of a DEM refers to the area covered by a cell on the ground. As a result, the higher the resolution, the more cells are per unit area of land, which results in a more detailed and accurate display of the topography of the area.

In models such as SLOSH (Sea, Lake, and Overland Surges from Hurricanes) developed by National Oceanic and Atmospheric Administration (NOAA), a set of differential equations describing the fluid motion are applied over a gridded area or "basin" representing the topography. For NYC, NY2 and NY3 (the most up to date) have been utilized. The following is a comparison of SLOSH runs on both NY2 and NY3 for Category 2 hurricanes under mean tide conditions. Probe flags were placed on corresponding locations so that inundation elevations for wastewater treatment plants could be compared. In order to do so, coordinates for treatment plant were obtained from online resources and manually located on the SLOSH maps as shown in both Figures 14.23 and 14.24. For more details on SLOSH refer to NOAA Technical Report NWS 48 –SLOSH (1984).

In all cases (except for Port Richmond), inundation depth increases when it is run on the NY3 basin as opposed to the NY2 basin. In some cases, the inundation differences are drastic. Disregarding Red Hook, which appears to be an outlier, nearly half of the treatment plants experience increased inundation depths of at least 3 ft. Because SLOSH calculates inundation for each grid in the basin, the values depicted are an average storm tide depth over the average land elevation for any given grid cell. The difference of which is the inundation depth as calculated by the SLOSH "subtract" function. Since the resolution of the grid cells in NY2 and NY3 are different (2 km × 2 km versus 1 km × 1 km, respectively), part of the difference in inundation depths between the two basins can be attributed to this. Additionally, various land use changes as well as natural and man-made geophysical characteristics may have been incorporated into the updated NY3 basin. The differences in the inundation elevations are shown in Table 14.3.

The evacuation zones were based on an outdated basin can lead to an underestimation of the flood risk during Superstorm Sandy in New York City. In addition, since SLOSH needs additional post processing in order to adequately display surge on a useful map, there is a need for a replicable methodology to do so. In this way, agencies or stakeholders may readily take the output of SLOSH and create useful evacuation zone maps or other emergency management tools.

Figure 14.25 shows the high-risk locations in the New York area. This image was updated after Hurricane Sandy due to urgency for utilizing the latest and updated base maps with much higher resolution.

In recent years, much attention has been paid to the kind of this map with high accuracy and quality. One of the tools for preparing these kinds of images is hydrological models with video outputs. For example, GSHHA model has the ability to display the amount of flooding in each cell separately after grading the area to the desired dimensions and performing calculations. This display is also numerical, by referring to the desired cell, the amount of flooding in that cell can be read. The model is also able to determine the flooding interval of each cell with changeable colors. This feature makes it possible to calculate the flooding for a specific area by using the flood height in

FIGURE 14.23 Inundation depths for a category 2 hurricane at mean tide level applied on the NY2 basin (Kokoszka, M. Reassessing vulnerability: Using a holistic approach and critical infrastructure to examine flooding, Master of Science Project, Polytechnic Institute of NYU, Brooklyn, NY, p. 44, (2013).).

certain cells and considering the dimensions of the cells. It can also provide a meaningful overview of flooding in the area, for example, showing areas whose flood height has exceeded the critical level in red. With the help of ArcMap software, it is one of the five main ArcGIS software. The obtained flood maps can be combined with hydrological models with land use maps, and high-risk points can be identified.

Figure 14.26 shows the output of the GSHHA model for floods with a return period of 100 years. As shown in the picture, initially the area is divided into $70 \times 70\,m^2$ cells, and by flooding, the flooding area in each cell is calculated. In this image, floods above 30 cm are omitted. Then the flood map is placed on the land use map of the area, and the flood in each section is clearly defined. For example, the treatment plant area is inundated/flooded. Figure 14.25 shows how to visualize a 100-year return flood in the Brooklyn area and the Coney Island treatment plant by utilizing the GSHHA model. The topographic grading of the area and some land use characteristics can be seen.

Output time series maps are very useful for gaining a visual sense of what is happening in the catchment at a given time. Using WMS to animate a series of maps provides the user with a series of animated times from the desired output. This allows the user to see how the variable spatial distribution (runoff) progresses over time. Also, due to the fact that the model offers flood maps for each hour (time is adjustable) as output, these maps can be displayed as movie loops using software

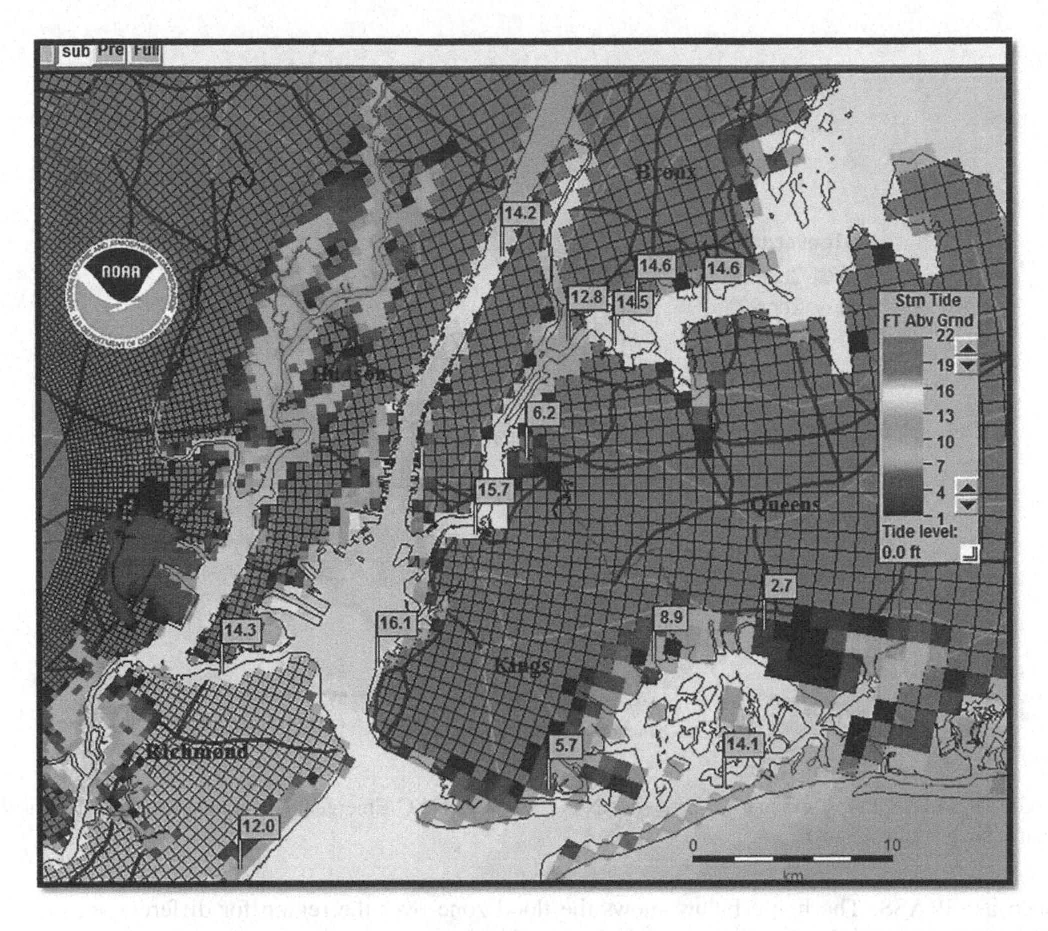

FIGURE 14.24 Inundation depths for a category 2 hurricane at mean tide level applied on the NY3 basin, Kokoszka (2013). (*Source*: NOAA, Technical Report NWS 48 – SLOSH, National Weather Service, Silver Spring, MD, 1984.)

TABLE 14.3
Inundation Comparison

Wastewater Treatment Plant	NY2 Inundation (ft)	NY3 Inundation (ft)	Delta (ft)
26th Ward	2.0	8.9	6.9
Bowery Bay	11.4	14.5	3.1
Coney Island	1.6	5.7	4.1
Hunts Point	11.5	14.6	3.1
Jamaica	1.7	2.7	1.0
Oakwood Beach	10.2	12.0	1.8
Owls Head	15.6	16.1	0.5
Newtown Creek	Dry	6.2	6.2
North River	13.5	14.2	0.7
Port Richmond	15.4	14.3	−1.1
Red Hook	Dry	15.7	15.7
Rockaway	13.7	14.1	0.4
Tallman Island	11.7	14.6	2.9
Wards Island	12.0	12.8	0.8

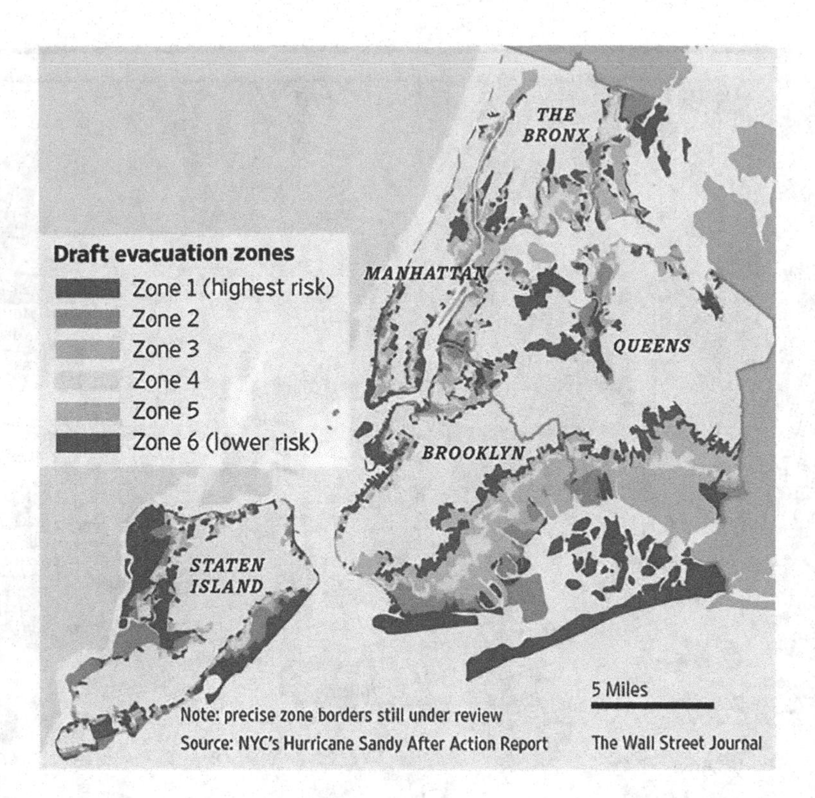

FIGURE 14.25 Latest evacuation zones for NYC. (*Source*: NYC Emergency Management, 2012, https://maps.nyc.gov/hurricane/.)

such as GRASS. The figure below shows the flood zone over the region for different scenarios. Figure 14.27 shows flood zoning with 20-year and 200-year return periods with and without protection strategies. These images show good information about the effectiveness or noneffectiveness of protection strategies.

14.5.3 Visualization to Control and Reduce Flood Risk

Compatible flood control strategies including best management strategies (BMPs) by RBD (2014) are divided into four groups: delay, discharge, storage, and resistance. The Office of Metropolitan Architecture (OMA) team proposes a four-pronged urban water strategy as shown in Figure 14.28 to use symbols and signs for better visual communication, including resistance (expansion of hard and soft landscape infrastructure for coastal protection) and delay (policy recommendations, guidelines, and urban infrastructure for reducing runoff from rainwater), storage (a cycle of green infrastructure to store and transport excess rainwater), and drainage (pumping water and creating alternative routes). To support the drainage system). This group excludes the risk of floods from overlapping the probability of flooding (which increases with increasing sea level) and the effects of flooding (which decreases with adaptation methods and increases with the expansion and development of urban development). This overlap has helped to identify high-risk areas to increase the focus of studies.

Karamouz et al. (2019) have introduced the fifth group called retreat. This group includes managed retreats that allow coastal areas that were not previously flooded to be flooded. This method emphasizes increasing compatibility and reliance on the natural response of the coast to floods. The resistance group includes solutions for flood protection structures such as coastal wall and torsion (gabion), which by green solutions (vegetation) direct effects of floods and storm surges along with increasing compatibility and maintaining the landscape. Delay strategies attempt to reduce the

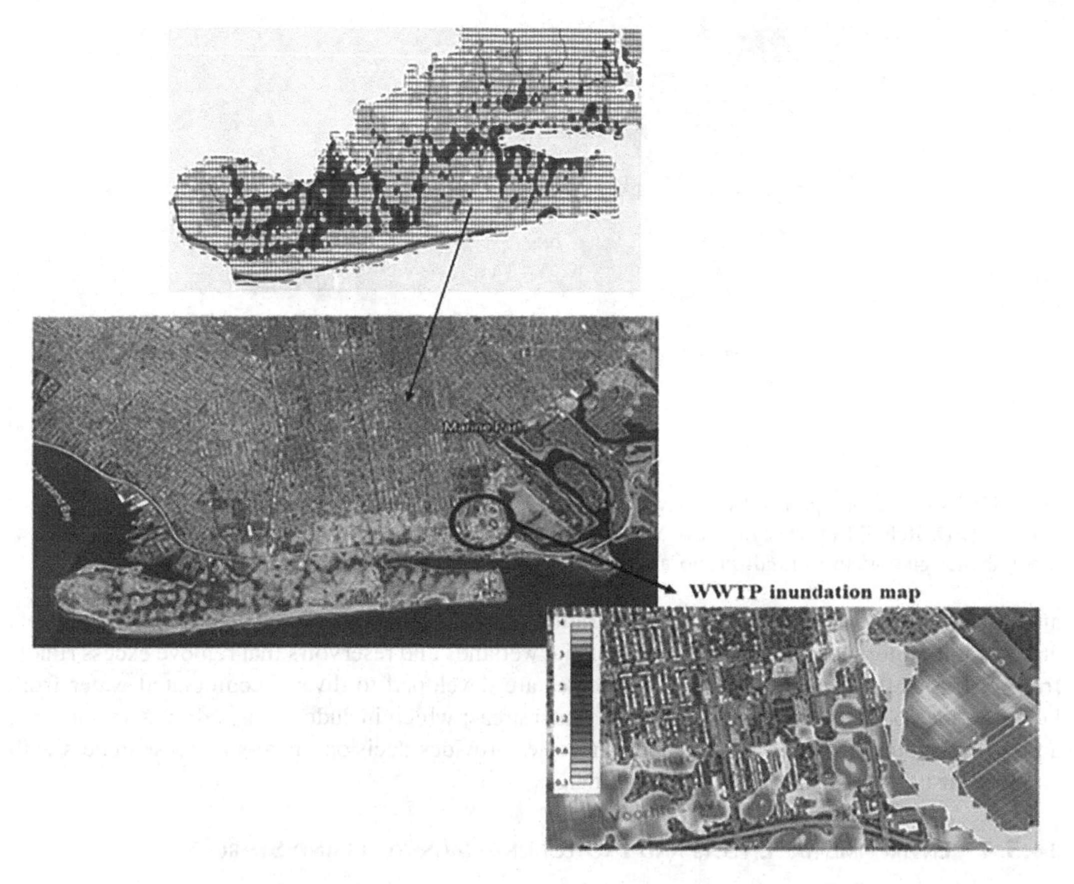

FIGURE 14.26 Brooklyn's shoreline's visualization of inundation map (GSSHA model). (*Source*: Karamouz, M. and Farzaneh, H., Water Res. Manage., 34(6), 2043–2059, 2020.)

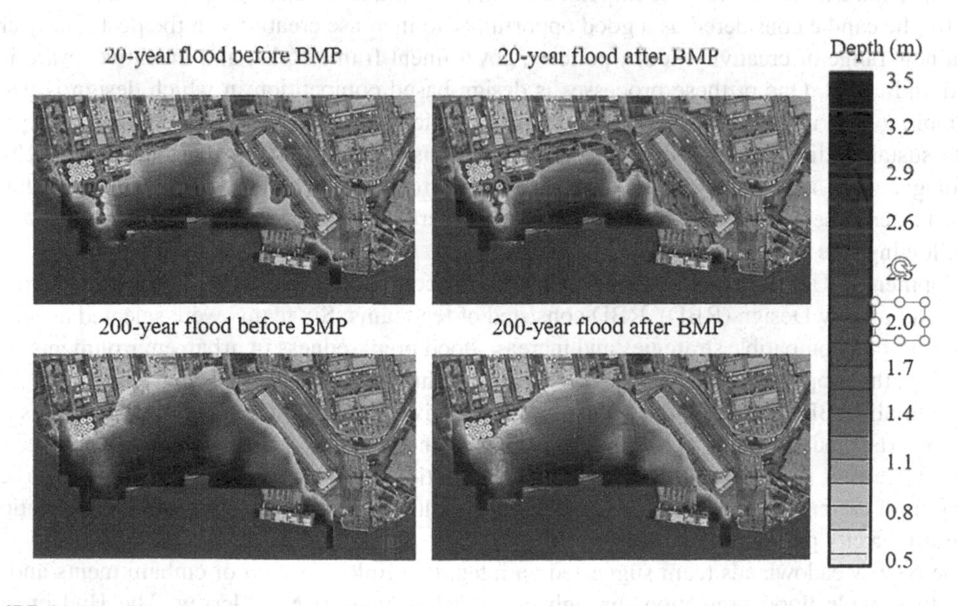

FIGURE 14.27 Flood zone imaging with a return period of 20 and 200 years with and without the application of flood control and mitigation (BMP) solutions in the Hunts Point treatment plant in Bronx, NY. (*Source*: Karamouz, M. and Farzaneh, H., *Water Res. Manage.*, 34(6), 2043–2059, 2020.)

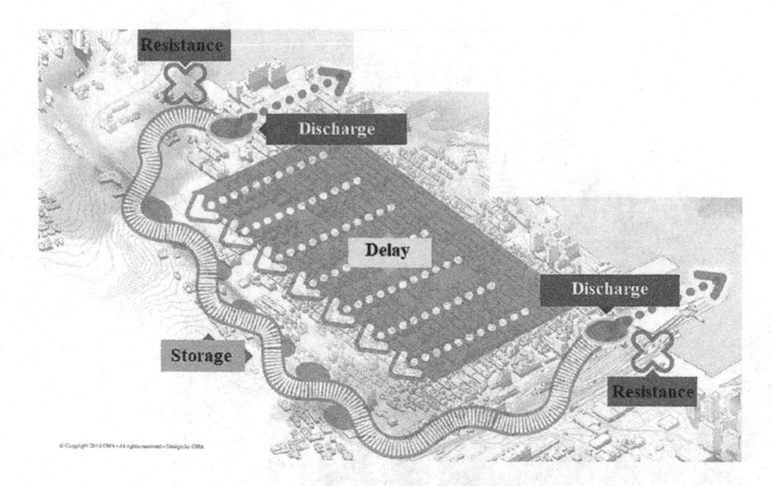

FIGURE 14.28 Flood-pronged urban water strategies. (From "YC mayor's office of recovery and resiliency." RBD, Rebuild by Design. New York City economic development corporation. (2014). http://www.rebuildbydesign.org/our-work/all-proposals/winning-projects/hunts-point-lifelines.)

amount of runoff peak by increasing the duration of the hydrograph. Bioswale and vegetated roofs are examples. Storage strategies include artificial wetlands and reservoirs that remove excess runoff from critical infrastructure. Drainage strategies are developed to divert accumulated water from floods in the coastal city to storage or open water areas; which include pumps, drainage channels, and artificial ducts. Visualization of such strategies provides decision makers and researchers with comprehensive information in the form of images.

14.5.4 Environmental Effects and Protection—Signatures and Symbols

Various activities have been carried out around the world to come up with ideas for environmentally friendly methods of flood control, one of the most significant of which was the aftermath of Hurricane Sandy, which was accompanied by severe climate change and rising sea levels. This catastrophe can be considered as a good opportunity to increase creativity in the post-Cindy era so that a new range of creativity, tools, policies, government frameworks, and incentives were introduced and tested. One of these processes is design-based competition in which design is used as a suitable tool to increase adaptation to complex problems and to produce integrated strategies to create sustainability and viability. A competition consists of experts in various fields including hydrology, economics and social sciences, landscape architecture, and civil engineers were formed in 2013 after Superstorm Sandy to generate ideas as further described below.

Following the presidential executive order, the US Department of Housing and Urban Development (HUD) initiated a competition with the collaboration of the Netherlands government called Rebuild by Design (RBD). RBD consisted of ten teams. Six teams were selected as winners in order to find compatible strategies and increase flood preparedness in urban environments. In the following, the proposals submitted by these teams of this competition are reviewed.

The finalists BIG U, New Meadowlands, Hudson River Project, and Hunts Point Lifelines made up of experts, landscape architects, and engineers to generate ideas and conceptual designs for flood risk management. The main objective was to find solutions suitable for flood control infrastructure during both extreme events and normal weather conditions. The BIG U represented the notion of integrating a city park with floodwalls.

The New Meadowlands team suggested an integrated linked system of embankments and wetlands to provide flood protection through the Meadowlands in New Jersey. The Hudson River Project proposed the green and gray infrastructure approach for reducing flood risk and achieving a more comprehensive flood management strategy such as landscaped-based and engineered-based coastal defenses. Hunts Point Lifelines' (RBD, 2014) proposal also included green and gray flood protection and measures to protect critical economic assets including transportation in the region. This competition provided a unique opportunity for engineers to explore the many ideas that were generated to rethink green solutions.

The BIG team (RBD, 2014) examined how flood protection structures were designed for Manhattan beaches in New York City with the goal of overcoming the problems of building a coastal wall that would separate the urban environment from the surrounding aquatic environment. Thus, as shown in Figure 14.29, communication between local communities, people, cultural and recreational activities, and flood control infrastructure were named as the most important features of protective structures. Flood control structures that protect some of Manhattan's shores in the event of a catastrophe meet this goal only in a short period of their useful life. It is essential that these structures be designed to improve the urban situation for the rest of their useful life. For this purpose, these structures should be primarily an active part of urban infrastructure. For example, by building sidewalks and bicycle paths on these structures, it reduced the density of urban traffic and increased the compatibility between the urban and aquatic environment. Also, the construction of recreational places in the space created in the shadow of these structures will increase the connection between different urban infrastructures and will maximize the use of financial investment created in the region. Figure 14.30 shows an example of these arrangements.

Inter bro team (RBD, 2014) concentrated on the long coasts of Long Island, NY where they are exposed to a variety of water hazards, including flooding due to rainfall inland areas, force caused by waves on coastal areas, and rising sea levels. Special measures have been devised. Due to the complexity of these challenges, they divided the region into five parts and offered integrated solutions.

In the coastal area, the group's focus has been on protecting coral reefs by raising hills and trapping sediments. In the defensive island region, the use of dykes has been suggested to protect critical infrastructure. The development and construction of swamp islands, in addition to reducing

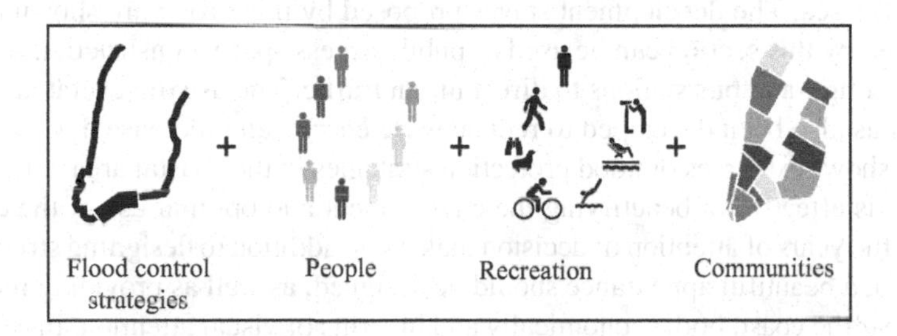

FIGURE 14.29 Symbol/Signage of main elements of flood control strategies. (From "YC mayor's office of recovery and resiliency." RBD, Rebuild by Design. New York City economic development corporation. (2014). http://www.rebuildbydesign.org/our-work/all-proposals/winning-projects/hunts-point-lifelines.)

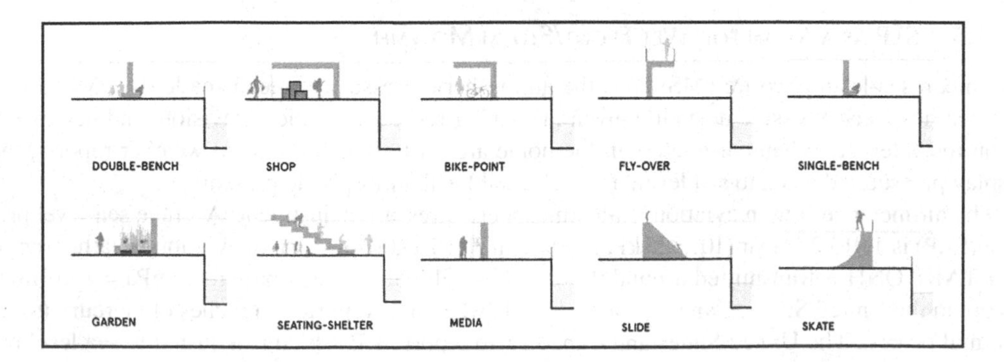

FIGURE 14.30 Signage of coastal measures in flood control. (From "YC mayor's office of recovery and resiliency." RBD, Rebuild by Design. New York City economic development corporation. (2014). http://www.rebuildbydesign.org/our-work/all-proposals/winning-projects/hunts-point-lifelines.)

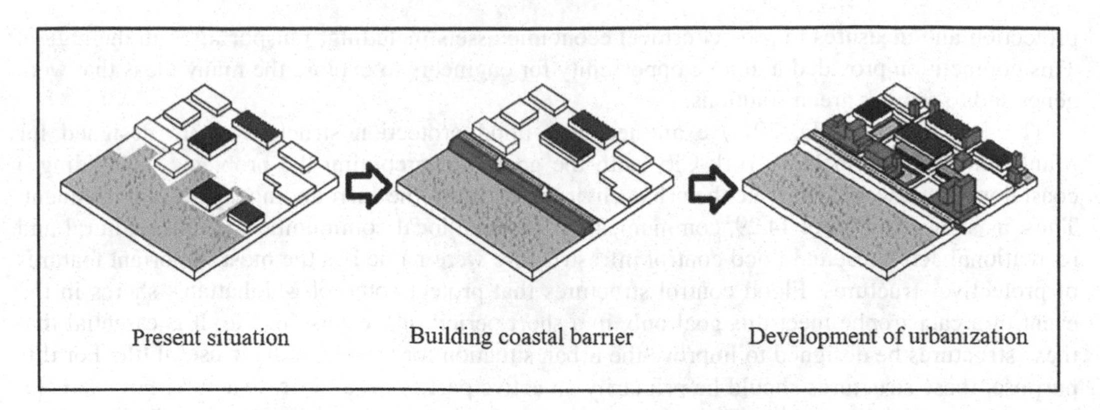

FIGURE 14.31 Coastal development stages. (From "YC mayor's office of recovery and resiliency." RBD, Rebuild by Design. New York City economic development corporation. (2014). http://www.rebuildbydesign.org/our-work/all-proposals/winning-projects/hunts-point-lifelines.)

the destructive effects of hurricane waves in order to provide more suitable places for recreation and improve the living environment, have been another suggestion of this team. Also to prevent pollution from entering the delta by streams and rivers, expansion of vegetation and increase of permeability of estuaries has been proposed, which also reduces the effects of storm waves.

The MIT CAU team (RBD, 2014) proposed three basic principles of protection, connection, and promotion consisting of a park and a coastal dam for the Midland area. The park connects the freshwater basin and existing lagoons and the urban area, which plays the role of protection against coastal floods and rainfall. The coastal dam is defined as the boundary between the developed urban area and the sea. The development stages proposed by this group are shown in Figure 14.31. The space created by this section can be used as public access space, construction of residential and commercial buildings, and bus stations to direct urban traffic. The use of vegetated areas in front of a levee or dam has also been developed to reduce wave energy and increase dyke stability.

Figure 14.32 shows examples of flood protection strategies in the coastal area, which in addition to flood prevention is effective in beautifying the environment and optimal use of the coastal environment. In fact, in the years of attention of decision makers, in addition to designing structures to prevent floods in the area, a beautiful appearance should be designed, as well as providing recreational areas in these areas near the coast, both economically and in terms of visual attention. In a simple example, instead of designing high walls against floods, using stairs will both break the wave and make the area more beautiful. Or in another example, we can mention the design of green space instead of giant concrete structures, which in addition to beautiful appearance, are effective against floods.

14.5.5 SLP as a Mean for Wet Front/Storm Movement

The mean sea-level pressure (MSLP) is the atmospheric pressure at mean sea level (PMSL). This is the atmospheric pressure normally given in weather reports on radio, television, and newspapers or on the Internet. When barometers in the home are set to match the local weather reports, they display pressure adjusted to sea level, not the actual local atmospheric pressure.

The altimeter setting in aviation is an atmospheric pressure adjustment. Average sea-level pressure (SLP) is 1,013.25 mbar (101.325 kPa; 29.921 in Hg; 760.00 mmHg). In aviation, weather reports (METAR), QNH is transmitted around the world in millibars or hectopascals (1 hPa = 1 millibar), except in the United States, Canada, and Colombia where it is reported in inches of mercury (to two decimal places). The United States and Canada also report SLP, which is adjusted to sea level by a different method, in the remarks section, not in the internationally transmitted part of the code, in hectopascals or millibars. However, in Canada's public weather reports, SLP is instead reported in kilopascals. Figure 14.33 shows showing atmospheric pressure.

FIGURE 14.32 Examples of coastal conservation strategies in coastal areas.

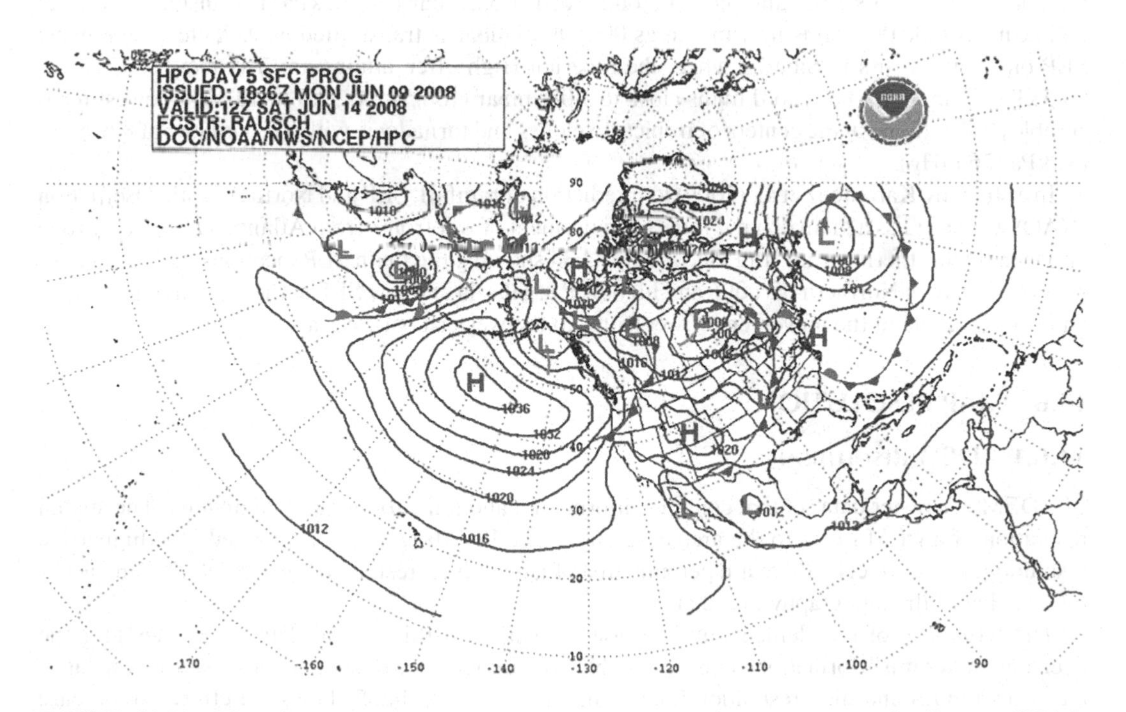

FIGURE 14.33 Map showing atmospheric pressure in mbar or hPa (hecto pascal). (*Source*: NOAA, Technical Report NWS 48 – SLOSH, National Weather Service, Silver Spring, MD, 1984.)

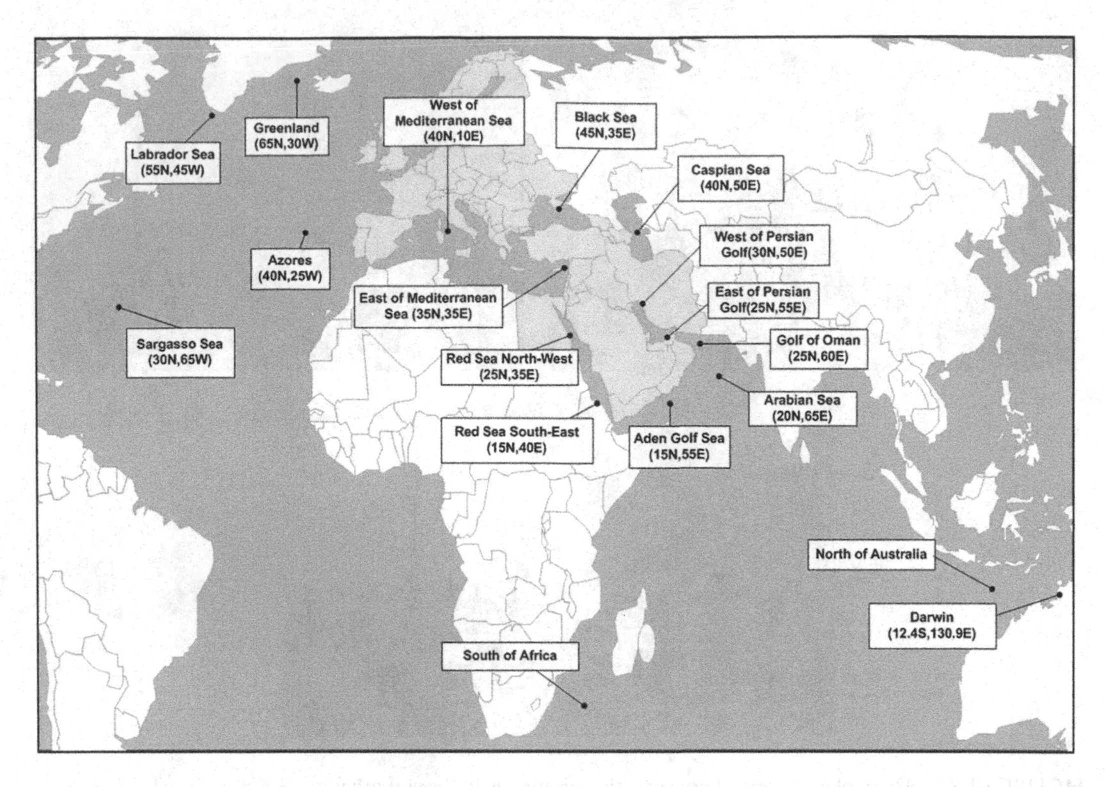

FIGURE 14.34 Effective signals (SLP teleconnection locations) on rainfall variations over Europe and the Middle East—North Atlantic Oscillation (NAO).

In the US weather code remarks, three digits are all that are transmitted; decimal points and the one or two most significant digits are omitted: 1,013.2 mbar (101.32 kPa) is transmitted as 132; 1000.0 mbar (100.00 kPa) is transmitted as 000; 998.7 mbar is transmitted as 987; etc. The highest SLP on Earth occurs in Siberia, where the Siberian High often attains an SLP above 1,050 mbar (105 kPa; 31 inHg), with record highs close to 1,085 mbar (108.5 kPa; 32.0 inHg). The lowest measurable SLP is found at the centers of tropical cyclones and tornadoes, with a record low of 870 mbar (87 kPa; 26 inHg).

In a study by Karamouz et al. (2009) for prediction of rainfall, SLP and North Atlantic Oscillation (NAO) were used as a main driving force for movement of wet fronts from Atlantic Ocean to Europe and then the northern part of Iran. See Figure 14.34 showing the main SLP locations. Other teleconnections between Northern part of Iran (high SLP at Siberia and low SLP at the red sea) have played an important role in the movement of wet fronts to the western part of Iran.

14.6 MAP RESOLUTION

14.6.1 DEM Resolution

The DEM is a raster data set that contains information about the topography of an area. The spatial resolution of a DEM refers to the area covered by a cell on the ground. As a result, the higher the resolution, the more cells there are per unit area of land, which results in a more detailed and accurate display of the topography of the area.

The large size of the elements and the low resolution of this spatial data sometimes give the decision maker misinformation. This has doubled the importance of spatial surveys using the latest advances in GIS and high-resolution images and maps. Figure 14.35 shows the efforts to increase the resolution of topographic maps in recent years.

1980's DMA 90 m
10^2 cells/km^2

1990's USGS DEM 30 m
10^3 cells/km^2

2000's NED 10-30 m
10^4 cells/km^2

2010's LIDAR ~1 m
10^6 cells/km^2

FIGURE 14.35 The challenge of increasing DEM resolution in recent years (Tarboton, D.G., Idaszak, R., Horsburgh, J.S., Ames, D.P., Goodall, J.L., Couch, A., Hooper, R., P., Dash, P.K., Stealey, M., Yi, H., Bandaragoda, C., and Castronova, A.M. The HydroShare collaborative repository for the hydrology community. *AGU Fall Meeting Abstracts*, IN12B-02. (2017)).

Any spatial data, even when it has a high resolution, is considered an approximation of reality and therefore has a level of error. This error is propagated in the modeling process, and the lower the resolution of the data, the higher the propagation of this error and the resulting uncertainty. Therefore, it is important to understand how this error is propagated.

The two main characteristics of DEMs namely the vertical accuracy and spatial resolution are directly related to the techniques of deriving data such as photogrammetric methods, terrestrial and airborne laser scanning, and radar interferometry (Vaze et al., 2010; Yunus et al., 2016). Satellite remote sensing techniques have provided DEMs for the entire world but with lower resolution and accuracy. However, the most detailed and accurate active remote sensing method called airborne Light Detection and Ranging (LiDAR) has been utilized to create DEMs of high resolutions (Liu, 2008). LiDAR is a technology similar to RADAR that can be used to create high-resolution DEMs with vertical accuracy as good as 10 cm.

LIDAR equipment, which includes a laser scanner, a Global Positioning System (GPS), and an Inertial Navigation System (INS), is generally mounted on a small aircraft. The laser scanner transmits brief laser pulses to the ground surface, from which they are reflected or scattered back to the laser scanner. Detecting the returning pulses, the equipment records the time that it took for them to go from the laser scanner to the ground and back. The distance between the laser scanner and the ground is then calculated based on the speed of light.

One of the main advantages of LiDAR data is that the pixel size of the DEM can be as low as 0.5 m or less which allows capturing many details about changes in elevation and the shape of a surface whether it is a structure or a natural landscape. Although many investigations reveal that the LiDAR datasets provide the most accurate outcome, it is not available in many places which limits its application.

Many studies have analyzed the impacts of DEM resolution on terrain characteristics (Gallant and Hutchinson, 1997; Holmes et al., 2000; Li and Wong, 2010; Vaze et al., 2010). Some investigators have specifically assessed the spatial resolution effects on flood modeling within a deterministic approach without incorporating uncertainty into the analysis (Werner, 2001; Omer et al., 2003; Raber et al., 2007; Chen and Hill, 2007; Kook and Merwade, 2009). Van de Sande et al. (2012) compared global DEMs and LiDAR DEM in terms of flood risk assessment for a coastal zone in

Nigeria using bathtub method, which assumes that flood inundation occurs only for locations with elevation below a given water level and concluded that the coastal flood risk is underestimated by global DEMs due to overestimating land elevation.

14.6.2 Digital Terrain Model

A digital terrain model (DTM) can be described as a three -dimensional representation of a terrain surface consisting of *X*, *Y*, *Z* coordinates stored in digital form. It includes not only heights and elevations but other geographical elements and natural features such as rivers and ridge lines . A DTM is effectively a DEM that has been augmented by elements such as break lines and observations *other* than the original data to correct for artifacts produced by using only the original data. With the increasing use of computers in engineering and the development of fast three-dimensional computer graphics, the DTM is becoming a powerful tool for a great number of applications in the earth and the engineering sciences.

14.6.3 Quality and Accuracy of DEM/DTM

The quality of a DEM/DTM is a measure of how accurate elevation is at each pixel (absolute accuracy) and how accurately is the morphology presented (relative accuracy). Several factors play important roles in the quality of DEM-derived products:

- Terrain roughness;
- Sampling density (elevation data collection method);
- Grid resolution or pixel size;
- Interpolation algorithm;
- Vertical resolution; and
- Terrain analysis algorithm.

Some 3D products derived from this matter include quality masks that give information on the coastline, lake, snow, clouds, correlation, etc.

14.6.4 Digital Surface Model (DSM)

Digital Surface Model (DSM) represents the MSL elevations of the reflective surfaces of trees, buildings, and other features elevated above the "Bare Earth." DSM includes ground surface, vegetation, and man-made objects. DSM demonstrates the natural and artificial features on the Earth's surface. DSM may be useful for RF planning, landscape modeling, city modeling, visualization applications, and more.

14.6.5 Common Uses of DEMs

A DEM is a digital representation of model (DTM). While the term can be used for any representation of terrain as GIS data, it is generally restricted to the use of a raster grid of elevation values. DEMs are commonly built using remote sensing techniques, but they may also be built from land surveying. DEMs are used often in GISs and are the most common basis for digitally produced relief maps of ground surface topography or terrain. It is also widely known as a digital terrain.

- Extracting terrain parameters.
- Modeling water flow or mass movement (e.g., landslides).
- Creation of relief maps.

- Rendering 3D visualizations
- Creation of physical models (including raised-relief maps).
- Rectification of aerial photography or satellite imagery.
- Reduction (terrain correction) of gravity measurements (gravimetry, physical geodesy).
- Terrain analyses in geomorphology and physical geography.

14.6.6 Effect of Map Resolution on Modeling

DEM data are used in flood models to represent a physical land surface. The spatial resolution of a DEM refers to the area of land being represented by a single grid cell. So, a spatial resolution of 10 m means one grid cell is representing a $10 \times 10\,m^2$ area of physical land. Low (or coarse) resolution and high (or fine) resolution are relative terms. "Higher" resolution implies comparatively greater preservation of land features, while a "lower" resolution dataset generally smooths over topographic details (Environmental Systems Research Institute Inc., 2016). DEM resolutions used in the hydrological and hydraulic modeling industry typically range from 1,000 to 2 m or less (Vaze et al., 2010).

Yet, although higher spatial resolution and accuracy often translate to improved results, these benefits cannot be considered in isolation, especially when we consider how numerical flood models work. Numerical models rely on multiple simulations to represent the physical processes of flooding, and these simulations can number into the hundreds of thousands or more depending on an area's extent. The more complex models are (i.e., the higher the spatial resolution), the greater the demands on computational power, processing time, file size, data storage, and cost (Environmental Systems Research Institute Inc., 2016). On the other hand, using comparatively lower- resolution data means lower feature spatial accuracy but faster processing and smaller file sizes. Thus, depending on modeling objectives and requirements, lower resolution may fit the computational difficulties.

14.6.7 Kriging Interpolation

In general, in spatial statistics, the interpolation of a quantity at a point with known coordinates is done using the value of the same quantity at other points with known coordinates. The most important estimator of spatial statistics is named kriging, in honor of one of the pioneers of geostatistics, D.G. Krige, a mining engineer in South Africa (Hengl et al., 2010). One of the most important features of kriging is that the error and, consequently, the confidence interval can be calculated for each estimate.

In the pure geostatistical approach, predictions are commonly made by calculating some weighted average of the observations:

$$\hat{z}_{OK}(s_0) = \sum_{i=1}^{n} \lambda_i \cdot z(s_i) \quad (14.62)$$

where $\hat{z}(s_0)$ is the predicted value of the target variable at an unvisited location s_0 given its map coordinates, the sample data $z(s_1), z(s_2), \ldots, z(s_n)$, and their coordinates. The weights λ_i are chosen such that the prediction error variance is minimized, yielding weights that depend on the spatial autocorrelation structure of the variable. This interpolation procedure is popularly known as ordinary kriging (OK).

Kriging can be seen as an enhanced mode of inverse distance weighting (IDW) interpolation. The main problem with IDW is that it does not care how important the neighbors used in interpolation are. In kriging, the weight assigned to each of the neighboring points reflects the true structure of spatial autocorrelation.

The kriging estimator is the best-unbiased estimator. Therefore, two conditions must be confirmed in the estimation. The first condition is the absence of systematic error in estimation (estimation must be unbiased). The average estimation error must be zero to meet this condition:

$$E\left[\hat{z}(s)-z(s)\right]=0 \tag{14.63}$$

Therefore, the condition $\sum_{i}^{n} w_i = 1$ (w_i is kriging weights) must be true, and the sum of the kriging coefficients must be equal to one. The second condition is that the variance of the estimate should be minimized. The variance of the estimate is calculated to establish this condition:

$$E\left[\left(\hat{z}(s)-z(s)\right)^2\right]=\bar{C}(s,s)-2\sum_{i=1}^{n} w_i\bar{C}(s,s_i)-\sum_{i=1}^{n}\sum_{j=1}^{n} w_i w_j \bar{C}(s_i,s_j) \tag{14.64}$$

By substituting semivariance ($\bar{\gamma}$) for covariance ($\bar{C}$) in the above relation and minimizing the variance in terms of kriging coefficients while maintaining the first condition, the final coefficients are obtained and estimation can be made anywhere in space.

14.6.8 Variogram Modeling

As mentioned, a variogram is needed to estimate the coefficients in kriging interpolation. Semivariance, first introduced by Matheron (1962) and Gandin (1963), is defined as follows between two points in each other's neighborhood:

$$\gamma(h)=\frac{1}{2}E\left[\left(z(s_i)-z(s_i+h)\right)^2\right] \tag{14.65}$$

where $z(s_i)$ is the value of the target variable at the location of estimation and $z(s_i+h)$ is the value of the neighbor at the distance s_i+h. Suppose there are n observation points that give rise to $n(n-1)/2$ pairs of points for which the semivariance is calculated. All of these semivariances can be plotted in terms of distance, which will result in a variogram cloud, as shown in Figure 14.36. These variogram clouds cannot be described visually, so values are averaged for standard distances, called (lag). If these mean values are displayed, an empirical variogram is obtained. What is clear is that the semivariance is smaller at shorter distances and stabilizes from a specific distance onward. This issue can be interpreted as follows.

The values of the target variable are more similar at closer distances, up to a certain distance where the difference between the pairs of points is more or less equal to the total variance. This is called the spatial autocorrelation effect.

It should also be noted that there is a difference between the factor of the range of impact and the practical range. The practical range is equal to the log(h) where $\gamma(h)=0.95\gamma(\infty)$, which means the semivariance interval is close to 95% of the sill. Figure 14.37 shows some basic concepts of the variogram.

Before using empirical variograms in estimation, it is necessary to fit the most appropriate theoretical model to them. The procedure is similar to fitting the probability distribution function to the data that shape the histogram. After calculating the empirical variogram, it can be fitted to a theoretical model such as linear, spherical, exponential, circular, and Gaussian.

Spherical, Gaussian, Exponential, and Matern theoretical models are commonly used to fit empirical data. Table 14.4 shows the equations of each of these models. It should be noted that the Matern model is a more general case than Exponential and Gaussian. All of these models are

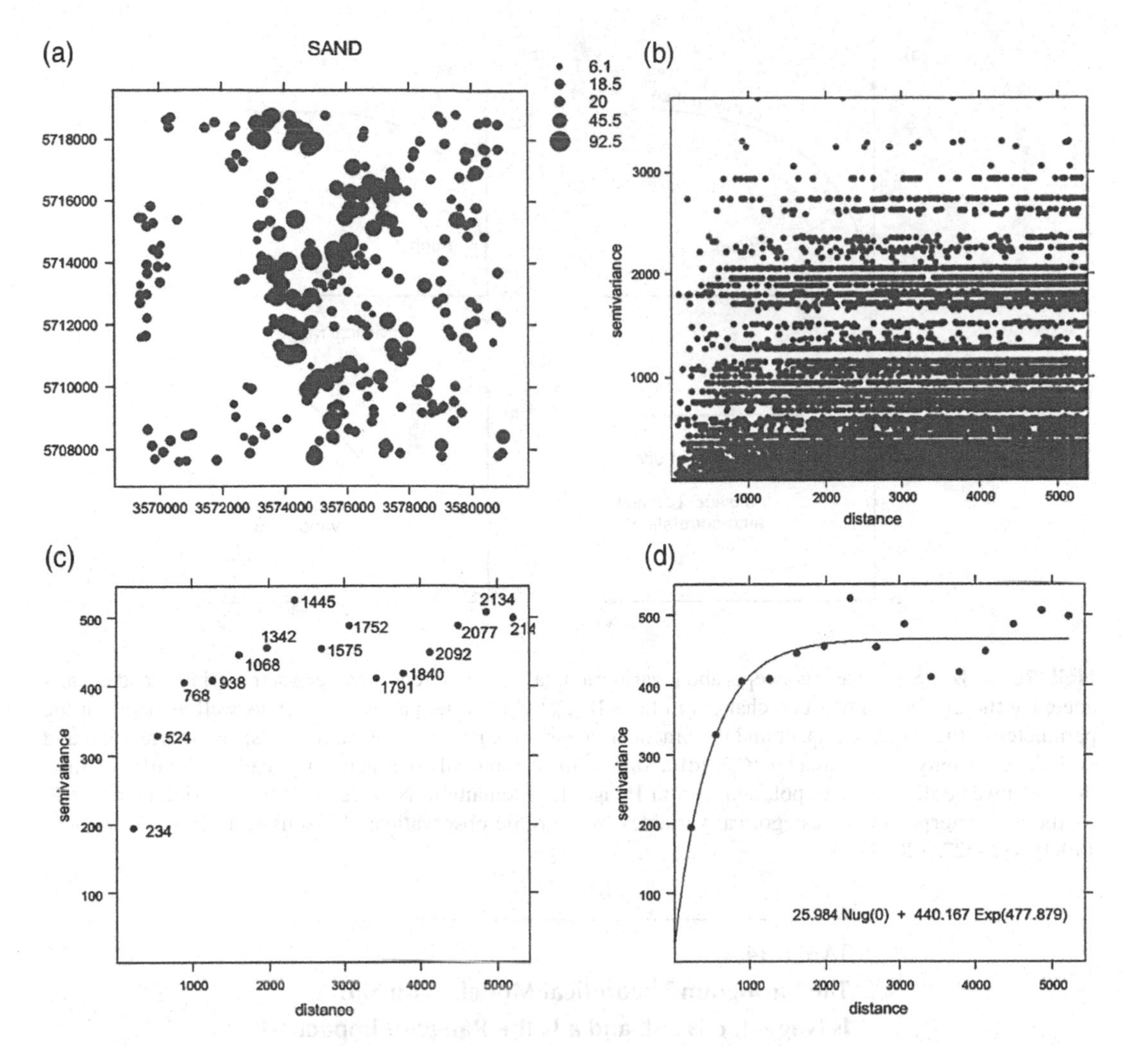

FIGURE 14.36 An example of a variogram modeling steps for mapping sand (%) in topsoil (Hengl et al., 2007), (a) location of observation points (300 points); (b) variogram cloud that shows semivariances for 44,850 pairs of points; (c) cumulative semivariances at intervals of about 300 meters; and (d) the final variogram (theory) model fitted to the data.

categorized as models with the sill. In this group of variograms, the value of $\gamma(h)$ increases with increasing h and then approaches the constant limit, which is the sill. One of the important features of this type of variogram is that the value of the sill is equal to the variance of the value of variable in the sample population. In all models with sill, the relationship between the semivariances and covariances is expressed as follows:

$$C(h) = \sigma^2 - \gamma(h) \tag{14.66}$$

This relationship states that in models with sill, the sum of the variogram and the covariogram is always equal to σ^2, same as the variogram sill.

The parameters mentioned in the table above should be estimated to fit to the experimental data. Nonlinear regression is used to fit the parameters. For this purpose, the sum of the weight of the squared errors $\sum_{j=1}^{p} w_j \left(\gamma(h) - \hat{\gamma}(h)\right)^2$ is minimized. Different models have been proposed for the

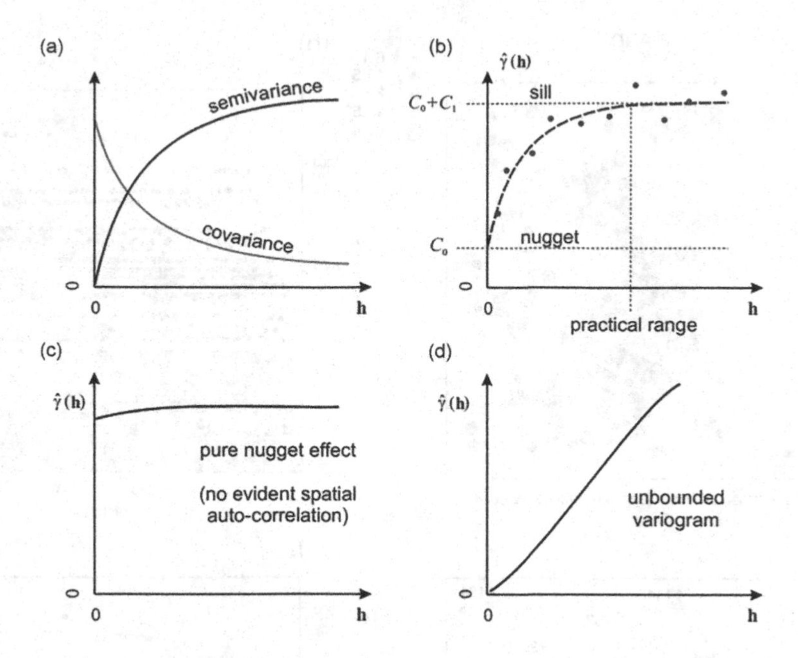

FIGURE 14.37 Some basic concepts about variogram: (a) The difference between semivariance and covariance; (b) the distinction between changes in the sill (C_0+C_1) and its parameter (C_1), as well as between the parameter of the range of impact and the practical range; (c) variogram that shows no spatial correlation and is defined by only one parameter (C_0); (d) a variogram without sill that generally leads to similar estimations of inverse distance interpolation. (From Hengl, T., Toomanian, N., Reuter, H.I., and Malakouti, M.J.. Methods to interpolate soil categorical variables from profile observations: Lessons from Iran. *Geoderma*, 140(4), 417–427. (2007).)

TABLE 14.4
The Variogram Theoretical Models with Sill; c_0 Is Nugget, c Is Sill, and a Is the Range of Impact

Equation	Model
$\gamma(h)=c_0+c\left[\frac{3h}{2a}-\frac{1}{2}\left(\frac{h}{a}\right)^3\right]$	Spherical
$\gamma(h)=c_0+c\left[1-\exp\left(-\frac{h}{a}\right)\right]$	Exponential
$\gamma(h)=c_0+c\left[1-\exp\left(-\frac{h^2}{a^2}\right)\right]$	Gaussian
$\gamma(h)=c_0+c\left[1-\frac{1}{2^{(v-1)}\Gamma(v)}\left(\frac{h}{a}\right)^a K_v\left(\frac{h}{a}\right)\right]$	Matern

weight w_j. A simple model can be defined as the relation N_j / h_j^2, where N_j is the number of pairs at certain lag and h_j is the distance (Bivand et al., 2013).

To compare the models and select the best one, their performance in interpolation to achieve the desirable output can be evaluated using the k-fold cross- validation method. In this method, the data is divided into k sets. From k sets, a single set is used for validation and the remaining $k-1$ set is used to train the model. Using this framework, the performance of variogram model is evaluated for reduction of estimation error. Interested readers are referred to gstat package in *R* programming language (Pebesma, 2004).

To test the performance of the models, two statistics proposed by Cressie (1993) can be used as follows:

$$\text{MPE} = \frac{1}{n}\sum_{i=1}^{n}\left\{\frac{e(v_i)-\hat{e}_{-i}(v_i)}{\sigma_{-i}(v_i)}\right\} \tag{14.67}$$

$$\text{RMSPE} = \left[\frac{1}{n}\sum_{i=1}^{n}\left\{\frac{e(v_i)-\hat{e}_{-i}(v_i)}{\sigma_{-i}(v_i)}\right\}^2\right]^{\frac{1}{2}} \tag{14.68}$$

where n is the number of control points used for interpolation, $e(v_i)$ is the observed variable at location v_i, $\hat{e}_{-i}(v_i)$ is the interpolated variable, and $\sigma_{-i}(v_i)$ is the standard deviation of the estimation at the location of v_i that is obtained by discarding $e(v_i)$ in the fitting process to the empirical variogram.

14.6.9 Resampling

DEM generation process is subject to uncertainty and different types of error (Mukherjee et al., 2013) in both vertical and horizontal coordinates, such as data collection errors (Rodriguez et al., 2006), errors due to excessive smoothing (Polidori et al., 1991), transformation (Carter, 1992), positioning (Triglav-Cekada et al., 2009) as well as unknown combinations of errors (Mukherjee et al., 2013). Grid spacing and interpolation techniques can be the other sources of errors (Mukherjee et al., 2011). Since flood analysis is directly related to the DEM accuracy and resolution, reliability of the results is significantly affected by these errors and uncertainties. Many investigators have analyzed DEM errors using statistical and visual analysis approaches (Hunter and Goodchild, 1995; Wechsler, 2007). Statistical approaches frequently employ root mean square error (RMSE) and standard deviation of the DEM error (Shearer, 1990; Wise, 1998; Carlisle, 2005; Chen and Yue, 2010). However, since the variability of DEM errors are not spatially homogeneous, they cannot be sufficiently assessed by a global metric such as RMSE (Carlisle, 2005). Therefore, to the extent by which each cell can represent a true ground elevation would lead to uncertainty in a DEM and the subsequent analysis. To evaluate this uncertainty, more information on the spatial structure of error is needed. This cannot be provided by the RMSE (Wechsler, 2007).

Representation technique is used to produce lower resolution DEMs. Representation is generally defined for raster data and refers to the process of interpolating the values of a cell while converting a raster set. The three most popular and used methods are Near Neighbor (NN), Bilingual (BL), and Cubic Estimation (CC), have the following characteristics, advantages, and disadvantages of each method:

- *Characteristics:*
  - *Nearest neighbor*: In this method, the value of the nearest pixel is assigned to each pixel.
  - *Bilinear interpolation method*: The values of four pixels are interpolated in a 2 × 2 window to calculate the value of each pixel.
  - *Cubic convolution method*: 16-pixel values are interpolated in a 4 × 4 window to calculate the value of each pixel.
- *Advantages:*
  - *Advantages of the NN method*: The original data is moved without interference. In other words, the new value of a pixel is selected from the same old values. This method is suitable for demarcation between phenomena and suitable for preclassification use. It is easier and faster to calculate than other methods.

    - *Advantages of the BL method*: Sudden values do not change and step effects on diagonal lines and curves are eliminated. Spatially, data is more accurate and is used when pixel dimension change.
    - *Advantages of CC method*: In this way, the borders become clearer and the noises become softer and more homogeneous. This method is recommended when the user changes the pixel size significantly.
- *Disadvantages:*
    - *Disadvantages of NN method*: When this method is used to restamp the image with larger pixel dimensions to smaller pixel dimensions, the lines and diagonal curves are stepped. Using this method for linear data such as road or river can cause breakage or distance in the network.
    - *Disadvantages of BL method*: The values of all the pixels are average, and therefore, like low-pass filters, they make the pixels smoother and less distinct and the boundaries between features more difficult to distinguish.
    - *Disadvantages of CC method*: The value of the pixels may change significantly. It is the most difficult method in terms of calculation and therefore the slowest method of sampling.

Fereshtepour and Karamouz (2018) explored the effects the DEM resolution on coastal flood assessment are investigated when taking DEM uncertainty into account. Considering 1-m LiDAR DEM as a baseline, DEMs of different resolutions produced by bilinear resampling method along with USGS National Elevation Dataset (NED) DEMs of 10 and 30m are used to obtain accuracy–efficiency trade-offs within the proposed probabilistic flood risk assessment framework. The results are then compared with a deterministic approach where DEMs are used without incorporating the associated errors into the analysis. A Monte Carlo-based method called Sequential Gaussian Simulation (SGS) is used to generate different realizations of the elevation for DEM dataset based on the spatial variability of errors. A hydrologically connected bathtub method is used for flood inundation modeling and then the Flood Damage Estimator (FDE) model is employed to determine the induced damages. The entire methodology is developed in *R* language and environment which provides more flexibility for statistical and spatial analysis. The study area is the coastal zones of the lower Manhattan in New York City. Figure 14.38 shows Monte Carlo simulation for DEM uncertainty analysis, using SGS.

14.6.10 DEM Error

Uncertainty in hydrological modeling is often associated with a model structure or calibration parameters, in which little consideration is given to the uncertainties associated with input or output data (Benke et al., 2011). DEM is an essential input among the needed data in assessing flood impacts (Fereshtehpour and Karamouz, 2018). DEM and its characteristics could be primary sources of errors in producing inundation maps (Bates et al., 2003). This error could be because of the difference of DEM data of low resolution from the actual values of spatial data. Therefore, reducing this error will be the most effective way to improve the results. In spatial data analysis, the actual value is unknown, or it is not usually available, which could be assessed through uncertainty analysis (Wechsler, 2003). Often high-resolution light detection and ranging (LIDAR) DEM data is considered as actual elevation data (Fereshtehpour and Karamouz, 2018).

Selecting a suitable approach for simulating DEM uncertainty is dependent on the availability of spatial structure of error. When this information is not available, RMSE provided with DEM metadata can be used to approximate random values (Hunter and Goodchild, 1997; Wechsler and Kroll, 2006). If the spatial structure of error is available from a high -accuracy ground control point (GCP) measurement or high-resolution DEM, other approaches such as kriging could also be integrated into random field generation (Aerts et al., 2003; Holmes et al., 2000; Oksanen, 2006). Because

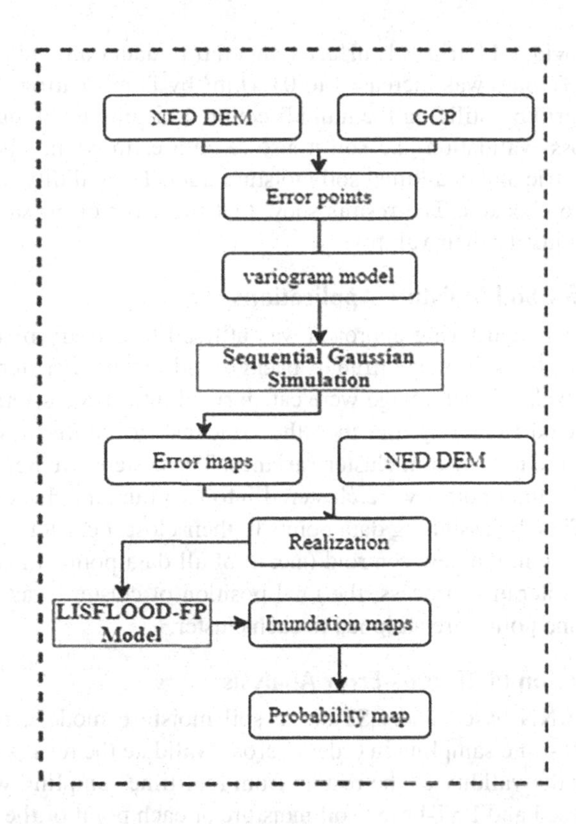

FIGURE 14.38 Flood probability map, developed by Sequential Gaussian Simulation (SGS) of DEM error analysis. (Adapted from Aerts, J.C. et al., Trans. GIS, 7(2), 211–230, 2003.)

DEM errors are spatially correlated, the second approach should lead to a better representation of uncertainties. As 1-m LiDAR DEM is available for certain regions, a large number of points are sampled from this DEM as the GCPs. In reality, field surveys for collecting GCPs are usually carried out along the accessible transects such as roads, beaches, and paths through the primary land cover (Leon et al., 2014). Therefore, land cover raster map is employed, and different probabilities are assigned to the existing land cover categories. Furthermore, probabilities for each land cover is needed to represent the location preferences for point sampling (the highest probability of sampling is assigned to the accessible transects). A set of spatially balanced points are then created using ArcMap toolbox in geostatistical analyst extension. It should be noted that these control points are considered as if they were obtained by actual observation.

14.7 CASE STUDIES

14.7.1 Case Study 1: A Satellite/Citizen Science-Based Soil Moisture Estimator

Karamouz et al. (2021) proposed a new framework and a platform to quantify citizen science-based and land surface temperature-based satellite data to estimate real- time soil moisture data. First, a user interface and a sampling process were designed to collect and analyze the subjective and gravimetric soil moisture data. Furthermore, explanatory moisture condition (EMC) from vegetation cover was analyzed as explanatory data. A statistical artificial neural network was used for quantifying subjective data, and soil moisture layouts were produced by utilizing the OK method. To investigate the relationship between the soil moisture and land surface temperature layouts, first, LST data was retrieved from the MODIS MOD11A1 product. MODIS sensor collects data in 36 spectral bands and various spatial resolutions. MOD11A1 product, which is developed based on the

Terra MODIS data, provides 1 km^2 land surface temperature data on a daily basis. Then after, the spatial resolution of LST data was increased to 0.04 km^2 by interpolating the original 1 km^2 data over a 200 m × 200 m grid by utilizing the multidirectional kriging technique. A second sampling was performed for cross- validation and soil moisture data estimation. The results show a close match between gravimetric and quantified soil moisture data. The validity of the results from real-time sampling was also checked. The results show that the error of all sample points was small enough (statistically) to justify their validity.

14.7.1.1 Clustering for Soil Moisture Applications

In the case study, k-means clustering approach was utilized to classify pixels of each photo into two clusters and separate the soil image from its background. At the first step, by utilizing k-means clustering algorithm, pixels of each image were categorized into two clusters. In the first iteration, clusters would be selected randomly and then the k-means procedure moves the cluster centers iteratively to minimize the total within cluster variance. These steps are iterated until convergence (Hastie et al., 2009). All data points were clustered into two random cluster centers selected during the first iteration. Then by assigning data points to their closest cluster centers according to the Euclidean distance function, the new centroid (mean) of all data points in each cluster was calculated. By repeating this iterative process, the final position of clusters was determined. The final position is when the same points are assigned to each cluster.

14.7.1.2 Data Estimation Platform—Error Analysis

After developing the ANN-based and LST-based soil moisture models, they should be implemented for any new real -time sampling in order to cross- validate the results from them. The cross-validation criterion for the validity of the results from real-time sampling was being the absolute error between ANN-based and LST-based soil moisture at each point of the real-time sampling in the range of mean plus/minus the standard deviation of the modeling stage in its EMC group. In the next step, the results from ANN-based and LST-based models for real-time sampling were modified and combined with each other. For this purpose, first, due to the importance of the location of points, the study area was divided into polygons by using the Thiessen method for the real-time sampling points.

Then two correction factors were calculated based on errors of the results. In order to calculate the error, first the difference between the field data and the ANN-based soil moisture in each point of first sampling in the polygon was calculated and then their mean was reported as $error_1$. Next $error_2$ was calculated as the difference between the field data and LST-based soil moisture in each point of first sampling in each polygon. After calculating $error_1$ and $error_2$, a correction factor was calculated based on the inverse ratio of errors in each polygon. Then the correction factor $\left(\frac{Error_2}{Error_1 + error_2}\right)$ related to the estimation from ANN-based model was multiplied by the ANN-based soil moisture and also the correction factor $\left(\frac{Error_2}{Error_1 + error_2}\right)$ related to the estimation from LST-based model was multiplied by the LST-based soil moisture (Table 14.5).

14.7.1.3 Study Area and Data Collection

The study area was located in the Shahriar County, Tehran province, in an area of 80 km^2 situated between E 50°58′ to 51°5′ and N 35°35′ to 35°41′. In selecting this region, factors such as soil type, land use diversity, availability of meteorological station, and ease of access were considered. The geomorphology of the site is relatively uniform according to the geological maps and is classified as Young Terrace. Furthermore, dominant soil texture of the area is clay marl texture (FAO, 2020) without gravels (about 80%) and about 10% of it contains sandy soils. The last remaining 10% contains volcanic rocks in the surrounding mountains. The 5-year average annual precipitation and temperature of the region are 199 mm and 16.9° C, respectively.

TABLE 14.5
Soil Moisture Classes and the Provided User Guide

	Classifications		
	Completely Dry	Relatively Dry	Moist
Visual Guide			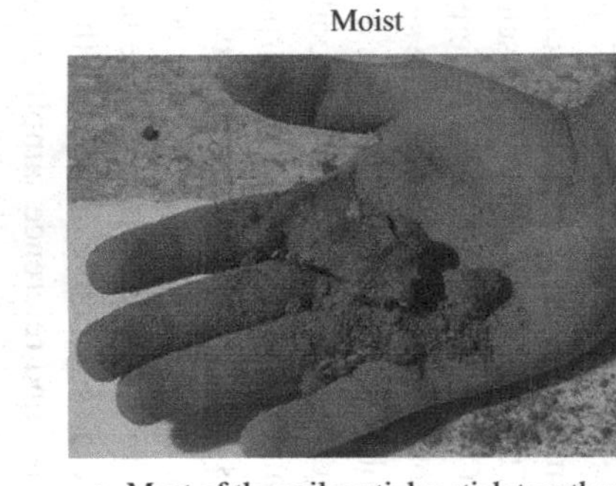
Text Guide	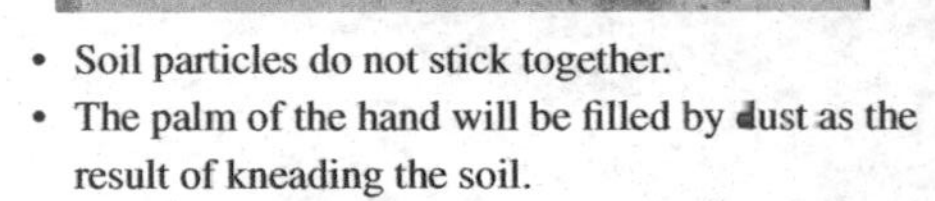 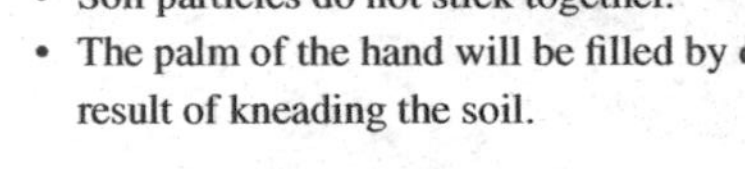 • Soil particles do not stick together. • The palm of the hand will be filled by dust as the result of kneading the soil.	• Some soil particles stick together. • The palm of the hand will be filled by dust as the result of kneading the soil.	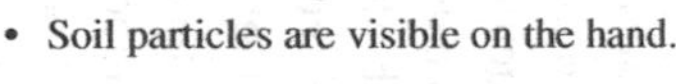 • Most of the soil particles stick together. • Soil mass is fragile. • Soil particles are visible on the hand.
	Relatively Wet	Completely Wet	
Visual Guide	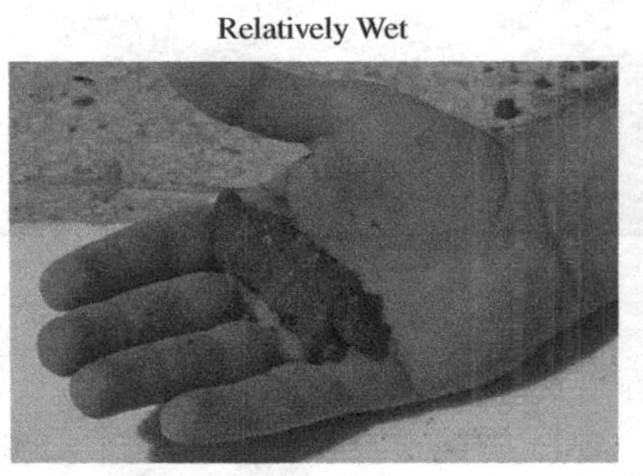	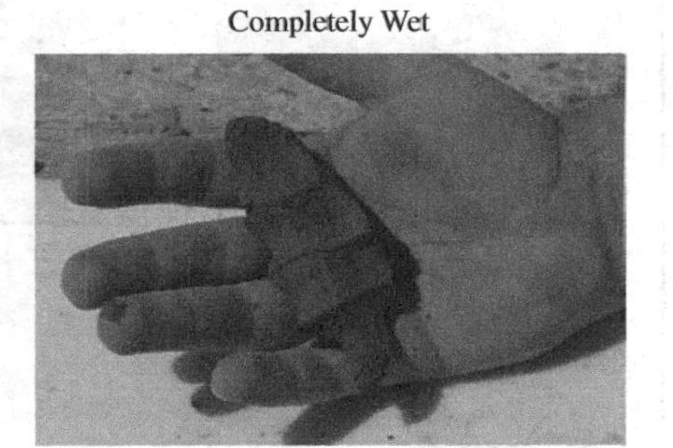	
Text Guide 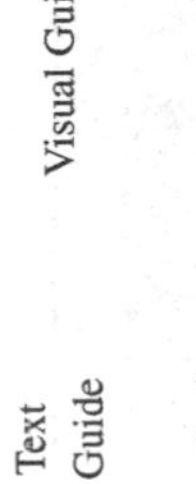	• Surface of the soil is rough. • Soil mass has some plasticity.	 • Surface of the soil is soft. • Soil mass has significant plasticity.	

To develop the methodology, two samplings were conducted as follows. The first sampling was conducted on May 2, 2019. Forty-two different locations were selected and subjective data collection process was carried out by 21 users. In order to select the sampling points, the study area was gridded into 1 km² cells and then, by considering limitations such as the existence of urban areas and inaccessibility to some of the private properties, the sampling points were selected.

In the second sampling, which was conducted on October 11, 2019, eight different locations were selected and the data collection process was carried out by ten users. Size of the second dataset was selected to be smaller than the first dataset (both number of the sampling points and users). The size reduction was due to the application of the second dataset. This dataset was collected to demonstrate the capability of the developed methodology and not for development purposes. Since in the modeling approaches the size of the testing dataset is usually set to about 30% of the dataset which is used for the developing stages (Xu and Goodacre, 2018), number of the second sampling points was selected to be 20% of the first sampling points. In addition, number of the users was reduced to check whether the models could quantify the gathered subjective data, when their size is limited. Figure 14.39 shows the study area and the first, second, and reference sampling points.

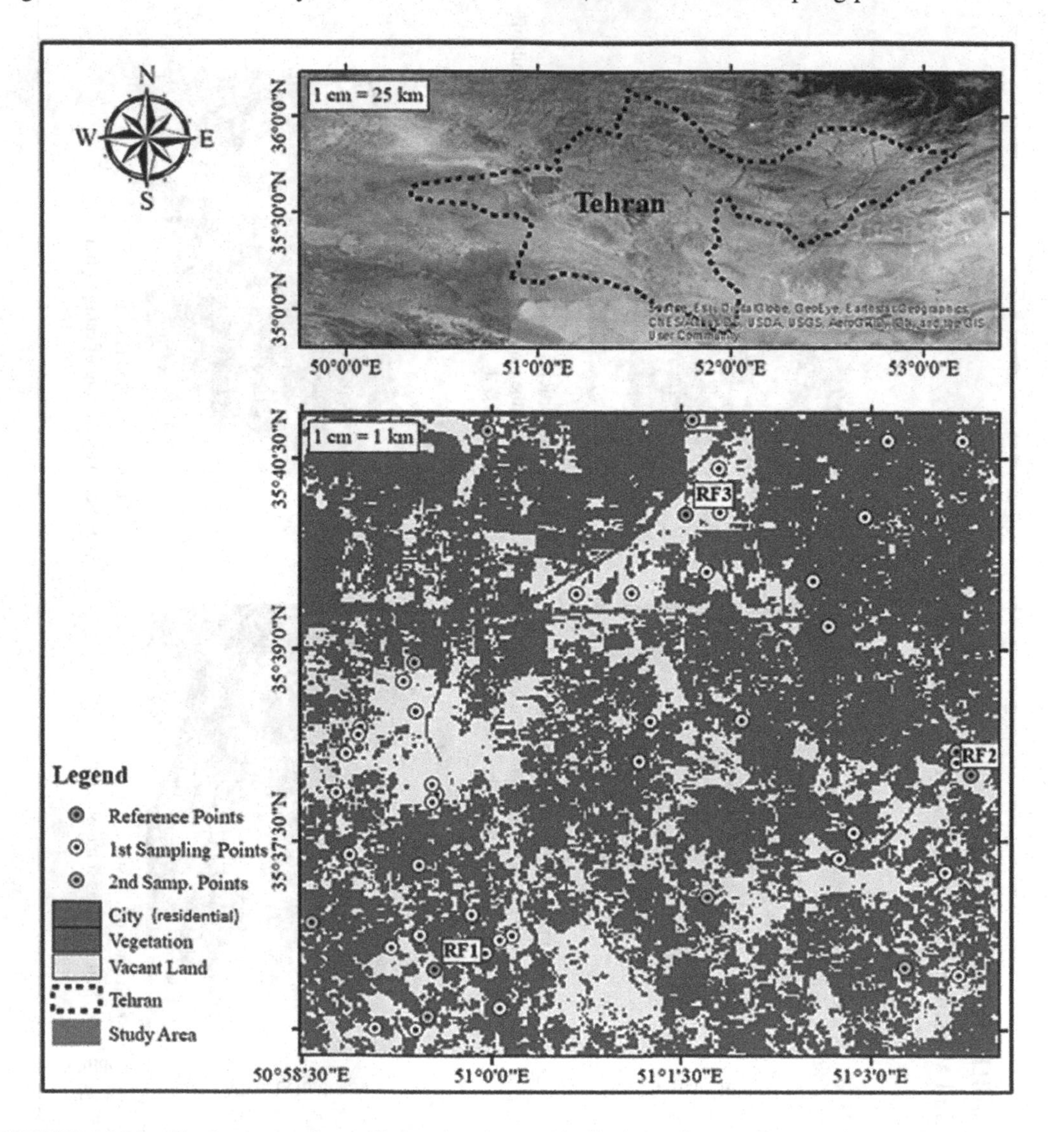

FIGURE 14.39 Study area and sampling points located in Shahriar County, Tehran, Iran.

The 5-day antecedent rainfall data before both sampling days was obtained from Tehran and Shahriar meteorological stations (www.irimo.ir). The 5-day antecedent precipitation in both sampling days was recorded as 0 mm even though there was a scattered rain in the area. So, antecedent moisture condition (AMC) of the sampling points was categorized as AMC I for both days.

14.7.1.4 Digital Image Processing

In this section, digital images regarded in the data collection are processed and the relationship between soil color and soil moisture is investigated. For this purpose, the image of each soil sample was analyzed by the Environment of Visualizing Images (ENVI) software. It should be noted that all of the images for communicating with user and soil classification were taken by a 12 Megapixel Canon DIGITAL IXUS 200 IS camera.

Digital images are most often represented by the RGB model. In this model, each pixel of the image contains three DNs. DNs are dimensionless values that represent the amount of light reflectance in the red, green, and blue bands. Combinations of these DNs are used to represent various colors. It should be noted that the bit depth concept is used to define the range of DNs. This concept refers to the color information stored in an image. If an image has a higher bit depth, it could store and represent more colors. For an N-bit digital image, each DN ranges from 0 to $2^N - 1$ (for an 8-bit image, the DN could be as high as 255). Because the light reflectance in "bright" colors is more than "dark" ones, the "bright" pixels have high DNs, while the "dark" pixels have low DNs. In other words, "0" means no reflection, and "$2^N - 1$" is maximum reflection in each color band. An 8-bit image could store and illustrate 16.7 million ($2^8 \times 2^8 \times 2^8 = 2^{24}$) different colors.

In this case study, k-means clustering approach was utilized to classify pixels of each photo into two clusters and separate the soil image from its background. Then, the soil cluster was analyzed to investigate the correlation between soil color and moisture.

14.7.1.5 Cross-Validation of Citizen Science with Satellite Data

To investigate the relationship between the soil moisture and land surface temperature layouts, first, LST data was retrieved from the MODIS MOD11A1 product. MODIS sensor collects data in 36 spectral bands and various spatial resolutions. MOD11A1 product, which is developed based on the Terra MODIS data, provides 1 km^2 land surface temperature data on a daily basis. Then after, the spatial resolution of LST data was increased to 0.04 km^2 by interpolating the original 1 km^2 data over a 200 m × 200 m grid by utilizing the multidirectional kriging technique. Figure 14.40 shows the high-resolution LST layout for the first sampling date (May 2, 2019). About 80% of the data were used to develop a power correlation between LST and soil moisture data and the rest of the data were used for testing the developed relationship. It is worthy to note that to prevent overfitting, ten different building-testing datasets were selected to develop and test the relationship. For this purpose, ten random datasets with size of 80% for the building stage and 20% for the testing stage were selected and the power relationship was developed/tested based on each of them. Finally, the average R^2 and RMSE indices were calculated based on all ten datasets and coefficients of the final power model (presented in Equation 11) were calculated by averaging the coefficients of the models which were developed based on ten datasets.

Several relationships such as linear, exponential, and power relationships were tested to investigate the correlation between the soil moisture and dimensionless land surface temperature (LST_{DL}), and finally, the power model was selected as the best model.

$$SM_{LST} = 4.296 \times LST_{DL}^{-178.67} \tag{14.69}$$

The power relationship shows a closed fit as shown in Figure 14.41 and yielded average R^2_{Train} (R^2_{Test}) of 0.69 (0.68) over ten building-testing datasets, and its average $RMSE_{Train}$ ($RMSE_{Test}$) was 2.31% (2.30%). The linear and exponential relationships yielded smaller coefficients of determination of 0.63 and 0.65, respectively.

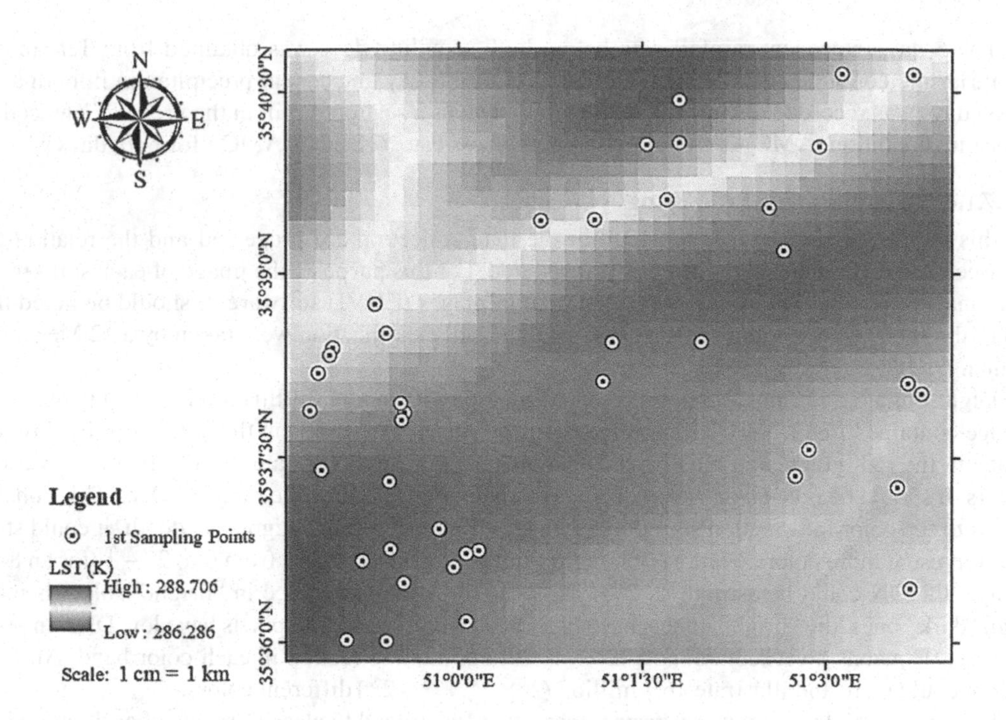

FIGURE 14.40 High- resolution land surface temperature (LST) layout for the first sampling date.

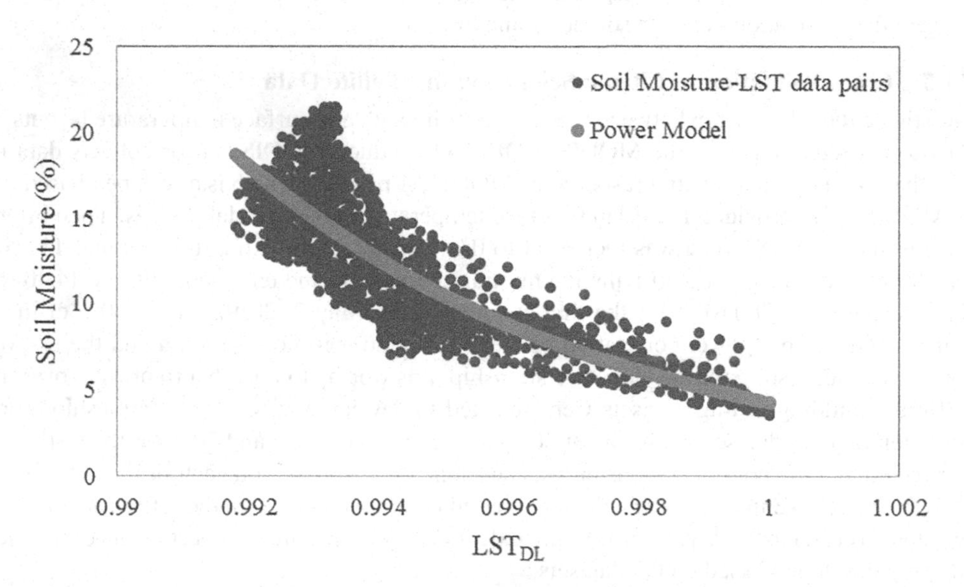

FIGURE 14.41 Fitted power model and the soil moisture-LST data pairs.

In the next, the results of ANN-based and LST- based models are used to obtain soil moisture at real time for a pilot. First, the result is verified and then merged with each other to obtain modified soil moisture.

14.7.2 Case Study 2: Data Assimilation for Flood Assessment

Adibfar (2021) used VIC together with the DA approach. The assimilation process was applied by using the EnKF to the NASA's SMAP satellite soil moisture data and VIC model inputs (such as temperature, pressure, long/short wave radiation). This study aims to develop an inundation model using precise antecedent soil moisture derived by the assimilation model. In order to emphasize on the effects of antecedent soil moisture on produced runoff, a lag time of 3 hours was applied to the soil moisture and inundation maps. The reason for this is that after measurement of soil moisture in 3 AM, water has been allowed to move around for 3 hours before the generation of the inundation maps in order to have a more accurate and more final inundation map of the study area. Assimilating the observational variability of the SMAP soil moisture data to the simulated VIC's simulated soil moisture leads to improved soil moisture updates and, hence, enhances initial conditions for runoff generation. Furthermore, VIC routing model (RVIC) is utilized to simulate runoff directions and produce inundation map. The proposed methodology and sampling stations can be seen in Figures 14.42 and 14.43. This methodology is tested on central and western parts of

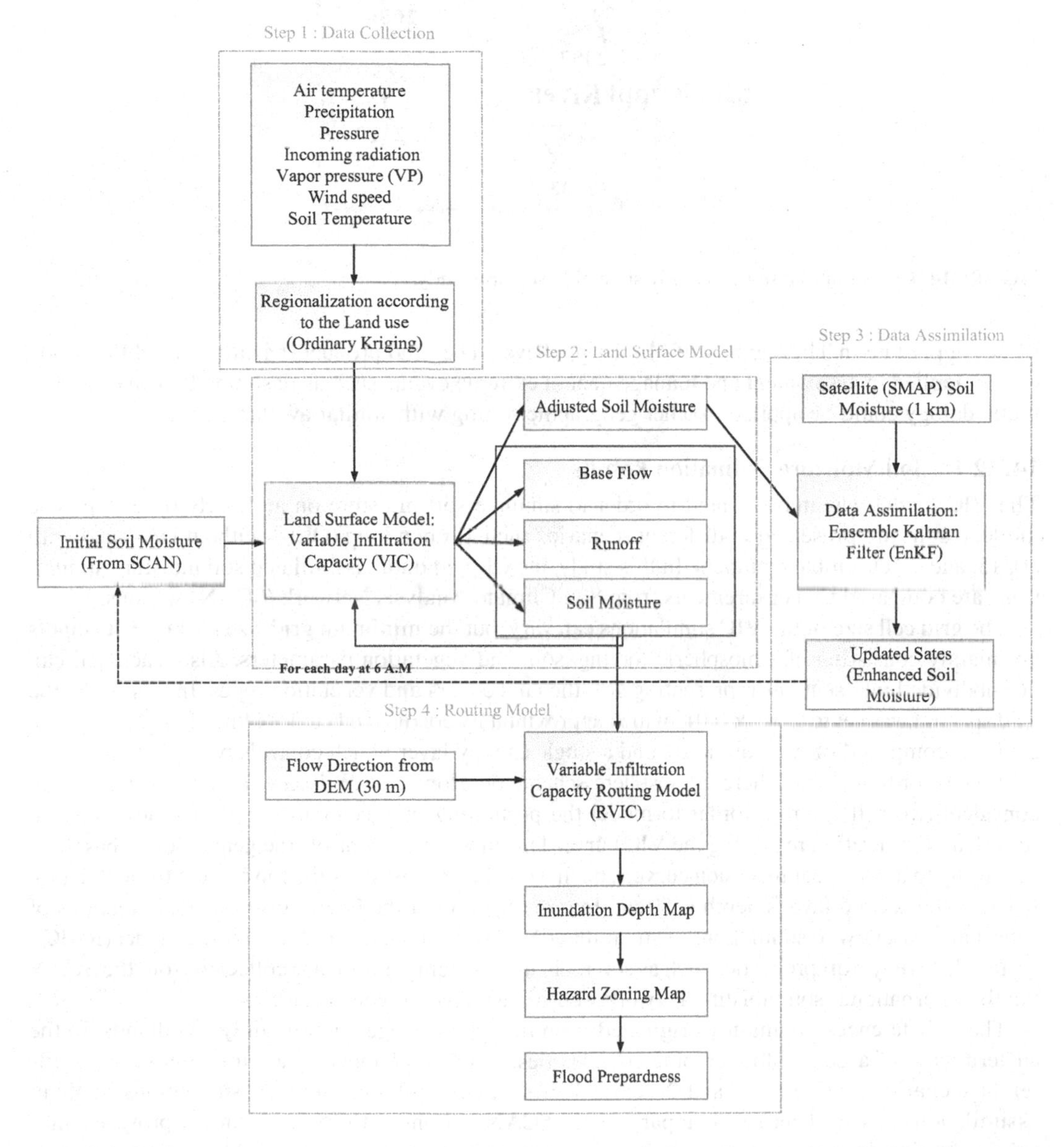

FIGURE 14.42 Proposed flood and damage assessment platform by enhanced soil moisture estimates for the study area.

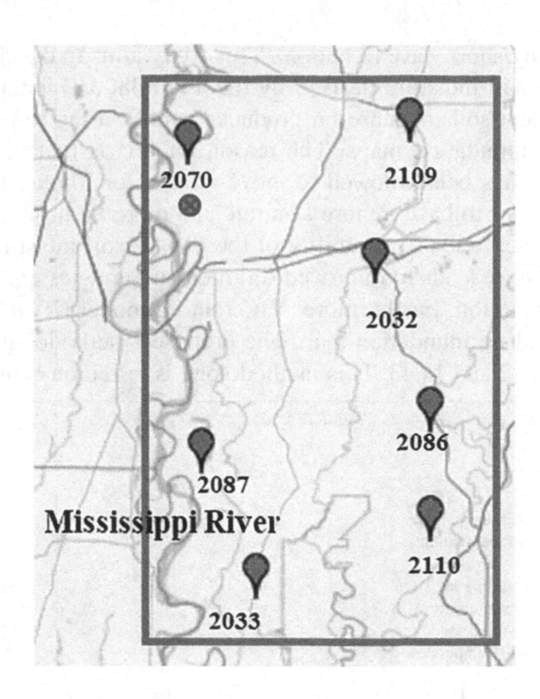

FIGURE 14.43 Sampling stations of Mississippi River case study.

Mississippi State in The Vicinity of Mississippi River. The final product (hazard map) of this study can be applied in assessment and management of extreme events such as flash floods. The proposed methodology could be applied to other geographic setting with similar available data.

14.7.2.1 Soil Moisture Estimation Results

The VIC model was implemented in order to simulate soil moisture on an hourly time step. The model results are presented in different scenarios such as open loop (OL)—without DA—and with 10, 15, and 20 ensemble members. In this study, the OL and data- assimilated soil moisture simulations are compared to measurements from Soil Climate Analysis Network (SCAN) stations.

The grid cell size of the VIC simulations can vary, but the minimum grid size is 1km. VIC inputs are mainly consisting of atmospheric forcing, soil, and vegetation parameters. Also, each cell can be subdivided into sections representing specific land covers and vegetation types. In this study, the grid spacing was set to $0.01° \times 0.01°$ which approximately corresponds to $1 \times 1\ km^2$. In addition, each grid cell composed of two soil layers and a single canopy layer which controls precipitation's interception according to biosphere–atmosphere scheme developed by Wilson et al. (1987). This study considers two soil layers in order to model the portioning of precipitation into baseflow, surface runoff, and infiltration regarding the VIC curve. This model also controls the generation of baseflow according to a nonlinear recession curve. The first soil layer represents the top 5 cm of the soil depth, whereas the second layer's depth is 50 cm. In order to convert the hourly grid-based simulations of runoff and baseflow to simulations of streamflow, VIC was coupled with VIC routing model (RVIC).

For this study, soil properties such as soil moisture and temperature are collected from the SCAN via the International soil moisture network website (ismn.geo.tuwien.ac.at).

These differences are mainly originated from the type of vegetation (mainly woodlands) in the eastern parts of area and topographic complexities. After implementing assimilation in three different scenarios (with 10, 15, and 20 ensemble members), all soil moisture simulations (without assimilation or assimilated) are compared to the SCAN stations. Also, SCAN stations provide atmospheric forcing data, such as air temperature, vapor pressure, and wind speed. This comparison was performed on SCAN stations within the study area. The DA scenario with 15 ensemble members demonstrated an RMSE of 63.25% lower in comparison to the open-loop scenario. Implementing the EnKF with 10 and 20 members showed enhanced performances compared to OL simulation in terms of simulating soil moisture in all seven SCAN stations. However, Using EnKF with 15 ensemble members not only showed better performance in estimating soil moisture in every SCAN station but also showed lower average computed errors.

14.7.2.2 RVIC Model Results

After implementing the best assimilation scenario on VIC model and improving precision of soil moisture prediction, a routing model should be run in order to track the produced runoff within study area grids. The RVIC streamflow routing model is a modified version of the streamflow routing model typically used as a postprocessor with the VIC model. The routing model is a source-to-sink model that solves a linearized version of the Saint-Venant equations. In this study, the RVIC model implemented on each cell grid and the results of simulated inundation depths are shown and the inundation depth in the most intense recorded precipitations from April 5th at 6 PM to April 7th at 6 AM. Also, as shown in the figure, the open-loop simulations predicted deeper inundation depths and more damage while DA simulations showed relatively lower inundation depths in the area (25% lower on average). This is because in DA simulations inclusion of soil moisture errors to propagate in time intervals is prevented, and thus, the simulated runoff and baseflow is predicted with better accuracy. As a result, the open-loop simulation overestimates predicted soil moisture. Therefore, the soil moisture estimates are corrected by implementing the EnKF. In addition, river banks and their adjacent areas including Cahoma, Quitman, Tallahatchie, Grenada, Leflore, Sunflower, Washington, Humphreys, Sharkey, and some parts of Issaquena counties are more prone to flash flood or river flood events. These areas are in the Mississippi river flood plains and have lower altitude in comparison with the eastern counties in the study area.

Inundation depth simulations are performed in order to estimate the experienced damages to structures and their contents. The direct building losses are the costs to repair or replace the damage caused to the structure and contents. The economic losses associated with the structural and content damages based on the HAZUS model functions for agricultural and urban land covers.

Regarding all the inundation depths simulations including DA and open-loop scenario from April 1–30, a flood damage estimation is offered based on flood depth-damage functions.

As shown in figure, near-river banks experienced severe inundation depths. These regions are typically used as agricultural or urban areas and are more susceptible to the flood damage in comparison to the eastern parts such as Sunflower and Carroll counties. The total estimated flood damage for agricultural and urban sections in both types of simulations calculated based on HAZUS damage functions corresponding with different land uses are shown. In near-river banks experienced severe inundation depths. These regions, typically used as agricultural or urban areas, are more susceptible to the flood damage in comparison to the eastern parts such as Sunflower and Carrol counties. Applying DA scheme to the LSM reduced the propagated soil moisture error which leads to the lower amount of produced runoff and baseflow. Therefore, the estimated flood damage in DA scenario is more reliable and accurate than the open-loop simulation by 20%–30% (different time steps). Analyzing and estimating the flood damages in least possible time and in any desirable region is prominent in the field of flood damage management. In addition, categorizing different infrastructures and considering their value is important in designing flood-control systems and it can be utilized in flood management systems. As can be seen in Figure 14.44, the assimilated model predicted lower soil moistures and, hence, lower flood volumes.

14.8 CONCLUDING REMARKS

The significance of visualization for information representation is presented in this chapter. Its functionality to serve as a storage mechanism, a processing and research instrument, and a communication tool is increasing with the past of advances in optic and imagining technologies. Visualization

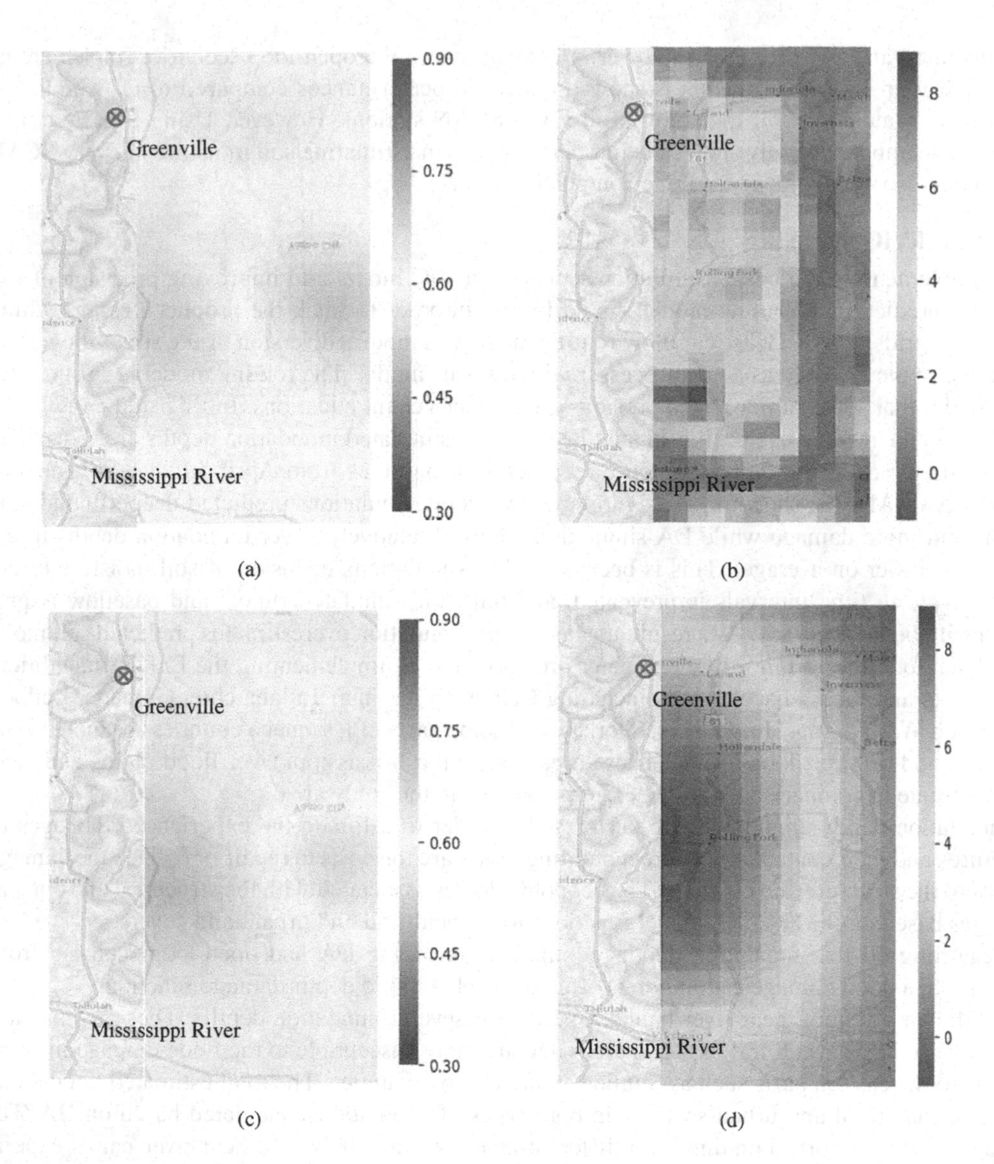

FIGURE 14.44 Soil moisture at 3 am and flood inundation at 6 am on April 5, 2019 with (a and b) and without (c and d) data assimilation.

technology is capable of transferring information into a simple image or animation. EV is one of the hottest areas that many imaging technologies have been realized and utilized with many more potential for ground-breaking new developments.

It can readily document many types of site data, including the ground surface, boring lithology and geologic strata, well construction information, groundwater conditions, and chemical concentrations of soil, water, and vapor. The generated image is typically a compilation of hundreds of pages of information from large and cumbersome reports. As a result, visualization acts as a data management tool that collates, organizes, and displays large volumes of information.

It has special application in disaster management and flood inundation that can be utilized by a variety of users with low to very high technical capabilities. The water movement and accumulation and water infrastructures need to be continuously monitored. With the materials presented in this chapter ranging from sensors, to pattern recognition tools and techniques, to satellite technology and data collections almost all in imaging forms, to development of citizen science applications, we have realized many immerging opportunities. The future of water recourses assessment, protection, water hazard prevention, and effective visual communication lie on the advancement of EV and how we are prepared to utilize it.

PROBLEMS

1. In case study 1, answer the following questions:

   a. According to the case study 1, which clustering approach was utilized to classify pixels of each photo into two clusters and separate the soil image from its background? Express how position of clusters was determined?

2. In case study 2, answer the following questions:

   a. What are the initial inputs in a land-surface model like VIC? How is this information obtained?
   b. How can the VIC software be installed? What is the prerequirement software for this package? The information is available at https://vic.readthedocs.io/en/master/ website.
   c. Provide information about cell resolution (also, explain what is the appropriate cell resolution for this type of modeling. Hint: different resolutions are used for different purposes).
   d. What are the inputs and outputs of VIC's model?

3. How are digital images represented by the RGB model?
4. Redo Example 14.5 and consider the following points:

$$x_1 = 1.5,\ x_2 = 2.5,\ x_3 = 3,\ x_4 = 0.5 \text{ and } x_5 = 7,$$

5. Suppose that there is an LSM that predicts soil moisture with ($\sigma_x^2 = 2$) and a measurement system with ($\sigma_y^2 = 1$) if the model's output and measurement is 0.24 and 0.27. Calculate the Kalman prediction in the next step.
6. According to the DEM error section of ArcMap toolbox in the geostatistical analyst extension how can we select random point with different weights? Investigate what algorithm this module uses.
7. Suppose that our data set is two-dimensional with two variables x, y, and that the eigenvectors and eigenvalues of the covariance matrix are as follows:

$$\vartheta_1 = \begin{bmatrix} 0.6778 \\ 0.7352 \end{bmatrix}, \quad \lambda_1 = 1.284$$

$$\vartheta_1 = \begin{bmatrix} -0.7352 \\ 0.6778 \end{bmatrix}, \quad \lambda_1 = 0.049$$

   Recast the data along the PC axes.
8. One of the biggest limitations of kriging is its main assumptions that the target variable is stationary. What procedures you can suggest to solve this problem and show nonstationary? (See Karamouz and Fereshtehpour's paper (2019) for guidance.)
9. Use the procedure in Section 14.2.4.1 and download and record four soil moisture readings for May 1, 2021, or the closest date that data is available, in the State of Colorado. Select your own 36-km pixels (cells) and show the coordination.

REFERENCES

Abdi, H. and Williams, L.J. (2010). Principal component analysis. *Wiley Interdisciplinary Reviews: Computational Statistics*, 2(4), 433–459.

Adibfar, A. (2021). Mapping knowledge of soil moisture driven method for flood vulnerability assessment. MS Thesis, University of Tehran, Tehran, IR.

Aerts, J.C., Goodchild, M.F., and Heuvelink, G.B. (2003). Accounting for spatial uncertainty in optimization with spatial decision support systems. *Transactions in GIS*, 7(2), 211–230.

Ailamaki, A., Faloutos, C., Fischbeck, P.S., Small, M.J., and VanBriesen, J. (2003). An environmental sensor network to determine drinking water quality and security. *ACM Sigmod Record,* 32(4), 47–52.

Ajisegiri, B., Andres, L.A., Bhatt, S., Dasgupta, B., Echenique, J.A., Gething, P.W., and Joseph, G. (2019). Geo-spatial modeling of access to water and sanitation in Nigeria. *Journal of Water, Sanitation and Hygiene for Development,* 9(2), 258–280.

Aloui, F. and Dincer, I. (eds.) (2018). *Exergy for a Better Environment and Improved Sustainability 2: Applications.* Springer, Berlin, Germany. ISBN: 978-3-319-62575-1.

Amitrano, D., Di Martino, G., Iodice, A., Riccio, D., and Ruello, G. (2014). A new framework for SAR multitemporal data RGB representation: Rationale and products. *IEEE Transactions on Geoscience and Remote Sensing*, 53(1), 117–133.

Andres, L., Boateng, K., Borja-Vega, C., and Thomas, E. (2018). A review of in-situ and remote sensing technologies to monitor water and sanitation interventions. *Water,* 10(6), 756.

Bates, P.D. and De Roo, A.P.J. (2000). A simple raster-based model for flood inundation simulation. *Journal of Hydrology*, 236(1–2), 54–77.

Benke, K.K., Pettit, C.J., and Lowell, K.E. (2011). Visualisation of spatial uncertainty in hydrological modelling. Journal of *Spatial Science*, 56(1), 73–88.

Bishop, C.M. (2006). *Pattern Recognition and Machine Learning.* Springer, Berlin, Germany.

Bivand, R.S., Pebesma, E.J., Gomez-Rubio, V., and Pebesma, E.J. (2013). *Applied Spatial Data Analysis with R* (Vol. 2). Springer, New York.

Boser, B.E., Guyon, I.M., and Vapnik, V.N. (1992, July). A training algorithm for optimal margin classifiers. In *Proceedings of the fifth annual workshop on Computational learning theory* (pp. 144–152).

Bouveyron, C. and Brunet, C. (2012). Probabilistic Fisher discriminant analysis: A robust and flexible alternative to Fisher discriminant analysis. *Neurocomputing*, 90, 12–22.

Cammalleri, C. and Ciraolo, G. (2012). State and parameter update in a coupled energy/hydrologic balance model using ensemble Kalman filtering. *Journal of Hydrology*, 416–417, 171–181. doi: 10.1016/j.jhydrol.2011.11.049.

Camps-Valls, G., Tuia, D., Gómez-Chova, L., Jiménez, S., and Malo, J. (2011). Remote sensing image processing. *Synthesis Lectures on Image, Video, and Multimedia Processing*, 5(1), 1–192.

Canadian Institute for Climate Studies (CICS), https://climatechoices.ca/.

Cao, F., Lu, Y., Dong, S., and Li, X. (2020). Evaluation of natural support capacity of water resources using principal component analysis method: A case study of Fuyang district, China. *Applied Water Science*, 10(8), 1–8.

Carlisle, B.H. (2005). Modelling the spatial distribution of DEM error. *Transactions in GIS*, 9(4), 521–540.

Carter, J.R. (1992). The effect of data precision on the calculation of slope and aspect using gridded DEMs. *Cartographica*, 29, 22–34.

Chen, J. and Hill, A. (2007). Modeling urban flood hazard: just how much does dem resolution matter? *Proceedings of applied geography conferences*, 30, 372–379.

Chen, S. and Zhang, D. (2004). Robust image segmentation using FCM with spatial constraints based on new kernel-induced distance measure. IEEE Transactions on Systems, Man, and Cybernetics, Part B (Cybernetics), 34(4), 1907–1916.

Chen, C. and Yue, T. (2010), A method of DEM construction and related error analysis. *Computers & Geosciences*, 36(6), 717–725.

Clark, M.P., Rupp, D.E., Woods, R.A., Zheng, X., Ibbitt, R.P., Slater, A.G., et al. (2008). Hydrological data assimilation with the ensemble Kalman filter: Use of streamflow observations to update states in a distributed hydrological model. Advances in *Water Resources*, 31(10), 1309–1324.

Cook, A., Merwade, V. (2009), Effect of topographic data, geometric configuration and modeling approach on flood inundation mapping. *Journal of Hydrology*, 377, 131–142.

Cressie, N. (1993). Statistics for Spatial Data, rev. edn. Wiley, New York (900 pp.). (Original edition, 1991. Paperback edition in the Wiley Classics Library: Wiley, Hoboken, NJ, 2015).

de Almeida, T.I.R., Penatti, N.C., Ferreira, L.G., Arantes, A.E., and do Amaral, C.H. (2015). Principal component analysis applied to a time series of MODIS images: The spatio-temporal variability of the Pantanal wetland, Brazil. *Wetlands Ecology and Management*, 23(4), 737–748.

Duda, R.O., Hart, P.E., and Stork, D.G. (2000). *Pattern Classification* (2nd ed.). Wiley, London, UK.

Dzombak, D., Vidic, R., and Landis, A. (2012). Use of treated municipal wastewater as power plant cooling system makeup water: Tertiary treatment versus expanded chemical regimen for recirculating water quality management. Carnegie Mellon University, Pittsburgh, PA.

Environmental Systems Research Institute (ESRI). (2016), ArcGIS Pro Tool Reference Resample. http://pro.arcgis.com/en/pro-app/tool-reference/data-management/resample.html

Evensen, G. (2003). The ensemble Kalman filter: Theoretical formulation and practical implementation. *Ocean Dynamics*, 53(4), 343–367. doi: 10.1007/s10236-003-0036-9.

Evensen, G. (1994a). Inverse methods and data assimilation in nonlinear ocean models. Physica D, 77, 108–129.

Evensen, G. (1994). Sequential data assimilation with a nonlinear quasi-geostrophic model using Monte Carlo methods to forecast error statistics. *Journal of Geophysical Research: Oceans*, 99(C5), 10143–10162.

Faloutsos, C. and Vanbriesen, J. (2006). Sensor mining at work: Principles and a water quality case-study. Knowledge Discovery in Databases (KDD).

FAO. (2020). The State of World Fisheries and Aquaculture 2020. Sustainability in action. Rome. https://doi.org/10.4060/ca9229en

Fazel, N., Berndtsson, R., Uvo, C.B., Madani, K., and Kløve, B. (2018). Regionalization of precipitation characteristics in Iran's Lake Urmia basin. *Theoretical and Applied Climatology*, 132(1), 363–373.

Fereshtehpour, M. and Karamouz, M. (2018). DEM resolution effects on coastal flood vulnerability assessment: Deterministic and probabilistic approach. *Water Resources Research*, 54(7), 4965–4982.

FEWS NET, the Famine Early Warning Systems Network, https://fews.net/.

Fukunaga, K. (1990). *Introduction to Statistical Pattern Recognition*. Elsevier Inc, Amsterdam, Netherlands. doi: 10.1016/C2009-0-27872-X.

Gallant, J.C. and Hutchinson, M.F. (1997). Scale dependence in terrain analysis. Mathematics and Computers in Simulation, 43(3–6), 313–321.

Gandin, L.S. (1963). *Objective Analysis of Meteorological Field*. Gidrometeorologicheskoe Izdate'stvo, Leningrad, 286 pp.

Gao, H., Tang, Q., Shi, X., Zhu, C., Bohn, T., Su, F., and Wood, E. (2010). Water budget record from Variable Infiltration Capacity (VIC) model.

García-Peñalvo, F.J. (2016). The WYRED project: A technological platform for a generative research and dialogue about youth perspectives and interests in digital society.

Grainger, J., and Dufau, S. (2012). The front-end of visual word recognition. In J.S. Adelman (ed.) *Visual Word Recognition: Models and Methods, Orthography and Phonology* (pp. 159–184). Psychology Press, New York, NY.

Guyon, I., Boser, B., and Vapnik, V. (1993). Automatic capacity tuning of very large VC-dimension classifiers. In Stephen Jose Hanson, Jack D. Cowan, and C. Lee Giles (eds.) *Advances in Neural Information Processing Systems* (pp. 147–155). Morgan Kaufmann Publishers Inc., San Francisco, CA.

Haak, L. and Pagilla, K. (2020). The water-economy nexus: a composite index approach to evaluate urban water vulnerability. Water Resources Management, 34(1), 409–423.

Hamman, J.J., Nijssen, B., Bohn, T.J., Gergel, D.R. and Mao, Y. (2018). The Variable Infiltration Capacity model version 5 (VIC-5): Infrastructure improvements for new applications and reproducibility. Geoscientific Model Development, 11(8), 3481–3496.

Harley, J., Donoughue, N., States, J., Ying, Y., Garrett, Jr., J.H., Jin, Y., Moura, J.M.F., Oppenheim, I.J., and Soibelman, L. (2009). Focusing of ultrasonic waves in cylindrical shells using time reversal. Proc., 7th International Workshop on Structural Health Monitoring, Stanford, CA.

Hastie, T., Tibshirani, R., and Friedman, J. (2009). Unsupervised learning. In The *Elements* of *Statistical Learning* (pp. 485–585). Springer, New York.

He, M., Hogue, T.S., Margulis, S.A., and Franz, K.J. (2012). An integrated uncertainty and ensemble -based data assimilation approach for improved operational streamflow predictions. *Hydrology and Earth System Sciences*, 16(3), 815.

Hengl, T., Toomanian, N., Reuter, H.I., and Malakouti, M.J. (2007). Methods to interpolate soil categorical variables from profile observations: Lessons from Iran. *Geoderma*, 140(4), 417–427.

Hengl, T., Heuvelink, G.B.M., and Van Loon, E.E. (2010). On the uncertainty of stream networks derived from elevation data: The error propagation approach. *Hydrology and Earth System Sciences*, 14(7), 1153–1165.

Holmes, K.W., Chadwick, O.A., and Kyriakidis, P.C. (2000). Error in a USGS 30-meter digital elevation model and its impact on terrain modeling. *Journal of Hydrology*, 233(1–4), 154–173.

Houser, P.R., De Lannoy, G.J., and Walker, J.P. (2005). Hydrologic data assimilation. In: A. Aswathanarayana, ed. Advances in Water Science Methodologies. AA Balkema, The Netherlands, p. 23.

Hu, Y.P. and Tsay, R.S. (2014). Principal volatility component analysis. *Journal of Business & Economic Statistics*, 32(2), 153–164.

Hunter, G.J., and Goodchild, M.F. (1997), Modeling the uncertainty of slope and aspect estimates derived from spatial databases. *Geographical Analysis*, 29(1), 35–49.

Hunter, G.J., and Goodchild, M.F. (1995), Dealing with error in a spatial database: A simple case study. *Photogrammetric Engineering and Remote Sensing*, 61(5), 529–537.

Isovitsch, S.L. and van Briesen, J.M. (2008). Sensor placement and optimization criteria dependencies in a water distribution system. *Journal of Water Resources Planning and Management*, 134(2), 186–196.

Jolliffe, I.T. (2002). Principal components in regression analysis. *Principal Component Analysis*, 167–198. doi:10.1007/0-387-22440-8_8.

Kanoun, O. and Trankler, H.R. (2004). Sensor technology advances and future trends. *IEEE Transactions on Instrumentation and Measurement*, 53(6), 1497–1501.

Karamouz, M. and Ebrahimi, E.S. (2022). A system dynamics-based evaluation of water resources carrying capacity. Proceedings of 2022 EWRI Congress, Atlanta, GA, June 5–8.

Karamouz, M. and Farzaneh, H. (2020). Margin of safety based flood reliability evaluation of wastewater treatment plants: Part 2-quantification of reliability attributes. *Water Resources Management, Water Resources Management*, 34(6), 2043–2059.

Karamouz, M. and Fereshtehpour, M. (2019). Modeling DEM errors in coastal flood inundation and damages: A spatial nonstationary approach. *Water Resources Research*, 55(8), 6606–6624.

Karamouz, M. and Fooladi Mahani, F. (2021). DEM uncertainty based coastal flood inundation modeling considering water quality impacts. Water Resource Management, 35(10.3), 3083–3103.

Karamouz, M., Nazif, S., Fallahi, M., and Imen, S. (2009). Rainfall prediction considering the climate signals using hybrid KNN-ANN Model, ASCE, Bangkok, Thailand.

Karamouz, M., Taheri, M., Khalili, P., and Chen, X. (2019). Building infrastructure resilience in coastal flood risk management. *Journal of Water Resources Planning and Management*, 145(4), 04019004.

Karamouz, M., Ebrahimi, E., and Ghomlaghi, A. (2021). Soil moisture data using citizen science technology cross-validated by satellite data. *Journal of Hydroinformatics*, doi: 10.2166/hydro.2021,029.

Kheirkhah, A., Azadeh, A., Saberi, M., Azaron, A., and Shakouri, H. (2013). Improved estimation of electricity demand function by using of artificial neural network, principal component analysis and data envelopment analysis. *Computers & Industrial Engineering*, 64(1), 425–441.

Kokoszka, M. (2013). Reassessing vulnerability: Using a holistic approach and critical infrastructure to examine flooding, Master of Science Project, Polytechnic Institute of NYU, Brooklyn, NY, p. 44.

Krause, A., Singh, A., and Guestrin, C. (2008). Near-optimal sensor placements in Gaussian processes: Theory, efficient algorithms and empirical studies. *Journal of Machine Learning Research*, 9(2), 235–284.

Leon, J.X., Heuvelink, G.B., and Phinn, S.R. (2014). Incorporating DEM uncertainty in coastal inundation mapping. *PLoS ONE*, 9(9), e108727.

Li, J. and Wong, D.W. (2010). Effects of DEM sources on hydrologic applications. Computers, Environment and *Urban Systems*, 34(3), 251–261.

Liang, X., Lettenmaier, D.P., Wood, E.F., and Burges, S.J. (1994). A simple hydrologically based model of land surface water and energy fluxes for general circulation models. Journal of Geophysical Research: Atmospheres, 99(D7), 14415–14428.

Liang, X., Wood, E.F., and Lettenmaier, D.P. (1999). Modeling ground heat flux in land surface parameterization schemes. *Journal of Geophysical Research: Atmospheres*, 104(D8), 9581–9600. doi: 10.1029/98JD02307.

Lievens, H., Tomer, S.K., Al Bitar, A., De Lannoy, G.J., Drusch, M., Dumedah, G. et al. (2015). SMOS soil moisture assimilation for improved hydrologic simulation in the Murray Darling Basin, Australia. Remote Sensing of Environment, 168, 146–162.

Ling, M. and Chen, J. (2014). Environmental visualization: Applications to site characterization, remedial programs, and litigation support. *Environmental Earth Sciences*, 72(10), 3839–3846.

Liu, Y. and Gupta, H.V. (2007). Uncertainty in hydrologic modeling: Toward an integrated data assimilation framework. *Water Resources Research*, 43(7), W07401.

Liu, Y., Weerts, A.H., Clark, M., Hendricks Franssen, H.J., Kumar, S., Moradkhani, H., et al. (2012). Advancing data assimilation in operational hydrologic forecasting: Progresses, challenges, and emerging opportunities. Hydrology and Earth System Sciences, 16(10), 3863–3887.

Liu, X. (2008). Airborne LiDAR for DEM generation: Some critical issues. Progress in *Physical Geography*, 32(1), 31–49.
Lu, Y. and Zhang, F. (2019). Toward ensemble assimilation of hyperspectral satellite observations with data compression and dimension reduction using principal component analysis. Monthly Weather Review, 147(10), 3505–3518.
Matheron, G. (1962). Precision of exploring a stratified formation by boreholes with rigid spacing—application to a bauxite deposit. In Mining Research (pp. 407–422), Pergamon.
McLaughlin, D. (1995). Recent developments in hydrologic data assimilation. Reviews of Geophysics, 33(S2), 977–984.
Moradkhani, H, Sorooshian, S., Gupta, H.V., and Houser, P.R. (2005). Dual state-parameter estimation of hydrological models using ensemble Kalman filter. *Advances in Water Resources*, 28(2), 135–147. doi: 10.1016/j.advwatres.2004.09.002.
Mukherjee, S., Garg, R.D., and Mukherjee, S. (2011). Effect of Systematic error on DEM and its derived attributes: a case study on Dehradun area using Cartosat-1 stereo data. *Indian Journal of Landscape System and Ecological Studies*, 34(1), 45–58.
Mukherjee, S., Joshi, P.K., Mukherjee, S., Ghosh, A., Garg, R.D., and Mukhopadhyay, A. (2013). Evaluation of vertical accuracy of open source Digital Elevation Model (DEM). *International Journal of Applied Earth Observation and Geoinformation*, 21, 205–217.
NOAA. (1984). Technical report NWS 48 – SLOSH, National Weather Service, Silver Spring, MD.
NYC Emergency Management. (2012). https://maps.nyc.gov/hurricane/.
Oksanen, J. (2006). *Digital Elevation Model Error in Terrain Analysis*, University of Helsinki, Helsinki, Finland, Ph.D., 51 p.
Omer, C.R., Nelson, E.J., and Zundel, A.K. (2003). Impact of varied data resolution on hydraulic modeling and floodplain delineation. *Journal of the American Water Resources Association*, 39(2): 467–475. doi:10.1111/j.1752- 1688.2003.tb04399.x.
Papadimitriou, S., Sun, J., and Faloutsos, C. (2005). Streaming pattern discovery in multiple time-series. Vancouver, British Columbia, Canada.
Parente, A. and Sutherland, J.C. (2013). Principal component analysis of turbulent combustion data: Data preprocessing and manifold sensitivity. *Combustion and Flame*, 160(2), 340–350.
Pauwels, V.R. and De Lannoy, G.J. (2009). Ensemble-based assimilation of discharge into rainfall-runoff models: A comparison of approaches to mapping observational information to state space. Water *Resources Research*, 45(8).
Pebesma, E.J. (2004). Multivariable geostatistics in S: The Gstat package. *Computers & Geosciences*, 30, 683–691.
Polidori, L., Chorowicz, J., and Guillande, R. (1991). Description of terrain as a fractal surface, and application to digital elevation model quality assessment. *Photogrammetric Engineering & Remote Sensing*, 57, 1329–1332.
Raber, G.T., Jensen, J.R., Hodgson, M.E., Tullis, J.A., Davis, B.A., and Berglund, J. (2007). Impact of LiDAR nominal post-spacing on DEM accuracy and flood zone delineation. *Photogrammetric Engineering & Remote Sensing*, 73(7), 793–804. doi: 10.14358/PERS.73.7.793.
Rahman, A.S. and Rahman, A. (2020). Application of principal component analysis and cluster analysis in regional flood frequency analysis: A case study in New South Wales, Australia. *Water*, 12(3), 781.
RBD, Rebuild by Design. (2014). YC mayor's office of recovery and resiliency, New York City economic development corporation. http://www.rebuildbydesign.org/our-work/all-proposals/winning-projects/hunts-point-lifelines.
Reich, B.J. and Ghosh, S.K. (2019). *Bayesian Statistical Methods*. Chapman & Hall/CRC Texts in Statistical Science Series, London, UK.
Robinson, A.R. and Lermusiaux, P.F. (2000). Overview of data assimilation. *Harvard Reports in Physical/Interdisciplinary Ocean Science*, 62(1), 1–13.
Rodriguez, E., Morris, C.S., and Belz, J.E. (2006). A Global assessment of the SRTM performance. *Photogrammetric Engineering & Remote Sensing*, 72, 249–260.
Sobrino, J.A., Jimenez-Munoz, J.C., and Paolini, L. (2004). Land surface temperature retrieval from Landsat TM 5. *Remote Sensing of Environment*, 90, 434–440.
Sun, J., Papadimitriou, S., and Faloutsos, C. (2006). Distributed pattern discovery in multiple streams. Proc., Pacific-Asia Conference on Knowledge Discovery and Data Mining (PAKDD), Singapore.
Santo, R.D.E. (2012). Principal component analysis applied to digital image compression. *Einstein (São Paulo)*, 10(2), 135–139.
Serneels, S. and Verdonck, T. (2008). Principal component analysis for data containing outliers and missing elements. *Computational Statistics & Data Analysis*, 52(3), 1712–1727.

Shearer, J.W. (1990). *The Accuracy of Digital Terrain Models. Terrain Modeling in Surveying and Engineering*. Whittles Publishing Services, Caithness, 315–336.

Sun, W. and Sun, J. (2017). Prediction of carbon dioxide emissions based on principal component analysis with regularized extreme learning machine: The case of China. *Environmental Engineering Research*, 22(3), 302–311.

Tarboton, D.G., Idaszak, R., Horsburgh, J.S., Ames, D.P., Goodall, J.L., Couch, A., Hooper, R., P., Dash, P.K., Stealey, M., Yi, H., Bandaragoda, C., and Castronova, A.M. (2017). The HydroShare collaborative repository for the hydrology community. AGU Fall Meeting Abstracts, IN12B-02.

The Moderate Resolution Imaging Spectroradiometer (MODIS), https://modis.gsfc.nasa.gov/.

The Multi-angle Imaging SpectroRadiometer, https://terra.nasa.gov/about/terra-instruments/misr.

Theodoridis, Y. (2003). Ten benchmark database queries for location-based services. The Computer Journal, 46(6), 713–725.

Triglav-Cekada, M., Crosilla, F., and Kosmatin-Fras, M. (2009), A simplified analytical model for a-priori LiDAR point-positioning error estimation and a review of LiDAR error sources. *Photogrammetric Engineering & Remote Sensing*, 75, 1425–1439.

USACE (U.S. Army Corps of Engineers). (1995). HEC–DSS, User's Guide and Utility Program Manual. Hydrologic Engineering Center, CPD-45, Davis, CA.

USEPA (U.S. Environmental Protection Agency). (2009a). Risk assessment guidance for superfund volume I human health evaluation manual. Available online at https://www.epa.gov/sites/default/files/2015-09/documents/partf_200901_final.pdf. Accessed on 12 November 2009.

USEPA (U.S. Environmental Protection Agency). (2009b). Exposure factors handbook. Available online at https://ofmpub.epa.gov/eims/eimscomm.getfile?p_download_id=492239. Accessed on 15 November 2009.

Vapnik, V., Golowich, S.E., and Smola, A. (1997). Support vector method for function approximation, regression estimation, and signal processing. Advances in *Neural Information Processing Systems*, 9, 281–287.

Vapnik, V. and Chervonenkis, A. (1974). *Theory of Pattern Recognition*. Nauka, Moscow.

Vapnik, V. (1995). *The Nature of Statistical Learning Theory*. Springer, New York.

Vapnik, V. (1982). *Estimation of Dependences Based on Empirical Data*. Springer, Berlin.

Van de Sande, B., Lansen, J., and Hoyng, C. (2012). Sensitivity of coastal flood risk assessments to digital elevation models. *Water*, 4(3), 568–579.

Vaze, J., Teng, J., and Spencer, G. (2010). Impact of DEM accuracy and resolution on topographic indices. *Environmental Modelling & Software*, 25(10), 1086–1098.

Vrugt, J.A., Gupta, H.V., Nualláin, B., and Bouten, W. (2006). Real-time data assimilation for operational ensemble streamflow forecasting. *Journal of Hydrometeorology*, 7(3), 548–565.

Wästberg, B., Billger, M., and Adelfio, M. (2020). A user-based look at visualization tools for environmental data and suggestions for improvement: An inventory among city planners in Gothenburg. *Sustainability*, 12(7), 2882.

Werner, M. (2001). Impact of grid size in GIS based flood extent mapping using a 1-D flow model. *Physics and Chemistry of the Earth, Part B:* Hydrology, Oceans and Atmosphere, 26, 517–522.

Wilson, M.F., Henderson-Sellers, A., Dickinson, R.E., and Kennedy, P.J. (1987). Sensitivity of the Biosphere-Atmosphere Transfer Scheme (BATS) to the inclusion of variable soil characteristics. *Journal of Climate & Applied Meteorology*, 26(3), 341–362. doi: 10.1175/1520-0450(1987)026<0341:SOTBTS>2.0.CO;2.

Wechsler, D. (2003). *Wechsler Intelligence Scale for Children–Fourth Edition: Technical and Interpretive Manual*. Psychological Corporation, San Antonio, TX.

Wechsler, S.P. (2007). Uncertainties associated with digital elevation models for hydrologic applications: A review. *Hydrology and Earth System Sciences*, 11(4), 1481–1500.

Wechsler, S.P., and Kroll, C.N. (2006). Quantifying DEM uncertainty and its effect on topographic parameters. *Photogrammetric Engineering & Remote Sensing*, 72(9), 1081–1090.

Weng, Q., Lu, D., and Schubring, J. (2004). Estimation of land surface temperature–vegetation abundance relationship for urban heat island studies. Remote sensing of Environment, 89(4), 467–483.

Wise, S. (1998). The effect of GIS interpolation errors on the use of digital elevation models in geomorphology. In S.N. Lane, K.S. Richards, and J.H. Chandler (eds.), *Landform Monitoring, Modelling and Analysis*, John Wiley and Sons, Chichester UK. 300 p.

World Bank. (2017). Nigeria WASH poverty diagnostic: Intermediate review meeting, World Bank Group, Washington, DC.

World Bank. (2018). Predicting deprivations in housing and basic services from space, World Bank Group, Washington, DC.

Wright, W.R. and Palkovits, R. (2012). Development of heterogeneous catalysts for the conversion of levulinic acid to γ-valerolactone. *ChemSusChem*, 5(9), 1657–1667. https://www.usgs.gov/media/images/gis-data-layers-visualization.

Xu, F. and Brambilla, G. (2008). Demonstration of a refractometric sensor based on optical microfiber coil resonator. *Applied Physics Letters*, 92(10), 101126.

Xu, Y. and Goodacre, R. (2018). On splitting training and validation set: A comparative study of cross-validation, bootstrap and systematic sampling for estimating the generalization performance of supervised learning. Journal of Analysis and Testing, 2(3), 249–262.

Yunus, A.P., Avtar, R., Kraines, S., Yamamuro, M., Lindberg, F., and Grimmond, C.S.B. (2016). Uncertainties in tidally adjusted estimates of sea level rise flooding (bathtub model) for the greater London. Remote Sensing, 8(5), 366.

Index

Note: **Bold** page numbers refer to tables and *Italic* page numbers refer to figures.

A

acceptable pressure range 333
acceptance probability 533
Accra, Ghana *344,* 344–345
activated carbon 378
activated sludge process 368–378, **369**, *369*
  aeration and mixing system 369
  bacterial kinetics *373,* 373–378, *374*
  biomass control 370
  bioreactor 369
  plant loading 370
  returned sludge 370
  sedimentation tank 370, 371
  sludge activity 370
  sludge settleability 370
  solids retention time 370
adaptive neuro fuzzy inference systems (ANFIS) 19
adaptive random sampling (ARS) 691
adaptive systems 16
Adibfar, A. 890
advanced first-order second-moment (AFOSM) method 648–649
advection 320
advection–dispersion transport 320
aeration 310, 366
aeration lagoons 367
aerobic treatment 366–367, *367*
Agenda21 378
agent-based modeling (ABM) 554–557, *556*
agricultural irrigation, water reuse 330
agricultural sector 543–544
Aharchay river basin 72, *72*
Ahmadi, A. 547
Ahmadi, B. 71
air stripping 366
Akaike, H. 159
Akaike Information Criterion (AIC) 159
Akan, A.O. 187
Alegre, H. 419
algorithms 20
Alipaz, S. 652
AM *see* asset management (AM)
American Society of Civil Engineers 451
American Water Works Association (AWWA) 314
Ampt, G.A. 106, 107
Amsterdam, Netherlands
  stormwater infrastructure
    improvement and future plans 273–274
    rainproof platform *274*
    recommendations 274
    system characteristics 273
  urban water supply infrastructures 342–344, *343*
  wastewater treatment 401
anaerobic baffled reactor (ABR) 379
anaerobic degradation 367
anaerobic digestion 367, *368*
anaerobic treatment 367–368, *368*
analytical hierarchy process (AHP) 456, 484
Anderson, T.L. 31
Ang, A.H.S. 598
ANNs *see* artificial neural networks (ANNs)
Annual Worth (AW) method 424
ant colony 533–534
anticipatory phase 676
aquaculture 331
aquatic ecosystems 96–97
aquifer storage and recovery (ASR) system 237, *238*
arch dams *293,* 293–294
architecture reaffirms 11
arid and semiarid areas 40, 41
Arnold, J.C. 493
artificial neural networks (ANNs) 19, 694
  architectures 494
  backpropagation algorithm 495, 496
  hard-limit transfer function 495
  linear neural networks 494
  linear transfer function 495
  log-sigmoid transfer function 495, *495*
  mean absolute error 496
  models 493–494
  multilayer perceptron network 496–500, **497–498**
  probabilistic neural network 694–696, **696**
  radial basis function 697–699, *697–699*
  root mean squared error 496
  sum squared error 496
  temporal neural networks 500
asset management (AM) 419
  attributes 430–431
  benefits 431, *431*
  condition assessment 440
  developing plans 442–443, **443**, *443*
  drivers 431–432
  governmental agencies 430
  infrastructure 443–445, *444*
  level of service 430, 432–433, **433**
  life cycle cost analysis 435–438, *436, 438*
  objectives 432
  program 430
  risk assessment *433,* 433–435, **434**
  software tools 441
  status and condition 432
  for stormwater and wastewater systems 451–458
  strategic 440
  sustainable service delivery *439,* 439–440
  tools and practices 440–441
  of urban water infrastructures 419
  water distribution system 446–451, **450**, *451*
  for water supply infrastructures 445–446, **446**, **447**
assets 428
Au, T. 421

autocorrelation function (ACF) 152–153, *162*
autoregressive (AR) models 154–157, **159**
autoregressive moving average (ARMA) model, time series 152–154, 158–162
  autocorrelation function 152–153, *162*
  characteristic behavior of **159**
  models considerations 160–162
  partial autocorrelation function 153–154, *162*
Au, T.P. 421
availability 649

B

backpropagation algorithm (BPA) 495, 496
backward elimination 841–842
balance sheet 428–429, **429**
Baltic Sea 401
Bandini, S. 555
Barrette Point Park 398
"baseline chemical" approach 602
basin area *(A)* 94
The Baxter Plant 338
Bayes decision theory 851
The Belmont Plant 338
Bender, M.J. 539
benefit–cost ratio method 425
best management practices (BMPs) 60, 93
  aquifer storage and recovery system 237, *238*
  bioretention basins 227
  bioretention swales 226–227, *227*
  constructed wetlands *231,* 231–233, *233*
  extended detention basin 233–235, **234**, *234,* **235**
  grass buffers 237
  infiltration systems 236, *236*
  ponds and lakes 236
  porous pavement 237, *238*
  retention pond 224–226, *225,* **226**
  sand filters 227–230, *228, 229*
  sediment basins 224, *224*
  swales and buffer strips 230–231
big data 44
BIG U 7
binomial distribution 580
biochemical oxygen demand (BOD) 331, 361, 365, 369, 390
biological treatment, of wastewater 364–366
bioreactor 364
bioretention basins 227
bioretention swales 226–227, *227*
biosphere 24
Bishop, C.M. 858
Bivand, R.S. 718
black box model 22–23, *23*
blackwater, separation of 328
Blaney–Criddle method 57
BMPs *see* best management practices (BMPs)
BOD *see* biochemical oxygen demand (BOD)
BOD_5 test 264
Bookan dam, curing *298,* 298–300, *299*
Booker, J.F. 549
Boussinesq equation 306
Bouveyron, C. 842
Brown, S.A. 244
Bruneau, M. 650
Brunet, C. 842
Budinger, T.F. 77
Building Information Tool (BIT) 606
bulk decay 321–322
buttress dam *293,* 294

C

calibration–validation process 21
capital investment 421
carbon dioxide (CO_2) 6, 43
carbon emissions 506
carbon:nitrogen:phosphorus (C:N:P) ratio of wastewater 366
Caribbean Sea 384
Caribbean wastewater treatment 384–386
Carroll, D.L. 527
cascade reservoirs 295, *296*
CATT *see* Centre for Advancement of Trenchless Technologies (CATT)
causal loop diagram (CLD) 540
causal systems 16
C-band radars 47
Centre for Advancement of Trenchless Technologies (CATT) 459
centrifugal pump 316
Chalkiadakis, G. 546
Chahar-Dangeh water distribution network, in Iran 436
channel flow 239
channel slope index 170
Chaves, H.M. 652
Check-Up Program for Small Systems (CUPSS) 459–460
chemical discharge 265
Chen, S. 40
Chen, Z. 10
Chervonenkis, A. 857
Chezy–Manning formula 325
chi-square test 253
chlorides 41
chlorination 284, 348, **391**
chlorine
  bulk decay 321–322
  sensors 824
  wall decay 321
Chow, V. 109
CHRS's Global IR data 49
classical risk estimation 584–585
clay 302
Clean Water Act (CWA) 29–31, 268, 361, 410
climate change 30, 260, 262–264
climatically appropriate for existing condition (CAFEC)
  evapotranspiration coefficient 61
  precipitation 61–62
climatic effects 5, 40–41
CMMS *see* Computerized Maintenance Management Systems (CMMS)
coagulation 309, 312
coastal floods 753, **754–755**, *756, 757*
  Hurricane Irene 760–761
  Hurricane Katrina 753–760
  Superstorm Sandy 762
  Tohoku earthquake and tsunami 760
Cobble, K. 539
codigestion 76
Colorado River 339
combination inlets *242,* 243

combined sewer overflow (CSO) 75, 79, 80, 248–249, 387
Commission for instruments and methods of observation (CIMO) 45
common probabilistic models
binomial distribution 580
exponential distribution 581
gamma distribution 582
log pearson type 3 distribution 582
normal distribution 580–581
complex systems 16
Computerized Maintenance Management Systems (CMMS) 441
cone of depression 303, *304*
confined aquifer 304, *305*
confined flow 304–306
conflict resolution 539
models 541–545, *545*
steps 539–540
systems approach 539–541, *540*
traditional approaches 539
consequence of failure (CoF) 435
constructed wetlands (CWs) *231*, 231–233, *233*, 380
consumer behavior theory (CBT) 397
continuously stirred-tank reactor (CSTR) 376
continuous simulation approach 256
continuous systems 16, *16*
conventional gravity sewers 382
conventional sanitation approach 359
conveyance tunnels 308–309
convolution integral 127–130, *128*
Conway, R.A. 602
cool coastal areas 41
cooperative game theory 546
cooperative water allocation 549–553
Copeland, C. 30
Copenhagen Climate Adaptation Plan 409
Copenhagen, Denmark
urban water supply infrastructures *341*, 341–342
water consumption 342
water supply plan 341–342
wastewater treatment 409
costs 421
Cressie, N. 883
criteria space 536, 538, 541
critical infrastructure 465–467, *466*
CSO *see* combined sewer overflow (CSO)
cultural and esthetic effects 6–7
cumulative distribution function (CDF) 574, 580, 611–612
CUPSS *see* Check-Up Program for Small Systems (CUPSS)
curb-opening inlets *242*, 243, 244, 278
Current Annual Real Losses (CARL) 333
curse of dimensionality 522
cut-set analysis 592
CWA *see* Clean Water Act (CWA)
CWs *see* constructed wetlands (CWs)
cyber dependencies 466
cylindrical rain gauges 46–47

D

damped systems 16
dams 290, 293, 445–446
area–volume–elevation curve of 294, *294*
curing 298–301
types of *293*, 293–294
Darcy's law 64, 106, 107, 302, 306
Darcy–Weisbach equation 193–196, 208
Darcy–Weisbach formula 325
data assimilation (DA) 862–863
for flood assessment 890–892, *891*, *892*
RVIC model 893
soil moisture estimation results 892–893
data collection platforms (DCPs) 642
data-driven mathematical models (DDM) 689
data preparation techniques 477
fuzzy inference system 488–489
fuzzy sets and parameter imprecision 486–488, *488*
multicriteria decision-making 484–486
regionalizing hydrologic data 477
cross-validation 483–484
fitting variogram 482–483, *483*
Kriging system 480–482
popular method 478
spatial prediction 478
theoretical semivariogram models 478–480, *479*, *480*
De Almeida, G.A. 708
De Bruijn, K.M. 650
defuzzification module 489
DeGarmo, E.P. 421, 424
De Lannoy, G.J. 865
Delaware River Watershed 403
Delaware Valley Early Warning System 338
Delft Dashboard (DDB) 713, *713*
Delft 3D-FLOW 711–713
Delft 3D-Wave module 714–715
Dellapenna, J.W. 30, 31
demand functions 427, *427*
density function estimation 852
Department of Environmental Protection (DEP) 79
depression storage 58–59, 99
design storm duration 257
detention time 363
"deterministic" approach 602
deterministic MCDM 485
devised systems 16
Dez reservoir basin, Iran 295, *296*
diffuse pollution 42
digital elevation map (DEM) 702
digital elevation model (DEM) 167, 177, 266–267, 617, 876–878, *877*, 884–885, *885*
digital image processing 889
digital surface model (DSM) 878
digital terrain model (DTM) 878
direct integration method 401, 596–597, *597*
direct runoff hydrograph (DRH) 135, **135**, 141, **141**
direct unit hydrograph (DUH) 132
disaster 25–26, *26*
indices 644
drought 652
reliability 644–649, *645*
resiliency 649–650
sustainability index 651–652
time-to-failure analysis 649
vulnerability 650–651
institutional roles 679
technology 678
training 678–679
disaster-resistant city (DRC) 635

discrete compounding interest factors, general cash flow elements using 421–422
discrete random variable 618
discrete systems 16, *16*
disinfection 310, 312
dispersion 320
dissolved-air flotation (DAF) systems 379
dissolved oxygen (DO) 264–265, 367
distributed hydrological models
GSSHA model 704–705, *705*
digital elevation model data collection 705
gridding 706
identify catchment area 706
infiltration 706–707
outputs 707, *707*
parameter estimation 706
roughness coefficient 707
LISFLOOD model 708–709
watershed modeling system 702–704, *704*
domestic/sanitary wastewater 40
domestic wastewater, untreated **361**
Dougherty, D.E. 533
downstream water supply 543
drainage
channel design 246–248
grass-lined channel 247–248
unlined channels 247, *247*
in urban watersheds 239
driveways 98
Driving force–Pressure–State–Impact–Response (DPSIR) approach 220, 604, 605
Driving force, State, and Response (DSR) model 221–222, *221*
dynamic strategy planning for sustainable urban land use management 222, *223*
drop manholes 245, *246*
drought *630*, 630–631, *631*, 641
drought severity index 62–63
Dublin Principles 378
Dufau, S. 892
dwelling unit (DU) 453
dynamic models 189
dynamic programming
curse of dimensionality 522
Markov chain 524–526, *526*
multireservoir system 521
recursive function 522
stochastic dynamic programming 524, *525*
successive approximation algorithms 523

E

earth observation technologies 826–828
earthquakes 632–633
Eastern Shore Regional GIS Cooperative (ESRGC) 604
Eckholm, E. 26
Eckstein, O. 780
ecoefficiency 325, 465
Ecological Sanitation (EcoSan) 378–379
economic consequences 435
economic leakage index (ELI) 333–335
economic optimum leakage reduction strategy 333, *335*
EDB *see* extended detention basin (EDB)
Eddy, M. 391
Egler, F.E. 22
electric power 465
electromagnetic inspection 458
electromagnetic meters 316
elevated tanks 307
ELI *see* economic leakage index (ELI)
elimination et choix traduisant la realité (ELECTRE) 484
embankment dams 293, *293*
energy head 180, *181*
energy production sector 544
energy retrofits, for buildings 83
Engineering and Environmental Services Department (EESD) 272
engineering economy 419
ensemble Kalman filter (EnKF) 864
Enterprise Asset Management (EAM) 441
entropy theory 618–620
environmental consequences 435
environmental economics 419
environmental management system (EMS) 411
environmental performance indicators (EPIs) 325, 463
Environmental Protection Agency (EPA) 319
environmental sector 543
environmental sensing 824–825, *825, 826*
remote sensing and earth observation technologies 826–828
soil moisture active passive 828–830
ubiquitous 825–826
environmental visualization (EV) 823
control and reduce flood risk 870–872, *872*
data assimilation 862–863
discrimination analysis
Fisher discriminant analysis 842–843
linear discriminant analysis 843–844
principal of component analysis 844–850, *846*
environmental effects and protection 872–874, *873, 874*
feature extraction and selection 841–842
image processing 859–860
RGB analysis *860*, 860–862, *861, 862*
intelligent visualization and image analysis systems 866–867
MAP resolution
common uses 878–879
DEM 876–878, *877*, 884–885, *885*
digital surface model 878
digital terrain model 878
effect of 879
Kriging interpolation 879–880
quality and accuracy 878
resampling 883–884
variogram modeling 880–883
mean sea-level pressure 874–876, *875, 876*
parameter estimation
maximum likelihood estimation 833–835
maximum posteriori 835–836
moments method 832–833
nonparametric density estimation 836–841, *837, 840, 841*
pattern recognition 830–832, *831*
sensed water infrastructure 823–824
supervised classification 851
Bayes decision theory 851
density function estimation 852

k-nearest neighbor estimation 853–856, *854, 855, 856*
min-mean distance classification 856
parametric density estimation 852–853
support vector machines 857
unsupervised classification 857, *858*
optimization-based clustering 858–859, *859*
sequential clustering 857–858
variable infiltration capacity model 863–866, *866*
water-related 867–870, *868*, **869**, *869, 870, 871*
environmental water demand 543
EPANET 321, 324–325, 715–717
equity 428
European Union Drinking Water Directive 98/83/EC 284
EV *see* environmental visualization (EV)
evaporation 37, 54–58, 103
evaluation
mass transfer method 55–56
pan evaporation 56
Penman equation 55
water budget method 55
measurement **87**
evaporation pan 56
evapotranspiration 54–58, 103
coefficient 61
estimation 67–68
measurement of 56–57
Thornthwaite method 57–58
Evenscn, G. 864
event-driven method 717–718
event tree 600–601
evolutionary algorithms (EAs)
ant colony 533–534
genetic algorithms 526–532, **528–530**, *531*
simulation annealing 532–533
Tabu search 534–536
exactions, and impact fee-funded systems 461
expected damage cost 781–787
exponential distribution 581
exponential family 511
exponential growth 502–503
exponential model 479
extended detention basin (EDB) 233–235
BMP characteristics 233, **235**
components 233, *234*
pond criteria **234**
extended detention shallow wetlands 231
external benefits 421
external costs 421
externalities 4
external rate of return (ERR) method 424

F

factor of safety (FS) 598
Falkland, A. 288
Famine Early Warning System Network (FEWS NET) 827
Farzaneh, H. 398, 486, 557
faster moving assets 428
fault tree analysis 389
fecal sludge 76
Federal Aviation Agency (FAA) equation 103
Federal Emergency Management Agency (FEMA) 267, 604, 631
Federal Water Pollution Control Act 410
Federal Water Pollution Control Act Amendments of 1972 *see* Clean Water Act (CWA)
feed-forward network *see* multilayer perceptron (MLP) network
Fereshtehpour, M. 501, 884
filtration 310, 312
financially sustainable management strategies 459
financial statements 427–429, **429**
finite memory 15
fire hydrant system **351**
Fisher discriminant analysis (FDA) 842–843
fixed asset 428
fixed costs 420
fixed grade nodes 315
flash floods 752
flocculation 309, 310, 312
flood(ing) 19
control 297
damage 178
assessment 724–729
control 543
inundation maps 617
management case 642
of urban drainage systems 260–264
Flood Damage Estimator (FDE) 782
Flood Information Tool (FIT) 606
flood mitigation performance index (FMPI) 727
floodplain 174
floodplain management 794
BMPs and flood control 795
conceptual design framework 800–803, *801, 802*
inland and coastal storms 795–798, *796, 797*
nonstationary-based framework 798–800, *799, 800*
urban drainage system 803–806, *804, 805*
insurance 807–809
structural and nonstructural measures 794–795
watershed flood early warning system 806–807, *808*
flood probability analysis 578–580, *579*
flood records 144–145, *146*
flood resiliency 751
analysis
partial frequency analysis 763–769, **768**
testing outliers 769–771, *770*
time series 762–763
coastal 753–762
control principles 773–774
damage 780
expected damage cost 781–787, *782*, **783**, **784**, *785*
stage–damage curve 780–781, *781*
evacuation zones 775–777, *776, 777*
hazards
climate change **774**, 774–775
sea level rise and storm surge 775
inland *753*, 752–753
recurrence interval 771
risk management *787*, 787–793, *789, 792, 793*
routing techniques *772*, 772–773
smart cities and livable cities 809
fighting climate change 811–812
infrastructure renewal 810–811

flood resiliency (*cont.*)
Mayor's office point 809
urban challenges *812,* 812–813
types and formations 752
water infrastructure performance 777–779, **779**
flow deflectors 245
flow splitters 245
flow and wave coupling 715
Fooladi, F. 692, 709, 718
Forgo, F. 542
forward selection 842
free chlorine, in water 312
frequency analysis 253–255
freshwater supply 286
friction factor 194–195
Froude number *(Fr)* 181, 246
Fujiwara, O. 691
full cooperation 546
Future Worth method 423–424
fuzzification module 488
fuzzy inference system (FIS) 19, 488–489, *489*
fuzzy sets 486–488

G

Galan, J.M. 555
game theory 546–553
gamma distribution 582
Gandin, L.S. 880
gas adsorption 366
Gaussian Semivariogram Model 478–479
Gaussian window function 696
generalized extreme value (GEV) distribution approach 798, 799
general meta-rationality (GMR) 549
genetic algorithms (GA) 526–532, *527*
geographical dependencies 466
geographic information system (GIS) 40, 266, 702
Ghadimi, S. 546
Ghana Water Company Limited (GWCL) 344, 345
Ghosh, S.K. 835
Global full-resolution IR composites 49
Global Positioning System (GPS) 877
global sensitivity analysis 398
global warming 43
urban heat islands 6
glucose 375
goal-seeking family 511–512
Goldberg, D.E. 526, 530
Gong, N. 267, 702
Grainger, J. 892
grass buffers 237
grass-lined channel design 247–248
grate inlet *242,* 243
gravity dam 293, *293*
gray box model 22–23, *23*
Gray, D.M. 134
gray infrastructure 78
Great Basin, digital snow maps for *52*
Green-Ampt model *106,* 106–108
Horton method 110–111
infiltration *106,* 111–112
ponding time 108–110
soil parameters for 108, **108**
greenhouse gas (GHG) 42, 81, 82
green infrastructures (GIs) 35–36, *79,* 79–80
green stormwater infrastructure (GSI) 269
Green, W.H. 106, 107
Gridded Surface-Subsurface Hydrological Analysis (GSSHA) hydrologic model 267, 704–705, 784
digital elevation model data collection 705
gridding 706
identify catchment area 706
infiltration 706–707
outputs 707
parameter estimation 706
roughness coefficient 707
gridded surface subsurface hydrologic analysis (GSSHA) software 390, 391
Grigg, N.S. 780
grit chamber 363
groundwater 63–64
recharge, water reuse 330
storage
confined flow 304–306
unconfined flow 306–307
well hydraulics 303–304
treatment plant 310, *310*
group rationality 547
Gumbel probability distribution 253–254, **254**
Guo, J.C.Y. 242
Gupta, R.S. 196

H

Hall, A.W. 28
Hamman, J.J. 865
Hamouda, M.A.A. 606
Hardin, G. 30
Hardy Cross method 208, 209, **209**, **211**
Hauger, M.B 258
Hazen–Williams formula 196–198, *197,* 325
HAZUS-MH 604, 606, 607
headwalls 250
Helweg, O.J. 780
Henze, M. 359, 360
heuristics 20
high-density polyethylene (HDPE) pipes 249
high-lift pumps 307
high-resolution picture transmissions (HRPTs) 51
Hipel, K.W. 549
Hirsch, F. 507
HOFOR 342
Hojjat-Ansari, A. 392
holistic attributes 430
Holland, G.J. 713
Holland, J. 526
Holling, C.S. 650
Holtan method 111
Holthuijsen, L.H. 711
Horton method 110–111
Horton, R.E. 58, 104, 110, 174
Houghtalen, R.J. 187
Housing and Urban Development (HUD) 220
Hudson River Project 7
Huffman, D. 539
human and institutional systems 25
Hunts Point Lifelines 7

Hunts Point plant 398–401, *399, 400*
Hurricane Irene 760–761
Hurricane Katrina 465, 753–760
Hurricane Sandy Rebuilding Task Force 7
Hwang, C.L. 484
hybrid drought index (HDI) 668–673, *669*, **670**, *671, 672*, **673**
Hybrid Sustainability Index (HSI) 70
hydraulic analysis, of open-channel flow 181–183
hydraulic-driven simulation models
  EPANET 715–717
  event-driven method 717–718
  QUALNET 717
hydraulic retention time (HRT) 367
hydraulic simulation, of water networks 324–325
hydrographs 5
hydrological risk (HR) 251–252
hydrological uncertainty 615
hydrologic cycle 18
  components 36
  defined 35
hydrologic data 573–574
  discrete and continuous random variables 574
  flood probability analysis 578–580, *579*
  moments of distribution 575–578, *575*, **577**, *577*
hydrologic effects 5, 41–42
hydrologic methods, of river routing 142–144
hydrologic modeling system (HEC-HMS) 18–19, 700–701, *701*
  empirical approach 21–22
  mathematical 19
  model resolution 22–23, *23*
  physical 19
  systems modeling 22
  theoretical approach 21–22
  tools and techniques 20–23, *21*
hydrologic system 21
hydrologic variability 17–18
hydrologic variables 18
hydrology 20, 611
  big data in 44
HydroNET 47
hydropower plant, element of 294, *295*
hyetographs 112

I

ICOLD *see* International Commission on Large Dams (ICOLD)
IHACRES 700
IIMM *see* International Infrastructure Management Manual (IIMM)
image analysis systems 866–867
impact population affluence technology (IPAT) identity 506
individual rationality 547
industrial sector 544
industrial wastewater 40
inference engine 489
infiltration 37, 59–60
  defined 105
  excess rainfall calculation 105–106
  stormwater 236, *236*
infiltration bed 236, *236*
infinite memory 15
influent, concentration 389–391
information technology 445
infrastructure asset management (IAM) 419
Infrastructure Leakage Index (ILI) 332, 333
infrastructure value index (IVI) 445
inland flooding 752–753, *753*
inlets
  efficiency of 243
  locations 243–244
  stormwater 240, 242
  types *242*, 243, *250*
in-line leak detection systems 458–459
input–output water balance 68–69, 71
input uncertainty 615
instantaneous unit hydrographs (IUH)
  convolution integral 127–130, *128*
  Laplace transformation model 138–142
  Nash model 133–137, **137**
  S-curve hydrograph 131
integrated attributes 431
integrated nonrevenue water 459
integrated water cycle management (IWCM) 28
Integrated water Resources Management (IWRM) 284
intensity–duration–frequency (IDF) curve 103, **113**, 113–114, 256
  for Kuala Lumpur *114*
  rainfall duration, selection of 113–114
Interagency Performance Evaluation Task Force (IPET) 757–759
Inter-American Development Bank (IDB) 827
interception storage 58–59, **59**
  excess rainfall calculation 104–105, *105*
interdependencies 25–26, *26*
  urban water cycle 4, *4*
interflow 39
Intergovernmental Panel on Climate Change (IPCC) 775
internal rate of return (IRR) method 424
International Commission on Large Dams (ICOLD) 445–446
International Hydrology Program (IHP 6) 93
International Infrastructure Management Manual (IIMM) 442
International Organization for Standardization (ISO) 410–411, *412*
International Water Association (IWA) 332, 442
interoperability 465
interstices 64
inundation probability map 617
Inventory Collection and Survey Tool (CAST) 606
inverse Laplace transformation 140
isohyetal method 50
IUH *see* instantaneous unit hydrographs (IUH)
Ivey, C.S. 83
IWA *see* International Water Association (IWA)

J

Jain, A.K. 196
James, L.D. 780
Jensen, M.B. 35
junction nodes 315
Juwana, I. 652

K

Kalan Power Plant 300
Kalman filter (KF) 863
Kapur, K.C. 597
Karageorgis, A.P. 604
Karamouz, M. 2, 24, 36, 51, 57, 62, 63, 64, 68, 70, 77, 103, 106, 110, 111, 124, 144, 145, 160, 161, 220, 239, 241, 248, 250, 262, 303, 307, 318, 319, 388, 392–394, 398, 437, 455, 475, 485, 486, 492, 500, 501, 511, 547, 557, 609, 620, 632, 646, 652, 661, 668, 690, 692, 702, 709, 718, 719, 724, 782, 783, 803, 870, 876, 884, 885
Kaya identity 506–507
Keffer, C. 325, 465
Kelly, R.A. 554
Kerachian, R. 528
Kerby-Hathaway formula 103
kernel density estimation 836
Ketabchi, H. 547
Kibler, D.F. 59, 104
Kinematic wave model **187**
Kirpich methods 103
Klijn, F. 650
K-means clustering algorithm 858–859
k-nearest neighbor (k-NN) estimation 853–856, *854, 855, 856*
knowledge base 488
Kresic, N. 29
Krige, D.G. 879
Kriging estimator 501
Kriging interpolation 879–880
Kriging system 50, 480–482
Krzysztofowicz, R. 615
kurtosis coefficient 578

L

Labi, S. 13, 477
Lake Bornsjön 345
Lake Malaren 345
lakes 65, 236
Lamberson, L.R. 597
land cover, on runoff volume 94–97
landfill leakages 5
land surface models (LSMs) 862
land use 3, 167
  planning 220–222
  runoff CNs for 117, **118–119**
  on runoff volume 94–97, *97*
Lan, T. 18
Laplace transformation model 138–142
Lar Dam, curing 300–301, *301*
large-scale centralized systems 328
LASAN 339
Latin America and Caribbean (LAC) 384–386, *385*
Lawrence, A.I. 35, 60
LCA *see* life cycle assessment (LCA)
leakage
  control 332
  detection 331
  management 331–335
    acceptable pressure range 333
    economic leakage index 333–335
Lee, W. 601
level of service (LOS) 430, 432–433
liabilities 428
Liang, X. 865
life cycle assessment (LCA) 8–9, 325–326
  urban water supply infrastructures 325–326
life cycle cost (LCC) 435–436, *436,* 437–438, *438,* 440–441
life cycle impact assessment (LCIA) 326
Light Detection and Ranging (LiDAR) 167, 177, 877
"Lindley's plan" 386
linear discriminant analysis (LDA) 843–844
linear growth 511
linear method 515–518, *518*
  simplex method 518–520, **519–520**
linear neural networks 494
linear quadratic estimation (LQE) *see* Kalman filter (KF)
LISFLOOD-FP 267
LISFLOOD model 708–709
Liu, L. 35
Liu, P.L. 712
load-resistance concept 595–596
load resistance method 389
Lodz combined sewerage system 386–387, **387**
Lodz's urban waterways 12
Loehman power 547
Loehman stability index 546
logical dependencies 466
logistic growth 503–504
log pearson type 3 distribution 582
Lohani, B.N. 360
London, England 272–273, *273*
LOS *see* level of service (LOS)
Los Angeles, California 269–271
Los Angeles Department of Water and Power (LADWP) 269
Loucks, D.P. 22, 77, 651, 652
lysimeter 56

M

Madani, K. 549
Mahmoodzadeh, D. 547
Mahmoudi, S. 719
manholes 244, *245*
Manning equation 239
Manning's coefficient 182, **182**, 184
Manning's equation 100–102
Manning's roughness coefficient 100, 175, **175–177**
MAP *see* maximum posteriori (MAP)
margin of safety (MOS) 401, 560, 598
Markov chain 524–526
Markov processes and Markov chains 690–691
Marryott, R.A. 533
Marsalek, J. 93
mass conservation 320
mass transfer method 55–56
matern semivariogram model 480
mathematical simulation techniques 689–690
  Markov processes and Markov chains 690–691
  Monte carlo technique/simulation 691–694, **693**
  stochastic simulation 690
mathematics of growth 501–502

demographic transition 505, *505*
environmental limits 505–506
Kaya identity 506–507, **507**
social limits 507–508
exponential growth *502,* 502–503
logistic growth 503–504
Matheron, G. 880
MATLAB® 19
MATLAB code 531
maximizing Nash product (MNP) 397
maximizing overall resilience (MOR) 397
maximum allowed velocity method 246
maximum likelihood method (MLE) 399, 833–835
maximum posteriori (MAP) 835–836
common uses 878–879
DEM 876–878, *877,* 884–885, *885*
digital surface model 878
digital terrain model 878
effect of 879
Kriging interpolation 879–880
quality and accuracy 878
resampling 883–884
variogram modeling 880–883, *881,* **882**, *882*
Mays, L.W. 317, 595, 596, 598, 647, 654
McCowan, J. 720
MCDM *see* multicriteria decision-making (MCDM)
McHarg, I. 11
McKee, R.H. 318
McPherson, C.B. 113
mean absolute error (MAE) 496
mean error (ME) 483
mean sea-level pressure (MSLP) 874–876, *875, 876*
mean square error (MSE) 20
mean square estimation error 485
mean value first-order second moment (MFOSM) method 648
"meat grinder" approach 22
membership function 486
Metropolis, N. 532
Metropolitan Water District (MWD) 339
MFOSM *see* mean value first-order second moment (MFOSM) method
microorganisms 364
Millennium Development Goals (MDGs) 317, 359
Millennium Development Task Force 359
Miller, D. 80
Milton, J.S. 493
minimized directly connected impervious areas (MDCIA) 258
minimum attractive rate of return (MARR) 424
minimum variance 481
min-mean distance classification 856
mixed liquor suspended solids (MLSS) 368, 369
mobile nonintrusive inspection (MNII) systems 336
model ecosystem approach 602
Moderate Resolution Imaging Spectroradiometer (MODIS) 827
moisture anomaly index *(Z)* 62
moisture loss coefficient 61
moments method 832–833
monetary indicators 463
Monod equation 374, *375*
Monod, J. 374
Monod model 371
Monte Carlo simulation (MCS) method 485, 500–501, 691–694, **693**
monthly average humidity 498
monthly average temperature 498
monthly precipitation 497
monthly water consumption (MCM) 497, 499
Moradkhani, H. 864
Movahhed, M. 455
moving average (MA) process 157, **159**
Multiangle Imaging Spectroradiometer (MISR) satellite 827
multicriteria decision-making (MCDM)
analytical hierarchy process 484
deterministic 485
ELECTRE 484
probabilistic method 485–486
PROMETHEE 484
TOPSIS 484
multijet meters 316
multilayer perceptron (MLP) network 496–500, **497–498**, *499, 500*
multiobjective optimization 536–538, *537, 538*
multi-objective water management 546–549
municipal wastewater 40, *360*
Muskingum-Cunge method 144
Muskingum method 142–144, 278
inflow and outflow hydrographs 189–191, *190*
urban channel routing 189–191

N

NASA's Earth Observing System 48
Nash, J.E. 133, 134
Nash model 133–137, **137**
Nash product 542
Nash theory 549
National Fire Protection Association (NFPA) 307
National Institute for Drinking Water Supply 343
National Oceanic and Atmospheric Administration (NOAA) 49, 51
high-resolution picture transmissions 51
Satellite Active Archive system 49
National Primary Drinking Water Regulations 29
National Secondary Drinking Water Standards 30
National Water and Wastewater Benchmarking Initiative (NWWBI) 442, 443
natural disasters 629
natural heat exchange 82–83
natural systems 16, 25
natural treatment systems 330–331
natural uncertainty 613
Nazari, S. 547
Nazif, S. 661
NDI *see* nondestructive testing (NDI)
near real-time risk assessment, of large diameter PCCP mains 458–459
negative-feedback mechanisms 15
Nelson, A.C. 460, 461
net future value (NFV) 424
net zero electricity consumption 82
New Meadowlands 7
Newton–Raphson method 208
New York City (NYC) 775–776
energy provision in 83–84, *84*

New York City Department of Environmental Protection (NYC-DEP) 455
New York City Department of Transportation (NYCDOT) 249
New York City's wastewater system 778–779, **779**
New York Harbor 248, 388
Nile, B.K. 18
nitrogen oxides (NO_x) 6, 43
non-cooperative game theory 546
non-cooperative stability 549–553
noncurrent asset *see* fixed asset
nondestructive evaluation (NDE) 335–336, **335–336**
nondestructive testing (NDI) 335–336, **335–336**
nonlinear method 520–521
nonparametric density estimation 836–841
Non-Revenue Water (NRW) 333, *334*
nonuniform flow 179
normal distribution 580–581
Normalized Differential Vegetation Index (NDVI) 827
normal ratio method 50
Noyola, A. 384
numerical modeling 710–715, *711*
NWWBI *see* National Water and Wastewater Benchmarking Initiative (NWWBI)
NYC-DEP *see* New York City Department of Environmental Protection (NYC-DEP)
NYC-DEP WWT system 392
NYC Green Infrastructure Plan 249

O

Office of Metropolitan Architecture (OMA) 870
Office of Watersheds (OOW) 269
Olyaei, M.A. 388, 394
open-channel flow
- channel, elements of *180*
- classification 181
- energy head 180, *181*
- hydraulic analysis of 181–183
- in watersheds 179–181

open-source models 711–715
operational uncertainty 615
optimal attributes 431
optimization techniques 514–515
- dynamic programming
  - curse of dimensionality 522
  - Markov chain 524–526
  - multireservoir system 521
  - recursive function 522
  - stochastic dynamic programming 524, *525*
  - successive approximation algorithms 523
- evolutionary algorithms
  - ant colony 533–534
  - genetic algorithms 526–532, **528–530**, *531*
  - simulation annealing 532–533
  - Tabu search 534–536
- linear method 515–518, *518*
  - simplex method 518–520, **519–520**
- multiobjective 536–538, *537, 538*
- nonlinear method 520–521

ordinary list square (OLS) method 389
ordinary rain gauges 47
oscillation 511
outliers 769–771
overland flow 100, 239
- effective Manning *n* values for **184**
- flow depth in 183–184
- GSSHA 267
- hydrographs *186*
- on impervious surfaces 184–187
- LISFLOOD-FP 267
- on pervious surfaces 187–189, *188*
- simulation 699–700
  - distributed hydrological models 702–709, *703, 704, 705, 707*
  - HBV 702
  - hydrologic modeling system 700–701, *701*
  - IHACRES 700
  - StormNET 701–702
- StormNET 267
- time of concentration for 184, 185

oxidation 361, 365
oxidation ponds *366,* 366–367
oxygen 39
ozonation 311

P

Palmer drought severity index (PDSI)
- agricultural drought indicators 60
- coefficients of water balance parameters 61
- drought severity index 62–63
- potential climatic values 61
- precipitation for climatically appropriate for existing condition 61–62

Palmer, W.C. 62
pan evaporation 56
pans 42
parallel pipe system *206,* 206–207, **207**, *207*
parallel reservoir 295–297, *296*
parametric density estimation 852–853
Paris, France 405–408
Paris Sanitation Section (SAP) 408
Parkinson, J. 218
partial autocorrelation functions (PACFs) 153–154, *162*
partial frequency analysis 763
- nonstationary case 765–766
- stationary analysis 763–765
- ungagged flood data **766**, 766–769, **767**

Parzen method 836–841
Parzen window 836, 837
path enumeration method *592,* 592–595, *594*
Pauwels, V.R. 865
pavement drainage inlets *242,* 242–244
Paydari Index (PI) 652
PDSI *see* Palmer drought severity index (PDSI)
peak discharge 114
peaks over threshold series 763
"Peaks over Threshold" (POT) series 763
Penman equation 55
perception 557
performance function W (X) 596
performance indicators (PIs) 69, 71, 325
Philadelphia, USA
- stormwater infrastructure
  - improvement and future plans 269
  - recommendations 269
  - sewer area and stormwater outfall *268*

system characteristics 267–268
urban water supply infrastructures 336–338, *337*
distribution system 338
supply system 337–338
wastewater treatment 403–405
Philadelphia Water Department (PWD) 269, 336–337
phosphate 378
physical–chemical separation 378
physical models (PBM) 689, 709–710
physical water system 476
Pimm, S.L. 650
pipe flow
energy equation of *192,* 192–193
head loss due to friction
cast iron pipe 196
compound pipeline *203,* 203–206, **204**, *204*
Darcy–Weisbach equation 193–196
Hazen–Williams equation 196–198, *197*
minor head loss 198–203, **199**, **200**
Moody diagram 194, 195, *195*
parallel or looping pipes *206,* 206–207, **207**, *207*
pipe networks 208–211, *210*
minor losses 192
PI-Plus 652
Planning for Sustainable Use Index (PSUI) 69, 71
Plastics Pipe Institute's Pipeline Analysis and Calculation Environment (PPI-PACE) 459
pocket wetlands 232
Poisson process 492
ponding/pluvial floods 752
ponds 236
pond/wetland systems 232
population 285, **285**
porous pavement 237, *238*
Porter-Cologne Act 269
Porto, R.L. 41
positive feedback mechanisms 15
potential moisture loss (PL) 61
potential recharge (PR) 61
potential runoff (PRO) 61
power generation 543
power model 479
PPI-PACE *see* Plastics Pipe Institute's Pipeline Analysis and Calculation Environment (PPI-PACE)
pre-chlorination 311
precipitation 37, 44–54
for climatically appropriate for existing condition 61–62
measurement
by satellite 48
by standard gauges 44–46, *46*
by weather radar 47–48
missing rainfall data, estimation of 49–51, **50**
PERSIANN system 48–49
rain gauges
cylindrical 46–47
cylindrical and ordinary 46
floating 45–46, *46*
ordinary 47
siphon 45, 47
tipping bucket 47
tipping bucket rain gauge recorder 47
snowmelt estimation 51–54, *52,* **52–54**
station average method 51
Precipitation Estimation from Remotely Sensed Information Using Artificial Neural Networks (PERSIANN) system 48–49
Predicted Environmental Concentrations (PEC) 265
Predicted No Effect Concentration (PNEC) 264
preference ranking organization method for enrichment evaluation (PROMETHEE) 484
Present Worth (PW) method 423
pressure-reducing valve (PRV) 315, 316
pressure sewer systems, using grinder pumps 383
Pressure, State, and Response (PSR) model 220
price elasticity, of water demand 323–325
Price, W.L. 691
principal of component analysis (PCA) 844–850
prism and wedge storage 142, *144*
private costs 421
probabilistic MCDM 485–486
probabilistic multicriteria decision-making 617–618
probabilistic neural network (PNN) 662, 694–696, **696**
probability
definition 573
theory 620–621, **621**
probability density function (PDF) 574
probability of failure (PoF) 434
production functions model 426
production possibility curve 426, *426*
Prommesberger, B. 258
proportional meters 316
public awareness 26–28
publicly owned treatment works (POTWs) 361
public–private partnership (PPP) 271
pumped pipeline systems 201, *201*

Q

quadratic discriminant analysis (QDA) 844
quality management system (QMS) 411
QUALNET 321, 717
quantized systems 16, *16*
The Queen Lane Plant 338
QUICKIN 720

R

radial basis function 697–699, *697–699*
radial basis function (RBF) 697–699
rainfall
abstractions 103
complex
hydrograph for 127–128, *129*
ordinates of 130, **130**
convective 5, 41
depth or intensity 18
design 255
selection 255–256
spatial and temporal distribution of 257
excess 103–112
Green–Ampt model *106,* 106–108, **108**
infiltration, estimation of 105–106
interception storage estimation 104–105, **105**
hyetograph **88**, *88,* 135, **135**, *163*
long periods of 5, 41
loss 99
measurement 112–114

rainfall (*cont.*)
intensity–duration–frequency curves **113**, 113–114, **114**
missing data, estimation of 49–51, **50**
probabilistic description of 251
in tropical regions 40–41
rainfall intensity 116
rainfall–runoff analysis 97–99
drainage area characteristics 98–99
rainfall losses 99
rainfall–runoff process 40
rainfall–runoff simulation 701
rain gauges **89**
cylindrical 46–47
cylindrical and ordinary 46
floating 45–46, *46*
ordinary 47
siphon 45, 47
solar-powered 45, *46*
tipping bucket 47
tipping bucket rain gauge recorder 47
rain shadow areas 41
rainwater collection system 288, *289*
rainwater harvesting 288
rapid depth filtration 312
rapid mixing 309
Rasoulkhani, K. 555
rational method 95
runoff volume, estimation of 114–117
raw sludge 363
RBD *see* Rebuild by Design (RBD)
readiness index (RI) 606
Rebuild by Design (RBD) 7, 220, 872
recarbonation 310
recharge coefficient 61
recycled water criteria 328, **328–329**
redundancy, in AM 433
regression analysis 319
rehabilitation phase 676
Reich, B.J. 835
reinforced concrete pipes (RCPs) 249
reliability
analysis 659
design by 572–573, *573*
difficulties 586
direct integration method 596–597, *597*
disaster indices 644–649, *645*
factor of safety 598
hazard rate 587
indices 647–648
load-resistance concept 595–596
margin of safety 598
parallel combinations 589, *589, 590*
path enumeration method *592,* 592–595, *594*
preparedness 673–674, *674*
probability 585
series combination of components 588, *588*
state enumeration method 590–592, *591*
wastewater treatment plants under coastal flooding 388–392, *390*
watershed 587
water supply indicators 599–600
environmental risk analysis 602–603
event tree 600–601, *601, 602, 603*
risk analysis methods and tools 600
Weibull distribution 586
remote sensing 826–828
reservoir(s) 290–292, *291,* 445–446
cascade 295, *296*
flood storage capacity of 297
operation 297
parallel 295–297, *296*
service 290
storage *295,* 307–308
storage capacity 65
reservoir routing method 259
final design 260
hydrological determination 259
initial configuration 259
initial sizing 259
outlets design 259
preliminary design 259–260
site selection 259
residence time 97
resiliency 2, 608–609
disaster indices 649–650
for flood risk management 790
Resource Conservation and Recovery Act 410
retention pond 224–226, **226**
basin shape 225
forebay 226
inlet 226
open water zone 226
outlet 226
overflow embankment 226
permanent pool 225–226
side slopes 226
site selection 225
structure 224, *225*
vegetation 226
water quality capture volume 224
return period 251, 582–584, *583*
design 256–257
"revealed societal preferences" approach 601
Reynolds number 181, 194
Richard equation 105, 106
Rinaldi, S.M. 466, 778
"riparian rights" 30–31
risk 571
climate change 572
components 571, *571*
hazards 571
vs. probability 585
vs. threat 585
vulnerability 571
risk-based attributes 431
risk of failure 571
river floods 752
river routing methods 142–144, *143*
Muskingum method 142–144
storage constant determination 144
river system components 562
Rogers, P. 28
rooftops 98
root mean squared error (RMSE) 20, 496
Rossman, L.A. 324
Ross, Sh.M. 493
roughness coefficient 100

routing techniques *772*, 772–773
Roy, B. 484
runoff 39
 concurrent 123
 curve number 95
 design, selection 255–256
 direct 123, **129**
 surface roughness 174–175
 volume
  estimation of 114–122
  hazard, vulnerability, and risk maps 95, *95*
  land use and cover impacts on 94–97, *97*
runoff coefficients 61, 95, 115, **115–116**
runoff curve number (CN) method 117–122, **118–120**

S

Saaty, T.L. 484
Sahara tethered acoustic sensor 458
St. Venant equations 142, 181–182, 189
Saint-Venant equations 239
Sakic Trogrlic, R. 220, 239, 241, 248, 250, 262
Salas, J.D. 161
salt crusts 42
sand filters 227–230
 basin geometry 228
 basin storage volume 228
 chambers 227
 inlet 228
 outlet 228
 site selection 228
 structure *228, 229*
 surcharge zone 227
 water quality capture volume, calculation of 229–230
Sandoval-Solis, S. 70, 652
Sandy 7
sanitary sewer system 276, **277**, *277*
Sarma, P.G.S. 145
satellite wastewater management 380–382
 decentralized systems 382
 extraction type 381
 infrastructure requirements 382
 interception type 381
 satellite water reclamation and reuse systems 380–381, *381*
 upstream type 382
Savenije, H.H.G. 28
scenario analysis 511
Schroeder, K. 248
Schumann, U. 40
Schuylkill Action Network 338
Schuylkill River Watershed 337, 338
SCOUT 47
screening, wastewater treatment 362
SCS *see* Soil Conservation Service (SCS)
S-curve 125
SD index (SDI) 325
sea-level pressure (SLP) 874–876, *875, 876*
seaward short-wave reproduction (SWAN) 714
secondary settling 309
sedimentation 309, 310, 312, 364
 basin 224, *224*, 363
semivariogram 478
sensing technology 823
sensitivity analysis 511
Sentinel-1 Program 827
septic tank effluent gravity (STEG) 382
septic tank effluent pumps (STEP) 382
sequential batch reactors (SBRs) 379
sequential Gaussian simulation (SGS) 500–501
sequential stability (SEQ) 549
service charge-funded systems 461
service reservoirs 290
SERVIR 827
sewage force 316
sewerage 74–75
sewershed 2
shallow wetlands 231
Shannon, C.E. 618
Shapley value 546, 553
S-hydrograph method 124–127, *126*
SIAAP 408
Simonovic, S.P. 476, 539
simple systems 16
simulation annealing (SA) 532–533
simulation systems 16
simulation techniques 489
 artificial neural networks
  architectures 494
  backpropagation algorithm 495, 496
  hard-limit transfer function 495
  linear neural networks 494
  linear transfer function 495
  log-sigmoid transfer function 495
  mean absolute error 496
  models 493–494
  multilayer perceptron network 496–500
  root mean squared error 496
  sum squared error 496
  temporal neural networks 500
 dynamics *508*, 508–514
 mathematics of growth 501–508
 Monte Carlo simulation 500–501
 probabilistic distribution 489–492, **492**
 stochastic process 492–493
sine hole effect model 479
siphon rain gauge 47
Skrzywan, S. 386
slope
 of channel 170–173, **173**, *173*
 of stream 174
 of watershed 170–171, **172**, **173**
SLOSH 867
slotted drain inlet *242*, 243
sludge 312
sludge processing 310
sludge volume index (SVI) 368
small-scale wastewater systems 330
smartball free-swimming acoustic sensor 458
smartball pipe wall assessment (PWA) tool 458
SMART grid 82, 84
smog 6
snow gauges 45
snow mapping and monitoring 51
snowmelt
 for catchment 53–54
 equations 52, **52**
 estimation 51–54, *52*, **52–54**

Snow Telemetry (SNOTEL) 44, *45*
Snyder, P. 31
social choice 426
social consequences 435
social costs 421
social responsibility (SR) 411, 634
Soil Climate Analysis Network (SCAN) 44, *45*
Soil Conservation Service (SCS) 95
  design rainfall 257
  runoff volume, estimation of 117–122
  unit hydrographs 122–123, *123*
soil moisture
  antecedent 119, **119**
  runoff CNs for 120, **120**
soil moisture active passive (SMAP) 44, **828**, 828–829, **829**, *829*, *830*
soil moisture deficit 60
soil moisture estimator 885–886
  clustering 886
  cross-validation 889–890, *890*
  data estimation platform 886, **887**
  digital image processing 889
  study area and data collection 886, *888*, 888–889
solid waste management 75–76
Source Water Assessments and Protection Plans 337
Spearman–Conley test 149–151, **150**, **151**
Spearman test 147–149, **148**, **149**
special assessment districts 461–463, **462**, *464*
specific retention 302
spherical model 479
SRI *see* system readiness index (SRI)
S-shaped family 511
stable systems 16
stage–damage curve 780–781
standards, on wastewater services 410–412, *411*, *412*
standpipes 307
Stankowski, S.J. 98
state enumeration method 590–592, *591*
station average method 51
Statistical Downscaling Model (SDSM) 803
Steel, R.K. 809
Sterman, J.D. 540
stochastic dynamic programming (SDP) 524, *525*
stochastic process 492–493
stochastic simulation 690
stochastic/statistical approach 602
Stockholm, Sweden
  stormwater infrastructure
    improvement and future plans 274–275
    recommendations 276
    sustainability factors *275*
    system characteristics 274
  urban water supply infrastructures 345–346
    supply system 345
    water distribution system 345–346
  wastewater treatment
    improvement and future plans 402–403
    system characteristics 401
storage constants, determination of 144
storage tanks 307–308
StormNET 267, 701–702
storm sewers 239
storm surge. 753
stormwater 218, 451–452
  cost 453–454, **454**
  gray infrastructure 219, *219*
  green infrastructure 219, *219*
  program funding 456–458
  scoring assets 452
  storage facilities 257–260
    conveyance or channel routing 258
    detention and retention 258, 259
    infiltration 258–259
    on-site runoff storage 257–258
    sizing of storage volumes 259–260
    types of 258–259
Stormwater Capture Master Plan (SCMP) 269
stormwater runoff 40, 42
stream channel 168
stream slopes 174
street gutters 239
  drainage capacity 241
structural integrity monitoring (SIM) 335, 347
subtropical areas 41
sum squared error (SSE) 20, 496
Superstorm Sandy 388, 392, 465, 762
support vector machines (SVM) 857
surface drainage channels
  design rainfall 255–256
  design return period 256–257
  design storm duration and depth 257
  probabilistic description of rainfall 251–255
  spatial and temporal distribution of design rainfall 257
surface reservoirs 307
surface runoff 42
surface sewer systems 244–246, *245*, *246*
Surface Water Treatment Rule (SWTR) 319
suspended growth 378
suspended solids (SS) 360
Sustainability Group Index (SGI) 69, 70, 609–610, *610*, 651–652
sustainability, Toronto's strategies for
  energy retrofits 83
  natural heat exchange 82–83
  net zero electricity consumption 82
  SMART grid 82
  transportation 83
sustainable attributes 431
sustainable development (SD) 286, 326–331
  adaptability 326
  awareness and promotion of sustainable behavior 327
  cost-effectiveness affordability 327
  demographics 327
  durability 326
  environmental protection 327
  flexibility 326
  goals 10
  noise, odor, and traffic 327
  personnel requirements 327
  recycling and reuse of resources 327
  reliability 326
  scale and degree of centralization 326–327
  social dimensions 327
  technical performance 326
  technology, selection of 327–331
sustainable development index (SDI) 463
sustainable sanitation 359–360
sustainable service delivery *439*, 439–440

Sustainable Urban Drainage Systems (SUDS) 409
sustainable urban land use management
Driving force, State, and Response framework 221–222, *221, 223*
dynamic strategy planning 220
identification of 220
Sustainable Water Management Improves Tomorrows Cities Health (SWITCH) 12
swales 239, 240
swales and buffer strips 230–231
Sweeney, L.B. 540
system 12–17
classifications 14–16
closed 14
defined 13
design problems 16, **17**
detection problems 16, **17**
general characteristics 13–15
identification problems 16, **17**
isolated 14
open 14, 15
properties 15–17
specifications of *17*
state of 14
stationary or steady state 14, *14*
synthetic problems 16, **17**
systematic attributes 430
system dynamics *508*, 508–509
modeling 509–511
time paths 511–514, *513–514*
system dynamics models 459, 689
systemic attributes 430
system readiness index (SRI) 661–663, **662**, *663*

T

Tabu search (TS) 534–536
tanks 56
tapped delay lines (TDLs) 694
tax-funded systems 460
Technical Committee (TC) 411
Tehran, Iran 288–290, *290, 291*
temperature, in urban areas 37
terrestrial ecosystems 96–97
Theim equation 306
theoretical semivariogram models 478–480, *479, 480*
Thiessen method 50
Thiessen polygon method *87*
Thomas model 66–68
Thompson, R.D. 41
Thornthwaite, C.W. 57
Thornthwaite method 57–58
T-hour unit hydrographs 125
tie-set analysis 593
time of concentration 99–100
time series analysis 151–162
Akaike Information Criterion 159
ARMA model 152–154, 158–162, **159**
autoregressive models 154–157, **159**
data preparation 151
goodness of fit of model 152
identification 152
model parameters, estimation of 152
moving average process 157, **159**
uncertainties, evaluation of 152
tipping bucket rain gauges 47, 112
Tohoku earthquake 760
TOPSIS 484
Toronto's strategies, for sustainability
energy retrofits 83
natural heat exchange 82–83
net zero electricity consumption 82
SMART grid 82
transportation 83
total actual renewable water resources (TARWR) 20
Total Coliform Rule (TCR) 319
total dissolved solids (TDS) 360
total suspended solids (TSS) 233
tournament selection 527, 528
tractive force approach 247
Transit City Light Rail 83
transpiration 37
transportation 76–77, 83
critical role of 810
trapezoidal open channel 246, *247*
travel time 99–103, 178
empirical formulas 103
sheet flow 102–103
time of concentration 99–100
time parameters 100–101, **101**
velocity method 101–102, **102**
trial-and-error procedure 195
TRMM level 2A (TRMM-2A 25) microwave imager data 49
TRMM Microwave Imager (TMI) 48
tropical areas 40
Tropic Rainfall Measuring Mission (TRMM) 48, 827
tsunami 760
Tucci, C.E.M. 41, 266, 773
tunnel boring machine (TBM) 309
turbine meters 316
turbulent flow 194
typical urban water dynamic model 563

U

ubiquitous environmental sensor technologies 825–826
UH *see* unit hydrographs (UH)
UHIs *see* urban heat islands (UHIs)
Ujang, Z. 359, 360
ultrasonic meters 316
unavailability 649
unavoidable annual real losses (UARL) 332, 333
unbiasedness 480–481
Uncertain Flood Inundation 616
uncertain soil moisture (SM) estimation 616
uncertainty *610–611*, 611–614, *613, 614*
analysis 653–656, **656**
of hydrological forecasting 615
implications 615, 653
measure 653
measurement 616–618
unconfined flow 306–307
ungagged coastal flooding 719–724, *721, 722, 723, 724*
United States 30–31
United States Agency for International Development (USAID) 827

United States Environmental Protection Agency (USEPA) 26, 29
United States Geological Survey (USGS) 318
unit hydrographs (UH) 122–127
 application of 123–124, **125**
 development 122
 ordinates of 130, **130**, 131
 SCS method 122–123, *123*
 S-hydrograph method 124–127, *126*
unlined channels, design of 247, *247*
up-flow anaerobic sludge blanket (UASB) reactors 379
urban channel routing 189–191
urban drainage systems
 characteristics 98–99
 risk issues in 260–265, **261**
 and solid waste management 75–76
 and wastewater treatment systems 74–75
urban floods 265–267, 752
urban heat islands (UHIs) 5–6, 43
 global warming 6, 43
 high energy use 6, 43
 issues in 43
 mitigation 6, 43
 poor air quality 6, 43
 reduction strategies 6, 43
 risks to public health 6, 43
urban hydrologic and hydrodynamic simulation
 artificial neural networks 694
  probabilistic neural network 694–696, **696**
  radial basis function 697–699, *697–699*
 characteristics 689
 DEM error realizations 718–719, *719*
 flood damage assessment 724–729
 hydraulic-driven simulation models
  EPANET 715–717
  event-driven method 717–718
  QUALNET 717
 infrastructure flood risk management 724–729, *726*, **727**, *728*, *729*
 inland BMPs 719–724
 mathematical simulation techniques 689–690
  Markov processes and Markov chains 690–691
  Monte carlo technique/simulation 691–694
  stochastic simulation 690
 numerical modeling 710–715, *711*
 overland flow simulation 699–700
  distributed hydrological models 702–709, *703*, *704*, *705*, *707*
  HBV 702
  hydrologic modeling system 700–701, *701*
  IHACRES 700
  StormNET 701–702
 physical model 709–710
 ungagged coastal flooding 719–724, *721*, *722*, *723*, *724*
urban irrigation, water reuse 330
urbanization 31, 93
 impacts on urban water cycle 4–5
 impacts on water cycle 39–40
 unit hydrograph
  before and after 39, *39*
  limitation 40
urban stormwater drainage systems
 Amsterdam, Netherlands
  improvement and future plans 273–274
  recommendations 274
  system characteristics 273
 best management practices
  aquifer storage and recovery system 237, *238*
  bioretention basins 227
  bioretention swales 226–227, *227*
  constructed wetlands *231*, 231–233, *233*
  extended detention basin 233–235, **234**, *234*, **235**
  grass buffers 237
  infiltration systems 236, *236*
  ponds and lakes 236
  porous pavement 237, *238*
  retention pond 224–226, *225*, **226**
  sand filters 227–230, *228*, *229*
  sediment basins 224, *224*
  swales and buffer strips 230–231
 Chongqing, China
  improvement and future plans 271–272
  recommendations 272
  system characteristics 271
 combined sewer overflow 248–249
 components
  flow in gutters *241*, 241–242
  pavement drainage inlets *242*, 242–244
  surface sewer systems 244–246, *245*, *246*
 culverts
  outlet control flow in 250, *251*
  outlet erosion protection 251, *252*
  sizing of 250
 design considerations 239–240, **240**
 drainage channel design 246–248
  grass-lined channel 247–248
  unlined channels 247, *247*
 drainage in urban watersheds
  channel flow 239
  overland flow 239
 land use planning 220–222
 London, England
  improvement and future plans 272–273
  system characteristics 272
 Los Angeles, California
  improvement and future plans 270–271
  recommendations 271
  system characteristics 269–270
 objectives of 217–218
 overland flow models 267
 Philadelphia, USA
  improvement and future plans 269
  recommendations 269
  system characteristics 267–268
 risk issues
  chemical discharge 265
  DO depletion in streams 264–265
  flooding 260–264
 Stockholm, Sweden
  improvement and future plans 274–275
  recommendations 276
  system characteristics 274
 stormwater storage facilities 257–260
  conveyance or channel routing 258
  detention and retention 259
  infiltration 258–259
  sizing of storage volumes 259–260
  types of 258–259

surface drainage channels
design rainfall 255–256
design return period 256–257
design storm duration and depth 257
probabilistic description of rainfall 251–255
spatial and temporal distribution of design rainfall 257
sustainable urban land use management
Driving force, State, and Response framework 221–222, *221, 223*
dynamic strategy planning 220
identification of 220
urban floods 265–267
urban transportation infrastructures 76–77
urban water cycle (UWC) 2–5
climatic effects 40–41
components 3, *3*, 37–39
cultural aspects 43–44
depression storage 58–59
greenhouse effect 42
groundwater 63–64
hydrologic effects 41–42
infiltration 59–60
interactions on urban components 40–44, 72–77
interception storage 58–59, **59**
interdependencies 4, *4*
land use 3
livable cities of future 77–86
Palmer drought severity index
agricultural drought indicators 60
coefficients of water balance parameters 61
drought severity index 62–63
potential climatic values 61
precipitation for climatically appropriate for existing condition 61–62
qualitative aspects 42
remotely sensed and satellite data 44
reservoirs and lakes 65
sustainable, community-based enabling systems for **10**
urban drainage systems
and solid waste management 75–76
and wastewater treatment systems 74–75
urban heat islands 43
urbanization, impact of 4–5
urbanization impacts 39–40
urban transportation infrastructures 76–**77**
urban water infrastructure 76–77
waste production 3
wastewater treatment system 73, *74*
urban drainage systems and 74–75
water and 73
water balance
evaporation 54–58
evapotranspiration 54–58
precipitation 44–54
Thomas model 66–68
water supply–demand sustainability 68–72, *69*
water demand 3
water supply and wastewater collection systems 74
urban water demand management (UWDM) 284, 317–324, **355**
urban water disaster management (UWDM)
attributes 677–679
comprehensive disaster management 635
planning cycle 635–636
disaster indices 644
drought 652
reliability 644–649
resiliency 649–650
sustainability index 651–652
time-to-failure analysis 649
vulnerability 650–651
drought 641
flood management case 642
governance perspective 637–638
initiation
community activities 640–641
dynamic process 638
multisectoral approach 638
policy implications 639
political and governmental commitment 638–639
public participation 640
stakeholder participation 638
planning process 634–635
policy, legal, and institutional framework 633
preparedness 660–661, 679–680
disaster and scale 673
evaluation 661–667, **662**, *663*, **665**, *665*, **666**, **667**, *667*
hybrid drought index 668–673, *669*, **670**, *671*, *672*, **673**
monitoring system 680
organization and institutional chart *680*, 680–683, *681, 682*
public issues of concern 675–676
reliability 673–674, *674*
water supply reliability indicators and metrics 674–675
risk analysis 656
creating incentives and constituencies 659–660
management and vulnerability 656–659, *657, 658*
of water resources systems 659
situation analysis 642–644
societal responsibilities 634
sources and kinds 629
drought *630*, 630–631, *631*
earthquakes 632–633
floods 631–632
system failures 632
widespread contamination 632
strategies 636–637
uncertainty 652–656, **656**
water resources disaster 676–677, *677*
urban water hydraulics
channel
cross section 174
geomorphology 167–178
law of stream slopes 174
length of 168–170, *168–170*
roughness 174–177, **175–177**
routing 189–191
slope of 170–173, **173**, *173*
urban morphology challenges 177–178
open-channel flow
classification 181
hydraulic analysis of 181–183
in watersheds 179–181
overland flow

urban water hydraulics (*cont.*)
effective Manning *n* values for **184**
hydrographs *186*
on impervious surfaces 184–187
on pervious surfaces 187–189, *188*
travel time 178
water distribution systems
energy equation of pipe flow *192,* 192–193
head loss due to friction, evaluation of 193–211
urban water hydrology
excess rainfall calculation 103–112
Green–Ampt model *106,* 106–108, **108**
infiltration, estimation of 105–106
interception storage estimation 104–105, **105**
instantaneous unit hydrographs
convolution integral 127–130, *128*
Laplace transformation model 138–142
Nash model 133–137, **137**
land use and cover impacts 94–97
rainfall measurement 112–114
intensity–duration–frequency curves **113**, 113–114, *114*
rainfall–runoff analysis 97–99
drainage area characteristics 98–99
rainfall loss 99
revisiting flood records 144–145
peak adjustment factor 145–146, *146*
peak discharge, urban effects on 145
river routing methods 142–144, *143*
runoff volume, estimation of 114–122
time series analysis 151–162
Akaike Information Criterion 159
ARMA model 152–154, 158–162, **159**
autoregressive models 154–157, **159**
moving average process 157, **159**
travel time 99–103
empirical formulas 103
sheet flow 102–103
time of concentration 99–100
time parameters 100–101, **101**
velocity method 101–102, **102**
unit hydrographs 122–127
application of 123–124, **125**
development 122
SCS method 122–123, *123*
S-hydrograph method 124–127, *126*
urban effect, test of significance of 146–151
Spearman–Conley test 149–151, **150**, **151**
Spearman test 147–149, **148**, **149**
watersheds 93–94
urban water infrastructure 76–77
urban water infrastructure management 7–12
community-based enabling systems **10**
environmental, economic, and social effects 9
life cycle assessment 8–9
urban landscape architecture 10–12
urban water management 167
urban water supply infrastructures
Accra, Ghana *344,* 344–345
Amsterdam, Netherland 342–344, *343*
water distribution system 344
water supply system 343–344
conveyance tunnels 308–309
Copenhagen, Denmark *341,* 341–342
water consumption 342
water supply plan 341–342
dams *293,* 293–294, *294*
curing 298–301
environmental performance assessment 325
EPANET 324–325
flood control 297
groundwater storage
confined flow 304–306
unconfined flow 306–307
well hydraulics 303–304
hydraulic simulation of water networks 324–325
leakage management 331–335
life cycle assessment 325–326
Los Angeles 339–341, *340*
nondestructive testing 335–336, **335–336**
parallel reservoir 295–297, *296*
Philadelphia, USA 336–338, *337*
planning issues 294–295
reservoirs 290–292, *291*
operation 297
urban storage 307–308
Stockholm, Sweden 345–346
sustainable development of 326–331
water demand management 317–324
water distribution system
components 314–316
hydraulic s of 316
wáter supply system challenges 317
water supply storage 290–292, *291*
water transfers 308–309
water treatment plants
infrastructure 309–311
unit operation 311–313
urban water system economics 419
analysis 423–425, **424**, *425*
analytical tools 419
asset management
attributes 430–431
developing plans 442–443
drivers 431–432
governmental agencies 430
infrastructure 443–445
level of service 432–433
levels of service 430
life cycle cost analysis 435–438, *436, 438*
objectives 432
program 430
risk assessment 433–435
select tools and practices 440–441
status and condition 432
for stormwater and wastewater systems 451–458
sustainable service delivery *439,* 439–440
water distribution system 446–451
for water supply infrastructures 445–446
benefit–cost ratio method 425–426
benefits 421
cash flow elements 421–422, **422**, *423*
Centre for Advancement of Trenchless Technologies 459
Check-Up Program for Small Systems 459–460
costs **420**, 420–421, **421**
critical infrastructure interdependencies 465–467, *466*
demand and production functions 426

demand functions 427
environmental performance 463–465
financial statements 427–429
financing methods 460–463
performance measures 442, 443
production possibility curve 426, *426*
resource-supply function 426
urine separation 328
US Department of Housing and Urban Development (HUD) 7
US Environmental Protection Agency (USEPA) 442
US-EPA landmark Acts 410
US Safe Drinking Water Act (SDWA) 284
UWC *see* urban water cycle (UWC)
UWDM *see* urban water demand management (UWDM); urban water disaster management (UWDM)

V

vacuum sewers 383
Valiantzas, J.D. 196
Van der Zaag, P.V. 28
Van de Sande, B. 877
Vapnik, V. 857
variable costs 420
variable infiltration capacity (VIC) model 863–866, *866*
variogram, fitting 482–483, *483*
variogram modeling 880–883, *881*, **882**, *882*
Vecht River 344
velocity method 101–102, **102**
Venturi/Dall tubes 332
Viessman, W. 59
Vision21 378
visual inspection, of internal condition 458
visualization technology 823
Vitry gare de Seine, urban project for 12
volatile organic compounds (VOCs) 6, 43
volatile solids (VS) 373, 374
volatile suspended solids (VSS) 373, 374
vulnerability
assessment tools 606–607, *607*
disaster indices 650–651
estimation 603–606, *605*
flood 780
risk management 656–659, *657*, *658*
risk reduction 607–608, *608*

W

wall decay 321
Wang, G.T. 40
Wang, L. Z. 549
Wang, M. 533
warning phase 676
waste production 3
wastewater collection network project 422–423
wastewater collection systems 40
wastewater linear assets 458–459
wastewater management 360
wastewater package plants 383–384
wastewater services, standards on 410–412, *411*, *412*
wastewater systems 359–360
asset management-based flood resiliency 455–456, **457**, *458*
costs of 452–453
program funding 454–455, **455**
wastewater treatment (WWT) 360–380
advanced 378
Amsterdam, Netherlands 401
Caribbean 384–386
Copenhagen, Denmark 409
developing region, technologies for 378–379
Lodz combined sewerage system 386–387
Paris, France 405–408
Philadelphia, USA 403–405
primary treatment *362*, 362–363
secondary treatment 363–378
activated sludge process 368–378, **369**, *369*
aerobic 366–367, *367*
anaerobic 367–368, *368*
biological process 364–366
suspended growth 378
Stockholm, Sweden
improvement and future plans 402–403
system characteristics 401
wetlands 380
Zaragoza, Spain 405
wastewater treatment plants (WWTPs) 73, 219, 248
in Brooklyn 388, **388**, *388*
under coastal flooding, reliability assessment 388–392, *390*
four criteria 393, **394**
margin of safety method 560
NYC service areas 392–393, *393*
probabilistic load and resistance reliability 559–560, *559*
safety-based flood reliability evaluation 398–401
water 35, 77, *77*
availability 284–286
economics of 28–29
free chlorine in 312
hierarchy of 23–26
management
agent based modeling (ABM) 555
price of 29
reclaimed 11
supplementary sources of 288
total systems approach 23–26
biosphere 24
built environment 25
disasters and interdependencies 25–26, *26*
human and institutional systems 25
natural systems 25
system-based thinking 24
transfers 308–309
water balance 2
evaporation 54–58
evapotranspiration 54–58
parameters, coefficients of 61
precipitation 44–54
Thomas model 66–68
water supply–demand sustainability 68–72, *69*
water budget method 55
water demand 3
forecasting 319
price elasticity of 323–325
water distribution system (WDS) 40, 420, 446–448
branch distribution system 314, *315*

water distribution system (WDS) (*cont.*)
  components 314–316
  design and construction 448–451
  hydraulics
    energy equation of pipe flow *192,* 192–193
    head loss due to friction, evaluation of 193–211
  hydraulic s of 316
  looped water distribution network 314, *315*
  pipes 314
  pumps 316
  tanks 316
  water supply system challenges 317
  water treatment and water mains 448
water diversion project 468
Water Eco-Nexus Cycle System 8, 10, *11*
water infrastructure asset management 443–445
water mains 448
water pressure 318
water quality 39
Water Quality Act (WQA) 29, 410
water quality capture volume (WQCV) 224, 226, 228–230
water quality model 319–321
water quality standards (WQSs) 29, 319
water quality volume (WQV) 235–236
water-related environmental visualization 867–870, *868,* **869**, *869, 870, 871*
water resources development projects 426
water resources disaster
  natural and man-induced disasters 676
  phases 676–677, *677*
water reuse 328, 330
water sensitive urban design (WSUD) 288
watershed 2, 14, 35
  average rainfall of 50–51
  characteristics **117**
  functions 93–94
  geomorphology 94
  length of 168–170, *168–170,* **172**
  peak adjustment factors for urbanizing 145, *146*
  slope of 170–171, **172, 173**
  snowmelt runoff from 51–52
watershed modeling system 702–704
watershed reliability 587
water supply
  demand sustainability 68–72, *69*
  development 284
  methods 287
  natural resources for 287
  share of water users 286–287
  storage 290–292, *291*
  and wastewater collection systems 74
water supply quantity standards 318–319
water supply sector 544
water systems analysis
  challenges 477
  conflict resolution 539
    models 541–545, *545*
    steps 539–540
    systems approach 539–541
    traditional approaches 539
  data preparation techniques 477
    fuzzy inference system 488–489, *489*
    fuzzy sets and parameter imprecision 486–488
    multicriteria decision-making 484–486
    regionalizing hydrologic data 477–484
  engineering applications 475
  optimization techniques 514–515
    dynamic programming 521–526
    evolutionary algorithms 526–536
    linear method 515–520
    multiobjective 536–538, *537, 538*
    nonlinear method 520–521
  prerequisite 477
  representation and domains 475–476, *476*
  sensitivity 477
  simulation techniques 475, 489
    artificial neural networks 493–500, **497–498,** *499, 500*
    dynamics 508–514
    mathematics of growth 501–508
    Monte Carlo simulation 500–501
    probabilistic distribution 489–492, **492**
    stochastic process 492–493
water tariff 221
water treatment
  coagulation 309, 312
  disinfection 310, 312
  filtration 310, 312
  flocculation 309, 312
  infrastructure 309–311
  rapid mixing 309
  screening and grit removal 309
  secondary settling 309
  sedimentation 309, 312
  sludge processing 310
  unit operation 311–313
water treatment plant (WTP) 333
WDS *see* water distribution system (WDS)
Weaver, W 618
Webster, P.J. 713
Weibull distribution 586
well drilling 303
well hydraulics 303–304
wetlands 380
  pocket 232
  runoff volume 96–97
  shallow 231
wet pond *see* retention pond
white box model 22–23, *23*
White, I.D. 13, 539
Wilson, M.F. 892
wing walls 250
Wolf, A. 539
Wolf, M. 557
World Bank 360
World Health Organization (WHO) 78
WQCV *see* water quality capture volume (WQCV)
Wu, C. 712
WWT *see* wastewater treatment (WWT)
WWTPs *see* wastewater treatment plants (WWTPs)

Y

Yen, B.C. 598
Yoon, K. 484
Young, P. 323
Young, R. A. 549
Yule Walker equations 153, 156

Z

Zaragoza, Spain
Almozara wastewater treatment plant in 405, *407*
Cartuja wastewater treatment in 405, *406*
wastewater treatment 405
Zayandeh Rud River Basin 62, 291, *292*
zero memory 15
Zheng, C. 533
Z-index series 62

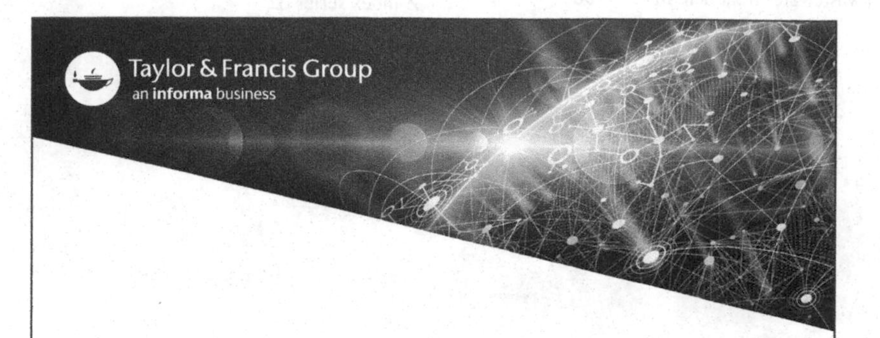

Taylor & Francis eBooks

www.taylorfrancis.com

A single destination for eBooks from Taylor & Francis with increased functionality and an improved user experience to meet the needs of our customers.

90,000+ eBooks of award-winning academic content in Humanities, Social Science, Science, Technology, Engineering, and Medical written by a global network of editors and authors.

TAYLOR & FRANCIS EBOOKS OFFERS:

A streamlined experience for our library customers

A single point of discovery for all of our eBook content

Improved search and discovery of content at both book and chapter level

REQUEST A FREE TRIAL

support@taylorfrancis.com

Printed in the United States
by Baker & Taylor Publisher Services